Chemical Processes in Soils

Soil Science Society of America Book Series

Books in the series are available from the Soil Science Society of America, 677 South Segoe Road, Madison, WI 53711 USA.

1. MINERALS IN SOIL ENVIRONMENTS. Second Edition. 1989.
   J. B. Dixon and S. B. Weed, *editors* R. C. Dinauer, *managing editor*
2. PESTICIDES IN THE SOIL ENVIRONMENT: PROCESSES, IMPACTS, AND MODELING. 1990.
   H. H. Cheng, *editor* S. H. Mickelson, *managing editor*
3. SOIL TESTING AND PLANT ANALYSIS. Third Edition. 1990.
   R. L. Westerman, *editor* S. H. Mickelson, *managing editor*
4. MICRONUTRIENTS IN AGRICULTURE. Second Edition. 1991.
   J. J. Mortvedt et al., *editors* S. H. Mickelson, *managing editor*
5. METHODS OF SOIL ANALYSIS: PHYSICAL AND MINERALOGICAL METHODS. Part 1. Second Edition. 1986.
   Arnold Klute, *editor* R. C. Dinauer, *managing editor*

   METHODS OF SOIL ANALYSIS: MICROBIOLOGICAL AND BIOCHEMICAL PROPERTIES. Part 2. 1994.
   R. W. Weaver et al., *editor* S. H. Mickelson, *managing editor*

   METHODS OF SOIL ANALYSIS: CHEMICAL METHODS. Part 3. 1996.
   D. L. Sparks, *editor* J. M. Bartels, *managing editor*

   METHODS OF SOIL ANALYSIS: PHYSICAL METHODS. Part 4. 2002.
   J. H. Dane and G. C. Topp, *editors* L. K. Al-Amoodi, *managing editor*
6. LAND APPLICATION OF AGRICULTURAL, INDUSTRIAL, AND MUNICIPAL BY-PRODUCTS. 2000.
   J. F. Power and W.A. Dick, *editors* J. M. Bartels, *managing editor*
7. SOIL MINERALOGY WITH ENVIRONMENTAL APPLICATIONS
   J. B. Dixon and D. G. Schulze, *editors* L. K. Al-Amoodi, *managing editor*
8. CHEMICAL PROCESSES IN SOILS
   M. A. Tabatabai and D. L. Sparks, *editors* L. K. Al-Amoodi, *managing editor*

Chemical Processes in Soils

Co-editors: M. A. Tabatabai and D. L. Sparks

Managing Editor: Lisa Al-Amoodi

Editor-in-Chief SSSA: Warren A. Dick

Number 8 in the Soil Science Society of America Book Series

Published by: Soil Science Society of America, Inc.
Madison, Wisconsin, USA

2005

Soil Science Society of America, Inc.
677 South Segoe Road, Madison, WI 53711-1086 USA

Library of Congress Control Number: 2005924447

Printed in the United States of America.

CONTENTS

DEDICATION

Vasilios Petrios (Bill) Evangelou 1947–2002

It is highly appropriate that *Chemical Processes in Soils* be dedicated to the memory of Dr. Vasilios Petrios (Bill) Evangelou, whose research career covered more than two decades in the area of soil chemistry.

He was born on January 8, 1947 in Stavros Greece, a small village in the northern prefecture of Halkidiki. His parents, Petros and Glykaria (Zotou) Evangelou, were ethnic Greek refugees who had escaped the religious persecution of communist Albania. Bill grew up in the neighboring seaside village of Olympiada, the youngest of four brothers. At age fourteen, he left his tiny village for the city of Thessaloniki to attend The American Farm School, a trade school that taught modern farming techniques to rural Greek boys. It was at the Farm School that he was given the name "Bill", which is the nickname used by most Greek boys named Vasilios who immigrate to English speaking countries.

At age eighteen, Bill completed his studies at the Farm School and was valedictorian of his senior class. It was the custom of the school to award its top student with livestock to help him begin his farming career. However, Bill did not have land on which to raise animals, so the school administrators asked him what he wanted. His answer: "I want to go to America." The school and several sponsors made this dream a reality. In December 1965, Bill arrived in Roseville, CA at the home of David and Dolly Fiddyment to live and work on their turkey ranch and attend Roseville High School. He graduated from Roseville High in June 1967.

He received his Associate of Arts Degree from Sierra Junior College in 1967 and from there transferred to California State University, Chico, where he received his Bachelor of Science in Agriculture (1972) and his Masters of Science in Plant and Soil Sciences (1974). Bill had been accepted to graduate school for his doctoral studies at the University of California, Davis upon his completion of his Masters, but there were no funds being offered to foreign students in those days so he accepted a position as an instructor of Agriculture at San Joaquin Delta Community College in Stockton, CA. He and his wife worked hard to save the money required to attend U.C. Davis in the fall of 1976. He completed all work required for his Ph.D. by July 1980 and accepted the position of Soils Extension Specialist at the University of Kentucky. Dr. Evangelou's career at the University of Kentucky spanned nearly 19 years and included a change in appointment to Research Professor of Soil/Water Chemistry.

In January 1999, Dr. Evangelou accepted the Iowa State University Agronomy Department's offer as Professor of Soil/Water Physical Chemistry. During his career, he published two advanced textbooks, 14 book chapters, and more than 150 scientific publications in refereed journals, national and international conference proceedings, and other professional publications. His areas of research included surface chemistry of clay minerals, reaction kinetics and thermodynamics, metal sulfide chemistry, and control mechanisms of acid drainage.

The depth and breadth of his research program allowed him the opportunity to publish two advanced college level textbooks since 1995. The first book, *Pyrite Oxidation and its Control* (CRC/Lewis Press, Boca Raton, FL), addresses acid mine drainage, surface chemistry of clay minerals and iron sulfides, oxidation mechanisms of pyrite, microbial role, kinetics, control, ameliorates, limitations, and microencapsulation. Because of his expertise on pyrite oxidation and its control acquired through his research program, he taught more than 30 short courses by invitation on the subject of environmental soil and water chemistry, pyrite chemistry, and acid mine drainage for federal, state, and private industry professionals in both North and South America. More than 2000 professionals from the USA, Canada, Europe, South America, and South Africa attended his short courses. He also conducted short courses, by invitation, for the U.S. Department of Energy and Department of Interior personnel; for personnel of the Department of Natural Resources from the states of Kentucky, Illinois, West Virginia, Utah, and New Mexico; and for coal mine industry affiliated members of the American Society of Surface Mine Reclamation.

The second book, *Environmental Soil & Water Chemistry: Principles and Application* (John, Wiley and Sons, New York), is a textbook on the fundamental principles and application of the chemistry of soil and water, behavior and treatment of soil water contaminants, agricultural chemicals, acid drainage, soil water restoration, revegetation, sodic soils, and brackish waters. Published in late 1998, it is an advanced undergraduate or graduate level textbook.

Dr. Evangelou taught two graduate level courses at Iowa State University. He was recognized for his scientific contributions with numerous awards, including the Marion L. and Chrystie M. Jackson Soil Science Award, Soil Science Society of America, for outstanding contributions in the areas of soil chemistry and mineralogy and graduate student education; Fellow, American Society of Agronomy; Fellow, Soil Science Society of America; a U.S. Patent on "Peroxide Induced Oxidation Proof Phosphate Surface Coating on Iron Sulfides"; a U.S. and Canadian Patent on "Oxidation Proof Silicate Surface Coating on Iron Sulfides"; Senior Fulbright Scholar Award; and Thomas Poe Cooper Award, University of Kentucky, College of Agriculture, 1994, for distinguished achievement in research. Dr. Evangelou was National Chair of the Soil Chemistry division of Soil Science Society of America (2001) and was in charge of organizing that division's national scientific meetings as well as a symposium on Surface Chemistry of (Bio)Organic Molecules in Soils for the national meetings of the American Society of Agronomy in Charlotte, NC. Dr. Evangelou also organized the 2002 International Bouyoucos Conference on (Bio)Availability held in Sani, Halkidiki, Greece.

FOREWORD

The Soil Science Society of America (SSSA) is dedicated to excellence in the acquisition of new knowledge, in the training of scientists, in the education of citizens, and in the applications of knowledge to challenges facing society. In addition, our scientific and professional society facilitates the teaching, research, and outreach programs that contribute to the long-term sustainability of food and fiber production systems and promote informed and wise stewardship of soil, water, and air resources. This mission is in no small part achieved through timely publication of books, monographs, journals, and other scholarship communicating current scientific knowledge in the diverse disciplines in soil science.

Chemical Processes in Soils represents a comprehensive and contemporary review of chemical behavior and reactions in soils. The authors have approached each topic with a focus on mineral solubility, surface exchange, and microbial processes influencing the availability and environmental fate of plant nutrients, heavy metals, and other elements. Several chapters are dedicated to the important role of soil organic matter, humic substances, and biogeochemical reactions in soils.

Understanding basic chemical and biological processes in soils is essential to sustaining agricultural productivity while protecting our vital natural resources. This thorough volume on soil chemical and biological processes will be a valuable asset to students, practitioners, educators, and researchers in soil, ecological, environmental, earth, and agricultural sciences.

JOHN HAVLIN
President
Soil Science Society of America

PREFACE

Chemical Processes in Soils provides an authoritative review of the principles governing some of the most important chemical reactions and behavior in soils. This volume is the result of serving on the editorial committee of the Soil Science Society of America Book Series 5, *Methods of Soil Analysis. Part 3. Chemical Methods*. In editing that book, we received several recommendations and proposals of book chapters to be included in the book on soil analysis. Many of those titles were not within the scope of a book on methods of analysis, but were in the area of soil chemistry. Some of those titles were selected to cover, in detail, the state of knowledge in the specific area of soil chemistry, with emphasis on the reactions, theories, and concepts involved. The authors were allowed considerable latitude in developing their chapters, resulting in both panoramic treatment of topics and detailed coverage of specific reactions.

This volume contains 15 chapters written by authorities in their fields. Soil organic matter is one the most complex and reactive fractions of soils. A major chapter on the chemistry of soil organic matter covers carbon in the environment, the genesis and fractionation of soil organic matter, isolation of humic substances, and considerations of their structural composition, soil saccharides, and soil peptides. The details of the reactions involved and the techniques and methods used are described. Other chapters explore in detail the chemistry of phosphorus, potassium, sulfur, and micronutrients in soils. Other important topics include the kinetics and mechanisms involved in biogeochemical processes, cation exchange reactions, soil acidity, chemistry of redox processes, equations and models describing adsorption processes, sorption and desorption rates for neutral organic compounds, metal complexation by soil humic substances, speciation of metals in soils, chemistry of speciation of trace elements in soil solution, and the chemistry of salt-affected soils. The literature accumulated in each of the topics is extensive, and exhaustive coverage of the literature was not always possible. Therefore, the editors and authors apologize for omission of any important work.

It is hoped that each chapter would serve as an independent source of information for scientists involved in teaching and research. References are listed at the end of each chapter that might help the reader in expanding the scope of interest.

We express our appreciation to the authors and the organization that they represent for cooperation and support. We acknowledge the assistance of the Headquarters staff, especially Lisa Al-Amoodi, for advice and assistance in editing and preparing the manuscripts for publication. We acknowledge Vasilios Petros (Bill) Evangelou and Ronald E. Phillips, co-authors of Chapter 7, whose deaths occurred while this book was in progress. The assistance of D. L. Sparks in proofreading and indexing Chapter 7 is gratefully acknowledged.

M. A. TABATABAI, co-editor
Department of Agronomy
Crop, Soil, and Environmental Sciences
Iowa State University
Ames, Iowa, USA

D. L. SPARKS, co-editor
Dep. of Plant and Soil Sciences
University of Delaware
Newark, DE, USA

CONTRIBUTORS

R. J. Bartlett	Department of Plant and Soil Science, University of Vermont, Hills Building, Burlington, VT 05405-0082
P. R. Bloom	Department of Soil, Water, and Climate, 439 Borlaug Hall, University of Minnesota, 1991 Upper Buford Circle, St. Paul, MN 55108
C. E. Clapp	USDA-ARS and Department of Soil, Water and Climate, University of Minnesota, 1991 Upper Buford Circle, St. Paul, MN 55108
V. P. Evangelou	Deceased (formerly Department of Agronomy, Iowa State University, Ames, IA)
S. Goldberg	USDA-ARS, George E. Brown, Jr., Salinity Laboratory, 450 West Big Springs Road, Riverside, CA 92507
M. H. B. Hayes	Department of Chemical and Environmental Sciences, University of Limerick, Limerick, Ireland
P. M. Huang	Department of Soil Science, University of Saskatchewan, Saskatoon, SK S7N 5A8, Canada
W. L. Kingery	Department of Plant and Soil Sciences, Mississippi State University, Mississippi State, MS 39762
E. Loffredo	Universita Degli Studi di Bari, Dipartimento di Biologia e Chimica Agroforestale i Ambientale, Via G. Amendola, 165/A, 70126 Bari, Italy
C. J. Matocha	Department of Agronomy, University of Kentucky, Lexington, KY 40546-0091
M. Nachtegaal	Paul Scherrer Institut, 5232 Villigen PSI, Switzerland
D. R. Parker	Department of Environmental Sciences, University of California, 2416 Geology Building, Riverside, CA 92521
R. E. Phillips	Deceased (formerly University of Kentucky, Lexington, KY)
G. M. Pierzynski	Department of Agronomy, Plant Sciences Center, 2040 Throckmorton Hall, Kansas State University, Manhattan, KS 66506-5501
D. Roberts	Water and Earth Science Associates Ltd., 3108 Carp Road, Ottawa, ON, Canada KOA1L0
D. S. Ross	Department of Plant and Soil Science, University of Vermont, Hills Building, Burlington, VT 05405-0082
S. Sauve	Department of Chemistry, University of Montreal, C.P. 6128, Succ. Centre-Ville, Montreal, QC H3C 3J7, Canada
K. G. Scheckel	USEPA, 5995 Center Hill Avenue, Cincinnati, OH 45224-4504
N. Senesi	Universita Degli Studi di Bari, Dipartimento di Biologia e Chimica Agroforestale i Ambientale, Via G. Amendola, 165/A, 70126 Bari, Italy
L. M. Shuman	Department of Crop and Soil Sciences, The University of Georgia, Griffin Campus, 1109 Experiment St., Griffin, GA 30223-1797

A. J. Simpson Department of Physical and Environmental Sciences, University of Toronto, 1255 Military Trail, Toronto, N1C 1A4, Canada

J. T. Sims Department of Plant and Soil Sciences, University of Delaware, 152 Townsend Hall, Newark, DE 19716-2170

U. L. Skyllberg Department of Forest Ecology, Swedish University of Agricultural Sciences, Umea, S-90183, Sweden

D. L. Sparks Department of Plant and Soil Sciences, University of Delaware, Newark, DE 19717-1303

D. L. Suarez USDA-ARS, U.S. Salinity Laboratory, 450 West Big Springs Road, Riverside, CA 92507

M. E. Sumner Department of Agronomy, University of Georgia, Watkinsville, GA 30677

M. A. Tabatabai Department of Agronomy, Iowa State University, Ames, IA 50011-1010

W. J. Weber, Jr. Department of Chemical Engineering, University of Michigan, Ann Arbor, MI 48109-1115

T. M. Young Department of Civil and Environmental Engineering, University of California, One Shields Avenue, Davis, CA 95616

Conversion Factors for SI and non-SI Units

Conversion Factors for SI and non-SI Units

To convert Column 1 into Column 2, multiply by	Column 1 SI Unit	Column 2 non-SI Units	To convert Column 2 into Column 1, multiply by
		Length	
0.621	kilometer, km (10^3 m)	mile, mi	1.609
1.094	meter, m	yard, yd	0.914
3.28	meter, m	foot, ft	0.304
1.0	micrometer, µm (10^{-6} m)	micron, µ	1.0
3.94×10^{-2}	millimeter, mm (10^{-3} m)	inch, in	25.4
10	nanometer, nm (10^{-9} m)	Angstrom, Å	0.1
		Area	
2.47	hectare, ha	acre	0.405
247	square kilometer, km^2 $(10^3\ m)^2$	acre	4.05×10^{-3}
0.386	square kilometer, km^2 $(10^3\ m)^2$	square mile, mi^2	2.590
2.47×10^{-4}	square meter, m^2	acre	4.05×10^{3}
10.76	square meter, m^2	square foot, ft^2	9.29×10^{-2}
1.55×10^{-3}	square millimeter, mm^2 $(10^{-3}\ m)^2$	square inch, in^2	645
		Volume	
9.73×10^{-3}	cubic meter, m^3	acre-inch	102.8
35.3	cubic meter, m^3	cubic foot, ft^3	2.83×10^{-2}
6.10×10^{4}	cubic meter, m^3	cubic inch, in^3	1.64×10^{-5}
2.84×10^{-2}	liter, L (10^{-3} m^3)	bushel, bu	35.24
1.057	liter, L (10^{-3} m^3)	quart (liquid), qt	0.946
3.53×10^{-2}	liter, L (10^{-3} m^3)	cubic foot, ft^3	28.3
0.265	liter, L (10^{-3} m^3)	gallon	3.78
33.78	liter, L (10^{-3} m^3)	ounce (fluid), oz	2.96×10^{-2}
2.11	liter, L (10^{-3} m^3)	pint (fluid), pt	0.473

	Mass		
2.20×10^{-3}	gram, g (10^{-3} kg)	pound, lb	454
3.52×10^{-2}	gram, g (10^{-3} kg)	ounce (avdp), oz	28.4
2.205	kilogram, kg	pound, lb	0.454
0.01	kilogram, kg	quintal (metric), q	100
1.10×10^{-3}	kilogram, kg	ton (2000 lb), ton	907
1.102	megagram, Mg (tonne)	ton (U.S.), ton	0.907
1.102	tonne, t	ton (U.S.), ton	0.907
	Yield and Rate		
0.893	kilogram per hectare, kg ha^{-1}	pound per acre, lb acre^{-1}	1.12
7.77×10^{-2}	kilogram per cubic meter, kg m^{-3}	pound per bushel, lb bu^{-1}	12.87
1.49×10^{-2}	kilogram per hectare, kg ha^{-1}	bushel per acre, 60 lb	67.19
1.59×10^{-2}	kilogram per hectare, kg ha^{-1}	bushel per acre, 56 lb	62.71
1.86×10^{-2}	kilogram per hectare, kg ha^{-1}	bushel per acre, 48 lb	53.75
0.107	liter per hectare, L ha^{-1}	gallon per acre	9.35
893	tonne per hectare, t ha^{-1}	pound per acre, lb acre^{-1}	1.12×10^{-3}
893	megagram per hectare, Mg ha^{-1}	pound per acre, lb acre^{-1}	1.12×10^{-3}
0.446	megagram per hectare, Mg ha^{-1}	ton (2000 lb) per acre, ton acre^{-1}	2.24
2.24	meter per second, m s^{-1}	mile per hour	0.447
	Specific Surface		
10	square meter per kilogram, m^2 kg^{-1}	square centimeter per gram, cm^2 g^{-1}	0.1
1000	square meter per kilogram, m^2 kg^{-1}	square millimeter per gram, mm^2 g^{-1}	0.001
	Density		
1.00	megagram per cubic meter, Mg m^{-3}	gram per cubic centimeter, g cm^{-3}	1.00
	Pressure		
9.90	megapascal, MPa (10^6 Pa)	atmosphere	0.101
10	megapascal, MPa (10^6 Pa)	bar	0.1
2.09×10^{-2}	pascal, Pa	pound per square foot, lb ft^{-2}	47.9
1.45×10^{-4}	pascal, Pa	pound per square inch, lb in^{-2}	6.90×10^3

(continued on next page)

Conversion Factors for SI and non-SI Units

To convert Column 1 into Column 2, multiply by	Column 1 SI Unit	Column 2 non-SI Units	To convert Column 2 into Column 1, multiply by
		Temperature	
1.00 (K – 273)	kelvin, K	Celsius, °C	1.00 (°C + 273)
(9/5 °C) + 32	Celsius, °C	Fahrenheit, °F	5/9 (°F – 32)
		Energy, Work, Quantity of Heat	
9.52×10^{-4}	joule, J	British thermal unit, Btu	1.05×10^{3}
0.239	joule, J	calorie, cal	4.19
10^{7}	joule, J	erg	10^{-7}
0.735	joule, J	foot-pound	1.36
2.387×10^{-5}	joule per square meter, J m^{-2}	calorie per square centimeter (langley)	4.19×10^{4}
10^{5}	newton, N	dyne	10^{-5}
1.43×10^{-3}	watt per square meter, W m^{-2}	calorie per square centimeter minute (irradiance), cal cm^{-2} min^{-1}	698
		Transpiration and Photosynthesis	
3.60×10^{-2}	milligram per square meter second, mg m^{-2} s^{-1}	gram per square decimeter hour, g dm^{-2} h^{-1}	27.8
5.56×10^{-3}	milligram (H_2O) per square meter second, mg m^{-2} s^{-1}	micromole (H_2O) per square centimeter second, μmol cm^{-2} s^{-1}	180
10^{-4}	milligram per square meter second, mg m^{-2} s^{-1}	milligram per square centimeter second, mg cm^{-2} s^{-1}	10^{4}
35.97	milligram per square meter second, mg m^{-2} s^{-1}	milligram per square decimeter hour, mg dm^{-2} h^{-1}	2.78×10^{-2}
		Plane Angle	
57.3	radian, rad	degrees (angle), °	1.75×10^{-2}

	Electrical Conductivity, Electricity, and Magnetism		
10	siemen per meter, S m^{-1}	millimho per centimeter, mmho cm^{-1}	0.1
10^4	tesla, T	gauss, G	10^{-4}
	Water Measurement		
9.73×10^{-3}	cubic meter, m^3	acre-inch, acre-in	102.8
9.81×10^{-3}	cubic meter per hour, $m^3 h^{-1}$	cubic foot per second, $ft^3 s^{-1}$	101.9
4.40	cubic meter per hour, $m^3 h^{-1}$	U.S. gallon per minute, gal min^{-1}	0.227
8.11	hectare meter, ha m	acre-foot, acre-ft	0.123
97.28	hectare meter, ha m	acre-inch, acre-in	1.03×10^{-2}
8.1×10^{-2}	hectare centimeter, ha cm	acre-foot, acre-ft	12.33
	Concentrations		
1	centimole per kilogram, cmol kg^{-1}	milliequivalent per 100 grams, meq 100 g^{-1}	1
0.1	gram per kilogram, g kg^{-1}	percent, %	10
1	milligram per kilogram, mg kg^{-1}	parts per million, ppm	1
	Radioactivity		
2.7×10^{-11}	becquerel, Bq	curie, Ci	3.7×10^{10}
2.7×10^{-2}	becquerel per kilogram, Bq kg^{-1}	picocurie per gram, pCi g^{-1}	37
100	gray, Gy (absorbed dose)	rad, rd	0.01
100	sievert, Sv (equivalent dose)	rem (roentgen equivalent man)	0.01
	Plant Nutrient Conversion		
	Elemental	***Oxide***	
2.29	P	P_2O_5	0.437
1.20	K	K_2O	0.830
1.39	Ca	CaO	0.715
1.66	Mg	MgO	0.602

Chapter 1

Chemistry of Soil Organic Matter

C. E. CLAPP, *USDA-ARS, University of Minnesota, St. Paul, Minnesota, USA*

M. H. B. HAYES, *University of Limerick, Limerick, Ireland*

A. J. SIMPSON, *University of Toronto, Toronto, Canada*

W. L. KINGERY, *Mississippi State University, Mississippi State, Mississippi, USA*

The term *soil organic matter* (SOM), according to Stevenson (1994), refers to the whole of the organic matter in soils, including the litter, the light fraction, the microbial biomass, the water-soluble organics, and the stabilized organic matter (humus). The term *natural organic matter* (NOM) is now widely used to describe the natural organic compounds in soils, sediments, and waters.

Hayes and Swift (1978) considered the complete soil organic fraction to be made up of live organisms and their partly undecomposed, partly decomposed, and completely transformed remains, as well as those of plants. However, they regarded SOM to be a more specific term for the nonliving components that may be described as a heterogeneous mixture composed largely of products resulting from microbial and chemical transformations of organic debris. The transformations are known collectively as the *humification* process, and the final product, or *humus*, is a mixture of substances that has some resistance to further degradation. This is similar to Stevenson's (1994) definition of humus as the total of organic compounds in soil exclusive of undecayed plant and animal tissues, their partial decomposition products, and the soil biomass. Stevenson refers to the partitioning of SOM into the active (or labile) and the stable pools. The *active fraction* contains the comminutive (pulverizable) nonliving plant matter (litter) that lies on the surface of the soil, the light fraction, the microbial biomass, and the nonhumified substances that are not bound to the soil minerals. Litter is important for the recycling of nutrients. The light fraction is incorporated in the soil but separable from it in liquids with density values in the range of 1.6 to 2.0 g cm^{-3}. It consists largely of plant residues in various stages of decomposition, has a rapid turnover rate, and hence provides a source of plant nutrients. There is an amazing number of microorganisms in the soil environment, perhaps 10^{10} g^{-1} of soil (Burns, 1990). These include viruses, bacteria (up to 10^9 g^{-1}), several hundred million actinomycetes, 10 to 20 × 10^6 fungi, 10^3 to 10^6 algae, up to 10^6 protozoa, and sometimes as many as 50 nematodes per gram of soil. Stevenson (1994) has quoted data for 570 and 1600 kg ha^{-1} for microbial C in temperate soils in England and Canada, respectively, and lesser amounts (460 kg) in tropical (Brazilian) soils. Microorganisms have a vital role for the turnover and transformation processes involving organic (and some inorganic) materials in the soil.

Chemical Processes in Soils. SSSA Book Series, no. 8.

These are therefore sources of enzymes in the soil environment, and (as well as plant root exudates) are responsible for water soluble organics in the soil solution.

The stable, or passive, humus pool of SOM was considered by Stevenson (1994) to be the humification products having some resistance to biodegradation processes. Hayes and Swift (1978), as referred to above, considered SOM to be the humified components of the soil organic fraction. De Saussure (1804) is credited with the term humus, the Latin equivalent of soil, to describe the dark-colored material in soil. Arguably, humus, or SOM, is one of the more complex materials in nature, containing most, if not all of the naturally occurring organic compounds. The major components are considered to be recalcitrant remains of plants and algae, including materials derived from lignins, tannins, sporopollenins, and large aliphatic molecules, such as algaenans, cutans, and suberans (Derenne and Largeau, 2001). There is increasing interest in the presence of black C (arising from the incomplete combustion of organic materials) in SOM. The major components of SOM can be considered to consist of humic substances (see Genesis of Soil Organic Matter below), saccharides and peptides, and products derived from recalcitrant materials such as those listed above.

Soil organic matter is fundamental to efforts to improve the environments for plants, animals, and people, including the protection from contaminants of air, water, and soil. Among the important functions that can be attributed to SOM are:

- formation and the stabilization of soil aggregate structures
- retention of plant nutrients attributable to its cation-exchange properties
- release of plant nutrients (especially N, P, S, and some trace elements) when the organic matter is degraded
- retention of soil moisture
- absorption of solar radiation (thereby increasing soil temperature)
- complexation of heavy metals
- retention of aromatic and sparingly soluble anthropogenic (synthetic) organic chemicals
- release of soluble and colored materials in drainage waters
- sequestration of C

More recently, the importance of SOM to considerations of environmental quality has been considered in relation to global warming, or the greenhouse effect. It is now accepted that human activity, both agricultural and industrial, has created an imbalance between global sinks and sources of C, giving rise to increases in atmospheric gases. Soil management is important in relation to the role of soils as sources and sinks of C in the environment, and hence it is important in terms of global warming issues (Lal et al., 1995). Of the five primary greenhouse gases, processes involving SOM account, to varying degrees, for the contributions of carbon dioxide (CO_2), methane (CH_4), and nitrous oxide (N_2O) (the other gases being chlorofluorocarbons and ozone) (Dale et al., 1993).

CARBON IN THE SOIL ENVIRONMENT

The soil degradative effects that accompany losses of SOM have long been recognized. Inevitably, the degradation of soil structure, which follows the depletion of SOM, leads to soil losses through erosion by wind and water. It is important to arrest the decline in SOM levels to establish ecologically and economically sustainable production systems.

Emphasis is being placed on the need to sequester into SOM-C from atmospheric CO_2 because of international concerns about greenhouse gas emissions and global warming. These concerns are highlighted by the Kyoto Protocol (Conference of the Parties, 1997), which seeks to limit the greenhouse gas emissions by the developed countries. This sequestration need should not diminish the fact that increasing SOM levels must be a priority to enhance the quality of our soils and improve soil fertility, soil structure, and nutrient cycling.

Carbon in Soils, Water, the Atmosphere, and Living Matter

The data in Table 1–1 show that, for soils in particular, there are considerable differences in the estimates of organic C. The amounts of C in the atmosphere can be estimated relatively accurately. There is not a uniform distribution of CO_2 throughout the atmosphere, but rapid and accurate analytical procedures allow frequent samplings to be made. Atmospheric scientists commonly observe a lowering in the atmospheric concentration of CO_2 at latitudes around 55°N.

In contrast to atmospheric measurements, difficulties are encountered when measuring values for soil organic carbon (SOC) at the regional, national, continental, or global levels (Swift, 2001). The spread of values in Table 1–1 is to be expected because of, among other things, the spatial variability of soils. Swift discussed the level of sampling detail that would be necessary to obtain reasonably accurate results, and he has referred to procedures that use modal profiles considered typical of a soil, vegetation, and the climate in question. In this way, modal or typical SOC values can be established for a range of soil types. These values can then be applied to the areas occupied by the given soil to build an inventory. He has referred to the fact that the great majority of estimates of SOC are taken from agricultural soils rather than from the natural ecosystems that cover large parts of the terrestrial surface. Soil depth also presents problems for estimations. The estimated reserves of SOC increase with the depth to which sampling is specified. Soil organic C is concentrated in the surface layers of most soils, but lesser amounts continue to be found at depth. For soils such as Vertisols and Histosols, substantial amounts of C are found

Table 1–1. Estimates of organic C in soil, water, atmosphere, and phytomass.

Location	Reference	C
		$\times 10^{15}$ g
Soil	Atjay et al. (1979), Bohn (1976)	2000–3000
	Eswaran et al. (1993)	1 555
Soil humic acids	Bazilevich (1974)	1000–1500
Oceans	Schlesinger (1995)	38 000
Dissolved organic C	Mopper and Degens (1979)	1 000
Particulate organic	Mopper and Degens (1979)	30
Plankton	Williams (1975)	3
Freshwater lakes and rivers	Schlesinger (1995)	0.4
Atmosphere	Schlesinger (1995)	720
Phytomass	Duvigneaud (1972)	592

at depth. Furthermore it is necessary to know the bulk density throughout the profile to calculate the amount of SOC on an area basis. Soil bulk density often changes under different land uses and management practices.

The data of Eswaran et al. (1993) suggest that the reserves of SOC in the soils of the world decrease in the order: Histosols 390 Pg (where 1 Pg = 10^{15} g) > Inceptisols (267 Pg) > Oxisols (150 Pg) > Alfisols (136 Pg) > Aridosols (110 Pg) > Entisols (106 Pg) > Ultisols (101 Pg). Aridisols (1044 Pg) have the major stores of soil inorganic C. The total SOC estimates are more than twice those for C in the atmosphere (720 Pg) and are estimated to be nearly three times the C in the biotic pool of all living matter (Schlesinger, 1995; Lal et al., 1998).

The Potential of Soils to Sequester Carbon

Under the terms of the Kyoto Protocol to the United Nations Framework Convention on Climate Change (Conference of the Parties, 1997), the developed countries agreed to several binding commitments to either decrease or to limit increases of their greenhouse gas emissions. The Protocol was essentially concerned with changes in C levels, rather than with stocks or amounts of C in any particular pool. Doubts were cast on the viability of the Protocol when the United States announced in 2001 that it did not intend to ratify the Kyoto Protocol. However, at the Conference of Parties in Bonn in July 2001, the other developed nations kept alive the possibility that it might still be implemented. Modifications were made then to the original Protocol.

In Article 3.3 in the original version of the Kyoto Protocol, soil is recognized as a significant terrestrial reservoir of C. Article 3.3 limits the allowable terrestrial sources and sinks of C to well-defined cases of afforestation, reforestation, and deforestation (Nabuurs et al., 1999). Fluxes of C to and from agricultural soils were not a part of the original Kyoto Protocol. However, at the Bonn meeting, it was decided to expand the land management systems eligible for inclusion as C sinks to include forest management, cropland management, and revegetation. This decision greatly increases the scope and opportunities for the sequestration of C by soils under a wide range of management and production systems to be incorporated legitimately under the articles of the Kyoto Protocol.

Major losses of C to the atmosphere have followed land use changes that transformed forests, grasslands, and wetlands into tillage operations where intensive management practices were followed (Jenkinson, 1990). Grassland and forest soils in the USA have tended to lose 20 to 50% of the SOM during 40 to 50 yr of cultivation (Mann, 1985, 1986; Johnson and Kern, 1991; Houghton, 1995). Kern (1994) estimated that the SOC content of the major U.S. field croplands was about 8.3 Pg before cultivation. The current SOC content of these soils is estimated to be approximately 7 Pg. Donigian et al. (1994) simulated the total C changes in the 0- to 20-cm depth between 1907 (when the soils were brought into long-term cultivation) and 1994 for soils of the central United States, which includes what is now the Corn Belt and a part of the Great Plains. Their data show a steep decline in SOC in the first 40 yr of cultivation, and in that time, 47% of the SOC was lost. A steady state was then achieved and maintained until management changes involving reduced tillage operations were introduced about 1970, and there has been a gradual accumulation of SOC since that time. Reduced tillage minimizes soil disturbance and plant residue return is an important component of the management system.

Figure 1–1 (Johnson, 1995) provides a diagrammatic representation of the changes with time to SOM levels that result from changes in management practices. Soils in long-term pasture will be in steady state with regard to SOM content. The organic matter content at steady state will be governed by the soil properties (e.g., mineralogy, drainage regime, fertility) and climate. When the management is changed, for example when the soils are brought into long-term cultivation, the SOM will be depleted through time until a new level is reached. When a new, more conservational management system is introduced, such as conservation tillage or reestablishing grassland, a reaccumulation of the SOM will take place, and a new level will eventually by reached. In this new steady state, the SOM content may be below, equal to, or above the original content, as indicated by A, B, and C in Fig. 1–1, depending on the management used.

The concept of utilizing soil to increase the sequestering of C is of great environmental significance. The annual emissions of C from soil, as the result of biological activity, is estimated to be of the order of 60 Pg, and that is balanced by the net fixation by land plants. Emissions from fossil fuels amount to about 5.5 Pg yr^{-1}. It is these emissions that are causing the increase in the levels of CO_2 in the atmosphere. The concentration of CO_2 in the atmosphere is estimated to have been about 250 ppmv (ppm by volume) in the CE 900 to 1200 era. The estimates for CE 1300 to 1850 era are 280 ppmv, and the concentrations in 1994 were 358 ppmv (Lal et al., 1998). The atmospheric levels of CO_2 are considered to be increasing at a rate of 0.5% yr^{-1}. The increases since 1850 are attributed to land use changes and to the increased combustion of fossil fuels since the industrial revolution. The annual increases in atmospheric CO_2 from these sources are estimated to be 1.5 and 3.3 ± 0.2 Pg C yr^{-1}, respectively (Post et al., 1990). These increases do not corre-

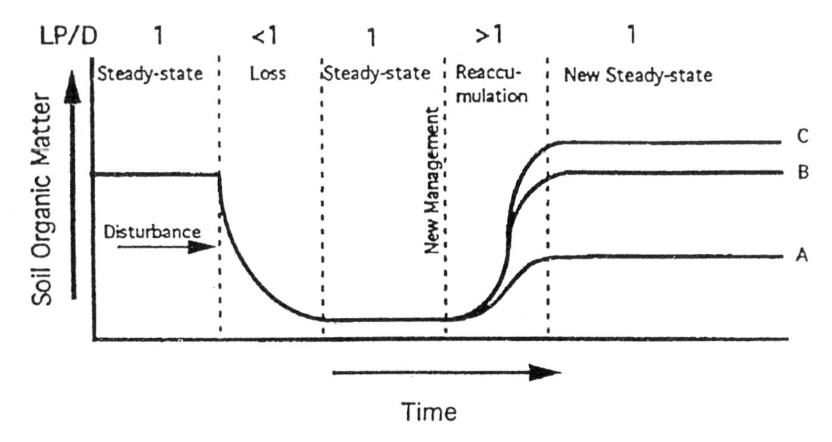

Fig. 1–1. A diagrammatic representation of the changes in soil organic matter (SOM) levels that can result from changes in management practices, such as long-term cultivation of native grassland. When a new management is introduced, such as the reintroduction of long-term grassland the SOM reaccumulates until a new steady state is reached. That new steady state will depend on the crop and management and may be below (A), equal to (B), or greater than (C) the SOM content in the original steady-state system. When LP/D < 1, SOM is in decline; when LP/D >1, SOM accumulates. LP = litter production and D = amount of decomposition (after Johnson, 1995).

spond to the emissions from anthropogenic activities referred to above, mainly because of the influences of sequestration processes.

Schlesinger (1995) showed the principal C pools and exchanges between them. Estimates for the fixation of C by land plants (120 Pg C yr^{-1}), for the emission of C by respiration from plants (60 Pg C yr^{-1}), for soil biological respiration processes (60 Pg C yr^{-1}), and for the net destruction of vegetation (2 Pg C yr^{-1}) are relevant to our topic. The net fixation by the oceans is estimated to be 2 Pg C yr^{-1} (107 Pg C yr^{-1} fixed from the atmosphere, 105 Pg C yr^{-1} emitted). Because there is an equilibrium between the CO_2 in the atmosphere and that sequestered by the oceans, increasing amounts of the CO_2 emitted are sequestered by the oceans. That is one of the reasons why the increases in atmospheric CO_2 are not in line with the emissions resulting from anthropogenic activities. In the absence of anthropogenic inputs, the processes referred to would be in steady state. However, the anthropogenic processes disturb these equilibria and result in increasing levels of atmospheric CO_2.

If the trend in increased C emissions continues, it is suggested that the concentration of CO_2 in the atmosphere will reach 600 (or perhaps 700) ppmv in the 21st century. The consequences to the environment of such increases are widely discussed. Thus, it is important to seek conservational procedures by which C can be sequestered from the atmosphere. Improved management of soils offers significant opportunities to enhance C sequestration.

Where soils are left to recover without applications of good management practices, it can take up to 100 yr for a new equilibrium to be established when SOM has been degraded to the steady-state level (Dick et al., 1998). The recovery is strongly influenced by temperature. Liski et al. (1998) estimated that it takes about 2000 yr to restore the SOC after forest fires in boreal forests. Batjes (1999) reviewed how management practices, organic amendments, and appropriate applications of fertilizers and water can enhance the C sequestration by soils.

On the basis of the data we have cited, the annual emissions of C from soils is of the order of 10 times greater than those from fossil fuels. Thus, any increase in sequestration of C by soils will decrease the accumulation of CO_2 in the atmosphere. According to Batjes (1999), restoration of degraded soils could increase C sequestration by 0.65 to 1.9 Pg C yr^{-1}. Improved management of nondegraded agricultural lands and improved management of extensive grasslands and forest regrowth could increase sequestration by 0.14 to 0.42 and 0.09 to 0.26 Pg C yr^{-1}, respectively. Swift (2001) pointed out that it will not be possible to make accurate estimates of the C sequestration potentials of soils until more is known about

- the ways in which organic debris is transformed in different soil environments
- the compositions and associations of the transformed products
- the mechanisms by which these are conserved in soil environments
- conservation practices that will decrease their susceptilibities to decomposition processes

Thus, the modern concerns about global warming are focusing attention on new aspects of the importance of SOM. It is now obvious that we need to know in detail the processes that lead to the conservation of C in the soil environment, and to obtain a better awareness of the structures and the interactions of the organic components of soils.

CLASSICAL DEFINITIONS

The term humus, as used above, refers to the components of SOM that have been so transformed that they bear no morphological resemblances to the materials of origin. Hayes and Swift (1978) subdivided these humified materials into:

- amorphous, polymeric, brown humic substances, which are differentiated on the basis of solubility properties into humic acids (HAs), fulvic acids (FAs), and humins
- compounds belonging to recognizable classes, such as polysaccharides, polypeptides, and altered lignins—these can be synthesized by microorganisms or can arise from modifications of similar compounds in the original debris

The more widely used definitions consider humic substances (HS) to be a series of relatively high molecular weight, yellow to black-colored substances formed by secondary synthesis reactions (Stevenson, 1994), and a category of naturally occurring, biogenic, heterogeneous organic substances that can generally be characterized as being yellow to black in color, of high molecular weight, and refractory (Aiken et al., 1985). These definitions are vague and, as will emerge during considerations of the material in this chapter, include statements that may not be considered valid at the present time. Nonetheless, the subdivision outlined is useful for purposes of classification.

Sprengel (1826) introduced fractionation on the basis of solubility characteristics in aqueous media. However, classification on the basis of solubility properties is purely operational and cannot be regarded as having any distinct compositional or structural significance. *Humic acids* are the components of HS that are soluble in base but precipitated in acidic media at pH 1 (Water chemists use pH 2 as the cut off, and effectively all of the precipitable components are precipitated at this pH). However, there are many nonhumic materials (e.g., some proteins) that are precipitated under similar conditions. In the classical definitions *fulvic acids* are the components of the alkaline extracts that remain in solution after the medium is acidified. Clearly that fraction would also contain many of the components of humus that belong to recognizable classes as mentioned above. Thus, the term *fulvic acid fraction* is now used to describe the materials that remain in solution in the acidified medium. The FA fractions from which the standard and reference FAs of the International Humic Substances Society (IHSS) were obtained were passed onto XAD-8 [(poly)methylmethacrylate] resin (see Fig. 1–5, **XXI**, p. 43) (Swift, 1996). The polar nonhumic (such as saccharides) components are considered to pass through the resin, and the FAs are retained by the resin (see Isolation of Humic Substances, p. 21, and Fractionation of Soil Organic Substances, p. 40). *Humin* is the component of HS that is not soluble in aqueous media at any pH value. It is a major component of the soil humic fraction and represents more than 50% of the organic

C of soil (Kononova, 1966) and more than 70% of the organic C in lithified sediments (Keil and Hedges, 1995). Often it is considered to be the component of HS in association with mineral colloids that is not extracted in aqueous base. However, lignin residues that have not been oxidized sufficiently to provide hydratable functionalities in the basic media, as well as algenans, cutans, and suberans (see p. 17), can contribute to the humin fraction (Derenne and Largeau, 2001). On the basis of considerations of solubility in aqueous media, any HS material that is extracted in an organic solvent, after exhaustive extraction in aqueous base, can be considered to be humin (see Isolation of Humic Substances, p. 21).

GENESIS OF SOIL ORGANIC MATTER

Organic residues, which enter the soil following the death of floral and faunal tissues and the applications of organic wastes, serve as a food source for the soil biomass. Under favorable environmental conditions, soil organisms begin to degrade macromolecular residues into their component monomers, and these are used as sources of tissue-building materials.

Several pathways, presented as theories, have been put forward to explain the genesis or formation of humus materials during the decay of plant and animal residues in soil. These can be categorized under two major headings. One involves humus formation as the result of the transformations of dead plant tissue, sometimes called the *degradative* approach. This is purely a biological process, and would involve modifications of relatively intractable plant components, such as lignins, suberins, and cutins (Hatcher and Spiker, 1988). The other involves biological processes followed by a chemical or *abiotic synthesis* process.

In a stable ecosystem, such as a forest or grassland, equilibrium exists between the formation and decomposition of humus (Hayes and Swift, 1990). If such an equilibrium did not exist, and if accumulation should exceed decay, the world would be deeply buried in humus residues (Jenkinson, 1981). Soil humus levels, and hence the rates of genesis and decay, are strongly influenced by the amounts and kinds of clay minerals and other inorganic colloids, and by the rainfall and temperature regimes. It is known that the constituents of different residues decompose at different rates, thereby giving rise to several identifiable stages of decomposition (Stevenson and Cole, 2000). Unless sterically protected, labile substances, such as simple sugars, amino acids, most proteins, and certain polysaccharides, decompose rapidly. Subsequent stages involve attack on organic intermediates and on newly formed biomass tissues. In the final stages, more resistant macromolecules, such as the polysaccharide cellulose, lignins, and some proteins, undergo more gradual decomposition, and other organic residues, such as cutins, long chain aliphatic hydrocarbons, fatty acids and esters, are more resistant still. During these biotransformation reactions, fully oxidized C (CO_2) is released. Reduced inorganic S and N are utilized as electron donors in metabolic synthesis, and thereby fully oxidized S and N are released. Organic P is hydrolyzed to yield oxidized inorganic P. This set of reactions forms the end-segment of the biogeochemical cycling of C, N, P, and S. These overall soil bioreactions are the basis on which bioremediation objectives are fashioned. Terrestrial ecosystems, to which contaminants enter, are managed in an attempt to integrate the contaminants into natural biogeochemical cycles where they are rendered innocuous by the ensuing decomposition and incorporation of by-products into target soil humus or biomass, and into plant tissue.

The relative abundances of ^{13}C and ^{15}N in SOM are useful indicators of the origins and the genesis of the materials. Differences in the natural abundance of ^{13}C between C_3 and C_4 plants have been used as in situ labeling of organic matter for the determination of organic C turnover, or of SOM dynamics (Balesdent et al., 1987, 1988; Martin et al., 1989). Higher plants that follow the C_3 pathway (Calvin cycle) discriminate against $^{13}CO_2$ during photosynthesis to a greater extent than do plants that follow the C_4 pathway (Hatch–Slack cycle). As the result of this, C_4 plants such as corn (*Zea mays* L.) and warm season grasses have $\delta^{13}C$ values between –9 and –17‰, with most values averaging –12‰. C_3 plants such as soybean [*Glycine max* L. (Merr.)], wheat (*Triticum aestivum* L.), cool season grasses, and forest trees have $\delta^{13}C$ values ranging from –23 to –34‰, and most have $\delta^{13}C$ values of the order of –26 ‰ (Deines, 1980).

Data for $\delta^{13}C$ values are obtained from the relationship:

$$\delta^{13}C = 1000(R_{sample}/R_{standard} - 1)$$

where R is the ratio of $^{13}C/^{12}C$. Pee Dee Belemnite (PDB) serves as an original standard, and urea, with a $\delta^{13}C$ value of –18.2‰ can serve as a working standard. The same formula applies for determinations of $\delta^{15}N$ values where R is the ratio of $^{15}N/^{14}N$. The standard for $\delta^{15}N$ is atmospheric N_2 ($\delta^{15}N = 0$), and acetanilide ($\delta^{15}N = -1.1$) can be used as a working standard.

Clapp et al. (1997) listed data for $\delta^{13}C$ values for several soils, coals, and for the HAs, FAs, and XAD-4 acids (see Modern Fractionation Procedures, p. 42) fractions isolated from soils and waters. The values could be related to the vegetation, and the data reflected changes from C_3 to C_4 plants in cropping systems (see also Spectroscopic Methods, p. 90). The fractions whose origins were microbial were slightly more enriched in ^{13}C, and that would suggest a slight preference for ^{13}C in synthesis processes involving soil microbes. Based on natural ^{13}C abundance data, Huggins et al. (1998) estimated soil organic C decay rates of 0.011 yr^{-1} for C_4 derived plants (corn) and 0.007 yr^{-1} for C_3 plants (soybean). Humification rates of 0.16 and 0.11 yr^{-1} were calculated for corn and soybean, respectively. Annual average C additions for maintenance of SOC in cultivated soils were estimated at 5.6 Mg C ha^{-1}.

Clapp et al. (2000) and Layese et al. (2002) used the natural abundance of ^{13}C to study how the dynamics of the soil C were influenced by different management practices. Corn was grown continuously for 13 yr followed by soybean for 4 yr, and the crop residues were either removed from or returned to the soil. Tillage operations were either no-till (NT), moldboard plow (MB), or chisel plow (CH). Based on the $\delta^{13}C$ data it became evident that, except at the surface (0–5 cm) in the case of the NT and CH with residue added, and at the 25- to 30-cm depth in the case of the MB with the residue removed, the SOC within the profile was predominantly relic C. Applications of N fertilizer could not sustain SOC without residue return.

A discrimination between lighter ^{14}N and the heavier ^{15}N isotopes occurs during biological and chemical processes (Delwiche and Steyn, 1970), and this leads to an enrichment of ^{15}N in the unreacted fraction of the substrate. Transformations of plant and soil N are accompanied by an isotope effect in which ^{14}N is mineralized preferentially (Sutherland et al., 1993). The mineralized N is susceptible to leaching loss and plant uptake. The unreacted portion of the substrate (i.e., organic soil N) is enriched in ^{15}N. Thus, crops can have a $\delta^{15}N$ value that is lower than that of the total N pool (Sutherland et al., 1991).

Létolle (1980) provided a chart that indicates the spread of abundances of $\delta^{15}N$ values for terrestrial materials. Of relevance to our considerations is the range for animal manures, from +5 to +15‰, that for organic matter in soils is in a range between +4 and +20‰, and commercial fertilizers range from –4 to +4‰. There are limitations, however, in interpretations of data because the range of $\delta^{15}N$ values do not have distinct boundaries. It is possible, for example, to have a sample with a $\delta^{15}N$ value of +6% that is made up of 50% N with a $\delta^{15}N$ of 0‰, and 50% with a $\delta^{15}N$ of 12‰. Such considerations are important where N inputs are mixed, for example mixtures of fertilizers and animal manures.

Hayes (1996) obtained $\delta^{15}N$ data for the HAs, FAs, and XAD-4 acids (see Modern Fractionation Procedures, p. 42) isolated from the drainage waters from two English soils, one of which, a Pelostagnogley (with 33% fine silt and 37% clay), was in long-term pasture and the other, also a Pelostagnogley (39% silt and 54% clay), was in long-term cultivation, under wheat. The $\delta^{15}N$ values for the drainage waters (which had passed through 0.85 m of the pasture soil) decreased in the order: XAD-4 acids > HA > FA, and the values for the HA and XAD-4 acids (2.8–3.2‰) were significantly greater than those for the FAs (1.2‰). That could suggest that the commercial fertilizers added to the grassland soil influenced the N in the FAs. The higher values for the HAs and XAD-4 acids could reflect inputs from grazing animal manure. There were differences for the fractions isolated from the runoff waters. The $\delta^{15}N$ values for the HA (1.6‰) and FA (0.8‰) would indicate significant contributions from commercial fertilizers. The XAD-4 acid (3.6‰) had inputs that diluted the influences of commercial fertilizers (applied at a rate of 400 kg N ha^{-1} yr^{-1}). The order for the samples isolated from the waters which had drained (≈0.8 m) through the cultivated soils was: XAD-4 (3.0‰) > FA (1.5‰) > HA (1.0‰). That data would suggest an influence from inputs from fertilizers. The order for samples isolated from the drainage waters from an imperfectly drained brown forest soil in a 3-yr grass–3-yr arable rotation was: HA (+4.7‰) > XAD-4 (+3.6‰) > FA (+3.5‰). These values were different than those for the arable and grassland soils, and are also considered to reflect inputs from fertilizers.

The data for the HAs, FAs, and XAD-4 acids isolated from the grassland and cultivated soils showed some differences from the similar samples isolated from the drainage waters. These fractions were isolated from the soils at pH levels of 7, 10.6, and 12.6. The XAD-4 acids had significantly higher $\delta^{15}N$ values than the HAs and FAs, but the highest values of all were for the hydrophobic extracts that were not recovered from the XAD-8 resins in NaOH (see Modern Fractionation Procedures, p. 42). These fractions were considered to be more closely related to the parent (plant or microbial) materials than the HA, FA, and XAD-4 acids isolated.

The limited data we have for $\delta^{13}C$ and $\delta^{15}N$ values for SOM fractions suggest that uses of isotope abundance techniques can have worthwhile applications in studies of the genesis of SOM components.

Genesis of the Humic Components of Humus

In his treatise "Humic Substances—in Search of a Paradigm," Wershaw (2000) stated that after 100 yr of research we are no closer to a single structural diagram for humic acid than we were at the beginning. Burdon (2001) reviewed structures proposed for HS and concluded that there is no biological or chemical logic to suggest that the structures would form in the soil environment. Burdon asked, "why would a microorganism expend energy and resources making a material that it has no use for? Any organism that did this would become extinct because of competition by organisms that did not waste energy and resources in this way." Then he went on to review the logic of the theories that have had support with regard to the genesis of humus substances. These variously suggest that humus materials could arise from:

- chemical reactions (in which case evolutionary pressures would not apply)
- biological processes that do not have a purpose in so far as microorganisms are concerned
- processes that involve degradations of the substrate materials but not the genesis of complex structures

His first two considerations can be related to those of Trusov (1917), who considered humification to take place by:

- decomposition of plant residues and the biosynthesis of simple aromatic compounds
- microbial oxidation of the latter to form hydroxyquinones
- condensation of the quinones into dark-colored (humic substances) products

It is appropriate to consider chemical reactions in tandem with biological processes. The biological contribution would release the interactive chemical species. The Maillard reaction (Maillard, 1912, 1916, 1917) is one example of how this might occur. It involves the reaction of amino acids with reducing sugars to give a product (melanoidin) with some properties similar to HAs. However, as Burdon (2001) pointed out, there would not be sufficient concentrations of sugars (or of amino acids) in the soil solution to allow the Maillard reaction to proceed to any significant extent (in terms of the residence time of the solution in most soil environments). Also, because the Maillard reaction proceeds best under conditions of high pH, alkaline soils should have more humic materials (than acid soils), but this is not the case. In addition, although, in terms of solubility and titration characteristics, humic-like products from the Maillard reaction can have properties similar to those of HS from soil, it is now known that the compositions of Maillard and humic products are significantly different. Early cross polarization magic angle spinning (CPMAS) ^{13}C-nuclear magnetic resonance (NMR) studies by Benzig-Purdie and Ripmeester (1983) suggested strong similarities between melanoidins and SOM, but subsequent ^{13}C-(Ikan et al., 1992) and ^{13}C- and ^{15}N-NMR spectroscopy studies (Benzig-Purdie et al., 1992; Knicker and Lüdemann, 1995; Knicker, 2000a; Kögel-Knaber, 2000) showed that the compositions of the melanoidin and organic matter materials are sufficiently different so that it is unlikely that melanoidin formation is an important contributor to humus genesis. It is possible, however, that melanoidin formation could contribute to the formation of humic-like substances in composting operations when the temperatures of the media are significantly raised. The work of Poirier et al. (2000, 2002) suggests that melanoidin-type substances were present in the humin fraction of a ferralitic soil from southwestern France and in the deep sample of a Congo soil where the C was stable (8300 yr old). In the latter instance the melanoidin-type substances compose 5% of the total soil C. However, we await convincing evidence that these materials were not artifacts of the drastic base and acid hydrolyses used in their isolation.

The ligno-protein concept of Waksman and Iyer (1933) is in the same category. It proposes that oxidized lignin when reacted with protein material will form a complex that has resistance to microbial decomposition and has properties similar to HAs. Hayes (1960) made an oxidized lignin–casein complex similar to that described by Waksman and Iyer, but when the complex was subjected to differential thermal analysis the thermogram had all the features of the oxidized lignin and of the casein, and was very different from that of a HA from a Sapric Histosol.

Lignin provides a major source of material for HS genesis. It has a degree of resistance to biodegradation, but white rot fungi, and to a lesser extent brown rot fungi, can degrade it. The degradation process is complex and not yet fully understood (Vicuna, 2000). Three types of oxidative enzymes (lignin peroxidase, manganese peroxidase, and laccase) are considered to be involved in the degradation process, and the same enzymes are involved in the synthesis. These enzymes will form polymers from lignin precursors in in vitro experiments. In the case of lignin peroxidase, at least, these synthetic polymers will degrade to small molecules (Nutsubidze et al., 1998). The question then arises, could humic-type molecules be synthesized in the soil environment from lignin-type precursors? Burdon (2001) was emphatic that this would not happen. Should polymerization take place, then the products formed would be lignin-like, just with different ratios of the types of bonds that occur in lignin, and significantly different at least from the HA and FA components of HS. That would not, however, exclude the synthetic polymers from being components of the humin fraction.

Wershaw (1993, 1994) outlined the process of enzymatic depolymerization by activated oxygen free radicals (e.g., superoxide radicals) of species such as lignins and tannins, and he noted how microorganisms also secrete oxidative enzymes that depolymerize plant polymers. Oxidative transformations will give rise to a variety of reactive functionalities, and especially carbonyl/carboxyl groups. In the case of lignin, for example, the unaltered parts are relatively nonpolar (hydrophobic) and the oxidized segments are polar and hydrophilic. Thus, the transformed lignin is an amphiphile (with separate hydrophobic and hydrophilic parts) and can exhibit surface active properties.

There is a logical basis to the suggestion that quinones could give rise to humic structures. There are many sources of di- and trihydroxy benzenes in plants that could, upon biological transformations, give rise to quinone structures. Lignin is one obvious source, and Flaig et al. (1975) discussed how polyphenols from lignin degradation could give rise to humic-type products. Lignin structures (Fig. 1–2) are based on the three cinnamyl building blocks: sinapyl alcohol (**I**), coniferyl alcohol (**II**), and *p*-coumaryl alcohol (**III**). In the hypothetical sketch of lignin structure shown (**IV**), cleavage of the ether linkage to the 3C side chain and demethylation would give a 1,2-dihydroxybenzene (catechol) structure. Sinapyl and coniferyl units are contained in angiosperm lignin, and the coumaryl unit is contained in the lignin

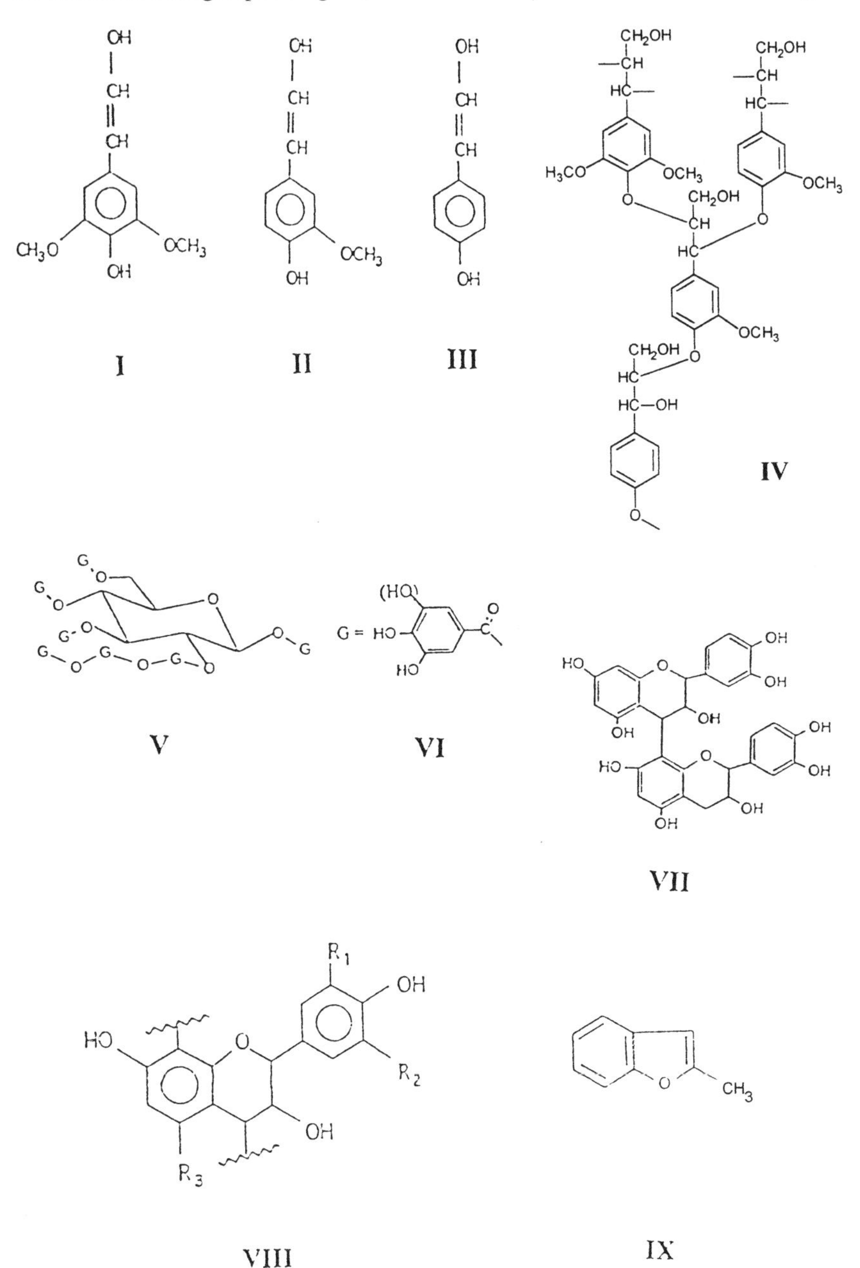

Fig. 1–2. Structures relating to precursors of humic substances from lignin and its degradation products (**I–IV**), hydrolyzable tannins (**V**, **VI**), condensed tannins (**VII**), proanthocyanidin (**VIII**), and phlorotannins (**IX**).

of nonwoody tissues. Another potential source of quinone structures is hydrolyzable tannin (**V**), which upon hydrolysis releases gallic acid (**VI**), for which the single bond from the carboxyl group is to OH, and this readily decarboxylates to 1,2,3-trihydroxybenzene. Compound **VII** is an example of a condensed tannin, and is an oligomeric proanthocyanidin (PAC), structure **VIII**, where R_1, R_2, R_3 is H or OH. One of the degradation products of the oligomer will have the 1,2-dihydroxybenzene structure, which can oxidize to a quinone.

Colored materials from catechol (Hänninen et al., 1987) have some properties that are similar to those for humic materials, but there are significant differences as well. The ^{13}C-NMR spectra reveal the differences. One of the differences is a deficiency in aliphatic components. However, as will be shown in the NMR Studies of Composition and Associations section (p. 97), we consider that humus and the HS constituents are associations of molecules, and so the absence of significant aliphatic parts should not rule out the products from catechol as components of humus. Perhaps more relevant to considerations of the contributions of polyphenols is the fact that the oxidation of the synthetic polymerization products does not give the range of benzenepolycarboxylic acids that is characteristic for humus substances subjected to the same degradation procedures (Glaser et al., 1998).

Soil inorganic components can catalyze the formation of humic-type substances from the phenolic compounds that can arise from the decomposition of organic residues. Clay-sized minerals (e.g., chlorites, smectites, vermiculite, kaolinite, halloysite, and dickite); primary minerals (e.g., olivines, pyroxenes, amphiboles, micas, and feldspars); and oxides or oxyhydroxides of Fe, Al, Si, and Mn differ in their catalytic capabilities to promote oxidative polymerization of phenolic compounds. Yields of humic-type substances from clay-catalyzed condensation (polymerization) of phenolic compounds, as a percentage of the total amount of compound added, ranged from 46.4% for Ca-montmorillonite to 8.9% for quartz (Wang et al., 1986). An abiotic polymer could form by the oxidation with air or with metallic cations (Fe^{3+}, Mn^{4+}) of 1,2,3-trihydroxybenzene structures (Shindo and Huang, 1984). The concentrations of functional groups in the humic-like substances produced in this way fall within the range of those found in natural humic substances. The process requires two reactions (degradation and synthesis), and the limiting factor could well be the concentration of polyphenols in the reaction medium. However, the fact that reactive components can concentrate on a soil mineral surface may well favor this process as a plausible mechanism for the genesis of some components of humus.

The consensus of evidence that is emerging would suggest that the humic fraction of SOM is a mixture of components of plant and microbial origins, and it seems certain that there will never be a single structure that is representative of a significant component of the humic fraction. This is implicit also in the treatises of Wershaw (2000) and MacCarthy (2001). Wershaw (2000) proposed that an appropriate approach to understanding humus genesis would be to study the chemical reactions that the chemical components of plant tissues undergo during and after senescence. In other words, emphasis might be placed on the humification process rather than on ill-defined intermediates in the continuum from well-characterized plant components to CO_2. Wershaw et al. (1996) showed that oxidative degradation of plant tissue leads to the formation of natural dissolved organic C (DOC) and humus. Their results indicated that lignin alteration products are major contributors to the genesis of humus. The alterations follow the sequence of O-demethylation and hydroxylation followed by ring fission, chain shortening, and oxidative removal of substituents. The extent to which oxidative depolymerization takes place will depend on the availability of O_2. It is not surprising, therefore, that humification is retarded under anaerobic conditions. Further studies by Wershaw et al. (1998a, 1998b) provided evidence for oxidation of lipids and cutins, hydrolysis of lignin methylether groups, lignin depolymerization, and hydrolysis of peptides. Leachates of the senescent leaves with water contained degradation products of lignin, hydrolizable tannins, nonhydrolyzable tannins, lipids, carbohydrates, and peptides. In essence, the argument of Wershaw indicates that humic components are mixtures of relatively refractive organic materials released during the transformation of organic residues. It is plausible to consider that these materials, which have a degree of resistance to further biological degradation, can sterically protect some more labile organic components. This concept is in line with that of Simpson et al. (2002), for which they provide evidence from NMR experiments, which will be discussed further (p. 97–102).

Senesi et al. (1996) reviewed and discussed the chemical and physicochemical data available for the evaluation on a molecular scale of the HS-like components in organic amendments of various origins and nature, and the effects of added HS-like materials on the composition, structure, and chemical reactivity of native soil HS.

Genesis of Nonhumic Components of Humus

The classical definitions explored above define soil HS on the basis of considerations of solubilities in aqueous acidic and basic media. It was pointed out that products such as saccharides and peptides would be contained in the FA fraction, and that proteins, as well as long chain fatty acids and esters, would be precipitated or coprecipitated in the HA fraction. Based on the classification outlined by Hayes and Swift (1978), such materials, formed in plants and in microbial tissues are components of humus, but not HS.

Soil Polysaccharides

Refer to the Soil Saccharides section later in the chapter (p. 103) for conformations, configurations, and structures of saccharide components. Soil polysaccharides are important components of SOM and have important roles in the soil environment. There is genetic or biological control of the synthesis of biological macromolecules such as polysaccharides, and there is convincing evidence to indicate that soil polysaccharides are mixtures of components with origins in plants and in microorganisms.

The glycosidic linkages (see Some Relevant Aspects of Sugar Chemistry, p. 103) in polysaccharides are readily hydrolyzed, enzymatically or chemically, and for that reason polysaccharides will have a transient existence in the soil unless protected. It is inevitable, however, that polysaccharides must have some steric protection in the soil, and this is likely to arise from associations with HS, with hydrophobic nonhumic components of humus, and with clays and (hydr)oxides.

Cheshire (1979) concluded that with the passage of time soil microorganisms will not produce a polysaccharide that has the same composition as that found in the soil. Thus, if the amounts of arabinose and of xylose are to be sustained in soil polysaccharides, there must be considerable inputs from plant sources. Hence, soil polysaccharides must be considered mixtures of materials from plant and microbial sources.

Peptide and Protein Materials

We still await the isolation of a peptide or protein material from SOM, although there is abundant evidence for the presence of peptide or protein components in humus. As yet, no peptide or protein structure has been isolated and characterized, although the amino acids detected in the acid hydrolysis digests of humus materials, and the evidence for enzymatic activities and from NMR studies (see Soil Peptides, p. 119), leave no doubt about their presence. The evidence we have suggests that these are mixtures of plant and of microbial compounds. However, the absence in soil hydrolyzates of hydroxyproline (Burdon, 2001), which accounts for up to 25% of the amino acids in plant cell walls (McNeil et al., 1979), favors microbial origins for the peptide components of humus.

Sporopollenins

Derenne and Largeau (2001) provided a succinct review of the origins, aspects of the morphologies and compositions, and resistances to biological and chemical degradations of sporopollenins. These are the macromolecular structural components of the outer walls (exines) of lower plant spores, and of higher plant pollen grains. Because of their persistence, they are among the most widely occurring plant fossils. However, some have greater resistance to biodegradation in the soil than others. These must be regarded as minor components of the SOM of most soils, and would be components of the humin fraction.

Tannins

We have tended to underestimate the probable contribution of tannins to SOM, and their possible role in the genesis of HS. These are fourth in order of abundance in terrestrial plants (after cellulose, hemicelluloses, and lignins), and may be placed in three structural categories, hydrolyzable tannins (**V**, Fig. 1–2), the PACs polymers (**VIII**), and the phlorotannins (PTs, **IX**). The ester linkages of the hydrolyzable tannins are readily cleaved by esterases, and are thought not to survive in the soil environment. There can be more than 100 units in the PACs, but for the most part there are 10 to 40 units in the structure. In most cases there are about 10 units in the structures of PTs, but molecular values of several hundreds have been reported (Ragan and Glombitza, 1986). Wilson and Hatcher (1988) referred to the large level of PACs in barks isolated from brown coals, and a communication by Joll et al. (2001), using tetramethylammonium hydroxide (TMAH) thermochemolysis, suggested a substantial tannin presence in aquatic HS.

Algenans, Cutans, and Suberans

The terms *algenan*, *cutan*, and *suberan* were introduced to describe the locations of the materials (i.e., in algal cell walls, cuticular envelopes, and suberized layers, respectively). These substances are insoluble in aqueous media, and are highly resistant to chemical degradation. Reference to Nip et al. (1989), Tegelaar et al. (1995), and Derenne and Largeau (2001) will introduce the reader to the distribution and relative importance of these materials in soils and sediments.

Algenans consist of a macromolecular network of long, generally unbranched hydrocarbon chains. Ether linkages might lead to crosslinking. It would appear that the steric protection provided by the hydrocarbon network inhibits attack on potentially labile units (cross) linking the structures. There is not, however, definite evidence of a contribution of algenans to the composition of SOM, but it would appear to be inevitable that these are present.

Cutans and suberans are widely distributed in the plant kingdom. Their structure is highly aliphatic (Poirier et al., 2000), but there is a possibility that these could contain some aromatic moieties (McKinney et al., 1996). These materials are widely preserved in fossils, and although there is no morphological evidence for their presence in SOM, it would appear to be certain that these are among its important components (in terms of reactivities). Such hydrocarbon materials have a high resistance to biological degradation in the soil environment. Baldock et al. (1997) reviewed NMR evidence that indicates increases in aliphaticity with increasing decomposition of SOM, and there is abundant evidence from pyrolysis studies for long chain hydrocarbon structures (e.g., van Bergen et al., 1997, 1998a, 1998b; Huang et al., 1998). Substantial amounts of such substances were detected in the insoluble and nonhydrolyzable components of the SOM in a forest soil.

Lipids

Lipid is a term used to describe a group of natural substances that are insoluble in water but soluble in hydrocarbons. It includes fats, waxes, and natural hydrocarbons. In general, biochemists reserve the term lipid for naturally occurring compounds that give fatty acids when hydrolyzed.

Lipids are outside of the HS classification, and their removal should be regarded as a prerequisite for studies of humic compositions. However, although some of the lipid contents are removable in nonpolar solvents, it is extremely difficult to remove all of the lipid substances because they can form strong associations with other components in the operationally defined humic fractions, and especially the HA and humin components.

Peats are especially rich in lipid materials, and it is these which give the oily appearance to the peat residues after extraction with NaOH. These are also present in mineral soils, and earlier studies by Meinschein and Kenny (1957) fractionated a benzene-alcohol soil extract on a silica gel column prewet with *n*-heptane. Four fractions were recovered by successive elutions with *n*-heptane, carbon tetrachloride, benzene, and methanol. The eluates in *n*-heptane and carbon tetrachloride were mainly saturated hydrocarbons. The wax, recovered in benzene, consisted of 5% of the extracted materials. The wax esters were converted to saturated hydrocarbons by high pressure hydrogenation, and mass spectrometric analyses. The waxes ranged from C_{36} to C_{52}, and the even C numbered products were 90% of the total.

When a neutral wax from an Australian soil (Butler et al., 1964) was saponified (hydrolyzed in base; see Information from Hydrolysis Processes, p. 57), acids (identified as esters, using gas chromatography) ranging from C_{12} to C_{30} were identified, with the C_{22} (13%), C_{24} (22%), and C_{26} (21%) acids in greatest abundance. The wax also contained a complex mixture of hydrocarbons.

A wax mixture was isolated from the surface organic mat of a soil in which biological activity had been inhibited by applications of copper (Himes and Bloomfield, 1967). A structure of the type $CH_3(CH_2)_xCOO(CH_2)_yCH_3$ was proposed, and the values of x and y suggested that the acid components were n-C_{14}, n-C_{16}, and n-C_{18}, with the primary alcohols as n-C_{26}, n-C_{28}, and n-C_{30}.

Reference was made to the presence of long chain hydrocarbons in the section above on algenans, cutans, and suberans. We were alerted to the possible origins in algae (Largeau et al., 1984) and in plant cuticles (Nip et al., 1986) of long chain parafinnic substances in soil. The reader is referred to Algenans, Cutans, and Suberans (p. 17) for further references to relevant work in this area.

Applications of pyrolysis (Py) analytical procedures (reviewed by Bracewell et al., 1989) have significantly advanced our awareness of long chain hydrocarbons, esters, acids, and alcohols in the soil environment. Alkanes, alkenes, fatty acid n-alkyl esters, and alkyl aromatics (aromatic rings linked covalently to aliphatic chains) were among the products identified by Py gas chromatography–mass spectrometry (GC–MS) in studies by Schnitzer and Schulten (1995) of whole soils and of different sized fractions from soils. Huang et al. (1998) have performed Py–GC–MS and in-source Py-electron impact (EI) MS studies on the dissolved organic matter (DOM) and on its parent SOM in the litter (L_f), as well as in the humified organic (O_h) horizons, and in the mineral soil of a peaty gley upland soil at the summit of the Great Dun Fell (845 m above sea level), Cumbria, England. Some of the samples were methylated using tetramethylammonium hydroxide (Hatcher et al., 1995). The major pyrolysis products included compounds derived from carbohydrates, lignin, and fatty acids (some with >50 C atoms). The fact that fatty acids were especially abundant in the O_h horizon is significant. Their abundance in the O_h horizon, and the corresponding decrease in lignin-derived components in that horizon was considered to reflect the extent of resistance of the acids to biological degradation. The C_{22} to C_{36} n-fatty acids were considered to arise from leaf waxes, and the C_{14} to C_{18} and unsaturated fatty acids could originate from both plants and microorganisms. The very long chain fatty acids (C_{43}–C_{53}) were considered to have likely origins in mycobacteria. The n-fatty acids in the DOM were shorter (maximum C_{22} with C_{16} dominant) indicating that bound lipids of longer chain lengths are less likely to be mobilized by water leaching.

Glomalin

The work by Wright and colleagues indicates that soil aggregation is largely dependent on “glues” called glomalin. Wright et al. (1996) showed that a protein-type material present in the range of 4.4 to 14.8 mg g^{-1} of soil had the characteristics of glomalin produced by the hyphae of arbuscular mycorrhizal fungi (AMF).

This proteinaceous material is insoluble in water, but it can be solubilized by 20 or 50 m*M* citrate at 121°C when present in both soil and in the hyphae of AMF (Wright et al., 1996, 1998; Wright and Upadhyaya, 1996). There was evidence to indicate that an oligosaccharide (a saccharide containing up to 10 sugar units) is linked to protein from the hyphae and the soil, and so the materials were considered to have properties of glycoprotein materials synthesized by the *Glomales* taxonomic order. The presence of N-linked oligosaccharides on the AMF hyphal protein confirmed that glomalin is a glycoprotein (Wright et al., 1998).

The AMF form a symbiotic relationship with 80% of vascular plants. These fungi adhere to the plant roots, and their hyphae extend far beyond the roots and penetrate to areas of the soils that the roots cannot reach, and extract P and other nutrients for the plants in return for C for growth of the fungus. It would appear also that the glomalin can protect the hyphae during dry periods. Also, the evidence that is available suggests that the glomalin enters the soil and attaches to minerals and the other SOM components. Growth of AMF occurs when there are active roots to colonize.

The work of the Wright Group has shown that when soil management is changed from continuous cultivation to no-till, significant increases in SOM, glomalin, and in stable aggregates are observed even in 3 yr (Wright et al., 1999; Wright, 2001). Glomalin has been found in significant concentrations in numerous soils (Wright and Upadhyaya, 1998), and Rillig et al. (2001) showed that AMF make a large contribution to soil C pools in tropical forest soils. Wright and Upadhyaya (1998) found that glomalin could be related to the organic matter of Scottish soils, and aggregate stabilities of these soils could be related to the glomalin contents.

Black Carbon or Char

Black carbon (BC) is the term used to describe forms of pyrogenic C, such as char, soot, charcoal, graphite and other products of incomplete combustion of organic substances (Goldberg, 1985). The term *char* is used to describe a material that forms at 370°C in the pores of spent C in the process of C generation. Black C can be visualized as stacked layers of fused aromatic structures, and it is thought that poorly oxygenated plant components, such as waxes, are major contributors to its genesis. The structures can be related to graphite (but not so ordered), in which C atoms are in fused hexagonal arrays with the layers 3.35 Å apart and held together by weak van der Waal forces. Because of its high resistance to thermal and chemical oxidation, BC can be considered to be an important sink for C in soils and sediments (Kuhlbusch and Crutzen, 1995; Skjemstad et al., 1996; Bird and Gröcke, 1997; Derenne and Largeau, 2001; Swift, 2001), and may be a significant contributor to the missing C sink in global C accounting (Kuhlbusch and Crutzen, 1995). However, it must not be thought that BC is inert to oxidation processes in the soil environment. Cookson (1978) outlined mechanisms by which oxidation of peripheral units in the network arrangements can take place to give rise to carboxyl groups, phenolic hydroxyls, and quinone type carbonyl groups. Thus, the presence of such functionalities can confer degrees of solubility to BC, and although for the most part BC might be considered a component of humin, it is also possible that it can contribute to the HA fraction.

Black C can contribute up to 45% of the total organic C in some soils (Skjemstad et al., 1996; Glaser et al., 1998; Schmidt et al., 2001). In general its concentrations increase with depth, and this is attributed to erosion of surficial particles, and burial by earthworm and termite activity (Saldarriaga and West, 1986; Glaser et al., 2000); ^{14}C data indicate that age increases with depth. Values obtained from $\delta^{13}C$ measurements have shown that the BC can be related to the plant materials of origin (Bird and Gröcke, 1997). Although some degradation of BC can take place in well-aerated soils (Bird et al., 1999), the incorporation of BC into soils and sediments can be considered an important mechanism of C stabilization (Schmidt and Novack, 2000).

The burning of vegetation and crop residues in the field has variously been used as an agronomic practice, and slash and burn techniques are still used in many tropical and subtropical areas. Sanford et al. (1985) found charcoal to be common in soils of mature rain forests in the north-central Amazon basin, and radiocarbon dating of the charcoal indicated that numerous fires had occurred since the mid-Holocene period. In a later study, Sanford and Horn (2000) described how large areas of lowland tropical rainforests have been burned in the past 6000 yr. The work of Skjemstad et al. (1996) suggests that the burning of vegetation by accident or by design contributed significantly to the C in many Australian soils. It is probable that char or BC contributes significantly to the black color in Mollisol soils in the midwestern United States.

Globally the burning of vegetation has been calculated to generate between 40 and 600×10^{12} g of BC, and between 7 and 24×10^{12} g are considered to be contributed from the burning of fossil fuels (Kuhlbusch and Crutzen, 1995). The small BC particles are readily transported by wind, and if not fixed in the soil, these can be transported in runoff water. Eventually, all of the BC will be fixed in soils and in sediments, and transport to the ocean sediments can take place in aerosols (Massiello and Druffel, 1998). Hedges and Keil (1995) estimated that the annual storage in ocean sediments can be up to 160×10^{12} g.

The quantitative estimation of BC in soils presents analytical difficulties, and hence the accuracy of the estimates referred to above may be questionable. In general BC is isolated from soil samples by using chemical (such as H_2O_2, $K_2Cr_2O_7/H_2SO_4$, hot concentrated HNO_3; Bird and Gröcke 1997; Verardo, 1997; Glaser et al., 1998), photochemical (using high-energy ultraviolet [UV] irradiation; Skjemstad et al., 1993, 1996), or thermal oxidation procedures in a stream of O_2 (Kuhlbusch, 1995; Gustaffson et al., 1997, 2001). These procedures seek to oxidize all organic residues other than the BC. Before oxidation, the soil samples are usually demineralized using HCl/HF treatments. Ongoing studies by Salloum and Hatcher (2004) suggest that such oxidative procedures lead to significant overestimations of BC and inaccuracies in the estimations of the contributions of BC to global C budgets. Salloum and Hatcher used sodium hypochlorite, which degrades to soluble products the lignin contribution at 130 ppm to the ^{13}C-NMR spectrum (see the Composition section under NMR Studies, p. 97). The highly condensed aromatic residues resist the treatment. Analyses of the results when this procedure and thermal oxidation and UV photooxidation methods were used on organic matter samples from peats, soils, marine sediments, shales, coals, and pure charcoal indicated estimates from the thermal and UV methods were significantly greater than those from the treatment with sodium hypochlorite. These thermal and UV oxidation procedures were considered to be unable to remove all nonpyrogenic C before analysis.

Mao and Schmidt-Rohr (2003) introduced a solid-state ^{13}C-NMR method that distinguishes various types of aromatic residues such as those of lignin from the fused rings of charcoal (see NMR Studies of Composition and Associations, p. 97).

ISOLATION OF HUMIC SUBSTANCES

Some soil scientists argue that the use of chemical procedures to isolate organic matter from soil alters the compositions of the organic materials, so subsequent studies are performed on artifacts. The counter argument is that the organic matter components of soils constitute such a gross mixture of materials that it is impossible to get an awareness of the composition of the mixtures without isolating the components from their natural environments. This argument is supported by the fact that the structures of the macromolecules of life, such as the proteins, nucleic acids, and polysaccharides, would not have been established without isolating these from their biological habitats. It is true, of course, that some damage can be caused where due attention is not given to the properties of the solvents and of the solutes. Those who believe SOM should be studied in its soil habitat are sometimes grouped as "lumpers", and those who use the isolation and fractionation approach are grouped as "splitters". Our approach is that of splitters. The advances being made in applications of NMR to studies of HS may well enable progress to be made in the future using the lumpers approach (see Spectroscopic Methods, p. 90).

Principles Involved in the Isolation of Humic Substances from Soil

Hayes (1985) discussed procedures and the principles involved in the isolation of HS from soils. To isolate HS from soil it is necessary to bring them into solution. Aqueous alkali, introduced by Achard (1786), are the traditional solvents used, but consideration is also given to applications of organic solvents, especially for attempts to isolate the humin components. To dissolve a solute, cavities or holes (lacunae) must form in the solvent to accommodate the solute molecules. Solubilization depends on how readily the cavities are formed and on the ability of the solvent molecules to solvate (i.e., surround) solute molecules. The self association, or attractive forces, between solvent molecules in the liquid phase must be overcome to create openings for the solute molecules. The solute dissolves when solute–solute attraction is of the same order as that between solvent molecules, but dissolution will not occur if solvent–solvent interactions are significantly greater than those of the solute (or vice versa). In addition, secondary forces (in contrast to ionic- and covalent-bond forces), which include in all cases multipole interactions and dispersion forces, and in some cases H bonding, must be overcome before dissolution will take place. The influences of the secondary forces increase as the size of the molecule increases.

It is appropriate now to consider the forces that are holding molecules together, and the properties that can disrupt these forces. These are discussed in some detail by Hayes (1985), and only brief mention is made here.

Interparticle Forces

There are in principle an infinite number of multipole interactions, but in principle only the monopoles, dipoles, and quadrupoles are important. When the electronegativities of two atoms that are bonded are different, the bonding electrons are held closer to the more electronegative atom and the bond is *polar covalent*. Should the difference in electronegativity be great, then an *ionic species* is formed as the electrons migrate to the highly electronegative atom. The *monopole* interaction force is equal to the total net charge on the molecule, and the interaction energy of the ionic species is dominated by the monopole interactions.

When two charges q and $-q$ are separated by a distance r, the magnitude of the dipole is qr. The dipole has magnitude and direction, and the dipole moment is expressed in *debye* (D) units where D = 3.338×10^{-30} coulomb•m.

In the absence of monopoles, molecules with large dipoles dominate the multipole interaction energies, and such molecules are polar. The term *polarity* does not refer directly to dipole moment, but polar molecules will solvate species that have regions of charge excess or of charge deficiency. Water, for example, is a highly polar molecule, but its dipole moment is relatively small compared with the dipolar aprotic solvents (discussed below) acetonitrile (**X**, Fig. 1–3), formamide (**XI**, Fig. 1–3), *N,N*-dimethylformamide (DMF, **XII**), and dimethylsulfoxide (DMSO, **XIII**).

When nonpolar molecules are exposed to an electric field, *induction forces* arise, resulting in dipolar properties. In effect, any polar molecule in the vicinity of another which is less polar will polarize the second molecule to give rise to *dipole-induced dipole* interactions.

Dispersion forces, known as *London* or *van der Waals forces* (although the latter are often considered to include multipole interactions and induction forces as well), are relevant to nonpolar molecules. Such molecules do not have permanent moments, but their fluctuating electron clouds give rise to what might be considered instantaneous dipoles. For example, the fluctuating dipole in Molecule A propagates a fluctuating electric field that travels outward with the velocity of light to induce a similar dipole in Molecule B. The second fluctuating dipole in B radiates a fluctuating field back to A, where it interacts with the original fluctuating dipole. The second then behaves in the same way with respect to the first. The two

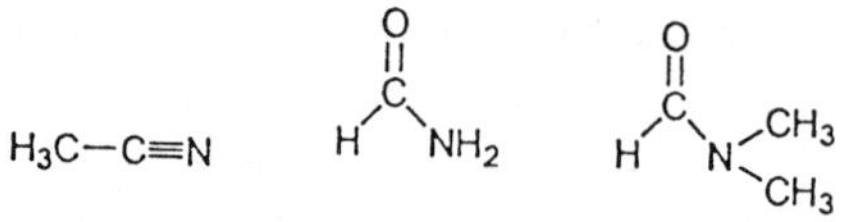

$H_3C{-}C{\equiv}N$ **X**

$HC(=O)NH_2$ **XI**

$HC(=O)N(CH_3)_2$ **XII**

$(H_3C)_2S{=}O$ **XIII**

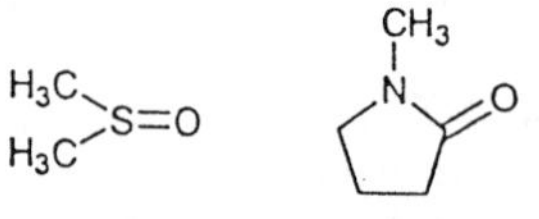

XIV

Fig. 1–3. The dipolar aprotic solvents acetonitrile (**X**), *N,N*-dimethylformamide (**XII**), dimethylsulfoxide (**XIII**), and *N*-methylpyrrolidone (**XIV**), and the dipolar protic solvent, formamide (**XI**).

dipoles give rise to an attractive interaction between the two molecules in what is called an *induced dipole-induced dipole* interaction.

Hydrogen bonding (both intermolecular and intramolecular) is a very important secondary interaction where HS, and in particular H^+-exchanged HS, are concerned. It is primarily responsible for the failure of H^+-exchanged HAs to dissolve in water. For two molecules to form an intermolecular hydrogen bond, it is necessary to have at least one electronegative atom, such as O or N on each molecule and one of these atoms must be bonded to H to constitute the donor group. The bond to the H atom in the donor group is strongly polarized. It is thus associated with a dipole moment directed along the bond pointing toward the H. The acceptor also has a dipole moment, which arises from the separation of the nonbonded electrons and points toward the nucleus of the electronegative atom. The two dipoles enter into a strong dipole–dipole interaction, which is the primary source of the energy of formation of the H bond.

Dipole–dipole interactions are strongest when the dipoles are aligned head to tail. Thus the strongest H bonds are formed when the two electronegative atoms and the H are colinear. Since the H atom has a very small van der Waals radius, the two dipoles can approach each other closely to give relatively strong bonds, having energies that can amount up to 20 kJ mol^{-1} when added to the much weaker contributions from the multipole, induction, and dispersion forces.

Structure and Some Solution Properties of Water

Because water is the major solvent used for the isolation of HS, it is important to be aware of the properties that make it a good solvent for humic preparations. Any consideration of water for use as a solvent or in a solvent system should take account of its structure and of some of its unique characteristics. Detailed information about the properties of water relevant to this topic is contained in the several volumes on *Water, A Comprehensive Treatise*, edited by Franks (1975) and referenced in Franks (1983).

Hayes (1985) reviewed aspects of water chemistry that are relevant to the topic under consideration. Calculations based on theory, where concepts of the structure of the isolated water molecule are used, suggest that the most stable form of interaction between water molecules involves the linear H bond. The estimated molar dissociation energies for such H-bonded structures range between 20 and 35 kJ, and the equilibrium O–O distances found are between 2.6 and 3.0 Å (Franks, 1983, p. 15).

Hydrogen bonding confers a considerable degree of structure to liquid water. In the "flickering" cluster model, the individual water molecules tend to form clustered structures by what is known as a cooperative interaction. It can be supposed that liquid water is composed of clusters of varying sizes in equilibrium with each other and with free unstructured water molecules. Local energy fluctuations allow the formation and degradation of H bonds and of clusters in a dynamic process in what is known as the *flicker effect*.

When soluble ions or highly polar compounds are introduced into water, solvation occurs because of associations, through electrostatic effects, between the solvent and solute species. The ordering of the solvent around the solute leads to a decrease in entropy, but this is compensated for by the increased entropy from the breakdown of the water cluster structure.

Nonpolar molecule–water interactions are different. When such molecules enter spaces between water clusters they interact more strongly with each other than with the nonclustered water molecules in the void spaces. However, the guest molecules disrupt the H bonds between the resident, nonclustered water molecules. This raises the energy of the nonbonded water, which then tends to associate with the clusters. Thus, the clusters grow at the expense of the free water, and the energy and entropy decrease as free water becomes localized in the clusters through H bonding. The result is that the nonpolar solutes tend to associate with themselves, and this effect has given rise to the concept of the hydrophobic bond between nonpolar molecules in the presence of water (Franks, 1975, 1983).

Properties of Organic Solvents

The acidity or basicity of a solvent system, the ability to hydrogen bond, the relative permittivity, and the dipole moment are important properties when considering the ability of a solvent to dissolve a solute. Peripheral properties, such as boiling point, viscosity, and density, do not necessarily affect separations, but it is very important to take these into account when designing a solvent system. A good solvent is of little value if the solute cannot be recovered from it, or recovered with compositional properties different from the material sought.

The various parameters of solvents that are relevant for the dissolution of HS are listed in Tables 1–2 and 1–3.

The relative permittivity (K_r), or the dielectric constant value, is important when considering the properties of a solvent for HS. The term *relative permittivity* refers to the ability of a solvent to decrease the coulombic field of an ion. For example, the value for water is 78.5, and this refers to the extent the field is decreased in the system, compared with a vacuum, at a distance r from the solute species. Thus, water is a very powerful solvent for salts because it greatly decreases the coulombic attractions between the oppositely charged species. Hence, these do not readily interact to form solid or crystalline salts.

In general, less polar solutes dissolve best in solvents with low K_r values, and higher values favor solvation of polar molecules. Definite trends cannot be established, however, because of the involvements of specific interactions, and especially H bonding (Snyder, 1978).

Consideration of dipole moment is important. The ability of a solvent to disrupt the ion or molecule associations in a compound depends on the extent to which it can solvate the component molecules or ions and decrease the interactions holding them together. One end of the dipole is attracted electrostatically to the ion of opposite charge, or to the region of the molecule having the appropriate charge deficit or excess. When specific solute–solvent interactions are not important, the dipole moment of the solvent largely determines the orientation of the solvent around the solute molecule, and this orientation is essential for the electrostatic solvation process. The formation of solvent shells around the molecule is essential to prevent the solute species from self associating and to allow solution to take place. Solvents (other than water) with high dielectric constants, and these include the dipolar aprotic solvents, dissolve ionic species by separating and solvating the ions.

Table 1–2. Boiling point, molar volume (V), refractive index (n), viscosity (η), density (ρ), relative permittivity (K_r), dipole moment (μ), electrostatic factor (EF), and base parameter (pK_{HB}) values for selected solvents.†

Solvent	Boiling point	V	n	η	ρ	K_r	μ	EF	pK_{HB}
	°C	$cm^3\ mol^{-1}$		cP 25°C					
n-Pentane	36	116.2	1.355	0.22	0.61	1.84	0	0	--
Diethylether	35	104.8	1.350	0.24	0.71	4.34	1.36	5.90	0.98
Formic acid	101	37.8	1.371	--	1.22	58.0	--	--	--
Ethanoic acid	118	57.1	1.370	1.1	1.04	6.13	0.83	5.09	--
Pyridine	115	80.9	1.507	0.88	0.98	12.4	--	--	1.88
Acetonitrile	82	52.6	1.341	0.34	0.78	37.5	3.84	144.0	1.05
Acetone	56	74.0	1.356	0.30	0.78	20.7	2.88	59.62	1.18
N-Methyl-2-pyrrolidone	202	96.5	1.468	1.67	1.03	32.0	--	--	2.37
Formamide	210	39.8	1.447	3.3	1.13	109.5	3.37	369.0	--
N,N-Dimethylformamide	153	77.0	1.428	0.80	--	36.7	3.82	140.2	2.06
Dimethylsulfoxide	189	71.3	1.477	2.00	1.10	46.6	4.49	209.2	2.53
Ethanol	78	58.5	1.359	1.08	0.79	24.3	1.68	40.82	--
Water	100	18.0	1.333	0.89	1.00	78.5	1.84	144.44	--

† From Taft et al. (1969), Barton (1975), Dack (1976), and Snyder (1978).

Dipolar aprotic solvents are defined by Parker (1962) as solvents that have dielectric constant (relative permittivity) values >15, and are incapable of donating H atoms to form strong hydrogen bonds. Acetone, acetonitrile (methylcyanide, **X**, Fig. 1–3), DMF (**XII**), DMSO (**XIII**), and N-methyl-2-pyrrolidone (**XIV**) are the dipolar solvents listed in Table 1–2. With the exception of acetonitrile, the solvents listed have exposed electronegative O atoms, as have the dipolar, but protic solvents formamide and N-methylformamide, which provide sites for solute and sol-

Table 1–3. Data for the Hildebrand (δ), total (δ_o), dispersive (δ_d), polar (δ_p), and H-bonding (δ_h) parameters†, and for proton donor (δ_a) and proton acceptor (δ_b) parameters‡.

Solvent	(δ)	δ_o	δ_d	δ_p	δ_h	δ_a	δ_b
n-Pentane	7.0	7.1	7.1	0.0	0.0	--	--
Diethylether	7.4	7.7	7.1	1.4	2.5	--	3.0
Formic acid	12.1	12.2	7.0	5.8	8.1	--	--
Ethanoic acid	10.1	10.5	7.1	3.9	6.6	--	--
Pyridine	10.7	10.7	9.3	4.3	2.9	--	4.9
Acetonitrile	11.9	12.0	7.5	8.8	3.0	--	3.8
Acetone	9.9	9.8	7.6	5.1	3.4	--	3.0
N-Methyl-2-pyrrolidone	11.3	11.2	8.8	6.0	3.5	--	--
Formamide	19.2	17.9	8.4	12.8	9.3	L§	L
N,N-Dimethylformamide	12.1	12.1	8.5	6.7	5.5	--	4.6
Dimethylsulfoxide	12.0	13.0	9.0	8.0	5.0	--	5.2
Ethanol	12.7	13.0	7.7	4.3	9.5	6.9	6.9
Water	23.4	23.4	7.6	7.8	20.7	L	L

† From Barton (1975).
‡ From Snyder (1978).
§ L = large.

vent H bonding. Because they are dipolar, such solvents will solvate polar organic molecules and in so doing will form tight solvent shells around the solute species.

Anions are less solvated in dipolar aprotic solvents than in water. In water and in protic solvents the anions will solvate by ion–dipole interactions on which strong H bonding is superimposed. In the cases of the dipolar aprotic solvents the anions will solvate also by ion–dipole interactions, but without the influence of H bonding. Solvation is aided by less energetic interactions arising from the mutual polarizability of the anions and the solvent molecules (Parker, 1962).

Electrostatic factor (EF) values arise from the product of the relative permittivity and the dipole moment; EF takes account of the influence of both of these properties on the electrostatic solvation of solutes.

Hydrogen bonding is an important property for considerations of solute–solvent interactions. Pimentel and McClellan (1960) classified H bonding solvents into proton donors, proton acceptors (such as keto and ether functionalities, and dipolar aprotics), proton donors and acceptors (such as alcohols, carboxylic acids, primary and secondary amines, and water), and the nonhydrogen bonding compounds (such as carbon disulfide and the paraffins).

Taft et al. (1969) defined a base parameter value, pK_{HB}, which measures the relative strength of the acceptor when a H-bonded complex is formed using any suitable hydroxyl reference acid. Values of pK_{HB} are not applicable to reference acids involved in intramolecular hydrogen bonding. The higher the pK_{HB} value (Table 1–2), the better is the compound as an acceptor in H bonding. This is taken into account in Extraction with Organic Solvents (p. 34) for the solubilization of HS.

The overall tendency of compounds to interact through dispersion forces is related to the refractive index values of the compound (Karger et al., 1973). The refractive index is the ratio of the speed of light in a medium to its speed in a vacuum, and it is related to the frequency of the light used. The greater the refractive index, the stronger the dispersion (London) interactions (The term *dispersion* derives from light propagation phenomena through the dielectric). Thus, the dipolar aprotic solvents and pyridine have the strongest influences on dispersion interactions of the compounds listed in Table 1–3. When refractive index is used in measurements of concentrations of solute, it is important to maximize the differences in the values between solvent and solute.

Low viscosity of the solvent is important for ease of handling. It is seen from Table 1–2 that, with the possible exception of DMF, there is a degree of matching of boiling points and viscosities. When mixed solvents are used, the viscosities of the mixtures are intermediate between the values for the pure components in the mixture (Snyder, 1978).

Consideration of solvent densities is important for separations using gravity. Again, mixtures can be used to regulate density because the density value for a mixture is close to the arithmetic average of the densities of the pure components in it.

Use of Solubility Parameters for Predicting Solubilities

Because solution takes place when the self attraction forces in solute and solvent molecules are of the same order of magnitude, a good solvent for a (nonelectrolyte) polymer solute will have, for example, a solubility parameter (δ) value close to that of the solute. It is often found that a mixture of two solvents, one having a δ value above and the other a δ value below that of the solute, can provide a better solvent for a particular solute than either solvent alone.

Hildebrand and coworkers introduced the concept of the one-component solubility parameter (δ) more than 50 yr ago (see Hildebrand and Scott, 1951, 1962; Hildebrand et al., 1970). It is defined as the square root of the cohesive energy density (CED) and may be expressed as:

$$\delta = (-E/V_1)^{0.5} = (\text{CED})^{0.5} \quad [1]$$

where V_1 is the molar volume of the liquid and $-E$ is the molar cohesive energy, or the molar energy of vaporization. The unit of solubility is the Hildebrand, expressed as $(\text{cal cm}^{-3})^{0.5}$, or $2.046\ (\text{J cm}^{-3})^{0.5}$.

The interactions between molecules which produce the cohesive energy characteristic of the liquid phase involve the dispersion forces, dipole–dipole and dipole-induced dipole interactions, and specific interactions, especially H bonding. If it is assumed that the intermolecular forces are the same in the vapor and liquid states, then $-E$ is the energy of a liquid relative to its ideal vapor at the same temperature. It can be described as the energy required to vaporize 1 mol of liquid to the saturated vapor phase plus the energy required for the isothermal expansion of the saturated vapor to infinite volume. Detailed discussion of the theory and derivations is given in the publications by Hildebrand and associates cited above.

The Hildebrand one-component solubility parameter, δ, is appropriate for solutions lacking in polarity and in specific interactions. This parameter, to some extent, is now replaced by the multicomponent solubility parameters that give values for each of the different interaction forces.

There are several excellent reviews of multicomponent solubility parameters, including those by Hansen (1967), Karger et al. (1973), Barton (1975), and Snyder (1978). Barton's review provides all the information necessary to familiarize the reader with the subject, and it contains a comprehensive reference list to the original work that developed the theory and techniques on which the multicomponent approach is founded.

Table 1–3 lists the Hildebrand solubility parameter δ, the total solubility parameter δ_0, and the multicomponent parameters for dispersion δ_d, polar δ_p, and H bonding δ_h forces for a number of solvents. These data are taken from the compilation by Barton (1975). He pointed out that the data become empirical when multicomponent parameters are used, and thus it is important to use a set of data that are self-consistent. Keller et al. (1971) and Karger et al. (1976) further subdivided the H bonding parameter into the acid or proton donor (δ_a) parameter and the base or proton acceptor (δ_b) parameter. Values for these are listed for some of the compounds in Table 1–3 from data provided by Snyder (1978). These data are not from the same source as those compiled by Barton (1975) and included in Table 1–3; hence, the values given for δ_h should not be compared directly with those for δ_a and δ_b.

It was pointed out that the overall δ values should be similar for solvent and solute to achieve maximum solubility. The same applies for the individual parameters δ_d, δ_p, and δ_h. Subdivision of δ in this way allows differences in solubility and solvent selectivity to be anticipated for solvents with similar polarities and similar overall values of δ.

The use of δ_a, and δ_b, as distinct from δ_h, can be a useful guide when selecting a solvent for a particular extraction process. Because maximum solubility is promoted by strong H bonding between solvent and solute molecules, solution should be greatest when the product of δ_a, (for solvent) and δ_b (for solute) (or vice versa) is a maximum, rather than when the values are equal for both solvent and solute. In the Extraction with Organic Solvents section (p. 34), consideration is given to solubility parameters in the dissolution of HS.

Solubilization of Macromolecules

The mechanisms in the dissolution of neutral polymers and polyelectrolytes are different. At low pH values (pH $\approx$ 4), H^+-exchanged HS have many of the physicochemical properties of neutral polymers having substantial amounts of H bonding. We are not suggesting that HS are polymers; we wish only to say that the ways in which the molecules associate, through H bonding for example, confers on the molecules some macromolecular properties that have similarities to neutral polymer systems. As the pH is raised, the acid groups in the HS dissociate and the molecules can assume the properties of polyelectrolytes. Therefore, it is appropriate to consider some of the features involved in dissolving neutral polymers and polyelectrolytes.

Solubilization of Neutral Polymers

The most widely used theory of polymer solutions is that initiated independently by Flory and by Huggins (Flory, 1953). This theory is based on the evaluation of the free energy change of mixing ($\delta_m G_p$) of a liquid and polymer to form a solution. Hayes (1985) reviewed aspects of the Flory and Huggins theory, and referred to deficiencies in this theory for solubilization of polymers as outlined by Barton (1975). Despite the shortcomings, evaluations of polymer-solvent parameters are still widely used.

Solubilization of Polyelectrolytes

Humic substances expand in water as the result of repulsions between conjugate bases (anions) formed after the release or dissociation of the counterions. The higher the charge density the greater the repulsion between the charges, and the greater the hydrodynamic volume.

Electrically charged, ionic species readily solvate or hydrate, and in so doing they bind solvent or water molecules because of strong monopole–dipole interactions. Smaller ions have stronger electric fields at their peripheries, and these bind solvent molecules more strongly. Such strongly bound molecules tend to be localized, which implies a loss of entropy. This loss, however, is compensated to some extent by the breakdown of water structure.

Richards (1980, p. 209) presented a description of the ion atmosphere for a polyelectrolyte in solution. He provided a clear interpretation of the electrical double-layer effects and of the mechanisms by which the presence of excess salts can depress the electrical potential and cause the highly charged polyanions to have many of the physicochemical properties of neutral molecules.

Isolation of Humic Substances from Soil

The HS components that have been isolated in significant yields are highly charged, and the bulk of the functionalities contributing to the charge characteristics are carboxyl and phenolic hydroxyl. Thus, charge-neutralizing hydrogen ions and divalent and polyvalent cations cause these materials to be sparingly soluble in water in their soil environment. Because such ions are undissociated, or only weakly dissociated from the anionic functional groups, a type of cross-linking effect takes place from inter- and intrastrand H bonding and from bridging in the cases of the divalent and polyvalent cations. The polyvalent cations can also form bridges between the humic molecules and the inorganic colloidal components of the soil, and this also complicates the processes of extraction. The net result is that we are dealing with a sparingly soluble pseudomacromolecular system. Most of the studies have used water as the solvent, but organic solvents have been used occasionally.

Criteria for Solvents and Methods for the Extraction of Soil Humic Substances

According to Whitehead and Tinsley (1964) effective solvents for the isolation of HS should have:

- a high polarity and a high dielectric (or permittivity) constant to assist the dispersion of the charged molecules
- a small molecular size to penetrate into the humic structures
- the ability to disrupt the existing H bonds and to provide alternative groups to form humic-solvent hydrogen bonds
- the ability to immobilize metallic cations

Four criteria for the ideal extraction method, as listed by Stevenson (1994) suggest that:

1. The method leads to the isolation of unaltered materials.
2. The extracted humic substances are free of inorganic contaminants, such as clays and polyvalent cations.
3. Extraction is complete, thereby ensuring representation of fractions from the entire molecular weight range.
4. The method is universally applicable to all soils.

Extraction of Humic Substances in Water and in Aqueous Salt Solutions

Fulvic acids, in the H^+-exchanged form, are, by definition, soluble in water, but HAs are not. As was pointed out, both acids can be regarded as neutral molecules with some macromolecular properties at pH values of 4 and below. (In the presence of strong acids, however, where protonation of appropriate functional groups can take place, the HS can be considered as polycations.) The easy expla-

Table 1–4. Yields and electron spin resonance (ESR) data for humic acids (HAs) and fulvic acids (FAs) extracted by different extractants from a H^+-exchanged Sapric Histosol soil.†

Solvent	Yield			ESR		
	HA	FA	Total	HA	FA	HA + FA
	%			spins $g^{-1} \times 10^{-16}$		
2.5 *M* EDA (pH 12.6)	49.0	14.0	63.0	15.0	27.5	--
EDA (anhydrous)	2.0	3.0	5.0	6.4	12.8	--
0.5 *M* NaOH	58.0	2.0	60.0	4.6	0.4	--
0.1 *M* $Na_4P_2O_7$ (pH 7)	13.7	0.8	14.5	4.5	1.9	--
1 *M* Na^+-EDTA	12.5	3.8	16.3	0.3	0.3	--
Pyridine‡	34.0	2.0	36.0	--	--	2.1
DMF‡	16.0	2.0	18.0	--	--	1.4
Sulfolane‡	10.0	12.0	22.0	--	--	1.0
DMSO‡	17.0	6.0	23.0	--	--	4.2

† From Hayes et al. (1975b).

‡ Extraction with the solvent was followed by exhaustive extraction with water. EDA = ethylenediamine; DMF = *N,N*-dimethylformamide; DMSO = dimethylsulfoxide; EDTA = ethylenediaminetetraacetic acid.

nation for the differences between the solubilities of the H^+-FAs and H^+-HAs is that the FAs contain more polar groups per unit of structure. This should be regarded only as a part of the explanation. The configurations and conformations of compositionally similar substances can be very important in considerations of their solubilities. For example, cellulose, a β-(1→4) linked polyglucose is not soluble in water whereas amylose, an α-(1→4) linked polyglucose is (see Soil Saccharides, p. 103). One reason for the difference is that the β-configuration allows the sugar monomer units in one polymer strand to have intimate contact with those on another strand to form a linear helix structure stabilized through H bonding. The α conformation does not allow such intermolecular association, and the polymer is soluble in water. It is interesting to note, though, that the β-D-glucopyranose molecule (see Soil Saccharides, p. 103) can replace almost exactly the chair conformation of water molecules inherent in the ice structure, giving rise to glucose–water H bonds instead of water–water H bonds (Suggett, 1975). This affinity, however, is not sufficient to overcome the extensive H bonding between the polymer strands of cellulose.

There is good reason to consider that FA molecules exhibit H bonding when dried. However, lack of regularity in the arrangements of the functional groups suggests that the molecules have insufficient order to allow regular and close sequences of inter- and intramolecular H bonding. Many FAs (especially when dealing with the FA fraction) contain sugar moieties, but it is evident that the molecular weights of these components and/or the conformations of the sugars and the configurations of their linkages do not allow sufficient H bonding between the molecules to prevent them from hydrating.

The data in Table 1–4 were obtained following water extraction of a H^+-exchanged Sapric Histosol that yielded 2.8% FA. The residual soil was extracted with the solvents systems listed. Extraction with each solvent was followed by exhaustive extraction with water. After extraction with sulfolane (**XV**, Fig. 1–4), DMSO, pyridine (**XVI**), and with ethylenediamine (EDA; **XVII**) or diaminoethane, additional amounts of FAs were isolated in the water extracts. This suggests that the FAs

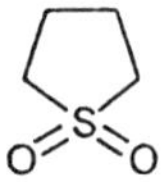

XV 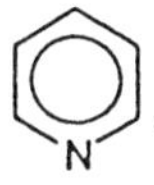 XVI 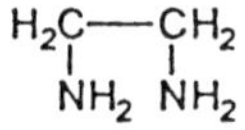 XVII

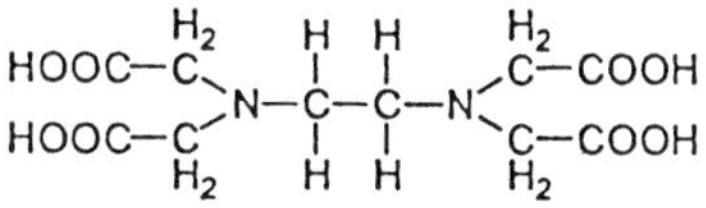

XVIII

Fig. 1–4. Organic solvents used in extracting soil organic matter (see text).

that were not extracted in water initially were adsorbed or in some way bound to insoluble or to the other soil components and were released by—though not necessarily dissolved in—the solvents used subsequently in the extraction processes. The effect of pH on the extraction processes is evidenced by the enhanced yields obtained when pyridine and EDA (aqueous) were used. The data for EDA, especially, suggests that the FAs were sorbed to (or trapped within) the HAs, or associated with other components that swelled in the alkaline medium releasing the FAs to solution. It is also possible, of course, that the FAs released were oxidized HA materials, as will be discussed below (p. 32).

In their search for nonalkaline solvents for soil HAs, Bremner and Lees (1949) showed that up to about 30% of SOM could be extracted as HS by sodium and potassium salts of inorganic and of organic acids. Of these, a 0.1 *M* solution of sodium pyrophosphate ($Na_4P_2O_7$) neutralized with phosphoric acid (H_3PO_4) was best, and was followed in order by solutions of the sodium salts of fluoride (NaF), hexametaphosphate $(NaPO_3)_6$, orthophosphate (Na_3PO_4), borate ($Na_2B_4O_7$), NaCl, NaBr, and NaI. The order of decreasing extraction efficiencies for the organic acid salts was: oxalate $(COONa)_2$, citrate[($NaOOCCH_2C(OH)COONaCH_2COONa$], tartarate [NaOOCCH(OH)CH(OH)COONa], malate [$NaOOCCH(OH)CH_2$–COONa], salicylate (*o*-OHC_6H_4COONa), benzoate (C_6H_4COONa), succinate ($NaOOCCH_2CH_2COONa$), sodium-4-hydroxybenzenecarboxylate, and ethanoate (CH_3COONa).

The best extractants in the above list of salts form complexes with the polyvalent metals that neutralize charges on the HS and/or link these to the inorganic soil colloids. These polyvalent ions are replaced by sodium ions from the salts. The efficiency of each solvent system will depend on the extent to which the resident cations are exchanged and removed from humic structures. Diffusion of the salts to the interior of solid HS is slow. Some channeling can take place, but extensive penetration would probably require the opening up from the outside of the molecular structures. It would be necessary for these structures to remain open to allow exchange from the interior to take place.

The evidence that we have suggests that the more highly oxidized HS and those in the lower molecular weight ranges are removed by the salt solutions. The greater the extent the substances are oxidized, the greater will be their charge densities, and the combination of high charge and relatively low hydrodynamic volume increases the possibilities for solution to take place.

The extent of dissolution is influenced by the amounts of low molecular weight electrolyte present. Excess electrolyte increases the ionic strength outside the humic structures relative to that inside, and it decreases the thickness of the electrical double layer, causing the molecules to contract. At high salt concentrations hydration is curtailed and dissolution may not take place.

It is not necessary to hydrogen-exchange HS before extraction with neutral salt solutions containing divalent and polyvalent metals, which can form complexes with the organic matter. Alexandrova (1960) showed that serosems (soils with high Ca contents) could be effectively extracted with $Na_4P_2O_7$ without prior exchange with H^+. In fact, the acidification process inherent in the H^+ exchange depresses the amounts extracted. For example, the low yields of HS obtained by Hayes et al. (1975b) when they exhaustively extracted a Sapric Histosol soil with neutralized $Na_4P_2O_7$ and ethylenediaminetetraacetic acid (EDTA, **XVIII**, Fig. 1–4) solutions probably arose because the extractants were not buffered and the acidity of the H^+-exchanged soil significantly lowered the pH of the solvents in contact with the soil. Hence, ionization of the acid groups on the HS was suppressed.

Extraction of Humic Substances under Basic Conditions

Since Achard (1786), aqueous basic solutions have been the predominant solvent system for the isolation of HS. He extracted peat with aqueous potassium hydroxide (KOH), and the supernatant gave a dark, amorphous precipitate upon acidification. This precipitate later became known as HA. Achard noted that larger amounts of dark-colored material could be extracted from the humified layers of the peat than from the less decomposed components. One of the earliest comprehensive studies of the origin and chemical characteristics of HS was performed by Sprengel (1826, 1837), and many of the procedures he introduced have become generally adopted. He concluded that the HAs were bound to minerals in soils rich in bases, and were considered more likely to be in solution in soils with low base contents.

The data in Table 1–4 show that the basic solvent systems, 0.5 *M* NaOH and aqueous 2.5 *M* EDA, extracted more HS from a H^+-exchanged organic soil than did the other solvents. Because of the highly alkaline conditions which prevailed (the soil suspension in the aqueous EDA system had a pH of 12.6) all the acid groups in the molecules were dissociated and the repulsion of charge gave the fully expanded (swollen) structure. Thus, the anionic and polar sites could be readily solvated with water molecules.

Use of basic solutions is likely to bring about some oxidation of the HS molecules. This was highlighted by Bremner (1950) who showed that solution in aqueous 0.5 *M* NaOH increased O_2 uptake by soil organic materials (Table 1–5), and subsequently Swift and Posner (1972) demonstrated that some breakdown of HAs can take place in alkaline conditions in the presence of O_2. It is evident from Table 1–5 that uptake of O_2 (oxidation) is small under mildly alkaline conditions. We do not have data for the uptake of O_2 by HS in NaOH solutions at pH values below 11, or in solution in EDA.

Table 1–5. The influence of pH on the uptake of O_2 by organic matter extracted for 7 h with solvents in media of different pH values.†

Reagent	O_2 uptake	
	Soil 1	Soil 2
	mm³ 0.2 g⁻¹	
0.5 *M* NaOH	896	712
0.5 *M* Na_2CO_3, pH 10.5	56	71
0.2 *M* Na citrate, pH 7.0	39	58
0.1 *M* $Na_4P_2O_7$, pH 7.0	7	37
pH 8.0	12	--
pH 9.0	31	52

† Stevenson (1994), from data by Bremner (1950).

Damage through oxidation can be partially avoided when work on humic solutions in alkaline media is performed under dinitrogen gas. Choudhri and Stevenson (1957) used this approach when they extracted their soil HAs with aqueous NaOH in an atmosphere of N_2 and in the presence of stannous chloride as an antioxidant. For the isolation of its collection of standard soil HS, the IHSS has recommended equilibrating the soils with 1 *M* hydrochloric acid, neutralizing to pH 7 with 1 *M* NaOH, and adding (under N_2 gas) 0.1 *M* NaOH to give a liquid/soil ratio of 10:1 (Swift, 1996).

The yields of HS (HAs + FAs) were slightly higher from the exhaustive extractions with aqueous EDA than from those with NaOH (Table 1–4). Data in this table suggest that less HA and more FA substances were extracted in the EDA than in the NaOH solutions. However, it was found that considerable amounts of the substances classified as FAs (components remaining in solution when the aqueous EDA extracts were acidified to pH 1) were precipitated in the bags during dialysis against distilled water. That suggests that some HAs were solubilized by EDA salts formed on acidification, and these acids were precipitated as the salts were lost during dialysis. (This observation is further discussed in the next section.)

The use of aqueous EDA, or indeed of any primary amine cannot be recommended as an extractant for HS. The C and N contents of the EDA-soluble substances were significantly higher than for those extracted with NaOH (Hayes et al., 1975b; Hayes, 1985). The reverse was true for the oxygen contents of these substances. EDA readily reacts with carbonyl groups to form Schiff base structures. Such reactions are rapid when the protonated carbonyl group reacts with the nonprotonated amine. Thus, the maximum reaction rate occurs when the product of the concentrations of the protonated carbonyl groups and free amines is maximum. The reaction continues, but at a rate that decreases in an exponential manner, as the pH is raised or lowered from the optimum value for the reaction.

In addition to Schiff base–type reactions, the amines react with C atoms alpha to the keto groups in quinone structures (see structure type **CCXXXVIII**, Fig. 1–29). Such bonds readily resist cleavage in the acid-wash procedure used by Hayes et al. (1975b) in their attempts to lower the N contents of the HS extracted by EDA. Dryden (1952) and Rybicka (1959) observed also that the N contents of coal materials extracted with EDA were raised. Incorporation of the solvent organic molecules into the humic structure would enhance the C and N contents at the expense of the O_2 lost in the water released in the condensation reactions.

The high free radical contents of the HAs and FAs isolated when aqueous EDA was used (Table 1–4), as measured by the electron spin resonance procedure, provides further evidence of denaturation in this solvent system.

Extraction with Organic Solvents

Except in the case of extraction with anhydrous EDA, data in Table 1–4 were obtained for HS isolated from an air-dried H^+-exchanged Sapric Histosol soil. For extractions with pyridine, DMF, DMSO, and sulfolane, soils (60 g) were thoroughly mixed with the appropriate solvent (250 mL). After centrifugation the residues were repeatedly extracted with water until the supernatants were only faintly colored. Supernatants for each of the solvent systems were combined, and the pH values of the solutions were adjusted to 1.0 using 6 *M* HCl. The HAs and FAs were separated by centrifugation.

For extraction with anhydrous EDA, H^+-exchanged soil was dried in vacuo at 75°C over phosphorus pentoxide (P_2O_5) and then thoroughly mixed with EDA (dried by distillation from solid NaOH). Extraction was repeated twice.

In the absence of water, EDA was a very poor solvent for HAs (Table 1–4). This suggests that the solute–solvent interactions were not sufficiently energetic and/or numerous to overcome the strong H bonding forces between the HA molecules. This is consistent with the notion that the more hydrophobic or less polar components of HS orientate toward the exteriors of the structures as shrinking takes place during drying. This effect would give rise to a shielding of the more polar and more readily solvated groups in the interiors of the molecules. The molecular orientations suggested would not be essential to explain the phenomena observed. The concept of micelle-type arrangements (see Sizes and Shapes of Humic Substances, p. 54) would suggest that the polar functionalities orientate to the outside in the associations of molecules in the humic structures. That arrangement would require strong H bonding involving the polar functionalities, and as suggested above, the solute–solvent interactions were not sufficiently energetic to overcome the attractive forces of the H bonding.

Inevitably some acid groups in the dry HS would be contacted by the EDA. These would protonate one or both of the amine groups of the solvent, and the conjugate acid structure ($EDAH^+$) would then be held by ion exchange to the conjugate base structures (e.g., $-COO^-$) of the dissociated acids. However, the size of the EDA molecule and of its protonated derivative would inhibit penetration into the tightly H-bonded humic matrix, and so extensive solvation would not take place. Thus, in the absence of water HAs cannot be expected to swell readily and to solvate in EDA or other organic amine solvents.

The yields of FAs (Table 1–4) from extraction with anhydrous EDA were similar to those for extraction with NaOH. Because FAs are considered to be smaller and are more polar than HAs, it seems likely that sufficient molecules that could be solvated were available to the EDA solvent to allow solution to take place.

Pyridine, followed by water, extracted more HS than did the DMF, DMSO, or sulfolane systems (Table 1–4). The enhanced solubilization by pyridine could be attributed partially to a pH effect. When pyridine (eight parts) was diluted with water (one part) to simulate the composition of the solvent in the air-dried soil, the pH of the mixture was 11.6. Because of the low buffering capacity of the solvent system, the pH of the extract was only 4.2 (Swift, 1968). Theory suggests that substantially more humic materials can be brought into solution by pyridine if the pH of the medium is maintained at 9.0 (the pK_b for pyridine is 8.96).

At pH values >9.0 pyridine molecules predominate in the medium, and these are then involved in solvating functional groups on the humic structures. There is a logarithmic increase in pyridinium ions as the pH falls, and the organocations formed are held by ion exchange to the conjugate bases of the acidic functional groups on the humic molecules. Although humate-pyridinium salts are less dissociated than those of the humate-monovalent inorganic cation species, they are significantly more dissociated than the H^+- and divalent and polyvalent cation-exchanged materials. This would explain the enhanced solubilization in water of HS from H^+-exchanged Histosol soils first extracted with pyridine (Hayes, 1985).

Electron spin resonance (ESR) data (Table 1–4) suggest that solution in pyridine does not significantly enhance the content of free radicals in HS. Pyridine does, however, significantly enhance their N contents, especially in the cases of the FAs, as was found also for extracts with EDA (Hayes et al., 1975b; Hayes, 1985). Sorption, by ion exchange and possibly by charge-transfer processes, might provide plausible explanations for the enrichments.

The amounts of HS extracted into the dipolar aprotic solvents DMF, DMSO, and sulfolane depend entirely on the extent to which these solvents can solvate the humic molecules. ESR data indicate that DMF and DMSO do not generate free radicals in the humic extracts. There is a slight enhancement of S in the HAs extracted in sulfolane, and this might be attributable to some adsorption of the solvent by the humic molecules. Otherwise the elemental analysis data of the HS extracted by these two solvents are similar (Hayes et al., 1975b), and the elemental data for the HAs isolated in the solvents and in 0.5 *M* NaOH are comparable. However, FAs from the basic solution have lower C and higher O contents, and this suggests that some uptake of O_2 occurred under the alkaline conditions. Because some colored materials precipitated during dialysis of the fulvic-type substances from the organic liquid extracts, it is probable that the true FAs are contaminated by HAs solvated by the residual solvents in the acidified mixtures. These acids precipitate when the solvating molecules are replaced with water during dialysis. An alternative explanation could be that molecular associations took place during the residence time of the FA substances in the dialysis bags.

In the experiments of Hayes et al. (1975b) DMSO was marginally better than DMF or sulfolane for dissolving HS (Table 1–4). Evidence from ESR shows a higher free radical concentration in DMSO than in either DMF or sulfolane. Because DMSO would not be expected to generate free radicals, it is reasonable to infer from the ESR data that humic components, which are insoluble in the DMF– and sulfolane–water systems, are dissolved in this solvent. Elemental contents were similar for the HAs and FAs of the DMSO extracts. That would infer that the major difference between the two fractions was one of molecular size, and more likely that could reflect higher extent of molecular associations in the cases of the HA fractions. Some FA materials were observed to precipitate during dialysis, as was

Table 1–6. Comparison of absorbance values for solutions obtained by mixing H^+-exchanged humic acids (0.2% w/w) with aqueous and organic solvents.†

Solvent	Absorbance value at 400 nm
Water	1.0
Dioxane	0.0
Acetonitrile	0.0
Ethanol	1.0
Formic acid (90%)	4.0
Pyridine	5.0
N,N-Dimethylformamide (DMF)	18.0
Formamide	19.0
Dimethyl sulfoxide (DMSO)	21.0
0.5 *M* NaOH, pH 9.2	23.0
0.5 *M* NaOH	24.0

† From Hayes (1985).

noted for the DMF and sulfolane systems, and that could infer reassociations of humic molecules during the dialysis process.

In a further set of experiments Hayes et al. (1975b) exhaustively extracted an H^+-exchanged sapric histosol with water, and then exhaustively with the series DMF–, sulfolane–, DMSO–, pyridine–, and EDA– water, 1:1 (v/v) mixtures in that order. The cumulative amounts of HS extracted were the same as the amounts isolated by the last solvent in the series without the aid of the others. This would suggest that the materials that dissolved in the less efficient solvents were dissolved also in the more efficient members of the series. The analytical data for the different fractions closely resembled those for the substances extracted by the single solvent systems (Hayes et al., 1975b; Hayes, 1985).

The pyridine and dipolar aprotic solvents' performances as humic solvents (Table 1–4) suggested that a comparison be made of the abilities of these and of other organic solvents to dissolve HAs. It was considered that any solvent and treatment that would dissolve HAs could be adapted for the extraction of HS from the soil. Samples of HA (isolated by the procedure recommended by the IHSS) from a Florida (Belle Glade) Sapric Histosol were extracted with the solvents (0.2% w/v) listed in Table 1–6. Swelling was allowed to take place overnight, and after centrifugation each supernatant solution was diluted with the solvent used for extraction until an absorbance reading against the solvent blank could be obtained at 400 nm in each case. The absorbance values quoted (Table 1–6) represent the product of the reading obtained and the dilution factor used.

Data in Table 1–6 show that acetonitrile, dioxane, ethanol, and water are very poor solvents for HAs. Humic acids are, by definition, insoluble in water, although traces of the H^+-exchanged substances are invariably dissolved in it. Similar trace amounts were dissolved in ethanol. Pyridine and formic acid (90%) were also poor solvents for HS. The extent of dissolution in pyridine was less than would be predicted from the data quoted for the H^+-exchanged soil in Table 1–4. However, the Florida peat HAs were significantly drier than the soil, and it is concluded that pyridine in the absence of water does not readily solvate HAs, and the results had similarities to those for anhydrous EDA.

Comparison of results with those obtained by Sinclair and Tinsley (1981) suggests that solvation by the 90% formic acid was disappointing. However, Sinclair and Tinsley used anhydrous formic acid. This acid has many of the desirable characteristics of a good solvent for HS (see Tables 1–2 and 1–3), as listed by Whitehead and Tinsley (1964) and as discussed in this chapter.

Formamide, DMSO, and DMF were good solvents for the Florida peat HA, and solvation by these approached that by 0.5 *M* NaOH. Swelling was slow in the organic solvents, and this contrasted with the behavior in the aqueous base.

Some of the properties listed for the solvents in Tables 1–2 and 1–3 help to explain the differences in solubilities of the peat HAs in the solvents listed in Table 1–6. For the purpose of the present discussion, solvents in Table 1–6 giving absorbance values <10 and >15 are regarded as poor and as good solvents, respectively, for the H^+-exchanged HA. Each of the good solvents has an electrostatic factor (EF) value >140, but so also has acetonitrile, a poor solvent. However, on the basis of the data in Tables 1–2 and 1–6, it is appropriate to consider that organic solvents with EF values >140 and with PK_{HB} (a measure of the strength of a solvent as an acceptor in hydrogen bonding) values >2 should be good solvents for HAs. The low pK_{HB} value for acetonitrile indicates that it is incapable of breaking the H bonds of the H^+-exchanged HAs. DMF and DMSO fulfill these requirements and are shown to be good solvents in Table 1–6. *N*-methyl-2-pyrolidone (**XIV**), has the required pK_{HB} value to be a good solvent, but the H-bonding parameter, δ_h, (Table 1–3 and discussed below) is unfavorable, and the high molar volume (Table 1–2) value suggests that steric constraints could hinder diffusion to solvation sites within the nonexpanded H^+-exchanged humic structures. Law (1988) showed that this compound is a poor solvent for HS in a H^+-exchanged Sapric Histosol, but it performs significantly better in the Ca^{2+}-exchanged soil.

Solubility parameter data (Table 1–3) provide less clear-cut differences between the good and poor solvents in Table 1–6. Mention was made of the need for δ, δ_d, δ_p, and δ_h parameters for solvent and solute to be similar for dissolution to take place. There are no measured solubility parameter data available for HS, and so accurate predictions cannot be made based on parameter values for solutes and solvents.

The dispersion force (δ_d) parameters are similar for the organic solvents used to provide the data in Table 1–6, but there are differences in the polar (δ_p) and H bonding (δ_h) parameters of these solvents. Comparison of data in Tables 1–3 and 1–6 suggests that the best solvents for H^+-exchanged HAs have δ_p, δ_h, and δ_b (proton acceptor) parameters on the order of, or greater than, 6, 5, and 5, respectively. Water satisfies all these criteria, although it is a poor solvent for H^+- and for divalent and polyvalent cation-exchanged HAs. Solution is greatest when the products δ_a (solvent) × δ_b (solute), or vice versa, are maximum (see Use of Solubility Parameters for Predicting Solubilities, p. 26). The very large values of δ_h, δ_a, and δ_b indicate the extent of self association through H bonding of water molecules. It is clear that the δ_a and δ_b values of the H^+-exchanged HAs are not sufficient to disrupt these attractive forces.

Formamide (**XI,** Fig. 1–3 and Table 1–6) was shown to be a good solvent for the H^+-exchanged HAs, and this might not readily be predicted from comparisons of its solubility parameter data with those for ethanol and water. However, although the values for δ_a and δ_b are large, it is clear that H bonding must take place between the solvent and the humic molecules. The value of δ_h is similar to that for ethanol, a poor solvent, but the high δ_p parameter, which distinguishes formamide from the other solvents in Table 1–3, is central to the performance of the solvent as an acceptor in H bonding with HAs.

The H^+-exchanged HAs studied were polydisperse. If it is assumed that size (in terms of molecules or aggregates of these) was the only major difference between the components, it is reasonable to suggest that the smaller materials (or associations of molecules) were solubilized preferentially. To apply effectively solubility parameter theory, it is desirable to work with humic fractions that are relatively homogeneous with regard to size and composition. That would involve an extensive fractionation program (see Fractionation of Soil Organic Substances, p. 40). Solubility parameter values could then be obtained for the humic fractions, and such data might suggest solvent systems for the isolation from soil of different fractions, or the design of solvent mixtures for more complete extraction of all soil humic components.

Applications of polymer solution theory to the studies of the dissolution of HAs and of their extraction from soils suffer most because interactions between each pair of components must be known. Unfractionated HAs, FAs, and HS in the soil are parts of multicomponent systems, and interactions between the different components are unknown.

Polymer solution theory was used by Chiou et al. (1983) to study the binding of small (see Flory, 1953) organic chemicals by soils. They considered the soil sorbent substances to be amorphous macromolecular HS, and they adapted the Flory–Huggins theory (see Flory, 1953) to a study of the sorbate species solubilized in the amorphous macromolecules. However, determination of the Flory–Huggins interaction parameter (χ) for solvent-polymer pairs again requires careful fractionation of the humic molecules. Barton (1975) cited shortcomings of this parameter as a practical criterion of solubility. In the view of the authors, much can be learned about the ways in which HS are associated through determination of Flory–Huggins and solubility parameters of carefully fractionated HS.

Extraction with Solvent Mixtures

Parker (1962) cited references for the solubilization of neutral polymers (such as polyacrylonitrile, nitrocellulose, cellulose acetate) and wood products in DMSO, DMF, and sulfolane. Unlike cellulose, these products are not strongly H bonded. The review by Suggett (1975) suggests that the addition of urea (10%) to DMSO effectively breaks the H bonds in cellulose and allows solution to take place. With this in mind, one of the authors (MHBH) conducted work in his laboratory, in which urea (10%, w/v) was added to DMSO, to 0.025 M $Na_2B_4O_7$ (pH 9.2), and to 0.1 M $Na_4P_2O_7$ (adjusted to pH 7 with H_3PO_4) solutions. Solution of the Histosol H^+-exchanged HAs in borate was depressed by the presence of urea, although urea did not influence extraction in the pyrophosphate. Aqueous 10% urea was a very poor solvent for the H^+-exchanged HAs, but solubilization was almost complete in 5 M urea. Complete solution of the HA was observed when water (5% of the total weight) was added to the urea (10%)–DMSO solution. However, solution was

Table 1–7. Solubilization of H^+-exchanged Fenland Sapric Histosol humic acid (H^+-HA) and humic substances from the H^+-exchanged Sapric Histosol soil (H^+-SHisl) in 0.1 *M* NaOH, DMSO, and acidified DMSO.†

Sample and weight	Solvent mixtures: DMSO	Solvent mixtures: Other solvents	Absorbance at 465 nm (E_4)	Absorbance at 565 nm (E_6)	E_4/E_6
mg	cm³	cm³			
H^+-HA (1.27)	0.0	0.1 *M* NaOH (5.2)	1.212	0.240	5.05
H^+-HA (1.27)	5.2	--	1.045	0.206	5.07
H^+-HA (1.27)	5.18	H_2O (0.02)	1.042	0.196	5.31
H^+-HA (1.27)	5.0	H_2O (0.2)	1.042	0.196	5.31
H^+-HA (1.27)	5.18	HCl (0.02)‡	0.910	0.110	8.27
H^+-HA (1.27)	5.0	HCl (0.20)‡	0.890	0.108	8.24
H^+-SHisl (11.7)‡	0.0	0.5 *M* NaOH (5.2)	2.06	0.348	5.92
H^+-SHisl (11.7)‡	11.7	--	1.71	0.328	5.21
H^+-SHisl (11.7)‡	11.0	H_2O (0.6); HC1(0. 10)§	1.90	0.226	8.41
H^+-SHisl (11.7)‡	11.6	HC1 (0.10)‡	1.98	0.232	8.53

† From Hayes (1985).
‡ On dry-weight basis.
§ Concentrated acid.

gradually depressed as additional increments of water were added. Although urea–water mixtures greatly improved the efficiency of dissolution and extraction of HS from soil, their uses cannot be recommended because the raised N contents of the dialyzed extracts suggest that urea interacts with the HS.

Table 1–7 presents data for HAs isolated from a Norfolk (England) Fenland soil (a Sapric Histosol) and for the HS isolated from the same soil using various combinations of DMSO, acid, and water. For comparison, some data are presented for solution in NaOH. Use is made of E_4/E_6 ratios to indicate differences in the solution conformations and/or compositions of the HS in the different solvent systems.

All the samples of the Fenland H^+-exchanged HAs were completely dissolved in the solvent systems used in Table 1–7. The data show that the absorbance value at 465 nm for the NaOH solution was greater than that for the same concentrations of solutes in the DMSO preparations. Although DMSO was stored over $CaCl_2$ to prevent excessive water uptake, no attempt was made to fully dry the solvent, and it is assumed that it contained some water. Additional water (up to about 4% of the total volume) did not significantly alter the solution properties of DMSO. Addition of concentrated HCl, even to the extent of only 0.4% of the total volume of the mixture, significantly altered the absorbances and the E_4/E_6 ratios.

Because all the HAs were dissolved in the DMSO, whether or not HCl was added to the DMSO, the changes in the E_4/E_6 ratios observed from the additions of acid might be attributed to alterations of the solution conformations of the humic molecules, as was implied in the work of Swift et al. (1970) and later suggested by Chen et al. (1977) and Ghosh and Schnitzer (1979). Decreases in the E_4 and E_6 absorbance values as a result of additions of acid and the increases in the E_4/E_6 ratios would therefore be consistent with shrinking and the suppressions of the hydrodynamic volumes of the HA molecules. An increase in the negative slope of the plot of log absorbance vs. wavelength (Chen et al., 1977) was observed when acid was added; this could be regarded as further evidence for changes of the solution conformations of the molecules in the presence of acid.

Table 1–7 also compares the organic extracts from a H^+-exchanged Fenland (Norfolk, England) Sapric Histosol soil. In the H^+-exchanged form this soil contained only about 5% ash. The data show that the E_4/E_6 ratios for extracts in 0.5 *M* NaOH and DMSO are similar, but that the E_4 and E_6 absorbance values are slightly higher for the base extract. This is in keeping with the trends observed for the H^+-exchanged HAs. The addition of acid predictably increased the E_4/E_6 ratio, and it can be concluded that it also increased the amount of organic matter extracted because of the increased E_4 values. It is not possible to compare quantitatively the amounts of organic matter extracted by the different solvent systems, but when the depressed absorbance values in DMSO (compared with NaOH) are taken into account, extraction with acidified DMSO nearly matches in efficiency that in the base.

Data from Law et al. (1985) and Fagbenro et al. (1985) showed that DMSO containing up to 6% of (12 *M*) HCl was a good solvent for Tropical Forest and Savanna soils. The addition of acid dispensed with the need to "pre-H^+-exchange" the soils. This can be explained by the fact that the HCl protonated the conjugated bases of the acidic functionalities neutralizing the charges of the exchangeable metal cations.

Anions are very sparingly solvated in DMSO, but cations are much more readily solvated (Martin and Hauthal, 1975, p. 131). Hence the exchangeable metal cations that neutralize the charges on the humic molecules in the divalent/polyvalent cation-exchanged tropical soils would be solvated by DMSO. The conjugate bases (carboxylates and phenolates) would not be solvated, however, until neutralized with H ions. Reaction Scheme [2] shows how the carboxyl and phenolic structures could hydrogen bond to DMSO, where Ø–OH represents phenolic constituents and R represents the remainder of the humic molecule:

```
        R
       /
  O=C          CH3  φ—R          R
     \         /   /              \
      O—H....O=S-----O        CH3   C=O
               \      \        /   /
               CH3    H....O=S------O
                              \      \
                              CH3    H...O—
```

[2]

FRACTIONATION OF SOIL ORGANIC SUBSTANCES

For the most part, the classical isolation procedures which use aqueous base solvents, as described in the previous section, will isolate solute molecules that have significant polar functionalities, and especially ionizable functionalities that give predominantly the conjugate bases of carboxyl and phenolic groups. However, to some extent nonpolar functionalities (such as some long chain hydrocarbons and the long chain esters of fatty acids and waxes) associated with the polar species are coextracted. Thus, the classical fractionation procedures made use of differences in solubility characteristics based on polarity and charge characteristics. In addition to procedures based on these characteristics, modern fractionation procedures also use differences in the sizes and shapes, as well as the adsorption characteristics of the molecules in the isolates.

Classical Fractionation Procedures

Fractionations based on solubility differences involve precipitations from solution in bases. In principle, humic fractions whose acidic functionalities are predominantly weakly dissociable should be first to precipitate as the pH of the alkaline medium is lowered, and the strongest acids would be last. This can be translated into extent of humification, with the less humified fractions precipitating first and the most humified last (lower in the pH scale). Use can also be made of additions of salt, which decreases the intra- and intermolecular charge repulsion effects. In the case of the intramolecular charges the salt causes solvent to be excluded and a shrinking of the molecules or of the molecular associations. Also, where a diffuse double layer of charges occurs, the salt causes a decrease in the extension or boundaries of the layers and molecules approach, giving rise to shrinking of the molecules or molecular aggregates by exclusion of water from the matrix. Further, the suppression of the electrical double layer also allows the molecules to approach more closely, promoting coagulation. This may be expressed as self association, or as the formation of pseudo micelle-type structures. Precipitation in this way is a salting out effect, and Theng et al. (1968) used ammonium sulfate for the fractional precipitation of soil HAs. A selective precipitation can be achieved using heavy metals [see MacCarthy and O'Cinneide (1974), who studied the interactions of Cu and Co with a FA fraction].

Uses can be made of the poor solvents listed in Table 1–6 (and of other solvents listed in Table 1–2 whose properties would classify these in the poor solvent category for HS) to fractionally precipitate HAs from alkaline solutions. Ethanol is in that category, and although ethanol has been used to dissolve the so-called *hymatomelanic* acid fraction from HAs (Stevenson, 1994), it also can be used to precipitate humic fractions from alkaline solutions of HAs (Kyuma, 1964; Kumada and Kawamura, 1968). Humic substances can also be partitioned between alkaline solutions and non- (or sparingly) miscible organic solvents to give a concentration of HS at the water–solvent interface. The principle involves association of the hydrophobic components of HS with the organic solvent at the interface. The HS do not cross the interface because the polar moieties are not solvated. An application of this principle by Rice and MacCarthy (1989) is discussed below. Eberle and Schweer (1974) developed a hydrophobic extractable ion pair by dissolving long chain tertiary or quaternary amines in chloroform and extracting HS and lignosulphonic acids from water at pH 5, then recovering the HS at pH 10. This suggested a potential for applications of counter-current distribution techniques, but such techniques are tedious, and emphasis is now given to fractionations on the basis of molecular size and charge density differences.

Some of the earlier (1950s) attempts to fractionate HS on the basis of charge density differences are referred to by Dawson et al. (1979) and by Hayes et al. (1975b). By means of continuous flow paper curtain (Durrum, 1951) and column electrophoresis experiments (Clapp, 1957), they showed that polysaccharides could be separated from colored HS. However, a Gaussian distribution of color components was obtained, and hence well-defined fractions of HS were not obtained.

Modern Fractionation Procedures

Modern fractionation procedures use the principles on which the classical procedures are based, and in addition apply procedures based on the molecular sizes and shapes and sorption characteristics of the organic molecules.

Successes in the uses of low pressure gel permeation chromatography (GPC) for separations on the basis of the molecular size differences of biological polymers has prompted applications of these techniques for the fractionation of HS. Until recently the majority viewpoint favored the concept that HS are polydisperse macromolecules, and so it was logical to attempt fractionations on the basis of molecular size differences using GPC techniques (e.g., Lindqvist, 1967; Swift and Posner, 1971; Cameron et al., 1972a, 1972b; Hayes and Swift, 1978; Wershaw and Aiken, 1985; DeNobili et al., 1989; Swift, 1989b, 1996; DeNobili and Chen, 1999; Perminova, 1999). In theory, when applied to a gel column a humic solution will be excluded from the gel pores (in the cases of molecules that are too large to enter the gel pores) or will enter the pores. For molecules that do enter the pores passage through the column will be retarded, and the extent of the retardation will depend on the size and shape of the molecule.

There are gels available that can separate fractions of biological molecules in the range of perhaps one thousand to millions of daltons. The gels that are in use most are composed of polysaccharides, such as the Sephadex (**XIX**, Fig. 1–5) gels [dextrans, or α-(1→6)-D-glucose] and (poly) acrylamide (**XX**) resins of different pore sizes. For separation to reflect size differences it is important that there should not be interactions between the gels and the solute materials, or that any such interactions be suppressed, should these occur. There are, for example, residual negative charges on dextran gels—charges can arise also in the (poly)acrylamide gels—and electrostatic repulsion between the charges can accelerate the passage of even small molecules through the gels to give effects that can be interpreted falsely in terms of large molecular sizes. Alternatively, hydrophobic interactions between the gel and the solute molecules can retard passage of the molecules through the gel columns and give rise to false small-size molecules. Mobile phases containing buffers of high ionic strengths have been recommended to counter the charge repulsion effects. However, as Specht and Frimmel (2000) and Piccolo (2001) pointed out, high ionic-strength buffers decrease the molar volume of HS in solution. This favors the hydrophobic associations of the molecules and enhances hydrophobic adsorption on the gel solid phase.

Swift (1996) gave a diagrammatic representation of the tedious procedure that can give fractions having a degree of molecular size homogeneity. Figure 1–6 gives a representation of how a gel column might be used to obtain organic fractions which should have, in theory, molecules of similar sizes. Consider the passage of polydisperse (range of molecular sizes) soil organic molecules through a gel column with discrete pore sizes. Molecules excluded from the gel pores are eluted in the void volume (V_0), and those that enter the gel are eluted between V_o and V_t (if true gel filtration is observed, the molecules eluted should decrease in

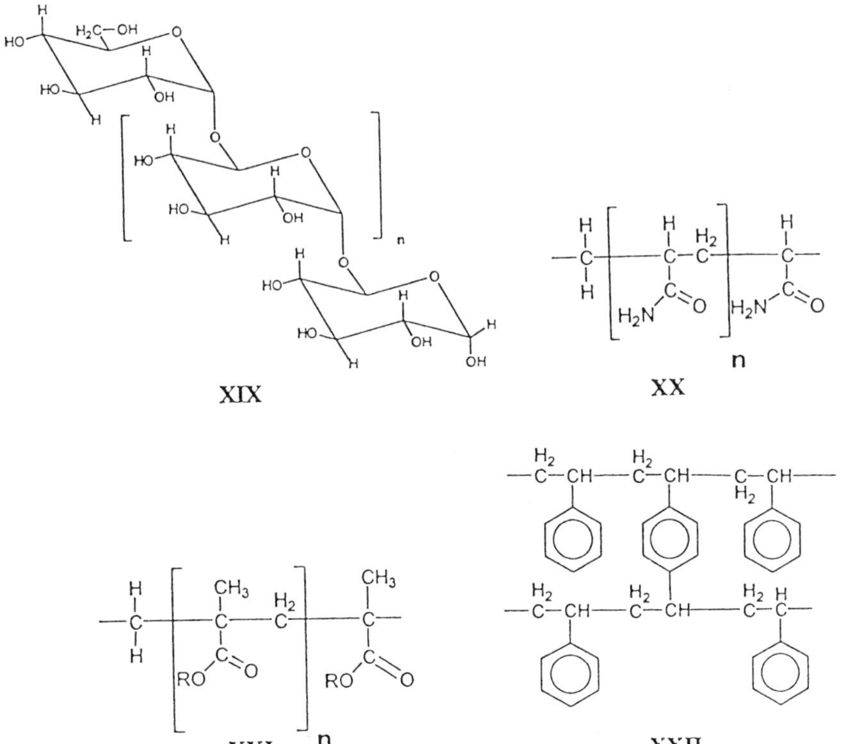

XIX

XX

XXI

XXII

Fig. 1–5. The gel polymers Sephadex (**XIX**) and (poly)acrylamide (**XX**), and the XAD-8 (**XXI**), and XAD-4 (**XXII**) used for fractionation.

size in the range of V_o to V_t). To obtain fractions of more discrete sizes, it would be appropriate to collect the fractions eluted between V_o and V_1, V_1 and V_2, V_2 and V_3, and so on. After several such runs, the fractions eluted in each discrete volume range might be combined, concentrated, and reprocessed through the same gel. This process might be repeated several times until the sample is effectively retained within the volume boundary. A variation of that procedure might be to isolate the material isolated in the V_1 and V_2 volume range and pass the concentrated fraction through a column of smaller exclusion volume size (e.g., Sephadex G-75 where Sephadex G-100 was used in the initial fractionation). Then the material eluted within the V_2 and V_3 boundary might be passed through Sephadex G-50, and so on. In theory, it should be possible to isolate several homogeneous fractions in that way. However, the possibility exists that molecular fractions could associate during the course of gel filtration to give retention properties indicating molecules with falsely high molecular weight (MW) values.

On the other hand, it has been noticed, for example in the work of Cameron et al. (1972a, 1972b), that some components eluted in a particular column volume gave rise to components that were of lower molecular sizes when reprocessed on the gel columns. That could be interpreted in terms of a breakdown of molecular associations (see Sizes and Shapes of Humic Substances, p. 48, and NMR Studies

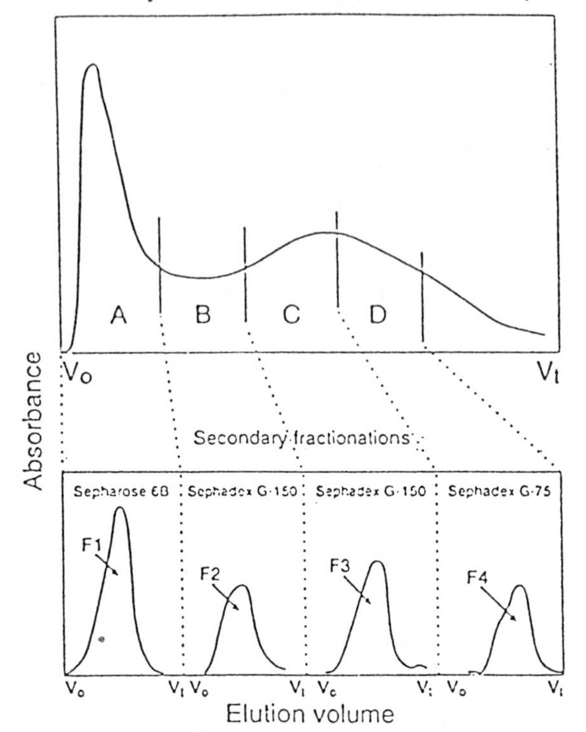

Fig. 1–6. Composite diagram showing initial elution pattern (absorbance at 400 nm vs. volume eluted) for whole humic acid (upper section of diagram) and patterns for the subsequent runs of the separate fractions, F1, F2, F3 and F4 (lower section of diagram) (from Swift, 1996).

of Composition and Associations, p. 97). Cameron et al. (1972b) were able to obtain fractions which, after extensive reprocessing, were eluted within discrete volume boundaries. The work of Cameron et al. (1972b) is now classic in its fractionation of HS utilizing gel filtration techniques. They also recovered fractions using the less tedious pressure filtration (through membranes of different pore sizes) technique. Tombácz (1999), employing a series of Amicon membranes and a stirred ultrafiltration cell, obtained six HA fractions from a HA isolate, and each was significantly different compositionally. Hence, a combination of techniques based on charge density and molecular size differences can give extensive fractionations of humic extracts. These types of fractionations are further discussed in Sizes and Shapes of Humic Substances (p. 48).

A procedure, based on that of Malcolm and MacCarthy (1992) and used by Ping et al. (1995) was adapted by Hayes (1996) and by Hayes et al. (1996) for the fractionation of HS on the basis of charge density differences. In this procedure soils are exhaustively extracted with distilled water, with 0.1 *M* $Na_4P_2O_7$, adjusted to pH 7 with H_3PO_4, with 0.1 *M* $Na_4P_2O_7$ (pH 10.6), and with 0.1 *M* $Na_4P_2O_7$ plus 0.1 *M* NaOH (pH 12.6). The aqueous solutions, recovered as described in Isolation of Humic Substances (p. 21), are treated in a manner similar to that for HS in waters.

In the cases of the alkaline solutions, the media are diluted and the pH adjusted to 7 to 8. Then the solutions are filtered through 0.2-μm pore-size filters, and the filtrates are diluted to <50 mg L^{-1}. (The filtrates are ideally diluted to <20 mg L^{-1}, or to a concentration in which there is no evidence for coagulation or precipitation when the pH is adjusted to 2. The situation when coagulation occurs is covered below. However, when the solutions are allowed to stand for prolonged periods molecular associations will take place and precipitation will occur.) This solution (at pH 2) is applied to XAD-8 [(poly) methylmethacrylate, **XXI**, Fig. 1–5] and XAD-4 (styrenedivinylbenzene, **XXII**) resins in tandem (each equilibrated with 0.01 *M* HCl). It is appropriate to desalt the materials held on the columns by passing distilled water through the columns until the conductivity is <100 μS cm^{-1}. The XAD-4 column is then back eluted with 0.1 *M* NaOH. The eluates are passed through IR-120 resin and the H^+-exchanged products are freeze-dried. This provides the XAD-4 acids fraction.

There are several possibilities for dealing with the HS sorbed to the XAD-8. One of these could involve desalting and back eluting (NaOH) as described for the XAD-4 acids. The pH of the back eluate when adjusted to 2 (6 *M* HCl) gives rise to the HA precipitate, and this is allowed to settle at 4°C, and the FAs are separated by centrifugation. The FA solution is then desalted by passing onto XAD-8 (at pH 2) and washing with distilled water until the conductivity is <100 μS cm^{-1}. Back elution (with NaOH) follows, and the eluate is passed through IR-120 resin and freeze-dried as described for the XAD-4 acids.

In an adaptation of that procedure, the retentates (HAs and FAs) on the XAD-8 resin are back eluted in NaOH (without desalting), the pH of the center cut is adjusted to 2, and the remainder of the procedure is the same as that described.

The HAs may be dialyzed to remove salt. This, however, leads to losses of dialyzable materials. Hayes and Graham (2000) described the use of XAD-8 to desalt and to further fractionate the HA fraction. The HAs are dissolved in dilute (0.1 *M*) NaOH, and the solution is diluted until the concentration is <20 ppm. The pH is adjusted to 2, and the solution is applied to the XAD-8 column. It was found that during the desalting process, in the cases of HAs from a fibric peat (of low charge densities), some coagulated materials were washed through. The materials retained on the column were recovered as described for the FAs. The nonsolubilized materials that eluted during desalting were redissolved in dilute base, diluted as before, and applied to the XAD-8 column at pH 2.5. Again some coagulated material was recovered during desalting, and the material sorbed by the resin was recovered as before. The process was repeated at pH 3, and again at pH 3.5. All of the remaining material was retained by the column at pH 3.5, and was recovered as described. In this way four HA fractions were obtained. Two fractions were obtained when the HA from a Mollisol soil was fractionated on XAD-8. The NMR spectra indicated that the fractions were compositionally different.

The procedure described was developed to fractionate extracts isolated at discrete pH values in a sequential exhaustive extraction process. The technique can be adapted to fractionate a single extract, such as that by 0.1 *M* NaOH. The extract may be adjusted to pH 7 to 8, filtered, diluted, adjusted to pH 2, and applied to XAD-8 and XAD-4 in tandem. The XAD-4 materials may be recovered as described. The XAD-8 retentates may be desalted, and back eluted sequentially with aqueous media at pH 7, and at discrete pH values between 7 and 12. This gives a fractionation on the basis of charge density differences.

Similar procedures may be applied for the fractionation of organic materials isolated in organic solvent systems when the solvents are soluble in water. We have experience only with DMSO systems (Clapp and Hayes, 1996; Häusler and Hayes, 1996). When a solution of HS in DMSO/HCl is applied to an XAD-8 column at pH 2, the DMSO/HCl solvent passes through the resin and the HS are sorbed to the resin. Coextracted saccharides and peptides that are not covalently linked to the humic molecules move with the DMSO/HCl. XAD-8 and XAD-4 resins in tandem have not been used for this work, so it is not known to what extent the materials that move in the DMSO/HCl system will be sorbed by XAD-4. To remove the DMSO from the system, the column is washed exhaustively with 0.01 *M* HCl, then with distilled water.

Duxbury (1989) reviewed applications of different electrophoretic separation methods, including zone electrophoresis, moving boundary electrophoresis, isotachophoresis, and isoelectric focusing (IEF). Polyacrylamide gel electrophoresis (PAGE) is also used, and this procedure can also provide fractionation based on size differences.

COMPOSITION AND STRUCTURES OF HUMIC SUBSTANCES

What Do We Mean by Compositions and Structures?

The compositions of HS refer to the component elements, the functional groups, and the different types of the molecules or "building blocks" that compose the actual HS molecules. The *elemental composition* is perhaps the most fundamental characteristic of a chemical compound. This specifies the types of atoms in the compound and their relative abundances. For pure compounds the abundances or the ratios are defined by integral values (a single molecule cannot contain fractional numbers of atoms). The *empirical formula* is the smallest integral value that defines the elemental composition. Because HS are complex nonstoichiometric mixtures, the elemental compositions are average values for large assemblies of molecules of different compositions.

The term *structure* describes connectivity in a molecule, but connectivity does not completely define the geometry of a molecule. To establish structure it is important to specify configuration. Configuration defines not only the composition and connectivity of the molecule but also the orientations of all bonds in space. To establish the molecular structure of a pure compound, the *molecular formula* follows the determination of the elemental composition and the empirical formula. The molecular formula establishes the exact number of each type of atom in the molecule and can be deduced when the MW values and the empirical formulas are known. They may be determined directly by mass spectrometry, in the cases of volatile compounds. An average molecular formula can be determined for HS fractions, but the value of such determinations is questionable. In the cases of pure substances, the determination of a *structural formula* follows logically that of the molecular formula. The structural formula describes the connectivity in the molecule, or the linkages of the atoms in the molecule, and the nature of the bonds (e.g., single, double). In modern organic chemistry structural formulas are determined by spectroscopic procedures or by combinations of chemical and spectroscopic processes. *Structural isomers* have the same molecular formulas but different structural formulas. *Configuration* indicates how atoms or groups are spatially distributed around dissymmetric centers (such as an asymmetric C), or about a rigid part of a molecule (double bonds or small saturated rings). In saturated or partly saturated acyclic systems where rotation around a C–C or other single bond can take place, the atoms can take up numerous positions in space for brief periods. The term *conformation* is used to denote any one of these numerous possible molecular arrangements.

The terms *primary*, *secondary*, *tertiary*, and *quaternary structures* are used to describe the structures of biological polymers and macromolecules, and sometimes (wrongly in our viewpoint) there is reference to humic structures in these terms (Refer to Considerations of Protein Structures, p. 123, for a brief outline of the meanings of the primary to quaternary structures of proteins). Humic substances lack the order that is characteristic of biological molecules whose genesis is genetically controlled, and they lack also the order that a predetermined design provides for most synthetic polymers. In our search for structure we can hope at best to reach an awareness of the types of molecules (or building blocks) that compose HS, of the molecular environments of the reactive functional groups which these contain, of the sizes and shapes of the component molecules, and of the extent and the mechanisms by which these interact or associate with each other, with other soil organic molecules, with water, metals, minerals, and with anthropogenic chemicals that enter the soil environment by accident or by design.

Structural Considerations as Applied to Humic Molecules

The elemental compositions of HS can be determined with reasonable precision. These, however, are average values for mixtures of molecules, so a meaningful empirical formula cannot be obtained for a particular HS. Also, the MW values we have are average values. As is evident from our discussion of fractionation, no discrete humic fraction has been isolated which approaches chemical purity. We know that there are contained in humic mixtures identifiable long chain acids, esters, and hydrocarbons, but these are not classed as humic, even though a case might be made for their inclusion as HS, as discussed earlier in Genesis of Soil Organic Matter (p. 8).

Because empirical, and consequently molecular formulas cannot be deduced for humic molecules it is obvious that concepts of configuration and accurate structural formulas cannot be applied. Thus, the basic concepts inherent in the definitions of the empirical formula, structural isomerism, and stereoisomerism that are applicable to discrete compounds do not apply for HS, at least on the basis of what we know at this time. To do so would require that there be biological or genetic control of the synthesis process, and there is no evidence for such. Humic substances must be considered as gross mixtures.

We have discussed the degradative and synthetic approaches in Genesis of Humic Components of Humus (p. 11). Products from the degradative approach will have compositional and some structural features of the parent molecules. Should molecules arise from random condensation processes, especially where there are numerous interacting molecular species, then there will not be regularity in the structures. Regardless of whether genesis is by degradation or by synthesis, it can be concluded that it will be impossible to isolate sufficient amounts of humic molecules from any batch that will have identical structures. The determination of primary structures for HS that would satisfy the criteria used, for example, for protein structures, would require regularities in the repeating units (which they do not have) to have secondary structures. Because of the likelihood of extensive H bonding, especially in the cases of the H^+-exchanged substances, it is plausible to visualize random three-dimensional arrangements.

In the classical arrangements, tertiary structures of HS would involve the folding of primary and secondary structures in regular three-dimensional arrangements, and the quaternary structures would involve specific associations between molecular species. Such associations might confer biological specificities to some molecules and not to others, and there is no evidence for such at this time. It would seem pointless, therefore, to set out to establish rigorous classical primary to quaternary structures for humic molecules. Nevertheless such concepts provide a useful frame of reference for considerations of the compositions, aspects of general structures, and especially sizes, shapes, and associations of humic molecules.

Sizes and Shapes of Humic Substances

The MW values of HS have an important bearing on their sizes and on the shapes or conformations that these can adopt. Thus, the MW values have an important role in the reactivities of HS in soil and water environments. There are several procedures that can be used for determinations of MW values, but it is evident that the values given by the different procedures vary greatly (Aiken and Gillam, 1989; Clapp et al., 1989; Swift, 1989a, 1989b; Stevenson, 1994). The variability in values has been attributed either to innate differences in the humic molecules or to the limitations of the methods when applied to polydisperse systems. We review below the principles behind the major procedures that have been used for determining the MW values of HS. Then we outline some of the criticisms of data obtained from applications of some of the procedures used.

The classical procedures for determining the MW values of HS used the ultracentrifuge. Swift (1989b) outlined the principles and the procedures employed in ultracentrifugation studies. The analytical ultracentrifuge can be operated in the sedimentation velocity mode and the sedimentation equilibrium mode. Consider a solution of macromolecules in a cell within a rotor and subjected to centrifugal force by rotation about the rotor axis. The fundamental equation for sedimentation of the macromolecule (Schachman, 1959) is

$$\frac{\delta c}{\delta t} = D\left(\frac{\delta^2 c}{\delta r^2} + \frac{1}{r} \bullet \frac{\delta c}{\delta r}\right) - \omega^2 S\left(r\frac{\delta}{\delta r} + 2c\right) \qquad [3]$$

where c is the solute concentration, t is the time, D is the diffusion coefficient of the solute, r is the distance from the axis of rotation, ω is the angular velocity, and

S is the sedimentation coefficient defined as the radial velocity (dr/dt) of the solute molecules divided by the centrifugal field ($r\omega^2$) and is expressed as

$$S = \frac{1}{r\omega^2} \bullet \frac{dr}{dt} = \frac{d \ln r}{\omega^2 dt} \qquad [4]$$

The working formulas for determining MW values from ultracentrifugation data can be derived from Eq. [3]. For the *sedimentation velocity* method high rotor speeds (up to 60 000 rpm) are used, and the velocity and shape of the sedimenting boundary are studied. A MW value (M) is obtained from

$$M = \frac{RTS}{(1 - \bar{V}\rho_s)D} \qquad [5]$$

where R is the gas constant, T is the absolute temperature, $\bar{V}$ is the partial specific volume, and ρ_s is the density of the solution. A number of assumptions and approximations are made in the derivation of this equation. Nevertheless, the method works well for single, well-defined molecules or with a limited number of components that separate during centrifugation into compounds which each give a single well-defined boundary. It is not appropriate to use the technique for highly polydisperse (i.e., wide range of MWs or sizes) because that leads to a rapid spreading of the sedimenting boundary.

Equilibrium ultracentrifugation can also be employed. Here lower rotor speeds (≈15 000–20 000 rpm) are employed. The rotor is spun for an extended period until the distribution of the molecules in solution is in the equilibrium state. In this state the concentration of molecules throughout the cell is such that at any point in the cell the sedimenting forces are equal and opposite to the diffusion forces. The MW values are obtained from Eq. [5], which is derived from Eq. [3]

$$M = \frac{2RT}{(1 - \bar{V}\rho_s)\omega^2} \bullet \frac{d \ln c}{dr^2} \qquad [6]$$

In this method MW values can be determined directly from measurements made in the ultracentrifuge. The method can operate when there is a limited amount of polydispersity in the sample, and it allows determinations of number-averaged ($\bar{M}_n$), weight-averaged ($\bar{M}_w$), and z-averaged ($\bar{M}_z$) MW values, which can be used to estimate the degree of polydispersity of the sample. These are given by

$$\bar{M}_n = \frac{\Sigma n_i M_i}{\Sigma n_i} \qquad [7]$$

where n_i is the number of ith molecules, and its MW is M_i

$$\bar{M}_w = \frac{\Sigma w_i M_i}{\Sigma w_i} \qquad [8]$$

where the weight of the ith kind of molecule is w_i and its MW is M_i. The weight of the ith kind of molecule (w_i) in the mixture can be written as

$$w_i = n_i M_i \quad [9]$$

and by substituting for w_i in Eq. [7], $\bar{M}_w$ can be written as

$$\bar{M}_w = \frac{\Sigma n_i M_i^2}{\Sigma n_i M_i} \quad [10]$$

The z-average MW ($\bar{M}_z$) is obtained from

$$\bar{M}_z = \frac{\Sigma n_i M_i^3}{\Sigma n_i M_i^2} \quad [11]$$

The sedimentation velocity depends on the frictional coefficient (f) of the sedimenting species. This is given, whatever the shape or degree of solvation of the sedimenting species, by

$$f = kT/D \quad [12]$$

where k is the Boltzman constant. For a rigid, condensed, solvated sphere the frictional coefficient (f_o) is obtained from Stokes' Law:

$$f_o = 6\pi\eta R_o \quad [13]$$

where R_o is the radius of the sphere.

The frictional ratio (f/f_{min}) is used to study molecular conformation, where f is the observed frictional coefficient and f_{min} is the hypothetical frictional coefficient of a condensed solid sphere having the same volume as the molecule under study. Should the molecule being studied have the conformation of a condensed sphere, then $f/f_{min} = 1$. Values greater than unity reflect distortions of the spherical shape, and changes in hydration. For highly solvated spherical molecules, or for rod- or disc-shaped molecules, f/f_{min} can range from 2 to 7 or more (see Tanford, 1961; Swift, 1989b). Tanford (1961) showed, using the frictional relationships and the sedimentation equations, that

$$f_{min} = 6\pi\eta \left(\frac{7M\bar{V}}{4\pi N} \right)^{1/3} \quad [14]$$

where N is Avagadro's number, and

$$\frac{f}{f_{min}} = \frac{1}{3\eta} \left[\frac{k^2T^2}{9\pi^2} \bullet \frac{(1 - \bar{V}_\rho)}{\bar{V}} \right]^{1/3} \frac{1}{S^{1/2}D^{2/3}} \quad [15]$$

and

$$\frac{f}{f_{min}} = \left(\frac{4}{3} \right)^{1/3} \frac{1}{6\eta(\pi N)^{2/3}} \left(\frac{1 - \bar{V}\rho}{\bar{V}^{1/3}} \right) \frac{M^{2/3}}{S} \quad [16]$$

These equations show the combinations of data from which frictional ratios can be calculated.

Aiken and Gillam (1989) discussed how colligative properties can be used for determinations of MW values. A colligative property is, by definition, a thermodynamic property that depends on the number of particles in solution, and not on the nature of the particles. The influences that a solute can have on a solvent can be used for MW determinations. Thus the effects of a solute on lowering the boiling point (vapor pressure osmometry) or freezing point (cryoscopy) of a solvent, or of raising the boiling point (ebulliometry) can be used for the determination of the MW value of the solute, so too can the osmotic effects that the solute (membrane osmometry) will have on the solvent.

Aiken and Gillam (1989) provided a very relevant review of determinations of the MWs of HS by colligative property measurements, and they outlined the theory behind the determinations. Vapor pressure osmometry and cryoscopy have been the techniques most widely used, and these procedures are useful for determining MW values up to about 50 000 Da (Glover, 1975).

Clapp et al. (1989) discussed applications of viscosity measurements for studying sizes and shapes of HS.

If a solvent of known viscosity (η_o), and density (d_o) takes t_o s to travel between the upper and lower marks of the viscometer, whereas the same volume of the sample under investigation, with density, d, takes t s to travel that same distance, the *viscosity* (η) is given by

$$\eta = \eta_o \frac{td}{t_o d_o} \qquad [17]$$

and the *relative viscosity* (η_{rel}) is given by

$$\eta_{rel} = \left(\frac{\eta}{\eta_o}\right) = \frac{td}{t_o d_o} \qquad [18]$$

The relative increase in the viscosity of the solution over that of the solvent is known as the *specific viscosity* (η_{sp}), where

$$\eta_{sp} = \frac{\eta - \eta_o}{\eta_o} = \eta_{rel} - 1 \qquad [19]$$

The specific viscosity is a direct measure of the increase in viscosity resulting from the presence of macromolecules. The specific viscosity divided by the concentration, c, of the solute, is known as the *reduced viscosity* (η_{red}), which is also referred to as the *viscosity number*. Values of reduced viscosity generally range from 0.02 to 0.05 for spherical molecules and from 0.5 to ≥5 for thread-like molecules.

The *intrinsic viscosity* ([η]), or the *limiting viscosity number*, is widely used for calculating MW values and molecular sizes. The relationship is given by

$$[\eta] = \left(\frac{\eta_{sp}}{c}\right)_{c\to 0} = \left(\frac{\eta_{rel} - 1}{c}\right)_{c\to 0} \qquad [20]$$

and the intrinsic viscosity (mL g^{-1}) is obtained by extrapolating to zero concentration in a plot of η_{sp}/c vs. sample concentration (c).

For a solution containing spherical molecules much larger than the solvent molecule, and at a concentration sufficiently small that interactions between the large spherical molecules may be neglected, Einstein (1906) showed that

$$\eta_{sp} = 2.5\ \phi \qquad [21]$$

where ϕ is the volume fraction occupied by the spherical molecules and can be expressed as

$$\phi = N v_h c/M \qquad [22]$$

where N is Avagadro's number, v_h is the hydrodynamic volume (i.e., the volume of the particle plus incorporated solvent) of the dissolved species, c is its concentration in g mL^{-1}, and M is its MW. From Eq. [19], [20], and [22], we obtain

$$[\eta] = 2.5(N v_h/M) \qquad [23]$$

which is independent of MW, and is assumed to be proportional to M.

The sedimentation velocity ultracentrifugation procedure used by Cameron et al. (1972b) referred to earlier (p. 42) has received most attention because of the efforts made, using gel chromatography and pressure filtration through graded porosity membranes, to obtain fractions that had relatively uniform molecular sizes for determinations of MW values. Values of 2400 and 4400 Da were obtained for fractions isolated in $Na_4P_2O_7$ and fractionated using gel chromatography. Values ranging from 12 800 to 412 000 Da were obtained for isolates in NaOH at 20°C, fractionated using gel chromatography and pressure filtration techniques. Values of 408 000 and 1 360 000 Da were obtained for fractions isolated in NaOH at 60°C (these latter samples were later found to contain silicate).

A linear relationship was obtained for the plot of the frictional ratio values (f/f_{min}) (Eq. [15] and [16]) obtained from ultracentrifugation data, against the log of MW values (Fig. 1–7). The equation for the straight line ($f/f_{min} = 0.30M^{1/6}$) fitted the relationship for nonbranched, randomly coiled macromolecules. The nonlinearity observed for the high MW samples was attributed to branching, and not to cross-linking, but now it is considered that the presence of silicates may have given rise to pseudo-high MW properties.

Stevenson et al. (1953) were the first to apply ultracentrifugation techniques (using the sedimentation velocity approach) for studies of the sizes and shapes of humic molecules. Sedimentation velocity ultracentrifugation studies by Flaig and Beutelspacher (1954, 1968) and by Piret et al. (1960) followed. Unlike the work of Cameron et al. (1972b), these earlier studies were performed on humic materials that had not been fractionated. Wershaw and Aiken (1985) pointed out that semi-empirical methods, such as sedimentation velocity ultracentrifugation, are not suitable for polydisperse systems because of the multiple diffusion coefficients and sedimentation constants for different size particles. The use of salt as background electrolyte is important to suppress the repulsive effects of the negative charges, but

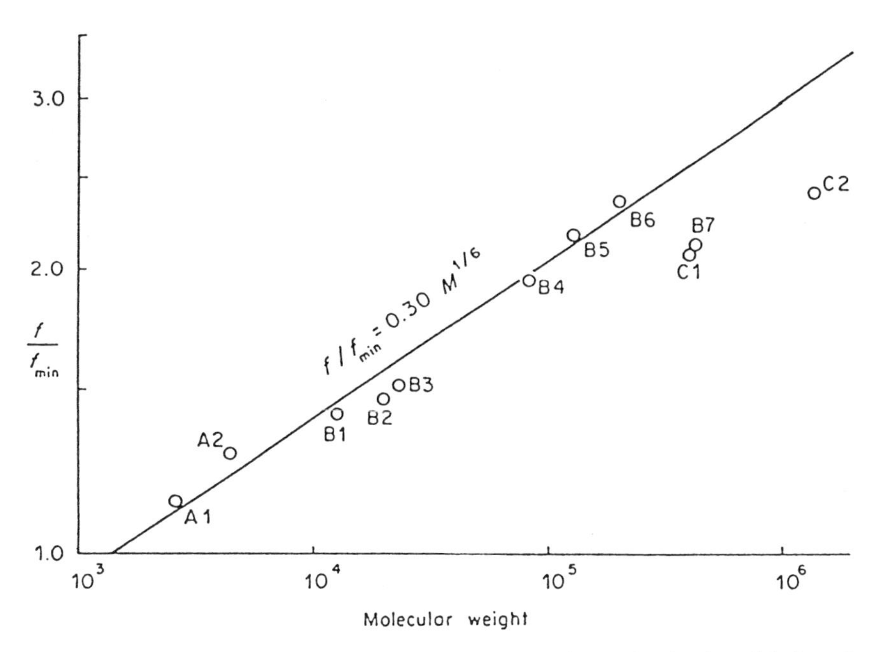

Fig. 1–7. Experimental relationship between the frictional ratio (f/f_{min}) and molecular weight for various humic acid fractions. The line is the theoretically derived relationship between frictional ratio and molecular weight for nonbranched, randomly coiled macromolecules (after Cameron et al., 1972b).

when Flaig and Beutelspacher (1968) used 0.2 *M* NaCl, they calculated a MW value of 77 000 Da for their polydisperse system, and a value of 2050 Da in the absence of the electrolyte. This huge increase could suggest aggregation in the presence of the electrolyte. Measurement of sedimentation coefficients of polydisperse materials, which include subunits, invariably leads to erroneous values of MW (Laue and Rodhes, 1990). Other studies using sedimentation velocity ultracentrifuge studies (Ritchie and Posner, 1982) and the more mathematically sound equilibrium centrifugation (Posner and Creeth, 1972; Reid ct al., 1990) also showed polydispersity in HS, and, for this very reason, confirmed the ambiguity of the MW values obtained by ultracentrifuge methods.

There are fewer discrepancies in the MW values of FAs isolated from waters, and values in the range of 614 to 1000 Da were quoted by Aiken and Gillam (1989) for a number of researches where cryoscopy and vapor phase osmometry procedures were used. The work of Thurman et al. (1982) showed how higher MW values were obtained when FAs were analyzed using ultrafiltration and gel filtration procedures than when osmometry and cryoscopy methods were applied. That might be explained by molecular associations occurring during the gel chromatography and pressure filtration procedures.

The concept of molecular associations is receiving wider acceptance since Wershaw (1986, 1993) proposed that HS consist of ordered aggregates of amphiphiles (compounds with hydrophobic stretches, as well as charged or polar centers), composed mainly of altered plant molecular structures that exhibit acidic functionality. The humic molecules would, in this model, be held together by hydrophobic bonding, charge transfer, and H bonding interactions. Piccolo (2001) considered that the molecular associations represent a self-assembling supramolecular association of HS arising from the "mutual affinities of certain molecules in aqueous solutions." The associations arise from intermolecular forces (Israelachvili, 1994), and the strengths of the associations depend on their molecular structures. The progressive associations of these structures can be observed in dilute solutions of HAs at pH values in which all the acidic groups are protonated. The hydrophobic components might also be contributed by nonhumic moieties, such as hydrocarbon and fatty acid/ester structures in the medium.

This model of Wershaw suggests micelle-type associations of molecules in contrast to the polydisperse random coil types of structures that had wide acceptance. However, the classical concept of an ordered micelle would not apply because of the heterogeneity of HS. A critical micelle concentration in the range of 1 to 10 g L^{-1}, reported by Hayase and Tsubota (1983), is much higher than that expected for surface active compounds having regular micellar structures. Although Engebretson and von Wandruszka (1994, 1997) and Engebretson et al. (1996) did not dismiss the concept of macromolecularity, they deduced, from observations of the quenching of pyrene (**XXIII**, Fig. 1–8) fluorescence by Br^- ions in the presence of HS, that humic molecules could spontaneously aggregate through hydrophobic associations and the pyrene that associated with the hydrophobic moieties in the interior structures would be protected from the Br^- ions. Kenworthy and Hayes (1997) also used the fluorescence quenching of pyrene by Br^- to investigate the nature of humic associations. They used HAs and FAs isolated in distilled water and at pH 7, 10.6, and 12.6 from the A_0 horizon of a Podzol soil and found that pyrene was more protected by the HAs than the FAs. The protection increased for both the HAs and FAs as the pH of the extraction media increased. The protection was lost, however, when ethanoic acid (AcOH), followed by a base, was added to the medium. The authors considered that hydrophobic associations of the humic mol-

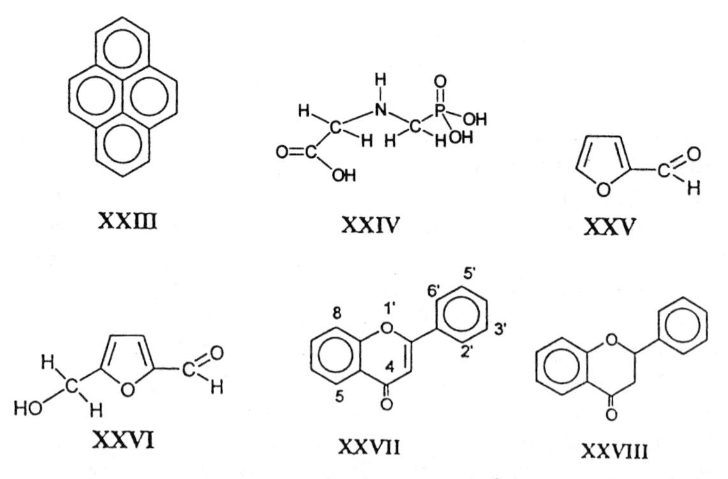

Fig. 1–8. Pyrene, (**XXIII**), glyphosate (**XXIV**), furfural (**XXV**), hydroxymethylfurfural (**XXVI**), chrysin (**XXVII**), and butein (**XXVIII**) mentioned in the text (after Miano et al., 1992; Engebretson and von Wandruszka, 1997; Clapp et al., 2001).

ecules protected the pyrene from the Br^-, when the pyrene had entered the molecular associations and associated with the hydrophobic sites in the humic structures. The extraction processes used were sequential (see Extraction of Humic Substances under Basic Conditions, p. 32); the more polar components were removed at the lower pH values, and the more hydrophobic moieties at the higher pH values. Treatment of the humic solution with AcOH was prompted by the proposal of Piccolo et al. (1996), which suggested that the AcOH treatment followed with base caused the humic aggregates to break up. Then the relatively low MW disaggregated HA and FA moieties would no longer provide protection for the pyrene fluorophore.

Fluorescence and infrared spectroscopy was used by Miano et al. (1992) to investigate the interactions of glyphosate [*N*-(phosphonomethyl)glycine, **XXIV**] herbicide with dissolved humic material. The pH of the solution was progressively lowered from 9. Increasing fluorescence quenching effects with increasing glyphosate contents at high wavelengths were observed in the excitation spectra. A decrease in the main peak intensity with increasing additions of herbicide was observed for the synchronous spectra. These results could be interpreted in terms of a disaggregation of the humic supramolecular association arising from the formation of multiple hydrogen bonds between glyphosate and the small humic molecules.

Kohl et al. (2000), using solid-state ^{19}F-NMR spectroscopy, found that the uptake of hexafluorobenzene by a peat containing its indigenous lipids was more rapid and complete than by the same peat when the lipids were removed.

In the study of Piccolo et al. (1996), the pH of a HA solution was lowered from 9.2 to 2.0 using either HCl or organic acids, applied to a gel column and eluted with 0.02 or 0.1 mol L^{-1} borate buffer. The elution profile from the gel chromatography column suggested to the authors that the organic acid treatment caused the HAs to be eluted in the low molecular size range, close to the column total volume (V_t). The process was reversible. To explain the results, the HAs were considered to be supramolecular associations of relatively small heterogeneous molecules held together by weak dispersive forces (see Piccolo, 2001). This thesis has been subjected to criticism.

Applications of gel chromatography procedures for determinations of molecular sizes and the concept proposed by Piccolo et al. (1996) were subjected to scrutiny in a series of articles in *Soil Science* (Vol. 166, No. 11, 1999). In that series DeNobili and Chen (1999) outlined some of the principles of gel chromatography that are important for the fractionation of HS. Perminova (1999) outlined the problems encountered in applications of size exclusion chromatography (SEC) for the fractionation of soluble polyelectrolytes using hydrophilic gels. She emphasized especially the problems that arise from ionic exclusion and specific adsorption and cited methods to counter these effects. She considered that the results obtained by Piccolo et al. (1996) could be explained by the high concentration of AcOH used, the salts generated, and the pH changes in the gel column.

In his challenge of the concepts of self-aggregation of smaller molecules into larger assemblies, as set out by Wershaw (1986, 1993) and by Piccolo et al. (1996), Swift (1999) emphasized the need to eliminate interferences caused by charge interactions, sorption reactions, electrolyte and nonelectrolyte concentrations, pH effects, and other factors. All of these can interfere with the fractionation process and change the elution pattern. Thus, the fractionation process and elution behavior would not be based on differences in molecular sizes. He expressed especial concern about the significant amount of colorless material eluted in the total column volume (V_t) in the work of Piccolo et al. (1996), as detected by refractive index (RI) measurements. He also referred to the fact that salts and simple organic molecules would be eluted at V_t. Although Swift agreed that accumulations of hydrophilic and hydrophobic moieties could take place in the solid phase, he considered that, because of the polyelectrolyte nature of HS, such would be unlikely to take place in solution because of charge repulsion effects. He was more amenable to the suggestion by von Wandruszka (1998) of the formation of intramolecular pseudomicelles. This idea could agree with the concept of flexible macromolecules capable of conformational rearrangements in solution to give accumulations of hydrophobic moieties within the molecular structure.

Despite the criticisms, Piccolo has provided strong new evidence to support his molecular associations concept (Piccolo, 2001; Piccolo et al., 2001). These later studies used high pressure (HP) SEC, avoided high concentrations of organic acids ($<0.5 \times 10^{-3}$ *M*), and avoided high ionic strengths. To a control eluent at pH 7 (0.05 *M* $NaNO_3$) was added 2.0×10^{-6} *M* of either methanol, HCl, or AcOH, which gave pH values of 6.97, 5.54, and 5.69, respectively. The ionic strength did not change ($I = 0.0506$ *M*). All of the results indicated that the treatments broke humic molecular associations into smaller component molecules. It is relevant to note that when standards of known compositions, such as a neutral polysaccharide and the negatively charged polyelectrolyte, (poly) styrenesulfonate, were subjected to the same procedure, elutions from the columns were as predicted.

One of the authors (MHBH) has been a firm supporter of the macromolecular concept for HS (Hayes and Swift, 1978, 1990; Hayes et al., 1989a, 1989b). He holds in high regard the work Swift led in fractionating HAs into discrete sizes using GLC and pressure filtration techniques (Cameron et al., 1972a, 1972b). He accepts the thoroughness of the work of Perminova (1999). He also finds it difficult to explain the question Swift raises with regard to the improbability that discrete fractions of HAs (isolated in the work of Cameron et al.) would reassemble into aggregates of the same molecular sizes after dispersion. These fractions were isolated from a Sapric Histosol, and we have observed that such soils are rich in aliphatic hydrophobic moieties. Such materials could be the glue that held the humic components together. However, whether or not specific glues would give rise to molecular associations of relatively discrete molecular sizes, such at those described by Cameron et al., is far from proven.

Despite the fact that there are pitfalls associated with the use of high concentrations of organic acids to break up associations of humic molecules, evidence from our work supports the concept of molecular associations in HS. Kenworthy and Hayes (1997) showed how pyrene is released when what might be considered to be HA aggregates or molecular associations are broken by additions of AcOH. By means of diffusion ordered spectroscopy (DOSY)-NMR spectroscopy Simpson et al. (2002) showed the types of associations that can constitute HA fractions (see NMR Studies of Composition and Associations, p. 97).

Degradation Studies and Composition

The traditional approach to studies of the molecules that compose HS uses identifications of compounds in digests of chemical degradation reactions. In most degradation reactions the compounds identified in the digests cannot be considered to be the exact component molecules or the building blocks in the humic structures. However, from awareness of the degradation mechanisms and the changes that can take place during residence in the digests of the structures released when bonds linking the component molecules are broken, it is possible to get indications of the types of component molecules and of their linkages in the structures. Hayes and Swift (1978) introduced considerations of such mechanistic approaches. More recently, advances in the techniques of Py–GC–MS are providing similar types of information, although we consider that the chemical degradation processes still provide the more interpretable data. Striking advances have been made of late in applications of NMR procedures, and these suggest that when better fractionation is achievable, it will be possible to make complete identifications of component humic molecules and of their linkages (see NMR Studies of Composition and Associations, p. 97).

Information from Hydrolysis Processes

Parsons (1989) reviewed the degradation mechanisms and the digest products formed when HS are hydrolyzed under basic and acidic conditions.

The components of the biological polymers, such as polysaccharides, peptides, proteins, glycoproteins, and nucleic acids have labile linkages that are readily hydrolyzed. Although it is possible that such molecules can be linked to humic structures, it seems likely that such does not happen to a significant extent. The polysaccharides and peptide structures, for example, are separate entities in the soil environment and are not covalently linked to the humic components (see Simpson et al., 2002). The information we have suggests that the component molecules of HS are relatively simple, but these molecules or units would appear to be linked by bonds that are, for the most part, difficult to cleave. For example, the "backbone" structures could be linked by ether linkages, by C–C bonds, or by other types of linkages that are degraded only when high inputs of energy are used.

Studies by Schnitzer and Neyroud (1975) showed that significant yields of fatty acids were obtained when FAs were hydrolyzed for 3 h at 170°C in 2 *M* NaOH. The most general mechanism involves cleavage of the acyl–oxygen bond, as shown in Reaction Scheme [24]. R and R′ can be aliphatic or aromatic substituents. Attack by OH^- takes place at the carbonyl C, and proceeds by a second-order mechanism (March, 1985, p. 290) in a reaction that is essentially irreversible:

$$\mathrm{R{-}C({=}O){-}OR'} \overset{OH^-\ (slow)}{\rightleftharpoons} \mathrm{R{-}C(O^-)(HO){-}OR'} \longrightarrow \mathrm{RC({=}O)OH} + \mathrm{RO^-} \longrightarrow \mathrm{RC({=}O)O^-} + \mathrm{R'OH} \qquad [24]$$

In the case of hydrolysis of peptides and proteins, the attack by OH^- takes place at the carbonyl C of the peptide bond, and the carboxylate and the amino functionalities are liberated in the amino acids released. Acid anhydrides and lactones will hydrolyze by similar mechanisms under basic conditions.

Hydrolysis and cleavage of double bonds will take place under alkaline conditions to release hydrocarbons and acids to the digest. Shemyakin and Shchukina (1956) have illustrated the reaction mechanisms, as outlined in Reaction Scheme [25], where again the Rs represent appropriate aliphatic or aromatic substituents. The mechanism also illustrates how ketonic functionalities can release carboxylate and hydrocarbon structures:

$$R_2C{=}CR_2 \xrightleftharpoons{H_2O} R_2C(OH){-}CR_2H \xrightleftharpoons{OH^-} R_2C(O^-){-}CR_2H \longrightarrow R_2C{=}O + H{-}CR_2{-}H$$

[25]

Under mildly alkaline conditions, β-keto esters decarboxylate as outlined in Reaction Scheme [26] to yield a ketone, CO_2, and an alcohol:

$$R^1{-}C(=O){-}C(R^2)(R^3){-}C(=O){-}OR^4 \xrightarrow[(OH^-,\ mild)]{H_2O} R^1{-}C(=O){-}C(R^2)(R^3){-}H + CO_2 + R^4OH$$

[26]

However, under strongly basic conditions, cleavage would take place at the keto functionality to give ester and acid products, as outlined in Reaction Scheme [27]:

$$R_1{-}C(=O){-}C(R_2)(R_3){-}C(=O){-}OR_4 \xrightarrow[(heat)]{OH^-} R_2(H)(R_3)C{-}C(=O){-}OR_4 + R_1{-}C(=O){-}O^-$$

[27]

Acids catalyze the hydrolysis of the functionalities mentioned above. Reaction Scheme [28] illustrates the acid catalysis of an ester involving the initial protonation of the carbonyl oxygen and leading to cleavage of the acyl–oxygen bond:

$$R{-}C(=O){-}OR' \xrightleftharpoons{H^+} R{-}\overset{+}{C}(OH){-}OR' \xrightleftharpoons{H_2O(slow)} R{-}C(OH)(H_2O^+){-}OR' \rightleftharpoons$$

$$(HO)_2C(R){-}\overset{+}{O}(H){-}R' \xrightleftharpoons{R'OH(slow)} HO{-}\overset{+}{C}(R){-}OH \xrightleftharpoons[H^+]{-H^+} R{-}C(=O){-}OH$$

[28]

In the case of acid hydrolysis of substituted amides, which includes peptides and proteins, protonation primarily takes place on the oxygen, although protonation of the N atom can also take place. The mechanisms are illustrated in Reaction Scheme [29]:

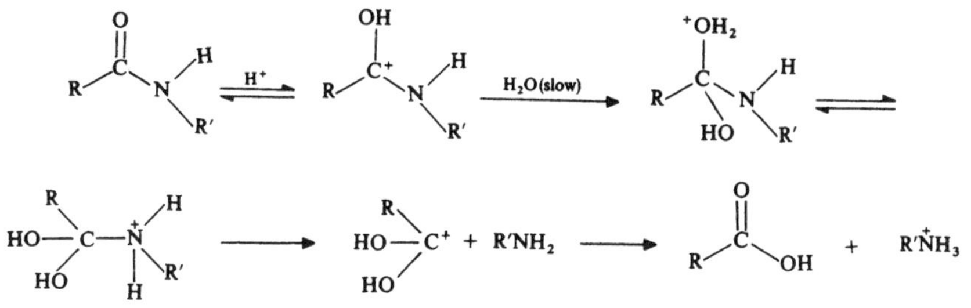

[29]

An understanding of the mechanisms of acid hydrolysis of oligo- and of polysaccharides can be obtained from a consideration of the hydrolysis of cellobiose, a disaccharide in which two β-D-glucose units are linked (1 → 4) and that has an acetal structure of the general type ROC(H) (OR″)R′, where R″ is the O–C-4 linked residue, as well as the hemiacetal functionality of the non-linked (to another sugar residue) C-1 (see Some Relevant Aspects of Sugar Chemistry, p. 103).

To facilitate hydrolysis, the alkoxy group (OR″) is converted to the conjugate acid, and the hydrolysis process then proceeds as outlined in the Reaction Scheme [30] to give the hemiacetal structure, which will be glucose in the case of the hydrolysis of cellobiose:

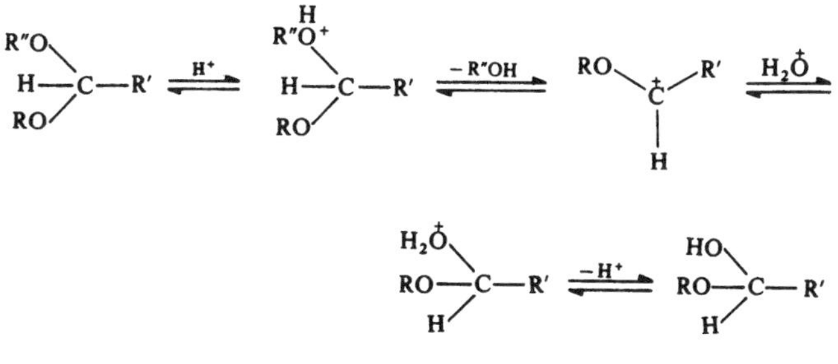

[30]

During the course of acid hydrolysis, up to 50% of the masses of soil HAs can be lost as CO_2 (Parsons, 1989), especially where carboxyl groups are activated by functionalities such as β-keto groups, and where two or three hydroxyl substituents are present in benzenecarboxylic acid structures [e.g., gallic acid (**VI**, Fig. 1–2), or 3,4,5-trihydroxybenzenecarboxylic acid, which decarboxylates even in hot water], and as soluble molecules (e.g., sugars, amino acids, small amounts of purine and pyrimidine bases, and phenolic substances). Decarboxylation of β-keto acids is well known under conditions of acid hydrolysis. The process, which gives rise to enol and keto species, as well as CO_2 is illustrated in Reaction Scheme [31], and this mechanism can be extended to decarboxylation of β-dicarboxylic acids:

[31]

Decarbonylation can also take place, and α-keto esters are decarbonylated on heating under acid conditions, as indicated in Reaction Scheme [32], and decarbonylation of β-hydroxy and of β-keto acids can also take place in acidic media:

[32]

Hydrolysis in 6 *M* HCl has been used as a pretreatment in studies of the degradation of HAs (Atherton et al., 1967; Riffaldi and Schnitzer, 1973). That is the classical procedure for liberating the amino acid components in proteins. However, pentose sugars dehydrate under these strongly acidic conditions and lose three molecules of water to give furfural (**XXV**, Fig. 1–8) and hexoses dehydrate to give hydroxymethylfurfural (**XXVI**). Polymerization of these, including interactions with the amino acids liberated to give Browning reaction (melanoidin) products, might account for the increase in the MW observed by Posner and Creeth (1972) when HAs were hydrolyzed in 6 *M* HCl. Even boiling in water can lead to a degree of decarboxylation, as was observed by Cranwell and Haworth (1975) when 30% of the mass was lost as CO_2. A further 15% was lost when the residue was subsequently boiled in 6 *M* HCl.

Acid hydrolysis leads to increases in the C contents and decreases in the N contents of HAs (Cheshire et al., 1967; Riffaldi and Schnitzer, 1973). By means of ^{13}C-NMR spectroscopy, Schnitzer and Preston (1983) observed that amino acid and sugar residues can effectively be removed during hydrolysis. They observed that aromaticity is increased, and this could reflect the losses in aliphatic hydrolyzable residues, or it could reflect the synthesis of new materials. Such observations were also reported by Preston and Ripmeester (1982). However, despite the losses of CO_2 observed, Riffaldi and Schnitzer (1973) found that the charge and the free radical contents of the hydrolyzed HAs were not altered significantly by the acid hydrolysis process. Similar observations with regard to total acidity were made by O'Callaghan (1980). Even when losses of residues released in the hydrolysis are taken into account, it would be necessary to have new acid groups formed (e.g., by the hydrolysis of esters, and lactones) to account for the contents of acidic functionalities detected in the hydrolyzed residues.

Information from Transesterification Processes

A mild boron trifluoride-methanol transesterification procedure was applied extensively for the partial depolymerization of wood by Browning (1997). Almendros and Sans (1989, 1991, 1992) introduced the procedure for studies of component molecules of HAs and humins. The rationale for this approach is that ester functionalities linked to the humic core would be cleaved to set free labile (derivatized) structures. The –COOH and –OH functionalities liberated are methylated in situ and recovered as the ester and methoxyl derivatives, respectively. The procedure is relatively mild, and artifact formation is minimized. In the cases of the studies of Almendros and Sans, yields of identifiable products were 30 to 35% of the mass of the starting material. A variety of monobasic straight chain and branched fatty acids, long chain dicarboxylic acids, di-hydroxymonobasic acids, tri-hydroxymonobasic acids, methoxybenzenecarboxylic acids, di-, tri-, and tetra-methoxybenzenecarboxylic acids, and a variety of other miscellaneous acids were identified in the humin digests. Although it may well be that these acids were present as esters in the humins, many (and especially the long chain aliphatic acids) are characteristic of cutins and suberins and of degradation products of plants. In view of the suggestions made by Rice and MacCarthy (1989, 1990) following their isolation of components of humin, it may well be that the long chain acids released from the humin in the transesterification process are not covalently linked to the humic components of humin.

As used by Simpson (1999), the procedure involved heating humic samples and BF_3 in methanol (20%) in a Teflon-lined stainless-steel bomb at 90°C for 16 h. Then, after addition of distilled water the mixture was extracted with chloroform and the extract was analyzed by gas chromatography–mass spectrometry (GC–MS). This transesterification procedure was the first step in the sequential degradation of HAs isolated from the A_h and B_h horizons of a Podsol soil that had been under oak forest for >3000 yr, from the B_h horizon of a neighboring Podzol from which the forest was cleared 400 yr BP, and for the FA from the B_h horizon of the forested soil. More than 30 esters, some of which also had ether functionalities, were identified in the digests of all of the fractions. The abundances of the compounds in the HA digests were similar, but different from these in the FA digest. Esters of aliphatic dicarboxylic acids were the most abundant species in the digests, and accounted for 50 to 60% of the digest products identified, and the methyl esters of octanedioic [$CH_3O_2C(CH_2)_6CO_2CH_3$] and nonanedioic [$CH_3O_2C(CH_2)_7CO_2CH_3$] acids accounted for 40% of the abundance of compounds identified in the HA digests. Strangely, these two esters were not present in the digests of the FA sample, where the most abundant species were the methyl ester of β-methoxy propanoic acid ($CH_3OCH_2CH_2CO_2CH_3$), the methyl diester of succinic (butanedioic, acid, [$CH_3O_2C(CH_2)_2CO_2CH_3$], the methyltriester of hexanetrioic (or propane-1,2,3 tricarboxylic) [$CH_3O_2CCH_2CH(CO_2CH_3)CH_2CO_2CH_3$] acid, the methyl ester of the β-dimethoxy propanoic acid [$(CH_3O)_2CHCH_2CO_2CH_3$], and the methyl ester of β-phenyl pentanoic [$CH_3CH_2CHØCH_2CO_2CH_3$, where Ø is the phenyl (C_6H_5) group] acid. These acids composed 22, 17, 17, 16, and 14% of the compounds identified in the FA digest. Strangely the heptanedioic, octanedioic, and nonanedioic esters were not detected in the digests of the FA.

The diesters identified are likely to have been derived from the methylation by means of the BF_3–MeOH treatment of dicarboxylic acids present in the starting materials. Short chain dicarboxylic acids occur in the soil as the result of a multistage breakdown of longer chain hydrocarbons. This is illustrated in Reaction Schemes [33], [34], and [35] (from Alexander, 1977). The process involves initial conversion to a fatty acid (Reaction Scheme [33]) followed by β-oxidation of the fatty acid (Reaction Scheme [34]), and finally oxidation of the terminal methyl group to give the dicarboxylic acid:

$$CH_3(CH_2)_nCH_2CH_2CH_3 \rightarrow CH_3(CH_2)_nCH_2CH_2CH_2OH \rightarrow$$
$$CH_3(CH_2)_nCH_2CH_2C(O)H \rightarrow CH_3(CH_2)_nCH_2\ CH_2CO_2H \qquad [33]$$

$$CH_3(CH_2)_nCH_2CH_2CO_2H \rightarrow CH_3(CH_2)_nC(O)CH_2CO_2H \rightarrow$$
$$CH_3(CH_2)_nCO_2H + CH_3CO_2H \qquad [34]$$

$$CH_3(CH_2)_nCO_2H \rightarrow HOOC(CH_2)_nCOOH \qquad [35]$$

Alexander (1977) discussed the capabilities of bacteria species *Mycobacterium, Nocardia, Pseudomonas, Streptomyces, Corynebacterium, Acetobacter,* and *Bacillus*; yeast species *Candida* and *Rhodoturta*; and several species of fungi to degrade long chain hydrocarbons.

Jones and Edington (1968) considered that 3 to 17% of organisms in the soil surface horizon will degrade hydrocarbon chains in the range of $CH_3(CH_2)_6CH_3$ to $CH_3(CH_2)_{14}CH_3$, but longer and shorter chain hydrocarbons were considered to be more resistant to attack. Thus, $CH_3(CH_2)_5CH_3$ is more stable than $CH_3(CH_2)_6CH_3$, and the abundance of $HOOC(CH_2)_6COOH$ might reflect the ease of oxidation of the $CH_3(CH_2)_6CH_3$ species, and the fact that about 30% of the diester of $HOOC(CH_2)_7COOH$ was present in the digest might reflect the relative resistance of C-7 carbon chains.

The fact that no diesters of acids with more than five C atoms were detected in the FA fraction can be attributed to the decrease in the solubilities of these acids as the C chain length increases. It is probable that the hydrophobicity of the longer chain acids (esters) at low pH values contributed to the precipitation of the components in the HA fraction. The significantly higher abundances of the hydrophilic acids in the FA fraction is further evidence for that concept.

Information from Reductive Degradation Processes

Reductive reactions take place when hydrogen is added to a compound and new covalent linkages are formed. The most widely used reducing agents are hydrogen, usually used in the presence of a solid catalyst utilizing metals such as Zn and Na which transfer electrons to the substrate, and metal hydrides, which transfer hydride (H^-) anions. Hydride transfer also takes place in basic media to give rise to oxidation and reduction products, as in the Canizzaro reaction (Reaction Scheme [36]), for example:

$$RHC{=}O + OH^- \longrightarrow R(HO)C(H)O^- + RCH{=}O \longrightarrow RCOOH + {}^-OCH_2R \longrightarrow RCOO^- + RCH_2OH \qquad [36]$$

The most widely used reagents in reduction reactions of HS have involved uses of Na amalgam, Zn dust distillation, and Zn dust fusion reactions. Of these, the Na amalgam process is relatively mild, whereas conditions used in Zn dust distillation and fusion processes are drastic. Reviews of reduction procedures are given by Hayes and Swift (1978) and by Stevenson (1989).

Reduction with Sodium Amalgam. The Na amalgam (Na/Hg) procedure for the degradation of HAs was introduced by Zetsche and Reinhart (1939), who observed that the dark color of HA disappeared on heating in the presence of Na amalgam. The dark color was restored when the digest was exposed to air.

Later Farmer and Morrison (1960) observed a similar effect. When they methylated the reduced product using dimethyl sulfate, they found that the methylated, reduced product had increased absorption bands in the aliphatic region of the infrared spectrum. Burges et al. (1963, 1964), using a Na amalgam degradation procedure, isolated yields of 30 to 35% of ether soluble phenolic products from the digests, and were first to identify compounds among these that could be related to meaningful components in humic molecules.

Model studies involving degradations of flavonoids with Na amalgam have helped our awareness of the degradation mechanisms involved. Chrysin (**XXVII**, Fig. 1–8), a flavone, and butein (**XXVIII**, a flavanone) are parent flavonoid structures, and the numbering system used is shown for chrysin. In Fig. 1–9 we show an outline of the cleavages brought about by Na amalgam in the case of the flavone apigenin (**XXIX**) and of the digest products identified by Hurst and Harborne (1967). Attack on the C–O–C linkage of apigenin gives phloretin (**XXX**), which degrades to phloroglucinol (1,3,5-trihydroxybenzene, **XXXI**) and resorcinol (1,3-dihydroxybenzene, **XXXII**) from the A ring, and 3-(4-hydroxyphenyl) propanoic acid (**XXXIII**), and 3-(4-hydroxyphenyl)propan-1-ol (**XXXIV**) from the B ring.

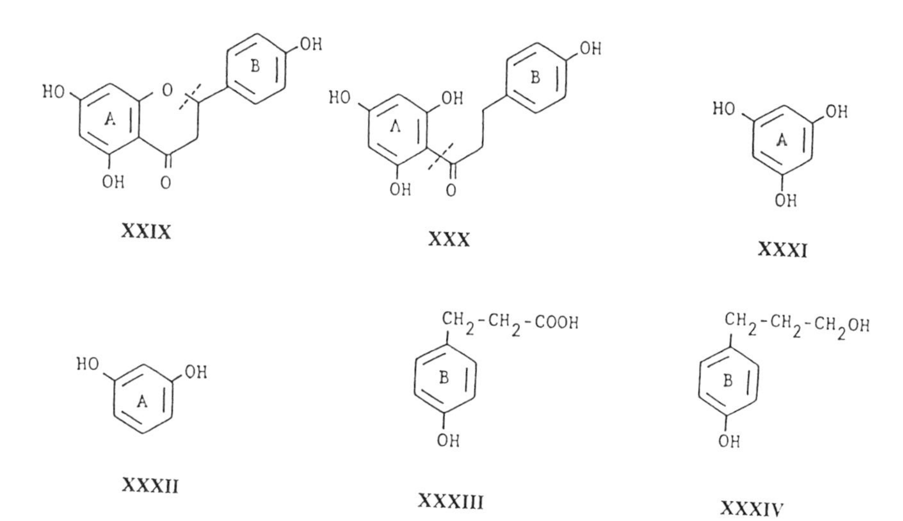

Fig. 1–9. Products from the reductive cleavage of apigenin (**XXIX**) by sodium amalgam under reductive conditions (after Hurst and Harborne, 1967).

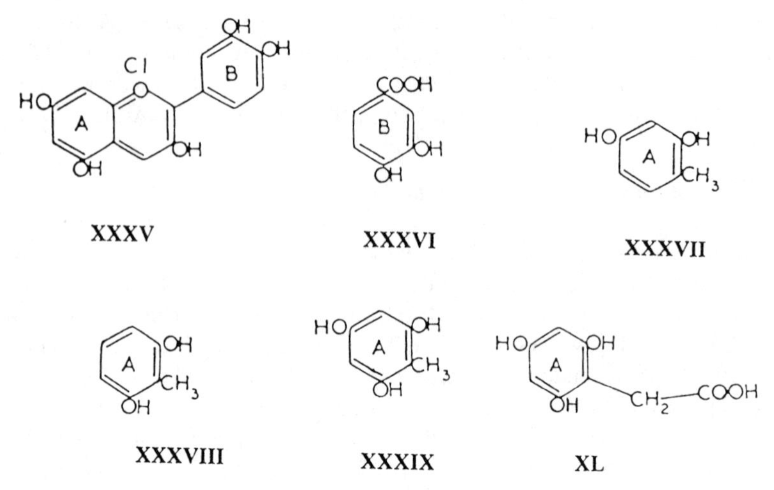

Fig. 1–10. Products identified in the digests of the reductive cleavage of cyanidin (**XXXV**) in sodium amalgam under alkaline conditions (after Burges et al., 1964).

Further model studies by Burges et al. (1964) identified an array of phenolic compounds formed in the digests of cyanidin (**XXXV**, Fig. 1–10) in Na amalgam under alkaline conditions. The phenols included phloroglucinol (**XXXI**, Fig. 1–9) and resorcinol (**XXXII**), 3,4-dihydroxybenzenecarboxylic acid (protocatechuic acid, **XXXVI**, Fig. 1–10), 2,4-dihydroxytoluene (**XXXVII**), 2,6-dihydroxytoluene (**XXXVIII**), 2,4,6-trihydroxytoluene (**XXXIX**), and 2,4,6-trihydroxyphenylethanoic acid (**XL**).

In their investigations of the mechanisms of degradations with Na amalgam, Piper and Posner (1972a, 1972b) considered that attack takes place, as can be deduced from considerations of the degradation products in the digests of apigenin (**XXIX**, Fig. 1–9) and of cyanidin (**XXXV**, Fig. 1–10), through atomic or nascent hydrogen ($H^{\bullet}$) on electron-rich areas of the substrate molecule. Because atomic hydrogen ($H^{\bullet}$) atoms readily combine to form hydrogen gas (H_2), it is necessary to use excess amalgam to provide the necessary supply of $H^{\bullet}$ to act as an electrophile. Biphenyl (**XLI**, Fig. 1–11) and 4-hydroxybiphenylmethane (**XLII**) were not de-

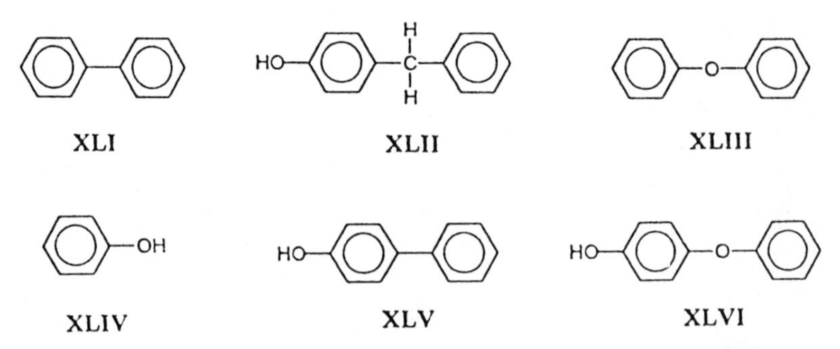

Fig. 1–11. Compounds used in model studies of degradations with sodium amalgam (after Piper and Posner, 1972a).

graded by Na amalgam, but there was some degradation of biphenylether (**XLIII**), and there also was some liberation of benzenol (phenol, **XLIV**, 2%) from 4-hydroxybiphenyl (**XLV**). However, activation by the 4-hydroxy substituent caused the release of 20% of phenol in the digest of 4-hydroxybiphenyl ether (**XLVI**).

About 34 different phenolic compounds have been identified in the Na amalgam digests of HS, although the maximum for any given study was 18. Compounds marked with symbol (‡) in Table 1–8 were methylated before identification by GC–MS and micro-infrared analyses (Schnitzer and Ortiz de Serra, 1973a). In general, the remaining compounds were identified by two-dimensional paper chromatography and cochromatography with standard compounds. Thus, these com-

Table 1–8. Compounds identified by various workers in sodium amalgam digests in alkaline conditions.†

Compound	Substituents
XLVII	$R_1 = R_2 = OH;\ R_3 = R_4 = R_5 = R_6 = H$
XXXII	$R_1 = R_3 = OH;\ R_2 = R_4 = R_5 = R_6 = H$
XLVIII	$R_1 = R_2 = R_6 = OH;\ R_3 = R_4 = R_5 = H$
XXXI	$R_1 = R_3 = R_5 = OH;\ R_2 = R_4 = R_6 = H$
XLIX	$R_1 = CH_3;\ R_3 = OH;\ R_2 = R_4 = R_5 = R_6 = H$
L	$R_1 = CH_3;\ R_4 = OH;\ R_2 = R_3 = R_5 = R_6 = H$
LI	$R_1 = CH_3;\ R_2 = R_4 = OH;\ R_3 = R_5 = R_6 = H$
LII	$R_1 = CH_3;\ R_3 = R_5 = OH;\ R_2 = R_4 = R_6 = H$
LIII	$R_1 = CH_3;\ R_2 = R_6 = OH;\ R_3 = R_4 = R_5 = H$
LIV	$R_1 = CH_3;\ R_2 = R_5 = OH;\ R_3 = R_4 = R_6 = H$
LV	$R_1 = CH_3;\ R_2 = R_4 = R_6 = OH;\ R_3 = R_5 = H$
LVI	$R_1 = R_3 = OH$; 2(CH_3) unassigned; 2(H) unassigned
LVII	$R_1 = COOH;\ R_3 = OH;\ R_2 = R_4 = R_5 = R_6 = H$
LVIII	$R_1 = COOH;\ R_4 = OH;\ R_2 = R_3 = R_4 = R_6 = H$
LIX	$R_1 = COOH;\ R_2 = R_4 = OH;\ R_3 = R_5 = R_6 = H$
LX	$R_1 = COOH;\ R_2 = R_6 = OH;\ R_3 = R_4 = R_5 = H$
LXI	$R_1 = COOH;\ R_3 = R_4 = OH;\ R_2 = R_5 = R_6 = H$
LXII	$R_1 = COOH;\ R_3 = R_5 = OH;\ R_2 = R_4 = R_6 = H$
LXIII	$R_1 = COOH;\ R_3 = R_4 = R_5 = OH;\ R_2 = R_6 = H$
LXIV	$R_1 = COOH;\ R_3 = OCH_3;\ R_4 = OH;\ R_2 = R_5 = R_6 = H$
LXV	$R_1 = COOH;\ R_3 = R_5 = OCH_3;\ R_4 = OH;\ R_2 = R_6 = H$
LXVI‡	$R_1 = COOH;\ R_3 = R_4 = OCH_3;\ R_2 = R_5 = R_6 = H$
LXVII‡	$R_1 = COOH;\ R_3 = R_5 = OCH_3;\ R_2 = R_4 = R_6 = H$
LXVIII‡	$R_1 = COOH;\ R_3 = R_4 = R_5 = OCH_3;\ R_2 = R_6 = H$
LXIX‡	$R_1 = CH(O);\ R_2 = R_4 = OCH_3;\ R_3 = R_5 = R_6 = H$
LXX‡	$R_1 = CH(O);\ R_3 = R_4 = R_5 = OCH_3;\ R_2 = R_6 = H$
LXXI‡	$R_1 = H_3CC(O);\ R_3 = R_4 = OCH_3;\ R_2 = R_5 = R_6 = H$
LXXII‡	$R_1 = H_3C(O);\ R_3 = R_4 = R_5 = OCH_3;\ R_2 = R_6 = H$
LXXIII‡	$R_1 = H_2CC(O)CH_3;\ R_3 = R_4 = R_5 = OCH_3;\ R_2 = R_6 = H$
LXXIV‡	$R_1 = H_2CCH_2COOH;\ R_3 = OCH_3;\ R_4 = OH;\ R_2 = R_5 = R_6 = H$
LXXV	$R_1 = CH_2CH_2COOH;\ R_3 = R_5 = OCH_3;\ R_4 = OH;\ R_2 = R_6 = H$
LXXVI	$R_1 = HC = CHCOOH;\ R_4 = OH;\ R_2 = R_3 = R_5 = R_6 = H$
LXXVII	$R_1 = HC = CHCOOH;\ R_3 = OCH_3;\ R_4 = OH;\ R_2 = R_5 = R_6 = H$

† From Burges et al. (1964), Stevenson and Mendez (1967), Martin and Haider (1969), Dormaar (1969), Piper and Posner (1972b), Schnitzer and Ortiz de Serra (1973b), Tate and Goh (1973), Martin et al. (1974), Matsui and Kumada (1977).

‡ Products that were methylated prior to identification using gas chromatography-mass spectrometry (GC-MS) and microinfrared (Schnitzer and Oritz de Serra, 1973a,b).

pounds were as formed in the digests. However, not all of the compounds can be considered to be the components of the humic molecules as released in the digests. Studies of the degradations of model compounds in the digest conditions suggest that in several instances the compounds released were altered under the digest reaction conditions. Model studies by Piper and Posner (1972b) suggested that some compounds (**XXXII**, Fig. 1–9; **XLVII**, **XLVIII**, **LXII**, **LXIII**, and **LXIV**, Table 1–8) did not degrade under the digest conditions. However, in model studies Martin et al. (1974) found that the compounds **XXXI**, Fig. 1–9, and **XLVIII**, **LIX**, **LX**, and **LXII**, Table 1–8, degraded to varying degrees under the digest conditions.

Decarboxylation of hydroxybenzene carboxylic acids will take place in the digest conditions, and the greater the extent of ring hydroxylation, the greater will be the decarboxylation. For example, 3,4,5-trihydroxybenzene carboxylic acid (**LXIII**, Table 1–8) will undergo decarboxylation even in water at 50°C, giving rise to 1,2,6-trihydroxybenzene (**XLVIII**), and decarboxylation could explain its origins. The same reasoning could explain the origins of catechol (**XLVII**) or 1,2-dihydroxybenzene from **LXI**, and of resorcinol (**XXXII**) or 1,3-dihydroxybenzene from **LIX**, **LX**, and **LXII**. Because compounds marked with (‡) in Table 1–8 were methylated before analyses, compounds **LXVI**, **LXVII**, and **LXVIII** would also have given dihydroxy- and trihydroxybenzene products.

The compositions of the digest compounds identified give indications with regard to their origins. Compounds shown in Fig. 1–2 (**I**, **II**, **III**, and **IV**, p. 13) give indications of the types of component molecules in lignin that could give rise to many of the aromatic structures listed in Table 1–8. Oxidation of the C_3 side chains would give rise to carbaldehyde and carboxyl functionalities on the aromatic nuclei, and cleavages of the ether linkages connecting the aromatic structures to the C_3 side chains would release 4-hydroxy-3-methoxy-4-hydroxy, and 3,5-dimethoxy-4-hydroxy benzene structures. Thus, it could be argued, on the basis of the information presented, that digest compounds **XLVII** and **XLVIII** (Table 1–8) could have arisen from lignin, provided that appropriate demethylation of the 3- and 5-methoxy substituents took place. Compounds **LVIII**, **LXI**, and **LXIII** (assuming demethylation of the 3- and 3,5-dimethoxy substituents), **LXIV**, the compounds identified after methylation (**LXV**, **LXVI**, **LXVIII**, and **LXX**), the compounds with C_3 side chains (**LXXIV**, **LXXV**, **LXXVI**, and **LXXVII**) have clear signals for lignin origins. The occurrence of unsaturated C_3 side chains (**LXXVI** and **LXXVII**) suggests incomplete hydrogenation under the reductive conditions used. A case can be made also for lignin origins for the keto structures, **LXXI** and **LXXII**.

The compounds with resorcinol-type (1,3-dihydroxybenzene, **XXXII**, Fig. 1–9) substituents [e.g., **LI**, **LII**, **LIII**, and **LVI**, Table 1–8) may be considered to have origins in fungi (Martin et al., 1967, 1972, 1973, 1974; Martin and Haider, 1969; Haider and Martin, 1970).

Reduction Using Zinc Dust Distillation and Fusion. The techniques involving the reduction of HS by Zn dust distillation and fusion were adapted from procedures used in studies of alkaloids, quinines, and phenols (see Valenta, 1963, 1973). The conditions used are drastic, strongly dehydrogenating, and can give rise to cleavages of C–C and of C–N bonds, and the formation of fused aromatic and of heteroaromatic structures.

The Zn dust distillation technique, as used by Hansen and Schnitzer (1969a), involved mixing HA (0.5 g) and Zn dust (12.5 g) in a Pyrex glass tube, covering with an additional layer of Zn dust (5 g), and heating at 510 to 530°C in a stream of inert gas for 15 min. The sublimed products were recovered in benzene, separated by thin layer chromatography (TLC) on silica gel and cellulose, and identified using ultraviolet spectroscopy and spectrofluorimetry.

Clar (1964) reviewed the Zn dust fusion process, which he introduced in 1939 for the reduction of high MW quinines. Samples were fused as melts of NaCl and $ZnCl_2$ at 200 to 300°C. In the techniques of Hansen and Schnitzer (1969a, 1969b) samples of HS (0.5 g) were ground with Zn dust (3 g), NaCl (1 g), and $ZnCl_2$ (5 g) and heated at 300°C for about 15 min. Samples were recovered and identified as for the Zn dust distillation. The products recovered were similar for the distillation and fusion methods.

The numbering system used for fused aromatic structures is illustrated for pyrene (**XXIII**, Fig. 1–12), and from that the numbering system for the other fused aromatic products listed can be deduced. Among the compounds isolated by Cheshire et al. (1967) were anthracene (**LXXVIII**) and 1,2-benzofluorene (**LXXXI**), isolated in crystalline forms. Other products, identified by GC–MS, were naphthalene (**LXXXII**), 1-methylnaphthalene (**LXXXIII**), 2-methylnaphthalene (**LXXXIV**) and higher homologs of methylnaphthalenes, pyrene (**XXIII**), 3,4-benzopyrene (**LXXXV**), 1,2-benzopyrene (**LXXXVI**), perylene (**LXXXVII**), triphenelene (**LXXXVIII**), chrysene (**LXXXIX**), 1,12-benzoperylene (**XC**), and coronene (**XCI**).

The major products of the Zn dust distillation and fusion of HAs identified by Hansen and Schnitzer (1969a) and their yields (as a percentage of the initial starting materials) were naphthalene (**LXXXII**) and 1,2,7-trimethylnaphthalene (**LXXXIV**, 0.06%), anthracene (**LXXVIII**) and its 1-methyl (**LXXIX**) and 9-methyl derivatives (**LXXX**, 0.06–0.07%), the 2-methyl (**XCII**) and the 3-methyl derivatives of phenanthrene (**XCIII**, 0.06–0.07%), pyrene (**XXIII**), and its 1-methyl and 4-methyl derivatives (0.12–0.13%), and perylene (**LXXXVII**, 0.15–0.20%). They identified as minor products fluoranthene (**XCIV**), 1,2-benzanthracene (**XCV**), and 2,3-benzofluorene (**XCVI**).

As indicated, where recoveries are listed, yields of products from Zn dust distillation and fusion processes are very low, and often amount to <1% of the starting material. Similar low yields are obtained for alkaloids and other polycyclic aromatic compounds for whose study the methods were devised. Despite the low yields Hansen and Schnitzer (1969a) extrapolated their findings to predict that polycyclic aromatic compounds might account for 12 to 25% of the mass of HS.

On the basis of the information from Zn dust distillation and additional data from other degradative and from spectroscopy techniques, Cheshire et al. (1967; see also Haworth, 1971) proposed that humic molecules have a polycyclic aromatic core to which polysaccharides, simple phenols, proteins or peptides, and metals are attached, as summarized diagrammatically in Fig. 1–13. This concept had its adherents for some time. However, it has become generally accepted that the drastic reaction conditions associated with Zn dust distillation and fusion processes can lead to excessive bond breaking and recombination of fragments to give fused aromatic structures. Polyhydroxyaromatics, quinones, and furfurals, for example, would

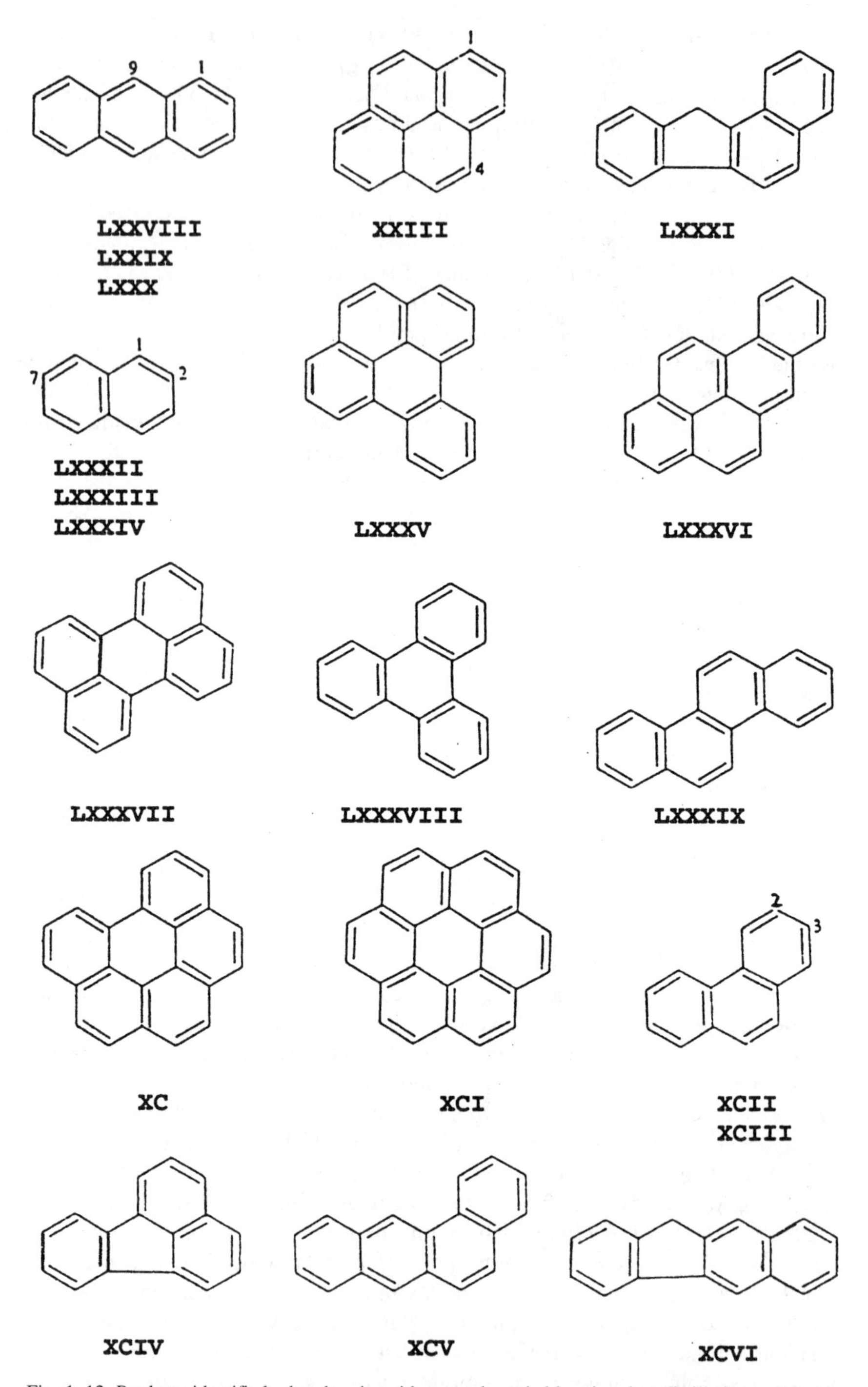

Fig. 1–12. Products identified when humic acids were degraded by zinc dust distillation and fusion processes (after Cheshire et al., 1967; Hansen and Schnitzer, 1969a, 1969b; Stevenson, 1989).

give rise to these under the reaction conditions applied, though Cheshire et al. (1968) considered that the yields were significantly less than those obtained for the degradation of HS. Long chain hydrocarbon products were also digest products, but these can be considered to have survived the degradation process.

Several of the polycyclic aromatic hydrocarbons listed above have been isolated from soils in organic solvents (Stevenson, 1994). Their presence in soils can be attributed to fires. It is appropriate to note that fused aromatic structures have also been identified among the pyrolysis degradation products (see Sequential Degradation Reactions, p. 83), and this can be attributed to the high temperatures used. It is worth noting that long chain aliphatic acids, esters, and hydrocarbons survive the pyrolysis processes, and these also survive the Zn dust distillation and fusion procedures. Also, the increasing awareness of char (see Information from Pyrolysis Degradation Processes, p. 85) in the soil environment focuses attention on the influences of fires in the formation of fused aromatic structures and the subsequent incorporation of these in SOM. However, on the basis of the evidence we have, we conclude that fused aromatic structures are unlikely to be significant components of naturally occurring SOM and are artifacts formed as the result of anthropogenic or naturally occurring burning.

Degradations in Sodium Sulfide Solutions

Mixtures of sodium sulfide (Na_2S) and of NaOH have been used for the delignification of wood in the paper industry since U.S. patents were awarded for the process in 1870 (Kleppe, 1970). The kraft process (from the Swedish word meaning strength) is considered to confer superior strength to the paper products. The process gives rise to useful byproducts, such as dimethyl sulfide, which can be oxidized to DMSO, ethanethiol, condensation polymers with methanal, hydroxy acids, and a variety of other aliphatic acids (Sjöström, 1981, p. 200–206).

The first application of Na_2S for the degradation of HAs was by Swift (1968; see also Hayes et al., 1972); these studies were later continued by Craggs (1972) and by O'Callaghan (1980). Hayes and O'Callaghan (1989) have given a detailed review of applications of the process.

Sodium sulfide hydrolyses in aqueous solutions (Marton, 1971; Sjöström, 1981, p. 125) are shown in Reaction Schemes [37] and [38]:

$$S^{2-} + H_2O \Leftrightarrow HS^- + HO^- \quad [37]$$

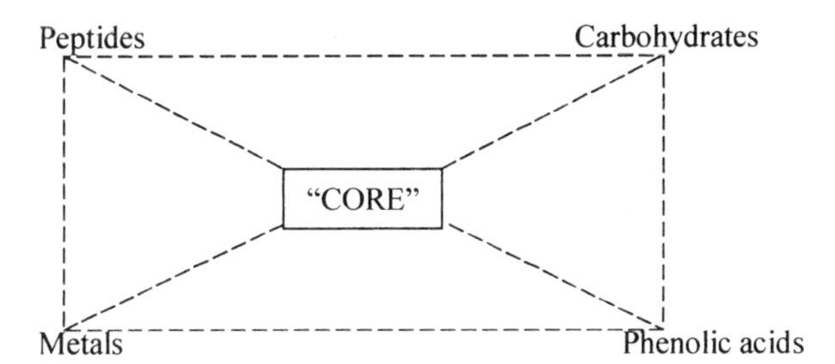

Fig. 1–13. A diagrammatic representation of humic acid (after Haworth, 1971).

$$HS^- + H_2O \Leftrightarrow H_2S + HO^- \quad [38]$$

The equilibrium constant K_1 for Reaction [37] is about 10, and the K_2 for Reaction [38] is about 10^{-7}:

$$K_1 = [HS^-][HO^-]/[S^{2-}] \quad [39]$$

$$K_2 = [H_2S][HO^-]/[HS^-] \quad [40]$$

Hence the formation of HS^- is strongly favored, and the sulfide anion has little importance. At elevated temperatures little S^{2-} and H_2S exist, and the reactive species are HS^- and HO^- (Ellis and Giggenback, 1971).

Nucleophilic strength increases in the order: $HO^- < HS^- < CH_3S^-$, but the basic strength, or the ability to extract a proton increases in the reverse order. The difference in polarizability is attributed to the larger HS^- compared with the HO^- species.

Hayes and O'Callaghan (1989) reviewed mechanisms of degradations with solutions of Na_2S at elevated temperatures, and referred to the model studies by Burdon et al. (1974). They then went on to describe the products formed when HAs were degraded, and have proposed origins in the HA structures for the digest products identified.

Most of their studies were performed using a saturated Na_2S solution (10%) at 250°C under autoclave conditions. The compounds identified in the digests are listed in Table 1–9.

"Acid boiled" (AB, for 24 h in 100-fold excess 6 *M* HCl in a stream of dinitrogen gas) and non-acid-boiled HAs were used in these studies. Up to 60% of the mass of the HAs in the non-AB materials in the Na_2S degradation digests were solvent soluble (Hayes et al., 1972; Craggs et al., 1974). The solid residue in the case of the AB material was 45% of the starting (prehydrolysis) material. This residue, when degraded in Na_2S at 250°C, yielded ether soluble materials that amounted to 40% of the mass of the starting materials. Thus, the acid hydrolysis, combined with the Na_2S degradation gave rise to >80% degradation of the starting HA material. Addition of sodium molybdate, a hydrodesulfurization catalyst commonly used in the petroleum industry (Mitchell, 1967), was shown by O'Callaghan (1980) to increase the yields of ether soluble digest products.

The digest products were methylated before identification by GC–MS. It is reasonable to assume that the methyl esters identified originated as carboxylic acid structures and are therefore listed as such in Table 1–9. We cannot state with certainty whether or not the methoxyl substituents were present as the methoxyl or hydroxyl functionalities.

The alcohols (**XCVII**, Table 1–9) and aliphatic dicarboxylic acids (**CIII**) were present in the digests of the AB HAs, and some of these alcohols and dicarboxylic acids, and the other aliphatic compounds listed (**XCVIII**, **XCIX**, **C**, **CI**, **CII**, **CIV**, **CV**, and **CVI**) were detected in the digests of the non-AB HAs. Ethanoic acid (**XCVIII**) was most abundant, and was likely to have had origins in carbohydrates, as did lactic acid, or 2-hydroxypropanoic acid (**C**), as discussed by Sjöström (1977, 1981) in his treatment of the mechanisms of alkaline degradation of polysaccha-

Table 1–9. Compounds identified in digests after humic acids were degraded in saturated aqueous sodium sulphide solutions at 250°C, under autoclave conditions.†

$CH_3(CH_2)_nOH$ (n =1 - 4) **(XCVII)**	CH_3COOH **(XCVIII)**	$CH_3(CH_2)_nCOOH$ (n = 1 - 4) **(XCIX)**
$H_3CCH(OH)COOH$ **(C)**	$H_3CC(O)(CH_2)_2COOH$ **(CI)**	$(COOH)_2$ **(CII)**
$HOOC(CH_2)_nCOOH$ **(CIII)**		$HOOCCH(CH_3)COOH$ **(CIV)**
$HOOCCH(CH_3)CH(CH_3)CH_2COOH$ **(CV)**		$CH_3C(O)C(CH_3)_2CH_2COOH$ **(CVI)**

CVII	$R_1 = HC = CHCH_2CH_3$; $R_2 = R_3 = R_4 = R_5 = R_6 = H$
CVIII	$R_1 = R_2 = COOH$; $R_3 = R_4 = R_5 = R_6 = H$
CIX	$R_1 = OCH_3$; $R_2 = R_3 = R_4 = R_5 = R_6 = H$
CX	$R_1 = CH_3$; $R_2 = OCH_3$; $R_3 = R_4 = R_5 = R_6 = H$
CXI	$R_1 = H_2CCH_3$; $R_4 = OCH_3$; $R_2 = R_3 = R_5 = R_6 = H$
CXII	$R_1 = CH_2OH$; $R_4 = OCH_3$; $R_2 = R_3 = R_5 = R_6 = H$
CXIII	$R_1 = H_2CCH_2COOH$; $R_4 = OCH_3$; $R_2 = R_3 = R_5 = R_6 = H$
CXIV	$R_1 = H_2CCH_2COOH$; 1 alkyl, 1 OCH_3 in ring; 3 Rs as H
CXV	$R_1 = H_2CC(O)CH_3$; $R_4 = OCH_3$; $R_2 = R_3 = R_5 = R_6 = H$
CXVI	$R_1 = C(O)CH_3$; $R_4 = OCH_3$; $R_2 = R_3 = R_5 = R_6 = H$
CXVII	$R_1 = COOH$; $R_3 = CH_3$; R_5 =Alkyl; $R_2 = R_4 = R_6 = H$
CXVIII	$R_1 = COOH$; $R_3 = CH_3$; $R_4 = OCH_3$; $R_2 = R_5 = R_6 = H$
CXIX	$R_1 = OCH_3$; $R_2 = COOH$; $R_6 = CH_3$; $R_3 = R_4 = R_5 = H$
CXX	$R_1 = R_2 = OCH_3$; $R_3 = R_4 = R_5 = R_6 = H$
CXXI	$R_1 = R_2 = OCH_3$; $R_6 = CH_3$; $R_3 = R_4 = R_5 = H$
CXXII	$R_1 = H_2CCH_3$; $R_2 = R_3 = OCH_3$; $R_4 = R_5 = R_6 = H$
CXXIII	$R_1 = H_2CCH_2COOH$; $R_2 = OCH_3$; $R_6 = C(O)CH_3$; $R_3 = R_4 = R_5 = H$
CXXIV	$R_1 = H_2CCH_2COOH$; $R_2 = R_6 = OCH_3$; $R_3 = CH_3$; $R_4 = R_5 = H$
CXXV	$R_1 = H_2CCH_2COOH$; $R_3 = R_4 = OCH_3$; $R_5 = CH_3$; $R_2 = R_6 = H$
CXXVI	$R_1 = H_2CCOOH$; $R_3 = R_4 = OCH_3$; $R_5 = H_2CCH_3$; $R_2 = R_6 = H$
CXXVII	$R_1 = COOH$; $R_3 = R_4 = OCH_3$; $R_5 = CH_3$; $R_2 = R_6 = H$

(Benzene ring with substituents R_1–R_6.)

† From Hayes and O'Callaghan (1989), Hayes and Swift (1990).

rides in the kraft process. Ethanoic acid could also arise from unsaturated aliphatic acids, and although these are not HS in the classical definitions, they are likely to be among the molecules in association in the mixtures that compose the humic media. The Verrentrapp reaction (Reaction Scheme [41]) illustrates one mechanism that could lead to the formation of ethanoic acid and other acids:

$$RCH_2{-}CH{=}CH{-}CH_2{-}CH_2{-}CO_2^- \underset{}{\overset{XH^-}{\rightleftharpoons}} RCH_2{-}CH{=}CH{-}\overset{-}{\underset{H}{C}}{-}CH_2CO_2^-$$

further migration to $RCH_2{-}CH_2CH_2CH{=}CHCO_2^-$

$$\xrightarrow{\text{fission}} RCH_2CH_2CH_2CO_2^- + CH_3CO_2^- \qquad [41]$$

Hayes and O'Callaghan (1989) discussed mechanisms that could give rise to aliphatic acids in the digests. Such acids could, for example, arise from cleavages of double bonds, and from cleavages of aliphatic side chains on aromatic nuclei when activating groups promote cleavages of the C–C bonds. The dicarboxylic acids could have origins in pairs of unsaturated bonds, and in carbohydrates. The numbers of methylene (CH_2) groups, as in structures depicted by **CIII** (Table 1–9), could show the separations of such bonds in parent structures.

The aromatic compounds identified in the degradation digests are different from those in digests of the classical oxidative degradation processes. The reader is referred to Hayes and O'Callaghan (1989) for details of the mechanisms involved in the release of aromatic components from humic core structures. However, one simple mechanism is outlined in Reaction Scheme [42]:

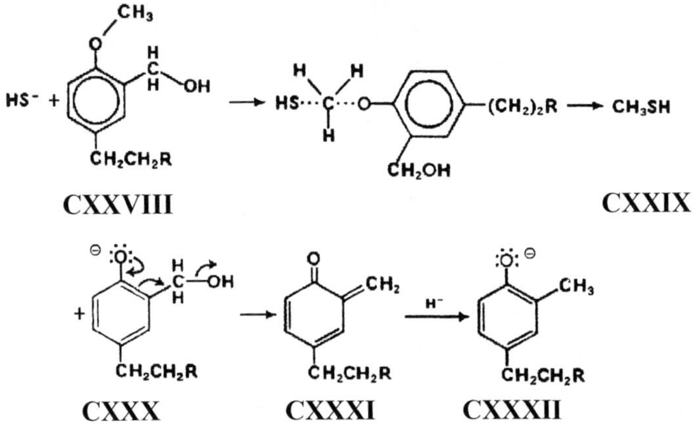

CXXVIII CXXIX

CXXX CXXXI CXXXII [42]

which can explain the origins of some of the compounds listed in Table 1–9. In Reaction Scheme [42], HS^- attack on the chemical species (**CXXVIII**) by an S_N2 mechanism leads to the formation of methyl mercaptan (**CXXIX**) and the anion species (**CXXX**). As indicated, the OH^- species is lost from the organic anion, and the quinone methide structure (**CXXXI**) is formed. This, when subjected to attack by the hydride anion (H^-) gives rise to the structure (**CXXXII**). There are a number of mechanisms that can give rise to the hydride species in the basic conditions, and one of these is the Dumas–Stass reaction shown in Reaction Scheme [43], where HX^- is HS^- or HO^-:

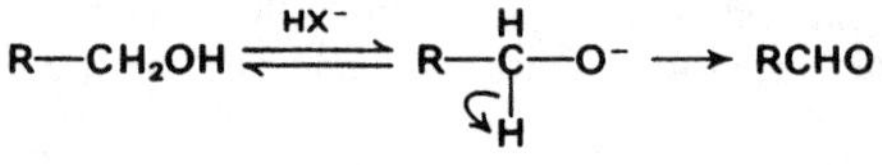

$$R{-}CH_2OH \xrightleftharpoons{HX^-} R{-}CH(H){-}O^- \longrightarrow RCHO \qquad [43]$$

The HS^- species can also accept H^- to give the S^{2-} species and H_2. Methyl mercaptan (Reaction Scheme [42], **CXXIX**) is ionized in the basic medium and the nucleophilic mercaptide can attack another phenylether group to give the phenate anion **CXXXII** and dimethylsulfide (CH_3SCH_3).

There are significant differences between the aromatic structures identified in the Na_2S (Table 1–9) digests and those in the digests of the oxidative degradation reactions (Table 1–10, p. 80). All of the di- and polycarboxylic acid derivatives of benzene have been identified in the oxidative digests, and with the exception of benzene-1,2-dicarboxylic acid (**CVIII**, Table 1–9), there was never more than one carboxyl substituent on the aromatic functionalities in the Na_2S digests.

Several of the aromatic structures shown in Table 1–9 have aliphatic substituents. It is very likely that the methyl substituents arose from hydroxy and/or ether functionalities on C atoms α to the aromatic ring and *ortho* or *para* to the phenolic hydroxyl substituent, as would apply for structures **CX**, **CXVIII**, **CXIX**, **CXXI**, and **CXXVII** (see Reaction Scheme [42]) (However, it is possible also that methyl substituents could be introduced during methylation when diazomethane is used). An ethyl group would form through the quinone methide intermediate in cases of secondary alcohol or ether groups, and such a mechanism could be involved in the genesis of structures shown as **CXI**, **CXXII**, and **CXXVI**.

In their model studies Burdon et al. (1974) showed how cinnamic acid (β-phenylpropenoic acid) when subjected to degradation with N_2S under autoclave conditions produced β-phenylpropanoic acid, benzenecarboxylic acid, benzyl alcohol, phenylmethyl ketone (acetophenone), and 1-phenylethanol. On the basis of mechanisms that led to the formation of these products it can be deduced that a propenoic acid substituent on the benzene ring could give rise to structure types (Table 1–9) **CVIII**, **CXIII**, **CXIV**, **CXVI**, **CXVIII**, **CXIX**, **CXXIII**, **CXXIV**, **CXXV**, **CXXVI**, and **CXXVII**. This information emphasizes the fact that a single precursor or parent product can give rise to several structures, and the large numbers of products identified in digests are multiples of the actual molecular components in the humic structures.

Most of the aromatic structures listed in Table 1–9 have methoxy substituents. Because the digest products were methylated it is not possible to say whether or not these (in the 3 and 5 ring positions) were present as methoxyl or as hydroxyl substituents (methoxyl groups would be demethylated in the Na_2S digests), but the likelihood is that these were present as methoxyls. A very strong case can be made for origins in lignin for structures such as **CXII**, **CXIII**, **CXIV**, **CXV**, **CXVI**, **CXX** (assuming that decarboxylation of the carboxyl group had taken place), and **CXXV**. On the basis of the component structures in lignin (**IV**, Fig. 1–2), a case might also be made to assign lignin origins to structures **CX**, **CXI**, **CXVIII**, **CXIX**, and **CXXIII**.

Oxidative Degradation Reactions

Oxidative processes involve the loss of electrons from the substrate leading to cleavages of C–H and C–C bonds. It requires very considerable energy inputs to rupture such bonds. The breaking of the C–H bond, for example, requires 300 to 400 kJ mol^{-1}.

Hayes and Swift (1978) outlined oxidative processes relevant for the degradation of HS. These include homolytic cleavages, which give rise to radical species, and heterolytic cleavages, which give rise to ionic species.

Oxidations with Permanganate. The classical procedures involve oxidations with permanganate. Manganese can have oxidation numbers of +1 to +7, but we are primarily concerned with Mn (VII) and Mn (V) for oxidations of HS. Mn (VII) is commonly the starting oxidant. It is relatively stable in acidic media. In 30 to 50% H_2SO_4 it forms highly reactive permanganic acid ($HMnO_4$), but in concentrated H_2SO_4 this decomposes to the less reactive soluble Mn (IV) species. The reactivities of the different species vary inversely with charge, and their oxidative effects follow the order: $HMnO_4 > MnO_4^- > MnO_4^{2-} > MnO_4^{4-}$.

$HMnO_4$ is too drastic for the oxidation of HS, and so the $KMnO_4$ salt is used. This gives a mildly alkaline aqueous solution that dissolves the HS. However, under such conditions, a slow decomposition of permanganate occurs. This is catalyzed by MnO_2, as indicated in Reaction Scheme [44]:

$$2MnO_4^- + H_2O \xrightarrow{MnO_2} 2MnO_2 + 2OH^- + 3/2O_2 \qquad [44]$$

Because MnO_2 is the final Mn product of oxidations by Mn (VII), it is essential to use excess permanganate. In strongly alkaline conditions permanganate decomposes slowly to give manganate (VI), the Mn_4^{2-} species. Manganate (VI) is an effective oxidizing agent, although less reactive than permanganate.

The compositions, stereochemistry, functionalities, and solubilities in the media govern the rates and extent of the oxidations of different substrates. In general anions are oxidized more readily than neutral molecules, and these in turn are more readily oxidized than cations.

Consider now reactions of permanganate with functionalities that can be important components of HS. Primary and secondary alcohols react more readily in basic than in neutral media because in these soluble alkoxide (anionic) species are formed. Reaction Scheme [45] shows the formation of the alkoxide anion from a primary alcohol,

$$RCH_2CH_2OH \xrightarrow{OH^-} RCH_2CH_2O^- \xrightarrow{Mn(VII)} RCH_2{-}C(=O)H \xrightarrow[Mn(VII)]{OH^-} (R)(H)C{=}C(OH)(H) \xrightarrow{Mn(VII)} R{-}CO_2H + CO_2$$

$$RCH_2CH_2O^- \xrightarrow{Mn(VII)} RCH_2CO_2H \qquad [45]$$

then oxidation with permanganate involving the abstraction of a hydride anion to give the carbaldehyde, which is further oxidized to the carboxylic acid. Enolizable aldehydes, and also ketones (from the oxidation of secondary alcohols) can lose CO_2 and form carboxylic acids having one C atom less than the alcohol (see Reaction Scheme [45]). Tertiary alcohols resist oxidation, except under drastic conditions.

Saturated straight chain aliphatic hydrocarbons have significant resistance to oxidation by permanganate, although the presence of carboxyl in the chain increases degradability. Branched chain hydrocarbons can oxidize to tertiary hydroxy acids (Kenyon and Symons, 1953).

Alkenes, on the other hand, are readily oxidized. In controlled reactions cyclic manganese esters are first formed (Stewart, 1965, p. 42), and these lead to the formation of *cis* glycols, as indicated in Reaction Scheme [46]:

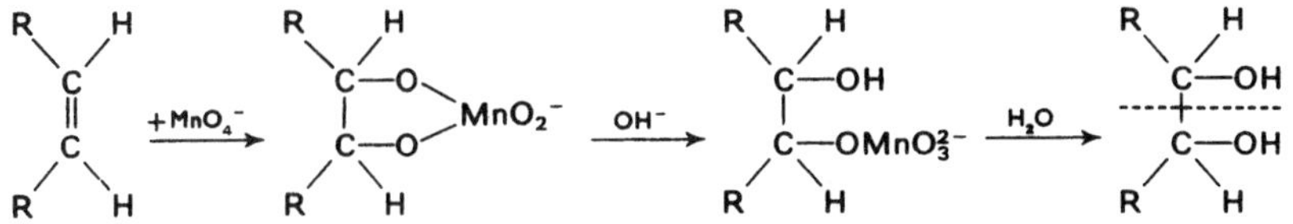

[46]

Further oxidation leads to cleavages to carboxylic acids, as indicated by the broken line in Reaction Scheme [46]. Such cleavages will result from applications of excess oxidant. Alkynes will form diketo and, more likely, carboxylic acid cleavage products.

The benzene ring has a resistance to oxidation with aqueous permanganate. Some degradation does take place when excess permanganate is used, and complete degradation will take place when amine and hydroxyl substituents are present. With the exception of tertiary C substituents, aliphatic hydrocarbon functionalities attached to the aromatic nuclei are oxidized to carboxyl. This is illustrated for compound **CXXXIII** in Reaction Scheme [47]:

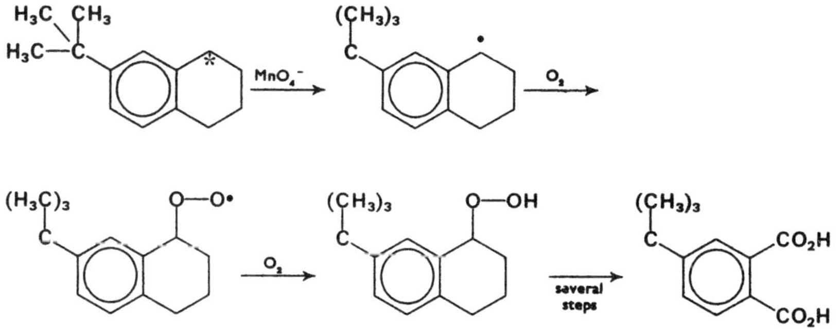

CXXXIII

[47]

(from Stewart, 1964, p. 67 and Lee 1969, p. 37), which summarizes the mechanisms involved in such oxidation processes. The tertiary C (with the absence of H on the C substituent α to the aromatic ring) resists oxidation, as stated. Attack on the benzylic C (marked with an asterisk) will give rise to the exclusion of a H atom, the radical species formed will react with O to give the peroxide species, and degradation will proceed to give the dicarboxylic acid.

Polycyclic aromatic hydrocarbons will oxidize to give a single ring with carboxylic substituents that can indicate the nature of the starting structure. Reaction Scheme [48],

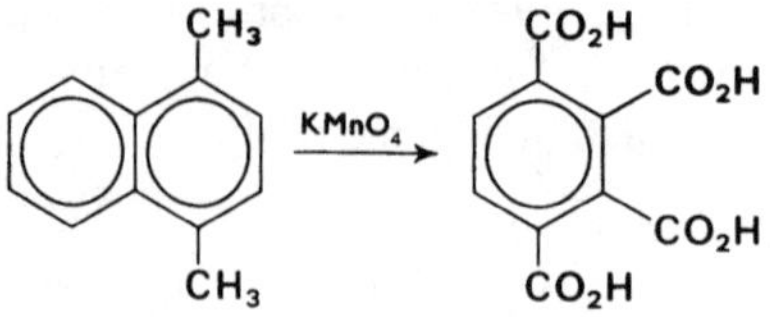

CXXXIV $\xrightarrow{KMnO_4}$ CXXXV

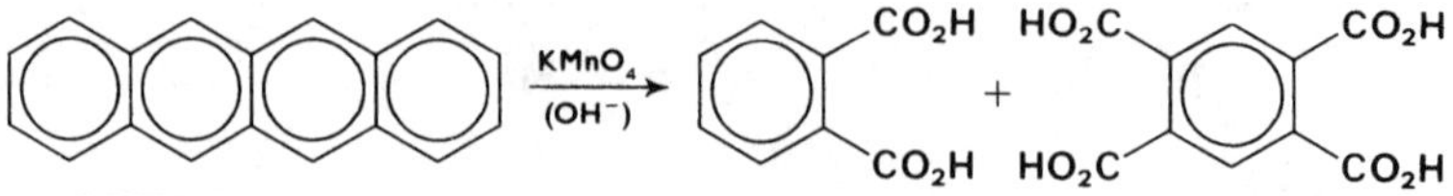

CXXXVI $\xrightarrow[(OH^-)]{KMnO_4}$ CXXXVII + CXXXVIII

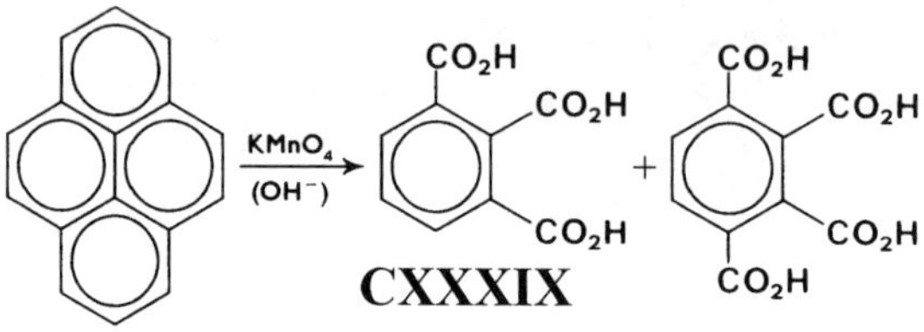

XXIII $\xrightarrow[(OH^-)]{KMnO_4}$ CXXXIX + CXXXV

[48]

indicates that oxidation of 1,4-dimethylnaphthalene (**CXXXIV**) gives 1,2,3,4-benzenetetracarboxylic acid (**CXXXV**), that of naphthacene (**CXXXVI**) gives 1,2-benzenedicarboxylic acid (**CXXXVII**) and 1,2,4,5-benzenetetracarboxylic acid (**CXXXVIII**), and that of pyrene (**XXIII**) gives 1,2,3-benzenetricarboxylic acid (**CXXXIX**) and (**CXXXV**).

Of relevance to our subject was the use of permanganate for the oxidative degradation of coal (Bone et al., 1930; Bone and Himus, 1936; Ward, 1947; van Krevalen, 1961, p. 225–231), and of lignin and wood (Chang and Allen, 1971, p. 452–457; Lai and Sarkanen, 1971, p. 217). Schnitzer and his colleagues, however, have focused humic scientists on the merits of permanganate for studies of the degradation of HS. In their earlier work, potassium permanganate in dilute potassium hydroxide was used to oxidize organic matter from the A and B_h horizons of an imperfectly drained podzol (Wright and Schnitzer, 1959; Schnitzer and Wright, 1960), and later to oxidize a HA preparation from the podzol B_h horizon (Hansen and Schnitzer, 1966). In subsequent work by the Schnitzer Group KOH was omitted and, to prevent degradation of –OH substituted aromatic nuclei, the HS were methylated using methyl iodide and silver oxide (Barton and Schnitzer, 1963), or with excess diazomethane (Schnitzer, 1974). Briggs and Lawson (1970) methylated using dimethyl sulfate and anhydrous potassium carbonate in acetone. The protective effect of converting phenolic to methoxyl substituents is evident from the work of Matsuda and Schnitzer (1972), who showed that yields of identifiable products from degradation of HS in permanganate solutions were 160 to 250% greater for methylated than for the non-methylated substrates when 1 g of HA was refluxed for 8 h in 250 mL of 0.25 *M* aqueous $KMnO_4$, pH 10.

Oxidations with Alkaline Cupric Oxide. The uses of alkaline cupric oxide arose from oxidations with alkali. Wallis (1971) showed how ether linkages in the aliphatic side chain of phenylpropane structures can be broken. Gierer and Noren (1962) worked with models for structural units of lignin and showed how ether linkages in the aliphatic side chain of phenylpropane structures can be broken in sodium hydroxide solutions at elevated temperatures. Compound **CXL** 1-(3,4-dimethoxyphenyl)-2-(2-methoxyphenoxy)propan-1,3-diol, when reacted for 2 h at 190°C in 2 *M* NaOH gave the products indicated in Reaction Scheme [49]:

H_2C—OH, HC—O—, HCOH, OCH_3, OCH_3, OCH_3 (OH⁻) → H_2C—OH, HC, O, HC, OCH_3, OCH_3 + OCH_3, OH

CXL **CXLI** **CXLII** [49]

This is β-ether cleavage and requires the presence of a hydroxyl group on the side chain carbons α- or γ- to the aromatic nucleus. The cleavage products are the epoxide structure (**CXLI**) and 2-methoxybenzenol (guaicol, **CXLII**) shown in Reaction Scheme [49]. The epoxide hydrolyzes to a triol.

The uses of alkaline cupric oxide media for studies of HS arose from applications in lignin chemistry. Alkaline solutions of cupric, mercuric, silver, and cobalt oxides were used to degrade lignins to aromatic carbaldehyde and carboxyls as the major products. The ratio of aldehydes to acids in the products is determined by the oxidizing potential of the oxidant. Silver oxide has the highest oxidizing potential of the four, and it produces mainly acids. Cupric oxide is a relatively weak oxidant, and under favorable conditions can be expected to produce aromatic aldehydes (Chang and Allen, 1971).

Cupric oxide is a one-electron transfer oxidant. Chang and Allen (1971, p. 444–446) suggest that, for the oxidation of lignin, the reaction is initiated by the extraction of an electron to give a phenoxy radical. A second electron is then transferred to another cupric oxide molecule, so coupling does not take place. The quinonemethide structure is formed, and the aldehyde, 3-methoxy-4-hydroxybenzenecarbaldehyde (**CXLIII**) is produced, as indicated in Reaction Scheme [50]:

·O—, $^-$OH, —CH_2OH, $-e^-$ → O=, =CHOH, $^-$OH → HO—, —CHO, OCH_3, OCH_3, OCH_3

CXLIII

[50]

It is important to use excess cupric oxide and the highest yields of compound **CXLIII** were obtained when 13.5 mol of reagent were used per lignin building unit (Chang and Allen, 1971).

Schnitzer and his colleagues also introduced the alkaline cupric oxide technique for the degradation of HS (Schnitzer, 1974, 1978; Griffith and Schnitzer, 1976, 1989). About one-third of the compounds in the cupric oxide digests were also contained in the permanganate digests. In more recent times ratios of digest products from alkaline cupric oxide degradations have been used to predict plant origins of HS. Hedges and Mann (1979) introduced (Fig. 1–14) syringyl/vanillyl (**CXLIV/CXLV**) or S/V ratios (where R_1 is CHO or COOH), and cinnamyl/vanillyl (**CXLVI/CXLV**) or CV ratios as useful indicators of lignin sources. Because *p*-hydroxybenzene structures are also found in the CuO oxidations of nonwoody angiosperm, and to a lesser extent from gymnosperm wood (Hedges and Ertel, 1982), the *p*-hydroxyphenol (**CXLVII**)/vanillyl (P/V) ratio is a useful indicator of non-woody angiosperm tissue origins (Cowie et al., 1992; Goni et al., 1993). Guggenberger and Zech (1994) effectively used such ratio values to trace the origins of the humic components from forest soils. Vanillyl phenols are found as oxidation products in all four categories of vascular plants (woody and nonwoody tissues of angiosperms, including hardwoods) and gymnosperms (including softwoods). Syringyl phenols, on the other hand, are produced only from angiosperm lignin, and cinnamyl units are formed from nonwoody tissue (Hedges and Parker, 1976; Hedges and Mann, 1979).

Simpson (1999) studied the C/V, S/V, and P/V ratios for HAs from the A_h and B_h horizons of a podzol soil under oak forest cover since 4000 BP. In addition, he determined the acid/aldehyde ratios in the syringyl $(Ad/Al)_s$ and vanillyl $(Ad/Al)_v$ families (Hedges et al., 1988). High acidic contents, and therefore high Ad/Al ratios, would be expected in lignin substrates that have undergone extensive oxidation. In intact vascular plants lignin ratios of 0.1 to 0.2 are seen, whereas in soil HS the ratios can vary from 0.6 to 2.5 (Ertel and Hedges, 1984; Ertel et al., 1986).

In his study Simpson (1999) exhaustively extracted soils from the A_h and B_h horizons at pH 7, 10.6, and 12.6. The C/V and P/V ratios suggested that the parent materials were predominantly woody in the case of the oak cover soil. Better information was obtained from the acid/aldehyde ratios. For example, the S/V ratios for the HAs at pH 7 and pH 12.6 from the A_h horizon were 0.33 and 0.59, respectively, suggesting contributions from angiosperm lignin. Low C/V (0.02 and 0.03) and P/V (0.04 and 0.13) ratios for these two isolates (pH 7 and 12.6) suggested that the source materials were predominantly woody. The $(Ad/Al)_v$ and $(Ad/Al)_s$ ratios were 11.4 and 4.2, respectively, for the pH 7 extracts, and 1.3 and 0.8, respectively, for the extracts at pH 12.6 This indicated that the components extracted at pH 12.6 were less oxidized, and were more indicative of the parent material. In the case of the HAs isolated at pH 7 from the B_h horizon, the $(Ad/Al)_v$ and $(Ad/Al)_s$ ratios were

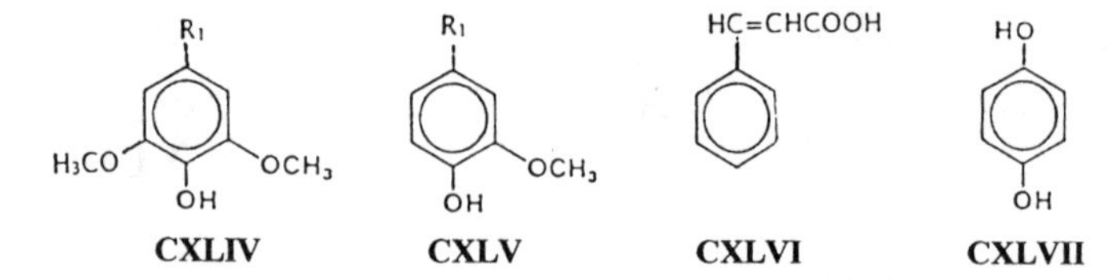

Fig. 1–14. Syringyl (**CXLIV**), vanillyl (**CXLV**), and cinnamyl (**CXLVI**) structures of compounds used in S/V and C/V ratios, and 1,4-dihydroxybenzene (**CXLVII**), indicators of origins in lignins (after Hedges and Mann, 1979).

60.7 and 5.5, respectively, which indicates that the lignin components were highly oxidized. That is as would be expected as the more highly oxidized components would be leached to lower levels in the soil profile.

Oxidations with Nitric Acid. Nitric acid has been used extensively for studies of lignin (Dence, 1971) and of coal (Van Krevelen, 1961, p. 220–224), and it has found limited use for degrations of HS. Digestions in nitric acid can give rise to a variety of aromatic substitution as well as oxidation products. Hence, considerable amounts of artifacts are present in the digests. We will not dwell on applications of nitric acid for oxidations of HS because the procedure is not being used to any significant extent at this time. Refer to Hayes and Swift (1978) for considerations of the mechanisms involved in oxidations with nitric acid and for indications of the types of products that can be formed.

Information from Oxidative Degradation Reactions

We list in Table 1–10 (p. 80) the variety of compounds that have been identified in the digests of the degradation of HS with alkaline permanganate and with alkaline cupric oxide (see also Hayes and Swift, 1978, 1990). Schnitzer and his colleagues obtained yields of 30 to 35% of the mass of the starting material as identifiable compounds in the digests of degradations with alkaline permanganate (see Griffith and Schnitzer, 1989). As indicated earlier, they also identified more than 60 compounds in the digests of alkaline permanganate degradations (Schnitzer, 1974, 1978; Griffith and Schnitzer, 1976, 1989), and about 25 of these were also contained in the permanganate digests.

It is essential to recognize that the digest compounds identified would rarely be the same as the structures in the HS from which they were released. The structural units undergo compositional alterations leading to their release from the parent molecules, and additional structural changes take place during the course of residence in the digests. Nevertheless, awareness of the degradation mechanisms that apply allows useful extrapolations to be made to the types of parent structures that can give rise to the degradation products.

The compounds listed in Table 1–10 contain a variety of aliphatic and aromatic structures. These were extracted in organic solvents, methylated, and fractionated and identified by gas liquid chromatography (GLC), GLC-MS, and microinfrared techniques. Thus, all of the carboxyls and hydroxyls (including phenols) were identified as the methyl esters and ethers, respectively. We have presented all of the methyl esters identified as the carboxylic acids parent compounds. The aliphatic di- and tricarboxylic acids would arise from cleavages of double bonds separated by the number of carbons linking the carboxyls. Long chain hydrocarbons and fatty acids are unlikely to be components of HS but were associated with these and were coextracted in the isolation processes. We have presented all of the methyl esters identified as the carboxylic acids parent compounds. Aliphatic di- and tricarboxylic acids (**CIII**, **CL**, **CLI**, and **CLII**) could arise from cleavages of double bonds separated by the number of carbons linking the carboxyls. The additional carboxyl groups (as in **CL** and **CLII**) could arise from olefinic side chains, and oxidizable functionalities, such as primary alcohol groups. Such dicarboxylic, tricarboxylic, and even tetracarboxylic acids can account for 20% of the digest products,

Table 1–10. Products identified in oxidative degradation reactions in alkaline permanganate and alkaline cupric oxide digests.

$HOOC(CH_2)_nCOOH$ (n = 2 - 8) **CIII**	$H_3C(CH_2)_nCH_3$ n = 12 - 38 **CIL**	$HOOCCH_2CH(COOH)CH_2COOH$ **CL**
$HOOCCH_2CH(OH)CH_2COOH$ **CLI**	$HOOCCH_2C(OH)(COOH)CH_2COOH$ **CLII**	H C COOH ‖ H C COOH **CLIII**
$H_3C(CH_2)_nCOOH$ **CLIV**	$H_3CCH(CH_3)CH_2COOH$ **CLV**	$H_3CCH(C_4H_9)CH_2COOH$ **CLVI**

Compound	Substituents
CLVII	$R_1 = R_3 = COOH; R_2 = R_4 = R_5 = R_6 = H$
CLVIII	$R_1 = R_4 = COOH; R_2 = R_3 = R_5 = R_6 = H$
CLIX	$R_1 = R_2 = R_3 = COOH; R_4 = R_5 = R_6 = H$
CLX	$R_1 = R_2 = R_4 = COOH; R_3 = R_5 = R_6 = H$
CLXI	$R_1 = R_2 = R_5 = COOH; R_3 = R_4 = R_6 = H$
CLXII	$R_1 = R_2 = R_3 = R_4 = COOH; R_5 = R_6 = H$
CLXIII	$R_1 = R_2 = R_3 = R_5 = COOH; R_4 = R_6 = H$
CLXIV	$R_1 = R_2 = R_4 = R_5 = COOH; R_3 = R_6 = H$
CLXV	$R_1 = R_2 = R_3 = R_4 = R_5 = COOH; R_6 = H$
CLXVI	$R_1 = R_2 = R_3 = R_4 = R_5 = R_6 = COOH$
CLXVII	$R_1 = COOH; R_3 = OCH_3; R_2 = R_4 = R_5 = R_6 = H$
CLXVIII	$R_1 = COOH; R_4 = OCH_3; R_2 = R_3 = R_5 = R_6 = H$
CLXIX	$R_1 = COOH; R_2 = R_3 = OCH_3; R_4 = R_5 = R_6 = H$
CLXX	$R_1 = COOH; R_3 = R_4 = OCH_3; R_2 = R_5 = R_6 = H$
CLXXI	$R_1 = COOH; R_3 = R_5 = OCH_3; R_2 = R_4 = R_6 = H$
CLXXII	$R_1 = COOH; R_3 = R_4 = R_5 = OCH_3; R_2 = R_6 = H$
CLXXIII	$R_1 = R_2 = COOH; R_3 = OCH_3; R_4 = R_5 = R_6 = H$
CLXXIV	$R_1 = R_2 = COOH; R_4 = OCH_3; R_3 = R_5 = R_6 = H$
CLXXV	$R_1 = R_3 = COOH; R_4 = OCH_3; R_2 = R_5 = R_6 = H$
CLXXVI	$R_1 = R_2 = R_3 = COOH; R_4 = OCH_3; R_5 = R_6 = H$
CLXXVII	$R_1 = R_2 = R_4 = COOH; R_3 = OCH_3; R_5 = R_6 = H$
CLXXVIII	$R_1 = R_3 = R_5 = COOH; R_2 = OCH_3; R_4 = R_6 = H$
CLXXIX	$R_1 = R_2 = R_3 = R_4 = COOH; R_5 = OCH_3; R_6 = H$
CLXXX	$R_1 = R_3 = R_4 = R_5 = COOH; R_2 = OCH_3; R_6 = H$
CLXXXI	$R_1 = R_2 = R_3 = R_4 = R_5 = COOH; R_6 = OCH_3$
CLXXXII	$R_1 = R_2 = COOH; R_3 = R_4 = OCH_3; R_5 = R_6 = H$
CLXXXIII	$R_1 = R_5 = COOH; R_3 = R_4 = OCH_3; R_2 = R_6 = H$
CLXXXIV	$R_1 = R_2 = R_3 = COOH; R_4 = R_5 = OCH_3; R_6 = H$
CLXXXV	$R_1 = CH_2COOH; R_2 = COOH; R_4 = OCH_3; R_3 = R_5 = R_6 = H$
CLXXXVI	$R_1 = CH_2COOH; R_3 = COOH; R_2 = R_4 = R_5 = R_6 = H$
CLXXXVII	$R_1 = CH_2COOH; R_3 = R_4 = COOH; R_2 = R_5 = R_6 = H$
CLXXXVIII	$R_1 = CH_2COOH; R_2 = R_3 = R_4 = R_5 = R_6 = COOH$
CLXXXIX	$R_1 = CH_2COOH; R_3 = R_4 = OCH_3; R_2 = R_5 = R_6 = COOH$
CXC	$R_1 = CH(O); R_3 = R_4 = OCH_3; R_2 = R_5 = R_6 = H$
CXCI	$R_1 = CH(O); R_3 = R_4 = R_5 = OCH_3; R_2 = R_6 = H$
CXCII	$R_1 = C(O)CH_3; R_3 = R_4 = OCH_3; R_2 = R_5 = R_6 = H$
CXCIII	$R_1 = C(O)CH_3; R_2 = R_3 = R_4 = OCH_3; R_5 = R_6 = H$
CXCIV	$R_1 = C(O)CH_3; R_3 = R_4 = R_5 = OCH_3; R_2 = R_6 = H$
CXCV	$R_1 = CH_2OCH_3; R_2 = R_3 = R_4 = R_5 = COOH; R_6 = H$
CXCVI	$R_1 = CH_2OCH_3; R_2 = R_3 = R_4 = R_6 = COOH; R_5 = H$
CXCVII	$R_1 = CH_2OCH_3; R_2 = R_3 = R_5 = R_6 = COOH; R_4 = H$
CXCVIII	$R_1 = COOC_4H_9; R_3 = R_4 = OCH_3; R_2 = R_5 = R_6 = H$
CXCIX	$R_1 = R_2 = COOCH_2CH(CH_3)_2; R_3 = R_4 = R_5 = R_6 = H$
CC	$R_1 = R_2 = OCH_3; R_3 = R_4 = R_5 = R_6 = H$
CCI	$R_1 = R_3 = OCH_3; R_2 = R_4 = R_5 = R_6 = H$

and pentanedioic (glutaric), hexanedioic (adipic), and heptanedioic acids are generally the most abundant of these in permanganate digests.

The same acids are also found in alkaline cupric oxide digests. Reaction Scheme [51] from Shemyakin and Shchukina (1956),

$$>C{=}C< \;\underset{}{\overset{H_2O}{\rightleftharpoons}}\; -\underset{HO}{C}-\underset{H}{C}- \;\rightleftharpoons\; >C{=}O + H_2C<$$

$$>\underset{O}{\overset{}{C}}-C< \;\rightleftharpoons\; -\overset{HO}{\underset{HO}{C}}-C- \;\longrightarrow\; -CO_2H + HC< \qquad [51]$$

indicates how, in the alkaline hydrolysis conditions that prevail in the alkaline cupric oxide degradation process, aliphatic carboxylic acids and hydrocarbons can form from olefinic structures in the digest medium. It is evident from that scheme how alcohol and ketone functionalities could also give rise to acids and hydrocarbons. It can readily be deduced how the positioning of such functionalities along the hydrocarbon chain can give dicarboxylic acids, as well as monocarboxylic acids and alkanes.

Long, straight chain hydrocarbons, represented by **CIL**, and fatty acids, represented by **CLIV**, are found especially in the digests of alkaline cupric oxide oxidations, and were shown in the work of Neyroud and Schnitzer (1974) and of Schnitzer and Neyroud (1975) to consist of >10% of the identified digest products. The majority of the alkanes were in the range of C_{18} to C_{36}, and had an odd/even C ratio of 1.0, suggesting microbial origins. Release of the long chain fatty acids required 3 h of hydrolysis at 170°C, and that would suggest saponification of esters, possibly phenolic esters. These digest products are unlikely to be components of HS, as defined earlier in the Classical Definitions (p. 7), but are likely to have been produced by microorganisms or by plant cuticles, associated with the humic molecules, and coextracted with these. The branched short chain acids (**CLV** and **CLVI**) are likely to be oxidized artifacts products produced in the digest conditions.

Perhaps most information can be obtained from the aromatic structures (**CLVII** to **CCI**). All of the benzenecarboxylic acids, with the exception of benzoic acid, were identified in the works referred to for the permanganate and alkaline cupric oxide digests. On the basis of mechanistic considerations, these could arise from oxidizable side chains on a single aromatic nucleus, such as straight chain and branched chain hydrocarbons, with or without various functionalities, and cyclic aliphatic substituents, such as those shown in Reaction Scheme [47]. Given what is now known about the components of HS, it is unlikely that extensive substitutions occur with functionalities of the types mentioned. However, there is general agreement that char materials contribute to the C in many soils. The char can arise from forest fires and from vegetation burning, which was a common agricultural practice in some environments. Fused aromatic structures can be models for char, and Reaction Scheme [48] shows how oxidation of fused aromatic structures can give rise to all of the benzenecarboxylic acids in the range depicted by compounds

CLVII to **CLXVI**. Alternatively, the possibilities for carbonylation reactions giving rise to the polycarboxylic acids in the digest conditions which operate cannot be overruled.

Compounds **CLXVII** to **CLXXXIV** contain one or more methoxy and one or more carboxyl substituents on the aromatic nucleus. It is tempting to infer that the compounds containing methoxyl in the 4-position, in the 3,4-positions, and in the 3,4,5-positions had origins in lignin, and that would include the benzenecarboxylic acid derivatives (**CLXVIII**, **CLXX**, **CLXXII**, and **CLXXIV**). On the basis of the reasoning outlined earlier (Degradations in Sodium Sulfide Solutions, p. 69), many of the structures in Table 1–10 can be considered to provide further proof of the contributions of lignin to the compositions of HS. The R_1 carboxyl substituent would arise from the oxidation of fragments of the C_3 lignin side chain. Even stronger proof is provided by the identification in alkaline cupric oxide digests of compounds **CXCI**, **CXCII**, **CXCIII**, and **CXCIV**, and these were found by Schnitzer and Ortiz de Serra (1973b) to contribute 26% of the mass of the structures identified in the CuO–NaOH digest of an unmethylated HA from the A_1 horizon of a Brunizem soil. Compounds **CLXX** and **CLXXII**, from the same digest, might be considered to be more advanced oxidation products from **CXCI** and **CXCII**, and these contributed an additional 14% of the mass of the compounds identified. When the HAs were methylated before digestion the amounts of these lignin-related products identified were 23% for the carbaldehydes plus the ketones, and 20% for the related acids. Thus, >40% of the mass of the products identified could be considered to be derived from lignin. Reaction Scheme [50] shows how benzene carbaldehyde structures can be formed from lignin fragments, and Reaction Scheme [52] from Wallis (1971),

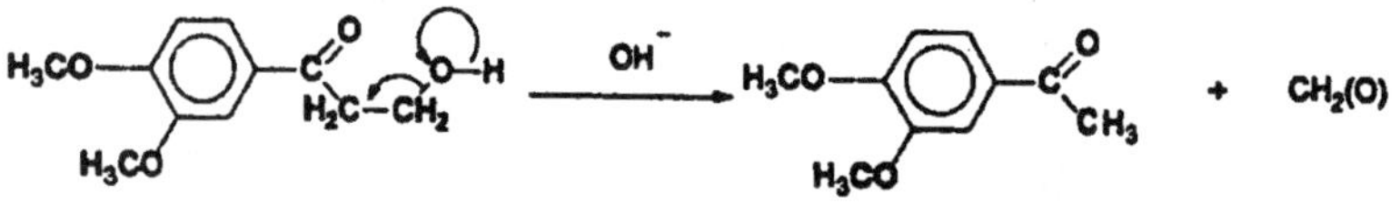

[52]

outlines a mechanism by which methyl ketones can form.

The identification among the alkaline cupric oxide digest degradation products of di- and tricarboxylic acid substituents in the methoxybenzene compounds shown as structures **CLXXIII** to **CLXXXIV** raises the possibility of carbonylation reactions giving rise to the carboxyls. Some of the methoxyl groups, but not all are in the ring positions expected for lignin structures, and the positions of others might suggest fungal origins. Haider and Martin (1967) showed how a variety of di- and trihydroxybenzenes, di- and trihydroxybenzenecarboxylic acids, and di- and trihydroxymethylbenzene structures were precursors to the formation of humic-type acids by fungi in culture media. Thus, substances synthesized by fungi might account for some of the digest compounds in the range of **CLXXIII** to **CLXXXIV**.

The phenylethanoic acid structures (**CLXXXV** to **CLXXXIX**) provide evidence for aliphatic side chain substituents on the aromatic nuclei. The ethanoic acid substituents could arise from incomplete oxidation of aliphatic side chains and from olefinic functionalities β to the aromatic nucleus. Such olefinic groups could be formed from lignin-type precursors using mechanisms based on that outlined in Reaction Scheme [49].

The benzylmethylether compounds identified (**CXCV** to **CXCVII**) are unlikely to have arisen from primary alcohol structures as the result of the methylation procedures used (diazomethane), but these could have arisen from the methylation of enols and subsequent cleavages. Because of the four carboxylic acid substituents on each aromatic nucleus, it is tempting to infer that the benzyl substituents arose from the incomplete oxidation of one of five aliphatic substituents on the aromatic nucleus. However, as referred to previously, there is no definite evidence for the presence of such structures in humic molecules.

Ester structures of the types represented by compounds **CXCVIII** and **CXCIX**, and detected in the digests of alkaline cupric oxide oxidations are unlikely to have been components of humic structures because the ester functionalities would have been hydrolyzed in the alkaline media. These could have formed as artifacts during the acidification of the digest products.

The 1,2-(**CC**) and 1,3-dimethoxybenzene (**CCI**) structures could have arisen from cleavages of the type shown in Reaction Scheme [49], and these could also have arisen from decarboxylation of compounds **CLXIX** and **CLXX**, in the case of **CC**, and **CLXXI** in the case of **CCI**.

The extensive studies of Schnitzer and his colleagues show that a broader range of identifiable compounds, and especially aliphatic and phenolic compounds (or perhaps more accurately methoxybenzene structures) are found in the digests of alkaline cupric oxide than in those of alkaline permanganate. Also, yields of solvent-soluble (ethyl acetate) digest products are greater in the cases of the alkaline cupric oxide digests. Thus, it is clear that a broader range of compounds survive in the digest conditions (than for permanganate), and therefore better comparisons can be made using alkaline cupric oxide (than permanganate) between the component molecules in HS from different sources.

Sequential Degradation Reactions

Hayes and Swift (1978) suggested the use of sequential degradation reactions for studies of the component molecules in HS, progressing from mild to increasingly reactive reagents. Simpson (1999) used the sequence BF_3 in methanol transesterification method (see Information from Transesterification Processes, p. 60), followed by Na_2S degradation (see Degradations in Sodium Sulfide Solutions, p. 69), followed by oxidation with permanganate (see Oxidations with Permanganate, p. 73) when studying the digest products from degradations of HA and FA from podzol soil profiles. Over 30 compounds were identified in the BF_3 in methanol digests in the cases of the HAs, and 15 of these were absent in the FA digests. More than 50% of the products identified in the HA digests were dicarboxylic esters, and the C_9 parent dicarboxylic acid accounted for 25 to 30% of these. The profile for the FA was different. No C_6–C_9 dicarboxylic acids were detected in the FA digests, and esters of C_4 and C_5 diacids accounted for about 20% of the products (compared with about 7% in the cases of the HAs). Short chain dicarboxylic acids are known to accumulate in the soil as products of multistage breakdown of longer chain hydrocarbons (Alexander, 1977). It is known that hydrocarbons with seven carbons in the chain have a degree of resistance to breakdown, and that might explain the relative abundance of $HOOC(CH_2)_7COOH$ in the HA digests of degradations with BF_3. Its absence in the FA digests might be attributed to solubility in the aqueous media. The importance of such acids in the HS of the soils used in this study has been emphasized in a separate study where NMR was used (Simpson et al., 2001a).

Bull et al. (2000) suggested that short chain (<C-20) *n*-alkylcarboxylic acids are derived from both aerial and subaerial vegetation. They identified suberin as the predominant source of longer-chain *n*-alkylcarboxylic acids, and considered that ω-hydroxycarboxylic acids and dicarboxylic acids are derived predominantly from inputs of free extractable polyesters and suberin intimately associated with plant roots. They stressed the importance of root material as a predominant source of aliphatic organic acids.

The aromatic compounds identified were contained in the HA and FA digests, and the esters of benzoic acid and of 3-phenylbutanoic acid were most abundant. The latter composed 14% of the products identified in the FA digests.

A total of 26 products were identified in the HA digests from the Na_2S degradation, and 14 of these were in the FA digest. The major products were aliphatic and similar to those liberated by BF_3 in methanol. About 50% of the identified HA digest products were the methyl esters of linear C_4 to C_9 dicarboxylic acids. Succinic acid (**CIII,** Table 1–10, where $n = 2$) accounted for about 18% of the identified HA digest products and about 37% of those from the FA digests, glutaric (**CIII**, where $n = 3$) and adipic acids (**CIII**, where $n = 4$) accounted for about 14 and 10%, respectively, of the products in the HA digests, but both were absent from the FA digest. Branched chain diesters, mainly the β-methyl derivatives of succinic acid (2-methylbutanedioic and 2-ethylbutanedioic acids) accounted for 24% of the HA and 40% of the product in the FA digest, but it is likely that these were formed from the succinic acid product in the digest. The shorter chain, C_4 to C_6 dicarboxylic acid esters dominated the digest products of the HAs, and not the C_7 to C_9 acids as in the case of the BF_3 in methanol digests. A number of longer chain fatty acids esters [$CH_3(CH_2)_nCOOCH_3$, where $n \geq 15$) were in the digests. It is likely that these were derived from free fatty acids associated with the soil HS and are known to make significant contributions to the SOM (Bol et al., 1996).

The esters of benzoic, phenylethanoic, and phenylpropanoic acids accounted for up to 30% of the digest products of the HAs and 18% of those in the FAs.

Only a small amount of residue was left for digestion with $KMnO_4$ following the Na_2S degradation. That would suggest that the first two steps were effective in releasing the component molecules from the podzol HA and FA samples studied. Thirty compounds were identified in the $KMnO_4$ digests of the HAs, but only six of these were contained in the FA digest. Long chain fatty acids were in all of the HA digests, and these were likely of plant origin. The major products identified were benzenecarboxylic acids. Benzene 1,2,4-tricarboxylic acid (**CLX**) composed 50 to 65% of the products identified in the HA and FA digests. Benzene 1,2,4,5-tetracarboxylic acid (**CLXIV**) composed 33% of the remainder of the compounds detected in the FA digest. Phthalic acid (benzene 1,2-dicarboxylic acid) was second in abundance in the HA digest components (28%). Other phthalate compounds were detected, and these might have arisen from plasticizers.

Information from Pyrolysis Degradation Processes

Pyrolysis (Py) procedures are increasingly used for studies of the component molecules of HS. Because of the large energy inputs in Py processes, most of the components of pyrolyzed materials are drastically altered from the structures that gave rise to these in the parent materials. Bracewell et al. (1989) have provided a comprehensive review of applications of Py to studies of the compositions of SOM and HS. The principles covered still hold, of course, but the techniques have advanced significantly in the last decade. More recently Hatcher et al. (2001) authored an excellent well-referenced review of Py techniques and results, and other relevant recent reviews were provided by Leinweber and Schulten (1999) and Kögel-Knabner (2000).

Pyrolysis–gas chromatography–mass spectrometry (Py–GC–MS) and Py field ionization mass spectrometry (FIMS) are the Py techniques most extensively used in studies of SOM and HS. In Py–GC–MS the Py products are separated chromatographically, the separated products are ionized (e.g., by electron impact), and detection is by MS. Py-FIMS uses soft electron impact ionization at low voltage to give molecular ions. Pyrolysis product identification is more difficult by this technique because of the lack of product separation.

Curie point Py involves heating the sample on a wire in an inert atmosphere to a temperature determined by the wire that supports the sample. In probe Py, for example, the sample is placed directly on a wire and the Py products are swept, using an inert gas, for analysis into the GC–MS system. In another system a chamber is used and calibrated Py temperatures are obtained using a metal wire coiled around a quartz tube (Saiz-Jiminez, 1994).

In our considerations of the products identified in the digests of chemical degradation reactions we stressed that the digest products were not the same as those that composed the HS and SOM molecules, and that, especially in cases where high energy inputs were used, these products were significantly different from those in the parent molecules. However, we pointed out that by taking account of the degradation mechanisms involved, useful predictions could be made of the precursors of the products identified. The same principles apply for considerations of products from Py processes. Pyrolysis products can give considerable broad-based information with regard to the possible origins of Py products.

Figure 1–15, from Hatcher et al. (2001), shows the chromatograms for the Py products from the organic horizon (a) and the mineral horizon (b) of a forest soil. The authors provided identifications of the numbered peaks shown. They considered that hydroxyl- and methoxybenzene structures had origins in lignin or lignin-derived precursors, although some of these could have also been generated from proteins. Thus, they assigned to origins in lignin-derived precursors the compounds represented by peaks 23 (2-hydroxy-1-methoxybenzene); 28 (2-hydroxy-1-methoxy-5-methylbenzene); 35 (2-hydroxy-1-methoxy-4-vinylbenzene, **CCII**, Fig. 1–16); 38 (4-allyl-2-methoxyphenol, or eugenol, **CCIII**); 44 (4-*trans*-allyl-2-methoxyphenol, or trans isoeuginol, **CCIV**); 46 (4-hydroxy-3-methoxybenzenecarbaldehyde, or vanillin); 51 (4-hydroxy-3-methoxyacetophenone, **CCV**); and 65 (*trans*-2,6-dimethoxy-4-propenyl phenol, **CCVI**). They considered that protein or peptide structures gave rise to peaks 5 (pyrrole, **CCVII**); 18 (benzonitrile,

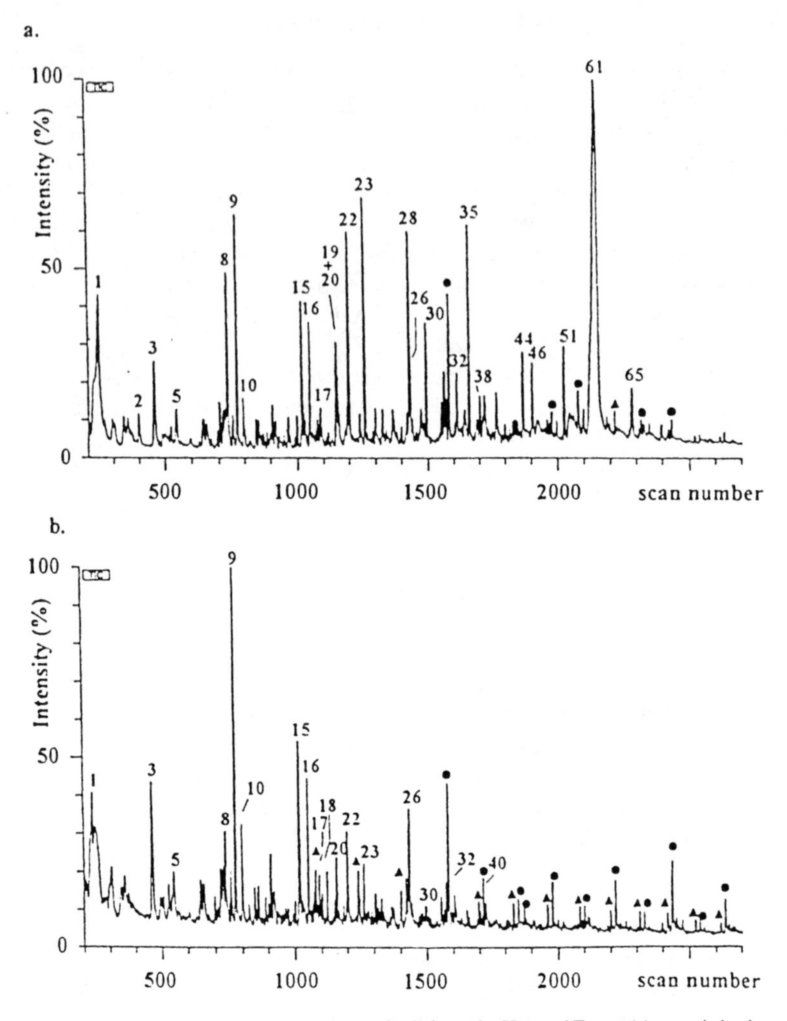

Fig. 1–15. Chromatogram of pyrolysis products of soil from the Harvard Forest (a) organic horizon and (b) mineral horizon. Peak assignments are given in text (after Hatcher et al., 2001).

CCVIII); and 40 (indole, **CCIX**), and that lignin + protein structures could have given rise to 15 (phenol, or hydroxybenzene, **XLIV**, Fig. 1–11); 20 (2-methyl phenol); 22 (3- and 4-methyl phenol); and 30 (4-vinylphenol, **CCX**).

Pyrolyzates of SOM and of HS are usually rich in compounds with origins in carbohydrates. Bracewell et al. (1989) found, particularly in FA fractions, compounds with origins in carbohydrates, motivating them to propose that *pseudopolysaccharides*, a term used to describe saccharide-type or -derived/related materials that give Py products such as furan (**CCXI**), methylfuran and dimethylfuran (with CH_3 substituents in the furan ring), furfural (**XXV**, Fig. 1–8, p. 54), and methylfurfural structures, are major components of the FA fraction. However, Bracewell et al. based their suggestions on Py data obtained for samples that had

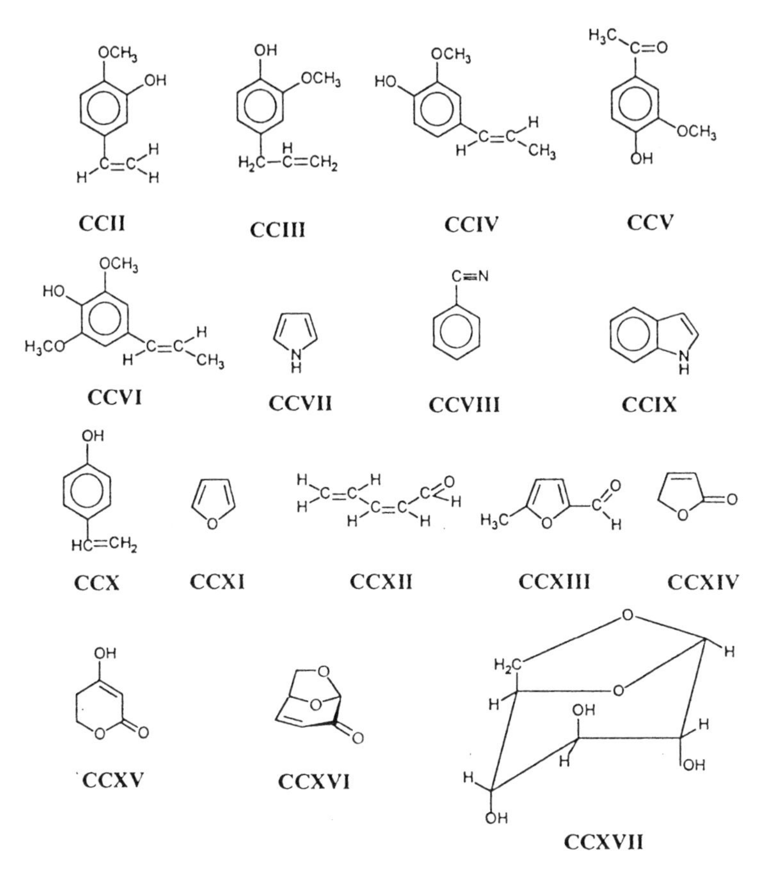

Fig. 1–16. Structures of some of the products from pyrolysis degradation processes (after Hatcher et al., 2001).

not been subjected to procedures that might remove noncovalently linked (to the HS) saccharide and peptide materials (such as the XAD-8 treatment described in Modern Fractionation Procedures, p. 42). Thus, it is likely that the substrates examined contained considerable amounts of saccharides and polysaccharides that may be associated with but are not now considered to be components of the FAs. There was less evidence for peptide- and carbohydrate-related structures when the samples were hydrolyzed in 6 *M* HCl before Py. Hydrolysis, however, increased the abundance of phenol Py products, with likely origins in lignins and in microbially synthesized polyphenol substances.

Hatcher et al. (2001) considered as carbohydrate-derived the compounds in the pyrograms in Fig. 1–15 labeled 1 (ethanoic or acetic acid), 3 (toluene or methylbenzene), 8 [(2H)-furan-2-one], 9 (furfural, or 2-furaldehyde), 10 (2,4-pentadienal, **CCXII**), 16 (5-methyl-2-furaldehyde, **CCXIII**), 17 [(5H)-furan-2-one, **CCXIV**], 19 [4-hydroxy-5,6-dihydro-(2H)-pyran-2-one, **CCXV**], 26 [(1s)-6,8-dioxabicyclo-3,2,1-oct-2 en-4-one, or levoglucosenone, **CCXVI**], and 61 (1,6-anhydro-β-D-glucapyranose, or levoglucosan, **CCXVII**).

The peaks represented by solid triangles in Fig. 1–15 were identified as alkanes, and those by solid circles as alkenes. The figure shows the very considerable abundance of alkanes and alkenes in the mineral horizon.

There is the risk, of course, that artifact products which cannot be related to structural components in HS arise from applications of Py procedures, especially when high temperatures are used. Py-FIMS, as used by Schnitzer and Schulten (1992) for studies on a whole podzol gave abundant evidence for carbohydrates through identifications of Py degradation products such as furfural (**XXV**, Fig. 1–8) from pentoses, and hydroxymethylfurfural (**XXVI**) from hexose sugars. Schnitzer (2000) has summarized data to show the dominance of carbohydrate fragments: (*m*/*z* = 60, 72, 82, 84, 96, 98, 110, 112, 114, 126, 132, 144, and 162; phenols (*m*/*z* = 94, 108, 110, 122, 124, 126, 138, and 154); lignin fragments (monolignins and dilignins), and long chain (suberin-derived) esters (*m*/*z* = 446, 474, 502, and 530); and n-C_{10} to n-C_{18} alkyl diesters. There was also evidence for alkylbenzenes ranging from *m*/*z* = 442 ($C_6H_5C_{26}H_{53}$) to *m*/*z* 470 ($C_6H_5C_{28}H_{57}$). Hatcher et al. (2001) stressed how the high temperature degradation process involved in Py can give an overabundance of alkanes and alkylbenzenes. The alkylbenzenes can arise from cyclization and aromatization processes involving an aliphatic group containing at least one functionality. Because Py decarboxylates fatty acids, alkanes and alkenes result. Nitrogen-containing compounds, with possible origins in peptide structures, were also identified.

Subsequent studies by Schnitzer and Schulten (1995) used Curie-point Py–GC–MS to investigate the Py products of soil HAs and FAs, humin, and the organic matter of whole soils and of the different size fractions of soils. They identified Py products of carbohydrates, phenols and lignins, fatty acid *n*-alkyl esters, N-containing compounds, alkanes and alkenes, and alkyl aromatics, including alkyl naphthalene and alkyl phenanthrene. There were surprising abundances of alkyl-aryl compounds consisting of aromatic rings linked covalently to aliphatic chains. The chain lengths were from C_1 to C_{22}. As indicated above, these could have arisen from aromatization processes.

Challinor (1995) introduced the tetramethylammonium hydroxide [$(H_3C)_4N^+OH^-$, TMAH) thermochemolysis GC–MS procedure for characterization of wood, and Hatcher et al. (1995) introduced it for studies of HS. Clifford et al. (1995), del Rio et al. (1998), Huang et al. (1998), Chefetz et al. (2000), and van Heemst et al. (2000) used the procedure for studies of HS and SOM. This is a one-step derivitization and degradation technique in which labile C–O bonds are cleaved. This is a base catalyzed reaction at high temperatures. Acidic protons in carboxyl and phenolic-OH functionalities are replaced by the CH_3 (to give methyl ester and methoxyl groups, respectively), esters are transesterified to the methyl ester (Filley et al., 1999), and the volatile products are separated and analyzed by GC–MS. The technique can be applied on-line (where the TMAH is added directly to a pyrolyzer linked to a GC–MS system) or off-line (where, for example, the reaction is performed in sealed ampoules).

The TMAH method has many advantages. It allows subpyrolysis temperatures to be used (250°C), and thus the products identified are closer to the parent structures in the native molecules than those from many of the harsher chemical degradation and Py procedures. The reader is referred to Hatcher et al. (2001) for a discussion of the process, a TMAH profile for a streamwater DOM, and a listing in the pyrogram of products of uncertain origins. They also list products from carbohydrate and lignin precursors, as well as N-containing compounds and alkanes.

Huang et al. (1998) performed Py–GC–MS and in-source Py-electron impact (EI) MS studies on the dissolved organic matter (DOM), on its parent SOM in the litter (L_f) and humified organic (O_h) horizons, and in the mineral soil of a peaty gley upland soil at the summit of the Great Dun Fell, Cumbria, England. Some samples were methylated using the TMAH procedure (Hatcher et al., 1995). The major pyrolyzates included compounds derived from carbohydrates, lignin, and fatty acids (some with >50 C atoms). Long chain alkylaromatics and fused aromatic structures were not listed; this suggests that artifact formation was greatly decreased in their procedures.

This well-structured study gives indications of the changes that take place to plant components during the humification process. Lignin components of the vegetation were clearly evident in the Py products from the litter material. There was an enrichment of the 4-hydroxybenzene components (of the phenylpropane structures) of lignins over the syringyl (3,5-dimethoxy-4-hydroxy) and guaiacyl (3-methoxy-4-hydroxy) units in the pyrolyzates of the material in the O_h horizons compared with the litter. That suggested a selective removal of the tri- and di-substituted aromatic structures, probably as the result of oxidation of the aromatic structures. There was evidence also for oxidative degradation of the C_3 side chains in O_h compared with the litter materials. Lignin signals were virtually absent from the Py products of the mineral soil, and that was considered to arise from the intensive oxidation of the aromatic rings during the transformation processes in the soil environment.

Saccharide-derived pyrolyzates from the L_f horizon had their origins primarily in plant materials. Polysaccharide signals decreased with depth, and there was preferential degradation of hemicellulose relative to cellulose. Contributions from fungal chitin and bacterial cell wall components in the O_h horizon were indicated by anhydroglucosamine pyrolyzates, and microbial polysaccharides synthesized in situ were the major contributors to the carbohydrates of the mineral soil.

The fact that fatty acids were especially abundant in the O_h horizon is significant. These substances are not, of course, humic, but are found in association with humic fractions, and it is difficult to separate these from the HS. Their abundance in the O_h horizon, and the corresponding decrease in lignin-derived components in that horizon, was considered to reflect the extent of resistance of the acids to biological degradation. The C_{22} to C_{36} *n*-fatty acids were attributed to inputs from leaf waxes, and the C_{14} to C_{18} and unsaturated fatty acids could originate from both plants and microorganisms. The very long chain fatty acids (C_{43} to C_{53}) were considered to have likely origins in mycobacteria.

The lignin-derived Py products in the DOM samples were more highly oxidized than those in the soil, and the products indicated that the saccharides present were significantly different from those in the soil horizons. The *n*-fatty acids were shorter (maximum C_{22} with C_{16} dominant), indicating that bound lipids of longer chain lengths are less likely to be mobilized by water leaching.

Spectroscopic Methods

Comprehensive reviews have been presented on the application of various spectroscopic techniques to the characterization of HS (see Part II, Spectroscopic Studies of the Structures of Humic Substances, in Hayes et al., 1989a). With the exception of nuclear magnetic resonance (NMR) spectroscopy, most of these will not be addressed in detail here. Hayes (1997) summarized some of the primary features of results obtained using various spectroscopic techniques other than NMR. General discussion of these analyses can be found in Silverstein et al. (1991), Skoog and Leary (1992), and Atkins (1994). In summary:

Infra-red. Infra-red (IR) spectra typically have broad bands characteristic of a variety of molecular structures. Although Diffuse Reflectance Fourier-Transform (DRIFT) IR spectroscopy provides better resolution, only limited information can be obtained from the IR technique. MacCarthy and Rice (1985) and Bloom and Leenheer (1989) reviewed applications of IR to studies of compositional aspects of HS.

Raman. Some of the earlier Raman spectra of HS were of limited value because of fluorescence problems. Improvements were realized when HS were sorbed to noble metal surfaces, which increased sensitivity. This technique is referred to as Surface-Enhanced Raman Spectroscopy (SERS). Results from SERS studies have emphasized that HS are of questionable homogeneity. Bloom and Leenheer (1989) reviewed aspects of the theory and general applications of Raman spectroscopy and outlined possible applications of the technique for studies of HS.

Ultraviolet-Visible. The utility of ultraviolet-visible (UV-Vis) spectroscopy is limited because spectral lines are too broad to provide structural details. Bloom and Leenheer (1989) reviewed aspects of the theory and the applications of UV-Vis spectroscopy to studies of HS.

Electron Paramagentic Resonance. Electron paramagentic resonance (EPR) involves the study of molecules containing unpaired electrons by determining the magnetic fields at which they come into resonance with monochromatic radiation. Applications of EPR to studies of HS have been described by Senesi and Steelink (1989). This approach has provided only limited information on molecular structures of HS.

X-ray Photoelectron Spectroscopy. X-ray photoelectron spectroscopy is used to study the electronic environment near an element by causing inner shell electrons to eject when a sample is subjected to X-ray photons. The kinetic energies of the ejected electrons are a function of their binding energies, which are in turn related to the atomic number of the element and to electron density around the inner shell electrons. This technique has not been shown to be sensitive to small differences in electron binding and is of limited value in determining humic structures (Bloom and Leenheer, 1989).

Mössbauer. Bloom and Leenheer (1989) stated that Mössbauer spectroscopy is not directly applicable to the study of the structure of organic compounds, but it can be used to study metal ion binding sites.

Fluorescence Spectroscopy. Molecular fluorescence is based on the emission of a photon resulting from the transition from the first excited-singlet state to the singlet ground state. The nature of the fluorescence stems from the structural components of the molecule. The diversity of molecular structures in HS makes direct identification of individual structural components responsible for fluorescence extremely difficult. This is partly due to the fact that indigenous fluorescent structures generally constitute a minor portion of HS. Observed fluorescence spectra have most often been poorly resolved (Senesi and Loffredo, 1999).

Nuclear Magnetic Resonance Spectroscopy. NMR is among the most powerful tools for determining molecular structure for both organic and inorganic compounds (Wilson, 1987). This method exploits the separation of nuclear-spin energy levels in a strong magnetic field (see reviews of NMR basics by Derome, 1987, and Macomber, 1998). The number and nature of chemical bonds, as well as the proximity of adjacent atoms in space all affect the energy splitting of target nuclei and thereby provide structural information. It is a nondestructive technique and under appropriate conditions can provide quantitative analyses of the NMR-active atoms in a sample. Nuclear magnetic resonance is the premier instrumental procedure for both compositional and structural descriptions of humic materials.

Solid-state NMR, in particular the experiment known as cross polarization magic angle spinning (CPMAS), has been the most widely applied NMR procedure in humic studies. The primary advantage it has over solution-state NMR is that some SOM is insoluble in the solvent systems that can be used and can only be analyzed effectively in the solid state (Malcolm, 1989; Kögel-Knabner, 1997). The CPMAS experiment works by transferring the magnetization of protons (^{1}H), which are 100% NMR active, to the NMR-active carbon (^{13}C) atoms that make up only about 1% of naturally occurring C nuclei. This leads to a sensitivity gain of about four times that of ^{13}C. In the solid state, interaction between atoms are dipolar and occur over long ranges. As a result, the transfer of ^{1}H magnetization is spread uniformly across the entire sample volume, and sensitivity enhancement occurs for all ^{13}C atoms in the sample. Furthermore, the time between magnetic pulses is proportional to the time required by the protons in a sample to reestablish equilibrium or relax. Therefore, since protons relax faster than ^{13}C atoms, there is less waiting between scans using CPMAS than conventional direct detection, which means that data can be collected faster. During the solid-state experiment a sample is spun at the magic angle (54.7°), which has the effect of decreasing anisotropy, that is, the nonspherical magnetic field around an atom causing broad NMR signals or lines. Although this produces sharper peaks, inevitably some line broadening results, giving what are known as spinning side bands. These side bands become impossible to remove from conventional CPMAS experiments when samples are spun at magnetic field strengths > 7.05 Tesla (300 MHz) using common commercially available instrumentation. For humics, therefore, CPMAS-NMR is generally limited to instruments operating at 300 MHz or less. When CPMAS was introduced in 1973 (Pines et al., 1973), only low-field NMR instruments (<200 MHz) were available. Thus, elimination of side bands was not a major concern. CPMAS, until recently, has provided the quickest means to collect quantitative NMR data on extracted materials and whole-soil samples.

It is possible to observe up to 97% of the C-containing structures in HS when CPMAS ^{13}C-NMR spectra are obtained under optimal conditions. The spectra can provide useful information about aliphatic/aromatic ratios, as well as the presence of certain classes of structures, such as carboxyl/ester/amide groups, phenols, ether-type functionalities (specifically methoxyl groups), and carbohydrates or carbohydrate-derived materials. In addition, it is an excellent "finger-printing" technique for comparing samples from different origins, or those obtained by different procedures and treatments (Clapp et al., 1993).

There have been a number of cases in the literature of overinterpretation of broad resonance peaks, especially in quantification (Clapp et al., 1993); therefore, it is important to optimize the NMR experiment. Several important factors have been identified with respect to quantifiable solid-state ^{13}C-NMR. These include:

- spinning speed at the magic angle to decrease side-bands
- sufficient high-power ^{1}H decoupling
- proper contact time to maximize cross-polarization for magnetization transfer from ^{1}H-spins to ^{13}C-spins, and appropriate delays for complete ^{1}H relaxation
- delay time that is at least three to five times longer than the longest ^{1}H relaxation time in the sample
- percentage of C in the sample
- amount of organic free radicals and/or inorganic paramagnetic species (e.g., Fe and Cu) in the sample, which may cause loss of the ^{13}C-signal in selective or in all parts of the spectrum due to influences on ^{13}C relaxation and cross-polarization efficiency.

The spectra in Fig. 1–17 (from Clapp and Hayes, 1999), for HAs isolated from the silt and clay fractions of a Mollisol to which maize (a C_4 plant) residues were added or were removed after harvesting during an 8-yr period, provide an example of the information that can be obtained by CPMAS ^{13}C-NMR. There are definite similarities in the spectra, but there are clear and informative differences as well. Resonance peaks in the 25- to 35-ppm range indicate aliphatic C, suggesting the presence of methylene groups in alkyl chains. The resonance centered around 56 ppm are likely to be methoxyl/ethoxyl groups, but amino functionalities, such as those in peptides, also resonate in this region. The distinct signals in the 65- to 70-ppm area could be aliphatic C moieties linked by O (ether and ester linkages), and/or C bonded to secondary alcohol-type structures, including saccharides. The small shoulder at about 100 to 105 ppm can represent anomeric C, or the C_1 of sugars in cyclic conformation (see Soil Saccharides, p. 103) or to saccharide-derived materials. Evidence for aromaticity is given by resonances between 110 and 140 ppm. The relative symmetry of that peak suggests that char materials contribute to the resonance. There is a distinct difference between the spectra for the silt- and clay-fraction HAs in the chemical shift region around 145 to 150 ppm. That resonance, characteristic of O-aromatic (e.g., –OH and/or –OCH_3) substituents, is not evident in the spectra of the clay HAs, and is indicative of a more pronounced lignin signal in the fraction from the silt-sized separates. This concept is supported by the sharper resonance at 56 ppm (indicating methoxyl, also from lignin components) in the cases of the HAs from silt. The 56-ppm resonance is more rounded in the cases

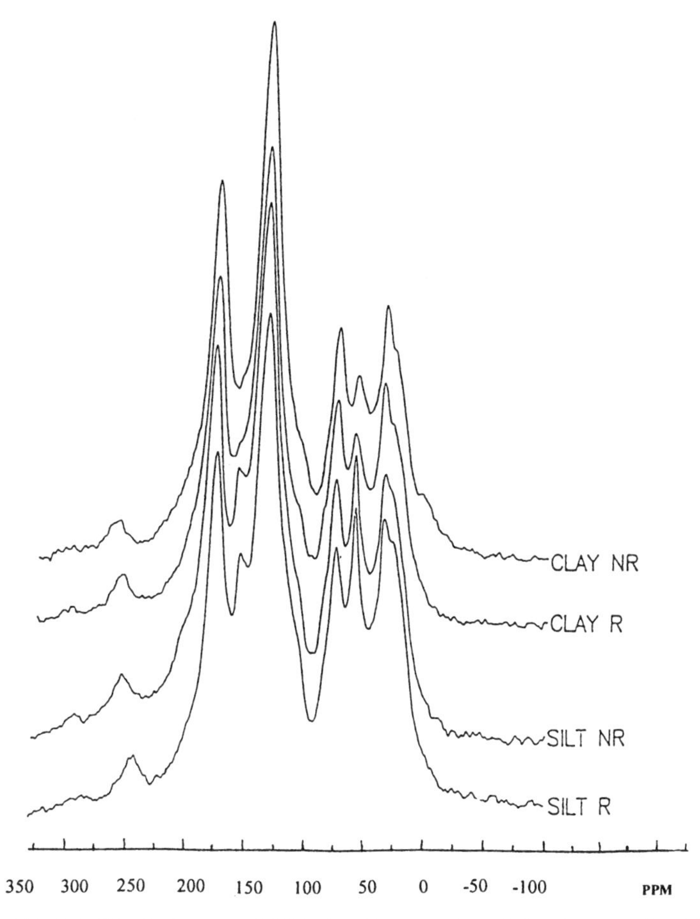

Fig. 1–17. CP-MAS ^{13}C-NMR spectra of HA isolated from the silt and clay fractions of a Mollisol to which maize residues had been added annually (R) or not added (NR) for 8 yr (from Clapp and Hayes, 1999).

of the clay HAs, and that would suggest contributions from amino/peptide functionalities. Delta ^{13}C measurements indicated that there was less contribution from the maize in the HAs from the silt-sized separates. It was concluded that the silt-sized fraction was made up of aggregates of clay particles that had sequestered and protected the organic matter before the introduction of maize cropping. That organic matter was less transformed (humified) than the material in contact with the clay-sized materials. The peaks at the 160- to 180-ppm resonances are due primarily to carboxyl, ester, and amide C, and those at 230- to 250-ppm are likely to be from carbonyl groups.

Solid-state ^{1}H-NMR is difficult with HS because spectral lines are often so broad that they provide little information. In cases where extracts from soils and waters are soluble in solvents such as D_2O, NaOD, or DMSO, solution-state NMR has significant advantages over solid-state NMR. In solution, the tumbling of individual molecules eliminates anisotropy, and spectra can be obtained with much greater resolution and without interferences from side bands. With the advent of higher field spectrometers (>300–900 MHz) and cryogenically cooled probes, direct detection via solution-state NMR can actually provide greater sensitivity than CPMAS, and with better spectral dispersion and resolution. A 900-MHz spectrometer will have three times the sensitivity of a 300-MHz instrument, while a cryoprobe is capable of providing four times the sensitivity of conventional probes. Both proton ^{1}H- and ^{13}C-NMR experiments can be employed in solution state to provide qualitative information on humic mixtures. Figure 1–18 shows examples of solution-state ^{1}H and ^{13}C spectra and a distortionless enhancement by polarization transfer (DEPT) spectrum for FA extracted from the B_h horizon of an oak forest soil and isolated on XAD-8 resin.

One-dimensional NMR datasets of soil extracts have considerable and often continuous overlap, and spectral matching of one-dimensional data are virtually impossible. Thus, to decrease spectral overlap it is essential to add a second dimension. Multidimensional experiments can easily be executed to correlate heteronuclear chemical shifts and to provide through-bond and interaction-through-space connectivity information. Most multidimensional NMR experiments are in two-dimensional as opposed to the earlier, more common one-dimensional spectra. Numerous recent studies using these techniques with humic materials have been performed successfully and have been reviewed by Simpson (2001).

With solution-state, multidimensional NMR spectroscopy it has been possible to observe the main structural components in FAs and HAs (Schmitt-Kopplin et al., 1998; Fan et al., 2000; Haiber et al., 2001; Simpson et al., 2001a). Chemical shift and coupling data from standard compounds can be used to identify the possible moieties present. For example, Simpson et al. (2001a) made a preliminary search using both library spectra and standard compounds from the laboratory to determine the chemical shifts and couplings in more than 1000 aromatic moieties that might be in FAs. A number of these had been identified in degradation digests. Others are derivatives of compounds in forms expected to be present in humic structures. From this, more than 30 aromatic compounds were found that gave similar or identical shift and coupling data to those observed in ^{1}H-NMR spectra (see Table 1–11, p. 96).

Simpson et al. (2001a) summarized the major components in an XAD-fractionated FA as follows:

- mono- and di-carboxylic acids in approximately equal amounts, to which can be assigned chain lengths of primarily 6, 7, or 10 C atoms—these are the major components in the sample
- smaller amounts, about 10 to 20% of the amounts of acids, comprised of esters and alcohols or ethers
- some carbohydrate and amino acid residues, most likely to be in the form of chains
- very small amounts of 1,2-, 1,4-, and 1,3,4-substituted benzenes and cinnamic acids

More recent studies using conventional multidimensional experiments coupled with liquid chromatography (which is a part of a newly expanding class of technology referred to as *hyphenated-NMR*), *diffusion ordered spectroscopy* (DOSY), and *three-dimensional NMR spectroscopy*, have shown that non-fractionated soil organic extracts are comprised mainly of lignin, carbohydrate, peptide, and aliphatic residues that are not cross-linked (Simpson et al., 2001b, 2002, 2003, 2004b). These findings have led to a description of humic materials as a mixture of extractable soil components, which accrue from the microbial and vegetative inputs of a specific locale.

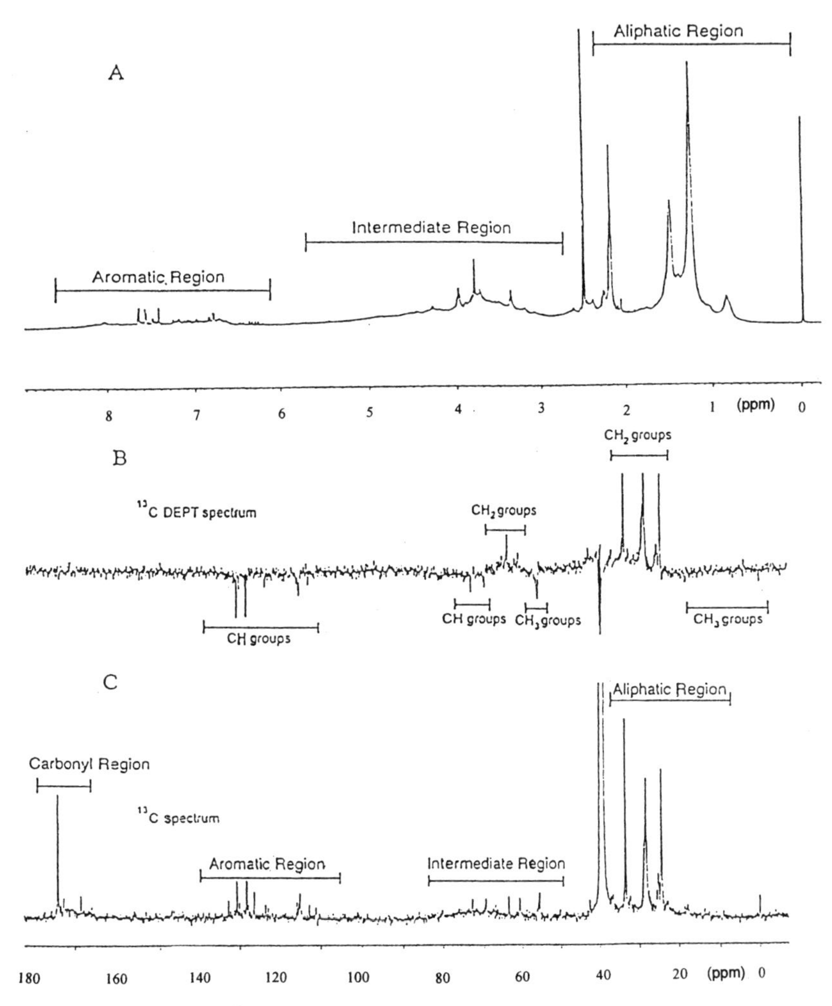

Fig. 1–18. (A) Solution-state ^{1}H-NMR spectrum of a fulvic acid (FA) dissolved in DMSO-d_6. The major spectral regions have been labeled as: aromatic region, which also contains amide signals; intermediate region containing a mixture of amino acid, sugar, and CH_2 groups adjacent to ester, and ether/hydroxyl; and aliphatic region of various units. (B) ^{13}C-DEPT and (C) ^{13}C one-dimensional spectra of a fractionated FA dissolved in DMSO-d_6. In (B) the CH_3 and CH groups point down and CH_2 groups up; quaternary carbons do not appear. In (C) the intermediate region contains signals from sugars, various amino acid carbons, and C adjacent to ether, ester, and hydroxyl groups; the aliphatic region includes amino acid side chains (after Simpson et al., 2001a).

Table 1–11. The chemical shifts of aromatic protons found in standard compounds and their correlation with the observed signals in the NMR spectra of a fulvic acid.†

R_1‡	R_2	Ring R_3	Substituents R_4	R_5	R_6	Standard Proton Shift		Observed Proton Shift		Zone (1–6)
						ppm				
$COCH_2CH_3$	H	H	OCH_3	H	H	$H_{5.3} = 6.90$	$H_{6.2} = 7.92$	$H_{5.3} = 6.89$	$H_{6.2} = 8.00$	1
COOH	H	H	$OCHCH_3CH_3$	H	H	$H_{5.3} = 6.90$	$H_{6.2} = 8.04$	$H_{5.3} = 6.89$	$H_{6.2} = 8.00$	1
COOH	H	H	$OCH_2CH_2CH_2CH_3$	H	H	$H_{5.3} = 6.93$	$H_{6.2} = 8.05$	$H_{5.3} = 6.89$	$H_{6.2} = 8.00$	1
$COOCH_2CH_3$	H	H	OCH_3	H	H	$H_{5.3} = 6.90$	$H_{6.2} = 7.99$	$H_{5.3} = 6.89$	$H_{6.2} = 8.00$	1
$COOCH_2CH_2CH_3$	H	H	OH	H	H	$H_{5.3} = 6.91$	$H_{6.2} = 7.95$	$H_{5.3} = 6.89$	$H_{6.2} = 8.00$	1
$COCH_2CH_2COOH$	H	H	OCH_3	H	H	$H_{5.3} = 7.04$	$H_{6.2} = 7.97$	$H_{5.3} = 7.04$	$H_{6.2} = 8.00$	1
$COOCH_2CH_2CHCH_3CH_3$	H	H	OH	H	H	$H_{5.3} = 6.90$	$H_{6.2} = 7.94$	$H_{5.3} = 6.89$	$H_{6.2} = 8.00$	1
$COOCH_2CHCH_3CH_3$	H	H	OH	H	H	$H_{5.3} = 6.92$	$H_{6.2} = 7.96$	$H_{5.3} = 6.89$	$H_{6.2} = 8.00$	1
$COOCH_3$	H	H	OH	H	H	$H_{5.3} = 6.88$	$H_{6.2} = 7.84$	$H_{5.3} = 6.83$	$H_{6.2} = 7.85$	2
COOH	H	OCH_3	OCH_3	H	H	$H_5 = 6.92$	$H_6 = 7.78$	$H_5 = 6.93$	$H_6 = 7.79$	2
COOH	H	OCH_2CH_3	OCH_2CH_3	H	H	$H_5 = 6.90$	$H_6 = 7.74$	$H_5 = 6.82$	$H_6 = 7.74$	2
COOH	H	H	OH	H	H	$H_{5.3} = 6.84$	$H_{6.2} = 7.80$	$H_{5.3} = 6.82$	$H_{6.2} = 7.74$	2
COOH	H	OH	OCH_3	H	H	$H_5 = 7.00$	$H_6 = 7.44$	$H_5 = 6.95$	$H_6 = 7.45$	3
$COCH_3$	H	OCH_3	OCH_3	H	H	$H_5 = 6.89$	$H_6 = 7.58$	$H_5 = 6.88$	$H_6 = 7.53$	3
$COOCH_2CH_3$	H	OH	OH	H	H	$H_5 = 6.83$	$H_6 = 7.34$	$H_5 = 6.84$	$H_6 = 7.42$	3
CHO	H	OCH_3	OH	H	H	$H_5 = 7.04$	$H_6 = 7.42$	$H_5 = 6.95$	$H_6 = 7.45$	3
CHO	H	OH	OCH_3	H	H	$H_5 = 6.96$	$H_6 = 7.42$	$H_5 = 6.84$	$H_6 = 7.42$	3
CHO	H	OCH_2CH_3	OCH_3	H	H	$H_5 = 6.98$	$H_6 = 7.45$	$H_5 = 6.84$	$H_6 = 7.42$	3
$COCH_3$	H	OCH_3	OCH_3	H	H	$H_5 = 6.84$	$H_6 = 7.54$	$H_5 = 6.80$	$H_6 = 7.50$	3
CHO	H	OCH_3	OCH_3	H	H	$H_5 = 6.99$	$H_6 = 7.47$	$H_5 = 7.08$	$H_6 = 7.45$	3
$COCH_3$	H	OCH_3	OH	H	H	$H_5 = 6.95$	$H_6 = 7.53$	$H_5 = 6.88$	$H_6 = 7.53$	3
$COCH_3$	H	OCH_3	OCH_2CH_3	H	H	$H_5 = 6.97$	$H_6 = 7.44$	$H_5 = 6.95$	$H_6 = 7.45$	3
CHO	H	OCH_2CH_3	OCH_2CH_3	H	H	$H_5 = 6.96$	$H_6 = 7.41$	$H_5 = 6.95$	$H_6 = 7.45$	3
CHO	H	OCH_3	H	OCH_3	H	$H_{6.2} = 6.70$	$H_5 = 7.00$	$H_{6.2} = 6.99$	$H_5 = 7.02$	4
COOH	H	OCH_3	H	OCH_3	H	$H_{6.2} = 6.70$	$H_5 = 7.26$	$H_{6.2} = 6.77$	$H_5 = 7.20$	4
CHCHCOOH	H	OH	OH	H	H	$H_6 = 6.78$	$H_6 = 7.00$	$H_5 = 6.78$	$H_6 = 7.03$	4
CHCHCOOH	H	OCH_3	OH	H	H	$H_6 = 6.80$	$H_6 = 7.11$	$H_5 = 6.80$	$H_{6.2} = 7.10$	4
CHCHCOOH	H	OH	OH	H	H	$H_{\alpha**} = 7.42$	$H_{\beta**} = 6.2$	$H_{\alpha**} = 7.45$	$H_{\beta**} = 6.25$	5
CHCHCOOH	H	OCH_3	OH	H	H	$H_{\alpha**} = 7.45$	$H_{\beta**} = 6.32$	$H_{\alpha**} = 7.46$	$H_{\beta**} = 6.32$	5
COOH	H	H	H	H	COOH	$H_{3.4} = 7.53$	$H_{5.2} = 7.73$	$H_{6.4} = 7.53$	$H_{5.2} = 7.70$	6

† From Simpson et al. (2001a).

‡ R_1 to R_6 represent the substituents on the aromatic nucleus.

There are concerns that extracting soil for the purpose of studying the structures and interactions of HS breaks the associations between the organic matter and clay mineral or oxide surfaces, alters conformations, and thereby alters the surface chemistry. The physicochemical processes at the solid–liquid interface determines the influence of soil colloids on the quantity and quality of our drinking water, C sequestration, food production, and the mobility and fate of anthropogenic chemicals and heavy metals. Approaches to the study involving solid-state NMR do not permit observations of specific structures at solvent–surface interfaces and, unlike solution-state experiments, cannot detect resonance signals arising from interactions through space via couplings through bonds, known as nuclear Overhausser effects (NOE). High resolution magic angle spinning (HRMAS) NMR spectroscopy is a newer technique that provides a means to study the structures, reactions, and interactions that occur at colloidal interfaces in samples that are unaltered from their natural setting. This technology allows for the analysis of materials that swell, become partially soluble, or form true solutions in a solvent, even when some solids are present. Solvent can be added to an analyte (e.g., intact soil particles), and after swelling the components in contact with the solvent become NMR-visible. Dipolar interactions that are predominant in the surface components in the solid state are decreased by the addition of solvent and are averaged by spinning the sample. A number of solution-state experiments that provide highly resolved structural information on the solvent accessible constituents may be employed using HRMAS (Simpson et al., 2001c). Figure 1–19 shows ^{1}H-HRMAS spectra for a whole soil sampled from the A_h horizon under an oak forest and swollen in either DMSO-d_6 or D_2O. Use of D_2O provides insights into the components that are water-accessible in nature and constitute the environmental–soil interface. DMSO, on the other hand, penetrates into the soil particles and provides information on subsurface structures and on the overall organization of components within the colloidal matrix. In the case of D_2O, HRMAS spectra indicate only those signals from components in a range of solvent-accessible environments, including those sorbed to mineral surfaces, in macromolecular domains, in association with other organic matter, or free in solution (Simpson et al., 2001c). Where DMSO is used, H bonds are broken, enabling the solvent to enter both hydrophobic and hydrophilic domains. Swelling the soil in DMSO can disrupt the soil aggregates and provide information on hydrophobic structures that may be physically protected by the arrangement of the organic matter under natural aqueous conditions. For example, compare spectra from soil swollen in DMSO-d_6 with those in D_2O wherein these signals are absent. The signals in DMSO-d_6 show a distinct band at ≈6.2 to 8.2 ppm.

NMR Studies of Composition and Associations

Composition

There are a variety of multidimensional NMR experiments that have been successfully applied to the study of HS (see the review by Simpson, 2001). These include total correlation spectroscopy (TOCSY), correlation spectroscopy (COSY), nuclear Overhauser effect spectroscopy (NOESY), heteronuclear single quantum coherence (HSQC), heteronuclear multiple bond connectivity (HMBC), and HSQC-

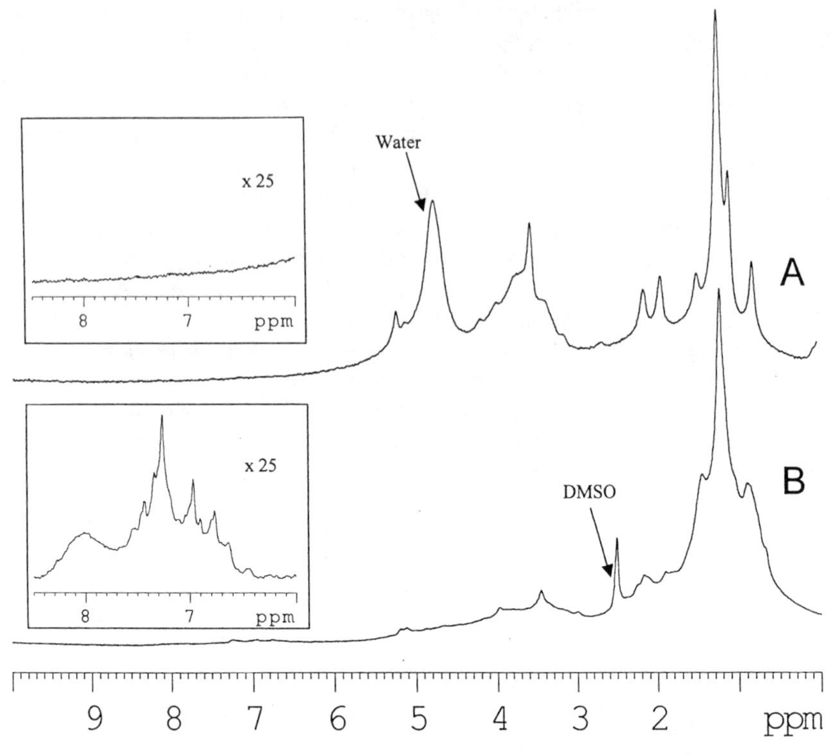

Fig. 1–19. ^{1}H-HRMAS spectra of intact soil particles swollen in D_2O (A) and in DMSO-d_6 (B). Inserts show the aromatic/amide region (adapted from Simpson et al., 2001c).

TOCSY. The interpretation of results from these techniques has been presented in far greater detail elsewhere and is largely beyond the scope of this chapter (Simpson, 2001; Simpson et al., 2001a). The collective evidence from these results allows the assignment of many of the structural connectivities in HS. Common structural components in many soil extracts identified by solution-state NMR include peptides (Simpson et al., 2002), carbohydrates (Simpson et al., 2001a), aliphatic species derived from cuticles (Simpson et al., 2003), lignin-derived, and aromatic species (Simpson et al., 2004a). These components combined represent the major constituents in many soil extracts (Simpson et al., 2002). However, knowledge is missing of the exact chemical composition of the different species present. An example of the TOCSY spectrum of a FA isolated from a pine forest is shown in Fig. 1–20. General assignments corresponding to labeled regions can be made as: (i) amino acids (amide-side chains); (ii) aromatics; (iii) methylene units adjacent to ethers, esters, and hydroxyls in aliphatic chains; amino acids (α-β-γ couplings); (iv) sugars; methine units bridging lignin aromatics; amino acids (α-β couplings); (v) methylene in aliphatic chains; and methyl units in amino acids and aliphatic chains. The major structural classes can in this case be summarized as components derived from plant and microbial biomass.

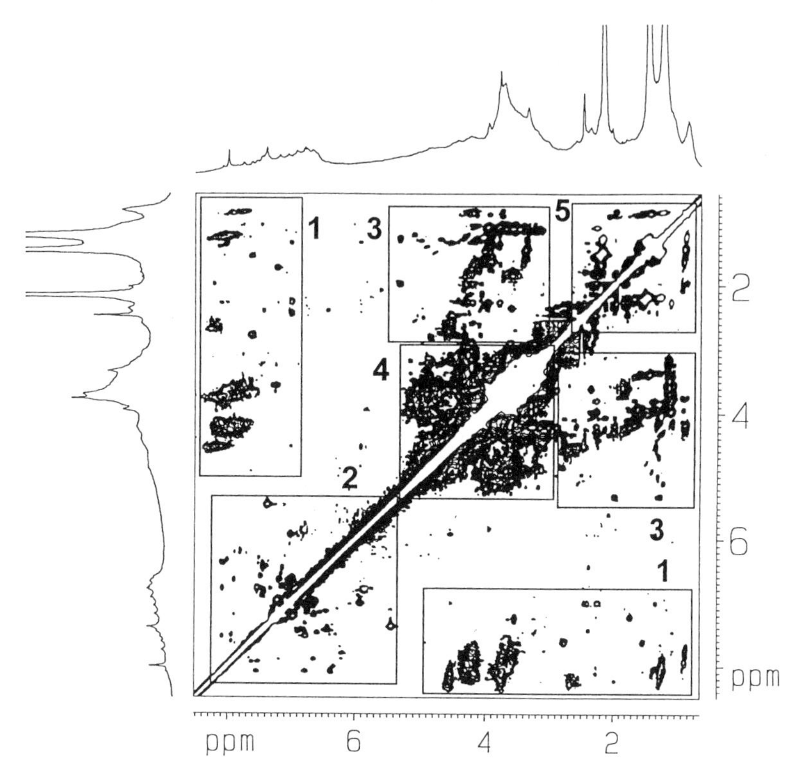

Fig. 1–20. TOCSY spectrum of a fulvic acid (FA) isolated from a pine forest soil. General assignments are included as follows: (1) amino acids (amide-side chains); (2) aromatics; (3) methylene units adjacent to ethers, esters, and hydroxyls in aliphatic chains; amino acids (α-β-γ couplings); (4) sugars; methine units bridging lignin aromatics; amino acids (α-β couplings); (5) methylene in aliphatic chains; and methyl units in amino acids and aliphatic chains (from Simpson, 2001).

It is logical to expect the major components in SOM to have plant and microbial origins because of the abundance and the influences of these in the soil environment. However, NMR findings to date do not rule out the presence of novel structural components formed by chemical and/or biological transformations of organic matter in the soil environment. If present, novel structures are likely to be at lower concentrations than those with origins directly attributable to parent materials that continually enter the soil and are readily identified in NMR studies. However, should novel structures be isolated and identified in the future, the awareness of these will have a significant impact on our knowledge of the humification process, and of terrestrial C storage.

Associations

Diffusion ordered spectroscopy is an NMR technique that employs the principle of molecular separation based on diffusion properties (Morris and Johnson, 1993). The resulting two-dimensional plot provides a direct correlation between diffusion coefficients (dc) and proton chemical shifts. Simpson (2002) employed DOSY to study the aggregation of the IHSS peat HA. In a pure aqueous solvent it was found that all the components displayed a single diffusion coefficient. This indicates that either all the components were covalently cross-linked as a macromolecule, or that they were present as a stable aggregate. With the addition of a trace amount of ethanoic acid (known to disaggregate proteins) separate diffusivities for the lignin, carbohydrate, and peptide components could then be clearly distinguished (see Fig. 1–21). Were these components bonded into a macromolecular assemblage, they would remain as a rigid unit and still display a more or less singular dc value. Since that is not what is observed, separation of the structures based on their diffusivities is definitive proof that they are not cross-linked or bound in the same complex, as suggested by Schulten and Schnitzer (1993). Furthermore, numerous DOSY experiments (Simpson, 2002) were able to monitor the varying degrees of HA aggregation at different concentrations, and demonstrate that weak acids could disrupt the aggregation process. At high concentration these aggregates display diffusivities that are consistent with those observed in large proteins (>66 000 Da). After dissaggregation and at low acid concentration, DOSY-NMR revealed that the major components (lignins, polysaccharides and peptides) in the peat HA had diffusivities consistent with MW compounds in the range of ≈2500, ≈1000, and 200 to 600 Da, respectively. In the same paper it was shown that the components in a FA did not tend to aggregate strongly, and these generally tended to be smaller than those found in the HA. This is consistent with earlier studies (Simpson et al., 2001b) in which the major components in a FA were easily separated in aqueous solvents and all components were consistent with species of average MW values <1500 Da. However, it should be noted that in a FA isolated from an Oak forest soil, large polysaccharides (maltodextrans) with average diffusivities of ≈6000 Da were identified. In a FA from an agricultural soil components consistent with ≈12 000 Da peptides/proteins were found. It is logical that an operationally defined extract of soil will result in a mixture of plant components at various stages of humifica-

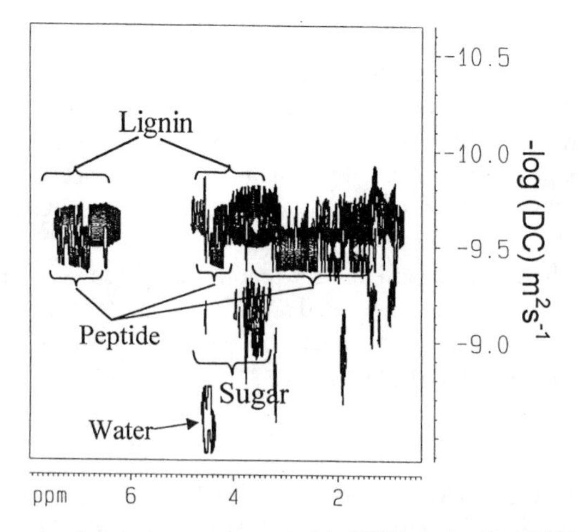

Fig. 1–21. Diffusion ordered spectroscopy spectrum of the IHSS peat humic acid (HA) standard in D_2O after the addition of 5 μL of acetic acid (adapted from Simpson, 2002).

tion with a range of molecular sizes and structures rather than macromolecules with undetermined structures.

Mao and Schmidt-Rohr (2003) introduced a solid-state ^{13}C-NMR method (mentioned earlier in Black Carbon or Char, p. 19) that distinguishes the fused rings of charcoal from aromatic structures, such as those derived from lignin which can be components of HS. The technique is based on long-range dipolar dephasing. In this the dephasing of unprotonated C signals is accelerated approximately three-fold compared with standard dipolar dephasing without recoupling. That gives a more efficient differential dephasing. Signals of unprotonated carbons with two or more protons at a two-bond distance dephase to <3% in <0.9 ms, and that is significantly faster than those of aromatic sites separated from the nearest proton by three or more bonds. Compared with lignin, slow dephasing is observed for the aromatic carbons in wood charcoal, and even slower for inorganic carbonate. Direct C-13 polarization is used on such structurally complex samples to prevent loss of the signals of interest, which by design originate from carbons that are distant from protons and therefore crosspolarize poorly. In natural organic matter such as HAs, this combination of recoupled dipolar dephasing and direct polarization at 7-kHz MAS enables selective observation of signals from fused rings that are characteristic of charcoal or char materials. It will be appropriate to apply this technique to HAs of Mollisols, where the method should make it possible to determine how much of the aromatic signals shown in Fig. 1–17 can be attributed to char in association with the HAs.

The traditional concept of HS as randomly coiled macromolecules is based on observations of apparent MW values, which can exceed 1 000 000 Da (Swift, 1989b; Schulten and Schnitzer, 1993). There has been recent speculation that the high MW values observed may be explained by the association of small components to form aggregates in aqueous solution with macromolecular-like properties (Piccolo et al., 1999, 2001; Wershaw, 1999). This new concept has been confirmed by the NMR experiments described above and by others (Simpson et al., 2002). That is, the major components found in alkali-extractable soil HS are in general of relatively low MW (<2500 Da), and associate in the presence of metals to form aggregates. This concept is illustrated in Fig. 1–22. The spheres represent generic metal cations and dashed lines show H bonds as well as hydrophobic associations, which are shown along with polysaccharides (the R group represents a hydroxyl or proton), polypeptides (R represents a side chain), aliphatic chains, and aromatic lignin fragments. It is important to note that humic fractions are complex mixtures of many components, which vary in ratio and structure, with sample origin, and with extraction procedure. Therefore, this concept should only be considered as an example of a humic aggregate and is not definitive. With respect to both the polysaccharides and polypeptides, the exact sequences of units have not been determined. Other studies of fractions from the same soils indicate glucose, galactose, mannose, arabinose, rhamnose, and xylose are the most abundant sugar monomers in acid digests, and aspartic acid, glutamic acid, arginine, glycine, leucine, threonine, and serine are the most abundant amino acid monomers (Simpson, 1999). Nuclear magnetic resonance spectroscopy shows these monomers to be linked into chains of up to 10 units. The aromatics are shown as portions of lignin that are cross-linked into structures with up to about eight units. The linkages between the rings of the struc-

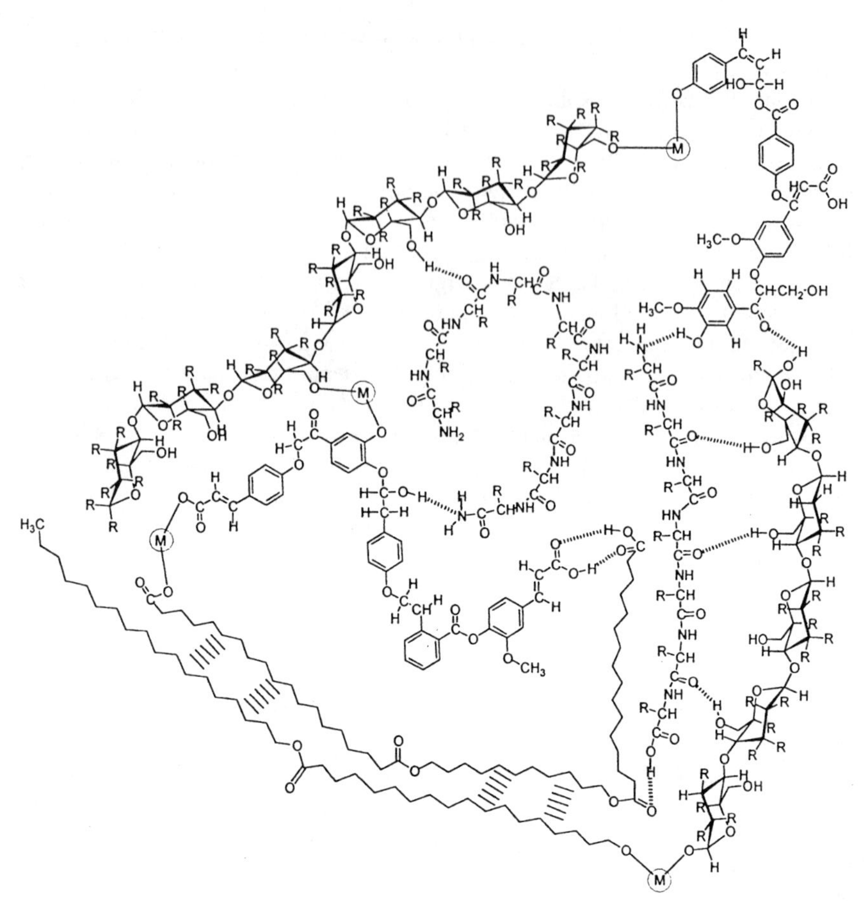

Fig. 1–22. An illustration of how major components (detected using advanced NMR techniques) in soil humic substances (HS) fractions might associate in aqueous environments in the presence of metals to form aggregates in aqueous environments. The spheres represent generic metal cations, the narrow dashed lines show H-bonds, and the wide dashed lines indicate hydrophobic associations. The primary structures depicted are polysaccharides, polypeptides, aliphatic components, and lignin fragments. The representation is based on evidence from Simpson et al. (2002).

tures cannot be determined exactly, but the nature of the rings and attached functionalities can be determined from long-range coupling NMR experiments. This information has been combined to suggest two lignin-derived units that are likely to exist in the humic materials. In all the above cases additional degradation and NMR studies on the fractions from the same and similar soils confirm these findings (Simpson, 1999; Simpson et al., 2001a, 2001b, 2002).

While aggregates will likely be held together through a complex combination of hydrophobic associations, charge interactions, H bonds, and metal bridging, results shown in this study suggest that metal ions play a crucial role in aggregate formation and stability.

SOIL SACCHARIDES

The roles of polysaccharides in soils are perhaps better understood than those of HS. Their compositions are better known, even well understood, but details of their structures are as vague as those for HS. Interest in soil polysaccharides was stimulated by Martin (1945, 1946), who established that the slimy bacterial products shown by Waksman and Martin (1939) to aggregate sand–clay mixtures were polysaccharides. At about the same time Stacey and colleagues at the University of Birmingham, England (Haworth et al., 1946), and Geoghegan and Brian (1946, 1948) of the ICI Research Station, Jealotts Hill, Berkshire, England (who were unknown to each other and to Martin) were investigating the uses of bacterial dextran and levan polysaccharides for the improvement of soil structure and for the retention of water in soil.

Statistical analyses (Rennie et al., 1954; Chesters et al., 1957) correlated good soil structure with microbial gums and polysaccharide substances. However, belief in this thesis was shaken when Mehta et al. (1960) showed that the aggregate structures were preserved when a Swiss Braunerde was treated with periodate ($NaIO_4$), a treatment that would degrade the saccharides. However, the work of Greenland et al. (1961, 1962) suggested that the stable periodate-treated soil crumbs were held together by fungal hyphae and myceliae (which resisted the periodate treatment). Also, the soils had free $CaCO_3$, which caused the soil crumbs in the soils studied by Clapp and Emerson (1965) to resist degradation after treatment with periodate.

To achieve an understanding of the interactive roles that saccharides and polysaccharides can have in the soil environment, it is appropriate to have an awareness of the configurations, conformations, and structures of saccharides. A brief review of the relevant chemistry follows.

Some Relevant Aspects of Sugar Chemistry

We begin with an outline of the compositions, configurations, and conformations of aldose (containing aldehyde functionalities) sugars. The compositions and configurations (orientations in space of the atoms in each sugar molecule) can be readily deduced by following the logic of the structures in Fig. 1–23. Compound **CCXVIII** is glyceraldehyde, a 3-C aldose shown in the D-configuration in the Fischer (open chain) projection. To obtain the Fischer structure we view each asymmetric C (in the zig-zag structure **CCXVIIIa**) along the C plane, but above the plane of carbons to which it is bonded. Thus, in the case of **CCXVIIIb**, the –OH on C-2, the asymmetric C, will be on the right-hand side and the H on the left. In this projection D-refers to the configuration of –OH on the asymmetric C. When the –OH is on the left, the structure is L-glyceraldehyde. In the "shorthand notation" used in depicting structures in the "sugar tree" (Fig. 1–23), the bottom spur of the vertical line represents the primary alcohol (H_2COH) functionality, the horizontal line represents the –OH of the secondary alcohol (HCOH) group, and the topmost spur is the aldehyde (HC=O) functionality. Now, consider lengthening the sugar chain by adding a HCOH group between the aldehyde functionality and the existing secondary alcohol group. We show addition first to the right (in the Fischer projection) in the sugar tree to give D-erythrose (**CCXIX**), then to the left to give D-threose

(**CCXX**). Structures **CCXIXa** and **CCXXa** represent erythrose and threose in the zig-zag forms. By following the same logic, D-erythrose will give two neutral sugars, D-ribose (**CCXXI**) and D-arabinose (**CCXXII**), and D-threose will give D-xylose (**CCXXIII**) and D-lyxose (**CCXXIV**). Following the same progression we derive two sugars from each pentose, and we obtain, as shown in Fig. 1–23, D-allose (**CCXXV**), D-altrose (**CCXXVI**), D-glucose (**CCXXVII**), D-mannose (**CCXXVIII**), D-gulose (**CCXXIX**), D-idose (**CCXXX**), D-galactose (**CCXXXI**), and D-talose (**CCXXXII**). It must be borne in mind that the structures shown are in Fischer projection, and based on the information given above the reader will be able to construct the zig-zag forms of the sugars. Without the aid of the sugar tree, and some pneumonics, the writers would have difficulty in remembering the structures of some of the sugars. They relate the tetroses to the film ET, the pentoses to RAXL, a primeval dance, perhaps, and then remember the hexoses in order by using the pneumonic "*all altruists gladly make gum in gallon tanks*." It makes it easier also to deduce the sugar structures by remembering the logic of placements of the –OH groups starting with the highest numbered asymmetric C and working down to the lowest. It is important to remember that the sugars will all have the L-configuration if built from L-glyceraldehyde. [D- and L- are not related to optical activity. Thus, a D-sugar that is dextrorotatory is depicted as D(+)- and when levorotatory by D(–)-].

The pentose sugars ribose, arabinose, and xylose are important in soils, and glucose, mannose, and galactose, and derivatives of these are the important hexoses. Fucose (**CCXXXIII**, or 6-deoxy-L-galactose), in which the –OH on C-6, the

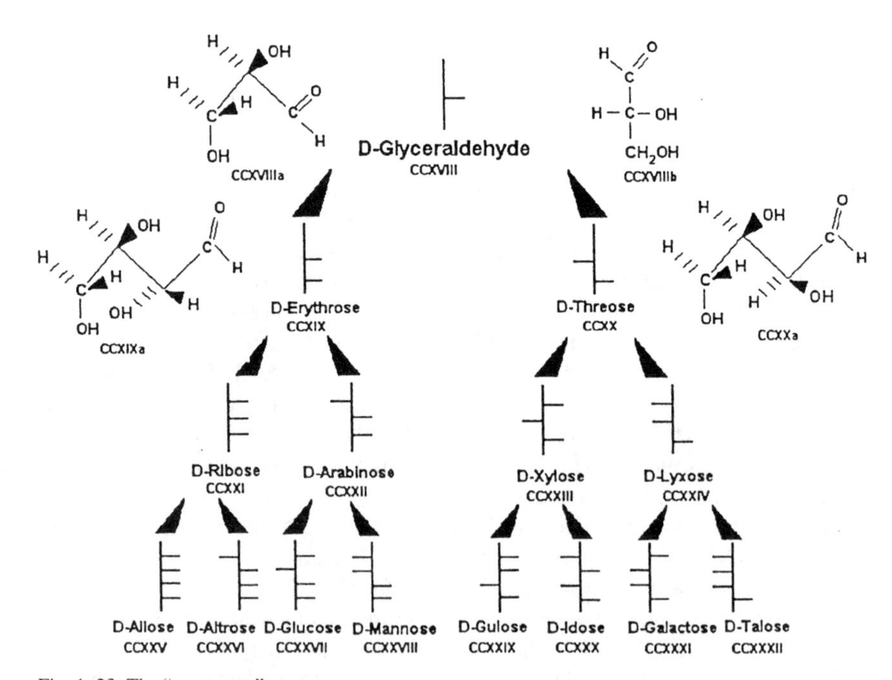

Fig. 1–23. The "sugar tree."

primary alcohol functionality, is replaced by H to give the methyl group and rhamnose (**CCXXXIV**, or 6-deoxy-L-mannose), in which the –OH group on C-6 is again replaced by H, are important deoxyhexose sugars in soils. Uronic acids, such as glucuronic, galacturonic, and mannuronic in which the primary alcohol groups are replaced by –COOH, are important components of acidic polysaccharides. Amino sugars in soils are important 2-deoxy sugars, in which the –OH on C-2 is replaced by an amino ($–NH_2$) group, or with an acetylated amino group [$NH(OCOCH_3)$].

Pentose and hexose sugars in nature are predominantly in ring forms, and these can be in furanose (5-membered) and pyranose (6-membered) ring structures. There is an equilibrium between the open chain and ring forms of pentoses and hexoses; however, more than 90% are in the cyclic (ring) forms in solution at any time. The open chain (Fischer projection), and ring structures (furanose, and pyranose) for D-arabinose and D-glucose are shown in Fig. 1–24. Note that the five-membered ring is formed between C-1 and C-4; C-1 then becomes an asymmetric C and part of a hemiacetal structure. In the pyranose structure the ring is formed between C-1 and C-5, and again C-1 becomes asymmetric and part of a hemiacetal functionality. The six-membered rings are shown in Haworth projection and in the more sterically realistic chair (C-1) conformations. The less relevant boat conformation is obtained by flipping the bonds between C-1 and C-2 and between C-1 and –O. The C-1 takes the 1-C conformation when the ring is "flipped." Then the equatorial substituents become axial and the axial substituents become equatorial. For the β-anomer the –OH on C-1 points down in the Haworth projection and is in the equatorial configuration in the chair (C-1) conformation; for the α-anomer the –OH points upwards and is in the axial configuration.

Five-membered rings are almost planar, but the six-membered rings are puckered and assume conformations, or positions in space, ranging from the "boat" to the "chair" extremes. Thus, the Haworth projection (Fig. 1–24) does not give a good representation of the conformations which 6-membered rings can adopt.

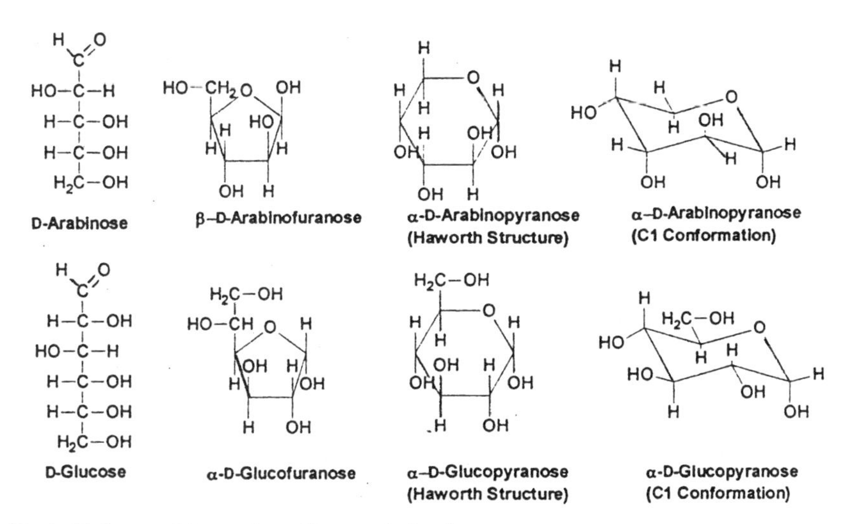

Fig. 1–24. Structural forms of D-arabinose and of D-glucose.

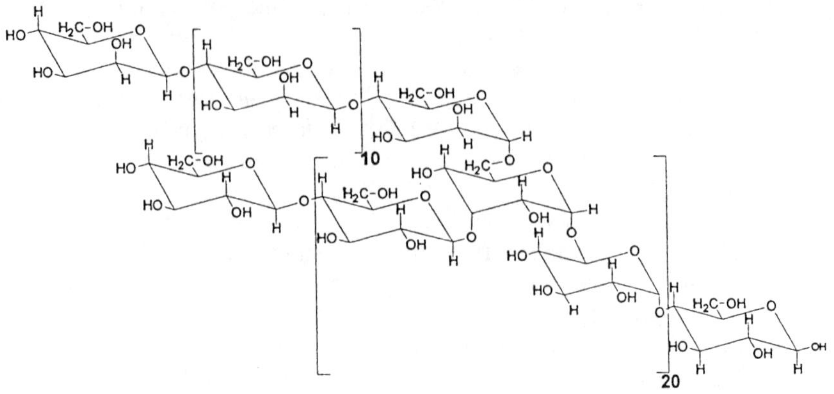

Fig. 1–25. A branched heteropolysaccharide.

When 8 to 10 sugar units are linked, an oligosaccharide is formed. When more than 10 units are linked, a polysaccharide results. Sugar units are linked through glycosidic linkages. These involve an –O– bridge between the C-1 of one sugar unit (in ring conformation) and the C (but not the C-1) of another sugar, and one H_2O molecule is eliminated for each linkage formed. Homopolysaccharides are polymers of one sugar, and heteropolysaccharides are composed of two or more sugars. Polysaccharides may be linear or branched. Figure 1–25 is an example of a branched heteropolysaccharide. In this figure the linear heteropolysaccharide chain is composed of, from the left in the linear chain, glucose, linked β-(1 → 4) to another glucose, linked β-(1 → 3) to galactose, linked α-(1 → 6) to glucose, linked α-(1 → 4) to β-D-glucose. The three sugars linked between the brackets repeat 20 times. For the branched chain, from the top, mannose is linked β-(1 → 4) to mannose (which is repeated 10 times), and the mannose side chain is linked α-(1 → 6) to the glucose in the horizontal chain.

Origins of Soil Saccharides

Carbohydrate residues added to soil tend to decompose relatively readily, although some decompose or are transformed faster than others. Water soluble carbohydrate components, such as starch, decompose much more readily than water insoluble materials, such as cellulose. Thus, for example amylose, the α-(1 → 4) linked polyglucose component of starch, is rapidly decomposed in soil whereas cellulose, a β-(1 → 4) linked polyglucose, has a considerable resistance to degradation. The reason for that is that the α-linkages provide a random-coil type of conformation, which allows very little H bonding to confer rigidity between the sugar molecules. The β-linkages, on the other hand, give a linear or linear helical structure to the cellulose. That allows the glucose components of the polymers to lie over each other, and the resulting H bonding gives a rigid structure, and hydrolysis of the glycosidic linkages by cellulase enzymes is impeded.

Hemicelluloses are found in the cell walls of plants. These are heteropolysaccharides (in which glucose may often be present in only small amounts) composed of 100 to about 200 hexose and pentose units. The principal constituent sugars in softwood hemicelluloses are, in decreasing abundances, mannose, xylose, glucose, galactose and arabinose. The order for hardwood hemicelluloses is xylose, mannose, glucose, and galactose, with minor amounts of arabinose and rhamnose. The backbone of the glucomannans is a chain of $(1 \rightarrow 4)$ linked β-D-glucopyranose and β-D-mannopyranose units, and the branches are contributed by $(1 \rightarrow 6)$ linked α-D-galactopyranose residues. The C-2 and C-3 positions in the mannose and glucose units are partially substituted by acetyl [$CH_3C(O)$] functionalities. In the cases of arabinoglucuronoxylans, for example, the backbone consists of a chain of $(1 \rightarrow 4)$ linked β-D-xylopyranose units partially substituted at C-2 by 4-*O*-methyl α-D-glucuronic acid, and at C-3 by α-L-arabinofuranose units (Goldstein, 1991).

Cheshire and Hayes (1990) pointed out that although a particular sugar might only be a small proportion of the sugars in plants, that sugar can with time become predominant in a residue if it has even a slightly greater resistance to degradation than the other sugars in the medium. Bacon and Cheshire (1971) illustrated that point by showing that 2-*O*-methylfucose and 2-*O*-methylxylose are prominent among the methylated soil saccharides, despite the fact that these sugars are minor components of the hemicelluloses of some plants. The *O*-methyl functionality is considered to block enzymatic processes leading to degradation of the sugars.

Glucose, galactose, and mannose are the major neutral sugars in animals (components of glycoproteins), and other sugars include glucuronic acid, glucosamine, and galactosamine (or their N-acetyl derivatives). Very small amounts of pentoses are found in animals, although of course ribose and deoxyribose (components of nucleic acids) will be present in all cells. Ribose is common in bacterial polysaccharides. Chitin, from insects, is a major source of acetylglucosamine.

Glucose, arabinose, and xylose are generally the major sugars in plant tissues, and the relatively high proportions of the pentose sugars arabinose and xylose in soil saccharides suggest plant origins for these. Arabinose in temperate zone plants is primarily in the L-form, and Cheshire and Thompson (1972) found that at least 90% of the arabinose in the soil studied was in the L-configuration. However, several yeast/fungal galactomannans, and some xylomannans, with xylose as the side chain, and an arabinoxylomannan, composed of D-mannose, D-xylose, and L-arabinose have been reported as cellular polysaccharides of yeast (Gorin and Barreto-Bergter, 1983). Moderate amounts of galactose are found in some plants (Stephen, 1983), but mannose and L-rhamnose are found in negligible amounts in plants. Mannose and galactose are more abundant in bacteria (Keene and Lindberg, 1983), in algae (Painter, 1983), and in fungi (Gorin and Barreto-Bergter, 1983). That implies that the latter two sugars primarily have microbial or animal origins. Rhamnose is usually present in the L-form in bacterial and fungal polysaccharides, and it composes the side chain of rhamnomannans. However, rhamnose can occur as a phenolic glycoside (in which the phenolic O is linked to the glycosidic, or anomeric C of the sugar) in tannin and flavonoid structures and is contained in the quercetin of oak bark. Fucose, which can occur in the D- and L-forms, is rare in plants but common in bacteria. There is a link with D-galactose in the biosynthesis of L-fucose, but the link with mannose is less clear in the synthesis of rhamnose.

Cheshire et al. (1969, 1971) showed that when readily degradable ^{14}C-labeled substrates, such as glucose and starch, were incubated under field conditions, the strongest label developed in the hexoses glucose, galactose, and mannose, as well as in the deoxyhexoses rhamnose and fucose. Labeling was very much less in the pentoses arabinose and xylose. That would suggest that microorganisms are largely responsible for the hexoses and deoxyhexoses in soil. When the experiments were extended for up to 2 yr there was evidence to show that high levels of activity developed in the deoxyhexoses, and especially in rhamnose. However, the compositions of the labeled sugars bore only a superficial resemblance to that of the total soil, or soil extract, and the authors considered that a sufficient time lapse might not have occurred to allow for differential degradation rates for the synthesized carbohydrates. It would seem, however, that such data would support the view that microbial synthesis is only a partial contributor to the origins of soil carbohydrates (Cheshire and Hayes, 1990).

Cheshire et al. (1974, 1979) incubated labeled plant material with soil to determine if plant carbohydrates can survive in the soil for a significant period. Some of the sugars in straw decomposed rapidly, but others appeared to decompose relatively slowly, and about 15% of these remained after 5 yr. The results suggested that little, if any, xylose or cellulose was synthesized by soil microorganisms. Observations by Cheshire and Anderson (1975) suggest that inputs of plant residues are essential to maintain the soil carbohydrate levels. Total carbohydrate contents of soils fallowed for 10 yr fell by as much as 50%. However, there was no significant change in the composition of the soil carbohydrate as the result of the fallow, and that would suggest that the sugars were equally susceptible to metabolism.

Oades (1984) used the ratio (mannose + galactose)/(xylose + arabinose) to suggest origins (plant or microbial) for sugars in soils. Ratio values <0.5 suggest origins in plants and values of the order of 2 suggest microbial sources. The same reasoning could be applied to the ratio (rhamnose + fucose)/(xylose + arabinose). It could also be meaningful to compare the ratios mannose/rhamnose and galactose/fucose for each sample.

Isolation and Fractionation of Soil Saccharides

Procedures for the isolation and fractionation of polysaccharides from soil have been reviewed by Mehta et al. (1961), Swincer et al. (1968), Greenland and Oades (1975), Hayes and Swift (1978), Cheshire (1979), Cheshire and Hayes (1990), and Stevenson (1994). To compare accurately the efficiencies of different procedures, it would be appropriate to have an estimate of the carbohydrate content of the soil before and after extraction using any particular system. A favored way for estimating sugars has been to pretreat soil samples with 72% H_2SO_4 followed by dilution to 0.5 *M*, then refluxing for a prescribed period (Cheshire and Mundie, 1966; Oades, 1967). Today hydrolysis is preferentially performed using trifluoroacetic acid, then the sugar hydrolyzates are reduced (e.g., with sodium borohydride in DMSO, **XIII**, p. 22) to the alditols, the alditol acetate derivatives are formed (using acetic anhydride), and the individual sugars are separated and quantified by GLC (Blakeney et al., 1983; Watt et al., 1996; Hayes et al., 1997).

Sodium hydroxide is probably the best of the aqueous solvents used. Its efficiency might be related to the varying extent of charge, arising predominantly from uronic acids, which are characteristic of many soil polysaccharides, and possibly from sulfonated polysaccharides (typically with SO_3H functionalities on C-6 of the hexose sugars). Swincer et al. (1968) found that pretreatment of the soil with 1 *M* HCl or HF (but not with 0.05 *M* H_2SO_4) significantly increased the saccharide yield in the aqueous base. That might be explained by a H^+-exchange (for divalent and polyvalent metals) effect involving the uronic acids and/or sulfonic acid groups (see Extraction of Humic Substances under Basic Conditions, p. 32). When the residue was acetylated (using acetic anhydride and concentrated H_2SO_4) at 60°C for 2 h and then extracted with chloroform, the carbohydrate yield was increased by 23%. The amounts of carbohydrate extracted using the sequence 1 *M* HCl, 0.5 *M* NaOH, and the acetylated material ranged from 57 to 74% of the sugar contents of the soil.

Some of the dipolar aprotic solvents (see Properties of Organic Solvents, p. 24), and especially DMSO (**XIII**, p. 22), can be expected to be good solvents for soil saccharides. DMSO has the ability to break H bonds and can even dissolve cellulose. Häusler and Hayes (1996) showed that considerable amounts of saccharides were removed from the HA fraction from a Sapric Histosol when that fraction was dissolved in DMSO/HCl and passed on to XAD-8 resin. The HA was sorbed by the resin, and saccharides were washed through. Thus, the use of DMSO/HCl and of XAD-8 (**XXI**, p. 43) and XAD-4 (**XXII**) resins in tandem may well provide a useful method for the isolation of soil saccharides (see Modern Fractionation Procedures, p. 42). The saccharides would be retained by the XAD-4 resins.

The bulk of the saccharides in soil aqueous extracts are considered to be contained in the FA fraction, or the organic fraction that remains in solution on acidification of soil extracts. However, the introduction of XAD resin technology for the fractionation of soil organic extracts has allowed the separation of saccharides from what are considered to be the true FAs. Swincer et al. (1968) used Polyclar-AT [a (poly)vinylpyrrolidone resin used for clarifying wine] to remove color (HS) from soil saccharide extracts, but XAD-8 has been found to be superior for this purpose. Barker et al. (1965, 1967) and Hayes et al. (1975a) used gel chromatography and ion exchange chromatography techniques to remove color successfully from their saccharide extracts. As mentioned above, polysaccharides can be associated with the HAs. These associations can involve physical or van der Waals forces, H bonding, physical (steric) entrapment, and it is possible that covalent linkages can form (e.g., through phenolic glycoside functionalities). DMSO would break the noncovalent associations (and it may or may not liberate the steric constraints).

The principles described for the fractionation of soil HS should apply also for soil saccharides, and these have been reviewed by Hayes and Swift (1978), Cheshire (1979), and Cheshire and Hayes (1990). Techniques that find most widespread applications in this area include gel chromatography for fractionation based on size differences, and anion-exchange chromatography for separations based on charge-density differences. Sephadex, an α-(1 $\rightarrow$ 6) polyglucose dextran (**XIX**, Fig. 1–5) variously cross-linked to give discrete pore sizes, and cross-linked poly(acrylamide) gels [-(CH_2–CH(C(O)NH_2)]$_n$ (**XX**) are widely used. However, since many of the components of soil polysaccharides are charged under the conditions employed, it is necessary to use dilute salt solutions to suppress the repulsion between the charges on the polysaccharides and the residual charges on the gels. Use can also be made of the complexes that sugars form with boric acid to achieve fractionation. Complexation, and the conferring to the complex of a borate charge, depends on the *cis-trans* relationship of hydroxyl groups on the sugars composing the polymers. Barker et al. (1967) and Finch et al. (1967) formed borate complexes of two polysaccharide components from a Sapric Histosol soil. Clapp (1957) also had made use of this principle to isolate a neutral polysaccharide from a New York muck soil using a Kirkwood electrophoresis cell (Kirkwood et al., 1950) and paper and column electrophoresis (Clapp and Davis, 1970; Clapp et al., 1979). Column electrophoresis and borate buffer systems were used to separate the component sugars in the hydrolyzate of the polysaccharide.

Finch et al. (1967) described the isolation of a neutral polysaccharide from the 0.3 *M* H_2SO_4 extract of a Sapric Histosol soil. The source of the first polysaccharide was the precipitate formed when the extract was neutralized with $NaHCO_3$. The material excluded from the Sephadex gel (hence the higher MW substances) had a MW value of 50 000 Da., determined by ultracentrifugation, and it moved as a single component in free boundary electrophoresis in borate (pH 9.1), phosphate (pH 7.0), and barbiturate (pH 9.0) buffers. These data indicated a high degree of homogeneity. A second polysaccharide, which remained in solution when the acid extract was neutralized, was also fractionated using Sephadex gels, and the material excluded from the Sephadex moved as a single component in barbiturate and phosphate buffers. However, four negatively charged components were evident when borate buffer was used, and that was conclusive evidence for inhomogeneity. When that polysaccharide was subjected by Barker et al. (1967) to ion exchange chromatography, using diethylaminoethyl (DEAE)-A50 Sephadex (an ion exchange gel), and eluting with a 0 to 2 *M* NaCl gradient, four fractions were obtained, as shown in Fig. 1–26.

All of the polysaccharide fractions isolated contained more than five sugar units (Hayes and Swift, 1978, p. 298); hence, it is concluded that we still await the isolation of a pure polysaccharide from soil. Among the six polysaccharide fractions isolated from a Sapric Histosol by Hayes et al. (1975a) two contained about 70% glucose (see Table 1–12). However, these polysaccharides contained six other

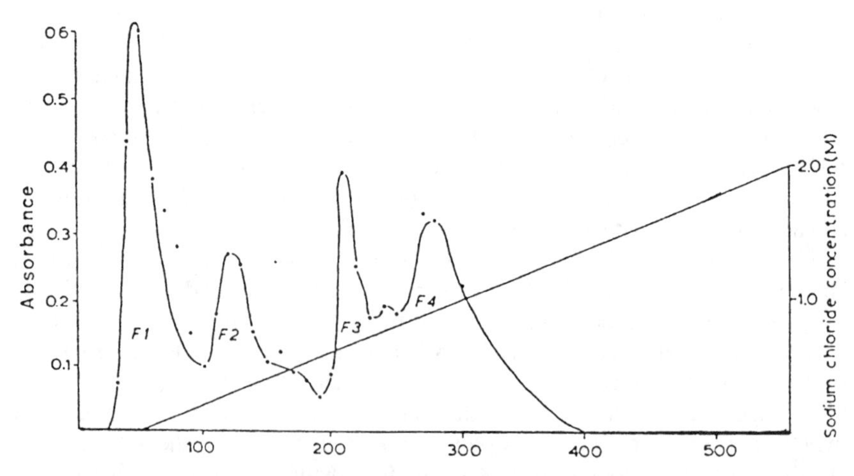

Fig. 1–26. Fractionation of a soil polysaccharide using gel ion-exchange chromatography (after Barker et al., 1967).

sugars in concentrations ranging from 2 to 9%, and hence must be considered to be impure.

Thus, it is clear that the purification of soil polysaccharides is a tedious process. Even the procedures that allow materials to be isolated having a degree of size and charge density homogeneities will not guarantee purity. The multitudes of microorganisms, plants, and processes that operate in the soil environment suggest that several polysaccharides having the same physicochemical characteristics can be present in the same soil sample.

Compositions and Structural Considerations of Soil Saccharides

A variety of analytical procedures are available for the quantitative estimations of the sugars in polysaccharides (Clapp et al., 1979; Cheshire et al., 1983a; Cheshire and Hayes, 1990; Watt et al., 1996). The glycosidic linkages between the sugar components in oligo- and polysaccharides are hydrolyzable, and the individual sugars can be identified. Also, because there is genetic control of their synthesis there will be a regularity in the sequences of the sugar units.

Table 1–12 gives the relative abundances of each neutral sugar in six polysaccharide components isolated by Hayes et al. (1975a) from a Sapric Histosol soil. Ribose is not included, nor are the contributions from unidentified sugars. Glucose is present in greatest abundance in Samples 1, 2, and 3, and galactose is in greatest abundance in Samples 4, 5, and 6. With the possible exception of saccharides numbered 2 and 3, the (mannose + galactose)/(arabinose + xylose) ratios indicate biological origins for the carbohydrates. This applies also for the (fucose + rhamnose)/(arabinose + xylose) ratios. Also, the ratio values for Samples 4 and 5 emphasize even more emphatically the (micro)biological origins of these carbohydrates. The high glucose contents of Samples 1 and 3 suggest relatively high contributions of altered cellulose components to these polysaccharides. It could well be, however, that any cellulosic components were separate entities.

High relative contents of the pentose sugars arabinose and xylose suggest plant origins for these sugars. As pointed out earlier, glucose, xylose, and arabinose usually constitute the major sugars of plant tissues. Microorganisms synthesize arabinose in the L- and D-configurations, whereas this sugar is synthesized only in the D-configuration by plants. Galactose can also be present in considerable abundance

Table 1–12. Sugar components† of saccharides isolated from a Histosol soil.‡

No.	Glu	Gal	Man	Xyl	Ara	Fuc	Rha	(Man + Gal)/(Ara + Xyl)	(Fuc + Rha)/(Ara + Xyl)
	%								
1	68.9	9.0	9.0	3.6	2.6	1.9	5.3	2.9	1.2
2	32.3	21.6	8.7	18.8	10.3	3.3	3.4	1.1	0.2
3	70.8	5.9	5.5	8.1	5.0	1.9	2.3	0.9	0.3
4	19.5	31.8	17.3	7.9	4.0	3.5	12.7	4.1	1.4
5	18.1	33.2	16.7	8.9	4.7	5.2	11.5	3.7	1.2
6	21.3	24.3	16.9	11.2	8.3	7.5	9.1	2.1	0.9

† Glu = glucose; Gal = galactose; Man = mannose; Xyl = xylose; Ara = arabinose; Fuc = fucose; Rha = rhamnose.

‡ From Hayes et al. (1975a), Hayes and Swift (1978).

in some plants, and L-fucose (or 6-deoxy-L-galactose) is present in lesser amounts. There usually are negligible amounts of mannose and of L-rhamnose (6-deoxy-L-mannose) in plants. The hexose sugars glucose, galactose, and mannose, the amino sugars glucosamine (from acetylglucosamine) and galactosamine (from acetylgalactosamine), and the two uronic acids, glucuronic acid and galacturonic acid, could be from animals. Chitin, from insects, would be an obvious major source of acetylglucosamine. Ribose, in small amounts, would be present in animal and plant tissues.

There are few accurate data for the sizes and shapes of soil polysaccharides. This arises from the difficulties faced in attempting to isolate from soil oligosaccharides and polysaccharides that are homogeneous with respect to size. Charge polydispersity is easier to deal with. It is clear, however, that some soil polysaccharides have high MW values, and there are reports of fractions with values ranging from a few thousand to several hundred thousand daltons (Finch et al., 1967; Swincer et al., 1968; Cheshire, 1979). The shapes of the polysaccharides will be determined primarily by the nature (α or β) of the glycosidic linkages (Cheshire and Hayes, 1990). For example, because the glucose units of cellulose are linked $\beta(1 \rightarrow 4)$, the chain of glucose molecules (in the chair conformation) assume linear helical structures. This leads to extensive H bonding between the component molecules in adjacent strands. Thus, cellulose does not dissolve in water, and it has some resistance to biological breakdown. In contrast, the $\alpha(1 \rightarrow 4)$ configuration of amylose gives a more open structure, and the glycosidic linkages are more readily cleaved. Structures of two microbially produced polysaccharides are shown in Fig. 1–27 (Olness and Clapp, 1975). Some characteristics of these polymers are given in Table 1–13 (Olness and Clapp, 1973).

In NMR Studies of Composition and Associations (p. 97) it was noted that the saccharides in association with the FA studied contained only three to eight sugar units. The procedures used to isolate that FA fraction would not have included polysaccharides, and these of course would be expected to be separable from the humic and fulvic acids in soils. It is relatively easy to visualize the structures of soil polysaccharides, even though no such structure has yet been elucidated. Assume that component pentose and hexose sugars are in the chair conformation; the shapes will then be governed by the positions and by the nature of the glycosidic linkages (α or β) between the sugars. When the linkages are predominantly β, the molecules will have a structure that is linear or a linear helix. When the linkages are α, the structures will be more compact, and spherical or globular. The presence of uronic acids will confer charge to the molecules. However, it is rare to find more than three or four sugars in a polysaccharide, and because of the range of sugars found so far in even the most highly purified soil polysaccharides, it has to be assumed that we have been dealing with gross saccharide mixtures.

Roles of Saccharides in the Soil

To be effective for the formation and stabilization of soil aggregates, a polysaccharide must bind to at least two soil particles. It is difficult to visualize how polysaccharide molecules could bridge two sand particles. However, bridging could readily take place in the cases of soil colloids such as clays and (hydr)oxides. Also, in

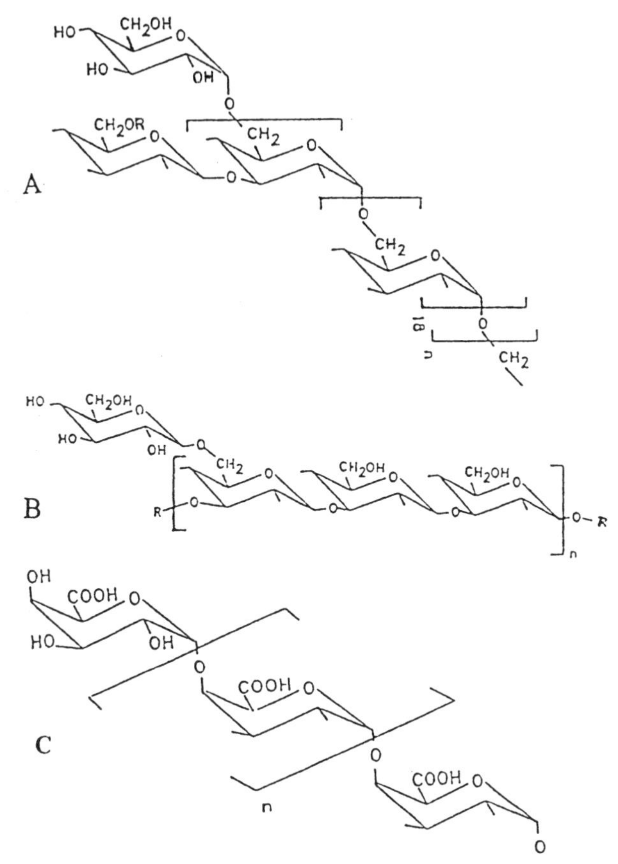

Fig. 1–27. Structures of (A) B-512F, a dextran composed of α-(1 → 6) and α-(1 → 3)-linked glucose units; (B) Polytran, a dextran composed of β-(1 → 3)- and β-(1 → 6)-linked glucose units; and (C) poly(galacturonic acid), an acidic polysaccharide composed of α-(1 → 4)-linked galacturonic acid units (after Olness and Clapp, 1975).

cases of clay domains polysaccharides could bridge across the edges of the particles composing the domains. Studies dealing with the adsorption of polysaccharides by clays have been reviewed by Theng (1979) and by Clapp et al. (1991).

Adsorption of Neutral Polysaccharides by Clays

Clapp et al. (1968) and Olness and Clapp (1973, 1975) studied the adsorption by Na^+-montmorillonite of extracellular rhizobial polysaccharides and the commercial dextrans B-512F and Polytran (Fig. 1–27). B-512F is a polyglucose from a strain of *Leuconostoc mesenteroides* (Northern Regional Research Laboratory, Peoria, Illinois) in which 95% of the linkages are α-(1 → 6) and 5% are α-(1 → 3). In contrast Polytran has 75% of the polyglucose units linked β-(1 → 3) and 25% β-(1 → 6). Both polymers have MW values of the order of 2×10^6. Isotherms for the adsorption of both polymers were of the high affinity (H)-type in the clas-

Table 1–13. Characteristics of the polymers Polytran and B-512F.†

Characteristic	Polymer	
	Polytran	B-512F
Molecular monomer	Glucose	Glucose
Structural monomer linkages		
Principal chain, % of total	β-1 → 3 (75%)	α-1 → 6 (95%)
Branches, % of total	β-1 → 3 (25%)	α-1 → 6 (5%)
Molecular weight	95% ≥ 2 × 10^6 5% 10^5 to 10^6	100% ≥ 2 × 10^6
Chemical composition		
C, %	39.0	41.8
H, %	5.8	6.4
N, %	<0.05	<0.05
Ignition residue, %	1.7	<0.05
Periodate consumption ratio‡	0.50	1.95
Hydroxyl ratio§		
Primary hydroxyl	0.75	0.05
Secondary hydroxyl	2.25	2.95

† From Olness and Clapp (1973).

‡ Periodate consumption ratios are reported as the numbers of moles of periodate (IO_4^-) reduced to iodate (IO_3^-) per mole of anhydroglucose unit (AGU).

§ Hydroxyl ratios were estimated from schematic representations of the polymers and represent maximum values. These are reported as the number of moles of R-OH per mole of anhydroglucose units (AGU).

sification of Giles et al. (1974), (i.e., effectively no, or minimal adsorptive, or polymer, was measurable in the equilibrium solution until the adsorbent surface was covered). However, after a plateau (leveling off of the isotherm) had been reached in the case of adsorption of Polytran, an inflection occurred and adsorption then continued to increase (Olness and Clapp, 1975). Thus, adsorption of Polytran was significantly greater than that of the B-512F dextran. The isotherms of Parfitt and Greenland (1970) for the adsorption of B-512F by Ca^{2+}- and Al^{3+}-montmorillonite were also H-type, but the fact that adsorption by the Na^+-clay was twice that by the Ca^{2+} and Al^{3+} preparations reflects the easier approach by the polymer to the dispersed clay. Parfitt and Greenland (1970) obtained similarly shaped isotherms for the adsorption of amylose [α-(1 → 4)-linked (poly)glucose]. X-ray diffraction studies in the work of Olness and Clapp (1973) showed that the basal spacings of Na^+-montmorillonite increased until the layer separation of the oven-dried Na^+-montmorillonite–Polytran and –B512F complexes reached 5.8 and 5.0 , respectively. That would suggest that one layer of polymer was adsorbed in the interlayer spaces. Based on the amounts adsorbed and the surface areas of the clays, it was clear that significantly more Polytran was adsorbed, and this could be accommodated by adsorption on the external surfaces.

Burchill et al. (1981) calculated that if the repeating unit in the β-(1 → 6)-linked polymer has a length of 0.5 to 0.6 nm, its total extended length would be in the range of 6.2 to 7.4 μm. The β-linkage would give an extended structure, but unlike the β-(1 → 4) structure of cellulose, there would not be the extensive H bonding that renders cellulose water insoluble. A more realistic conformation thus would be intermediate between extended linear and random coil. Since all –OH and bulky substituents are equatorial, the pyranose groups could make intimate contact with the clay surface. The α-glycosidic linkage would, on the other hand, give rise to a more compact random coil solution conformation that would decrease the contacts between the component sugars and the clay surfaces.

When a random coil makes contact with an adsorbing surface, the three-dimensional distribution in space is lost as each segment seeks to make contact with the surface. The more intimate the contact, the greater is the energy of adsorption. As adsorption progresses the amounts of surface available is decreased and loops of the adsorptive (polymer) extend into the bulk solution. Thus, as adsorption proceeds the energy of adsorption per segment of the molecule will decrease.

Microcalorimetry provides an appropriate tool for investigating this thesis (Burchill et al., 1981), and without that data it is difficult to provide unambiguous explanations for the differences between the adsorption of B-512F and Polytran by Na^+-montmorillonite. On the basis of the above discussion, we conclude that Polytran anchored to the external clay surfaces and then extended as loops into the bulk solution. In this way more polysaccharide would be adsorbed than the more globular, random coiled B-512F.

Adsorption of Charged Polysaccharides by Clays

Solution conformations of polyanions are influenced by the extent of dissociation in the medium of the anionizable groups. Thus, based on the discussion in Solubilization of Polyelectrolytes (p. 28), water insoluble anionizable polysaccharides would be expected to become water soluble when sufficient amounts of the conjugated bases of the dissociated functionalities are solvated. We have discussed reasons for the insolubility of cellulose in water. Thus, when sufficient ionizable groups are introduced into the cellulose molecule, the energy of salvation of the conjugated bases formed on dissociation will break the H bonds, and solution will take place.

Parfitt and Greenland (1970) found that (poly)galacturonic acid (PGA, Fig. 1–27) was only weakly adsorbed by Na^+- and Ca^{2+}-exchanged montmorillonite clays when the pH values of the media were 6, and it did not adsorb at all to Al^{3+}-montmorillonite at that pH. Adsorption was observed at lower pH values and uptake at a particular pH (<6) decreased in the order: Al^{3+}-montmorillonite > Ca^{2+}-montmorillonite > Na^+-montmorillonite. A plateau was observed when 300 mg g^{-1} of clay was held by Al^{3+}-montmorillonite at pH 3.6 and 275 K. X-ray diffraction indicated that interlayer adsorption did not take place. The amount adsorbed would correspond to a surface coverage of 450 to 500 m^2 g^{-1}. It must be concluded that only a fraction of the polymer segments made contact with the outer surface of the clay, and the bulk of the polymer extended as loops into the bulk solution.

In seeking an explanation for the observed behavior, we must consider the composition of PGA. It is α-($1 \rightarrow 4$)-linked, and the configuration of the –OH on C-4 is axial. Because the carboxyl groups will be dissociated at pH 6, repulsion between the charges will cause the polymer to assume conformations in which repulsion between the charges is least. At lower pH values, the uronic acids will be undissociated and the polymer will effectively behave like a neutral molecule. This gives opportunities for H bonding to the siloxane surface, and more realistically to coordination with the resident cations, as shown in infrared data by Parfitt (1972), or through H bonding between the carboxyl groups and the water molecules (Theng, 1979, p. 255–257). In separate studies involving one of the authors (Burchill and Hayes, unpublished data, 1981), data from microcalorimetry indicated that the energy of adsorption of PGA by Al^{3+}-montmorillonite was 37.5 kJ mol^{-1} for each segment of the anhydrogalacturonic acid. That relatively high interaction energy would be consistent with the water–bridge interaction mechanism.

The axial configuration of the C-4 hydroxyl would inhibit maximum contact between the sugar units in the polymer and the clay surface. The fact that interlayer adsorption did not take place suggests that the PGA polymer was less flexible than Polytran or B-512F. Hence it was unable to enter the narrow gaps between the layers, especially in the cases of the Ca^{2+}- and Al^{3+}-exchanged clays. Binding to the external surface would involve coordination to the charge neutralizing cation, and that would allow the major part of each polymer to extend into the bulk solution.

Data by Moavad et al. (1974) for the adsorption of anionic extracellular microbial polysaccharides by monovalent-, divalent-, and trivalent-cation exchanged kaolinite provide further evidence for the bridging influence of cations for the binding of acidic polysaccharides. Burchill et al. (1981) calculated the areas that the adsorbed molecules would cover if extended to flattened conformations. Surface coverage for the Na^+-, and K^+-kaolinite would be 60 to 70 m^2 g^{-1}. For the Ba^{2+}-, Ca^{2+}-, and Mn^{2+}-clay the range would be 100 to140 m^2 g^{-1}. For the Al^{3+}- and Fe^{3+}-exchanged clays the values would be of the order of 500 and 2000 m^2 g^{-1}, respectively. Corresponding values for H^+-exchanged clay (more realistically, perhaps a H^+/Al^{3+} system) were 170 to 180 m^2 g^{-1}.

Finch et al. (1967) calculated the plateau adsorption of a soil polysaccharide by Georgia Na^+-exchanged kaolinite to correspond to a surface coverage of 80 m^2 g^{-1}. Surface area measurements by standard techniques gave a value of 96 m^2 g^{-1} for the clay, and that would suggest contamination by a clay of higher surface area (montmorillonite). Periodate oxidation of the adsorbed polysaccharide was greatly retarded, and that would suggest that the adsorbed material was held close to the surface. Less protection would be afforded to loops extending away from the surface. The adsorption data are in line with those of Moavad et al. (1974) for their Na^+- and K^+-kaolinites. Coordination provides one explanation for the vast increases in adsorption, especially by the Fe^+ clay, but we do not know the extent to which (hydr)oxides were involved in the adsorption mechanisms.

Adsorption of Polysaccharides by (Hydr)oxides

Acidic polysaccharides can be expected to be held by coulombic attraction to (hydr)oxides at pH values below the point of zero charge (PZC) of the minerals. We still await detailed studies of interactions of polysaccharides by (hydr)oxides, and especially those of Fe and Al, which can be important in soils.

Adsorption of Soil Polysaccharides by Clays

Finch et al. (1967) reacted a polysaccharide material isolated from a Sapric Histosol soil with a H^+-exchanged clay. The clay was prepared by passing Na^+-montmorillonite through a H^+-exchanged ion-exchange resin and immediately freeze

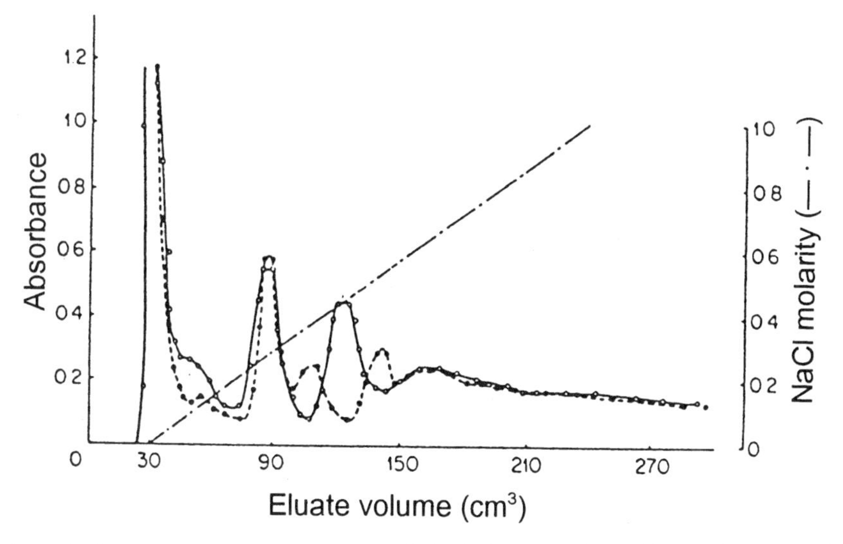

Fig. 1–28. Anion-exchange chromatography on DEAE-A50 Sephadex in pH 6.0 phosphate buffer of a soil polysaccharide mixture before (open circles) and after (filled circles) interaction with an H^+- or H^+/Al^{3+}-montmorillonite preparation (after Finch et al., 1967).

dried. The polysaccharide mixture, before and after reaction with the clay, was eluted from DEAE-A50 Sephadex at pH 6.0 using a 0 to 1 *M* NaCl gradient. The results in Fig. 1–28 show that the saccharide eluted from the column at about 120 mL was effectively removed by the clay preparation. Subsequently, Barker et al. (1967) showed that the component that was adsorbed contained one uronic acid per six to seven sugar units. The isotherm was Langmuirian (in contrast to the high affinity isotherms that generally apply for the adsorption of polymers by clays). The linear Langmuir plot is obtained using Eq. [53]:

$$C_2^{\mathrm{l}}/n_2^{\sigma(v)} = 1/K_{\mathrm{L}}(n^{\sigma})_{\mathrm{m}} + C_2^{\mathrm{l}}/(n^{\sigma})_{\mathrm{m}} \quad [53]$$

where C_2^{l} is the bulk phase concentration of the adsorptive (polysaccharide); $n_2^{\sigma(v)}$ is the Gibbs adsorption, which closely approximated to the amount adsorbed at the surface; K_{L} is the adsorption isotherm constant; and $(n^{\sigma})_{\mathrm{m}}$ is the number of surface sites, or the surface phase capacity. The linear plot when extrapolated gives the saturation monolayer coverage, assuming that the adsorbed polysaccharide took up the flattened conformation on the clay surface. The value of 730 $m^2\ g^{-1}$ obtained is a good approximation of the surface area of montmorillonite clay. However, the actual maximum uptake measured would correspond to a coverage of only 370 m^2. That would suggest that a monolayer was adsorbed between the clay layers, as was observed by Olness and Clapp (1973) for the uptake of Polytran and B-512F. Because the pH of the clay-polysaccharide medium was 2.1, it is safe to assume that the sorbed polysaccharide was not ionized in the medium. The electrophoretic mobility was decreased 10-fold when the pH was lowered from 7 to 4.4.

Finch et al. (1967) proposed H bonding between the clay and the saccharide, but it is possible that coordination to Al^{3+}, released in the H^+-exchange process, or to water coordinated with the Al^{3+}, was involved.

Evidence provided by electron microscopy of polysaccharide–clay interactions should be interpreted with caution. When soil particles are examined at high magnification, polysaccharides can appear as strands and loops stretching from one particle to another. These strands may represent the shrunken residues of an extensive gel structure that, before drying, was spread over a much larger area of the clay surfaces. Alternatively, they could represent strands with limited contact with the surface that, in the course of drying, collapsed on the surface.

Polysaccharides and Soil Aggregates

We have referred to the importance of the β-glycosidic linkage for the adsorption of polysaccharides on planar surfaces. Pagliai et al. (1979) showed that the stabilities of aggregates incorporating β-glycosidic-linked (poly)glucose are proportional to the MW of the polymers. Also, the stabilities could be related to the viscosities and the extent of adsorption. Pagliai et al. (1979) also showed that the relationship between MW and biodegradation is not straightforward. For example, polymers that extend as loops into the bulk solution would be liable to biodegradation, although extensive loops might occur only when there is extensive surface coverage by the polymer.

Cheshire et al. (1983a, 1983b) used oxidation with sodium periodate to study the physical stabilities of soil aggregates. They found that in some cases the stability of the aggregates was maintained for long periods, and even when aggregates were degraded, a significant proportion of the carbohydrate remained. Their data provided convincing evidence for a direct relationship between carbohydrates and soil structure (Cheshire et al., 1983a). The carbohydrate that remained after periodate treatment was rich in glucose, arabinose, and xylose, suggesting plant origins. However, in studies by Cheshire et al. (1985), ^{14}C-labeled glucose was incubated with soil and after subsequent treatment with periodate, the microbially synthesized sugars with the greatest resistance to degradation were glucose, arabinose, and xylose. A similar pattern was observed when ^{14}C-labeled plant material was incubated with soil. Thus, it would appear that the residues that persist after periodate treatment are independent of substrate and related to the structure of the polysaccharide.

To be able to stabilize preformed aggregates the polysaccharide should have sufficient solubility in water to diffuse to the adsorption site in the internal surfaces. Cellulose is not an effective binding agent when added to soil. Mention has been made of the strong H bonds between the β-linked (poly)glucose strands. However, should these bonds be broken, the polymer would be soluble and would be expected to be a good aggregate stabilizer. Page (1980) found that cellulose xanthate (in which the –OH group on C-6 is replaced by $-OCS_2^-$), a water-soluble cellulosic polymer, was an excellent stabilizer of soil aggregates. Harrison (1982) showed that when the xanthate is adsorbed to clays, CS_2 is given off and the cellulose remained sorbed to the clay (as a neutral polymer).

Clapp and Emerson (1972) used a series of homo- and heteropolysaccharides of varying intrinsic viscosity, some of which were charged (containing uronic acids), in studies of the effects in binding to and stabilizing aggregates of Ca^{2+}-montmorillonite preparations (see also Clapp et al., 1991). Sorption of the neutral polysaccharides increased with their intrinsic viscosity, [η], values. When Na^{+}-saturation was used (which involved a sequence of treatments with different concentrations of NaCl) to disperse the aggregates it was found that samples which adsorbed high levels of polysaccharide could be dispersed only when prolonged treatments with 0.05 *M* periodate were used to break the polysaccharide structures. However, the anionic polysaccharides, as expected adsorbed least, even when these had high [η] values. Also, three of the four anionic polysaccharides that had high [η] values were dispersed when treated with 0.1 *M* sodium pyrophosphate ($Na_4P_2O_7$), and that indicated that interactions of the anionic polymers involved complexations with the Ca^{2+} ion. Polysaccharide mixtures extracted from soils, although adsorbed to varying extents, failed to stabilize the aggregates. This was ascribed to their lower (than required) [η] and MW values, and hence their inabilities to bridge between clay particles.

Aggregate formation may be visualized as the interaction of mobile clay particles with mucigel from plant roots and extracellular polysaccharides from bacteria (see Oades, 1990). There is, however, a contradiction in the results obtained for growing plants. During growth many plants appear to give rise to a breakdown of microaggregates in the rhizosphere. This could result from a priming action of the indigenous organic matter. Although it appears inevitable that polysaccharides are important for the formation of microaggregates, there is enough evidence to indicate that fungal hyphae and plant rootlets are important for the stabilization of macroaggregates. The fact that macroaggregates will be broken down as the result of extended treatments with periodate indicates the fundamental role of polysaccharides in the formation and maintenance of good soil structure.

General Conclusions Regarding Saccharides

Polysaccharides are major aggregating agents in soils. We have a good awareness of the compositions and aspects of the structures of polysaccharides that give rise to aggregate formation and stabilization. However, we have only limited awareness of the compositions and the structures of the indigenous soil polysaccharides that promote the formation of soil aggregates. The study by Finch et al. (1967) shows how polysaccharides that interact with clays can be isolated. There must be the will to carry out such studies, and although there was considerable emphasis on soil polysaccharides during the period between the late 1940s and mid 1980s, that interest has not been maintained. This is an area of study that can greatly advance our awareness of the roles of organic matter in the soil environment.

SOIL PEPTIDES

Amino acids, amino groups, amino sugars, and nucleic acid derivatives usually account for >95% of the organic N in soils (Anderson et al., 1989), and many other N-containing compounds have been reported in trace amounts (Stevenson, 1994). Traces of L-phosphatidic acid, choline, and ethanolamine have also been found, and there is evidence for the presence of uric acid, the endproduct of N metabolism of many animals. It can be oxidized to allantoin, cyanuric acid, and urea (see Anderson et al., 1989). Amino acids, the components of peptides and proteins, are by far the most important sources of organic N in the soil, and amino sugars, and nucleic acid derivatives are significantly lesser contributors.

In this section we are mainly concerned with the presence of proteins, peptides, and amino acids in soil. It is inevitable that there will be considerable amounts of these in the soil at any time because of the inputs from plant roots, from plant materials directly added to the soil, or entering the soil as senescent matter from vegetative cover (introduced by organisms such as earthworms). Additions also can come from the microfauna (composed of bacteria, fungi, and viruses which can number in the region of 1×10^7 to 1×10^{10} g^{-1} of dry weight of soil (Burns, 1990). Estimates of living tissue microbial biomass range from 17 to 22 g m^{-2} of soil (Jenkinson and Ladd, 1981). There will be significant contributions also from protozoa, eelworms, earthworms, soil insects, and burrowing animals. Most of the proteins and peptide materials in dead organisms will be recycled by the soil biota, but inevitably some will survive, protected in soil aggregates and in HS. McLaren (1954) and McLaren et al. (1958) were first to show how enzymes could interact with clays and enter the interlamellar spaces. McLaren et al. (1975) also showed how the activity of enzymes is preserved in associations with humus.

Compositions and Some Considerations of Structures

The reader is referred to standard texts in biochemistry for details about the techniques of establishing the structures of protein and peptide materials. The discussion provided here is intended to present the structures and groupings of the amino acids commonly found in soil hydrolyzates, and to provide concepts of shapes that are relevant in considerations of interactions of proteins and peptides with other soil components.

Amino Acids

More than 100 amino acids have been isolated from natural sources, but only about 20 of these are naturally occurring components of proteins. The great majority of the naturally occurring amino acids have the amino group attached to the C α to the carboxyl, and with few exceptions the α-C also bears a H atom. The fourth bond of the α-C is joined to a group with more than 100 variations. Thus, most of the naturally occurring amino acids differ only in the structure of the residue R (**CCXXXV**, Fig. 1–29, shown in Fischer projection) attached to the α-C. All of the amino acids isolated from proteins have the L-configuration.

Because the carboxylic group is acidic and the amino group is basic, the amino acids exist as *dipolar ions* (*zwitterions*), as indicated by **CCXXXVI**. Thus, the charge will be influenced by the pH of the medium. When there are no ionizable groups on R, the amino acid will have two ionization constants, as shown in Reaction Schemes [54] and [55]:

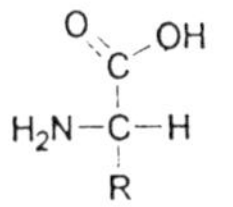

CCXXXV 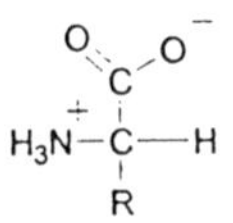 **CCXXXVI**

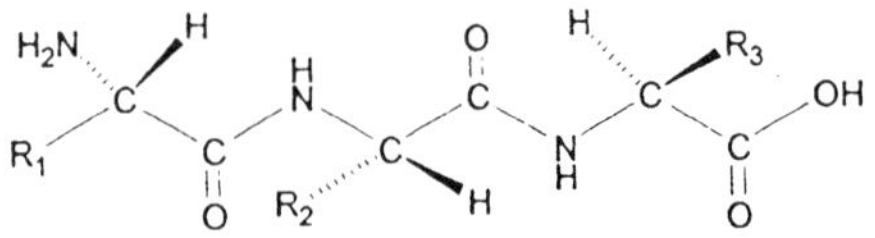

CCXXXVII 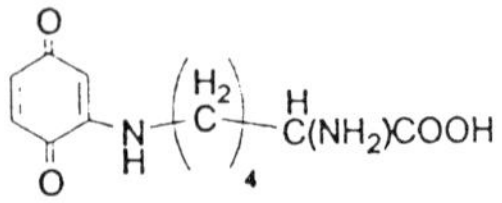 **CCXXXVIII**

Fig. 1–29. Structures of amino acids (**CCXXXV** and **CCXXXVI**) and a peptide (**CCXXXVII**), and a product-type from the reaction of a basic amino acid with hydroquinone (**CCXXXVIII**).

$$\underset{\substack{|\\ ^{+}NH_3}}{R\text{–}CH}\text{–}COOH + H_2O \leftrightarrow \underset{\substack{|\\ ^{+}NH_3}}{R\text{–}CH}\text{–}COO^- + H_3O^+ \qquad K_a \approx 10^{-2} \qquad [54]$$

$$\underset{\substack{|\\ ^{+}NH_3}}{R\text{–}CH}\text{–}COO^- + H_2O \leftrightarrow \underset{\substack{|\\ NH_2}}{R\text{–}CH}\text{–}COO^- + H_3O^+ \qquad K_a \approx 10^{-9} \qquad [55]$$

When an electrical potential is introduced across two electrodes in a solution of amino acids, migration to the anode or to the cathode will depend on the charge on the amino acid, and that in turn will depend on pH. There is no net migration at the *isoelectric point* characteristic for each amino acid because the concentration of the anion is the same as that of the cation, as shown in Reaction Scheme [56]:

$$\left[\underset{\substack{|\\ NH_2}}{R\text{–}CH}\text{–}COO^-\right] = \left[\underset{\substack{|\\ ^{+}NH_3}}{R\text{–}CH}\text{–}COOH\right] \qquad [56]$$

Peptides are formed by interaction between the α-amino functionality in one amino acid and the carboxyl in another, with the elimination of water. Structural type **CCXXXVII**, Fig. 1–29, is a diagrammatic representation of amino acids in a peptide sequence where R_1, R_2, etc. represent specific amino acid residues.

The peptide bond is readily cleaved by hydrolysis, and acid catalysis is usually employed in chemical degradations, as shown in Reaction Scheme [29], p. 59. Enzymatic catalysis (by peptidases/proteases) processes are the major mechanisms for hydrolysis in the soil environment.

Table 1–14 lists the amino acids that may be found in soil hydrolyzates. These are grouped into acidic, basic, neutral hydrophilic (NHi), neutral hydrophobic (NHo), S-containing, and labile heterocyclic. Cystine is, of course a dimer of cysteine, and it readily cleaves at the S–S bond during chemical degradations to give

Table 1–14. Amino acids which can occur in soil hydrolyzates.

Number	Structure	Name	Abbreviation
	Acidic		
1	$HOOCCH_2CH(NH_2)COOH$	L-Aspartic acid	Asp
2	$HOOC(CH_2)_2CH(NH_2)COOH$	L-Glutamic acid	Glu
	Basic		
3	$HN{=}C(NH_2)NH(CH_2)_3CH(NH_2)COOH$	L-Arginine	Arg
4	$CH_2CH(NH_2)COOH$ on imidazole ring (N, NH)	L-Histidine	His
5	$H_2N(CH_2)_4CH(NH_2)COOH$	L-Lysine	Lys
	Neutral Hydrophilic (NHi)		
6	$CH_3CH(NH_2)COOH$	L-Alanine	Ala
7	H_2NCH_2COOH	Glycine	Gly
8	$HOCH_2CH(NH_2)COOH$	L-Serine	Ser
9	$CH_3CH(OH)CH(NH_2)COOH$	L-Threonine	Thr
10	HO–(benzene ring)–$CH_2CH(NH_2)COOH$	L-Tyrosine	Tyr
	Neutral Hydrophobic (NHo)		
11	$CH_3CH_2CH(CH_3)CH(NH_2)COOH$	L-Isoleucine	Ile
12	$(CH_3)_2CHCH_2CH(NH_2)COOH$	L-Leucine	Leu
13	(benzene ring)–$CH_2CH(NH_2)COOH$	L-Phenylalanine	Phe
14	$CH_3CH(CH_3)CH(NH_2)COOH$	L-Valine	Val
	S-Containing		
15	$HSCH_2CH(NH_2)COOH$	L-Cysteine	CySH
16	$HOOCHC(H_2N)H_2CS\text{-}SCH_2CH(NH_2)COOH$	L-Cystine	CyS-SCy
	Labile Heterocyclic		
17	$H_2C{-}CH_2$, H_2C, $CHCOOH$, N, H (pyrrolidine ring)	L-Proline	Pro
18	$CH_2CH(NH_2)COOH$ on indole ring (N, H)	L-Tryptophan	Trp

cysteine. Hence, it can be classified among the labile amino acids. Tryptophan is highly labile in acid digests and is rarely detected in soil hydrolyzates.

Considerations of Protein Structures

In protein chemistry, *primary structure* refers to the types and the quantities of amino acids, and their sequences, as linked covalently through peptide bonds. The primary structure of a protein specifies the structural formulas and the configurations of the amino acid residues. In theory a large polypeptide molecule could be expected to take several possible conformations. This does not happen in proteins in their natural environment because of the influences of secondary interactions that stabilize specific conformations.

Secondary structure describes the first hierarchical level of conformational shapes that polypeptide molecules may adopt. A regular coiling of the polypeptide chain along one dimension, as in fibrous proteins, is one example. There is some double bond character in the C–N bonds of polypeptides, and thus rotation about the bond is restricted; the –C(O)NH segments tend to be planar. The sizes, shapes, and the various functionalities of the R substituents also lead to constraints in the rotation about the C to C bonds, and that gives further restrictions to the numbers of conformations that the molecule can adopt. In the α-helix, the simplest protein structure, the molecules take the form of a helical coil, with the single complete turn of the helix extending to about 5.4 Å along the polypeptide chain. This structure allows intramolecular H bonding between the H attached to the N (in the peptide bond) in one amino acid residue and the carbonyl of the third amino acid residue beyond it. Thus, the extensive H bonding along the polypeptide chain stabilizes the helical conformation. Breaking the H bonding disrupts the structure and denatures the protein.

Tertiary structure describes another level of conformational shapes that the molecule can adopt. It refers to a folding of the shapes described by secondary structure. In this the polypeptide chains are tightly folded into component three-dimensional structures, such as those that are typical of globular proteins. Tertiary structures are stabilized by H bonds, as well as by coulombic attraction between protonated basic amino acids (which have a second, non-α amino substituent). Ionized carboxyls in the acidic amino acids (which contain a second carboxyl functionality) are also possible.

The *quaternary structure* of proteins refers to associations, without the formation of covalent bonds, between the protein molecule and other macromolecules, some of which include carbohydrates and other nonpeptide structures. In the cases of proteins, each will have its own primary, secondary, and tertiary structure. Thus, the quaternary structure describes how different folded polypeptide units are associated in oligomeric proteins. Because the different units may have different conformations, the three-dimensional conformations of the quaternary structures can be highly complex.

Considerations of Structure and Interactions

The shapes of the protein or peptide molecules will significantly influence their interactions with soil components, especially with the soil colloidal constituents. Similar considerations will apply with regard to size, charge, and shape, as was discussed for polysaccharides. From our considerations of the ionizable functionalities in amino acids it is evident that the charges on these will be pH dependent. The relevant concepts extend to proteins. The amounts of terminal carboxyl and amino groups, as well as acidic and basic amino acids (see Table 1–14), in the structures will greatly influence the charges that these will carry at a particular pH. The charges will, in turn, influence physicochemical properties, as well as shapes, and the discussion with regard to charged polysaccharides outlined in Adsorption of Charged Polysaccharides by Clays (p. 115) can be extended to considerations of the adsorption of proteins and peptides, at least by clays.

The α-helix would be expected to be the peptide or protein structure most extensively adsorbed by clays, and the conditions outlined for the adsorption of random-coil/globular-shaped polysaccharides would apply also for the adsorption of similar shaped proteins by clays. Shapes are likely to be less determinative for interactions with HS. There are possibilities for the formation of ester linkages between terminal carboxyl groups as well as the free carboxyl in the acidic amino acids (Asp and Glu) and the –OH of alcohols and phenols in the humic structures. The free amino groups in Arg and Lys can form Schiff base functionalities with aldehyde and keto groups. Such amino functional groups can also form covalent linkages with quinones in humic structures, as indicated for **CCXXXVIII**, Fig. 1–29. However, the formation of such covalent linkages would not necessarily protect the protein or peptide from microbial and enzymatic attack. The protection is most likely to arise from an engulfing of the proteinaceous or peptide material within the humic matrix, possibly involving associations with hydrophobic stretches of the humic molecule, or with hydrophobic moieties that are part of the mixtures of organic components of SOM.

Peptide and Other Forms of Organic Nitrogen in Soils

Without protection from microbial (enzymatic) digestion and degradation processes, peptides and proteins would have very transitory existences in the soil. The isolation from soil of a chemically pure protein presents a daunting problem. This is not surprising in view of the numerous enzymes, each in very small amounts, from microbial processes that exist in the soil. Furthermore, for these to have other than transitory existences, they must be protected. Protection is considered usually to involve adsorption onto clays, associations with the hydrophobic components of SOM (especially the HAs and the humin materials), as referred to above, and seclusion in the pores of microaggregates and soil aggregates. These protective mechanisms can place the biodegradable material out of the reaches of microorganisms and their extracellular enzymes (Burns, 1990).

Knicker (2000a) reviewed her NMR evidence that shows that >80% of the organic N in soils is in peptide-like structures. In a study of mineral N utilization in the composting of wheat straw, Knicker and Lüdemann (1996) found, using ^{15}N-NMR (solid state), that after 8 d of incubation >80% of the total ^{15}N intensity was found in the chemical shift range between –220 and –285 ppm (amide-N). The composting medium had been amended with $(^{15}NH_4)_2SO_4$. No evidence for heterocyclic aromatic N has been found in normal natural organic matter (Knicker, 2000b), but it has been observed in the HAs of soil incubated with ^{15}N-labeled trinitrotoluene. It may, however, be contained in organic matter undergoing transition to coal. Knicker et al. (2002) did detect a clear shoulder in the chemical shift region for pyrrole- or indole-heteroaromatic N (−145 to −220 ppm) in the ^{15}N-NMR spectrum of the deepest layer of a peat that was at least 10 000 yr old. That peat could be considered to be at the beginning of the coalification stage, and there is abundant evidence for heterocyclic N in coal.

Peptide-N (−220 to −285 ppm) was the major source of organic N evident in the ^{15}N-NMR spectra for the humic isolates from different depths of an organic rich Sapropel from Mangrove Lake, Bermuda (Knicker and Hatcher, 2001). Amide-N was also found in the residues of 6 *M* HCl hydrolysis of the humin materials. The resistance to hydrolysis was considered to be related to encapsulation in the paraffinic network of algenans. Such encapsulation, in hydrophobic media, was considered also to provide protection from attack by hydrophilic enzymes. In another study along these lines, Knicker et al. (2001), using TMAH thermochemolysis/GC–MS techniques (see Information from Pyrolysis Degradation Processes, p. 85), identified protein remnants in base-insoluble geopolymers. Hydrolysis had failed to release amino acids entrapped in the hydrophobic humin residues.

We are unaware of accurate determinations of the overall contribution of peptidic components in humic mixtures. This is because of the difficulty of separating these components from the humic and other components in the mixtures. The uses of DMSO/HCl and XAD-8 resin technology (see Modern Fractionation Procedures, p. 42) enabled Häusler and Hayes (1996) to obtain an average decrease of 10% in the amino acid content in the HAs from a Sapric Histosol. The changes ranged from 0% for six of the amino acids identified to 40% in the case of His. It may well be, of course, that the losses represented specific peptides that were set free from the HAs by the DMSO/HCl treatment. Appelqvist et al. (1996) dissolved in DMSO + 1% 12 *M* HCl the HAs isolated using O.1 *M* NaOH from a H^+-exchanged Sapric Histosol and recovered the HAs using the XAD-8 resin technique (see p. 46). The DMSO/HCl procedure decreased the amino acid content of the HA by 23%. The decrease was uniform for the different amino acid groupings.

Peptide resonances are readily observed in two-dimensional TOCSY experiments. The couplings from amide protons to those on the α-C and side chains appear as a characteristic trail (see Fig. 1–30A). If D_2O is added to the sample the N–H groups exchange to become N–D, and the trail disappears. For comparison, the spectrum of a protein, bovine albumin, is shown in Fig. 1–30B. In HA and FA mixtures the amides display a characteristic resonance peak that often has a center at about 8.1 ppm. This amide resonance can be irradiated using a selective one-dimensional TOCSY experiment in which the magnetization is passed from the amide protons to protons that are in close proximity, i.e., the protons on the α-C, and on the side chains. In this case the only signals in the resulting one-dimensional spectrum are from peptides. Figure 1–31 compares a conventional ^{1}H spectrum of the IHSS peat HA (Fig. 1–31A) with that of the one-dimensional selective TOCSY experiment (Fig. 1–31B). The results from the selective irradiation TOCSY are not quantitative, but they do clearly highlight the peptide signals. These can be identified in the complete one-dimensional proton spectrum of the material (Fig. 1–31A). Here the one-dimensional proton spectrum is quantitative. In this sample it is easy to see that the peptide signals are clearly abundant in the mixture. Indeed, in many humic materials peptides can be seen as major components (Fan et al., 2000; Kingery et al., 2000; Simpson et al., 2001a), and should not be underestimated when considering the structural components of humic materials. Yet, despite their prevalence in HS, the precise structures and recalcitrant behavior of these components are far from resolved.

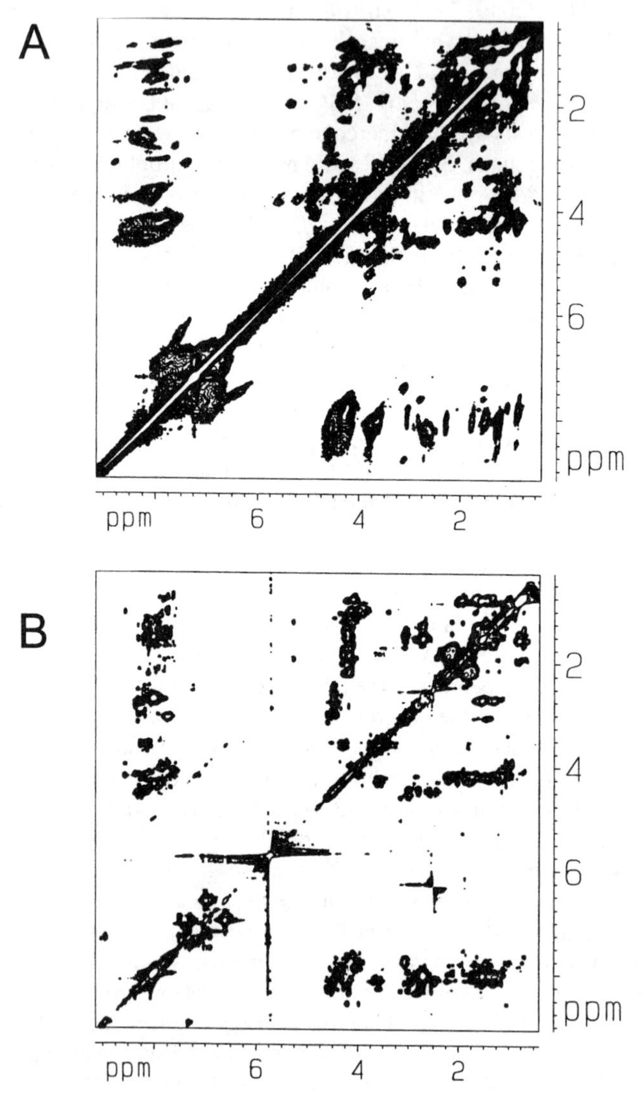

Fig. 1–30. Comparison of the TOCSY-NMR spectrum of (A) the IHSS peat humic acid (HA) with (B) the 0.1 *M* NaOH digest of bovine albumin (a pure protein). Both samples are dissolved in DMSO-d6 for NMR. The boxes highlight couplings between amide protons and protons on the α-carbon on peptides (as mentioned in the text). Note the similarities of the spectra indicate peptide structures are abundant in the IHSS peat HA (from Simpson, unpublished data).

The possibilities for the identification of forms of N other than peptide N in soil fractions have been greatly advanced by recent developments in solid-state NMR pulse sequences. Carbon directly bonded to N can now be observed through saturation-pulse induced dipolar exchange with recoupling (SPIDER), as described by Schmidt-Rohr and Mao (2002b). This procedure involves the recoupling of ^{13}C–^{14}N dipolar couplings and effective pulsed saturation of ^{14}N. MAS is employed at spinning frequencies that allow detection of aromatic C residues. This technique will allow detection of N substituents on aromatic nuclei. Such considerations are very important in view of the growing evidence for organically immobilized N in SOM. Yields of lowland rice (*Oryza sativa* L.), for example, have been shown to decrease by more than 35% during 20 to 30 yr of double and triple cropping. The total soil N did not decrease, and it is believed that the unavailable organic N was bonded to lignin residues which accumulate in the anaerobic conditions (Olk et al., 1996, 2000; Schmidt-Rohr and Mao, 2002a, 2002b).

Amino Acid Residues in Soil Hydrolyzates

There have been numerous measurements of amino acids in soil hydrolyzates, but these do not give indications about the origins in the soil of the amino acid pre-

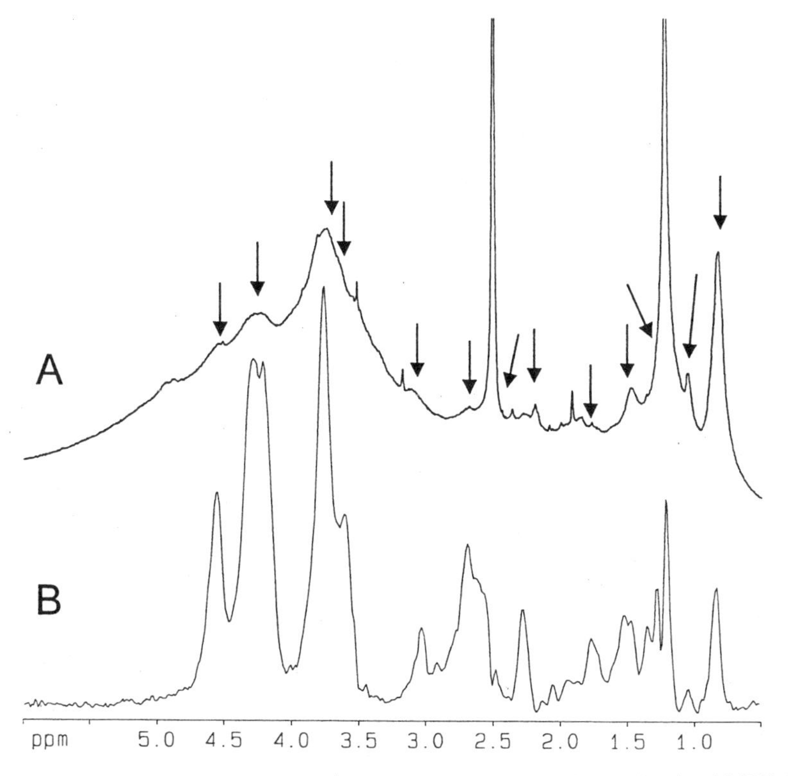

Fig. 1–31. Comparison of the conventional ^{1}H-NMR spectrum of the IHSS peat humic acid (HA) (A) with the one-dimensional selective TOCSY NMR spectrum (B). Arrows indicate the signals from the peptide structures in the conventional ^{1}H spectrum (from Simpson, unpublished data).

cursors. Because of the difficulties in isolating protein and peptide materials work has focused on examinations of the amino acid contents in different fractions of SOM.

Hayes et al. (1999) summarized data for the amino acids in the HAs, FAs, and XAD-4 acids (recovered from the XAD-4 resin when the XAD-8 and XAD-4 resin in tandem procedure was used) isolated at pH 7.0, pH 10.6, and pH 12.6 from a clay, noncalcareous grassland soil. The amino acids were also analyzed in the HAs, FAs, and XAD-4 acids from the drainage waters from the soil. The soil HAs, FAs, and XAD-4 acids had, on average, five, seven, and five times more amino acids, respectively, than did the amino acids from the corresponding water fractions. The concentration of amino acids in the soil isolates at any particular pH tended to decrease in the same order as the samples from the drainage waters: HAs > XAD-4 acids > FA. For the most part the relative abundances of the acid groupings followed the order: total NHi >> total acidic > total NHo >> total basic. In general the abundances of the individual amino acids decreased in the order: Gly > Asp > Ala > Glu > Thr > Val Leu > Ser > Ile > Lys > Phe > Tyr > Arg > His > Met.

Clapp and Hayes (1996) exhaustively extracted a Waukegan Mollisol soil with 0.1 *M* $Na_4P_2O_7$ (Pyro) at pH 7 (Extracts 1–7), then with Pyro, pH 10.6 (Extracts 8–12), then with Pyro + 0.1 *M* NaOH, pH 12.6 (Extracts 13–15), then with 0.1 *M* NaOH (Extracts 16–17), and finally with 94% DMSO/6% 12 *M* HCl. The HAs were recovered from Extracts 1, 8, 13, 16, and 18 (i.e., is the first extract from each of the solvent systems used). Recovery from the DMSO/HCl system used the XAD-8 resin technique (see Extraction with Organic Solvents, p. 34).

The quantities of the 17 amino acids detected in the hydrolyzates of the HAs are given in Table 1–15. Again, the order of abundances of the amino acid groups followed the same order as observed by Hayes et al. (1999) for the grassland soil, and glycine was most abundant, or in two samples equally abundant, with Asp. Overall, the relative abundances of the individual amino acids in the samples were similar. However, a feature of these results is the fact that the amounts of amino acids in the HAs isolated in pyrophosphate at both pH 7 and pH 10.6 were less than one-half the amounts in the HAs isolated in NaOH. (It should be borne in mind that extraction with one set of solvents was discontinued when the absorbance of the extract at 400 nm was <0.5; invariably there was a big increase in absorbance when the next solvent in the sequence was introduced.) Pyrophosphate at pH 7 isolates the most highly oxidized HS and is considered to do so by complexing with metals (divalent/polyvalent) that immobilize these in the soils. More peptide materials were released when the pH of the pyrophosphate solution was raised (Extract 8) to pH 10.6, but the amounts extracted were still less than one-half those extracted in Pyro + NaOH at pH 12.6. The surprising feature was that most peptides were extracted at pH 12.6 in the absence of pyrophosphate.

The Mollisol soil was largely Ca^{2+}-exchanged. We presume that the highly humified HA extracted in Pyro at pH 7 had associated the more highly transformed peptide materials. That would mean that the peptides were likely to have microbial origins. The materials extracted at pH 12.6 were less humified and were likely to contain more plant related-structures.

Table 1–16 gives data for the amino acids in the HAs and FAs isolated from the A_h and B_h horizons of an oak forested Podzol soil (Simpson, 1999). Extraction

Table 1–15. Amino acid contents of humic acids isolated from a Waukegan soil using different extracting solutions.†

Amino acid	Extract number‡				
	1	8	13	16	18
	nmol mg^{-1}§				
Acidic					
Asp	63	58	140	150	74
Glu	51	52	110	110	72
Basic					
Arg	10	13	38	38	9
His	25	32	55	54	38
Lys	39	22	59	50	8
NHo¶					
Pro	22	23	58	47	26
Val	23	26	78	85	28
Ile	11	14	42	49	16
Leu	15	20	66	74	11
Tyr	36	74	82	90	64
Phe	11	14	45	50	18
NHi¶					
Thr	31	33	79	84	30
Ser	38	43	89	110	42
Gly	66	73	160	180	78
Ala	43	52	110	130	52
S-containing					
Met	3	4	10	10	4
CySH	2	2	4	4	2
Total	480	555	1225	1317	600

† From Clapp and Hayes (1996).
‡ (1) 0.1 *M* pyrophosphate (Pyro), pH 7.0; (8) 0.1 *M* Pyro, pH 10.6; (13) 0.1 *M* Pyro + 0.1 *M* NaOH, pH 12.6; (16) 0.1 *M* NaOH, pH 12.6; and (18) DMSO 94% + 6% 12 *M* HCl.
§ Corrections made for ash contents.
¶ NHo = Neutral hydrophobic; NHi = Neutral hydrophilic.

was exhaustive with water (in the case of the A_h horizon); Pyro, pH 7; Pyro, pH 10.6; and Pyro + NaOH, pH 12.6. The yield pattern was different from that observed for the Mollisol. The significant increases in peptide yields in the pH 12.6 seen for Mollisols were absent for the HAs, and also for the FAs. There was an extraordinary enrichment of peptide materials in the water extract (pH 5.5) from the A_h horizon. However, that enrichment was not evident in the extracts from the B_h horizon. There was a big enrichment also in amino acids in the extract at pH 10.6 from the A_h horizon, but the abundances of the HAs isolated at the different pH values were similar in the case of the B_h HAs. That emphasizes that the enrichments of peptides in the A_o horizon would be susceptible to transport in the soil profile.

The reader will observe that the order of abundances of the amino acid groups are not predictable, as seen for the Mollisol and grassland clay soil referred to above. Note the considerable abundances of the acidic group amino acids in the FAs and HAs isolated at pH 7 from the A_h and B_h horizons. Predictably the neutral hydrophilic (NHi) group dominated, but not highly significantly in the cases of extracts at pH 10.6 and 12.6.

Table 2–16. Amino acids (AA), expressed as percentage of the total amino acids in the fulvic acids (FAs), humic acids (HAs), and XAD-4 acids isolated in water, pH 5.5; 0.1 *M* sodium pyrophosphate (Pyro), pH 7; Pyro, pH 10.6; and Pyro + 0.1 *M* NaOH, pH 12.6, from the A_h and B_h horizons of a Podzol soil.†

	Acidic			Basic				Neutral hydrophilic (NHi)					Neutral hydrophilic (NHo)					Other		
Sample	Asp	Glu	Total acidic	His	Lys	Arg	Total basic	Gly	Thr	Ser	Ala	Total NHi	Tyr	Val	Phe	Ileu	Total NHo	Met	Total AA	Total AA by weight
	%																		nmol mg^{-1}	%
A_h FA (5.5)	17.0	10.6	27.6	1.7	1.7	10.6	14.0	17.4	8.5	7.6	1.3	34.8	1.4	7.6	2.9	4.7	23.0	0.2	235.8	2.4
A_h FA (7.0)	20.8	14.6	35.4	1.1	3.5	9.4	14.0	16.4	7.6	7.6	1.4	33.1	0.9	6.6	1.7	3.7	17.3	0.1	457.8	4.7
A_h FA (10.6)	16.8	11.1	28.0	1.0	3.0	10.3	14.3	13.8	8.8	8.1	1.5	32.2	1.7	8.4	3.4	5.1	24.9	0.5	593.7	5.9
A_h FA (12.6)	15.9	11.5	27.5	1.1	2.4	9.8	13.4	15.8	8.6	8.4	1.6	34.3	1.9	8.1	3.4	5.1	24.5	0.1	515.6	5.2
A_h HA (5.5)	10.7	10.2	20.9	1.9	4.8	12.6	19.3	9.0	6.6	5.6	4.7	25.9	3.0	8.7	4.6	6.4	32.6	1.3	2681.0	26.4
A_h HA (7.0)	18.3	12.3	30.5	2.3	4.8	9.2	16.3	11.8	7.1	7.1	3.8	29.8	1.5	7.5	3.1	4.6	22.9	0.4	651.6	6.6
A_h HA (10.6)	2.7	3.8	6.4	1.8	13.6	10.1	25.5	34.7	3.9	5.8	5.4	49.8	1.7	4.6	3.0	3.2	17.8	0.5	1884.5	16.5
A_h HA (12.6)	14.3	10.5	24.7	2.0	4.8	9.9	16.7	13.4	7.2	6.8	3.7	31.1	2.2	8.1	3.9	5.4	27.1	0.2	736.8	7.2
A_h XAD-4(5.5)	19.9	12.5	32.3	1.3	2.4	12.1	15.8	18.2	8.1	8.4	1.0	35.7	1.0	6.1	1.7	3.0	15.5	0.3	297.0	3.0
A_h XAD-4(12.6)	20.4	6.0	26.4	1.3	1.3	25.5	28.1	7.2	5.1	1.7	3.6	17.6	1.7	10.6	2.3	5.5	27.4	0.6	235.2	2.2
B_h FA (7.0)	28.8	20.9	49.7	0.8	2.0	10.2	13.1	14.6	4.7	4.7	1.2	25.2	0.3	4.7	1.2	2.5	11.5	0.2	253.8	2.7
B_h FA (10.6)	15.7	10.0	25.7	0.6	3.3	16.9	20.7	17.2	4.4	2.1	3.3	27.0	2.9	12.8	1.0	1.7	25.7	0.7	521.9	5.2
B_h FA (12.6)	14.6	11.2	25.8	1.3	3.8	11.8	16.9	12.2	7.3	6.5	3.4	29.4	2.2	8.3	4.0	5.3	27.2	0.5	756.0	7.4
B_h HA (7.0)	16.9	13.0	29.9	1.4	4.5	11.2	17.1	11.6	7.0	6.5	3.0	28.1	1.5	7.9	3.4	5.4	24.3	0.5	756.9	7.5
B_h HA (10.6)	15.0	12.0	27.0	1.4	3.8	12.6	17.8	6.5	6.9	6.1	4.0	23.5	2.5	9.2	4.4	6.4	31.0	0.6	707.0	6.9
B_h HA (12.6)	14.6	11.2	25.8	1.3	3.8	11.8	16.9	12.2	7.3	6.5	3.4	29.4	2.2	8.3	4.0	5.3	27.2	0.5	756.0	7.4
B_h XAD-4 (7.0)	22.6	9.9	32.5	0.4	2.9	13.6	16.8	28.0	6.6	9.9	0.8	45.3	0.2	2.9	0.2	0.8	5.0	0.2	485.7	5.0
B_h XAD-4 (10.6)	35.5	14.9	50.4	0.8	1.9	10.4	13.1	10.9	5.8	7.0	1.2	24.8	0.9	3.9	1.2	2.1	11.1	0.3	430.9	4.6

† From Simpson (1999).

Another feature of the data in Table 1–16 is the order in relative abundance of Asp and the relatively low contribution of the amino acids to the mass of the XAD-4 acids. We consider that the XAD-4 acids are the most likely of all the fractions to have had origins in microbiological processes. Hence, the reader is invited to consider correlations and ratios between the abundances of the amino acids groups and the individual acids for the different samples. We have done this for Arg/Ala ratios; these are highest for the XAD-4 acids, followed by those for the FAs, and the lowest ratios are for the HAs. That decrease in ratio values can be correlated with extent of oxidation, which we consider to be biologically initiated.

General Conclusions

It is accepted that peptide materials are important sources of organic N in the soil environment. Their existence in the soil is transitory, unless protected through associations with the colloidal inorganic and organic soil components, and in soil microaggregates and aggregates to which degradative enzymes and microorganisms do not have access. We know that soil proteins exist because of the enormous variety of enzymes in every fertile soil. Many of these proteins may well be denatured to peptide structures and these are also adsorbed by soil colloids.

It is highly unlikely, because of the intimate associations between protein and peptide materials and the humic and nonhumic components of SOM, that it will be possible to isolate a soil protein in meaningful yield. This should not be a matter for concern because assay procedures show that specific enzymes are present in soils. Because the chemistry of proteins and peptides is well understood, we can make predictions of the physicochemical reactions with which these can be involved in the soil environment.

It will be of interest to know more about the mechanisms of interactions between proteins and peptides and the other organic components of SOM. It will be of interest also to know more about the diagenesis of the soil proteins and peptides, and how these can be related to the soil plant and microbial biota in a particular soil environment. Progress can be made by relating the compositions of those in soil with those in the vegetation, and with the kinds of microflora that are present in a particular soil environment.

SUMMARY

It is evident from our discussion of the genesis of soil organic matter that the component molecules of HS are still not known unambiguously. So far the best indications of the natures of these can be deduced from identifications of products of degradation reactions, and from an awareness of the mechanisms involved in the degradation processes. Awareness of likely component molecules, and of functionality, provided by techniques such as wet chemical methods, NMR, FTIR, titration data, and developments in Raman spectroscopy (which may provide valuable information in the near future) has not changed dramatically in the last 10 yr. We knew then, as now, that 25 to 45% of the components of soil HAs are aromatic (Malcolm, 1989), and that value may be higher in some instances (e.g., for samples from the B_h horizons of some podzols). That information was derived from

CPMAS ^{13}C-NMR data, but it is considered that such NMR techniques might underestimate the aromatic functionality content. Ten years ago the consensus favored single ring aromatic components (Hayes et al., 1989b) having three to five ring substituents. That view has not changed, although five-ring substituents now seem less likely. On the basis of products identified by pyrolysis procedures, some hold the view that fused aromatic compounds could be components of humic structures. We do not consider that such structures are covalently linked in humic structures, and are likely to be artifacts of the thermal degradation procedures. It is possible that such structures can be present in soils where burning has taken place. In such cases the compounds can be considered to be associated (e.g., by van der Waals forces) with the humic materials, but not components of the humic molecules.

Humic acids in soil aqueous systems are less aromatic and more highly oxidized than those that are components of the solid HS. The HS in drainage waters and in the soil solution have significantly fewer sugars and amino acids than those in soils. That would suggest that the sugars and amino acids are sterically protected from enzymatic attack in their humic associations. These could involve hydrophobic constituents, such as long chain hydrocarbons and fatty acids and waxes, or physically protected through associations with mineral colloids and soil aggregates. The NMR data of Simpson et al. (2002) emphasize how various organic components can be in association to give a conglomerate of materials that come under the broad classification of HS, based on the operational definitions that apply at this time. Conformational changes in HS in the soil environment, or the disruption of aggregates, could break up associations with sugars and peptides, and expose glycosidic and peptide linkages to microbial attack. Some of the core humic structures released in this way would dissolve in the soil solution, and that could explain some of the properties of the HS in the soil solution.

Hydrogens in three to five of the aromatic ring positions may be replaced by substituents consisting of hydroxyl, methoxyl, and less likely, by aliphatic hydrocarbon structures. Some of these may be involved in linking aromatic structures. There is evidence also for aldehyde and keto functional groups attached to some of the aromatic nuclei, for phenylpropane (3-carbon chains attached to the aromatic rings) units, and for hydroxyl and methoxyl substituents. The phenylpropane structures, and hydroxy/methoxy substituents in the 4-, 3- and 4-, and in the 3-, 4-, and 5-ring positions would suggest origins in lignins, whereas the presence of these substituents in the 3- and 5-ring positions suggests origins in the skeletal structures, or the metabolic products of microorganisms.

Ether functionalities can link aromatic structures, and it is logical to assume that aromatic-aliphatic ethers are also present, such as those arising from lignin structures. There is not convincing evidence for the linking of aromatic units by hydrocarbons, and it seems likely that long chain hydrocarbons, with origins in plant cuticles and in algal species could be "impurities" held to the humic structures by van der Waals forces. Some of the hydrocarbon structures are olefinic, as suggested from the presence of mono-, di-, and sometimes tri-carboxylic acids in the digests of oxidative degradation reactions (see Oxidative Degration Reactions, p. 73).

Fatty acids in degradation digests may be released from esters of phenols and other hydroxyl groups in the "backbone" structures. Such components could arise from waxes, suberins, and microbial products associated with HS. These could contribute to the hydrophobic properties which HS can display, and have a role in self-association effects.

Titration data show that the acid groups in HS provide a continuum of dissociable protons, and acid strengths range from strong to very weak. The strongest acids are carboxylic, and some are activated by appropriate adjacent functional groups. Phenolic hydroxyls also contribute to the acidity, and these are most abundant in the HA fraction. It would seem that their contribution to the total acidity is greatest in newly formed HS, and especially in those with origins in the lignified components of plants. As oxidation takes place, the phenols are oxidized, and eventually carboxylic acids are formed. Some of the phenolic substituents can have enhanced acidity because of the influences of other substituents on the aromatic structures. Enols and other weakly dissociable groups also contribute to the charge characteristics under alkaline conditions.

Fulvic acids have compositional characteristics similar to HAs, but they have differences as well. The FAs are less aromatic (i.e., they can contain <25% aromatic components in some instances), and are smaller, more polar, and more highly charged. Thus, FAs would be less self-associated than HAs. Because FAs are not precipitated under acid conditions, or by low concentrations of divalent metals, they might be expected to be readily removed from soils in drainage waters. That this does not happen is an indication of interactions (steric and physical bonding) between FAs and insoluble components, especially HAs, of soils.

Soil carbohydrates have an important role in the formation of soil aggregates, but these would be unlikely to survive in the soil environment in the absence of physical protection by the humic components, by the hydrophobic constituents of SOM, and by the soil inorganic colloids. Carbohydrates are likely to be independent of core HS, although there are possibilities for covalent linkages through glycosidic bonds, such as phenolic glycoside structures.

Soil peptides are major sources of soil organic N, and these also survive in the soil through associations with the soil organic and mineral colloids, or through protection from associations with hydrophobic organic molecules. They are likely to exist independently of HS, but there are possibilities also for covalent linkages through peptide bonds between free amino and carboxyl groups. Such linkages can also occur through interactions between the free amino groups and quinone structures.

Glomalin, now perceived as a mucopolysaccharide-type substance, can form strong associations with the mineral soil colloids and may have an important role in soil aggregate formation and stabilization. The survival of these compounds in the soil is likely to be attributable to the associations they form with the soil mineral colloids and with the hydrophobic soil colloids.

The soil organic components isolated from XAD-4 resins when soil organic solutions are passed through XAD-8 and XAD-4 resins in tandem are mainly composed of polar carbohydrate and peptide substances, and with some highly polar humic constituents. It can be predicted that further separations of this fraction will be possible.

Many attempts have been made in the past to present structural representations of soil humic molecules. However, as Burdon (2001) has pointed out, there is little chemical logic to support any of these hypothetical structures. We consider that, despite the significant advances that are being made, it would be pointless to attempt to depict humic structures in terms of molecular formulas. The representation in Fig. 1–22 is intended to give indication of some of the types of associations between organic molecules that can occur in soil organic matter, and should not be considered as an attempt to represent an accurate structural concept. It is clear that HS are not polymers, and HAs and FAs are unlikely even to be macromolecules. The current awareness would suggest that HS are associations of molecules formed from the biological oxidations of lignin components, from the products of microbial metabolism, and from the cell walls, especially those of recalcitrant plant components and fungi. There is real hope that soon there will be a realistic awareness of the compositions and associations of the molecules that give rise to the humic fractions. But, because HS are gross mixtures, it will be pointless to try to establish accurate structures. When we know more about the compositions of the component molecules, and of the ways in which these are associated, it will be possible to reach a better understanding of the vital reactions and interactions that involve the organic materials in the soil environment.

REFERENCES

Aachard, F.K. 1786. Chemische Utersuchung des Torfs. Crell's Chem. Ann. 2:391–403.

Aiken, G.R., and A.H. Gillam. 1989. Determination of molecular weights of humic substances by colligative property measurements. p. 515–544. *In* M.H.B. Hayes et al. (ed.) Humic substances II: In search of structure. John Wiley and Sons, Chichester, UK.

Aiken, G.R., D.M. McKnight, R.L. Wershaw, and P. MacCarthy. 1985. An introduction to humic substances in soil, sediment, and water. p. 1–9. *In* G.R. Aiken et al. (ed.) Humic substances in soil, sediment, and water: Geochemistry, isolation and characterization. John Wiley and Sons, New York.

Alexander, M. 1977. The carbon cycle. p.115–203. *In* Introduction to soil microbiology. John Wiley and Sons, New York.

Alexandrova, L.N. 1960. The use of sodium pyrophosphate for isolating free humic substances and their organic-mineral compounds from the soil. Soviet Soil Sci. 2:190–197.

Almendros, G., and J. Sans. 1989. Compounds released from humic acid upon BF_3–MeOH transesterification. Sci. Total Environ. 81/82:51–60.

Almendros, G., and J. Sans. 1991. Structural study on the soil humin fraction-boron trifluoride-methanol transesterification of soil humin preparations. Soil Biol. Biochem. 23:1147–1154.

Almendros, G., and J. Sans. 1992. A structural study of alkyl polymers in soil after perborate degradation of humin. Geoderma 53:79–95.

Anderson, H.A., W. Bick, A. Hepburn, and M. Stewart. 1989. Nitrogen in humic substances. p. 223–253. *In* M.H.B. Hayes et al. (ed.) Humic substances II: In search of structure. John Wiley and Sons, Chichester, UK.

Appelqvist, I.A.M., C.L. Graham, and M.H.B. Hayes. 1996. Isolation, fractionation, and characterization of humic acids from a Sapric Histosol. p. 33–39. *In* C.E. Clapp et al. (ed.) Humic substances and organic matter in soil and water environments: Characterization, transformations and interactions. IHSS, University of Minnesota, St. Paul.

Atherton, N.M., P.A. Cranwell, A.J. Floyd, and R.D. Haworth. 1967. Humic acid—1. ESR spectra of humic acids. Tetrahedron 23:1653–1667.

Atjay, G.L., P. Ketner, and P. Duvigneaud. 1979. Terrestrial primary production and phytomass. p. 129–181. *In* B. Bolin et al. (ed.) The global carbon cycle. John Wiley and Sons, Chichester, UK.

Atkins, P.W. 1994. Physical chemistry. 5th ed. W.H. Freeman, New York.

Bacon, J.S.D., and M.V. Cheshire. 1971. Apiose and mono-ethyl sugars as minor constituents of the leaves of deciduous trees and various other species. Biochem. J. 124:555–562.

Baldock, J.A., J.M. Oades, P.N. Nelson, T.M. Skene, A. Golchin, and P. Clarke. 1997. Assessing the extent of decomposition of natural organic materials using solid-state ^{13}C NMR spectroscopy. Aust. J. Soil Res. 35:1061–1083.

Balesdent, J., A. Mariotti, and B. Guillet. 1987. Natural ^{13}C abundance as a tracer for studies of soil organic matter dynamics. Soil Biol. Biochem. 19:25–30.

Balesdent, J., G.H. Wagner, and A. Mariotti. 1988. Soil organic matter turnover in long term field experiments as revealed by carbon-13 natural abundance. Soil Sci. Soc. Am. J. 52:118–124.

Barker, S.A., P. Finch, M.H.B. Hayes, R.G. Simmonds, and M. Stacey. 1965. Isolation and preliminary characterization of soil polysaccharides. Nature (London) 205:68–69.

Barker, S.A., M.H.B. Hayes, R.G. Simmonds, and M. Stacey. 1967. Studies on soil polysaccharides. 1. Carbohydr. Res. 5:13–24.

Barton, A.F.M. 1975. Solubility parameters. Chem. Rev. 75:731–753.

Barton, D.H.R., and M. Schnitzer. 1963. A new experimental approach to the humic acid problem. Nature (London) 198:217–218.

Batjes, N.H. 1999. Management options for reducing CO_2–concentrations in the atmosphere by increasing carbon sequestration in the soil. NRP Rep. 4102 000031. ISRIC Technical Paper 30. International Soil Reference and Information Centre, Wageningen, The Netherlands.

Bazilevich, N.I. 1974. Soil-forming role of substance and energy exchange in the soil–plant system. Trans. 10th Int. Congr. Soil Sci. (Moscow) 6:17–27.

Benzig-Purdie, L., M.V. Cheshire, B.I. Williams, C.I. Ratcliffe, J.A. Ripmeester, and B.A. Goodman. 1992. Interactions between peat and sodium acetate, ammonium sulphate, urea or wheat straw during incubation studied by ^{13}C and ^{15}N NMR spectroscopy. J. Soil Sci. 43:113–125.

Benzig-Purdie, L., and J.A. Ripmeester. 1983. Melanoidins and soil organic matter: Evidence of strong similarities revealed by ^{13}C CP-MAS NMR. Soil Sci. Soc. Am. J. 47:56–61.

Bird, M.I., and D.R. Gröcke. 1997. Determination of the abundance and carbon isotope composition of elemental carbon in sediments. Geochim. Cosmochim. Acta 61:3413–3414.

Bird, M.I., C. Moyo, E.M. Veenendaal, J. Lloyd, and P. Frost. 1999. Stability of elemental carbon in a savanna soil. Global Biogeochem. Cycles 13:923–932.

Blakeney, A.B., P.J. Harris, R.J. Henry, and B.A. Stone. 1983. A simple and rapid preparation of alditol acetates for monosaccharide analysis. Carbohyd. Res. 113:291–299.

Bloom, P.R., and J.A. Leenheer. 1989. Vibrational, electronic, and high-energy spectroscopic methods for characterizing humic substances. p. 409–446. *In* M.H.B. Hayes et al. (ed.) Humic substances II: In search of structure. John Wiley and Sons, Chichester, UK.

Bohn, H.L. 1976. Estimate of organic carbon in world soils. Soil Sci. Soc. Am. J. 40:468–470.

Bol, R., Y. Huang, J.A. Medrith, G. Eglinton, D.D. Harkness, and P. Ineson. 1996. The ^{14}C age of organic matter and its lipid constituents in a stagnohumic gley soil. Eur. J. Soil Sci. 47:215–222.

Bone, W.A., and G.W. Himus. 1936. Coal, its constitution and uses. Longman Green, London.

Bone, W.A., L. Horton, and S.G. Ward. 1930. Researches on the chemistry of coal. VI. Its benzenoid constitution as shown by its oxidation with alkaline permanganate. Proc. R. Soc. London, Ser. A 127:480–510.

Bracewell, J.M., K. Haider, S.R. Larter, and H.-R. Schulten. 1989. Thermal degradation relevant to structural studies of humic substances. p. 181–222. *In* M.H.B. Hayes et al. (ed.) Humic substances II: In search of structure. John Wiley and Sons, Chichester, UK.

Bremner, J.M. 1950. Some observations on the oxidation of soil organic matter in the presence of alkali. J. Soil Sci. 1:198–204.

Bremner, J.M., and H. Lees. 1949. Studies of soil organic matter: II. The extraction of organic matter from soil by neutral reagents. J. Agric. Sci. 39:274–279.

Briggs, G.C., and G.J. Lawson. 1970. Chemical constitution of coal: 16. Methylation studies on humic acid. Fuel 49:39–48.

Browning, B.L. 1997. Methods in wood chemistry. Vol. 2. Interscience, New York.

Bull, I.D., P.F. van Bergen, C.J. Nott, P.R. Poulton, and R.P. Evershed. 2000. Organic geochemical studies of soils from the Rothamsted classical experiments. V. The fate of lipids in different long-term experiments. Org. Geochem. 31:389–408.

Burchill, S., M.H.B. Hayes, and D.J. Greenland. 1981. Adsorption. p. 221–400. *In* D.J. Greenland and M.H.B. Hayes (ed.) The chemistry of soil processes. John Wiley and Sons, Chichester, UK.

Burdon, J. 2001. Are the traditional concepts of the structures of humic substances realistic? Soil Sci. 166:752–769.

Burdon, J., J.D. Craggs, M.H.B. Hayes, and M. Stacey. 1974. Reactions of sodium sulphide: 1. With compounds containing hydroxyl groups. Tetrahedron 30:2729–2733.

Burges, N.A., H.M. Hurst, and S.B. Walkden. 1964. The phenolic constituents of humic acid and their relation to the lignin of the plant cover. Geochim. Cosmochim. Acta 28:1547–1555.

Burges, N.A., H.M. Hurst, S.B. Walkden, F.M. Dean, and M. Hurst. 1963. Nature of humic acids. Nature (London) 199:696.

Burns, R.G. 1990. Microorganisms, enzymes and soil colloid surfaces. p. 337–361. *In* M.F. De Boodt et al. (ed.) Soil colloids and their associations in aggregates. Plenum, New York.

Butler, J.H.A., D.T. Downing, and R.J. Swaby. 1964. Isolation of chlorinated pigment from green soil. Aust. J. Chem. 17:817–819.

Cameron, R.S., R.S. Swift, B.K. Thornton, and A.M. Posner. 1972a. Calibration of gel permeation chromatography materials for use with humic acid. J. Soil Sci. 23:342–349.

Cameron, R.S., B.K. Thornton, R.S. Swift, and A.M. Posner. 1972b. Molecular weight and shape of humic acid from sedimentation and diffusion measurements on fractionated extracts. J. Soil Sci. 23:394–408.

Challinor, J.M. 1995. Characterization of wood by pyrolysis derivitization–gas chromatography/mass spectrometry. J. Anal. Appl. Pyrolysis 35:93–107.

Chang, H.-M., and G.G. Allan. 1971. Oxidation. p. 433–485. *In* K.V. Sarkanen and C.H. Ludwig (ed.) Lignins. Wiley, New York.

Chefetz, B., Y. Chen, C.E. Clapp, and P.G. Hatcher. 2000. Characterization of organic matter in soils by thermochemolysis using tetramethylammonium hydroxide (TMAH). Soil Sci. Soc. Am. J. 64:583–589.

Chen, Y., N. Senesi, and M. Schnitzer. 1977. Information provided on humic substances by E_4/E_6 ratios. Soil Sci. Soc. Am. J. 41:352–358.

Cheshire, M.V. 1979. Nature and origin of carbohydrates in soils. Academic Press, London.

Cheshire, M.V., and G. Anderson. 1975. Soil polysaccharides and carbohydrate phosphates. Soil Sci. 119:356–372.

Cheshire, M.V., P.A. Cranwell, C.P. Falshaw, A.J. Floyd, and R.D. Haworth. 1967. Humic acid: II. Structure of humic acids. Tetrahedron 23:1669–1682.

Cheshire, M.V., P.A. Cranwell, and R.D. Haworth. 1968. Humic acid. III. Tetrahedron 24:5155–5167.

Cheshire, M.V., and M.H.B. Hayes. 1990. Composition, origins, structures, and reactivities of soil polysaccharides. p. 307–336. *In* M.F. DeBoodt et al. (ed.) Soil colloids and their associations with aggregates. Plenum, New York.

Cheshire, M.V., and C.M. Mundie. 1966. The hydrolytic extraction of carbohydrates from soil by sulphuric acid. J. Soil Sci. 17:372–381.

Cheshire, M.V., C.M. Mundie, J.M. Bracewell, G.W. Robertson, J.D. Russell, and A.R. Fraser. 1983a. The extraction and characterization of soil polysaccharides by whole soil methylation. J. Soil Sci. 34:539–554.

Cheshire, M.V., C.M. Mundie, and H. Shepherd. 1969. Transformation of (^{14}C) glucose and starch in soil. Soil Biol. Biochem. 1:117–130.

Cheshire, M.V., C.M. Mundie, and H. Shepherd. 1971. The origin of the pentose fraction of soil polysaccharide. J. Soil Sci. 22:222–236.

Cheshire, M.V., G.P. Sparling, and R.W.E. Inkson. 1979. The decomposition of straw in soil. p. 65–71. *In* E. Grossard (ed.) Straw decay and its effect on disposal and utilization.

Cheshire, M.V., G.P. Sparling, and C.M. Mundie. 1983b. Effects of periodate treatment of soil on carbohydrate constituents and soil aggregation. J. Soil Sci. 34:105–112.

Cheshire, M.V., G.P. Sparling, and C.M. Mundie. 1985. The effect of oxidation by periodate on carbohydrate derived from plants and microorganisms. J. Soil Sci. 36:351–356.

Cheshire, M.V., G.P. Sparling, and H. Shepherd. 1974. Transformation of sugars when rye hemicellulose labelled with (^{14}C) decomposes in soil. J. Soil Sci. 25:90–98.

Cheshire, M.V., and S.J. Thompson. 1972. Configuration of soil arabinose. Biochem. J. 129:19.

Chesters, G., O.J. Attoe, and O.N. Allen. 1957. Soil aggregation in relation to various soil constituents. Soil Sci. Soc. Am. Proc. 21:272–277.

Chiou, C.T., P.E. Porter, and D.W. Schmedding. 1983. Partition equilibria of nonionic organic compounds between soil organic matter and water. Environ. Sci. Technol. 17:227–231.

Choudhri, M.B., and F.J. Stevenson. 1957. Chemical and physicochemical properties of soil colloids. III. Extraction of organic matter from soils. Soil Sci. Soc. Am. Proc. 21:508–513.

Clapp, C.E. 1957. High molecular weight water-soluble muck: Isolation and determination of constituent sugars of a borate complex-forming polysaccharide employing electrophoretic techniques. Ph.D. diss. Cornell University, Ithaca, NY.

Clapp, C.E., R.R. Allmaras, M.F. Layese, D.R. Linden, and R.H. Dowdy. 2000. Soil organic carbon and ^{13}C abundance as related to tillage, crop residue, and nitrogen fertilization under continuous corn management in Minnesota. Soil Tillage Res. 55:127–142.

Clapp, C.E., and R.J. Davis. 1970. Properties of extracellular polysaccharides from Rhizobium. Soil Biol. Biochem. 2:109–117.

Clapp, C.E., J.E. Dawson, and M.H.B. Hayes. 1979. Composition and properties of a purified polysaccharide isolated from an organic soil. p. 153–167. *In* K.M. Schallinger (ed.) Proc. Int. Symp. Peat in Agric. and Hort. Spec. Publ. 205. Agric. Res. Org., Bet Dagan, Israel.

Clapp, C.E., and W.W. Emerson. 1965. The effect of periodate oxidation on the strength of soil crumbs: I. Qualitative studies. II. Quantitative studies. Soil Sci. Soc. Am. Proc. 29:127–134.

Clapp, C.E., and W.W. Emerson. 1972. Reactions between Ca-montmorillonite and polysaccharides. Soil Sci. 114:210–216.

Clapp, C.E., W.W. Emerson, and A.E. Olness. 1989. Sizes and shapes of humic substances by viscosity measurements. p. 497–514. *In* M.H.B. Hayes et al. (ed.) Humic substances II. In search of structure. John Wiley and Sons, New York.

Clapp, C.E., R. Harrison, and M.H.B. Hayes. 1991. Interactions between organic macro-molecules and soil inorganic colloids and soils. p. 409–468. *In* G.H. Bolt et al. (ed.) Interactions at the soil colloid–soil solution interface. Kluwer, Dordrecht, the Netherlands.

Clapp, C.E., and M.H.B. Hayes. 1996. Isolation of humic substances from an agricultural soil using a sequential and exhaustive extraction process. p. 3–11. *In* C.E. Clapp et al. (ed.) Humic substances and organic matter in soil and water environments: Characterization, transformations, and interactions. IHSS, University of Minnesota, St. Paul.

Clapp, C.E., and M.H.B. Hayes. 1999. Characterization of humic substances isolated from clay- and silt-sized fractions of a corn residue-amended agricultural soil. Soil Sci. 164:899–913.

Clapp, C.E., M.H.B. Hayes, and U. Mingelgrin. 2001. Measurements of sorption–desorption and isotherm analyses. p. 205–240. *In* C.E. Clapp et al. (ed.) Humic substances and chemical contaminants. SSSSA, Madison, WI.

Clapp, C.E., M.H.B. Hayes, and R.S. Swift. 1993. Isolation, fractionation, functionalities, and concepts of structures of soil organic macro molecules. p. 31–69. *In* A.J. Beck et al. (ed.) Organic substances in soil and water: Natural constituents and their influence on contaminant behaviour. Roy. Soc. Chem., Cambridge, UK.

Clapp, C.E., M.F. Layese, M.H.B. Hayes, D.R. Huggins, and R.R. Allmaras. 1997. Natural abundances of ^{13}C in soils and waters. p. 158–175. *In* M.H.B. Hayes and W.S. Wilson (ed.) Humic substances, peats and sludges: Health and environmental aspects. Roy. Soc. Chem., Cambridge, UK.

Clapp, C.E., A.E. Olness, and D.J. Hoffmann. 1968. Adsorption studies of a dextran by montmorillonite. Trans. 9th Int. Congr. Soil Sci. (Adelaide) 1:627–637.

Clar, E. 1964. Polycyclic hydrocarbons. Vol 1. Academic Press, New York.

Clifford, D.J., D.M. Carson, D.E. McKinney, J.M. Bortiatynski, and P.G. Hatcher. 1995. A new rapid technique for the characterization of lignin in vascular plants: Thermochemolysis with tetramethylammonium hydroxide (TMAH). Org. Geochem. 23:169–175.

Conference of the Parties. 1997. Kyoto Protocol to the Framework Convention on Climate Change. Kyoto, Japan. Dec. 1997. United Nations, New York.

Cookson, J.T., Jr. 1978. Adsorption mechanisms: The chemistry of organic adsorption on activated carbon. p. 241–279. *In* P.N. Cheremisinoff and F. Ellerbusch (ed.) Carbon adsorption handbook. Ann Arbor Science, Ann Arbor, MI.

Cowie, G.L., J.I. Hedges, and S.E. Calvert. 1992. Sources and relative reactivities of amino acids, neutral sugars, and lignin in an intermittently anoxic marine environment. Geochim. Cosmochim. Acta 56:1963–1978.

Craggs, J.D. 1972. Sodium sulphide reactions with humic acid and model compounds. Ph.D. diss. University of Birmingham, UK.

Craggs, J.D., M.H.B. Hayes, and M. Stacey. 1974. Sodium sulphide reactions with humic acid and model compounds. Trans. 10th Int. Congr. Soil Sci. (Moscow) 2:318–324.

Cranwell, P.A., and R.D. Haworth. 1975. The chemical nature of humic acids. p. 13–18. *In* D. Povoledo and K. Gotterman (ed.) Proceedings, International Meeting of Humic Substances, Nieuwesluis. PUDOC, Wageningen, the Netherlands.

Dack, M.R.J. 1976. The influence of solvent on chemical reactivity. p. 95–157. *In* M.R.J. Dack (ed.) Solutions and solubilities. Techniques of chemistry. Vol. VIII. Part II. Wiley, New York.

Dale, V.H., R.A. Houghton, A. Grainger, A.E. Lugo, and S. Brown. 1993. Emissions of greenhouse gases from tropical deforestation and subsequent uses of the land. p. 215–260. *In* Sustainable agriculture and the environment in the humid tropics. Nat. Res. Coun., Nat. Acad. Press, Washington, DC.

Dawson, J.E., C.E. Clapp, and M.H.B. Hayes. 1979. Studies of the physical and physico-chemical properties of extracts from organic soils: I. Electrophoretic characteristics of water soluble extracts. p. 278–281. *In* K.M. Schallinger (ed.) Proc. Int. Symp. Peat in Agric. and Hort. Spec. Publ. 205. Agric. Res. Org., Bet Dagan, Israel.

Deines, P. 1980. The isotopic composition of reduced organic carbon. p. 329–406. *In* P. Fritz and J.Ch. Fontes (ed.) Handbook of environmental isotope geochemistry. Vol. 1. Elsevier, New York.

Del Rio, J., D.E. McKinney, H. Knicker, M.A. Nanny, R.D. Minard, and P.G. Hatcher. 1998. Structural characterization of bio- and geo-macromolecules by off-line thermochemolysis with tetramethylammoniun hydroxide. J. Chromatogr. A 823:433–438.

Delwiche, C.C., and P.L. Steyn. 1970. Nitrogen fractionation in soils and microbial reactions. Environ. Sci. Technol. 4:929–935.

Dence, C.W. 1971. Halogenation and nitration. p. 373–402. *In* K.V. Sarkanen and C.H. Ludwig (ed.) Lignins. Wiley, New York.

De Nobili, M., and Y. Chen. 1999. Size exclusion chromatography of humic substances: Limits, perspectives and prospectives. Soil Sci. 164:825–833.

De Nobili, M., E. Gjessing, and P. Sequi. 1989. Sizes and shapes of humic substances by gel chromatography. p. 561–591. *In* M.H.B. Hayes et al. (ed.) Humic substances II: In search of structure. John Wiley and Sons, Chichester, UK.

Derenne, S., and C. Largeau. 2001. A review of some important families of refractory macromolecules: Composition, origin, and fate in soils and sediments. Soil Sci. 166:833–847.

Derome, A.E. 1987. Modern NMR techniques for chemistry research. Pergamon Press, Oxford, UK.

de Saussure, Th. 1804. Recherches chemiques sur la végétation. Paris, Ann. 12:162.

Dick, W.A., R.L. Blevins, W.W. Frye, S.E. Peters, D.R. Christensen, F.J. Pierce, and M.L. Vitosk. 1998. Impacts of agricultural management practices on C sequestration in forest-derived soils of the eastern Corn Belt. Soil Tillage Res. 47:235–244.

Donigian, A.S., Jr., T.D. Barnwell, R.B. Jackson, A.S. Patwardham, K.B. Weinreich, A.L. Rowell, R.V. Chinnaswanny, and C.V. Cole. 1994. Assessment of alternative management practices and policies affecting soil carbon in agroecosystems of the central United States. EPA/600/R-94/064. USEPA, Athens, GA.

Dormaar, J.F. 1969. Reductive cleavage of humic acids of chernozemic soils. Plant Soil 31:182–184.

Dryden, J.G.C. 1952. Solvent power for coals at room temperature. Chemical and physical factors. Chem. Ind. (London), p. 502–508.

Durrum, E.L. 1951. Continuous electrophoresis and ionophoresis on filter paper. J. Am. Chem. Soc. 73:4875–4880.

Duvigneaud, P. 1972. Morale et ècologie. La Pensée et Les Hommes 7:286–303.

Duxbury, J.M. 1989. Studies of the molecular size and charge of humic stubstances by electrophoresis. p. 593–620. *In* M.H.B. Hayes et al. (ed.) Humic substances II: In search of structure. John Wiley and Sons, Chichester, UK.

Eberle, S.H., and K.H. Schweer. 1974. Bestimmung von Huminsaure und Ligninsulfonsaure im Wasser durch Flussig-Flussingextraktion. Vom Wasser 41:27–44.

Einstein, A. 1906. Eine neue Bestimmung der Molekuldimensionen. Ann. Phys. 19:289–306.

Ellis, A.J., and W. Giggenback. 1971. Hydrogen sulphide ionization and sulphur hydrolysis in high temperature solution. Geochim. Cosmochim. Acta 35:247–260.

Engebretson, R.R., T. Amos, and R. Von Wandruszka. 1996. Quantitative approach to humic acid associations. Environ. Sci. Technol. 30:990–997.

Engebretson, R.R., and R. von Wandruszka. 1994. Micro-organization of dissolved humic acids. Environ. Sci. Technol. 28:1934–1941.

Engebretson, R.R., and R. von Wandruszka. 1997. The effect of molecular size on humic acid associations. Org. Geochem. 26:759–767.

Ertel, J.R., and J.I. Hedges. 1984. The lignin component of humic substances: Distribution among soil and sedimentary humic, fulvic, and base-insoluble fractions. Geochim. Cosmochim. Acta 48:2065–2074.

Ertel, J.R., J.I. Hedges, A.H. Devol, and E.R. Jeffrey. 1986. Dissolved humic substances of the Amazon River System. Limnol. Oceanogr. 31:739–754.

Eswaran, H., E. van den Berg, and P. Reich. 1993. Organic carbon in soils of the world. Soil Sci. Soc. Am. J. 57:192–194.

Fagbenro, J.M., M.H.B. Hayes, I.A. Law, and A.A. Agboola. 1985. Extraction of soil organic matter and humic substances from two Nigerian soils using three solvent mixtures. p. 22–26. *In* M.H.B. Hayes and R.S. Swift (ed.) Volunteered Papers, 2nd Int. Conf. IHSS, University of Birmingham, Birmingham, UK.

Fan, T.W.-M., R.M. Higashi, and A.N. Lane. 2000. Chemical characterization of a chelator-treated soil humate by solution-state multinuclear two-dimensional NMR with FTIR and pyrolysis-GC-MS. Environ. Sci. Technol. 34:636–646.

Farmer, V.C., and R.I. Morrison. 1960. Chemical and infrared studies on phragmites peat and its humic acid. Sci. Proc. R. Dublin Soc. Ser. A. 1:85–104.

Filley, T.R., R.D. Minard, and P.G. Hatcher. 1999. Tetramethylammonium hydroxide (TMAH) thermochemolysis: Proposed mechanisms based upon the application of ^{13}C-labeled TMAH to a synthetic model lignin dimer. Org. Geochem. 30:607–621.

Finch, P., M.H.B. Hayes, and M. Stacy. 1967. Studies of soil polysaccharides and on their interactions with clay preparations. p. 19–32. *In* Int. Soc. Soil Sci. Trans., Comm. IV and VI, Aberdeen. 1966. Aberdeen Univ. Press, Aberdeen, UK.

Flaig, W., and H. Beutelspacher. 1954. Physikalische Chemie der Huminsären. Landbouwkd. Tijdschr. 66:306–336.

Flaig, W., and H. Beutelspacher. 1968. Investigations of humic acids with the analytical ultracentrifuge. p. 23–30. *In* Isotopes and radiation in soil organic matter studies. IAEA, Vienna.

Flaig, W., H. Beutelspacher, and E. Reitz. 1975. Chemical composition and physical properties of humic substances. p. 1–211. *In* J.E. Gieseking (ed.) Soil components: Vol. 1. Organic components. Springer-Verlag, New York.

Flory, P.J. 1953. Principles of polymer chemistry. Cornell University Press, Ithaca, NY.

Franks, F. 1975. The hydrophobic interaction. p. 1–94. *In* F. Franks (ed.) Water, a comprehensive treatise. Aquous solutions of amphiphiles and macromolecules. Vol. 4. Plenum, New York.

Franks, F. 1983. Water. Roy. Soc. Chem., London.

Geoghegan, M.J., and R.C. Brian. 1946. Influence of bacterial polysaccharides on aggregate formation in soils. Nature (London) 158:837.

Geoghegan, M.J., and R.C. Brian. 1948. Aggregate formation in soil. 2. Influence of various carbohydrates and proteins in aggregation of soil particles. Biochem. J. 43:14.

Ghosh, K., and M. Schnitzer. 1979. UV and visible adsorption spectroscopic investigations in relation to macromolecular characteristics in humic substances. J. Soil Sci. 30:735–745.

Gierer, J., and I. Noren. 1962. Reactions of lignin on sulfate digestion. II. Model experiments on the cleavage of aryl alkyl ethers by alkali. Acta Chem. Scand. A 16:1713–1729.

Giles, C.H., D. Smith, and A. Huitson. 1974. A general treatment and classification of the solute adsorption isotherm. I. Theoretical. J. Colloid Interface Sci. 47:755–765.

Glaser, B., E. Ballashov, L. Haumaier, G. Guggenberger, and W. Zech. 2000. Black carbon in density fractions of anthropogenic soils of the Brazilian Amazon region. Org. Geochem. 31:669–678.

Glaser, B., L. Haumaier, G. Guggenberger, and W. Zech. 1998. Black carbon in soils: The use of benzenecarboxylic acids as specific markers. Org. Geochem. 29:811–819.

Glover, C.A. 1975. Absolute colligative property measurements. p. 79–159. *In* P.E. Slade, Jr. (ed.) Polymer molecular weights. Part I. Marcel Dekker, New York.

Goldberg, E.D. 1985. Black carbon in the environment. John Wiley and Sons, New York.

Goldstein, I.S. 1991. Overview of the chemical composition of wood. p. 1–5. *In* M. Levin and I.S. Goldstein (ed.) Wood structure and composition. Dekker, New York.

Goni, M.A., B. Nelson, R.A. Blanchette, and J.I. Hedges. 1993. Fungal degradation of wood lignins: Geochemical perspectives from CuO-derived phenolic dimers and monomers. Geochim. Cosmochim. Acta 57:3985–4002.

Gorin, P.A.J., and E. Barreto-Bergter. 1983. The chemistry of polysaccharides of fungi and lichens. p. 365–409. *In* G.O. Aspinall (ed.) The polysaccharides. Vol. 2. Academic Press, Orlando.

Greenland, D.J., G.R. Lindstrom, and J.P. Quirk. 1961. Role of polysaccharides in stabilisation of natural soil aggregates. Nature (London) 191:1283–1284.

Greenland, D.J., G.R. Lindstrom, and J.P. Quirk. 1962. Organic materials which stabilize natural soil aggregates. Soil Sci. Soc. Am. Proc. 26:366–371.

Greenland, D.J., and J.M. Oades. 1975. Saccharides. p. 213–257. *In* J.E. Gieseking (ed.) Soil components. Vol 1: Organic components. Springer-Verlag, Berlin.

Griffith, S.M., and M. Schnitzer. 1976. The alkaline cupric oxide oxidation of humic and fulvic acids extracted from tropical volcanic soils. Soil Sci. 122:191–201.

Griffith, S.M., and M. Schnitzer. 1989. Oxidative degradation of soil humic substances. p. 69–98. *In* M.H.B. Hayes et al. (ed.) Humic substances II: In search of structure. John Wiley and Sons, Chichester, UK.

Guggenberger, G., and W. Zech. 1994. Dissolved organic carbon in forest floor leachates: Simple degradation products or humic substances? Sci. Total Environ. 152:37–47.

Gustaffson, O., T.D. Bucheli, Z. Kukulska, M. Anderson, C. Largeau, J.-N. Rouzaud, C.M. Reddy, and T.I. Eglinton. 2001. Evaluation of a protocol for the quantification of black carbon in sediments, soils, and aquatic particles. Global Biogeochem. Cycles 15:881–890.

Gustaffson, O., F. Haghseta, C. Chan, J. Macfarlane, and P.M. Gschwend. 1997. Quantification of the dilute sedimentary soot phase: Implications for PAH speciation and bioavailability. Environ. Sci. Technol. 31:203–209.

Haiber, S., H. Herzog, P. Burba, B. Gosciniak, and J. Lambert. 2001. Two-dimensional NMR studies of size fractionated Suwannee river fulvic and humic acid reference. Environ. Sci. Technol. 35:4289–4294.

Haider, K., and J.P. Martin. 1967. Synthesis and transformation of phenolic compounds by *Epicoccum nigrun* in relation to humic acid formation. Soil Sci. Soc. Am. Proc. 31:766–772.

Haider, K., and J.P. Martin. 1970. Humic acid-type phenolic polymers from *Aspergillus sydowi* culture medium, *Stachybotrys* spp. cells and autooxidized phenol mixtures. Soil Biol. Biochem. 2:145–156.

Hänninen, K.I., R. Klöcking, and B. Helbig. 1987. Synthesis and characterization of humic acid-like polymers. Sci. Total Environ. 62:201–210.

Hansen, C.M. 1967. Three-dimensional solubility parameter and solvent diffusion coefficient. Danish Technical Press, Copenhagen, Denmark.

Hansen, E.H., and M. Schnitzer. 1966. The alkaline permanganate oxidation of Danish alluvial organic matter. Soil Sci. Soc. Am. Proc. 30:745–748.

Hansen, E.H., and M. Schnitzer. 1969a. Zinc dust distillation and fusion of a soil humic and fulvic acid. Soil Sci. Soc. Am. Proc. 33:29–36.

Hansen, E.H., and M. Schnitzer. 1969b. Zinc dust distillation of soil humic compounds. Fuel 48:41–46.

Harrison, R. 1982. A study of some montmorillonite-organic complexes. Ph.D. diss. University of Birmingham, UK.

Hatcher, P.G., K.J. Dria, S. Kim, and S.W. Frazier. 2001. Modern analytical studies of humic substances. Soil Sci. 166:770–794.

Hatcher, P.G., M.A. Nanny, R.D. Minard, S.C. Dible, and D.M. Carson. 1995. Comparison of two thermochemolytic methods for the analysis of lignin in decomposing wood: The CuO oxidation method and the method of thermochemolysis with TMAH. Org. Geochem. 23:881–888.

Hatcher, P.G., and E.C. Spiker. 1988. Selective degradation of plant biomolecules. p. 59–74. *In* F.H. Frimmel and R.F. Christman (ed.) Humic substances and their role in the environment. Wiley, Chichester, UK.

Häusler, M.J., and M.H.B. Hayes. 1996. Uses of the XAD-8 resin and acidified dimethylsulfoxide in studies of humic acids. p. 25–32. *In* C.E. Clapp et al. (ed.) Humic substances and organic matter in soil and water environments: Characterization, transformations and interactions. IHSS, University of Minnesota, St. Paul.

Haworth, R.D. 1971. The chemical nature of humic acids. Soil Sci. 111:71–79.

Haworth, W.N., F.W. Pinkard, and M. Stacey. 1946. Function of bacterial polysaccharides in soil. Nature (London) 158:836–837.

Hayase, K., and H. Tsubota. 1983. Sedimentary humic acid and fulvic acid as surface active substances. Geochim. Cosmochim. Acta 47:947–952.

Hayes, M.H.B. 1960. Subsidence and humification in peats. Ph.D. diss. The Ohio State University, Columbus.

Hayes, M.H.B. 1985. Extraction of humic substances from soils. p. 329–362. *In* G. Aiken et al. (ed.) Humic substances in soil, sediment, and water: Geochemstry, isolation and characterization. Wiley, New York.

Hayes, M.H.B. 1997. Emerging concepts of the composition and structures of humic substances. p. 3–30. *In* M.H.B. Hayes and W.S. Wilson (ed.) Humic substances, peats, and sludges: Health and environmental aspects. Roy. Soc. Chem., Cambridge, UK.

Hayes, M.H.B., and C.L. Graham. 2000. Procedures for the isolation and fractionation of humic substances. p. 91–109. *In* G. Davies and E.A. Ghabbour (ed.) Humic substances. Versatile components of plants, soils and waters. Roy. Soc. Chem., Cambridge, UK.

Hayes, M.H.B., P. MacCarthy, R.L. Malcolm, and R.S. Swift. 1989a. Humic substances II. *In* search of structure. John Wiley and Sons, Chichester, UK.

Hayes, M.H.B., P. MacCarthy, R.L. Malcolm, and R.S. Swift. 1989b. Structures of humic substances: The emergence of forms. p. 687–733. *In* M.H.B. Hayes et al. (ed.) Humic substances II: In search of structure. John Wiley and Sons,Chichester, UK.

Hayes, M.H.B., and M.R. O'Callaghan. 1989. Degradations with sodium sulfide and with phenol. p. 143–180. *In* M.H.B. Hayes et al. (ed.) Humic substances II: In search of structure. John Wiley and Sons, Chichester, UK.

Hayes, M.H.B., M. Stacey, and R.S. Swift. 1972. Degradation of humic acid in a sodium sulphide solution. Fuel 51:211–213.

Hayes, M.H.B., M. Stacey, and R.S. Swift. 1975a. Techniques for fractionating soil polysaccharides. p. 75–81. *In* Trans. 10th Int. Congr. Soil Sci. (Moscow), Suppl. Vol.

Hayes, M.H.B., and R.S. Swift. 1978. The chemistry of soil organic colloids. p. 179–320. *In* D.J. Greenland and M.H.B. Hayes (ed.) The chemistry of soil constituents. John Wiley and Sons, Chichester, UK.

Hayes, M.H.B., and R.S. Swift. 1990. Genesis, isolation, composition and structures of soil humic substances. p. 245–305. *In* M.F. DeBoodt et al. (ed.) Soil colloids and their associations in aggregates. Plenum, New York.

Hayes, M.H.B., R.S. Swift, R.E. Wardle, and J.K. Brown. 1975b. Humic materials from an organic soil: A comparison of extractants and of properties of extracts. Geoderma 13:231–245.

Hayes, T.M. 1996. Study of the humic substances from soils and waters and their interactions with anthropogenic organic chemicals. Ph.D. diss. University of Birmingham, UK.

Hayes, T.M., M.H.B. Hayes, and A.J. Simpson. 1999. Considerations of the amino nitrogen in humic substances. p. 206–227. *In* W.S. Wilson et al. (ed.) Managing risks of nitrates to humans and the environment. Roy. Soc. Chem., Cambridge, UK.

Hayes, T.M., M.H.B. Hayes, J.O. Skjemstad, R.S. Swift, and R.L. Malcolm. 1996. Isolation of humic substances from soil using aqueous extractants of different pH and XAD resins, and their characterization by ^{13}C-NMR. p. 13–24. *In* C.E. Clapp et al. (ed.) Humic substances and organic matter in soil and water environments: Characterization, transformations and interactions. IHSS, University of Minnesota, St. Paul.

Hayes, T.M., B.E. Watt, M.H.B. Hayes, C.E. Clapp, D. Scholfield, R.S. Swift, and J.O. Skjemstad. 1997. Dissolved humic substances in waters from drained and undrained grazed grassland in SW England. p. 107–120. *In* M.H.B. Hayes and W.S. Wilson (ed.) Humic substances, peats, and sludges: Health and environmental aspects. Roy. Soc. Chem., Cambridge, UK.

Hedges, J.I., W.A. Clark, and G.L. Cowie. 1988. Fluxes and reactivities of organic matter in a coastal marine bay. Limnol. Oceanogr. 33:1137–1152.

Hedges, J.I., and J.R. Ertel. 1982. Characterization of lignin by gas capillary chromatography of cupric oxide oxidation products. Anal. Chem. 54:174–178.

Hedges, J.I., and R.G. Keil. 1995. Sedimentary organic matter preservation: An assessment and speculative synthesis. Mar. Chem. 49:81–115.

Hedges, J.I., and D.C. Mann. 1979. The lignin geochemistry of marine sediments from the southern Washington coast. Geochim. Cosmochim. Acta 43:1809–1818.

Hedges, J.I., and P.L. Parker. 1976. Land-derived organic matter in surface sediments from the gulf of Mexico. Geochim. Cosmochim. Acta 40:1019–1029.

Hildebrand, J.H., J.N. Prausnitz, and R.L. Scott. 1970. Regular and related solutions. Van Nostrand-Reinhold, New York.

Hildebrand, J.H., and R.L. Scott. 1951. Solubility of non-electrolytes. 3rd ed. Reinhold, New York.

Hildebrand, J.H., and R.L. Scott. 1962. Regular solutions. Prentice Hall, Englewood Cliffs, NJ.

Himes, F.L., and C. Bloomfield. 1967. Extraction of triacontyl sterarate from a soil. Plant Soil 26:383.

Houghton, R.A. 1995. Changes in storage of terrestrial carbon since 1950. p. 45–65. *In* R. Lal et al. (ed.) Soils and global change. CRC/Lewis Publishers, Boca Raton, FL.

Huang, Y., G. Eglinton, E.R.E. Van Der Hage, J.J. Boon, R. Bol, and P. Ineson. 1998. Dissolved organic matter in grass upland soil horizons studied by analytical pyrolysis techniques. Eur. J. Soil Sci. 49:1–15.

Huggins, D.R., C.E. Clapp, R.R. Allmaras, J.A. Lamb, and M.F. Layese. 1998. Carbon dynamics in corn-soybean sequences as estimated from natural carbon-13 abundance. Soil Sci. Soc. Am. J. 62:195–203.

Hurst, H.M., and J.B. Harborne. 1967. Plant polyphenols. 16. Identification of flavonoids by reductive cleavage. Phytochemistry 6:1111–1118.

Ikan, R., Y. Rubinsztain, and Z. Aizenshtat. 1992. Chemical, isototopic, spectroscopic and geochemical aspects of natural and synthetic humic substances. Sci. Total Environ. 118:1–12.

Israelachvili, J.N. 1994. Intermolecular and surface forces. 2nd ed. Academic Press. London.

Jenkinson, D.S. 1981. The fate of plant and animal residues in soil. p. 505–561. *In* D.J. Greenland and M.H.B. Hayes (ed.) The chemistry of soil processes. John Wiley and Sons, Chichester, UK.

Jenkinson, D.S. 1990. The turnover of organic carbon and nitrogen in the soil. Philos. Trans. R. Soc. London, Ser. B. 329:361–368.

Jenkinson, D.S., and J.N. Ladd. 1981. Microbial biomass in soil: Measurement and turnover. p. 415–417. *In* E.A. Paul and J.N. Ladd (ed.) Soil biochemistry. Vol 5. Marcel Dekker, New York.

Johnson, M.G. 1995. The role of soil management in sequestration of soil carbon. p. 351–364. *In* R. Lal et al. (ed.) Soil management and the greenhouse effect. Lewis Publ., Boca Raton, FL.

Johnson, M.G., and J.S. Kern. 1991. Sequestering C in soils. A workshop to explore the potential for mitigating global climate change. USEPA Rep. 600/3–91–031, USEPA Environ. Res. Lab., Corvallis, OR.

Joll, C., A. Heitz, T. Huynh, and R.I. Kagi. 2001. Thermochemolysis reactions of tetramethylammonium hydroxide with tannins: Origins of methoxybenezenes and methoxytoluenes. *In* Abstracts, 221st ACS National Meeting, San Diego, CA. ACS, Washington, DC.

Jones, J.G., and M.A. Edington. 1968. An ecological survey of hydrocarbon-oxidizing micro-organisms. J. Gen. Microbiol. 52:381–390.

Karger, B.L., L.R. Snyder, and C. Eon. 1976. An expanded solubility parameter treatment for classification and use of chromatographic solvents and adsorbents. Parameters for dispersion, dipole and hydrogen bonding interactions. J. Chromatogr. 125:71–88.

Karger, B.L., L.R. Snyder, and C. Horvath. 1973. An introduction to separation science. Wiley-Interscience, New York.

Keene, L., and B. Lindberg. 1983. Bacterial polysaccharides. p. 287–363. *In* G.O. Aspinall (ed.) The polysaccharides. Vol. 2. Academic Press, Orlando, FL.

Keil, R.G., and J.I. Hedges. 1995. Sedimentary organic matter preservation: An assessment and speculative synthesis. Mar. Chem. 49:81–115.

Keller, R.A., B.L. Karger, and L.R. Snyder. 1971. Use of the solubility parameter in predicting chromatographic retention and eluotropic strength. p. 125–140. *In* R. Stock (ed.) Gas chromatography. Institute of Petroleum, London.

Kenworthy, I.P., and M.H.B. Hayes. 1997. Investigations of some structural properties of humic substances by fluorescence quenching. p. 39–45. *In* M.H.B. Hayes and W.S. Wilson (ed.) Humic substances, peats, and sludges. Health and environmental aspects. Roy. Soc. Chem., Cambridge, UK.

Kenyon, J., and M.C.R. Symons. 1953. The oxidation of carboxylic acids containing a tertiary carbon atom. J. Chem. Soc., p. 2129–2132.

Kern, J.S. 1994. Spatial patterns of soil organic carbon in contiguous United States. Soil Sci. Soc. Am. J. 58:439–455.

Kingery, W.L., A.J. Simpson, M.H.B. Hayes, M.A. Locke, and R.P. Hicks. 2000. The application of multidimensional NMR to the study of soil humic substances. Soil Sci. 165:483–494.

Kirkwood, J.G., J.R. Cann, and R.A. Brown. 1950. The theory of electrophoresis-convection. Biochim. Biophys. Acta 5:301–314.

Kleppe, P.J. 1970. Kraft pulping. Tappi 53:35–47.

Knicker, H. 2000a. Biogenic nitrogen in soils as revealed by solid-state carbon-13 and nitrogen-15 nuclear magnetic resonance spectroscopy. J. Environ. Qual. 29:715–723.

Knicker, H. 2000b. Double cross polarization magic angle spinning ^{15}N ^{13}C NMR spectroscopic studies for characterization of immobilized nitrogen in soils. p. 1105–1108. *In* Proc. IHSS, Toulouse, France.

Knicker, H., J.C. del Rio, P.G. Hatcher, and R.D. Murad. 2001. Identification of protein remnants in insoluble geopolymers using TMAH thermochemolysis/GC-MS. Org. Geochem. 32:397–409.

Knicker, H., and P.G. Hatcher. 2001. Sequestration of organic nitrogen in sapropel from mangrove Lake, Bermuda. Org. Geochem. 32:733–744.

Knicker, H., P.G. Hatcher, and F.J. Gonzales-Vila. 2002. Formation of heteroaromatic nitrogen after prolonged humification of vascular plant remains as revealed by nuclear resonance spectroscopy. J. Environ. Qual. 31:444–449.

Knicker, H., and H.D. Lüdemann. 1995. N-15 and C-13 CPMAS and solution NMR-studies of N-15 enriched plant-material during 600 days of microbial degradation. Org. Geochem. 23:329–341.

Knicker, H., and H.-D Lüdemann. 1996. CPMAS ^{13}C- and ^{15}N-NMR studies on the stabilization of inorganic nitrogen during composting. p. 425–431. *In* C.E. Clapp (ed.) Humic substances and organic matter in soil and water environments: Characterization, transformations and interactions. IHSS, University of Minnesota, St. Paul.

Kögel-Knaber, I. 1997. 13-C and 15-N NMR spectroscopy as a tool in soil organic matter studies. Geoderma 80:243–270.

Kögel-Knaber, I. 2000. Analytical approaches for characterizing soil organic matter. Org. Geochem. 31:1023–1028.

Kohl, S.D., P.J. Toscano, W. Hou, and J.A. Rice. 2000. Solid-state ^{19}F NMR investigation of hexofluorobenzene sorption to soil organic matter. Environ. Sci. Technol. 34:204–210.

Kononova, M.M. 1966. Soil organic matter: Its nature, its role in soil formation and in soil fertility. 2nd English ed. Pergamon Press, Oxford, UK.

Kuhlbusch, T.A.J. 1995. Method for determining black carbon in residues of vegetation fires. Environ. Sci. Technol. 29:2695–2702.

Kuhlbusch, T.A.J., and P.J. Crutzen. 1995. Toward a global estimate of black carbon in residues of vegetation fires representing a sink of atmospheric CO_2 and a source of O_2. Global Biogeochem. Cycles 9:491–501.

Kumada, K., and Y. Kawamura. 1968. On the fractionation of humic acids by a fractional precipitation technique. Soil Sci. Plant Nutr. 14:198–200.

Kyuma, K. 1964. A fractional precipitation technique applied to soil humic substances. Soil Sci. Plant Nutr. 10:33–35.

Lai, Y.Z., and K.V. Sarkanen. 1971. Isolation and structural studies. p. 165–240. *In* K.V. Sarkanen and C.H. Ludwig (ed.) Lignins. Wiley, New York.

Lal, R., J.M. Kimble, R.F. Follett, and C.V. Cole. 1998. The potential of the United States cropland to sequester carbon and to mitigate the greenhouse effect. Ann Arbor Press, Chelsea, MI.

Lal, R., J.M. Kimble, E. Levine, and B.A. Stewart. (ed.) 1995. Soil management and greenhouse effect. Lewis Publishers, Boca Raton, FL.

Largeau, C., F. Casadevall, A. Kadouri, and P. Metgger. 1984. Formation of *Botryococcus*-derived kerogens. Comparative study of immature Torbanite and of the extant alga *Botryococcus braunii.* Org. Geochem. 6:327–332.

Laue, T.M., and D.G. Rodhes. 1990. Determination of size, molecular weight and presence of subunits. p. 566–587. *In* M.P. Deutscher (ed.) Guide to protein purification. Methods of Enzymology. Vol. 182. Academic Press, San Diego, CA.

Law, I.A. 1988. The extraction, fractionation and characterization of humic substances, and their sorption behaviour towards metal cations. Ph.D. diss. University of Birmingham, UK.

Law, I.A., M.H.B. Hayes, and J.J. Tuck. 1985. Extraction of humic substances from soils using acidified dimethylsulphoxide. p. 18–21. *In* M.H.B. Hayes and R.S. Swift (ed.) Volunteered Papers, 2nd Int. Conf. IHSS, Birmingham. 1984. IHSS, Univ. of Minnesota, St. Paul.

Layese, M.F., C.E. Clapp, R.R. Allmaras, D.R. Linden, S.M. Copeland, J.A.E. Molina, and R.H. Dowdy. 2002. Current and relic carbon using natural abundance carbon-13. Soil Sci. 167:315–326.

Lee, D.G. 1969. Hydrocarbon oxidation using transition metal compounds. p. 1–51. *In* R.L. Augustine (ed.) Oxidation: Techniques and applications in organic synthesis. Vol. 1. Marcel Dekker, New York.

Leinweber, P., and H.-R. Schulten. 1999. Advances in analytical pyrolysis of soil organic matter. J. Anal. Appl. Pyrolysis 49:359–383.

Létolle, R. 1980. Nitrogen-15 in the natural environment. p. 407–433. *In* P. Fritz and J. Ch. Fontes (ed.) Handbook of environmental isotope geochemistry, Vol. 1. Elsevier, Amsterdam.

Lindqvist, I. 1967. Adsorption effects in gel filtration of humic acid. Acta Chem. Scand. A 21:2564–2566.

Liski, J., H. Ilvesniemi, A. Makela, and M. Starr. 1998. Model analysis of the effects of soil age, fires and harvesting on the carbon storage of boreal forest soils. Eur. J. Soil Sci. 49:406–416.

MacCarthy, P. 2001. The principles of humic substances. Soil Sci. 166:738–751.

MacCarthy, P., and S. O'Cinneide. 1974. Fulvic acid. II. Interaction with metal ions. J. Soil Sci. 25:429–437.

MacCarthy, P., and J.A. Rice. 1985. Spectroscopic methods (other than NMR) for determining functionality in humic substances. p. 527–559. *In* G.R. Aiken et al. (ed.) Humic substances in soil, sediment and water. John Wiley and Sons, New York.

Macomber, R.S. 1998. A complete introduction to modern NMR spectroscopy. John Wiley and Sons, New York.

Maillard, L.C. 1912. Action des acides amines sur les sucre; formation des melanoidinens par voie methodique. C.R. Acad. Sci. Paris 154:66–68.

Maillard, L.C. 1916. Synthese des materies humiques par action des acides amines sur les sucres reductours. Ann. Chim. 5:258–317.

Maillard, L.C. 1917. Identite des materies humiques de syntheses avec les materies humiques naturelles. Ann. Chim. 7:113–152.

Malcolm, R.L. 1989. Application of solid-state ^{13}C N.M.R. spectroscopy to geochemical studies of humic substances. p. 339–372. *In* M.H.B. Hayes et al. (ed.) Humic substances II. In search of structure. John Wiley and Sons, Chichester, UK.

Malcolm, R.L., and P. MacCarthy. 1992. Quantitative evaluation of XAD-8 and XAD-4 resins used in tandem for removing organic solutes from water. Environ. Int. 18:597–607.

Mann, L.K. 1985. A regional comparison of C in cultivated and uncultivated Alfisols and Mollisols in central United States. Geoderma 36:241–253.

Mann, L.K. 1986. Changes in soil C storage after cultivation. Soil Sci. 142:279–288.

Mao, J.-D., and K. Schmidt-Rohr. 2003. Recoupled long-range C–H dipolar dephasing in solid-state NMR, and its use for spectral selection of fused aromatic rings. J. Magn. Reson. 162:217–227.

March, J. 1985. Advanced organic chemistry: Reactions, mechanisms, and structure. 3rd ed. John Wiley and Sons, New York.

Martin, A., J. Mariotti, J. Balesdent, P. Lavelle, and R. Vuattoux. 1989. Estimate of organic matter turnover rate in a savanna soil by ^{13}C natural abundance measurements. Soil Biol. Biochem. 22:517–523.

Martin, D., and H.G. Hauthal. 1975. Dimethyl sulphoxide. E.S. Halberstadt (transl.). Van Nostrand-Reinhold, New York.

Martin, J.P. 1945. Microorganisms and soil aggregation. I. Origin and nature of some of the aggregating substances. Soil Sci. 59:163–174.

Martin, J.P. 1946. Microorganisms and soil aggregation. II. Influence of bacterial polysaccharides on soil structure. Soil Sci. 61:157–166.

Martin, J.P., and K. Haider. 1969. Phenolic polymers of *Stachybotrys atra*, *Stachybotrys chartarum* and *Epicoccum nigrum* in relation to humic acid formation. Soil Sci. 107:260–270.

Martin, J.P., K. Haider, and E. Bondietti. 1973. Properties of model humic acids synthesized by phenoloxidase and autooxidation of phenols and other compounds formed by soil fungi. p. 171–185. *In* Proc. of Internat. Meeting on Humic Substances (Nieuwersluis). PUDOC, Wageningen, the Netherlands.

Martin, J.P., K. Haider, and C. Saiz-Jimenez. 1974. Sodium amalgam reductive degradation of fungal and model phenolic polymers, soil humic acids and simple phenolic compounds. Soil Sci. Soc. Am. Proc. 38:760–765.

Martin, J.P., K. Haider, and D. Wolf. 1972. Synthesis of phenols and phenolic polymers by *Hendersonula toruloidea* in relation to humic acid formation. Soil Sci. Soc. Am. Proc. 36:311–315.

Martin, J.P., S.J. Richards, and K. Haider. 1967. Properties and decomposition and binding action in soil "humic acid" synthesized by *Epicoccum nigrum*. Soil Sci. Soc. Am. Proc. 31:657–662.

Marton, J. 1971. Reactions in alkaline pulping. p. 639–694. *In* K.V. Sarkanen and C.H. Ludwig (ed.) Lignins. John Wiley and Sons, New York.

Massiello, C.A., and E.R.M. Druffel. 1998. Black carbon in deep-sea sediments. Science 280:1911–1913.

Matsuda, K., and M. Schnitzer. 1972. The permanganate oxidation of humic acids extracted from acid soils. Soil Sci. 114:185–193.

Matsui, Y., and K. Kumada. 1977. Aromatic constituents of humic acids released by Na-amalgam reductive cleavage, KOH fusion and zinc dust fusion. Soil Sci. Plant Nutr. 23:491–501.

McKinney, D.E., J.M. Bortiatynski, D.M. Carson, D.J. Clifford, J.W. de Leeuw, and P.G. Hatcher. 1996. Tetramethylammonium hydroxyde (TMAH) thermochemolysis of the aliphatic biopolymer cutan: Insights into the chemical structure. Org. Geochem. 24:641–650.

McLaren, A.D. 1954. The adsorption and reactions of enzymes and proteins on kaolinite. J. Phys. Chem. 58:129–137.

McLaren, A.D., G.H. Peterson, and I. Barshad. 1958. The adsorption reactions of enzymes and proteins on clay minerals. IV. Kaolinite and montmorillonite. Soil Sci. Soc. Am. Proc. 22:239–244.

McLaren, A.D., A.H. Puktie, and I. Barshad. 1975. Isolation of humus with enzymatic activity from soil. Soil Sci. 119:178–180.

McNeil, M., A.G. Darvill, and P. Albersheim. 1979. The structural polymers of primary cell walls of dicots. Org. Nat. Prod. 37:191–249.

Mehta, N.C., P. Dubach, and H. Deuel. 1961. Carbohydrates in the soil. Adv. Carbohydr. Chem. 16:335–355.

Mehta, N.C., H. Streuli, M. Müller, and H. Deuel. 1960. Role of polysaccharides in soil aggregation. J. Sci. Food Agric. 11:40–47.

Meinschein, W.G., and G.S. Kenny. 1957. Analyses of a chromatographic fraction of organic extracts of soils. Anal. Chem. 29:1153–1159.

Miano, T.M., A. Piccolo, G. Celano, and N. Senesi. 1992. Infrared and fluorescence spectroscopy of glyphosate-humic acid complexes. Sci. Total Environ. 123:83–92.

Mitchell, P.C.H. 1967. The chemistry of some hydrodesulphurisation catalysts containing molybdenum. Climax Molybdenum Co., London.

Moavad, H., V.S. Guzev, I.P. Babyeva, and D.G. Zuyagintsev. 1974. Adsorption of the extracellular polysaccharide of the yeast *Lipomyces lipofer* on kaolinite. Pochvovedenie 11:79–84.

Mopper, K., and E.T. Degens. 1979. Organic carbon in the ocean: Nature and cycling. p. 293–316. *In* B. Bolin et al. (ed.) The global carbon cycle. John Wiley and Sons, Chichester, UK.

Morris, K.F., and C.S. Johnson. 1993. Resolution of discrete and continuous molecular size distributions by means of diffusion-ordered 2D NMR spectroscopy. J. Am. Chem. Soc. 115:4291–4299.

Nabuurs, G.J., A.J. Dolman, E. Verkaik, A.P. Whitmore, W.P. Deamen, O. Denema, P. Kabat, and G.M.J. Mobren. 1999. Resolving issues on terrestrial biosphere sinks in the Kyoto Protocol. Dutch National Research Program on Global Air Pollution and Climate Change, Bilthoven, The Netherlands.

Neyroud, J.A., and M. Schnitzer. 1974. The exhaustive alkaline cupric oxide oxidation of humic and fulvic acid. Soil Sci. Soc. Am. Proc. 38:907–913.

Nip, J., J.W. de Leeuw, P.A. Schenck, W. Winding, H.L.C. Meuzelaar, and J.C. Crelling. 1989. A flash pyrolysis and petrographic study of cutinite from the Indiana paper coal. Geochim. Cosmochim. Acta 53:671–683.

Nip, M., E.W. Tegelaer, H. Brinkhuis, J.W. de Leeuw, P.A. Schenck, and P.J. Holloway. 1986. Analysis of modern and fossil plant cuticles by Curie Point pyrolysis-gas chromatography-mass spectrometry: Recognition of a new highly aliphatic and resistant biopolymer. Org. Geochem. 10:769–778.

Nutsubidze, N.N., S. Sarkanen, E.L. Schmidt, and S. Shashikanth. 1998. Consecutive polymerization and depolymerization of Kraft lignin by *Trametes cingulata.* Phytochemistry 49:1203–1212.

Oades, J.M. 1967. Carbohydrates in some Australian soils. Aust. J. Soils Res. 5:103–115.

Oades, J.M. 1984. Soil organic matter and structural stability: Mechanisms and implications for management. Plant Soil 76:319–337.

Oades, J.M. 1990. Associations of colloids in soil aggregates. p. 463–483. *In* M.F. De Boodt et al. (ed.) Soil colloids and their associations in aggregates. Plenum Press, New York.

O'Callaghan, M.R. 1980. Some studies in soil chemistry. Ph.D. diss. University of Birmingham, UK.

Olk, D.C., G. Brunetti, and N. Senesi. 2000. Decrease in humification of organic matter with intensified lowland rice cropping: A wet chemical and spectroscopic investigation. Soil Sci. Soc. Am. J. 64:1337–1347.

Olk, D.C., K.G. Cassman, E.W. Randall, P. Kinchesh, L.J. Sanger, and J.M. Anderson. 1996. Changes in chemical properties of organic matter with intensified rice cropping in tropical lowland soil. Eur. J. Soil Sci. 47:293–303.

Olness, A.E., and C.E. Clapp. 1973. Occurrence of collapsed and expanded crystals in montmorillonite-dextran complexes. Clays Clay Miner. 21:289–293.

Olness, A.E., and C.E. Clapp. 1975. Influence of polysaccharide structure on dextran adsorption by montmorillonite. Soil Biol. Biochem. 7:113–118.

Page, E.R. 1980. Cellulose xanthate as a soil conditioner: Laboratory experiments. J. Sci. Food Agric. 31:1–6.

Pagliai, H., G. Guidi, and G. Petruzzelli. 1979. Effect of molecular weight on dextran-soil interactions. p. 175–180. *In* W.W. Emerson et al. (ed.) Modifications of soil structure. Wiley, New York.

Painter, T.J. 1983. Algal polysaccharides. p. 195–285. *In* G.O. Aspinall (ed.) The polysaccharides. Vol. 2. Academic Press, Orlando, FL.

Parfitt, R.L. 1972. Adsorption of charged sugars by montmorillonite. Soil Sci. 113:417–421.

Parfitt, R.L., and D.J. Greenland. 1970. Adsorption of polysaccharides by montmorillonite. Soil Sci. Soc. Am. Proc. 34:862–866.

Parker, A.J. 1962. The effects of solvation on the properties of anions in dipolar aprotic solvents. Q. Rev. Chem. Soc. 16:163–187.

Parsons, J.W. 1989. Hydrolytic degradation of humic substances. p. 99–120. *In* M.H.B. Hayes et al. (ed.) Humic substances II: In search of structure. John Wiley and Sons, Chichester, UK.

Perminova, I.V. 1999. Size exclusion chromatography of humic substances: Complexities of data interpretation attributable to non-size exclusion effects. Soil Sci. 164:834–840.

Piccolo, A. 2001. The supramolecular structure of humic substances. Soil Sci. 166:810–832.

Piccolo, A., P. Conte, and A. Cozzolino. 1999. Effects of mineral and monocarboxylic acids on the molecular association of dissolved humic substances. Eur. J. Soil Sci. 50:687–694.

Piccolo, A., P. Conte, A. Cozzolino, and R. Spaccini. 2001. p. 89–118. *In* C.E. Clapp et al. (ed.) Humic substances and chemical contaminants. SSSA, Madison, WI.

Piccolo, A., S. Nardi, and G. Concheri. 1996. Macromolecular changes in humic substances induced by interaction with organic acids. Eur. J. Soil Sci. 47:319–328.

Pimentel, G.C., and A.L. McClellan. 1960. The hydrogen bond. Freeman, San Francisco, CA.

Pines, A., M.G. Gibby, and J.S. Waugh. 1973. Proton-enhanced NMR of dilute spins in solids. J. Chem. Phys. 59:569–590.

Ping, C.L., G.J. Michaelson, and R.L. Malcolm. 1995. Fractionation and carbon balance of soil organic matter in selected cryic soils in Alaska. p. 307–314. *In* R. Lal et al. (ed.) Soils and global change. CRC Lewis, Boca Raton, FL.

Piper, T.J., and A.M. Posner. 1972a. Sodium amalgam reduction of humic acid—I. Evaluation of the method. Soil Biol. Biochem. 4:513–523.

Piper, T.J., and A.M. Posner. 1972b. Sodium amalgam reduction of humic acid—II. Application of the method. Soil Biol. Biochem. 4:525–531.

Piret, E.L., R.G. White, H.C. Walther, Jr., and A.J. Madden, Jr. 1960. Some physicochemical properties of peat humic acids. Sci. Proc. R. Dublin Soc., Ser. 1A, p. 69–79.

Poirier, N., S. Derenne, J. Balesdent, J.-N. Rouzaud, A. Mariotti, and C. Largeau. 2002. Abundance and composition of the refractory organic fraction of an ancient, tropical soil (Pointe Noire, Congo). Org. Geochem. 33:383–391.

Poirier, N., S. Derenne, J.-N. Rouzaud, C. Largeau, A. Mariotti, J. Balesdent, and J. Maquet. 2000. Chemical structure and sources of the macromolecular, resistant, organic fraction isolated from a forest soil (Lacadée, south-west France). Org. Geochem. 31:813–827.

Posner, A.M., and J.M. Creeth. 1972. A study of humic acid by equilibrium ultracentrifugation. J. Soil Sci. 23:333–341.

Post, W.M., T.H. Peng, W.R. Emanuel, A.W. King, V.H. Dale, and D.L. DeAngelis. 1990. The global carbon cycle. Am. Sci. 78:310–326.

Preston, C.M., and J.A. Ripmeester. 1982. Application of solution and solid-state ^{13}C NMR to four organic soils, their humic acids, fulvic acids, humins and hydrolysis residues. Can. J. Spectrosc. 27:99–105.

Ragan, M.A., and K.W. Glombitza. 1986. Phlorotannins brown algal polyphenols. Progr. Phycol. Res. 4:129–241.

Reid, P.M., A.E. Wilkinson, E. Tipping, and N.J. Malcolm. 1990. Determination of molecular weights of humic substances by analytical (UV scanning) ultracentrifugation. Geochim. Cosmochim. Acta 54:131–138.

Rennie, D.A., E. Truog, and O.N. Allen. 1954. Soil aggregation as influenced by microbial gums, level of fertility, and kind of crop. Soil Sci. Soc. Am. Proc. 18:399–403.

Rice, J., and P. MacCarthy. 1989. Isolation of humin by liquid-liquid partitioning. Sci. Total Environ. 81/82:61–69.

Rice, J.A., and P. MacCarthy. 1990. A model of humin. Environ. Sci. Technol. 24:1875–1877.

Richards, E.G. 1980. An introduction to the physical properties of large molecules in solution. Cambridge University Press, Cambridge, UK.

Riffaldi, R., and M. Schnitzer. 1973. Effects of 6N HCl hydrolysis on the analytical characteristics and chemical structure of humic acids. Soil Sci. 115:349–356.

Rillig, M.C., S.F. Wright, K.A. Nichols, W.F. Schmidt, and M.S. Torn. 2001. Large contribution of arbuscular mycorrhizal fungi to soil carbon pools in tropical forest soils. Plant Soil 233:167–177.

Ritchie, G.S.P., and A.M. Posner. 1982. The effect of pH and metal binding on the transport properties of humic acids. J. Soil Sci. 33:233–247.

Rybicka, S.M. 1959. The solvent extraction of low rank vitrain. Fuel 38:45–54.

Saiz-Jiminez, C. 1994. Analytical pyrolysis of humic substances: Pitfalls, limitations, and possible solutions. Environ. Sci. Technol. 28:1773–1780.

Saldarriaga, J.G., and D.C. West. 1986. Holocene fires in the northern Amazon basin. Quat. Res. 26:358–366.

Salloum, M.J., and P.G. Hatcher. 2004. Determination of black carbon in natural organic matter by chemical oxidation and solid-state ^{13}C nuclear magnetic resonance spectroscopy. Org. Geochem. 35:923–935.

Sanford, R.L., Jr., and S.P. Horn. 2000. Holocene rain-forest wilderness: A neotropical perspective on humans as an exotic, invasive species. *In* S.F. McCool et al. (ed.) Wilderness science in a time of change conference, Proc. RMRS-P-15-VOl-3. USDA, Ogden.

Sanford, R.L., J. Saldarriaga, K.E. Clark, C. Uhl, and R. Herrera. 1985. Amazon rain-forest fires. Science 227:53–55.

Schachman, H.K. 1959. Ultracentrifugation in biochemistry. Academic Press, New York.

Schlesinger, W.H. 1995. An overview of the carbon cycle. p. 9–25. *In* R. Lal et al. (ed.) Soils and global change. Lewis Publ., Boca Raton, FL.

Schmidt, M.W.I., and A.G. Novack. 2000. Black carbon in soils and sediments: Analysis, distribution, implications and current challenges. Global Biogeochem. Cycles 14:777–793.

Schmidt, M.W.I., J.O. Skjemstad, C.I. Czimczik, B. Glaser, K.M. Prentice, Y. Gelinas, and T.A.J. Kulbusch. 2001. Comparative analysis of black carbon in soils. Global Biogeochem. Cycles 15:163–167.

Schmidt-Rohr, K., and J.-D. Mao. 2002a. Efficient CH-group selection and identification in C-13 solid-state NMR by dipolar DEPT and H-1 chemical-shift filtering. J. Am. Chem. Soc. 124:13938–13948.

Schmidt-Rohr, K., and J.-D. Mao. 2002b. Selective observation of nitrogen-bonded carbons in solid-state NMR by saturation-pulse induced dipolar exchange with recoupling. Chem. Phys. Lett. 359:403–411.

Schmitt-Kopplin, P., N. Hertkorn, H.-R. Schulten, and A. Kettrup. 1998. Structural changes in a dissolved soil humic acid during photochemical degradation processes under O_2 and N_2 atmospheres. Environ. Sci. Technol. 32:2531–2541.

Schnitzer, M. 1974. Alkaline cupric oxide oxidation of methylated fulvic acid. Soil Biol. Biochem. 6:1–6.

Schnitzer, M. 1978. Humic substances: Chemistry and reactions. p. 1–64. *In* M. Schnitzer and S.U Khan (ed.) Soil organic matter. Elsevier North-Holland, New York.

Schnitzer, M. 2000. A lifetime perspective on the chemistry of soil organic matter. Adv. Agron. 68:1–58.

Schnitzer, M., and J.A. Neyroud. 1975. Further investigations on the chemistry of fungal 'humic acids.' Soil Biol. Biochem. 7:365–371.

Schnitzer, M., and M.I. Ortiz de Serra. 1973a. The chemical degradation of a humic acid. Can. J. Chem. 51:1554–1566.

Schnitzer, M., and M.I. Ortiz de Serra. 1973b. The sodium-amalgam reduction of soil and fungal humic substances. Geoderma 9:119–128.

Schnitzer, M., and C.M. Preston. 1983. Effects of acid hydrolysis on the ^{13}C NMR spectra of humic substances. Plant Soil 75:201–211.

Schnitzer, M., and H.-R. Schulten. 1992. The analysis of soil organic matter by pyrolysis-field ionization mass spectrometry. Soil Sci. Soc. Am. J. 56:1811–1817.

Schnitzer, M., and H.-R. Schulten. 1995. Analysis of organic matter in soil extracts and whole soils by pyrolysis-mass spectrometry. Adv. Agron. 55:168–217.

Schnitzer, M., and J.R. Wright. 1960. Studies of the oxidation of organic matter of the A_o and B_h horizons of a podzol. Trans. 7th Int. Congr. Soil Sci 2:112–119.

Schulten, H.R., and M. Schnitzer. 1993. A state of the art structural concept for humic substances. Naturwissenschaften 80:29–30.

Senesi, N., and E. Loffredo. 1999. The chemistry of soil organic matter. p. 239–370. *In* D.L. Sparks (ed.) Soil physical chemistry. 2nd ed. CRC Press, Boca Raton, FL.

Senesi, N., T.M. Miano, and G. Brunetti. 1996. Humic-like substances in organic amendments and effects on native soil humic substances. p. 531–593. *In* A. Piccolo (ed.) Humic substances in terrestrial ecosystems. Elsevier, New York.

Senesi, N., and C. Steelink. 1989. Application of ESR spectroscopy to the study of humic substances. p. 373–408. *In* M.H.B. Hayes et al. (ed.) Humic substances II: In search of structure. John Wiley and Sons, Chichester, UK.

Shemyakin, M.M., and L.A. Shchukina. 1956. Oxidative-hydrolytic splitting of carbon-carbon bonds of organic molecules. Q. Rev. Chem. Soc. 10:261–282.

Shindo, H., and P.M. Huang. 1984. Catalytic effects of manganese (IV), iron (III), aluminium and silicon oxides on the formation of phenolic polymers. Soil Sci. Soc. Am. J. 48:927–934.

Silverstein, R.M., G.C. Bassler, and T.C. Morril. 1991. Spectrometric identification of organic compounds. 5th ed. John Wiley and Sons, New York.

Simpson, A.J. 1999. The structural interpretations of humic substances isolated from podzols under varying vegetation. Ph.D. diss. University of Birmingham, UK.

Simpson, A.J. 2001. Multidimensional solution-state NMR of humic substances: A practical guide and review. Soil Sci. 166:795–809.

Simpson, A.J. 2002. Determining the molecular weight, aggregation, structures and interactions of natural organic matter using diffusion ordered spectroscopy. Magn. Reson. Chem. 40:S72–S82.

Simpson, A.J., J. Burdon, C.L. Graham, N. Spencer, M.H.B. Hayes, and W.L. Kingery. 2001a. Interpretation of heteronuclear and multidimensional NMR spectroscopy as applied to humic substances. Eur. J. Soil Sci. 35:4421–4425.

Simpson, A.J., W.L. Kingery, and P.G. Hatcher. 2003. The identification of plant derived structures in humic materials using 3-dimensional NMR spectroscopy. Environ. Sci. Technol. 37:337–342.

Simpson, A.J., W.L. Kingery, M.H.B. Hayes, M.E. Spraul, E. Humpfer, P. Dvortsak, R. Kerssebaum, M. Godejohann, and M. Hofmann. 2001b. The separation of structural components in natural organic matter by diffusion ordered spectroscopy. Environ. Sci. Technol. 35:4421–4425.

Simpson, A.J., W.L. Kingery, M.H.B. Hayes, M. Spraul, E. Humpfer, P. Dvortsak, R. Kerssebaum, M. Godejohann, and M. Hofmann. 2002. The structures and associations of organic molecules in the terrestrial environment. Naturwissenschaften 89:84–88.

Simpson, A.J., W.L. Kingery, M.E. Spraul, E. Humpfer, and P. Dvortsak. 2001c. The application of ^{1}H Hr-MAS NMR spectroscopy for the study of structures and associations of organic components at the solid-aqueous interface of whole soil. Environ. Sci. Technol. 35:3321–3325.

Simpson, A.J., B. Lefebvre, A. Moser, A. Williams, N. Larin, M. Kvasha, W.L. Kingery, and B.K. Kelleher. 2004a. Identifying residues in natural organic matter through spectral prediction and pattern matching of 2-D NMR datasets. Magn. Reson. Chem. 22:14–22.

Simpson, A.J., L.H. Tseng, M.J. Simpson, M. Spraul, U. Braumann, W.L. Kingery, B.P. Kelleher, and M.H.B. Hayes. 2004b. The application of LC-NMR and LC-SPE-NMR to compositional studies of natural organic matter. Analyst 129:1216–1222.

Sinclair, A.H., and J. Tinsley. 1981. Studies of soil organic matter using formic acid solvents: Factors affecting composition and degree of formylation of organic isolates. J. Soil Sci. 32:103–117.

Sjöström, E. 1977. The behaviour of wood polysaccharides during alkaline pulping processes. Tappi 60:151–154.

Sjöström, E. 1981. Wood chemistry. Fundamentals and applications. Academic Press, New York.

Skjemstad, J.O., P. Clark, J.A. Taylor, J.M. Oades, and S.G. McClure. 1996. Chemistry and nature of protected carbon in soil. Aust. J. Soil Res. 34:251–271.

Skjemstad, J.O., L.J. Janik, M.J. Head, and S.G. McClure. 1993. High energy ultraviolet photo-oxidation: A novel technique for studying physically protected organic matter in clay- and silt-sized aggregates. J. Soil Sci. 44:485–499.

Skoog, D.A., and J.J. Leary. 1992. Principles of instrumental analysis. 4th ed. Saunders College Publishing, Fort Worth, TX.

Snyder, L.R. 1978. Solvent selection for separation processes. p. 25–75. *In* E.S. Perry and A. Weissberger (ed.) Separation and purification. Techniques of chemistry. Vol. XII. 3rd ed. Wiley, New York.

Specht, C.H., and F.H. Frimmel. 2000. Specific interactions of organic substances in size-exclusion chromatography. Environ. Sci. Technol. 34:2361–2366.

Sprengel, C. 1826. Über Pflanenzhumus, Humussäure and humussaure Slaze. Kastne's Arch. Ges. Naturlehre 8:145–220.

Sprengel, C. 1837. Die Bodenkunde oder die Lehre vom Boden. Immanuel Müller Publ. Co., Leipzig, Germany.

Stephen, A.M. 1983. Other plant polysaccharides. p. 97–193. *In* G.O. Aspinall (ed.) The polysaccharides. Vol. 2. Academic Press, Orlando, FL.

Stevenson, F.J. 1989. Reductive cleavage of humic substances. p. 121–142. *In* M.H.B. Hayes et al. (ed.) Humic substances II: In search of structure. John Wiley and Sons, Chichester, UK.

Stevenson, F.J. 1994. Humus chemistry: Genesis, compositions, reactions. 2nd ed. John Wiley and Sons, New York.

Stevenson, F.J., and M. Cole. 2000. Cycles of soil: Carbon, nitrogen, phosphorus, sulfur, micronutrients. 2nd ed. John Wiley and Sons, New York.

Stevenson, F.J., and J. Mendez. 1967. Reductive cleavage products of soil humic acids. Soil Sci. 103:383–388.

Stevenson, F.J., Q. van Winkle, and W.P. Martin. 1953. Physicochemical investigations of clay-adsorbed organic colloids: II. Soil Sci. Soc. Am. Proc. 17:31–34.

Stewart, R. 1964. Oxidation mechanisms. W.A. Benjamin, New York.

Stewart, R. 1965. Oxidation by permanganate. p. 1–68. *In* K.B. Wiberg (ed.) Oxidation in organic chemistry. Part A. Academic Press, New York.

Suggett, A. 1975. Polysaccharides. p. 519–567. *In* F. Franks (ed.) Water, a comprehensive treatise: Vol. 4. Aqueous solutions of amphiphiles and macromolecules. Plenum, New York.

Sutherland, R.A., C. van Kessel, R.E. Farrell, and D.J. Pennock. 1993. Landscape-scale variations in plant and soil nitrogen-15 natural abundance. Soil Sci. Soc. Am. J. 57:169–178.

Sutherland, R.A., C. van Kessel, and D.J. Pennock. 1991. Spatial variability of nitrogen-15 abundance. Soil Sci. Soc. Am. J. 55:1339–1347.

Swift, R.S. 1968. Physico-chemical studies on soil organic matter. Ph.D. diss. University of Birmingham, UK.

Swift, R.S. 1985. Fractionation of soil humic substances. p. 387–408. *In* G.R. Aiken et al. (ed.) Humic substances in soil, sediment, and water. John Wiley and Sons, New York.

Swift, R.S. 1989a. Molecular weight, size, shape, and charge characteristics of humic substances: Some basic considerations. p. 449–466. *In* M.H.B. Hayes et al. (ed.) Humic substances II: In search of structure. John Wiley and Sons, Chichester, UK.

Swift, R.S. 1989b. Molecular weight, shape, and size of humic substances by ultracentrifugation. p. 467–495. *In* M.H.B. Hayes et al. (ed.) Humic substances II: In search of structure. John Wiley and Sons, Chichester, UK.

Swift, R.S. 1996. Organic matter characterization. p. 1011–1069. *In* D.L. Sparks (ed.) Methods of soil analysis. Part 3. SSSA Book Ser. 5. SSSA., Madison, WI.

Swift, R.S. 1999. Macromolecular properties of soil humic substances: Fact, fiction and opinion. Soil Sci. 164:790–802.

Swift, R.S. 2001. Humic substances and carbon sequestration. Soil Sci. 166:858–871.

Swift, R.S., and A.M. Posner. 1971. Gel chromatography of humic acid. J. Soil Sci. 22:237–249.

Swift, R.S., and A.M. Posner. 1972. Autoxidation of humic acid under alkaline conditions. J. Soil Sci. 23:381–393.

Swift, R.S., B.K. Thornton, and A.M. Posner. 1970. Spectral characteristics of a humic acid fractionated with respect to molecular weight using an agar gel. Soil Sci. 110:93–99.

Swincer, G.D., J.M. Oades, and D.J. Greenland. 1968. Studies on soil polysaccharides. I. The isolation of polysaccharides from soil. Aust. J. Soil Res. 6:211–224.

Taft, R.W., D. Gurka, L. Joris, V.R. Schleyer, and J.W. Rakshys. 1969. Studies of hydrogen-bonded complex formation with *p*-fluorophenol. V. Linear free energy relationships with OH reference acids. J. Am. Chem. Soc. 91:4801–4808.

Tanford, C. 1961. Physical chemistry of macromolecules. Wiley, New York.

Tate, K.R., and K.M. Goh. 1973. Reductive degradation of humic acids from three New Zealand soils. N.Z. J. Sci. 16:59–69.

Tegelaar, E.W., G. Hollman, P. van der Vegt, J.W. de Leeuw, and P.J. Holloway. 1995. Chemical characterization of periderm tissue of some angiosperm species: Recognition of an insoluble, nonhydrolyzable, aliphatic biomacromolecule (Suberan). Org. Geochem. 23:239–250.

Theng, B.K.G. 1979. Formation and properties of clay-polymer complexes. Developments in Soil Science 9. Elsevier, Amsterdam.

Theng, B.K.G., J.R.H. Wake, and A.M. Posner. 1968. The fractional precipitation of soil humic acid by ammonium sulphate. Plant Soil 29:305–316.

Thurman, E.M., R.L. Wershaw, R.L. Malcolm, and D.J. Pinckney. 1982. Molecular size of aquatic humic substances. Org. Geochem. 4:27–35.

Tombácz, E. 1999. Collodial properties of humic acids and spontaneous changes of their colloidal state under variable solution conditions. Soil Sci. 164:814–824.

Trusov. 1917. Contributions to the study of soil humus. I. Processes of the formation of "humic acid." Contrib. Study Russ. Soils. 26–27:1–210.

Valenta, Z. 1963. Zinc dust distillation. p. 644–753. *In* K.W. Bentley (ed.) Elucidation of structures by chemical methods. Part 2. John Wiley and Sons, New York.

Valenta, Z. 1973. Dehydrogenation and zinc dust distillation. p. 1–76. *In* K.W. Bentley et al. (ed.) Elucidation of organic structures by physical and chemical methods. Techniques of organic chemistry. IV. 2nd ed. Wiley-Interscience, New York.

van Bergen, P.F., I.A. Bull, P.R. Poulton, and R.P. Evershed. 1997. Organic geochemical studies of soils from the Rothamsted Classical Experiments. I. Total lipid extracts, solvent insoluble residues and humic acids from Broadbalk Wilderness. Org. Geochem. 26:117–135.

van Bergen, P.F., M.B. Flannery, P.R. Poulton, and R.P. Evershed. 1998a. Organic geochemical studies of soils from the Rothamsted Classical Experiments. III. Nitrogen-containing macromolecular moieties in soil organic matter from Geescroft Wilderness. p. 321–338. *In* B.A. Stankiewicz and P.F. van Bergen (ed.) Nitrogen-containing macromolecules in the bio- and geosphere. ACS Symp. Ser. 707. ACS, Washington, DC.

van Bergen, P.F., C.J. Nott, I.A. Bull, P.R. Poulton, and R.P. Evershed. 1998b. Organic geochemical studies of soils from the Rothamsted Classical Experiments. IV. Preliminary results from a study of the effect of soil pH on organic matter decay. Org. Geochem. 29:1779–1795.

van Heemst, J.D.H., J.C. del Rio, P.G. Hatcher, and J.W. de Leeuw. 2000. Characterization of estuarine and fluvial dissolved organic matter by thermochemolysis using tetramethylammonium hydroxide. Acta Hydrochim. Hydrobiol. 28:69–76.

van Krevelen, D.W. 1961. Coal. Elsevier, Amsterdam.

Verardo, D.J. 1997. Charcoal analysis in marine sediments. Limnol. Oceanogr. 42:192–197.

von Wandruszka, R. 1998. The micellar model of humic acid: Evidence from pyrene fluorescence measurements. Soil Sci. 163:921–930.

Vicuna, R. 2000. Ligninolysis—A very peculiar microbial process. Mol. Biotechnol. 14:173–176.

Waksman, S.A., and K.R.N. Iyer. 1933. Contribution to our knowledge of the chemical nature and origin of humus: IV. Fixation of proteins by lignins and formation of complexes resistant to microbial attack. Soil Sci. 36:69–82.

Waksman, S.A., and J.P. Martin. 1939. The role of microorganisms in the conservation of the soil. Science 90:304–305.

Wallis, A.F.A. 1971. Solvolysis by acids and bases. p. 345–362. *In* K.V. Sarkanen and C.H. Ludwig (ed.) Lignins. Wiley, New York.

Wang, T.S.C., P.M. Huang, C.-H. Chou, and J.-H. Chen. 1986. The role of soil minerals in the abiotic polymerization of phenolic compounds and formation of humic substances. p. 251–281. *In* P.M. Huang and M. Schnitzer (ed.) Interactions of soil minerals with natural organics and microbes. SSSA Spec. Publ. 17. SSSA, Madison, WI.

Ward, S.G. 1947. Coal—Its constitution and utilisation as a chemical and as a raw material. J. Inst. Fuel 21:67–70.

Watt, B.E., R.L. Malcolm, M.H.B. Hayes, N.W.E. Clark, and J.K. Chipman. 1996. The chemistry and potential mutagenicity of humic substances in waters from different watersheds in Britain and Ireland. Water Res. 6:1502–1516.

Wershaw, R.L. 1986. A new model for humic materials and their interactions with hydrophobic chemicals in soil-water or sediment-water systems. J. Contam. Hydrol. 1:29–45.

Wershaw, R.L. 1993. Model for humus in soils and sediments. Environ. Sci. Technol. 27:814–816.

Wershaw, R.L. 1994. Membrane-micelle model for humus in soils and sediments and its relation to humification. USGS Water-Supply paper 2410. USGS, Reston, VA.

Wershaw, R.L. 1999. Molecular aggregation of humic substances. Soil Sci. 164:803–813.

Wershaw, R.L. 2000. The study of humic substances—In search of a paradigm. p. 1–7. *In* E.A. Ghabbour and G. Davies (ed.) Humic substances. Versatile components of plants, soil and water. Roy. Soc. Chem., Cambridge, UK.

Wershaw, R.L., and G.R. Aiken. 1985. Molecular size and weight measurements of humic substances. p. 477–492. *In* G.R. Aiken et al. (ed.) Humic substances in soil, sediment, and water. John Wiley and Sons, New York.

Wershaw, R.L., K.R. Kennedy, and J.E. Henrich. 1998a. Use of ^{13}C NMR and FTIR for elucidation of degradation pathways during natural litter decomposition and composting. II. Changes in leaf composition after senescence. p. 29–46. *In* G. Davies and E.A. Ghabbour (ed.) Humic substances: Structures, properties and uses. Roy. Soc. Chem., Cambridge, UK.

Wershaw, R.L., J.A. Leenheer, and K.R. Kennedy. 1998b. Use of ^{13}C NMR and FTIR for elucidation of degradation pathways during natural litter decomposition and composting. III. Characterization of leachate from different types of leaves. p. 47–68. *In* G. Davies and E.A. Ghabbour (ed.) Humic substances: Structures, properties and uses. Roy. Soc. Chem., Cambridge, UK.

Wershaw, R.L., J.A. Leenheer, K.R. Kennedy, and T.I. Noyes. 1996. Use of ^{13}C NMR and FTIR for elucidation of degradation pathways during natural litter decomposition and composting. I. Early stage leaf degradation. Soil Sci. 161:667–679.

Whitehead, D.C., and J. Tinsley. 1964. Extraction of soil organic matter with dimethylformamide. Soil Sci. 97:34–42.

Williams, P.J. LeB. 1975. Biological and chemical aspects of dissolved organic material in sea-water. p. 301–363. *In* J.P. Riley and G. Skirrow (ed.) Chemical oceanography. 2nd ed. Vol II. Academic Press, London.

Wilson, M.A. 1987. NMR techniques and applications in geochemistry and soil chemistry. Pergamon Press, Oxford, UK.

Wilson, M.A., and P.G. Hatcher. 1988. Detection of tannins in modern and fossil barks and plant residues by high-resolution solid-state ^{13}C nuclear magnetic resonance. Org. Geochem. 12:539–546.

Wright, J.R., and M. Schnitzer. 1959. Alkaline permanganate oxidation of the organic matter of the A_0 and B_2 horizons of a podzol. Can. J. Soil Sci. 39:44–53.

Wright, S.F. 2001. Glomalin, a manageable glue. USDA-ARS Brochure. USDA-ARS Soils Microbial Systems Laboratory, Beltsville, MD.

Wright, S.F., M. Franke-Snyder, J.B. Morton, and A. Upadhyaya. 1996. Time-course study and partial characterization of a protein on hyphae of arbuscular mycorrhizal fungi during active colonization of roots. Plant Soil 181:193–203.

Wright, S.F., J.L. Starr, and I.C. Paltinenu. 1999. Changes in aggregate stability and concentration of glomalin during tillage management transition. Soil Sci. Soc. Am. J. 63:1825–1829.

Wright, S.F., and A. Upadhyaya. 1996. Extraction of an abundant and unusual protein from soil and comparison with hyphal protein of arbuscular mycorrhizal fungi. Soil Sci. 161:575–586.

Wright, S.F., and A. Upadhyaya. 1998. A survey of soils for aggregate stability and glomalin, a glycoprotein produced by hyphae of arbuscular mycorrhizal fungi. Plant Soil 198:97–107.

Wright, S.F., A. Upadhyaya, and J.S. Buyer. 1998. Comparison of N-linked oligosaccharides of glomalin from arbuscular mycorrhizal fungi and soils by capillary electrophoresis. Soil Biol. Biochem. 30:1853–1857.

Zetsche, F., and H. Reinhart. 1939. Beitrag zur Reduktion de Huminsauren. Brennst.-Chem. 20:84–87.

Chapter 2

Chemistry of Phosphorus in Soils

J. THOMAS SIMS, *University of Delaware, Newark, Delaware, USA*

GARY M. PIERZYNSKI, *Kansas State University, Manhattan, Kansas, USA*

Phosphorus is essential to all forms of life on earth and is primarily conserved in soils and sediments. Understanding the chemical, biological, and physical processes in soils that affect the availability of P to terrestrial plants, and ultimately to animals and humans, is therefore of fundamental importance to the development of sustainable management practices for all agroecosystems. Similarly, knowledge of processes controlling the transport of P from soils to waters, and the availability of P to aquatic biota is of considerable ecological significance because of the well-known role of P in *eutrophication* ("...enrichment of surface waters by plants nutrients...a form of pollution that restricts the potential uses of impacted water bodies"; Foy and Withers, 1995).

Much modern soil chemistry and soil management now addresses this balance between sustainable agricultural practices and protection of the environment. Achieving production and environmental goals for soil P management begins with an understanding of how these goals are related to the soil P cycle, the chemical and biochemical processes that control the forms, biological availability, and mobility of soil P (Fig. 2–1). By understanding the fundamental principles that underlie P cycling between soils, biological organisms, and aquatic ecosystems, we can advance our ability to manage agroecosystems in a profitable, environmentally sustainable manner. The major components of the soil P cycle include: *dissolution–precipitation* (mineral equilibria), *sorption–desorption* (interactions between P in solution and soil solid phases), and *mineralization–immobilization* (biologically mediated conversions of P between inorganic and organic forms). All affect the concentration of P in the soil solution and soil solid phases and thus the amount of P that is available for uptake by plants and microorganisms and for transport to surface and groundwaters. Phosphorus moves within soils (e.g., to plant roots) primarily by *diffusion* (movement through soil pores in response to a concentration gradient) because the low concentrations of dissolved P in the soil solution preclude significant movement by *mass flow* (movement in the convective flow of water induced by evapotranspiration). Losses of P from soils occur mainly by *erosion* and *surface runoff*, (particulate and dissolved P) because *gaseous losses* of P do not occur and *leaching* or *subsurface runoff* of P (primarily as dissolved P) are rare in most cases.

Chemical Processes in Soils. SSSA Book Series, no. 8.

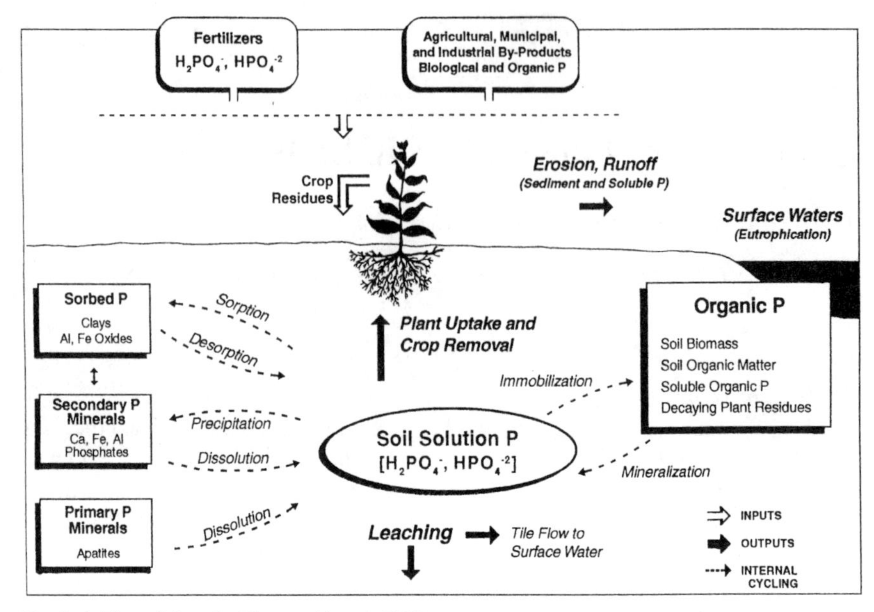

Fig. 2–1. The soil P cycle (Pierzynski et al, 2000).

Total soil P in topsoil horizons (0–15 cm) ranges from 50 to 3000 mg kg^{-1}, depending on parent material, soil type, and soil management practices (e.g., fertilization, manuring, cropping system; Foth and Ellis, 1997; Frossard et al., 1995). Most (50–75%) of the P in mineral soils is inorganic in nature and is predominantly associated with Al and Fe in acidic soils and Ca in alkaline, calcareous soils. Inorganic P occurs as primary minerals (those derived directly from weathered parent material) and secondary minerals (those formed by precipitation of P with Al, Ca, and Fe), P adsorbed onto the surface of clay minerals, Fe, and Al oxyhydroxides or carbonates, and P physically occluded within secondary minerals. The major minerals that release P into the soil solution as they weather differ between soils, mainly as a function of time and extent of soil development. In unweathered or moderately weathered soils, apatites [$Ca_{10}(X)(PO_4)_6$, where X is either F^-, Cl^-, OH^-, or CO_3^{2-}] are the primary mineral sources of P. As soils become progressively more weathered Al and Fe phosphates and organic forms of P predominate (Fig. 2–2). Organic P comprises between 30 and 65% of total P in mineral soils, depending on soil type, and is found in two broadly defined pools, that which is susceptible to microbial decomposition and is thus biologically available (e.g., inositol phosphates, phospholipids, nucleic acids and their derivatives) and that which is stable and highly resistant to microbial activity (the humic acid fraction) (Stevenson, 1982). In organic soils (>20–30% organic matter [OM] by weight) from 60 to 90% of the total P can be found in organic forms (Harrison, 1987). The inositol phosphates are the largest fraction of organic P in most soils and can be as much as 80% of total organic P. Microbial biomass is also an important form of soil organic P, represent-

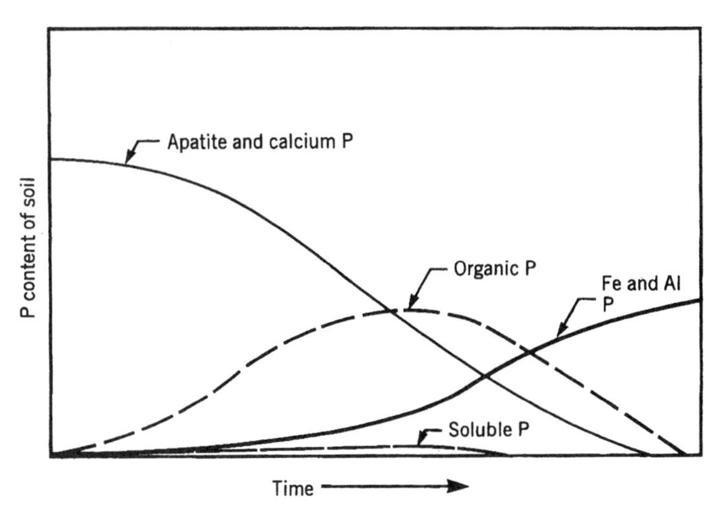

Fig. 2–2. Relative distribution of the major forms of soil P as related to time of soil development (Foth and Ellis, 1997).

ing from 2 to 5% of organic P in arable soils and up to 25% of organic P in grassland soils (Brookes et al., 1982, 1984; Turner et al., 2001). Studies have shown varying contributions of organic P to plant-available P and also investigated the means by which organic P is retained in soils. However, in general, the mechanisms by which organic P is converted to forms that are available to plants or aquatic organisms or that move in runoff and leaching waters remain poorly understood.

The inorganic and organic forms of soil P are in dynamic equilibria with dissolved P in the soil solution, which is mainly present as primary (PO_4^{3-}) or secondary (HPO_4^{2-}, $H_2PO_4^-$) orthophosphates, varying in relative concentrations as a function of soil pH. Most agricultural soils range in pH from 4.0 to 9.0. At pH 4.0 to 5.5 $H_2PO_4^-$ predominates, while at pH > 8.0 HPO_4^{2-} is the major orthophosphate species present. Ion pairs (e.g., $CaHPO_4^0$, $MgHPO_4^0$, or $CaPO_4^-$) may also be present in the soil solution at higher pH values (pH > 7.0), but in most cases chemical dissolution of these soluble complexes rapidly converts them to orthophosphate. Less is known about the soluble organic P species present in soils or their contribution to soil P bioavailability, but it is believed that microbial hydrolysis of dissolved organic P to orthophosphate is rapid (Stevenson, 1986). Concentrations of soil solution P typically range from <0.01 mg P L^{-1} in very infertile soils to 1 mg P L^{-1} in well-fertilized soils, but can sometimes be as high as 7 to 8 mg P L^{-1} in soils recently amended with fertilizers or organic by-products. A soil solution P concentration of 0.2 mg P L^{-1} is often reported as the value required for optimum plant growth; however, studies have shown that some plants can attain optimum yields when soil solution P is as low as 0.03 mg P L^{-1}. Phosphorus concentrations associated with eutrophication of freshwater lakes (0.01–0.03 dissolved P L^{-1} and 0.035–0.10 mg total P L^{-1}) are in about the same range as those required by terrestrial plants (Correll, 1998). The quantity of P in the soil solution at a given time generally represents <1% of the total quantity of P in the soil, on the order of <1 kg ha^{-1} (Pierzynski, 1991). To meet plant needs, P in the soil solution must be replenished many times over the life of the plant by release from the organic and inorganic solid phases in soils to the solution phase. Soil P in solid phases that rapidly equilibrate with the soil solution or runoff and surface waters is often referred to as *labile P* while forms of soil P that slowly release P to solution are termed *nonlabile.*

Anthropogenic factors can markedly influence P concentrations in the solution and solid phases of soils. Major inputs and outputs of P associated with human activity are fertilization, use of animal and green manures, addition of soil amendments such as lime or municipal and industrial by-products (e.g., biosolids [sewage sludge], composts), and plant uptake and P removal in harvested crops. Much of P management consists of making decisions about the need to add fertilizers and by-products to soils at rates and in manners that will sustain economically optimum plant production and minimize the likelihood of P losses to water. Fertilization is an essential part of soil management because as plants absorb P from the soil solution or soluble P is lost in runoff, labile P dissolves or desorbs from the solid phase to solution. If soils are left unfertilized or are underfertilized, labile P concentrations will progressively decrease, through crop removal or P loss, to the point that the soil can no longer adequately meet plant P needs. New additions of P are then necessary to ensure that sufficient labile P is present to meet plant needs. The greater the concentration of labile P, the longer the soil will be able to maintain plant growth at desired levels. However, consistent overfertilization, to the point that labile P increases to concentrations well beyond that needed for optimum plant growth, should be avoided, as excessively fertilized soils have been shown to have greater potential for environmentally significant losses of P to surface waters and shallow groundwaters (Barberis et al., 1996; Breeuwsma et al., 1995; De Smet et al., 1996; Sharpley et al., 1994; Simard et al., 1995; Sims et al., 2000; Whalen and Chang, 2001).

The intent of this chapter is to clearly link the fundamental principles of soil P chemistry with the applied aspects of soil P management. Increasing global concerns about how to feed a growing world population and the impacts of nonpoint P pollution on water quality make the need for well-designed best management practices (BMPs) for P more important than ever. To be most effective, BMPs must be based on a sound knowledge of soil P chemistry and reflect the influence of natural and anthropogenic factors (e.g., soil physical and chemical properties, climate, hydrology, cropping systems, soil management) on the soil P cycle. There are many examples of how we have applied our understanding of the interactions between soil P chemistry and soil management to develop BMPs that are widely used today, such as soil P testing, the selection and management of fertilizers and organic P sources, the use of soil conservation practices (e.g., tillage, buffer strips, constructed wetlands) to prevent P loss, and the computer models that predict P availability to crops and transport to surface and groundwaters. Despite successes such as these, P management remains a major agrienvironmental challenge. Future innovations in P management depend on our ability to advance the understanding of soil P chemistry and creatively apply this knowledge to practical measures that can increase agricultural productivity and not only protect, but improve water quality.

SOIL PHOSPHORUS CHEMISTRY: FUNDAMENTAL PROCESSES

Mineral Equilibria: Dissolution and Precipitation Reactions

The major primary P minerals in soils are reasonably well known (Table 2–1; Lindsay, 1979; Lindsay et al., 1989). In soils that have not been extensively weathered, P mineralogy is dominated by sparingly soluble apatites [$Ca_{10}X(PO_4)_6$, where X can be OH^-, F^-, Cl^-, or CO_3^{2-}] and other more soluble calcium phosphates [octacalcium phosphate ($Ca_8H(PO_4)_6 \bullet H_2O)_5$, monetite ($CaHPO_4$), brushite ($CaHPO_4 \bullet 2H_2O$)]. As soils weather, apatites dissolve, Ca^{2+} is lost by leaching or crop uptake, and Al or Fe forms of P become more important. In highly weathered, acidic soils, Al phosphates [berlinite ($AlPO_4$), variscite ($AlPO_4 \cdot 2H_2O$), sterretite ($Al(OH_2)_3HPO_4 \bullet H_2PO_4$), tarakanite ($H_6K_3Al_5(PO_4)_8 \bullet 18H_2O$)] and Fe phosphates [strengite ($FePO_4 \bullet 2H_2O$), tinticite ($Fe_6(PO_4)_4(OH)_6 \bullet 7H_2O$), vivianite ($Fe_3(PO_4)_2 \bullet 8H_2O$)] predominate. Caution is needed when using the term *mineral* since this implies an ordered, three-dimensional atomic structure, which may not always be the case in a complex matrix such as soil. For instance, secondary solid phases, formed by the coprecipitation of P added to soils in organic and inorganic fertilizers with soluble Al, Ca, and Fe in soils, are often amorphous in nature and mixed in composition.

Knowledge of soil P mineralogy and soil properties controlling mineral solubility provides important insight into the P concentrations expected in soil solutions at equilibrium with different P minerals. However, because more than one solid phase of P is usually present in any soil and because chemical and biological processes continuously change the nature of the soil solution and soil solid phases, true chemical equilibria between P minerals and soil solution P are rarely, if ever, achieved. The kinetics of dissolution and precipitation of P minerals must also be considered because the rates of some chemical reactions are so slow that equilibrium may never occur. Thus, while P minerals undoubtedly play a major role in controlling the solubility of P in most soils, nonequilibrium conditions and kinetic factors preclude more than a qualitative use of information on mineral solubility to predict soil solution P concentrations.

Despite these difficulties a considerable body of research has been directed towards characterizing soil P mineralogy, and advances continue today. There are basically two approaches that have most often been used to identify and study solid phases of soil P. The first relies on solubility equilibrium experiments using extracted soil solutions or aqueous solutions that have been brought into equilibrium with the soil. Both types of these experiments require the assumption that the soil solids are either in equilibrium with the aqueous phase or that a steady-state condition (pseudo-equilibrium) exists in which the indicators of equilibrium are time invariant within the experiment. The second approach uses a wide range of instrumental methods by which the presence of a particular solid phase can be directly determined, such as X-ray diffraction (XRD), electron microscopy and diffraction, optical microscopy, nuclear magnetic resonance spectroscopy (NMR), and X-ray absorption fine structure spectroscopy (EXAFS).

Sposito (1989) presented a generalized precipitation–dissolution reaction that can be used to illustrate the solubility equilibrium approach:

Table 2–1. Phosphate minerals thought to be important in soils. Data from Boyle and Lindsay (1985, 1986), Lindsay (1979), and Veith and Sposito (1977).

Mineral name	Formula	Equilibrium reaction	Log K
Variscite	$AlPO_4{\bullet}2H_2O$	$AlPO_4{\bullet}2H_2O + 2H+ \rightleftharpoons Al^{3+} + H_2PO_4^- + 2H_2O$	−2.50
Amorphous variscite	$Al(OH)_2H_2PO_4$	$Al(OH)_2H_2PO_4 \rightleftharpoons A1^{3+} + 2OH^- + H_2PO_4^-$	−27 to −29
Strengite	$FePO_4{\bullet}2H_2O$	$FePO_4{\bullet}2H_2O + 2H^+ \rightleftharpoons Fe^{3+} + H_2PO_4^- + 2H_2O$	−6.85
Vivianite	$Fe_3(PO_4)_2{\bullet}8H_2O$	$Fe_3(PO_4)_2{\bullet}8H_2O + 4H^+ \rightleftharpoons 3Fe^{2+} + 2H_2PO_4^- + 8H_2O$	3.11
Monocalcium phosphate	$Ca(H_2PO_4)_2{\bullet}H_2O$	$Ca(H_2PO_4)_2{\bullet}H_2O \rightleftharpoons Ca^{2+} + 2H_2PO_4^- + H_2O$	−1.15
Brushite (dicalcium phosphate dihydrate)	$CaHPO_4{\bullet}2H_2O$	$CaHPO_4{\bullet}2H_2O + H^+ \rightleftharpoons Ca^{2+} + H_2PO_4^- + 2H_2O$	0.63
Monetite (dicalcium phosphate)	$CaHPO_4$	$CaHPO_4 + H^+ \rightleftharpoons Ca^{2+} + H_2PO_4^-$	0.30
Octacalcium phosphate	$Ca_4H(PO_4)_3{\bullet}2.5H_2O$	$Ca_4H(PO_4)_3{\bullet}2.5H_2O + 5H^+ \rightleftharpoons 4Ca^{2+} + 3H_2PO_4^- + 2.5H_2O$	11.76
β-tricalcium phosphate	β-$Ca_3(PO_4)_2(c)$	β-$Ca_3(PO_4)_2(c) + 4H^+ \rightleftharpoons 3Ca^{2+} + 2H_2PO_4^-$	10.18
Hydroxyapatite	$Ca_5(PO_4)_3OH$	$Ca_5(PO_4)_3OH + 7H^+ \rightleftharpoons 5Ca^{2+} + 3H_2PO_4^- + H_2O$	14.46
Fluorapatite	$Ca_5(PO_4)_3F$	$Ca_5(PO_4)_3F + 6H^+ \rightleftharpoons 5Ca^{2+} + 3H_2PO_4^- + F^-$	−0.21
Hydroxypyromorphite	$Pb_5(PO_4)_3OH(c)$	$Pb_5(PO_4)3OH(c) + 7H+ \rightleftharpoons 5Pb^{2+} + 3H_2PO_4^- + H_2O$	−4.14
Chloropyromorphite	$Pb_5(PO_4)3Cl(c)$	$Pb_5(PO_4)3Cl(c) + 6H^+ \rightleftharpoons 5Pb^2 + 3H_2PO_4^- + Cl^-$	−25.05
None	$MnPO_4{\bullet}1.5H_2O$	$MnPO_4{\bullet}1.5H_2O + 2H^+ + e^- \rightleftharpoons Mn^{2+} + H_2PO_4^-$	10.28
None	$MnHPO_4{\bullet}3H_2O$	$MnHPO_4{\bullet}3H_2O + H^+ \rightleftharpoons Ma^{2+} + H_2PO_4 - 3H_2O$	−0.24
None	$Mn_5H_2(PO_4)_4{\bullet}4H_2O$	$Mn_5H_2(PO_4)_4{\bullet}4H_2O + 6H^+ \rightleftharpoons 5Mn^{2+} + 4H_2PO_4^- + 4H_2O$	4.76

$$M_aL_b(s) \leftrightarrow aM^{m+}(aq) + bL^{l-}(aq) \quad [1]$$

where M_aL_b is any solid phase that dissolves to release a cation (M^+) and an anion (L^-) in an aqueous solution (aq) and m and l are the ionic charges. A specific example for P is the equilibrium reaction for hydroxyapatite (see Table 2–1 for equilibrium reactions of the major P minerals present in soils):

$$Ca_5OH(PO_4)_3(H_2O)_5 \leftrightarrow 5Ca^{2+} + OH^- + 3PO_4^{3-}$$

From this one can calculate the ion activity product (IAP):

$$IAP = (M^{m+})^a(L^{l-})^b \quad [2]$$

where parentheses denote ion activities. After the appropriate determination of the necessary ion activities in the soil solution (e.g., Ca^{2+}, OH^-, PO_4^{3-}), one can compare calculated values of IAP to that expected for a particular solid phase. If the calculated IAP compares favorably, then indirect evidence indicating that the solution is in equilibrium with that solid has been obtained. If the IAP does not compare favorably, then the system may not have reached equilibrium, the system may be in equilibrium with a different solid, or the system may be in equilibrium with the solid but the solid may not be in the standard state selected by the researcher (Sposito, 1981).

As noted above, use of the solubility equilibrium approach requires knowledge of the minerals or solid phases that potentially might be in equilibrium with the aqueous solution being studied. Under high pH conditions the known soil minerals most often considered to control P solubility in soils are fluoroapatite, hydroxyapatite, tricalcium phosphate, octacalcium phosphate, dicalcium phosphate, and dicalcium phosphate dihydrate, listed in order of increasing solubility (Lindsay, 1979). Under low pH conditions variscite and strengite are likely possibilities for control of P solubility. Some work has suggested the formation of an amorphous analog of variscite (Pierzynski et al., 1990a; Veith and Sposito, 1977). Manganese phosphates have also been investigated (Boyle and Lindsay, 1986). Mineral equilibria reactions (Table 2–1) can be used to characterize P dissolution and precipitation in soils (looking ahead to Fig. 2–5). The lines in this figure represent the concentration of $H_2PO_4^-$ or HPO_4^{2-} expected if the soil solution is at equilibrium with that mineral phase (Lindsay, 1979). If the actual P concentration is above the solubility line for any P mineral, the soil solution is considered supersaturated with P and precipitation of that mineral should eventually occur. If the P concentration is below a mineral's solubility line, the solution is undersaturated and that mineral phase should eventually dissolve and release P into solution.

Use of the solubility equilibria approach to predict soil solution P concentrations has had limited success. For Ca phosphates, typical results are that the IAP falls between standard values for common mineral species, with the soil solution typically being supersaturated with respect to the apatites and undersaturated with respect to dicalcium phosphate dihydrate (Harrison and Adams, 1987; O'Connor et al., 1986; Pierzynski et al., 1990a). Similar results have been found for Al phosphates (Karanthanasis, 1991). A variety of reasons have been given for the apparent discrepancy between thermodynamically predicted and experimental results. Harrison and Adams (1987) speculated that apatites had formed in an Ultisol, but the crystals were encapsulated by more soluble Ca phosphate precipitates; thus, the solution concentrations reflected the equilibrium reactions of more than one solid phase of P. Inhibition of apatite formation by organic acids has also been demonstrated (Inskeep and Silvertooth, 1988). In the case of variscite, an amorphous analog has been described which may crystallize very slowly, if at all, thus at least partially explaining why few studies have demonstrated equilibrium with the crystalline form (Hsu, 1982a; Veith and Sposito, 1977). Similar conclusions were reported for Fe (III) phosphate (Hsu, 1982b). A solid solution between $AlPO_4$ and $Al(OH)_3$ in soils has also been hypothesized (Blanchar and Stearman, 1985).

Some studies have had more success with the solubility equilibrium approach. Hettiarachchi and Pierzynski (1996) evaluated the influence of P additions on Pb solubility in a Pb-contaminated soil. This was done to assess the feasibility of using P to reduce Pb solubility and thus bioavailability in contaminated soils, a recently proposed alternative for the currently accepted remediation strategy of soil excavation and replacement. Prior to P additions the soil appeared to be in equilibrium with cerrusite ($PbCO_3$), but after addition of a soluble P source (KH_2PO_4) the equilibrium shifted to one controlled by hydroxypryromorphite [$Pb_5(PO_4)_3OH$], a less soluble Pb mineral phase (Fig. 2–3).

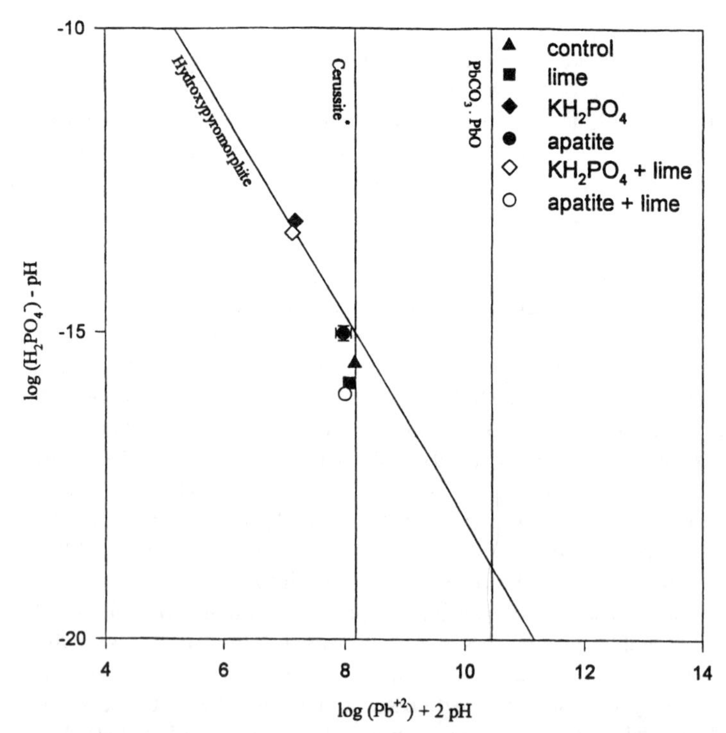

Fig. 2–3. Lead phosphate solubility diagram showing the shift in equilibrium from cerrusite ($PbCO_3$) to hydroxypyromorphite ($Pb_5(PO_4)_3OH$) upon addition of KH_2PO_4 to a Pb-contaminated soil (Hettiarachchi and Pierzynski, 1996).

The use of any direct methodology to determine solid phases of P present in soils has always been complicated by the fact that the relatively low concentration of P in most soils can be at or near the detection limits of many of the instruments used in this approach (Lindsay et al., 1989; Sample et al., 1980). For example, in most soils there would not be sufficient P for detection by X-ray diffraction (XRD), even if soil P was very high and all P was present as a crystalline mineral. Exceptions do exist, such as studies of phosphatic soils that were able to utilize XRD; however, these soils had total P concentrations as high as 51 000 mg kg^{-1}, relative to average total P values in normal soils of 500 to 1000 mg kg^{-1} (Karanthanasis, 1991). In this case, apatites were detected in whole soil samples having total P concentrations as low as 2600 mg kg^{-1} but no crystalline P minerals were found in soils with <1750 mg P kg^{-1}. Similarly, using XRD and optical and electron microscopy, Harris et al. (1994) detected poorly crystalline apatite and vivianite [$Fe_3(PO_4)_2 \cdot 8H_2O$] in a stream sediment having about 1700 mg P kg^{-1}. The stream sediments had been heavily impacted by nearby dairy operations. These studies suggest that direct methods may be of most value in soils that have been enriched in P by anthropogenic activities (i.e., overapplications of fertilizers, manures, biosolids) or that are naturally high in P because of geologic reasons.

Numerous studies have used direct methods to detect the presence of P minerals upon reaction of model systems (e.g., pure minerals such as calcite, gibbsite, and goethite) with P (Martin et al., 1988), but relatively few have been successful with whole soils or fractions thereof. Three studies have used a combination of solubility equilibria experiments and a direct methodology for detecting P solid phases in soils.

Hinedi and Chang (1989) studied a sludge-amended Domino soil (fine-loamy, mixed, superactive, thermic Xeric Petrocalcid) and found the equilibrium solution was slightly undersaturated with respect to hydroxyapatite and suggested that the P solid phase was carbonated apatite formed through coprecipitation. This suggestion was supported by NMR analyses that indicated the P was present as a poorly crystalline Ca phosphate with a chemical shift similar to that of a highly carbonated apatite.

Pierzynski et al. (1990a, 1990b, 1990c) studied P-rich particles in 11 excessively fertilized soils in conjunction with solubility equilibrium studies with the same soils. They separated the soils based on particle size and used density separation to isolate the P-rich particles, which were then examined by transmission electron microscopy (TEM), energy dispersive X-ray analysis (EDS), XRD, and Fourier-transformed infrared (FTIR) spectroscopy. Soils that had equilibrium solutions with a pH >7.0 generally were supersaturated with respect to hydroxyapatite and undersaturated with respect to octacalcium phosphate. Only one sample provided strong evidence for control of P solubility by a particular Ca phosphate solid phase. Soils that had equilibrium solutions with pH values <7.0 generally indicated equilibrium with an amorphous analog of variscite. A detailed examination of 260 P-rich particles showed that 240 particles had detectable quantities of both Al and Si, with an additional 14 particles having detectable quantities of just Al. An example spectra from EDS analysis of a P-rich particle is given in Fig. 2–4. Calcium did not dominate the elemental composition of particles separated from high pH soils, as thermodynamics would suggest. Four particles examined in more detail were amor-

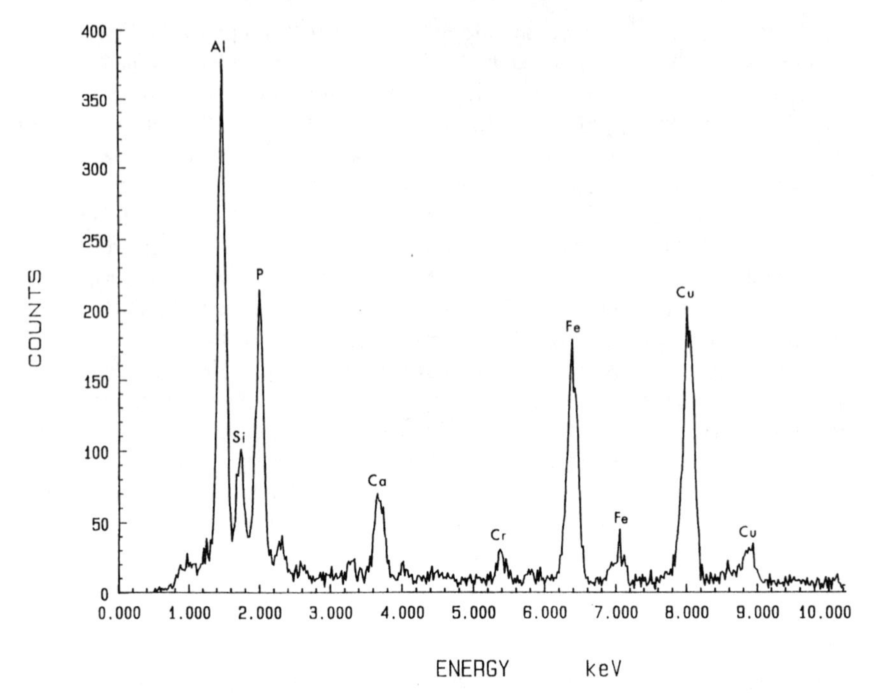

Fig. 2–4. Energy dispersive X-ray spectra from a P-rich particle found in the clay-sized fraction of the surface horizon of a Blount silt loam (Aeric Ochraqualf) (Pierzynski et al., 1990c).

phous by electron diffraction. Overall, the elemental composition of the P-rich particles did not support conclusions drawn from solubility equilibrium work, with the exception of the soils with a low pH where the particles were amorphous and contained Al. Further, P-rich particles isolated from one soil were highly variable in composition.

Karanthanasis (1991) studied two phosphatic soils from Kentucky with solubility equilibrium experiments and XRD, differential scanning calorimetry, and scanning electron microscopy (SEM). The XRD results suggested apatite was present in the relatively unweathered BC and C horizons and that wavellite was present in the BC horizon of one soil. The SEM results also supported these observations. The DSC results were less definitive, merely suggesting the presence of Al or Fe phosphates in the BC and C horizons. Solubility equilibrium studies were not useful in identifying the presence of a particular mineral in any soil horizon. Rather, the results suggested most horizons were supersaturated with respect to apatite and undersaturated with respect to more stable Ca, Al, and Fe phosphates. Overall, the results supported the hypothesis of apatite weathering to more stable secondary P solid phases.

In summary, significant amounts of P exist in a variety of solid phases in soils. Thermodynamics suggest the formation of crystalline products of relatively simple composition; however, the utility of the thermodynamic approach is limited by the database of potential solid phases that are likely to exist in natural soils. In particular, the direct examination of P solid phases in soils suggests a complicated elemental composition in an amorphous matrix. Thermodynamic data are not available for such materials. Further, the elemental composition of the particles varies considerably within a soil. This situation would suggest that solubility equilibrium experiments may not be the best approach for studying P solid phases and that an equilibrium between the soil solution and a single, stable solid phase may be difficult to reach. The kinetics of precipitation–dissolution reactions and the diffusion of reactants and products further complicate interpretation of this approach to understanding the relationship between solid and solution phases of soil P. Further advances in direct methods to identify solid phases of P alone, or in combination with solubility experiments, are clearly needed today.

The importance of advancing our understanding of mineral dissolution and precipitation to P bioavailability and mobility in soils can be easily illustrated. Traditionally, the study of soil P chemistry has focused on the need to correct P deficiencies for crops. In this case a soil solution P concentration of 0.2 mg L^{-1} (6.45×10^{-6} *M*) is often used as a sufficiency level (Sanchez and Uehara, 1980). When excess soil P and offsite movement of soluble P is a concern, a soil solution concentration of 1.0 mg L^{-1} (3.2×10^{-5} *M*) is occasionally used as a benchmark. This concentration often reflects the goal for wastewater discharges to rivers and streams and has been applied to soils on the premise that the discharge of P from soils to water should be held to the same standard. It is relatively straightforward to predict which mineral species should maintain solution P concentrations of 0.2 and 1.0 mg L^{-1}. For example, using the P phase diagram of Lindsay (1979), it is apparent that between pH 5 and 7 only apatites and tri-calcium phosphate can maintain solution P concentrations below 1.0 mg L^{-1}, and only apatites would maintain P concentrations below 0.2 mg L^{-1} (Fig. 2–5). If the soil pH range was expanded to 8.0, octacalcium phosphate could also maintain P concentrations near 1.0 mg L^{-1}. In theory, one could predict whether the soil could adequately supply P to crops or represented a risk for offsite movement of soluble P based on knowledge of the P solid phase in equilibrium with the soil solution. Of course there are several shortcomings to this approach. As mentioned above, there are no thermodynamic data on the complex P solid phases that can occur in soils, so it is not possible to predict equilibrium solution P concentrations. In addition, equilibrium P concentrations are a measurement of intensity, not quantity, and no information is obtained on whether the soil could replenish the solution with P if it were depleted. Similarly, if the soil solution were simply displaced or diluted, the kinetics of reestablishing the equilibrium condition are not known. It is also difficult to separate the effects of P desorption from mineral dissolution.

Despite these potential shortcomings, several studies have shed some light on the role of mineral dissolution in P bioavailability. In the case of crop removal of P, a relatively gradual process compared with the possible interaction between P solid phases in soils and surface runoff, one would expect an IAP to remain constant as P was removed from the soil solution via plant uptake and the solid phase dissolved to replenish the P that was absorbed. Such evidence has been obtained for calcareous soils over a 3-yr period in a field study with alfalfa (*Medicago sativa* L.) in which soil solutions maintained equilibrium with either tricalcium phosphate or octacalcium phosphate (Fixen et al., 1983). Less direct evidence of this same

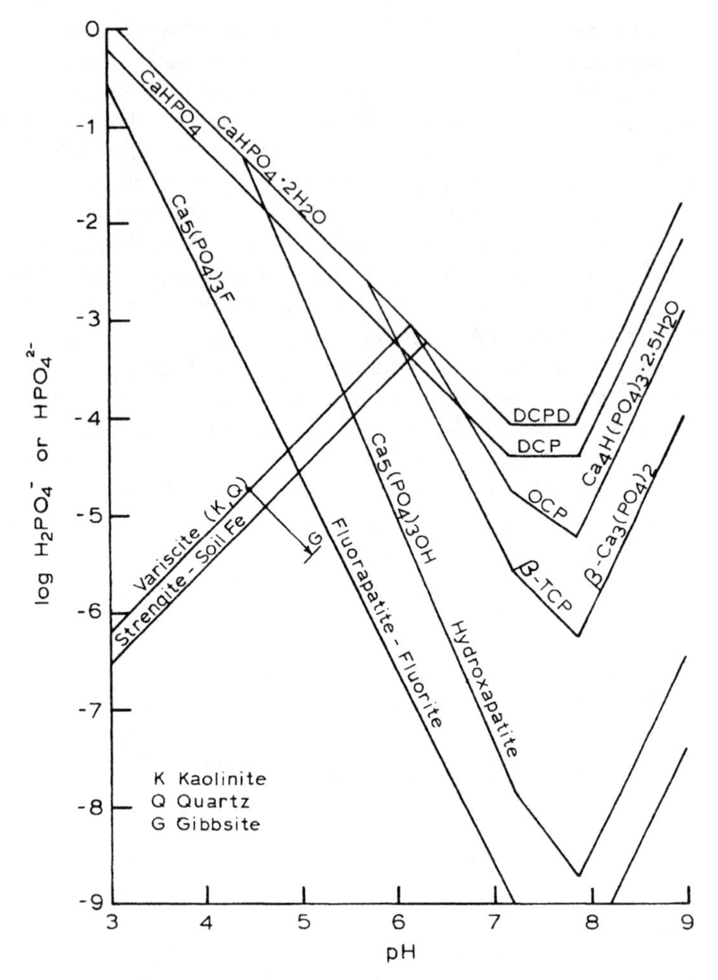

Fig. 2–5. Generalized phase diagram illustrating the effect of pH on P solubility in soils. The activity of Ca^{2+} is assumed to be $10^{-2.5}$ *M* and CO_2 is 0.000 303 MPa (0.0003 atmospheres).

phenomenon takes the form of soil test P levels that do not decline over extended periods of time in response to crop removal of P, suggesting that a P solid phase is maintaining bioavailable P levels (McCollum, 1991; Selles et al., 1995). Conversely, studies have shown that crystalline P minerals did not dissolve readily in soils, implying that such minerals could not easily interact with the soil solution to contribute to P bioavailability (Murmann and Peach, 1969).

It is clear that it is difficult to elucidate the role of mineral precipitation and dissolution in P biovavailability. Thermodynamics implies that P solid phases should serve as a sink for P applied to soils and as a source for P when P is removed from the soil solution. Direct and indirect evidence supporting the presence of P solid phases in soils exists in the literature and, in some cases, how the identity of those solid phases might change with addition or removal of P. However, with regard to long-term reaction products, no definitive studies exist showing increases or decreases in the amount of P solid phases in a soil upon addition or removal of P.

The difficulties that have been described have led to the use of more empirical estimates of the amount of P that can be released from soil solid phases to the soil solution. Generally these have taken the form of exposing the soil to a strong P sink, such as iron oxide (FeO) coated filter papers or anion exchange resins, or repeated extractions of the soil with water or 0.01 *M* $CaCl_2$. The source of the P removed by these rapid soil test methods is not clear and undoubtedly includes both soil solution P, desorbed P, and P that has dissolved from P solid phases. These methods are discussed in more detail later in this chapter.

Sorption and Desorption

Sorption is a general term that refers to several chemical processes that result in the retention of a *sorbate* (e.g., orthophosphate) at the surface of a *sorbent* (e.g., a soil colloid), including *adsorption*, *surface precipitation*, and *polymerization* (Sparks, 1995). Sorption is the preferred term to use when the exact mechanism of retention of a sorbate by a sorbent is unknown, which is often the case in studies of the chemical reactions of P with soils. Adsorption was defined by Sposito (1989) as the "...net accumulation of matter at an interface between a solid phase and an aqueous solution phase. It differs from precipitation because it does not include the development of a three-dimensional molecular structure even if that structure grows on a surface." Adsorption can be caused by physical processes, such as van der Waals forces of attraction and electrostatic outer sphere complexes (anion exchange, referred to as *nonspecific adsorption*) and by chemical processes, such as inner-sphere *ligand exchange* (covalent bonding, referred to as *specific adsorption* or *chemisorption*) and H bonding (Sparks, 1995). Adsorption is usually considered a two-dimensional process, unlike precipitation reactions, which result in the formation of three-dimensional molecular structures.

Adsorption of P in acid soils is believed to occur primarily by the formation of an inner-sphere complex between an orthophosphate anion (e.g., $H_2PO_4^-$) and a metal cation that is a constituent of a soil solid, such as a clay mineral (e.g., kaolinite) or a crystalline or amorphous metal oxide (e.g., an Al or Fe oxide). In this process, the orthophosphate anion undergoes an exchange reaction with a functional group on the colloidal surface, either a hydroxyl ion (OH^-) or a water molecule (OH_2), resulting in the formation of a covalent bond between orthophosphate and the metal cation and a lowering of the point of zero charge (PZC) of the surface due to its increased negative charge density. Inner-sphere complexes are characterized by very strong adsorptive forces, and adsorption reactions by this mechanism are not readily reversible. Ligand exchange can result in monodentate, bidentate, or binuclear forms of adsorbed P, with the monodentate form considerably more reversible than other adsorbed species. Generalized representations of common adsorption processes are shown in Fig. 2–6a and a specific example illustrating the formation of an inner-sphere complex between $H_2PO_4^-$ and an Al (or Fe) oxide surface in Fig. 2–6b. Note that formation of a single Al–O–P bond results in a more labile form of soil P than formation of bidentate complex. In this chapter we use the term ad-

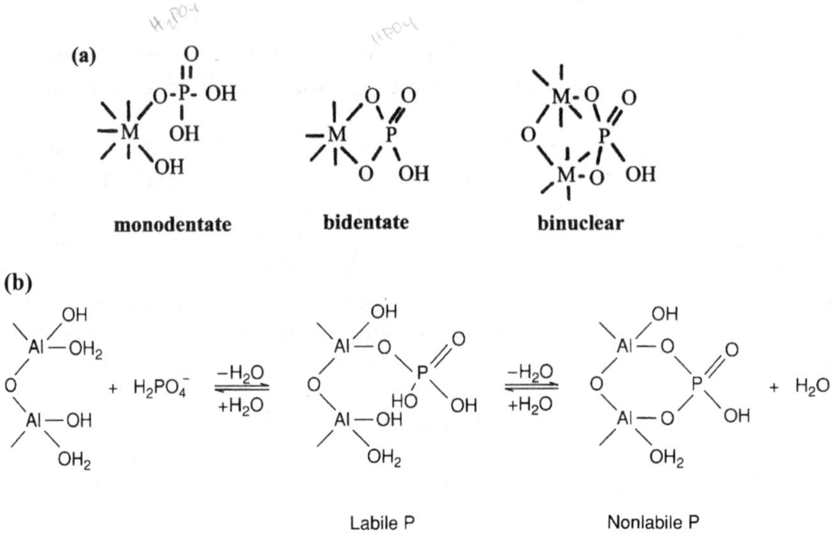

Fig. 2–6. Illustration of (a) monodentate, bidentate, and binuclear P sorption by a model metal oxide and (b) P sorption reactions with an Al oxide, illustrating mechanisms that result in labile and nonlabile P (Tisdale et al., 1993).

sorption only when the mechanism of P retention is known; otherwise we use the more inclusive and less specific term, sorption. We avoid use of the overly broad, but widely used term P *fixation* ("...processes in a soil by which certain chemical elements essential for plant growth are converted from a soluble or exchangeable form to a much less soluble or nonexchangeable form"; Brady and Weil, 2000) since it has no mechanistic connotations.

Desorption has not been defined as rigorously or studied in as mechanistic a manner as adsorption. In general, desorption refers to the release of a sorbate from a sorbent into the solution phase, without reference to any specific mechanism. For many sorbates, including P, the desorption process is poorly reversible, a condition referred to as *hysteresis,* and the bulk of the sorbed material remains tightly bound to the sorbent, even when subjected to repeated extractions. The reversibility (or lack thereof) of the P sorption process is of considerable importance because it affects not only the availability of native or added P for plant nutrition but the potential for P loss to waters in surface or subsurface runoff. Many studies have shown that P sorption is highly hysteretic, a phenomenon that contributes to the retention of P in soils in forms that are only slowly available for plant uptake and that facilitates P transport to surface waters in association with eroded soil particles.

Hundreds of scientific papers have been published on the macroscopic aspects of sorption and desorption of P by soils, and several excellent review chapters are available on this subject (Barrow, 1985; Frossard et al., 1995; Iyamuremye and Dick, 1996; Sample et al., 1980; Sanyal and De Datta, 1991; Syers and Curtin, 1989). Fewer studies have reported detailed mechanistic information on the specific chemical processes by which P is sorbed to soils. Even fewer still have examined the principles and causes of P desorption from soils or soil constituents. While it is well beyond the scope of this chapter to completely review the literature on P sorption and desorption, several studies are cited to illustrate the factors controlling sorption and desorption and the importance of these processes to plant nutrition and environmental quality.

Most studies of P sorption (and desorption) have used short-term laboratory sorption isotherm experiments in which a series of solutions containing increasing concentrations of P is allowed to react with air-dried soils or model soil constituents (e.g., calcite, gibbsite, goethite) for a defined time period, typically 24 to 48 h, at a constant temperature and a wide solution to soil ratio (e.g., 25:1) (Nair et al., 1984; Olsen and Watanabe, 1957). The solid and solution phases are then separated by centrifugation and/or filtration and the P removed from solution is said to have been sorbed by the solid phase. The intent of studies such as these is usually to compare the magnitude of P sorption between soils as a function of soil properties or soil management practices, not to deduce actual sorption mechanisms.

Several mathematical approaches have been used with data from sorption isotherms to calculate parameters related to P availability, such as sorption maxima and bonding energies between P and the solid phase. A common example is the widely used Langmuir equation:

$$P_{ads} = (kP_{eq}b)/(1 + kP_{eq}) \quad [3]$$

where P_{ads} is the amount of P sorbed by the soil, P_{eq} is the concentration of P in solution at equilibrium, and k and b are parameters calculated from the linearized version of this equation

$$P_{eq}/P_{ads} = (1/kb + P_{eq}/b) \quad [4]$$

and presumably represent estimates of the bonding energy of P with the soil (k) and the maximum amount of P that can be sorbed by the soil (b) (Fixen and Grove, 1990). An example of this approach is illustrated in Fig. 2–7, which shows the varying effects of P source (fertilizer vs. dairy manure) on P sorption by three soils varying in clay content from 15 to 50% (Sharpley, 2000). Calculations using the Langmuir equation showed that P sorption maxima (b) for the Hagerstown (15% clay; fine, mixed, semiactive, mesic Typic Hapludalfs), Opequon (25% clay; clayey, mixed, active, mesic Lithic Hapludalfs), and Kingsbury (50% clay; very-fine, illitic, mesic Aeric Epiaqualfs) soils were 172, 303, and 476 mg P kg^{-1} when fertilizer P was added and 245, 405, and 909 mg P kg^{-1} when manure was the P source.

Other methods used to quantify P sorption reactions with soils include the Freundlich equation ($P_{ads} = kP_{eq}^{1/n}$) and the Temkin equation ($P_{ads} = a\ln C + b$), where k, n, a, and b are empirical constants derived from fitting P sorption data to these equations (Olsen and Khasawneh, 1980). The assumptions and interpretation of these three equations have been questioned on theoretical and experimental grounds (Barrow, 1989; Harter and Baker, 1977; Larsen, 1967; Sparks, 1995). For example, the Langmuir equation assumes that the system is at equilibrium, that adsorption is reversible, that only monolayer coverage of the surface occurs, that the adsorption energy is the same for all sites and is independent of surface coverage, and that there is no interaction between sorbed species. Most of these assumptions are unlikely to be true in a complex, heterogenous matrix such as the soil. The general

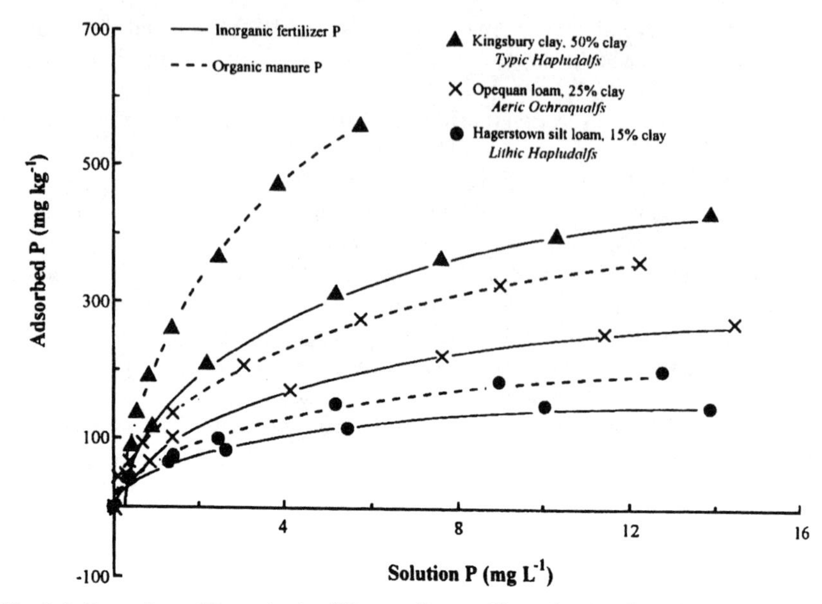

Fig. 2–7. Comparison of P sorption by different soil types, illustrating the effect of soil properties and manuring on P sorption (Sharpley, 2000).

consensus today is that these types of equations are useful in qualitatively describing and comparing the effects of soil properties and soil management on P sorption, but cannot be used to provide mechanistic information on the processes by which P is adsorbed to soil constituents or the energies of adsorption for P (Sparks, 1995). Encouragingly, recent advances in spectroscopic analysis of soils or soil constituents, such as XAFS and X-ray absorption near-edge spectroscopy (XANES), are now providing more insight into the mechanisms of P sorption by soils.

While short-term sorption isotherm type studies cannot answer fundamental questions about the exact chemical processes involved in P retention or the long-term sorption of P that occurs at low soil moisture contents, they have elucidated several aspects of P sorption by soils that are useful in our efforts to improve P management. First, the major factors that influence P sorption and desorption by soils are now well known and include clay content and mineralogy; the amount of amorphous Al and Fe oxyhydroxides, carbonates, and OM present in the soil; and the effects of soil solution chemistry (pH, ionic strength, competing anions, oxidation–reduction status) on the solid phases and forms of P in solution. Second, P sorption reactions are kinetically biphasic; that is, they are typified by a rapid initial reaction, typically one day or less, followed by a much longer, slow reaction that can last for weeks or longer (Enfield et al., 1981; Munns and Fox, 1976; Parfitt, 1978; van der Zee and van Riemsdijk, 1988). The initial reaction represents nonspecific adsorption and ligand exchange on mineral edges or by amorphous oxides and carbonates. The second, slow reaction may involve surface precipitation or polymerization on mineral surfaces and/or diffusion of adsorbed P into the interior of the solid phase. The time-dependent nature of P sorption is clearly illustrated in Fig. 2–8 for seven acidic Irish soils that varied widely in amorphous Al and Fe contents

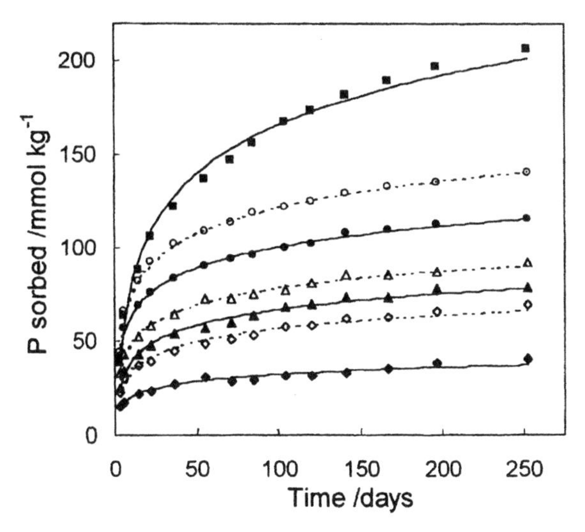

Fig. 2–8. Illustration of slow and fast P sorption reactions occurring in seven acidic agricultural soils from Northern Ireland (Maguire et al., 2001).

(Maguire et al., 2001). Desorption of P also proceeds in a temporally biphasic manner, as shown in Fig. 2–9 where the fitted desorption curves for six Belgian soils are partitioned into "fast" and "slow" desorbing pools (Lookman et al., 1995). The "fast desorbing pool" presumably represents P bound to reactive surfaces in direct contact with the aqueous phase, soluble P resulting from desorption or dissolution of recent additions of fertilizer or manure, physically adsorbed orthophosphate, and

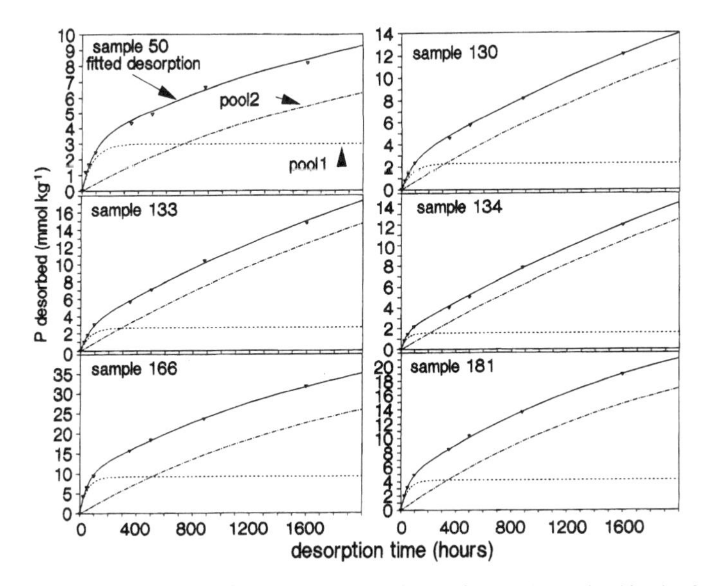

Fig. 2–9. Fitted desorption curves for six Belgian soils illustrating the desorption kinetics for the fast and slow desorbing pools of soil P (Lookman et al.,1995).

P complexed by OM. Soil P in the "slow desorbing pool" most likely originates from diffusion of P from interior sites inside soil solid phases or aggregates or slow dissolution of amorphous or crystalline solid phases of P.

The amount, type, and combination of inorganic and organic constituents present in a soil plays a major role in determining the extent of P sorption and the nature and rate of P desorption. In general, as shown in Fig. 2–10, P sorption increases in proportion to the amount of clay and reactive oxides or carbonates present in the soil, and is influenced by the physical and chemical characteristics of these solid phases (e.g., mineralogy, crystallinity). In highly weathered, acidic, mineral soils (e.g., Ultisols, Oxisols), P sorption is primarily controlled by 1:1 clay minerals, such as kaolinite, and amorphous oxyhydroxides of Al (e.g, gibbsite) and Fe (e.g., goethite). The Al and Fe oxyhydroxides may exist as discrete phases, as coatings on soil particles, or in complexes with soil OM. However, under conditions of extensive weathering, initially amorphous Al and Fe oxides can crystallize and P sorption capacity can decrease. Less weathered mineral soils that develop in more temperate climates (e.g., Mollisols, Vertisols) are dominated by 2:1 clay minerals (e.g., montmorillonite) and have lower P sorption capacities. Organic soils and very sandy soils have the lowest P sorption capacities. Lower P sorption capacities in organic soils result from several factors, including the small amount of clay minerals and metal oxides present, the coating of clays and oxides by humic molecules which can reduce the affinity of these solid phases for P, and the presence in the soil solution of organic anions from the decomposition of OM that compete with

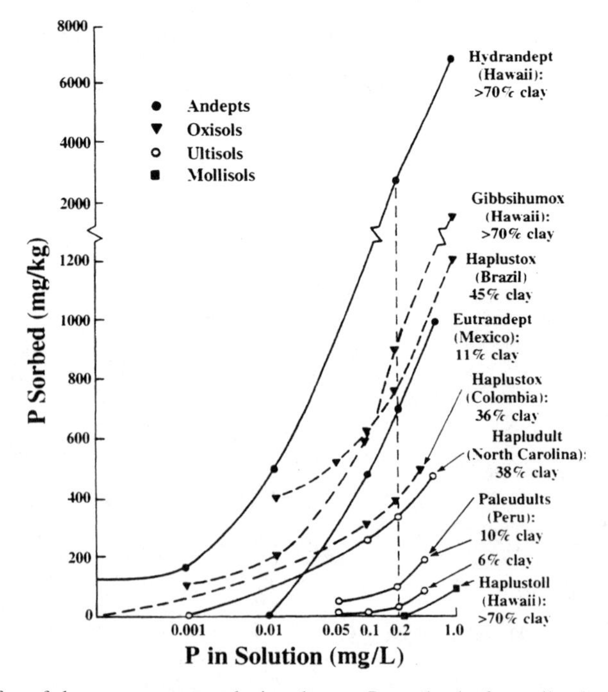

Fig. 2–10. Effect of clay type, content, and mineralogy on P sorption by four soil orders (Sanchez and Uehara, 1980).

P for sorption sites on mineral phases. In very sandy soils, P sorption is reduced because of the low quantity of reactive mineral phases; most sorption is due to coatings of Al and Fe oxides on sand grains. In alkaline soils, P sorption is predominantly due to reactions of orthophosphate with Ca carbonates and hydrous Fe oxides.

Soil pH affects the reactivity of soil constituents for orthophosphate, and other anions, by changing the nature of the functional groups and the electronegativity of the surface charge. The result is a well-defined relationship between anion adsorption capacity and soil pH, commonly referred to as an *adsorption envelope* (Fig. 2–11a; Sparks, 1995). As soil pH increases surface functional groups become de-

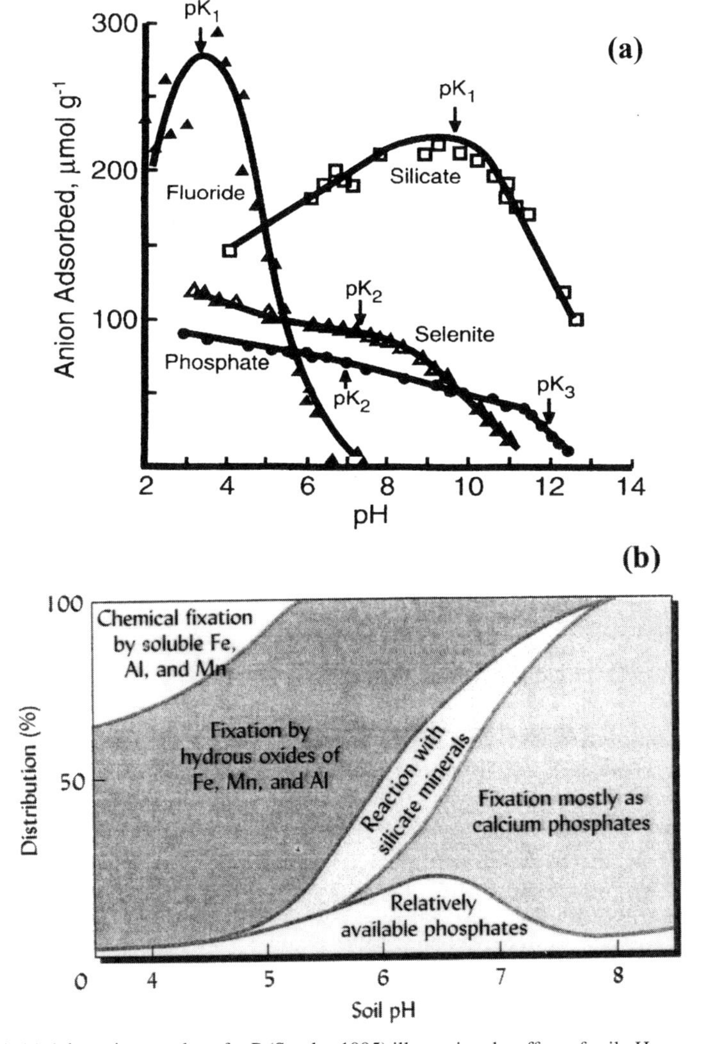

Fig. 2–11. (a) Adsorption envelope for P (Sparks, 1995) illustrating the effect of soil pH on sorption of P and other anions; (b) generalized relationship between soil pH and the forms and availability of P in soils (Brady and Weil, 2002).

protonated (e.g. $OH_2^+ \rightarrow OH_2 \rightarrow OH^-$), the colloidal surface becomes progressively more negatively charged, and P adsorption decreases. In the case of polyprotic conjugate acids such as H_3PO_4, clear breaks can be distinguished at each pK_a value in the slope of the relationship between the amount adsorbed and soil pH (pK_a = negative log of the acid dissociation constants. For H_3PO_4, $pK_{a1} = 2.1$; for $H_2PO_4^{-1}$, $pK_{a2} = 7.2$; for HPO_4^{-2}, $pK_{a3} = 12.3$). Phosphorus availability, then, is generally favored by liming acid soils because this decreases the adsorption of P by clay minerals and metal oxides as well as the potential for precipitation of Al and Fe secondary P minerals. Overliming acid soils (pH > 7.0), however, can reduce P availability by promoting the precipitation of solid phases of Ca–P. For most plants, the optimum pH range for plant availability of P is about pH 5.5 to 6.5 (Fig. 2–11b). Sorption of P is also influenced by other factors such as the presence of competing anions, the oxidation–reduction (redox) status of the soil, and the extent to which a soil's sorption capacity is saturated with P. Some key aspects of each of these factors are summarized next.

Inorganic and organic anions can compete with orthophosphate for sorption sites on soil colloids, resulting in a decrease in P sorption. The major competing inorganic anions in most soils are sulfate (SO_4^{2-}), hydroxide (OH^-), silicate ($H_3SiO_4^-$), and molybdate (MoO_4^{2-}). Nitrate (NO_3^-) and chloride (Cl^-) can also be present but are weakly held by soils and do not compete well with P for sorption sites. In general, orthophosphate is bound to soil colloids more tightly than the other major inorganic anions present in most soil solutions. Hence, competitive effects should be minimal except at high solution concentrations of these anions. A wide range of organic anions (e.g., carboxylic acids, oxalate, citrate) that originate from the decomposition of native and added OM (crop residues, manures, biosolids) can also compete with orthophosphate for sorption sites on clays and metal oxides. The importance of these competing anions at reducing P sorption and thus increasing P availability to plants and mobility in soils depends on their relative affinity for the colloidal surface, which can depend on soil pH, and their relative concentration in the soil solution.

Iyamuremye and Dick (1996) reviewed the effects of organic amendments on P sorption by soils in both aerobic and anaerobic environments and identified four main processes that could affect P availability. First, organic acids can complex with metal ions (Al, Fe) in solid phases, reducing their reactivity for P and inhibiting P sorption. Second, organic acids can directly compete with orthophosphate for sorption site on clays and metal oxides. Third, organic acids can affect surface charge, generally increasing the electronegative nature of soil colloids which then inhibits sorption of negatively charged orthophosphate ions. Fourth, organic amendments can dissolve P solid phases, through solubilization reactions between the carboxyl and hydroxyl functional groups of organic acids and metal ions in solid phases, thus reducing the surface areas for P sorption.

Studies of the effects of adding pure organic compounds, crop residues, manures, and biosolids to soils on P sorption have produced contradictory results that appear to depend on the nature of the organic material used, the type of soil, and soil pH (Hue, 1991, 1992; Ohno and Crannel, 1996; Ohno and Erich, 1997; Singh and Jones, 1976; Violante et al., 1991). Sibanda and Young (1986) and Violante and Gianfreda (1993) studied P sorption in the presence of organic acids (e.g., humic and fulvic acids, citrate, malate, and oxalate) and showed that these organic compounds could inhibit P sorption by soils (Fig. 2–12a). The concentration of P in organic amendments will also affect P availability as some of this P will be released into the soil solution by mineralization and dissolution reactions. Singh and Jones (1976) reported that a wide range of organic materials all inhibited P sorption by a Typic Vitrandept in the near term (<30 d), but that in the longer term (30–150 d) only amendments with a P content >0.31% (alfalfa hay, bean straw, barley [*Hordeum vulgare* L.] straw, and poultry manure) decreased P sorption and increased labile P. Materials with lower P contents (sawdust, cornstalks [*Zea mays* L.], and wheat

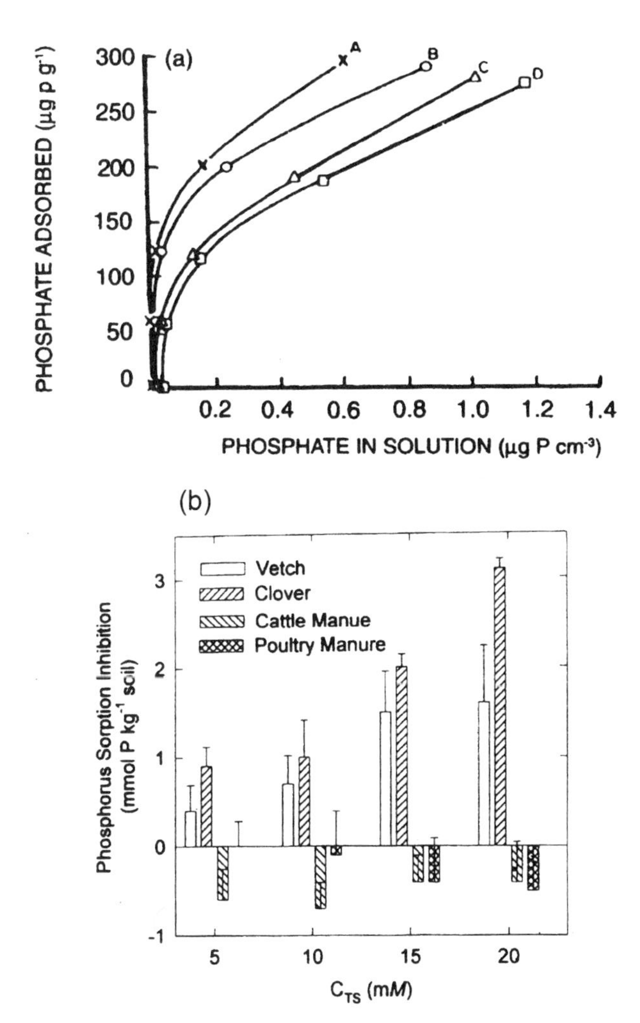

Fig. 2–12. (a) Effect of humic acid (HA) on P sorption by a tropical soil: no HA (A), 0.4% HA (B), 1.6% HA (C); 3.0% HA (D) (Sibanda and Young, 1986). (b) Comparison of the effectiveness of organic ligands isolated from aqueous extracts of green manures and animal manures on the inhibition of P sorption by an Aquic Haplorthod; C_{TS} is the concentration of total soluble C in each extract (Ohno and Crannell, 1996).

straw [*Triticum aestivum* L.]) increased P sorption and decreased labile P between 30 and 150 d.

Only a few studies have provided direct evidence of the mechanisms by which organic amendments affect P sorption. Ohno and Crannell (1996) found that dissolved OM (DOM) in aqueous extracts from plant residues (hairy vetch [*Vicia villosa* Roth subsp. *villosa*] and crimson clover [*Trifolium incarnatum* L.]) inhibited P sorption in an acidic Aquic Haplorthod, but that DOM in extracts of dairy and poultry manure did not have the same effect (Fig. 2–12b). They concluded that the chemical nature of the DOM from the plant residues caused it to be more effective at complexing with solid phase Al in this soil than the DOM in the manure extracts and that the formation of organic ligand–Al complexes inhibited P sorption. A possible reason for the greater effectiveness of the plant DOM was its lower molecular weight (≈710–850 vs. ≈2000–2800 for manures) and less rigid molecular structure. These factors may have allowed for more extensive bonding between DOM from plant residues and the soil surfaces responsible for P sorption, thus decreasing the number of sites capable of sorbing P. Their results were similar to those reported by Traina et al. (1986), who used model systems and found that at higher organic acid concentrations, citric acid complexed and then solubilized solid phase Al as an Al–OM complex, decreasing P sorption capacity. Clearly, given the widespread and increasing use of green manures, animal manures, and municipal biosolids as soil amendments, basic and applied research on the effect of organic compounds on P sorption and desorption is badly needed. Iyamuremye and Dick (1996) summarized their review of the effects of organic amendments on P sorption by soils by concluding, "Although there is considerable evidence that organic amendments can decrease P sorption in acid soils, we were struck by the dearth of field-based research to determine whether the results of more basic studies can be applied under field conditions." They identified several key research needs in this area, including (i) efforts to the effects of P contributions (inorganic or organic) vs. the effect of organic acids on P sorption when organic residues are added to soils, (ii) more in situ studies of the effects of organic amendments on P sorption in flooded soils, (iii) more field studies of the potential for organic amendments to increase plant P availability and fertilizer P efficiency by decreasing P sorption, and (iv) development of predictive models of the effect of organic amendments on P sorption based on amendment composition and soil properties.

Sorption and desorption of P can also be markedly affected by soil oxidation–reduction (redox) status. This is of particular importance in soils that are flooded, either naturally (e.g., riparian zones, wetlands) or for agricultural or environmental purposes (e.g., rice [*Oryza sativa* L.] production, constructed wetlands). The most common effects on P sorption are those associated with changes in the solid phases of Fe that are susceptible to dissolution as soils become anaerobic under flooded conditions and to reprecipitation as soils are drained and aerobic conditions are reestablished. Similar redox changes can affect Fe and P solubility in sediments in fresh waters and estuaries where oxic zones can exist at the sediment–water interface, overlying and interacting with anaerobic zones in deeper sediments.

The major changes that occur when soils are flooded were reviewed by Sanyal and De Datta (1991) and Iyamuremye and Dick (1996). In brief, the driving force for soil reduction reactions under anaerobic conditions is the microbial decomposition of soil or added OM, which generates electrons that under aerobic conditions would be accepted by O_2, forming H_2O. Flooding soils decreases the O_2 concentration in soil pores and results in a lower redox potential (Eh). In this situation, alternative electron acceptors, such as NO_3^-, $MnO_2(s)$, amorphous and crystalline Fe oxides, Fe^{3+}–bound phosphate minerals, SO_4^{2-}, and CO_2 accept the electrons generated by microbial respiration and are reduced, resulting in changes in both soil solution composition and the nature of soil solid phases. Soil pH also changes when soils become reduced, usually increasing in acidic soils as H^+ ions are consumed in reduction reactions and decreasing slightly in alkaline soils as CO_2 produced during anaerobic respiration is converted to carbonic acid (H_2CO_3; Ponnamperuma, 1972). Soil reduction affects P concentrations in the soil solution by three main processes. First, pH effects affect P concentrations. In response to the increased pH values that occur when acidic soils are reduced, hydrolysis of Al and Fe phosphates and desorption of P by anion exchange reactions with OH^- are enhanced. In calcareous soils the pH decreases caused by H_2CO_3 acid formation can also dissolve Ca–P minerals. Second, solubilizing Fe^{3+}–P minerals which slowly dissolve as soils become reduced, releasing Fe^{2+} and P into solution. Third, altering the surface reactivity of Fe oxides for P, by dissolving crystalline $Fe(OH)_3$ type minerals upon reduction (e.g., flooded soils), thus decreasing P sorption capacity and bonding strength for P. Note that upon reoxidation (e.g., drained soils), amorphous $Fe(OH)_2$ solid phases with greater surface areas and higher P sorption capacities may reprecipitate and restore or increase P sorption capacity (Iyamuremye and Dick, 1996). Any external factors that accelerate soil reduction, such as addition of readily decomposable OM or higher soil temperatures, both of which stimulate microbial activity, will increase the extent to which P sorption and desorption change upon flooding.

Changes in soil P chemistry that occur upon flooding and draining are complex and do not always follow clearly defined patterns. Numerous studies have investigated the effects of changing redox conditions on P solubility and bioavailability (Holford and Patrick, 1979; Khalid et al., 1977; Nair et al., 1999; Sah and Mikkelsen, 1986, 1989a, 1989b, 1989c; Vadas and Sims, 1999; Yu, 1985). Two studies, one involving soils and one with lake sediments, illustrate the importance of redox changes to soil P chemistry, plant P availability, and water quality.

Sah and Mikkelsen (1989a) evaluated the effects of flooding, OM addition, and temperature on P sorption by 10 flooded-drained California soils. Soils were flooded for periods up to 90 d, drained, air-dried, and used in P sorption isotherm experiments. Results showed that flooding and draining only increased P sorption in 5 of the 10 soils investigated. Soils where P sorption did not increase were either low in organic C, low in Fe content, or high in $CaCO_3$. Addition of OM (cellulose + starch) to soils prior to flooding and draining and incubation at higher temperatures increased P sorption, even in soils where flooding and draining alone had no effect on sorption (Fig. 2–13a). A related study (Sah and Mikkelsen, 1989b) showed that OM additions also increased the concentration of amorphous Fe in these soils, which, in turn, was highly correlated with P sorption, a phenomenon also reported by Khalid et al. (1977) and Roy and De Datta (1985). A third study by Sah and Mikkelsen (1989c) showed that flooding and draining soils that contained easily decomposable OM and reducible Fe induced P deficiency in corn by increas-

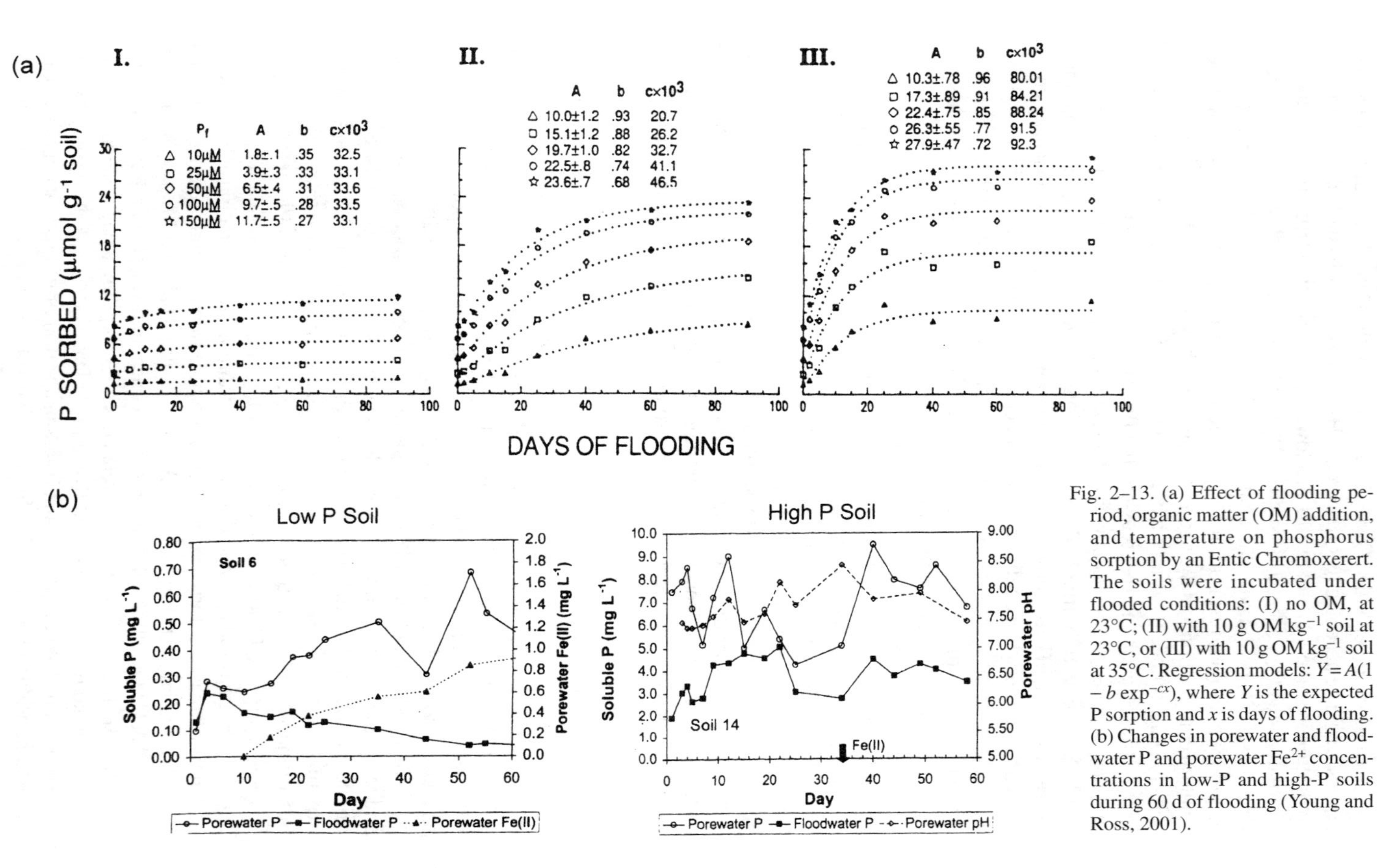

Fig. 2–13. (a) Effect of flooding period, organic matter (OM) addition, and temperature on phosphorus sorption by an Entic Chromoxererert. The soils were incubated under flooded conditions: (I) no OM, at 23°C; (II) with 10 g OM kg^{-1} soil at 23°C, or (III) with 10 g OM kg^{-1} soil at 35°C. Regression models: $Y = A(1 - b \exp^{-cx})$, where Y is the expected P sorption and x is days of flooding. (b) Changes in porewater and floodwater P and porewater Fe^{2+} concentrations in low-P and high-P soils during 60 d of flooding (Young and Ross, 2001).

ing P sorption and decreasing P desorption. Sah and Mikkelsen (1989a) concluded that predicting changes in P sorption and bioavailability upon flooding and draining must include an assessment of the interactions between factors that affect microbial activity, and thus the extent of soil reduction that will occur (e.g., OM, temperature) and the forms and concentrations of Fe in soils.

Young and Ross (2001) characterized the effect of flooding on changes in P sorption and solubility in 14 soils from Vermont using a 60-d laboratory microcosm study that allowed them to separately measure dissolved P concentrations in porewaters of the flooded soils and in the overlying waters. Flooding soils increased P concentrations in porewaters from 2- to 27-fold; however, increases in dissolved P in the overlying waters were much lower (about four times the initial concentration). Porewater Fe^{2+} concentrations also increased with time of flooding, but Fe^{2+} concentrations in the overlying floodwater were always below detection limits and Eh measurements showed that the floodwater remained oxidized throughout the study (Fig. 2–13b). Porewater and floodwater P concentrations also increased in proportion to the concentration of available soil P and the degree of soil P saturation. Average P concentrations in porewater and floodwater from low P soils ranged from 0.04 to 0.34 and 0.02 to 0.06 mg P L^{-1}, respectively. In high P soils, P concentrations ranged from 0.94 to 7.0 mg P L^{-1} (porewater) and 0.17 to 1.85 mg P L^{-1} (floodwater). Young and Ross (2001) concluded that oxidizing conditions forming in a redox interface between the flooded soils and the floodwater caused reprecipitation of Fe^{2+} dissolved from the soils and simultaneous sorption of P released from the reduced soils. This could potentially mitigate the movement of P from reduced soils or sediments to overlying surface waters. Moore and Reddy (1994) observed similar trends for the effect of a redox interface on Fe and P solubility in sediments from Lake Okeechobee in Florida. However, they also stated that the sorption capacity of freshly precipitated Fe oxides could be exceeded if soils are highly saturated with P, resulting in increased concentrations of dissolved P in floodwaters, an environmental concern.

Organic Phosphorus in Soils: Forms and Transformations

Organic P transformations are critical components of P cycling in managed and natural ecosystems and can have major impacts on soil productivity. As mentioned above, from 30 to 65% of total soil P in mineral soils and 90% in organic soils can be in organic forms (Table 2–2; Harrison, 1987). Natural (climate, vegetation, soil type) and anthropogenic (use of green and animal manures, crop rotation, fertilization) factors can affect the amount of organic P in soils. In a review of the literature, Harrison (1987) identified total P content, soil pH, organic C content, and sampling depth as the most significant soil properties for predicting organic P content. However, significant interactions with other parameters were found, and in some instances the method of determination (ignition vs. extraction), soil texture, geographical region, land class, or soil order influenced the ability to predict organic P. Many studies have shown organic P can provide a significant amount of available P to plants and much interest exists today in better incorporating organic P dynamics into nutrient management recommendations and computer models that describe P cycling (Blair and Boland, 1978; Dalal, 1979; Magid et al., 1996;

Table 2–2. Range in organic P content for selected soil orders as summarized by Harrison (1987).

Soil type	Range in organic P content
	mg kg^{-1}
Alfisols	0–1105
Aridisols	0–1775
Entisols	0–1090
Histisols	420–845
Inceptisols	2–1410
Mollisols	0–1510
Oxisols	1–840
Spodosols	0–1360
Ultisols	0–955
Vertisols	6–1035

McLaughlin et al., 1988; Oehl et al., 2001; Sharpley, 1992; Sharpley and Smith, 1989; Stewart and Tiessen, 1987; Tate, 1984; White and Ayoub, 1983).

Understanding the chemical nature of organic P in soils is a prerequisite to efficient P management of this important component of the P cycle. The ultimate sources of soil organic P are plant and animal residues. Microorganisms decomposing these residues can transform the organic P compounds found in them to other forms of varying resistance to further degradation. Inositol phosphates, being relatively resistant to degradation, generally comprise the majority of the identifiable organic P in soils. The basic inositol group is the six C ring structure, hexahydrohexahydroxybenzene. Inositol phosphates are monoesters with the hexaphosphate ester, phytic acid, being relatively common in soils (Stevenson, 1986). The myo-inositol form of the hexaphosphate form is shown in Fig. 2–14. Mono-, di-, tri-, tetra-, and pentaphosphates are also found and may be degradation products of the hexaphosphate form of inositol phosphate. Phospholipids are the second most abundant identifiable form of organic P in soils. This fraction is thought to contain phosphatidyl inositol, lecithin, serine, and ethanolamine (Stevenson, 1986). These groups contain glycerol, fatty acids, and phosphate, as shown in Fig. 2–14, with R and R′ being fatty acids and the R″ group being inositol, lecithin, serine, or ethanolamine. The P in the structure is a diester, which is more susceptible to degradation in soils compared with monoesters. Nucleic acids represent a small portion

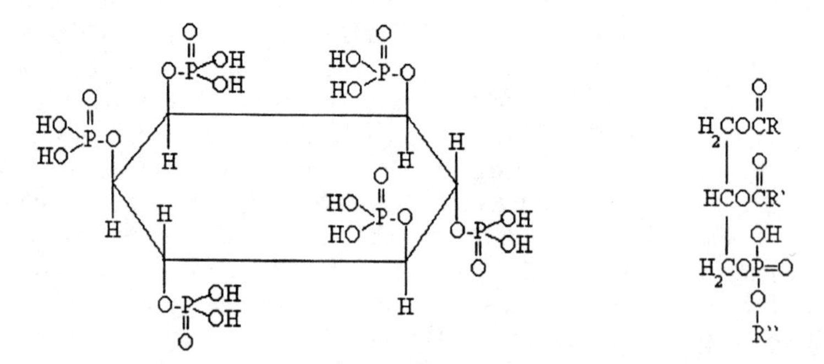

Fig. 2–14. Myo-inositol hexaphosphate and a generalized structure for phospholipids, with R and R′ representing fatty acids and R″ representing inositol, lecithin, serine, or ethanolamine.

of identifiable organic P in soils. They are readily degraded and their presence is related more to the continued production by soil microorganisms than to their persistence in the soil environment. Phosphonates are organic P compounds with direct C–P bonding, as opposed to the C–O–P bonding in most other organic P molecules commonly found in soils. Phosphonates were only discovered in soils in the 1980s (Newman and Tate, 1980), and the main compound reported to date has been 2-aminoethyl phosphonic acid (AEP).

Methods to measure and identify soil organic P have evolved slowly and, despite some recent advances, a large proportion of soil organic P remains difficult to precisely identify or place into a specific category. This may be related to the incorporation of organic P into soil OM or the formation of insoluble organic P compounds such as phytin, the Ca or Mg salt of phytic acid. In the former case complex entities may be formed that are difficult to identify or characterize. In both cases, the organic P may be in forms that are not easily extracted or detected or that change during the process of extracting and separating organic and inorganic P.

Much of the research investigating soil organic P has relied on ignition or sequential chemical extraction methods to estimate the total organic P concentration in soils or to partition organic P into pools of varying bioavailability (Bowman and Moir, 1993; Hedley et al., 1982; Kuo, 1996; Tiessen and Moir, 1993). The ignition (heating soil at high temperature) method, which involves acid extraction of unignited and ignited samples of the same soil, has been widely used to measure total organic P. This method assumes that the increase in acid-extractable inorganic P after ignition is caused by the release of P from organic P compounds oxidized during ignition. The details and limitations of methods to characterize organic P have been reported (Agbenin et al., 1999; Condron et al., 1990; Williams et al., 1970). For example, Condron et al. (1990) found the ignition method could overestimate organic P in tropical soils because heating the soil at high temperatures increased the solubility of inorganic P compounds. Sequential extraction methods have also been widely used. However, they have been criticized because the reagents used cannot be shown to be selective for specific compounds (inorganic or organic) or to accurately measure the bioavailability of P, and because of experimental concerns (e.g., reagents may hydrolyze organic P to inorganic P, or readsorption of P may occur between extraction steps) (Barbanti et al., 1994; Magid et al., 1996).

Advances in the application of analytical techniques are improving our ability to identify the forms of organic P in soils and how they change with time and soil management practices. Examples include NMR spectroscopy and various types of chromatography (gas, high pressure liquid, thin-layer). However, the equipment and analytical costs, time involved, and the complex nature of interpretation of analytical results do not make these methods suitable for routine characterization of soils for organic P. Despite these problems, use of ^{31}P NMR, both solid state and liquid, has been particularly helpful in increasing our knowledge of the nature and forms of organic P in soils and sediments (Adams and Byrne, 1989; Cade-Menun and Preston, 1996; Hupfer and Gachter, 1995; Mahieu et al., 2000; Robinson et al., 1998; Wilson, 1987). The use of NMR alters our classification of organic P somewhat, as it may only be able to distinguish monoester, diester, or phosphonate forms. Inositol phosphate, the phosphorylated sugars in nucleic acids, and some phospholipids are monoesters while other phospholipids and nucleic acids are diesters. Identification of individual moieties within the mono- and diester groups, and how they change with soil management, is sometimes possible. A study of soils under native vegetation or cultivation indicated that the undisturbed soils had orthophosphate mono- and diester forms or organic P while cultivated soils had only monoester forms (Condron et al., 1990). Young and poorly drained soils, where mineralization was suppressed, on a toposequence had higher proportions of orthophosphate diesters than older soils in the same sequence (Gil-Sortes et al., 1990). Diester forms of organic P added to a soil with biosolids disappeared within 28 d in an incubation study, presumably due to microbial degradation, while monoester forms persisted for at least 140 d (Hinedi and Chang, 1989).

The contribution of organic P to plant-available P and the potential for the loss of dissolved or particulate organic P by leaching or runoff is not fully understood and can be difficult to study. The balance between mineralization (biological process where enzymes produced by plants and microorganisms hydrolyze organic compounds releasing inorganic P into solution) and immobilization (conversion of soil inorganic P to organic P in microbial or plant biomass) ultimately controls the concentration of P in the soil solution. Mineralization and immobilization are influenced by factors that affect biological activity (e.g., temperature, moisture, energy and nutrient availability, redox, pH) and thus can vary spatially and temporally as a function of soil type, climate, and soil management practices. Further, studying P mineralization–immobilization reactions in soils or from organic soil amendments is a complicated endeavor. The fate of inorganic P released by mineralization is the same as that of any other inorganic P already in or added to the soil system: fairly rapid sorption by soil constituents or absorption by plants or microbes. Thus, unlike studies of the mineralization of organic N where the end-product (NO_3–N) is very soluble and easy to measure, the rapid disappearance from solution of the end-products of organic P mineralization makes it difficult to quantify the rate and extent of conversion of organic P to inorganic P. Some researchers have used changes in measured soil organic P concentrations, changes in operationally defined soil P pools (i.e., P fractions defined by sequential extraction procedures), and radiolabeled isotopes (^{32}P and ^{33}P) to study P mineralization. Organic P compounds can also be sorbed by soils, which can affect their availability for mineralization or immobilization, their diffusion in soils, and their loss by leaching and runoff. Research has shown that orthophosphate mono- and diesters can be sorbed by similar mechanisms as inorganic orthophosphate (e.g., ligand exchange) and that the rate and extent of sorption depends on the nature of the soil solid phase and the ester (Anderson et al., 1974; Celi et al., 1999; Shang et al., 1990). Lower molecular weight monoesters (e.g., inositol hexaphosphates) are sorbed more strongly than diesters, which may contribute to the predominance of inositol phosphates in soils, as sorption of organic P to soil mineral phases can cause it to be more resistant to microbial degradation.

Despite the difficulties in studying organic P transformations, it is clear that under some circumstances inorganic P released via mineralization can supply much of a crop's P requirements and that some soil management practices can seriously deplete organic P. In temperate soils mineralization can release from 5 to 20 kg P ha^{-1} yr^{-1}, similar to the annual uptake of P by most crops and native vegetation (Brookes et al., 1984). In highly weathered soils, such as those in the trop-

Table 2–3. Effect of soil type and crop residue type on the amount of inorganic and organic P mineralized and leached during an 84-d laboratory incubation study (Sharpley and Smith, 1989).

		Alfalfa		Corn		Soybeans		Wheat	
Soil type	Control	Incorp.	Surface	Incorp.	Surface	Incorp.	Surface	Incorp.	Surface
		mg inorganic P L^{-1}							
Bowie loam	0.54	0.48	0.72	0.56	0.66	0.38	0.56	0.38	0.52
Claremore clay	0.27	0.35	0.56	0.37	0.46	0.27	0.49	0.31	0.46
Ruston sandy loam	0.65	0.72	0.94	0.89	0.96	0.68	0.84	0.85	0.96
San Saba clay	0.35	0.26	0.55	0.48	0.55	0.33	0.45	0.35	0.43
Teller loam	3.84	4.14	5.42	3.08	3.58	4.23	5.04	3.78	4.05
Yahola silty clay	1.90	2.20	3.01	1.64	2.07	2.45	3.07	1.91	2.25
Avg.									
Inorganic P	1.13	1.24	1.73	1.05	1.27	1.25	1.60	1.14	1.33
Organic P	0.54	1.50	1.01	0.91	0.59	1.56	0.76	0.58	0.37

ics, mineralization of organic P is particularly important given the low solubility of inorganic P due to sorption and precipitation reactions with Al and Fe. The C/P ratio of crop residues and manures and the P content of these organic materials have also been shown to affect plant-available P in soils. In general, if residue P content is <0.2% or the C/P ratio is >300:1, immobilization of P can be expected. Mineralization is likely if residue P is >0.3% and C/P is <200:1 (Tisdale et al., 1993). Fertilization may also affect P mineralization. Sharpley (1985) found that net P mineralization was greater in unfertilized than in fertilized soils at 20 sites in Oklahoma and Texas, that most mineralized P came from a moderately labile pool of organic P ($NaHCO_3$ extractable), and that P mineralization was highly correlated with soil phosphatase activity. Sharpley and Smith (1989) found that type of residue, soil type, and whether or not the residue was incorporated by tillage could affect mineralization and leaching of P from residue-amended soils. More inorganic P was mineralized and leached when residues were surface applied, and more organic P was leached when residues were incorporated (Table 2–3). They concluded that leaving crop residues on the surface, instead of incorporation by tillage, could provide available P during early plant growth, but could also increase the loss of dissolved P in runoff, unless infiltrating water from rainfall leached the soluble P into the soil.

Depletion of organic P pools can cause significant changes in the management of agroecosystems. For example, several studies have documented declines in soil OM, organic P, and labile P in natural soils in the Great Plains that have been brought under cultivation (Lamb et al., 1985; Tiessen et al., 1983). In the initial years of conversion of prairie soils to cropland, mineralization of organic P provided adequate available P for crop production. However, with time, the mineralization of organic P could no longer supply sufficient plant-available P to meet crop requirements and fertilizer use was required to sustain yields. The decline in labile organic P reflects both the loss of P by crop harvest and the precipitation of mineralized organic P above crop requirements as sparingly soluble Ca–P minerals (apatite) (Tiessen et al., 1994). Similar trends were reported by Tiessen et al. (1992) for a Brazilian Oxisol, where cultivation with minimal fertilization reduced soil C, N, and P by 30% in 6 yr. However, the high P sorption capacity of this soil resulted in retention of mineralized organic P in forms that could not sustain crop production.

SOIL CHEMISTRY AND SOIL PHOSPHORUS MANAGEMENT

Decades of research and practical evaluations of soil management practices have resulted in the development of best management practices (BMPs) for P that, if implemented, should ensure agricultural productivity and protect water quality (Beegle et al., 2000; Chambers et al., 2000; Sharpley and Tunney, 2000; Steinhilber and Weismiller, 2000; Uusi-Kamppa et al., 2000). These BMPs are based on our knowledge of the soil P cycle and soil P chemistry. They reflect our understanding of when and how P in fertilizers and organic by-products must be applied to agricultural lands to sustain production of food, fiber, and other plant products and to establish and maintain the vegetation needed to reduce water, soil, and nutrient losses in erosion and runoff. These BMPs also recognize that, while the magnitude of P losses from soils to waters (<1 to 2 kg P ha^{-1} yr^{-1}) is generally not economically important to farmers, these losses do carry off-farm economic impacts that are clearly associated with water quality deterioration. Remediating impaired surface waters is lengthy and expensive. Therefore, preventing the nonpoint pollution of surface waters by P is the most economically efficient way for agriculture to protect and improve water quality.

In agricultural ecosystems this means identifying and implementing P management practices that meet agronomic and environmental goals. To do this, soil P fertility must be sustained and soil P losses must be controlled. Today, the overall guiding principles for sound P management include two fundamental concepts: (i) nutrient balance, at all scales, from field to farm to watershed and (ii) selection and efficient management of P sources in a manner that minimizes the transport of dissolved and particulate P to surface and shallow groundwaters. As illustrated in the following examples, each of these concepts is firmly linked to our understanding of the chemistry of P in soils and how this affects plant availability of P and the potential for P movement from land to water.

Nutrient Balance: Relationship to Soil Phosphorus Chemistry

An efficient P management program for any agroecosystem begins with a nutrient budget designed to calculate nutrient inputs to the system in feed, fertilizers, and by-products and nutrient outputs in crops, animal products, and losses by erosion, runoff, and leaching (Beegle et al., 2002). The dual goals of nutrient budgets are to (i) balance nutrient inputs and outputs in a manner that sustains soil fertility at a level appropriate to the intended use of the land and (ii) prevent the accumulation of nutrients in soils to values that enhance the risk of nonpoint pollution of surface and groundwaters. Systems with P deficits are usually those with inadequate resources to replenish P removed from soils by crop harvest, animal grazing, and soil erosion. With time, P depletions lead to a loss of soil fertility and productivity and eventually to soil degradation. Systems with P surpluses typically are those involving intensive animal production, where nutrient inputs in feed (fodder, mineral concentrates) often exceed nutrient outputs in animal products (meat, milk, eggs). Feed nutrients in excess of those exported in animal products accumulate in animal manures, which are usually applied to fields as a nutrient source for crops, substituting partially or wholly for commercial fertilizers. With time, P surpluses in the

Table 2–4. Effect of 16 yr of annual manure applications on total P and Olsen P in an irrigated Typic Haploboroll used for barley production in Canada (Whalen and Chang, 2001).

	Total P				Available P (Olsen P)†			
	Annual manure application rate (Mg ha^{-1} wet wt. basis)							
Soil depth	0	60	120	180	0	60	120	180
cm	g kg^{-1}				mg kg^{-1}			
0–15	0.92	1.76	2.8	3.75	95	452	736	964
15–30	0.64	1.11	1.76	2.02	21	229	517	612
30–60	0.57	0.67	0.8	0.93	4	30	99	176
60–90	0.6	0.54	0.63	0.64	3	5	18	26
90–120	0.61	0.6	0.61	0.64	3	6	20	21
120–150	0.6	0.6	0.66	0.67	5	10	32	48

† Olsen P value recommended for optimum crop production ranges from 20 to 30 mg kg^{-1}.

system lead to the buildup of soil P to values well above those needed for crop production, increasing the risk of environmentally significant losses of P to waters. This is clearly illustrated in Table 2–4, which shows the effects of 16 yr of cattle feedlot manure application on total P and plant-available P (Olsen P) in an irrigated Typic Haploboroll used for barley production in Canada (Whalen and Chang, 2001). The authors constructed a P mass balance for this site and reported that, at the 60, 120, and 180 Mg ha^{-1} yr^{-1} rates, 93, 88, and 85% of the P applied could be accounted for by crop removal or soil P. They speculated that unrecovered P could have been lost by leaching to groundwater; water tables at this site ranged from 50 to 250 cm and are always >150 cm. At a similar, nonirrigated site, approximately 100% of applied P was recovered in the crop and soil pools at all manure rates, suggesting that P losses were minimal without irrigation.

Our understanding of soil P chemistry plays a key role in the development of balanced nutrient budgets for agricultural systems. The most important example of this is the widespread use of soil testing to assess the P fertility status of soils or, more recently, the potential environmental impacts of P. Soil P testing to identify the need for P fertilization is a well-established practice, supported by decades of research and field-scale validation (Fixen and Grove, 1990; Kamprath and Watson, 1980; Sharpley et al., 1994; Sims, 2000). Most routinely used agronomic soil P tests (e.g., Bray P_1, Mehlich 1, Mehlich 3, Morgan, Olsen P) are mixtures of several chemical reagents (acids, bases, chelates) that were selected because our knowledge of P speciation and sorption–desorption reactions in soils provided evidence that they would extract labile P from inorganic solid phases (e.g., mineral and amorphous forms of Al–P, Ca–P, and Fe–P). For example, the Bray P_1 soil test (0.025 M HCl + 0.03 M NH_4F) was developed for use in acidic soils where Al–P is the primary solid phase affecting P solubility and desorption (Tisdale et al., 1993). This soil test was designed to use an appropriate concentration of F^-, well known to complex strongly with Al^{3+} in solution, to promote dissolution of Al–P, thus providing an indication of the ability of the soil to buffer depletions in solution P caused by plant uptake. The HCl in the extractant also dissolves Ca–P minerals that can be present in slightly acid to neutral soils. The soil P test used varies geographically as well, reflecting the physiographic differences in soil properties (e.g., acidic soils in humid regions vs. calcareous soils in arid climates) that soil chemistry studies have shown to affect P availability to plants. One of the limitations of soil P testing, and a key research need in this area, is the fact that most agronomic soil P tests were not designed to measure labile organic P in soils. This is particularly important for both P deficit farms, where crop residues may be a major, or the only, source of P, and P surplus farms, where overapplications of organic P sources, such as manures, regularly occur. However, given the complexity of organic P cycling, it seems unlikely that a simple, rapid, chemical extraction method for labile organic P can be developed. Tiessen et al. (1994) suggested that it is more likely that supplemental information obtained about the soil and soil management practices (e.g., OM content, pH, soil type, tillage, manure use) could be used to simulate the turnover of organic P and then be used in conjunction with chemical soil tests, which currently provide adequate estimates of available inorganic P, to estimate the total pool of labile P in soils.

In recent years there has been growing interest in using agronomic soil P tests, or other rapid soil tests, to characterize the risk of P loss from soil to water (Chardon and van Faassen, 1999; Sibbesen and Sharpley, 1997; Sims et al., 2000). In this case, the objectives are to determine if a soil is sufficiently enriched with P such that (i) high concentrations of soluble P are continually present and susceptible to loss by runoff and leaching, or (ii) erodible soil particles are enriched in forms of P that would be biologically available when they are deposited in surface waters. Because agronomic soil P tests measure biologically available P, including soluble P, and are used to identify the extractable P value in soils needed for economically optimum crop yields, they have been suggested as a rapid means to identify soils that are sufficiently overfertilized to be of concern for nonpoint P pollution of ground and surface waters. A number of recent studies have shown that agronomic soil test P is well correlated with P concentrations in runoff and leaching, suggesting that this approach may have some merit despite the fact that these tests were not developed specifically for this purpose (Heckrath et al., 1995; McDowell and Sharpley, 2001; Pote et al., 1996; Simard et al., 2000; Sims et al., 1998). Other approaches to environmental soil testing for P are direct measurement of water soluble P, estimating potentially biologically available and desorbable P with Fe oxide–impregnated filter paper strips, and characterizing the degree of P saturation of soils. Each of these methods has a foundation in soil P chemistry. Water soluble P directly measures P in the soil solution, the most biologically available and mobile form in soils. Short-term reactions of soils with an Fe oxide strip that acts as a sorption sink for P measures P in an "easily desorbable" pool, which should be most susceptible to losses in runoff and leaching (Chardon et al., 1996). Studies have also shown that Fe oxide P is well correlated with algal growth and plant P uptake, suggesting that it is also measuring bioavailable forms of soil P, important when assessing the impact of particulate P losses by erosion (Sharpley, 1993). Soil P saturation tests determine the ratio of the amount of P sorbed relative to the total P sorption capacity of soils, an index that has been shown to be well correlated with soluble P and P transport in runoff and by leaching (Sims et al., 2000). For instance, the original P saturation method developed in the Netherlands used acid ammonium oxalate extraction to measure total sorbed P and estimated P sorption capacity as the sum of oxalate-extractable Al and Fe. This was possible because short and long-term studies of P, Al, and Fe chemistry in Dutch soils showed that amorphous Al and Fe were the major solid phases responsible for P sorption. Recent studies have also shown that P saturation can be estimated using the Mehlich 3 soil test, as the ratio of M3-P to (M3-Al + M3-Fe) (Khiari et al., 2000; Sims et al., 2002). This allows for rapid, inexpensive determination of plant-available P (M3-P) and an environmental measure of soil P (M3 P saturation) with the same soil sample.

Phosphorus Source and Transport Management

In agricultural settings where soil testing, or some other approach, has shown that P inputs are needed to achieve the desired level of crop productivity, understanding how the many potential sources of fertilizer P (inorganic and organic) or crop residues interact with soils is critical and has been the focus of decades of soil P chemistry research. Key goals in this research have been determining the most effective application rates for various P sources and how application method (broadcasting, banding, injection) can affect the potential for plant P uptake and P loss to water. Similarly, in settings where P inputs are not needed, but surplus P exists (e.g., manures, biosolids) and land application is the only option, an understanding of soil P chemistry, along with an assessment of site hydrology and P management practices, can help to identify sites where P loss to water is most likely. This allows for a more effective prioritization of BMPs that can help reduce P transport, such as use of buffer strips, grassed waterways, and the like—another longstanding area of P research.

It is beyond the scope of this chapter to describe all the P sources used in agriculture, their chemical and biological reactions with soils, and how our understanding of these reactions has improved P management. Many reviews and textbooks provide this information (Foth and Ellis, 1997; Mengel and Rehm, 2000; Power and Dick, 2000; Sample et al., 1980; Tisdale et al., 1993). A few examples, however, illustrate how our knowledge of soil P chemistry can be used to maximize the efficient management of P sources. Most P fertilizers are derived from calcium phosphate rich ores (e.g., "rock phosphates" such as hydroxyapatite), which, as we know from mineral equilibrium studies, have inherently low P solubilities in most soils. Soil chemistry research has shown the agronomic efficiency of rock phosphates can be increased by (i) using them in warm, moist, acid soils with low concentrations of Ca^{2+}, conditions that promote apatite dissolution, and (ii) producing, through industrial processes, superphosphate fertilizers by reacting rock P with H_2SO_4 or H_3PO_4, which dissolves apatite, increasing the water solubility of these fertilizers in soils. Our knowledge of the solubility of P minerals has also been applied to reduce the environmental impact of manures and biosolids. Studies have shown that treating poultry litters with alum $[(Al)_2(SO_4)_3]$ results in formation of amorphous Al hydroxides that sorb soluble P in the litters, reducing runoff of dissolved P from pastures fertilized with these litters (Moore et al., 2000; Sims and Luka-McCafferty, 2002). Other studies have shown that biosolids produced at wastewater treatment plants that add Fe or Al salts to remove soluble P from wastewaters have lower P solubility in soils because of the formation of sparingly soluble forms of Al–P and Fe–P. This can result in both lower P availability to plants

(Frossard et al., 1996; McCoy et al, 1986) and reduced losses of dissolved P in runoff (Penn and Sims, 2001).

Reducing P loss to water requires BMPs that incorporate soil P chemistry in their design and mode of action. Phosphorus is lost from agricultural fields as either dissolved P (inorganic or organic) in runoff (surface and subsurface) or leaching waters, or as particulate P bound to eroded soil particles such as silt, clay, and OM. As rainfall interacts with the soil, P dissolves from soil particles, crop residues, fertilizers, and manures and then, depending on site hydrology, moves with surface runoff, laterally flowing subsurface water, or by leaching through the soil profile. However, as dissolved P moves along the surface of soil or through soil, it may be resorbed by soil or runoff sediment and become particulate P. The eroded sediment load leaving a field in surface runoff will determine the extent of particulate P losses. Losses of dissolved P and particulate P are influenced by the amount and forms of P present in soils, the physicochemical properties of any P sources used, and the rate, time, and method of P application. The most P will be lost from fields that have (alone or in combination): high soil P contents, highly erodible soils, artificial drainage systems (e.g., tile drains) that facilitate subsurface runoff, and high rates of organic or inorganic P applied. Losses can be exacerbated when P sources are broadcast and unincorporated, and when intense runoff events occur soon after P application. The least P will be lost from fields with agronomically optimum P values (low soil P can lead to losses of vegetation and enhance soil erosion), where little P is applied, where P is knifed into the soil in bands, and when intense runoff events occur well after application.

Based on our understanding of soil P chemistry and the processes by which P is transported from soil to water, it is relatively straightforward to outline strategies for reducing P losses from the landscape. In general, practices that reduce soil erosion, reduce the amount of surface runoff, or reduce the amount of P in the soil that interacts with runoff (the "mixing zone", approximately the uppermost 2.5 cm of soil) will minimize P losses. Past studies provide examples of these approaches.

Andraski et al. (1985a, 1985b) used a rainfall simulator to study soil, surface runoff, and P losses from simulated single storm events during a 3-yr period. For this soil the conventional tillage (moldboard plow) had consistently higher surface runoff than chisel, till–plant, and no-till systems. Soil and total P losses were also highest for the conventional tillage. Tillage effects were less consistent for dissolved and algal-available P losses, but, when significant differences occurred the losses were generally higher for the conventional tilled treatments than the remaining tillage systems.

The influence of tillage on runoff losses is not always consistent. Another rainfall simulator study found numerically higher surface runoff losses for no-till compared with conventional and chisel tillage systems for each of five simulations (Mueller et al., 1984a, 1984b). Manure applications had significant interactions with tillage for total, dissolved, and algal-available P losses. In the first year of the study there were no effects of tillage on total P losses. In the second year total P losses were higher for conventional tillage than for the chisel tillage or no-till systems without manure. Results varied when manure was applied. Soluble and algal-available P losses tended to be higher for no-till with manure applied compared with other tillage–manure combinations.

Blevins et al. (1990) reported significantly higher runoff volumes and sediment losses from conventional tilled treatments compared with chisel and no-till systems. Soluble P losses from the three tillage systems were somewhat variable, and consequently, no significant differences due to tillage were found. Unfortunately, total P losses were not reported.

Sharpley et al. (1992) demonstrated the relationship between soil cover, soil erosion, and, consequently, total P losses from natural runoff events in Texas and Oklahoma (Fig. 2–15). This study also brought forth another important conclusion in that as soil cover increased the percentage of the total P that was present in bioavailable forms increased, raising the question of the net effect of reduced tillage systems in protecting surface water quality.

Many other studies have shown increases in P concentrations in runoff or in P losses as P is added to soils or as soil P levels increase (Mueller et al., 1984b; Reddy et al., 1978; Romkens and Nelson, 1974; Sharpley et al., 1992). Less work has been done to examine the effects of P placement method on P losses. Romkens et al. (1973) found significantly less soluble P loss when P was incorporated as compared with surface broadcast applications. Similar results were reported by Baker and Laflen (1982) and Mueller et al. (1984b). Subsurface band applications of P have been shown to significantly reduce soluble and bioavailable P losses compared with broadcast applications for ridge-till and no-till, with less consistent effects on total P losses (Fig. 2–15b; Janssen et al., 1996).

In summary, our knowledge of the basic principles of soil P chemistry has proven invaluable in the development of BMPs that can sustain agricultural productivity and protect water quality. Yet, we know that significant challenges remain and that more basic and applied research on P chemistry is needed. Many soils in the world are severely depleted in P and cannot sustain food production. How can we use our knowledge of P cycling to restore the productivity of these soils? Other soils are saturated with P to the point that normal runoff and leaching are enriching surface and groundwaters and contributing to water quality degradation. How can we manage P saturated soils to prevent these P losses? In many cases, simple adoption of current recommendations based on our current understanding of P cycling and transport can markedly improve P management. However, in other cases we are in need of a greater understanding of soil P chemistry. For example, there is great interest in the role of organic P sources (crop residues, manures, composts) as soil amendments for severely P-depleted soils, especially in areas where fertilizer P is limited or not available. There is also interest in the long-term chemistry of P-saturated soils. If we stop further inputs of P to these soils, will the chemical species of P present gradually convert to more stable forms and result in a lower risk of soluble P loss? Or, will they remain slow-release sources of P to ground and surface waters for decades? Encouragingly, our ability to conduct research in areas such as these is improving, with the advent of advanced analytical techniques for inorganic and organic soil P and the availability of increasingly sophisticated field monitoring systems that measure P transport via leaching, erosion, and runoff. These approaches will provide more specific information about the forms of P (inorganic and organic) in soils and the mechanisms by which these different forms of P are retained, lost, and become biologically available. From this should come new BMPs that will further advance the agricultural and environmental management of P.

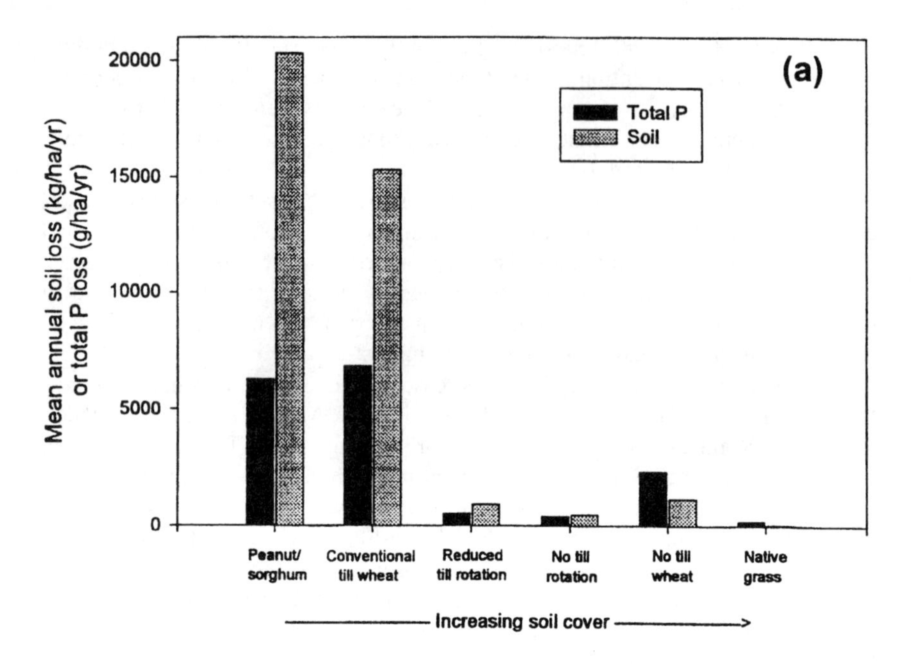

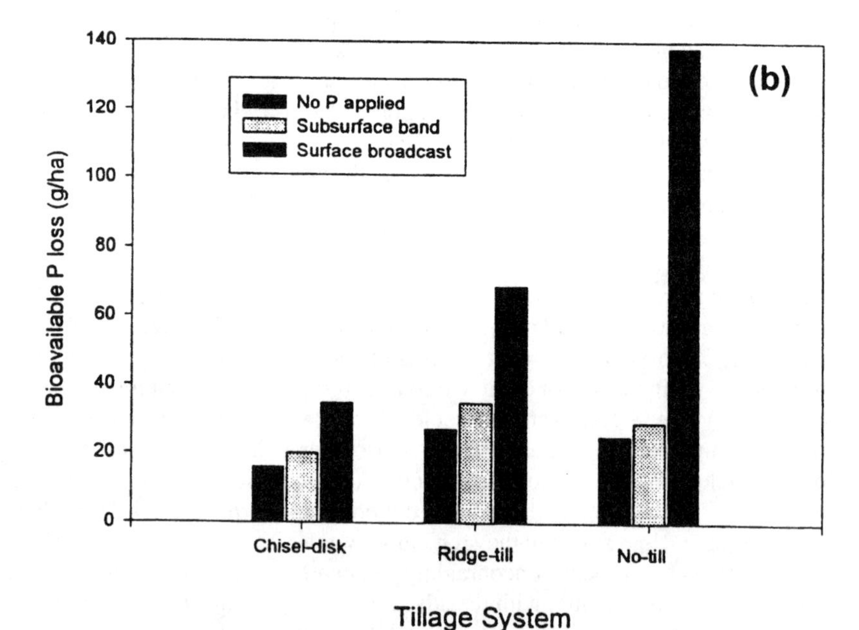

Fig. 2–15. (a) Generalized relationship between soil loss (erosion) and total P losses as influenced by soil cover (Sharpley et al.,1992). (b) Influence of P use, P application method, and tillage on bioavailable P losses in runoff (Janssen et al., 1996).

REFERENCES

Adams, M.A., and L.T. Byrne. 1989. ^{31}P-NMR analysis of phosphorus compounds in extracts of surface soils from selected karri (*Eucalyptus diversicolor* F. Muell.) forests. Soil Biol. Biochem. 21:523–528.

Agbenin, J.O., E.N.O. Iwuafor, and B. Ayuba. 1999. A critical assessment of methods for determining organic phosphorus in savanna soils. Biol. Fertil. Soils 28:177–181.

Anderson, G., E.G. Williams, and J.O. Moir. 1974. A comparison of the sorption of inorganic orthophosphate and inositol hexaphosphate by six acid soils. J. Soil Sci. 25:51–62.

Andraski, B.J., D.H. Mueller, and T.C. Daniel. 1985a. Phosphorus losses in runoff as affected by tillage. Soil Sci. Soc. Am. J. 49:1523–1527.

Andraski, B.J., D.H. Mueller, and T.C. Daniel. 1985b. Effects of tillage and rainfall simulation date on water and soil losses. Soil Sci. Soc. Am. J. 49:1512–1517.

Baker, J.L., and J.M. Laflen. 1982. Effect of crop residue and fertilizer management on soluble nutrient runoff losses. Trans. ASAE 25:344–348.

Barbanti, A., M.C. Bergamini, F. Frascari, S. Miserocchi, and G. Rosso. 1994. Critical aspects of sedimentary phosphorus chemical fractionation. J. Environ. Qual. 23:1093–1102.

Barberis, E., F.A. Marsan, R. Scalenghe, A. Lammers, U. Schwertmann, A. C. Edwards, R. Maguire, M.J. Wilson, A. Delgado, and J. Torrent. 1996. European soils overfertilized with phosphorus: I. Basic properties. Fert. Res. 45:199–207.

Barrow, N.J. 1985. Reactions of anions and cations with variable charge soils. Adv. Agron. 38:183–230.

Barrow, N.J. 1989. Surface reactions of phosphate in soils. Agric. Sci. 2:33–37.

Beegle, D., L.E. Lanyon, and J.T. Sims. 2002. Nutrient balance. p. 171–192. *In* P. Haygarth and S. Jarvis (ed.) Agriculture, hydrology, and water quality. CAB International, Oxon, UK.

Beegle, D.B., O.T. Carton, and J.S. Bailey. 2000. Nutrient management planning: Theory, justification, and practice. J. Environ. Qual. 29:72–79.

Blair, G.J., and O.W. Boland. 1978. The release of phosphorus from plant material added to soil. Aust. J. Soil Res. 16:101–111.

Blanchar, R.W., and G.K. Stearman. 1985. Prediction of phosphate sorption in soils from regular solid-solution theory. Soil Sci. Soc. Am. J. 49:578–583.

Blevins, R.L., W.W. Frye, P.L. Baldwin, and S.D. Robertson. 1990. Tillage effects on sediment and soluble nutrient losses from a Maury silt loam soil. J. Environ. Qual. 19:683–686.

Bowman, R.A., and J.O. Moir. 1993. Basic EDTA as an extractant for soil organic phosphorus. Soil Sci. Soc. Am. J. 57:1516–1518.

Boyle, F.W., and W.L. Lindsay. 1985. Preparation, x-ray diffraction pattern, and solubility product of manganese (III) phosphate hydrate. Soil Sci. Soc. Am. J. 49:758–766.

Boyle, F.W., and W.L. Lindsay. 1986. Manganese phosphate equilibrium relationships in soils. Soil Sci. Soc. Am. J. 50:588–593.

Brady, N.C., and R.R. Weil. 2002. Nature and property of soils. 13th ed. MacMillan, New York.

Breeuwsma, A., J.G.A. Reijerink, and O.F. Schoumans. 1995. Impact of manure on accumulation and leaching of phosphate in areas of intensive livestock farming. p. 239–249. *In* K. Steele (ed.) Animal waste and the land–water interface. Lewis Publ., New York.

Brookes, P.C., D.S. Powlson, and D.S. Jenkinson. 1982. Measurement of microbial biomass phosphorus in soil. Soil Biol. Biochem. 14:319–329.

Brookes, P.C., D.S. Powlson, and D.S. Jenkinson. 1984. Phosphorus in the soil microbial biomass. Soil Biol. Biochem. 16:169–175.

Cade-Menun, B.J., and C.M. Preston. 1996. A comparison of soil extraction procedures for ^{31}P NMR spectroscopy. Soil Sci. 161:770–785.

Celi, L., S. Lamacchia, T.A. Marsan, and E. Barberis. 1999. Interaction of inositol phosphate on clays: Adsorption and charging phenomena. Soil Sci. 164:574–585.

Chambers, B.J., T.W.D. Garwood, and R.J. Unwin. 2000. Controlling soil water erosion and phosphorus losses from arable land in England and Wales. J. Environ. Qual. 29:145–150.

Chardon, W.J., R.G. Menon, and S.H. Chien. 1996. Iron oxide impregnated filter paper (Pi test): A review of its development and methodological research. Nutr. Cycl. Agroecosyst. 46:41–51.

Chardon, W.J., and H.G. van Faassen. 1999. Soil indicators for critical source areas of phosphorus leaching. Rep. 22. Programmabureau Geintegreerd Bodemondezoek, Wageningen, the Netherlands.

Condron, L.M., E. Frossard, H. Tiessen, R.H. Newman, and J.W.B. Stewart. 1990. Chemical nature of organic phosphorus in cultivated and uncultivated soils under different environmental conditions. J. Soil Sci. 41:41–50.

Correl, D.L. 1998. The role of phosphorus in the eutrophication of receiving waters: A review. J. Environ. Qual. 27:261–266.

Dalal, R.C. 1979. Mineralization of carbon and phosphorus from carbon-14 and phosphorus-32 labelled plant material added to soil. Soil Sci. Soc. Am. J. 43:913–916.

De Smet, J., G. Hofman, J. Vanderdeelen, M. van Meirvenne, and L. Baert. 1996. Phosphate enrichment in the sandy loam soils of West Flanders, Belgium. Fert. Res. 43:209–215.

Enfield, C.G., T. Phan, D.M. Wolters, and R. Ellis, Jr. 1981. Kinetic model for phosphate transport and transformation in calcareous soils. Soil Sci. Soc. Am. J. 45:105–1064.

Fixen, P.E., and J.H. Grove. 1990. Testing soils for phosphorus. p. 141–180. *In* R.L. Westerman (ed.) Soil testing and plant analysis. 3rd ed. SSSA Book Ser. 3. SSSA, Madison, WI.

Fixen, P.E., A.E. Ludwick, and S.R. Olsen. 1983. Phosphorus and potassium fertilization of irrigated alfalfa on calcareous soils: II. Soil phosphorus solubility relationships. Soil Sci. Soc. Am. J. 47:112–117.

Foth, H.D., and B.G. Ellis. 1997. Soil fertility. 2nd ed. Lewis Publ., Boca Raton, FL.

Foy, R.H., and P.J.A. Withers. 1995. The contribution of agricultural phosphorus to eutrophication. The Fertilizer Society. Proc. No. 365. Greenhill House, Thorpe Wood, Petersborough, UK.

Frossard, E., M. Brossard, M.J. Hedley, and A. Metherell. 1995. Reactions controlling the cycling of P in soils. p. 107–137. *In* H. Tiessen (ed.) Phosphorus in the global environment. John Wiley & Sons, New York.

Frossard, E., S. Sinaj, L.-M. Zhang, and J.L. Morel. 1996. The fate of sludge phosphorus in soil–plant systems. Soil Sci. Soc. Am. J. 60:1248–1253.

Gil-Sotres, F., W. Zech, and H.G. Alt. 1990. Characterization of phosphorus fractions in surface horizons of soils from Galicia (N.W. Spain) by ^{31}P NMR spectroscopy. Soil Biol. Biochem. 22:75–79.

Harris, W.G., H.D. Wang, and K.R. Reddy. 1994. Dairy manure influence on soil and sediment composition: Implications for phosphorus retention. J. Environ. Qual. 23:1071–1081.

Harrison, A.F. 1987. Soil organic phosphorus—A review of world literature. CAB International, Wallingford, UK.

Harrison, R.B., and F. Adams. 1987. Solubility characteristics of residual phosphate in a fertilized and limed Ultisol. Soil Sci. Soc. Am. J. 51:963–969.

Harter, R.D., and D.E. Baker. 1977. Applications and misapplications of the Langmuir equation to soil adsorption phenomena. Soil Sci. Soc. Am. J. 41:1077–1080.

Heckrath, G., P.C. Brookes, P.R. Poulton, and K.W.T. Goulding. 1995. Phosphorus leaching from soils containing different phosphorus concentrations in the Broadbalk experiment. J. Environ. Qual. 24:904–910.

Hedley, M.J., J.W.B. Stewart, and B.S. Chauhan. 1982. Changes in inorganic and organic soil phosphorus fractions induced by cultivation and by laboratory incubations. Soil Sci. Soc. Am. J. 46:970–976.

Hettiarachchi, G.M., and G.M. Pierzynski. 1996. The influence of phosphorus source and rate on soil solution cadmium and lead activities. p. 333. *In* Agrononomy abstracts. ASA, CSSA, and SSSA, Madison, WI.

Hinedi, Z.R., and A.C. Chang. 1989. Solubility and phosphorus-31 magic angle spinning nuclear magnetic resonance of phosphorus in sludge-amended soils. Soil Sci. Soc. Am. J. 53:1057–1061.

Holford, I.C.R., and W.H. Patrick, Jr. 1979. Effects of reduction and pH changes on phosphate sorption and mobility in an acid soil. Soil Sci. Soc. Am. J. 43:292–297.

Hsu, P.H. 1982a. Crystallization of variscite at room temperature. Soil Sci. 133:305–313.

Hsu, P.H. 1982b. Crystallization of iron (III) phosphate at room temperature. Soil Sci. Soc. Am. J. 46:928–932.

Hue, N.V. 1991. Effects of organic acids/anions on P sorption and phytoavailability in soils with different methodologies. Soil Sci. 152:463–471.

Hue, N.V. 1992. Correcting soil acidity of a highly weathered Ultisol with chicken manure and sewage sludge. Commun. Soil Sci. Plant Anal. 23:241–264.

Hupfer, M., and R. Gachter. 1995. Polyphosphate in lake sediments: ^{31}P NMR spectroscopy as a tool for its identification. Limnol. Oceanogr. 40:610–617.

Inskeep, W.P., and J.C. Silvertooth. 1988. Inhibition of hydroxyapatite precipitation in the presence of fulvic, humic, and tannic acids. Soil Sci. Soc. Am. J. 52:941–946.

Iyamuremye, F., and R.P. Dick. 1996. Organic amendments and phosphorus sorption by soils. Adv. Agron. 56:139–185.

Janssen, K.A., G.M. Pierzynski, and P.L. Barnes. 1996. Phosphorus looses in runoff water as influenced by tillage and methods of phosphorus fertilizer application. p. 283. *In* Agronomy abstracts. ASA, CSSA, SSSA, Madison, WI.

Kamprath, E.J., and M.E. Watson. 1980. Conventional soil and tissue tests for assessing the phosphorus status of soils. p. 433–469. *In* F.E. Khasawneh et al. (ed.) The role of phosphorus in agriculture. ASA, CSSA, and SSSA, Madison, WI.

Karathanasis, A.D. 1991. Phosphate mineralogy and equilibria in two Kentucky alfisols derived from ordivician limestones. Soil Sci. Soc. Am. J. 55:1774–1782.

Khalid, R.A., W.H. Patrick, Jr., and R.D. DeLaune. 1977. Phosphorus sorption characteristics of flooded soils. Soil Sci. Soc. Am. J. 41:305–310.

Khiari, L., L.E. Parent, A. Pellerin, A.R.A. Alimi, C. Tremblay, R.R. Simard, and J. Fortin. 2000. An agri-environmental phosphorus saturation index for acid coarse-textured soils. J. Environ. Qual. 29:1561–1567.

Kuo, S. 1996. Phosphorus. p. 869–919. *In* D.L. Sparks (ed.) Methods of soil analysis. Part 3. SSSA Book Ser. 5. SSSA, Madison, WI.

Lamb, J.A., G.A. Peterson, and C.R. Fenster. 1985. Wheat fallow tillage systems' effect on a newly cultivated grassland soils' nitrogen budget. Soil Sci. Soc. Am. J. 49:352–356.

Larsen, S. 1967. Soil phosphorus. Adv. Agron. 19:151–210.

Lindsay, W.L. 1979. Chemical equilibria in soils. Wiley-Interscience, New York.

Lindsay, W.L., P.L.G. Vlek, and S.H. Chien. 1989. Phosphate minerals. p. 1089–1130. *In* J.B. Dixon and S.B. Weed (ed.) Minerals in soil environments. SSSA Book Ser. 1. ASA, CSSA, and SSSA, Madison, WI.

Lookman, R., D. Freese, R. Merckx, K. Vlassak, and W. H. van Reimsdijk. 1995. Long-term kinetics of phosphate release from soil. Environ. Sci. Technol. 29:1569–1575.

Magid, J., H. Tiessen, and L.M. Condron. 1996. Dynamics of organic phosphorus in soil under natural and agricultural ecosystems. p. 429–466. *In* A. Piccolo (ed.) Humic substances in terrestrial ecosystems. Elsevier Science, Amsterdam.

Maguire, R.O., R.H. Foy, J.S. Bailey, and J.T. Sims. 2001. Estimation of the phosphorus sorption capacity of acidic soils in Ireland. Eur. J. Soil Sci. 52:479–487.

Mahieu, N, D.C. Olk, and E.W. Randall. 2000. Analysis of phosphorus in two humic acid fractions of intensively cropped lowland rice soils by ^{31}P NMR. Eur. J. Soil Sci. 51:391–402.

Martin, R.R., R.S.C. Smart, and K. Tazaki. 1988. Direct observation of phosphate precipitation in the goethite/phosphate system. Soil Sci. Soc. Am. J. 52:1492–1500.

McCollum, R.E. 1991. Buildup and decline in soil phosphorus: 30-year trend on a Typic Umbraquult. Agron. J. 83:77–85.

McCoy, J.L., L.J. Sikora, and R.R. Weil. 1986. Plant availability of phosphorus in sewage sludge compost. J. Environ. Qual. 15:403–409.

McDowell, R.W., and A.N. Sharpley. 2001. Approximating phosphorus release from soils to surface runoff and subsurface drainage. J. Environ. Qual. 30:508–520.

McLaughlin, M.J., A.M. Alston, and J.K. Martin.1988. Phosphorus cycling in wheat–pasture rotations. III. Organic phosphorus turnover and phosphorus cycling. Aust. J. Soil Res. 26:343–353.

Mengel, D., and G. Rehm. 2000. Fundamentals of fertilizer application. p. D-155–D-174. *In* M.E. Sumner (ed.) Handbook of soil science. CRC Press, Boca Raton, FL.

Moore, P.A., Jr., T.C. Daniel, and D.R. Edwards. 2000. Reducing phosphorus runoff and inhibiting ammonia loss from poultry manure with aluminum sulfate. J. Environ. Qual. 29:37–49.

Moore, P.A., Jr., and K.R. Reddy. 1994. Role of Eh and pH on phosphorus geochemistry in sediments of Lake Okeechobee, Florida. J. Environ. Qual. 23:955–964.

Mueller, D.H., R.C. Wendt, and T.C. Daniel. 1984a. Phosphorus losses as affected by tillage and manure application. Soil Sci. Soc. Am. J. 48:901–905.

Mueller, D.H., R.C. Wendt, and T.C. Daniel. 1984b. Soil and water losses as affected by tillage and manure application. Soil Sci. Soc. Am. J. 48:90896–900.

Munns, D.N., and R. L. Fox. 1976. The slow reaction which continues after phosphate adsorption. Soil Sci. Soc. Am. J. 40:46–51.

Murmann, R.P., and M. Peech. 1969. Relative significance of labile and crystalline phosphates in soil. Soil Sci. 107:249–255.

Nair, P.S., T.J. Logan, A.N. Sharpley, L.E. Sommers, M.A. Tabatabai, and T.L. Yuan. 1984. Interlaboratory comparison of a standardized phosphorus adsorption procedure. J. Environ. Qual. 13:591–595.

Nair, V.D., R.R. Villapando, and D.A. Graetz. 1999. Phosphorus retention capacity of the spodic horizon under varying environmental conditions. J. Environ. Qual. 28:1308–1313.

Newman, R.H., and K.R. Tate. 1980. Soil phosphorus characterization by ^{31}P nuclear magnetic resonance. Commun. Soil Sci. Plant. Anal. 11:835–842.

O'Connor, G.A., K.L. Knudsten, and G.A. Connell. 1986. Phosphorus solubility in sludge-amended calcareous soil. J. Environ. Qual. 15:308–312.

Oehl, F., A. Oberson, S. Sinaj, and E. Frossard. 2001. Organic phosphorus mineralization studies using isotopic dilution techniques. Soil Sci. Soc. Am. J. 65:780–787.

Ohno, T., and B.S. Crannell 1996. Green and animal manure-derived dissolved organic matter effects on phosphorus sorption. J. Environ. Qual. 25:1137–1143.

Ohno, T., and M.S. Erich. 1997. Inhibitory effects of crop residue-derived organic ligands on phosphate adsorption kinetics. J. Environ. Qual. 26:889–895.

Olsen, S.R., and F.E. Khasawneh. 1980. Use and limitations of physical-chemical criteria for assessing the status of phosphorus in soils. p. 361–410. *In* F.E. Khasawneh et al. (ed.) The role of phosphorus in agriculture. ASA, Madison, WI.

Olsen, S.R., and F.S. Watanabe. 1957. A method to determine phosphorus adsorption maximum in soils as measured by the Langmuir isotherm. Soil Sci. Soc. Am. Proc. 21:144–149.

Parfitt, R.L. 1978. Anion adsorption by soils and soil materials. Adv. Agron. 30:1–50.

Penn, C.J., and J.T. Sims. 2002. Phosphorus forms in biosolids-amended soils and losses in runoff: Effects of wastewater treatment process. J. Environ. Qual. 31:1349–1361.

Pierzynski, G.M. 1991. The chemistry and mineralogy of phosphorus in excessively fertilized soils. Crit. Rev. Environ. Control 21:265–295

Pierzynski, G.M., T.J. Logan, S.J. Traina, and J.M. Bigham. 1990a. Phosphorus chemistry and mineralogy in excessively fertilized soils: Quantitative analysis of phosphorus-rich particles. Soil Sci. Soc. Am. J. 54:1576–1583.

Pierzynski, G.M., T.J. Logan, S.J. Traina, and J.M. Bigham. 1990b. Phosphorus chemistry and mineralogy in excessively fertilized soils: Descriptions of phosphorus-rich particles. Soil Sci. Soc. Am. J. 54:1583–1589.

Pierzynski, G.M., T.J. Logan, and S.J. Traina. 1990c. Phosphorus chemistry and mineralogy in excessively fertilized soils: Solubility equilibria. Soil Sci. Soc. Am. J. 54:1589–1595.

Pierzynski, G.M., J.T. Sims, and G.F. Vance. 2000. Soils and environmental quality. 2nd ed. Lewis Publ., Chelsea, MI.

Ponnamperuma, F.N. 1972. The chemistry of submerged soils. Adv. Agron. 24:29–96.

Pote, D.H., T.C. Daniel, A.N. Sharpley, P.A. Moore, Jr., D.R. Edwards, and D.J. Nichols. 1996. Relating extractable phosphorus to phosphorus losses in runoff. Soil Sci. Soc. Am. J. 60:855–859.

Power, J.F., and W.A. Dick (ed.) 2000. Land application of agricultural, industrial, and municipal by-products. SSSA Book Ser. 6. SSSA, Madison, WI.

Reddy, G.Y., E.O. McLean, G.D. Hoyt, and T.J. Logan. 1978. Effects of soil, cover crop, and nutrient source on amounts and forms of phosphorus movement under simulated rainfall conditions. J. Environ. Qual. 7:50–54.

Robinson, J.S., C.T. Johnston, and K.R. Reddy. 1998. Combined chemical and ^{31}P-NMR spectroscopic analysis of phosphorus in wetland organic soils. Soil Sci. 163:705–713.

Romkens, M.J.M., and D.W. Nelson. 1974. Phosphorus relationships in runoff from fertilized soils. J. Environ. Qual. 3:10–13.

Romkens, M.J.M., D.W. Nelson, and J.V. Mannering. 1973. Nitrogen and phosphorus composition of surface runoff as affected by tillage method. J. Environ. Qual. 2:292–295.

Roy, A.C., and S. K. De Datta. 1985. Phosphate sorption isotherm for evaluating phosphorus requirement of wetland rice soils. Plant Soil 94:185–186.

Sah, R.N., and D.S. Mikkelsen. 1986. Effects of temperature and prior flooding on intensity and sorption of phosphorus in soil. Plant Soil 95:163–171.

Sah, R.N., and D.S. Mikkelsen. 1989a. Phosphorus behavior in flooded-drained soils: I. Effects on phosphorus sorption. Soil Sci. Soc. Am. J. 53:1718–1722.

Sah, R.N., and D.S. Mikkelsen. 1989b. Phosphorus behavior in flooded-drained soils: II. Iron transformations and phosphorus sorption. Soil Sci. Soc. Am. J. 53:1723–1729.

Sah, R.N., and D.S. Mikkelsen. 1989c. Phosphorus behavior in flooded-drained soils: III. Phosphorus desorption and availability. Soil Sci. Soc. Am. J. 53:1729–1732.

Sample, E.C., R.J. Soper, and G.J. Racz. 1980. Reactions of phosphate fertilizers in soils. p. 263–310. *In* F.E. Khasawneh et al. (ed.) The role of phosphorus in agriculture. ASA, CSSA, and SSSA, Madison, WI.

Sanchez, P.A., and G. Uehara. 1980. Management considerations for acid soils with high phosphorus fixation capacity. p. 471–514. *In* F.E. Khasawneh et al. (ed.) The role of phosphorus in agriculture. ASA, Madison, WI.

Sanyal, S.K., and S.K. De Datta. 1992. Chemistry of phosphorus transformations in soil. Adv. Soil Sci. 16:1–120.

Selles, F., C.A. Campbell, and R.P. Zenter. 1995. Effect of cropping and fertilization on plant and soil phosphorus. Soil Sci. Soc. Am. J. 59:140–144.

Shang, C., P.M. Huang, and J.W.B. Stewart. 1990. Kinetics of adsorption of organic and inorganic phosphates by short-range ordered precipitate of aluminum. Can. J. Soil Sci. 70:461–470.

Sharpley, A.N. 1985. Phosphorus cycling in unfertilized and fertilized agricultural soils. Soil Sci. Soc. Am. J. 49:905–911.

Sharpley, A.N. 1992. Mineralization of organic phosphorus in agricultural soils: A manageable resource? p. 25–30. *In* F.J. Sikora (ed.) Future directions for agricultural phosphorus research. TVA Bull. Y-224. TVA, Muscle Shoals, AL.

Sharpley, A.N. 1993. Estimating phosphorus in agricultural runoff available to several algae using iron-oxide impregnated filter strips. J. Environ. Qual. 22:678–680.

Sharpley, A.N. 2000. Phosphorus availability. p. D-18–D-38. *In* M.E. Sumner (ed.) Handbook of soil science. CRC Press, Boca Raton, FL.

Sharpley, A.N., S.C. Chapra, R. Wedepohl, J.T. Sims, T.C. Daniel, and K.R. Reddy. 1994. Managing agricultural phosphorus for protection of surface waters: Issues and options. J. Environ. Qual. 23:437–451.

Sharpley, A.N., and S.J. Smith. 1989. Mineralization and leaching of phosphorus from soil incubated with surface-applied and incorporated corp residues. J. Environ. Qual. 18:101–105.

Sharpley A.N., S.J. Smith, O.R. Jones, W.A. Berg, and G.A. Coleman. 1992. The transport of bioavailable phosphorus in agricultural runoff. J. Environ Qual. 21:30–35.

Sharpley, A.N., and H. Tunney. 2000. Phosphorus research strategies to meet agricultural and environmental challenges of the 21st century. J. Environ. Qual. 29:176–181.

Sibanda, H.M., and S.D. Young. 1986. Competitive adsorption of humus acids and phosphate on goethite, gibbsite, and two tropical soils. J. Soil Sci. 37:197–204.

Sibbesen, E., and A.N. Sharpley. 1997. Setting and justifying upper critical limits for phosphorus in soils. p. 151–176. *In* H. Tunney et al. (ed.) Phosphorus loss from soil to water. CAB International, London.

Simard, R., S. Beaucheim, and P.M. Haygarth. 2000. Potential for preferential pathways for phosphorus transport. J. Environ. Qual. 29:97–105.

Simard, R.R., D. Cluis, G. Gangbazo, and S. Beauchemin. 1995. Phosphorus status of forest and agricultural soils from a watershed of high animal density. J. Environ. Qual. 24:1010–1017.

Sims, J.T. 2000. Soil fertility evaluation. p. D-113–D-154. *In* M.E. Sumner (ed.) Handbook of soil science. CRC Press, Boca Raton, FL.

Sims, J.T., A.C. Edwards, O.F. Schoumans, and R.R. Simard. 2000. Integrating soil phosphorus testing into environmentally based agricultural management practices. J. Environ. Qual. 29:60–72.

Sims, J.T., and N.J. Luka-McCafferty. 2002. On-farm evaluation of aluminum sulfate (alum) as a poultry litter amendment: Effects on litter properties. J. Environ. Qual. 31:2066–2073.

Sims, J.T., R.O. Maguire, A.B. Leytem, K.L. Gartley, and M.C. Pautler. 2002. Evaluation of Mehlich 3 as an agri-environmental soil phosphorus test for the Mid-Atlantic United States of America. Soil Sci. Soc. Am. J. 66:301–319.

Sims, J.T., R.R. Simard, and B.C. Joern. 1998. Phosphorus losses in agricultural drainage: Historical perspective and current research. J. Environ. Qual. 27:277–293.

Singh, B.B., and J.P. Jones. 1976. Phosphorus sorption and desorption characteristics of soil as affected by organic residues. Soil Sci. Soc. Am. J. 40:389–394.

Sparks, D.L. 1995. Environmental soil chemistry. Academic Press, New York.

Sposito, G. 1981. The thermodynamics of soil solutions. Oxford University Press, New York.

Sposito, G. 1989. The chemistry of soils. Oxford University Press, New York.

Steinhilber, P., and R. Weismiller. 2000. On-farm management options for controlling phosphorus inputs to the Chesapeake Bay. p. 169–178. *In* A.N. Sharpley (ed.) Agriculture and phosphorus management: The Chesapeake Bay. Lewis Publ., Boca Raton, FL.

Stevenson, F.J. 1982. Humus chemistry. Wiley Interscience, New York.

Stevenson, F.J. 1986. Cycles of soil. Wiley Interscience, New York.

Stewart, J.W.B., and H. Tiessen. 1987. Dynamics of soil organic phosphorus. Biogeochemistry 4:41–60.

Syers, J.K., and D. Curtin. 1989. Inorganic reactions controlling phosphorus cycling. p. 17–29. *In* H. Tiessen (ed.) Phosphorus cycles in terrestrial and aquatic ecosystems. Saksatchewan Inst. Pedology, Saskatoon, Canada.

Tate, K.R. 1984. The biological transformation of P in soil. Plant Soil 76:245–256.

Tiessen, H., and J.O. Moir. 1993. Characterization of available P by sequential extraction. p. 75–86. *In* M.R. Carter (ed.) Soil sampling and methods of analysis. Lewis Publ., Ann Arbor, MI.

Tiessen, H., I.H. Salcedo, and E.V.S.B. Sampaio. 1992. Nutrient and soil organic matter dynamics under shifting cultivation in semi-arid northeastern Brazil. Agric. Ecosyst. Environ. 38:139–151.

Tiessen, H., J.W.B. Stewart, and J.R. Bettany. 1982. Cultivation effect on the amounts and concentrations of carbon, nitrogen and phosphorus in grassland soils. Agron. J. 74:831–835.

Tiessen, H., J.W.B. Stewart, and J.O. Moir. 1983. Changes in organic and inorganic phosphorus composition of two grassland soils and their particle size fractions during 60–90 years of cultivation. J. Soil Sci. 34:815–823.

Tiessen, H., J.W.B. Stewart, and A. Oberson. 1994. Innovative soil phosphorus availability indices: Assessing organic phosphorus. p. 143–162. *In* J.L. Havlin and J.S. Jacobsen (ed.) Soil testing: Prospects for improving nutrient recommendations. SSSA Spec. Publ. 40. Am. Soc. of Agron., Madison, WI.

Tisdale, S.L., W.L. Nelson, J.D. Beaton, and J.L. Havlin. 1993. Soil fertility and fertilizers. 5th ed. Macmillan, New York.

Traina, S. J., G. Sposoito, D. Hesterberg, and U. Kafkafi. 1986. Effects of pH and organic acids on orthophosphate solubility in an acidic, montmorillonitic soil. Soil Sci. Soc. Am. J. 41:870–876.

Turner, B.L., A.W. Bristow, and P.M. Haygarth. 2001. Rapid estimation of microbial biomass in grassland soils by ultra-violet absorbance. Soil Biol. Biochem. 33:913–919.

Uusi-Kamppa, J., B. Braskerud, H. Jansson, N. Syversen, and R. Uusitalo. 2000. Buffer zones and constructed wetlands as filters for agricultural phosphorus. J. Environ. Qual. 29:151–157.

Vadas, P.A., and J.T. Sims. 1998. Redox status, poultry litter, and phosphorus solubility in Atlantic Coastal Plain soils. Soil Sci. Soc. Am. J. 62:1025–1034.

van der Zee, S.E.A.T.M., and W. H. van Riemsdijk. 1988. Model for long-term phosphate reactions in soil. J. Environ. Qual. 17:35–41.

Veith, J.A., and G. Sposito. 1977. Reactions of alumino-silicates, aluminum hydrous oxides, and aluminum oxide with *o*-phosphate: The formation of x-ray amorphous analogs of variscite and montebrasite. Soil Sci. Soc. Am. J. 41:870–876.

Violante, A., C. Colombo, and A. Buondonno. 1991. Competitive adsorption of phosphate and oxalate by aluminum oxides. Soil Sci. Soc. Am. J. 55:65–70.

Violante, A., and L. Gianfreda. 1993. Comparison in adsorption between phosphate and oxalate on an aluminum hydroxide montmorillonite sample. Soil Sci. Soc. Am. J. 57:1235–1241.

Whalen, J.K., and C. Chang. 2001. Phosphorus accumulation in cultivated soils from long-term annual applications of cattle feedlot manure. J. Environ. Qual. 30:229–237.

White, R.E., and A.T. Ayoub. 1983. Decomposition of plant residues of variable C:P ratio and the effect on soil phosphate availability. Plant Soil 74:163–173.

Williams, J.D.H., J.K. Syers, T.W. Walker, and R.W. Rex. 1970. A comparison of methods for the determination of soil organic phosphorus. Soil Sci. 110:3–18.

Wilson, M.A. 1987. NMR techniques and applications in geochemistry and soil chemistry. Pergamon, New York.

Young, E.O., and D.S. Ross. 2001. Phosphate release from seasonally flooded soils: A laboratory microcosm study. J. Environ. Qual. 30:91–101.

Yu, T. 1985. Physical chemistry of paddy rice soils. Springer-Verlag, Beijing.

Chapter 3

Chemistry of Sulfur in Soils

M. ALI TABATABAI, *Iowa State University, Ames, Iowa, USA*

The average S content of the earth's crust is estimated to be between 0.06 and 0.10%. It is usually ranked as the 13th most abundant element in nature. Sulfur occurs in soils in organic and inorganic forms, with organic S accounting for more than 95% of the total S in most soils from the humid and semihumid regions. The proportion of organic and inorganic S in soils samples, however, varies widely according to soil type and depth of sampling. It is usually somewhat lower in subsurface than in surface soils (Tabatabai and Bremner, 1972a, 1972b).

Although it is well known that S in soils is present mainly in organic combinations, very little is kwon about the identities of these S compounds. The inorganic S fraction in soils may occur as SO_4^{2-} and compounds of lower oxidation state such as sulfide (S^{2-}), thiosulfate ($S_2O_3^{2-}$), tetrathionate ($S_4O_6^{2-}$), polysulfides (S_n^{2-}, where $n>10$), sulfite (SO_3^{2-}), and elemental S (S°). The last four are detected in soils treated with elemental S or certain pollutants. In well-drained, aerated soils, most of the inorganic S normally occurs as SO_4^{2-}, and the concentrations of reduced S compounds are generally 1%. There are several states of SO_4^{2-} in soils. These include easily soluble SO_4^{2-}, adsorbed SO_4^{2-}, insoluble SO_4^{2-}, and SO_4^{2-} coprecipitated (cocrystallized) with $CaCO_3$. Under anaerobic conditions, particularly in tidal swamps and poorly drained or waterlogged soils, the main form of inorganic S in soils is sulfide and, often, elemental S. This chapter covers the chemistry of organic and inorganic S in soils. The biochemical processes involved in S cycling in soils was reported in review articles by Freney (1967) and Germida and Gupta (1992), and the role of S in agriculture is covered in a monograph edited by the author (Tabatabai, 1986).

CARBON–NITROGEN–PHOSPHORUS–SULFUR RELATIONSHIPS

Few attempts have been made to determine the organic S contents of soils. The results obtained by many researchers indicate that most of the S in surface soils of humid and semihumid regions is in the organic form (Tabatabai and Bremner, 1972a). This has been suggested from the close relationship between total S and total C and between total S and total N in those soils. Significant information is now available on the relationships among C, N, P, and S in soils around the world. Because, unlike P, which can be present in significant proportions as organic and in-

Chemical Processes in Soils. SSSA Book Series, no. 8.

Table 3–1. Carbon, N, P, and S relationships in some Brazilian and Iowa surface soils.†

	Ratio				
Soil no.	Organic C/ total N	Total N/ organic P	Total N/ total S	Organic P/ total S	Organic C/total N/ organic P/ total S (organic S)
			Brazilian soils		
1	21.0	11.8	3.4	0.3	210:10:0.9:3.0(2.5)
2	21.3	9.7	7.0	0.9	213:10:1.0:1.4(1.1)
3	12.8	13.3	8.3	0.7	128:10:0.8:1.2(1.1)
4	13.9	3.3	6.1	2.0	139:10:3.1:1.6(1.6)
5	23.6	18.7	12.0	0.7	136:10:0.5:0.8(0.8)
6	23.8	10.2	6.3	0.7	238:10:1.0:1.6(1.5)
Avg.	19.4	11.2	7.2	0.9	194:10:1.2:1.6(1.4)
			Iowa soils		
7	9.4	6.2	9.1	1.6	94:10:1.6:1.1(1.0)
8	10.8	7.0	9.2	1.3	108:10:1.4:1.1(1.0)
9	10.4	7.0	6.9	1.0	104:10:1.4:1.4(1.4)
10	10.3	8.0	7.4	0.9	103:10:1.2:1.3(1.3)
11	12.4	8.2	8.0	1.0	124:10:1.2:1.3(1.2)
12	13.1	7.6	6.9	0.9	131:10:1.3:1.5(1.4)
Avg.	11.1	7.3	7.9	1.1	110:10:1.4:1.3(1.2)

† Neptune et al. (1975).

organic combinations, inorganic N and S values are very small relative to organic forms of these elements in soils. Therefore, often the relationship among organic C, total N (instead of organic N), organic P, and total S (instead of organic S) are reported. Significant variation can occur in the C/N/P/S ratios of individual soils, but the mean ratios for groups of soils from different regions are similar. Agricultural soils, in general, have a mean C/N/P/S ratio of about 130:10:1.3:1.3. Soils under native grass have ratios of the order of 200:10:1:1. Peat and organic soils have intermediate ratios of approximately 160:10:1.2:1.2. These ratios have been illustrated for six Brazilian surface soils and six Iowa surface soils in Table 3–1.

SOURCES OF SULFUR IN SOILS

Mineral Sources

Many S-containing minerals occur in nature. The main S-bearing minerals in rocks and soils are present two states: (i) as SO_4^{2-} such as in gypsum ($CaSO_4 \cdot 2H_2O$), anhydrite ($CaSO_4$), epsomite ($MgSO_4 \cdot 7H_2O$), and mirabilite ($Na_2SO_4 \cdot 10H_2O$), or (ii) as sulfide such as pyrite and marcasite (FeS_2), sphalerite (ZnS), chalcopyrite ($CuFeS_2$), cobaltite (CoAsS), pyrrhotite ($Fe_{11}S_{12}$), galena (PbS), arsenopyrite ($FeS_2 \cdot FeAs_2$), and pentlandite $(Fe,Ni)_9S_8$.

Fertilizer Sources

There are many fertilizer materials, both liquid and solid, which are used to supply S to growing crops. The S source selected for any particular situation is determined by the crop to be grown, the S level of the soil, the cost of the material, and the ease with which it can be used in a particular fertilizer program.

Atmospheric Sources

Rainfall and the atmosphere constitute a third important source of S in soils. It is estimated that in the United States more than 25 million tonnes of SO_2, or 13 million tonnes of S, are emitted annually into the atmosphere. Most of these amounts are derived from combustion of fossil fuels, but industrial processes as ore smelting, petroleum refining operations, and other such sources contribute about 20% of the total S emitted into the atmosphere. Another major source is volcanic activities around the world. For review of the literature about the effect of acid rain on soils, see Tabatabai (1983, 1985).

CHEMICAL NATURE OF ORGANIC SULFUR IN SOILS

Many S compounds are produced by organisms in soils; some of these are listed in Table 2. Few organic S compounds in free state have been isolated from soils. Trithiobenzaldehyde $(SCHC_6H_5)_3$ was extracted from soils by Shorey (1913), who postulated that it is formed in soils by the reaction of H_2S released from bacteria with benzaldehyde produced from the decomposition of lignin. Traces of free cystine (16 mg kg^{-1} soil) have been extracted by ethanol from of a nonrhizosphere soil (Putnam and Schmidt, 1959), and methionine sulfoxide, cystine, and methionine were found in soils incubated with glucose and potassium nitrate (Paul and Schmidt, 1961). Other recent studies using sulfur K-edge X-ray near-edge structure spectroscopy (XANES) has been used to identify multiple organic S oxidation states in aquatic and soil humic substances (Xia et al., 1998). The XANES results suggested that the S in humic substances exists in four major oxidation groups similar to sulfate ester, sulfonate, sulfoxide, and thiol-sulfide. It is well known that most of the cystine and methionine in soils occurs in combined form. This is discussed below for its determination with Raney Ni.

Understanding the nature and properties of the organic S fractions in soils is important because these compounds govern the release of plant-available S. Even though much of the organic S compounds in soils remain unidentified, three broad groups of S compounds are recognized. These groups have been classified according to the nature of the reagents used or according to the groups of S compounds attacked by the reagents. Thus, three distinct groups of S-containing compounds have been identified (Tabatabai, 1996) (Fig. 3–1):

1. Organic S that is not directly bonded to C and is reduced to H_2S by hydriodic acid (HI). This fraction is believed to be largely in the form of sulfate ester with C–O–S linkages. Examples of substances that contain this linkage include arylsulfate, alkylsulfates, phenolic sulfate, sulfated polysuccharides, choline sulfate, and sulfated lipids. Other organic sulfates could be present as sulfamates (C–N–S) and sulfated thioglycosides (N–O–S). On average, about 50% of the total organic S in humid and semihumid regions is present in this form, but it can range from 30 to 60%

Table 3–2. Sulfur compounds isolated from plants, animals, and microorganisms.†

Compound	Formula	Source
	Aminoacids	
L-Methionine	$CH_3SCH_2CH_2CH(NH_2)COOH$	proteins
L-Cystine	$S\ CH_2CH(NH_2)COOH$ | $S\ CH_2CH(NH_2)COOH$	proteins bacitracin
L-Cysteine	$HSCH_2CH(NH_2)COOH$	proteins
Substituted cysteines (e.g., S-methyl lanthionine, cystathionine, felinine)	$RSCH_2CH(NH_2)COOH$	plants proteins antibiotics hair locust muscle cattle tissue cat urine microorganisms
L-Cysteine sulfinic acid	$HOOSCH_2CH(NH_2)COOH$	rat brain lugworm
L-Cysteic acid	$HOO_2SCH_2CH(NH_2)COOH$	fleece of sheep proteins of red algae
Taurine	$HOO_2SCH_2CH_2NH_2$	bile plants
	Sulphonium compounds	
S-Methyl methionine	$(CH_3)_2S^+CH_2CH_2CH(NH_2)COOH$	animals plants microorganisms
S-adenosyl methionine	NH_2 (adenine ring) $-CHCH(OH)CH(OH)CHCH_2$ (O ring) $HOOCCH(NH_2)CH_2CH_2S^+CH_3$	animal tissues
Dimethyl-β-propiothetin	$(CH_3)_2S^+CH_2CH_2COOH$	red algae green algae
	Sulfate esters	
Phenolic sulfates e.g., tyrosine-O-sulfate	$HO_3SO-C_6H_4-CH_2CH(NH_2)COOH$	urine from dogs, rates, and pregnant mares
Choline sulfate	$(CH_3)_3N^+CH_2CH_2OSO_3^-$	*Aspergillus sydowi* lichens *Penicillium chrysogenum*
Fucoidan	Possibly L-fucose units sulfated in the 4 positions	brown seaweeds (*Fucus vesiculosus*) giant kelp (*Macrocystis pyrifera*) *Laminaria digitata*
Chondroitin sulfate B (β-heparin)	Suggested structure: CH_2OH, $NaSO_3O$, O, O, $NHCOCH_3$, O, COONa, OH, OH, O	animal tissues

(continued on next page)

Table 3–2. Continued.

Compound	Formula	Source
	Vitamins	
Thiamine	CH_3, N, N, NH_2, CH_2, CH_3, CH_2CH_2OH, S, N^+	*Neurospora*
Biotin	HN, CO, NH, HC, CH, H_2C, S, $CH(CH_2)_4COOH$	yeast liver, kidney, and microorganisms
	Antibiotics	
Penicillins	RCONHCH—CH, S, $C(CH_3)_2$, CO—N, CHCOOH	*Penicillium* spp. *Aspergillus* spp.
Thiolutin	S, S, N, CO, $NHCOCH_3$	*Streptomyces albus*
Gliotoxin	OH, N, S, S, O, O, N—CH_3, CH_2OH	*Trichoderma viride*
Actithiazic acid	O, NH, S, $(CH_2)_5COOH$	*Streptomyces* sp.
Aureothricin	S, S, N, CO, $NHCOCH_2CH_3$, CH_3	*Streptomyces* sp.
	Sulfides	
Dimethyl sulfide	CH_3SCH_3	common bracken (*Pteridium aquilinum*) Fern (*Athyrium filix-foemina*) asparagus
Divinyl sulfide	$CH_2{=}CHSCH{=}CH_2$	*Allium ursinum*
Methyl 3-methylthio-propionate	$CH_3SCH_2CH_2COOCH_3$	pineapple (*Ananas sativus*)
	Sulfoxides	
Sulforaphene	$CH_3SOCH{=}CHCH_2CH_2NCS$	*Raphanus sativus*
Glucoiberin	$CH_3SOCH_2CH_2CH_2C$(S-glucose)($=NOSO_2O^-$)	*Iberis amara*

(continued on next page)

Table 3–2. Continued.

Compound	Formula	Source
Biotin-1-sulfoxide	CO / HN NH / HC—CH / H_2C $CH(CH_2)_4COOH$ / SO	*Aspergillus niger*
	Isothiocynates	
Erysolin	$CH_3SO_2(CH_2)_4NCS$	*Erysimum perofskianum*
Allyl isothiocyanate	$CH_2{=}CHCH_2NCS$	black mustard (*Brassica nigra*) and other crucifers
Mustard oil glucosides (e.g., sinalbin)	R—C(S-D-glucosyl)(=$NOSO_2O^-$)	*Cruciferae* *Resedaceae* *Tropaeolaceae*
	Miscellaneous compounds	
α-Lipoic acid	$CH_2CH_2CH(CH_2)_4COOH$ / S—S	microorganisms plants, animals
Coenzyme A	$HSCH_2CH_2NHCOCH_2CH_2$ / $OP(OH)OCH_2C(CH_3)_2CH(OH)CONH$ / O / $OP(OH)OCH_2CHCHCH(OH)CH$ (—O— ring) / O / $OP(OH)_2$ / adenine (NH_2)	pigeon liver *Streptomyces fradiae* higher plants
Ergothioneine	$CH_2CH(COO^-)N^+(CH_3)_3$ / N NH / SH	rye ergot animal tissue *Neurospora crassa*
Glutathione	$HOOCCH(NH_2)CH_2CH_2CONHCHCH_2SH$ / $HOOCCH_2NHCO$	microorganisms, plants, animals
"Plant sulfolipid"	$CH_2SO_2O^-$ / HO —O $OCH_2CHOHCH_2O$ / OH / $C_{17}H_{33}CO$ / OH	green plants

† Freney, 1967.

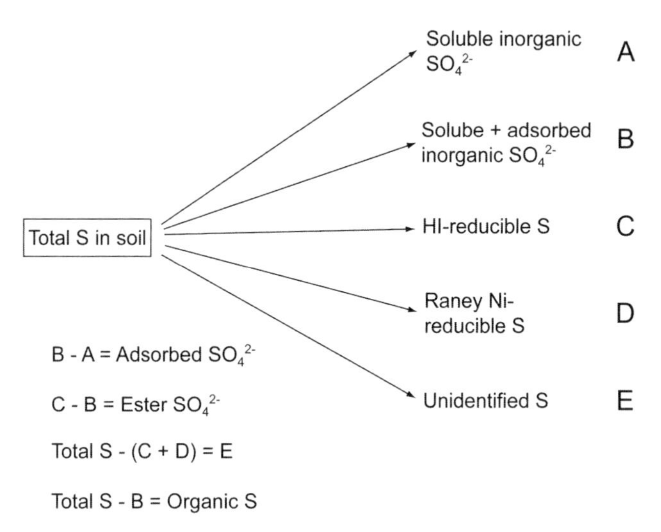

Fig. 3–1. Organic and inorganic S fractions in soils.

(Table 3–3). Values as high as 95% have been reported for Iowa subsoils. Unless otherwise indicated, the HI-reducible S includes the inorganic SO_4^{2-} fraction (Table 3–4).

2. Organic S that is directly bonded to C (C–S) and is reduced to inorganic sulfide by Raney Ni (50% each of Ni and Al powder) in an alkaline medium (NaOH). This fraction is believed to consist largely of S in the form of S-containing amino acids, such as methionine, cystine, and cysteine, but not alkyl sulfones or ester sulfates, which do not liberate sulfide when treated with Raney Ni. Its concentration in soils ranges from 10 to 30% of the total organic S (Tables 3–3 and 3–4), but values as high as 58% have been reported for some Quebec soils (Lowe and DeLong, 1963).
3. Organic S that is not reduced by either of the reagents employed in estimation of Fractions 1 and 2. This fraction is inert to hydriodic acid and Raney Ni. This unidentified fraction is generally in the range of 30 to 40% of the total organic S (Tables 3–3 and 3–4). It is very stable because it re-

Table 3–3. Fractionation of S in surface soils from different regions.†

	Percentage of total S in form specified					
	HI-reducible		C-bonded‡		Unidentified	
Location§	Range	Mean	Range	Mean	Range	Mean
Alberta, Canada (15)	25–74	49	12–32	21	7–45	30
Australia (15)	32–63	47	22–54	30	3–31	23
Brazil (6)	36–70	51	5–12	7	24–59	42
Iowa, USA (34)	36–66	52	5–20	11	21–53	37
Quebec, Canada (3)	44–78	65	12–32	24	0–44	11

† Tabatabai (1984).
‡ Determined by reduction with Raney Ni.
§ Numbers in parentheses indicate number of samples.

Table 3–4. Total S and percentage distribution of various forms of S in some Brazilian and Iowa surface soils.†

Soil no.	Total S	Percentage of total soil S in form specified						
		Inorganic SO_4^{2-}–S		HI-reducible S	Ester sulfate S	C-bonded S	Unidentified organic S‡	Total organic S
		LiCl	$Ca(H_2PO_4)_2$					
	mg kg^{-1} soil	%						
				Brazilian soils				
1	59	5.1	15.3	35.6	20.3	5.1	59.3	84.7
2	43	9.3	23.3	51.2	27.9	4.7	44.1	76.7
3	72	6.9	12.5	52.8	40.3	11.1	36.1	87.5
4	214	3.7	5.1	70.1	65.0	5.6	24.3	94.9
5	209	1.4	4.8	53.6	48.8	12.4	34.0	95.2
6	398	2.0	6.3	43.0	36.7	5.0	52.0	93.7
Avg.	166	4.7	11.2	51.1	39.8	7.3	41.7	88.8
				Iowa soils				
7	55	7.3	7.5	50.9	43.4	18.2	30.9	92.5
8	174	5.8	4.9	42.9	47.0	9.2	37.9	94.1
9	331	5.4	5.5	57.7	52.2	12.1	30.2	94.5
10	338	2.4	2.5	55.3	52.8	8.9	35.8	97.5
11	438	1.8	1.8	61.6	59.8	7.3	31.1	98.2
12	580	3.5	3.5	48.6	45.1	12.1	39.3	96.5
Avg.	319	4.4	4.5	54.5	50.1	11.3	34.2	95.6

† Neptune et al. (1975).
‡ Unidentified organic S was calculated from total S – (HI-reducible S + C-bonded S).

sists degradation by costic chemical reagents. Therefore, this fraction is of little importance as potential source of S for plants.

Work by Pirela and Tabatabai (1988) showed that Sn–H_3PO_4 reagent reduces a number of S model compounds (Table 3–5) and organic S in soils (Table 3–6). Phosphoric acid becomes anhydrous at 150°C, gradually changes to pyrophosphoric acid ($H_4P_2O_7$) at about 200°C, and changes to metaphosphoric acid [$(HPO_3)_n$] when heated above 300°C (Windholz, 1976). Evidently, the metaphosphoric acid produced upon heating in combination with Sn is the reagent responsible for hydrolysis and/reduction of certain organic and inorganic S compounds to H_2S.

Expressed as a percentage of total organic S in 13 Iowa surface soils, the amounts of Sn–H_3PO_4–reducible S ranged from 15 to 64% (avg. = 45%) and from 42 to 91% (avg. = 60%) for 1 and 10 h of distillation, respectively. The corresponding percentages of seven Chilean surface soils were from 8 to 75 (avg. =25%) and from 29 to 100% (avg. = 55%). These values are compared with those reduced with HI or Raney Ni (Table 3–6). The parabolic curves obtained for cumulative amounts of S reduced in surface and subsurface soils as a function of distillation time suggested that the chemical nature of organic S is dissimilar among some of the surface soils (Fig. 3–2) and that it differs with depth of sampling (Fig. 3–3). The amounts of organic S potentially reducible (S_r) with this reagent ranged from 35 to 86% (avg. = 61%) of organic S in Iowa surface soils. The corresponding values for seven Chilean soils ranged from 33 to 100% (avg. = 66%). The time required to distill

Table 3–5. Recovery of S (32 μg) from organic and inorganic compounds by tin-phosphoric acid and hydriodic acid methods.†

Compound	Nature of grouping	Recovery	
		$Sn–H_3PO_4$+‡	HI
		%	
	Organic		
Allyl isothiocyanate	–N=C=S	0	8.7
Cysteine hydrochloride	–C–SH	20.5 (40.2)	0
Cystine	–C–S–S–C–	85.9 (100.5)	0
Methionine	–C–S–C–	0	0
Methionine sulfone	$–C–SO_2–C–$	1.9 (1.9)	0
Methionine sulfoxide	–C–SO–C–	0	0
Sodium diethyldithiocarbamate	–C(=S)–S$^-$	0	0
Thioacetamide	$–C(=S)–NH_2$	100.7	0
Thiourea	$H_2N–C(=S)–NH_2$	10.9 (21.0)	0
Dimethylsulfamoyl chloride	NSO_3Cl	53.1 (68.8)	5.0
Ethyl dichlorothiophosphate	–O–P(=S)(Cl)(Cl)	100.0	75.0
p-nitrophenyl sulfate	$–O–SO_3^-$	92.8	93.6
Sinigrin monohydrate	$–S–C=NOSO_3^-$	91.6 (100.0)	95.0
Cysteic acid	$–C–SO_2OH$	0	0
Sulfanilamide	$–C–SO_2–NH_2$	0	0
Sulfanilic acid	$–C–SO_2OH$	0	0
Taurine	$–C–SO_2OH$	0	0
	Inorganic		
Magnesium sulfate	SO_4^{2-}	101.5	100.0
Potassium sulfate	SO_4^{2-}	99.6	98.4
Sodium sulfate	SO_4^{2-}	100.0	100.0
Zinc sulfate	SO_4^{2-}	96.6	95.2
Sodium thiosulfate	$S_2O_3^{2-}$	97.3	95.8
Sodium sulfate	SO_3^{2-}	97.8	97.8
Sodium thiocyanate	–S–C≡N	0	3.5
Potassium thiocyanate	–S–C≡N	0	6.1
Sodium sulfide	S^{2-}	100.0	100.0
Elemental sulfur	S^0	47.6 (70.5)	87.5

† Pirela and Tabatabai (1988).
‡ 30 min of digestion. Numerals in parentheses are those obtained after 4 h of digestion.

50% of S_r ranged from 0.3 to 1.7 h (avg. = 0.8 h) for the Iowa surface soils and from 0.4 to 9.2 h (avg. = 3 h) for the Chilean soils. Other studies with five Iowa soil profiles showed that the percentages of organic S reduced with Sn and H_3PO_4 in 1 and 10 h of distillation varied with depth of sampling and that these values were similar, lower, or greater than those obtained with the HI–reducing mixture (Pirela and Tabatabai, 1988).

Table 3–6. Total S and distribution of various S fractions in Iowa and Chilean soils.†

Soil	Total S	Inorg. S‡	Percentage of organic S reduced by reagent specified			Organic S (% of total S)			Total††	S_r‡‡	K_t§§
			Sn–H_3PO_4§	Sn–HCl	HI	Ester¶	C-bonded	Unid.#			
	mg kg^{-1} soil		%								
Iowa soils											
Ida	140	2	64 (91)	11	64	62.9	6.2	27.9	98.6	86	0.28
Hayden	128	2	49 (72)	5	44	43.8	37.5	17.2	98.4	67	0.29
Downs	196	8	30 (65)	4	51	48.5	12.2	35.2	95.9	70	1.14
Luther	125	3	26 (52)	7	58	56.8	13.6	27.2	97.6	48	0.70
Fayette	226	5	35 (59)	3	47	45.6	11.9	40.3	97.8	57	0.47
Tama	240	8	21 (46)	3	69	67.1	17.1	12.5	96.7	53	1.72
Lester	401	4	15 (42)	3	38	37.2	11.7	50.1	99.0	35	1.10
Clarion	319	3	29 (47)	4	66	64.9	8.2	26.0	99.1	46	0.51
Muscatine	347	3	28 (64)	10	57	56.8	11.5	30.8	99.1	66	1.23
Nicollet	390	3	48 (68)	3	49	48.7	5.1	45.4	99.2	73	0.64
Harps	470	4	41 (55)	4	67	66.4	8.3	24.5	99.1	55	0.34
Okobojo	437	5	32 (61)	3	55	54.2	14.0	30.7	98.9	73	1.39
Canisteo	478	5	34 (59)	4	59	58.2	9.8	31.0	99.0	67	1.12
Avg.			35 (60)	5	56	54.7	13.0	30.7	98.3	61	0.84
Chilean soils											
Alhue	358	7	73 (100)	40	85	83.2	8.7	6.1	98.0	100	0.38
Constitucion	159	5	12 (29)	1	32	31.4	20.8	44.7	96.9	33	1.99
Maipo	1692	70	21 (81)	8	75	71.8	2.1	22.0	95.9	83	1.84
Agua del Gato	626	141	18 (45)	6	62	48.1	10.5	18.8	77.5	47	1.59
Collipulli	366	3	34 (61)	2	69	68.6	7.7	23.0	99.2	75	1.63
Santa Barbara	648	12	11 (36)	3	39	38.7	12.3	47.1	98.1	54	4.57
Osorno	893	7	8 (35)	2	55	54.6	12.0	32.7	99.2	69	9.23
Avg.			25 (55)	9	60	56.6	10.6	27.8	95.0	66	3.03

† Pirela and Tabatabi (1988).
‡ Inorganic S extracted with 500 mg P L^{-1} as $Ca(H_2PO_4)_2$ and determined by reduction with HI.
§ Distilled for 1 h. Figures in parentheses are values obtained for distillation time of 10 h.
¶ Ester S was calculated from [HI-reducible S]–[$Ca(H_2PO_4)_2$]–extractable SO_4–S].
# Unidentified organic S was calcualted from [total S]–[(HI-reducible + C-bonded)–S].
†† Total organic S was calculated from [total S]–[inorganic S].
‡‡ Percentage of organic S potentially reducible with Sn and H_3PO_4.
§§ Digestion time required to reduce 50% of S_r.

INORGANIC SULFUR IN SOILS

Inorganic S in soils may occur as SO_4^{2-} and as compounds of lower oxidation states, such as sulfide, polysulfides, sulfite, thiosulfate, and elemental S. In well-drained, well-aerated soils most of the inorganic S normally occurs as SO_4^{2-}, and the concentrations of reduced S compounds are generally trace, if any. Among the various forms of S in the biosphere, SO_4^{2-} has received the most attention because of its mobility, availability to plants and microorganisms, its precipitation reactions, and its reactivity with positively charged surfaces in soils. Sulfate occupies the center of the S cycle, with a significant portion of the cycle being in soils. Soil as medium for plant growth receives SO_4^{2-} from a variety of sources, and some of it is then is used by plants, incorporated into microbial biomass, retained in soils by chemical or physicochemical reactions, or lost from the soil system by leaching. Sulfate is found in soils in concentrations ranging from <2 mg of S kg^{-1} soil in humid to semi-humid cultivated soils to >1000 S kg^{-1} in soils of the arid and semiarid regions. The concentration of soluble SO_4^{2-} in soils is estimated after extraction with 0.01 *M* LiCl or 0.15% $CaCl_2$ solution, and the concentration of soluble plus the adsorbed fraction is estimated after extraction with $Ca(H_2PO_4)_2$ solution containing 500 mg P

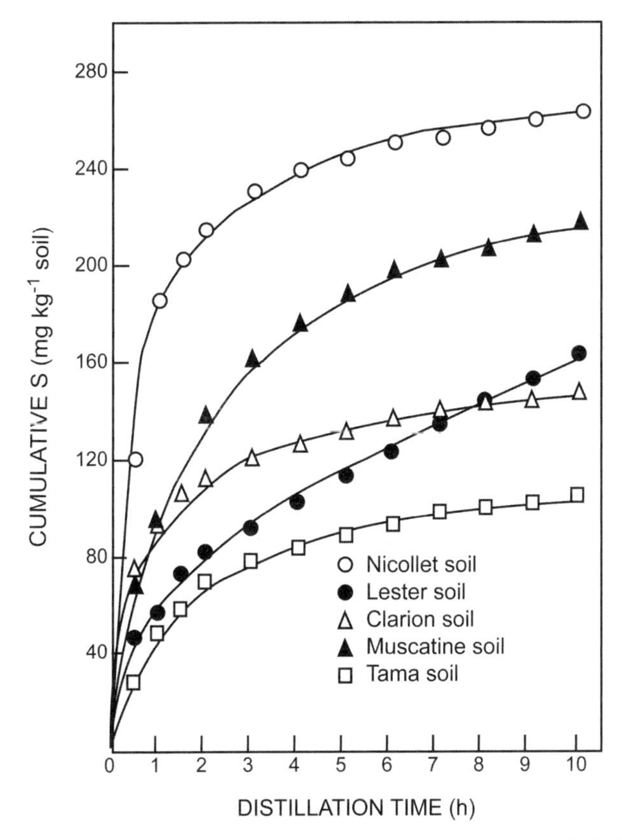

Fig. 3–2. Effect of distillation time on the reduction of organic S with Sn and H_3PO_4 in five Iowa surface soils (Pirela and Tabatabai, 1988).

L^{-1}. The amount of adsorbed SO_4^{2-} per unit mass of soil is then calculated from the difference between the amounts of the SO_4^{2-} extracted by the two reagents (Table 3–4).

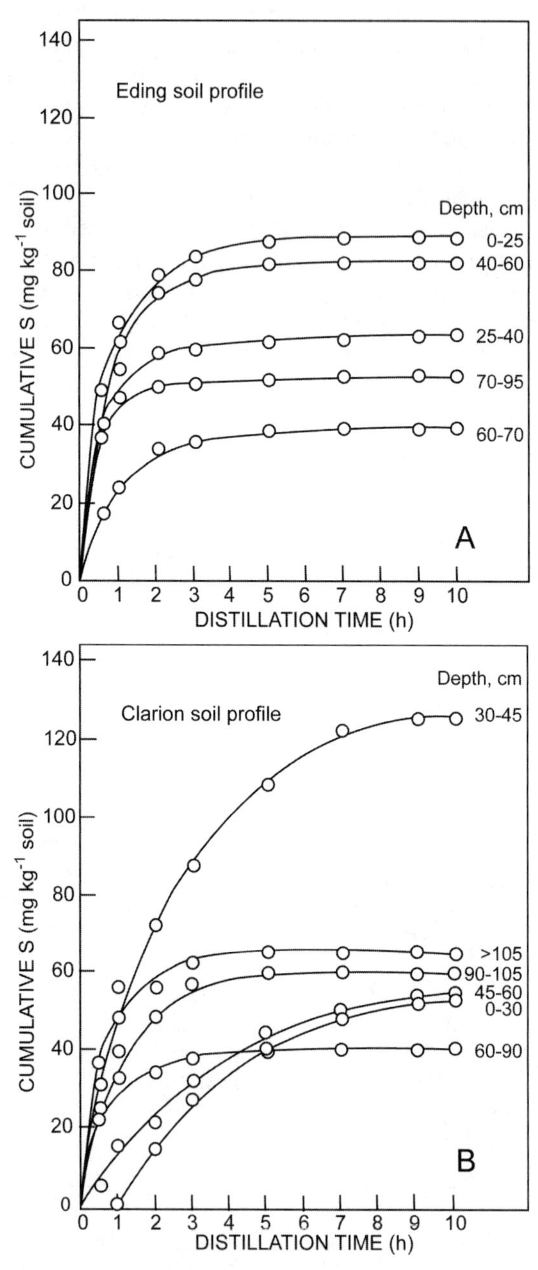

Fig. 3–3. Effect of distillation time on the reduction of organic S with Sn and H_3PO_4 in samples of (A) Edina soil profile and (B) Clarion soil profile.

Fate of Inorganic Sulfate in Soils

The inorganic SO_4^{2-} in soils may occur as water-soluble salts that can be leached from soils or adsorbed by soil colloids, and it may occur as insoluble forms. Soil chemical properties, such as pH, type of clay, and presence of cations, are important factors in governing the leaching and adsorption of SO_4^{2-} in soils. The mechanism of SO_4^{2-} adsorption by soils involves coordination with hydrous oxides, exchange on edges of silicate clays, and molecular adsorption. Both the water-soluble and adsorption SO_4^{2-} are available to plants. Under anaerobic (waterlogged) conditions SO_4^{2-} is reduced to S^{2-} and precipitated as metal sulfide. This process is biological in nature.

Leaching Losses

The movement of SO_4^{2-} in soils determines the magnitude of losses of S in drainage water. Leaching may have a marked influence on the distribution of the plant-available form of S within a soil profile. Sulfate movement with water is the major mechanism by which SO_4^{2-} is supplied to plant roots. Transport of SO_4^{2-} within a soil profile is influenced by its concentration in soil solution, its reaction with the solid phase components, and movement, velocity, and pattern of water movement within the soil. The relative magnitude of these factors and interactions among them determine the physicochemical fate of the SO_4^{2-} released from mineralization of organic S or added to soils as fertilizer, crop residue decomposition, irrigation waters, and atmospheric deposition.

Several lysimeter studies have been conducted to determine the SO_4^{2-} concentration of percolating water (Joffe, 1933; Driebelbis, 1947; Lunt, 1937; Mckell and Williams, 1960). The information obtained from such studies, however, are site specific. Losses of SO_4^{2-} by leaching vary widely. Some drainage water contain more S than the rain supplies even though little or none is added in fertilizer. The extra quantity may be deposited directly on plants and soils from the atmosphere or released from soil organic matter or minerals. Expressed in kilograms of S per hectare, thc annual losses from unfertilized fields by drainage water in the state of Illinois ranged from 1.5 to 65 (Whitehead, 1964), in Western Germany soils averaged 33, in Europe and North America averaged 15, in South America averaged 4.5, and in some areas of Australia is <1 (Freney et al., 1962). It has been estimated that between 3 and 32 kg S ha^{-1} is lost by tile drainage in the state of Iowa (Baker et al., 1975). In general, the annual loss from soils by leaching varies from insignificant amounts to as much as 320 kg S ha^{-1} from soils treated with S fertilizers (Harward and Reisenauer, 1966). From the information available on SO_4^{2-} losses with percolating water, the following conclusions can be made:

1. The SO_4^{2-} concentration of percolating water increases when soils are amended with S.
2. Losses of SO_4^{2-} are reduced by cropping and are less with rooting and perennial crops than with annual crops.
3. Leaching losses of SO_4^{2-} are greatest when monovalent ions such as K^+ predominate; next in order are the divalent ions such as Ca^{2+} and Mg^{2+}.

Leaching losses are the least when soils are acid and appreciable concentrations of Fe and Al hydrous oxides are present.

4. Under comparable soil and cropping conditions, the amount of SO_4^{2-} removed from the soil profile is generally directly related to the amount of leachate.
5. Sulfate adsorption leads to more rapid and complete removal of Cl^- from acid soils.
6. Sulfate losses increase with liming or amendment with phosphate.
7. Sulfate losses are less when the S fertilizer is banded than when broadcasted.

Both physical and chemical soil properties affect the movement of SO_4^{2-}. Within a soil profile the physical properties determine the rate and pattern of water movement and diffusion of SO_4^{2-} ions, whereas the chemical properties determine the exchange and reactions of SO_4^{2-} ions with soil constituents. The combined effect of these properties determines the final distribution of SO_4^{2-} within the soil profile. Complete separation of the influence of physical and chemical soil properties on transport of an anion such as SO_4^{2-}, however, is not possible because the nature of the anion influences the physical properties of soils (Longenecker, 1960). Several factors affect the distribution of SO_4^{2-} ions in the soil pores. These include (i) the electric field surrounding the individual soil particles and (ii) the magnitude of the repulsive (or attraction) forces, which are dependent on the mineralogy and chemical composition of the solid phase and the pH and salt content of the aqueous phase (Chao and Harward, 1963; Ensminger, 1954; Kamprath et al. 1956).

Sulfate Adsorption by Soils

Soils vary widely in their capacity to adsorb SO_4^{2-}. Because SO_4^{2-} adsorption occurs at low pH values ($<<6$), its adsorption is negligible in most agricultural soils (pH values >6). Its adsorption in subsurface acid horizons plays an important part in contributing to the S requirement of crops, conserving S from excessive leaching, and in determining S distribution in soil profiles. Sulfate adsorption is a reversible process and is influenced by a number of soil properties. These include:

1. Clay content and type of clay mineral. Sulfate adsorption usually increases with clay content of soils. Kaolin minerals retain more SO_4^{2-} than the montmorillonite clays. Capacities of H^+-saturated clays for SO_4^{2-} adsorption are: kaolinite > illite >bentonite. When saturated with Al, SO_4^{2-} adsorption is about the same for kaolinite and illite, but much lower for bentonite.
2. Hydrous oxide of Al and Fe. Hydrous oxides of Al, and to a lesser extent of Fe, show marked tendencies to retain SO_4^{2-}. These compounds seem to be responsible for most of the SO_4^{2-} adsorption in many soils.
3. Soil horizon or depth. Most soils have some capacities to adsorb SO_4^{2-}. The amounts of SO_4^{2-} adsorption in surface horizons may be low but are often greater in lower soil horizons due to the presence of more clay and Fe and Al hydrous oxides.

4. Soil pH. Adsorption of SO_4^{2-} in soils is favored by strong acid conditions. It becomes almost negligible at pH values >6. That is because the positive change in soils increases with decreasing pH, thus attracting the negatively charged SO_4^{2-} ions.
5. Sulfate concentration and temperature. The amount of SO_4^{2-} ions adsorbed depends on concentration and temperature. The adsorbed SO_4^{2-} is in kinetic equilibrium with that in solution. Adsorption maxima have not been reached in many laboratory studies, particularly under acid conditions. Temperature has a relatively small effect on SO_4^{2-} adsorption by soils.
6. Effect of time. Adsorption of SO_4^{2} increases with the length of time it is in contact with soil.
7. Presence of other anions. It is generally considered that SO_4^{2-} is weakly held by soils. Relative to other anions, the retention decreases in the following order: hydroxyl > phosphate > SO_4^{2-} = acetate > nitrate = chloride. Phosphate will displace or reduce the adsorption of SO_4^{2-}, but SO_4^{2-} ions have little effect on phosphate adsorption. There is little, if any effect of Cl^- on SO_4^{2-} adsorption. Molybdate ions will depress SO_4^{2-} adsorption.
8. Effect of cations. The amount of SO_4^{2-} adsorbed is affected by the associated cation of the salt or by the exchangeable cation. This effect follows the lyotropic series; that is, $H^+ > Sr^{2+} > Ba^{2+} > Ca^{2+} > Mg^{2+} > Rb^+ > K^+ > NH_4^+ > Na^+ > Li^+$. Both the cation and the SO_4^{2-} from the salt are retained, but the tendency of adsorption of SO_4^{2-} is different from that of the associated cation.

MECHANISMS OF SULFATE ADSORPTION BY SOILS

Several mechanisms have been proposed for adsorption of SO_4^{2-} by soils. These are discussed in the following sections.

Coordination with Hydrous Oxides

In acid soils SO_4^{2-} adsorption essentially involves the chemistry of Fe and Al. For this reason, Fe and Al hydrous sols have been used as model compounds in studies of the mechanisms of SO_4^{2-} adsorption by soils. The hydrous Fe and Al oxides tend to form coordination complexes due to the donor properties of oxygen. These are polymeric compounds with various proportions of aquo, hydroxo, ol, and oxo groups as shown:

(—M—OH$_2$) (—M—OH) (—M<(OH)(OH)>M—) (—M<(O)(O)>M—)

Aquo Hydroxo Ol Oxo

The proportion of each of these compounds and, consequently, the charge are determined by the equilibrium pH of the system. The formation and composition of colloidally dispersed hydrous oxides and precipitation may be explained on the basis of olation, oxolation, and anion penetration (Rollinson, 1956). As an oxyanion, SO_4^{2-} also has donor properties and can replace aquo, hydroxo, or ol groups in the coordination complex (Bailar and Busch, 1956). Sulfate is a moderately strong coordinator with Al (Marion and Thomas, 1946) and can occupy either 1 or 2 coordination position, but the number 2 position is not well understood (Bailar and Busch, 1956). When SO_4^{2-} salts are added to Al oxide sols, anion penetration occurs with an increase in pH due to displacement of hydroxo or ol group (Rollinson, 1956). This type of reaction has been demonstrated for a number of Oregon acid soils by measuring the differences in pH of soil–K_2SO_4 or –KCl solution mixtures (Chao et al., 1965). This work demonstrated that equilibrating soils with K_2SO_4 solution gave greater pH value than those obtained when the soils were equilibrated with KCl solution, with the difference being greater in soils high in Fe and Al hydrous oxides.

The reactions of acids with hydrous Al oxides were discussed by Graham and Thomas (1947). These reactions involve two steps. The first step involves the rapid conversion of hydroxo groups into aquo groups. The second step involves anion penetration and coordination of the anion to Al by breaking ol and oxo linkages and neutralization of the OH^- groups produced. The reactivity of some of acids are given as: $H_3PO_4 > H_2SO_4 > HNO_3$. By equilibrating acid soils with K_2SO_4 solutions for various times, Chang and Thomas (1963) reported an initial rapid rate of SO_4^{2-} adsorption followed by a slow rate. It has been suggested by Harward and Reisenauer (1966) that the initial rate of SO_4^{2-} adsorption observed by Chang and Thomas (1963) was due to reaction with surface groups whereas the slow rate was due to rate-limiting rupture of ol and oxo linkages and anion penetration.

Infrared spectroscopic studies by Parfitt and Smart (1978) showed four bands in the ν S–O stretching region for the adsorption of SO_4^{2-} on Fe oxides under acid conditions, which led them to develop a structural model for the reaction. This involves the replacement of two surface OH groups (or OH_2^+) by one SO_4^{2-}. The two O atoms of the SO_4^{2-} are coordinated each to a different Fe^{3+}, resulting in binuclear bridging surface complex as follows:

$$\begin{matrix} FeOH \\ FeOH_2^+ \end{matrix} + SO_4^{2-} \longrightarrow \begin{matrix} Fe{-}O \\ Fe{-}O \end{matrix} \!\!> S(=O)_2 + OH^- + H_2O \qquad [1]$$

The complex is formed on the surfaces of goethite (α-FeOOH), akaganeite (β-FeOOH), lepidocrocite (γ-FeOOH), hematitie (α-Fe_2O_3), and amorphous $Fe(OH)_3$. Because soils contain Fe oxides, it assumed that a similar reaction takes place in soils under acid conditions.

The effect of pH on SO_4^{2-} adsorption by hydrous oxides in soil must consider the zero point of charge (the pH at which the change on the surface is zero). Using Al oxide as an example, the potential determining role of H^+ and OH^- involves a charge transfer across the Al–water interface by equilibration with Al^{3+} or O^{2-} in the lattice (Jackson, 1963). Deviation from this pH value involves protonation or deprotonation of Al oxides as follows:

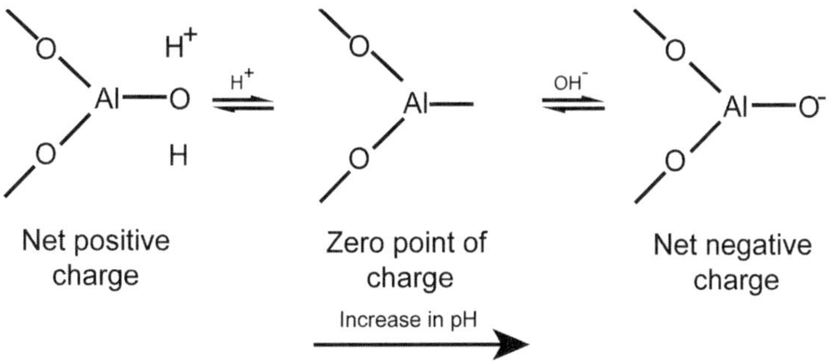

$$\text{Al–OH}_2^{+} \underset{H^+}{\rightleftharpoons} \text{Al–} \underset{OH^-}{\rightleftharpoons} \text{Al–O}^-$$

Net positive charge — Zero point of charge — Net negative charge

Increase in pH → [2]

A similar protonation or deprotonation of Fe oxides is possible:

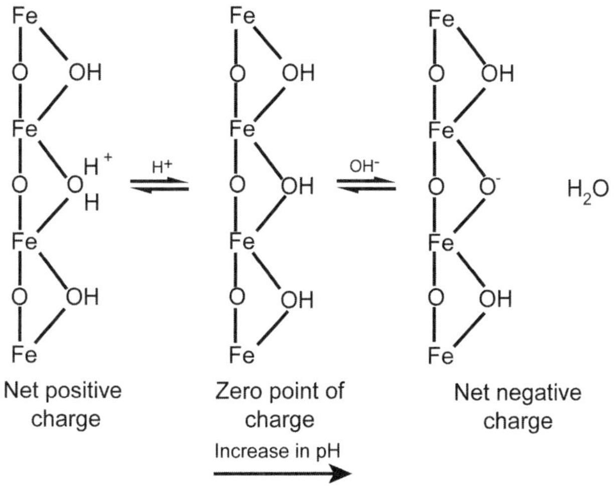

$$\text{Fe–OH}_2^{+} \underset{H^+}{\rightleftharpoons} \text{Fe–OH} \underset{OH^-}{\rightleftharpoons} \text{Fe–O}^- + H_2O$$

Net positive charge — Zero point of charge — Net negative charge

Increase in pH → [3]

The positively charged Al and Fe hydrous oxides adsorb the SO_4^{2-} ions.

In addition to the mechanisms described above, other reactions may account for the adsorption of SO_4^{2-} by soils (Chang and Thomas, 1963). One such mechanism assumes a homoionic Al-saturated clay coated with hydrated oxides R (Fe and Al) adsorb SO_4^{2-} ions as follows:

$$yK^{+} + Al_x\text{—clay} + yH_2O \rightarrow Al_x\,(OH)_y{}^{Ky}\text{— clay} + yH^{+} \qquad [4]$$

$$SO_4{}^{2-} + R_x(OH)_y\text{—clay} \rightarrow R_x[(OH)_{y-z}(SO_4)_z]\text{ — clay} + zOH^{-} \qquad [5]$$

In this mechanism, it is assumed that K^+ ions adsorption sites developed from the exchange and/or hydrolysis of Al on the clay surface. As a result of this hydrolysis, some Al ions are released into the solution. At the same time SO_4^{2-} ions replace the OH ions from R(OH) coating on clay and substitutes for them. The replaced OH ions in turn react with H ions. According to this mechanism, whether the pH of the system increases or decreases depends on the relative rates of the two reac-

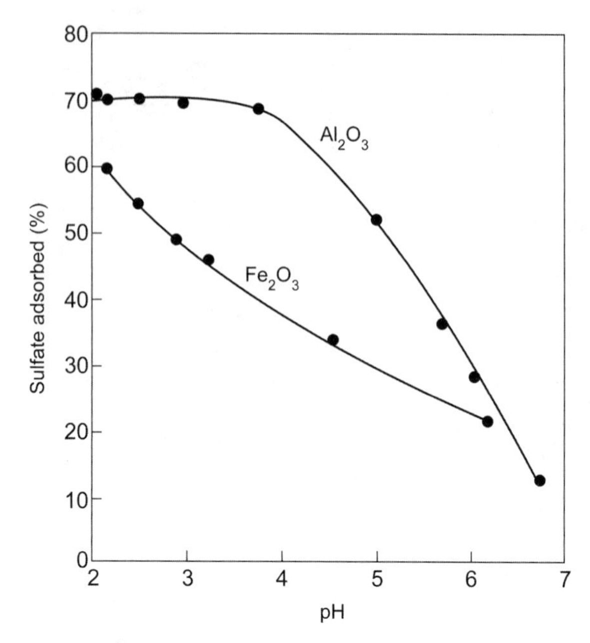

Fig. 3–4. Effect of pH on SO_4^{2-} adsorption from 2.5 m*M* K_2SO_4 solution by reagent-grade (Baker analyzed) Al_2O_3 and Fe_2O_3 (Chao et al., 1964).

tions, Al hydrolysis and OH^- ion exchange. The effect of pH on adsorption of SO_4^{2-} by reagent grade Al_2O_3 and Fe_2O_3 is shown in Fig. 3–4.

Coating soils with Fe increases the adsorption of SO_4^{2-} under acid conditions. Studies on Oregon soils (Chao et al., 1964) showed that SO_4^{2-} adsorption increased when surface-adsorbing soils were coated with Fe and Al oxides (Fig. 3–5, 3–6, and 3–7). Using Virginia soils, other workers (Berg and Thomas, 1959) suggested that there are no permanent sites that hold anions, but that such sites are formed in increasing number as the pH decreased. Although the information available indicates that the zero point of charge of Al compounds is at a pH of about 9, the results obtained with soils indicate that no significant amounts of SO_4^{2-} are adsorbed by soils at pH values >6.

The adsorbed SO_4^{2-} may be replaced by other anions of greater penetration (coordination) ability, such as PO_4^{3-}. This has been demonstrated by displacing the adsorbed SO_4^{2-} from surface soils by application of phosphate fertilizers (Chao et al., 1962; Ensminger, 1954; Kamprath et al., 1956).

Exchange on Edges of Silicate Clays

This exchange presumably involves replacement of OH^- by SO_4^{2-} in terminal octahedral coordination with Al. The general effects and mechanism are similar to those discussed above for Al and Fe hydrous oxides. The following type of reaction can occur with both hydrated Al oxides and the Al layer kaolinite:

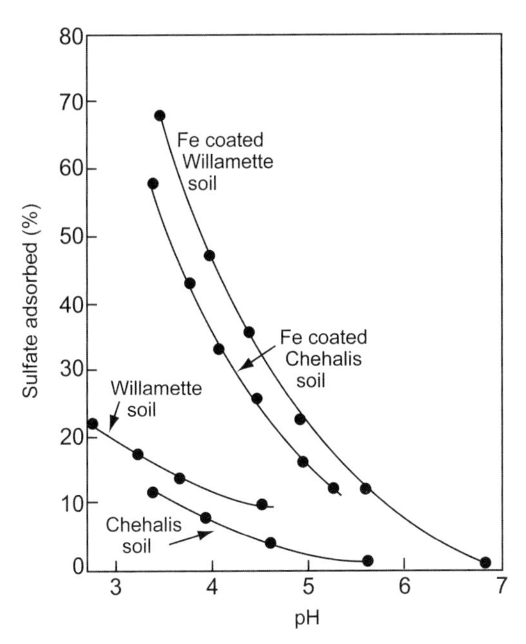

Fig. 3–5. Sulfate adsorption from a 2.5 mM K_2SO_4 solution by Fe-coated Willamette and Chehalis soils (Chao et al., 1964).

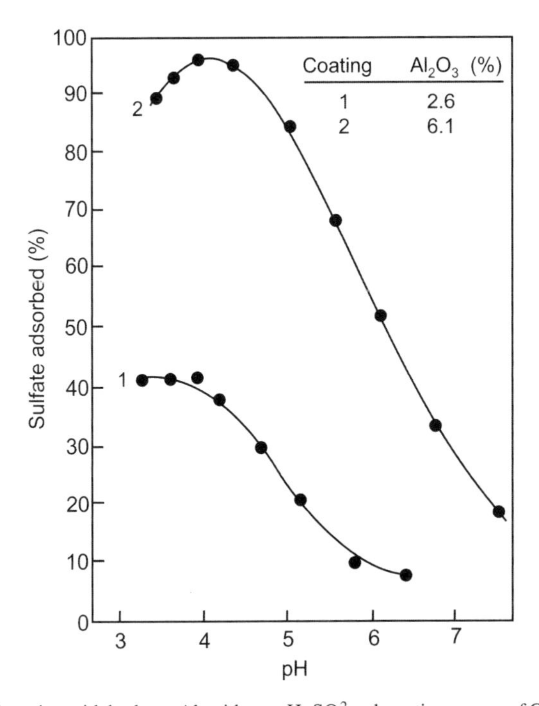

Fig. 3–6. Effect of coating with hydrous Al oxide on pH–SO_4^{2-} adsorption curves of Chehalis soil (Chao et al., 1964).

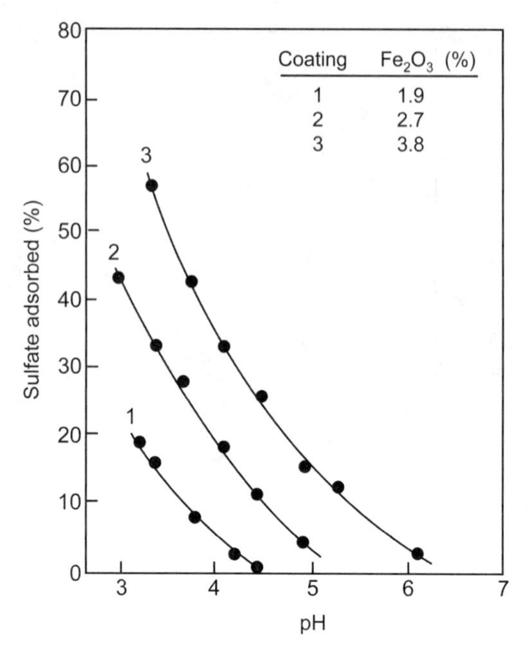

Fig. 3–7. Effect of coating with hydrous Fe oxide on pH–SO_4^{2-} adsorption curves of Chelalis soil (Chao et al., 1964).

$$XAl\begin{matrix} \diagup OH \diagdown \\ \diagdown OH \diagup \end{matrix}AlX + KHSO_4 \rightleftharpoons XAl\begin{matrix} \diagup OH \diagdown \\ \diagdown SO_4 \diagup \\ K \diagup \end{matrix}AlX + H_2O \qquad [6]$$

Molecular Adsorption (Metal-Induced Sulfate Adsorption)

This mechanism is less understood than the others described above. It implies that SO_4^{2-} is adsorbed by some mechanism by which the associated cation is retained by soils. This mechanism has been referred to as "salt" adsorption, "molecular" adsorption, and imbibition.

Sulfate is present in soils with various metal ions, and the different metal ions associated with SO_4^{2-} may significantly affect its reactions on soil colloidal surfaces and, consequently, its mobility in soils. Early work by Chao et al. (1963) on the effect of cations on SO_4^{2-} adsorption by soils showed that the amounts of SO_4^{2-} adsorbed from different salt solutions were in the order: $CaSO_4 > K_2SO_4 > (NH_4)_2SO_4 > Na_2SO_4$. They found that the amounts of SO_4^{2-} adsorbed by soils saturated with different cations followed the order of chemical valence of the saturated cations ($Al^{3+} > Ca^{2+} > K^+$). Other studies of simultaneous retention of Ca^{2+} and SO_4^{2-} by variable-charge soils showed that SO_4^{2-} adsorption increased with increasing Ca^{2+}

(Bolan et al., 1993; Marcano-Martinez and McBride, 1989). Several mechanisms have been proposed to explain the positive effect of Ca^{2+} on SO_4^{2-} adsorption. These include:

1. Precipitation reactions at high cation concentration and pH values >7. This mechanism has been ruled out because the concentrations of Ca^{2+} and SO_4^{2-} were deliberately kept undersaturated with respect to solubility of gypsum.
2. Specific sorption of Ca^{2+} by soil containing Fe and Al hydrous oxides. This increases the positive charge on the surface and thereby increases the adsorption of SO_4^{2-}. It has been estimated that from 75 to 98% of the increases in the adsorption of SO_4^{2-} due to Ca^{2+} adsorption could be attributed to increase in positive charge (Bolan et al., 1993).
3. Surface complexation reactions by which a metal ion (e.g., Ca^{2+}) coordinates to two adsorbed SO_4^{2-} groups and thereby enhancing further adsorption. This mechanism does not account for the changes in surface charge with SO_4^{2}.
4. Formation of ion pair such as CaSO4° that adsorbs to mineral surfaces (Bolan et al., 1993; Marcano-Martinez and McBride, 1989). The formation of the ion pair is enhanced by increasing the concentration of the metal ion (e.g., Ca^{2+}), leading to increases in SO_4^{2-} adsorption.

Recent work on metal-induced SO_4^{2-} adsorption by soil indicates that it is affected by pH, ionic strength, metal type, valence, and concentration of the associated cation (i.e., whether the metal ion is present at equivalent concentration to SO_4^{2} or the metal ion concentration is kept constant) (Ajwa and Tabatabai, 1995a, 1995b, 1997). The effect of pH on SO_4^{2-} adsorption and its associated cation (K^+, Cs^+, NH_4^+, Ca^{2+}, Mg^{2+}, Al^{3+}, or In^{3+}) by four surface soils (two from Iowa and one each from Chile and Costa Rica) showed that, for all the metal ions tested, SO_4^{2-} adsorption was the greatest at the lower pH values (4–5) and decreased with increasing pH of the equilibrium solution (up to pH 7) (Ajwa and Tabatabai, 1995a). This decrease in SO_4^{2-} adsorption is caused by the decrease in positive charges of the surfaces with increasing pH. This was supported by results obtained from potentiometric titration curves. Other studies of SO_4^{2-} adsorption have shown that maximum adsorption occurs at low pH values (3.5–4) and decreases with increasing pH values (Bolan et al., 1986; Courchesne, 1991; Tabatabai, 1987). Because of their greater contents of Al and Fe oxides, SO_4^{2-} adsorption by the soils from Chile and Costa Rica were about four times those of the Iowa soils. In addition to pH, the type and valence of the metal ion and the concentration of the electrolyte (NaCl) in the equilibrium solution considerably affect the surface charge of soils (Bolan et al., 1986; Bowden et al., 1980) and, therefore, alter SO_4^{2-} adsorption behavior. Increasing the concentration of the background electrolyte always decreases the amount of SO_4^{2-} adsorbed. This is the result of increased Cl^- ion competition for the adsorption sites.

The amounts of SO_4^{2-} adsorbed from solutions containing different metal ions is affected by the valence of the metal ion and, in general, follows the order: $In_2(SO_4)_3 > Al_2(SO_4)_3 > CaSO_4 > MgSO_4 > Cs_2SO_4 > K_2SO_4 > (NH_4)_2SO_4$ [because $(NH_4)_2SO_4$ is used as a fertilizer, it was used for comparison]. Also, for any metal ion, SO_4^{2-} adsorption decreased by increasing the concentration of the background electrolyte (NaCl) from 1 to 10 mmol L^{-1}. Sulfate adsorbed per unit decrease in pH, as calculated from the slope of the linear segment of the adsorption curves for the various SO_4^{2-} salts, showed that within a pH range of 3.4 and 4.6 (or 5.0), SO_4^{2-} adsorption generally is not affected by the type and valence of the associated metal ion, suggesting that the SO_4^{2-} adsorption is mainly caused by its retention on the positively charged oxide surfaces, and the metal ion has no role in this process. To support this conclusion, use of a solubility product relationship, possible precipitation of SO_4^{2-} and Ca^{2+} or Al^{3+}, has been ruled out (Ajwa and Tabatabai, 1995a).

The effects of the metal ions on SO_4^{2-} adsorption by soils when the SO_4^{2-} and its metal ions were added at equivalent concentrations have shown that the order of SO_4^{2-} adsorption followed the sequence listed above for the metal salts (Fig. 3–8). For metal ions of the same valence, SO_4^{2-} adsorption increased with increasing ionic radius of the metal ion; in 10^{-8} cm these were: 0.80, 0.54, 0.99, 0.65, 1.69, and 1.33 for In^{3+}, Al^{3+}, Ca^{2+}, Mg^{2+}, Cs^{+}, and K^{+}, respectively (Ajwa and Tabatabai, 1995b). Studies by Chao et al. (1963) on the effect of some of the metal ions listed above and NH_4^{+} on SO_4^{2-} adsorption showed that the magnitude of the SO_4^{2-} adsorption followed the order: $Ca^{2+} > K^{+} > NH_4^{+} > Na^{+}$, and this was affected by the type and valence of the associated cation. Their explanation of this order was based on the ability of the metal ions or cations to form bridges between SO_4^{2-} and the soil surface. Increasing the charge or size of the specifically adsorbed hydrated metal ions decreases the zeta potential of the soil colloids and, therefore, increases SO_4^{2-} adsorption. Experiments conducted by Ajwa and Tabatabai (1995b) showed adsorption of divalent and trivalent metals always increased with increasing SO_4^{2-} adsorption. They also found that the adsorption of the monovalent metal ions was not affected by the increase in SO_4^{2-} adsorption, suggesting that SO_4^{2-} is adsorbed by a ligand-exchange or anion penetration, mechanisms that render the surface of the Al or Fe oxides more negative, and therefore, enhances metal adsorption.

Other studies on the effect of the type and valence of the metal ions on the relationship between metal ions and SO_4^{2-} adsorption by soils when the metal ions were maintained at a constant concentration (12 $mmol_c$ L^{-1}) were reported by Ajwa and Tabatabai (1995b). They showed that SO_4^{2-} adsorption under those conditions varied significantly among the four soils tested, the metal ions present, and the valence of the metal ions, and that the adsorption of monovalent metal ions was not affected by the increase in SO_4^{2-} adsorption (Fig. 3–9). They reported that their results suggest that SO_4^{2-} adsorption under those conditions involves a ligand-exchange, or anion penetration, mechanism that renders the surface of the Al or Fe oxides more negative and, therefore, enhances metal ion adsorption. If this is the case, the increased negative charges should also increase the adsorption of the monovalent ions (Curtin and Syers, 1990). This has been observed when the SO_4^{2-} and metal ions were added at equivalent concentration, but not when the metal ion was maintained at a constant concentration. The cooperative adsorption of metal ions and SO_4^{2} as ion pairs, however, could better explain the results when the metal ion concentration kept constant. Because SO_4^{2-} and monovalent metal ions do not form ion pairs, the increase in SO_4^{2-} adsorption did not affect their adsorption. However, the formation of divalent and trivalent metal SO_4^{2-} ion pairs is enhanced with increasing SO_4^{2-} concentration in the equilibrium solution. Metal ion adsorption is then increased as a result of the simultaneous adsorption of the metal and SO_4^{2-} as ion pairs. This was supported by the results obtained by using the MINTEQA2 speciation model (Allison et al., 1990) showing that percentages of $CaSO_4^\circ$ and $MgSO_4^\circ$ ion pairs were less than the percentage of $AlSO_4^+$ (Ajwa and Tabatabai, 1995b). At pH values between 3.5 and 7.0, ion pair formation was constant, whereas $AlSO_4^+$ formation increased with decreasing pH.

It is known that the pH-dependent charges of soils are attributed to (i) the reactions of the potential determining ions (i.e., H^+ and OH^-) with surface hydroxyl

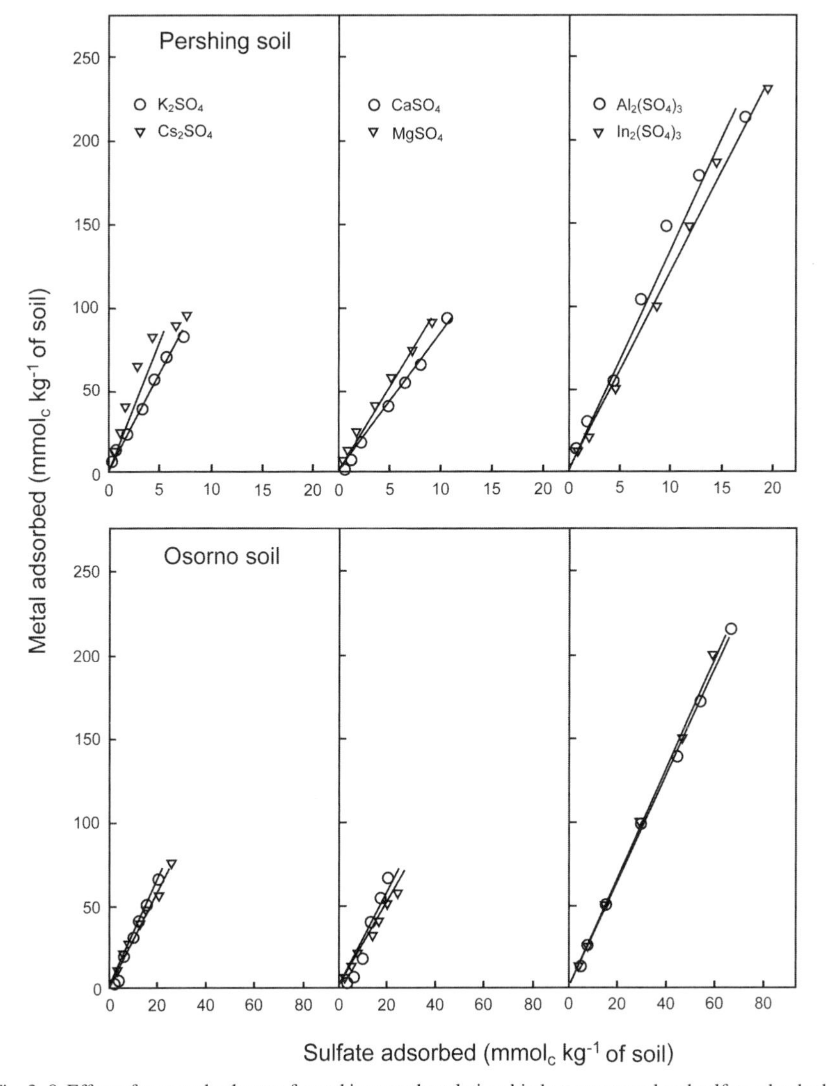

Fig. 3–8. Effect of type and valence of metal ions on the relationship between metal and sulfate adsorbed by Pershing and Osorno soils. Sulfate and metal ions were added at equivalent concentrations (Ajwa and Tabatabai, 1995b).

groups such as XOH and (ii) the interaction of electrolyte ions with surface hydroxyl groups (Sposito, 1984). The adsorption of SO_4^{2-} ions may involve the formation of surface complexes in the protonated form (i.e., HSO_4^-) as follows (Davis and Leckie, 1980):

$$XOH + H^+ + HSO_4^- = XOH_2^+\text{–}HSO_4 \quad [7]$$

Or adsorption of SO_4^{2-} ions may involve the formation of bidentate or bridging surface sites (Hingston et al., 1972):

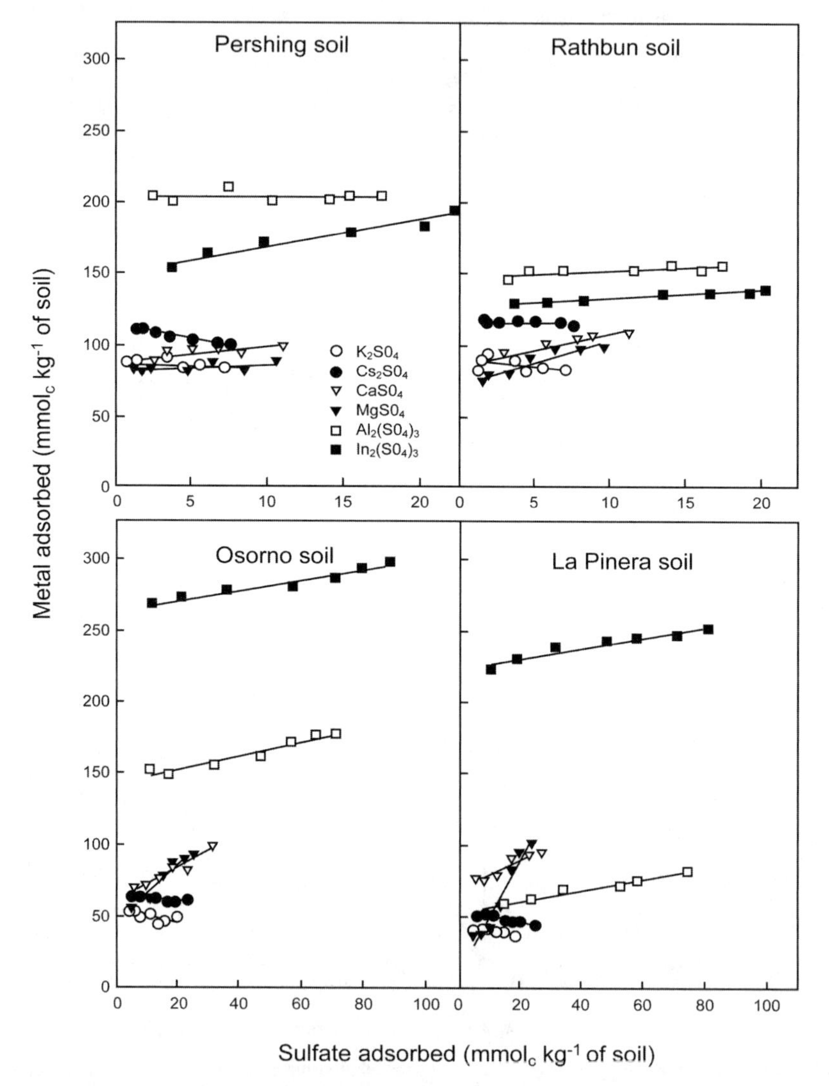

Fig. 3–9. Effect of the type and valence of metal ions on the relationship between metal and sulfate adsorbed by soils. Metal ions were maintained at a constant concentration (12.0 $mmol_c$ L^{-1}) (Ajwa and Tabatabai, 1995b).

$$(XOH)_2 + 2H^+ + SO_4^{2-} = \begin{matrix} XOH_2^+ \\ \quad \searrow \\ \quad\quad SO_4^{2-} \\ \quad \nearrow \\ XOH_2^+ \end{matrix} \qquad [8]$$

The stoichiometry of the reactions involved in cation adsorption can be estimated by measuring the amount of H^+ released, whereas the extent of SO_4^{2-} adsorption can be estimated from the amount of H^+ consumed (or OH^- released) during the adsorption process. Sulfate adsorption by ligand exchange results in the release of OH^-.

DISTRIBUTION COEFFICIENT OF ADSORPTION

An ion exchange model has been used to explain the effect of pH on SO_4^{2-} adsorption, where SO_4^{2-} is adsorbed by replacing OH^- (or OH_2) groups from the surfaces as follows:

$$X\text{-}OH_\alpha + SO_4^{2-} = X\text{-}SO_4 + \alpha OH^- \qquad [9]$$

where X is the surface containing adsorption sites, $X\text{-}SO_4$ is the SO_4^{2-} adsorbed by the surface and α is the (hydroxyl) coefficient. Then, by defining $K_{d\alpha}$, the SO_4^{2-} distribution coefficient, as the ratio of $X\text{-}SO_4$ to SO_4^{2-} in the equilibrium concentration and by substituting into the equilibrium constant, the expression obtained is

$$\log K_{d\alpha} = \log[(K)(X\text{-}OH_\alpha)] + \alpha pOH \qquad [10]$$

By replacing pOH by 14 – pH and by assuming α to be constant, the expression obtained is

$$\log K_{d\alpha} = \{\log[(K)(X\text{-}OH_\alpha)] + 14\alpha\} - \alpha pH \qquad [11]$$

For four soils, a relatively linear relationship was obtained between $\log K_{d\alpha}$ and pH of the equilibrium solution when the metal ions were monovalents, regardless of the concentration of the NaCl electrolyte. Similarly, when the metal ions are divalent or trivalent, a linear relationship is obtained between those variables for some soils, but a curvilinear relationship is obtained for other soils. The linear relationship supports the assumption that α is a constant. The calculated α values vary among soils, metal ions, background electrolyte concentration, and pH (Ajwa and Tabatabai, 1995a).

A similar ion exchange model has been used to explain the effect of pH on metal adsorption by soils (Basta and Tabatabai, 1992). In this model, competition of proton with metal ions for cation exchange sites is used to explain metal ion adsorption. The exchange reaction of a metal ion (M) present in the equilibrium solution can be presented as

$$X\text{-}H_\beta + M = X\text{-}M = \beta H^+ \qquad [12]$$

where X represents the surface containing the adsorption sites, X-M is the metal adsorbed by the soils, and β is the proton coefficient. By defining $K_{d\beta}$, the metal distribution coefficient, as the ratio of X-M to M concentration in the equilibrium solution and by substituting into the equilibrium equation for the reaction, the expression obtained is:

$$\log K_{d\beta} = \log[(K)(X\text{-}H_{\beta})] + \beta \text{pH} \quad [13]$$

Relatively linear relationships are obtained for some soils between $K_{d\beta}$ and pH for monovalent metal ions regardless of the NaCl concentration, but slightly curvilinear relationships are obtained for adsorption of divalent metal ions, and a parabolic relationship was obtained for the adsorption of Al^{3+} and In^{3+} by four soils (Ajwa and Tabatabai, 1995a). The calculated β values, however, varied among the four soils and the six metal ions studied. Even though Eq. [13] can be cautiously applied to the linear slopes obtained for some of the mono- and divalent metal ions, it failed to describe the relationship for the adsorption of trivalent metal ions. Basta and Tabatabai (1992) identified a limitation of this approach to describing the pH-dependency of metal adsorption: metal ion adsorption does not occur only by simple exchange reaction. Other mechanisms, such as complexation, chelation, dissolution, or precipitation reactions could be involved in metal adsorption and, consequently, alter the apparent SO_4^{2-} adsorption behavior.

SULFUR TRANSFORMATIONS IN SOILS

The transformations of S in soils are many and varied (e.g., oxidation, reduction, volatilization, decomposition and mineralization of plant and microbial residues), and often the changes are cyclic as S changes from inorganic to organic forms (immobilization) and back again by living organisms.

Mineralization

The conversion of an element from organic form to inorganic state as a result of microbial activity is termed *mineralization*. As is the case with C and N, organic S in soils is mineralized to inorganic forms, mainly SO_4^{2-}, the form taken up by plant roots. The mechanisms involved in this transformation, however, are not clear. It appears that microorganisms are involved in this process as they obtain their energy from the oxidation of carbonaceous materials in soils. During this process organic S is mineralized. Some of the mineralized S is used for synthesis of new microbial cell materials (immobilization) and the portion not required for synthesis is released as inorganic S. Mineralization and immobilization occur simultaneously in soils whenever organic material is undergoing microbial decomposition. The effect of temperature on the rate of N and S mineralization in 12 Iowa surface soils is shown in Table 3–7.

Sources of Mineralizable Sulfur

It is believed that ester sulfates in soils are main sources of S mineralization in soils. However, C-bonded S (C–S) cannot be excluded because this fraction con-

Table 3–7. Comparison of rates of N and S mineralization in Iowa surface soils.†

Soil	Rate of mineralization					
	at 20°C			at 35°C		
	N	S	N/S	N	S	N/S
	kg ha^{-1} wk^{-1}					
Lester	6.7	1.6	4.2	22.6	4.9	4.9
Ackmore	4.9	1.6	3.1	27.6	4.9	5.6
Fayette	5.6	1.8	3.1	23.5	5.2	4.5
Downs	8.1	2.7	3.0	35.9	7.0	5.1
Clarion	6.7	1.8	3.7	26.7	7.8	3.4
Muscatine	7.4	1.8	4.1	33.2	6.7	5.0
Nicollet	3.8	1.3	2.9	17.3	4.0	4.3
Tama	9.0	2.2	4.1	34.3	7.2	4.8
Webster	9.4	2.2	4.3	38.1	6.5	5.9
Canisteo	4.9	1.3	3.8	20.9	4.0	5.2
Harps	4.3	1.2	3.6	21.7	3.6	6.0
Okoboji	6.3	1.8	3.5	34.3	6.1	5.6
Avg.	6.4	1.8	3.6	28.0	5.7	5.0

† Soil–glass beads columns were incubated at 20 or 35°C, and the quantities of mineral N and S produced were determined after leaching every 2 wk with 0.01 *M* KCl for a total of 26 wk (Tabatabai, 1984).

tains the amino acids methionine, cystine, and cystiene, which can be converted to inorganic SO_4^{2-} under aerobic conditions. The information available suggests that this form of S in soils can be a source for plant uptake.

Role of Arylsulfatase in Sulfur Mineralization

Because a large proportion of the organic S in soils appears to be present as ester sulfate, it seems reasonable to expect that some organic S is mineralized by the action of arylsulfatase enzyme (EC 3.1.6.1). This enzyme was detected in soils by Tabatabai and Bremner (1970). It catalyzes the hydrolysis of ester sulfate releasing SO_4^{2-} for plant uptake. The reaction is as follows:

$$\text{R–C–O–SO}_3^- + \text{H}_2\text{O} \xrightarrow{\text{Arylsulfatase}} \text{R–C–OH} + \text{SO}_4^{2-} + 2\text{H}^+ \qquad [14]$$

For recent reviews about this enzyme, see Tabatabai (1994) and Tabatabai and Dick (2002).

Pattern of Sulfate Release

Because the opposing reactions of mineralization and immobilization can occur simultaneously, different patterns of SO_4^{2-} release have been observed, depending on the energy materials available for the microorganisms (Fig. 3–10). These patterns include (i) immobilization of S during the initial stages of incubation followed by SO_4^{2-} release, (ii) steady linear release of SO_4^{2-} over the whole period of incubation, and (iii) a rate of release that decreases with incubation time.

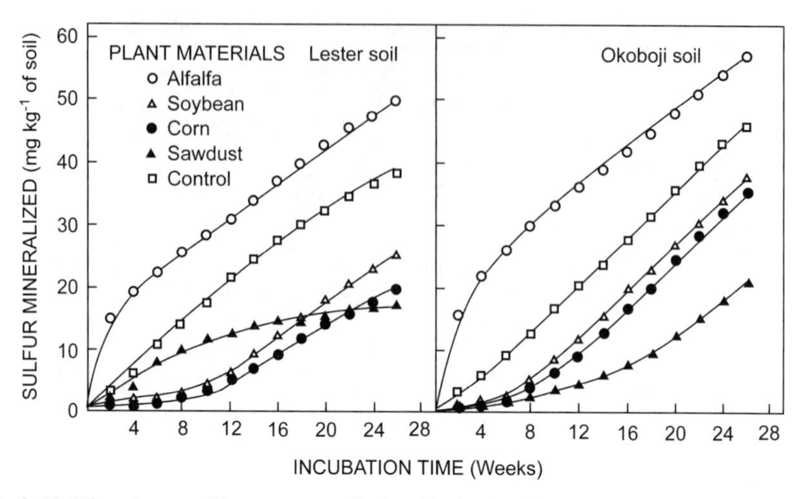

Fig. 3–10. Effect of crop-residue treatment on S mineralization in soils (Tabatabai and Al-Khafaji, 1980).

The pattern of SO_4^{2-} release is not related to any specific soil properties and apparently are caused by adjustment of the microbial populations to the incubation conditions and to the availability of the initial substrates.

Factors Affecting Sulfur Mineralization

Because the S mineralization is microbiological in nature, any variable that affects microbial growth should affect S mineralization. Therefore, temperature, moisture, pH, and the availability of nutrients are the most important.

Oxidation

Elemental S is one of the main sources of S added to soils. Before it can be utilized by crops, however, elemental S has to be oxidized to SO_4^{2-}. Elemental S is oxidized in soils by chemical and biochemical processes, and a number of factors affect these processes. Microbial reactions dominate the processes. The microorganisms involved in oxidation of elemental S in soils are of three groups: (i) chemolithotrophs (e.g., members of the genus *Thiobacillus*), (ii) photoautrotrophs (e.g., species of purple and green S bacteria), and (iii) heterotrophs, including wide ranges of bacteria and fungi. Chemolithotrophs and photoautrotrophs are mainly responsible for oxidation of reduced S compounds in aerobic soils. Phototrophic bacteria are the predominant organisms responsible for oxidizing S^{2-} at the soil–water interface in flooded soils and in the rhizosphere of rice (*Oryza sativa* L.) plants. The major reduced forms of inorganic S found in elemental S-treated soils are S°, S^{2-}, and the oxyanions $S_2O_3^{2-}$, $S_4O_6^{2-}$, and SO_3^{2-}. These anion are oxidized, ultimately to SO_4^{2-} (Fig. 3–11). The reactions involved in oxidation of elemental S in soils seem to be as follows:

$$S^\circ \rightarrow \overbrace{S_2O_3^{2-} \rightarrow S_4O_6^{2-} \rightarrow SO_3^{2-}}^{\text{Rhodanese}} \rightarrow SO_4^{2-} \qquad [15]$$

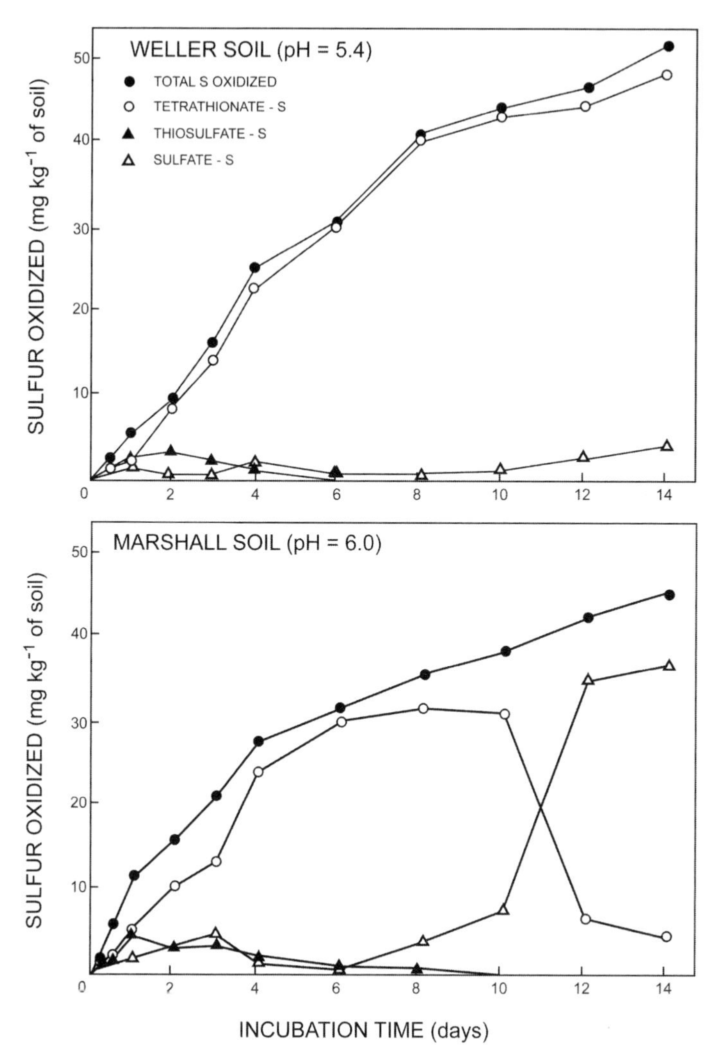

Fig. 3–11. Amounts of thiosulfate-, tetrathionate-, and sulfate-S produced in two Iowa surface soils (Weller and Marshall soils) amended with elemental S (200 mg S kg^{-1} soil) and incubated at 30°C for various times (Nor and Tabatabai, 1977).

It is not clear whether these reactions occur directly by biochemical reactions (microbial processes) in soils or whether some of the intermediate products formed are the results of abiotic reactions. This is especially important in the case of the intermediates $S_2O_3^{2-}$ and $S_4O_6^{2-}$.

The enzyme rhodanese (thiosulfate-cyanide sulfotransferase, EC 2.8.1.1) was detected in soils by Tabatabai and Singh (1976). It catalyzes the conversion of the intermediate $S_2O_3^{2-}$ to SO_3^{2-} as follows:

$$S_2O_3^{2-} + CN^- \xrightarrow{\text{Rhodanese}} SO_3^{2-} + SCN^- \quad [16]$$

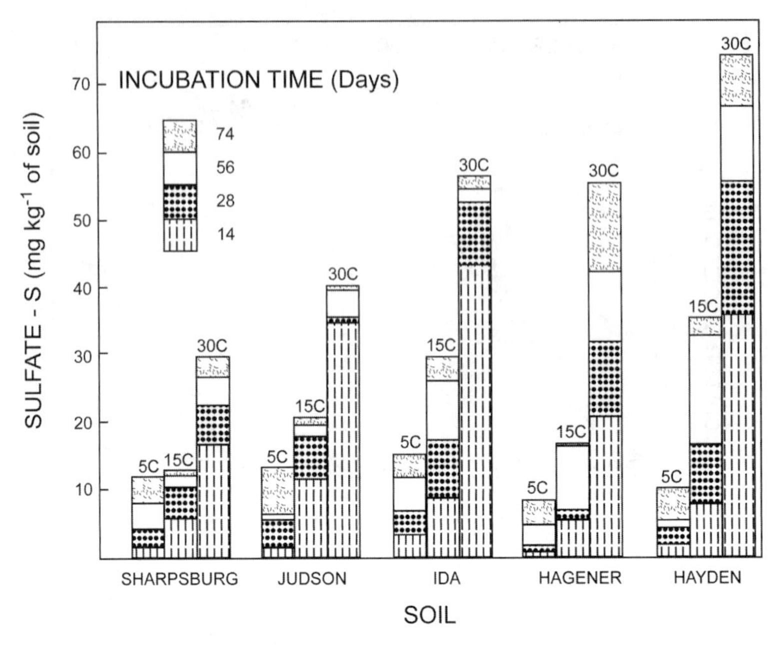

Fig. 3–12. Effects of temperature and time of incubation on oxidation of elemental S (100 mg S kg^{-1} soil) in soils (Nor and Tabatabai, 1977).

Several factors affect the rate of oxidation of elemental S in soils (Tabatabai, 1994). These include (i) particle size—the finer the particles, the faster is the reaction because of the increase of surface area with decreasing particle size; (ii) temperature—the higher the temperature, the greater the reaction rate (Fig. 3–12; this is true between 10 and 40°C); (iii) time of contact with soil—the longer the reaction, the more oxidation (Fig. 3–13); and (iv) effect of pH—the oxidation appears to be faster in alkaline than in acid soils.

Reduction

The reduction of SO_4^{2-} to H_2S is a process that occurs mainly by anaerobic bacteria. Thus, it occurs only in anaerobic soils. This process is not important in aerobic agricultural soils, except perhaps in anaerobic microsites in soil aggregates. This process, however, is a major reaction of the S cycling in soils that are waterlogged or undergo periodic flooding, especially in soils containing readily decomposable plant residues, such as alfalfa (*Medicago sativa* L.). Bacterial reduction of SO_4^{2-} involves either assimilation or dissimilation processes. In assimilation, SO_4^{2-} is reduced to the thiol (–SH) group of organic compounds for protein synthesis. In dissimilation, the reduction leads to production of H_2S under very low redox potential (Eh) values (Tabatabai, 1994). Under normal conditions, however, H_2S is not volatilized from soils because it precipitates with Fe^{2+}, Mn^{2+}, Cu^{2+}, Cu^{+}, and/or Zn^{2+} present in soils. In the case of Fe^{2+} ferrous sulfide (FeS) and pyrite (FeS_2) are formed

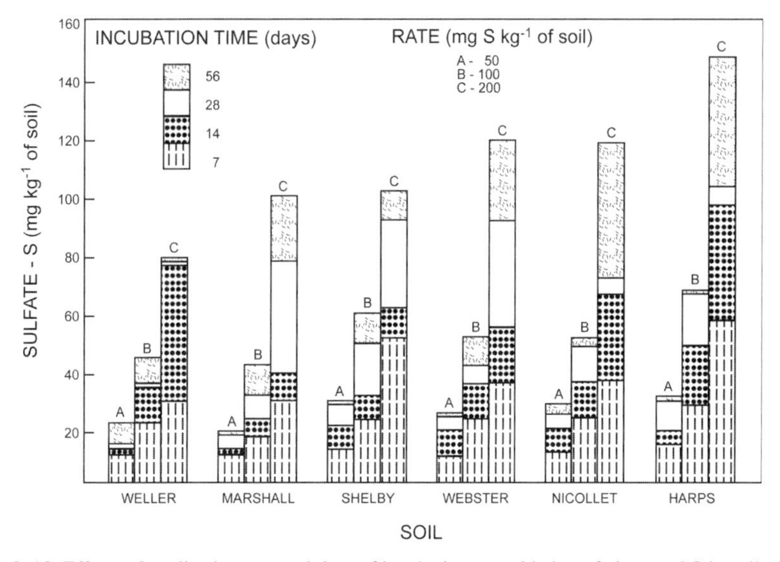

Fig. 3–13. Effects of application rate and time of incubation on oxidation of elemental S in soils. The values for Weller and Marshall soils treated with 200 mg S kg^{-1} soil include the amounts of thiosulfate- and tetrathionate-S produced during 7 and 14 d of incubation (Nor and Tabatabai, 1977).

in severely reducing conditions by the reduction of SO_4^{2-} to S^{2-} by the bacteria *Desulfovibrio desulfuricans,* which reacts with FeS, producing FeS_2 (for reviews, see Germida and Gupta, 1992; Evangelou, 1998; Bigham and Nordstrom, 2000).

Volatilization

Relatively small amounts, if any, of S-containing gases, including H_2S, are released from aerobic, agricultural soils, even when such soils are under waterlogged conditions. Substantial amounts of H_2S are released, however, from salt marsh soils. Several gases are released from soils when treated with animal manures, sewage sludge, and protein-rich plant materials, such as alfalfa, especially under waterlogged conditions. These include carbon disulfide (CS_2), which results from decomposition of the amino acids cystine and methionine; carbonyl sulfide (COS), released during decomposition of thiocyanate and isothiocyanate; and methyl mercaptan (CH_3SH), dimethyl sulfide (CH_3SCH_3), and dimethyl disulfide (CH_3SSCH_3), which result from decomposition of methionine-containing materials (Bremner and Steele, 1978).

ACKNOWLEDGMENTS

This work was supported, in part, by the Iowa Agriculture and Home Economics Experiment Station and the Biotechnology By-product Consortium of Iowa.

REFERENCES

Ajwa, H.A., and M.A. Tabatabai. 1995a. Metal-induced sulfate adsorption by soils: I. Effect of pH and ionic strength. Soil Sci. 159:32–42.

Ajwa, H.A., and M.A. Tabatabai. 1995b. Metal-induced sulfate adsorption by soils: II. Effect of metal type, valence, and concentration. Soil Sci. 160:281–290.

Ajwa, H.A., and M.A. Tabatabai. 1997. Metal-induced sulfate adsorption by soils: III. Application of Langmuir equations. Soil Sci. 162:169–180.

Allison, J.D., D.S. Brown, and K.J. Novo-Gradac. 1990. MINTEQA2/PRODEFA2, a geochemical assessment model for environmental systems. Version 3.0 user's manual. USEPA, Environ. Res. Lab., Athens, GA.

Bailar, J.C, Jr., and D.H. Busch. 1956. General survey of the coordination compounds. p. 1–99. *In* J.C. Bailar, Jr. (ed.) Chemistry of coordination compounds. Reinhold Publishing Corp., New York.

Baker, J.L., J.M. Laflen, H.P. Johnson, and J.J. Hanway. 1975. Nitrate, phosphorus, and sulfate in subsurface drainage water. J. Environ. Qual. 4:406–412.

Basta, N.T., and M.A. Tabatabai. 1992. Effect of cropping systems on adsorption of metals by soils. II. Effect of pH. Soil Sci. 153:195–204.

Berg, W.A., and G.W. Thomas. 1959. Anion elution patterns from soils and soil clays. Soil Sci. Soc. Am. Proc. 23:348–350.

Bigham, J.M., and D.K. Nordstrom. 2000. Iron- and aluminum-hydroxysulfate minerals from acid sulfate waters. p. 1–52. *In* C.N Alpers et al. (ed.) Sulfate minerals: Crystallography, geochemistry, and environmental significance. Reviews in Mineralogy and Geochemistry. Vol. 40. Mineralogical Society of America, Washington, DC.

Bolan, N.S., J.K. Syers, and M.E. Sumner. 1993. Calcium-induced sulfate adsorption by soils. Soil Sci. Soc. Am. J. 57:691–696.

Bolan, N.S., J.K. Syers, and R.W. Tillman. 1986. Ionic strength effects on surface charge and adsorption of phosphate and sulfate by soils. J. Soil Sci. 37:379–388.

Bowden, J.W., A.M. Posner, and J.P. Quirk. 1980. Adsorption and charging phenomena in variable change soils. p. 147–166. *In* B.K.G Theng (ed.) Soils with variable charge. N Z. Soil Sci., Palmerston, New Zealand.

Bremner, J.M., and C.G. Steele. 1978. Role of microorganisms in the atmospheric sulfur cycle. p. 155–201. *In* M. Alexander (ed.) Advances in microbial ecology. Plenum Press, New York.

Chang, M.L., and G.W. Thomas. 1963. A suggested mechanism for sulfate adsorption by soils. Soil Sci. Soc. Am. Proc. 27:281–283.

Chao, T.T., M.E. Harward, and S.C. Fang. 1962. Movement of S^{35} tagged sulfate through soil column. Soil Sci. Soc. Am. Proc. 26:27–37.

Chao, T.T., M.E. Harward, and S.C. Fang. 1963. Cationic effects on sulfate adsorption by soils. Soil Sci. Soc. Am. Proc. 27:35–38.

Chao, T.T., M.E. Harward, and S.C. Fang. 1964. Iron and aluminum coatings in relation to sulfate adsorption characteristics of soils. Soil Sci. Soc. Am. Proc. 28:632–635.

Chao, T.T., M.E. Harward, and S.C. Fang. 1965. Exchange reactions between hydroxyl and sulfate ions in soils. Soil Sci. 99:104–108.

Courchesne, F. 1991. Electrolyte concentration and composition effects on sulfate sorption by two spodosols. Soil Sci. Soc. Am. J. 55:1576–1581.

Curtin, D., and J.K. Syers. 1990. Mechanism of sulphate adsorption by two tropical soils. J. Soil Sci. 41:295–304.

Davis, F.L., and J.O. Leckie. 1980. Surface ionization and complexation at the oxide/water interface. J. Colloid Interface Sci. 63:480–499.

Driebelbis, F.R. 1947. Some plant nutrient losses in gravitational water from monolith lysimeters at Coshocton, Ohio. Soil Sci. Soc. Am. Proc. 11:182–188.

Ensminger. 1954. Some factors affecting the adsorption of sulfate by Alabama soils. Soil Sci. Soc. Am. Proc. 18:259–264.

Evangelou, V.P. 1998. Acid mine drainage. p. 1–17. *In* R.A. Meyers (ed.) Encyclopedia of environmental analysis and remediation. Wiley, New York.

Freney, J.R. 1967. Sulfur–containing organics. p. 229–259 *In* A.D. McLaren and G.H. Peterson (ed.) Soil biochemistry. Vol. 1. Marcel Dekker, New York.

Freney, J.R., N.J. Barrow, and K. Spencer. 1962. A review of certain aspects of sulphur as a soil constituent and plant nutrient. Plant Soil 17:295–308.

Germida, J.J., and V.V.S.R. Gupta. 1992. Biochemistry of sulfur cycling in soil. P.:1–53. *In*: G. Stotzky and J.-M. Bollag (ed.) Soil biochemistry. Vol. 7. Marcel Dekker, New York.

Graham, R.P., and A.W. Thomas. 1947. The reactivity of hydrous ammonia towards acids. Am. Chem. Soc. 69:816–821.

Harward, M.E., and H.M. Reisenauer. 1966. Reactions and movement of inorganic soil sulfur. Soil Sci.101:326–335.

Hingston, R.J., A.M. Posner, and J.P. Quirk. 1972. Anion adsorption by goethite and gibbsite. I. The role of the proton in determining adsorption envelopes. J. Soil Sci. 23:177–193.

Jackson, M.L. 1963. Aluminum bonding in soils: A unifying principle in soil science. Soil Sci. Soc. Am. Proc. 27:1–10.

Joffe, J.S. 1933. Lysimeter studies: II. Soil Sci. 35:239–257.

Kamprath, E.J., W.L. Nelson, and J.W. Fitts. 1956. The effect of pH, sulfate and phosphate concentrations on the adsorption of sulfate by soils. Soil Sci. Soc. Am. Proc. 20:463–466.

Longenecker, D.E. 1960. Influence of soluble anion on some physical and physico-chemical properties of soils. Soil Sci. 90:185–191.

Lowe, L.E., and W.A. DeLong. 1963. Carbon bonded sulfur in selected Quebec soils. Can. J. Soil Sci. 43:151–155.

Lunt, A.H. 1937. Forest lysimeter studies under Red Pine. Connecticut Agric. Exp. Stn. Bull. 394.

Marcano-Martinez, E., and M.B. McBride. 1989. Calcium and sulfate retention by two oxisols of the Brazilian Cerrado. Soil Sci. Soc. Am. J. 53:63–69.

Marion, S.P., and A.W. Thomas. 1946. Effect of diverse anion on the pH of maximum precipitation of aluminum hydroxide. Colloid Sci. 1:221–234.

McKell, C.M., and W.A. Williams. 1960. A lysimeter study of sulfur fertilization of an annual-range soil. J. Range Manage. 13:113–117.

Neptune, A.M.L., M.A. Tabatabai, and J.J. Hanway. 1975. Sulfur fractions and carbon–nitrogen–phosphorus–sulfur relationships in some Brazilian and Iowa soils. Soil Sci. Soc. Am. Proc. 39:51–55.

Nor, Y.M., and M.A. Tabatabai. 1977. Oxidation of elemental sulfur in soils. Soil Sci. Soc. Am. J. 41:736–741.

Parfitt, R.L., and R.S.C. Smart. 1978. The mechanism of sulfate adsorption on iron oxides. Soil Sci. Soc. Am. Proc. 42:48–50.

Paul, E.A., and E.L. Schmidt. 1961. Formation of free amino acids in rhizosphere and nonrhizosphere soil. Soil Sci. Soc. Proc. 25:359–362.

Pirela, H.J., and M. A. Tabatabai. 1988. Reduction of organic sulfur in soils with tin and phosphoric acid. Soil Sci. Soc. Am. J. 52:959–964.

Putnam, H.D., and E.L. Schmidt. 1959. Studies of free amino acid fraction in soils. Soil Sci. 87:22–27.

Rollinson, C.L. 1956. Olation and related chemical processes. p. 448–471. *In* J.C. Bailar, Jr. (ed.) Chemistry of coordination compounds. Reinhold Publishing, New York.

Shorey, E.C. 1913. Some organic soil constituents. USDA Bur. of Soils, Bull. 88.

Sposito, G. 1984. The surface chemistry of soils. Oxford Univ. Press, New York.

Tabatabai, M.A. 1983. Atmospheric deposition of nutrients and pesticides. p. 92–108. *In* F.W. Schaller and G.W. Bailey (ed.). Agricultural management and water quality. Iowa State University Press, Ames.

Tabatabai, M.A. 1984. Importance of sulphur in crop production. Biogeochemistry 1:45–62.

Tabatabai, M.A. 1985. Effect of acid rain on soils. Crit. Rev. Environ. Control. 15:65–110.

Tabatabai, M.A. (ed.) 1986. Sulfur in agriculture. Agron. Monogr. 27. ASA, CSSA, and SSSA, Madison, WI.

Tabatabai, M.A. 1987. Physicochemical fate of sulfate in soils. J. Air Pollut. Control Assoc. 37:34–38.

Tabatabai, M.A. 1994. Soil enzymes. p. 775–833. *In* R.W. Weaver et al. (ed.) Methods of soil analysis. Part 2. SSSA Book Ser. 5. SSSA, Madison, WI.

Tabatabai, M.A. 1996. Sulfur. p. 921–960. *In* D.L. Sparks (ed.) Methods of soil analysis. Part 3. SSSA Book Ser. 5. SSSA, Madison, WI.

Tabatabai, M.A., and A.A. Al-Khafaji. 1980. Comparison of nitrogen and sulfur mineralization in soils. Soil Sci. Soc. Am. J. 44:1000–1006.

Tabatabai, M.A., and J.M. Bremner. 1970. Arylsulfatase activity of soils. Soil Sci. Soc. Am. Proc. 34:225–229.

Tabatabai, M.A., and J.M. Bremner. 1972a. Distribution of total and available sulfur in selected soils and soil profiles. Agron. J. 65:40–44.

Tabatabai, M.A., and J.M. Bremner. 1972b. Forms of sulfur, and carbon, nitrogen, and sulfur relationships, in Iowa soils. Soil Sci. 114:380–386.

Tabatabai, M.A., and W.A. Dick. 2002. Soil enzymes: research and developments in measuring activities. p. 567–596. *In* R.G. Burns and R.P. Dick (ed.) Enzymes in the environment: Activity, ecology, and applications. Marcel Dekker, New York.

Tabatabai, M.A., and B.B. Singh. 1976. Rhodanese activity of soils. Soil Sci. Soc. Am. J. 40:381–385.

Whitehead, D.C. 1964. Soil and plant nutrient aspects of the sulphur cycle. Soils Fert. 27:1–8.

Windholz, M. 1976. The Merck index. 9th ed. Merck and Co., Raway, NJ.

Xia, K., F. Weesner, W.F. Bleam, P.R. Bloom, U.L. Skyllberg, and P.A. Helmke. 1998. XANES studies of oxidation states of sulfur in aquatic and soil humic substances. Soil Sci. Soc. Am. J. 62:1240–1246.

Chapter 4

Chemistry of Potassium in Soils

P. M. HUANG, *University of Saskatchewan, Saskatoon, Saskatchewan, Canada*

Potassium is a major component of the earth's crust, soils, and plants. In the earth's crust, K is the seventh most abundant element. The lithosphere contains an average of about 25.9 g K kg^{-1} (Hurlbut and Klein, 1977). In the soil, the common range of K content is 0.4 to 30 g K kg^{-1} (Jackson, 1964; Helmke, 2000). Of the major and secondary nutrient elements, K is generally the most abundant in the soil (Rich, 1968a; Sparks and Huang, 1985).

Relative abundance and some chemical characteristics of K and certain elements common in the earth's crust are given in Table 4–1. Among the mineral cations essential for plants, K is the largest in size. Therefore, the number of oxygen ions coordinating K in mineral structures is high. Consequently, the strength of each K–O bond is relatively weak. Potassium has a lower polarizability than NH_4^+, Rb^+, Cs^+, and Ba^{2+}. By contrast, K, relative to Ca^{2+}, Mg^{2+}, Li^+, and Na^+ ions, has a higher polarizability. Ions with higher polarizability would be preferentially selected in ion exchange reactions. Compared with Li^+, Na^+, Mg^{2+}, and Ca^{2+}, K^+ has a lower hydration energy (Helfferich, 1962) and thus would cause little swelling in the interlayer space.

The role of K in soils is prodigious; of the many plant nutrient–soil mineral relationships, those involving K are of major significance (Sparks, 2000b). Potassium plays a very important role in enzyme catalysis, photosynthesis and respiration, assimilation and transport, protein and oil metabolism, legume dinitrogen fixation, disease reduction, and interactions with other nutrients and with crop varieties or hybrids (Munson, 1985). Therefore, K is vital in sustaining plant growth, yield formation and crop quality. Furthermore, K is essential in human and animal nutrition and is closely related to certain medical and health aspects. The availability of soil K to plants is related to the nature of soil K reserves, the chemistry of the structural configurations and surface properties of soil components involved, and the dynamics and equilibria of K in soil environments. The objective of this chapter is to integrate the current knowledge and discuss future prospects on this subject.

FORMS OF SOIL POTASSIUM

Forms of soil K include solution, exchangeable, fixed, and structural K, as shown in Fig. 4–1. Solution and exchangeable K generally account for relatively

Chemical Processes in Soils. SSSA Book Series, no. 8.

Table 4–1. Relative abundance and chemical characteristics of K and certain elements common in earth's crust.

Ion	Crustal avg.†	Crystalline radii‡	Hydrated radii‡	Coordination number‡	Polarizability‡	Debye–Hückel parameter‡
	mg kg $^{-1}$	——— nm	———		nm^3	nm
Li^+	30	0.060	0.382	6	0.0079	0.432
Na^+	28 300	0.095	0.358	6, 8	0.0196	0.397
K^+	25 900	0.133	0.331	8–12	0.0876	0.363
Rb^+	120	0.148	0.329	8–12	0.1407	0.349
Cs^+	1	0.169	0.329	12	0.2452	--
Mg^{2+}	20 900	0.065	0.428	6	0.0110	0.502
Ca^{2+}	36 300	0.099	0.412	6, 8	0.0523	0.473
Sr^{2+}	450	0.113	0.412	8	0.0880	0.461
Ba^{2+}	400	0.135	0.404	8–12	0.1682	0.445
Al^{3+}	81 300	0.050		4, 6		
Si^{4+}	277 200	0.041				

† Berry and Mason (1959).
‡ Gast (1977).

small proportions of the total K in soils. Potassium-bearing micas and feldspars are the major K reserves of the soil (Fanning et al., 1989; Huang, 1989). Muscovite, biotite, microcline, and orthoclase are the main K-bearing primary minerals. Other micas and feldspars and other minerals may contain substantial amounts of K. Upon reaction with weathered mica, vermiculite (Rich, 1968a; Douglas, 1989; Fanning et al., 1989), allophane, and zeolites (Stitcher, 1972; Wada, 1989; Harsh, 2000), K is also present in fixed form in soils. In the immediate vicinity of the fertilized zone, K can react with aluminum hydroxides and acid phosphate solution to form taranakite (Taylor et al., 1963; Taylor and Gurney, 1965). Recent studies showed that the nucleation and crystallization of taranakite can be perturbed in soils with high Fe content, especially under reduced and acidic conditions (Zhou et al., 2000; Liu et al., 2002). In acid soils K also could be coprecipitated with Al and sulfate to

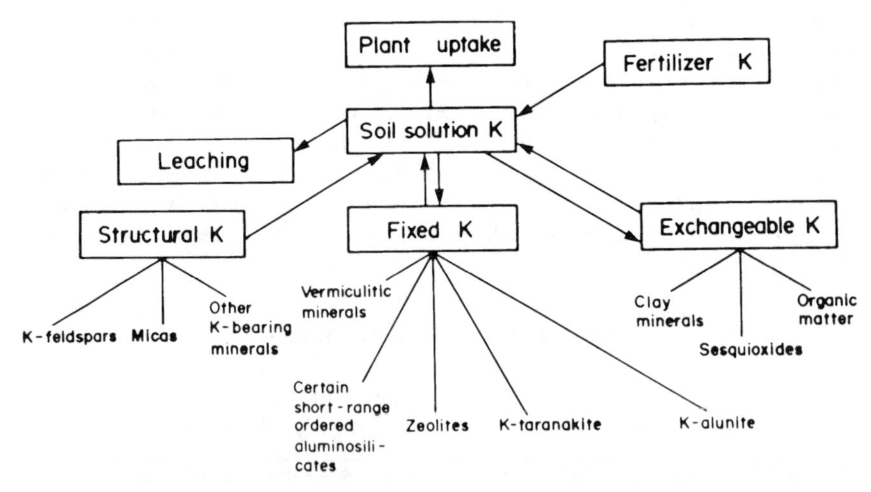

Fig. 4–1. Interrelationships of various forms of soil K (Sparks and Huang, 1985).

form alunite (Adams and Rawajfih, 1977). The forms of soil K in the order of their bioavailability are: solution > exchangeable > fixed (nonexchangeable) > structural K (Sparks, 2000b). Most K in soils is transported to the roots from the soil solution through diffusion and mass flow.

Solution Potassium

Soil solution K is the form of K that is directly taken up by plants and microorganisms and also is the form most subject to leaching in soils. The concentration of K in the soil solution varies from 2 to 5 mg K L^{-1} for normal agricultural soils of humid regions and is an order of magnitude higher in arid region soils (Haby et al., 1990). The amount of soil solution K is generally too low to meet the requirement of K by a crop during a growing season. Levels of solution K are governed by the dynamics and equilibria that occur between the forms of soil K under the influence of soil moisture, cations in solution and on the exchanger phase, solution inorganic and organic ligands, plant removal, microbial activity, fertilization, and leaching, as discussed below.

Exchangeable Potassium

Exchangeable K is the fraction of soil K that is electrostatically bound as an outer-sphere complex to the surfaces of clay minerals, sesquioxides, and organic matter. Therefore, it is readily exchanged with other cations and bioavailable. For optimum K nutrition of a crop, the replenishment of a K-depleted soil solution is affected by the release of exchangeable K from soils.

Fixed Potassium

Potassium becomes fixed because the binding forces between K and the mineral surfaces are greater than the hydration forces between individual K^+ ions. This results in a partial collapse of the crystal structures, and the K^+ ions are physically trapped to varying degrees, making K release a slow, diffusion-controlled process (Sparks, 1987). Fixed K can be found in a series of soil components (Rich, 1972; Sparks and Huang, 1985). It is also present in wedge zones of weathered micas and vermiculites (Rich, 1964, 1972). Cations with a size similar to K^+ (e.g., NH_4^+ and H_3O^+) can exchange K from wedge zones. Large hydrated cations, such as Ca^{2+} and Mg^{2+}, cannot fit into the wedge zones and thus cannot release fixed K. Fixed K is moderately to sparingly available to plants (Mengel, 1985). Release of fixed K to the exchangeable form can take place when levels of exchangeable and solution K are decreased by plant removal, microbial activity, and leaching (Sparks, 2000b).

Structural Potassium

The vast majority of the total soil K is in the form of structural K, mainly as K-bearing primary minerals such as feldspars and micas (Sparks and Huang, 1985). Structural K is generally assumed to be only slowly available to plants (Sparks,

2000b). However, the dynamics of K release from these primary minerals is related to their crystal structure and atomic bonding (Huang et al., 1968; Sparks, 1987). The availability of structural K depends on the dynamics of K release from K-bearing micas and feldspars (Scott and Reed, 1962a, 1962b; Reed and Scott, 1962, 1966; Scott, 1968; Huang et al., 1968; Song and Huang, 1988; Zhou, 1995), the degree of weathering of these minerals, and the level of K in the other forms (Sparks, 2000b).

SOIL COMPONENTS INVOLVED IN POTASSIUM DYNAMICS AND EQUILIBRIA

Potassium-Bearing Micas

Micas are 2:1 phyllosilicates having a charge imbalance that is satisfied by a tightly held, nonhydrated interlayer cation. The interlayer cation is mainly the K^+ ion in the K-bearing micas. The 2:1 layer is composed of an octahedral sheet between two tetrahedral sheets. In trioctahedral micas, such as biotite, all three octahedral positions are filled. In dioctahedral micas, such as muscovite, only two out of three octahedral cation positions are filled. Bailey (1984) presented a comprehensive review of micas.

The double sheets of the hexagonal network of linked tetrahedra are placed together, with the vertices of their tetrahedra pointing inward. These vertices are cross-linked by Al atoms in muscovite (dioctahedral mica) or Mg and Fe atoms in phlogopite and biotite (trioctahedral micas). Hydroxyl groups are incorporated, linked to Al, Mg, or Fe alone. There is a firm bound double sheet, with bases of the tetrahedra on each outer side. The mica structure is a succession of such double sheets with K^+ placed between them. Figure 4–2a shows the hexagonal network of linked tetrahedral groups, with the tetrahedra pointing downward. The lengths of the *a* and *b* axes of micas are determined by the dimensions of this network and shown by dotted lines. The double sheet is shown in detail in Fig. 4–2b. The structure of a double sheet is such that the hexagonal Si–O rings of a single upper sheet are not directly above the corresponding rings of the lower sheet, the amount of "stagger" being $a/3$ or 0.17 nm.

The 2:1 mica layer must superimpose so the hexagonal rings of adjacent basal surfaces line up and enclose the interlayer K^+ ions. In most species, adjacent layers are propped slightly apart by the interlayer cations. This is because tetrahedral rotation moves every other basal oxygen toward the center of each ring and thereby reduces the size of the opening.

The layer charge of K-bearing micas ideally is -1.0 mol_c per formula unit. It may originate entirely from substitution of R^{3+} (primarily Al, Fe, or Cr) for Si^{4+} in tetrahedral positions, or may originate entirely from substitution of R^+ or R^{2+} for R^{2+} or R^{3+} in octahedral positions. The layer charge also may come partly from both.

Micas occur in almost any geological environment and are abundant in many rocks, including shales, slates, phyllites, schists, gneiss, granites, and sediments derived from these rocks (Olson et al., 2000). Biotite is the common trioctahedral mica. Muscovite is the abundant dioctahedral mica. Muscovite and biotite are the most

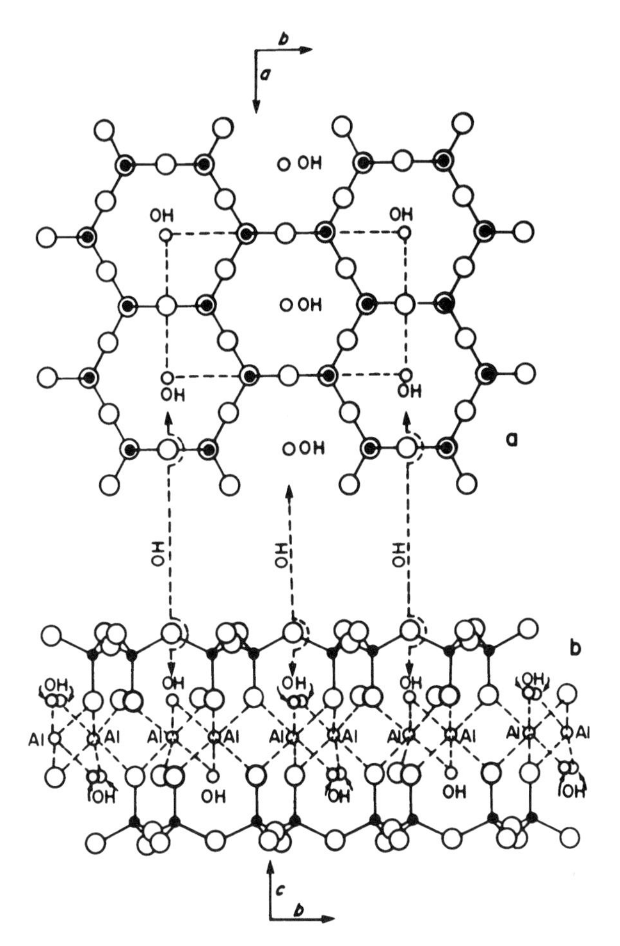

Fig. 4–2. Portions of mica structure. (a) Single layer of the hexagonal network of Si–O tetrahedra with OH located in the plane of their vertices at the centers of every sixfold ring. (b) Edge-on view of two of these layers with inward pointing vertices showing their relative orientations and the locations of the Al (or Mg) atoms between them (Bragg and Claringbull, 1965).

extensive micas in igneous and metamorphic rocks. Phlogopite is a trioctahedral mica and is a product of metamorphism of magnesian limestones or dolomitic limestones and also occurs in serpentinitic rocks.

Micas in soil are primarily inherited from parent material (Fanning et al., 1989; Olson et al., 2000). The most common mica groups in rocks and soils are muscovite, biotite, and phlogopite. Dioctahedral micas such as muscovite are more resistant to weathering than trioctahedral micas such as biotite and phlogopite. Therefore, the clay-sized mica in most soils is predominantly dioctahedral.

Illite was a term originally proposed in the 1930s as a group name for clay-sized, micaceous minerals (Grim et al., 1937). Illite is not accepted as species name or end member of a mineral composition series (Moore and Reynolds, 1997). Although illite is usually dioctahedral, it differs from muscovite in that the octahedral sheet of illite contains more Fe^{2+} and Mg^{2+} in place of Al^{3+} than muscovite. The overall layer-charge density of illite is –0.6 to –0.8 mol_c per formula unit, which is less than that of biotite or muscovite. In the tetrahedral sheet of illite, Al^{3+} substitutes less for Si. The K content of illite is also lower due to the lower negative layer charge that must be countered. Illite is the most abundant clay mineral in sedimentary rock. It also forms during surficial weathering, as well as in hydrothermal and in metamorphic environments. Illite is found in the clay fraction of soils and sediments.

Potassium-Bearing Feldspars

Potassium-feldspars are tectosilicates having a three-dimensional framework of linked SiO_4 and Al_2O_3 tetrahedra, with sufficient opening in the framework to accommodate K to maintain electroneutrality (Smith, 1974). Four-membered rings of tetrahedra are the basic units in building up the framework. These rings are linked together to form a honeycomb type of arrangement, as illustrated by the projection of the Si/Al positions on the (001) plane in Fig. 4–3. If Si and Al occur with no preference of site, the resulting symmetry is monoclinic. If Al takes up a preferred site, a triclinic structure results. This relationship is referred to as the *order–disorder relationship*. The ordered K-feldspar has a triclinic structure; the disordered K-feldspar has a monoclinic structure.

The K-feldspar polymorphs—sanidine, orthoclase, microcline, and adularia—have identical chemical composition (van der Plas, 1966; Barth, 1969). *Sanidine* is a monoclinic alkali feldspar with small optical axial angle (2*V*) and commonly occurs in volcanic rocks. The monoclinic alkali feldspars, that have the larger optic axial angle (2*V*), look homogeneous, and do not show cross-hatched twinning, are referred to as *orthoclase*. *Microcline* is triclinic, exhibits the typical cross-hatched twinning (Fig. 4–4A) and has the larger optic axial angle (2*V*). The alkali feldspars that may be either monoclinic or triclinic but have special crystal habit and occur in low-temperature hydrothermal veins are referred to as *adularia*.

One out of every four Si atoms in the framework of the K-feldspar structure is replaced by Al. This substitution imparts a negative charge to the framework which is neutralized by incorporation of other positively charged ions, such as K^+. Only rarely do K-feldspars of ideal chemical composition ($KAlSi_3O_8$) occur in nature (Huang, 1989). The K-feldspars contain foreign cations either in fourfold coordination replacing Al or Si or in higher coordination replacing K^+ ions. Lithium and

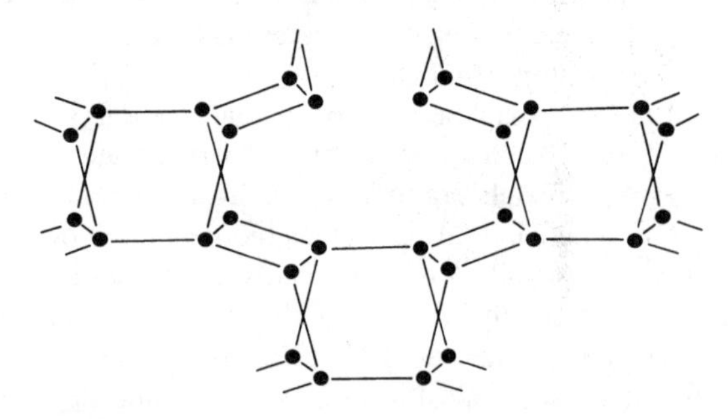

Fig. 4–3. Projection of the Al/Si positions of the $AlSi_3O_8$ framework on the (001) plane (Laves, 1960).

Be have the suitable size relationship to occupy the tetrahedral position. Frequently Na substitutes for K in the K-feldspar structure, and a complete solid solution series exists between the end members of composition $KAlSi_3O_8$ and $NaAlSi_3O_8$. The K- and Na-rich members of the group with a small amount of Ca are known as *alkali feldspars*. The lamallar aggregate, which is composed of a large amount of alkali feldspar and a subordinate amount of albite ($NaAlSi_3O_8$), is referred to as *perthite*. Most naturally occurring alkali feldspars, except authigenic K-feldspars, usually contain varying amounts of Na in the structure and are thus more or less perthitic (Barth, 1969). The morphology of soil perthitic crystals, which appear to contain a second phase, most likely albite, in a host crystal of K-feldspar is illustrated in Fig. 4–4B.

The total K-bearing feldspars including alkali feldspars make up nearly 31% of the earth's crust (Barth, 1969). Microcline is usually formed at lower temperature than is orthoclase. It is the common K-feldspar of pegmatites and hydrothermal veins and also occurs in metamorphic rocks. Orthoclase is the characteristic K-feldspar of igneous rocks and occurs alone and in perthitic intergrowth with albite. Orthoclase also occurs in metamorphic rocks. Sanidine is present in K-rich volcanic rocks, such as rhyolite and trachyte. Adularia occurs in low temperature hydrothermal veins. A major portion of K-feldspars in sedimentary rocks is of igneous

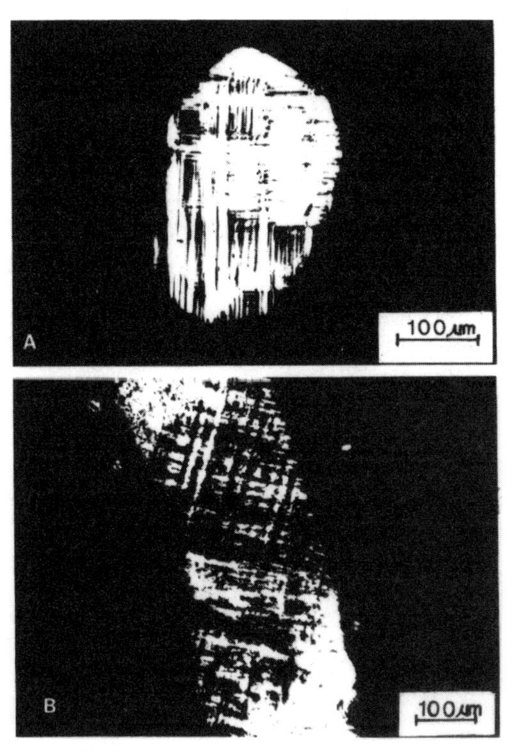

Fig. 4–4. Photomicrographs of K-feldspar crystals from the Ap horizon of a Haverhill soil (Aridic Haploboroll) in Saskatchewan, Canada. (A) K-feldspar crystal with the cross-hatched pattern of maximum microcline, fine sand (50–250 µm) fraction. (B) K-feldspar crystal showing two phases, coarse to very coarse sand (500–2000 µm) fraction (Somasiri and Huang, 1973).

Table 4–2. Distribution of feldspar K in various size fractions of a Haverhill soil profile (Aridic Haploboroll) in Saskatchewan, Canada (Somasiri et al., 1971).

Horizon	Particle size	Total K	Feldspar K	Mica K	% fraction of K	
					Feldspar	Mica
	μm	g kg^{-1}				
Ap	<0.2	21.1	--	21.1	--	100
	0.2–2	26.1	0.8	5.3	3	97
	2–5	20.1	6.1	14.0	30	70
	5–20	17.3	8.7	8.6	50	50
	20–50	15.4	9.7	5.7	63	37
	50–500	12.7	7.6	5.1	60	40
	500–2000	17.3	17.3	--	100	--
	>2000†	16.9	6.8	10.1	40	60
Bm	<0.2	18.3	--	18.3	--	100
	0.2–2	29.9	0.0	29.9	--	100
	2–5	24.6	3.6	21.0	15	85
	5–20	20.4	7.7	12.7	38	62
	20–50	14.7	8.5	6.2	58	42
	50–500	14.2	8.4	5.8	59	41
	500–2000	18.2	18.2	--	100	--
	<2000	17.6	6.5	11.1	38	63
Ck	<0.2	16.8	--	16.8	--	100
	0.2–2	27.4	2.1	25.3	8	92
	2–5	26.2	8.7	17.5	33	67
	5–20	22.7	9.2	13.5	41	59
	20–50	15.6	8.3	7.3	53	47
	50–500	11.9	9.2	2.7	77	23
	500–2000	20.7	20.7	--	100	--
	<2000	17.7	6.9	10.8	39	61

† Whole mineral fraction.

origin. The authigenic feldspars formed at low temperatures and pressures near the earth surface are present in sedimentary rocks, such as limestones, sandstones, siltstones, and shales (Huang, 1989).

The frequency distribution of K-bearing feldspars varies with the capacity and intensity factors of weathering reactions (Huang, 1989). In strongly weathered soils, K-feldspars are present in only small quantities or completely absent, though the parent material contains considerable quantities of those minerals. However, some feldspars may be found in humid tropical soils, which contain relatively fresh rock materials due to erosional and depositional processes. The K-feldspars and K-micas are the major K reserves of available K. The distribution of K-feldspars and micas in a series of particle-size fractions of moderately weathered soil profiles is illustrated in Table 4–2. The fraction of K from feldspars increases with increasing particle size; the opposite trend is true for the fraction of K from micas.

Vermiculites

The structure of vermiculites consists of 2:1 layers and exchangeable cations that have a plane of water molecules on each side. Water molecules are tetrahedrally bonded to oxygens of 2:1 layer surfaces through H bonding. Vermiculites are separated from smectites on the basis of a higher layer charge, tentatively set as greater than −0.6 mol_c per formula unit (Bailey, 1980). However, vermiculites and smectites display a continuum of chemical properties (Olson et al., 2000). Soil vermiculites have been identified with some characteristics of both smectites and vermiculites. These have been termed *high-charge smectites* or *low-charge vermiculites*.

Vermiculites are present in all soil orders of U.S. soil taxonomy (Soil Survey Staff, 1998) but are most often found in soils of temperate and subtropical climates. Vermiculites are present in all particle-size fractions ranging from fine clay to coarse sand. Soil vermiculites are nearly always reported to accompany muscovite, biotite, and chlorite. All vermiculites are believed to be alteration products of micas and chlorites (Douglas, 1989). Dioctahedral vermiculite is more common then trioctahedral vermiculite in soils. Trioctahedral vermiculite may comprise a significant amount of the sand and silt fractions of soils, whereas dioctahedral vermiculite is seldom observed as discrete crystallite >5 μm. In intensive weathering environments, vermiculites invariably have a hydroxy-Al interlayer.

Smectites

Smectites resemble the mica and vermiculite structures in that all three have an octahedral sheet that share oxygen atoms with two tetrahedral sheets (Borchardt, 1989) Cationic substitution exists in octahedral or tetrahedral sheets. The dioctahedral smectites—montmorillonite, beidellite, and nontronite—may form as a result of weathering of micas and vermiculites, whereas the trioctahedral smectites—hectorite (Li-rich), saponite (Mg-rich), and sauconite (Zn-rich)—appear to be inherited from the parent material and seldom found in soils.

More than 2 350 000 km^2 of clay soils high in smectites are distributed over the world (Buol et al., 1980). The terminology of almost all soils classified as Vertisols is dominated by smectites (Soil Survey Staff, 1998). Smectites, along with vermiculites are substantially responsible for the high cation exchange capacity (CEC) of soils in which they are present (Olson et al., 2000). Smectites are likely to have residual mica cores surrounded by "wedge sites" that may be primarily responsible for K fixation. The layer silicate clay complex is depicted in Fig. 4–5.

Hydroxy Interlayered Expansible Phyllosilicates

Hydroxy-Al interlayered smectites and vermiculites are weathering products derived from chlorite weathering or more commonly from deposition of hydroxy-Al polymers within the interlayer spaces of these expansible layer silicates (Barnhisel and Bertsch, 1989; Olson et al., 2000). These minerals form solid solution series with pure end members smectites or vermiculites at one end and pedogenic or aluminous chlorite at the other. Their chemical composition is entirely dependent on the basic 2:1 clay mineral plus the type and amount of interlayer material. Under intense weathering environments such as acid soil conditions, vermiculites almost always contain a hydroxy-Al interlayer (Rich and Obenshain, 1955; Rich, 1958). Hydroxy-interlayered vermiculites and hydroxy-interlayered smectites are most abundant in the upper solumn of acid soils. They have a wide geographical

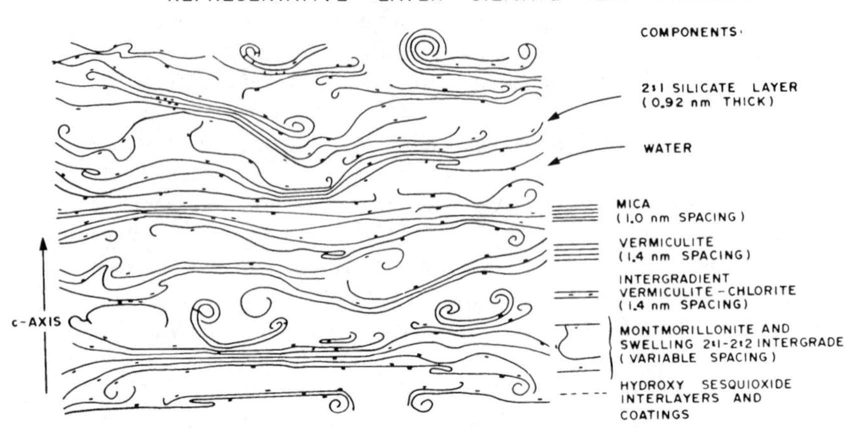

Fig. 4–5. Principal features of a dominantly montmorillonite layer silicate clay complex formed by weathering of mica. Different proportions of the various components occur in mica-derived layer-silicate clays of different soils and other sediments (Jackson, 1964) (illustration by G.A. Borchardt).

distribution and can be found in several soil orders. These minerals are most abundant in Ultisols, Alfisols, and Spodosols (Soil Survey Staff, 1998).

The hydroxy-Al components affect the physicochemical properties of soils (Barnhisel and Bertsch, 1989). Hydroxy-interlayered vermiculites and hydroxy-interlayered smectites have a wide range of CEC values. Positively charged hydroxy-Al polymers satisfy charge imbalance, but the polymers are not exchangeable. Therefore, the CEC decreases in proportion to how completely the interlayer space is filled. Hydroxy interlayers can inhibit K fixation by preventing layer collapse and thus increasing its selectivity.

Short-Range Ordered Aluminosilicates

Allophane is a group name for noncrystalline clay minerals consisting of silica, alumina, and water in chemical combination (Farmer and Russell, 1990). There are three major types of allophane in natural environments. They have been tentatively named proto-imogolite allophane, halloysite-like (or defect kaolin) allophane, and hydrous feldspathoid allophane (Harsh, 2000). Proto-imogolite allophane is characterized by an Al/Si ratio close to 2, a gibbsite-like sheet of octahedrally coordinated Al with orthosilicate groups sharing three O atoms with Al. Therefore, proto-imogolite allophane has the same short-range order as imogolite, but does not exhibit the tubular morphology. Allophane with an Al/Si ratio closer to one probably has a structure that is closer to kaolinite or halloysite with defects in the tetrahedral sheet. The hydrous feldspathoid allophane (also Si rich) contains no imogolite units and a significant, if not dominant, amount of tetrahedrally coordinated Al. The basic structure appears to be that of a framework silicate with 1:3 Al for Si substitution.

Allophane is often associated with soils derived from volcanic debris, because the rapid release of Al and Si from materials such as volcanic glass results in the precipitation of noncrystalline aluminosilicates. Noncrystalline aluminosilicates have also been identified in soils derived from many parent materials, including sandstone, gneiss, granite, and basalt (Harsh, 2000).

Allophane has a definite affinity for K because of its structural configuration. Wada and Harada (1969) obtained unusually high CEC values for Japanese soil allophanes when determined with K salts. The K fixation by synthetic aluminosilica gels is sterically controlled; channels in the gels impose a sieve action on the entry and passage of certain counterions (van Reeuwijk and de Villiers, 1968). Adsorption of exchangeable cations on allophanic soils appears to be similar to that on soils dominated by smectites (Nakahara and Wada, 1994). There have been reports of highly selective K exchange on Andisols. In addition to reactions with allophane or imogolite, this could result from trace amounts of illitic minerals or alunite formation (Harsh, 2000).

Zeolites

Zeolites are crystalline, hydrated aluminosilicates of alkali and alkaline earth cations that possess infinite, three-dimensional crystal structures (Ming and Mumpton, 1989). They are further characterized by their abilities to hydrate and dehydrate reversibly and to exchange some of their constituent cations, both without major change of structure. The negative charge of zeolites arising from the isomorphous substitution of Al^{3+} for Si^{4+} is balanced by the presence of alkali and alkaline earth cations, principally Na^+, K^+, Ca^{2+}, and Mg^{2+}, within the existing pores. The structural channels of various zeolites are formed by different combinations of linked tetrahedral rings. Each ring is composed of either 4, 5, 6, 8, or 12 tetrahedra with specific combinations present within a given zeolite, thereby providing for channels or restrictions of known size. Greater numbers of tetrahedra per ring result in wider channels. The size of the cation that can be introduced into the structure is dependent on the width of the channel at its narrowest restriction. Zeolites can act as ionic sieves with characteristic upper limits on the size of the ion that can permeate these structural channels. Zeolites have a strong affinity for K^+ than for Na^+, Ca^{2+}, and Mg^{2+}. Potassium ions may occupy positions in the structural channel of zeolites from which K^+ ion cannot be removed without destroying the crystal structure (Peterson et al., 1965)

Zeolites occur in soils most commonly where the parent rocks are zeolitic, suggesting inheritance (Ming and Mumpton, 1989). However, it also appears likely that they have been neoformed from strongly alkaline solutions in salt-affected soils (Churchman, 2000).

Potassium Taranakite

Taranakites, including potassium taranakite [$K_3Al_5H_6(PO_4)_8 \cdot 18H_2O$] and ammonium taranakite [$(NH_4)_3 Al_5H_6(PO_4)_8 \cdot 18H_2O$] have the same structure (Frazier and Taylor, 1965), although their composition is different. Natural K-taranakite was first found from trachytic rocks of the Sugarloaves, Taranaki, New Zealand in 1865 (Bannister and Hutchinson, 1947). Potassium-taranakite is formed in soils through the reaction of concentrated K and acid phosphate solutions with Al hydroxides (Taylor and Gurney, 1965). This compound can be hydrolyzed (Taylor et al., 1963) upon dilution and is thus considered to be a slow-releasing K source.

Application of fertilizer that contains phosphate, ammonium, and potassium may result in the formation of taranakites as reaction products in the immediate vicinity of fertilizers (Lindsay et al., 1962; Sarkar et al., 1977; Prabhudesai and Kadrekar, 1984; Zhou and Huang, 1995). Recently, the data of Zhou et al. (2000) and Liu et al. (2002) showed that NH_4-taranakite formation is perturbed with high Fe content, especially under reduced and acidic conditions. Similar perturbation of K-taranakite formation by Fe may occur in soils.

Potassium Alunite

Alunite [$KAl_3(SO_4)_2(OH)_6$] occasionally occurs as a component of soil surface crusts. It also has been described in the karst bauxite deposit of Europe (Bardossy, 1982). Potassium-alunite can be formed in acid soils when the ionic activities of K^+, Al^{3+}, and SO_4^{2-} are sufficiently high. The p*K*sp of alunite is 85.4 (Adams and Rawajfih, 1977). The solubility data indicate the feasibility of simultaneous retention of K^+ and SO_4^{2-} in the form of alunite in acid soils. Alunite formation may be partially responsible for highly selective K exchange in Andisols (Harsh, 2000).

Kaolinite and Halloysite

The kaolinite structure consists of one octahedral and one tetrahedral sheet. Halloysite is similar in structure to kaolinite except for a layer of water molecules intercalated between the 1:1 layer.

Kaolinite is commonly found in weathered surface soils and volcanic soils where environments of formation have been somewhat acidic. In U.S. soil taxonomy, kaolinite-rich soils are classified as Ultisols and Oxisols, and occasionally Andisols (Soil Survey Staff, 1998). Kaolinite is a common clay mineral of buried soils. Halloysite is found primarily in soils of volcanic origin (Olson et al., 2000).

Permanent negative charge on kaolinite from isomorphous substitution has been postulated because exchangeable cations are retained under acidic conditions (Schofield and Samson, 1954). More recent findings show smectite interstratification with kaolinite and the presence of mica zones in kaolinite particles (Atschuler et al., 1963; Lee et al., 1975a, 1975b). Furthermore, the differential heats of K^+–Ca^{2+} exchange, coupled with entropies of exchange (Talibudeen and Goulding, 1983b), suggest that 0.1 to 10% vermiculites, micaceous, and smectitic layers are present in kaolins. The permanent negative charge previously attributed to kaolinite is probably due to these impurities.

Metal Oxides

Oxide minerals comprise the oxides, hydroxides, oxyhydroxides, and hydrated oxides of Al, Fe, Mn, Si, and Ti (Kämpf et al., 2000). These minerals commonly occur in soils, particularly those in advanced stages of weathering. The surface charge of these minerals are pH dependent. The competitive adsorption of K and Ca by Mn oxide has been shown (Wang et al., 1995, 1996). The role of other metal oxides in K adsorption still remains obscure.

Organic Matter

The organic fraction of soils often accounts for a small but variable proportion of total soil mass. Despite its often minor contribution to the total mass of mineral soils, the organic matter can exert a profound influence on soil processes (Baldock and Nelson, 2000). Organic matter contributes 25 to 90% of the CEC of surface layers of mineral soils (Stevenson, 1994). The contribution is greatest for soils with low clay content or where the clay fraction is dominated by minerals with a low charge density, such as kaolinite, and is lowest for soils with high contents of highly charged minerals, such as vermiculites and smectites. However, studies on the influence of organic matter on K dynamics in soils are very limited (Wang and Huang, 2001).

CHEMICAL EQUILIBRIA OF POTASSIUM IN SOIL ENVIRONMENTS

Ion Selectivity

Potassium selectivity of soil particles is affected by mineralogical and chemical factors as described below:

Mineralogical Properties

Layer Charge. The site of negative charge and the charge density of minerals may affect K^+ selectivity (Rich, 1968a). The high K^+ selectivity of clay-sized muscovite is attributed to the negative charge arising in the tetrahedral sheet (Schwertmann, 1962a). However, the negative sites of vermiculite, a mineral which has a low K selectivity compared with muscovite (Schwertmann, 1962a; Rich and Black, 1964), are also primarily in the tetrahedral sheet.

The K^+/Ca^{2+} cation exchange selectivity (CES) values, expressed as the equivalent ratio of exchangeable K^+ to Ca^{2+}, ranged from 0.1 for montmorillonite to 1.7 for muscovite (Table 4–3). The K^+/Ca^{2+} equivalent ratios generally increase

Table 4–3. Potassium/calcium cation exchange selectivity (CES) of Na-saturated exchange materials (Dolcater et al., 1968).†

Exchange materials	K*X*‡	Ca*X*	K*X* + Ca*X* (CES)	K*X*/Ca*X* (CES)
	mmol kg^{-1}			
Montmorillonite, 0.2–0.8 µm	93	816	910	0.11
Vermiculite, <5 µm	326	1410	1740	0.23
Harpster B3, <0.2 µm	182	723	910	0.25
Harpster B3, 2–0.2 µm	89	274	360	0.32
Triangle B2tb, <5 µm	142	464	610	0.31
Biotite, 2–0.2 µm	67	72	140	0.93
Biotite, 5–2 µm	24	26	50	0.95
Muscovite, 0.2–0.08 µm	200	156	360	1.3
Muscovite, 2–0.2 µm	66	39	110	1.7

† Expressed as the equivalent ratio of K^+ to Ca^{2+} exchangeable to 0.5 *M* $Mg(OAc)_2$.
‡ *X* denotes exchange materials.

in the order: montmorillonite < vermiculite < Harpster < Triangle < biotite < muscovite. This corresponds to the order of increasing layer charge and surface charge density of the cation exchangers.

Particle Size. Potassium selectivity varies with particle size. In some soils, K selectivity increases with particle size (Schwertmann, 1962a; Rich and Black, 1964). Selectivity may decrease with particle size at later stages of weathering. Much of the K selectivity apparently occurs in the interlamellar regions.

Configuration of Interlayers and Wedge zones. A diagram of three weathered mica particles, each with about 50% expansible layers, is shown in Fig. 4–6. In Fig. 4–6a and 4–6b, the expanded interlayers are not continuous but have internal terminations forming wedge zones. In Fig. 4–6c, the interlayer spaces are either expanded or nonexpanded uniformly through each interlayer. Exchangeable K^+ is not distributed over the exchange surface uniformly, but rather a portion of the K^+ is present in the wedge zones. Only such cations as H_3O^+, K^+, NH_4^+, Rb^+, and Cs^+ can fit into these wedge zones because of the geometry of the zones.

Position of Wedge Zones in Interlayers. Potassium selectivity is influenced by the position of the wedge zone in interlayers. Instead of large hydrated ions such as Ca^{2+} and Mg^{2+}, K^+ is selected due to the space limitations for diffusion of these large hydrated cations in the wedge zones. If the wedge zone is near the edge of the particle, it can be effective in only a small amount of K selection. If the wedge zone is deep within the particle, much more K can be selected. A new wedge zone is formed as silicate layers close around the selected K^+ ions (Rich, 1968a).

Wedge zones increase as more interlayers are partly opened and decrease as the interlayers are opened all the way through the particle (Jackson, 1963; Le Roux

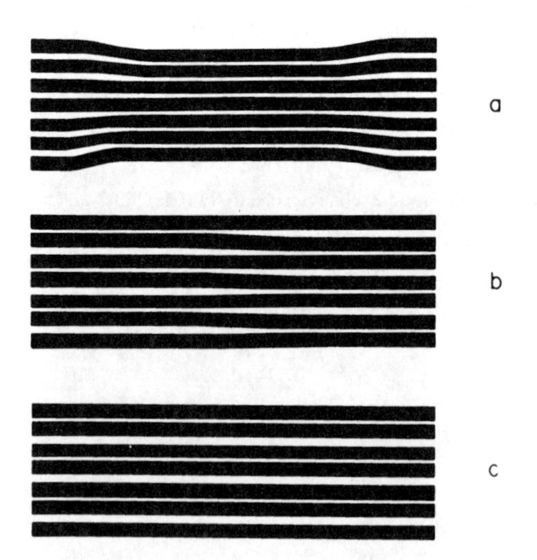

Fig. 4–6. Diagram of a weathered mica particle containing about 50% expanded (vermiculite) layer. (a) "Frayed edge" and mica core; (b) alternate layers open half-way through interlayer; and (c) regularly interstratified mica–vermiculite (no wedge zones) (Rich, 1972).

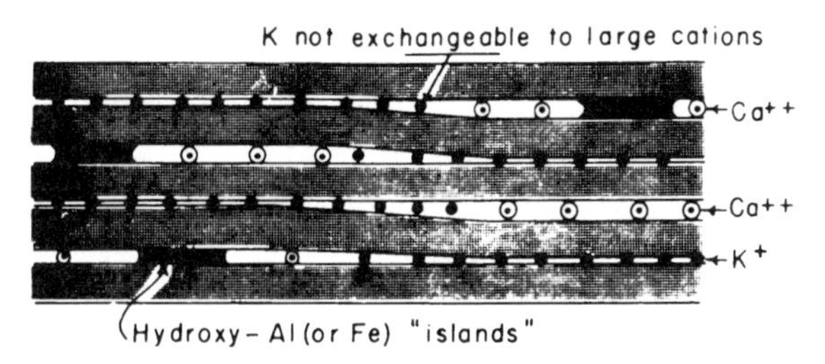

Fig. 4–7. Proposed model of an expansible layer silicate with interlayers indicating effect on K fixation (Rich, 1968a).

and Rich, 1969). This mechanism is limited if the interlayer space collapses at the particle edge (Le Roux et al., 1970).

Hydroxy Interlayers. Potassium selectivity is enhanced by the presence of hydroxy-Al and hydroxy-Fe^{3+} polymers in the interlayer space (Rich and Black, 1964; Rich, 1968b). These interlayer materials act as props to hinder the collapse of the layers about the K^+ ions. Apparently K^+ ions are able to move more easily in propped open interlayer spaces than in collapsed interlayers (Fig. 4–7). Besides the propping effect, the preferential occupation by the Al polymers of exchange sites that normally would adsorb Ca may also affect K selectivity (Kozak and Huang, 1971). Because of their size the polymers normally would not be adsorbed at wedge zones but rather in the more expanded interlayer position (Fig. 4–8). Therefore, sites that normally would adsorb Ca^{2+} ions would be occupied by Al polymers, whereas wedge sites would be relatively unaffected upon adsorption of Al polymers.

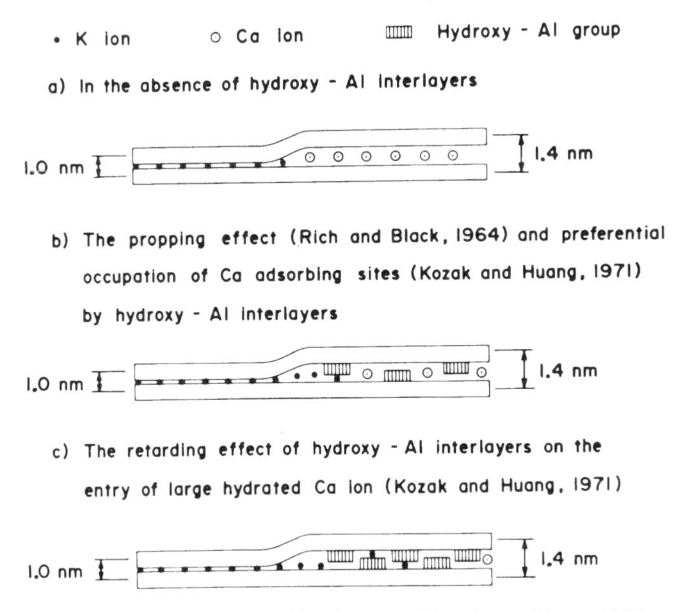

Fig. 4–8. Model of the effects of hydroxy-Al interlayers on K^+ exchange (Huang, 1987).

Further, hydroxy-Al interlayers may also create a retarding effect on the entry of the more hydrated Ca^{2+} ions, thereby increasing the selectivity toward the less hydrated K ions (Fig. 4–8).

Variable Charge Minerals. Besides layer silicates, variable charge minerals may contribute to adsorption and selectivity of K ions, since the negative charge of these minerals are pH dependent. Wang et al. (1995) showed that both cationic and anionic environments have a profound influence on the adsorption of K^+ and Ca^{2+} in a Mn oxide (birnessite) system. This is of fundamental significance in understanding the chemistry of K in soils, particularly those with variable charge.

Chemical Properties

Ion Size and Ion Valence. Potassium selectivity is greatly affected by the hydrated ion size (Rich, 1964; Rich and Black, 1964; Dolcater et al., 1968; Murdock and Rich, 1972). In soils containing partially weathered micas and partially closed vermiculites where wedge zones exist, NH_4OAc removes more exchangeable K than does $Mg(OAc)_2$, although Mg^{2+} usually is a better replacer of K^+ than $NH_4{}^+$. The wedge zones are able to screen out the hydrated Mg^{2+} ion and select the K^+ ion that could fit into the wedge zone. Magnesium chloride can remove more K^+ than can $Mg(OAc)_2$. This is attributed to the depression of soil pH by $MgCl_2$ and the formation of H_3O^+ by exchange or hydrolysis of exchangeable and polymeric Al. Being similar in size to K^+, the H_3O^+ ion is able to fit into the wedge zone and exchange K^+. A combination of NH_4OAc and $MgCl_2$ is more effective in removing exchangeable K^+ than either electrolyte alone. This is apparently due to Mg^{2+} opening up the interlayer space and facilitating the exchange of K^+ by NH_4^+.

Solution pH. Potassium exchange is enhanced by low pH (Rich, 1964; Rich and Black, 1964). Ions do not migrate rapidly enough to account for the reaction rate involving H_3O^+ ions. The lifetime of an individual H_3O^+ is exceedingly short, about 10^{-13} s, since all of the protons are undergoing rapid exchange (Cotton and Wilkinson, 1980). As long as there is a continuous film of H_2O to a K^+ ion, the exchange of H_3O^+ for K^+ could proceed rapidly.

The effect of pH on ion selectivity varies with the associated cations (Fig. 4–9). pH has little effect for NH_4^+; NH_4^+ saturation collapses the interlayer structure and thus excludes H_2O, which is the medium for transport of protons to and from the exchange sites. By contrast, the effect of pH for Li^+ and Mg^{2+} is pronounced; the increased efficiency of Li^+ and Mg^{2+} with a decrease in pH is ascribed to an expansion of the interlayer structure. Li^+ and Mg^{2+} apparently pry open the interlayer structure and facilitate the entry of water and protons, resulting in the exchange of K^+ ions by H_3O^+.

Potassium Concentration. The concentration of K^+ is critical in K selection in the wedge zones (Sparks and Huang, 1985). At low K^+ concentration, there is K selection at the wedge site. As K^+ concentration increases, the silicate layer collapses at the wedge site, resulting in an entrapment of ions deeper in the interlayer space. In the absence of wedge zones, a high K^+ concentration is necessary to initiate collapse of the interlayer space of vermiculite and to enhance K selectivity.

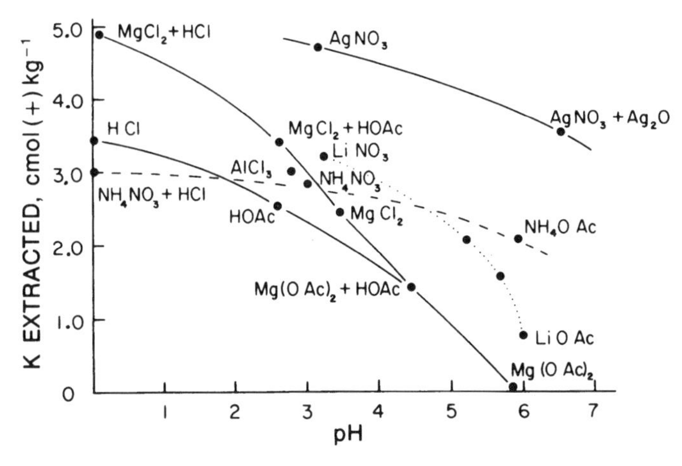

Fig. 4–9. Displacement of K from Nason (fine, mixed, semiactive, thermic Typic Hapludults) B3 soil as affected by pH and extracting solution. All solutions are normal with respect to the component indicated (Rich, 1964).

Temperature. Potassium selectivity of Delaware soils decreases with a rise in temperature (Sparks and Liebhardt, 1982), indicating decreased K sorption with increasing temperature. This is attributed to the presence of wedge zones and hydroxy-Al interlayers. The Delaware soils contain wedge zones that would selectively screen out large hydrated cations such as Ca^{2+} and Mg^{2+} and favor selection of smaller cations such as NH_4^+ and K^+. As temperature increases, the hydration of Ca^{2+} is decreased relative to that of K. Therefore, size is not as important in K selectivity. Further, these soils contain chloritized vermiculite (Rich, 1960). Increased temperature could cause collapse of the interlayer and decrease K selectivity.

Thermodynamics of Potassium Exchange Reactions

Thermodynamics for K^+ exchange reactions in soil and clay systems has been studied by using three principal approaches: equilibrium approaches of Argersinger et al. (1950) and Gaines and Thomas (1953) and a kinetic approach of Sparks and Jardine (1981). Goulding (1983) and Sparks and Huang (1985) reviewed these approaches. The theories of Argersinger et al. (1950) and Sparks and Jardine (1981) are briefly summarized below.

Theory of Argersinger et al. (1950). The thermodynamic equilibrium constant, K_{eq}, for the following reversible K^+–Ca^{2+} exchange reaction as shown in Eq. [1] can be expressed by Eq. [2]:

$$Ca^{2+}_{(ad)} + 2K^+ \Leftrightarrow 2K^+_{(ad)} + Ca^{2+} \quad [1]$$

$$a^2_{K_{(ad)}} a_{Ca} / a_{Ca_{(ad)}} a^2_K = K_{eq} \quad [2]$$

where $a_{(ad)}$ is the activity of the cation on the exchanger phase and a is the activity of the cation in the solution phase.

If we consider the exchanger as a nonideal solid solution and the activity of adsorbed H_2O to be negligible (Högfeldt, 1953; Gaines and Thomas, 1953; Sposito, 1981), the activity of the cations on the exchanger and in solution can be expressed in Eq. [3] and [4], respectively:

$$a^2_{K_{(ad)}} = f^2_K \bar{N}^2_K \text{ and } a_{Ca_{(ad)}} = f_{Ca} \bar{N}_{Ca} \quad [3]$$

$$a^2_K = \delta^4_{KCl} m^2_K \text{ and } a_{Ca} = \delta^3_{CaCl_2} m_{Ca} \quad [4]$$

where f is the activity coefficient of the cation on the exchanger phase, $\bar{N}$ is the mole fraction of the cation on the exchanger phase, δ is the mean activity coefficient of the salt in the solution phase, and m is the molality of the cation in the solution phase.

Therefore Eq. [2] can be converted to

$$f^2_K \bar{N}^2_K \delta^3_{CaCl_2} m_{Ca} / f_{Ca} \bar{N}_{Ca} \delta^4_{KCl} m^2_K = (f^2_K / f_{Ca}) k_v = K_{eq} \quad [5]$$

where k_v is the Vanselow selectivity coefficient.

If the Gibbs–Duhem equation is applied to this binary system and combined with Eq. [5], the following equation is obtained using a mole fraction of unity as the standard state for adsorbed cations (Argersinger et al., 1950; Sposito, 1981):

$$\ln K_{eq} = \int_0^1 \ln k_v \, d\bar{X}_K \quad [6]$$

where $\bar{X}_K$ is the equivalent fraction of K on the exchanger phase.

The equilibrium constant, K_{eq}, can then be calculated by plotting $\ln k_v$ as a function of $\bar{X}_K$ and determining the area under the curve by applying the trapezoidal rule. Knowing K_{eq}, one can calculate other thermodynamic parameters of K^+ exchange reactions, including $\Delta G°$, the standard Gibbs free energy of exchange, $\Delta H°$, the standard enthalpy of exchange, and $\Delta S°$, the standard entropy of exchange as shown below:

$$\Delta G° = -RT \ln K_{eq} \quad [7]$$

$$(d \ln K_{eq})/dT = \Delta H° / RT^2 \quad [8]$$

$$\Delta S° = (\Delta H° - \Delta G°)/T \quad [9]$$

Theory of Sparks and Jardine (1981). Although a relationship between chemical equilibrium and kinetics of completely reversible reactions was established many decades ago (Glasstone et al., 1941; Laidler, 1965), this relationship was first introduced by Sparks and Jardine (1981) to K^+ exchange in soil and clay systems. Potassium adsorption–desorption in soil and clay system is shown in Eq. [10]:

$$\text{K adsorption} \underset{k_d}{\overset{k_a}{\rightleftharpoons}} \text{K desorption} \quad [10]$$

where k_a and k_d are apparent adsorption and desorption coefficients, respectively.

Table 4–4. Standard free energies ($\Delta G°$), enthalpies ($\Delta H°$), and entropies ($\Delta S°$) of K^+–Ca^{2+} exchange on soils calculated from equilibrium and kinetic theories (Sparks and Jardine, 1981).

Method of calculation	$\Delta G°$	$\Delta H°$	$\Delta S°$
	J mol^{-1}		J mol^{-1} K^{-1}
Kinetics data	–5 338	–7 081	–5.9
Eyring's absolute reaction rate theory	–5405	–7 039	–5.5
Equilibrium data†	–4 400 to –14 330	–3 813 to –35 573	–12.1 to –92.6

† Deist and Talibudeen (1967).

The thermodynamic equilibrium constant for a completely reversible reaction is given by Denbigh (1966) and shown in Eq. [11]:

$$K_{eq} = k_a/k_d \quad [11]$$

If k_a and k_d can be determined, K_{eq} can be calculated from Eq. [11]. Knowing K_{eq}, $\Delta G°$ can be calculated from Eq. [7].

The energy of activation for K^+ adsorption (E_a) and desorption (E_d) can be calculated from the Arrhenius equation:

$$\mathrm{d}\ln k/\mathrm{d}T = E_i/RT^2 \quad [12]$$

where i is "a" or "d". The standard enthalpy of exchange is obtained from Eq. [13]:

$$\Delta H° = E_a - E_d \quad [13]$$

where E_a is the activation energy of the forward reaction and E_d is the activation energy of the reverse reaction. The standard entropy of exchange can then be calculated from Eq. [9].

The $\Delta G°$, $\Delta H°$, and $\Delta S°$ of K^+–Ca^{2+} exchange on soils calculated from equilibrium and kinetic approaches are shown in Table 4–4. The agreement in the values of these thermodynamic parameters obtained from the methods is excellent. Further, Ogwada and Sparks (1986a) made a critical evaluation on the use of kinetics for determining thermodynamics of ion exchange in soils. Therefore, thermodynamic parameters can be calculated from a kinetic approach.

Significance of Thermodynamic Parameters

Gibbs Free Energy. The standard Gibbs free energy, $\Delta G°$, may be depicted as a measure of the tendency of a system to undergo a particular reaction or the drive of a system to react. Therefore, small values for the $\Delta G°$ result from the restricted molecular motion of ions involved in exchange reactions.

The $\Delta G°$ values of K^+–Ca^{2+} exchange on a kaolinitic soil clay are –9875 and –6891 J mol^{-1} at 298 and 303 K, respectively, indicating the preference of the clay for K ions (Udo, 1978). Jensen (1973a, 1973b) also reported that the $\Delta G°$ values of K^+–Ca^{2+} exchange on kaolinite are negative, whereas $\Delta G°_C$ values, the standard free energy change considering only coulombic interactions, are positive. The affinity of kaolinite for K^+ ions was ascribed to specific interactions of K^+ ions at edge sites on the kaolinite (Jensen, 1973a, 1973b; Udo, 1978). However, smectite interstratification with kaolinite and mica zones in kaolinite particles have been reported (see Kaolinite and Halloysite above in the Soil Components Involved in Potassium Dynamics and Equlibria section). Kaolinite often contains 2:1 mineral impurities (Goulding and Talibudeen, 1980). Virtually no montmorillonite and kaolinite are completely free of micaceous impurities (Talibudeen and Goulding, 1983a, 1983b). Therefore, the preference of kaolinite for K^+ over Ca^{2+} ions is attributable to the wedge sites of these micaceous impurities.

Negative $\Delta G°$ values for K^+–Ca^{2+} exchange were also reported for some British soils (Deist and Talibudeen, 1967) and for the Ca^{2+} saturated silt loam soil in Delaware (Sparks and Jardine, 1981). The negative values indicate a strong preference of K^+ ions relative to Ca^{2+} on the soils. In the soil studied by Sparks and Jardine (1981), the clay mineralogy is dominated by vermiculite type minerals.

Jardine and Sparks (1984b) investigated the thermodynamics of K^+–Ca^{2+} exchanges on an Evesboro soil from Delaware. At 283 and 298 K, the soil has a strong preference for K^+ at low values of N_K (mole fraction of K^+ in the solution phase) and for Ca^{2+} at higher values. This selectivity reversal is apparently due to the multireactive nature of exchange sites of the soil for K^+ and Ca^{2+}. The selectivity reversal no longer exists at 313 K and is related to the loss of biphasic kinetics at the higher temperature. Although K^+ is selectively bound at low N_K, the positive $\Delta G°$ values indicate that the Evesboro soil exhibits Ca^{2+} preference at all temperatures. The overall Ca^{2+} preference of the soil is attributed to the organic matter of the soil (Naylor and Overstreet, 1969; Van Bladel and Gheyi, 1980; Goulding, 1981).

Enthalpy. Enthalpy, $\Delta H°$, may be considered a measure of molecular motion or an indication of the binding strength with which a particular cation is held (Dikerson, 1969). The $\Delta H°$ can be determined directly by calorimetry and indirectly by the van't Hoff equation.

Heats of reaction for adsorption of the preferred ion are generally exothermic. Therefore, heat is liberated upon chemical bond breakage during the exchange process. Exothermic reactions most likely result in spontaneous reactions in the standard state. Positive $\Delta H°$ values indicate a process that absorbs heat during a chemical reaction. They may be attributed to factors such as molecular diffusion to and in a particular absorbent, lateral interactions or repulsive forces between sorbed molecules, and desorption of solvent molecules from the surface of the exchanger to the bulk solution (Moreale and Van Bladel, 1979). The $\Delta H°$ values of K^+–Ca^{2+} exchange reactions in soil and clay systems are negative, indicating stronger binding for K^+ (Hutcheon, 1966; Deist and Talibudeen, 1967; Goulding and Talibudeen, 1979, 1980; Sparks and Jardine, 1981).

Entropy. Entropy, $\Delta S°$, is a measure of molecular arrangement; positive $\Delta S°$ values infer a system that is randomly arranged and negative $\Delta S°$ values depict a system that is ordered in its molecular arrangement. The total entropy change of an aqueous system may be considered as the sum of the configuration entropy and the ion hydration entropy. The entropy associated with ion hydration likely plays the predominant role in determining the total entropy change for reactions involving the exchange of two different-sized cations (Laudelout et al., 1968).

The ΔS° values for K^+–Ca^{2+} exchange reactions in soil and clay systems are negative(Hutcheon, 1966; Deist and Talibudsen, 1967; Sparks and Jardine, 1981; Jardine and Sparks, 1984b). The negative ΔS° is attributed to a decrease in both the configuration entropy of the clay and the ion hydration entropy (Sparks and Jardine, 1981). However, Goulding and Talibudeen (1980) suggested that rearrangements within the solid phase contribute most to entropy changes. They reported that the ΔS° values for K^+–Ca^{2+} exchange on clay minerals are negative for the expanded 2:1 layer silicates vermiculite, illite, and montmorillonite but positive for the muscovite.

Fixation of Potassium

Potassium fixation in soil and clay systems was first recognized by Dyer in 1893 (Kunze and Jeffries, 1953). Truog and Jones (1938) and Volk (1934) conducted pioneer studies on K fixation in soils.

Potassium fixation has been interpreted in terms of the good fit of K^+ (the crystal radius and coordination number are ideal) in an area created by holes and adjacent oxygen layers (Barshad, 1951, 1954). The fixation process may be enhanced by van der Waal's forces resulting from the high polarizability of K^+ (van der Marel, 1954). The important forces involved in the interlayer reactions in mineral colloids are electrostatic attractions between the negatively charged layers and the interlayer cations, and expansive forces due to ion hydration. Potassium fixation occurs in vermiculite because the electrostatic forces of attraction exceed those of hydration (Kittrick, 1966; Sawhney, 1966).

Potassium fixation in soils and clays has been studied extensively. The degree of K fixation is related to the kind of clay mineral and its charge density, the degree of hydroxy interlayering, the concentrations of K^+ ions and competing ions, the moisture content, and the solution pH (Rich, 1968a; Thomas and Hipp, 1968; Khan et al., 1994; Olk et al., 1995; Saha and Inoue, 1998).

The expanding 2:1 layer silicates, such as weathered micas, vermiculites, and smectites, are the major clay minerals responsible for K fixation in soils (Rich, 1968a). Weathered micas and vermiculites fix K under moist as well as dry conditions, whereas smectites fix K only under dry conditions. In the case of smectites, the amount of fixation is very small unless the charge density is high (Weir, 1965). A low charge montmorillonite (Wyoming) maintains a 1.5-nm *d* value when K saturated, unless it is heated (Laffer et al., 1966). Some soil smectites have a greater capacity to fix K than many specimen type smectites (Schwertmann, 1962a, 1962b). These soil smectites have a higher charge density and likely have wedge sites near micalike zones where K selectivity is high and K fixation can take place (Rich, 1968a).

In acid soils the principal mineral responsible for K fixation probably is dioctahedral vermiculite (Brown, 1953). Counteracting the effect of high charge density on K fixation in many acid soils is the presence of hydroxy-Al and -Fe(III) interlayer groups (Rich, 1968b). These groups act as props between the unit silicate layers and inhibit collapse of the layers about the K^+ ions. The introduction of hydroxy-Al groups into vermiculite increases the Gapon coefficient k_G (K/Ca) from 5.7 to 11.1 × 10^{-2} L(mmol)$^{1/2}$ (Rich and Black, 1964). However, K fixation is reduced markedly. The relationships of maximum K fixing capacities of the hydroxyaluminum/hydroxyaluminosilicate–vermiculite complexes with the amount of Al or Al + Si fixed in vermiculite are exponential and negative (Saha and Inoue, 1998).

Freezing and thawing contribute to fixation or release of K^+ (Fine et al., 1940). Since this pioneer research, the significance of wetting and drying and freezing and thawing on K fixation has been recognized by numerous researchers (Attoe, 1947; Mitra and Prekash, 1955; Luebs et al., 1956; Hanway and Scott, 1957, 1959; Scott et al., 1957; McLean and Simon, 1958; Cook and Hutcheson, 1960; Scott and Hanway, 1960). Wetting and drying cycles promote interstratification in K-smectites (Miklos and Cicel, 1993), enhance K fixation (Sucha and Siranova, 1991), and reduce soil K mobility and plant K uptake (Zeng and Brown, 2000). The degree of K fixation or release on wetting or drying depends on the type of colloid present and the level of K^+ in the soil solution.

Potassium fixation by clay minerals may be strongly influenced by the kind of adsorbed cations or anions. Montmorillonite clays dried with K_2SiO_3 are altered in their swelling properties and fix large amounts of K (Mortland and Gieseking, 1951). Hydrous mica clays also fix large amounts of K that could not be removed with boiling HNO_3.

The effect of pH on K fixation in soil and clay systems is related to (i) the concentration of H_3O^+ and the competing cations and (ii) the extent of interlayering and the resultant charge neutralization and propping effect. A marked increase in K fixation in soils was observed when pH was raised to about 9 or 10 with Na_2CO_3 (Volk, 1934). At pH values up to 2.5 there was no K fixation, and between pH 2.5 and 5.5 the amount of K fixation increased very rapidly (Martin et al., 1946). Fixation increased more slowly above pH 5.5. At low pH, the lack of K fixation is attributed to high concentration of H_3O^+ and its ability to replace K^+ (Rich, 1964; Rich and Black, 1964). At pH >5.5, Al^{3+} cations precipitate as hydroxy polycations, which increase in the number of OH groups as pH increases until a discrete gibbsite phase is formed (Hsu, 1989). At pH about 8, Al^{3+} does not neutralize the charge on the clay and cannot prevent K fixation. The increase in K fixation between pH 5.5 and 7.0 is attributable to the decreased number of hydroxy interlayer material that decreases K fixation (Rich, 1964; Rich and Black, 1964; Murdock and Rich, 1972).

KINETICS AND MECHANISMS OF POTASSIUM RELEASE AND EXCHANGE IN SOIL ENVIRONMENTS

Kinetics is a part of the science of motion. It deals with the rate of chemical reaction, all factors that influence the rate of reaction, and the explanation of the rate in terms of the reaction mechanisms. To understand chemical reaction rates and reaction pathways, it is essential to have a knowledge of the kinetics involved.

The study of chemical kinetics, even in homogeneous systems, is complex and often arduous. When one attempts to study the kinetics of reactions in heterogeneous systems, such as soils, and even soil components, such as clay minerals, hydrous oxides, and humic substances, the difficulties are greatly magnified (Sparks, 2000a). This is largely because of the complexity of soils, which are composed of a series of inorganic and organic components. These components often interact with each other and display different types of sites. Furthermore, the variety of particle sizes and porosities in soils adds to their heterogeneity. In most cases, both chemical kinetics and multiple transport processes are occurring simultaneously. Therefore, the determination of chemical kinetics, which can be defined as the investigation of rates of chemical reactions of molecular processes by which reactions occur where transport is not limiting (Gardiner, 1969), is extremely difficult in soil systems. Therefore, in these systems, kinetics is a generic term referring to time-dependent or nonequilibrium processes.

Soils rarely reach equilibrium, if ever. Therefore, kinetics should be a major leitmotif in soil chemistry research for decades to come. Although much uncertainty remains, major progress has been made in better understanding the kinetics of soil chemical processes. Agricultural soils are nearly always in a state of disequilibrium with regard to K transformation because of cropping, fertilization, and leaching processes. The kinetics and mechanisms of K release and exchange are thus fundamental to understanding the chemistry of soil K.

Equations Describing Rate Processes of Potassium

The kinetic equations commonly used in soil chemistry include zero-order, first-order, second-order, Elovich, parabolic diffusion, and power function models. For more complete details and applications of these models, one may consult Sparks (1989, 1995, 1998, 1999, 2000a).

The zero-order, first-order, Elovich, and parabolic diffusion equations have been used to describe K kinetics in clays and soils (Table 4–5).

First-Order Equation

The first-order equation often describes K reaction kinetics in clay and soil systems (Mortland and Ellis, 1959; Mortland, 1961; Reed and Scott, 1962; Huang et al., 1968; Scott, 1968; Sivasubramaniam and Talibudeen, 1972; Sparks et al., 1980b, 1980c; Feigenbaum et al., 1981; Sparks and Jardine, 1981, 1984; Martin and Sparks, 1983; Jardinc and Sparks, 1984a). Sparks and Jardine (1984) reported that

Table 4–5. Kinetic equations commonly used to describe K kinetics in soils and soil materials.

Zero order†	$[A]_t = [A]_0 - kt$
First order†	$\log[A]_t = \log [A]_0 - kt/2.303$
Elovich equation‡	$(K_0 - K_t) = (1/\beta)\ln(1 + \alpha\beta t)$
Parabolic diffusion equation§	$(K_t)/(K_\alpha) = Rt^{1/2}$ + constant

† For K^+ adsorption, $[A]_t$ is the concentration of K^+ in the solution at time t, $[A]_0$ is the initial concentration of K^+ added at time zero, and k is the adsorption rate constant. For K^+ release, $[A]_t$ is the concentration of K^+ in the soil or soil material at time t, $[A]_0$ is the initial concentration of K^+ in the soil or the soil material at time zero, and k is the release rate constant.

‡ $(K_0 - K_t)$ represents the net amount of K^+ sorbed or released by the soil or soil material at time t, and α and β are constants during any one experiment.

§ K_t is the quantity of K^+ adsorbed or desorbed at time t, K_α is the amount of K^+ adsorbed at equilibrium, and R is an overall diffusion coefficient.

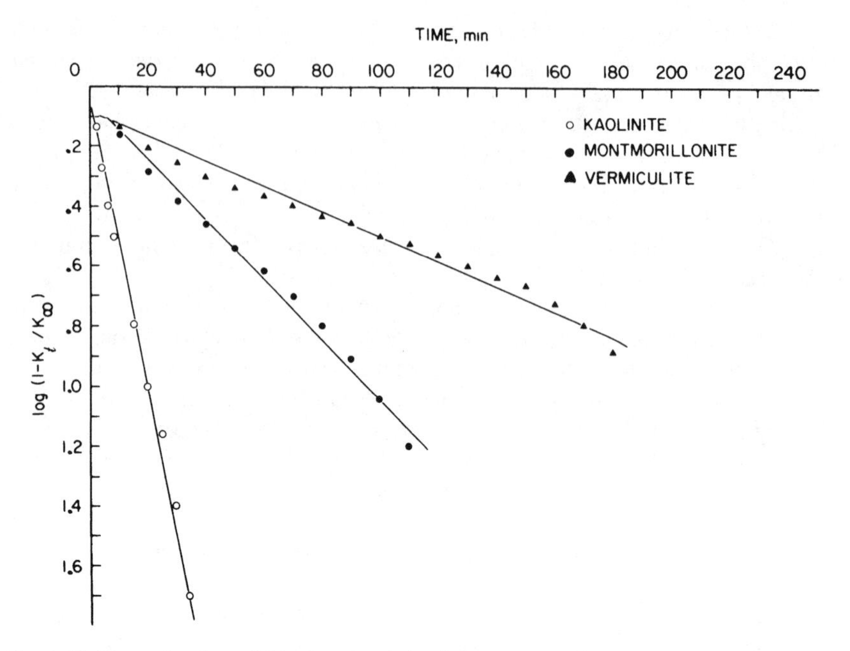

Fig. 4–10. First-order plots of K^+ adsorption in kaolinite, montmorillonite, and vermiculite systems (Sparks and Jardine, 1984).

the first-order equation describes K adsorption on kaolinite, montmorillonite, and vermiculite extremely well (Fig. 4–10).

Zero-Order Equation

The zero-order equation has been applied to describe K reaction kinetics in soil and clay systems by a few soil chemists and mineralogists (Mortland, 1958; Burns and Barber, 1961, Zhou and Huang, 1995; Zhou, 1995) In the study of the effect of temperature and moisture on exchangeable K in some silt loam soils, Burns and Barber (1961) reported that the rate of K release during an extensive time conformed initially to the first-order kinetics, whereas subsequent release followed the zero-order kinetics. Mortland (1958) found that the initial rate of K release from biotite is zero-order. Zhou and Huang (1995) reported that the monoammonium phosphate-induced K release from soils obeyed the zero-order kinetics.

Elovich Equation

The Elovich equation has been successfully used to describe the kinetics of sorption and desorption of various inorganic materials on soils (Sparks, 1989, 1995, 1998, 1999, 2000a). This equation has a relatively limited application in describing K reaction kinetics in soil and clay systems (Martin and Sparks, 1983; Sparks and Jardine, 1984; Sparks and Huang, 1985; Havlin et al., 1985; Rahmatulla and Mengel, 2000).

Parabolic Diffusion Equation

The parabolic diffusion equation is often used to suggest that diffusion-controlled phenomena are rate limiting (Sparks, 2000a). The parabolic diffusion law has been used to describe the release of nonexchangeable and structural K from clays (Barshad, 1954; Chute and Quirk, 1967; Feigenbaum et al., 1981). However, this equation does not always adequately describe kinetics of exchange between solution and exchangeable K (Sivasubramaniam and Talibudeen, 1972; Sparks et al., 1980c; Jardine and Sparks, 1984a). Nonlinearity with the parabolic diffusion equation for the initial minutes of K desorption in clays and soils is attributed to film diffusion-controlled exchange in the early minutes of K exchange (Chute and Quirk, 1967; Sparks et al., 1980c).

Kinetics and Mechanisms of Potassium Release from Primary Potassium-Bearing Minerals

Micas

Potassium release from micas may proceed by two processes, transformation of K-bearing micas to expansible 2:1 layer silicates by exchanging the K^+ with hydrated cations and the dissolution of micas resulting in the formation of weathering products. Both types of reactions occur in soil environments.

The transformation of micas to 2:1 expansible layer silicates through K^+ release proceeds by (i) edge weathering (Mortland, 1958; Rausell-Colom et al., 1965; Scott and Smith, 1967) and (ii) layer weathering (preferential weathering planes of Jackson et al., 1952). These weathering mechanisms are depicted in Fig. 4–6. In Fig. 4–6a and 4–6b, the expanded interlayers are not continuous but have internal terminations forming wedge zones (Jackson, 1963; Rich and Black, 1964). In Fig. 4–6c, interlayer spaces are either expanded all the way through the particle or nonexpanded uniformly through each layer. Edge weathering is most common in larger mica particles; in contrast, layer weathering is most common in smaller particles (Scott, 1968; Huff, 1972; Norrish, 1973). The mechanisms of K^+ release from micas by dissolution (Huang et al., 1968; Kittrick, 1973: Song and Huang, 1988; Zhou, 1995) are more complex than transformation of micas to 2:1 expansible layer silicates. However, K^+ release through mineral dissolution occurs in soils, especially in the rhizosphere and the immediate vicinity of fertilizer zones. The main factors that influence the release of K^+ from micas by both cation exchange and dissolution reactions are discussed below.

Tetrahedral Rotation and Cell Dimensions. The K would be in 12 coordination with 6 oxygens above and 6 oxygens below in the ideal hexagonal arrangement of basal oxygens of layer silicates (Fig. 4–11a). However, the interlayer K in muscovite and other micas is close to six basal oxygens, three above and three below (Radoslovich, 1960; Güven, 1971; Rich, 1972). The other six oxygens are at a greater distance from the K. This is caused by the ditrigonal configuration of the oxygens in each mica layer (Fig. 4–11b). Therefore, the coordination number of oxygen about K is reduced from 12 to 6, and the K–O bond is shortened.

A large rotation exists in dioctahedral micas, whereas the rotation is substantially less in trioctahedral micas (Radoslovich and Norrish, 1962). As micas take

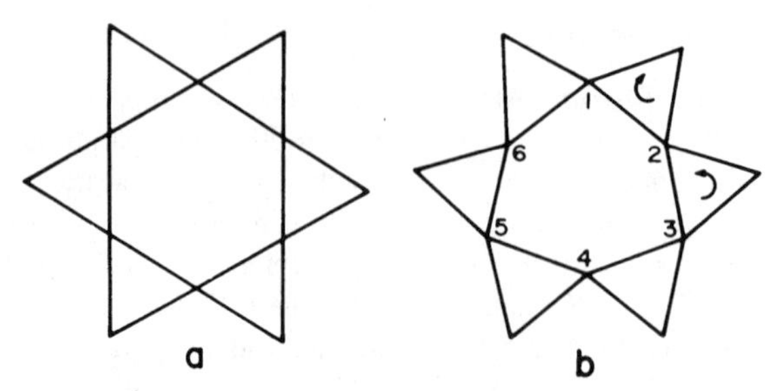

Fig. 4–11. Ideal hexagonal arrangement (a) of tetrahedra in layer silicates and (b) actual ditrigonal arrangement. The degree of tetrahedral rotation varies and generally is greater for dioctahedral than for trioctahedral layer silicates (Rich, 1968a).

in more Mg^{2+} and Fe^{2+} in the octahedral sheet and become more trioctahedral, the *b* dimension increases and the rotation becomes less (Radoslovich, 1962). Decreases in the degree of tetrahedral rotation result in longer K–O bond lengths. The bonding strength in trioctahedral micas is thus weaker than in dioctahedral micas. This results in the easier release of K from the trioctahedral micas than the dioctahedral micas. When K^+ is removed from the interlayer, the *b* dimension is decreased (Burns and White, 1963). Therefore, K release from micas would increase the tetrahedral rotation and strengthen the K–O bond.

Tetrahedral Tilting. Tilting of tetrahedra of layer silicates produces corrugations that may or may not mesh from one layer to the next, depending on the stacking sequence of 1.0-nm layers (Rich, 1968a). The relations of corrugations in one layer to the next may influence the diffusion path and diffusion rate of interlayer K. The tilting of tetrahedra and the subsequent corrugations in the trioctahedral micas are not nearly as pronounced as in the dioctahedral micas. Therefore, the tetrahedral tilting of micas can partially account for the slower release of K^+ from dioctahedral micas than trioctahedral micas.

Hydroxyl Orientation. Hydroxyl orientation of layer silicates is related to the stability of K in micas (Bassett, 1960). In trioctahedral micas, hydroxyl is perpendicular to the sheet plane and the K is thus repelled by hydroxyl. By contrast, the dipole moment of the hydroxyl is not directed toward K in dioctahedral micas, the interlayer K is more stable. Various populations of octahedral cations give different hydroxyl stretching frequencies and directions (Farmer and Russell, 1967; Wilkins, 1967). Weathering reactions may modify the octahedral population by release of octahedral cations and, thus, affect the orientation of hydroxyl dipole and the K–O bond strength.

Chemical Composition. Tetrahedral rotation and tilting and hydroxyl orientation of micas are governed by the chemical composition of their octahedral sheets. The replacement of hydroxyl by F^- can increase the retention of K^+ by micas because the repulsion of K^+ by the proton of a directed OH group would not occur (Rausell-Colom et al., 1965; Leonard and Weed, 1970). The chemical composition of micas is also an important factor governing the release of K^+ from muscovite,

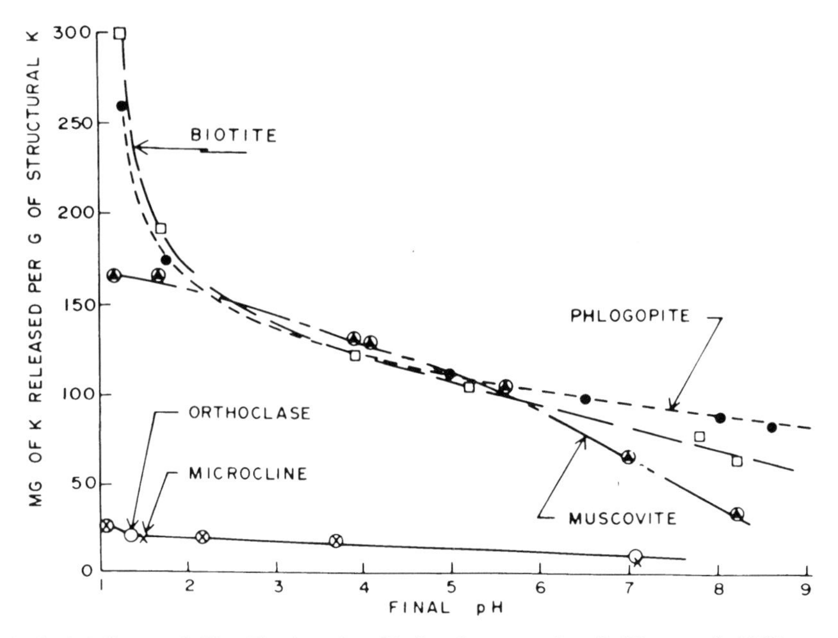

Fig. 4–12. Influence of pH on K^+ release from K minerals common in soils (Huang et al., 1968).

biotite, and phlogopite (Huang et al., 1968). The release of K^+ from the micas with different chemical composition as a function of pH is illustrated in Fig. 4–12.

Particle Size. A definite relationship between particle size and diffusion rate is prevented by irregularities in particle shape and structural imperfection (Sparks and Huang, 1985). Cox and Joern (1997) reported that K release rates of illite increase as particle size decreases. However, the mica particles of small diameter generally release K^+ through cation exchange reactions at a slower rate than coarser particles (Mortland and Lawton, 1961; Smith et al., 1968; von Reichenbach and Rich, 1969). This trend is shown in Fig. 4–13. The influence of particle size on the K^+ release is apparently governed by frequency of discontinuities and the degree of bending of the silicate layers (Rich, 1972). As particle size of large-grained micas is reduced, splitting would likely first occur along planes of discontinuity. The proportion of these weak planes may be reduced with the decrease of particle size, and K^+ release may thus decrease. Further, the silicate layer must bend to accommodate the larger hydrated cations that replace K^+ at the edges of mica layers. The larger and thicker the particle, the more it bends. Bending would tend to loosen K–O bonds (von Reichenbach and Rich, 1969). However, compared with large particles, the distance necessary to split a particle all the way through one side to the other is less in small particles. The opening of an interlayer in small particles may thus continue to completion. This would strengthen the bonding of K in the adjacent interlayers (Bassett, 1959), resulting in the formation of 1:1 regular mica–vermiculite interstratification (Reynolds and Hower, 1970).

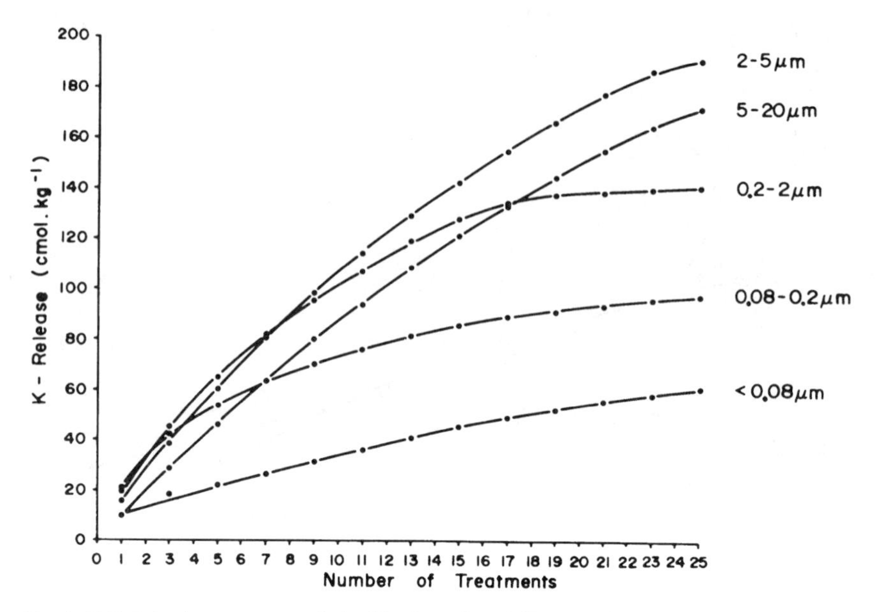

Fig. 4–13. Relation between accumulative K^+ release from different particle size fractions of muscovite and number of 5-d treatments (von Reichenbach and Rich, 1969).

Structural Imperfections. Muscovite frequently has discontinuities along the basal plane (Caslovsky and Vadam, 1970a, 1970b). If bending promotes K^+ release, these planes of discontinuity may initially increase the weathering rate. However, once they are opened all the way through a particle, the strain caused by bending is released and further release of K^+ is thus restricted.

Portions of the basal surfaces of micas show openings into the structure from which the layers roll back and form scrolls (Fig. 4–14). Such entries into interlayers should increase the K^+ release. Scrolls apparently form at borders of crystals of different crystallographic orientation and are present in the same particle (Le Roux et al., 1970).

Structural imperfections enable penetration of exchanging cations into the interlayers not only through lateral edges but also through fissures and cracks normal to the *a–b* plane (Wells and Norrish, 1968). Such structural imperfections may also be promoted by weathering dissolution or by inclusion of nonmica minerals (von Reichenbach, 1972).

Layer Charge Alterations and Related Reactions. The release of K^+ from micas and the fixation of K by phyllosilicates depend in part on their layer charge characteristics and density. A multiple regression analysis demonstrated that both components of total CEC (i.e., tetrahedral and octahedral) have a simultaneous effect on K fixation (Bouabid et al., 1991). Tetrahedral CEC explains more of the variation (64%) than octahedral CEC (36%). This greater effect is attributed mainly to the proximity of tetrahedral charge to the interlayer space of 2:1 phyllosilicates. The mechanisms by which the negative charge is changed are also significant. Charge

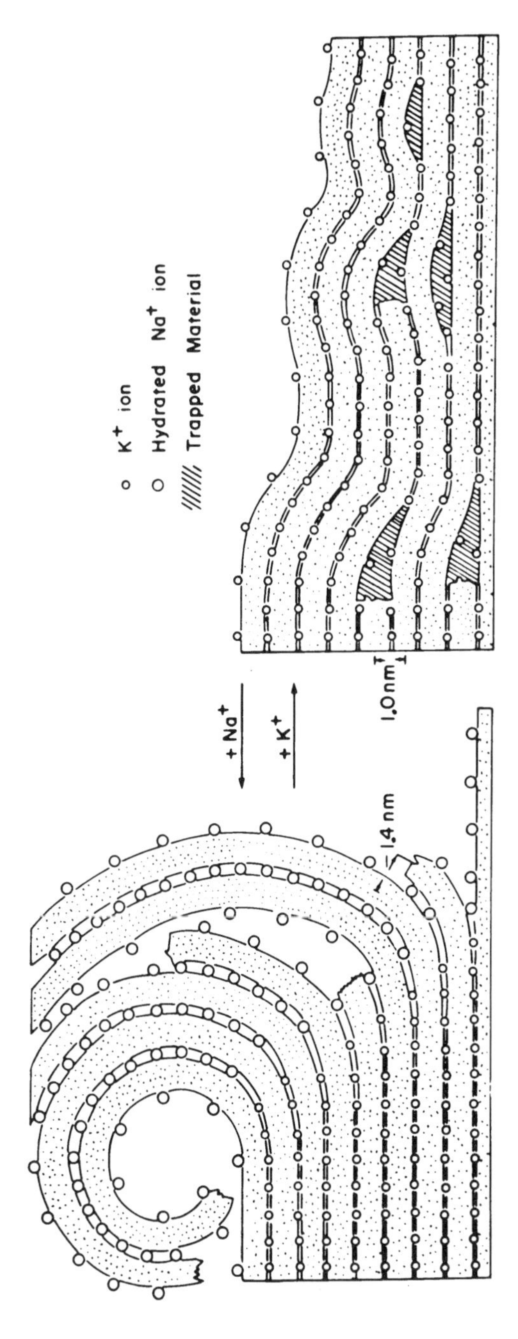

Fig. 4–14. Curling of layers by exchange of hydrated Na^+ ions for K^+ ions at the mica cleavage surface (Jackson, 1964) as revealed by the electron microscope (Raman and Jackson, 1964) (illustration by S.B. Jackson).

reduction of micas is associated with K^+ release by neutral salt solution. Micas containing larger amounts of Fe^{2+} seem most susceptible to those alterations.

Oxidation of Fe^{2+} is related to K^+ depletion (Robert, 1971). Oxidation of Fe^{2+} occurs only in the expanded portion of the mica particles. Therefore, K^+ release within the exchange front should remain unaffected by oxidation. However, several observations cast some doubt on the coordination of these processes (von Reichenbach, 1972). Boettcher (1966) suggested that K ions are expelled from the interlayers by Fe^{2+} oxidation to maintain electroneutrality. Potassium fixation is enhanced by the reduction of octahedral ferric Fe in smectites (Chen et al., 1987). By contrast, K fixation capacity of vermiculite is increased with the oxidation of Fe^{2+} to Fe^{3+}. The difficulty of replacing interlayer K in biotite also increases with the increase of oxidation of octahedral Fe^{2+} to Fe^{3+} (Barshad and Kishk, 1968). After Br_2 treatment, substantially less K is extracted from biotites (Dreher and Niederbudde, 1994).

The formation of regular interstratification in hydrobiotites begins by K^+ release around the edge of a surface layer of the biotite and then proceeds by attack on successive alternate interlayers (Farmer and Wilson, 1970). This process strengthens the bonding of K^+ in adjacent layers by Fe oxidation. Therefore, even if it is assumed that the release of K^+ from the exchange front of the still contracted zone of mica remains unaffected by the secondary process of layer charge variations, oxidation of the Fe^{2+} and associated reactions may influence the release of K^+ from micas.

Ionic Factors. The hydronium ion, H_3O^+, is one of the dominant ions in acidic soil solutions. The H_3O^+ binds three more water molecules fairly strongly, presumably attracting an O atom to each of its rather positive H atoms (Cotton and Wilkinson, 1980). Therefore, the aqueous "hydrogen ion" is best represented as $(H_9O_4)^+$. Reaction rates involving the H^+ ion are not restricted by diffusion rates of H_3O^+ (Bernal and Fowler, 1933). Therefore, proton transfer may be rapid to the wedge zone at the junction of "open" to "closed" mica. The rate of H_3O^+–K^+ exchange may be limited by the rate of K^+ diffusion.

There are many sources of H_3O^+ in soils. These include water through its dissociation, the dissolution of CO_2 from the air and soil atmosphere, strong mineral acids originated from weathering reactions, and low molecular weight organic acids. Hydronium ions are probably the cause of major release of structural K^+ by weathering reactions. This mechanism is especially important in the soil rhizosphere, which is the most important plant feeding zone. Therefore, K^+ release from soil minerals by H_3O^+ ions is significant in the study of the chemistry of soil K. The extraction of soils with boiling 1 M HNO_3, dilute HCl solutions and other acid dissolution techniques have been proposed for measuring the K-supplying power of soils (McLean and Watson, 1985; Sparks, 2000b).

The release of K^+ from micas by cation exchange is greatly influenced by the activity of K^+ ions in soil solution. When the K level is less than the critical value, K^+ is released from the interlayer by other cations from the solution. By contrast, when the K level is higher than the critical value, the mica-expansible 2:1 minerals adsorb K^+ from the solution. The critical K level is highly mineral dependent, being much lower for muscovite than for trioctahedral minerals (Scott and Smith,

1966; Newman, 1969; von Reichenbach, 1973; Henderson et al., 1976). The nature and concentration of the replacing cations also influences the critical K level in the solution (Rausell-Colom et al., 1965). For example, the critical K levels decrease in the order: $Ba^{2+} > Mg^{2+} > Ca^{2+} \cong Sr^{2+}$ for the same concentration of these ions and with a constant mica particle. For significant K^+ release to occur, the activity of all these replacing ions in the solution must be much higher than that of the K^+.

Besides cations, many organic and inorganic ligands in the soil solution also affect K release from minerals. The influence of different types of organic acids on weathering of rock-forming minerals have been studied (Huang and Keller, 1970, 1972; Tan, 1986; Robert and Berthelin, 1986). The kinetics of K release from muscovite and biotite is greatly influenced by the nature of organic acids (Song and Huang, 1988), which are common in soil environments, especially in localized zones where biological activity is intense such as in the rhizosphere and near decomposing plant residues.

Besides organic ligands, some strong complexing inorganic ligands such as phosphate also can influence the kinetics of K release from micas. In the immediate vicinity of the phosphate fertilizer zone, the concentration of phosphate can be as high as 2.9 to 4.5 *M* (Khasawneh et al., 1974; Tisdale et al., 1993). Phosphate at such high concentrations should induce K release from a series of K-bearing minerals of soils. Zhou (1995) reported that the rates of K release from muscovite, biotite, and illite are significantly enhanced by 1 *M* $NH_4H_2PO_4$ (pH 4.0). NH_4-taranakite is formed when $NH_4H_2PO_4$ is reacted with muscovite (Fig. 4–15B). However, NH_4-taranakite was not observed in the $NH_4H_2PO_4$–biotite system (Fig. 4–15A). This is attributed to the perturbation of taranakite formation by Fe (Zhou et al., 2000; Liu et al., 2002) released from biotite during the reaction.

Other Factors. Wetting and drying influence the release of K from soils. The release of K^+ upon drying of a soil is related to the clay fraction (Scott and Hanway, 1960). When a soil is dried, the degree of rotation of weathered micas may be changed and the K–O bond may be modified. Dehydration of interlayer cations may permit a redistribution of interlayer cations, because Ca^{2+} could now compete with K^+ for wedge sites. This apparently explains the release of K from soils upon drying. Air drying increases the potential buffering capacity of K (PBC^K) of a Vertisol; results suggest that soil water management schemes that include a dry period minimize the loss of available K (Assimakopoulos et al., 1994). The release of K^+ from a soil, which releases large amounts of K^+ during drying, is inhibited by nonvolatile liquids (alcohols with a high boiling temperature) and sucrose (Bates and Scott, 1964). Soils, which contain hydroxy-Al interlayers and appreciable amounts of K, do not release K upon drying (Rich, 1972). The presence of these organic and inorganic materials in the interlayer of weathered micas may change the *b* dimension of micas, the degree of tetrahedral rotation, and the length and strength of the K–O bond.

Temperature affects the rate of K^+ release from micas. Increasing temperature has been shown to increase the rate of K^+ release from biotite (Mortland, 1958; Rausell-Colom et al., 1965). Preheating of micas to high temperatures (1273 K) prior to tetraphenyl boron extraction increases the rate of K^+ release from muscovite, de-

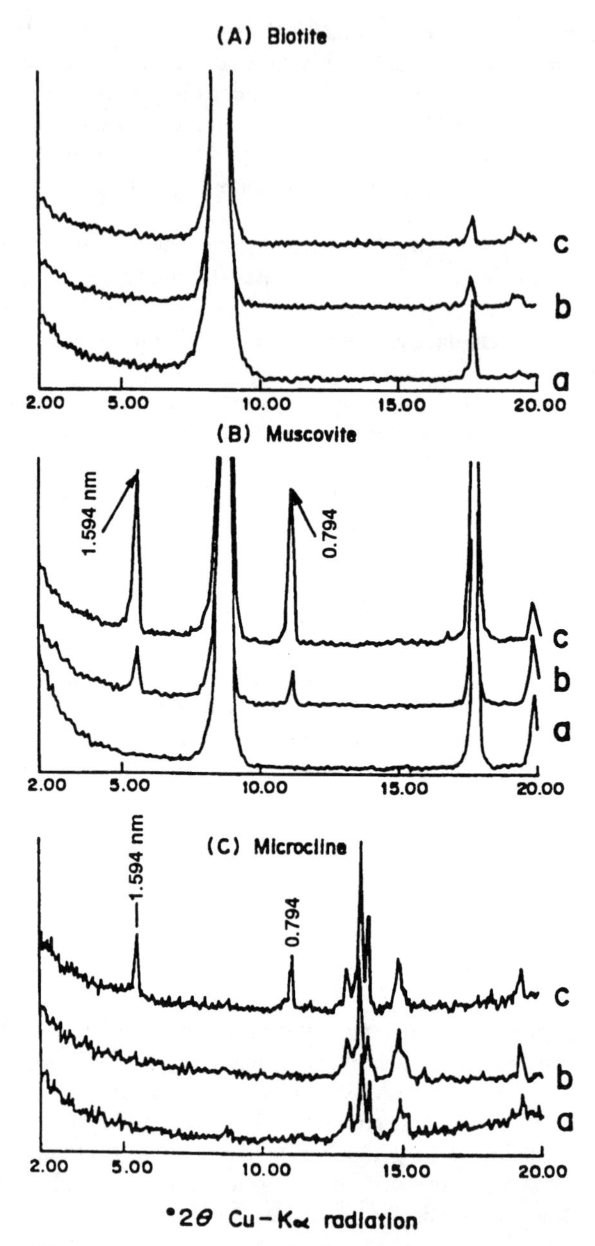

Fig. 4–15. X-ray diffractograms of untreated and $NH_4H_2PO_4$-treated K-bearing minerals (2–5 μm) (A) biotite, (B) muscovite, and (C) microcline: untreated minerals (a), the minerals treated with 1 *M* $NH_4H_2PO_4$ (pH 4.0) at 25°C for 4 wk (b), and the minerals treated with 1 *M* $NH_4H_2PO_4$ (pH 4.0) at 45°C for 4 wk (c) (Zhou, 1995). The unlabeled peaks in the diffractograms are those of the K-bearing minerals studied (biotite: 1.000, 0.500, and 0.460 nm; muscovite: 1.000, 0.499, and 0.448 nm; and microcline: 0.675, 0.651, 0.641, 0.594, and 0.461 nm).

creases the rate for biotite, and has little effect on phlogopite, except at very high temperatures. The decrease of K^+ release from biotite by preheating is attributable to oxidation of Fe(II); the mechanism causing the more rapid rate of K^+ release from muscovite upon heating remains obscure.

Redox potential of soils influences the K^+ release from micas because the interlayer K is more tenaciously held after oxidation of the structural Fe(II) (Sparks and Huang, 1985). Other factors being equal, the extent of K^+ release appears to be less in oxidized soil environments than in reduced soil environments.

Leaching enhances the K^+ release from micas through the removal of the reaction products. It accelerates the transformation of micas to vermiculites and other weathering products.

Potassium-Bearing Feldspars

Besides micas, K-bearing feldspars are important K reserves of soils. Therefore, the release of K from feldspars through weathering reactions is an important process in replenishing labile pool of soil K.

Structural Properties. The weathering rate of K-bearing feldspars appears to be affected by their structural properties (Sparks and Huang, 1985), including chemical composition, principally the inclusion of Na in the structure; the order of Si and Al in the structure; crystal and particle size; crystallization of more than one mineral within one particle; and the nature of twinning (Barth, 1969). The weathering rate of K-bearing feldspars probably increases with Si/Al disorder and Na content. A perthitic feldspar would likely weather more rapidly than a homogeneous particle. Albite would be weathered more rapidly in perthitic structures than the K-feldspar. Upon the removal of the lamellae of Na-feldspar, more internal surface would be opened and the K-feldspar phase would then be weathered more rapidly than a homogeneous crystal. It is likely that weathering is also more rapid at the junction of twins because of disorder at these planes. Therefore, authigenic maximum microcline should be most resistant to weathering by virtue of its Si/Al order, low Na content, lack of perthitic structure, and crystal size.

Ionic Factors. Hydronium ions are an important chemical weathering agent of K-bearing feldspars (Churchman, 2000). The initial weathering of K-bearing feldspars proceeds by surface reactions in which K^+ is displaced by H_3O^+ (Tamm, 1929). The substitution of K^+ by H_3O^+ weakens the feldspar structure. The attack by H_3O^+ is accompanied by a slight expansion of the lattice in feldspar surface layers (Bondam, 1967). The infrared absorption evidence indicates the presence of H bonds between H_3O^+ and surrounding lattice-bound oxygen atoms. The adsorbed H_3O^+ ions may, thus, facilitate the breakdown of the Al–O bonds. The Al is thereby changed from fourfold coordination to sixfold coordination and expelled from the feldspar structure. At the same time, SiOH groups are believed to be formed.

Potassium-bearing feldspars dissolve incongruently at the initial stage of weathering (Tamm, 1929; Correns and Engelhardt, 1939). The apparent incongruity of such reactions can be attributed to the precipitation of neoformed products (De-

Table 4–6. Apparent rate constants and Arrhenius heats of activation for the release of structural K from K minerals (Huang et al., 1968).

Mineral	Rate constant 301 K	311 K	Arrhenius heat of activation
	h^{-1}		kJ mol^{-1}
Biotite	1.46×10^{-2}	3.09×10^{-2}	58.66
Phlogopite	9.01×10^{-4}	2.44×10^{-3}	77.81
Muscovite	1.39×10^{-4}	4.15×10^{-4}	85.52
Microcline	7.67×10^{-5}	2.63×10^{-4}	96.24

carreau, 1977; Holdren and Berner, 1979). There have been intensive studies on the artificial weathering of feldspars (Huang, 1989; Blum and Stillings, 1995). Much attention has been paid to the occurrence, nature, and role of leached layers on feldspar surfaces. Leached layers, which may form on weathering of feldspars, might not be thick enough to slow diffusion of K^+. More recent studies, using secondary ion mass spectrometry (SIMS) (Muir and Nesbitt, 1997), elastic recoil detection (ERD) and Rutherford backscattering (RBS) (Casey et al., 1989), and also XPS (Muir et al., 1990; Hellmann et al., 1990) indicated formation of a dealkalized leached layer that was as deep as 100 nm and was especially thick at low pH. Nonetheless, studies of the alteration of feldspars at pHs between 5 and 8, comparable to pHs encountered in most soils, have generally shown only thin leached layers forming (Blum and Stillings, 1995).

The kinetics of dissolution of K^+ from K-bearing minerals by 1 *M* HNO_3 indicated that for the temperature range studied, the rates of K^+ release from muscovite, phlogopite, and biotite are about 2, 9 to 12, and 118 to 190 times, respectively, greater than the rate of K^+ release from microline (Table 4–6). The release of K^+ from a tectosilicate is more difficult than from the interlayers of micas. The Arrhenius heat of activation for the release of K^+ from microcline is greater than for the micas. The heat of activation is interpreted as the energy level the structural K must acquire to be able to react. As the heat of activation increases, the rate constant decreases, since the mineral with higher heat of activation for the release of structural K would release less K per unit time. The heat of activation for the release of K^+ from the K-bearing minerals (Table 4–6) is on the order of those obtained for the K^+ release from Indiana soils (Burns and Barber, 1961). The Arrhenius heat of activation and apparent rate constant for K^+ release from the K-bearing minerals (Table 4–6) are fundamental to understanding the availability of K^+ in feldspars and micas to plants (Reitemeier, 1951; Rich, 1968a; Smith et al., 1968). The release of K^+ from K-feldspars is generally more hindered than from micas, but the role of K-feldspars in supplying K to plants cannot be ignored. The K-feldspars are the largest natural reserve of K in many soils.

Besides H_3O^+ ions in soil solutions, ligands of complexing organic acids also have a significant role in the release of K from the feldspar structure. The dissolution rates of rock-forming minerals are dependent on the complexing properties of organic acids (Huang and Keller, 1970). Humic and fulvic acids are also capably of dissolving very small amount of K, Si, and Al from soil minerals at pH 7.0 (Tan, 1980). The dissolution of the elements is increased considerably at pH 2.5. Low mol-

Table 4–7. Potassium released from K-bearing primary minerals by oxalic or citric acid or sodium tetraphenylboron (NaTPB) solution at the end of a 10-d reaction period (Song and Huang, 1988).

Mineral	Organic acid†		NaTPB‡
	Oxalic	Citric	
	g K released kg^{-1} structural K		
Biotite	44.1	10.5	1000
Muscovite	1.1	0.9	83.0
Microcline	2.3	1.4	1.6
Orthoclase	2.2	1.2	1.3

† The concentration of organic acid solutions was 0.01 *M*.
‡ One *M* NaCl–0.2 *M* NaTPB–0.01 *M* EDTA.

ecular weight organic acids are very effective in releasing K^+ from K-bearing minerals (Song and Huang, 1988). The comparison between K released by organic acids and sodium tetraphenyl boron (NaTPB) solution (Table 4–7) shows that the sequence of K release from K-bearing minerals in organic acids is: biotite > microcline ≅ orthoclase > muscovite, whereas in the NaTPB solution, it becomes: biotite > muscovite > microcline ≅ orthoclase. This indicates that the mechanism of K release by organic acids differs from that by the NaTPB solution. The NaTPB solution has been used to investigate the exchangeability of K in minerals and soils (Scott and Reed, 1962a, 1962b; Reed and Scott, 1966; Scott, 1968; Smith et al., 1968). The K in the mineral structure is replaced by Na in the NaTPB solution through a cation exchange reaction, and the TPB combines with released K and forms precipitates to ensure continuation of the exchange reaction. Exchange of Na for K in the interlayers of micas is much easier than in the interstices of the three-dimensional framework structure of K-feldspars. Therefore, more K is released through cation exchange from biotite and muscovite than from microcline and orthoclase (Table 4–7). By contrast, microcline and orthoclase release more K than muscovite does in organic acid solutions. The K in microline and orthoclase is more susceptible to the attack of organic acids than the K in muscovite. This is especially important in the soil rhizosphere (see Potassium Chemistry in the Soil Rhizosphere below in the Chemistry of Labile Pool of Soil Potassium section). The sequence of rates of K release from K-bearing minerals varies with the mechanisms of K release.

In the immediate vicinity of the phosphate fertilizer zone, high concentration of phosphate in soil solution (Tisdale et al., 1993) may also induce K release from K-feldspars. Zhou (1995) reported that 1 *M* $NH_4H_2PO_4$ (pH 4.0) substantially enhances the rate of K release from microcline. Further, NH_4-taranakite is formed in the $NH_4H_2PO_4$–microcline system (Fig. 4–15C). The formation of the taranakite provides evidence for dissolution of microcline in the $NH_4H_2PO_4$ solution.

Other Factors. Climate, topography, degree of leaching, and Eh also influence the stability of K-bearing feldspars in soils (Huang, 1989; Churchman, 2000). All experimental evidence shows that chemical breakdown of K-feldspars accelerates with raising temperature (Rasmussen, 1972). Warm climate and high precipitation in tropical soils may have depleted all of their K-feldspars. However, these minerals may persist in temperate soils, especially in sand and silt fractions. The prevailing Eh of a soil may not be of direct concern to chemical weathering of K-bearing feldspars because their major elements do not exist in more than one valence state. However, complexing organic acids, which can be degraded by oxidation, can affect the weathering transformation of K-bearing minerals. Therefore, the prevailing Eh of a soil may be indirectly related to the stability of K-bearing feldspars.

Kinetics and Mechanisms of Potassium Release from Natural Soils

A number of chemical methods have been used to extract nonexchangeable K of soils. These include boiling HNO_3, H_2SO_4, hot HCl, electroultrafiltration, Na tetraphenylboron with EDTA, and ion exchange resins such as H and Ca saturated resins (Hunter and Pratt, 1957, Scott and Reed, 1962a, 1962b; Smith et al., 1968; Martin and Sparks, 1985; Helmke and Sparks, 1996). However, limited studies have been conducted on the kinetics of the release of nonexchangeable K from soils. Ion exchange resins are the most practical techniques for this purpose because time and K concentration in solution can be carefully controlled (Sparks and Huang, 1985).

Talibudeen et al. (1978) reported that soil K released to Ca resin could be divided graphically into simultaneous rates of K release from the surface of the soil complex, the weathered periphery, and the micaceous matrix. The kinetics of K release were analyzed mathematically to define the three simultaneous rate processes. They found that diffusion coefficient (3×10^{-19} cm^2 s^{-1}) calculated for the slowest process, which is suggested to be from the mineral matrix, is of the same magnitude as that for illitic clay minerals. Munn et al. (1976) extracted the K from different soil fractions of four Ohio soils with 0.01 M $CaCl_2$. Their results indicated that the K release from all soil separates can be described by the first-order kinetics. The apparent rate constants for K release from the silt fraction are generally lower than those calculated for the sand and clay fractions, and the rate constants for the clay and sand fractions are similar across four soils. Comparing the values of rate constants for K release from soils with those for K release from minerals (Huang et al., 1968), Munn et al. (1976) concluded that the rate constants of K release in their study are within the range expected for K release from the mixtures of micas and K-feldspars. It has been shown that K^+ can be removed more easily from coarser particles than finer particles of the Marshal soil (Fig. 4–16). Bending and deformation of mica layers during K^+ release increase with particle thickness. (Ross and Rich, 1973) (see also the section on mica particle size in Kinetics and Mechanisms of Potassium Release from Primary Potassium-Bearing Minerals).

Havlin et al. (1985) determined K release from different fractions of six Great Plain soils by successive extraction with Ca-saturated cation exchange resin. They found that the Elovich, power function, and parabolic diffusion equations adequately described the cumulative K release, whereas the first-order rate equation did not. The rate constants were highly correlated with mica contents of the soils, cumulative K uptake, and cumulative alfalfa (*Medicago sativa* L.) yield (Havlin et al., 1985; Havlin and Westfall, 1985).

Martin and Sparks (1983) investigated the kinetics of the release of nonexchangeable K from two Coastal Plain soils from Delaware using a H resin. They reported that the first-order rate equation best described the kinetics of K release

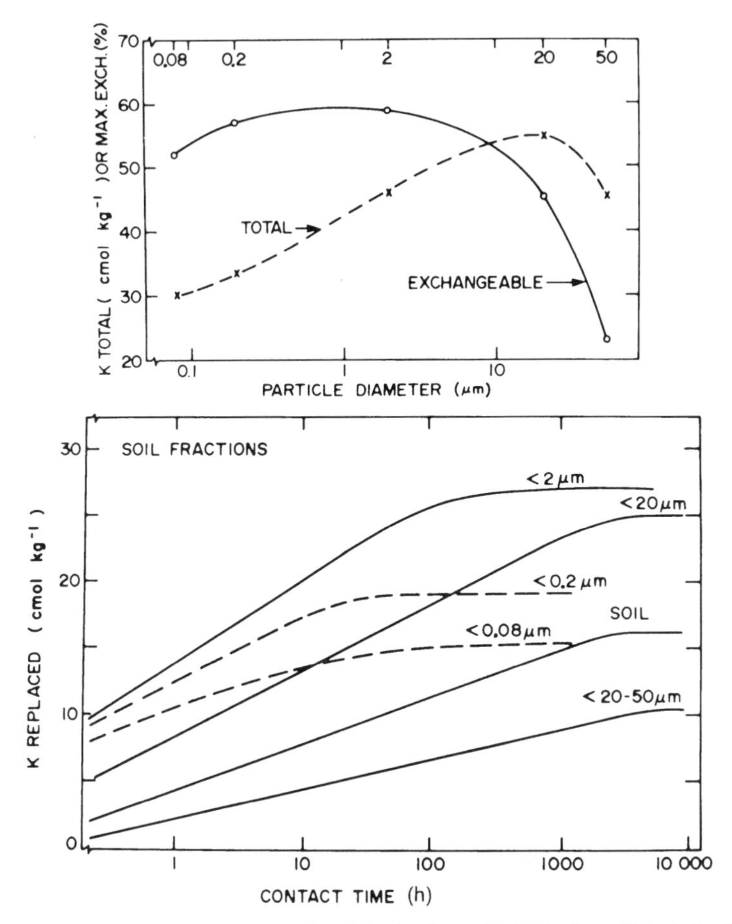

Fig. 4–16. Potassium replaced from whole soil and size fractions of the B2 horizon of Marshall soil (fine-silty, mixed, mesic Typic Hapludolls) by different contact periods with sodium tetraphenylboron solution (Smith et al., 1968).

from the soils studied. The parabolic diffusion law also described the data satisfactorily, indicating diffusion-controlled exchange, whereas the zero-order and Elovich equations did not describe the data well. The rate coefficients of the release of nonexchangeable K ranged from 1.5 to 2.2×10^{-3} h^{-1} in the Kalmia soil and from 1.6 to 2.9×10^{-3} h^{-1} in the Kennansville soil (Arenic Hapludults), indicating low rates of the release of nonexchangeable K from the two soils (Table 4–8). The magnitudes of the rate coefficient values differed little between depths in the two soils. This is attributed to the similar clay contents and clay mineralogy of the two soils.

Sadusky et al. (1987) investigated the kinetics and mechanisms of K release from sandy soils from the Delaware Coastal Plain. A number of simple kinetics models including zero- and first-order equations were applied to the K release data. All these equations described the data well, and it was difficult to conclude whether a simple zero- or first-order equation best described the K release data from the soils.

Table 4–8. First-order nonexchangeable K release rate coefficients (k) of Kalmia and Kennansville soils (Martin and Sparks, 1983).

Depth	$k \times 10^3$
m	h^{-1}
Kalmia soil	
0–0.15	1.9
0.15–0.30	1.9
0.30–0.45	2.1
0.45–0.60	1.5
0.60–0.75	1.8
0.75–0.90	2.2
Kennansville soil	
0–0.15	1.8
0.15–0.30	1.6
0.30–0.45	1.7
0.45–0.60	2.3
0.60–0.75	2.9
0.75–0.90	2.5

However, overall K release from the sand fractions was best described by the zero-order rate equation. The results from different studies in the literature showed that several equations often describe one reaction satisfactorily based on linear regression analysis, but no single equation best describes every case of K release from soils. This may be attributed to the nature and properties of soils and experimental conditions. Conformity of data to a certain kinetic model does not yield mechanistic information, nor can one definitely conclude that it is the best model (Sparks, 1989, 1998, 1999, 2000a).

The release of nonexchangeable K from soils is not only from clay-sized K-bearing minerals but also from micas and feldspars in silt and sand fractions (Sadusky et al., 1987; Fanning et al., 1989; Simard et al., 1992). Rahmatulla and Mengel (2000) investigated the kinetics of K release by using a device similar to that developed by Schubert et al. (1990), which simulates plant roots in extruding H^+ particularly at their tips (Hauter and Mengel, 1988). Nonexchangeable K was released from the separated sand, silt, and clay fractions by a H^+-saturated ion exchanger in a K^+ exchange device (Fig. 4–17). A semipermeable dialyzing membrane of 20-μm thickness was used to separate sample and ion exchanger. The amount of K^+ extracted during almost 1000 h from the sand, silt, and clay fractions varied considerably in each fraction. The extracted K ranged from 389 to 489 mg K^+ kg^{-1} in sand, 801 to 1010 mg K^+ kg^{-1} in silt and 899 to 1105 mg K^+ kg^{-1} in clay fraction. It is interesting to notice that maximum quantities of K were released from feldspar-rich sand fraction of sandy soils from the Delaware Coastal Plain in the USA (Sadusky et al., 1987) and from silt fraction of some Canadian soils (Simard et al., 1992). The H^+-saturated ion exchanger initiated a substantial release of nonexchangeable K^+ apparently by lowering K^+ concentration in the solution rather than by breaking down the mineral structure by low pH (Feigenbaum et al., 1981). The release of K^+ by the H^+-saturated ion exchanger from the sand, silt, and clay fractions as a function of time followed the Elovich, power function, first-order, and

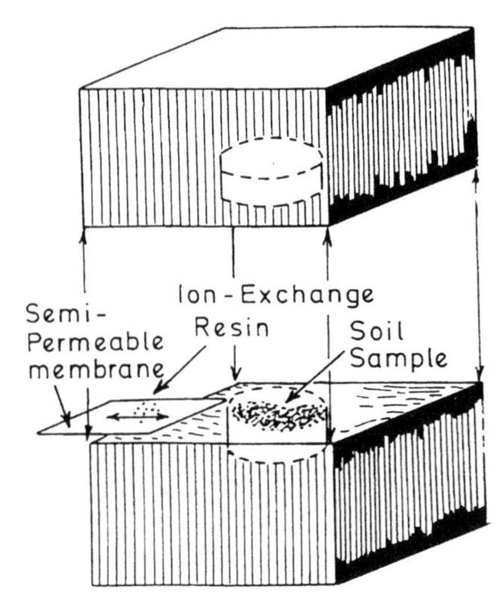

Fig. 4–17. Diffusion cell used in the study of K release kinetics from soils (Rahmatulla and Mengel, 2000).

parabolic diffusion equations (Rahmatulla and Mengel, 2000). The results presented in Fig. 4–18 are based on the Elovich equation. The variation in the slope of the Elovich equation correlated significantly with the concentration of micaceous minerals in the sand, silt, and clay fractions and may have an important impact on K for plant uptake.

The surface coatings may affect the rate of K release by weathering (Courchesne et al., 1996). Removal of surface coatings, which are composed of Al, Fe, Si, and organic C with traces of Ca, from the <2-mm fractions of B horizons of the Spodosols enhances their K release rates. In contrast, coating removal has no effect on their K release rates from the C horizon, indicating that mineral surfaces are apparently not altered by treatment with dithionite-citrate-bircarbonate and H_2O_2. The accumulation of coatings appears to be a process isolating primary mineral surfaces, thus reducing the extent of reactive surfaces available for K release.

In all of the studies discussed above, except for Sadusky et al. (1987), K^+ in soils was released by cations in the solutions. Few studies investigated the effect of anions on the kinetics of K^+ release from soils. The work of Zhou and Huang (1995) was the first report on the effect of orthophosphate on K release from soils. The K^+ released from the Xuwen (Oxisol), Wuxi (Alfisol), and Fenqiu (Entisol) soils in 1 *M* $NH_4H_2PO_4$ solution (pH 4.0) for 4 wk was 4.77, 114.0, and 207.8 mg kg^{-1}, respectively (Fig. 4–19). The results also clearly indicated that the K^+ release in 1 *M* $NH_4H_2PO_4$ solution at pH 4.0 from the three soils was 2.3 to 2.9 times higher than that in 1 *M* NH_4Cl solution at the same initial pH. Compared with the NH_4Cl systems, the excess K released from the soils in the $NH_4H_2PO_4$ systems is attributed to the phosphate-induced alteration of K-bearing minerals of the soils. Zhou (1995) showed that 1 *M* $NH_4H_2PO_4$ induced alteration of micas and feldspars and the subsequent K release. The kinetics of the $NH_4H_2PO_4$-induced K^+ release from the soils can be described by the zero-order rate equation. The rate coefficients of

K release from the soils in 1 *M* $NH_4H_2PO_4$ and NH_4Cl solutions are shown in Table 4–9. In the Xuwen soil, K-bearing minerals were virtually nonexistent and any K^+ release was probably from the interlayered vermiculite and muscovite, which could be present in a trace amount. The young Fenqiu soil contained the highest amount of K-bearing minerals (micas and feldspars), which was related to the high rate of K^+ release. The Wuxi soil was intermediate in the rate of K^+ release. The rate coefficients of K release from the three soils in 1 *M* $NH_4H_2PO_4$ solution (pH 4.0) were much higher than those in 1 *M* NH_4Cl solution (pH 4.). The amount and rate of the

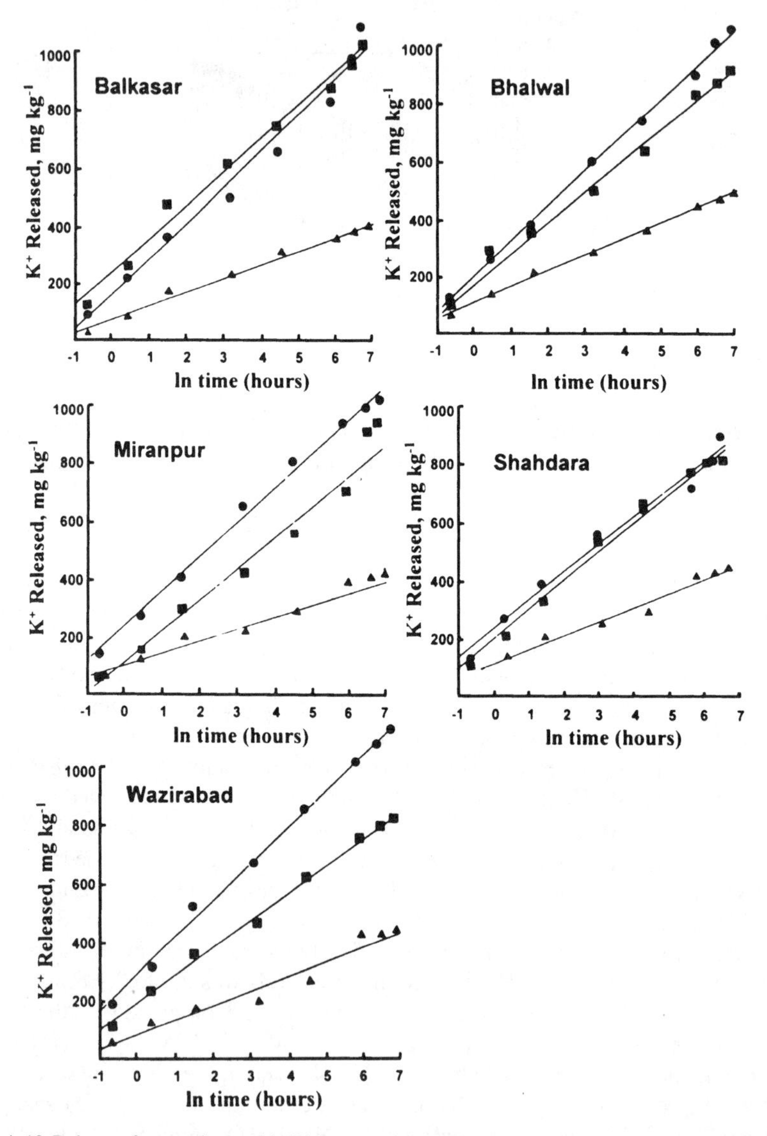

Fig. 4–18. Release of nonexchangeable K^+ from sand (triangle), silt (square), and clay (circle) fractions (Rahmatulla and Mengel, 2000).

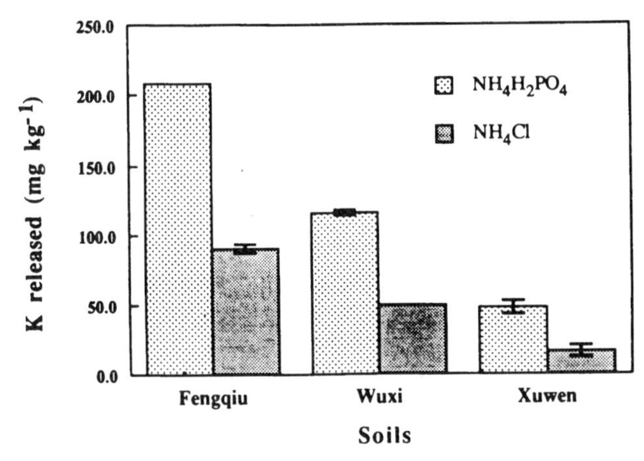

Fig. 4–19. Potassium released from the soils in 1 *M* $NH_4H_2PO_4$ and 1 *M* NH_4Cl solutions (pH 4.0) at 25°C in 4 wk. The standard errors of the K release in $NH_4H_2PO_4$-Fengqiu and NH_4Cl-Wuxi soil systems were too small to be shown in the figure (Zhou and Huang, 1995).

$NH_4H_2PO_4$-induced K^+ release from the soils were related to the degree of weathering of the soils. The combined effect of phosphate and proton on the alteration of K-bearing minerals and the formation of NH_4-taranakite were the major mechanisms of K^+ release from the soils in the $NH_4H_2PO_4$ solution. Application of $NH_4H_2PO_4$ fertilizer to soils may promote K^+ release from the soils, thus increasing their K-supplying rate.

As nonexchangeable K of soils contributes a significant proportion of K to plant nutrition, a measurement of buffer power that takes into account the contributions of exchangeable and nonexchangeable K and soil mineralogy is necessary for accurate modeling of soil K fertility and practical recommendations (Sparks and Huang, 1985; Schneider, 1997a, 1997b; Ghosh and Singh, 2001). The kinetics of release of nonexchangeable K are inversely related to the degree of weathering of soils (Dhillon ad Dhillon, 1990; Zhou and Huang, 1995). The levels of exchangeable K and K fixation capacities are influenced by long-term K fertility management (Liu et al., 1997). Continuous cropping without K inputs through fertilizers or manure cause a decline in nonexchangeable K reserves and release rate while application of recommended NPK plus farmyard manure maintain higher release

Table 4–9. Rate coefficients of K release from the soils in 1 *M* $NH_4H_2PO_4$ and 1 *M* NH_4Cl solutions (pH 4.0) at 25°C (Zhou and Huang, 1995).

	Rate coefficient (μg kg^{-1} h^{-1})	
Soil†	NH_4Cl	$NH_4H_2PO_4$
Xuwen (Oxisol)	4	72
Wuxi (Alfisol)	27	102
Fengqiu (Entisol)	37	170

† The soil samples were taken from the top 15 cm of the fields from tropical (Xuwen), subtropical (Wuxi), and temperate (Fenqiu) regions of China.

rates (Rao et al., 1999). Therefore, long-term cropping, fertilization, and manuring influence K release from the nonexchangeable K fraction of soils.

Kinetics and Mechanisms of Potassium Adsorption and Desorption Reactions

Kinetics of reactions existing between solution, exchangeable, nonexchangeable, and mineral phases of K profoundly influence K chemistry of soil. The rate and direction of these reactions determines whether applied K will be leached into lower horizons, taken by plants, converted into unavailable forms, or released into available forms. A substantial amount of research has been conducted on ion exchange with K^+, but only a meager amount of research on kinetics of K^+ reactions in soils has appeared in the literature.

The kinetics of K^+ reaction between soil solution and exchangeable phases is very much dependent on the kind of clay minerals present (Barshad, 1951; Sivasubramaniam and Talibudeen, 1972; Sparks, 1980; Sparks et al., 1980b, 1980c; Sparks and Jardine, 1981, 1984; Sparks and Rechcigl, 1982; Jardine and Sparks, 1984a). The K^+ reaction rate also strongly depends on the method to measure kinetics of K^+ exchange (Sparks and Rechcigl, 1982). Kinetics of K exchange on kaolinite, montmorillonite, and vermiculite are illustrated in Fig. 4–20. In the case of montmorillonite, the interlayer is able to swell with adequate hydration and thus allows for rapid passage of ions. With kaolinite, only edge sites are available for K^+ exchange. Kinetics of K^+ exchange on vermiculite tend to be extremely slow. Compared with montmorillonite the interlayer space of vermiculite is rather restricted and thus impedes many ion exchange reactions. Bolt et al. (1963) theorized the existence of three types of binding sites for K^+ exchange on a "hydrous mica." They hypothesized that rapid kinetics are due to external planar sites, slow kinetics to interlattice exchange sites, and intermediate kinetics to interlayer edge sites.

Only limited studies have been conducted on the kinetics between solution and exchangeable forms of K^+ in soil systems (Talibudeen and Dey, 1968; Siva-

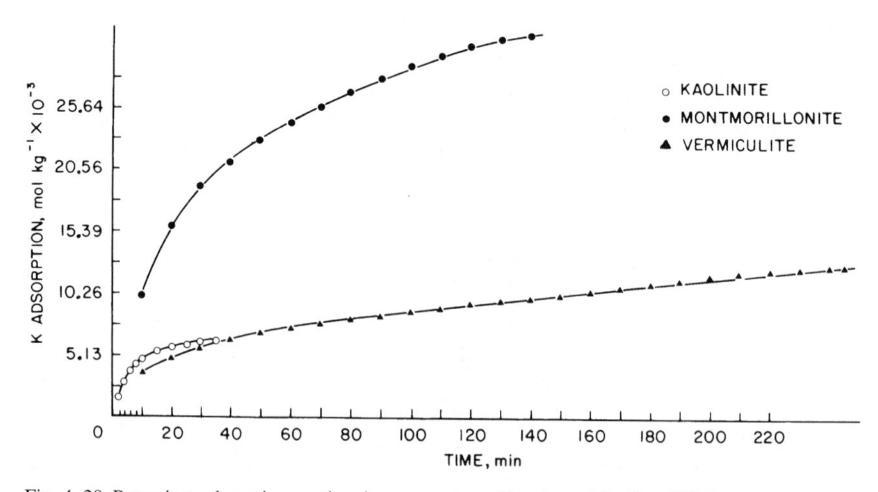

Fig. 4–20. Potassium adsorption vs. time in pure systems (Sparks and Jardine, 1984).

subramaniam and Talibudeen, 1972; Goulding and Talibudeen, 1979; Goulding, 1980, 1981; Sparks et al. 1980b, 1980c; Sparks and Jardine, 1981, 1984; Sparks and Rechcigl, 1982: Jardine and Sparks, 1984a; Wang and Huang, 1990a, 1990b, 1993, 2001). Sparks et al. (1980b) reported that the rate coefficient of K adsorption in two Dothan soils (fine-loamy, kaolinitic, thermic Plinthic Kandiudults) from Virginia ranged from 1 to 20 h^{-1}. This suggests slow rates of K adsorption compared with values of 81 to 216 h^{-1} calculated for Florida soils (Selim et al., 1976). The slow rate of K^+ exchange in these Virginia soils was attributed to intraparticle transport and diffusion processes that reflected the relatively large quantities of vermiculitic minerals present in the soils. The rapid rate of K adsorption in the Florida soils can be ascribed to the predominance of kaolinite.

The study on the kinetics of K^+ adsorption and desorption in a Matapeake soil (fine-silty, mixed, semiactive, mesic Typic Hapludults) from Delaware showed that K^+ desorption is slower than for K^+ adsorption (Sparks and Jardine, 1981). This is because of the difficulty in desorbing K^+ from the partially collapsed interlayer space of vermiculitic minerals. The Coloumbic attraction between K^+ ions and silicate layers would be greater than the hydration forces between the individual K^+ ions (Kittrick, 1966). This would result in a partial interlayer collapse. Therefore, K kinetics are nonsingular, and hysteresis could occur (Ardakani and McLaren, 1977; Rao and Davidson, 1978; Sparks et al., 1980c). Slower rates for K^+ desorption than for adsorption have also been reported by others (e.g., Talibudeen and Dey, 1968; Feigenbaum and Levy, 1977).

As with general ion exchange reactions in soil (Talibudeen, 1981; Sparks, 1989, 2000a), it is well accepted that at least five steps are involved in the K exchange reaction. They are (i) diffusion of K ions from bulk soil through the liquid film that surrounds the particles of soil exchange materials (film diffusion), (ii) diffusion of K ions through a hydrated interlayer surface of the particles (particle or intraparticle diffusion), (iii) exchange of K ions with counter ions (e.g., Ca ions) on exchange sites of the particle surface (simple chemical reaction), (iv) diffusion of displaced counter ions through the hydrated interlayer space of the particles (particle diffusion), and (v) diffusion of the displaced counter ions through the liquid film away from the particle (film diffusion). Therefore, K exchange kinetics on soil particles involve mass transfer (particle diffusion and film diffusion) and chemical reaction processes. For an exchange reaction to occur, ions must be transported to the exchange sites of the particles. Initially, both the film of liquid adhering to and surrounding the particles and the hydrated interlayer spaces in the particle are zones of low concentration of the ions. The ions in these zones are constantly depleted by ion adsorption to the sites. The decrease in concentration of the ions in these interfacial zones then is compensated by the ion diffusion from the bulk solution. The rate of the reaction will be governed by the slowest step in these processes. It is generally accepted that for most ion exchange reactions in a porous solid exchanger system, either particle diffusion or film diffusion is rate limiting (Helfferich, 1983; Sparks, 1989).

A method using combined apparent rate constants (rate coefficients) under static, stirred, and vortex mixing conditions was developed to quantify resistances of film diffusion, particle diffusion, and chemical reaction for K–Ca exchange under normal static state in kaolinite and vermiculite systems (Ogwada and Sparks,

Table 4–10. Apparent rate constants, activation energies and preexponential factor of the fast K adsorption reactions for the three soils based on the first-order kinetics (Wang and Huang, 1990a).

Soil†	Apparent rate constant				Activation energy	Preexponential factor
	278 K	288 K	298 K	308 K		
	$\times 10^5$ s^{-1}				kJ mol^{-1}	s^{-1}
Jinghua	8.94	16.7	18.4	23.6	21.6	1.14
Fuyang	9.88	17.4	19.3	23.5	19.4	0.49
Saoxin	11.9	21.9	26.6	34.0	24.0	4.18

† The soil samples were taken from the top 15 cm of Jinghua (Ferric Ultisol), Fuyang (Orthic Ultisol), and Saoxin (Vertisol) in Zhejiang, China.

1986b). Since the resistances are reciprocals of the rate constants of the processes, the individual rate constant for each process may also be quantified. Therefore, the rate-limiting step can be established.

The Arrhenius equation has been frequently used in soil K research (Sparks and Huang, 1985). Theoretically, the magnitude of the rate, as expressed by the Arrhenius equation, is a function of activation energy and the pre-exponential factor (collision number). Although attention was paid to the activation energy of K exchange reaction, the pre-exponential factor was not studied. Wang and Huang (1990a) reported that the magnitude of the rate constant of K adsorption by the soils is governed not only by the energy of activation but also by the pre-exponential factor. This relationship is illustrated in Table 4–10. The activation energies of K adsorption by these soils (19.4–24.0 kJ mol^{-1}) indicated that diffusion-controlled exchange (Glasstone et al., 1941) is the rate-limiting process. The activation energies of K exchange reactions of soils were reported to vary inversely with the rate of the reaction (Sparks and Huang, 1985). However, this was not the case in the data presented in Table 4–10. At any temperatures studied, both the rate constant and the activation energy of K adsorption for the Saoxin soil was larger than those for the other two soils. This was evidently because of the influence of the pre-exponential factor on the rate constant. Among the three soils, the Saoxin soil had the highest pre-exponential factor for the K adsorption. According to classic collision theory of a gas reaction, the pre-exponential factor in the Arrhenius equation refers to the frequency factor (Laidler, 1965). If the theory is applied to a reaction in a solution, particularly for those diffusion-controlled reactions (Collision-encounter theory), the pre-exponential factor is treated as the collision number (Adamson, 1979), which is a function of diffusion rate of the reactant. The diffusion rate of an ion in soils is affected by the tortuosity factor (Nye, 1966; Wild, 1981). The passage of K ion transport in the Saoxin soil is apparently more straightforward than in the other two soils. The differences in the magnitude of the pre-exponential factor of K adsorption for the soils are probably related to their mineralogical and chemical properties.

In studying K desorption from soils, Wang and Huang (1993) reported that desorption of K obeyed a polyphasic first-order kinetics. The activation energy of K desorption from the soils did not consistently vary inversely with their reaction rates. This is attributed to the effect of the pre-exponential factor on the desorption rate of K. The nature and association of soil components in relation to the pre-ex-

Table 4–11. The influence of soil organic matter on the amounts of K adsorbed by the soils in the reaction period of 0 to 30s (Wang and Huang, 2001).

Soil	Amount of K adsorbed (mg kg^{-1})	
	Before removal of organic matter	After removal of organic matter†
Jinghua	210	35
Fuyang	158	0.5
Saoxin	363	47

† Based on the weight after the H_2O_2 treatment.

ponential factor and activation energy of K desorption from soils merit increasing attention.

Besides mineral colloids, organic matter is a major contributor of CEC in soil systems. The average contribution of organic matter to the CEC of A horizons ranges from 14 to 50% (Thompson et al., 1989). Although selectivity of organic matter for divalent cations generally is higher than that for monovalent ones, unusual selectivity is displaced by certain organic compounds (e.g., valinomycin) for monovalent cations such as the K^+ ion (Talibudeen, 1981). Strong adsorption of K by a specific fraction of soils organic matter could account for the high-affinity K adsorption attributed to soil organic matter by Evangelou and Belvins (1988) among others. Soil organic matter may assume three-dimensional configurations that are K selective, as do enzymes (Evans and Wildes, 1971) and ionophores (Bruening et al., 1990). Evidence suggesting structural configurations selective for K^+ was found for fulvic acids (Gamble, 1973) and for humic acids (Bonn and Fish, 1993). However, little was known on the influence of organic matter on kinetics and mechanisms of K dynamics in $^+$ soils. Wang and Huang (2001) reported that at the end of the 30-s reaction, in which the adsorption is too fast for one to determine the rate coefficients, the amount of K adsorbed by the H_2O_2-treated soils was 0.5 to 47 mg kg^{-1}, compared with 158 to 363 mg kg^{-1} for the untreated soils (Table 4–11). The data indicated that the removal of organic matter results in the decrease of K adsorption by the soils by 6 to 316 times in the first 30 s. The rate coefficients of K adsorption of the soils in the reaction period of 30 to 120 s decreased by 1.5 to more than three times after the removal of the organic matter; the decrease in magnitude of the rate coefficient in 120 to 160 s was by 36% at most (Wang and Huang, 2001). The accessibility for K ions is expected to decrease after the removal of organic matter since the majority of the exchange sites of the soils after the removal of organic matter are located in the internal surface of smectites and vermiculites in the system and are relatively difficult for K to reach. Rich (1964) indicated that the diffusion path is tortuous for monovalent ions towards the exchange sites in the internal surface of these minerals. In contrast, the organic matter model of Stevenson (1994) suggested that ion exchange sites on the surface of organic matter are relatively easy to access because they are basically external sites. The fact that the organic matter affects the rate of K adsorption in the earlier reaction period more than during the later period supports the above reasoning, since initial K adsorption should take place on those sites that are relatively more accessible to K. Possible side effects brought about by the H_2O_2 treatment of soils tend to increase the rate of K adsorption (Wang and Huang, 2001). However, the data show the opposite trend. This provides further evidence that compared with mineral components soil organic matter has a faster rate of K adsorption.

On the other hand, the mobile humic acid (MHA) may reduce K fixation and result in greater total extractable K and highly labile K as well as greater plant K uptake (Olk and Cassman, 1995). Young N-rich soil organic matter fractions such as MHA may play an important role in governing K availability in soils with high K-fixing potential.

MOVEMENT OF POTASSIUM IN SOIL ENVIRONMENTS

Potassium in the soil solution can be leached, adsorbed and fixed by soil particles, or taken up by plants (Fig. 4–1). Physical, chemical, and biological processes may thus affect the movement of soil K^+. Factors that influence the movement of K^+ in soils include CEC, soil acidity and liming, mode and rate of application of potash fertilizers, and K^+ uptake by plants (Terry and McCants, 1968; Sparks, 1980, 2000b).

The retention of applied K by a soil is very much dependent on the CEC of the soil. Therefore, the extent of leaching of K^+ is strongly influenced by the nature and amount of mineral colloids and organic matter of the soil. Soils with higher CEC have a greater ability to retain applied K; in contrast, leaching of K^+ is often a problem in sandy soils (Sparks, 2000b).

Application of lime often enhances K^+ retention in sandy, Atlantic Coastal Plain soils because the CEC of such variable charge soils is increased as soil pH is increased. Liming a Lakeland fine sand (thermic, coated Typic Quartzipsamment) to pH 6 to 6.5 resulted in maximum retention of applied K (Nolan and Pritchett, 1960). At higher rates of limestone application, K^+ was replaced by Ca^{2+} on the cation exchange sites of the soil. At pH 6 to 6.5, less leaching of K^+ occurred because of enhanced substitution of K^+ for Ca^{2+} than for Al^{3+}, which was more abundant at low pH. Leaching of K^+ occurred on unlimed areas but not on limed areas when 112 to 224 kg of K ha^{-1} was applied on a Eustis loamy fine sand (sandy, siliceous, thermic Psammentic Paleudult) (Lutrick, 1963).

Movement of applied K in soil is related to the mode of potash application. Nolan and Pritchett (1960) investigated banded and broadcast placement of KCl applied at several rates to an Arredondo fine sand (loamy, siliceous, hyperthermic Grossarenic Paleudult) in lysimeters under winter and summer crops. Cumulative K^+ removal for both placements was only about 5.0 kg K ha^{-1} for the low rate of application.

The relationship of crop uptake and rate of K application to leaching of K has been determined in several studies (Sparks, 2000b). In a field experiment conducted by Jackson and Thomas (1960), K was applied at several rates prior to planting sweet potatoes (*Ipomoea batatas* L.) on a Norfolk sandy loam (fine-loamy, siliceous, thermic Typic Paleudult). At the 131 and 262 kg of K ha^{-1}, soil and plant K exceeded applied K at harvest time. However, 38 kg of K was unaccounted for by soil and plant K at the 524 kg of K ha^{-1} rate. This loss of K was apparently due to leaching below sampling depths. Sparks et al. (1980a) reported that at the 83 and 249 kg of applied K ha^{-1} rate during a 2-yr study with corn (*Zea mays* L.) on two Dothan soils of Virginia, the exchangeable K^+ in the E and B21t horizons of the two soils was increased. This was also attributed to leaching of applied K in soil environments. The extent of leaching is related to the rate of potash application.

CHEMISTRY OF LABILE POOL OF SOIL POTASSIUM

Quantity/Intensity Relationships of Soil Potassium

Schofield (1947) proposed that the ratio of the activity of cations such as K^+ and Ca^{2+} was defined by the relation $a_K/(a_{Ca})^{1/2}$ where a is the ion activity. He appears to be the first person who has applied the concepts of quantity (Q) and intensity (I) to the mineral nutrient status of soils (Schofield, 1955). Following a consideration of the Ratio Law (Schofield, 1947), Beckett (1964a) suggested that the I of K in a soil at equilibrium with its soil solution could best be defined by the ratio $a_K/(a_{Ca} + a_{Mg})^{1/2}$ of the soil solution. This equilibrium activity ratio for K is referred to as AR^K (Beckett, 1964a) and has often been used as a measure of K^+ availability (Sumner and Marques, 1966; Le Roux and Sumner, 1968a; Evangelou et al., 1994; Sparks, 2000b).

The exchangeable K^+ appears to be held by two distinct mechanisms (Beckett, 1964c). The majority of exchangeable K^+ is held by general force fields comparable to those that retain exchangeable Ca^{2+} or Na^+. A small fraction of exchangeable K^+ is retained at sites that have a specific binding for K^+ but not for Ca^{2+} and Mg^{2+}. The electrochemical potential of exchangeable K^+ in the diffuse double layer dictates the chemical potential of K^+ in the soil solution. The K^+ activity is also affected by the difference in electrical potential across the diffuse double layer surrounding the exchange complex. Therefore, there is no simple relationships between the activity of K^+ in the soil solution and quantity of K^+ on the exchange complex (San Valentin et al., 1973). A soil with a given complement of exchangeable K^+, Ca^{2+}, and Mg^{2+} gives rise to an equilibrium activity ratio for K(AR^K) in the soil solution that is characteristic of that soil and independent of the soil/solution ratio and total electrolyte concentration. The ratio depends only on K^+ saturation and strength of adsorption of cations (Moss, 1967).

Soils showing the same value of AR^K may not possess the same capacity for maintaining AR^K while K is removed by plant uptake (Beckett, 1964b). For this reason, one must include not only the current potential of K in the labide pool but also the way in which the potential depends on quantity of labile K present. These findings brought about the classic Q/I relationships where the activity ratio AR^K value is related to the change in exchangeable K to obtain the effect of quantity of labile K (exchangeable) on intensity of labile K (AR^K). The Q/I concept has been widely disseminated in the soil literature in studying soil K status (Evangelou et al., 1994, Sparks, 2000b). For the Q/I relationships to be valid in indicating the amount of soil K available for plant uptake during the growing period, they must be unaffected by the amount of K normally released, fixed, or added during the growing season. These assumptions have been proved to be valid (Matthews and Beckett, 1962; Beckett et al., 1966; Beckett and Nafady, 1967; Nafady and Lamm, 1973).

The traditional method for constructing a typical *Q*/*I* curve involves equilibrating a soil with solutions containing a constant amount of $CaCl_2$ and increasing the amount of KCl (Beckett, 1964a). The soil gains or loses K to achieve the characteristic AR^K of the soil or remains unchanged if its AR^K is the same as the equilibrating solution. The AR^K values are calculated from the measured concentrations of K^+, Ca^{2+}, and Mg^{2+} converted to their chemical activities by application of the extended Debye–Hückel equation (Sparks and Liebhardt, 1981). The AR^K values are then plotted vs. the gain or loss of K to form the characteristic *Q*/*I* curve (Fig. 4–21). Several parameters can be obtained from the *Q*/*I* plot to characterize soil K status. The AR^K when the *Q* factor (ΔK) equals zero is a measure of the degree of K availability at equilibrium or AR^K_e. The value of ΔK when AR^K equals zero is a measure of labile or exchangeable K in soils or ΔK°. The slope of the linear portion of the curve gives the potential buffering capacity of K (PBC^K) and is proportional to the CEC of the soil. The number of specific sites for K (K_x) is the difference between the intercept of the curve and the linear portion of the *Q*/*I* plot at $AR^K = 0$ (Beckett, 1964b; Sparks and Leibhardt, 1981; Evangelou et al., 1994).

Advances in ion selective electrode (ISE) technology have allowed for more rapid *Q*/*I* analysis (Evangelou et al., 1994). An ISE simplified *Q*/*I* method whereby a single K-ISE in an electrochemical cell with liquid junction was used to determine the K concentration (C_K) in equilibrated soil suspensions (Para and Torrent, 1983). The values of AR^K were estimated based on the expression $AR^K = (11.5 - 0.3b)C_K + 22 \times 10^{-6}$, where b is the CEC ($cmol_c kg^{-1}$) based on the weight of the soil samples used. The AR^K values calculated by this method were comparable to those determined with the traditional *Q*/*I* method. Wang et al. (1988) modified the procedure of Parra and Torrent (1983) by making direct determinations of CR^K [concentration ratio = $C_K/(C_{Ca+Mg})^{1/2}$] values with Ca- and K-ISEs in an electrochemical cell with and without liquid junction.

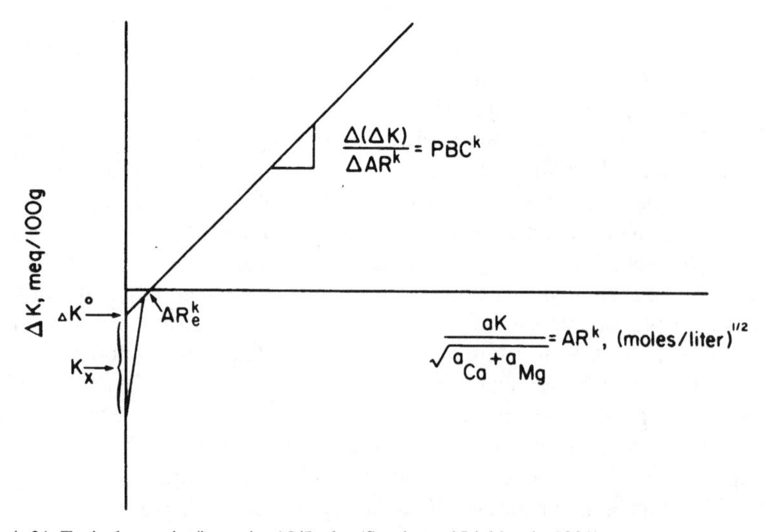

Fig. 4–21. Typical quantity/intensity (*Q*/*I*) plot (Sparks and Liebhardt, 1981).

Various interpretations and applications have been made on the parameters that can be derived from the *Q/I* plot. Below an AR^K of 0.001 $(mol\ L^{-1})^{1/2}$, the bulk of K is adsorbed at interlattice positions, between AR^K values of 0.001 and 0.01 $(mol\ L^{-1})^{1/2}$ at edge positions, and >0.01 $(mol\ L^{-1})^{1/2}$ at planar positions (van Schouwenburg and Schuffelen, 1963; Sparks and Liebhardt, 1981). A small percentage of edge positions may be filled at AR^K values up to 0.1 $(mol\ L^{-1})^{1/2}$.

The linear portion of the *Q/I* curve is a particular case of the Gapon equation (Le Roux and Sumner, 1968a; Sparks and Huang, 1985). The linear section of the *Q/I* relationship gives a Gapon selectivity coefficient (k_G) value exhibiting a preference for K over Ca plus Mg of 2.7; the curved portion, converted also into a Gapon-type relationship, gives a k_G value of 600 (Bolt, 1964). The former value is similar to that found for planar site exchange (van Schouwenburg and Schuffelen, 1963) whereas the latter is similar to that for edge–interlattice exchange (Bolt et al., 1963).

Different soils having the same AR_e^K do not have the same capacity for maintaining AR_e^K while K is being removed by plant roots (Le Roux, 1966). Soils can have the same AR_e^K but contain different amounts of labile K. Potassium fertilization, liming, and cropping can change AR_e^K values (Beckett, 1964b; Le Roux, 1966; Le Roux and Sumner, 1968b; San Valentin et al., 1973; Sparks and Liebhardt, 1981).

The ΔK° values are a better estimate of soil labile K than normal exchangeable K (Le Roux, 1966). Higher values of labile K ($-\Delta K^\circ$) indicate a greater K^+ release into the soil solution, resulting in a larger pool of labile K. The ΔK° becomes more negative or the labile K pool increases with K fertilization (Le Roux and Sumner, 1968b; San Valentin et al., 1973; Sparks and Liebhardt, 1981) and with liming on cropped soils (San Valentin et al., 1973; Sparks and Liebhardt, 1981).

The PBC^K indicates the ability of a soil to maintain the intensity of K^+ in the soil solution. A high value of PBC^K is indicative of constant availability of K^+ in the soil for a long period. In contrast, a low PBC^K would indicate the need for frequent fertilization. However, Sparks and Leibhardt (1981) reported that crops grown on a Kalmia soil in the Delaware Coastal Plain had not responded to potash fertilization (Liebhardt et al., 1976), even though this soil had low PBC^K values. This lack of response to applied K was attributed to the presence of the large amount of K-bearing minerals in the soil that with time could release K^+ to exchangeable and solution forms (Sparks, 1980; Sparks et al., 1980a). Cropping and liming can change the PBC^K values of soils (Bradfield, 1969; San Valentin et al., 1973).

Specific sites for K(K_x) in a Kalmia soil from the Delaware Coastal Plain tended to increase with increasing K fertilization and liming (Sparks and Liebhardt, 1981). The Kalmia soil contains hydroxy-Al interlayered vermiculite. With liming, the size or total number of interlayer islands could increase. Therefore, K_x could be influenced by liming due to steric factors and cation distribution.

Q/I parameters are affected by temperature (Sparks and Liebhardt, 1982). The amount of K^+ adsorbed by three Delaware soils decreased as temperature increased. Therefore, the K^+ adsorption process in the soils appeared to be exothermic. As temperature increased, ΔK decreased while AR^K increased for similar initial electrolyte concentrations. The ΔK° values changed little with temperature, whereas the PBC^K values decreased as temperature increased.

Potassium Chemistry in the Soil Rhizosphere

The term *rhizosphere* was first used by Hiltner (1904) but has since been modified and redefined. It is the narrow zone of soil influenced by the root and exudates. The extent of the rhizosphere may vary with soil type, plant species, age, and many other factors (Curl and Truelove, 1986), but it is usually considered to extend from the root surface out into the soil for a few millimeters. The rhizosphere can be divided into the ectorhizosphere, which is the soil region, and the endorhizosphere, which includes the rhizosphere and the epidermis and cortical cells of the root that are invaded by microorganisms (Fig. 4–22). The ectorhizosphere can extend a substantial distance from the root with the development of mycorrhizal fungal associations (Lynch, 1990).

Microbial populations in the rhizosphere can be 10 to 100 times larger than the populations in bulk soil (Sposito and Reginato, 1992). Therefore, the rhizosphere is bathed in root exudates and microbial metabolites. Chemistry of soil K in the rhizosphere should thus differ significantly from that in bulk soil.

Plant roots release a wide variety and considerable amounts of organic compounds into the rhizosphere. On average, between 30 and 60% of the net photosynthetic C is allocated to the roots in annual species; of this C, an appreciable portion (4–70%) is released as organic compounds into the rhizosphere (Lynch and Whipps, 1990; Liljeroth et al., 1994). Over the whole vegetation period in soil-grown crop species, more than twice as much organic C is released into the rhizosphere than retained in the root system at harvest (Sauerbeck et al., 1981). The major components of rhizodeposition are sloughed-off cells and their lysates, mucilage, and low molecular weight organic compounds. Sloughed-off cells and their lysates are primarily a C substrate for rhizosphere microorganisms. Mucilage consists primarily of polysaccharides, such as polygalacturonic acids (Rovira et al., 1979). Low molecular weight organic compounds are organic acids, amino acids, sugars, and phy-

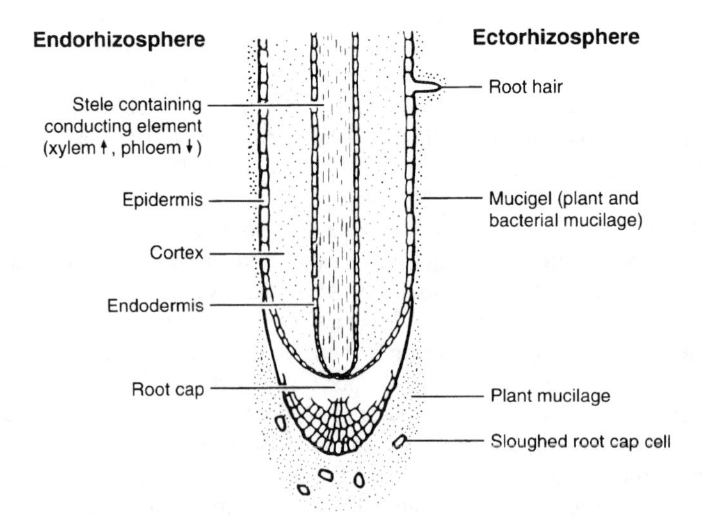

Fig. 4–22. Root region (Lynch, 1983).

Table 4–12. Amount of low molecular weight organic acids in rhizosphere soils of durum wheat (*Triticum durum* Desf.) cv. Kyle grown in three different soils as determined by gas chromatography (Szmigielska et al., 1997).

Acid	Soil†		
	Yorkton	Sutherland	Waitville
	μg kg^{-1} soil		
Malonic	99a‡	56a	68a
Succinic	22a	35476c	10826b
Fumaric	12a	150b	71ab
Malic	45a	898c	370b
Tartaric	ND§	665b	214a
trans-Aconitic	ND	13a	3a
Citric	ND	195b	81a
Acetic	865a	29245c	12240b
Propionic	ND	499a	ND
Butyric	ND	7604b	2127a
Total	1043a	74801c	26000b

† The soil samples were taken from rhizosphere soils of the Yorkton (Mollisol), Sutherland (Mollisol), and Waitville (Luvisol) soils in Saskatchewan, Canada.
‡ The data within the same row having the same letter are not significantly different ($p \leq 0.05$).
§ ND = not detected.

tosiderophores (Marschner, 1998). Both the amounts and proportion of the various organic compounds of root exudates vary substantially with plant species and cultivars. Further, the same plant cultivar grown in different soils varies in the kind and amount of low molecular weight organic acids present in the rhizosphere (Table 4–12). The impact of plant species and cultivars and their interactions with soil properties on the chemistry of biomolecules in the rhizosphere and the dynamics of soil K remains to be uncovered.

Root exudation is increased by various forms of stress, such as mechanical impedance and nutrient deficiency (Lynch and Whipps, 1990; Marschner, 1998). Schematic presentation of root exudation as affected by nutrient deficiency and mechanical impedance is shown in Fig. 4–23. Root exudation is much higher from roots growing in solid substrates, such as quartz sand, glass beads, or soils than in nutrient solution. Root length is usually strongly negatively related to the bulk density of the substrate. Therefore, allocation of photosynthates to roots is enhanced by mechanical impedance; the consumption of photosynthates for respiration and exudation per unit root length may increase by a factor of two in soils with high bulk density as compared with low bulk density (Sauerbeck and Helal, 1986). Mechanical impedance increases root exudation of sugars by a factor of 3 (Schönwitz and Ziegler, 1982) and of phenolics by a factor of 10 (D'Arcy, 1982). However, the impact of various forms of stress on the relationship between root exudation and K dynamics remains obscure.

Biological activity promotes K^+ release from micas and the concomitant formation of vermiculites (Mortland et al., 1956; Boyle et al., 1967; Weed et al., 1969; Sawhney and Voight, 1969; Hinsinger and Jaillard, 1993; Marschner, 1995). Plant roots deplete the K of the soil solution, and their action may be linked to that of tetraphenylboron in artificial weathering of micas. Furthermore, the overall action

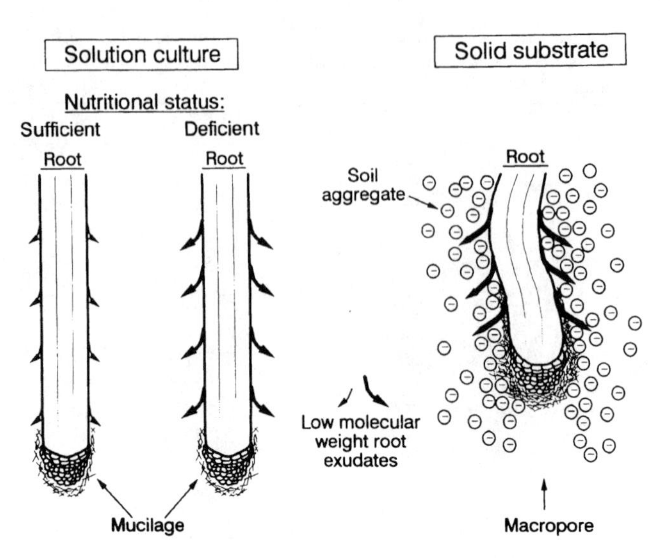

Fig. 4–23. Schematic presentation of root exudation as affected by mineral nutrient deficiency and mechanical impedance (Marschner, 1995). The size of the arrows indicates the extent of exudation.

of biological activity is more complex when organic acids are produced in root exudates and microbial metabolites in the rhizosphere. The influence of organic acids on the dynamics of K^+ release from K-bearing minerals has been observed (Song and Huang, 1988). The effect of organic acids in releasing K is attributed to the dissociated H^+ ions and complexing organic ligands. The weakening of the metal–O bonds by protonation of surface OH groups and the formation of inner sphere surface complexes with organic ligands (Stumm et al., 1985; Stumm and Wallast, 1990) certainly enhances mineral dissolution. The release of K is thus affected by the kinds of organic acids. In the study of Song and Huang (1988), concentrations of dissociated organic ligands and H^+ ions in 0.01 *M* oxalic acid solution were about four and three times, respectively, higher than those in 0.01 *M* citric acid solution because of the difference of their p*K* values. Therefore, at the same concentrations, oxalic acid releases much more K^+ from the K-bearing minerals than citric acid (Fig. 4–24). The organic ligands and H^+ ions are consumed during the reaction processes. At the same time, the concentrations of cations released from the mineral structure to the solution is increased. The renewal of the extracting solution replenishes organic ligands and H^+ ions as well as decreases the concentration of the structural ions released in the solution. Therefore, the rate of K^+ release is increased and sustained upon renewal of the extracting solution. The role of organic acids in the soil rhizosphere in affecting the chemistry of labile pool of soil K, thus, merits close attention.

SUMMARY AND CONCLUSIONS

The chemistry of soil K is related to forms of soil K, the nature of soil components involved, chemical equilibria and kinetics, and the labile pool of soil K.

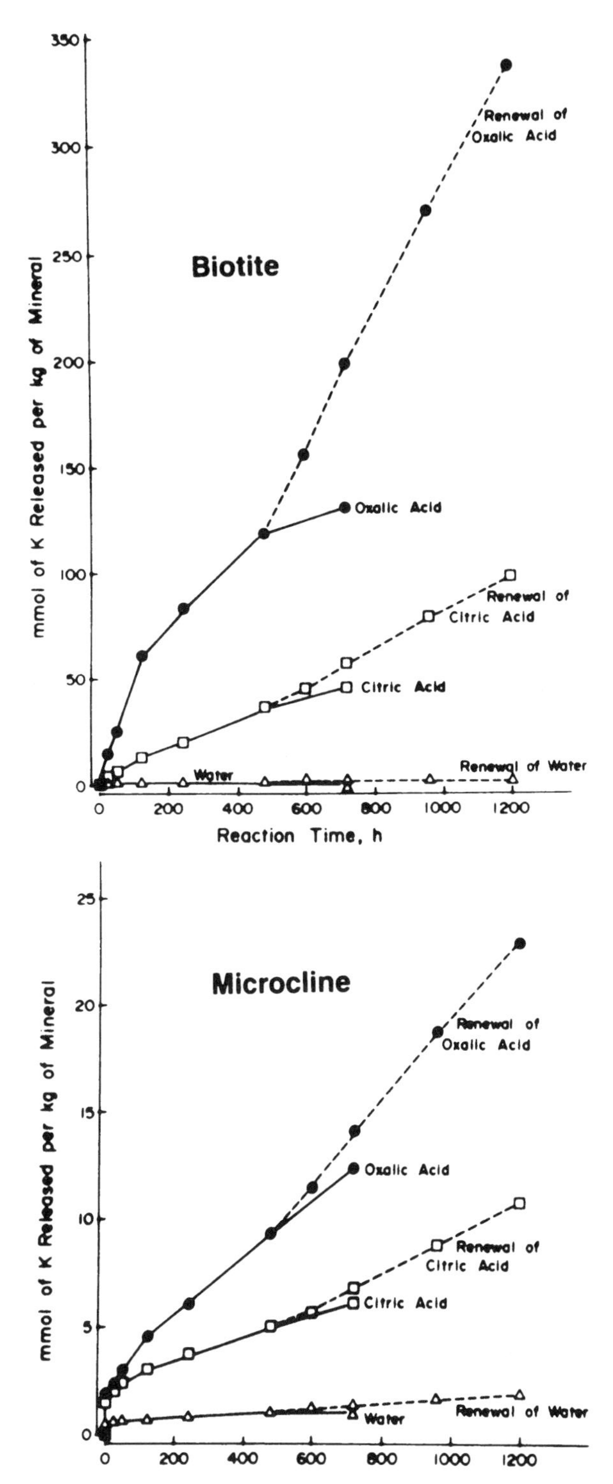

Fig. 4–24. Release of K from K-bearing minerals to water and 0.01 *M* oxalic or citric acid solution during renewal treatments (Song and Huang, 1988). (Continued on next page.)

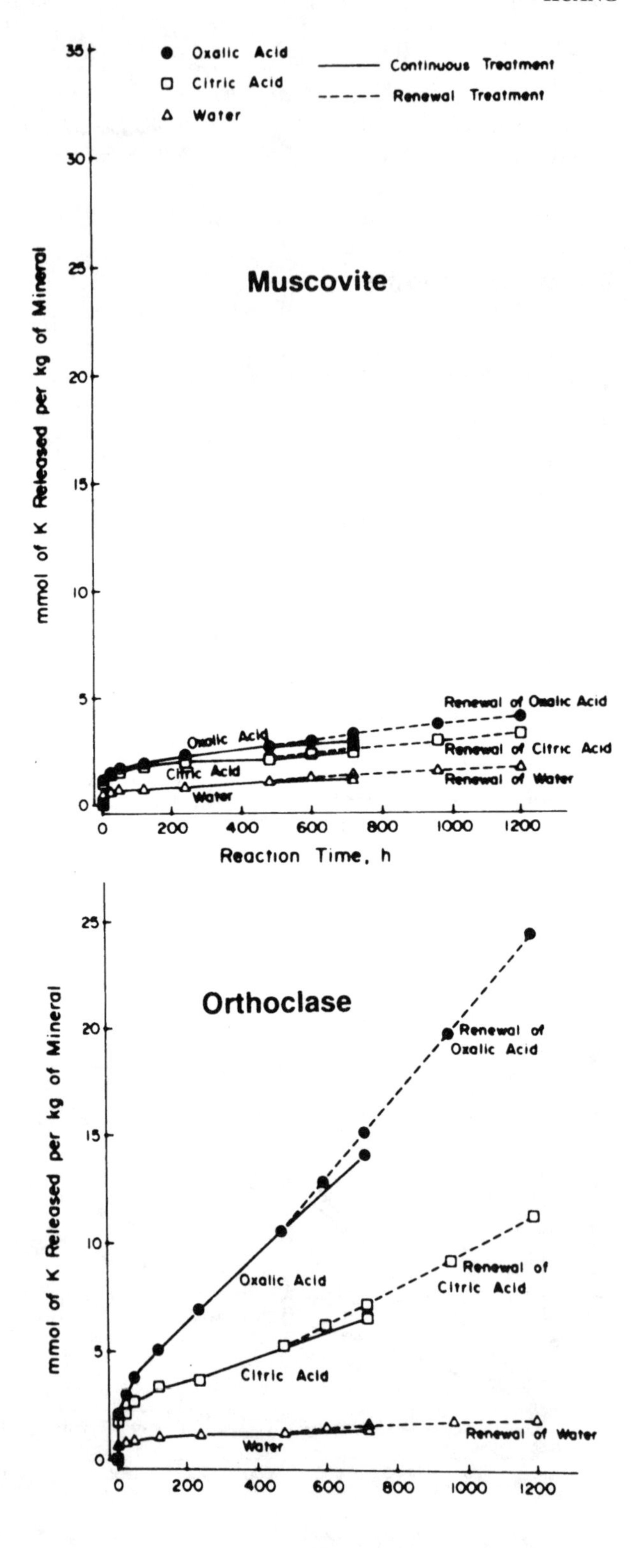

Fig. 4–24. Continued.

Much fundamental research work has been conducted on pure clay mineral systems. More research should be conducted on the effects of the degree of weathering of K-bearing minerals and the resultant alteration of structural and surface properties of those minerals on the kinetics and mechanisms of the transformation and labile pool of soil K.

Scant attention has been paid to the role of organic matter and organomineral complexes in influencing soil K behavior. Evidence suggesting structural configurations selective for K^+ has been shown for fulvic acids and humic acids. Strong adsorption of K by a specific fraction of soil organic matter, which apparently has three-dimensional configurations that are K selective, would account for the high-affinity K adsorption attributed to organic matter of some soils. In contrast, the mobile humic fraction may reduce K fixation and thus enhance labile K. The impact of cropping systems on the nature of soil organic matter and related soil surface chemistry pertaining to the dynamics and mechanisms of soil K transformation merits in-depth research. More research effort should be devoted to investigate K reactions in natural field conditions in a wide range of pedogenic environments under the influence of natural vegetation and cropping systems.

Little is known about rhizosphere chemistry of soil K. The rhizosphere is the bottleneck of K supply from soil to plant. The transformations, dynamics, and bioavailability of soil K should be substantially influenced by chemistry of soil rhizosphere. The rhizosphere chemistry of soil K as influenced by plant species and cultivars, soil properties, and land management practices merits close attention.

ACKNOWLEDGMENTS

This study was supported by the Natural Sciences and Engineering Research Council of Canada and the Potash and Phosphate Institute of Canada.

REFERENCES

Adams, F., and Z. Rawajfih. 1977. Basaluminite and alunite: A possible cause of sulfate retention by acid soils. Soil Sci. Soc. Am. J. 41:686–692.

Adamson, A.W. 1979. A textbook of physical chemistry. Academic Press, New York.

Ardakani, M.S., and A.D. McLaren. 1977. Absence of local equilibrium during ammonium transport in a soil column. Soil Sci. Soc. Am. J. 41:877–879.

Argersinger, W.J., Jr., A.W. Davidson, and O.D. Bonner. 1950. Thermodynamics and ion exchange phenomena. Trans. Kans. Acad. Sci. 53:404–410.

Assimakopoulos, J.H., N.J. Yassoglou, and C.P. Bovis. 1994. Effects of incubation at different water contents, air-drying and K-additions on potassium availability of a Vertisol sample. Geoderma 61:223–236.

Atschuler, Z.S., E.J. Dworrik, and H. Kramer. 1963. Transformation of montmorillonite to kaolinite during weathering. Science 141:148–152.

Attoe, O.J. 1947. Potassium fixation and release in soils occurring under moist and dry conditions. Soil Sci. Soc. Am. Proc. 11:145–149.

Bailey, S.W. 1980. Structures of layer silicates. p. 2–124. *In* G.W. Brindley and G. Brown (ed.) Crystal structures of clay minerals and their x-ray identification. Monogr. 5. Mineralogical Society, London.

Bailey, S.W. (ed.) 1984. Micas. Reviews in Mineralogy Vol. 13. Mineralogical Society of America, Washington, DC.

Baldock, J.A., and P.N. Nelson. 2000. Soil organic matter. p. B25–B84. *In* M.E. Sumner (ed.) Handbook of soil science. CRC Press, Boca Raton, FL.

Bannister, F.A., and G.E. Hutchinson. 1947. The identity of minervite and palmerite with taranakite. Min. Mag. J. Min. Soc. 28:31–35.

Bardossy, G. 1982. Karst bauxites. Elsevier Sci. Publ. Co., Amsterdam.

Barnhisel, R.I., and P.M. Bertsch. 1989. Chlorites and hydroxy-interlayered vermiculite and smectite. p. 729–788. *In* J.B. Dixon and S.B. Weed (ed.) Minerals in soil environments. 2nd ed. SSSA Book Ser. 1. SSSA, Madison, WI.

Barshad, I. 1951. Cation exchange in soils: I. Ammonium fixation and its relation to potassium fixation and to determination of ammonium exchange capacity. Soil Sci. 77:463–472.

Barshad, I. 1954. Cation exchange in micaeous minerals: II. Replaceability of ammonium and potassium from vermiculite, biotite, and montmorillonite. Soil Sci. 78:57–76.

Barshad, I., and F.M. Kishk. 1968. Oxidation of ferrous iron in vermiculite and biotite alters fixation and replaceability of potassium. Science 162:1401–1402.

Barth, T.W.F. 1969. Feldspars. Wiley-Interscience, New York.

Bassett, W.A. 1959. The origin of the vermiculite deposit at Libby, Montana. Am. Mineral. 44:282–299.

Bassett, W.A. 1960. Role of hydroxyl orientation in mica alteration. Geol. Soc. Am. Bull. 71:449–456.

Bates, T.E., and A.D. Scott. 1964. Changes in exchangeable potassium observed on drying soils after treatment with organic compounds: I. Release. Soil Sci. Soc. Am. Proc. 28:769–772.

Beckett, P.H.T. 1964a. Studies of soil potassium: I. Confirmation of the ratio law: Measurement of potassium potential. J. Soil Sci. 15:1–8.

Beckett, P.H.T. 1964b. Studies on soil potassium: II. The "immediate" *Q/I* relations of labile potassium in the soil. J. Soil Sci. 15:9–23.

Beckett, P.H.T. 1964c. K–Ca exchange equilibria in soils: Specific sorption sites for K. Soil Sci. 97:376–383.

Beckett, P.H.T., and M.H. Nafady. 1967. Studies on soil potassium: VI. The effect of K-fixation and release on the form of the K: (Ca + Mg) exchange isotherm. J. Soil Sci. 18:244–262.

Beckett, P.H.T., J.B. Craig, M.H.M. Nafady, and J.P. Watson. 1966. Studies in soil potassium: V. The stability of *Q/I* relations. Plant Soil 25:435–455.

Bernal, J.D., and R.H. Fowler. 1933. A theory of water and ionic solution, with particular reference to hydrogen and hydroxyl ions. Chem. Phys. 1:515–548.

Berry, L.G., and B. Mason. 1959. Mineralogy. W.H. Freeman and Co., San Francisco, CA.

Blum, A.E., and L.L. Stillings. 1995. Feldspar dissolution kinetics. p. 291–351. *In* A.F. White and S.L. Brantley (ed.) Chemical weathering rates of silicate minerals. Reviews in Mineralogy Vol. 31. Mineralogical Society of America, Washington, DC.

Boettcher, A.L. 1966. Vermiculite, hydrobiotite and biotite in the Rainy Creek igneous complex near Libby, Montana. Clay Miner. 6:283–296.

Bolt, G.H. 1964. Potassium–calcium exchange in soils: Specific adsorption sites of potassium. Soil Sci. 97:376–383.

Bolt, G.H., M.E. Sumner, and A. Kamphorst. 1963. A study between three categories of potassium in an illitic soil. Soil Sci. Soc. Am. Proc. 27:294–299.

Bondam, J. 1967. Structural changes in adularia in hydrolytic environments. Bull. Geol. Soc. Den. 17:357–370.

Bonn, B.A., and W. Fish. 1993. Measurement of electrostatic and site-specific associations of alkali metal cations with humic acid. J. Soil Sci. 44:335–345.

Borchardt, G. 1989. Smectites. p. 675–727. *In* J.B. Dixon and S.B. Weeds (ed.) Minerals in soil environments. 2nd ed. SSSA Book Ser. 1. SSSA, Madison, WI.

Bouabid, R., M. Badraoui, and P.R. Bloom. 1991. Potassium fixation and charge characteristics of soil clays. Soil Sci. Soc. Am. J. 55:1493–1498.

Boyle, J.R., G.K. Voight, and B.L. Sawhney. 1967. Biotite flakes: Alteration by chemical and biological treatment. Science 155:193–195.

Bradfield, E.G. 1969. Quantity/intensity relations in soils and the potassium nutrition of the strawberry plant (*Fragoria* sp.). J. Sci. Food Agric. 20:32–38.

Bragg, L., and G.F. Claringbull. 1965. The crystalline state. Vol. IV. Crystal structures of minerals. G. Bell & Sons, London.

Brown, G. 1953. The dioctahedral analogue of vermiculite. Clay Miner. Bull 2:64–69.

Bruening, R.L., R.M. Izatt, and J.S. Bradshaw. 1990. Understanding cation-macrocycle binding selectivity in single-solvent, extraction, and liquid membrane systems by quantifying thermodynamic interactions. p. 111–132. *In* Y. Inoue and G.W. Gokel (ed.) Cation binding by macrocycles: Complexation of cationic species by crown ethers. Marcel Dekker, New York.

Buol, S.W., F.D. Hole, and R.J. McCracken. 1980. Soil genesis and classification. 2nd ed. Iowa State University Press, Ames.

Burns, A.F., and S.A. Barber. 1961. Effect of temperature and moisture on exchangeable potassium. Soil Sci. Soc. Am. Proc. 25:349–352.

Burns, A.F., and J.L. White. 1963. The effect of potassium removal in the b-dimension of muscovite and dioctahedral soil micas. p. 9–16. *In* Proc. Int. Clay Conf., Stockholm. MacMillan, New York.

Casey, W.H., H.R. Westrich, G.W. Arnold, and J.F. Banfield. 1989. The surface chemistry of dissolving labradorite feldspar. Geochim. Cosmochim. Acta 53:2795–2807.

Caslovsky, J.L., and K. Vadam. 1970a. Examination of imperfect muscovite crystals by X-ray. J. Appl. Phys. 4:50–53.

Caslovsky, J.L., and K. Vadam. 1970b. Muscovite with isotropic and anisotropic elasticity in the basal plane. Am. Mineral. 55:1633–1638.

Chen, S.Z., P.F. Low, and C.B. Roth. 1987. Relation between potassium fixation and the oxidation state of octahedral iron. Soil Sci. Soc. Am. J. 51:82–86.

Churchman, G.J. 2000. The alteration and formation of soil minerals by weathering. p. F1–F76. *In* M.E. Sumner (ed.) Handbook of soil science. CRC Press, Boca Raton, FL.

Chute, J.H., and J.P. Quirk. 1967. Diffusion of potassium from mica-like materials. Nature (London) 213:1156–1157.

Cook, M.G., and T.B. Hutcheson, Jr. 1960. Soil potassium reactions as related to clay mineralogy of selected Kentucky soils. Soil Sci. Soc. Am. Proc. 24:252–256.

Correns, C.W., and W.V. Engelhardt. 1939. Neue Untersuchungen uber die Verwitterung des Kalifeldspates. Chem. Erde 12:1–22.

Cotton, F.A., and G. Wilkinson. 1980. Advanced inorganic chemistry. 4th ed. Interscience, New York.

Courchesne, F., M.-C. Turmel, and P. Beauchemin. 1996. Magnesium and potassium release by weathering in Spodosols: Grain surface coating effects. Soil Sci. Soc. Am. J. 60:1188–1196.

Cox, A.E., and B.C. Joern. 1997. Release kinetics of nonexchangeable potassium in soils using sodium tetraphenylboron. Soil Sci. 162:588–598.

Curl, E.A., and B. Truelove. 1986. The rhizosphere. Spinger-Verlag, Berlin.

D'Arcy, L.A. 1982. Etude des exsudats racinaires de soja et de lentils: I. Cinétique d'exsudation des composés phénoliques, des amino acides at des sucres, au cours de premiers jours de la vie des plantules. Plant Soil 68:399–403.

Decarreau, A. 1977. Etudes experimentales d'alteration en systeme ouvert de materiaux pologiques naturels. Approache experimentale des mechanismes d'alteration. Comportement geochimique des elements majeurs et en traces. Bull. Soc. Fr. Mineral. Cristallogr. 100:289–301.

Denbigh, K.G. 1966. The principles of chemical equilibrium with applications in chemistry and chemical engineering. 2nd ed. Cambridge University Press, New York.

Dickerson, R.E. 1969. Molecular thermodynamics. W.A. Benjamin, Menlo Park, CA.

Deist, J., and O. Talibudeen. 1967. Ion exchange in soils from ion pairs K–Ca, K–Rb, K–Na. J. Soil Sci. 18:125–137.

Dhillon, S.K., and K.S. Dhillon. 1990. Kinetics of release of non-exchangeable potassium by cation-saturated resins from red (Alfisols), black (Vertisols) and alluvial (Inceptisols) soils of India. Geoderma 47:283–300.

Dolcater, D.L., E.G. Lotse, J.K. Syers, and M.L. Jackson. 1968. Cation exchange selectivity of some clay-size minerals and soil materials. Soil Sci. Soc. Am. Proc. 32:795–798.

Douglas, L.A. 1989. Vermiculite. p. 635–674. *In* J.B. Dixon and S.B. Weed (ed.) Minerals in soil environments. 2nd ed. SSSA Book Ser. 1. SSSA, Madison, WI.

Dreher, P., and E.-A. Niederbudde. 1994. Potassium release from micas and characterization of the alteration products. Clay Miner. 29:77–85.

Evangelou, V.P., and R.L. Belvins. 1988. Effect of long-term tillage systems and nitrogen addition on potassium quantity–intensity relationships. Soil Sci. Soc. Am. J. 52:1047–1054.

Evangelou, V.P., J. Wang, and R.E. Phillips. 1994. New developments and perspectives on soil potassium quantity/intensity relationships. Adv. Agron. 52:173–227.

Evans, H.J., and R.A. Wildes. 1971. Potassium and its role in enzyme activation. p. 13–39. *In* potassium in biochemistry and physiology, 8th Colloq., Uppsala, Sweden. 14–17 June 1971. Int. Potash Inst., Berne, Switzerland.

Fanning, D.S., V.Z. Keramidas, and M.A. El-Desoky. 1989. Micas. p. 551–634. *In* J.B. Dixon and S.B. Weed (ed.) Minerals in soil environments. 2nd ed. SSSA Book Ser. 1. SSSA, Madison, WI.

Farmer, V.C., and J.D. Russell. 1967. Infrared absorption spectrometry in clay studies. Clays Clay Miner. 15:121–142.

Farmer, V.C., and J.D. Russell. 1990. Structures and genesis of allophanes and imogolite and their distribution in non-volcanic soils. p. 165–178. *In* M.F. De Boodt et al. (ed.) Soil colloids and their associations in aggregates. Plenum Press, New York.

Farmer, V.C., and M.J. Wilson. 1970. Experimental conversion of biotite to hydrobiotite. Nature (London) 226:841–842.

Feigenbaum, S., R. Edelstein, and I. Shainberg. 1981. Release rate of potassium and structural cations from micas to ion exchangers in dilute solutions. Soil Sci. Soc. Am. J. 45:501–506.

Feigenbaum, S., and R. Levy. 1977. Potassium release in some saline soils of Israel. Geoderma 19:159–169.

Fine, L.O., L.A. Bailey, and E. Truog. 1940. Availability of fixed potassium as influenced by freezing and thawing. Soil Sci. Soc. Am. Proc. 5:183–186.

Frazier, A.W., and A.W. Taylor. 1965. Characterization of taranakites and ammonium aluminum phosphates. Soil Sci. Soc. Am. Proc. 29:545–547.

Gaines, G.L., and H.C. Thomas. 1953. Adsorption studies on clay minerals: II. A formulation of the thermodynamics of exchange adsorption. J. Chem. Phys. 21:714–718.

Gamble, D.S. 1973. Na^+ and K^+ binding by fulvic acid. Can. J. Chem. 51:3217–3222.

Gardiner, W.C., Jr. 1969. Rates and mechanisms of chemical reactions. Benjamin, New York.

Gast, R.G. 1977. Surface and colloid chemistry. p. 27–73. *In* J.B. Dixon and S.B. Weed (ed.) Minerals in soil environments. SSSA Book Ser. 1. SSSA, Madison, WI.

Ghosh, B.N., and R.D. Singh. 2001. Potassium release characteristics of some soils of Uttar Pradesh hills varying in altitude and their relationship with forms of soil K and clay mineralogy. Geoderma 104:135–144.

Glasstone, S., K.J. Laider, and H. Eyring. 1941. Theory of rate processes. McGraw-Hill, New York.

Goulding, K.W.T. 1980. The thermodynamics of ion exchange adsorption in soils and soil clay minerals. Ph.D. diss. University of London, London.

Goulding, K.W.T. 1981. Potassium retention and release in Rothamsted and Saxmundhum soils. J. Sci. Food Agric. 32:617–670.

Goulding, K.W.T 1983. Adsorbed ion activities and other thermodynamic parameters of ion exchange defined by mole or equivalent fractions. J. Soil Sci. 34:69–74.

Goulding, K.W.T., and O. Talibudeen. 1979. Potassium reserves in a sandy clay soil from the Saxmundhum experiment: Kinetics and equilibrium thermodynamics. J. Soil Sci. 30:291–302.

Goulding, K.W.T., and O. Talibudeen. 1980. Heterogeneity of cation-exchange site for K–Ca exchange in aluminosilicates. J. Colloid Interface Sci. 78:15–24.

Grim, R.E., R.H. Bray, and W.F. Bradley. 1937. The mica in argillaceous sediments. Am. Minerol. 22:813–829.

Güven, N. 1971. Structural factors controlling stacking sequences in dioctahedral micas. Clays Clay Miner. 19:159–166.

Haby, V.A., M.P. Russelle, and E.O. Skogley. 1990. Testing soil for potassium, calcium, and magnesium. p. 181–228. *In* R.L. Westerman (ed.) Soil testing and plant analysis. 3rd ed. SSSA Book Ser. 3. SSSA, Madison, WI.

Hanway, J.J., and A.D. Scott. 1957. Soil potassium–moisture relations: II. Profile distribution of exchangeable K in Iowa soils as influenced by drying and rewetting. Soil Sci. Soc. Am. Proc. 20:501–504.

Hanway, J.J., and A.D. Scott. 1959. Soil potassium–moisture relations: III. Determining the release in exchangeable soil potassium on drying soils. Soil Sci. Soc. Am. Proc. 23:22–24.

Harsh, J. 2000. Poorly crystalline aluminosilicate clays. p. F169–F182. *In* M.E. Sumner (ed.) Handbook of soil science. CRC Press, Boca Raton, FL.

Hauter, R., and K. Mengel. 1988. Measurement of pH at the root surface of red clover (*Trifolium pratense*) grown in soils differing in proton buffer capacity. Biol. Fertil. Soil. 5:295–298.

Havlin, J.L., and D.G. Westfall. 1985. Potassium release kinetics and plant response in calcareous soils. Soil Sci. Soc. Am. J. 49:366–370.

Havlin, J.L., D.G. Westfal, and S.R. Olsen. 1985. Mathematical models for potassium release kinetics in calcareous soils. Soil Sci. Soc. Am. J. 49:371–376.

Helfferich, F. 1962. Ion exchange. McGraw-Hill Book Co., New York.

Helfferich, F. 1983. Ion exchange kinetics—Evolution of a theory. p. 159–179. *In* L. Liberti and F. Helfferich (ed.) Mass transfer and kinetics of ion exchange. NATO AS1 Series, Series E: Appl. Sci. 71. Kluwer, Amsterdam.

Hellman, R., C.H. Egglestone, H.F. Hochelle, Jr., and D.A. Crerar. 1990 The formation of leached layers on albite surfaces during dissolution under hydrothermal conditions. Geochim. Cosmochim. Acta 54:1267–1282.

Helmke, P.A. 2000. The chemical composition of soils. p. B3–B24. *In* M.E. Sumner (ed.) Handbook of soil science. CRC Press, Boca Raton, FL.

Helmke, P.A., and D.L. Sparks. 1996. Lithium, sodium, potassium, rubidium, and cesium. p. 551–574. *In* D.L. Sparks (ed.) Method of soil analysis. Part 3. SSSA Book Ser. 5. SSSA, Madison, WI.

Henderson, J.H., H.E. Doner, R.M. Weaver, J.K. Syers, and M.L. Jackson. 1976. Cation and silica relationships of mica weathering to vermiculite in calcareous Harp soils. Clays Clay Miner. 24:93–100.

Hiltner, L. 1904. Uber neuere Erfahrungen und Probleme auf dem Gebiet der Bodenbacteriologie und unter besonderer Berucksichtigung und Brache. Arb. Dtsch. Landwirt. Ges. 98:59–78.

Hinsinger, P., and B. Jaillard. 1993. Root-induced release of interlayer potassium and vermiculitization of phlogopite as related to potassium depletion in the rhizosphere of ryegrass. J. Soil Sci. 44:525–534.

Högfeldt, E. 1953. On ion exchange equilibria: II. Activities of the components in ion exchangers. Ark. Kemi. 5:147–171.

Holdren, G.R., and R.A. Berner. 1979. Mechanism of feldspar weathering: I. Experimental studies. Geochim. Cosmochim. Acta 43:1161–1171.

Hsu, P.H. 1989. Aluminum oxides and oxyhydroxides. p. 331–378. *In* J.B. Dixon and S.B. Weed (ed.) Minerals in soil environments. 2nd ed. SSSA Book Ser. 1. SSSA, Madison, WI.

Huang, P.M. 1987. Aluminum and the fate of nutrients and toxic substances in terrestrial and freshwater environments. p. 262–268. *In* M. Singh (ed.) Encyclopedia of systems and control. Pergamon Press, Oxford, England.

Huang, P.M. 1989. Feldspars, olivines, pyroxenes, and amphiboles. p. 975–1050. *In* J.B. Dixon and S.B. Weed (ed.) Minerals in soil environments. 2nd ed. SSSA Book Ser. 1. SSSA, Madison, WI.

Huang, P.M., L.S. Crosson, and D.A. Rennie. 1968. Chemical dynamics of K-release from potassium minerals common in soils. Trans. Int. Congr. Soil Sci., 9th 2:705–712.

Huang, W.H., and W.D. Keller. 1970. Dissolution of rock forming minerals in organic acids. Am. Mineral. 55:2076–2094.

Huang, W.H., and W.D. Keller. 1972. Organic acids as agents of chemical weathering of silicate minerals. Nature (London) 239:149–151.

Huff, W.D. 1972. Morphological effects as a result of potassium depletion. Clays Clay Miner. 20:295–301.

Hunter, A.H., and P.F. Pratt. 1957. Extraction of potassium from soils by sulfuric acid. Soil Sci. Soc. Am. Proc. 21:595–598.

Hurlbut, C.S., Jr., and C. Klein. 1977. Manual of mineralogy: After James D. Dana. 19th ed. John Wiley & Sons, New York.

Hutcheon, A.T. 1966. Thermodynamics of cation exchange on clay: Ca-K-montmorillonite. J. Soil Sci. 17:339–355.

Jackson, M.L. 1963. Interlayering of expansible layer silicates in soils by chemical weathering. Clays Clay Miner. 11:29–46.

Jackson, M.L. 1964. Chemical composition of soils. p. 71–141. *In* F.E. Bear (ed.) Chemistry of the Soil. Van Nostrand Reinhold Co., New York.

Jackson, M.L., Y. Hseung, R.B. Corey, E.J. Evans, and R.C. Vanden Heuvel. 1952. Weathering of clay-size minerals in soils and sediments: II. Chemical weathering of layer silicates. Soil Sci. Soc. Am. Proc. 16:3–6.

Jackson, W.A. and G. W. Thomas. 1960. Effects of KCl and dolomitic limestones on growth and ion uptake of the sweet potato. Soil Sci. 89:347–352.

Jardine, P.M., and D.L. Sparks. 1984a. Potassium–calcium exchange in a multireactive soil system: I. Kinetics. Soil Sci. Soc. Am. J. 47:39–45.

Jardine, P.M., and D.L. Sparks. 1984b. Potassium–calcium exchange in a multireactive soil system: II. Thermodynamics. Soil Sci. Soc. Am. J. 47:45–50.

Jensen, H.E. 1973a. Potassium–calcium exchange equilibria on a montmorillonite and a kaolinite clay: I.A test on the Argersinger thermodynamic approach. Agrochimica 17:181–190.

Jensen, H.E. 1973b. Potassium–calcium exchange equilibria on a montmorillonite and a koalinite clay: II. Application of double-layer theory. Agrochimica 17:191–198.

Kämpf, N., A.C. Scheinost, and D.G. Schulze. 2000. Oxide minerals. p. F125–F168. *In* M.E. Sumner (ed.) Handbook of soil science. CRC Press, Boca Raton, FL.

Khan, H.R., S.F. Elahi, M.S. Hussain, and T. Adachi. 1994. Soil characteristics and behavior of potassium under various moisture regimes. Soil Sci. Plant Nutr. 40:243–254.

Khasawneh, F.E., E.C. Sample, and I. Hashimoto. 1974. Reactions of ammonium ortho- and polyphosphate fertilizers in soil: I. Mobility of phosphate. Soil Sci. Soc. Am. Proc. 38:446–451.

Kittrick, J.A. 1966. Forces involved in ion fixation by vermiculite. Soil Sci. Soc. Am. Proc. 30:801–803.

Kittrick, J.A. 1973. Mica-derived vermiculites as unstable intermediates. Clays Clay Miner. 21:479–488.

Kozak, L.M., and P.M. Huang. 1971. Adsorption of hydroxy-Al by certain phyllosilicates and its relation to K/Ca cation exchange selectivity. Clays Clay Miner. 19:95–102.

Kunze, G.W., and C.D. Jeffries. 1953. X-ray characteristics of clay minerals as related to potassium fixation. Soil Sci. Soc. Am. Proc. 17:242–244.

Laffer, B.G., A.M. Posner, and J.P. Quirk. 1966. Hysteresis in the crystal swelling of montmorillonite. Clay Miner. 6:311–321.

Laidler, K.J. 1965. Chemical kinetics. 2nd ed. McGraw-Hill, New York.

Laudelout, H., R. Van Bladel, G.H. Bolt, and A.L. Page. 1968. Thermodynamics of heterovalent cation exchange reactions in a montmorillonite clay. Trans. Faraday Soc. 64:1477–1488.

Laves, F. 1960. Al/Si Verteilungen, Phasen-Transformation und Namen der Alkalifeldspate. Z. Kristallogr. Kristallgeom. Kristallphys. Kristallichem. 113 (Lave Festschrift):265–296.

Le Roux, J. 1966. Studies on ionic equilibria in Natal Soils. Ph.D. diss., University of Natal, Pistermaritzburg, Republic of South Africa.

Le Roux, J., and C.I. Rich. 1969. Ion selectivity of micas as influenced by the degree of potassium depletion. Soil Sci. Soc. Am. Proc. 33:684–690.

Le Roux, J., C.I. Rich, and P.H. Ribbe. 1970. Ion selectivity by weathered micas as determined by electron probe analysis. Clays Clay Miner. 18:333–338.

Le Roux, J., and M.E. Sumner. 1968a. Labile potassium in soils: I. Factors affecting the quantity–intensity (*Q/I*) parameters. Soil Sci. 106:35–41.

Le Roux, J., and M.E. Sumner. 1968b. Labile potassium in soils: II. Effect of fertilization and nutrient uptake on the potassium status of soils. Soil Sci. 106:331–337.

Lee, S.Y., M.L. Jackson, and J.L. Brown. 1975a. Micaceous occlusions in kaolinite observed by ultramicrotomy and high resolution electron microscopy. Clays Clay Miner. 23:125–129.

Lee, S.Y., M.L. Jackson, and J.L. Brown. 1975b. Micaceous vermiculite, glauconite, and mixed-layered kaolinite–montmorillonite examination by ultramicrotomy and high resolution electron microscopy. Soil Sci. Soc. Am. Proc. 39:793–800.

Leonard, R.A., and S.B. Weed. 1970. Mica weathering rates as related to mica type and composition. Clays Clay Miner. 15:149–161.

Liebhardt, W.C., L.V. Svec, and M.R. Teel. 1976. Yield of corn as affected by potassium on a Coastal Plain soil. Commun. Soil Sci. Plant Anal. 7:265–277.

Liljeroth, E., P. Kuikman, and J.A. Van Veen. 1994. Carbon translocation to the rhizosphere of maize and wheat and influence on the turnover of native soil organic matter at different soil nitrogen levels. Plant Soil 161:233–240.

Lindsay, W.L., A.W. Frazier, and H.F. Stephenson. 1962. Identification of reaction products from phosphate fertilizers in soils. Soil Sci. Soc. Am. Proc. 26:446–452.

Liu, C., P.M. Huang, and J.M. Zhou. 2002. Residence time effect on iron perturbation of taranakite formation. Soil Sci. Soc. Am. J. 66:109–116.

Liu, Y.J., D.A. Laird, and P. Barak. 1997. Release and fixation of ammonium and potassium under long-term fertility management. Soil Sci. Soc. Am. J. 61:310–314.

Luebs, R.E., G. Stanford, and A.D. Scott. 1956. Relation of available potassium to soil moisture. Soil Sci. Soc. Am. Proc. 20:45–50.

Lutrick, M.C. 1963. The effect of lime and phosphate on downward movement of potassium in Red Bay fine sandy loam. Proc. Soil Crop Sci. Soc. Fla. 23:90–94.

Lynch, J.M. 1983. Soil biotechnology. Microbiological factors in crop productivity. Blackwell Scientific Publications, Oxford, UK.

Lynch, J.M. 1990. Introduction: Some consequences of microbial rhizosphere competence for plant and soil. p. 1–10. *In* J.M. Lynch (ed.) The rhizosphere. John Wiley and Sons, Chichester, England.

Lynch, J.M., and J.M. Whipps. 1990. Substrate flow in the rhizosphere. Plant Soil 129:1–10.

Marschner, H. 1995. Mineral nutrition of higher plants. 2nd ed. Academic Press, London.

Marschner, H. 1998. Soil–root interface: Biological and biochemical processes. p. 191–231. *In* P.M. Huang et al. (ed.) Soil chemistry and ecosystem health. SSSA Spec. Publ. 52. SSSA, Madison, WI.

Martin, H.W., and D.L. Sparks. 1983. Kinetics of nonexchangeable potassium release from two Coastal Plain soils. Soil Sci. Soc. Am. J. 47:883–887.

Martin, H.W., and D.L. Sparks. 1985. On the behavior of nonexchangeable potassium in soils. Commun. Soil Sci. Plant Anal. 16:133–162.

Martin, J.C., R. Overstreet, and D.R. Hoagland. 1946. Potassium fixation in soils in replaceable and nonreplaceable forms in relation to chemical reactions in the soil. Soil Sci. Soc. Am. Proc. 10:94–101.

Matthews, B.C., and P.H.T. Beckett. 1962. A new procedure for studying the release and fixation of potassium ions in soils. J. Agric. Sci. 58:59–64.

McLean, E.O., and R.H. Simon. 1958. Potassium status of some Ohio soils as revealed by greenhouse and laboratory studies. Soil Sci. 85:324–332.

McLean, E.O., and M.E. Watson. 1985. Soil measurements of plant-available potassium. p. 277–308. *In* R.D. Munson (ed.) Potassium in agriculture. ASA, CSSA, and SSSA, Madison, WI.

Mengel, K. 1985. Dynamics and availability of major nutrients in soils. Adv. Soil Sci. 2:65–131.

Miklos, D., and B. Cicel. 1993. Development of interstratification in K- and NH_4-smectite from Jelsovy Potok (Slovakia) treated by wetting and drying. Clay Miner. 28:435–443.

Ming, D.W., and F.A. Mumpton. 1989. Zeolites in soils. p. 873–911. *In* J.B. Dixon and S.B. Weed (ed.) Minerals in soil environments. 2nd ed. SSSA Book Ser. 1. SSSA, Madison, WI.

Mitra, S.P., and D. Prekash. 1955. Potassium fixation under wet and alternate wet and dry conditions by clay minerals. Proc. Natl. Acad. Sci. India Sect. A 24(2):182–186.

Moore, D.M., and R.C. Reynolds, Jr. 1997. X-ray diffraction and the identification and analysis of clay minerals. 2nd ed. Oxford University Press, New York.

Moreale, A., and R. Van Bladel. 1979. Soil interactions of herbicide-derived aniline residues: A thermodynamic appraoch. Soil Sci. 120:1–9.

Mortland, M.M. 1958. Kinetics of potassium release from biotite. Soil Sci. Soc. Am. Proc. 22:503–508.

Mortland, M.M. 1961. The dynamic character of potassium release and fixation. Soil Sci. 91:11–13.

Mortland, M.M., and B.G. Ellis. 1959. Release of fixed potassium as a diffusion-controlled process. Soil Sci. Soc. Am. Proc. 23:363–364.

Mortland, M.M., and J.E. Gieseking. 1951. Influence of the silicate ion on potassium fixation. Soil Sci. 71:381–385.

Mortland, M.M., and K. Lawton. 1961. Relationship between particle size and potassium release from biotite and its analogues. Soil Sci. Soc. Am. Proc. 25:473–476.

Mortland, M.M., K. Lawton, and G. Uehara. 1956. Alteration of biotite to vermiculite by plant growth. Soil Sci. 82:477–481.

Moss, P. 1967. Independence of soil quantity/intensity relationships to changes in exchangeable potassium: Similar potassium exchange constants for soils within a soil type. Soil Sci. 103:196–201.

Muir, I.J., and H.W. Nesbit. 1997. Reactions of aqueous anions and cations at the labradorite–water interface: Coupled effects of surface processes and diffusion. Geochim. Cosmochim. Acta 61:265–274.

Muir, I.J., G.M. Bancroft, W. Shotyk, and H.W. Nesbit. 1990. A SIMS and XPS study of dissolving plagioclase. Geochim. Cosmochim. Acta 54:2247–2256.

Munn, D.A., L.P. Wilding, and E.O. McLean. 1976. Potassium release from sand, silt, and clay soil separates. Soil Sci. Soc. Am. J. 40:363–366.

Munson, R.D. 1985. Potassium in agriculture. ASA, CSSA, and SSSA, Madison, WI.

Murdock, L.W., and C.I. Rich. 1972. Ion selectivity in three soil profiles as influenced by mineralogical characteristics. Soil Sci. Soc. Am. Proc. 36:167–171.

Nafady, M.H., and C.A. Lamm. 1973. Plant availability in soils: IV. The effect of K-fixation on the quantity/intensity relationship, with reference to factors affecting the rate of capacity of potassium fixation and the corresponding changes in soil cation exchange capacity. Agrichimica 17:307–315.

Nakahara, O., and S.-I. Wada. 1994. Ca^{2+} and Mg^{2+} adsorption by an allophanic and a humic Andisol. Geoderma 61:203–212.

Naylor, D.V., and R. Overstreet. 1969. Sodium–calcium exchange behavior in organic soils. Soil Sci. Soc. Am. Proc. 33:848–851.

Newman, A.C.D. 1969. Cation exchange properties of micas: I. The relation between mica composition and potassium exchange in solutions of different pH. J. Soil Sci. 20:357–373.

Nolan, C.N., and W.L. Pritchett. 1960. Certain factors affecting the leaching of potassium from sandy soils. Proc. Soil Crop Sci. Soc. Fla. 20:139–145.

Norrish, K. 1973. Factors in the weathering of mica to vermiculite. p. 417–432. *In* J.M. Serratosa (ed.) 1972 Proc. 14th Int. Clay Conf., Madrid. Division de Ciencias C.S.I.C, Madrid, Spain.

Nye, P.H. 1966. The effect of the nutrient intensity and buffering power of a soil and the absorbing power, size and root hairs of a root, on nutrient absorption by diffusion. Plant Soil 5:81–105.

Ogwada, R.A., and D.L. Sparks. 1986a. A critical evaluation on the use of kinetics for determining thermodynamics of ion exchange in soils. Soil Sci. Soc. Am. J. 50:300–305.

Ogwada, R.A., and D.L. Sparks. 1986b. Kinetics of ion exchange on clay minerals and soils. II. Elucidation of rate-limiting steps. Soil Sci. Soc. Am. J. 50:1162–1164.

Olk, D.C., and K.G. Cassman. 1995. Reduction of potassium fixation by two humic acid fractions in vermiculitic soils. Soil Sci. Soc. Am. J. 59:1250–1259.

Olk, D.C., K.G. Cassman, and R.M. Carlson. 1995. Kinetics of potassium fixation in vermiculitic soils under different moisture regimes. Soil Sci. Soc. Am. J. 59:423–429.

Olson, C.G., M.L. Thompson, and M.A. Wilson. 2000. Phyllosilicates. p. F77–F123. *In* M.E. Sumner (ed.) Handbook of soil science. CRC Press, Boca Raton, FL.

Para, M.A., and J. Torrent. 1983. Rapid determination of potassium quantity–intensity relationships using a potassium-selective ion electrode. Soil Sci. Soc. Am. J. 47:335–337.

Peterson, D.L., H. Helfferich, and G.C. Blytas. 1965. Sorption and ion exchange in sedimentary zeolites. J. Phys. Chem. Solids 26:835–848.

Prabhudesai, S.S., and S.B. Kadrekar. 1984. Reaction products from fertilizer phosphorus in medium black soil of Konkan region. J. Indian Soc. Soil Sci. 32:52–56.

Radoslovich, E.W. 1960. The structure of muscovite, $KAl_2(Si_3Al)O_{10}(OH)_2$. Acta Crystallogr. 13:919–932.

Radoslovich, E.W. 1962. The cell dimensions and symmetry of layer–lattice silicates: II. Regression relations. Am. Miner. 46:617–636.

Radoslovich, E.W., and K. Norrish. 1962. The cell dimensions and symmetry of layer-lattice silicates: I. Some structural considerations. Am. Miner. 47:599–616.

Rahmatullah and K. Mengel. 2000. Potassium release from mineral structures by H^+ ion resin. Geoderma 96:291–305.

Raman, K.V., and M.L. Jackson. 1964. Vermiculite surface morphology. Clays Clay Miner. 12:423–429.

Rao, Ch.S., A. Swarup, A.S. Rao, and V.R. Gopal. 1999. Kinetics of nonexchangeable potassium release from a Tropaquept as influenced by long-term cropping, fertilization, and manuring. Aust. J. Soil Res. 37:317–328.

Rao, P.S.C., and J.M. Davidson. 1978. Nonequilibrium conditions for ammonium-adsorption-desorption during flow in soils. Soil Sci. Soc. Am. J. 42:668.

Rasmussen, K. 1972. Potash in feldspars. Proc. Colloq. Int. Potash Inst. 9:57–60.

Rausell-Colom, J.A., T.R. Sweetman, L.B. Wells, and K. Norrish. 1965. Studies in the artificial weathering of micas. p. 40–70. In E.G. Hallsworth and D.V. Crawford (ed.) Experimental pedology. Butterworths, London.

Reed, M.G., and A.D. Scott. 1962. Kinetics of potassium release from biotite and muscovite in sodium tetraphenylboron solutions. Soil Sci. Soc. Am. Proc. 26:437–440.

Reed, M.G., and A.D. Scott. 1966. Chemical extraction of potassium from soils and micaceous minerals with solutions containing sodium tetraphenylboron: IV. Muscovite. Soil Sci. Soc. Am. Proc. 30:185–188.

Reitemeier, R.F. 1951. The chemistry of soil potassium. Adv. Agron. 3:113–164.

Reynolds, R.C., and J. Hower. 1970. The nature of interlayering in mixed layer illite–montmorillonite. Clays Clay Miner. 18:25–36.

Rich, C.I. 1958. Muscovite weathering in a soil developed in the Virginia piedmont. Clays Clay Miner. 5:203–212.

Rich, C.I. 1960. Aluminum in interlayers of vermiculite. Soil Sci. Soc. Am. Proc. 24:26–32.

Rich, C.I. 1964. Effect of cation size and pH on potassium exchange in Nason soil. Soil Sci. 98:100–106.

Rich, C.I. 1968a. Mineralogy of soil potassium. p. 79–96. *In* V.J. Kilmer et al. (ed.) The role of potassium in agriculture. ASA, CSSA, and SSSA, Madison, WI.

Rich, C.I. 1968b. Hydroxy interlayers in expansible layer silicates. Clays Clay Miner. 16:15–30.

Rich, C.I. 1972. Potassium in minerals. Proc. Colloq. Int. Potash Inst. 9:15–31.

Rich, C.I., and W.R. Black. 1964. Potassium exchange as affected by cation size, pH, and mineral structure. Soil Sci. 97:384–390.

Rich, C.I., and S.S. Obenshain. 1955. Chemical and clay mineral properties of a Red-yellow Podzolic soil derived from mica schist. Soil Sci. Soc. Am. Proc. 19:334–339.

Robert, M. 1971. Etude experimentale de l'evolution de micas (biotites) Les aspects due processus de vermiculitisation. Ann. Agron. 22:43–93.

Robert, M., and J. Berthelin. 1986. Role of biological and biochemical factors in soil mineral weathering. p. 453–495. *In* P.M. Huang and M. Schnitzer (ed.) Interactions of soil minerals with natural organics and microbes. SSSA Spec. Publ. 17. SSSA, Madison, W.I.

Ross, G.J., and C.I. Rich. 1973. Effect of particle thickness on potassium exchange from phlogopite. Clays Clay Miner. 21:77–81.

Rovira, A.D., R.C. Foster, and J.K. Martin. 1979. Origin, nature, and nomenclature of the organic materials in the rhizosphere. p. 1–4. *In* J.L. Harley and R.S. Russell (ed.) The soil–root Interface. Academic Press, London.

Sadusky, M.C., D.L. Sparks, M.R. Noll, and G.J. Hendricks. 1987. Kinetics and mechanisms of potassium release from sandy Middle Atlantic Coastal Plain soils. Soil Sci. Soc. Am. J. 51:1460–1465.

Saha, U.K. and K. Inoue. 1998. Hydroxy-interlayers in expansible layer silicates and their relation to potassium fixation. Clays Clay Miner. 46:556–566.

San Valentin, G.O., L.W. Zelazny, and W.K. Robertson. 1973. Potassium exchange characteristics of a Rhodic Paleudult. Proc. Soil Crop Sci. Soc. Fla. 32:128–132.

Sarkar, D., M.C. Sarkar, and S.K. Ghosh. 1977. Reaction products in red soils of West Bengal. J. Indian Soc. Soil Sci. 25:141–149.

Sauerbeck, D., and H.M. Helal. 1986. Plant root development and photosynthate consumption depending on soil compaction. p. 948–949. *In* Trans. 13th Congr. Int. Soc. Soil Sci., Hamburg, Germany.

Sauerbeck, D., S. Nonnen, and J.L. Allard. 1981. Assimilateverbrauch und-umsatz im Wurzelraum in Abhängigkeit von Pflanzenart und-anzucht. Landwirtsch. Forsch. Sonderh. 37:207–216.

Sawhney, B.L. 1966. Kinetics of cesium sorption by clay minerals. Soil Sci. Soc. Am. Proc. 30:565–569.

Sawhney, B.L., and G.K. Voight. 1969. Chemical and biological weathering in vermiculite from Transvaal. Soil Sci. Soc. Am. Proc. 33:625–629.

Schneider, A. 1997a. Release and fixation of potassium by a loamy soil as affected by initial water content and potassium status of soil samples. Eur. J. Soil Sci. 48:263–271.

Schneider, A. 1997b. Short-term release and fixation of K in calcareous clay soils. Consequence for K buffer power prediction. Eur. J. Soil Sci. 48:499–512.

Schofield, R.K. 1947. A ratio law governing the equilibrium of cations in the soil solution. Proc. Int. Congr. Pure Appl. Chem. 11(3):257–261.

Schofield, R.K. 1955. Can a precise meaning be given to "available" soil phosphorus. Soils Fert. 18:373–375.

Schofield, R.K., and H.R. Samson. 1954. Flocculation of kaolinite due to the attraction of oppositely charged crystal faces. Discuss. Faraday. Soc. 18:135–145.

Schönwitz, R., and H. Ziegler. 1982. Exudation of water-soluble vitamins and of some carbohydrates by intact roots of maize seedlings (*Zea mays* L.) into a mineral nutrient solution. Z. Pflanzenphysiol. 107:7–14.

Schubert, S., R. Paul, and K. Uhlenbecker. 1990. Characterisierung des nachlieferbaren kaliums aus der nicht austauschbaren Fraction von acht Boden mittels einer Austanschermethode und EUF. VD-LUFA-Shriftenr. 30:324–329.

Schwertmann, U. 1962a. Die selecktive kationensorption der Tonfraktion einiger Boden aus Sedimenten. Z. Pflanzenernaehr. Dueng. Bodenk. 97:9–25.

Schwertmann, U. 1962b. Eigenschaften und beldung aufwertbarer (quellbarer) Dreischicht-tonminerale in Böden aus sedimenten. Beitr. Mineral. Petrogr. 8:199–209.

Scott, A.D. 1968. Effect of particle size on interlayer potassium exchange in micas. Trans. Int. Congr. Soil Sci. 2:649–660.

Scott, A.D., and J.J. Hanway. 1960. Factors influencing the change in exchangeable soil K observed on drying. Trans. Int. Congr. Soil Sci. 7th 4:72–79.

Scott, A.D., and M.G. Reed. 1962a. Chemical extraction of potassium from soils and micaceous minerals with solutions containing sodium tetraphenylboron: II. Biotite. Soil Sci. Soc. Am. Proc. 26:41–45.

Scott, A.D., and M.G. Reed. 1962b. Chemical extraction of potassium from soils and micaceous minerals with solutions containing sodium tetraphenylboron: III. Illite. Soil Sci. Soc. Am. Proc. 26:45–48.

Scott, A.D., and S.J. Smith, 1966. Susceptibility of interlayer potassium in micas to exchange with sodium. Clays Clay Miner. 14:69–81.

Scott, A.D., and S.J. Smith. 1967. Visible changes in macro mica particles that occur with potassium depletion. Clays Clay Miner. 15:357–373.

Scott, A.D., J.J. Hanway, and E.M. Stickney. 1957. Soil potassium–moisture relations: I. Potassium release observed on drying Iowa soils with added salts or HCl. Soil Sci. Soc. Am. Proc. 21:498–501.

Selim, H.M., R.S. Mansell, and L.W. Zelazny. 1976. Modeling reactions and transport of potassium in soils. Soil Sci. 122:77–84.

Simard, R.R., C.R. De Kimpe, and J. Zizka. 1992. Release of potassium and magnesium from soil fractions and its kinetics. Soil Sci. Soc. Am. J. 56:1421–1428.

Sivasubramaniam, S., and O. Talibudeen. 1972. Potassium–aluminum exchange in acid soils. I. Kinetics. J. Soil Sci. 23:163–176.

Smith, J.V. 1974. Feldspars minerals: Crystal structure and physical properties, Vol. 1. Chemical and textural properties, Vol. 2. Springer-Verlag, New York.

Smith, S.J., L.J. Clark, and A.D. Scott. 1968. Exchangeability of potassium in soils. Trans. Int. Congr. Soil Sci. 9th 2:661–669.

Soil Survey Staff. 1998. Keys to soil taxonomy. 8th ed. USDA-NRCS. U.S. Gov. Print. Office, Washington, DC.

Somasiri, S., and P.M. Huang. 1973. The nature of K-feldspars of selected soils in the Canadian Prairies. Soil Sci. Soc. Am. Proc. 37:461–464.

Somasiri, S., S.Y. Lee, and P.M. Huang. 1971. Influence of certain pedogenic factors on potassium reserves of selected Canadian Prairie soils. Soil Sci. Soc. Am. Proc. 35:500–505.

Song, S.K., and P.M. Huang. 1988. Dynamics of potassium release from potassium-bearing minerals as influenced by oxalic and citric acids. Soil Sci. Soc. Am. J. 52:383–390.

Sparks, D.L. 1980. Chemistry of soil potassium in Atlantic Coastal Plain soils: A review. Commun. Soil Sci. Plant Anal. 11:435–449.

Sparks, D.L. 1987. Potassium dynamics in soils. Adv. Soil Sci. 6:1–63.

Sparks, D.L. 1989. Kinetics of soil chemical processes. Academic Press, San Diego, CA.

Sparks, D.L. 1995. Environmental soil chemistry. Academic Press, San Diego, CA.

Sparks, D.L. 1998. Kinetics of soil chemical phenomena: Future directions. p. 81–102. *In* P.M. Huang et al. (ed.) Future prospects for soil chemistry. SSSA Spec. Publ. 55. SSSA, Madison, WI.

Sparks, D.L. 1999. Kinetics of sorption/release processes on natural surfaces. p. 413–448. *In* P.M. Huang et al. (ed.) Structure and surface reactions of soil particles. Vol. 4. John Wiley & Sons, New York.

Sparks, D.L. 2000a. Kinetics and mechanisms of soil chemical reactions. p. B123–167. *In* M.E. Sumner (ed.) Handbook of soil science. CRC Press, Boca Raton, FL.

Sparks, D.L. 2000b. Bioavailability of soil potassium. p. D-38–D53. *In* M.E. Sumner (ed.) Handbook of soil science. CRC Press, Boca Raton, FL.

Sparks, D.L., and P.M. Huang. 1985. Physical chemistry of soil potassium. p. 201–276. *In* R.D. Munson (ed.) Potassium in agriculture. ASA, CSSA, and SSSA, Madison, WI.

Sparks, D.L., and P.M. Jardine. 1981. Thermodynamics of potassium exchange in soil using a kinetic approach. Soil Sci. Soc. Am. J. 45:1094–1099.

Sparks, D.L., and P.M. Jardine. 1984. Comparison of kinetic equations to describe K–Ca exchange in pure and in mixed systems. Soil Sci. 138:115–122.

Sparks, D.L., and W.C. Liebhardt. 1981. Effect of long-term lime and potassium application on quantity–intensity (*Q/I*) relationships in sandy soil. Soil Sci. Soc. Am. J. 45:786–790.

Sparks, D.L., and W.C. Liebhardt. 1982. Temperature effects on potassium exchange and selectivity in Delaware soils. Soil Sci. 133:10–17.

Sparks, D.L., D.C. Martens, and L.W. Zelazny. 1980a. Plant uptake and leaching of applied and indigenous potassium in Dothan soils. Agron. J. 72:551–555.

Sparks, D.L., and J.E. Rechcigl. 1982. Comparison of batch and miscible displacement techniques to describe potassium adsorption kinetics in Delaware soils. Soil Sci. Soc. Am. J. 46:875–877.

Sparks, D.L., L.W. Zelazny, and D.C. Martens. 1980b. Kinetics of potassium exchange in a Paleudult from the Coastal Plain of Virginia. Soil Sci. Soc. Am. J. 44:37–40.

Sparks, D.L., L.W. Zelazny, and D.C. Martens. 1980c. Kinetics of potassium desorption in soil using miscible displacement. Soil Sci. Soc. Am. J. 44:1205–1208.

Sposito, G. 1981. The thermodynamics of soil solutions. Clarendon Press, Oxford, England.

Sposito, G., and R.J. Reginato. 1992. Opportunities in basic soil science research. SSSA, Madison, WI.

Stevenson, F.J. 1994. Humus chemistry. Genesis, composition, reactions. 2nd ed. John Wiley and Sons, New York.

Stitcher, H. 1972. Potassium in allophane and zeolites. Proc. Colloq. Int. Potash Inst. 9:43–51.

Stumm, W., G. Furrer, E. Wieland, and B. Zinder. 1985. The effects of complex-forming ligands on the dissolution of oxides and aluminosilicates. p. 55–74. *In* J.I. Drever (ed.) The chemistry of weathering. D. Reidel, Boston.

Stumm, W., and R. Wallast. 1990. Coordination chemistry of weathering. Kinetics of the surface-controlled dissolution of oxide minerals. Rev. Geophys. 28:53–69.

Sucha, V., and V. Siranova. 1991. Ammonium and potassium fixation in smectite by wetting and drying. Clays Clay Miner. 39:556–559.

Sumner, M.E., and J.M. Marques. 1966. Ionic equilibria in a ferrallitic clay: Specific adsorption sites for K. Soil Sci. 102:187–192.

Szmigielska, A.M., K.C.J. Van Rees, G. Cieslinski, and P.M. Huang. 1997. Comparison of liquid and gas chromatography for analysis of low molecular weight organic acids in rhizosphere soil. Commun. Soil Sci. Plant Anal. 28:99–111.

Talibudeen, O. 1981. Cation exchange in soils. p. 115–177. *In* D.J. Greenland and M.H.B. Hayes (ed.) The chemistry of soil processes. John Wiley and Sons, New York.

Talibudeen, O., J.D. Beasley, P. Lane, and N. Rajendran. 1978. Assessment of soil potassium reserves available to plant root. J. Soil Sci. 29:207–218.

Talibudeen, O., and S.K. Dey. 1968. Potassium reserves in British soils. Parts I and II. J. Agric. Sci. 71:95–104, 405–411.

Talibudeen, O., and K.W.T. Goulding. 1983a. Charge heterogeneity in smectites. Clays Clay Miner. 31:37–42.

Talibudeen, O., and K.W.T. Goulding. 1983b. Apparent charge heterogeneity in kaolins in relation to their 2:1 phyllosilicate content. Clays Clay Miner. 31:137–142.

Tamm, O. 1929. An experimental study on clay formation and weathering of feldspars. Medd. Statens Skogsforsoksanstalt 25:1–28.

Tan, K.H. 1980. The release of silicon, aluminum, and potassium during decomposition of soil minerals by humic acid. Soil Sci. 129:5–11.

Tan, K.M. 1986. Degradation of soil minerals by organic acids. p. 1–27. *In* P.M. Huang and M. Schnitzer (ed.) Interactions of soil minerals with natural organics and microbes. SSSA Spec. Publ. 17. SSSA, Madison, WI.

Taylor, A.W., and E.L. Gurney. 1965. Precipitation of phosphate by iron oxide and aluminum hydroxide from solutions containing calcium and potassium. Soil Sci. Soc. Am. Proc. 29:18–22.

Taylor, A.W., W.L. Lindsay, E.O. Huffman, and E.L. Gurney. 1963. Potassium and ammonium taranakite, amorphous aluminum phosphate and variscite as sources of phosphate for plants. Soil Sci. Soc. Am. Proc. 27:145–151.

Terry, D.L., and C.B. McCants. 1968. The leaching of ions in soils. North Carolina Agric. Exp. Stn. Tech. Bull. 184. North Carolina State Univ., Raleigh, NC.

Thomas, G.W., and B.W. Hipp. 1968. Soil factors affecting potassium availability. p. 269–291. *In* V.J. Kilmer et al. (ed.) The role of potassium in agriculture. ASA, CSSA, and SSSA, Madison, WI.

Thompson, M.L., H. Zhang, M. Kazemi, and J.A. Sandor. 1989. Contribution of organic matter to cation exchange capacity and specific surface area of fractionated soil materials. Soil Sci. 148:250–257.

Tisdale, S.L., W.J. Nelson, J.D. Beaton, and J.L. Havlin. 1993. Soil fertility and fertilizer. McMillan Publishing Co., New York.

Truog, E., and R.J. Jones. 1938. Fate of soluble potash applied to soils. J. End. Eng. Chem. 30:882–885.

Udo, E.J. 1978. Thermodynamics of potassium–calcium and magnesium–calcium exchange reactions on a kaolinitic soil clay. Soil Sci. Soc. Am. J. 42:556–560.

Van Bladel, R., and H.R. Gheyi. 1980. Thermodynamics of calcium–sodium and calcium–magnesium exchange in calcareous soils. Soil Sci. Soc. Am. J. 44:938–942.

van der Marel, H.W. 1954. Potassium fixation in Dutch soils: Mineralogical analyses. Soil Sci. 78:163–179.

van der Plas, L. 1966. The identification of detrital feldspars. Elsevier North-Holland, New York.

Van Reeuwijk, L.P., and J.M. de Villiers. 1968. Potassium fixation by amorphous aluminosilicate gels. Soil Sci. Soc. Am. Proc. 32:238–240.

van Schouwenburg, J.Ch., and A.C. Schuffelen. 1963. Potassium-exchange behavior of an illite. Neth. J. Agric. Sci. 11:13–22.

Volk, N.J., 1934. The fixation of potash in difficulty available forms in soils. Soil Sci. 37:267–287.

von Reichenbach, H.G. 1972. Factors of mica transformation. Proc. Colloq. Int. Potash Inst. 9:33–42.

von Reinchenbach, H.G. 1973. Exchange equilibria of interlayer cations in different particle size fractions of biotite and phlogopite. p. 480–487. In J.M. Serratosa (ed.) 1972 Proc. 14th Int. Clay Conf., Madrid.

von Reichenbach, H.G., and C.I. Rich. 1969. Potassium release from muscovite as influenced by particle size. Clays Clay Miner. 17:23–29.

Wada, K. 1989. Allophane and imogolite. p. 1051–1087. *In* J.B. Dixon and S.B. Weeds (ed.) Minerals in soil environments. 2nd ed. SSSA Book Ser. 1. SSSA, Madison, WI.

Wada, K., and Y. Harada. 1969. Effects of salt concentration and cation species on the measured cation-exchange capacity of soils and clays. Proc. Int. Clay. Conf., Tokyo 1:561–571.

Wang, J., R.E. Farrell, and A.D. Scott. 1988. Potentiometric determination of potassium Q/I relationships. Soil Sci. Soc. Am. J. 52:657–662.

Wang, F.L., and P.M. Huang. 1990a. Kinetics of K adsorption by selected soils. Trans. 14th Int. Congr. Soil Sci., Kyoto, Japan II:10–15.

Wang, F.L., and P.M. Huang. 1990b. Ion-selective electrode determination of solution K in soil suspensions and its significance in kinetic studies. Can. J. Soil Sci. 70:411–424.

Wang, F.L., and P.M. Huang. 1993. Kinetics of potassium desorption from soils. p. 171–176. *In* Proc. Int. Conf. Physical Chemistry and Mass Exchange Processes in Soils, Pushchino, Russia. 12–16 Oct. 1992. Academy of Sciences of Russia, Pushchino, Russia.

Wang, F.L., and P.M. Huang. 2001. Effects of organic matter on the rate of potassium adsorption by soils. Can. J. Soil Sci. 81:325–330.

Wang, F.L., T.R. Yu, and P.M. Huang. 1995. Effects of selected cations and anions on pK –0.5 pCa values for a Mn oxide system. Z. Pflanzenernähr. Bodenk. 161:173–178.

Wang, F.L., T.R. Yu, P.M. Huang, G.S.R. Krishnamurti, S.X. Li, and D. Fairhurst. 1996. Competitive adsorption of potassium and calcium on manganese oxide as studied with two ion-selective electrodes. Z. Pflanzenernähr. Bodenk. 159:93–99.

Weed, S.B., C.B. Davey, and M.G. Cook. 1969. Weathering of mica by fungi. Soil Sci. Soc. Am. Proc. 33:702–706.

Weir, A.H. 1965. Potassium retention in montmorillonite. Clay Miner. 6:17–22.

Wells, C.B., and K. Norrish. 1968. Accelerated rates of release of interlayer potassium from micas. Trans. Int. Congr. Soil Sci., 9th 2:683–694.

Wild, A. 1981. Mass flow and diffusion. p. 37–80. *In* D.J. Greenland and M.H.B. Hayes (ed.) The chemistry of soil processes. John Wiley and Sons, New York.

Wilkins, R.W.T. 1967. The hydroxyl-stretching region of biotite mica spectrum. Mineral. Mag. 36:325–333.

Zeng, Q., and P.H. Brown. 2000. Soil potassium mobility and uptake by corn under differential soil moisture regimes. Plant Soil 221:121–134.

Zhou, J.M. 1995. Kinetics and mechanisms of phosphate-induced potassium release from selected K-bearing minerals and soils. Ph.D. diss. University of Saskatchewan, Saskatoon.

Zhou, J.M., and P.M. Huang. 1995. Kinetics of monoammonium phosphate-induced potassium release from selected soils. Can. J. Soil Sci. 75:197–203.

Zhou, J.M., C. Liu, and P.M. Huang. 2000. Perturbation of taranakite formation by ferrous and ferric iron under acidic conditions. Soil Sci. Soc. Am. J. 64:885–892.

Chapter 5

Chemistry of Micronutrients in Soils

LARRY M. SHUMAN, *University of Georgia, Griffin, Georgia, USA*

Micronutrients are defined as the elements that plants need in very small quantities in contrast to the macronutrients used in larger quantities, such as N. As such they are mostly found in soils in low quantities, with the exception of Fe, which is fourth in abundance after O_2, Si, and Al. Iron is a micronutrient because of low plant requirements. Despite its abundance, Fe can become deficient due to insolubility at higher pH values. Thus, the micronutrients considered here are Zn, Cu, Mn, Fe, B, and Mo. There are some other elements that have been shown to be needed by plants in trace amounts that are generally ubiquitous (like Cl) and will not be considered here.

The elements will be grouped in pairs for discussion because of some similarities of each pair. Zinc and Cu are the first pair discussed. They are transition metals (all those here are transition elements, except B, a Group III element) that are next to each other on the periodic table with Zn having one more electron in the 4s layer (Table 5–1). The second pair is Mn and Fe, which are also next to each other in the periodic table and are the ones that display the most oxidation–reduction transitions in soils. The last pair is B and Mo, which are quite different from the others in that they form oxyanions in soil environments. They are unrelated to each other in the periodic table, representing the lightest and heaviest of the elements discussed (Table 5–1).

Zinc is 100 times more abundant than Cu and found in nature mostly as the sulfide. It is present nearly always as the divalent species and has little chance for covalent bonding. Zinc forms stable complexes, which is important in soil systems, since both inorganic and organic complexes can contribute to solvation and transfer. Like Zn, Cu is usually mined as the sulfide and is most often found as the divalent species. However, under reducing conditions it does form a monovalent species. In aqueous solution, though, the monovalent species is auto-oxidized to the divalent species by

$$2Cu^{+} \rightarrow 2Cu^{2+} + Cu(s) \qquad [1]$$

Manganese can be found in soils in several oxidation states, the most common of which are Mn^{2+}, Mn^{3+}, and Mn^{4+}. The minerals of Mn found in well-aerated soils are forms of MnO_2; however, these solids are always a mixture, with the relative number of oxygen atoms to Mn being <2. Iron is the most abundant of the mi-

Chemical Processes in Soils. SSSA Book Series, no. 8.

Table 5–1. General properties.

Element	Atom. no.†	Atom. wt.‡	Outside electrons	Atom. rad.§	Ionization potential	Electro negativity	O/R potential¶
		G-atoms		Å	eV		V
Zn	30	65.38	$...3d_{10}4s_2$	1.25	9.39	1.6	+0.76 to Zn^{2+}
Cu	29	53.54	$...3d_{10}4s_1$	1.17	7.72	1.9	−0.34 to Cu^{2+} −0.52 to Cu^{+}
Mn	25	54.94	$...3d_5 4s_2$	1.17	7.43	1.5	+1.18 to Mn^{2+}
Fe	26	55.85	$...3d_6 4s_2$	1.17	7.90	1.8	+0.44 to Fe^{3+}
B	51	0.82	$1s_2 2s_2 2p_1$	0.80	8.30	2.0	+0.87 to H_3BO_3
Mo	42	95.95	$...4d_5 4f_0 5s_1$	1.29	7.18	1.8	+0.20 to Mo^{3+}

† Atomic number.
‡ Atomic weight.
§ Atomic radius.
¶ Oxidation–reduction potential.

cronutrients, being fourth of all elements after O, Si, and Al (5% of the earth's crust). In its natural form Fe is usually found combined with O, Si, or S. The common oxidation states are Fe^{2+} and Fe^{3+}, both of which form complex ions. The reduced form is easily oxidized upon exposure to air, so the ferrous form is only found in wet soils that lack O_2. Boron is a semimetal, and the small atom size is responsible for its properties. As mentioned, B is the only element considered that is not a transition metal, but rather a Group III element and as such would normally be found in the B^{3+} state. However, it holds its three valence electrons too strongly and forms only covalent bonds. It is usually assigned a +3 valence because it combines with more electronegative elements. It hydrolyzes to a great extent forming boric acid, $B(OH)_3$. Boron occurs in nature as borates (oxyborate anions) such as borax ($Na_2B_4O_7 \cdot 10H_2O$). Molybdenum is needed by plants in the smallest quantities of the six considered. It is found as the sulfide (molybdenite), and its most important oxidation state is +6 as in MoO_3. In aqueous solution it forms a series of oxyanions called molybdates, of which MoO_4^{2-} is the most important.

PRECIPITATION–DISSOLUTION

The solubility of solids is the first consideration in determining the concentration of a micronutrient in soil solution. For all the elements considered, soil pH is usually the overriding factor in solubility, although oxidation–reduction can be important for Mn and Fe, and soil moisture always plays a part. Pure minerals of the elements considered are seldom found in soils, with the exception of Fe and Mn oxides. Thus, the solubility of solids with which the elements are associated becomes important.

The oxides and hydroxides of Zn, along with the carbonate, are too soluble to exist in soils, but the Fe–Zn mineral franklinite is less soluble than measured Zn solubility in soils (Lindsay, 1979). Lindsay defined "soil-Zn" solubility as

$$\text{soil-Zn} + 2H^+ \rightleftarrows Zn^{2+} \qquad \log K = 5.8 \qquad [2]$$

Table 5–2. Solid forms of micronutrients.†

Element	Principal modes of occurrence in primary minerals and secondary soil minerals and organic matter
Zn	Sulfide inclusions in silicates; isomorphic substitution for Fe in olivines, pyroxenes, amphiboles, and for Fe or Mn in oxides Coprecipitated with Fe and Al oxides, Mn oxides, illites, smectites, organic matter
Cu	Sulfide inclusions in silicates; isomorphic substitution for Fe and Mg in olivines, pyroxines, amphiboles, and micas and for Ca, K, or Na in feldspars Coprecipitated with Fe and Al oxides, illites, smectites, and organic matter
Mn	Birnessite; lithiophorite; isomorphous substitution and inclusion in Fe oxides, Ca carbonates, vermiculites, smectites
Fe	Goethite; hematite; ferrihydrite; isomorphous substitution and inclusion in Ca carbonates, vermiculites smectites
B	Tourmaline [$NaMg_3Al_6B_3Si_6O_{27}(OH,F)_4$]; isomorphic substitution for Si in micas Coprecipitated with Fe and Al oxides, illites, smectites
Mo	Molybdenite (MoS_2); isomorphic substitution for Fe in oxides Coprecipitated with Mn oxides, illites

† Adapted from Sposito (1989, Table 1.4, Table 1.5, p. 11, 12; not Mn and Fe) and Sposito (1984, Table 1.4 and elsewhere, p. 8; for Fe and Mn).

This equation is based on the solubility of Zn in many soils adjusted to various pH values. The solubility does not coincide with any known mineral. The solubility of Zn in soils is controlled by solid phases other than pure minerals. It is found as sulfide inclusions and in isomorphic substitutions for Fe in various minerals (Table 5–2) and as coprecipitates with oxides, secondary clay minerals, and soil organic matter. The solubility, then, is controlled by the solubility of these solids and by the lability of the Zn associated with organic matter. One way to approach the solubility of elements in soils is to sequentially extract them with solutions of increasing strength to dissolve less and less labile forms of the element. Attempts are made to link these operationally defined fractions with some of the above-mentioned phases such as organic matter, Fe and Mn oxides, and solid minerals (Shuman, 1991). Using these schemes Zn is found to reside principally in primary and secondary minerals (residual form), and to a lesser degree in oxides and organic matter (Fig. 5–1).

Copper forms and solubility follow Zn to a great extent, but both pH and redox can play a part. Copper becomes more soluble at lower pH values and under reducing conditions. As for Zn, the oxides, hydroxides, and carbonates are too soluble to exist in soils. Lindsay (1979) gives an equilibrium reaction for "soil-Cu" as

$$\text{soil-Cu} + 2H^+ \rightleftharpoons Cu^{2+} \qquad \log K \text{ of } 2.8 \qquad [3]$$

Copper, then, is associated with minerals other than pure Cu minerals as sulfide inclusions in silicates and as isomorphic substitution for Fe and Mg in several minerals (Table 5–2). It coprecipitates with similar soil materials with which Zn associates. Since divalent Cu can be reduced as

$$Cu^{2+} + e^- \rightleftharpoons Cu^+ \qquad \log K = 2.62 \qquad [4]$$

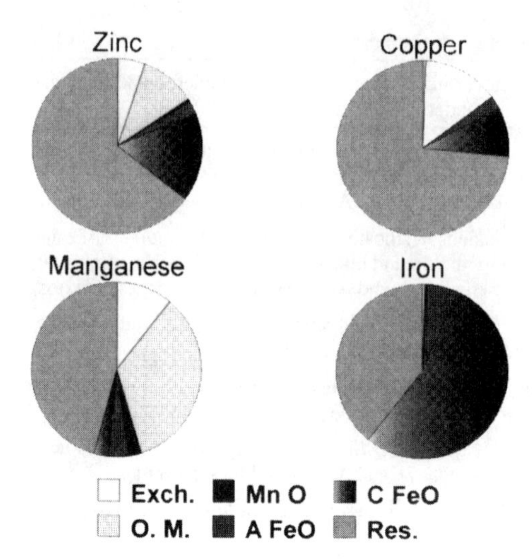

Fig. 5–1. Metals in soil fractions for a Cecil sl topsoil. Soil pH: 5.4. Organic matter content: 25.5 g kg^{-1}. CEC: 6.64 $cmol^{+}$ kg^{-1}. (Exch., exchangeable; O.M., organic matter; MnO, manganese oxides; AFeO, amorphous iron oxides; CFeO, crystalline iron oxides; Res., residual.) (Shuman, 1985).

there are minerals that could control Cu solubility in reduced soil. The soil-Cu equation (Eq. [3]) controls Cu solubility to $pe + pH = 14.89$, whereas below this redox cuprous ferrite could control Cu solubility that is in turn dependent on the solubility of Fe. The sequential extraction of Cu to look at various solid phases reveals that more is found in the organic fraction than for Zn which comes from its propensity to be complexed by organic ligands (Fig. 5–1).

Manganese solubility is controlled as much by redox as by pH. Pyroluscite and birnessite are two of the less soluble Mn minerals under well oxidized conditions. Below $pe + pH = 16.62$ the stable mineral is manganite, an oxyhydroxide, and at very low redox, rhodochrosite, a carbonate, would be stable (Fig. 5–2, Lindsay, 1979). Of course, Mn is also found in association with other soil minerals as an isomorphous species (Table 5–2). If a soil is high in Mn, then a larger proportion of it will be found in the Mn oxide form, as determined by sequential extraction. The extractant for this form is a mild reducing agent, hydroxylamine hydrochloride. The object is to reduce the Mn oxides without solubilizing the amorphous Fe oxides, which is often difficult. In high-Mn soils, wet, reducing conditions can cause an elevation of Mn in the soil solution such that when the soil begins to dry out, Mn becomes toxic to crops.

As for Mn, Fe solubility is controlled both by pH and redox. Iron becomes more soluble at lower pH values and under reducing conditions. Iron is not a micro*element*, since it is found in large quantities in soils. However, it is a micro*nutrient*, since plants need it in small amounts and it can become so insoluble that it is deficient, especially under high pH conditions. Unlike Zn and Cu the "soil-Fe" designation by Lindsay (1979) is more soluble than known oxide minerals such as goethite and hematite. Even in well-drained soils, Fe solids and coatings are continuously going through redox cycles such that most of the time some of the Fe is

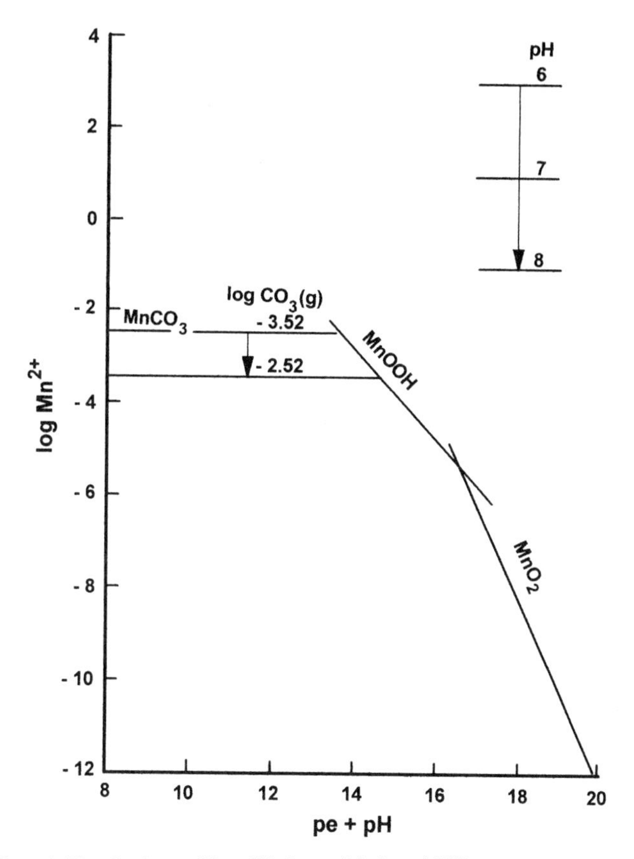

Fig. 5–2. Effect of pH and redox on Mn solid phases (Lindsay, 1979).

a freshly precipitated oxide that is very soluble. Within a short time (several days) some structuring occurs such that the solubility lowers to that of soil Fe given as

$$Fe(OH)_3(soil) + 3H^+ \rightleftharpoons Fe^{3+} + 3H_2O \qquad \log K = 2.70 \qquad [5]$$

Hydrolysis of the oxides is important, since it tends to raise the concentration of Fe in soil solution. The oxides are more prevalent than Fe^{3+} in the normal pH range of soils. The solid phases and the isomorphous substitution forms of Fe are shown in Table 5–2. Sequential extraction schemes extract the Fe oxides specifically, and some divide the fraction into amorphous and crystalline forms. Although much of the Fe in soil resides in residual, nonreactive forms, the reason to emphasize the oxide fraction is that many micronutrients and microelements are associated with soil Fe oxides, both within the structure and as adsorbents on the surface (Fig. 5–1).

The inorganic phases controlling B solubility in soils are largely unknown. It appears in very insoluble minerals like tourmaline (Table 5–2) and in isomorphic substitution for Si in micas. It is found in association with Fe and Al oxides in soils, which may be a significant source for plant B. In soil solution B is in the form of

H_3BO_3. In selective extraction studies where successive extractants are used on the same sample to determine amounts of an element in different fractions, B is found principally in residual minerals, but in high organic matter soils, a significant amount is found in the organic fraction.

Molybdenum occurs at very low levels in soil solution, indicating low solubility in soils. Unlike other micronutrients, Mo becomes less available under acid soil conditions. Lindsay (1979) showed that "soil-Mo" is less soluble than any of the Mo minerals that may be found in soils. The "soil-Mo" was given as

$$\text{soil-Mo} \rightleftharpoons MoO_4^{2-} + 0.8H^+ \qquad \log K = -12.40 \qquad [6]$$

Redox has little bearing on Mo solubility, since the minerals that are sensitive to redox are much more soluble than soil Mo. Molybdenum is found in association with Fe oxides and Mn oxides (Table 5–2). A common soil extractant for Mo is oxalate, which dissolves soil Fe oxides.

ADSORPTION–DESORPTION

Much of the dynamic movement of micronutrients in soils between solid and solution forms occurs via adsorption and desorption reactions, which are faster than those of precipitation and dissolution. Ions are held mainly on the colloidal surfaces of clays; oxides of Al, Fe, and Mn; and organic matter both by exchange sites (outer sphere) and by specific adsorption sites (inner sphere). This section will cover briefly some of the commonly used adsorption isotherm equations and adsorption mechanisms. Other chapters address these in detail along with the kinetics of adsorption reactions. Most of the discussion will concern adsorption of individual micronutrients. Since adsorption plays a minor role in the reactions of Fe and Mn, they will not be discussed. The reactions controlling these ions are principally precipitation and dissolution driven by pH and redox.

Adsorption isotherms are a convenient method to summarize adsorption data generated by a batch method of equilibrating soil or soil components with solutions of various target ion concentrations and measuring the amount adsorbed. Two of the older equations used in soil science are the Langmuir and the Freundlich equations, and a newer approach is that of the constant capacitance model.

The Langmuir equation was originally devised to describe gas molecule adsorption on a flat surface, but was eventually expanded to use with adsorbents on colloidal surfaces in solution. The form is

$$x/m = mkC/(1 + kC) \qquad [7]$$

where x/m is the amount of ion adsorbed per mass of adsorbent, m is the maximum adsorption capacity, C is the concentration of the remaining ion in solution after adsorption, and k is a constant that has been related to the bonding energy of adsorption (Fig. 5–3). The constants m and k are compared for different adsorbents as well as changing equilibrium conditions, such as pH and the ionic strength of the solution. For example, soils high in clay and/or organic matter usually have a

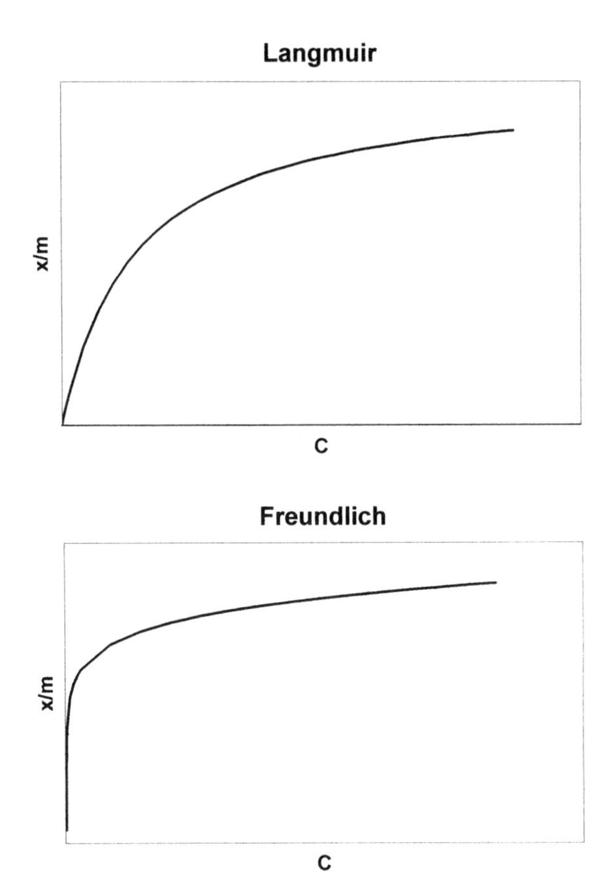

Fig. 5–3. Langmuir and Freundlich adsorption isotherms.

higher adsorption capacity than sandy soils that are low in organic matter. A reciprocal relation often exists between the m and k values. Fitting experimental data to the curve says nothing about adsorption mechanisms, but does allow the types of comparisons among adsorbents mentioned. To fit data to the curve, a linear form of the equation is used and the fit of the data to the line is tested statistically. In some cases two straight lines best describe the data (Shuman, 1975). This situation is described by a two-surface Langmuir equation with two m and two k values. Again, even though it is tempting to associate this equation with two types of surfaces (e.g., organic matter and clay), it has been shown that any data constructed in a smooth convex or concave line with a maximum can be modeled by any four-parameter equation (Sposito, 1982).

The Freundlich equation is not often used, but does fit many adsorption data sets. It is a power equation for data that rise rapidly over a narrow range as the concentration of the adsorbing ion is increased above zero. It has the form

$$x/m = KC^{1/n} \qquad [8]$$

where x/m is the amount adsorbed per unit mass of adsorbent, C is the concentration of the remaining adsorbing ion after adsorption, and K and n are constants (Fig. 5–3).

The constant capacitance model is somewhat less empirical than the ones described above in that it attempts to put the adsorption model on a chemical basis. It is much more complex, requiring the use of equilibrium computer models such as GEOCHEM or MINTEQA2 to actually calculate the predicted amounts adsorbed. One must first of all devise chemical equations to describe the surface reactions of the ions in question that involve inner-sphere complexation, not only for the target ion, but for all major ions in the system. The conditional equilibrium constants must then be given for each equation, and then a series of intrinsic equilibrium constants for each reaction are required. Thus, there are many constants to be given, and errors in these would lead to a less than perfect modeling of the data.

Zinc and Cu are adsorbed as the divalent species at the usual soil pH values of 6 and below. Only at high pH values do they hydrolyze to $ZnOH^+$, and $Cu(OH)_2^0$, which has an effect on the adsorption. As divalent species, much of the adsorption is by ion exchange, and it has been found that they can often be adsorbed in excess of the exchange capacity. Specific adsorption is also important, especially for adsorption on oxide and organic surfaces (Fig. 5–4). Soil clays have a great capacity for adsorption of Zn and Cu, but adsorption is a complex phenomenon and often does not follow the exchange capacity of the soil clays present. Research on adsorption vs. clay type has had mixed results, indicating that other parameters come into play. Soils that have a large proportion of their clay-sized material in 1:1 clays or hydrous oxides where hydroxyl groups are exposed will demonstrate variable

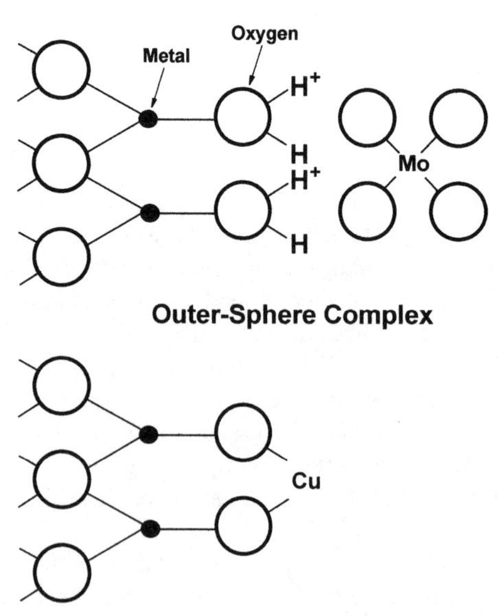

Fig. 5–4. Surface complexes between inorganic ions and hydroxyl groups of an oxide surface (Hayes, 1987).

charge that changes with pH. Soils with predominantly 2:1 clays that have charge originating from isomorphous substitution within the clay lattice will have permanent charge that does not vary with pH. Of course, soils with clays and hydrous oxides with variable charge are expected to have more variation in adsorption capacity with pH than soils with predominately permanent charge clays. Adsorption of Zn and Cu by Fe oxides is dependent on pH and the crystallinity of the oxides. Amorphous Fe oxides have about 10 times the surface area of crystalline Fe oxides and not surprisingly adsorb more Zn (Fig. 5–5) (Shuman, 1977). Note the large difference in the *y* axis scales for the crystalline and amorphous Fe oxides in Fig. 5–5. Organic matter is important in adsorption of Zn and Cu, but more so with Cu, which forms surface complexes. Removing organic matter from whole soils lowers the adsorption capacity (Shuman, 1988). In contrast, complexing of Zn and Cu by soluble organic matter in soil solution can reduce adsorption, although whether complexes are more or less adsorbed is sometimes dependent on pH. The adsorption of both Zn and Cu vary little in the presence of competing ions such as Ca and Mg

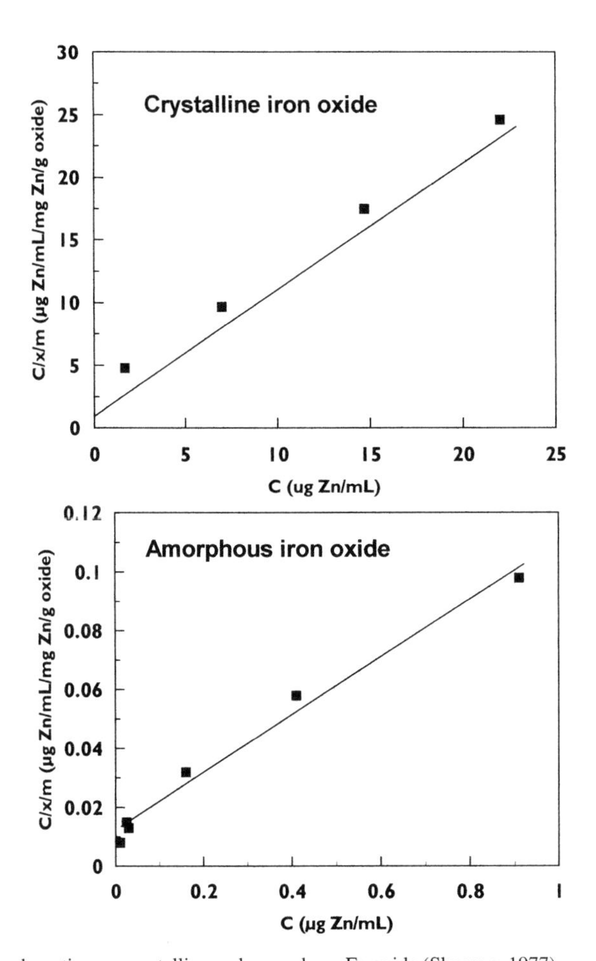

Fig. 5–5. Zinc adsorption on crystalline and amorphous Fe oxide (Shuman, 1977).

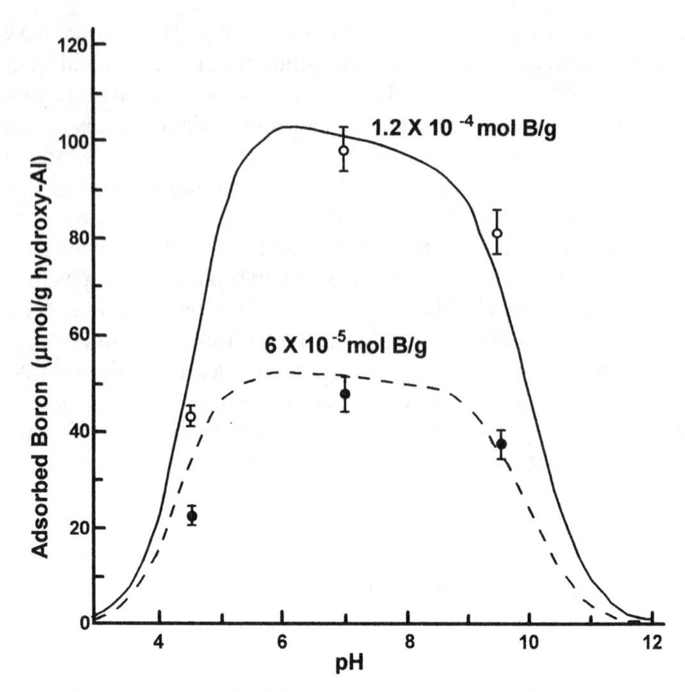

Fig. 5–6. Boron adsorbed on hydroxyl Al vs. pH for two concentrations of B in solution. Line computed by a model (Keren and Gast, 1983).

in soil solution. However, the preference for the metal decreases as adsorption increases.

As mentioned above, B is found in solution as boric acid, $B(OH)_3$, and Mo as the molybdate ion, MoO_4^{2-}. However, both tend to polymerize, which could add to the adsorption mechanisms involved. Adsorption of both B and Mo is very pH dependent, with maximum B retention occurring at a pH range of about 6 to 8 (Fig. 5–6) and maximum Mo adsorption below pH 5 (Fig. 5–7). As for the metals, many soil properties influence adsorption of these anions including clay type and content, oxide content, and organic matter as well as soil pH. Also the crystallinity of the Fe oxides are very important with adsorption decreasing dramatically as the ox-

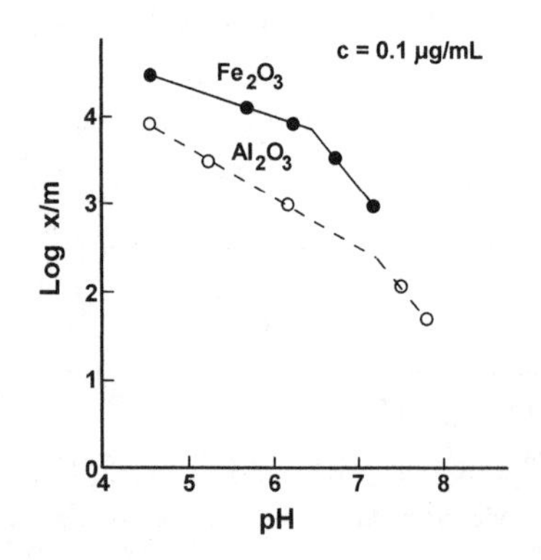

Fig. 5–7. Molybdate adsorbed on Fe and Al oxide vs. pH (Reisenauer et al., 1962).

ides age and become more crystalline. Since the anions are attracted to positive sites, the amount of edge area of a clay displayed that is dependent on the fineness of the particles will influence the adsorption of B and Mo. Phosphate is an anion that is analogous to molybdate and has been found to interfere with Mo adsorption. However, B adsorption is specific and nearly independent of the presence of other anions in the soil solution.

COMPLEXATION

A central ion surrounded by one or more ligands in solution is termed a *complex*. Although micronutrient ions form inorganic complexes that change the charge characteristics of the ion, more important are the soluble organic complexes. Synthetic chelates have more than one position to bind the central ion and are termed *polydentate*. Chelates, both synthetic and natural, have the effect of usually reducing adsorption on soil surfaces, and thus increase the total amount of metal ion in soil solution and plant availability. Inorganic complexation by anions such as Cl^- and SO_4^{2-} have some influence on adsorption, but usually less than organic complexes. The sulfate complex of Zn has a higher adsorption maximum for soil than the chloride complex, presumably because the sulfate is a stronger complex with Zn and has more effect on the charge characteristics. The sulfate complex of Zn may contribute to the solubility of Zn in soils. Copper complexes rather strongly with phosphate and sulfate, especially at lower pH values. At higher pH the Cu carbonate complex becomes important in Cu solubility (Lindsay, 1979). Equilibrium models are used to describe the relation between free uncomplexed ions in solution and those bound by complexing agents using conditional equilibrium constants.

Zinc and Cu are especially prone to complexation by organics, which can have both beneficial and negative effects. One of the benefits is to raise the amounts of metal in solution causing transfer from solids or adsorption–exchange sites to plant roots, which helps Zn, Cu, and Fe plant nutrition under potentially deficient conditions. Complexation is difficult to account for in chemical equilibrium models in that they can predict the complexes formed when solution concentrations of metal and ligands are given, but cannot give the effects of this complexation on ion exchange or adsorption–desorption within the same model. These dynamics are extremely important to soil chemistry, and the interrelationships are an active research topic. Metals such as Zn and Cu are held by chelating sites in organic matter mainly at oxygen-containing functional groups such as COOH, phenolic-OH, and C=O groups (Adriano, 1986). A great percentage of the Cu in soil solution is complexed by soluble organics. Surface waters that contain soluble organics may have a heavier total metal load due to the organic complexation of the metals. Since organics are generally more soluble in basic solutions, organic complexation has the effect of making metals more soluble at higher pH, when inorganic equilibrium would predict precipitation as hydroxides.

OXIDATION–REDUCTION

The solubility of Fe and Mn can be greatly affected by oxidation–reduction reactions in soils. The oxidized forms accept electrons to become reduced under low O_2 conditions or when fresh organic matter is undergoing decay, which donates electrons. The reduced forms of these metals are more soluble than the oxidized forms; thus, they are more plant available and can even become toxic. Other cations, such as Zn and Cu are affected indirectly as the adsorbing surfaces of Fe and Mn oxides are altered by oxidation–reduction conditions.

The oxidation state of Zn does not change in soils, but oxidation–reduction does indirectly affect Zn availability and solubility through effects on the oxides of Fe and Mn, which strongly adsorb Zn. Thus, reducing conditions brought about by flooding or by bacterial oxidation of fresh organic matter can release Zn, which was held on oxides surfaces or within the amorphous oxide solid. Copper solubility can be affected by the same mechanisms, but also can actually be reduced itself in soils under rather severe reducing conditions. As mentioned above, the cuperous form of the Fe–Cu compound, cuperous ferrite, is much more soluble than the ferric form. However, the effects of the adsorption of Cu by oxides seems to be much more important relative to oxidation–reduction effects than the reduction of Cu.

As mentioned above, Mn solubility is largely determined by the oxidation status of soils. Manganese (IV) is more easily reduced than Fe (III) according to the reduction potentials. Under reducing conditions, the soil pH level often decreases, which makes Mn more soluble. Under flooding conditions, Mn moves into more available forms. When the soil starts to dry out, the Mn can become so high in the soil solution that it is toxic to plants. Manganese has been implicated in coupled oxidation–reduction reactions in soils with specific metals and organics.

Of course, Fe solubility hinges to a great extent on oxidation–reduction. Even short-term reducing conditions in soils after a rainfall can reduce Fe and help keep Fe oxide coatings in an amorphous state. However, continuous waterlogging does have more effect on Fe redox status than short-term wet conditions. As mentioned above, the amorphous forms of Fe oxides have about 10 times the surface area of the more crystallized forms. Amorphous Fe oxides, which have a large adsorption capacity for metals, are produced under fluctuating oxidation–reduction conditions.

MASS FLOW, DIFFUSION, AND TRANSPORT

Micronutrient ions, except B, are considered to be nonmobile in soils, but can move slowly by diffusion. Boron is water soluble and moves in soluble form. Molybdenum is considered mobile under neutral or alkaline conditions. At low pH the molybdate ion is strongly chemisorbed, analogous to the phosphate anion. Under some conditions, such as low pH or when organically complexed at higher pH, metals can move in the profile in soluble forms. If there is movement of soil colloids, sorbed metals can move by colloid transport.

Zinc movement in soils is often through diffusion, which is influenced by soil pH, bulk density, moisture content, and organic matter content. Both Zn and Cu are strongly hydrating cations that are considered to be exchangeable and somewhat mobile. The mobility of Zn under low pH is enhanced by redistribution of Zn from less to more soluble forms. Another major mechanism for movement of Zn is the same mechanism affecting Cu movement, that of organic complexation. As noted above, these metals are held in solution by being complexed by natural organics and as such move by mass flow with the water. Of course, cracks in the soil or root channels greatly exacerbate the transport of these complexed metals to groundwater giving direct "pipelines" through the soil profile. This mode of water movement is usually termed *preferential flow*. Copper in soil solution at low pH is dominated by the divalent species followed by inorganic complexes, such as copper sulfate. At high pH values the natural organic matter complexes dominate the solution species. Thus, Cu moves in soil profiles both as free cations at lower pH and as soluble organic complexes at higher pH values. Manganese and Fe mobility are influenced by redox in that as divalent ions they are about the same as Zn and Cu in mobility. When Mn is oxidized to +4 or Fe is oxidized to +3, they become amphoteric oxides and hydroxides and are immobile in solution because they are insoluble. The elements discussed are shown on Fig. 5–8 grouped by ionic charge (valence) and radius according to their solubility. Note that Mn appears three times due to the different oxidation states found in soil.

A third mechanism for transport affecting Fe greatly, and also operative for Zn and Cu, is that of colloid transport. The presence of colloids can increase metal transport by several fold. Especially under low flow conditions, Fe moves through the soil profile as colloidal oxides, and these act as carriers of metals. However, organic C is even more effective as a metal carrier than inorganic surfaces. Particles that can transport metals are clay minerals, Fe oxides, Fe sulfides, and quartz par-

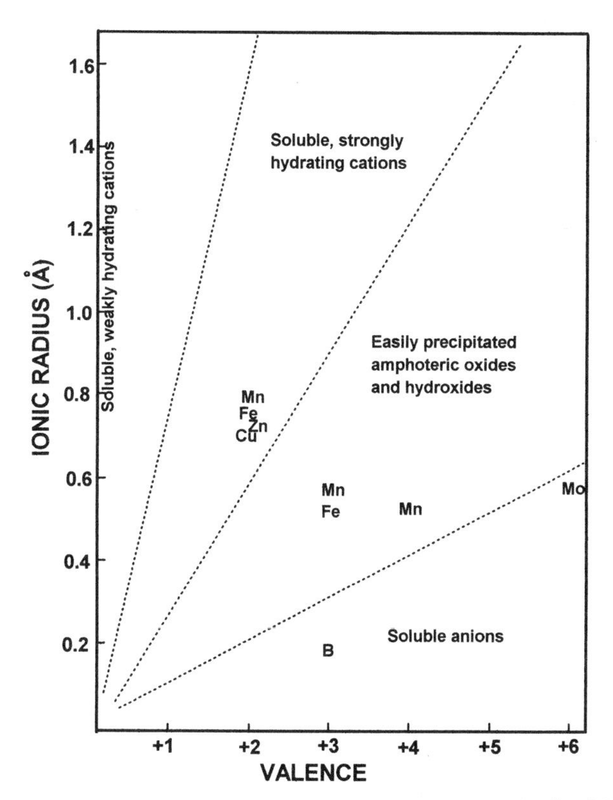

Fig. 5–8. Classification of elements into four groups based in ionic charge and radius (McBride, 1994).

ticles with high amounts of organic C (Villhoth, 1999). Adsorption of organic matter by Fe oxides can actually reverse the surface charge from positive to negative (Kretzschmer and Sticher, 1997), giving the necessary charge characteristics for metal transport. Colloids with high negative surface charge and organic C content are the ones that enhance metal transport the most, with less transport by larger particles with low negative surface charge and high contents of Fe and Al oxides (Karathanasis, 1999).

BIOLOGICAL PROCESSES

Plant roots alter soil conditions so as to affect some of the processes discussed above. Roots can exude protons to lower the pH and exude organic materials, which act as complexing agents for the micronutrient ions, and even O_2, which oxidizes Fe and Mn. The organics produced are used by soil bacteria as a C source, which builds the populations near the root. This increased bacterial activity can also have effects on micronutrients by the same mechanisms as plant roots.

Certain plants can cause as much as a threefold increase in metals in the soil solution by the action of biologically produced chelates in the rhizosphere. These materials alter the flux to the roots by mass flow and diffusion processes. These chelates have been found to be high in soils with actively growing plants whereas they are missing in fallow soil. Plants are induced to produce increased amounts of chelates when metals become deficient. These materials are effective in mobilizing all the metal cations discussed. Plants exhibit two strategies for obtaining Fe from soils of low Fe availability. The grasses produce organic phytosiderophores to chelate the Fe, while other species produce protons to acidify the soil and mobilize Fe. Mugineic acid is a representative compound of the phytosiderophores. These have to be constantly produced to be effective due to rapid decomposition by bacteria. As for the general case, more phytosiderophores are produced under low Fe conditions, but also Fe-insensitive (to deficiency) varieties produce more than Fe-sensitive varieties. Since amorphous Fe oxides are more soluble, these chelates are more effective in providing the plant with Fe when the amorphous forms are present than when only the more crystalline forms are present in the soil.

Rice (*Oryza sativa* L.) plants have the ability to exude O_2 into the rhizosphere, which has the effect of keeping the copious amounts of Fe in solution under flooding (reduced) conditions from being toxic to the plant roots. This O_2 precipitates Fe as hydrous oxides, making it unavailable to the plant. The oxidation of Fe also produces protons, which can result in a drop of more than 0.2 pH units next to the root. The reactions is

$$4Fe^{2+} + O_2 + H_2O \rightleftharpoons 4Fe(OH)_3 + 8H^+ \quad [9]$$

Figure 5– 9 shows that as the pH decreases near the plant root, there is a buildup of Fe, with the Fe(III) increasing due to the oxidation. The effect causes a plaque of Fe oxide to form on root surfaces. This Fe oxide layer can have the effect of adsorbing metals and phosphate, causing deficiencies or alleviating toxic conditions.

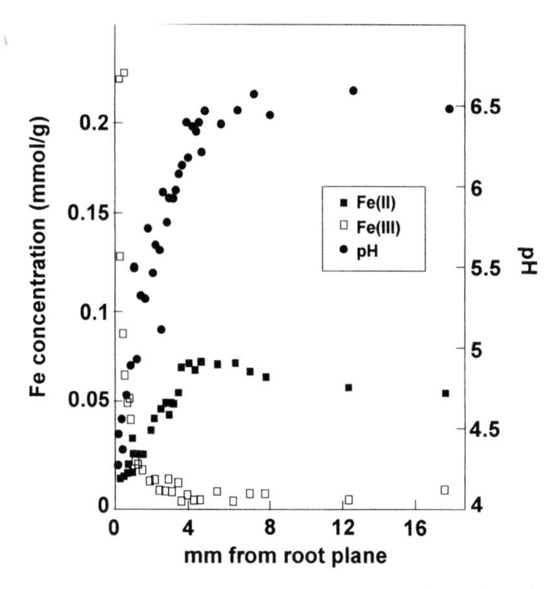

Fig. 5–9. Profiles of Fe (II), Fe(III), and pH in flooded soil exposed to a planer layer of rice roots for 10 d (Kirk et al., 1993).

REFERENCES

Adriano, D.C. 1986. Trace elements in the terrestrial environment. Springer-Verlag, New York.

Hayes, K.F. 1987. Equilibrium, spectroscopic, and kinetic studies on ion adsorption at the oxide/aqueous interface. Stanford University, Stanford, CA.

Karathanasis, A.D. 1999. Subsurface migration of copper and zinc mediated by soil colloids. Soil Sci. Soc. Am. J. 63:830–838.

Keren, R., and R.G. Gast. 1983. pH-dependent boron adsorption by montmorillonite hydroxy-aluminum complexes. Soil Sci. Soc. Am. J. 47:1116–1121.

Kirk, G.J.D., C.B.M. Begg, and J.L. Solivas. 1993. The chemistry of lowland rice rhizosphere. Plant Soil 155/156:83–86.

Kretzschmar, R., and H. Sticher. 1997. Transport of humic-coated iron oxide colloids in a sandy soil: Influence of Ca^{2+} and trace metals. Environ. Sci. Technol. 31:3497–3504.

Lindsay, W.L. 1979. Chemical equilibria in soils. John Wiley and Sons, New York.

McBride, M.B. 1994. Environmental chemistry of soils. Oxford University Press, New York.

Reisenauer, H.M., A.A. Tabikh, and P.R. Stout. 1962. Molybdenum reactions with soils and the hydrous oxides of iron, aluminum, and titanium. Soil Sci. Soc. Am. Proc. 26:23–27.

Shuman, L.M. 1975. The effect of soil properties on zinc adsorption by soils. Soil Sci. Soc. Am. Proc. 39:454–458.

Shuman, L.M. 1977. Adsorption of Zn by Fe and Al hydrous oxides as influenced by aging and pH. Soil Sci. Soc. Am. J. 41:703–706.

Shuman, L.M. 1985. Fractionation method for soil microelements. Soil Sci. 140:11–22.

Shuman, L.M. 1988. Effect of removal of organic matter and iron- or manganese-oxides on zinc adsorption by soil. Soil Sci. 146:248–254.

Shuman, L.M. 1991. Chemical forms of micronutrients in soil. p. 113–44. *In* J.J. Mortvedt et al. (ed.) Micronutrients in agriculture. 2nd ed. SSSA Book Ser. 4. SSSA, Madison, WI.

Sposito, G. 1982. On the use of the Langmuir equation in the interpretation of "adsorption" phenomena: II. The "two-surface" Langmuir equation. Soil Sci. Soc. Am. J. 46:1147–1152.

Sposito, G. 1984. The surface chemistry of soils. Oxford University Press, New York.

Sposito, G. 1989. The chemistry of soils. Oxford University Press, New York.

Villholth, K. G. 1999. Colloid characterization and colloidal phase partitioning of polycyclic tic hydrocarbons in two creosote-contaminated aquifers in Denmark. Environ. Sci. Technol. 33:691–699.

Chapter 6

Kinetics and Mechanisms of Soil Biogeochemical Processes

C. J. MATOCHA, *University of Kentucky, Lexington, Kentucky, USA*

K. G. SCHECKEL, *USEPA, Cincinnati, Ohio, USA*

D. L. SPARKS, *University of Delaware, Newark, Delaware, USA*

The application of kinetic studies to soil chemistry is useful to determine reaction mechanisms and fate of nutrients and environmental contaminants. How deeply one wishes to query the mechanism depends on the detail sought. Reactions that involve chemical species in more than one phase are termed *heterogeneous* and occur in soil and geochemical environments (Lasaga, 1981; Sparks, 1999). The mixture of inorganic and organic components with a range of reactivities that vary spatially makes the study of chemical kinetics in soils difficult. Separating the rates of biologically mediated processes from abiotic processes is an important consideration in modeling fate of plant nutrients, trace metals, and environmental contaminants in soils. A combination of kinetic and spectroscopic tools is essential to elucidate reaction mechanisms (Sposito, 1994; Bertsch and Hunter, 1998; Sparks, 1999). Analogies between surface complexes and their dissolved counterparts can be drawn to assist in proposing reaction mechanisms (Phillips et al., 1997, 1998).

Equilibrium models have provided valuable insight into predicting the distribution of chemical species likely to develop in soils under typical environmental conditions. However, most soil systems are in disequilibrium, and using only equilibrium constants for modeling purposes would be misleading because they provide no information about reaction rates (Sparks, 1989; Sparks et al., 1996). Typical soil chemical reactions have time scales that range from microseconds to days for some aqueous complexation and sorption phenomena to years for mineral precipitation processes (Fig. 6–1). The choice of experimental methodology depends on the time scale of the reaction kinetics (Amacher, 1991; Sparks, 1999). The focus of this chapter is to develop the principles of kinetics into a framework to solve problems related to soil chemistry. For additional discussions on topics of soil chemical kinetics, the reader should consult recent books and monographs (Amacher, 1991; Sposito, 1994; Sparks, 1989, 1998a, 1998b, 1999).

Chemical Processes in Soils. SSSA Book Series, no. 8.

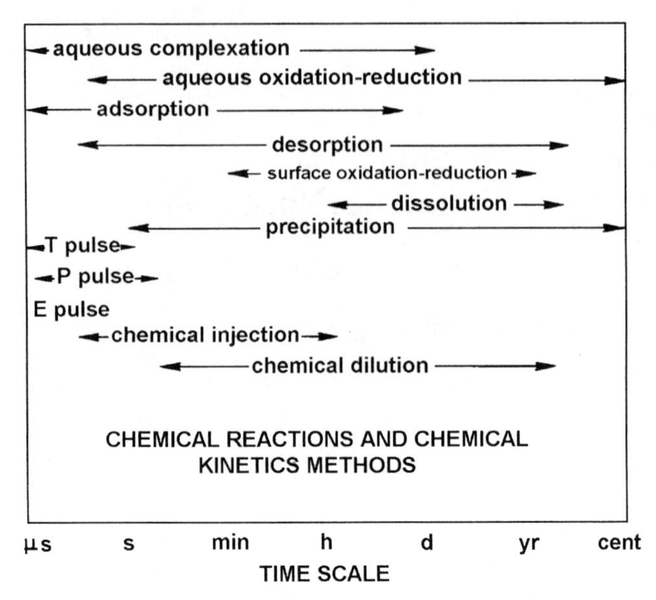

Fig. 6–1. Characteristic time scales for soil chemical reactions and for some typical experimental kinetic techniques (from Sposito, 1994).

FUNDAMENTALS OF KINETICS APPLIED TO HETEROGENEOUS SYSTEMS

Elementary and Overall Reactions

It is important to first distinguish between overall and elementary reactions. An elementary reaction occurs in a single molecular process with no intermediate species appearing before product formation and describes an exact reaction mechanism or pathway (Lasaga, 1981; Espenson, 1995). Elementary reactions are usually uni- or bimolecular. An example of an elementary reaction would be the self-ionization of water:

$$H_2O \rightarrow H^+ + OH^- \quad [1]$$

This reaction has a molecularity of one (unimolecular). Molecularity is identical to the overall order of the rate expression for elementary reactions. The same is not true of an overall reaction, where the order of the rate expression reflects the net result of a series of elementary reactions. Fenton's reaction for Fe^{2+} oxidation by H_2O_2 was outlined by Lasaga (1981) as an example of an overall reaction:

$$2Fe^{2+} + H_2O_2 + 2H^+ \rightarrow 2Fe^{3+} + 2H_2O \quad [2]$$

The likelihood of simultaneous collision of 1 mol of H_2O_2, 2 mol of Fe^{2+}, and 2 mol of H^+ to produce two Fe^{3+} and two H_2O is small. In fact, several stepwise reactions involving important radical intermediates $^{\bullet}OH$ and HO_2 comprise the overall reaction and do not appear in Eq. [2]. These intermediates are important in determining the overall rate.

The overall reaction stoichiometry is a form of chemical bookkeeping and does not describe a reaction mechanism (Brezonik, 1994). However, it is necessary to verify prior to experimentally determining a rate expression because changes in solution conditions such as pH can affect product distribution and surface loading and eventually reaction rates (Lasaga, 1981). For example, the reduction of Cr(VI) by Fe(II) in solution (Fendorf and Li, 1996; Buerge and Hug, 1997):

$$3Fe^{2+} + HCrO_4^- + 7H^+ \rightarrow 3Fe^{3+} + Cr^{3+} + 4H_2O \qquad [3]$$

$$3Fe^{2+} + HCrO_4^- + 8H_2O \rightarrow Fe_3Cr(OH)_{12}(s) + 5H^+ \qquad [4]$$

is acid consuming at pH < 4 (Eq. [3]) and acid producing at pH > 4 (Eq. [4]) because of formation of a mixed Fe–Cr precipitate.

Rate Expressions and Equilibrium Constants

One of the most useful concepts is the rate law or rate expression, which describes the time-dependent velocity or rate at which a reaction proceeds in terms of the concentrations of all species that affect the rate (Espenson, 1995). The kinetic approach is useful because it expresses the rate as changes in reactant and product concentration with time and it empirically determines the reaction orders. Consider the following simple forward reaction:

$$A \xrightarrow{k_1} P \qquad \text{rate} = -d[A]/dt = d[P]/dt = k_1[A]^a \qquad [5]$$

where a is the power dependency or reaction order and k_1 is the rate constant, assuming the reaction stoichiometry is known and constant for the time period investigated. The value of k_1 should depend only on temperature; thus, it should be specified and time invariant (Sparks, 1989, 1995; Brezonik, 1994). The rate law is defined once the value of a is determined. If unity, then the reaction is first order in A and first order overall. Fractional order rate expressions are not uncommon (Sparks, 1989). The rate, $-d[A]/dt$, is usually given in terms of molarity per second, and the units of k for a first-order reaction are per second, minute, or hour.

If A and P in Eq. [5] are at equilibrium, then a thermodynamic equilibrium constant K can be defined (neglecting activity corrections):

$$A \underset{k_{-1}}{\overset{k_1}{\rightleftharpoons}} P \qquad K = [P]/[A] = k_1/k_{-1} \qquad [6]$$

where k_1 is the rate constant for the forward reaction and k_{-1} is the rate constant for the reverse reaction. The chemical thermodynamic approach defines the relationship governing the concentrations at equilibrium, and the exponents are raised to the values in the stoichiometric equation. This approach provides no information on the reaction rates, and the dependence on concentration does not necessarily have to follow Eq. [6]. Some reactions may be thermodynamically favorable but kinetically inert. An example is the solution complexation reaction of the Co(III) cation in octahedral coordination with NH_3 ligands (Stumm and Morgan, 1996):

$$Co(NH_3)_6^{3+} + 6H_3O^+ \rightarrow Co(H_2O)_6^{3+} + 6NH_4^+ \qquad K \sim 10^{25} \qquad [7]$$

This reaction requires several days to be complete despite the large thermodynamic driving force. Some redox reactions at room temperature only proceed at measurable rates in the presence of a catalyst. Chemodenitrification of NO_3^- to N_2 by aqueous Fe(II):

$$10Fe^{2+} + 2NO_3^- + 24H_2O$$

$$\rightleftharpoons 10Fe(OH)_3(s) + N_2(g) + 18H^+ \; \Delta G^\circ = -18 \text{ kJ mol}^{-1} \qquad [8]$$

is thermodynamically favorable, but rate controlled because a catalyst such as a solid phase or dissolved Cu^{2+} is required for the reaction to proceed at room temperature (Buresh and Moraghan, 1976).

Experimental Determination of Elementary and Complex Rate Expressions

Rate expressions can be determined using integrated equations, nonlinear least square fits of the raw data, and initial rates with the method of isolation (Sparks et al., 1996; Sparks, 1998b). It is generally assumed that reaction rates depend on concentration to conform to the law of mass action (Brezonik, 1994). Zero-order reactions are a clear violation of this law in that there is no dependence on concentration (Fig. 6–2):

$$A \xrightarrow{k_0} P \qquad -d[A]/dt = k_0[A]^0 = k_0 \qquad [9]$$

After integration, a plot of $[A]_t$ vs. time will be linear:

$$[A]_t = [A]_0 - k_0 t \qquad [10]$$

Zero-order dependence usually implies that heterogeneous or homogeneous catalysis is involved in the reaction and excess substrate is available for reacting (Brezonik, 1994). The integrated form of Eq. [5] with boundary conditions of $[A] = [A]_0$ at $t = 0$, would be

$$\ln[A]_t/[A]_0 = -k_1 t \qquad [11]$$

A plot of $\ln[A]_t/[A]_0$ against t should be linear if the decay of A is first order (Sparks, 1998b), and $-k_1$ will be the slope of the line. The determination of half-life ($t_{1/2}$), the time required for one-half of the reactant A to be consumed, is a useful quantity and can be evaluated at $t = t_{1/2}$, and $[A]_t = 1/2[A]_0$ to give

$$t_{1/2} = \ln 2/k_1 = 0.693/k_1 \qquad [12]$$

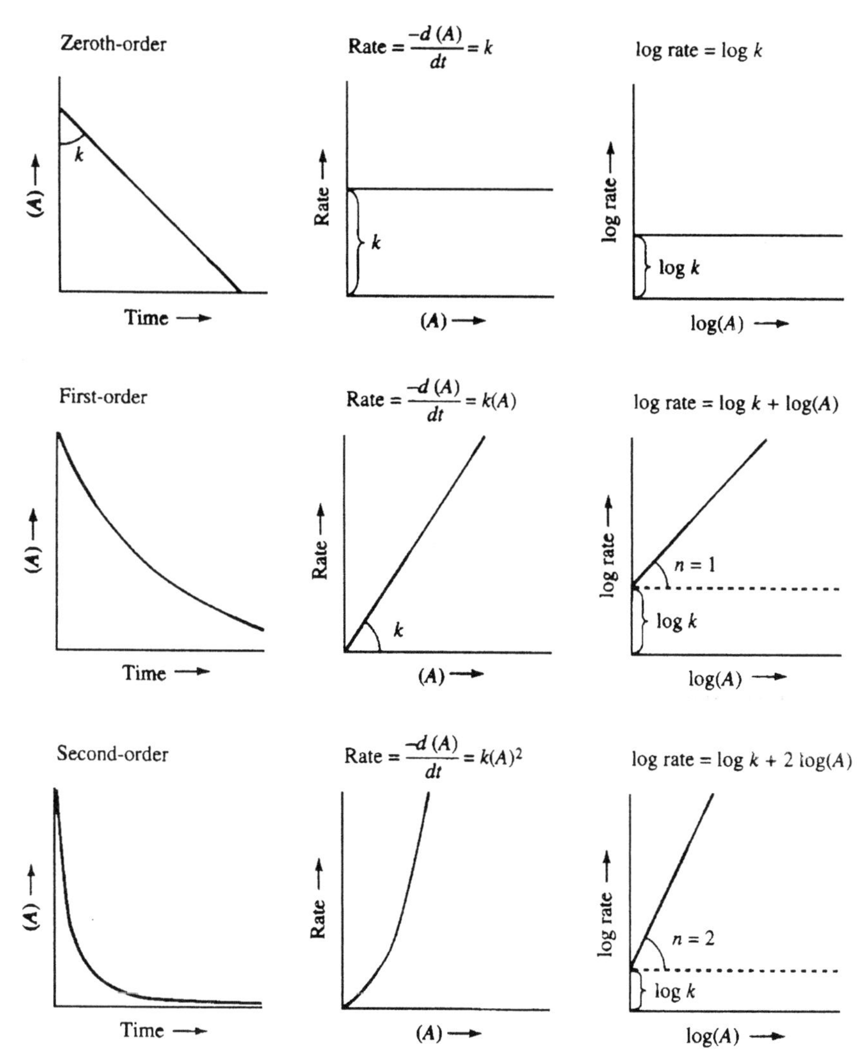

Fig. 6–2. The general appearance of concentration–time, rate–concentration, and log rate–log concentration profiles for simple rate expressions involving reactant A (from Appelo and Postma, 1993).

For first-order reactions, $t_{1/2}$ is inversely dependent on k_1 and independent of initial concentration.

Elementary bimolecular reactions follow second-order kinetics as shown:

$$2A \xrightarrow{k_2} P \qquad d[A]/dt = -k_2[A]^2 \qquad [13]$$

$$A + B \xrightarrow{k_2} P \qquad dx/dt = k_2[A][B] \qquad [14]$$

For Eq. [13], assuming an irreversible reaction, integration and linearization lead to

$$1/[A]_t = 1/[A]_0 + k_2t \quad [15]$$

The general appearance of simple rate laws involving reactant A differ predictively for each reaction order (Fig. 6–2).

Assuming that $[A]_0$ and $[B]_0$ are initial concentrations not equal to one another, and x is the amount reacted at time t for an irreversible reaction with boundary conditions $x = 0$ at $t = 0$, the integrated form of Eq. [14] is

$$\frac{1}{[A]_0 - [B]_0} \ln \frac{[B]_0[A]_t}{[A]_0[B]_t} = k_2t \quad [16]$$

To obtain $t_{1/2}$ for a second-order reaction, it must be assumed that $[A]_0 = [B]_0$ or that Eq. [13] holds, resulting in

$$t_{1/2} = 1/[A]_0k_2 \quad [17]$$

which indicates that $t_{1/2}$ is inversely proportional to initial concentration (Sparks, 1989).

In the case where back reactions are operative, Eq. [5] now becomes

$$A \underset{k_{-1}}{\overset{k_1}{\rightleftarrows}} P \qquad \mathrm{d}[A]/\mathrm{d}t = -k_1[A] + k_{-1}[P] \quad [18]$$

where k_1 and k_{-1} are the first-order rate constants for the forward and reverse reactions (Sparks, 1989). To solve the reversible first-order equation, two other expressions are needed. The stoichiometry of the reaction and the thermodynamic equilibrium condition (subscripted e) (Eq. [19] and [20]) must be defined:

$$[A]_0 + [P]_0 = [A]_e + [P]_e = [A]_t + [P]_t \quad [19]$$

$$k_1[A]_e = k_{-1}[P]_e \quad [20]$$

After rearrangement one obtains

$$[P]_e/[A]_e = K = k_1/k_{-1} \quad [21]$$

An expression in which $[A]$ is the only concentration variable can be derived and integration leads to

$$-\mathrm{d}[A]/\mathrm{d}t = (k_1 + k_{-1})([A] - [A]_e) \quad [22]$$

$$\ln \frac{[A]_t - [A]_e}{[A]_0 - [A]_e} = -(k_1 + k_{-1})t \quad [23]$$

A plot of the left side of Eq. [23] against time will be linear, and the negative of its slope is $-(k_1 + k_{-1})$ (Espenson, 1995).

Often, processes in soils will not cease after one step, but may continue. If each step is first order, then consecutive first-order reactions can be represented in general as

$$A \xrightarrow{k_1} B$$

$$B \xrightarrow{k_2} C$$

$$-\mathrm{d}[A]/\mathrm{d}t = k_1[A], \quad \mathrm{d}[B]/\mathrm{d}t = k_1[A] - k_2[B], \quad \mathrm{d}[C]/\mathrm{d}t = k_2[B] \qquad [24]$$

At any time, $[A]_0 = [A] + [B] + [C]$ and at $t = 0$, $[A] = [A]_0$, $[B] = [C] = 0$, which allows one to integrate as follows:

$$[A]_\mathrm{t} = [A]_0\, \mathrm{e}^{-k_1 t} \qquad [25]$$

$$[B]_t = \frac{[A]_0 k_1}{k_2 - k_1}(\mathrm{e}^{-k_1 t} - \mathrm{e}^{-k_2 t}) \qquad [26]$$

$$[C]_t = [A]_0\left[1 - \frac{k_2}{k_2 - k_1}\,\mathrm{e}^{-k_1 t} + \frac{k_1}{k_2 - k_1}\,\mathrm{e}^{-k_2 t}\right] \qquad [27]$$

The steady-state approximation can be employed to simplify the expressions by assuming that $[B]_t$ is small and does not change much during the reaction ($k_1 < k_2$):

$$\mathrm{d}[B]/\mathrm{d}t \cong 0 \text{ and } \mathrm{d}[B]/\mathrm{d}t = k_1[A] - k_2[B] \cong 0 \qquad [28]$$

This reduces the expressions to

$$[B]_\mathrm{t} = \left(\frac{k_1}{k_2}\right)[A] = \left(\frac{k_1}{k_2}\right)[A]_0\, \mathrm{e}^{-k_1 t} \qquad [29]$$

$$[C]_t = [A]_0 - [A] - [B] \qquad [30]$$

$$\cong [A]_0 (1 - \mathrm{e}^{-k_1 t}) \qquad [31]$$

These simplified expressions are useful in that they allow one to predict the concentration of all three species as a function of time by knowing only $[A]_0$, k_1, and k_2.

In deciding whether integrated and linearized forms of the elementary rate expressions are correct, nonlinear least squares programs can be used to fit data directly to various differential equations to directly calculate k_1 and k_2, thus minimizing the error associated with data linearization (Espenson, 1995).

The method of initial rates is particularly suited for rapid heterogeneous systems because one measures the concentration of reactant or product over a short reaction time to preclude back reactions and simplify rate expressions (Lasaga, 1981; Walker et al., 1988; Espenson, 1995; Sparks, 1998b). For very rapid reactions, one must employ an experimental technique capable of rapid quantification of reactant or product concentration such as relaxation or stopped-flow techniques to measure initial rates (Sparks et al., 1996). The method of isolation is often used in measuring the initial rate by analyzing initial rate of reactant consumption or product formation in the presence of a large excess of the other reactant. Fixing concentrations of all but one reactant and varying that reactant concentration will allow the determination of reaction order with respect to that reactant. The rate then simplifies to a pseudo-first-order expression, and the initial rate can be plotted vs. the initial concentration of the variable reactant (Brezonik, 1994). For example:

$$A + B \xrightarrow{k_2} P \qquad \text{rate} = -\mathrm{d}[A]/\mathrm{d}t = k_2[A]^a[B]^b \qquad [32]$$

determination of the reaction order a for reactant A could be achieved by measuring the initial rate of decrease in $[A]$ as a function of $[A]_0$ in a large excess (10-fold or more) of B. Initial rate data can then be reexpressed as

$$\text{initial rate} = -\mathrm{d}[A]/\mathrm{d}t = k'_2[A]^a \qquad [33]$$

where $k_2' = k_2[B]^b$, B is the fixed reactant, and k_2' is the apparent or pseudo-first-order rate constant. The slope, a, of the log-log plot:

$$\log \text{rate} = \log k_2' + a\log[A] \qquad [34]$$

is equal to reaction order for reactant A, and the y intercept provides the pseudo-first-order rate constant. The process is repeated for B, leading to the overall rate constant k_2.

Other Kinetic Models

The Elovich equation was originally derived to describe heterogeneous chemisorption of gases on solid surfaces (Low, 1960) and has been adopted in soil chemistry to describe P, B, K, and As sorption and desorption kinetics on soils and soil minerals (Sposito, 1984; Sparks, 1989, and references therein). More recently, Wang et al. (2000) described salicylate adsorption on Al oxide to follow an Elovich rate law. In its usual form,

$$\mathrm{d}q/\mathrm{d}t = X\exp(-Yq) \qquad [35]$$

the rate of adsorption, $\mathrm{d}q/\mathrm{d}t$, exponentially decreases with increasing surface coverage, q, X is the reaction rate at zero coverage, and Y is the coverage scale factor. The integrated form is

$$q(t) = (1/Y)\ln(XYt_0) + (1/Y)\ln(1 + t/t_0) \qquad [36]$$

where t_0 is a fitting parameter and a plot of $q(t)$ vs. $\ln(1 + t/t_0)$ yields the slope Y. The Elovich equation is generally regarded as an empirical description of rate data (Sposito, 1984; Sparks, 1989).

The parabolic diffusion equation has often been used to indicate that diffusion-controlled phenomena are controlling the reaction rate (Sparks, 1989). Other models include the power function model and the $Z(t)$ and diffusion models. Recognition of the continuum between many soil chemical processes and transport phenomena has stimulated the derivation of chemical and physical nonequilibrium models that consider multiple components and sites. For more complete details and applications of these models, one may consult other sources (Sparks, 1989, 1995, 1999).

Activation Parameters

The requirement that temperature be specified in determining the rate expression is manifested in the exponential temperature dependence of k based on the Arrhenius equation (Sparks, 1989):

$$k = A\exp(-E_a/RT) \quad [37]$$

where A is a preexponential factor, E_a is the energy of activation (kJ mol^{-1}), R is the molar gas constant (0.008 314 kJ mol^{-1} K^{-1}), and T is absolute temperature (K). A plot of $\ln k$ vs. $1/T$ gives a slope of $-E_a/R$ and an intercept of $\ln A$. In soil chemistry, values of E_a >42 kJ mol^{-1} are indicative of surface chemical–controlled reactions, while E_a values <42 kJ mol^{-1} are diffusion-controlled reactions (Sparks, 1989). The overall activation energy will be the net result of a combination of activation energies for elementary reactions that comprise the reaction mechanism.

The rate constant is related to the free energy of activation, $\Delta G^{\ddagger}$, by the Eyring expression (Atwood, 1985):

$$k = (k'\mathrm{T}/h)\exp(-\Delta G^{\ddagger}/RT) \quad [38]$$

where h is Planck's constant (6.626×10^{-34} J s) and k' is Boltzmann's constant (1.381×10^{-23} J K^{-1}). Substituting $\Delta G^{\ddagger} = \Delta H^{\ddagger} - T\Delta S^{\ddagger}$ into Eq. [38] and taking the logarithm gives

$$\ln(k/T) = -\Delta H^{\ddagger}/RT + \ln(k'/h) + \Delta S^{\ddagger}/R \quad [39]$$

A plot of $\ln(k/T)$ vs. T^{-1} yields a straight line giving $\Delta H^{\ddagger}$ from the slope and $\Delta S^{\ddagger}$ from the intercept. The values of $\Delta S^{\ddagger}$ can be useful in determining the reaction mechanism. Values of $\Delta S^{\ddagger}$ less than −10 J mol^{-1} K^{-1} indicate an associative reaction and those reactions >10 J mol^{-1} K^{-1} are dissociative. An associative reaction predicts that an intermediate is formed with a higher coordination number than the original complex. A dissociative reaction mechanism predicts that an intermediate with a reduced coordination number is formed subsequent to the cleavage of the bond to the leaving group (Shriver et al., 1994). For dissociative reactions, $\Delta H^{\ddagger}$ can provide information about bond strength (Atwood, 1985).

REACTION MECHANISMS

The rate equation mathematically describes the rate of the rate-controlling step in a process and is expressed in terms of measurable quantities, such as concentrations of original reactants or products (Espenson, 1995). Equipped with the rate equation, a mechanism can be proposed that is consistent with the experimental data. It is important to keep in mind that a rate equation alone cannot give a unique mechanism, but several mechanisms often can be written for the same rate equation (Brezonik, 1994). The previous example of Cr(VI) reduction by Fe(II) (see Eq. [3] and [4]) was reported to follow a second-order rate law (Buerge and Hug, 1997):

$$-d[Cr(VI)]/dt = k_{65}[Fe^{2+}][Cr(VI)] \quad [40]$$

where $k_{65} = k_{obs}(pH)$. The rate equation requires that the reaction occurs by a multistep process because the one step reaction would be a higher order rate law based on the third-order dependence in $[Fe^{2+}]$. The reaction mechanism was written as successive one-electron transfers, with the first transfer representing the rate-limiting step:

$$Cr(VI) + Fe(II) \underset{k_{56}}{\overset{k_{65}}{\rightleftharpoons}} Cr(V) + Fe(III) \quad [41]$$

$$Cr(V) + Fe(II) \underset{k_{45}}{\overset{k_{54}}{\rightleftharpoons}} Cr(IV) + Fe(III) \quad [42]$$

$$Cr(IV) + Fe(II) \underset{k_{34}}{\overset{k_{43}}{\rightleftharpoons}} Cr(III) + Fe(III) \quad [43]$$

That Eq. [41] is the rate-limiting step is internally consistent with the empirical rate expression (Eq. [40]), where both Fe(II) and Cr(VI) appear.

Effects of reaction conditions (reactant concentrations, pH, ionic strength, temperature, and flow rate) on the observed rate constant can provide valuable information on the rate-controlling step and nature of the transition state, and can support or rule out hypothesized mechanisms. Observed rate constants can be expressed at any pH for each acid–base species times the fraction of reactant in each of its various protonated forms:

$$k_{obs} = \sum_{i=0}^{n} k_i \alpha_i \quad [44]$$

where k_{obs} is the observed rate constant, i represents the degree of deprotonation, and k_0 and α_0 refer to the fully protonated species (Brezonik, 1994). For a monoprotic acid, HA:

$$HA \rightleftharpoons H^+ + A^- \qquad K_a = [H^+][A^-]/[HA] \quad [45]$$

$\alpha_{H\alpha}$ can be calculated:

$$\alpha_{HA} = 1/(1 + K_a/[H^+]) \quad [46]$$

and then evaluated as shown in Eq. [44].

Direct spectroscopic and microscopic confirmation of heterogeneous reaction products such as sorbed species or solid phase transformations can provide additional support for proposed mechanisms (Sparks, 1995). In solution chemistry where homogeneous systems are studied, diffusion-limited reactions are rapid ($>10^9$ M^{-1} s^{-1}) (Espenson, 1995). Rates of soil chemical processes, in contrast, are often controlled by slower intraaggregate and interaggregate diffusion because of the presence of a porous solid phase (Sparks, 1989). Transport processes include transport in the soil solution, film diffusion, interaggregate (interparticle) diffusion, and intraagregate diffusion (Fig. 6–3). Good tests of whether one is measuring chemical kinetics or transport are to run reactions at different flow rates or initial concentrations of reactant. If the rate constant changes, then it is likely that transport processes are being measured (Sparks, 1989, 1998a). Experimentally determined E_a values indicate transport-controlled vs. surface chemical reaction-controlled processes. In soil systems, rate constants are generally conditional or apparent because of the influence of transport processes (Skopp, 1985; Sparks, 1999).

Another useful method of determining mechanistic information, particularly for organic reactions, would be to compare reaction rates using radioisotopes as tracers. By substituting D for H at a suspected reactive site in a reactant, the difference in mass of the elements gives rise to a greater dissociation energy for the C–D bond when compared with the C–H bond making the C–D bond more difficult to cleave (Adamson, 1979). Thus, for rate-limiting C–H bond cleavage, a primary kinetic isotope effect would be observed where $k_{C-H}/k_{C-D} > 1$. Determination of where bonds are broken and formed can be made by characterizing products as a function of time using other isotopes of carbon (^{14}C, ^{13}C) and oxygen (^{17}O, ^{18}O) (Brezonik, 1994).

The correlation of rate constants with thermodynamic data can be useful in predicting contaminant fate. The most common relationships are the correlations between rate and equilibrium constants for related reactions, known as linear free-energy relationships (LFERs). Kinetic LFERs are important because they assist in

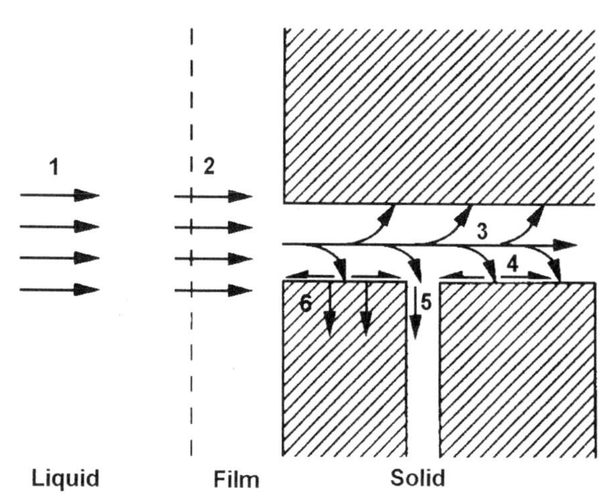

Fig. 6–3. Transport processes in solid–liquid soil reactions: transport in the soil solution (1), transport across a liquid film at the solid–liquid interface (2), transport in a liquid-filled micropore (3), diffusion of a sorbate at the surface of the solid (4), diffusion of a sorbate occluded in a micropore (5), and diffusion in the bulk of the solid (6) (from Aharoni and Sparks, 1991).

elucidating reaction mechanisms and allow the prediction of reaction rates from more available equilibrium properties (Brezonik, 1994). However, without careful study, one should not assume a priori that these relationships exist.

KINETICS OF IMPORTANT SOIL PROCESSES

Sorption–Desorption and Biotransformation Kinetics

The kinetics of sorption and desorption are extremely important in regulating the solubility and fate of many plant nutrients, trace metals, and organic compounds in soil environments (Sparks, 1995). Trace metal sorption reactions can occur at time scales ranging from less than milliseconds to greater than months, depending on the reaction environment of the sorbate–sorbent system. The availability, types, and accessibility of sorption sites on the sorbent, the reactive nature of the metal cation (sorptive), and solute conditions influence the sorption kinetics. Nonmetal cations typically sorb onto mineral surfaces by ion exchange processes (outer-sphere complexation) and metal cations and oxyanions by outer-sphere, inner-sphere, or surface precipitation processes (Sparks, 1999). It is important to acknowledge that outer- and inner-sphere complexes often evolve simultaneously on a surface. Inner-sphere complexation is often not reversible, and adsorption by this mechanism is weakly affected by the ionic strength of the aqueous solution. Inner-sphere complexation can increase, reduce, neutralize, or reverse the charge on the surface regardless of the original charge. Adsorption of ions via inner-sphere complexation can occur on a surface indifferent of the surface charge and involves a ligand exchange process where the metal or metalloid exchanges with OH^- or H_2O and becomes directly bound to the metal in the mineral's crystalline structure (Sparks, 1995). Biotransformation of inorganic and organic contaminants further complicates the effects of sorption and desorption, and needs to be considered in field-scale predictions of contaminant fate.

Ion Exchange

Ion exchange reactions are crucial in nutrient dynamics in soils and involve mixtures of inorganic and organic components (McBride, 1994; Sposito, 1994; Sparks, 1995). Ion exchange kinetics greatly depend on the sorbent and ion, involve long-range electrostatic bonds of low energy (outer sphere surface complexes), and are typically rapid (Sparks, 1995). Pressure-jump relaxation indicated that Ca–Na exchange on montmorillonite:

$$\text{2Na-montmorillonite} + Ca^{2+} \underset{k_{-1} = 643\ M^{-2}\ s^{-1}}{\overset{k_1 = 953\ M^{-2}\ s^{-1}}{\rightleftarrows}} \text{Ca-montmorillonite}_2 + 2Na^+ \qquad [47]$$

was complete in <100 ms (Fig. 6–4), with forward (k_1) and reverse (k_{-1}) rate constants greater than those measured for Ca–K exchange ($k_1 = 385$ M^{-2} s^{-1}, $k_{-1} = 432$ M^{-2} s^{-1}). This kinetic approach supported previous findings that have indicated preference of K over Ca in soil minerals (Tang and Sparks, 1993).

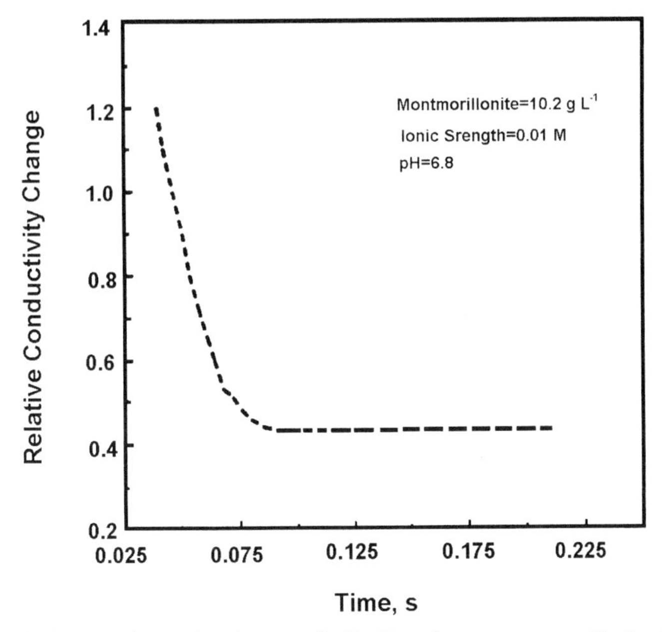

Fig. 6–4. Typical pressure jump relaxation curve for Ca–Na exchange on montmorillonite showing relative change in conductivity vs. time (from Tang and Sparks, 1993).

Chemisorption

The tendency for bonding of a metal ion with surface oxo groups that represent σ- and π-donors ligands (OH^-) and coordination of the same metal ion with OH^- in solution has been documented and can be represented for the metal M^{n+}(McBride, 1994):

$$>S\text{–}OH + M^{n+} \rightarrow >S\text{–}O\text{–}M^{(n-1)+} + H^+ \qquad [48]$$

$$H\text{–}OH + M^{n+} \rightarrow H\text{–}O\text{–}M^{(n-1)+} + H^+ \qquad [49]$$

where >S–OH represents a metal surface site. These inner-sphere surface complexes are more stable because the metal ion binds directly with the surface functional group (Sposito, 1984). Inner-sphere surface complexation and outer-sphere complexation can occur simultaneously (Sparks, 1995). In general, sorption behavior of transition metals is initially rapid followed by a slow reaction (Strawn and Sparks, 1999; Sparks, 1999). Transition metal ions with d electrons have initial sorption rates (k_{int}) on Fe and Al oxides that scale linearly with rates of water exchange, k_{H2O}, indicating formation of monodentate inner sphere surface complexes (Fig. 6–5 and 6–6). Manganese(II) adsorption on birnessite (δ-MnO_2) was complete within milliseconds, and initial first-order rate plots yielded a logk_{int} of –0.071 M^{-1} s^{-1} (Fig. 6–7). This value falls just below the line in Fig. 6–6, but still demonstrates the reasonable agreement between k_{H2O} and k_{int} on a variety of metal oxide minerals.

Some anions are adsorbed as outer-sphere complexes on positively charged surfaces; however, oxyanions, such as sulfate, phosphate, and arsenate, can be sorbed both as inner-sphere and outer-sphere complexes (Zhang and Sparks, 1990; Hayes et al., 1987; Peak et al., 1999). Dissolved $H_2AsO_4^-$ and $HCrO_4^-$ formed inner sphere

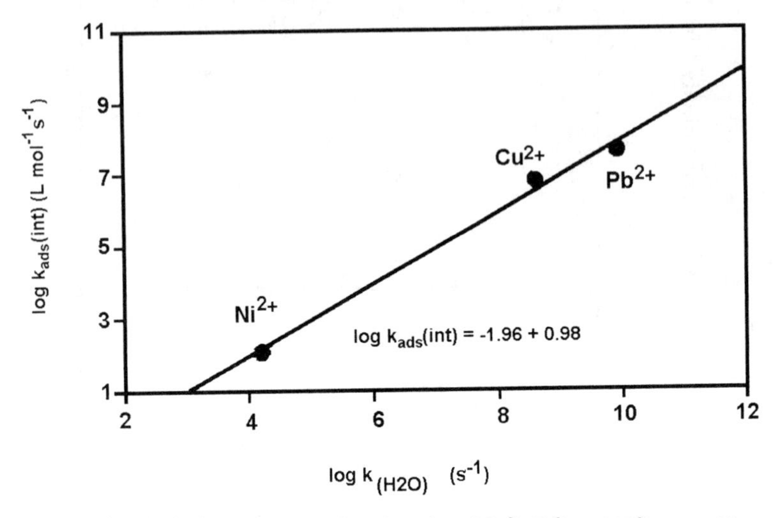

Fig. 6–5. Log of the intrinsic rate constants for adsorption of Cu^{2+}, Pb^{2+}, and Ni^{2+} on goethite vs. log of the rate constants for water exchange from the primary hydration spheres of these divalent metal cations (from Grossl et al., 1994).

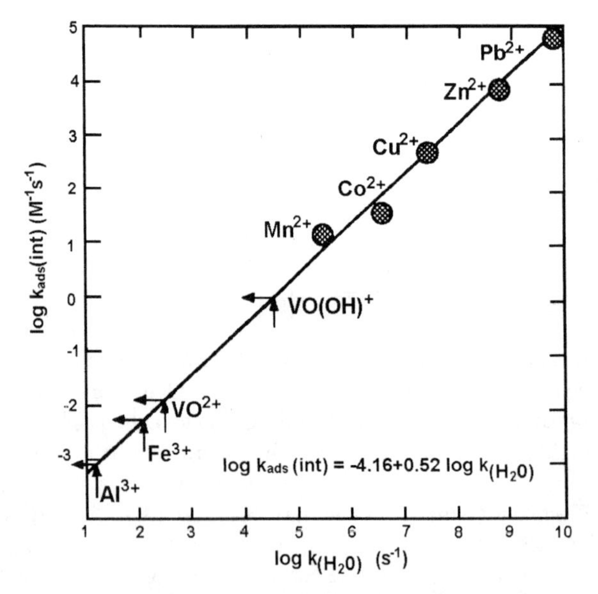

Fig. 6–6. Linear free energy relation between the rate constants for water exchange k_{-w} (s^{-1}) and the intrinsic adsorption rate constants $k_{ads(int)}$ (M^{-1} s^{-1}) from the pressure jump experiments of Hachiya et al. (1984). The intrinsic constants refer to an uncharged surface. The linear-free energy relations based on the experimental points are extended to some ions with lower k_{-w} (s^{-1}) in order to predict adsorption rates (from Wehrli, 1990).

surface complexes on goethite by a two-step sequence (Fendorf et al., 1997; Grossl et al., 1997). Kinetic evidence derived from pressure jump experiments and X-ray absorption fine structure (XAFS) spectroscopy revealed the presence of a monodentate inner sphere surface complex formed by ligand exchange in the first step ($k_1 = 10^{5.8}$ and $10^{6.3}$ L mol^{-1} s^{-1} for $HCrO_4^-$ and $H_2AsO_4^-$), followed by a second slower step ($k_2 = 16$ and 15 s^{-1} for CrO_4^- and $H_2AsO_4^-$) of bidentate surface complexation (Fig. 6–8).

Heterogeneous adsorption reactions of organic ligands at the metal oxide–water interface can resemble dissolved solution (homogeneous) complexes in structure and reactivity based on in situ spectroscopic results (Ainsworth et al., 1998). For example, the bidentate surface complex formed when reacting salicylate on Al oxide to form a six-membered ring could be modeled using the aqueous phase Eigen–Wilkens–Werner mechanism, where fast formation of an outer sphere ion pair is followed by a first-order ligand substitution process (Sposito, 1994; Wang et al., 2000). Wang et al. (2000) recently observed two pathways of salicylate adsorption on colloidal δ-Al_2O_3: a rapid reaction characterized by Elovich kinetic behavior followed by a slower reaction that obeyed pseudo-first-order kinetics (Fig. 6–9 and 6–10). The overall reaction mechanism was controlled by the leaving group lability at the Al–OH_2^+ and Al–OH surface sites.

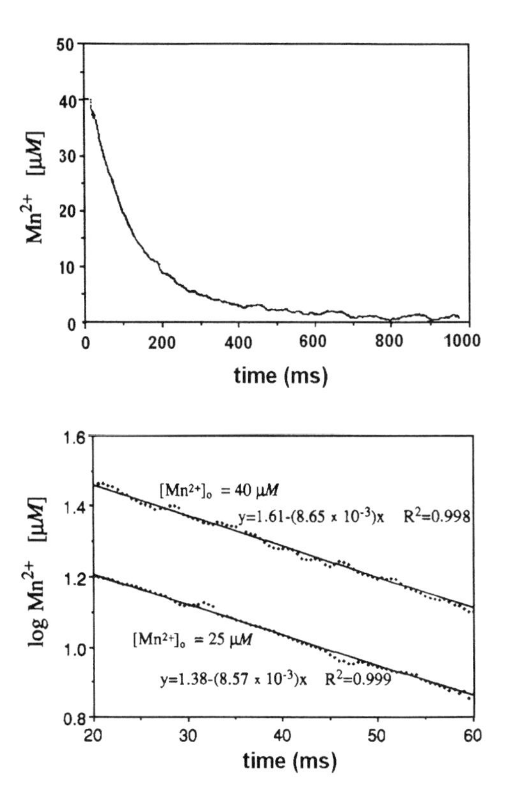

Fig. 6–7. Initial first-order reaction rate plots describing Mn(II) adsorption on δ-MnO_2 (from Fendorf et al., 1993).

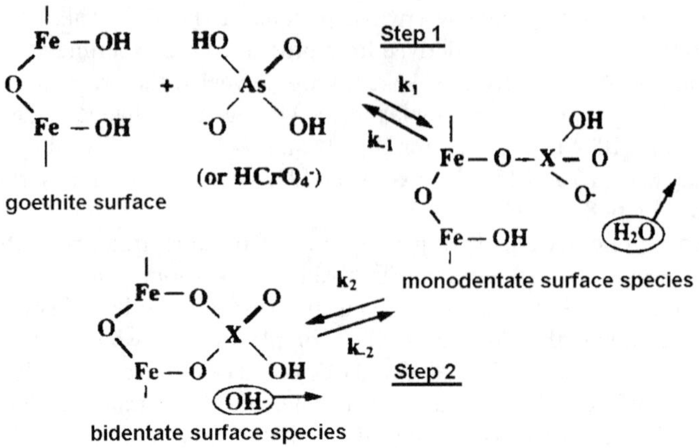

Fe — OH
O
Fe — OH
goethite surface
HO
As
O
OH
(or HCrO₄⁻)
Step 1
k1
k-1
Fe — O — X — O
OH
O
Fe — OH
H₂O
monodentate surface species
k2
k-2
Step 2
Fe — O
X
O
OH
Fe — O
OH-
bidentate surface species

Fig. 6–8. Proposed mechanism for oxyanion adsorption–desorption on goethite. The X represents either As(V) or Cr (VI) (from Grossl et al., 1997).

Desorption

While it is enormously important to understand sorption mechanisms of metals in soils and natural environments, it is equally necessary to examine potential desorption reactions because these systems are seldom at equilibrium. This is particularly true for soils that are already contaminated. Research on desorption will allow scientists to make better judgments regarding the fate and mobility of metal contaminants in soils and to develop environmentally sound, cost-effective remediation strategies. An important factor affecting the degree of desorption of metals is the time period the sorbate has been in contact with the sorbent (residence time). A number of researchers have noted that desorption trends are not usually the in-

7x10⁻⁶
6
5
4
3
2
1
0
[Al-Sal]
-1.0
-0.5
0.0
0.5
1.0
1.5
Log(t+t₀)(Second)

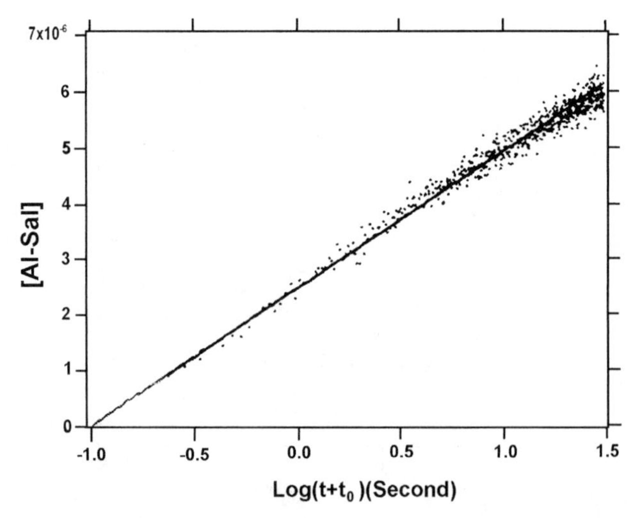

Fig. 6–9. Elovich plot of salicyclate adsorption kinetics on Al oxide at reaction time <30 s (from Wang et al., 2000).

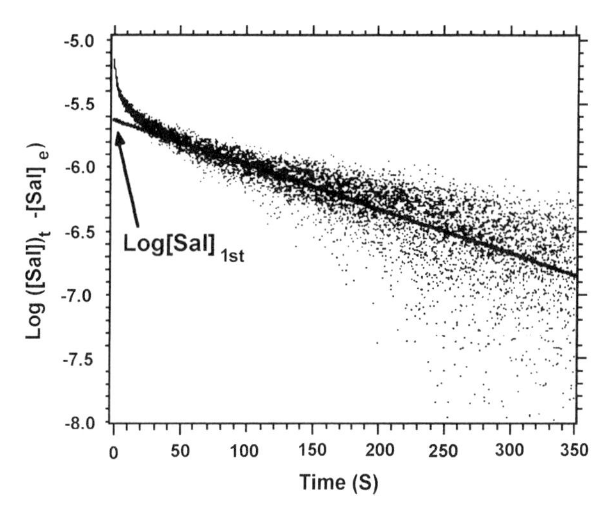

Fig. 6–10. Pseudo-first-order plot of salicyclate adsorption kinetics on Al oxide at reaction time >30 s and approximately 70% of the adsorption (from Wang et al., 2000).

verse of sorption processes, indicating that desorption tends to be irreversible because not all of the adsorbate is desorbed. This apparent irreversibility is often referred to as hysteresis. Ainsworth et al. (1994) studied the desorption of Co, Cd, and Pb on hydrous ferric oxide (HFO) as a function of oxide aging and metal sorption residence time. They noted that oxide aging did not cause hysteresis of metal cation desorption. However, increasing the residence time between the oxide and sorbed cations resulted in hysteresis with Cd and Co, but little hysteresis was observed with Pb. The relationship of reversibility with aging time for Co, Cd, and Pb was inversely proportional to the ionic radii of the metal ions (i.e., Co < Cd < Pb) and was attributed to isomorphic substitution into a crystallizing sorbent phase.

Precipitation–Dissolution

The slower uptake step in metal sorption studies has been ascribed to interparticle diffusion, surface precipitation, or sorption onto high activation energy sites (Steinberg et al., 1987; Brümmer et al., 1988; Sposito, 1984; Farley et al., 1985; Chisholm-Brause et al., 1990; Charlet and Manceau, 1992; Fendorf et al., 1994; Comans and Hockley, 1992; Benjamin and Leckie, 1981; Strawn and Sparks, 2000; Scheidegger et al., 1996a, 1996b, 1997, 1998; Towle et al., 1997; Xia et al., 1997; Thompson, 1998; Ford et al., 1999; Scheinost et al., 1999; Scheckel and Sparks, 2000). While adsorption has been characterized as a two-dimensional surface feature, surface precipitation has been depicted as three-dimensional growth. According to Sposito (1986), precipitation is the formation of a solid mixture either by inclusion or by coprecipitation on the surface of a preexisting solid phase (surface precipitate).

Pioneering work by Scheidegger et al. (1996a, 1996b, 1997, 1998) helped to explain the kinetic reactions of Ni(II) sorption on the surfaces of pyrophyllite and metal oxides by showing the formation of a mixed cation (Ni–Al) hydroxide phase.

They demonstrated that surface precipitation can occur at (i) low sorbate concentrations, much less than monolayer coverage; (ii) short time scales, within 15 min; and (iii) pH values lower than where one would expect the precipitation of a metal's hydroxide phase according to its thermodynamic solubility product. It appears that Al plays an important part in the formation of these surface precipitates. In the case where the sorbent phase contained Al, the precipitates were predominantly Al-containing layered double-hydroxide (LDH) phases (Scheidegger et al., 1997; Scheinost et al., 1999). Metal adsorption on Al-free sorbents resulted in α-$Ni(OH)_2$ precipitates (Scheinost et al., 1999; Scheinost and Sparks, 2000). In fact, O'Day et al. (1994), observing Co (II) reactions on kaolinite; Charlet and Manceau (1992), studying Cr (III) interactions with hydrous ferric oxides; Roe et al. (1991), examining Pb(II) reactions on goethite; and Chisholm-Brause et al. (1990), working with Pb on γ-Al_2O_3, also determined that the metals they studied formed multinuclear complexes on the surface of clay minerals and metal oxides. However, additional XAFS Pb studies on γ-Al_2O_3 by Strawn et al. (1998) and on α-Al_2O_3 by Bargar et al. (1997) demonstrated that surface precipitation is not the dominant sorption mechanism.

Although the mechanism(s) for the formation of these surface precipitates has not been clearly identified, it has been postulated that they cannot form without the presence of a surface interface in solution (James and Healy, 1972). O'Day et al. (1994) theorized that the mechanism(s) for surface complex formation could be related to (i) cluster size and structural misregistry, assuming the formation of critical-sized clusters of atoms from supersaturated solutions as the first step in the growth of a new phase; (ii) enhanced surface concentration of the sorbed metal contained within a perturbed interfacial volume at the surface; and (iii) influences of the reaction surface reducing the dielectric constant of water, which then results in a decrease in ΔG (James and Healy, 1972; Sposito, 1984).

Scheidegger and Sparks (1996) examined the dissolution of polynuclear Ni(II) surface complexes from pyrophyllite via proton-promoted dissolution with HNO_3 at pH = 4 and pH = 6. Proton-promoted dissolution occurs when protons bind to surface oxide ions, which causes the bonds to weaken. Thereafter, the metal is detached into the solution phase (Sparks, 1995). This study (Scheidegger and Sparks, 1996) was unique in that it examined the kinetics of precipitation and dissolution of the mixed Ni–Al hydroxide phases from the surface. Nickel detachment from surface complexes was rapid initially at both pH values (with <10% of total Ni released) and was attributable to desorption of specifically adsorbed, mononuclear bound Ni. Dissolution then slowed tremendously, primarily due to the gradual dissolution of the multinuclear surface precipitates. A reference compound, crystalline $Ni(OH)_2$, was also examined for its dissolution potential. Compared with the dissolution of crystalline $Ni(OH)_2$ (≈96% dissolved), Ni release from pyrophyllite was extremely slow.

Dissolution kinetics of surface precipitates based on Ni release from Ni–Al LDH phases formed on pyrophyllite drastically decreased with aging time from 1 h to 1 yr. The aging effect was due to Al-for-Ni substitution in the octahedral sheets and silicate-for-nitrate exchange in the interlayers of the precipitates, which transformed the initial Ni–Al LDH into a Ni–Al phyllosilicate precursor (Ford et al., 1999; Scheckel and Sparks, 2000). A similar increase in precipitate stability with aging time (Fig. 6–11) was paralleled by solid-state transformation to a Ni-phyllosilicate phase in a gibbsite–silica mixed system (Fig. 6–12). The above research demonstrates the necessity to recognize that surface precipitates form on clay mineral and metal oxide surfaces during short reaction times. Separating the kinetics of inner sphere adsorption and surface precipitation, both of which can occur at early stages of the reaction, is an active area of research.

Organic Contaminant Reactivity

The behavior of organic contaminants in heterogeneous soil environments is exceedingly complex because abiotic and biotic processes interact in an incom-

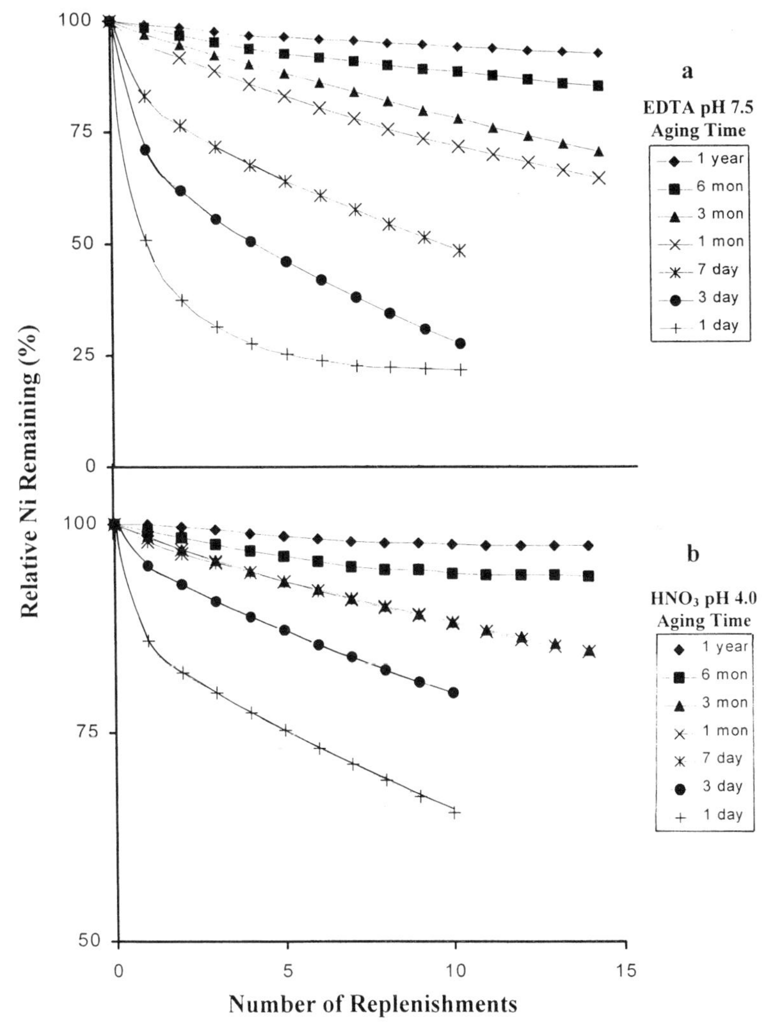

Fig. 6–11. Dissolution behavior of aged Ni precipitates on a gibbsite–silica mixture showing relative amount of Ni remaining on the mixture surface following extraction with (a) 1 m*M* EDTA at pH 7.5 and (b) HNO_3 at pH 4.0 plotted against the total number of replenishments. The stability of the Ni surface precipitates increases with aging time (from Scheckel and Sparks, 2000).

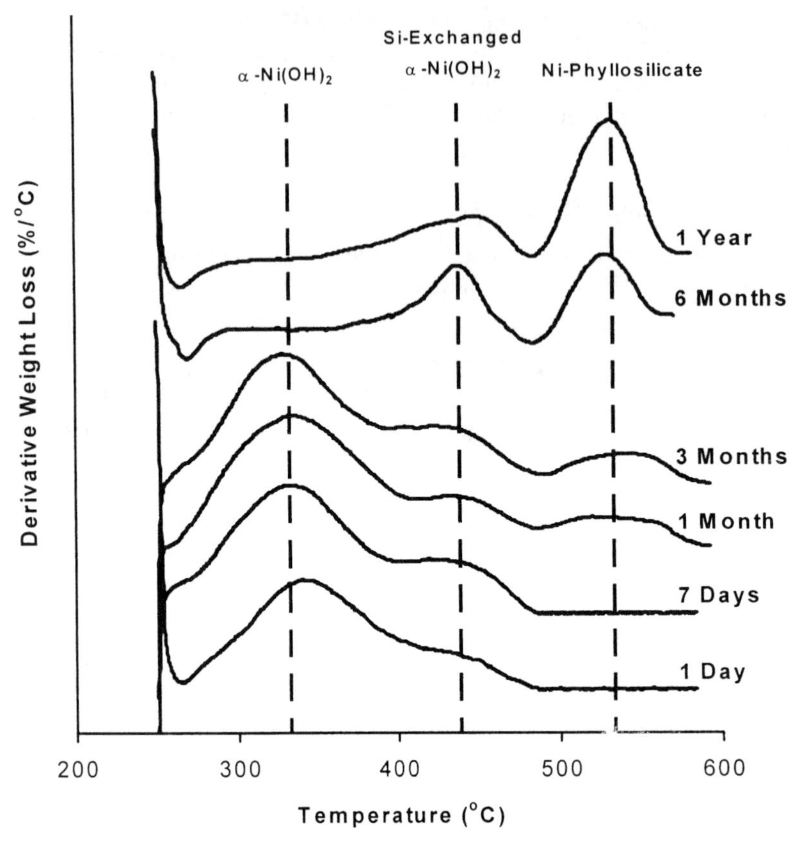

Fig. 6–12. Changes in the thermal stability of the Ni surface precipitates on the gibbsite–silica mixture showing conversion from α-$Ni(OH)_2$ to a Ni phyllosilicate with aging. The derivative weight loss curves are background subtracted from an unreacted sample to show only weight loss events associated with the precipitate phases. The identification markers were derived from reference compounds (from Scheckel and Sparks, 2000).

pletely understood fashion. Rates of sorption and desorption are biphasic, characterized by a rapid, reversible reaction followed by a slower, irreversible stage (Pignatello, 2000, and references therein). The fraction that is rapidly sorbed is easily desorbed and bioavailable, whereas the slow sorbing fraction is sequestered, unavailable for microbial attack, and difficult to desorb (Sparks, 1999). The fact that bioavailability can be rate limited by desorption would suggest that microbes can only access dissolved organic contaminants, a finding that past investigators have reported (Ogram et al., 1985). However, some microbes are able to mineralize sequestered fractions indirectly, likely by accelerating the desorption rates via a concentration gradient created by uptake from solution and biosurfactant production (Guerin and Boyd, 1992; Pignatello, 2000).

The importance of intraparticle and interparticle diffusion of the organic compound into soil organic matter fraction in governing the slow sorption–desorption rates is increasingly recognized (Steinberg et al., 1987). A coupled sorp-

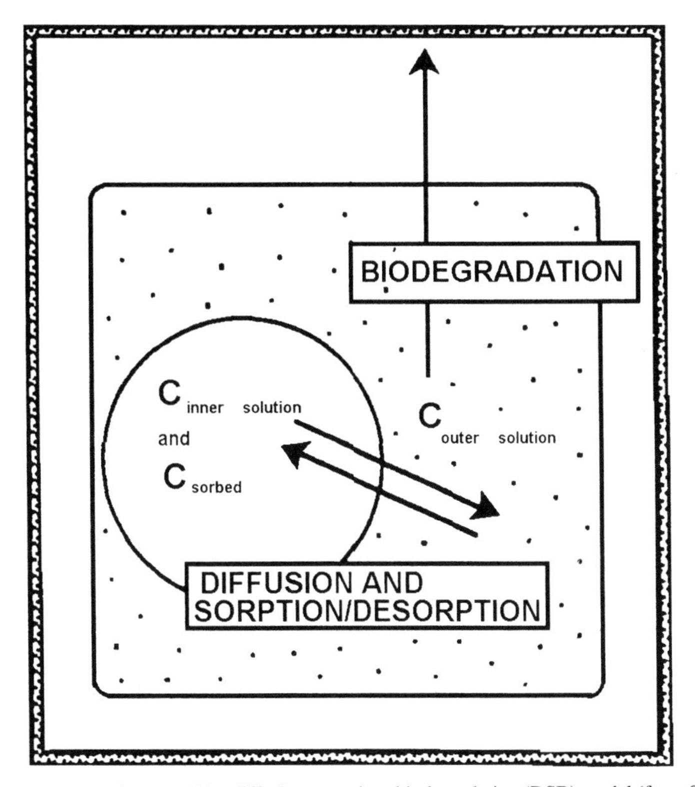

Fig. 6–13. Schematic diagram of the diffusion–sorption–biodegradation (DSB) model (from Scow and Hutson, 1992).

tion–biodegradation model called both the diffusion–sorption–biodegradation (DSB) model and the two-compartment model treats the system as an inner compartment, the soil aggregate, and an outer compartment, the soil solution (Fig. 6–13). This model assumes that organic contaminants follow linear sorption isotherms and diffuse via Fick's second law into the inner compartment and biodegradation follows Monod kinetics reduced to first order in the outer compartment. Without accounting for diffusion and sorption, the first-order biodegradation model underestimated the persistence of the contaminant when compared with the DSB model. Biodegradation of phenol by *Pseudomonas* was described well by the DSB model at different aggregate sizes (Fig. 6–14). Diffusive entry of the organic contaminant into nanopores was recently verified by Nam and Alexander (1998). Reduced bioavailability occurred when hydrophobic compounds entered hydrophobic nanopores. For a more complete description of the models describing organic contaminant reactivity, the reader is referred to Pignatello (2000).

Redox Kinetics

In soil environments, the most important chemical elements to undergo redox reactions are C, N, O, S, Mn, and Fe. In contaminated soils, As, Se, Cr, U, Hg, and

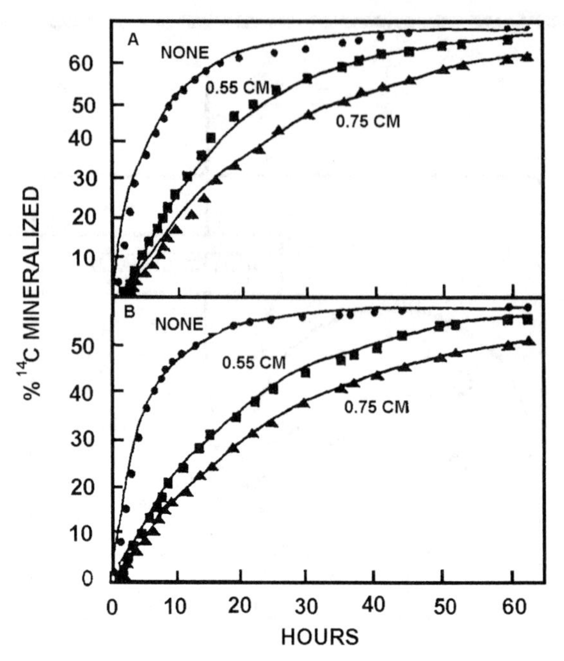

Fig. 6–14. Mineralization of ^{14}C-labeled phenol by *Pseudomonas* spp. in the presence of aggregates with radii of 0.55 or 0.75 cm and in the absence of clay aggregates. Solid lines show the simulation of the DSB model. The inoculum was (A) 5.0×10^7 or (B) 5.0×10^6 cells mL^{-1} (from Scow and Alexander, 1992).

Co are redox active (Sposito, 1994). Traditionally, the soil organic matter decomposition or diagenesis model in soils and sediments (Table 6–1) ranks the sequence of oxidants as O_2 reduction, NO_3^- reduction, MnO_2 reduction, $Fe(OH)_3$ reduction, SO_4^{2-} reduction, and methanogenesis based on the oxidant yielding the most free energy change per mole of organic C (as CH_2O) oxidized (Froelich et al., 1979; Postma and Jakobsen, 1996). The sequence proceeds until all oxidants or all oxidizable organic matter is consumed. In natural settings, this model is not always followed. Sulfate reduction preceded Fe oxide reduction in sediment porewater samples in some cases, a reverse of the normal redox sequence (Fig. 6–15). Thus, equilibrium behavior between dissolved redox species and with minerals cannot be assumed (Lindberg and Runnels, 1984), which underscores the need for the study of redox reaction rates (Sparks, 1989). Most of these reactions are heterogeneous and can involve complex multistep processes and competing biotic and abiotic pathways (Chapelle, 1993; Lovley, 1995).

Nutrient Cycling

The fate of missing N in soil is a long-standing enigma in agriculture. Transformations and losses of N make soil test recommendations challenging. It has long been assumed that the global N cycle involves coupled reactions mediated by bac-

Table 6–1. Some important reduction half-reactions in soils (from Postma and Jakobsen, 1996).

Reaction	Energy
	kJ mol^{-1} CH_2O
$CH_2O + O_2 \rightarrow CO_2 + H_2O$	–475
$5CH_2O + 4NO_3^- \rightarrow 2N_2 + 4HCO_3^- + CO_2 + 3H_2O$	–448
$CH_2O + 3CO_2 + 2MnO_2(s) + H_2O \rightarrow 2Mn^{2+} + 4HCO_3^-$	–349
$CH_2O + 7CO_2 + 4Fe(OH)_3(s) \rightarrow 4Fe^{2+} + 4HCO_3^- + 3H_2O$	–114
$2CH_2O + SO_4^{2-} \rightarrow H_2S + 2HCO_3^-$	–77
$2CH_2O \rightarrow CH_4 + CO_2$	–58

teria (Averill and Tiedje, 1982). Microbial nitrification in soil can be described by the following reactions:

$$NH_4^+ + 1.5O_2 \underset{k_2}{\overset{k_1}{\rightarrow}} NO_2^- + H_2O + 2H^+ \quad [50]$$

$$NO_2^- + 0.5O_2 \rightarrow NO_3^- \quad [51]$$

where O_2 oxidation of NH_4^+ is catalyzed by enzymes of the autotrophic bacteria *Nitrosomonas* and NO_2^- oxidation is ascribed to *Nitrobacter* (Alexander, 1977). This sequence of reactions can be described by a consecutive, first-order process (Brezonik, 1994; Sparks, 1989). Assuming that Step 1 is rate-limiting with $k_1 = 0.2$ d^{-1} and applying the steady-state approximation to the NO_2^- intermediate with $k_2 = 1.0$ d^{-1}, then the concentrations of the three N species can be modeled over time for an initial NH_4^+ concentration of 2 m*M* and a continuous excess supply of O_2 from the atmosphere (Fig. 6–16):

$$[NH_4^+]_t = [NH_4^+]_0 e^{-k_1 t} \quad [52]$$

$$[NO_2^-]_t = \frac{k_1}{k_2}[NH_4^+]_0 e^{-k_1 t} \quad [53]$$

$$[NO_3^-]_t = [NH_4^+]_0(1 - e^{-k_1 t}) \quad [54]$$

Conversion of NH_4^+ to NO_2^- controls the overall formation of plant available NO_3^- because $k_2 > k_1$.

Microbial nitrification may compete with abiotic pathways in some soil environments. Oxidation of NO_2^- to NO_3^- by reactive Mn oxides was reported in the presence and absence of O_2 and may compete effectively with the biological pathway to help explain the lack of NO_2^- accumulation in soils (Bartlett, 1981). Recent evidence indicated that when NO_2^- accumulated, MnO_2 was reduced to Mn(II) (Vandenabeele et al., 1995).

Under anoxic conditions, a variety of microorganisms couple oxidation of reduced organic substrates with the reduction of NO_3^- or NO_2^- to gaseous NO, N_2O, or N_2 in a process called denitrification (Chapelle, 1993). Various scientists have reported that denitrification follows both zero- and first-order kinetics (Patrick, 1960;

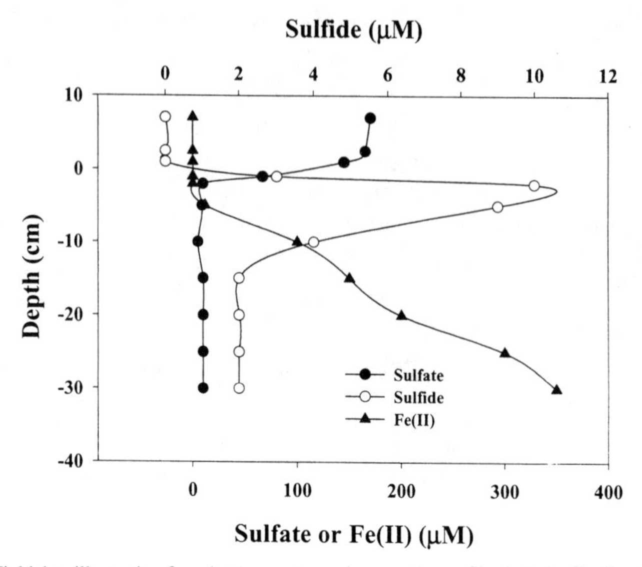

Fig. 6–15. Field data illustrating from bottom water and porewater profiles in Lake Greifen (from Wersin et al., 1991, as cited by Postma and Jakobsen, 1996).

Stanford et al., 1975). The occurrence of N_2 production and NO_3^- respiration in oxic surface soils and sediments has been documented (Seitzinger, 1988; Carter et al., 1995). These findings are surprising given the commonly held assumption that biological denitrification does not occur below the rooting zone for lack of carbon (Sorensen, 1987) and is repressed by O_2. Chemodenitrification by naturally occurring reductants in soils may also cause some gaseous N loss.

There are potential chemodenitrification pathways that could occur in soils containing Fe(II)-bearing minerals and low in organic matter. Nitrite (NO_2^-) re-

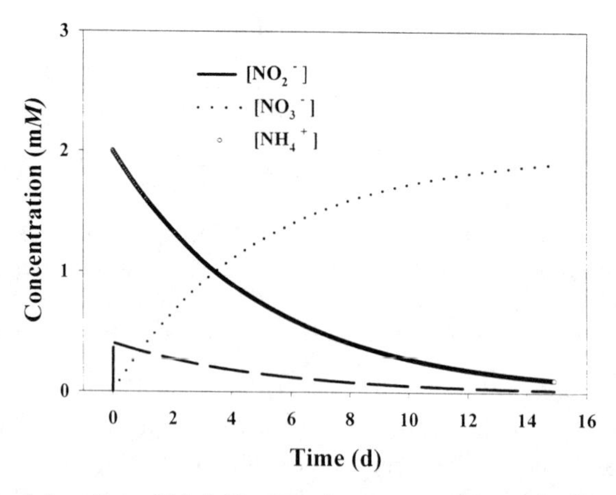

Fig. 6–16. Simulation of microbial nitrification using a two-step, consecutive first-order process with initial NH_4^+ concentration of 2 mM and $k_1 = 0.2$ d^{-1} and $k_2 = 1.0$ d^{-1}.

duction to N_2O by Fe(II) sorbed to lepidocrocite was rapid and the maximum rate of 10 μmol h^{-1} occurred at pH 8.5 (Sorensen and Thorling, 1991). The authors explained that Fe(II) formed a bidentate complex with surface oxo groups, which provided an ideal environment for NO_2^- adsorption and subsequent N–N double bond formation, similar to the function of the Fe metal center in nitrite reductase enzyme (Averill and Tiedje, 1982).

Green rust [$Fe_4^{II}Fe_2^{III}(OH)_{12}SO_4 \cdot yH_2O$], a naturally occurring Fe(II)/Fe(III) oxide mineral, promoted the rapid dissimilatory conversion of NO_3^- to NH_4^+ with a pseudo-first-order rate constant describing the NH_4^+ production rate of 10^{-5} s^{-1} (Koch and Hansen, 1997). Further studies indicated that NO_3^- was electrostatically bound prior to electron transfer at positively charged sites on green rust (Hansen and Koch, 1998). The rate-limiting step in the mechanism was postulated to be the initial two-electron transfer because NO_2^- reduction by green rust was more rapid (Hansen et al., 1994). In contrast, the maximum rate of NO_3^- reduction by Fe(II)-bearing silicates occurred at pH 4 (Postma, 1990). Other clay minerals with structural Fe(II) that have been implicated in NO_3^- reduction, particularly at lower depths in the soil profile, are chlorite and pyrite (Postma et al., 1991). The N_2O produced may take part in global warming and atmospheric ozone destruction because it is a more potent greenhouse gas than CO_2 and represents an irreversible loss of N from soil. Abiotic dissimilatory conversion of NO_3^- preserves N in the soil. More work is necessary to understand the role of soil Fe and Mn in the N cycle.

Soil organic C is a key indicator of soil quality and is important for agricultural production. To predict changes in soil C, information on key processes controlling soil C stabilization is necessary (Percival et al., 2000). Investigations in the last two decades have reported that transition metal oxides and transition metal-bearing phyllosilicates common to soil environments can be effective oxidants of oxidizable organic ligands resembling organic matter (Stone and Morgan, 1984a; Hering and Stumm, 1990; Sparks, 1995). In solution chemistry, the mechanisms for redox reactions are classified as inner sphere electron transfer, where the coordination sphere of the transition state is markedly changed, and outer sphere electron transfer, where the coordination spheres of the reactants remain intact (Atwood, 1985). In some cases, inner sphere electron transfer involves a bridging ligand that has available lone-pair electrons. For heterogeneous redox processes in soils and geochemical environments, the surface complexation model has been used to describe reductive dissolution of metal oxides as a function of solution variables (Stone and Morgan, 1984a).

The general mechanism of abiotic reductive dissolution of metal oxides includes the following steps (Stone and Morgan, 1984a, 1984b; Stone, 1991): (i) diffusion of soluble oxidizable ligands and metals to the mineral surface, (ii) sorption forming inner sphere or outer sphere precursor surface complexes, (iii) electron transfer to lower metal oxidation state and weaken metal–oxygen bonds, (iv) detachment of products (oxidized reductant and reduced metal) from the mineral surface, and (v) diffusion of products into bulk solution. Manganese (III,IV) oxides are reduced by organic ligands several orders of magnitude more rapidly than Fe(III) oxides (Stone, 1991).

Pivotal early investigations into oxidizable organic ligand reactions with solid Mn(III,IV) (hydr)oxides postulated that inner-sphere electron transfer or re-

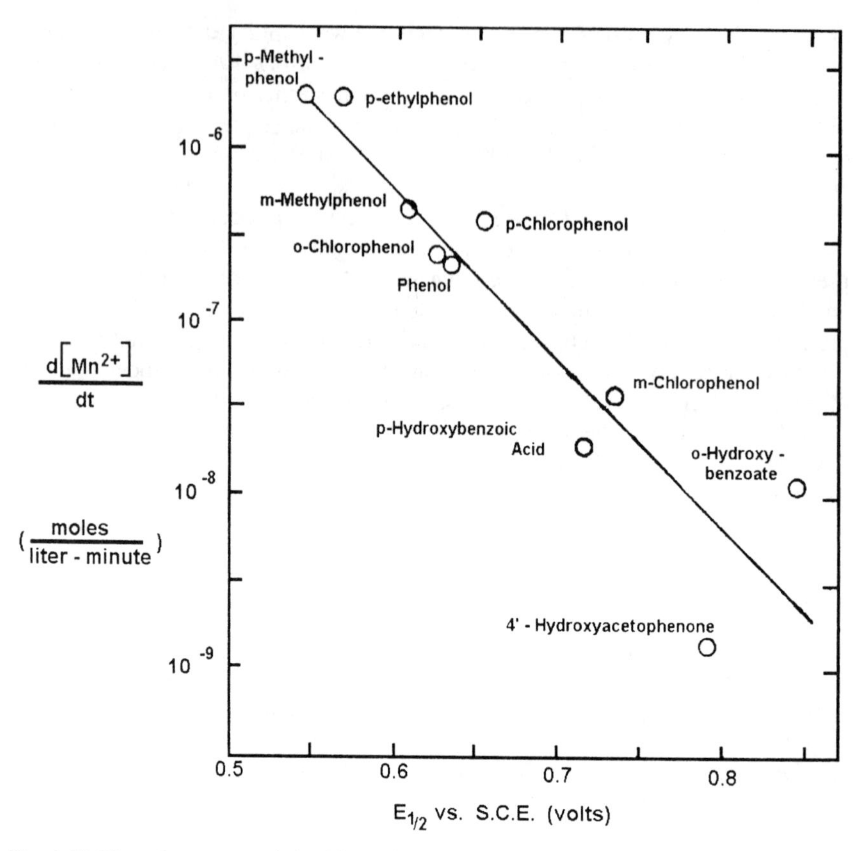

Fig. 6–17. Linear free energy relationship (LFER) describing Mn oxide reduction rates by substituted phenols (from Stone, 1987).

actions involving surface species were rate-limiting (Stone and Morgan, 1984a, 1984b). Adsorbed Ca^{2+} and PO_4^{3-} strongly inhibited dissolution rates of feitknechtite (β-MnOOH), and small changes in reductant structure caused large rate changes, providing strong support for the importance of surface complexation in reductive dissolution. Reduction rates were inversely related to the electron withdrawing properties of substituted phenols based on Hammett constants, indicative of a LFER (Fig. 6–17) with electron transfer as the rate-limiting step (Stone, 1987).

The most common Mn oxide controlling solubility of plant available Mn(II), birnessite (δ-$MnO_{1.7}$), was reduced rapidly to dissolved Mn(II) by catechol and required use of an electron paramagnetic resonance stopped-flow (EPR-SF) technique to measure reaction kinetics (Matocha et al., 2001). Initial rate plots indicated the reaction was first order in catechol and surface area and zero order in pH (Fig. 6–18) with a second-order rate constant $k = 4 \times 10^{-3}$ L m^{-2} s^{-1}. The E_a of 59 kJ mol^{-1} and $\Delta S^{\ddagger}$ of −78 J mol^{-1} K^{-1} show the reaction to be associative and surface-chemical controlled. It was proposed that precursor surface complex formation (Step ii) was rate-limiting.

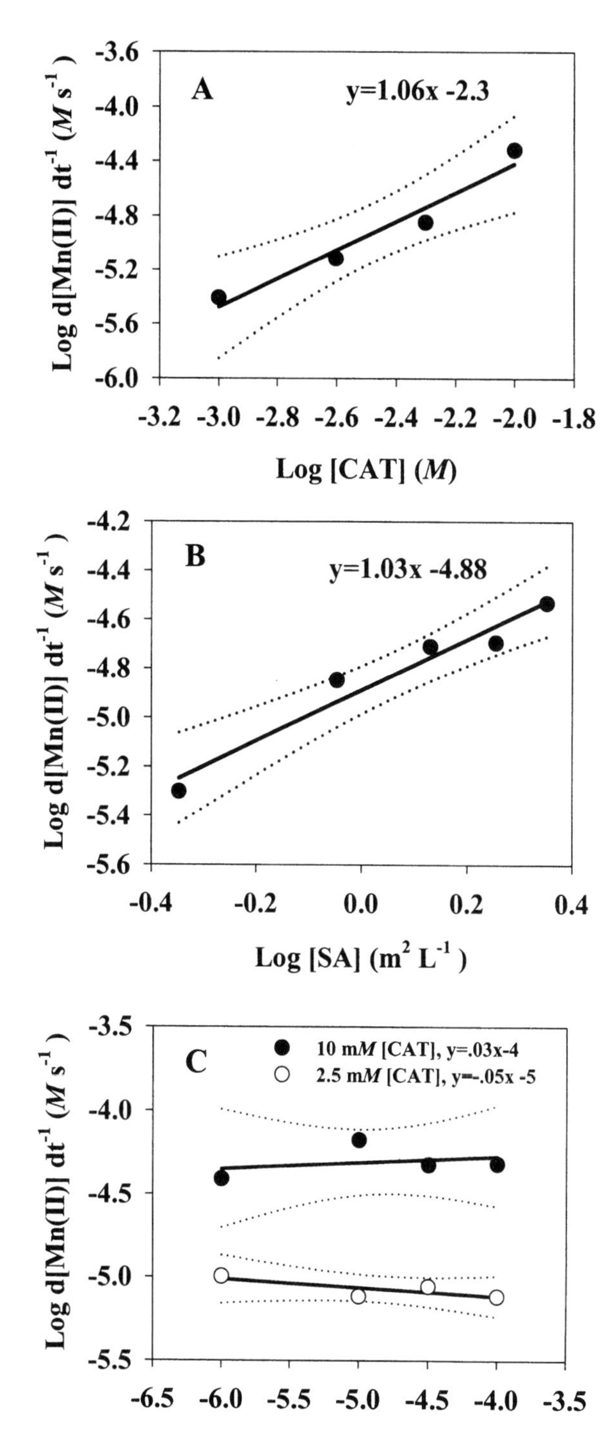

Fig. 6–18. Representative initial reaction rate plots measured by the EPR-SF technique as a function of: (A) [CAT]; (B) [SA]; and (C) [H^+]. The dotted lines represent the 95% confidence interval bands (from Matocha et al., 2001).

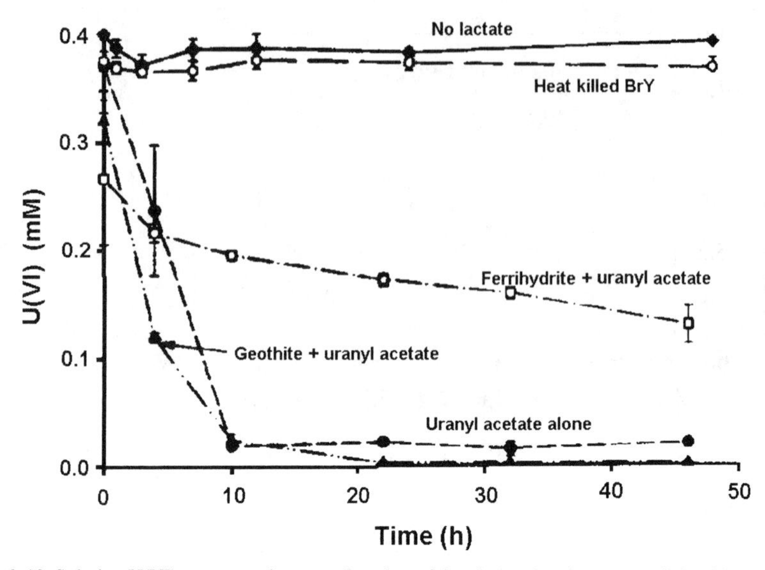

Fig. 6–19. Solution U(VI) concentrations as a function of time in batch cultures containing Shewanella alga strain Br Y. Both goethite and ferrihydrite were present at Fe concentrations of approximately 20 m*M* (from Wielinga et al., 2000).

Contaminant Cycling

In the last 15 yr, researchers have isolated dissimilatory metal-reducing bacteria that can couple organic matter oxidation to metal contaminant reduction (Lovley, 1993). Rates of these reactions depend on the reduction potential of the solid- or solution-phase metal, the surface area, and the presence of competing terminal electron acceptors (TEAs). This was illustrated in a study where the presence of ferrihydrite suppressed the rate of uranium [U(VI)] reduction by *Shewanella alga* (Wielinga et al., 2000). Ostensibly, the structural Fe(III) in the amorphous ferrihydrite phase was more available as a TEA than crystalline Fe(III)-goethite based on the large decrease in U(VI) reduction rates (Fig. 6–19). Direct microbial cell–metal oxide contact has been proposed to be requisite for Fe(III) and Mn(IV) reduction, but the mechanisms are still unclear at this point (Lovley, 1993). Solid Mn(III,IV) and Fe(III) oxide reduction rates by microorganisms have been reported to be highly dependent on mineral surface area (Burdige et al., 1992; Roden and Zachara, 1996; Larsen et al., 1998).

Manganese oxides are also important in controlling speciation of contaminants and can directly oxidize trace and transition metals such as Co(II), Fe(II), Cr(III), As(III), Se(IV), and Pu(IV) (Sparks, 1999). The adsorption step was rate-controlling for As(III) and Cr(III) oxidation on birnessite (Fendorf et al., 1992; Scott and Morgan, 1995). Formation of surface precipitates at higher pHs and Cr(III) concentrations was found to inhibit Cr(III) oxidation rates. The slowest step for Se(IV) oxidation by birnessite was postulated to be electron transfer based on the unfavorable thermodynamic driving force (Scott and Morgan, 1996).

The coupling of the Mn cycle with N, C, and contaminants can be explained by frontier molecular orbital theory (FMOT) (Luther, 1990). Oxides and hydrox-

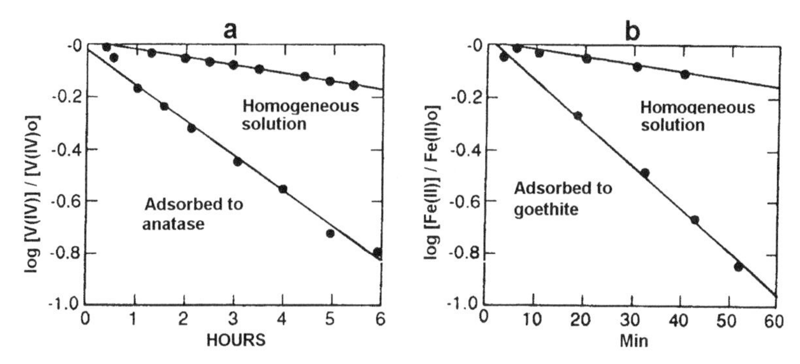

Fig. 6–20. First-order plots demonstrating the catalysis of (A) vanadium (IV) and (B) Fe(II) oxygenation by oxide surfaces (taken from Wehrli et al., 1989).

ides of Mn(III) and Mn(IV) are kinetically better oxidants than O_2 despite their position on the thermodynamic sequence of oxidants (Table 6–1). This is because the lowest unoccupied molecular orbitals (LUMO) for Mn(III) and Mn(IV) are the e_g orbitals, which are of sigma (σ*) symmetry and available to accept electrons from σ- and π-donor organic ligands or contaminant metals. For Mn(IV), with a $d^3(t_{2g}^3\ e_g^0)$ electronic configuration, both e_g orbitals are vacant and a direct two-electron transfer can occur, without unpairing electrons from the reductant ligand. In contrast, the LUMO for O_2 is of π* symmetry ($\pi_{2px}^{1*}\ \pi_{2py}^{1*}$), singly occupied, and requires unpairing of electrons from the incoming reductant (Luther, 1990).

Surface oxygen functional groups on metal oxides can catalyze redox reactions due to inner-sphere complexation (Wehrli et al., 1989). Enhancement of the oxygenation rate over the homogeneous rate was shown to result from inner-sphere adsorption of vanadyl (VO^{2+}) and Fe(II) on TiO_2 and FeOOH (Fig. 6–20). Dissolved oxo ligands or surface functional oxo groups that bind Fe(II) enable it to become a better reductant because the oxo group can transfer electron density through the σ and π systems with proper electron overlap to O_2 in an outer sphere process (Luther, 1990). Reactivity of both surface and solution complexed Fe(II) species decreases in the order: $Fe(OH)_2^0 > FeOH^+ \sim Fe(OM=)_2 > Fe^{2+}$ (Wehrli et al., 1989).

Reduction of U(VI), Tc(VII), and nitroaromatic compounds by Fe(II)-surface species have been reported (Klausen et al., 1995; Cui and Eriksen, 1996; Liger et al., 1999). No reaction was observed between U(VI) and Fe(II) in homogeneous solution at pH 7.5 despite the favorable thermodynamics for electron transfer. However, in the presence of Fe(II)-hematite, U(VI) was reduced at minute time scales and followed a second-order rate law:

$$d[U(VI)]/dt = -k[{=}Fe^{III}OFe^{II}OH^\circ][U(VI)]_{ads} \quad [55]$$

where [$=Fe^{III}OFe^{II}OH^\circ$] was the hydroxo surface complex term, Fe(II) adsorbed on hematite, that controlled the reduction rate, $[U(VI)]_{ads}$ represented the amount of adsorbed U(VI), and the value of k was 399 M^{-1} min^{-1} (Liger et al., 1999). Four-nitrobenzene reduction by [$=Fe^{III}OFe^{II}OH^\circ$] species was slower ($k = 22\ M^{-1}$ min^{-1}) and these differences were postulated to reflect differences in surface complexes of the oxidants (Charlet et al., 1998). Rates of abiotic U(VI) reduction were comparable to rates of biotic reduction and both pathways may compete in natural geochemical environments (Liger et al., 1999).

CONCLUSION

It is clear that the study of reaction rates of soil chemical processes will continue because of recognition of the fact that these processes are seldom at equilibrium. There is a need to assess the role of biological and abiological processes in the cycling of nutrients and contaminants. Linkage of mechanistic data obtained via kinetic and spectroscopic studies to field-scale transport processes will provide a better prediction of contaminant and nutrient fate in natural soil environments.

REFERENCES

Adamson, A.W. 1979. A textbook of physical chemistry. Academic Press, New York.

Aharoni, C., and D.L. Sparks. 1991. Kinetics of soil chemical reactions—A theoretical treatment. p. 1–18. *In* D. Sparks and D.L. Suarez (ed.) Rates of soil chemical processes. SSSA Spec. Publ. 27. SSSA, Madison, WI.

Ainsworth, C.C., D.M. Friedrich, P.L. Gassman, Z. Wang, and A.G. Joly. 1998. Characterization of salicylate–alumina surface complexes by polarized fluorescence spectroscopy. Geochim. Cosmochim. Acta 62:595–612.

Ainsworth, C.C., J.L. Pilon, P.L. Gassman, and W.G. Van Der Sluys. 1994. Cobalt, cadmium, and lead sorption to hydrous ferric oxide: Residence time effect. Soil Sci. Soc. Am. J. 58:1615–1623.

Alexander, M. 1977. Nitrification. p. 251–271. *In* Introduction to soil microbiology. 2nd ed. John Wiley and Sons, New York.

Amacher, M.C. 1991. Methods of obtaining and analyzing kinetic data. p. 19–59. *In* D.L. Sparks and D.L. Suarez (ed.) Rates of soil chemical processes. SSSA Spec. Publ. 27. SSSA, Madison, WI.

Appelo, C.A.J., and D. Postma. 1993. Geochemistry, groundwater, and pollution. Balkema, Rotterdam, The Netherlands.

Atwood, J.D. 1985. Inorganic and organometallic reaction mechanisms. VCH Publishers, New York.

Averill, B.A., and J.M. Tiedje. 1982. The chemical mechanism of microbial denitrification. FEBS Lett. 138:8–12.

Bargar, J.R., S.N. Towle, G.E. Brown, Jr., and G.A. Parks. 1997. XAFS and bond-valence determination of the structures and compositions of surface functional groups and Pb(II) and Co(II) sorption products on single-crystal α-Al_2O_3. J. Colloid Interface Sci. 185:473–492.

Bartlett, R.J. 1981. Nonmicrobial nitrite-to-nitrate transformation in soils. Soil Sci. Soc. Am. J. 45:1054–1058.

Benjamin, M.M., and J.O. Leckie. 1981. Multi-site adsorption of Cd, Co, Zn, and Pb on amorphous iron oxyhydroxide. J. Colloid Interface Sci. 79:209–221.

Bertsch, P.M., and D.B. Hunter. 1998. Elucidating fundamental mechanisms in soil and environmental chemistry: The role of advanced analytical, spectroscopic, and microscopic methods. p. 103–122. *In* P.M. Huang et al. (ed.) Future prospects for soil chemistry. SSSA Spec. Publ. 55. SSSA, Madison, WI.

Brezonik, P.L. 1994. Chemical kinetics and process dynamics in aquatic systems. Lewis Publ., Boca Raton, FL.

Brümmer, G.W., J. Gerth, and K.G. Tiller. 1988. Reaction kinetics of the adsorption and desorption of nickel, zinc, and cadmium by goethite. I. Adsorption and diffusion of metals. J. Soil Sci. 39:37–52.

Buerge, I.J., and S.J. Hug. 1997. Kinetics and pH dependence of chromium (VI) reduction by iron(II). Environ. Sci. Technol. 31:1426–1432.

Burdige, D.J., S.P. Dhakar, and K.H. Nealson. 1992. Effects of manganese oxide mineralogy on microbial and chemical manganese reduction. Geomicrobiol. J. 10:27–48.

Buresh, R.J., and J.T. Moraghan. 1976. Chemical reduction of nitrate by ferrous iron. J. Environ. Qual. 5: 320–324.

Carter, J.P., Y.H. Hsiao, S. Spiro, and D.J. Richardson. 1995. Soil and sediment bacteria capable of aerobic nitrate respiration. Appl. Environ. Microbiol. 61:2852–2858.

Chapelle, F.H. 1993. Ground-water microbiology and geochemistry. Wiley and Sons, New York.

Charlet, L., and A. Manceau. 1992. X-ray absorption spectroscopic study of the sorption of Cr(III) at the oxide–water interface. II: Adsorption, co-precipitation and surface precipitation on ferric hydrous oxides. J. Colloid Interface Sci. 148:443–458.

Charlet, L., E. Silvester, and E. Liger. 1998. N-compound reduction and actinide immobilization in surficial fluids by Fe(II): the surface $=Fe^{III}OFe^{II}OH^{\circ}$ species, as major reductant. Chem. Geol. 151:85–93.

Chisholm-Brause, C.J., P.A. O'Day, G.E. Brown, Jr., and G.A. Parks. 1990. Evidence for multinuclear metal–ion complexes at solid/water interfaces from x-ray absorption spectroscopy. Nature 348:528–531.

Comans, R.N.J., and D.E. Hockney. 1992. Kinetics of cesium sorption on illite. Geochim. Cosmochim. Acta 56:1157–1164.

Cui, D., and T.E. Ericksen. 1996. Reduction of pertechnetate in solution by heterogeneous electron transfer from Fe(II)-containing geological material. Environ. Sci. Technol. 30:2263–2269.

Espenson, J.H. 1995. Chemical kinetics and reaction mechanisms. 2nd ed. McGraw-Hill, New York.

Farley, K.J., D.A. Dzombak, and F.M. Morel. 1985. A surface precipitation model for the sorption of cations on metal oxides. J. Colloid Interface Sci. 106:226–242.

Fendorf, S.E., M.J. Eick, P.R. Grossl, and D.L. Sparks. 1997. Arsenate and chromate retention mechanisms on goethite. 1. Surface structure. Environ. Sci. Technol. 31:315–320.

Fendorf, S.E., M. Fendorf, D.L. Sparks, and R. Gronsky. 1992. Inhibitory mechanisms of Cr(III) oxidation by δ-MnO_2. J. Colloid Interface Sci. 153:37–54.

Fendorf, S.E., G.M. Lamble, M.G. Stapleton, M.J. Kelley, and D.L. Sparks. 1994. Mechanisms of chromium (III) sorption on silica. I. Cr(III) surface structure derived by extended x-ray absorption fine structure spectroscopy. Environ. Sci. Technol. 28:284–289.

Fendorf, S.E., and G. Li. 1996. Kinetics of chromate reduction by ferrous iron. Environ. Sci. Technol. 30:1614–1617.

Fendorf, S.E., D.L. Sparks, J.A. Franz, and D.M. Camaioni. 1993. Electron paramagnetic resonance stopped-flow kinetic study of manganese(II) sorption–desorption on birnessite. Soil Sci. Soc. Am. J. 57:57–62.

Ford, R.G., A.C. Scheinost, K.G. Scheckel, and D.L. Sparks. 1999. The link between clay mineral weathering and structural transformation in Ni surface precipitates. Environ. Sci. Technol. 33:3140–3144.

Froelich, P.N., G.P. Klinkhammer, M.L. Bender, N.A. Luedtke, G.R. Heath, D. Cullen, P. Dauphin, D. Hammond, B. Hartman, and V. Maynard. 1979. Early oxidation of organic matter in pelagic sediments of the eastern equatorial Atlantic: suboxic diagenesis. Geochim. Cosmochim. Acta 43:1075–1090.

Grossl, P.R., M.J. Eick, D.L. Sparks, S. Goldberg, and C.C. Ainsworth. 1997. Arsenate and chromate retention mechanisms on goethite. 2. Kinetic evaluation using a pressure-jump relaxation technique. Environ. Sci. Technol. 31:321–326.

Grossl, P.R., D.L. Sparks, and C.C. Ainsworth. 1994. Rapid kinetics of Cu(II) adsorption/desorption of goethite. Environ. Sci. Technol. 28:1422–1429.

Guerin, W.R., and S.A. Boyd. 1992. Differential bioavailability of soil-sorbed naphthalene to two bacterial species. Appl. Environ. Microbiol. 58:1142–1152.

Hachiya, K., M. Sasaki, T. Ikeda, N. Mikami, and T. Yasunaga. 1984. Static and kinetic studies of adsorption–desorption of metal ions on a γ-Al_2O_3 surface. 2. Kinetic study by means of pressure-jump technique. J. Phys. Chem. 88:27–31.

Hansen, H.C.B., O.K. Borggaard, and J. Sorensen. 1994. Evaluation of the free energy of formation of Fe(II)-Fe(III) hydroxide-sulphate (green rust) and its reduction of nitrite. Geochim. Cosmochim. Acta 58:2599–2608.

Hansen, H.C.B., and C.B. Koch. 1998. Reduction of nitrate to ammonium by sulphate green rust: Activation energy and reaction mechanism. Clay Miner. 33:87–101.

Hayes, K.F., A.L. Roe, G.E. Brown, Jr., K.O. Hodgson, J.O. Leckie, and G.A. Parks. 1987. In situ x-ray absorption study of surface complexes: Selenium oxyanions on α-FeOOH. Science 238:783–786.

Hering, J.G., and W. Stumm. 1990. Oxidative and reductive dissolution of minerals. Rev. Mineral. 23:427–465.

James, R.O., and T.W. Healy. 1972. Adsorption of hydrolyzable metal ions at the oxide–water interface. I. Co(II) adsorption on SiO_2 and TiO_2 as model systems. J. Colloid Interface Sci. 40:42–52.

Klausen, J.W., S.P. Trober, S.B. Haderlein, and R.P. Schwarzenbach. 1995. Reduction of substituted nitrobenzenes by Fe(II) in aqueous mineral suspensions. Environ. Sci. Technol. 29:2396–2404.

Koch, C.B., and H.C.B. Hansen. 1997. Reduction of nitrate to ammonium by sulphate green rust. Adv. GeoEcol. 30:373–393.

Larsen, I., B. Little, K.H. Nealson, R. Ray, A.T. Stone, and J. Tian. 1998. Manganite reduction by *Shewanella putrefaciens* MR-4. Am. Mineral. 83:1564–1572.

Lasaga, A.C. 1981. Rate laws of chemical reactions. p. 1–68. *In* A.C. Lasaga and R.J. Kirkpatrick (ed.) Kinetics of geochemical processes. Reviews in Mineralogy, Vol. 8. Mineralogical Society of America, Washington, DC.

Liger, E., L. Charlet, and P. Van-Cappellen. 1999. Surface catalysis of uranium(VI) reduction by iron(II). Geochim. Cosmochim. Acta 63:2939–2955.

Lindberg, R.D., and D.D. Runnells. 1984. Ground water redox reactions: An analysis of equilibrium state applied to E_h measurements and geochemical modeling. Science 225:925–927.

Lovley, D.R. 1995. Microbial reduction of iron, manganese, and other metals. Adv. Agron. 54:175–231.

Low, M.J.D. 1960. Kinetics of chemisorption of gases on solids. Chem. Rev. 60:267–312.

Luther, G.W. III. 1990. Frontier molecular orbital theory in geochemical processes. *In* W. Stumm (ed.) Aquatic chemical kinetics: Reaction rates of processes in natural water. Wiley-Interscience, New York.

Matocha, C.J., D.L. Sparks, J.E. Amonette, and R.K. Kukkadapu. 2001. Kinetics and mechanism of birnessite reduction by catechol. Soil Sci. Soc. Am. J. 65:58–66.

McBride, M.B. 1994. Environmental chemistry of soils. Oxford University Press, New York.

Nam, K., and M. Alexander. 1998. Role of nanoporosity and hydrophobicity in sequestration and bioavailability: Tests with model soils. Environ. Sci. Technol. 32:71–74.

O'Day, P.A., G.E. Brown, Jr., and G.A. Parks. 1994. X-ray absorption spectroscopy of cobalt (II) multinuclear surface complexes and surface precipitates on kaolinite. J. Colloid Interface Sci. 165:269–289.

Ogram, A.V., R.E. Jessup, L.T. Ou, and P.S.C. Rao. 1985. Effects of sorption on biological degradation rates of 2,4-D in soils. Appl. Environ. Microbiol. 49:582–587.

Patrick, W.H. 1960. Nitrate reduction rates in a submerged soil as affected by redox potential. 7th Int. Congr. Soil Sci. Trans. II:494–500.

Peak, D., R.G. Ford, and D.L. Sparks. 1999. An in-situ ATR-FTIR investigation of sulfate bonding mechanisms on goethite. J. Colloid Interface Sci. 218:289–299.

Percival, H.J., R.L. Parfitt, and N.A. Scott. 2000. Factors controlling soil carbon levels in New Zealand grasslands: Is clay content important? Soil Sci. Soc. Am. J. 64:1623–1630.

Phillips, B.L., S.N. Crawford, and W.H. Casey. 1997. Rate of water exchange between $Al(C_2O_4)(H_2O)_4^+(aq)$ complexes and aqueous solutions determined by ^{17}O-NMR spectroscopy. Geochim. Cosmochim. Acta 61:4965–4973.

Phillips, B.L., J.A. Tossell, and W.H. Casey. 1998. Experimental and theoretical treatment of elementary ligand exchange reactions in aluminum complexes. Environ. Sci. Technol. 32:2865–2870.

Pignatello, J.J. 2000. The measurement and interpretation of sorption and desorption rates for organic compounds in soil media. Adv. Agron. 69:1–73.

Postma, D. 1990. Kinetics of nitrate reduction by detrital Fe(II)-silicates. Geochim. Cosmochim. Acta 54:903–908.

Postma, D., C. Boesen, H. Kristiansen, and F. Larsen. 1991. Nitrate reduction in an unconfined sandy aquifer: Water chemistry, reduction processes, and geochemical modeling. Water Resour. Res. 27:2027–2045.

Postma, D., and R. Jakobsen. 1996. Redox zonation: Equilibrium constraints on the Fe(III)/SO_4 reduction interface. Geochim. Cosmochim. Acta 60:3169–3175.

Roden, E.E., and J. Zachara. 1996. Microbial reduction of crystalline Fe^{3+} oxides: Influence of oxide surface area and potential for cell growth. Environ. Sci. Technol. 30:1618–1628.

Roe, A.L., K.F. Hayes, C.J. Chisholm-Brause, G.E. Brown, Jr., G.A. Parks, K.O. Hodgson, and J.O. Leckie. 1991. In-situ X-ray absorption study of lead ion surface complexes at the goethite/water interface. Langmuir 7:367–373.

Scheckel, K.G., and D.L. Sparks. 2000. Kinetics of the formation and dissolution of Ni precipitates in a gibbsite/amorphous silica mixture. J. Colloid Interface Sci. 229:222–229.

Scheidegger, A.M., M. Fendorf, and D.L. Sparks. 1996a. Mechanisms of nickel sorption on pyrophyllite: Macroscopic and microscopic approaches. Soil Sci. Soc. Am. J. 60:1763–1772.

Scheidegger, A.M., G.M. Lamble, and D. L. Sparks. 1996b. Investigation of Ni sorption on pyrophyllite: An XAFS study. Environ. Sci. Technol. 30:548–554.

Scheidegger, A.M., G.M. Lamble, and D.L. Sparks. 1997. Spectroscopic evidence for the formation of mixed-cation hydroxide phases upon metal sorption on clays and aluminum oxides. J. Colloid Interface Sci. 186:118–128.

Scheidegger, A.M., G.M. Lamble, and D.L. Sparks. 1998. The kinetics of nickel sorption on pyrophyllite as monitored by x-ray absorption fine structure (XAFS) spectroscopy. Geochim. Cosmochim. Acta 62:2233–2245.

Scheidegger, A.M., and D.L. Sparks. 1996. Kinetics of the formation and the dissolution of nickel surface precipitates on pyrophyllite. Chem. Geol. 132:157–164.

Scheinost, A.C., R.G. Ford, and D.L. Sparks. 1999. The role of Al in the formation of secondary Ni precipitates on pyrophyllite, gibbsite, talc, and amorphous silica: A DRS study. Geochim. Cosmochim. Acta 63:3193–3203.

Scheinost, A.C., and D.L. Sparks. 2000. Formation of layered single- and double-metal hydroxide precipitates at the mineral/water interface: A multiple-scattering XAFS analysis. J. Colloid Interface Sci. 223:167–178.

Scott, M.J., and J.J. Morgan. 1995. Reactions at oxide surfaces. 1. Oxidation of As(III) by synthetic birnessite. Environ. Sci. Technol. 29:1898–1905.

Scott, M.J., and J.J. Morgan. 1996. Reactions at oxide surfaces. 1. Oxidation of Se(IV) by synthetic birnessite. Environ. Sci. Technol. 30:1990–1996.

Scow, K.M., and J. Hutson. 1992. Effect of diffusion and sorption on the kinetics of biodegradation: Theoretical considerations. Soil. Sci. Soc. Am. J. 56:119–127.

Scow, K.M., and M. Alexander. 1992. Effect of diffusion on the kinetics of biodegradation: experimental results with synthetic aggregates. Soil Sci. Soc. Am. J. 56:128–134.

Seitzinger, S.P. 1988. Denitrification in freshwater and marine ecosystems: Ecological and geochemical significance. Limnol. Oceanogr. 33:702–724.

Shriver, D.F., P. Atkins, and C.H. Langford. 1994. Inorganic chemistry. 2nd ed. W.H. Freeman and Co., New York.

Skopp, J. 1985. Analysis of time-dependent chemical processes in soil. J. Environ. Qual. 15:205–213.

Sorensen, J. 1987. Nitrate reduction in marine sediment: Pathways and interactions with iron and sulfur cycling. Geomicrobiol. J. 5:401–421.

Sorensen, J., and L. Thorling. 1991. Stimulation by lepidocrocite (γ-FeOOH) of Fe(II)-dependent nitrite reduction. Geochim. Cosmochim. Acta 55:1289–1294.

Sparks, D.L. 1989. Kinetics of soil chemical processes. Academic Press, San Diego, CA.

Sparks, D.L. 1995. Environmental soil chemistry. Academic Press, San Diego, CA.

Sparks, D.L. 1998a. Kinetics and mechanisms of chemical reactions at the soil mineral/water interface. p. 135–192. *In* D.L. Sparks (ed). Soil physical chemistry. 2nd ed. CRC Press, Boca Raton, FL.

Sparks, D.L. 1998b. Kinetics of sorption/release reactions on natural particles. p. 413–448. *In* P.M. Huang et al. (ed). Structure and surface reactions of soil particles. Vol. 4. John Wiley and Sons, New York.

Sparks, D.L. 1999. Kinetics and mechanisms of soil chemical reactions. p. B-123–167. *In* M.E. Sumner (ed.) Handbook of soil science. CRC Press, Boca Raton, FL.

Sparks, D.L., S.E. Fendorf, C.V. Toner IV, and T.H. Carski. 1996. Kinetic methods and measurements. p. 1275–1307. *In* D.L. Sparks et al. (ed.) Methods of soil analysis. Part 3. SSSA Book Ser. 5. SSSA, Madison, WI.

Sposito, G. 1994. Chemical equilibria and kinetics in soils. Oxford Univ. Press, New York.

Sposito, G. 1984. The surface chemistry of soils. Oxford University Press, New York.

Sposito, G. 1986. Distinguishing adsorption from surface precipitation. p. 217–228. *In* J.A. Davis and K.F. Hayes (ed.). Geochemical processes at mineral surfaces. ACS Symp. Ser. 323. ACS, Washington, DC.

Stanford, G., R.A. Vander Pol, and S. Dzienia. 1975. Denitrification rates in relation to total and extractable soil carbon. Soil Sci. Soc. Am. Proc. 39:284–289.

Steinberg, S.M., J.J. Pignatello, and B.L. Sawhney. 1987. Persistence of 1,2-dibromoethane in soils: Entrapment in intraparticle micropores. Environ. Sci. Technol. 21:1201–1208.

Stone, A.T. 1987. Reductive dissolution of manganese (III/IV) oxides by substituted phenols. Environ. Sci. Technol 21:979–988.

Stone, A.T. 1991. Oxidation and hydrolysis of ionizable organic pollutants at hydrous metal oxide surfaces. p. 231–254. *In* D.L. Sparks and D.L. Suarez (ed.) Rates of soil chemical processes. SSSA Spec. Publ. 27. SSSA, Madison, WI.

Stone, A.T., and J.J. Morgan. 1984a. Reduction and dissolution of manganese (III) and manganese (IV) oxides by organics. 1. Reaction with hydroquinone. Environ. Sci. Technol. 18:450–456.

Stone, A.T., and J. J. Morgan. 1984b. Reduction and dissolution of manganese (III) and manganese (IV) oxides by organics. 2. Survey of the reactivity of organics. Environ. Sci. Technol. 18:617–624.

Strawn, D.G., A.M. Scheidegger, and D.L. Sparks. 1998. Kinetics and mechanisms of Pb(II) sorption and desorption at the aluminum oxide water interface. Environ. Sci. Technol. 32:2596–2601.

Strawn, D.G., and D.L. Sparks. 1999. Sorption kinetics of trace elements in soils and soil materials. p. 1–28. *In* H.M. Selim and I. Iskandar (ed.) Fate and transport of heavy metals in the vadose zone. Lewis Publ., Chelsea, MI.

Strawn, D.G., and D.L. Sparks. 2000. Effects of soil organic matter on the kinetics and mechanisms of Pb(II) sorption and desorption in soil. Soil Sci. Soc. Am. J. 64:144–156.

Stumm, W., and J. J. Morgan. 1996. Aquatic chemistry. Wiley, New York.

Tang, L., and D.L. Sparks. 1993. Cation-exchange kinetics on montmorillonite using pressure-jump relaxation. Soil Sci. Soc. Am. J. 57:42–46.

Thompson, H.A. 1998. Dynamic ion partitioning among dissolved, adsorbed, and precipitated phases in aging cobalt(II)/kaolinite/water systems. Ph.D. diss. Stanford University, Stanford, CA.

Towle, S.N., J.R. Bargar, G.E. Brown, Jr., and G.A. Parks. 1997. Surface precipitation of $Co(II)_{(aq)}$ on Al_2O_3. J. Colloid Interface Sci. 187:62–68.

Vandenabeele, J., M.V. Woestyne, F. Houwen, R. Germonpre, D. Vandesande, and W. Verstraete. 1995. Role of autotrophic nitrifiers in biological manganese removal from groundwater containing manganese and ammonium. Microb. Ecol. 29:83–98.

Walker, W.J., C.S. Cronan, and H.J. Patterson. 1988. A kinetic study of aluminum adsorption by aluminosilicate clay minerals. Geochim. Cosmochim. Acta 52:55–62.

Wang, Z., C.C. Ainsworth, D.M. Friedrich, P.L. Gassman, and A.G. Joly. 2000. Kinetics and mechanism of surface reaction of salicylate on alumina in colloidal aqueous suspension. Geochim. Cosmochim. Acta 64:1159–1172.

Wehrli, B. 1990. Redox reactions of metal ions at mineral surfaces. p. 311–336. *In* W. Stumm (ed.) Aquatic chemical kinetics. John Wiley and Sons, New York.

Wehrli, B., B. Sulzberger, and W. Stumm. 1989. Redox processes catalyzed by hydrous oxide surfaces. Chem. Geol. 78:167–179.

Wersin, P., P. Hohener, R. Giovanoli, and W. Stumm. 1991. Early diagenetic influences on iron transformations in a freshwater lake sediment. Chem. Geol. 90:233–252.

Wielinga, B., B. Bostick, C.M. Hansel, R.F. Rosenweig, and S. Fendorf. 2000. Inhibition of bacterially promoted uranium reduction: Ferric (hydr)oxides as competitive electron acceptors. Environ. Sci. Technol. 34:2190–2195.

Xia, K., A. Mehadi, R.W. Taylor, and W.F. Bleam. 1997. X-ray absorption and electron paramagnetic resonance studies of Cu(II) sorbed to silica: Surface-induced precipitation at low surface coverages. J. Colloid Interface Sci. 185:252–257.

Zhang, P.C., and D.L. Sparks. 1990. Kinetics and mechanisms of sulfate adsorption/desorption on goethite using pressure-jump relaxation. Soil Sci. Soc. Am. J. 54:1266–1273.

Chapter 7

Cation Exchange in Soils

V. P. EVANGELOU, *(deceased) formerly Iowa State University, Ames, Iowa, USA*

R. E. PHILLIPS, *(deceased) formerly University of Kentucky, Lexington, Kentucky, USA*

Soils are multicomponent systems consisting of solid (inorganic and organic components), liquid (soil solution), and gaseous phases. These three phases are in a constant state of flux, trying to maintain an equilibrium state. Any externally induced change in one phase influences the other two phases, and a new equilibrium state is established. Cation exchange equilibrium represents a type of equilibrium affected mostly by the distribution of cations between the soil mineral surface phase and the solution phase.

Cation exchange reactions result primarily from excess negative charge on soil colloids. There are mainly two types of negative charge found in soil systems, permanent negative charge and variable, or pH-dependent, negative charge (McBride, 1989; Sposito, 1984b; White and Zelazny, 1986 and references therein). Permanent negative charge on soil clay mineral surfaces is generated through isomorphic substitution of coordinating cations with a given positive charge by cations with lower positive charge. Isomorphic substitution can occur mostly in tetrahedral sheets (e.g., vermiculite) or mostly in octahedral sheets (e.g., smectite). The magnitude of charge generated by isomorphic substitution is generally unaffected by pH. On the other hand, variable negative charge on mineral surfaces results from organic matter functional groups, such as carboxyls, and/or surface hydroxyls of inorganic minerals (Talibudeen, 1981 and references therein). The magnitude of the variable negative charge is influenced by pH and ionic strength. Any increase in pH and/or ionic strength is followed by an increase in negative charge (Gillman, 1984; Singh and Uehara, 1986; Uehara and Gillman, 1980).

Soils are mostly a mixture of various types of clay minerals, and because more than two cations are present at any one time (e.g., Ca^{2+}, Mg^{2+}, K^+, Na^+, NH_4^+), a rigorous theoretical and experimental description of ionic distribution is difficult and time consuming. For this reason, most often soil minerals are studied as binary cation exchange systems. Several theoretical approaches have been used to derive binary exchange equations. Most often mentioned in the literature are the mass-action and the double layer approaches. A number of equations have been proposed for the mass-action approach involving empirical and thermodynamic justification. They include the Vanselow, the Gaines and Thomas, the Gapon, the Rothmund and Kornfeld, the Kerr, and the Krishnamoorthy–Overstreet (Table 7–1) (Sposito, 1977, 1981a, 1981b, 1984b; Sparks, 1995; McBride, 1994, Evangelou et al., 1994 and ref-

Chemical Processes in Soils. SSSA Book Series, no. 8.

Table 7–1. Cation exchange selectivity coefficients for homovalent (K–Na) and heterovalent (K–Ca) exchange.†

Selectivity coefficient	Homovalent exchange‡	Heterovalent exchange§
Kerr	$K_k = \frac{\{K\text{-soil}\}[Na^+]¶}{\{Na\text{-soil}\}[K^+]}$	$K_k = \frac{\{K\text{-soil}\}^2[Ca^{2+}]}{\{K\text{-soil}\}^n[K^+]}$
Rothmund and Kornfeld#	$K_f = \frac{\{K\text{-soil}\}^n[Na^+]}{\{Na\text{-soil}\}[K^+]}$	$K_f = \frac{\{K\text{-soil}\}^n[Ca^{2+}]}{\{Ca\text{-soil}\}[K^+]^2}$
Vanselow††	$Kv = \frac{\{K\text{-soil}\}[Na^+]}{\{Na\text{-soil}\}[K^+]}$,	$Kv = \left[\frac{\{K\text{-soil}\}^2[Ca^{2+}]}{\{Ca\text{-soil}\}\ [K+]^2}\right]$
or	$K_v = K_k\left[\frac{1}{\{K\text{-soil}\} + \{Ca\text{-soil}\}}\right]$	
or	$K_k \quad \left[\frac{1}{\{K\text{-soil}\} + \{Ca\text{-soil}\}}\right]$	
Krishnamoorthy-Overstreet	$Kk_o = \frac{\{K\text{-soil}\}[Na+]}{\{Na\text{-soil}\}[K^+]}$,	$Kk_o = \left[\frac{\{K\text{-soil}\}^2[Ca^{2+}]}{\{Ca\text{-soil}\}[K^+]^2}\right]$
or	$Kk_o = K_k\left[\frac{1}{\{K\text{-soil}\} + 1.5\{Ca\text{-soil}\}}\right]$	
Gaines-Thomas††	$K_{GT} = \frac{\{K\text{-soil}\}[Na+]}{\{Na\text{-soil}\}[K^+]}$,	$K_{GT} = \left[\frac{\{K\text{-soil}\}^2[Ca^{2+}}{\{Ca_{1/2}\text{-soil}\}[K^+]^2}\right]$
or	$K_{GT} = K_k\left[\frac{1}{2[2\{Ca\text{-soil}\} + \{K\text{-soil}\}]}\right]$	
Gapon	$K_G = \frac{\{K\text{-soil}\}[Na^+]}{\{Na\text{-soil}\}[K^+]}$,	$K_G = \frac{\{K\text{-soil}\}[Ca^{2+}]^{1/2}}{\{Ca_{1/2}\text{-soil}\}[K^+]}$
or	$K_G = K_k$	

† Table adapted from Sparks (1995).
‡ The homovalent exchange reaction (K–Na exchange) is Na-soil + $K^+ \Leftrightarrow$ K-soil + Na^+.
§ The heterovalent exchange reaction (K–Ca exchange) is Ca-soil + $2K^+ \Leftrightarrow$ 2K-soil + Ca^{2+}, except for the Gapon convention, where it would be $Ca_{1/2}$-soil + $K^+ \Leftrightarrow$ K-soil + $1/2Ca^{2+}$.
¶ Brackets denote concentration in the solution phase (mol L^{-1}); braces denote concentration in the exchanger phase (mol kg^{-1}).
# Walton (1949) units expressed as concentration.
†† Vanselow (1932) and Gaines and Thomas (1953) originally expressed both solution and exchanger phases as activity. For simplicity, in this table they represent concentrations.

erences therein). The double layer approach (Bolt, 1955; Erickson, 1952) is rarely used to evaluate cation exchange reactions in soil and will not be discussed in any detail in this chapter. The formal thermodynamic approach, based on the mass action principle, gives no direct information about the molecular mechanisms and the forces operating in such systems. On the other hand, the diffuse double layer approach provides a description of Coulumbic forces operating in ion exchange processes (Van Bladel et al., 1972; Shainberg and Letey, 1984, and references therein).

There are a number of limitations in binary cation exchange equations as applied to soils. These limitations are (Bohn et al., 1985):

- Binary ion exchange is frequently considered, but the presence of additional ions is rarely acknowledged, even in highly acidic systems.
- Ion exchangers are assumed to possess constant exchange capacity, but it often varies with the nature of exchanging ions, solution concentration, and pH.
- Simple stoichiometric (1 for 1) ion exchange is generally assumed, but apparent deviations from 1:1 stoichiometry are usually explained in terms of simultaneous adsorption of anions and cations or in terms of the formation of complex ions.
- Complete reversibility is usually taken for granted.

Cation exchange reactions can be used for various applications in soils for the purpose of predicting cation release and removal in soil solution, nutrient availability to plants, and soil physical behavior, such as soil dispersion/flocculation.

The most widely studied binary cation exchange reactions in soils are those involving heterovalent cations, namely Na^+–Ca^{2+} and K^+–Ca^{2+}. A large number of studies involving Na^+–Ca^{2+} exchange have been conducted with respect to influence of ionic strength and solution composition. The reason for this great interest in Na^+–Ca^{2+} exchange is to enable us to predict surface composition and hydraulic properties of saline-sodic soils for the purpose of reclaiming and/or managing such soils. Soil hydraulic properties appear to be related to colloid dispersion and flocculation phenomena induced by Na^+ as affected by ionic strength, pH, and mineralogy (Marci and Evangelou, 1991a, 1991b, 1991c, and references therein). The soil colloid dispersion effect induced by exchangeable Na^+ is related to the highly hydrated nature of this ion. Soils commonly disperse when in equilibrium with an electrolyte solution under the "flocculation value." This flocculation value depends on solution composition (Na adsorption ratio, SAR), solution ionic strength, and clay mineralogy (El-Swaify, 1976; van Olphen, 1977; Greene et al., 1978), and pH (Tama and El-Swaify, 1978; Keren et al., 1988). For example, flocculation values for Na/Ca-montmorillonite are 3.0, 4.0, and 7.00 $mmol_c\ L^{-1}$, and 6.0, 10.0 and 18.0 $mmol_c\ L^{-1}$ for Na/Ca illite with exchangeable Na percentage (ESP) values of 5, 10, and 20, respectively (Shainberg and Letey, 1984).

Clay dispersion causes modification of soil-pore distribution, affecting soil hydraulic conductivity and gas diffusion. An increase in Na^+ levels in the soil solution or on the exchange phase (ESP) causes soil saturated hydraulic conductivity and gas diffusion to decrease (Russo and Bresler, 1977; Agassi et al., 1981; Kazman et al., 1983). The magnitude of ESP is related to the relative ratio of Na^+ to Ca^{2+} in the solution phase, known as SAR. An empirical exchange equation involving the relationship between ESP and SAR representing soils of the arid west was first developed by the U.S. Salinity Laboratory Staff (1954). It shows that when SAR is approximately 10 to 15, the ESP is also in the range of 10 to 15. In this ESP range, soils of the arid west will undergo dispersion. This relationship, however, does not necessarily apply to all soils (Marsi and Evangelou, 1991a).

Extensive K^+–Ca^{2+} exchange studies have also been conducted to predict plant available K^+ (Beckett, 1964, 1972; Evangelou et al., 1994, and references therein). The amount of "available" K in soil is conventionally determined by its quantity, that is, the amount extracted by neutral ammonium acetate solution. Woodruff (1955), emphasizing the "Intensity Factor," claimed that available K in soils could be defined in a rigorous thermodynamic manner by the potential of the solid phase to supply K to the soil solution. This potential is expressed as the difference between the electrochemical potential of K in the solid phase and that of any competing ion, such as Ca^{2+}. The difference in electrochemical potential can be approximated experimentally by measurements in the soil solution and expressed as the change in free energy (ΔG) accompanying the exchange of the two competing ions, namely $\Delta G = RT\ln[(\alpha_K)/(\alpha_{Ca})^{1/2}]$, where R is the universal gas constant, T is temperature, and α_K and α_{Ca} are the single-ion activities of K^+ and Ca^{2+}, respectively, in the soil solution.

Although Na^+–Ca^{2+} and K^+–Ca^{2+} exchange reactions are widely studied, such reactions are not limited to the above cations. Exchange reactions involve charged species, including metals and nonmetals, and many exchange studies on metals have been reported in the literature (Deist and Talibudeen, 1967a, 1957b; Faucher and Thomas, 1954; Hunsacker and Pratt, 1971; Maes et al., 1976; Pleysier et al., 1979; Sposito and Mattigod, 1979; Sposito, 1981a).

In the past two decades a number of review articles on exchange reactions have been written. Thomas (1977) reviewed the development of knowledge of cation reaction in soils and focused mainly on experimentation dealing with the subject. Sposito (1981a) presented the historical development of the mathematical equations and theoretical concepts describing cation exchange in soils. Sposito expanded his 1981 review article on exchange to a textbook (Sposito, 1984b) that describes in some detail mineral surface chemistry and the various mechanisms accounting for cation adsorption and exchange. Evangelou et al (1994) presented cation exchange equations with respect to thermodynamics and their potential in releasing cations to the soil solution. In a book chapter on cation exchange, Sparks (1995) presented the history and its basis to chemical thermodynamics.

The purpose of this chapter is to introduce the most widely used exchange equations in soil chemistry research, including the Vanselow (1932), the Gapon (1933), and Gaines and Thomas (1953) equations. We demonstrate the mathematical and thermodynamic interrelationships between these three equations and carry out a sensitivity analysis on these equations.

CATION EXCHANGE REACTIONS

Cation exchange in soils or clay minerals involves replacement of a given cation from a given surface by another cation. There are basically two major types of cation exchange reactions in soil systems: (i) homovalent cation exchange and (ii) heterovalent cation exchange. Below, details of such reactions will be discussed, invoking the Ca^{2+}–Mg^{2+} system as an example of a homovalent exchange reaction, and the Na^+–Ca^{2+} system as an example of a heterovalent exchange system.

Homovalent Cation Exchange

Homovalent cation exchange refers to exchanging cations with similar valence. For example, in the case of Ca^{2+}–Mg^{2+} exchange the reaction can be expressed as follows:

$$\text{ExMg} + \text{Ca}^{2+} \Leftrightarrow \text{ExCa} + \text{Mg}^{2+} \quad [1]$$

Based on Reaction [1], the cation exchange selectivity coefficient (K_{Ca-Mg}) can be expressed as

$$K_{Ca-Mg} = [\text{ExCa}/\text{ExMg}][(\text{Mg}^{2+})/(\text{Ca}^{2+})] \quad [2]$$

where ExMg and ExCa denote exchangeable cations (cmol kg^{-1}), Ex denotes the soil exchanger with a charge of 2–, and Mg^{2+} and Ca^{2+} denote solution concentration (mol L^{-1} or mmol L^{-1}). Equation [2] can be used to solve for $ExCa_{1/2}$ ($ExCa_{1/2}$ = $cmol_c$ kg):

$$\text{ExCa}_{1/2} = K_{Ca-Mg}(\text{CEC})(\text{CR}_{Ca})[1 + \text{CR}_{Ca}K_{Ca-Mg}]^{-1} \quad [3]$$

where CEC is cation exchange capacity and CR_{Ca} denotes Ca adsorption ratio and is expressed by

$$\text{CR}_{Ca} = [\text{Ca}^{2+}]/[\text{Mg}^{2+}] \quad [4]$$

where brackets denote concentration (mol L^{-1} or mmol L^{-1}) in the solution phase, and

$$\text{CEC} = \text{ExCa}_{1/2} + \text{ExMg}_{1/2} \quad [5]$$

A plot of $ExCa_{1/2}$ vs. CR_{Ca} will produce a curvilinear line asymptotically approaching CEC. The pathway of such line from $ExCa_{1/2} = 0$ to $ExCa_{1/2}$ = CEC depends on K_{Ca-Mg} and CEC (Fig. 7–1).

An important component of homovalent exchange is the magnitude of the exchange selectivity coefficient. Commonly, homovalent cation exchange reactions in soils or soil minerals exhibit a selectivity coefficient somewhere around 1, with some exceptions (Table 7–2). This value signifies that the soil mineral surface does not show any particular adsorption preference for either of the two cations. How-

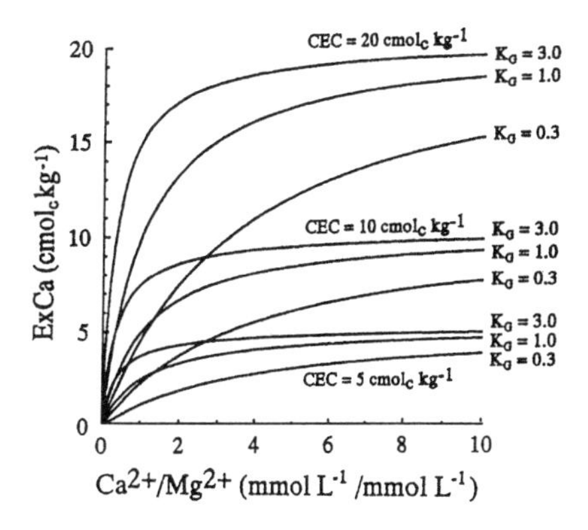

Fig. 7–1. Predicted relationship between Ca^{2+}/Mg^{2+} solution ratio and exchangeable Ca^{2+} for charged surfaces with various selectivity coefficients and CEC values.

Table 7–2. Thermodynamic equilibrium constants of exchange (K_{eq}) for binary exchange processes on soils and soil components.†

Exchange process	Exchanger	K_{eq}‡	Reference
Ca–Mg	calcareous soils	0.89–0.75	Van Bladel and Gheyi (1980)
Ca–Mg	Camp Berteau montmorillonite	0.95	Van Bladel and Gheyi (1980)
Ca–Cu	Wyoming bentonite	0.96	El-Sayed et al. (1970)
Na–Li	World vermiculite	11.42	Gast and Klobe (1971)
Na–Li	Wyoming bentonite	1.08	Gast and Klobe (1971)
Na–Li	Chambers montmorillonite	1.15	Gast and Klobe (1971)
Mg–Ca	soil	0.61	Jensen and Babcock (1973)
Mg–Ca	kaolinitic soil clay (303 K)	0.65	Udo (1978)
K–Na	soils	4.48–6.24	Deist and Talibudeen (1967a)
K–Na	Wyoming bentonite	1.67	Gast (1972)
K–Na	Chambers montmorillonite	3.41	Gast (1972)

† Table adapted from Sparks (1995).
‡ The exchange studies were conducted at 298K; exceptions are noted.

ever, for a mineral with a $K_{Ca-Mg} > 1$, the surface prefers Ca^{2+}, and for a mineral with $K_{Ca-Mg} < 1$, the surface prefers Mg^{2+} (Beckett, 1965; Sposito and LeVesque, 1985; Sposito et al., 1986).

Homovalent Cation Exchange Preference Isotherms

Cation preference is commonly demonstrated through fractional isotherms. Fractional isotherms are plots of equivalent fraction on the exchange phase vs. equivalent fraction in the solution phase. For a homovalent ion exchange reaction (e.g., Ca^{2+}–Mg^{2+}) when $K_{Ca-Mg} = 1$, it follows from Eq. [5] that

$$[ExCa_{1/2}/CEC] = [Ca^{2+}]/[Ca^{2+} + Mg^{2+}] = [Ca^{2+}]/[Cl^{-}/2] = [Ca^{2+}]/[SO_4^{2-}] \quad [6]$$

and therefore

$$[Ca^{2+}]/[Ca^{2+} + Mg^{2+}] = [Ca^{2+}/Mg^{2+}]/[1 + Ca^{2+}/Mg^{2+}] \quad [7]$$

Furthermore, upon introducing solution single-ion activities, ϕ_i using the Davies equation (Davies, 1962) (Fig. 7–2)

$$\log\phi_i = -Az_i^2\,[I^{1/2}/(1 + I^{1/2}) - 0.3I] \quad [8]$$

where A is a constant that equals 0.512 at 25°C, z denotes ion valence, and I ionic strength, and

$$[Ca^{2+}]\phi_{Ca}/[Ca^{2+} + Mg^{2+}]\phi_{Ca,Mg} = [Ca^{2+}]\phi_{Ca}/Mg^{2+}]\gamma_{Mg}/\{1 + [Ca^{2+}]\phi_{Ca}/[Mg^{2+}]\phi_{Mg}\} \quad [9]$$

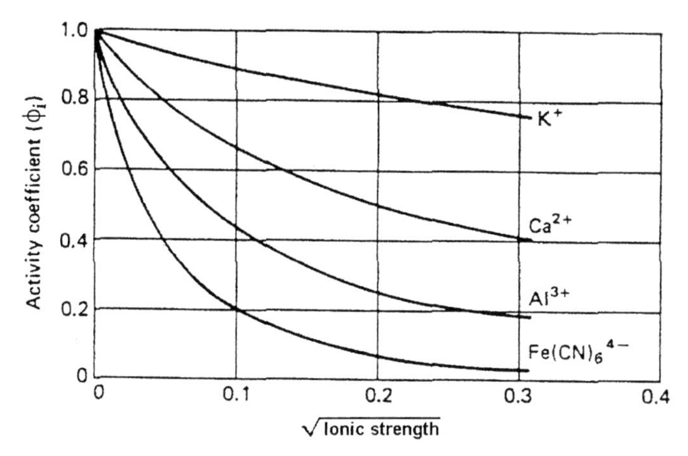

Fig. 7–2. Relationship between solution single-ion activity coefficients and ionic strength for various cations.

It appears that since ϕ_{Ca} equals ϕ_{Mg} the relationship of $ExCa_{1/2}$/CEC vs. $[Ca^{2+}]/[Ca^{2+} + Mg^{2+}]$ is shown to be independent of ionic strength. Such an isotherm is shown in Fig. 7–3. The diagonal line shows no preference ($K_{Ca\text{–}Mg} = 1$). A line above the diagonal line reveals that the surface prefers Ca^{2+}, while any line below the non-preference line reveals that the surface prefers Mg^{2+}.

Heterovalent Cation Exchange

Heterovalent cation exchange reactions in soils or soil minerals commonly involve exchange of a monovalent cation with a divalent cation or vice versa. Cations with valence higher than 2 (i.e., Al^{3+}) are also involved, but to a lesser ex-

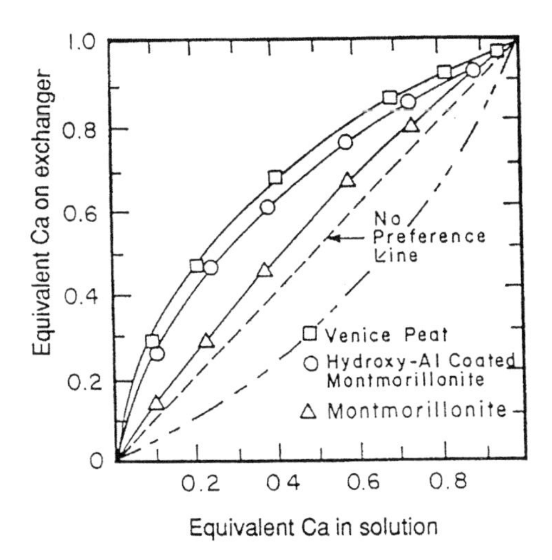

Fig. 7–3. Relationships between equivalent fraction of solution phase vs. equivalent fraction of exchange phase for Ca^{2+}–Mg^{2+} exchange using various soil minerals (Hunsaker and Pratt, 1972).

tent because of the low pH needed for the Al^{3+} species to persist in solution and exchange phase. The equation that is most commonly used to describe heterovalent cation exchange is the Gapon exchange equation (Gapon, 1933). For example, for Na^+–Ca^{2+} exchange:

$$ExCa_{1/2} + Na^+ \Leftrightarrow ExNa + 1/2Ca^{2+} \qquad [10]$$

Based on Reaction [10], the Gapon exchange selectivity coefficient (K_G) can be expressed as

$$K_G = [ExNa/ExCa_{1/2}][Ca^{2+}]^{1/2}/[Na^+] \qquad [11]$$

where ExNa and $ExCa_{1/2}$ denote exchangeable cations ($cmol_c\ kg^{-1}$), Ex denotes the soil exchanger with a charge of 1–, and Na^+ and Ca^{2+} denote solution cations in terms of concentration ($mol\ L^{-1}$ or $mmol\ L^{-1}$). Equation [11] can be rearranged to solve for ExNa

$$ExNa = (K_G CEC)(SAR)[1 + SAR K_G]^{-1} \qquad [12]$$

where

$$SAR = [Na^+]/[Ca^{2+}]^{1/2} \qquad [13]$$

where brackets denote concentration ($mol\ L^{-1}$ or $mmol\ L^{-1}$) in the solution phase, and

$$CEC = ExCa_{1/2} + ExNa \qquad [14]$$

A plot of SAR vs. ExNa will produce a curvilinear line, asymptotically approaching CEC. Rearranging Eq. [14] gives

$$ExNa/ExCa_{1/2} = K_G[Na^+]/[Ca^{2+}]^{1/2} \qquad [15]$$

Theoretically, a plot of $ExNa/ExCa_{1/2}$ or exchangeable Na ratio (ESR) vs. SAR will produce a straight line with slope = K_G. The average magnitude of K_G for soils of the arid west is approximately 0.015 $(mmol\ L^{-1})^{-1/2}$. However, experimental K_G appears to depend on pH, salt concentration, and clay mineralogy. Furthermore, experimental K_G does not appear to be constant across the various Na loads. Commonly, as Na load increases, K_G also increases (Oster and Sposito, 1980). Furthermore, as pH increases in variable charge soils, K_G decreases (Pratt et al., 1962). Note also that the K_G of a heterovalent exchange depends on solution concentration units, whereas the K_G of a homovalent exchange is independent of solution units. A heterovalent K_G obtained (with solution units in $mmol\ L^{-1}$) when multiplied by $(1000)^{1/2}$ gives the K_G $[(mol\ L^{-1})^{-1/2}]$. Commonly, the K_G of Na^+–Ca^{2+} exchange may vary from 0.50 to 2 $(mol\ L^{-1})^{-1/2}$, whereas the K_G of K^+–Ca^{2+} exchange may vary from 2 to 30 $(mol\ L^{-1})^{-1/2}$, depending on mineralogy and pH (Evangelou and Blevins, 1988, and references therein). The greatest variation in K_G (e.g., K^+–Ca^{2+}) is due to mineralogy in the order: illite > smectite > organic matter (Fig. 7–4).

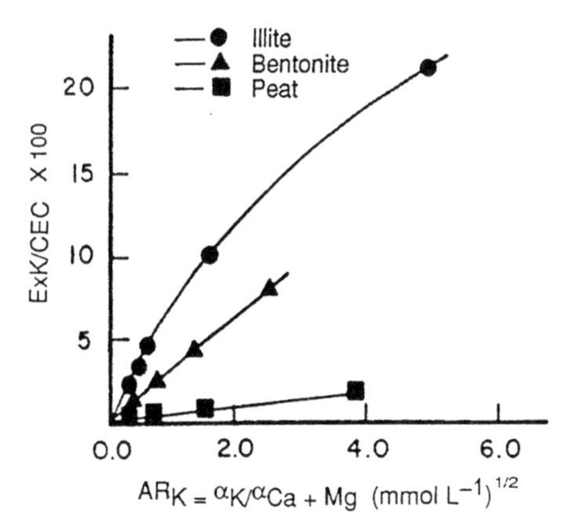

Fig. 7–4. Relationship between AR_K and percentage CEC occupied by K^+ for illite, bentonite, and peat (Salmon, 1964).

Heterovalent Cation Exchange Preference Isotherms

Cation preference in heterovalent exchange can also be demonstrated through fractional isotherms. Fractional heterovalent isotherms are plots of equivalent fractions in the solution phase vs. equivalent fractions in the exchange phase (e.g., in the case of Na^+–Ca^{2+}, $Na^+/[Na^++2Ca^{2+}]$ vs. E_{Na} = ExNa/CEC or $2Ca^{2+}/[Na^++2Ca^{2+}]$ vs. $ExCa_{1/2}$/CEC). The nonpreference line for a heterovalent exchange shown in Fig. 7–5 is the solid line. A line above the nonpreference line reveals that the surface prefers the monovalent cation, while a line below the nonpreference line reveals that the surface prefers the divalent cation. Heterovalent fractional cation preference isotherms differ from homovalent cation preference isotherms in that (i) the nonpreference line in heterovalent exchange is not the diagonal line and (ii) preference

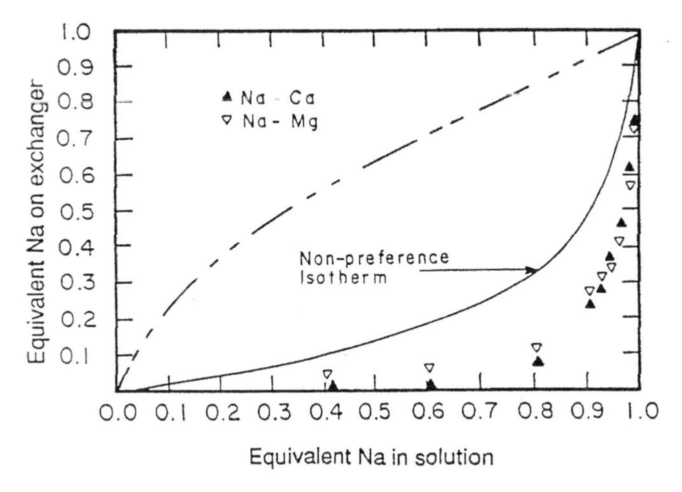

Fig. 7–5. Relationships between Na load on the exchange phase vs. Na-equivalent fraction in the solution phase for Na^+–Ca^{2+} and Na^+–Mg^{2+} exchange in montmorillonite and a hypothetical Na-preference isotherm (Sposito and LeVesque, 1985).

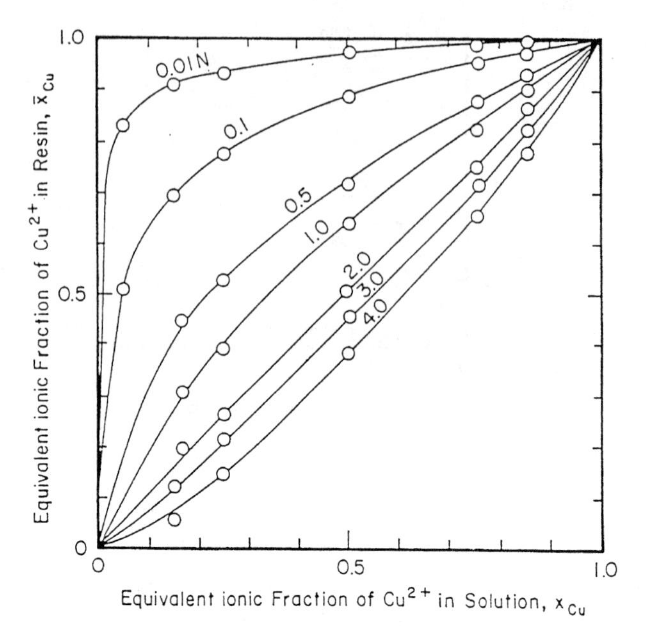

Fig. 7–6. Relationships between Cu^{2+} load on the exchanger phase vs. Cu^{2+}-equivalent fraction in the solution phase for Na^+–Cu^{2+} exchange on cation exchange resin.

in heterovalent exchange depends on ionic strength. The latter is also known as the "square root effect" (Fig. 7–6).

The exchange mathematical expression used to produce the nonpreference isotherm is the Vanselow equation. It is suggested that the Vanselow equation is consistent with thermodynamics of chemical reactions because it employs units of solution and exchange phases in moles (for detailed discussion see the section on thermodynamics of cation exchange). In the case of Na^+–Ca^{2+} exchange

$$K_v = [M_{Na}/M_{Ca}]^{1/2}[(\alpha_{Ca})^{1/2}/\alpha_{Na}] \qquad [16]$$

where $\alpha_{Na}/\alpha_{Ca}^{1/2} = AR_{Na}$. Upon rearranging and solving for ExNa

$$ExNa = CECK_vAR_{Na}/[4 + (K_VAR_{Na})2]^{1/2} \qquad [17]$$

Considering that for the nonpreference isotherm as $I \rightarrow 0$, $K_v = 1$ and $\Delta G^\circ = 0$, then

$$ExNa/CEC = AR_{Na}/[4 + (AR_{Na})2]^{1/2} \qquad [18]$$

Using a number of AR_{Na} values, as $I \rightarrow 0$, to cover the entire exchange isotherm, equivalent fractional loads for Na^+ can be estimated using Eq. [18]. A plot of ExNa/CEC vs. $(Na^+)/[(Na^+) + (2Ca^{2+})]$ will produce a curvilinear line representing the nonpreference isotherm. Upon introducing ϕ_i values in the $Na^+/[Na^+ + 2Ca^{2+}]$ term such that

Table 7–3. Calculated exchange equivalent ratios using Eq. [18].

	Solution			
Equiv. ratio $[Na^+/(Na^+ + 2Ca^{2+})]$	Activity ratio AR_{Na}	Total salt	K_v	Exchange eq. ratio [ExNa/CEC]
			$mol_c\ L^{-1}$	
0.5	2.236×10^{-2}	5×10^{-4}	1	0.01
0.5	2.69×10^{-1}	5×10^{-2}	1	0.13
0.5	8.95×10^{-1}	5×10^{-1}	1	0.41
	Conc. ratio CR_{Na}			
0.5	2.236×10^{-2}	5×10^{-4}	1	0.01
0.5	2.236×10^{-1}	5×10^{-2}	1	0.11
0.5	7.071×10^{-1}	5×10^{-1}	1	0.33

$$[(Na^+)/\phi_{Na}]/\{(Na^+)/\phi_{Na} + 2[(Ca^{2+})/\phi_{Ca}]\} \quad [19]$$

It can be shown that as I increases, ϕ_{Na} and ϕ_{Ca} decrease disproportionally, with ϕ_{Ca} decreasing significantly more than ϕ_{Na}. Therefore, as I increases, the term $[(Na^+)/\phi_{Na}]/\{(Na^+)/\phi_{Na} + 2[(Ca^{2+})/\phi_{Ca}]\}$ decreases, suggesting that the nonpreference isotherm, ExNa/CEC vs. $(Na^+)/[(Na^+) + (2Ca^{2+})]$, is shifted towards higher apparent preference for the monovalent cation, making it ionic strength dependent (Fig. 7–6). The influence of ϕ_i, induced by increasing I, on apparent ion preference, however, is very small in comparison with the square root effect on apparent ion preference induced also by increasing I. This is demonstrated in Table 7–3. It shows that, based on Eq. [18], by assuming similar solution equivalent ratios but different ionic strength, for the same solution equivalent ratios as I increases from 5×10^{-4} to 5×10^{-1} M, AR_{Na} increases from 2.23×10^{-2} to 8.95×10^{-1}. Hence, exchange equivalent ratios increase from 0.01 to 0.41. When solution ions are introduced as concentration, exchange equivalent ratios increase from 0.01 to 0.33. Therefore, it is clear that the square root effect has a much larger impact on the isotherm than solution activity.

Sposito (1981a) derived a nonpreference heterovalent (monovalent-divalent) exchange isotherm showing analytically its dependency on ionic strength. The equation is

$$E_{Ca} = 1 - \{1 + 2/b\mathrm{TMC}[1/(1 - E_{Cas})2 - (1 - E_{Cas})]\}^{-1/2} \quad [20]$$

where E_{Ca} is the equivalent fraction of Ca^{2+} on the solid phase, TMC is the total metal concentration (TMC = $Na^+ + 2Ca^{2+}$), E_{Cas} is the equivalent Ca^{2+} concentration in the solution phase, and b is ϕ^2_{Na}/ϕ_{Ca}, where ϕ_{Na} and ϕ_{Ca} represent single-ion activity coefficients for Na^+ and Ca^{2+}, respectively. Note that Na^+–Ca^{2+} here is only used as an example since the above equations apply to all heterovalent ion exchange reactions.

Ion Preference

Ion preference or selectivity is defined as the potential of a charged surface to demonstrate preferential adsorption of one ion over another ion. Such ion preference is described by the lyotropic series and an example, from highest to lowest preference, follows: $Ba^{2+} > Pb^{2+} > Sr^{2+} > Ca^{2+} > Ni^{2+} > Cd^{2+} > Cu^{2+} > Co^{2+} > Mg^{2+} > Ag^{+} > Cs^{+} > Rb^{+} > K^{+} > NH_4^{+} > Na^{+} > Li^{+}$ (Helfferich, 1972). Note, however, the above lyotropic series is not universally applicable. The series often depends on the nature of the adsorbing surface. For example, the relative replacing order of a number of hard cations on montmorillonite is: $K^{+} < Mg^{2+} < Ca^{2+} < Sr^{2+} < Ba^{2+}$; on vermiculite is: $K^{+} < Ba^{2+} < Sr^{2+} < Ca^{2+} < Mg^{2+}$; on biotite is: $Mg^{2+} < K^{+} < Ca^{2+} < Sr^{2+} < Ba^{2+}$; on muscovite is: $Mg^{2+} < Ca^{2+} < Sr^{2+} < K^{+} < Ba^{2+}$ (Sullivan, 1977, and references therein). Various models for predicting ion selectivity have been reported in the literature (Eisenman, 1962; Sullivan, 1977; Xu and Harsh, 1990a, 1990b, 1992; Auboiroux et al., 1998), but with generally limiting satisfaction. Ion selectivity reflects the outcome of many interacting factors between cations and surfaces, making it difficult to incorporate all these factors quantitatively into any given model. A number of these cation selectivity factors are listed below.

Selectivity Factors Based on Physical Chemistry of Cations

1. Generally, given two cations with the same valence (e.g., Na^{+} vs. K^{+}), the cation with the smaller hydrated radius, or least negative heat of hydration (Tables 7–4 and 7–5), is preferred (K^{+}).
2. Generally, given two cations with different valence (e.g., Na^{+} and Ca^{2+}), the cation with the higher valence (Ca^{2+}) is preferred because of its greater polarization potential (polarization is the distortion of the electron cloud around the adsorbing cation due to the electric field of the charged surface). Note, larger anions show greater polarization potential than smaller anions. For this reason, larger anions would be preferred by a positively charged surface.
3. Given two cations with the same valence but one being a stronger acid than the other (e.g., Cu^{2+} vs. Ca^{2+}), the cation with stronger acid behavior or higher hydrolytic constant would be preferred if the surface behaved as a relatively strong base.
4. For any cation with any valence, when in the presence of two different anions, the anion with the highest potential to form neutral pairs with the cation controls the latter's adsorption potential, assuming that the anions do not react with the surface. For example, Ca^{2+} in the presence of Cl^{-} exhibits greater adsorption potential than Ca^{2+} in the presence of SO_4^{2-} because of the latter's greater potential to form neutral $CaSO_4^{\circ}$ pairs.
5. For any heavy metal cation in the presence of two different anions, the anion with the highest potential to form surface complexes controls the metal's adsorption potential. For example, Ni^{2+} in the presence of $Ca(NO_3)_2$ exhibits greater adsorption potential than Ni^{2+} in the presence of $CaSO_4$ because of the sulfate's potential to react with the surface and produce sites with high specificity for Ca^{2+}.

Selectivity Factors Based on Physical Chemistry of Surfaces

1. A surface that has the potential to form inner-sphere complexes with certain monovalent cations (e.g., K^{+} or NH_4^{+}) shows stronger preference for such cations than any other cation (e.g., vermiculite-K^{+} or vermiculite-NH_4^{+} vs. vermiculite-Na^{+} or vermiculite-Ca^{2+}).

Table 7–4. Crystallographic radii and heats and entropies of ion hydration at 25°C.

Ion	Crystallographic ion radius	Heat of hydration, ΔH	Entropy of hydration, ΔS
	nm	$kJ\ mol^{-1}$	$J\ mol^{-1}\ K^{-1}$
H^+	—	−10	109
Li^+	0.060	−506	117
Na^+	0.095	−397	87.4
K^+	0.133	−314	51.9
Rb^+	0.148	−289	40.2
Cs^+	0.169	−225	36.8
Be^{2+}	0.031	−2470	—
Mg^{2+}	0.065	−1910	268
Ca^{2+}	0.099	−1580	309
Ba^{2+}	0.135	−1290	159
Mn^{2+}	0.080	−1830	243
Fe^{2+}	0.076	−1910	272
Cd^{2+}	0.097	−1790	230
Hg^{2+}	0.110	−1780	180
Pb^{2+}	0.050	−4640	464
Fe^{3+}	0.064	−4360	460
La^{3+}	0.115	−3260	368
F^-	0.136	−506	151
Cl^-	0.181	−377	98.3
Br^-	0.195	−343	82.8
I^-	0.216	−297	59.8
S^{2-}	0.184	−1380	130

2. A surface that does not have the potential to form inner-sphere complexes prefers cations with higher valence. This preference depends on the magnitude of the surface electrical potential. For example, a surface with high electrical potential shows lower preference for monovalent cations (e.g., Na^+) in the presence of a divalent cation (e.g., Ca^{2+}) than a surface with low electrical potential.
3. Surfaces that exhibit pH-dependent electrical potential show various degrees of selectivity for the same cation. For example, kaolinite or kaolinitic soils at high pH (high negative surface electrical potential) show increasing preference for divalent cations than monovalent cations, while at low pH (low negative electri-

Table 7–5. Ion sizes and ionic hydration.

Ion	Ionic radii	
	Not hydrated	Hydrated
	Å	Å
Li^+	0.68	10.03
Na^+	0.98	7.90
K^+	1.33	5.32
NH^{4+}	1.43	5.37
Rb^+	1.49	5.09
Cs^+	1.65	5.05
Mg^{2+}	0.89	10.80
Ca^{2+}	1.17	9.6
Sr^{2+}	1.34	9.6
Ba^{2+}	1.49	8.8

cal potential), kaolinite, or kaolinitic soils show increasing preference for monovalent cations rather than divalent cations.

4. Surfaces that have the potential to undergo conformational changes (e.g., humic acids) prefer higher valence cations (e.g., Ca^{2+}) than lower valence cations (e.g., K^+).
5. Surfaces that exhibit weaker acid behavior (high pK_α) show stronger preference for heavy metals than hard metals in comparison to surfaces with stronger acid behavior (low pK_α). For example, illite or kaolinite shows stronger preference for Cu^{2+} or Cd^{2+} than does montmorillonite.

THERMODYNAMICS OF CATION EXCHANGE IN SOILS

A soil binary exchange reaction at equilibrium involving for example Na^+ and Ca^{2+} is commonly written as

$$Ex_2Ca + 2Na^+ \Leftrightarrow ExNa_2 + Ca^{2+} \quad [21]$$

However, for the purpose of directly comparing K_G and K_v later in this chapter, the Na^+–Ca^{2+}exchange expression is presented as (Evangelou and Phillips, 1987, 1989, and references therein)

$$1/2Ex_2Ca + Na^+ \Leftrightarrow ExNa + 1/2Ca^{2+} \quad [22]$$

where Ex denotes an exchanger phase taken to have a charge of negative one (–1), and Na^+ and Ca^{2+} denote solution species. Note that according to the Vanselow expression the difference between the exchange selectivity coefficients of the above two equations is that the K_v of Reaction Expression [21] is equal to the K_V of Reaction Expression [22] raised to the second power.

A criterion of chemical-reaction equilibrium is (Smith and Van Ness, 1987)

$$\Sigma v_i \mu_i = 0 \quad [23]$$

where v_i stoichiometric coefficient for species i, and μ_i chemical potential for species i. The chemical potential μ_i of species i in solution is the same as the partial molar Gibbs energy, G_i

$$d\mu_i = dG_i = RT d\ln f_i \quad [24]$$

where R is the universal gas constant. Under constant temperature, T, Eq. [24] relates the quantities μ_i and G_i to the fugacity (f_i) in solution. Integration of Eq. [24] from the standard state of species i to any other state of species i in solution produces

$$\mu_i - G_i^\circ = RT\ln(f_i/f_i^\circ) \quad [25]$$

where G_i° is molar Gibbs energy for species i. The ratio f_i/f_i° is defined as the activity α_i in solution. For solids and liquids the standard state f_i° is the pure solid or liquid at 1 atm and the systems temperature; thus, for solid and liquid phase reactions $\alpha_i = f_i/f_i^\circ$.

Based on the above definitions and equations

$$\mu_i = G_i^\circ + RT\ln\alpha_i \quad [26]$$

and for a chemical reaction at the thermodynamic equilibrium point

$$\Sigma v_i(G_i^\circ + RT\ln\alpha_i) = 0 \quad [27]$$

Therefore,

$$\pi(\alpha_i)^{v}i = \exp -\left(\frac{\Sigma v_i G_i^\circ}{RT}\right) = K_{eq} \quad [28]$$

where π signifies the overall product of species i for a given chemical reaction and K_{eq} is the equilibrium constant. Note also that

$$-RT\ln K_{eq} = \Sigma v_i^\circ G_i^\circ = \Delta G^\circ \quad [29]$$

The Gibbs energy, G_i°, of the pure component is a property of pure species i in its standard state and fixed pressure and depends only on temperature. According to Eq. [29], K_{eq} depends also only on temperature, and ΔG° is the standard Gibbs energy change of reaction. For any reaction at equilibrium

$$K_{eq} = \pi(\alpha_i^\circ)^{v}i = \pi(f_i/f_i^\circ)^{v}i \quad [30]$$

Thus, from a thermodynamic equilibrium point of view the activities α_i are completely defined only when the pure component reference states f_i° and G_i° are also defined.

The thermodynamic exchange equilibrium constant K_{eq} for Reaction [22] at room temperature (22°C) and 1 atm pressure is represented by

$$K_{eq} = (\alpha_{Ca})^{1/2}(\alpha_{ExNa})/(\alpha_{Na})(\alpha_{Ex2Ca})^{1/2} \quad [31]$$

where α_{Na} and α_{Ca} are the activities of solution phase Na^+ and Ca^{2+}, respectively, and α_{ExNa} and α_{Ex2Ca} are the activities of exchange phase Na^+ and Ca^{2+}, respectively.

Activity, α_i, is defined by

$$\alpha_i = \phi_i c_i \quad [32]$$

where ϕ_i is the activity coefficient of species i and c_i is the concentration of species i.

When solution ionic strength, I, approaches zero, solution phase α_i is set to 1; hence, $\phi_i = 1$. For mixed electrolyte solutions when $I > 0$, the single ion activity concept introduced by Davies (1962) is employed to estimate ϕ_i.

Activity of the adsorbed or solid phase species is defined by employing the mole fraction concept (Vanselow, 1932). For a heterovalent binary exchange reaction (e.g., Na^+–Ca^{2+}) assuming that the system obeys ideal solid–solution theory (Sposito, 1981b, and references therein), the activity term ($^{\alpha}$Exi) is defined by

$$^{\alpha}\mathrm{ExNa} = M_{\mathrm{Na}} = \mathrm{ExNa}/(\mathrm{ExNa} + \mathrm{Ex_2Ca}) \quad [33]$$

and

$$^{\alpha}\mathrm{Ex_2Ca} = M_{\mathrm{Ca}} = \mathrm{Ex_2Ca}/(\mathrm{ExNa} + \mathrm{Ex_2Ca}) \quad [34]$$

where M_{Na} and M_{Ca} denote mole fraction of Na^+ and Ca^{2+}, respectively, and Ex denotes exchange phase with a valence of 1−. When ideal solid–solution behavior is not obeyed

$$^{\alpha}\mathrm{Ex}_i = \gamma_i M_i \quad [35]$$

where γ_i denotes adsorbed-ion activity coefficient. In the mole fraction concept the sum of exchangeable Na^+ (ExNa) and exchangeable Ca^{2+} (Ex_2Ca) is expressed in moles per kilogram exchanger, making the denominator of Eq. [33] and [34] for the entire isotherm a variable. When exchangeable Na^+ and exchangeable Ca^{2+} are expressed in charge equivalents (E_i), their sum across the entire isotherm is constant. Equivalent fractions for Na^+ and Ca^{2+} are given by

$$E_{\mathrm{Na}} = \mathrm{ExNa}/(\mathrm{ExNa} + 2\mathrm{Ex_2Ca}) \quad [36]$$

and

$$E_{\mathrm{Ca}} = 2\mathrm{Ex_2Ca}/(\mathrm{ExNa} + 2\mathrm{Ex_2Ca}) \quad [37]$$

Note, Eq. [36] is used to estimate ESP by multiplying E_{Na} by 100. For the above binary exchange system, CEC is taken to be

$$\mathrm{CEC} = \mathrm{ExNa} + 2\mathrm{Ex_2Ca} \quad [38]$$

and it is assumed that any other cations (e.g., exchangeable K^+ and/or H^+) are present in negligible quantities and do not interfere with Na^+–Ca^{2+} exchange, or H^+ is tightly bound to the charged surface, giving rise to pH-dependent charge (Fletcher et al., 1984).

On the basis of the above equations and definitions, an equilibrium exchange expression for Reaction [22] can be given by

$$K_{\mathrm{v}} = (M_{\mathrm{Na}})(\alpha_{\mathrm{Ca}})^{1/2}/(M_{\mathrm{Ca}})^{1/2}\,(\alpha_{\mathrm{Na}}) \quad [39]$$

where $\alpha_{\mathrm{Na}}/\alpha_{\mathrm{Ca}}^{1/2}$ is known as the Na activity ratio ($\mathrm{AR_{Na}}$), and K_{v} is the Vanselow exchange selectivity coefficient. The magnitude of K_{v} is taken to represent relative affinity of Na^+ with respect to Ca^{2+} by a charged surface (Sposito, 1981a; Shainberg et al, 1980). When K_{v} equals 1 at a given level of exchangeable Na^+, the exchanger at that level of Na^+ load shows no preference for either Na^+ or Ca^{2+}. On the other hand, when $K_{\mathrm{v}} > 1$ at a given level of exchangeable Na^+, it signifies exchanger preference for Na^+, and when $K_{\mathrm{v}} < 1$ at a given level of exchangeable Na^+, it signifies preference for Ca^{2+}.

A variable K_v with respect to exchangeable Na^+ load can be transformed to the thermodynamic exchange constant (K_{eq}) by

$$K_{eq} = K_v(\gamma_{Na}/(\gamma_{Ca})^{1/2} \quad [40]$$

where γ_{Na} and γ_{Ca} denote adsorbed-ion activity coefficient for Na^+ or Ca^{2+}, respectively. It follows from Eq. [40] that any variation in K_v with respect to exchange phase composition is followed by a variation in the solid phase activity coefficients (γ_i) such that

$$\mathrm{d}\ln K_v = -\mathrm{d}\ln\gamma_{Na} + 1/2\mathrm{d}\ln\gamma_{Ca} \quad [41]$$

Furthermore, any variation in γ_{Na} must be compensated for by a variation in γ_{Ca} such that

$$M_{Na}\mathrm{d}\ln\gamma_{Na} + M_{Ca}\mathrm{d}\ln\gamma_{Ca} = 0 \quad [42]$$

The solutions of Eq. [41] and [42] with respect to γ_{Na} and γ_{Ca} are

$$1/2\mathrm{d}\ln\gamma_{Ca} = -E_{Na}\mathrm{d}\ln K_v \quad [43]$$

and

$$\mathrm{d}\ln\gamma_{Na} = (1 - E_{Na})\mathrm{d}\ln K_v \quad [44]$$

where E_{Na} is the equivalent charge fraction of adsorbed Na^+ (Argersinger et al., 1950). Integration of Eq. [43] and [44] generates two new equations, allowing us to quantify $\ln\gamma_{Na}$ and $\ln\gamma_{Ca}$ at any value of E_{Na}. These equations are (Argersinger et al., 1950)

$$\ln\gamma_{Na} = (1 - E_{Na})\ln K_v - \int_{E_{Na}}^{1} \ln K_v \mathrm{d}E_{Na} \quad [45]$$

and

$$1/2\ln\gamma_{Ca} = -E_{Na}\ln K_v + \int_{0}^{E_{Na}} \ln K_v \mathrm{d}E_{Na} \quad [46]$$

Additional information on solid phase activity coefficients will be given later on in this chapter. Thermodynamic exchange constants (K_{eq}) for a number of heterovalent exchange reactions are given in Table 7–6.

Exchange reactions of Na^+ with trace metal cations (Cd^{2+}, Co^{2+}, Cu^{2+}, Ni^{2+}, and Zn^{2+}) on Camp Berteau montmorillonite (Sposito and Mattigod, 1979) showed that K_v was constant and independent of exchanger composition, suggesting ideal solid–solution behavior, up to an equivalent fraction of trace metal cations of 0.70. Sodium–calcium exchange studies on similar clay minerals showed that there was a more pronounced selectivity of clay for Ca^{2+} at the Ca-rich end of the isotherm (Van Bladel et al., 1972; Levy and Hillel, 1968). In general, the selectivity coefficient (K_v) for a binary exchange reaction depends primarily on the ionic strength of the solution, on the proportion of cations in the soil absorbing complex, and the proportion of the cations in the soil solution phase (Shainberg et al., 1980; Morgan and Pachepskiy, 1986).

Table 7–6. Thermodynamic equilibrium constants of exchange (K_{eq}) for binary exchange processes on soils and soil components.†

Exchange process	Exchanger‡	K_{eq}§	Reference
Ca–Na	Soils	0.42–0.043	Mehta et al. (1983)
Ca–Na	calcareous soils	0.38–0.09	Van Bladel and Gheyi (1980)
Ca–Na	World vermiculite	0.98	Wild and Keay (1964)
Ca–Na	Camp Berteau montmorillonite	0.72	
Ca–K	Chambers montmorillonite	0.045	Hutcheon (1966)
Ca–NH_4	Camp Berteau montmorillonite	0.035	Laudelout et al. (1967)
Na–Ca	soils (304 K)	0.82–0.16	Gupta et al. (1984)
Mg–Na	soils	0.75–0.053	Mehta et al. (1983)
Mg–Na	World vermiculite	1.73	Wild and Keay (1964)
K–Ca	soils	5.89–323.3	Deist and Talibudeen (1967a)
K–Ca	soil	12.09	Jensen and Babcock (1973)
K–Ca	soils	19.92–323.3	Deist and Talibudeen (1967b)
K–Ca	soil	0.46	Jardine and Sparks (1984b)
K–Ca	soils	0.64–6.65	Goulding and Talibudeen (1984)
K–Ca	soils	6.42–6.76	Ogwada and Sparks (1986b)
K–Ca	soil silt	0.86	Jardine and Sparks (1984b)
K–Ca	soil clay	3.14	Jardine and Sparks (1984b)
K–Ca	kaolinitic soil clay (303 K)	16.16	Udo (1978)
K–Ca	Clarsol montmorillonite	12.48	Jensen (1972)
K–Ca	Danish kaolinite	32.46	Jensen (1972)
K–Mg	soil	5.14	Jensen and Babcock (1973)
K–Na	soils	4.48–6.24	Deist and Talibudeen (1967a)

† Table adapted from Sparks (1995).
‡ The exchange studies were conducted at 298K; exceptions are noted.
§ $K_{eq}/2$ approximately equals K_G.

On the basis of double layer theory (van Olphen, 1977) clay surfaces, with respect to cation selectivity, can be separated into two major classes (Shainberg et al., 1980): (i) internal surfaces and (ii) external surfaces. Internal clay surfaces show (i) high preference for divalent cations (e.g., Ca^{2+}) because of their high surface electrical potential and enhancement by double layer overlap; (ii) decreasing preference for divalent cations as a result of increasing ionic strength in the bulk solution, causing double layer suppression and lowering surface electrical potential; and (iii) lowering preference for divalent cations as surface coverage is increased, suppressing the surface electrical potential. External clay surface areas show (i) high preference for monovalent cations (e.g., Na^+) because of their low surface electrical potential, (ii) increasing preference for divalent cations because increasing ionic strength causes greater specific interactions between surface and divalent cations, and (iii) increasing preference for monovalent coverage cations as surface coverage by divalent cations is increasing due to suppressing of external surface electrical potential.

Experimental evidence on Na^+–Ca^{2+} exchange on soil clay minerals for internal and external surfaces does not always agree with the predictions made based on the double layer model. The reason for this discrepancy is perhaps that the assumptions made with respect to the double layer model are not valid (Shainberg et al., 1980). These assumptions include uniformly charged surfaces, ions as point charges, and nonspecific interactions of the ions with the charged surfaces. For example, Shainberg et al. (1980) reported that the affinity of illite for Ca^{2+} in the presence of Na^+ decreased with an increase in the fraction of exchangeable Ca^{2+}; however, the reverse was true for montmorillonite.

For a number of soils, the exchange selectivity coefficient of Na^+–Ca^{2+} exchange increased as pH decreased (Pratt et al., 1962). This increase in the selectivity coefficient signified increase in affinity of the Na^+ by the clay surface through decreasing surface charge density or decreasing surface electrical potential. Stumm and Bilinski (1973) pointed out that deprotonating clay edge surfaces have greater affinity for a monovalent cation than a divalent cation because the monovalent cation requires much less free energy to desolvate and thus come closer to the adsorbing surface. Soils of humid regions exhibit a much higher affinity for Na^+ than the average salt-affected soil in the western USA (U.S. Salinity Laboratory Staff, 1954; Marsi and Evangelou, 1991a). A possible explanation for this behavior is that soils of humid regions commonly exhibit low pH, and the interlayers of 2:1 clay minerals are occupied by Al-hydroxy polymers. Hence, low charge density and/or cation adsorption is taking place by deprotonating clay edge surfaces.

SENSITIVITY ANALYSIS ON THE GAPON AND VANSELOW EXCHANGE COEFFICIENTS

Basis of the Exchangeable Sodium Percentage–Sodium Adsorption Ratio Relationship

Gapon Expression

The Gapon exchange reaction for a Na^+–Ca^{2+} system can be formulated as follows:

$$\mathrm{ExCa_{1/2}} + \mathrm{Na^+} \Leftrightarrow \mathrm{ExNa} + 1/2\mathrm{Ca^{2+}} \qquad [47]$$

From Eq. [47] the Gapon exchange selectivity coefficient (K_G) can be expressed as

$$K_G = [\mathrm{ExNa}/\mathrm{ExCa_{1/2}}]\{[\mathrm{Ca^{2+}}]^{1/2}/[\mathrm{Na^+}]\} \qquad [48]$$

where ExNa and $ExCa_{1/2}$ denote exchangeable cations ($cmol_c\ kg^{-1}$), Ex denotes the soil exchanger with a charge of 1–, and Na^+ and Ca^{2+} denote solution cations in terms of total cation concentration (mmol L^{-1}).

Equation [48] can be used to solve for ExNa:

$$\mathrm{ExNa} = K_G(\mathrm{CEC})(\mathrm{SAR})[1 + \mathrm{SAR}K_G]^{-1} \qquad [49]$$

where

$$\mathrm{SAR} = [\mathrm{Na^+}]/[\mathrm{Ca^{2+}}]^{1/2} \qquad [50]$$

Taking the limits for Eq. [49] at SAR approaching zero and SAR approaching infinity:

$$\underset{SAR \to 0}{\text{limit}}\ \text{ExNa} = 0; \qquad \underset{SAR \to \infty}{\text{limit}}\ \text{ExNa} = \text{CEC}$$

The derivative of Eq. [49] with respect to SAR is

$$\text{dExNa/dSAR} = [K_G\text{CEC}][1 + K_G\text{SAR}]^{-2} \qquad [51]$$

and

$$\underset{SAR \to 0}{\text{limit}}\ \text{dExNa/dSAR} = K_G\text{CEC}; \qquad \underset{SAR \to \infty}{\text{limit}}\ \text{dEXNa/dSAR} = 0$$

The limits of Eq. [49] and [51] at zero and infinity imply that a plot of Eq. [49] in terms of SAR vs. ExNa (with ExNa being the dependent variable) at a constant CEC and K_G will give a curvilinear function approaching the CEC asymptotically (Evangelou and Phillips, 1987).

Vanselow Expression

The Vanselow expression (Vanselow, 1932) for Na^+–Ca^{2+} exchange is given as shown by Eq. [16]; mole fractions are described by Eq. [33] and [34]. The assumption in the Vanselow exchange expression is that the mole fractions for Na^+ and Ca^{2+} (M_{Na}, M_{Ca}) in the exchanger represent the ideal solid–solution state. Therefore, mole fractions represent activities of the adsorbed cations. Furthermore, one has to assume that the single-ion activity concept in the solution phase is representative of the activity of the ions in that phase (Sposito, 1981a, 1981b). Based on these assumptions, the K_v is considered to be a thermodynamic constant. Single-ion activities in solution are assigned units of moles per liter or millimoles per liter.

Equation [39] can be used to solve for ExNa by substituting into it

$$M_{Na} = \text{ExNa}[(1/2\text{CEC} + 1/2\text{ExNa})]^{-1} \qquad [52]$$

and

$$M_{Ca} = [1/2\text{CEC} - 1/2\text{ExNa}][1/2\text{CEC} + 1/2\text{ExNa}]^{-1} \qquad [53]$$

Substituting Eq. [52] and [53] into Eq. [39] and rearranging, Eq. [39] takes the form of a quadratic equation:

$$(\text{ExNa})^2 = [K_v(\text{SAR}')(\text{CEC})]^2[(K_v\text{SAR})^2 + 4]^{-1} \qquad [54]$$

where SAR′ denotes Na activity ratio. Take the positive root of Eq. [54] to get

$$\text{ExNa} = \text{CEC}K_v\text{SAR}'[4 + (K_v\text{SAR}')2]^{-1/2} \qquad [55]$$

Taking the limits of Eq. [55] as SAR′ approaches zero (0) and SAR′ approaches infinity (∞)

$$\underset{SAR' \to 0}{\text{limit}}\ \text{ExNa} = 0; \qquad \underset{SAR' \to \infty}{\text{limit}}\ \text{ExNa} = \text{CEC}.$$

The derivative of Eq. [55] with respect to SAR′ is

$$\mathrm{dExNa/dSAR'} = 4K_v\mathrm{CEC}[(K_v\mathrm{SAR'})^2 + 4]^{-3/2} \quad [56]$$

Take the limits of Eq. [56] as SAR approaches zero (0) and SAR approaches infinity (∞) to get

$$\lim_{\mathrm{SAR'} \to 0} \mathrm{dExNa/dSAR'} = (\mathrm{CEC}K_v/2)$$

and

$$\lim_{\mathrm{SAR'} \to \infty} \mathrm{dExNa/dSAR'} = 0$$

The limits for Eq. [55] and [56] at zero and infinity imply that a plot of Eq. [55] in terms of ExNa vs. SAR′ with a constant CEC and K_v will give a curvilinear function asymptotically approaching the CEC. Recall that the same is said for the Gapon equation (Eq. [49]). As expected, the derivatives of ExNa with respect to SAR and SAR′ for the Gapon and Vanselow equations, respectively, have identical limits as SAR and SAR′ approach zero and as SAR and SAR′ approach infinity. The two equations, however, are expected to differ in the pathway between the two limits.

Interrelationships between the Gapon and Vanselow Expressions

Evangelou and Phillips (1987) generated values of exchangeable cations for a wide range of E_{Na} values for different soils with various K_v and various CEC values by employing the Vanselow equation (Eq. [55]). They plotted these values, ExNa = 0 to ExNa = CEC, as a function of SAR′ and showed that the limits were approached by differing pathways (Fig. 7–7). The pathways were CEC and K_v dependent. When the same values were plotted as SAR′ vs. E_{Na} (Fig. 7–8), the plot revealed that, as expected from the theoretical treatment, the pathway by which the limit of $E_{Na} \rightarrow 1$ was met was dependent on the magnitude of K_v.

When values of ExNa and $\mathrm{ExCa}_{1/2}$, obtained from the Vanselow equation (Eq. [55]) for each assumed SAR′ value but a constant CEC and K_v, were reintroduced into the Gapon equation (Eq. [48]), a corresponding K_G value for each K_v as a function of E_{Na} was estimated. These values, shown in Fig. 7–9, clearly demonstrate that the K_v and K_G are equal only at E_{Na} of 0.60. The same point is demonstrated using published experimental data (Fig. 7–10). This crossover point is independent of the magnitude of CEC and K_v. It can also be demonstrated as follows.

Rearrange Eq. [48], divide ExNa by CEC, and solve to get (see Appendix II)

$$\mathrm{SAR'} = E_{Na}[K_G(1 - E_{Na})]^{-1} \quad [57]$$

Also, solve from Eq. [39] for SAR′ to get

$$\mathrm{SAR'} = 2E_{Na}[K_v(1 - E_{Na}^2)^{1/2}]^{-1} \quad [58]$$

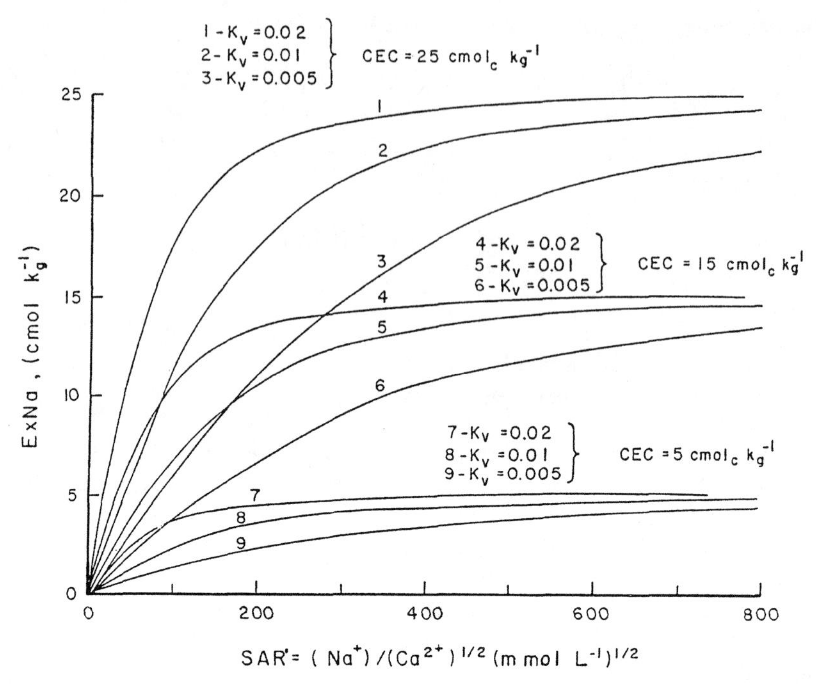

Fig. 7–7. Relationship between Na adsorption ratio (SAR) based on ion activity and exchangeable Na (ExNa) for systems of various CEC and K_v values (Evangelou and Phillips, 1987).

Equate Eq. [57] and [58] by assuming that SAR′ and K_G in Eq. [57] were obtained by using solution activity to get

$$E_{Na}[K_G(1 - E_{Na})]^{-1} = 2E_{Na}[K_v(1 - E_{Na}{}^2)^{1/2}]^{-1}$$

$$= 2E_{Na}\,[K_v(1 - E_{Na})^{1/2}(1 + E_{Na})^{1/2}]^{-1} \qquad [59]$$

Set $K_G = K_V$ and solve for the E_{Na} value where $K_G = K_v$ to find $E_{Na} = 0.60$ is the only E_{Na} value where $K_G = K_v$.

The traditional Gapon exchange selectivity coefficient (K_G) (Gapon, 1933) is often employed in the literature when investigating Na^+–Ca^{2+} exchange reactions in soils (Shainberg et al., 1980) and/or when reclaiming sodic soils (U.S. Salinity Laboratory Staff, 1954). As pointed out previously, the traditional Gapon exchange selectivity coefficient, K_G, has no direct molecular interpretation because it considers that cations react with an exchanger in equivalents (Sposito, 1977). The Vaneslow exchange selectivity coefficient (K_v) (Vanselow, 1932), however, does have molecular interpretation because it considers that cations react with an exchanger in moles. A corrected expression for the Gapon coefficient (K'_G) employing equivalents by satisfying the relationship $K_v = K'_G$ was derived by Sposito (1977). For example, for the reaction listed in Eq. [47]

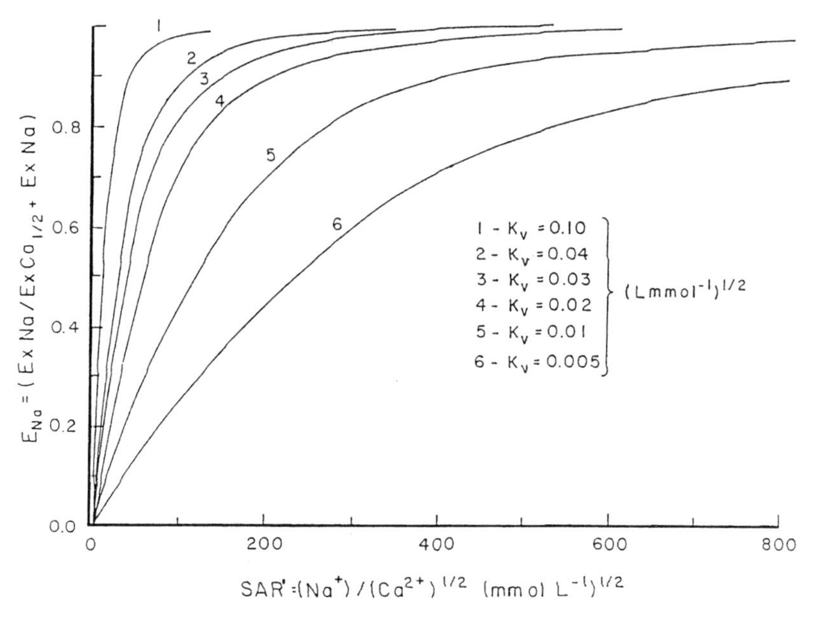

Fig. 7–8. Relationship between Na adsorption ratio (SAR) based on ion activity and Na equivalent fraction (E_{Na}) in the exchanger for systems of various CEC and K_v values. Note that Curve 4 (K_v = 0.02) corresponds to Curves 1, 4, and 7 of Fig. 7–7; Curve 5 (K_v = 0.01) corresponds to Curves 2, 5, and 8 of Fig. 7–7; Curve 6 (K_v = 0.005) corresponds to Curves 3, 6, and 9 of Fig. 7–7. (From Evangelou and Phillips, 1987.)

$$K'_G = (2E_{Na}/\mathrm{SAR}')(1 - E_{Na}^2)^{-1/2} = K_v = (M_{Na}/M_{Ca}^{1/2})[\mathrm{Ca}^{2+})^{1/2}]/(\mathrm{Na}^+) \quad [60]$$

Sposito (1977) demonstrated that the traditional Gapon exchange selectivity coefficient, K_G, for Na^+–Ca^{2+} exchange (Eq. [48]) is indistinguishable from the corrected thermodynamic Gapon exchange selectivity coefficient (K'_G or K_v) if the $E_{Na} < 0.20$. This can be demonstrated by rearranging the left-hand side of Eq. [60] into

$$(4/K'^2_G)(\mathrm{SAR}')^{-2} = (1/E^2_{Na}) - 1 \quad [61]$$

For E_{Na} values <0.20, the term $1/E_{Na}^2$>>>1 and therefore the numeral 1 on the right-hand of Eq. [61] may be dropped; thus, a linear relationship between E_{Na} and SAR is attained. This justifies the conclusion that for $E_{Na} < 0.20$ the traditional K_G is approximately equal to the corrected thermodynamic Gapon cation exchange selectivity coefficient K'_G or K_v. For a particular group of soils, Oster and Sposito (1980) showed $K_G \approx K_v$ for E_{Na} values <0.40. This limiting condition for which $K_G \approx K_v$ is not expected to hold up for all soils, regardless of the magnitude of K_v. Since E_{Na} = 0.20 can be attained at various SAR′ values, depending on the magnitude of K_v, and because mineralogically different soils may exhibit different K_v (Pratt et al., 1962; Shainberg et al., 1980), its magnitude is expected to control the apparent de-

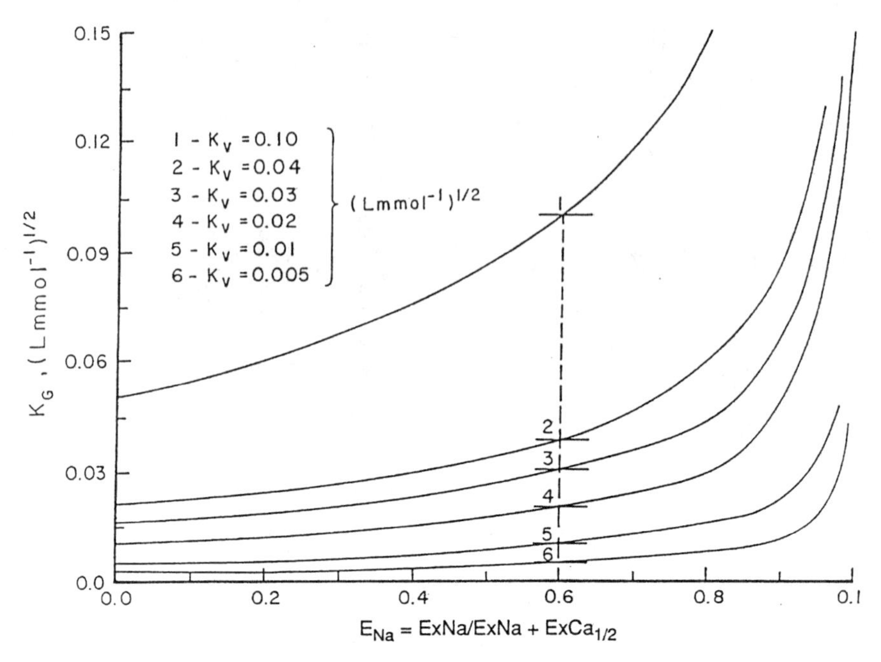

Fig. 7–9. Relationship between Na-equivalent fraction (E_{Na}) in the exchanger and Gapon exchange coefficient (K_G) for soil systems of various K_v values (data were generated by employing the Vanselow equation; the short lines intercepting the K_G plots represent the corresponding K_v values). (From Evangelou and Phillips, 1987.)

gree of curvilinearity between SAR′ and all possible E_{Na} values. Therefore, the condition $K_G \approx K_v$ would appear to be dependent on the magnitude of K_v and E_{Na}.

The plots in Fig. 7–9 demonstrate that the condition $K_v \approx K_G$ is met only for low (<0.015) K_G values. It has been reported that soils vary in K_G from approximately 0.008 to approximately 0.03 (L mmol^{-1})$^{1/2}$ (Evangelou and Coale, 1987). The heterogeneous group of soils used by Oster and Sposito (1980) to compare the K_G and K_v possessed a K_G of approximately 0.015 (L mmol^{-1})$^{1/2}$. Figure 7–9 shows that the K_G is never equal to the K_v for the E_{Na} value that Sposito (1977) considered in his theoretical analyses. However, there is enough noise in the experimental data so that at E_{Na} values <0.40, as shown by Oster and Sposito (1980) for K_v values <0.015, the K_G is indistinguishable from the K_v because of their small absolute difference. The plots in Fig. 7–11 demonstrate the precise relationship between K_v and K_G at the E_{Na} values of 0.05, 0.20, 0.50, and 0.90.

The information in Fig. 7–8 reveals that the E_{Na} range for which a linear relationship between SAR′ and E_{Na} is observed appears to be dependent on the magnitude of K_v. The larger the K_v, the larger the apparent E_{Na} range for which the E_{Na}–SAR′ relationship is linear. For these conditions, however, this larger E_{Na} range is represented by a narrow range in SAR′ values. On the other hand, the smaller the K_v the smaller the E_{Na} range is for which the E_{Na}–SAR′ relationship exhibits apparent linearity. This narrow E_{Na} range, however, is represented by a wide SAR range. The plots in Fig. 7–12 also demonstrate that the degree of the apparent linearity between exchangeable Na ratio [ESR = ExNa/(CEC – $ExCa_{1/2}$)] and SAR′ also depends on the magnitude of K_v. Note that the data in Fig. 7–12 was produced using the K_v equation (Eq. [55]) for a number of assumed K_v values and a wide range of assumed SAR′ values for each K_v. The larger the K_v, the larger the apparent ESR range for which the slope of the line (slope = K_G), starting at the origin, appears linear. It is interesting to note, however, that experimental data do not necessarily support this conclusion (Fig. 7–13). This may be due to the fact that soil minerals do not obey ideal solid–solution theory.

The overall information in Fig. 7–12 demonstrates that the relationship between K_v and K_G as a function of E_{Na} is curvilinear. They are interrelated by Eq. [59], which is reproduced below after simplification by writing $[1 - E_{Na}{}^2]^{1/2} = [1 - E_{Na}]^{1/2}[1 + E_{Na}]^{1/2}$ to get (see Appendix II)

$$K_G = K_v[1 + E_{Na}]^{1/2}[2(1 - E_{Na})^{1/2}]^{-1} \qquad [62]$$

The equation shows that the limit $K_G = 0.5K_v$ when $E_{Na} \rightarrow 0$, and the limit $K_G = \infty$ when $E_{Na} \rightarrow 1$. The curvilinearity can better be shown by taking the derivative of K_G with respect to K_v from Eq. [62] and plotting (dK_G/dK_v) or slope, as a function of E_{Na}. The derivative K_G with respect to K_v (dK_G/dK_v) or the slope at any E_{Na} value is

$$dK_G/dK_v = [1 + E_{Na}]^{1/2}[2(1 - E_{Na})^{1/2}]^{-1} \qquad [63]$$

The curvilinearity of the slope is shown in Fig. 7–14. It is almost linear up to E_{Na} values of 0.7. There is enough noise or experimental error in experimental data that one could not distinguish the curvilinearity up to E_{Na} values of 0.8 and maybe 0.9. The slope increases exponentially at E_{Na} values of about 0.9. The value of dK_G/dK_v at $E_{Na} = 0.9999$ is 71 and is infinite at $E_{Na} = 1.00$.

The information presented above demonstrates that the traditional K_G is indistinguishable from the thermodynamic K_v for a limited range of E_{Na} values (E_{Na} range 0–0.60) and for K_v values <0.015 (L mmol^{-1})$^{1/2}$. However, at low K_v values, any small absolute change in the corresponding K_G has significant impact on the degree of the E_{Na} range for which a linear relationship between E_{Na} and SAR′ is observed. On the other hand, at larger K_v values, large absolute changes in the magnitude of the K_G have relatively small influences on the E_{Na} range for which a linear relationship between E_{Na} and SAR′ is observed. At the larger K_v values, the E_{Na} range is larger than that of the lower K_v values for which a linear relationship between E_{Na} and SAR is observed. The extent of the apparent linearity between E_{Na} ($E_{Na} \times 100$ = ESP) and SAR′ as first introduced in the literature for a heterogeneous group of soils by the U.S. Salinity Laboratory Staff (1954) varies for individual soils depending on the magnitude of their respective K_v. There are many factors in soil that control the magnitude of the K_v, such as pH (Pratt et al., 1962), ionic strength and ESP (Shainberg et al., 1980) and during reclamation of a soil with an E_{Na} value >0.60, all three parameters would most likely be in a state of flux. Because of this flux, the E_{Na}–SAR′ relationship is expected to change slope and degree of curvilinearity.

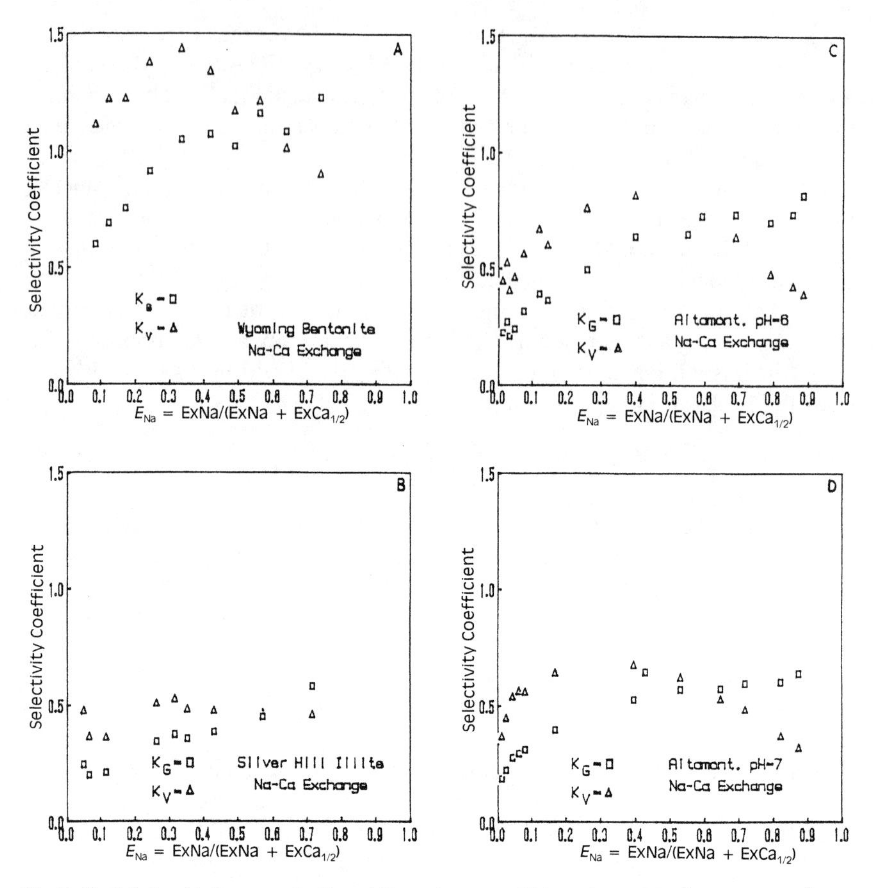

Fig. 7–10. Relationship between the K_v and K_G exchange coefficients for Na^+–Ca^{2+} and Na^+–Mg^{2+} systems as a function of Na-equivalent fraction (E_{Na}). Raw data were taken from Sposito et al. (1983), Sposito and LeVesque (1985), and Fletcher et al. (1984). (From Evangelou and Phillips, 1988.)

The theoretical comparisons between K_v and K_G presented above are based on the assumption that the mole fraction (M_X) of a cation X in the exchanger represents the ideal solid–solution state, and therefore it is equated to activity. Therefore, the degree of application of this information to a heterogeneous group of soils depends on how well and to what extent this assumption holds (Sposito and Mattigod, 1979; Shainberg et al., 1980).

COMPARISON OF THE ARGERSINGER AND GAINES AND THOMAS EXCHANGE CONVENTIONS

Fundamental Differences

Cation exchange thermodynamics were introduced in the literature by Vanselow (1932), Argersinger et al. (1950), Hogfeldt et al. (1953), and Gaines and

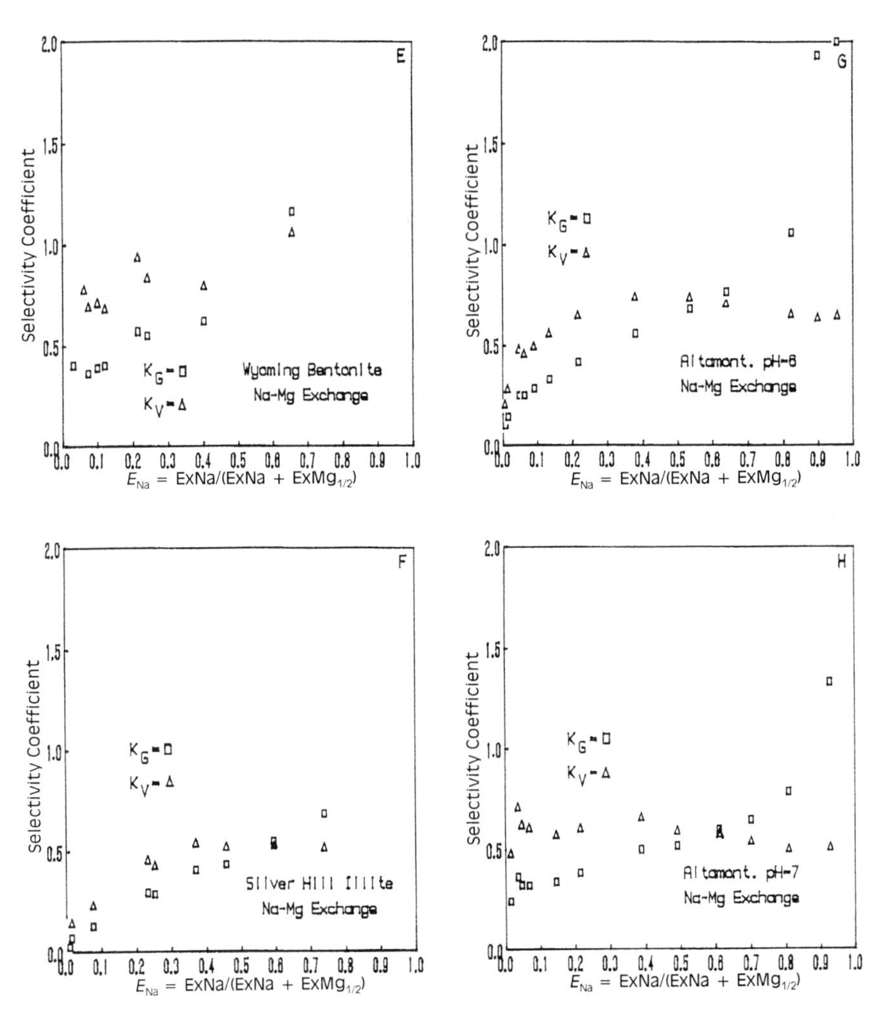

Fig. 7–10. Continued.

Thomas (1953). The different approaches for describing components of exchange expressions thermodynamically were reexamined by Sposito (1977, 1981a, 1981b, 1984a, 1984b), Sposito and Mattigod (1979), Sparks and Huang (1985), Ogwada and Sparks (1986a, 1986b), and Evangelou and Phillips (1989).

Two ways of describing adsorbed cations are (i) the mole fraction and (ii) the equivalent fraction (Sposito, 1977; Sposito and Mattigod; 1979, Goulding, 1983). The mole fraction is generally associated with the Vanselow (1932) exchange selectivity coefficient (K_v) while the equivalent fraction is associated with the Gaines and Thomas exchange selectivity coefficient (K_{GT}). In order for an exchange coefficient to be of thermodynamic value, the exchange phase should be represented as the mole fraction (Sposito, 1977; 1981a, 1981b; Sposito and Mattigod, 1979). The solution phase can be expressed as activity by the single-ion activity concept

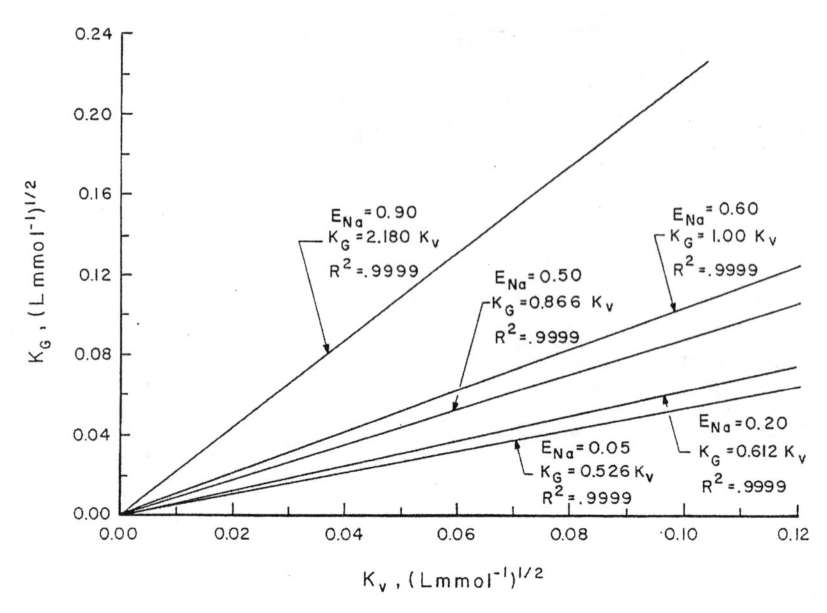

Fig. 7–11. Relationships between the Vanselow exchange coefficient (K_v) and the Gapon exchange coefficient (K_G) at various E_{Na} values. Data were generated by the Vanselow equation. (From Evangelou and Phillips, 1987.)

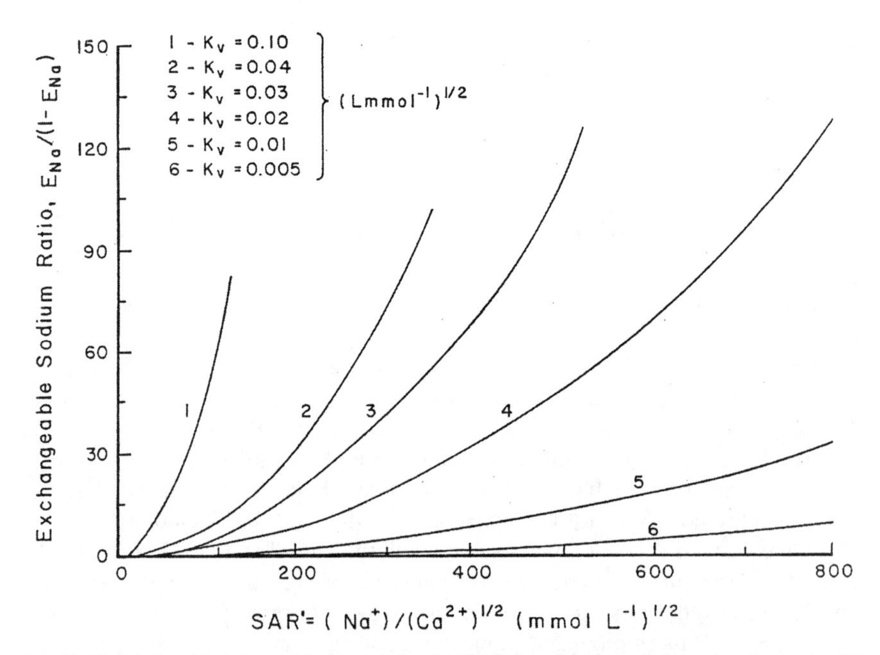

Fig. 7–12. Relationship between Na adsorption ratio (SAR) based on ion activity and exchangeable Na ratio (ESR) ($E_{Na}/(1 - E_{Na})$) for systems of various K_v and CEC values (data points are calculated employing the Vanselow equation). (From Evangelou and Phillips, 1987.)

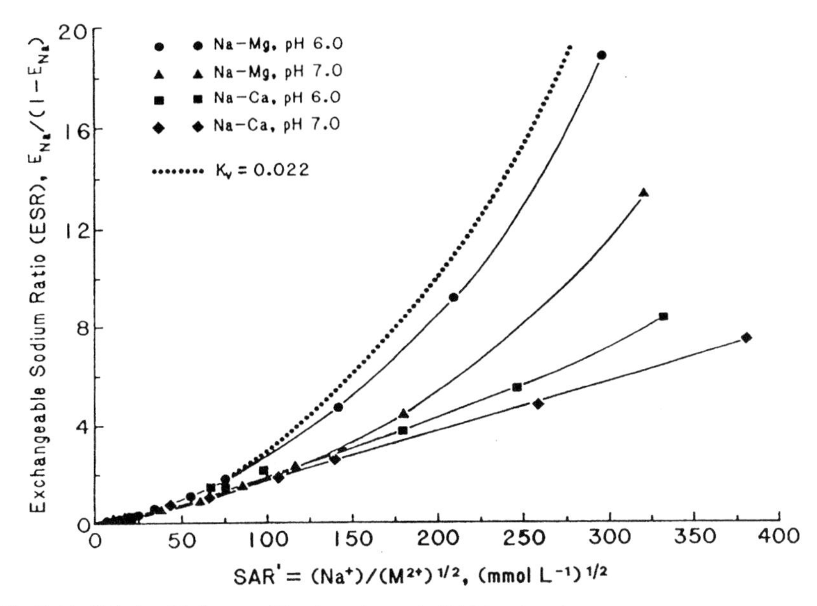

Fig. 7–13. Relationship between Na adsorption ratio (SAR) and exchangeable Na ratio (ESR) for various exchange systems in Silver Hill illite. Raw data were taken from Fletcher et al. 1984).

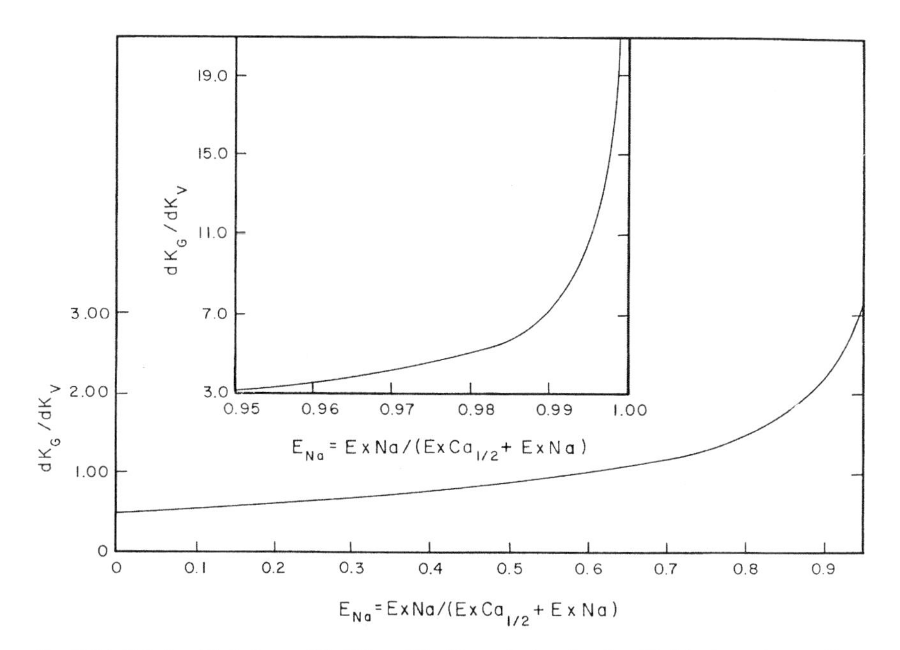

Fig. 7–14. Relationship between dK_G/dK_v and the Na-equivalent fraction (E_{Na}) in the exchanger (from Evangelou and Phillips, 1987).

(Sposito, 1984a). The mole fraction equals activity if the mixture is ideal; if not, activity coefficients are needed (Sposito and Mattigod, 1979; Sposito, 1981a). The equivalent fraction cannot be assumed to represent the ideal solid–solution state, and it is only a formal parameter (Sposito, 1977, 1981a, 1981b).

The Gaines and Thomas (1953) exchange expression, K_{GT}, is often used in the literature to estimate thermodynamic exchange parameters and/or demonstrate ideal solid–solution exchange behavior (Maes et al., 1976; Udo, 1978; Goulding, 1983; Ogwada and Sparks, 1986a; Pleysier et al., 1979). Maes et al. (1976) used the K_{GT} expression to demonstrate that for the exchange reaction of Na^+ with the cations Cd^{2+}, Co^{2+}, Cu^{2+}, Ni^{2+}, and Zn^{2+} on Camp Berteau montmorillonite, the systems do not obey ideal solid–solution theory. Sposito and Mattigod (1979) reexamined the data of Maes et al. (1976) using mole fractions. Mathematical findings showed that K_v and K_{GT} are not equal, nor are they analogous with respect to the entire exchange isotherm. The same authors also showed that when ideal solid–solution behavior is met, the magnitude of K_{GT} depends on the metal equivalent fraction on the exchanger. By reevaluating Maes et al.'s (1976) data using mole fractions on the exchanger, Sposito and Mattigod (1979) found that the K_v remains constant up to a heavy metal equivalent fraction on the exchanger of 0.7.

Ogwada and Sparks (1986a) evaluated K^+–Ca^{2+} exchange data for two soils by the K_v and K_{GT}. They concluded that these systems do not obey solid–solution theory. However, even though the adsorbed-ion activity coefficients and the exchange selectivity coefficients by the two conventions differ in magnitude, they showed an analogous trend. Similar conclusions were drawn by Goulding (1983). Therefore, these authors concluded that similar inferences of ion behavior would be made by either of the two conventions (mole fractions or equivalent fractions).

Gaines and Thomas Exchange Convention

For the Na^+–M^{2+} (M^{2+} could represent any divalent cation, i.e., Ca^{2+} or Mg^{2+}) exchange reaction (Eq. [11]) and the equivalent fractions, E_i, defined by Eq. [36] and [37], the Gaines and Thomas (1953) exchange expression is given by

$$K_{GT} = [E_{Na}/E_M{}^{1/2}][(M^{2+})^{1/2}/(Na^+)] \qquad [64]$$

where the units of Na^+ and M^{2+} are in moles per liter. Note, Eq. [64] differs from that given by Sposito and Mattigod (1979) and Ogwada and Sparks (1986a) with respect to stoichiometry and direction of the reaction. These authors evaluated Reaction [11] by considering 1 mol of a divalent cation exchanging with 2 mol of Na^+ (Reaction [22]). Also, Sposito and Mattigod (1979) considered Reaction [22] from right to left. The direction and stoichiometry of the exchange reaction in this manuscript is consistent with the theoretical analysis in the above sections.

It was demonstrated that a plot of SAR′ vs. ExNa (with ExNa being the dependent variable) for a constant CEC and K_v will result in a curvilinear function approaching the CEC asymptotically. Furthermore, it was shown that the relationship between SAR′ and Na equivalent fraction in the exchanger (E_{Na}) is dependent on K_v, but it is independent of CEC. Similar findings can be demonstrated with the

Gaines and Thomas exchange expression (Eq. [64]) (see Appendix II). Rearranging Eq. [64] yields

$$(\text{ExNa})^2 + [(K_{\text{GT}})(\text{SAR}')]^2\text{CECExNa} - [(K_{\text{GT}})(\text{CEC})(\text{SAR}')]^2 = 0 \quad [65]$$

The positive root of Eq. [65] of the quadratic equation in ExNa is

$$\text{ExNa} = (\text{CEC}/2)\{-(\text{SAR}'K_{\text{GT}})^2 + \text{SAR}'K_{\text{GT}}\,[(\text{SAR}'K_{\text{GT}})^2 + 4]^{1/2}\} \quad [66]$$

The limits of Eq. [66] as SAR′ approaches zero and as SAR′ approaching infinity are

$$\lim_{\text{SAR}' \to 0} \text{ExNa} = \text{O}; \qquad \lim_{\text{SAR}' \to \infty} \text{ExNa} = \text{CEC}$$

The derivative of Eq. [66] with respect to SAR′ is

$$\text{dExNa/dSAR}' = K_{\text{GT}}\text{CEC}\{-K_{\text{GT}}\text{SAR}' + (1/2)[(K_{\text{GT}}\text{SAR}')^2 + 4]^{1/2} + [(K_{\text{GT}}\text{SAR}')^2/2][(K_{\text{GT}}\text{SAR}')^2 + 4]^{-1/2}\} \quad [67]$$

and

$$\lim_{\text{SAR}' \to 0} \text{dExNa/dSAR}' = K_{\text{GT}}\text{CEC}; \qquad \lim_{\text{SAR}' \to \infty} \text{dExNa/dSAR}' = 0$$

The limits of Eq. [66] and [67] as SAR′ approaches zero and infinity imply that a plot of SAR′ vs. ExNa (with ExNa being the dependent variable) for a constant CEC and K_{GT} will result in a curvilinear function approaching the CEC asymptotically (Fig. 7–15). Furthermore, Eq. [67] shows that the relationship between SAR′ and equivalent fraction in the exchanger (E_{Na}) is dependent on K_{GT}, but it is independent of CEC (Fig. 7–16). For the Vanselow (1932) equation, the limits of ExNa and dExNa/dSAR at SAR′ approaching zero (0) and SAR′ approaching infinity are identical to those of the Gaines and Thomas (1953) equation. The two equations, however, differ in the pathway between these two limits.

Interrelationships between the Gaines and Thomas and Vanselow Exchange Conventions

For an exchange reaction between two competing cations, we can set the condition

$$K_{\text{GT}} = FK_{\text{v}} \quad [68]$$

where F is a function of E_{Na} and K_{v} is the Vanselow (1932) exchange coefficient. For homovalent exchange reactions, F is equal to 1 and $K_{\text{GT}} = K_{\text{v}}$. For the heterovalent Na^+–M^{2+} exchange, F can be determined from (see Appendix II)

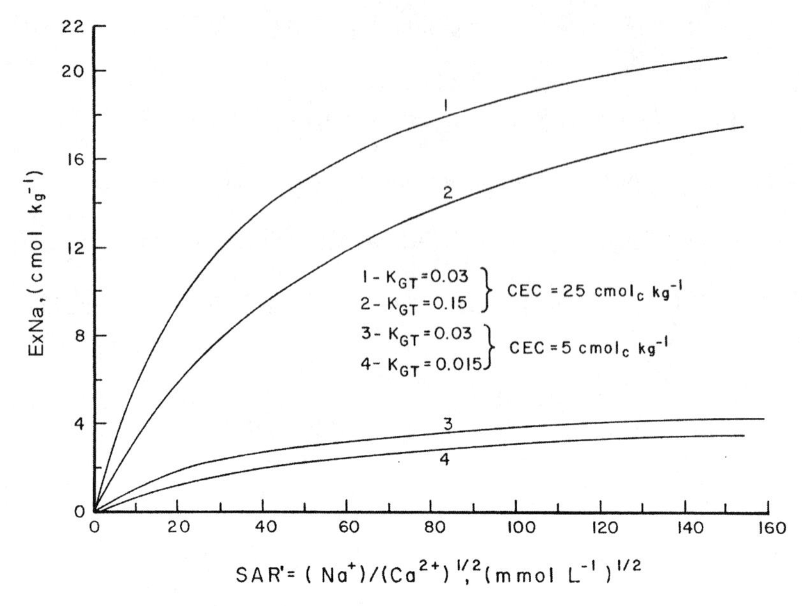

Fig. 7–15. Relationship between exchangeable Na (ExNa) and Na adsorption ratio (SAR) soils or exchange systems of various CEC and K_{GT} values.

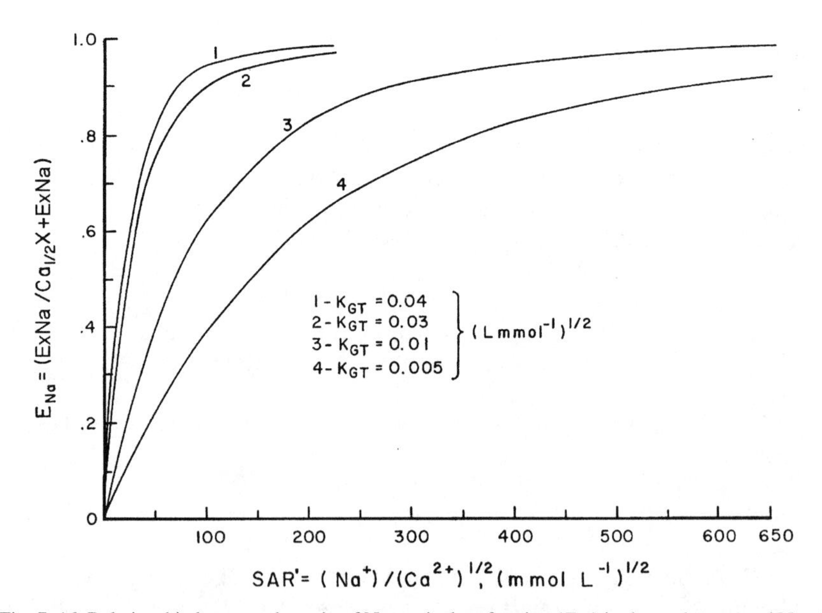

Fig. 7–16. Relationship between the ratio of Na-equivalent fraction (E_{Na}) in the exchanger and Na adsorption ratio (SAR) for soils and exchange systems of various CEC and K_{GT} values.

$$\mathrm{SAR}' = E_{\mathrm{Na}}/[K_{\mathrm{GT}}(1 - E_{\mathrm{Na}})^{1/2}] \qquad [69]$$

and from Evangelou and Phillips (1987)

$$\mathrm{SAR}' = 2E_{\mathrm{Na}}/[K_{\mathrm{v}}(1 - E_{\mathrm{Na}}{}^2)^{1/2}] = 2E_{\mathrm{Na}}/K_{\mathrm{GT}}(1 - E_{\mathrm{Na}})^{1/2}(1 + E_{\mathrm{Na}})^{1/2}] \qquad [70]$$

Equating Eq. [69] and Eq. [70] and solving for K_{GT} yields

$$K_{\mathrm{GT}} = (K_{\mathrm{v}}/2)[1 + E_{\mathrm{Na}}]^{1/2} \qquad [71]$$

Thus, $F = (1/2)[1 + E_{\mathrm{Na}}]^{1/2}$ for the heterovalent Na^+–M^{2+} exchange reaction. Output data obtained from the Vanselow equation (Eq. [55]) by solving for exchangeable Na^+ (ExNa) are shown in Fig. 7–17. The slope of the lines for the different plots of various soils or exchangers is representative of the magnitude of the K_{GT} for each corresponding K_{v}. The data indicate that for all practical purposes the K_{GT} is a constant with respect to SAR′, and consequently of E_{Na}. Note however, that the *y* intercepts of the plots in Fig. 7–17 are not zero. A *y* intercept is obtained, and the magnitude of its negative value is dependent on the magnitude of the slope of the lines. This suggests that the relationship between $E_{\mathrm{Na}}/(1 - E_{\mathrm{Na}})^{1/2}$ vs. SAR′ is a slightly curvilinear one. For all practical purposes, however, it is considered linear.

The data in Fig. 7–17 were also obtained by solving the Vanselow equation. For each set of ExNa, $ExCa_{1/2}$, and SAR obtained from the Vanselow equation (Eq.

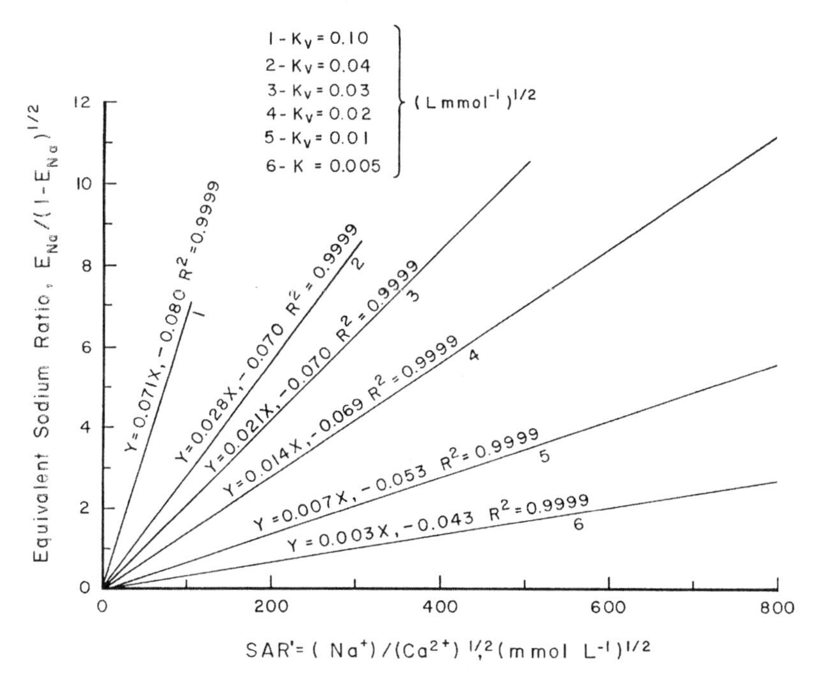

Fig. 7–17. Relationship between equivalent Na ratio in the exchanger and Na adsorption ratio (SAR) for soils or exchangers of various CEC and various K_{v} values (data were generated by employing the Vanselow equation).

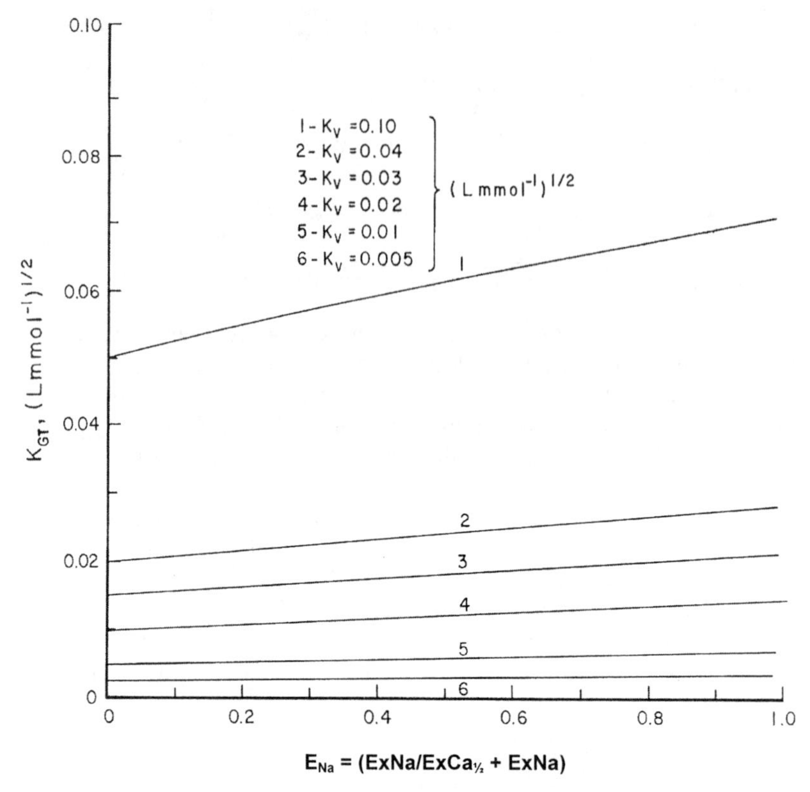

Fig. 7–18. Relationship between the Gaines and Thomas exchange coefficient (K_{GT}) and Na-equivalent fraction (E_{Na}) in the exchanger (data for estimating K_{GT} values were generated by the Vanselow equation.

[55]), a K_{GT} value was estimated with Eq. [64]. These K_{GT} values for each soil or exchanger, represented by a different K_v, are plotted as a function of E_{Na} (Fig. 7–18). The data reveal that the relationship between K_{GT} and E_{Na} is for all practical purposes linear. At values <0.01 the K_{GT} and K_v could become indistinguishable when estimated from experimental data due to the noise in such data. The difference between K_v and K_{GT} becomes wider as K_v values become larger.

The derivative of Eq. [71] with respect to K_v is

$$dK_{GT}/dK_v = (1/2)[1 + E_{Na}]^{1/2} \qquad [72]$$

Equation [72] demonstrates that the K_{GT} is dependent on the magnitude of E_{Na}. The limits of Eq. [72] as E_{Na} approaches zero (0) and as E_{Na} approaches 1 demonstrate that the limit $K_{GT} = 0.50K_v$, when $E_{Na} \rightarrow 0$ and that the limit $K_{GT} = (2^{-1/2})K_v = 0.707K_v$ when $E_{Na} \rightarrow 1$. These relationships are also approached when experimental data are plotted in a similar manner (Fig. 7–19). The K_{GT} data contained in Fig. 7–18 is replotted in Fig. 7–20 as a function of K_v for several E_{Na} values. Figure 7–20 demonstrates that the relationship between K_v and K_{GT} is linear for a constant E_{Na}. The plot in Fig. 7–21 where dK_{GT}/dK_v (slope) is plotted vs. E_{Na} graphically illustrates the minimum and maximum differences between K_{GT} and K_v. The minimum difference in the slope is 0.5 while the maximum difference is $(2)^{-1/2}$ (0.707); that is,

$$\lim_{E_{Na} \to 0} dK_{GT}/dK_v = 1^{1/2}(1/2) = 0.5$$

$$\lim_{E_{Na} \to 1} dK_{GT}/dK_v = [1 + 1]^{1/2}2^{-1} = [2]^{-1/2} = 0.707$$

Equation [72] shows that the pathway of K_{GT} between the E_{Na} limits of 0 and 1 for data obeying ideal solid–solution theory is curvilinear due to $(1 + E_{Na})^{1/2}$.

The general conclusions that we can draw from the above is that when the mole fraction in the exchanger represents the ideal solid–solution state, the values of K_{GT} across the entire CEC cannot be constant and analogous to K_v. The K_{GT} is linearly related to E_{Na} values. The noise, however, in the experimental data is expected to mask the slope of the K_{GT} as a function of E_{Na}.

Adsorbed-Ion Activity Coefficients

An example of K_{eq} and adsorbed-ion activity coefficients for homovalent exchange is given below (see Appendix I). Consider the homovalent cation exchange reaction

$$\text{ExCa} + Mg^{2+} \Leftrightarrow \text{ExMg} + Ca^{2+} \quad [73]$$

where ExCa is exchangeable Ca^{2+} (mol kg^{-1}), ExMg is exchangeable Mg^{2+} (mol kg^{-1}), and Ex is the exchange phase with charge of 2–.

$$K_{eq} = \left(\frac{\text{ExMg}}{\text{ExCa} + \text{ExMg}}\right)\lambda_{Mg}\,(Ca^{2+}) \Big/ \left(\frac{\text{ExCa}}{\text{ExCa} + \text{ExMg}}\right)\lambda_{Ca}\,(Mg^{2+}) \quad [74]$$

where K_{eq} is the thermodynamic exchange constant, λ_{Mg} and λ_{Ca} are the rationale activity coefficients for Ca^{2+} and Mg^{2+} on the solid phase, and the activities of Ca^{2+} and Mg^{2+} in the solution phase are in parentheses:

$$\text{Mole fraction } Mg^{2+} = M_{Mg} = \text{ExMg}/(\text{ExCa} + \text{ExMg}) \quad [75]$$

$$\text{Mole fraction } Ca^{2+} = M_{Ca} = \text{ExCa}/(\text{ExCa} + \text{ExMg}) \quad [76]$$

Substituting Eq. [75] and [76] into Eq. [74] gives

$$K_{eq} = \frac{[M_{Mg}]\,\lambda_{Mg}\,(Ca^{2+})}{[M_{Ca}]\,\lambda_{Mg}\,(Mg^{2+})} \quad [77]$$

The Vanselow exchange selectivity coefficient is given by

$$K_v = \frac{[M_{Mg}]\,(Ca^{2+})}{[M_{Ca}]\,(Mg^{2+})} \quad [78]$$

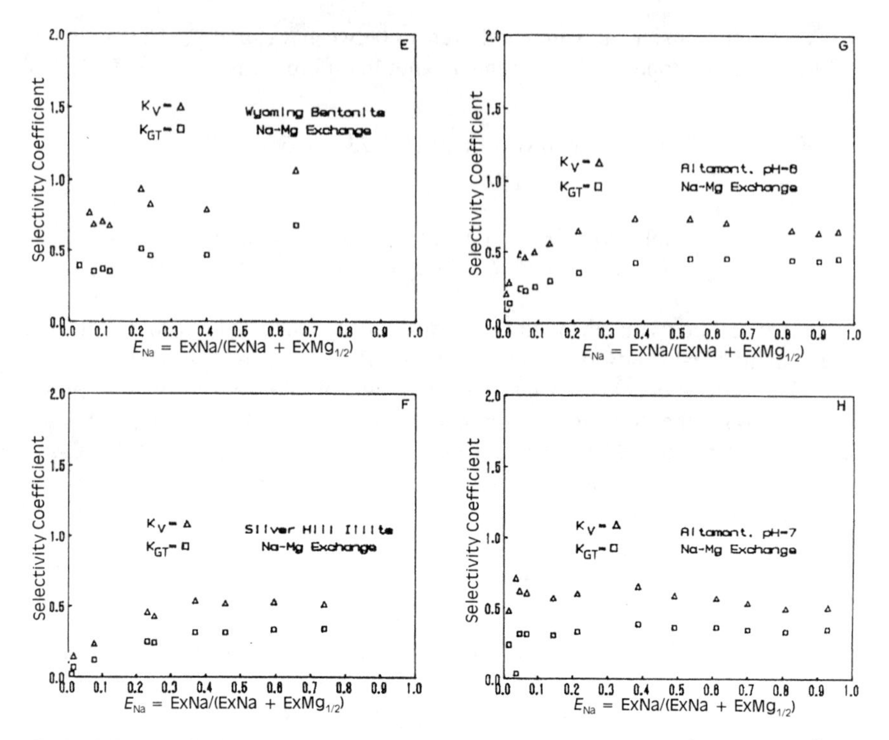

Fig. 7–19. Relationship between the K_v and K_{GT} exchange coefficients for Na^+–Ca^{2+} and Na^+–Mg^{2+} systems as a function of Na-equivalent fraction (E_{Na}). Raw data were taken from Sposito et al. (1983), Sposito and LeVesque (1985), and Fletcher et al. (1984). (From Evangelou and Phillips, 1989.)

Therefore,

$$K_{eq} = K_v \frac{\lambda_{Mg}}{\lambda_{Ca}} \quad [79]$$

Equation [79] indicates that in order for K_{eq} to remain constant across the entire exchange isotherm, upon meeting the condition

$$M_{Ca} + M_{Mg} = 1 \quad [80]$$

any variation in K_v must be followed by a variation in λ_{Ca} and λ_{Mg}. The equation that relates the K_{eq} to K_v is given by (see Appendix I)

$$\ln K_{eq} = \int_0^1 \ln K_v dE_{Mg} \quad [81]$$

Below we give a hypothetical homovalent example involving Mg^{2+}–Ca^{2+} exchange to demonstrate how one might treat these experimental data to produce values for K_{eq}, λ_{Mg}, and λ_{Ca}. A plot of E_{Mg} vs. hypothetical experimental K_v is given in Fig. 7–22. Regression analysis of the data gives the following equation:

$$\ln K_v = -2E_{Mg} + 1 \quad [82]$$

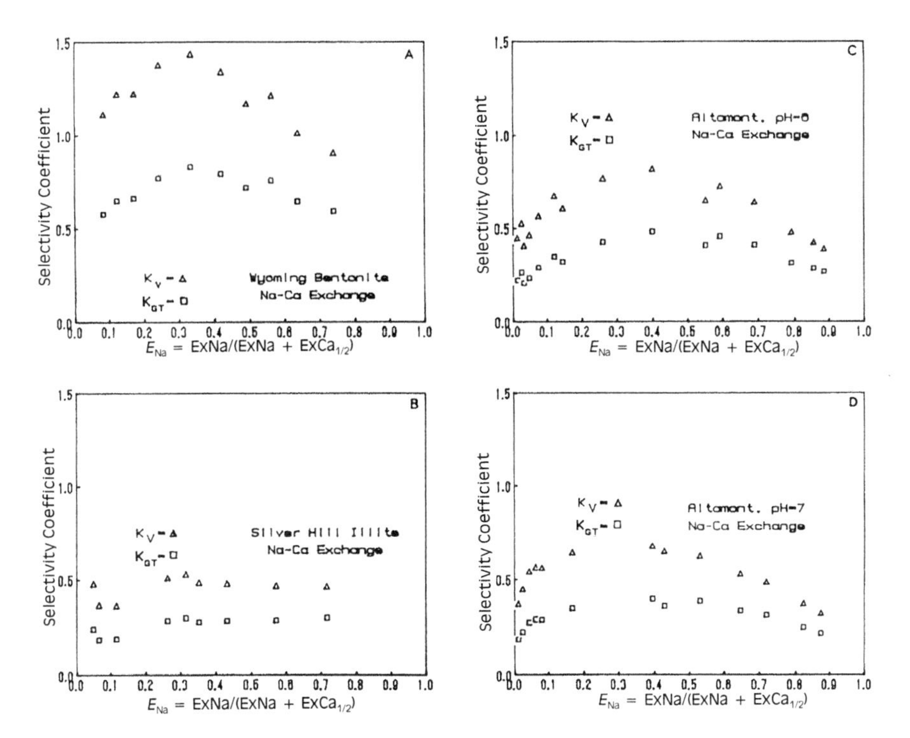

Fig. 7–19. Continued.

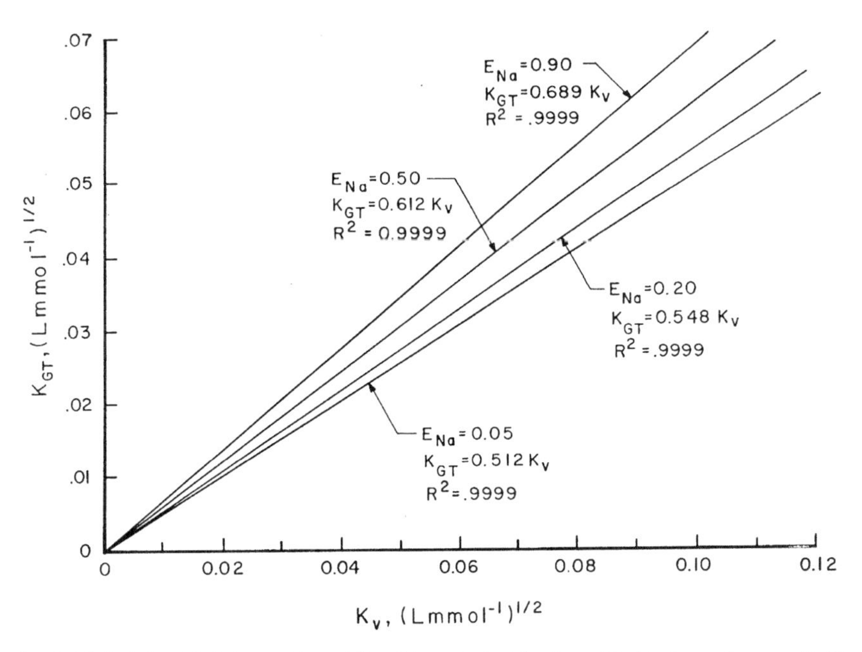

Fig. 7–20. Relationship between K_G and K_v at Na-equivalent fractions (E_{Na}) in the exchanger of 0.05, 0.20, 0.50, and 0.90 (data for estimating K_{GT} values were generated by the Vanselow equation).

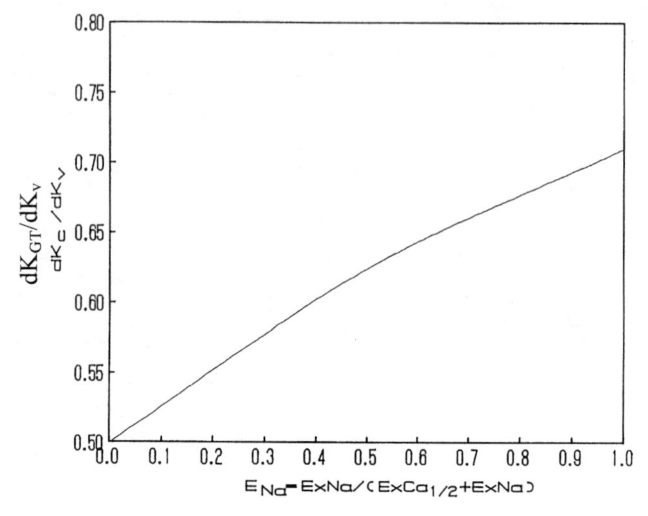

Fig. 7–21. Relationship between dK_{GT}/dK_v and Na-equivalent fraction (E_{Na}) in the exchanger (from Evangelou and Phillips, 1989).

From Eq. [81]:

$$\ln K_{eq} = \int_{E_{Mg}=0}^{E_{Mg}=1} (-2E_{Mg} + 1)dE_{Mg} \qquad [83]$$

and

$$\ln K_{eq} = -E_{Mg}^2 + E_{Mg}\int_0^1 = -1^2 + 1 - 0 \qquad [84]$$

Therefore,

$$\ln K_{eq} = 0, \text{ and raising it to base } e \text{ gives } K_{eq} = 1 \qquad [85]$$

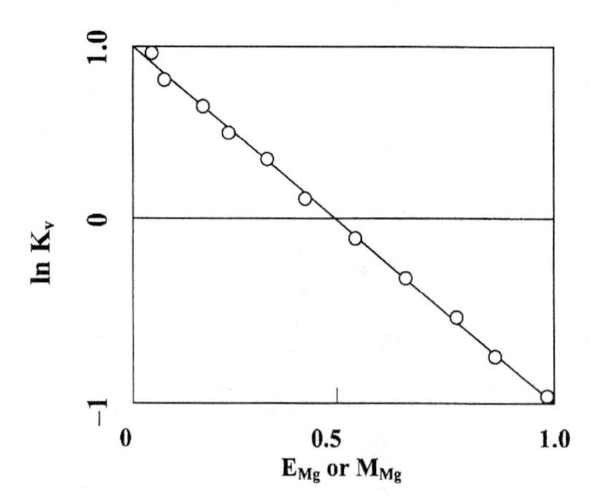

Fig. 7–22. Hypothetical Ca–Mg exchange data plotted as Mg-equivalent fraction on the exchange vs. $\ln K_V$.

Calculating λ_g and λ_{Ca} at $E_{Mg} = 0.25$ (see Appendix I)

Considering that

$$-\ln\lambda_{Mg} = (1 - E_{Mg})\ln K_v - \int_{E_{Mg}=0.25}^{E_{Mg}=1} (-2E_{Mg} + 1)dE_{Mg} \quad [86]$$

and

$$-\ln\lambda_{Mg} = (1 - E_{Mg})\ln K_v - [(-E_{Mg}^2 + E_{Mg})]_{0.25}^{1} \quad [87]$$

at $E_{Mg} = 0.25 \ln K_v = 0.50$; therefore,

$$-\ln\lambda_{Mg} = (1 - 0.25)(0.5) - [-(1)^2 + 0.25^2 + 1 - 0.25] \quad [88]$$

$$-\ln\lambda_{Mg} = (1 - 0.25)(0.50) - [-0.1875] \quad [89]$$

and

$$\ln\lambda_{Mg} = -0.5625, \text{ or } \lambda_{Mg} = 0.57 \quad [90]$$

Continuing for $\ln\lambda_{Ca}$

$$-\ln\lambda_{Ca} = -(E_{Mg})(\ln K_v) + \int_{E_{Mg}=0}^{E_{Mg}=0.25} (-2E_{Mg} + 1)dE_{Mg} \quad [91]$$

$$-\ln\lambda_{Ca} = -(0.25)(0.50) + \int_{E_{Mg}=0}^{E_{Mg}=0.25} (-2E_{Mg} + 1)dE_{Mg} \quad [92]$$

$$-\ln\lambda_{Ca} = -0.125 + [-(E_{Mg})^2 + E_{Mg}]_0^{0.25} \quad [93]$$

$$-\ln\lambda_{Ca} = -0.125 + [-(0.25)^2 + 0 + 0.25 - 0] \quad [94]$$

$$-\ln\lambda_{Ca} = -0.125 + 0.1875 \quad [95]$$

and

$$\ln\lambda_{Ca} = -0.0625, \text{ or } \lambda_{Ca} = 0.939 \quad [96]$$

Substituting values from the above example into Eq. [79] and considering that at $E_{Mg} = 0.25$, $K_v = 1.65$

$$K_{eq} = K_v(-\lambda_{Mg}/\lambda_{Ca}) = 1.65(0.57/0.939) = 1.00 \quad [97]$$

or

$$\ln K_{eq} = \ln K_v + \ln\lambda_g - \ln\lambda_{Ca} \quad [98]$$

Substituting values from the above example into Eq. [98] and considering that at $E_{Mg} = 0.25 \ln K_v = 0.50$

$$\ln K_{eq} = 0.50 - 0.5625 + 0.0625 = 0 \text{ as expected} \quad [99]$$

Example of K_{eq} and Adsorbed-Ion Activity Coefficients, γ_i for Heterovalent Exchange

The equation employed to calculate the thermodynamic exchange constant (K_{eq}) from K_v values across the entire exchange isotherm of Na^+–Ca^{2+} (see Sensitivity Analysis on the Gapon and Vanselow Exchange Coefficients above) is as follows:

$$\ln K_{eqV} = \int_0^1 \ln K_v dE_{Na} = \ln K_{eq} \quad [100]$$

The $\ln K_{eq}$ constant can be calculated from Eq. [100] by first producing a plot of E_{Na} vs. $\ln K_v$ as shown in Fig. 7–23. Recall that K_v can be determined from Eq. [33], [34], and [39] employing experimental solution and exchange data for Na^+ and Ca^{2+} covering the entire exchange isotherm. The plot in Fig. 7–23 is then separated into five "trapezoids." Summation of the areas ($\ln K_v E_{Na}$) of all five trapezoids produces $\ln K_{eq}$.

Technically, $\ln K_{eq}$ can be estimated through Eq. [100] by integrating under the entire $\ln K_v$ curve in Fig. 7–23 as demonstrated for the Mg^{2+}–Ca^{2+} exchange system in Fig. 7–22. However, since there is no mathematical function describing the relationship between $\ln K_v$ and E_{Na}, the process becomes time-consuming and requires a practical approach. This approach involves carefully cutting up the entire figure (Fig. 7–23) along its borders and then taking the weight of this paper sample. After obtaining the relationship of paper weight per unit surface area ($\ln K_v$ vs. E_{Na}), the area under any fraction of the plot, or the entire $\ln K_v$ plot, can be estimated. Using this approach the area in Fig. 7–23 was estimated to be 2.035 or $\ln K_{eq} = 2.035$. Additional approaches for integrating under the curve (E_i vs. $\ln K_v$) of irregular exchange data include computer graphics or analytical approaches (Sparks, 1995).

For the direction and stoichiometry of Na^+–Ca^{2+} exchange in Reaction [23], the equations for estimating adsorbed-ion activity coefficients according to Argersinger et al. (1950) are

$$-\ln \gamma_{Na} = (1 - E_{Na})\ln K_v - \int_{E_{Na}}^1 \ln K_v dE_{Na} \quad [101]$$

and

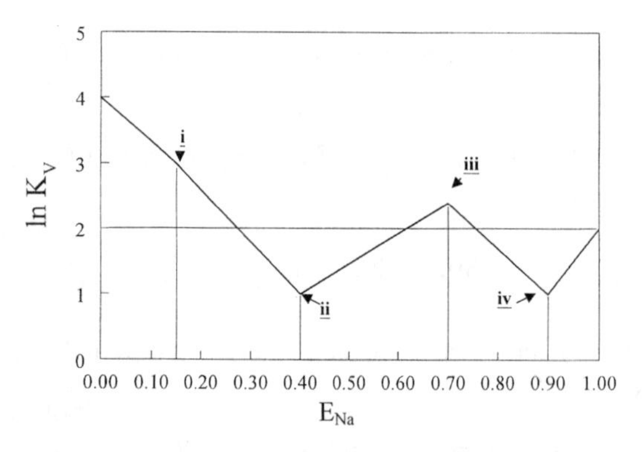

Fig. 7–23. Hypothetical relationship between K-equivalent fraction on the exchange phase (E_K) and $\ln K_v$.

$$-1/2\ln\gamma_{Ca} = -(E_{Na})\ln K_v + \int_0^{E_{Na}} \ln K_v dE_{Na} \quad [102]$$

The process outlined below demonstrates how one estimates adsorbed-ion activity coefficients for Points i, ii, iii, and iv in Fig. 7–23. For Point i, integrating under $E_{Na} = 0$ to $E_{Na} = 0.14$ and solving Eq. [102] gives $\gamma_{Ca} = 0.86$, whereas for adsorbed Na^+ on Point i, integrating under $E_{Na} = 0.14$ to $E_{Na} = 1$ and solving Eq. [101] gives $\gamma_{Na} = 0.36$. For Point ii, integrating under $E_{Na} = 0$ to $E_{Na} = 0.4$ and solving Eq. [102] gives $\gamma_{Ca} = 0.04$ whereas for adsorbed Na^+ on Point ii, integrating under $E_{Na} = 0.40$ to $E_{Na} = 1$ and solving Eq. [101] gives $\gamma_{Na} = 1.53$. For Point iii, integrating under $E_{Na} = 0$ to $E_{Na} = 0.7$ and solving Eq. [102] gives $\gamma_{Ca} = 1.537$ while for adsorbed Na^+ on Point iii, integrating under $E_{Na} = 0.70$ to $E_{Na} = 1$ and solving Eq. [101] gives $\gamma_{Na} = 0.78$. Finally, for Point iv, integrating under $E_{Na} = 0$ to $E_{Na} = 0.9$ and solving Eq. [102] gives $\gamma_{Ca} = 0.14$, while for adsorbed Na^+ on Point iv, integrating under $E_{Na} = 0.90$ to $E_{Na} = 1$ and solving Eq. [101] gives $\gamma_{Na} = 1.05$.

Multiplying the K_v corresponding to Points i, ii, iii, and iv in Fig. 7–23 with the ratio of $\gamma_{Na}/\gamma_{Ca}{}^{1/2}$ at each corresponding point produces the K_{eq} obtained by solving Eq. [100]. For example, $\ln K_v$ at Point i is 3. Taking the antilog $K_v = 20.08$, multiplying it by $\gamma_{Na}/\gamma_{Ca}{}^{1/2}$ at Point i, 0.36/0.93, gives K_{eq} 7.77 vs. 7.65 obtained by graphically solving Eq. [100]. The small difference between the two K_{eq} values is possibly due to error in estimating graphically the area under the curve in Fig. 7–23. The same calculation at Point ii where $\ln K_v = 1$ or $K_v = 2.72$, multiplying it by $\gamma_{Na}/\gamma_{Ca}{}^{1/2}$ at Point ii, 1.53/0.53, gives $\ln K_{eq} = 2.061$ or $K_{eq} = 7.85$.

Interrelationships between the Gaines and Thomas and Vanselow Adsorbed-Ion Activity Coefficients

The thermodynamic exchange constant (K_{eq}) is related to the K_v or K_{GT} by the equations (Sposito, 1981b; Sposito and Mattigod, 1979; Ogwada and Sparks, 1986a)

$$K_{eqV} = K_v(\gamma_{Na}/\gamma_M^{1/2}) \quad [103]$$

or

$$K_{eqGT} = K_{GT}(f_{Na}/f_M^{1/2}) \quad [104]$$

where γ_i and f_i are the adsorbed-ion activity coefficients for K_v and K_{GT}, respectively. Note, however, that for the ideal solid–solution state only γ_{Na} and γ_{Ca} are equal to 1 (Sposito, 1981b). The equation employed to calculate the thermodynamic exchange constant (K_{eq}) from K_v (referred to from here on as K_{eqV}) is

$$\ln K_{eqV} = \int_0^1 \ln K_v dE_{Na} \quad [105]$$

From the thermodynamic exchange equilibrium constant, one may calculate Gibbs free energy using classical thermodynamic relationships. For example, Gibbs free energy (ΔG°) of exchange for any given two cations is given by

$$\Delta G^\circ_{EX} = -RT\ln K_{eqV} \quad [106]$$

The parameters for calculating K_v include the mole fraction on the exchange, which is estimated by Eq. [33] and [34], and the single-ion activity in solution, which can be calculated from the ionic strength and the Davies (1962) equations. The ionic strength (I) equation is given by $I = 1/2\Sigma m_i z_i^2$ where m_i is molarity of species i in solution and z_i is valence of solution species i. The single-ion solution activity coefficient (ϕ) is calculated by the Davies (Davies, 1962) equation (Eq. [8]). The solution of these two equations is now commonly done by ion association models, such as GEOCHEM and MINTEQ (Allison et al., 1991; Sposito and Mattigod, 1979; Parker et al., 1994).

For the direction and stoichiometry of the exchange reaction used in this chapter (Reaction [22]), the equations for estimating adsorbed-ion activity coefficients according to the Argersinger et al. (1950) approach, which employs the K_v, are

$$-\ln\gamma_{Na} = (1 - E_{Na})\ln K_v - \int_{E_{Na}}^{1} \ln K_v dE_{Na} \quad [107]$$

and

$$-1/2\ln\gamma_M = -(E_{Na})\ln K_v + \int_0^{E_{Na}} \ln K_v dE_{Na} \quad [108]$$

According to the Gaines and Thomas approach (Ogwada and Sparks, 1986a; Hutcheon, 1966) adsorbed-ion activity coefficients, considering the direction and stoichiometry of Reaction [23], are given by the equations

$$-\ln f_{Na} = (1 - E_{Na})(\ln K_{GT} - 1/2) - \int_{E_{Na}}^{1} \ln K_{GT} dE_{Na} \quad [109]$$

and

$$-1/2\ln f_M = -E_{Na}(\ln K_{GT} - 1/2) + \int_0^{E_{Na}} \ln K_{GT} dE_{Na} \quad [110]$$

By subtracting Eq. [109] from Eq. [107], substituting the term K_v as a function of K_{GT} of Eq. [71] and carrying out the integration gives (see Appendix II)

$$f_{Na}/\gamma_{Na} = 2/(1 + E_{Na}) \quad [111]$$

Similarly, from Eq. [110], [108], and [71]

$$f_M/\gamma_M = 1/(1 + E_{Na}) \quad [112]$$

Equations [111] and [112] can also be obtained from the generalized equations, relating the two adsorbed-ion activity coefficients, reported by Sposito and Mattigod (1979). Equations [111] and [112] indicate that the interrelationship of the adsorbed-ion activity coefficients obtained by the Argersinger and Gaines and Thomas conventions are E_{Na} dependent. Because of this E_{Na} dependency values of f_{Na} and γ_{Na} or f_M or γ_M cannot be truly considered to be similar. From the above relationships one can also show that (see Appendix II)

$$\ln K_{eqV} = 1/2 + \ln K_{eqGT} \quad [113]$$

Considering that according to Gaines and Thomas

$$\ln K_{eqGT} = \int_0^1 \ln K_{GT} dE_{Na} \qquad [114]$$

Subtracting Eq. [113] from Eq. [114], substituting K_{GT} from the right-hand side of Eq. [113] as a function of K_v from Eq. [71], and carrying out the integration, gives (Evangelou and Phillips, 1989)

$$\ln K_{eqV} - \ln K_{eqGT} = 1/2\int_0^1 [\ln(1 + E_{Na}) - 2\ln 2]dE_{Na}$$

$$= -1/2[(1 + E_{Na})\ln(1 + E_{Na}) - E_{Na} - 2E_{Na}\ln 2]\Big|_0^1$$

$$= -1/2[2\ln 2 - 1 - 2\ln 2] = +1/2 \qquad [115]$$

Therefore, Eq. [113] ($\ln K_{eqV} = 1/2 + \ln K_{eqGT}$) is demonstrated. This illustrates that the thermodynamic exchange equilibrium constant (K_{eq}) differs form the Gaines and Thomas mean exchange equilibrium coefficient (K_{eqGT}) by a constant factor of exp(1/2) for the heterovalent Na^+–M^{2+} exchange reaction. In this discussion, we chose the convention of one sodium replacing half a calcium or any other divalent cation on the exchanger. For the case where two sodium replace one calcium or any other divalent cation in the exchanger, the relationship is

$$\ln K_{eqV} = \ln K_{eqGT} + 1 \qquad [116]$$

The 1 in the equation above replaces the 1/2 from Eq. [113]. If the direction of the reaction is reversed so that the calcium is replacing sodium on the exchanger, then the plus sign becomes minus. The change in stoichiometry will also change the magnitude of the K_{eqGT} and consequently the magnitude of the K_{eqV}, but the relationships will not change the magnitude of f_i or γ_i, nor the relationships described by Eq. [111] and [112].

Using experimental data, we estimated the ratio of K_V/K_{GT} at several E_{Na} values (Table 7–7). From the theoretical analysis section it is shown that the $K_{GT} = (K_v/2)[1 + E_{Na}]^{1/2}$ and that

$$\lim_{E_{Na} \to 0} K_{GT} = 0.5K_v, \text{ and } \lim_{E_{Na} \to 1} K_{GT} = 0.71K_v$$

The experimental data (K_{eqV}/K_{eqGT}) shown in Table 7–7 fall within one standard deviation of the value of the above theoretical ratio of K_{eqV}/K_{eqGT}. The data in Table 7–8 show values of K_{eqGT}, K_{eqV}, and K'_{eqV} ($K'_{eqV} = 1/2 + \ln K_{eqGT}$). In the theoretical analyses presented here, it was shown that for the stoichiometry and direction of the heterovalent binary exchange considered $\ln K_{eqV}' = \ln K_{GT} + 1/2$ (Eq. [113]). The experimental results shown in Table 7–8 are also in agreement with the interrelationships of K_{eqV} and K_{eqGT}.

Sposito and Mattigod (1979) pointed out that ideality in cation exchanger phases must be studied in terms of the Vanselow selectivity coefficient (K_v). They also showed that for ideal cation exchangers (K_v is constant across the entire ex-

Table 7–7. Calculated and experimental ratios of K_v/K_{GT} at several equivalent fractions in the exchanger (E_{Na}) values (K_{GT} is the Gaines and Thomas selectivity coefficient; K_v is the Vanselow selectivity coefficient).

Soil or clay	Exchange system	CEC†	K_v/K_{GT} at E_{Na} of 0.1	0.2	0.3	0.4	0.5	0.6	0.7	0.8	0.9
		$cmol_c\ kg^{-1}$									
Silver Hill illite	Na–Ca	14.2	1.90	1.83	1.76	1.69	1.64	1.59	1.54	1.49	1.45
Altamont soil, pH 7.0	Na–Ca	55.3	1.90	1.82	1.75	1.69	1.64	1.59	1.54	1.49	1.45
Altamont soil, pH 6.0	Na–Ca	38.5	1.90	1.83	1.76	1.69	1.62	1.57	1.52	1.48	1.45
Wyoming bentonite	Na–Ca	103.3	1.91	1.84	1.76	1.70	1.64	1.58	1.53	1.49	1.45
Silver Hill illite	Na–Mg	12.7	1.90	1.83	1.75	1.69	1.63	1.58	1.53	1.49	1.44
Altamont soil, pH 7.0	Na–Mg	65.6	1.90	1.82	1.75	1.69	1.63	1.58	1.53	1.48	1.44
Altamont soil, pH 6.0	Na–Mg	51.1	1.91	1.82	1.75	1.69	1.63	1.58	1.54	1.49	1.45
Wyoming bentonite	Na–Mg	88.5	1.91	1.84	1.76	1.69	1.63	1.58	1.53	--	--
Mean			1.90	1.83	1.76	1.69	1.63	1.58	1.53	1.49	1.45
($K_V/K_{GT} = 2/[1 + E_{Na}]^{1/2}$) (see Eq. [71])			1.907	1.826	1.754	1.690	1.633	1.581	1.534	1.491	1.451
SD, ±			0.005	0.008	0.005	0.004	0.007	0.006	0.007	0.005	0.005
SEM			0.002	0.003	0.002	0.001	0.002	0.002	0.002	0.002	0.002
CV, %			0.27	0.46	0.30	0.20	0.43	0.41	0.46	0.33	0.34

† CEC = ExNa + $ExM_{1/2}$.

Table 7–8. Thermodynamic exchange constants (K_{eqV}) and calculated K'_{eqV} and K_{eqGT}.†

Sample identification	Exchange system	K_{eqV}	K_{eqGT}	K'_{eqV}	K_{eqGT}/K_{eqV}	K'_{eqV}/K_{eqV}
Silver Hill illite	Na–Ca	0.462	0.266	0.440	.575	0.952
Altamont soil, pH 7	Na–Ca	0.538	0.310	0.512	.577	0.953
Altamont soil, pH 6	Na–Ca	0.613	0.370	0.610	.603	0.995
Wyoming bentonite	Na–Ca	1.088	0.645	1.066	.593	0.980
Silver Hill illite	Na–Mg	0.379	0.259	0.427	.683	1.125
Altamont soil, pH 7	Na–Mg	0.579	0.357	0.591	.618	1.022
Altamont soil, pH 6	Na–Mg	0.607	0.389	0.639	.641	1.052
Wyoming bentonite	Na–Mg	0.882	0.689	1.135	.781	1.287
Mean	0.634	1.046				
SD	+0.231	+0.165	+0.273	+.069	+0.113	
CV, %	35.96	40.27	40.24	10.94	10.81	
SEM	0.082	0.058	0.096	.025	.040	
Ratio‡						0.606+ = exp(−1/2)

† K'_{eqV} calculated by Eq. [113], K_{eqV} calculated by integration from the experimental data plot using Eq. [106], and K_{eqGT} calculated by integration of the experimental data plot using Eq. [114].
‡ Ratio calculated from Eq. [113].

change isotherm) the K_{GT} is dependent on E_{Na} levels. Our studies (Evangelou and Phillips, 1989) of the interrelationship between K_v and K_{GT} for ideal exchangers demonstrate that when $E_{Na} \rightarrow 0$, $K_{GT} = 0.5K_v$, and when $E_{Na} \rightarrow 1$, $K_{GT} = 0.71K_v$. These relationships between K_v and K_{GT} should also hold regardless of whether ideal solid–solution theory is obeyed or not. The overall relationship with respect to the entire exchange isotherm for ideal exchangers is $K_v = 2K_{GT}(1 + E_{Na})^{-1/2}$. Therefore, the pathway between the two limits 0 and 1 is slightly curvilinear. The experimental results are in general agreement with the equations of the interrelationships (Table 7–7).

The data in Fig. 7–24 and 7–25 show the ratio of the adsorbed-ion activity coefficients by the Argersinger and Gaines and Thomas conventions for all the clay minerals. The data in Table 7–9 show the regression equations representing plots of experimental data vs. theoretical values. These data demonstrate the dependency of the interrelationship of the two types of adsorbed-ion activity coefficients obtained by the Argersinger (γ_i) and Gaines and Thomas (f_i) conventions as a function of E_{Na}. These relationships are in general agreement with Eq. [111] and [112], which are curvilinear. Because of the interrelationship shown in Fig. 7–24 and 7–25 (also Eq. [71]) the two exchange coefficients (K_v and K_{GT}) when plotted as a function of E_{Na}, regardless of whether ideal solid–solution theory is obeyed or not, do not compare in magnitude but appear to show analogous trends. However, these two lines, representing K_v and K_{GT}, are always converging as $E_{Na} \rightarrow 1$ (Fig. 7–26). Also, there is no apparent limiting condition for which K_v equals K_{GT} or K_{eqV} equals K_{eqGT}. The above serve as reasons, in addition to chemical thermodynamic considerations (Sposito, 1981a, 1981b; Sposito and Mattigod, 1979), for employing only the Vanselow convention to demonstrate ideal solid–solution behavior by an exchanger.

Theoretical analysis showed that $f_{Na}/\gamma_{Na} = 2/(1 + E_{Na})$ and $f_M/\gamma_M = 1/(1 + E_{Na})$. From this f_{Na}/γ_{Na} relationship it is also shown that when $E_{Na} \rightarrow 0$, $f_{Na}/\gamma_{Na} = 2$ and when $E_{Na} \rightarrow 1$, $f_{Na}/\gamma_{Na} = 1$. The regression analyses shown in Table 7–9 represent-

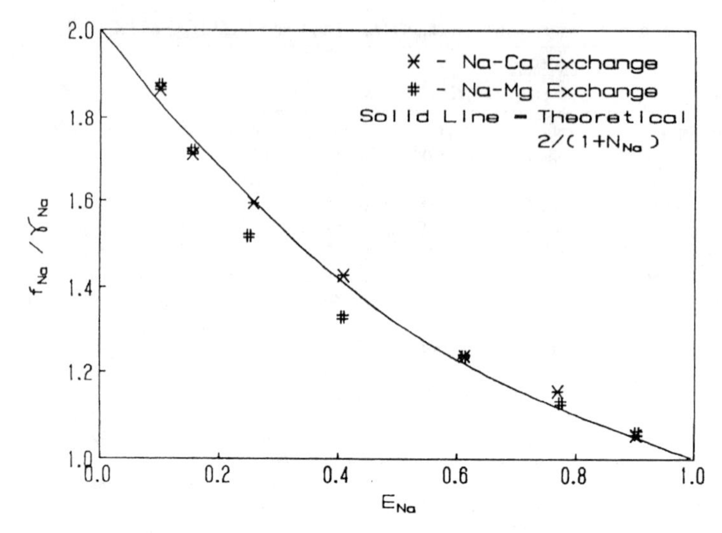

Fig. 7–24. Relationship between ratio of adsorbed-Na activity coefficients (f_{Na}/γ_{Na}) obtained by the Gaines and Thomas (f_i) and Argersinger (γ_i) conventions, and Na-equivalent fraction on the exchanger (E_{Na}) for the Na^+–Ca^{2+} and Na^+–Mg^{2+} exchanger systems. Raw data were taken from Sposito (1983), Sposito and LeVesque (1985), and Fletcher et al. (1984) (from Evangelou and Phillips, 1989).

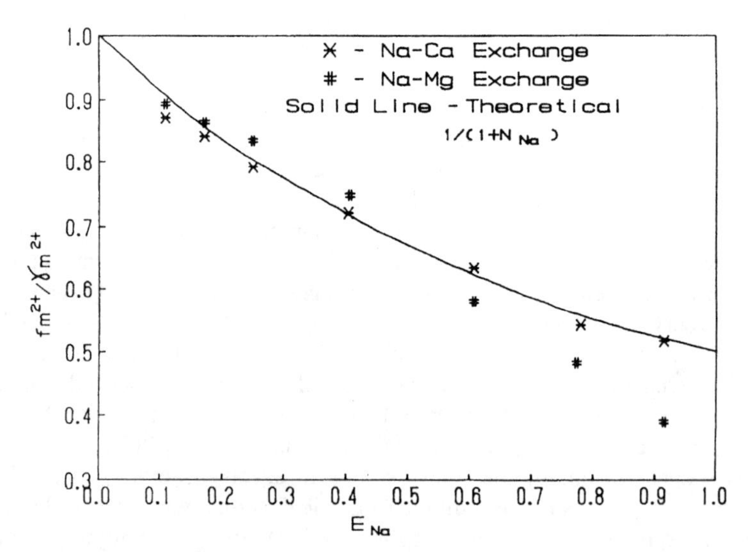

Fig. 7–25. Relationship between ratio of adsorbed-Ca or adsorbed-Mg activity coefficients (f_{Ca}/γ_{Ca} or f_{Mg}/γ_{Mg}), obtained by the Gaines and Thomas (f_i) and Argersinger (γ_i) conventions, and Na-equivalent fraction on the exchanger (E_{Na}) for the Na^+–Ca^{2+} and Na^+–Mg^{2+} exchange systems. Raw data were taken from Sposito et al. (1983), Sposito and LeVesque (1985), and Fletcher et al. (1984) (from Evangelou and Phillips, 1989).

Table 7–9. Correlation and regression equations for Na^+–M^{2+} exchange, relating theoretical ratios of adsorbed-ion activity coefficients f_i/γ_i, to experimental f_i/γ_i ratios (f_i = adsorbed-ion activity coefficients by the Gaines and Thomas convention:, γ_i = adsorbed-ion activity coefficients by the Argersinger convention).

Exchange system	f/γ	Regression equations†	R^2	Hypothesis that slope = 1.0
Na–Ca	f_{Na}/γ_{Na}	$Y = 0.0000 + 1.0057X$	0.9942	NS
Na–Ca	f_{Ca}/γ_{Ca}	$Y = 0.0169 + 0.9985X$	0.9915	NS
Na–Mg	f_{Na}/γ_{Na}	$Y = 0.0436 + 1.0026X$	0.9747	NS
Na–Mg	f_{Mg}/γ_{Mg}	$Y = 0.2182 + 0.7325X$	0.9674	‡

† Each regression equation represents the mean of four clay and soil systems. $X = f_i/\gamma_i$ experimental. $Y = f_i/\gamma_i$ theoretical. Theoretical $= f_i/\gamma_i$ for the Na^+, Ca^{2+}, or Mg^{2+} were estimated with Eq. [111] and [112], respectively.

‡ Denotes slope is significantly different from 1 at the 0.05 probability level (t test).

ing the Na^+–Ca^{2+} and Na^+–Mg^{2+} systems are in agreement with these theoretical findings. These experimental results and findings are also in agreement with the experimental results presented on K^+–Ca^{2+} exchange by Goulding (1983). Goulding (1983) showed that when $E_K \rightarrow 0, f_K/\gamma_K > 1$; and when $E_K \rightarrow 1, f_K/\gamma K = 1$. For the f_M/γ_M relationship, it can be shown that when $E_{Na} \rightarrow 0, f_M/\gamma_M = 1.0$, and when $E_{Na} \rightarrow 1, f_M/\gamma_M = 1/2$. The results reported by Goulding (1983) and Ogwada and Sparks (1986a) for the K^+–Ca^{2+} system in Cl^- background ionic media with respect to f_{Ca}/γ_{Ca} are in agreement with our theoretical findings. These authors reported that when $E_K \rightarrow 0, f_{Ca}/\gamma_{Ca} = 1$, and when $E_K \rightarrow 1, f_{Ca}/\gamma_{Ca} < 1$.

The dependency of the adsorbed-ion activity coefficients by the Gaines and Thomas convention on E_{Na} for a heterovalent-binary exchange is expected, since as $E_{Na} \rightarrow 1$ the K_v and K_{GT} converge (Fig. 7–26). Because the dependency of f_i/γ_i

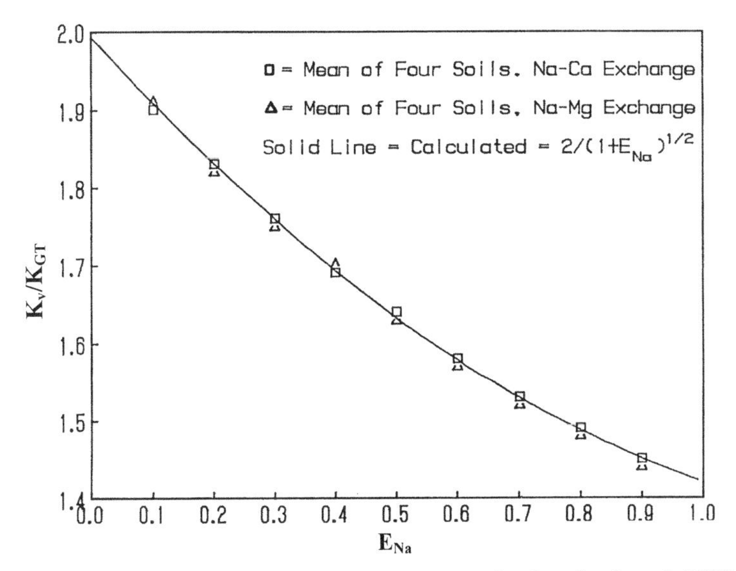

Fig. 7–26. Relationship between K_V/K_{GT} and E_{Na}. Raw data were taken from Sposito et al. (1983), Sposito and LeVesque (1985), and Fletcher et al. (1984) (from Evangelou and Phillips, 1989).

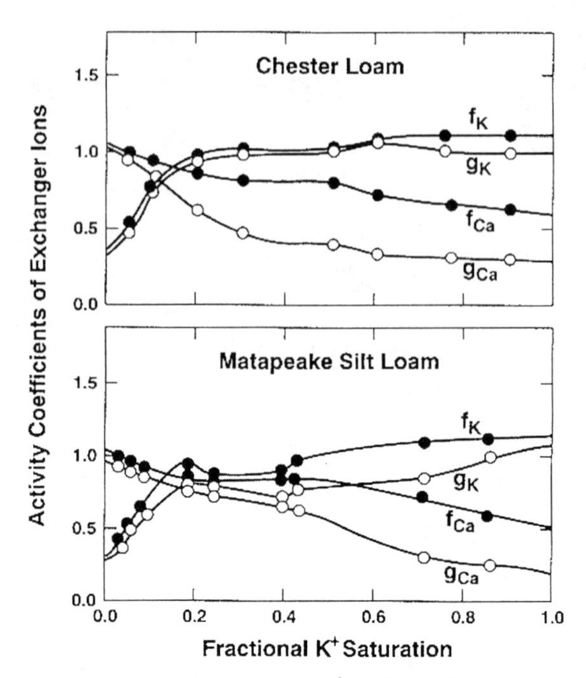

Fig. 7–27. Adsorbed-ion activity coefficients for K^+–Ca^{2+} exchange obtained by the Argersinger et al. (1950) approach (γ_i)and the Gaines and Thomas approach (f_i) (from Ogwada and Sparks, 1986).

on E_{Na} is described by nearly linear functions (Fig. 7–24 and 7–25), one may conclude that for all heterovalent binary exchange systems the same conclusions or inferences can be made by adsorbed-ion activity coefficients obtained by the Argersinger or Gaines and Thomas conventions (Fig. 7–27).

Influence of Anions on Exchange Reactions

Ion exchange equilibria are affected by the type of anions used (Babcock and Schultz, 1963; Rao et al., 1968) because of the pairing properties of divalent ions (Tanji, 1969; Adams, 1971). Tanji (1969) demonstrated that approximately one-third of the dissolved $CaSO_4^\circ$ of a solution in equilibrium with gypsum ($CaSO_42H_2O$) is in the $CaSO_4^\circ$ pair form. Magnesium behavior is similar to Ca^{2+} since Mg^{2+} pairing ability with SO_4^{2-} is approximately equal to that of Ca (Adams, 1971). By investigating Na^+–Ca^{2+} exchange reactions in Cl^- or SO_4^{2-} solutions, Babcock and Schultz (1963) and Rao et al. (1968) found that exchange selectivity coefficients for the two cations were different, but they demonstrated that this difference was nearly eliminated when solution phase activities and $CaSO_4^\circ$ were considered.

Evangelou (1986) investigated the influence of Cl^- or SO_4^{2-} anions on K^+–Ca^{2+} exchange in several soils. The data from these studies demonstrated that correcting the solution data for single-ion activity considering ion pairing and single-ion solution activity did not compensate for the experimental differences in K^+–Ca^{2+} exchange between the Cl^- and SO_4^{2-} systems. These differences may be due to com-

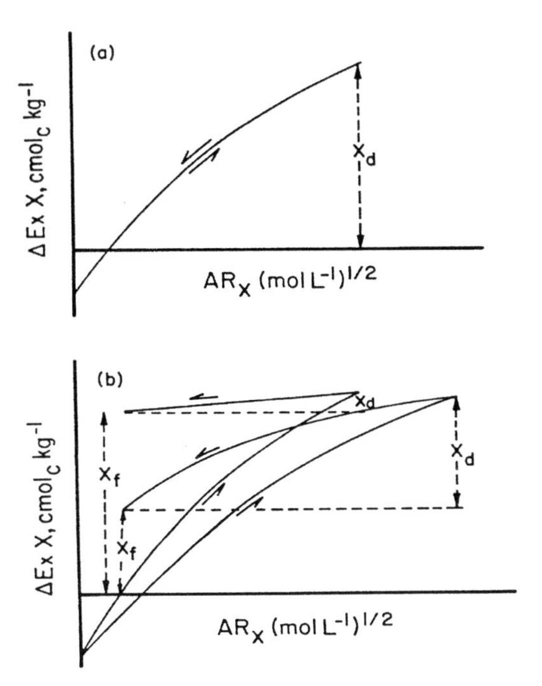

Fig. 7–28. Relationship between AR_X and ΔExX for the forward and reverse X^+–Ca^{2+} exchange reaction.

petitive interactions of $CaCl^+$ with Ca^{2+} and/or KSO_4^- with SO_4^{2-} for exchange sites and possibly SO_4^{2-} specific adsorption onto the soil's solid surfaces. Sposito et al. (1983) showed that external clay surfaces adsorbed charged pairs of $CaCl^+$ and $MgCl^+$ in Na^+–Ca^{2+}–Cl^- and Na^+–Mg^{2+}–Cl^- clay suspension systems.

The influence of anion on K^+ rate of adsorption, mobility, and retention were investigated by Sadusky and Sparks (1991) on two Atlantic Coastal Plain soils. They reported that the type of anion had little effect (if any) on the rate of K^+ adsorption but had an effect on the amount of K^+ adsorbed. They found that K^+ adsorbed in the presence of a particular accompanying anion was of the order: $SiO_3 > PO_4 > SO_4 > Cl > ClO_4$. These findings strongly imply that exchange reactions in soils are affected by the anion present in the soil. For a major review of the above, see also Evangelou et al. (1994).

Exchange Reversibility

It is assumed that the shape and form of cation exchange is identical between the adsorption and desorption mode (Fig. 7–28a). Nye and Tinker (1977) suggested that the desorption process may be affected by hysteresis (Fig. 7–28b). The graphical relationship presented in Fig. 7–28b shows that the difference between the quantity of adsorbed X^+ and the quantity of desorbed $X^+(X_d)$ is the quantity of X^+ fixed, X_f, (Arnold, 1970). In this case, the term *fixed* denotes exchange reaction nonreversibility under the conditions described by the experiment. The term fixed as often seen in the literature denotes nonextractability of the cation by concentrated salt solutions (Quirk and Chute, 1968; Barshad, 1954; Lurtz, 1966; Opuwaribo and Odu, 1974). Arnold (1970) pointed out that an isotherm demonstrating a hysteresis effect can exhibit two different shapes (linear and curvilinear). If the desorption isotherm is linear, it indicates that the (X^+) is being desorbed from low affinity sites only. If the desorption isotherm is curvilinear, the cation (X^+) is being desorbed from high (stern layer) and low (diffuse layer) affinity sites.

The hysteresis effect appears to be affected by length of equilibration period. Nye and Tinker (1977) pointed out that a true hysteresis effect would persist no matter what length of time was given for a true equilibrium to establish. In contrast, if the difference between an adsorption and a desorption isotherm is eliminated by extending the equilibration time, this would be considered a relaxation effect (Everett and Whitton, 1952).

KINETICS AND THERMODYNAMIC RELATIONSHIPS: FACTORS CONTROLLING REACTION RATES

Reaction rates are determined by the number of successful collisions between reactants, which are controlled by the product of the total number of collisions of the reactants, the number of collisions that have sufficient energy to cause a reaction event (energy factor), and the fraction of collisions that have the proper orientation (probability factor). The theoretical reaction rate (k) is given by

$$k = PZ\exp(-E_\alpha/RT) \qquad [117]$$

where P is the probability factor, Z is the collision frequency, E_α is activation energy, T is temperature, R is the universal gas constant, and $\exp(-E_\alpha/RT)$ is the fraction of collisions with energy sufficient to cause a reaction event.

The product PZ is related to the pre-exponential factor (A) of the Arrhenius reaction and the entropy (S) of the reaction by

$$PZ = A\exp(S/R) \qquad [118]$$

Based on Reactions [117] and [118], the reaction rate constant is inversely related to the activation energy (E_α) of the reaction and directly related to the entropy of activation.

Temperature Influence

For heterogeneous reactions, the observed reaction rate is determined by the amount of surface covered by reacting molecules and by the specific velocity of the surface reaction. The influence of temperature on the rate therefore must include two factors, the effect on the surface area covered and the effect on the surface reaction itself.

As for homogeneous reactions, the influence of temperature on the rate constant of heterogeneous reactions is given by the Arrhenius equation:

$$k = A\exp(-E_{\alpha}^{*}/RT) \quad [119]$$

where k is the specific rate constant (time^{-1}), A is the frequency factor or preexponential factor (a constant), E_{α}^{*} is the activation energy (kJ mol^{-1}), R is the universal gas constant, and T is absolute temperature.

E_{α}^{*} is often referred to as the "apparent energy of activation." This energy of activation evaluated from the observed rate constants is a composite term and is not necessarily the energy required to activate the reactants on the surface, which is the true activation energy. The apparent energy of activation includes not only the true activation energy of the surface reaction, but also heats of adsorption of reactants, or reactants and products, to yield an apparent activation energy that may be quite different from the true one.

Integrating Eq. [119] considering that E_{α} is not itself temperature dependent, and hence a constant, gives

$$\ln k = -E_{\alpha}^{*}/RT + \ln A \quad [120]$$

where $\ln A$ is the constant of integration. In terms of logarithms to the base 10, Eq. [120] may be written as

$$\log k = -E_{\alpha}^{*}/2.303R(1/T) + \log A \quad [121]$$

From Eq. [121], it follows that a plot of the logarithm of the rate constant against the reciprocal of the absolute temperature should be a straight line with

$$\text{slope} = E_{\alpha}^{*}/2.303R = E_{\alpha}^{*}/4.58 \quad [122]$$

and y intercept equals $\log_{10}A$. By taking the slope of the line, E_{α}^{*} may be calculated readily from Eq. [120].

$$E_{\alpha}^{*} = -4.58(\text{slope}) \quad [123]$$

Activation energies <42 kJ mol^{-1} indicate diffusion controlled reactions, whereas reactions with E_{α} values >42 kJ mol^{-1} indicate chemical reactions or surface controlled processes.

The rate of the forward reaction of K^{+}–Ca^{2+} exchange (see Reaction [11]) can be expressed by

$$-\mathrm{dExCa}_{1/2}/\mathrm{d}t = k[K^{+}][\mathrm{ExCa}_{1/2}] \quad [124]$$

where k is the rate constant (time^{-1}). Assuming that during cation exchange K^{+} is kept constant, rearranging Eq. [124] gives

$$-\mathrm{dExCa}_{1/2}/\mathrm{ExCa}_{1/2} = k'\mathrm{d}t \quad [125]$$

where $k' = k[K^{+}]$. Setting the appropriate boundary conditions and integrating

$$\ln[\mathrm{ExCa}_{1/2}/\mathrm{ExCa}_{1/2}\mathrm{o}] = -k't \quad [126]$$

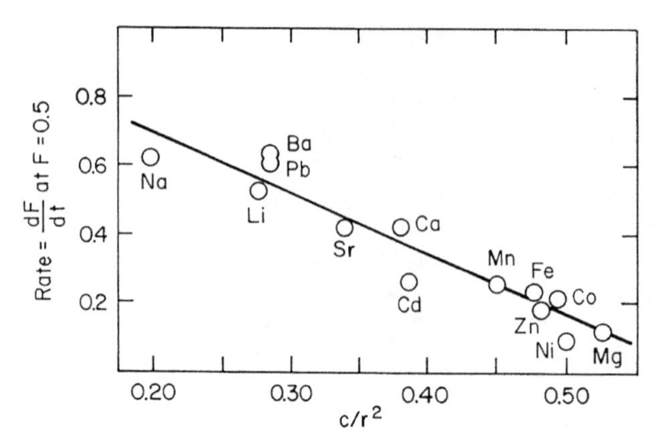

Fig. 7–29. Relationship between rate of exchange and c/r^2 (c is ion valence, and r is ionic radius) (from Keay and Wild, 1961).

or

$$[\mathrm{ExCa}_{1/2}] = [\mathrm{ExCa}_{1/2}\mathrm{o}]\exp(-k't) \qquad [127]$$

and

$$\log[\mathrm{ExCa}_{1/2}/\mathrm{ExCa}_{1/2}\mathrm{o}] = -k't/2.303 \qquad [128]$$

A plot of log[$\mathrm{ExCa}_{1/2}/\mathrm{ExCa}_{1/2}\mathrm{o}$ vs. t would produce a straight line with slope –k/2.303.

In 2:1 clay minerals (e.g., vermiculite) the rate of cation exchange depends on their ionic potential (c/r^2, where c denotes charge of the cation and r denotes ionic radius) of the cations involved, as long as interlayer spacing remains constant. Cation exchange data show that c/r^2 is inversely related to the rate of cation exchange (Fig. 7–29). The reason for this relationship is that the higher the ionic potential of an ion, the lower its entropy of activation, or the higher the energy by which hydration-sphere water is held (Bohn et al., 1985). Generally, cations that hold water tightly exhibit low diffusion potential in clay interlayer water.

From the K_{eq} constant, estimated graphically from the Vanselow exchange selectivity coefficient, a number of thermodynamic parameters can be estimated. For example,

$$\Delta G^\circ = -RT\ln K_{eq} \qquad [129]$$

where ΔG° is the standard Gibbs free energy of exchange and all other terms are as previously defined. Substituting values for the constants in Eq. [129] converts to

$$\Delta G^\circ = -\,1.364\log K_{eq} \qquad [130]$$

For any given reaction, where reactants and products are at the standard state, a negative ΔG° denotes that the reaction will move spontaneously from left to right. When

Table 7–10. Gibb's free energy (ΔG°_{ex}) of homovalent exchange and standard enthalpy of exchange (ΔH°_{ex}) values for binary exchange processes on soils and soil components.†

Exchange process	Exchanger	ΔG°_{ex}	ΔH°_{ex}	Reference
		kJ mol^{-1}		
Ca–Cu	Wyoming bentonite	0.11	−18.02	El-Sayed et al. (1970)
Na–Li	World vermiculite	−6.04	−23.15	Gast and Klobe (1971)
Na–Li	Wyoming bentonite	−0.20	−0.63	Gast and Klobe (1971)
Na–Li	Chambers montmorillonite	−0.34	−0.47	Gast and Klobe (1971)
K–Na	Wyoming bentonite	−1.28	−2.53	Gast (1972)
K–Na	Chambers montmorillonite	−3.04	−4.86	Gast (1972)

† Table adapted from Sparks (1995). The exchange studies were conducted at 298 K.

the ΔG° is zero, it implies that $K_{eq} = 1$. Finally, for any given reaction, where reactants and products are at the standard state, a positive ΔG° denotes that the reaction will not move spontaneously from left to right. Such a reaction, however, could become spontaneous by changing the concentrations of the reactants and/or products, thus producing a negative ΔG° for the reaction. Finally, the more negative the ΔG°, the greater the K_{eq}—greater adsorption potential, greater solubility, greater ion preference. In the case of exchange reactions, ΔG° denotes ion preference between two competing cations for the same surface. In the case of K^+–Ca^{2+}, it signifies that Ca^{2+} is the surface cation and K^+ is the solution cation. If ΔG° is negative the surface prefers K^+, but if ΔG° is positive, the surface prefers Ca^{2+}. Note that this preference is independent of ionic strength. The data in Tables 7–10 and 7–11 present ΔG° and ΔH° values for various homovalent and heterovalents exchange reactions, respectively.

Also, from the K_{eq} one can calculate the standard enthalpy of exchange, ΔH°, using the van't Hoff equation:

Table 7–11. Gibbs free energy (ΔG°) of heterovalent exchange and standard enthalpy of exchange (ΔH°_{ex}) values for binary exchange processes on soils and soil components.†

Exchange process	Exchanger‡	ΔG°_{ex}	ΔH°_{ex}	Reference
		kJ mol^{-1}		
Ca–Na	World vermiculite	0.06	39.38	Wild and Keay (1964)
Ca–K	Chambers montmorillonite	7.67	16.22	Hutcheon (1966)
Ca–NH_4	Camp Berteau montmorillonite	8.34	23.38	Laudelout et al. (1967)
Mg–Na	World vermiculite	−1.36	40.22	Wild and Keay (1964)
Mg–NH_4	Camp Berteau montmorillonite	8.58	23.38	Landelout et al. (1967)
K–Ca	soils	−7.42 to −14.33	−3.25 to −5.40	Deist and Talibudeen (1967b)
K–Ca	soil	1.93	−15.90	Jardine and Sparks (1984b)
K–Ca	soils	1.10 to −4.70	−3.25 to −5.40	Goulding and Talibudeen (1984)
K–Ca	soils	−4.61 to −4.74	−16.28	Ogwada and Sparks (1986b)
K–Ca	kaolinitic soil clay (303 K)	−6.90	−54.48	Udo (1978)

† Table adapted from Sparks (1995).
‡ The exchange studies were conducted at 298 K; exceptions are noted.

$$\ln(K_{eq}T_1/K_{eq}T_2) = (\Delta H^\circ/R) - [(1/T_2) - (1/T_1)] \quad [131]$$

where T_1 and T_2 denote two different temperatures. From the classical thermodynamic relationship

$$\Delta G^\circ = \Delta H^\circ - T\Delta S^\circ \quad [132]$$

the standard entropy of exchange, ΔS° can be calculated by rearranging Eq. [132]:

$$\Delta S^\circ = (\Delta H^\circ - \Delta G^\circ)/T \quad [133]$$

On the basis of Reaction [133] one may observe the requirements of ΔS° and ΔH° in order for $K_{eq} > 1$ or for the reactants in their standard state to be converted spontaneously to their products in their standard state. If ΔH° is negative and ΔS° is positive, the reaction will be spontaneous. If, on the other hand, the ΔH° is positive and ΔS° is negative, the reaction will not be spontaneous. Finally, if ΔH° and ΔS° have the same sign, the reaction may or may not be spontaneous, depending on the magnitude of ΔS°, ΔH°, and temperature. Enthalpy, ΔH°, is defined as the heat given out or taken up during a reaction. If heat is given out during a reaction, it is referred to as an exothermic reaction, and the ΔH° would be negative. If, on the other hand, heat is taken up during a reaction, it is referred to as an endothermic reaction, and the ΔH° would be positive. A negative ΔH° denotes stronger bonds within the products whereas a positive ΔH° denotes stronger bonds within the reactants.

Entropy, ΔS°, is defined as the tendency of a given system to come apart. For any given ΔH°, the greater the ΔS°, the greater the tendency of a reaction to become spontaneous because of the greater tendency of the system to come apart or go to disarray. Exchange reactions on clay surfaces involving hard cations are known to be mostly electrostatic processes, and entropy of exchange plays a significant role in determining cation preference with respect to size and valence.

Relationships between Kinetics and Thermodynamics of Exchange

An equilibrium state is defined as the point where the forward reaction equals the reverse reaction, thus

$$K_{eq} = k_f/k_r \quad [134]$$

where K_{eq} is the thermodynamic exchange constant, k_f is the rate constant of the forward reaction, and k_r is the rate constant of the reverse reaction. The standard enthalpy of exchange (ΔH°) can be calculated using the van't Hoff equation or from the activation energies (E_α) of the forward and reverse reactions by

$$E_{\alpha f} - E_{\alpha r} = \Delta H^\circ \quad [135]$$

From the classical thermodynamic relationship

$$\Delta G^\circ = \Delta H^\circ - T\Delta S^\circ \quad [136]$$

The standard entropy of exchange, ΔS°, can now be calculated by rearranging Eq. [136]:

$$\Delta S^\circ = (\Delta H^\circ - \Delta G^\circ)/T \quad [137]$$

and

$$\Delta G^\circ = -1.364 \log K_{eq} \quad [138]$$

Note, when using activation energies to estimate ΔH° one needs first to demonstrate that the reactions, forward and reverse, are not limited by mass-transfer or diffusion processes. This basically limits the application of this concept to external surfaces only under rapid solute–solid mixing. For additional information on kinetics and thermodynamics see Jardine and Sparks (1984a, 1984b), Ogwada and Sparks (1986b), and Sparks (1989).

APPENDIX I

Derivation of K_{eq} and Solid Phase Activity Coefficients, λ

Consider the homovalent cation exchange reaction

$$\text{ExCa} + \text{Mg}^{2+} \Leftrightarrow \text{ExMg} + \text{Ca}^{2+} \quad [1]$$

where ExCa is exchangeable Ca2+ (mol kg^{-1}), ExMg is exchangeable Mg^{2+} (mol kg^{-1}), and Ex is exchange phase with charge 2–.

$$K_{eq} = \left(\frac{\text{ExMg}}{\text{ExCa} + \text{ExMg}}\right) \lambda_{Mg} (\text{Ca}^{2+}) \Big/ \left(\frac{\text{ExCa}}{\text{ExCa} + \text{ExMg})}\right) \lambda_{Ca} (\text{Mg}^{2+}) \quad [2]$$

where K_{eq} is the thermodynamic exchange constant, λ_{Mg} and λ_{Ca} are activity coefficients for Ca^{2+} and Mg^{2+} on the solid phase-rationale activity coefficients, and parentheses denote the activities of Ca^{2+} and Mg^{2+} in the solution phase.

$$\text{Mole fraction Mg}^{2+} = M_{Mg} = \text{ExMg}/(\text{ExCa} + \text{ExMg}) \quad [3]$$

$$\text{Mole fraction Ca}^{2+} = M_{Ca} = \text{ExCa}/(\text{ExCa} + \text{ExMg}) \quad [4]$$

Substituting Eq. [3] and [4] into Eq. [2] gives

$$K_{eq} = \frac{[M_{Mg}]\lambda_{Mg}(\text{Ca}^{2+})}{[M_{Ca}]\lambda_{Ca}(\text{Mg}^{2+})} \quad [5]$$

The Vanselow exchange selectivity coefficient is given by

$$K_v = \frac{[M_{Mg}](\text{Ca}^{2+})}{[M_{Ca}](\text{Mg}^{2+})} \quad [6]$$

Therefore,

$$K_{eq} = K_v(\lambda_{Mg}/\lambda_{Ca}) \quad [7]$$

Equation [7] indicates that in order for K_{eq} to remain constant across the entire exchange isotherm (defined by $M_{Ca} + M_{Mg} = 1$), any variation in K_v must be followed by a variation in λ_{Ca} and λ_{Mg}.

Taking natural logs on both sides of Eq. [7] gives

$$\ln K_{eq} = \ln K_v + \ln\lambda_{Mg} - \ln\lambda_{Ca} \quad [8]$$

Taking the derivative of Eq. [8] with respect to fractional metal load, M_x or E_x, where x is any cation (Ca^{2+} or Mg^{2+}) (E denotes equivalent fraction, but in homovalent exchange $M = E$) gives

$$+d\ln K_v = -d\ln\lambda_{Mg} + d\ln\lambda_{Ca} \quad [9]$$

From classical thermodynamics (Gibbs–Duhem equation)

$$M_{Ca}d\ln\lambda_{Ca} + M_{Mg}d\ln\lambda_{Mg} = 0 \quad [10]$$

Rearranging:

$$d\ln\lambda_{Ca} = -M_{Mg}d\ln\lambda_{Mg}/M_{Ca} \quad [11]$$

and

$$d\ln\lambda_{Mg} = -M_{Ca}d\ln\lambda_{Ca}/M_{Mg} \quad [12]$$

Substituting Eq. [11] and [12] into Eq. [9] gives

$$d\ln\lambda_{Ca} + (M_{Ca}d\ln\lambda_{Ca}/M_{Mg}) = d\ln K_v \quad [13]$$

$$-d\ln\lambda_{Mg} + (-M_{Mg}\, d\ln\lambda_{Mg}/M_{Ca}) = d\ln K_v \quad [14]$$

Rearranging Eq. [13] and [14] :

$$\frac{M_{Mg}d\ln\lambda_{Ca} + M_{Ca}d\ln\lambda_{Ca}}{M_{Mg}} = d\ln K_v \quad [15]$$

$$\frac{-[M_{Ca}d\ln\lambda_{Mg} + M_{Mg}d\ln\lambda_{Mg}]}{M_{Ca}} = d\ln K_v \quad [16]$$

Rearranging Eq. [15]:

$$d\ln\lambda_{Ca} = \frac{M_{Mg}}{M_{Ca} + M_{Mg}}\, d\ln K_v \quad [17]$$

Rearranging Eq. [16]:

$$d\ln\lambda_{Mg} = \frac{M_{Ca}}{M_{Ca} + M_{Mg}}\, d\ln K_v \quad [18]$$

Rearranging and substituting Eq. [17] and [18] into Eq. [7] gives

$$+\mathrm{d}\ln\lambda_{Mg} = -(1 - M_{Mg})\mathrm{d}\ln K_v \quad [19]$$

$$\mathrm{d}\ln\lambda_{Ca} = M_{Mg}\mathrm{d}\ln K_v \quad [20]$$

where $[1 - M_{Mg}] = M_{Ca}$.

Integrate Eq. [19] and [20] by parts (Note: $\ln\lambda_{Ca} = 0$ at $M_{Ca} = 1$ and $\ln\lambda_{Mg} = 0$ at $M_{Ca} = 0$).

$$-\ln\lambda_{Mg} = (1 - M_{Mg})\ln K_v - \int_{M_{Mg}}^{1}\ln K_v\,\mathrm{d}M_{Mg} \quad [21]$$

and

$$-\ln\lambda_{Ca} = -M_{Mg}\ln K_v + \int_{0}^{M_{Mg}}\ln K_v\,\mathrm{d}M_{Mg} \quad [22]$$

Substituting Eq. [21] and [22] into Eq. [8], $\ln K_{eq} = \ln K_v + \ln\lambda_{Mg} - \ln\lambda_{Ca}$ gives

$$\ln K_{eq} = \int_{0}^{1}\ln K_v \mathrm{d}M_{Mg} \quad [23]$$

APPENDIX II

Derivation of Algebraic Interrelationships between Selectivity Exchange Coefficients

We will consider Na^+ ions replacing Ca^{2+} ions on an exchanger (soil); for example, the reaction is given in Eq. [1]

$$1/2Ex_2Ca^{2+} + Na^+ \Leftrightarrow ExNa + 1/2Ca^{2+} \quad [1]$$

We will then derive the interrelationships of the Gapon, Gaines and Thomas, and Vanselow selectivity coefficients. These interrelationships apply to any divalent and monovalent ions in soil or soil minerals.

Gapon Selectivity Coefficient

The Gapon selectivity coefficient of the exchange reaction of Eq. [1] is defined to be

$$K_G = [E_{Na}/E_{Ca}][(Ca^{2+})^{1/2}/(Na^+)] \quad [2]$$

where K_G is the Gapon exchange selectivity coefficient, E_{Na} is the equivalent fraction of Na^+ on the exchanger (= ExNa/[1/2Ex_2Ca + ExNa]), E_{ca} is the equivalent fraction of Ca^{2+} on the exchanger (=Ex_2Ca/[1/2Ex_2Ca + ExNa]), Ex is the soil exchanger with a charge of 1−, parentheses denote activities, Ca^{2+} and Na^+ are the solution phase cations (mmol L^{-1}), and ExNa and 1/2Ex_2Ca are exchangeable cations (cmol kg^{-1}).

Cation exchange capacity in units of equivalents of exchangeable cation is

$$1/2Ex_2Ca + ExNa = CEC \quad [3]$$

Solving for 1/2Ex_2Ca from Eq. [3], replacing it in Eq. [2], and solving for ExNa we get

$$ExNa = K_G(SAR)CEC[1 + K_G(SAR)]^{-1} \quad [4]$$

where SAR is the Na adsorption ration [= $(Na^+)/(Ca^{2+})^{1/2}$].

It will become apparent below (see also text) why Eq. [4] is solved for ExNa. The limits of Eq. [4] as SAR approached 0 and infinity are

$$\lim_{SAR \to 0} ExNa = 0; \qquad \lim_{SAR \to \infty} ExNa = CEC$$

The limit ExNa = 0/1. The limit ExNa is an indeterminate form of ∞/∞

$$\lim_{SAR \to 0} \qquad \lim_{SAR \to \infty}$$

Differentiate the numerator and denominator of Eq. [4] to get

$$\lim_{SAR \to \infty} ExNa = K_G(CEC)/K_G = CEC$$

The derivative of Eq. [4] with respect to SAR is

$$dExNa/dSAR = K_G(CEC)[1 + K_G(SAR)]^{-2} \qquad [4A]$$

and

$$\lim_{SAR \to 0} dExNa/dSAR = K_G(CEC)[1 + 0]^{-2} = K_G CEC \qquad [4B]$$

$$\lim_{SAR \to \infty} dExNa/dSAR = K_G(CEC)/[1 + \infty]^{+2} = 0 \qquad [4C]$$

Note that a derivative of a dependent variable (ExNa above) with respect to an independent variable (SAR above) is the slope of the curve resulting from a plot of the dependent variable as a function of the independent variable at any value of the independent variable.

The limits of ExNa from the Gapon equation and its derivative as SAR approaches zero and infinity imply that a plot of ExNa vs. SAR with SAR, the independent variable, and where CEC and K_G remain constant, will result in a curvilinear function approaching the CEC asymptotically.

Vanselow Selectivity Coefficient

The Vanselow selectivity coefficient of the exchange reaction of Eq. [1] is defined to be

$$K_V = [M_{Na}/(M_{Ca})^{1/2}][(Ca^{2+})^{1/2}/(Na^+)] \qquad [5]$$

where K_V is the Vanselow selectivity coefficient, M_{Na} (= ExNa $[Ex_2Ca + ExNa]^{-1}$) is the mole fraction of Na^+ on the exchanger, M_{ca} (= $Ex_2Ca[Ex_2Ca + ExNa]^{-1}$) is the mole fraction of Ca^{2+} on the exchanger, SAR = $(Na^+)/(Ca^{2+})^{1/2}$, parentheses denote activity, and Ex is soil exchange with a charge of 1−.

We assume that the mole fractions of Na and Ca, M_{Na} and M_{Ca}, in the exchanger represent the ideal solid–solution state, and, therefore, mole fractions represent activities of the adsorbed cations. Equation [5] can be solved for ExNa by substituting it into Eq. [6] and [7].

$$M_{Na} = ExNa[1/2CEC - 1/2ExNa]^{-1} \qquad [6]$$

and

$$M_{ca} = [1/2\text{CEC} - 1/2\text{ExNa}][1/2\text{CEC} + 1/2\text{ExNa}]^{-1} \quad [7]$$

Equations [6] and [7] convert molar fraction, M_{Na} and M_{Ca}, into equivalent fractions, E_{Na} and E_{Ca}. To show that

$$M_{Ca} = [1/2\text{CEC} - 1/2\text{Ex}_{Na}][1/2\text{CEC} + 1/2\text{ExNa}]^{-1}$$

Replace CEC with its equivalent from Eq. [3]:

$$M_{Ca}[(1/2\text{Ex}_2\text{Ca} + 1/2\text{ExNa}) - 1/2\text{ExNa}][(1/2\text{Ex}_2\text{Ca} + 1/2\text{ExNa}) + 1/2\text{ExNa}]^{-1}$$

$$= 1/2\text{Ex}_2\text{Ca} + 1/2\text{ExNa} - 1/2\text{ExNa}][1/2\text{Ex}_2\text{Ca} + 1/2\text{ExNa} + 1/2\text{ExNa}]^{-1}$$

$$= 1/2\text{Ex}_2\text{Ca}[1/2\text{Ex}_2\text{Ca} + \text{ExNa}]^{-1}$$

The last expression is equal to the equivalent fractions of Ca on the exchanger, E_{Ca}.

Substituting Eq. [6] and [7] into Eq. [5] and rearranging Eq. [7] takes the form of a quadratic equation in ExNa.

$$\text{ExNa}^2 = [K_V(\text{SAR})\text{CEC}]^2[4 + (K_V\text{SAR})^2]^{-1} \quad [8]$$

Take the positive root of Eq. [8] to get

$$\text{ExNa} = [K_V(\text{SAR})\text{CEC}][4 + K_V(\text{SAR})^2]^{-1/2} \quad [9]$$

The limits of Eq. [9] as SAR approaches 0 (becomes very small) and as SAR approaches infinity (becomes very large) are as follows:

$$\lim_{\text{SAR} \to 0} \text{ExNa} = 0, \quad \text{and} \quad \lim_{\text{SAR} \to \infty} \text{ExNa} = \text{CEC}$$

$$\lim_{\text{SAR} \to 0} \text{ExNa} = [\text{CEC}K_V(\text{SAR})][4 + (K_V\text{SAR})^2]^{-1/2} = 0/(4)^{1/2} = 0 \quad [9A]$$

$$\lim_{\text{SAR} \to \infty} \text{ExNa} = K_V(\text{CEC})\text{SAR}[4 + (K_V\text{SAR}^2)]^{-1/2} \quad [9B]$$

Equation [9B] is an indeterminate form (∞/∞). Divide the numerator and denominator by $(K_V\text{SAR})$ to get

$$\lim_{\text{SAR} \to \infty} \text{ExNa} = [\text{CEC}(K_V\text{SAR})/(K_V\text{SAR})][4/(K_V\text{SAR})^2 + (K_V\text{SAR})^2/(K_V\text{SAR})^2]^{-1/2} \quad [9C]$$

$$\lim_{\text{SAR} \to \infty} \text{ExNa} = [\text{CEC}][0 + 1]^{-1/2} = \text{CEC} \quad [9D]$$

The derivative of Eq. [9] with respect to SAR is

$$\text{dExNa/dSAR} = 4K_V\text{CEC}[4 + (K_V\text{SAR})^2]^{-3/2} \quad [10]$$

$$\lim_{\text{SAR} \to 0} \text{dExNa/dSAR} = 4K_V\text{CEC}/(4)^{3/2} = (K_V)\text{CEC}/2 \quad [10A]$$

and

$$\lim_{SAR \to \infty} dExNa/dSAR = 4K_V CEC/\infty = 0 \quad [10B]$$

The limits of Eq. [9] and [10] as SAR approaches zero and infinity mean that a plot of ExNa vs. SAR with constant K_V and CEC will result in a curvilinear function approaching the CEC asymptotically. Recall the same was said about the Gapon equation, Eq. [4]. The limits of the derivatives of the Gapon equation, Eq. [4B] and the Vanselow equation, Eq. [10A] as SAR approaches zero means that slope of a plot of ExNa vs. SAR of the Vanselow coefficient at the origin will be equal to $2K_G$, where the CEC, K_G, and K_V remain constant; that is, K_GCEC $= K_V$CEC/2 or $K_V = 2K_G$.

Gaines and Thomas Selectivity Coefficient

The Gaines and Thomas exchange coefficient is defined in terms of equivalent fractions, as is the Gapon coefficient. However, the Gaines and Thomas coefficient is defined in terms of the square root of the divalent cation equivalent fraction on the exchanger while the Gapon coefficient is defined in terms of the linear term of the divalent cation on the exchanger.

The Gaines and Thomas exchange selectivity coefficient, K_{GT}, is expressed as follows:

$$K_{GT} = [E_{Na}/(E_{Ca})^{1/2}][(Ca^{2+})^{1/2}/Na^+] \quad [11]$$

where all quantities on the right-hand side of Eq. [11] have been defined above. Expressing E_{Na} and E_{Ca} in terms of ExNa and Ex_2Ca from Eq. [2A] and [2B], solving for Ex_2Ca from Eq. [3] and substituting Ex_2Ca back into Eq. [1], $(Na^+)/(Ca^{2+})^{1/2}$ = SAR. Rearrange to get a quadratic equation in ExNa:

$$(ExNa)^2 + [K_{GT}(SAR)]^2 CECExNa - [K_{GT}(CEC)SAR]^2 = 0 \quad [12]$$

The positive root of the quadratic equation, Eq. [12], is ExNa $= (-2c/[b + (b^2 - 4ac)^{1/2}]^{-1}$, where a, b, and c are the coefficients of the quadratic, linear, and constant term of Eq. [12], respectively.

$$ExNa = 2CEC(SAR)K_{GT}[K_{GT}SAR + \sqrt{4 + (K_{GT}SAR)^2}]^{-1} \quad [13]$$

The limits of Eq. [13] as SAR approaches zero and infinity are

$$\lim_{SAR \to 0} ExNa = 0, \quad \text{and} \quad \lim_{SAR \to \infty} ExNa = CEC$$

$$\lim_{SAR \to 0} ExNa = [0/(0 + \sqrt{4})] = 0 \quad [13A]$$

In order to obtain the limit of ExNa of Eq. [13] as SAR approaches infinity, divide the numerator and denominator by K_{GT}SAR to get

$$\lim_{SAR \to \infty} ExNa = 2CEC/[(1 + \sqrt{1})] = 2CEC/2 = CEC \quad [13B]$$

The derivative of Eq. [13] with respect to SAR is

$$\Delta \text{ExNa/dSAR} = 2K_{GT}\text{CEC}[4 - (K_{GT}\text{SAR})^2]/ [4 + (K_{GT}\text{SAR})^2]^{1/2}[K_{GT}\text{SAR} + [\sqrt{4 + (K_{GT}\text{SAR})^2}]^2 \quad [13C]$$

The limits of dExNa/dSAR as SAR approaches zero and infinity are

$$\lim_{\text{SAR} \to 0} \text{dExNa/dSAR} = K_{GT}\text{CEC}, \quad \text{and} \quad \lim_{\text{SAR} \to \infty} \text{dExNa/dSAR} = 0$$

$$\lim_{\text{SAR} \to 0} \text{dExNa/dSAR} = 8K_{GT}\text{CEC}/[4\sqrt{4}] = K_{GT}\text{CEC} \quad [14A]$$

Equation [14A] as SAR approaches zero means that the slope of a plot ExNa vs. SAR of the Vanselow coefficient at the origin will be equal to $2K_C/(\text{CEC})^{2,}$ where CEC, K_V and K_{GT} remain constant; that is, $K_V(\text{CEC})/2 = K_{GT}/\text{CEC}$ or $K_V = 2K_{GT}/\text{CEC}^2$.

$$\lim_{\text{SAR} \to \infty} \text{dExNa/dSAR is an indeterminate form of } \infty/\infty$$

Divide the numerator and denominator by $(K_{GT}\text{SAR})^2$ to get

$$\lim_{\text{SAR} \to \infty} \text{dExNa/dSAR} = -2K_{GT}\text{CEC}/[\sqrt{1}][2K_{GT}(\text{CEC})\text{SAR} + 4 + 2\sqrt{4 + (K_{GT}\text{SAR})^2}] = 0 \quad [14B]$$

Obtaining the Interrelationships of K_G and K_V, and K_{GT} and K_V in Terms of ExNa

The interrelationships can be obtained in several different ways. We have chosen to obtain them by solving Eq. [4], [9], and [13] for SAR in terms of, or a function of, E_{Na} and then equating Eq. [4] and [9] and Eq. [9] and [13].

Recopying Eq. [4] and solving for SAR:

$$K_G(\text{SAR})[1 + K_G(\text{SAR})]^{-1} = \text{ExNa/CEC} = E_{Na}$$

$$\text{SAR} = E_{Na}[K_G(1 - E_{Na})]^{-1} \quad [15]$$

Recopying Eq. [9] and solving for SAR:

$$K_V(\text{SAR})[4 + (K_V\text{SAR})^2]^{-1/2} = \text{ExNa/CEC} = E_{Na}$$

$$\text{SAR} = 2E_{Na}[K_V(1 - E_{Na}{}^2)]^{-1} \quad [16]$$

Equate Eq. [15] and [16] to get

$$\text{SAR} = E_{Na}[K_G\,(1 - E_{Na})]^{-1} = 2E_{Na}[K_V(1 - E_{Na}{}^2)]^{-1}.$$

Solve for K_G to get

$$K_G = (K_V/2)[1 + E_{Na}]^{1/2}[1 - E_{Na}]^{-1/2} \quad [17]$$

Recopy Eq. [13] and solve for SAR to get

$$\text{SAR} = E_{Na}[K_C(1 - E_{Na})^{1/2}]^{-1} \quad [18]$$

Equate SAR from Eq. [16] and [18] to get

$$SAR = 2E_{Na}[K_V(1 - E_{Na}{}^2)^{1/2}]^{-1} = E_{Na}[K_{GT}(1 - E_{Na})^{1/2}]^{-1}$$

$$K_{GT} = (K_V/2)[1 + E_{Na}]^{1/2} \quad [19]$$

The smallest K_{GT} can be as a function of K_V is $0.5(K_C = 0.5K_V)$; this occurs when E_{Na} approaches zero. The largest K_{GT} can be as a function of K_V is $0.707(K_C = 0.707K_V)$; this occurs when E_{Na} approaches one.

There is no reason to obtain the relationship of K_{GT} and K_G since neither has thermodynamic meaning. However, if one had a reason to calculate the relationship as a function of E_{Na}, it is

$$K_G = K_{GT}[1 - E_{Na}]^{-1/2}$$

or

$$K_{GT} = K_G[1 - E_{Na}]^{1/2} \quad [19A]$$

Adsorbed Activity Coefficients

Absorbed activity coefficients correct the equivalent and mole fraction terms for departures from the ideal solid–solution state in exchange reactions. The equations of the Vanselow selectivity coefficient and Gaines and Thomas selectivity coefficient for the exchange reaction for ideal solid–solution exchange are

$$K_V = M_{Na}/(M_{Ca})^{1/2}[(Ca^{2+})^{1/2}/(Na^+)] \quad [20]$$

$$K_{GT} = [E_{Na}/(E_{Ca})^{1/2}][(Ca^{2+})^{1/2}/(Na^+)] \quad [21]$$

Solid and/or clay exchange reactions are seldom if ever ideal. Therefore, adsorbed activity coefficients correct for this nonideality. The selectivity coefficients of Eq. [20] and [21] above are then modified as follows:

$$K_{eq} = K_{eqV} = K_V(\gamma_{Ca})^{1/2}/\gamma_{Na} \quad [22]$$

and

$$K_{eq} = K_{eqGT} = K_C(f_{Ca})^{1/2}/f_{Na} \quad [23]$$

where K_{eqV} and K_{eqGT} are thermodynamic selectivity coefficients, γ_{Ca} and γ_{Na} are adsorbed activity coefficients associated with the Vanselow coefficient and each are equal to unity if ideal soil-solution state is met, f_{Ca} and f_{Na} are adsorbed activity coefficients associated with the Gaines and Thomas coefficient and are not equal to unity even if ideal solid–solution state is met.

For the direction and stoichemistry of the exchange reaction given in Eq. [1], the equations for estimating the adsorbed-ion activity coefficients according to Argersinger et al. (1950) approach, which employs the K_V, are

$$-\ln\gamma_{Na} = (1 - E_{Na})\ln K_v - \int_{E_{Na}}^{1}(\ln K_V)dE_{Na} \quad [24]$$

and

$$-1/2\ln\gamma_{Ca} = -E_{Na}\ln K_V + \int_{0}^{E_{Na}}(\ln K_V)dE_{Na} \quad [25]$$

According to the Gaines and Thomas approach (Hutcheson, 1966), the equations are

$$-\ln f_{Na} = (1 - E_{Na})(\ln K_{GT} - 1/2) - \int_{E_{Na}}^{1}(\ln K_V)dE_{Na} \quad [26]$$

and

$$-1/2\ln f_{Ca} = -E_{Na}(\ln K_{GT} - 1/2) + \int_0^{E_{Na}}(\ln K_{GT})dE_{Na} \quad [27]$$

and

$$f_{Na}/\gamma_{Na} = 2/(1 + E_{Na}) \quad [28]$$

To show that Eq. [28] is correct, subtract Eq. [26] from Eq. [24] to get

$$\ln f_{Na} - \ln\gamma_{Na} = (1 - E_{Na})[1/2 - \ln(K_{GT}/K_V) + \int_{E_{Na}}^{1}\ln(K_{GT}/K_V)dE_{Na} \quad [29]$$

Substitute K_V in terms of K_{GT} from Eq. [19] to get

$$\ln f_{Na} - \ln\gamma_{Na} = (1 - E_{Na})[1/2 + \ln 2 - 1/2\ln(1 + E_{Na})] + \int_{E_{Na}}^{1}[1/2\ln(1 + E_{Na}) - \ln 2]dE_{Na} \quad [30]$$

Perform the integration in Eq. [30] and simplify to get

$$\ln f_{Na} - \ln\gamma_{Ca} = \ln 2 - \ln(1 + E_{Na}) \quad [31]$$

Therefore,

$$f_{Na}/\gamma_{Na} = 2/(1 + E_{Na}) \quad [32]$$

Note that the integral $\int\ln(1 + E_{Na})dE_{Na} = (1 + E_{Na})\ln(1 + E_{Na}) + C$, where C is the constant of integration.

To find the ratio of f_{Ca}/γ_{Ca}, proceed as for finding the ratio of f_{Na}/f_{Ca} except begin by subtracting Eq. [27] from Eq. [25]. After simplifying, find

$$\ln f_{Ca} - \ln\gamma_{Ca} = -\ln(1 + E_{Na}) = \ln[1/(1 + E_{Na})]$$

$$f_{Ca}/\gamma_{Ca} = [1 + E_{Na}]^{-1} \quad [33]$$

The authors found that in obtaining Eq. [32] and [33] it is easier to first work through the nonintegral portions before working through the integral portions. Then combine the non-integral and integral portions.

The equation employed to calculate the thermodynamic K_{eqV} (K_{eq}) from the Vanselow coefficient is

$$\ln K_{eqV} = \int_0^1 (\ln K_V)dE_{Na} = \ln K_{eq} \quad [34]$$

Equation [34] is mathematically the mean value theorem that states the mean value or average value v_m of a variable v with respect to the second variable μ for the internal $a < \mu < b$ is defined by the equation

$$v_m = [\int_a^b v d\mu]/[\int_a^b d\mu]$$

where $v_m = \ln K_{eqV}$, $v = \ln K_V$, $\mu = E_{Na}$, $a = 0$, and $b = 1$ [35]

The denominator of Eq. [35] is $\int_0^1 dE_{Na} = 1$. To get the value of $\int_0^1 K_V dE_{Na}$ from experimental data, plot $\ln K_V$ vs. E_{Na} and then graphically evaluate the integral.

The K_{GT} can be used to estimate K_{eqGT} from experimental data by plotting $\ln K_{GT}$ vs. E_{Na} and then graphically evaluating the integral; that is,

$$\ln K_{eqGT} = \int_0^1 \ln K_{GT} dE_{Na} \quad [36]$$

If Eq. [1] is an ideal solid–solution exchange reaction, $K_V = K_{eqV} = K_{eq}$ for the direction and stoichiometry of Eq. [1].

$$\ln K_{eqV} = 1/2 + \ln K_{eqGT} = \ln K_{eq} \quad [37]$$

To show this, write down the interrelationship of K_{GT} and K_V, $K_{GT} = (K_V/2)(1 + E_{Na})^{1/2}$, and take ln of both sides of the equation:

$$\ln K_{GT} = \ln K_V - \ln 2 + (1/2)\ln(1 + E_{Na})$$

Rearrange to get

$$\ln K_V - \ln K_C = \ln 2 - (1/2)\ln(1 + E_{Na})$$

Integrate both sides of the equation with respect to E_{Na} with lower limit equal to zero and the upper limit equal to one.

$$\int_0^1 [\ln K_V - \ln K_{GT}]dE_{Na} = \int[\ln 2 - (1/2)\ln(1 + E_{Na})dE_{Na}$$

$$\int_0^1 (\ln K_V)dE_{Na} - \int_0^1 \ln K_{GT} dE_{Na} = 1/2\int_0^1 [2\ln 2 - \ln(1 + E_{Na})]dE_{Na}$$

Replace the integrals on the left hand side by their equivalents from Eq. [34] and [36] and integrate the right-hand side to get

$$\ln K_{eqV} - \ln K_{eqGT} = 1/2[(2\ln 2)E_{Na} - (1 + E_{Na})\ln(1 + E_{Na}) + E_{Na}\Big]_0^1$$

$$\ln K_{eqV} - \ln K_{eqGT} = 1/2[2\ln 2 - 2\ln 2 + 1]$$

$$\ln K_{eqV} - \ln K_{eqGT} = 1/2 \quad [38]$$

Thus, Eq. [37] is demonstrated.

For the case of two sodium ions replacing one calcium ion, Eq. [38] becomes

$$\ln K_{eqV} = 1/2 + \ln K_{eqGT} = \ln K_{eq}$$

If the direction of the reaction is reversed so that Ca is replacing Na on the exchangers, the +1/2 and +1 become −1/2 and −1, respectively.

REFERENCES

Adams, F. 1971. Ionic concentration and activities in soil solutions. Soil Sci. Soc. Am. Proc. 35:420–426.

Agassi, M., I. Shainberg, and J. Morin. 1981. The effect of electrolyte concentration and soil sodicity on infiltration and crust formation. Soil Sci. Soc. Am. J. 45:848–851.

Allison, J.D., D.S. Brown, and K.J. Novo-Gradac. 1991. MINTEQA2/prodefa2, A geochemical assessment model for environmental systems. EPA 600/3-91/021. USEPA Office of Research and Development, Athens, GA.

Argersinger, W.J., Jr., A.W. Davison, and O.D. Bonner. 1950. Thermodynamics and ion exchange phenomena. Trans. Kansas Acad. Sci. 53:404–410.

Arnold, P.W. 1970. The behavior of potassium in soils. Fertil. Soc. Proc. 115:3–30.

Auboiroux, M., F. Melou, F. Bergaya, and J.C. Touray. 1998. Hard and soft acid–base model applied to bivalent cation selectivity on a 2:1 clay mineral. Clays Clay Miner. 46:546–555.

Babcock, K.L., and R.K. Schultz. 1963. Effect on anion on the sodium–calcium exchange in soils. Soil Sci. Soc. Am. Proc. 27:630–632.

Barshad, I. 1954. Cation exchange in micaceous minerals. II. Replaceability of ammonium and potassium from vermiculite, biotite, and montmorillonite. Soil Sci. 78:57–76.

Beckett, P.H.T. 1964. Studies on soil potassium. I. Confirmation of the ratio law: Measurement of potassium potential. J. Soil Sci. 15:1–8.

Beckett, P.H.T. 1965. The cation-exchange equilibria of calcium and magnesium. Soil Sci. 100:118–122.

Beckett, P.H.T. 1972. Critical cation activity ratios. Adv. Agron. 24:376–412.

Bohn, H., B. McNeal, and G. O'Connor. 1985. Soil chemistry. 2nd ed. John Wiley and Sons, New York.

Bolt, G.H. 1955. Ion adsorption by clays. Soil Sci. 79:267–276.

Davies, C.W. 1962. Ion association. Butterworth, Washington, DC.

Deist, J., and O. Talibudeen. 1967a. Ion exchange in soils from the ion pairs K–Ca, K–Rb, and K–Na. J. Soil Sci. 18:1225–1237.

Deist, J., and O. Talibudeen. 1967b. Thermodynamics of K–Ca exchange in soils. J. Soil Sci. 18:1238–1248.

Eisenman, G. 1962. Cation selective glass electrodes and their mode of operation. Biophys. J. 2:259–323.

El-Swaify, S.A. 1976. Changes in physical properties of soil clays due to precipitated Al and Fe hydroxides. II. Colloidal interaction in the absence of drying. Soil Sci. Soc. Am. Proc. 40:516–520.

El Sayed, M.H., R.G. Burau, and K.L. Babcock. 1970. Thermodynamics of copper (II)–calcium exchange on bentonite clay. Soil Sci. Soc. Am. Proc. 34:397–400.

Erickson, E. 1952. Cation-exchange equilibria on clay minerals. Clay Miner. 74:103–113.

Evangelou, V.P. 1986. The influence of anions on potassium quantity–intensity relationships. Soil Sci. Soc. Am. J. 50:1182–1188.

Evangelou, V.P., and F.J. Coale. 1987. Dependence of the Gapon coefficient on exchangeable sodium for mineralogically different soils. Soil Sci. Soc. Am. J. 51:68–72.

Evangelou, V.P., and R.E. Phillips. 1987. Sensitivity analysis on the comparison between the Gapon and Vanselow exchange coefficients. Soil Sci. Soc. Am. J. 51:1473–1479.

Evangelou, V.P., and R.E. Phillips. 1988. Comparison between the Vanselow and Gapon exchange selectivity coefficients. Soil Sci. Soc. Am. J. 52:379–382.

Evangelou, V.P., and R.E. Phillips. 1989. Theoretical and experimental interrelationships of thermodynamic exchange parameters obtained by the Argersinger and Gaines and Thomas conventions. Soil Sci. 148:311–321.

Evangelou, V.P., and R.L. Blevins. 1988. Effect of long-term tillage systems and nitrogen addition on quantity–intensity relationships (Q/I) in long-term tillage systems. Soil Sci. Soc. Am. J. 52:1047–1054.

Evangelou, V.P., J. Wang, and R.E. Phillips. 1994. New developments and perspectives in characterization of soil potassium by quantity–intensity (Q/I) relationships. Adv. Agron. 52:173–227.

Everett, D.H., and W.I. Whitton. 1952. A general approach to hysteresis. I. Trans. Faraday Soc. 48:749–757.

Faucher, J.A., Jr., and H.C. Thomas. 1954. Adsorption studies on clay minerals. IV. The system montmorillonite cesium-potassium. Ibid. 22:258–261.

Fletcher, I., G. Sposito, and C.S. LeVesque. 1984. Sodium–calcium–magnesium exchange reactions on a montmorillonitic soil: I. Binary exchange reactions. Soil Sci. Soc. Am. J. 48:1016–1021.

Gaines, G.L., and H.C. Thomas. 1953. Adsorption studies on clay minerals. II. A formulation of the thermodynamics of exchange adsorption. J. Chem. Phys. 21:714–718.

Gapon, E.N. 1933. On the theory of exchange adsorption in soils. J. Gen. Chem. (U.S.S.R.) #:144–163. (Abstract in Chem. Abs. 28:4149, 1934).

Gast, R.G. 1972. Alkali metal cation exchange on Chambers montmorillonite. Soil Sci. Soc. Am. Proc. 36:14–19.

Gast, R.G., and W.D. Klobe. 1971. Sodium–lithium exchange equilibria on vermiculite at 25° and 50°C. Clays Clay Miner. 19:311–319.

Gillman, G.P. 1984. Using variable charge characteristics to understand the exchange cation status of toxic soils. Aust. J. Soil Res. 22:71–80.

Goulding, K.W.T. 1983. Adsorbed ion activities and other thermodynamic parameters of ion exchange defined by mole or equivalent fractions. J. Soil Sci. 34:69–74.

Goulding, K.W.T., and O. Talibudeen. 1984. Thermodynamics of K–Ca exchange in soils. II. Effects of mineralogy, residual K and pH in soils from long-term ADAS experiments. J. Soil Sci. 35:409–420.

Greene, R.S.B., A.M. Posner, and J.P. Quirk. 1978. A study of the coagulation of montmorillonite and illite suspension by $CaCl_2$ using electron microscope. p. 35–40. *In* W.W. Emerson et al. (ed.) Modification of soil structure. John Wiley and Sons, New York.

Helfferich, F.J. 1972. Ion exchange. McGraw Hill, New York.

Hunsaker, V.E., and P.F. Pratt. 1971. Calcium–magnesium exchange equilibrium in soils. Soil Sci. Soc. Am. Proc. 35:151–152.

Hutcheon, A.I. 1966. Thermodynamics of cation exchange on clay; Ca-K montmorillonite. J. Soil Sci. 17:339–355.

Jardine, P.M., and D.L. Sparks. 1984a. Potassium–calcium exchange in a multi-reactive soil system. I. Kinetics. Soil Sci. Soc. Am. J. 48:39–45.

Jardine, P.M., and D.L. Sparks. 1984b. Potassium–calcium exchange in a multi-reactive soil system. II. Thermodynamics. Soil Sci. Soc. Am. J. 48:45–50.

Jensen, H.E. 1972. Potassium–calcium exchange on a montmorillonite and a kaolinite clay. I. A test on the Argersinger thermodynamic approach. Agrochimica 17:181–190.

Jensen, H.E., and K.L. Babcock. 1973. Cation exchange equilibria on a Yolo loam. Hilgardia 41:475–487.

Kazman, Z., I. Shainberg, and M. Gal. 1983. Effect of low levels of exchangeable Na and applied phosphogypsum on the infiltration rate of various soils. Soil Sci. 135:184–192.

Keay, J., and A. Wild. 1961. The kinetics of cation exchange in vermiculite. Soil Sci. 92:49–59.

Keren, R., I. Shainberg, and Eva Klein. 1988. Settling and flocculation value of sodium-montmorillonite particles in aqueous media. Soil Sci. Soc. Am. J. 52:76–80.

Laudelout, H., R. van Bladel, G.H. Bolt, and A.L. Page. 1967. Thermodynamics of heterovalent cation exchange reactions in a montmorillonite clay. Trans. Faraday Soc. 64:1477–1488.

Levy, R., and D. Hillel. 1968. Thermodynamic equilibrium constants of sodium–calcium exchange in some Israel soils. Soil Sci. 106:393–398.

Lurtz, J.A., Jr. 1966. Ammonium and potassium fixation and release in selected soils of south eastern United States. Soil Sci. 102:366–372.

Maes, A., P. Peigneur, and A. Creners. 1976. Thermodynamics of transition metal ion exchange in montmorillonite. p. 319–329. *In* Proc. Int. Clay Conf., Mexico City. 16—23 July 1975. Applied Publ. Ltd., Wilmette, IL.

Marsi, M., and V.P. Evangelou. 1991a. Chemical and physical behavior of two Kentucky soils: I. Sodium–calcium exchange. J. Environ. Sci. Health A26:1147–1176.

Marsi, M., and V.P. Evangelou. 1991b. Chemical and physical behavior of two Kentucky soils: II. Saturated hydraulic conductivity–exchangeable sodium relationships. J. Environ. Sci. Health A26:1177–1194.

Marsi, M., and V.P. Evangelou. 1991c. Chemical and physical behavior of two Kentucky soils: III. Saturated hydraulic conductivity–Imhoff cone test relationships. J. Environ. Sci. Health A26:1195–1215.

McBride, M.B. 1989. Surface chemistry of soil minerals. p. 35–160. *In* J.D. Dixon and S.B. Weed (ed.) Minerals in soil environments. SSSA Book Ser.1. 2nd ed. SSSA, Madison, WI.

McBride, M.B. 1994. Environmental chemistry of soils. Oxford University Press, New York.

Mehta, S.C., S.R. Poonia, and R. Pal. 1983. Sodium–calcium and sodium–magnesium exchange equilibria in soil for chloride- and sulfate-dominated systems. Soil Sci. 136:339–346.

Morgun, Y.G., and Y.A. Pachepskiy. 1986. Selectivity of ion exchange sorption in $CaCl_2$–$MgCl_2$–NaCl–H_2O soil system. Soviet Soil Sci. 19:1–10.

Nye, P.H., and P.B. Tinker. 1977. Solute movement in the soil–root system. Univ. of California Press, Berkeley.

Ogwada, R.A., and D.L. Sparks. 1986a. Use of mole or equivalent fractions in determining thermodynamic parameters for potassium exchange in soils. Soil Sci. 141:268–273.

Ogwada, R.A., and D.L. Sparks. 1986b. A critical evaluation on the use of kinetics for determining thermodynamics of ion exchange in soils. Soil Sci. Soc. Am. J. 50:300–305.

Opuwaribo, E., and C.T.I. Odu. 1974. Fixed ammonium in Nigerian soils. I. Selection of a method and amounts of native fixed ammonium. J. Soil Sci. 25:256–264.

Oster, J.D., and G. Sposito. 1980. The Gapon coefficient and the exchangeable sodium percentage–sodium adsorption ratio relation. Soil Sci. Soc. Am. J. 44:258–260.

Parker, D.R. W.A. Novell, and R.L. Chaney. 1994. GEOCHEM-PC: A chemical speciation program for IBM and compatible personal computers. p. 253–269. *In* R.H. Leoppert et al. (ed.) Chemical equilibrium and reaction models. SSSA Spec. Publ. 42. SSSA and ASA, Madison, WI.

Pleysier, J.L., A.S.R. Juo, and A.J. Herbillion. 1979. Ion exchange equilibria involving aluminum in a kaolinitic ultisol. Soil Sci. Soc. Am. J. 43:875–880.

Pratt, P.F., L.D. Whittig, and B.L. Grover. 1962. Effect of pH on the sodium–calcium exchange equilibria in soils. Soil Sci. Soc. Am. Proc. 26:227–230.

Quirk, J.P., and J.H. Chute. 1968. Potassium release from mica-like clay minerals. Trans. Inst. Congr. Soil Sci. 9th 3:671–681.

Rao, T.S., A.L. Page, and N.T. Colemam. 1968. The influence of ionic strength and ion-par formation between alkaline-earth metals and sulfate on Na-divalent cation-exchange equilibria. Soil Sci. Soc. Proc. 32:543–639.

Russo, D., and E. Bresler. 1977. Effect of mixed Na/Ca solution on the hydraulic properties of unsaturated soils. Soil Sci. Soc. Am. J. 41:713–717.

Sadusky, M.C., and D.L. Sparks. 1991. Anionic effects on potassium reactions in variable-charge Atlantic Coastal Plain soils. Soil Sci. Soc. Am. J. 55:371–375.

Salmon, R.C. 1964. Cation exchange reactions. J. Soil Sci. 15:273–283.

Shainberg, I., and J. Letey. 1984. Response of soils to sodic and saline conditions. Hilgardia 52:1–57.

Shainberg, I., J. Oster, and J.D. Wood. 1980. Sodium/calcium exchange in montmorillonite and illite suspensions. Soil Sci. Soc. Am. J. 44:960–964.

Singh, U., and G. Ueharea. 1986. The electrochemistry of the double layer: Principles and applications to soils. p. 1–38. *In* D. L. Sparks (ed.) Soil physical chemistry. CRC Press, Boca Raton, FL.

Smith, J.M., and H.C. Van Ness. 1987. Introduction to chemical engineering thermodynamics. 4th ed. McGraw-Hill, New York.

Sparks, D.L. 1995. Environmental soil chemistry. Academic Press, San Diego, CA.

Sparks, D.L., and P.M. Huang. 1985. Physical chemistry of soil potassium. p. 201–266. *In* R.D. Munson (ed.) Potassium in agriculture. ASA, CSSA, SSSA, Madison, WI.

Sparks, D.L. 1989. Kinetics of soil chemical processes. Academic Press, San Diego, CA.

Sposito, G. 1977. The Gapon and the Vanselow selectivity coefficients. Soil Sci. Soc. Am. J. 41:1205–1206.

Sposito, G. 1981a. Cation exchange in soils: A historical and theoretical perspective. p. 13–30. *In* R.H. Dowdy (ed.) Chemistry in the soil environment. ASA Spec. Publ. 40. ASA and SSSA, Madison, WI.

Sposito, G. 1981b. The thermodynamics of soil solutions. Clarendon Press, Oxford, UK.

Sposito, G. 1984a. The future of an illusion: Ion activities in soil solutions. Soil Sci. Soc. Am. J. 48:451–531.

Sposito, G. 1984b. The surface chemistry of soils. Oxford University Press, New York.

Sposito, G., and C.S. LeVesque. 1985. Sodium–calcium–magnesium exchange on Silver Hill illite. Soil Sci. Soc. Am. J. 49:1153–1159.

Sposito, G., and S.V. Mattigod. 1979. GEOCHEM: A computer program for calculation of chemical equilibria in soil solutions and other natural water systems. The Kearney Foundation of Soil Science, Univ. of California, Riverside, CA.

Sposito, G., and S.V. Mattigod. 1979. Ideal behavior in Na-trace metal cation exchange on Camp Berteau montmorillonite. Clays Clay Miner. 27:125–128.

Sposito, G., C.S. LeVesque, and D. Hesterberg. 1986. Calcium–magnesium exchange on illite in the presence of adsorbed sodium. Soil Sci. Soc. Am. J. 50:905–909.

Sposito, G., K.M. Holtzclaw, L. Charlet, C. Jouany, and A.L. Page. 1983. Sodium–calcium and sodium–magnesium exchange on Wyoming bentonite in perchlorate and chloride background ionic media. Soil Sci. Soc. Am. J. 47:51–56.

Stumm, W., and H. Bilinski. 1973. Trace metals in natural waters: Difficulties of interpretation arising from our ignorance on their speciation. Adv. Water Pollut. Res. 6:39–49.

Suarez, D.L., J.D. Rhoades, R. Lavado, and C.M. Grieve. 1984. Effect of pH on saturated hydraulic conductivity and soil dispersion. Soil Sci. Soc. Am. J. 48:50–55.

Sullivan, P.J. 1977. The principle of hard and soft acids and bases as applied to exchangeable cation selectivity in soils. Soil Sci. 124:117–121.

Talibudeen, O. 1981. Cation exchange in soils. p. 115–177. *In* D.J. Greenland and M.A.B. Hays (ed.) The chemistry of soil and processes. John Wiley and Sons, New York,

Tama K., and S.A. El-Swaify. 1978. Charge, colloidal, and structural stability interrelationships for oxidic soils. p. 40–49. *In* W.W. Emerson et al. (ed.) Modification of soil structure. John Wiley and Sons, New York.

Tanji, K.K. 1969. Solubility of gypsum in aqueous electrolytes as affected by ion association and ionic strengths up to 0.15M and at 25° C. Environ. Sci. Technol. 3:656–661.

Thomas, G.W. 1977. Historical developments in soil chemistry: Ion exchange. Soil Sci. Soc. Am. J. 41:230–238.

U.S. Salinity Laboratory Staff. 1954. Diagnosis and improvement of saline and alkali soils. USDA Agric. Handb. 60. U.S. Gov. Print. Office, Washington, DC.

Udo, E.J. 1978. Thermodynamics of potassium–calcium and magnesium–calcium exchange reactions on a kaolinitic soil clay. Soil Sci. Soc. Am. J. 42:556–560.

Uehara, G., and G.P. Gillman. 1980. Charge characteristics of soils with variable and permanent charge minerals: I. Theory. Soil Sci. Soc. Am. J. 44:250–252.

U.S. Salinity Laboratory Staff. 1954. Diagnosis and improvement of saline and alkali soils. USDA Agric. Handb. 50. U.S. Gov. Print. Office, Washington, DC.

Van Bladel, R., and H.R. Gheyi. 1980. Thermodynamic study of calcium–sodium and calcium–magnesium exchange in calcareous soils. Soil Sci. Soc. Am. J. 44:938–942.

Van Bladel, R., G. Gavira, and H. Laudelout. 1972. A comparison of the thermodynamic, double-layer theory and empirical studies of the Na–Ca exchange equilibria in clay water systems. p. 385–398. *In* Proc. Int. Clay Conf.

van Olphen. 1977. An introduction to clay colloid chemistry. 2nd ed. John Wiley and Sons, New York.

Vanselow, A.P. 1932. Equilibria of the base-exchange reactions of bentonites, permutities, soil colloids, and zeolites. Soil Sci. 33:95–113.

White, G.N., and L.W. Zelazny. 1986. Charge properties of soil colloids. p. 39–81. *In* D.L. Sparks (ed.) Soil physical chemistry. CRC Press, Boca Raton, FL.

Wild, A., and J. Keay. 1964. Cation-exchange equilibria with vermiculite. J. Soil Sci. 15:135–144.

Woodruff, C.M. 1955. The energies of replacement of calcium by potassium in soils. Soil Sci. Soc. Am. Proc. 19:167–171.

Xu, S., and J.B. Harsh. 1990a. Hard and soft acid–base model verified for monovalent cation selectivity. Soil Sci. Soc. Am. J. 54:1596–1601.

Xu, S., and J.B. Harsh. 1990b. Monovalent cation selectivity quantitatively modeled according to Hard/Soft Acid/Base theory. Soil Sci. Soc. Am. J. 54:357–363.

Xu, S., and J.B. Harsh. 1992. Alkali cation selectivity and surface charge of 2:1 clay minerals. Clays Clay Miner. 40:567–574.

Chapter 8

Soil Acidity

PAUL R. BLOOM, *University of Minnesota, St. Paul, Minnesota, USA*

ULF L. SKYLLBERG, *Swedish University of Agricultural Sciences, Umea, Sweden*

MALCOLM E. SUMNER, *University of Georgia, Watkinsville, Georgia, USA*

The pH value is probably the single most important chemical characteristic of a soil. The pH is a "master variable" (McBride, 1994), and knowledge of pH in soil is needed to understand important chemical processes, such as ion mobility, metal ion equilibria, and rate of precipitation and dissolution reactions. Knowledge of soil pH is also needed to understand nutrient availability to plants, toxicity of trace metals, and the negative response of many plant species to soil acidity.

Soils are made up of a complex mixture of solid phase components that can react with additions of acid or base. Reactions of soil components control soil pH and buffer soils against pH changes. Knowledge of the capacity and intensity of soil acidity and pH buffering processes is necessary to understand the response of a soil to liming, acid-forming N fertilizers, acid mine wastes, and acid rain.

In this chapter we will use Brönsted's definition of acids and bases where an acid is a proton (H^+) donor and a base is an proton acceptor. The strength of an acid is defined by its ability to donate protons (Stumm and Morgan, 1996). According to this definition, ionization of an acid form produces a conjugate base, which can accept a proton if the reaction is reversed. Conversely, when a base accepts a proton, a conjugate acid is produced.

MEASURES OF ACIDITY IN SOIL

Soil pH

Because pH is a term that is only defined for solutions , it can only with difficulty be applied to a solid phase–dominated material like soil. However, the pH in water in equilibrium with a soil can easily be measured. Differences in measured pH reflect differences in soil solution (water in the soil pores) pH. The activity of H^+ in the soil greatly affects the equilibrium and kinetics of biotic and abiotic reactions in soils.

Soil pH is commonly determined after equilibrating soil with distilled or deionized water or a salt solution, as in the definition in the *Glossary of Soil Science Terms* (SSSA, 2001):

Chemical Processes in Soils. SSSA Book Series, no. 8.

pH, soil—The pH of a solution in equilibrium with soil. It is determined by means of a glass, quinhydrone, or other suitable electrode or indicator at a specified soil-solution ratio in a specified solution, usually distilled water, 0.01 *M* $CaCl_2$, or 1 *M* KCl.

The soil pH is a measure of the *intensity* of acidity in a soil. In a modern laboratory it is measured using an H^+ ion-specific electrode constructed with a glass membrane that is sensitive to proton activity. Use of pH-sensitive dyes is still common for quick determination of soil pH in the field. For more practical details on how to perform pH measurements in soil see Thomas (1996).

The pH in soil suspensions is greatly affected by the displacement of H^+ and Al^{3+} ions from soil surfaces into solution by other cations (e.g., Ca^{2+}, Mg^{2+}, Na^+, and K^+). The displacement of Al^{3+} affects pH because Al^{3+} ion can hydrolyze to produce H^+ ions. The dilution of soil solution salts generally results in greater measured pH value in distilled water (pH_{H2O}) than in soil solution at natural soil water content. This effect is illustrated by a 0.4 increase in pH obtained for dilution of a soil suspension from 1:1 to 1:10 (w/v) (Thomas, 1996). The measured pH_{H2O} is also influenced by the natural seasonal variation in soil solution salt concentrations (Skyllberg, 1991) and the effects of soil management (e.g., recent fertilization).

To counteract the effect of varying concentrations of salts, an ionic medium such as 0.01 *M* $CaCl_2$ or 1 *M* KCl can be added to the soil prior to pH measurement (Thomas, 1996). The pH values obtained in $CaCl_2$ and KCl are designated as pH_{CaCl2} and pH_{KCl}. Fixing the salt concentration greatly decreases the effect of soil/solution ratio on measured pH (Skyllberg, 1995). Calcium is used in the determination of pH because it is the predominant soil solution cation in most soils of temperate regions. For soils that are dominated by soil organic matter and/or permanently charged clay minerals, the pH values in 0.01 *M* $CaCl_2$ and 1 *M* KCl are substantially lower than measured in water. Lathwell and Peech (1964) reported that for soils of in the state of New York pH_{CaCl2} values are generally 0.5 units lower than pH_{H2O}. More recent research results for 236 soils in Western Australia show that the differences between the two pH measurements is described by: $pH_{CaCl2} = 0.918pH_{H2O} - 0.356$ ($R^2 = 0.94$) (Brennan and Bolland, 1998). This expression yields a difference of 0.69 pH units at pH 4.0, with an increasing difference in measured pH at higher pH. A concentration of 0.01 *M* is somewhat higher than the Ca concentration found in most soil solutions, and thus the pH values obtained by this method are usually lower than pH measured in soil solution extracted by centrifugation (discussed below). Some investigators have advocated using 0.002 *M* $CaCl_2$ to better approximate the concentration of Ca in soils solutions (e.g., Aitken and Moody, 1991). In very acid, highly weathered subsoils with low soil organic matter (SOM) that are dominated by kaolinite clays and hydrous oxides of Al and Fe, pH in $CaCl_2$ and KCl can exceed pH in water (van Raij and Peech, 1972). In these soils the added Cl^- ion will release hydroxyl ions from oxide surfaces and raise pH (McBride, 1994).

The pH of soil solutions can be determined directly if the solution phase is separated from soil. One method is to take field moist soils directly to the laboratory and extract the soil solution by high-speed centrifugation (Giesler and Lundström, 1993). A second method is to collect soil solutions using tension or zero-tension lysimeters installed in the field. Comparison of zero-tension lysimeter data with centrifugation data shows both higher and lower pH for the lysimeter solutions for a Spodosol in a forest in Sweden (Giesler et al., 1996). Lysimetry works best in low pH soils because in higher pH soils, where bicarbonate ions form a significant fraction of the anions, the loss of dissolved CO_2 by degassing in the lysimeter can result in increased solution pH (Suarez, 1987).

Drying and storage can significantly change soil pH values, but it is not easy to predict the results of drying and storage. Bartlett and James (1980) reported a 0.6 unit increase after drying a pH 5.0 mineral soil. However, after liming to pH 6.7, drying produced a 0.5 unit decrease in pH. Slatery and Burnett (1992) reported that 7 yr of storage of soils dried at 40°C resulted in a 0.23 unit increase in pH_{CaCl2} and a 0.55 increase in pH_{H2O}. Van Lierop and Mackenzie (1977) reported an average decrease of 0.5 for pH_{H2O} and 0.2 for pH_{CaCl2} (0.015 *M*) upon drying 10 organic soils.

Quantity of Acidity in Acid Soils

Soils with similar pH can contain vastly different quantities of acidity. The practical effect of this is that the quantity of agricultural lime (ground limestone) needed to yield the same increase in pH may differ greatly for acid soils with similar pH values. Thus, there is a need to determine the quantity of acidity in acid soils. The *Glossary of Soil Science Terms* (SSSA, 2001) defines three measures of the quantity of soil acidity:

> acidity, total—The total acidity including residual and exchangeable acidity. Often it is calculated by subtraction of exchangeable bases from the cation exchange capacity determined by ammonium exchange at pH 7.0. It can be determined directly using pH buffer–salt mixtures (e.g. $BaCl_2$ plus triethanolamine, pH 8.0 or 8.2) and titrating the basicity neutralized after reaction with a soil.
>
> acidity, residual—Soil acidity that is neutralized by lime or a buffered salt solution to raise the pH to a specified value (usually 7.0 or 8.0) but which cannot be replaced by an unbuffered salt solution. It can be calculated by subtraction of salt replaceable acidity from total acidity.
>
> acidity, salt-replaceable—The aluminum and hydrogen that can be replaced from an acid soil by an unbuffered salt solution such as KCl or NaCl.

Total acidity can be determined by the titration of a soil suspension in a salt solution to a reference pH using a strong base. This produces a plot of the pH vs. the quantity of base consumed. However, because of the slow reaction of some of the residual acidity, this is a very slow procedure. A more rapid standard method for the determination of total acidity is to react a soil for several hours or overnight with a solution containing 0.5 *M* $BaCl_2$ plus a triethanolamine (TEA) buffer adjusted to pH 8.2 (Thomas, 1982). Triethanolamine is well buffered at pH 8.2. The Ba^{2+} is included to displace acidity from soil components. A reference pH of 8.2 is chosen to represent the pH attained when a soil is treated with excess ground limestone. From a more fundamental point of view pH 8 is a good choice because at this pH organic carboxyl acidity, as well as the acidity of Al^{3+} bound to clays, has been neutralized. An alternative standard method for calculating total acidity is the determination of the difference between the cation exchange capacity (CEC) determined at pH 7.0 with ammonium acetate and the sum of exchangeable cations (Evangelou and Phillips, 2005, this publication). In this method soil is reacted with an ammonium acetate solution at pH 7.0, the excess ammonium acetate is leached from the soil, and the NH_4^+ ions bound to cation exchange sites are extracted with a salt solution and quantified (Sumner and Miller, 1996). Subtraction of the exchangeable bases in the leachate represents the quantity of acidity that must be neutralized to raise the pH to 7.0. Because total acidity by this method is determined at a lower pH value, the measured acidity is lower than determined by the $BaCl_2$–TEA method.

Total acidity is operationally defined as consisting of two components, salt-replaceable and residual (nonextractable) acidity. Salt-replaceable acidity, also known as extractable or exchangeable acidity, is the H^+ and Al^{3+} replaceable with a concentrated nonbuffered salt solution, usually 1 *M* KCl, while the residual acidity is the acidity that is titratable, but is not easily exchangeable (Bertsch and Bloom, 1996). In the calculation of exchangeable acidity the molar concentration of extracted Al^{3+} is multiplied by 3. In practice the residual acidity is determined by the difference between the total acidity neutralized by raising the pH of a soil to 7.0 or 8.2 and the salt-extractable acidity.

Total Acidity and Cation Exchange Capacity

Total and salt replaceable acidity values are important components of the CEC in acid soils. The *Glossary of Soil Science Terms* (SSSA, 2001) defines CEC as follows:

> cation exchange capacity (CEC)—The sum of exchangeable bases plus total soil acidity at a specific pH, values, usually 7.0 or 8.0. When acidity is expressed as salt-extractable acidity, the cation exchange capacity is called the effective cation exchange capacity (ECEC) because this is considered to be the CEC of the exchanger at the native pH value. It is usually expressed in centimoles of charge per kilogram of exchanger ($cmol_c\ kg^{-1}$) or millimoles of charge per kilogram of exchanger.

For acid soils, CEC at pH 7 (CEC_7) and at pH 8.2 (CEC_8) include the pH-dependent charges that are created by neutralizing weak acid sites, and these CEC values represent the potential cation exchange sites that can be created with liming. In acid soils high in pH dependent charge components like SOM and oxides and hydroxides of Al and Fe(III), CEC_7 and CEC_8 greatly exceed ECEC.

SOIL ACIDIFICATION

Many processes in soils produce acidity, and acidification is a natural process in all nonacid soils in which leaching occurs. In addition, human activities can have a very significant impact on soil acidification; for example, fertilizer amendments can produce acidity. Acid rain and other pollutants can also acidify soils. Soil acidification occurs when there is a net donation of protons to soil components with a loss of bases by leaching or removal by harvest of plant materials. This can be expressed as a decrease in the acid neutralization capacity (ANC) of soils (van Breemen et al. 1983).

Natural Acidification Processes

Natural acidification of soils under most conditions is a result of processes linked to plants and their ability to assimilate carbon dioxide. The proton transfer reactions that occur in a soil with growing plants and decaying plant debris are shown in Table 8–1. Of these reactions the most important for natural acidification of soils are the production of CO_2 by microbial and root respiration and the production of organic acids by plants and soil microbes.

When CO_2 in is produced by respiration it can to dissolve to produce carbonic acid (Stumm and Morgan, 1996):

$$CO_2(g) + H_2O(l) = H_2CO_3^* \qquad \log K = -1.51 \qquad [1]$$

where the units of K are in moles per liter atmospheres. The dissolved CO_2 in Eq. [1] is designated as $H_2CO_3^*$ because this so-called carbonic acid, is really mostly hydrated molecular CO_2 and not molecular H_2CO_3.

In nonacid and weakly acid soils carbonic acid ionizes to produce bicarbonate and H^+ (Stumm and Morgan, 1996):

$$H_2CO_3^* = HCO_3^- + H^+ \qquad \log K = -6.55 \qquad [2]$$

Because of respiration, the soil atmosphere has a partial pressure of $CO_2(g)$ well above the value of 0.00038 atm in the earth atmosphere. In well-drained mineral forest soils P_{CO2} normally varies between 0.002 and 0.001 atm (Fernandez and Kosian, 1987; Magnusson, 1992). In agricultural prairie soils, Inskeep and Bloom

Table 8–1. Schematic H^+ transfer reactions in soils.†

Acidification (process from left to right)	Reaction	Proton consumption (process from right to left)
Respiration and CO_2 dissociation	$CH_2O + O_2 \rightarrow CO_2 + H_2O = H_2CO_3^*$ $H_2CO_3^* = HCO_3^- + H^+$	NA
Organic acid production and dissociation	$RCH_3 + 3/2O_2 \rightarrow RCOOH + H_2O$ $RCOOH = RCOO^- + H^+$	NA
Plant uptake and assimilation of NH_4–N	$ROH + NH_4^+ \leftrightarrow RNH_2 + H_2O + H^+$	Ammonification
Nitrification	$NH_4^+ + 2O_2 \rightarrow NO_3^- + 2H^+ + H_2O$	NA
Microbial oxidation of amino N in SOM	$RNH_2 + 2CO_2 + 2H_2O \leftrightarrow ROH + NO_3^- + H^+ + 2CH_2O$	Plant uptake and assimilation of NO_3–N
Plant uptake of Me^{n+} and complex formation internal to plants	$nRCOOH + Me^{n+} \leftrightarrow (RCOO)_nMe + nH^+$	Release of metals from plant debris and SOM
NA	$CO_2 + H_2O + Me^{n+} \leftarrow (RCOO)_nMe + nH^+$	Microbial oxidation of metal complexes
Plant uptake and assimilation of SO_4–S	$ROH + SO_4^{2-} + 2H^+ + 2CH_2O \leftrightarrow RSH + 2CO_2 + 3H_2O$	Microbial oxidation of thiol S in SOM

† The double arrow (↔) means that the reaction occurs in both directions but does not mean it is reversible. The forward and reverse reactions occur by different mechanisms. The = sign is used to designate reversible reactions. NA = Not Applicable. Table is based on van Breemen et al. (1983) and De Vries and Breeuwsma (1987).

(1986) found that filling of soil pores with water greatly decreased CO_2 exchange with the air and increased P_{CO2}. In pots with soybean [*Glycine max* (L.) Merr.] plants the P_{CO2} ranged from 0.0006 atm in drier soils to 0.031 atm at moisture contents approaching saturation. In waterlogged soils P_{CO2} can be much higher (Magnusson, 1994; Inskeep and Bloom, 1986). In reality, P_{CO2} is never constant and varies substantially within the soil profile and over the year (Magnusson, 1992, 1994).

Under conditions of good drainage and rainfall sufficient to produce leaching, CO_2(g) is the predominant acidifying agent when soil pH is greater than approximately 5.0 to 5.5. Carbonic acid can donate protons (Eq. [1]) to CEC sites releasing base cations (e.g., Ca^{2+}) and bicarbonate ions become charge balancing anions in the leachate. As pH decreases to lower values, ionization of H_2CO_3* occurs to only a minor extent and CO_2(g) is no longer an effective proton donor. In calcareous soils, soils with $CaCO_3$(s), H_2CO_3* donates protons, to $CaCO_3$(s) resulting in dissolution to form calcium bicarbonate in solution. As long as solid phase $CaCO_3$(s) is present, pH is maintained. In noncalcareous soils, the very slow weathering of primary silicate minerals can partially counteract acidification by H_2CO_3*. These processes are discussed from a pH buffering perspective in the section on pH buffering below.

At pH values <5.0 to 5.5 organic acids are more important in soil acidification than CO_2(g). Organic acids in soils originate from plants, either as litter, degradation products of litter, or as exudates from plant roots. Some of these acids are low in molecular weight and are highly soluble whereas others are low solubility macromolecular components formed through humification processes in soil. These acids contain weak acid carboxyl and phenolic groups. Carboxylic acid functional groups have pK_a values in the range of 2 to 7, and phenolic acid functional groups have pK_a values in the range of 6 to 10. Organic acids exuded from plants and microbes are generally low-molecular weight mono-, di-, and tricarboxylic acids, with pK_a values in the range of 3 to 5 (Strobel et al., 1999; van Hees et al., 1999). Dissolved organic acids can donate protons to CEC sites replacing base cations, and organic anions (conjugate bases of the organic acids) become charge balancing anions for base cations in leachate. Because organic acids have lower pK_a values than H_2CO_3*, the acidification by organic acids can ultimately result in lower pH values than acidification by carbonic acid. Organic acids can lower soil pH to <4, which is the case in many acidic topsoils in forests, such as O horizons of Spodosols and Inceptisols, as well as in acidic peat soils. Low molecular weight acids generally have a short lifetime in soils because they can be consumed by soil microbes and oxidized to CO_2(g) and water. Oxidation of organic acids to CO_2(g) removes the potential for proton donations except by the dissolution of the CO_2(g) produced by oxidation.

In addition to the acidification by CO_2(g) and organic acids there are other reactions in the soil–plant system that produce and consume protons (Table 8–1). However, these additional reactions have little net long-term effect in a natural system where no soil amendments are added and where plant material is not harvested. When plants are harvested the natural cycle is broken and soil pH changes can occur. For example, in forests, long-term tree harvest acidifies soils. When a plant root takes up anions and cations either bicarbonate or protons are excreted in order to maintain the charge balance in the plant. In Table 8–1 the release of alkalinity in the form of bicarbonate ions is described as the consumption of a proton. The acidity produced at the surface of roots can be calculated as the difference between uptake of cation and anion charges. Trees in forests generally take up more cations (mostly base cations and NH_4^+–N) than anions (mostly nitrate and sulfate). Thus, the effect near the surface of roots is soil acidification. Decay of forest litter, however, reverses this acidification by producing a biocarbonate anion. In contrast, anion uptake causes a release of alkalinity, which is reversed during plant decay. Root uptake of a sulfate anion will result in the efflux of two bicarbonate anions, but when dead plant material decays two protons are produced by the oxidation of plant S to SO_4^-. For NO_3^-–N, plant uptake results in the release of one proton and decomposition of plant debris, with mineralization of organic amino N to ammonium and subsequent oxidation of N to nitric acid, results in the release of one bicarbonate anion.

Acid-Producing Fertilizers

Nitrogen fertilizers that produce NH_4^+ in soils (ammonium sulfate, ammonium nitrate, anhydrous ammonia, and urea) acidify soils. In soil, bacteria oxidize NH_4^+ to NO_3^-, producing acidity (Tisdale et al., 1985):

$$NH_4^+ + 2O_2(g) \rightarrow 2H^+ + NO_3^- + 2H_2O \qquad [3]$$

This reaction can produce sufficient acidification to require periodic addition of lime to cropland (Adams, 1984).

Acidification by Sulfur Compounds

Sulfuric acid produced in acid mine wastes or when the soils of coastal swamps are drained can have a large impact on soils. When pyrite, $Fe(II)S_2(s)$, or iron sulfide, $Fe(II)S(s)$, associated with a seam of coal or a coastal organic soil is exposed to air, bacterial oxidation results in the production of sulfuric acid (Sanchez, 1976):

$$\underset{\text{pyrite}}{FeS_2(s)} + 15/4O_2(g) + 7/2H_2O \rightarrow Fe(OH)_3 + 2H_2SO_4 \qquad [4]$$

Under these conditions soil pH values of <3 are not uncommon. This reaction has produced large amounts of acid release after drainage and cultivation of acid sulfate soils in Vietnam, Malaysia, and Indonesia (Sanchez, 1976).

Acidification of nonacid soil to enhance the growth of acidophilic plants (plants adapted to acid soils) can be accomplished using elemental S. Bacterial oxidation causes a similar effect as for the oxidation of pyrite:

$$S + 3/2O_2(g) + H_2O \rightarrow H_2SO_4 \qquad [5]$$

Growers using elemental S must be very careful because pH values of 3.5 are easily attainable when adding excess elemental S to mineral soils.

An alternative method is to add alum, $KAl(SO_4)_2$, which produces acidity according to the reaction:

$$2KAl(SO_4)_2(s) + 6H_2O \rightarrow 2Al(OH)_3(s) + 3H_2SO_4 + K_2SO_4 \qquad [6]$$

Acidification by alum in mineral soils is limited by the solubility of $Al(OH)_3$, and the pH will not be decreased below about 4. This makes this more expensive option preferable for applications to very small areas.

pH BUFFERING IN SOILS AND THE CONTRIBUTION OF SOIL COMPONENTS TO TOTAL ACIDITY IN ACID SOILS

Overview of pH Buffering Processes in Soils

Soils are complex weak acid–weak base systems that resist pH changes upon addition of acid or base. This is illustrated by the data of Magdoff and Bartlett (1985) for acid and lime additions to soil (Fig. 8–1 and 8–2). Magdoff and Bartlett investigated the pH buffering in 51 surface soils of Vermont by reacting these soils with H_2SO_4 or $CaCO_3$ for 30 d and determining the final pH in 0.01 *M* $CaCl_2$. Figure 8–1 presents the measured data for 4 of the soils. Figure 8–2 is a plot of all of the data with zero addition points transposed along the *x* axis to correspond to the plot of a reference soil. The acid or lime added is reported per unit organic matter. Figures 8–1 and 8–2 can be considered acid–base titration plots for these soils and the pH buffer intensity of the soils at any given point is inversely proportional to the slope of the plot at any point (Stumm and Morgan,1996). The data in Fig. 8–2 show that the minimal buffer intensity of these soils is at about pH 5.5 (maximal slope)

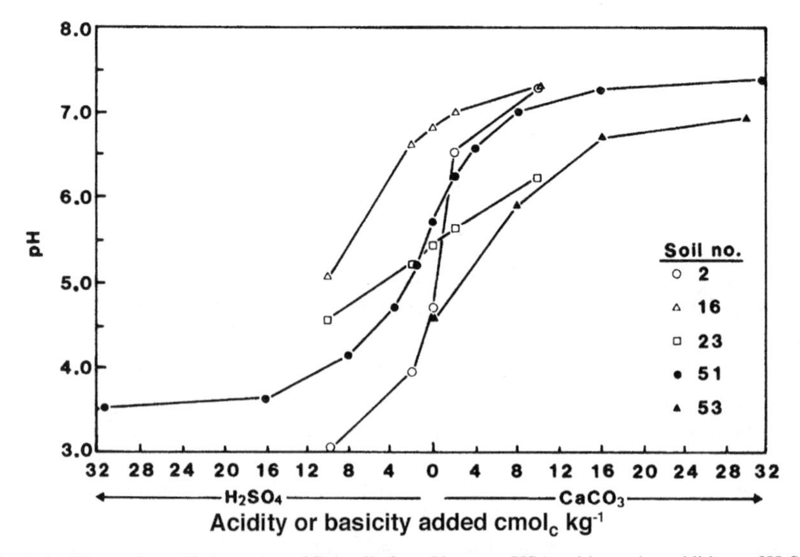

Fig. 8–1. pH_{CaCl2} after a 30-d reaction of five soils from Vermont, USA, with varying additions of H_2SO_4 and $CaCO_3$ (Magdoff and Bartlett, 1985).

while maximal buffering occurs at pH <3.5 and pH >7.5. In the pH range of 4.5 to 6.5, a range that encompasses most acid agricultural soils and many forest soils, the buffer plots are approximately linear.

The pH buffering reactions in soils include proton desorption and adsorption reactions on minerals and soil organic matter as well as ion exchange, dissolution, and precipitation reactions. Some soil components are effective in buffering over a wide range of pH values, whereas others are effective over limited pH range (Table 8–2). The buffer ranges in Table 8–2 are somewhat similar to those suggested by Ulrich (1991) with major modifications and additions. For example, Ulrich did not include soil organic matter and surface reactions of oxides and hydrous oxides.

Most of the reactions in Table 8–2 are reversible for proton uptake or proton release. Other reactions, like the decomposition and weathering of primary silicates, are irreversible. The rates of many proton adsorption and desorption reactions vary from almost instantaneous to very slow. Thus, the time chosen for equilibration significantly affects the laboratory determination of pH buffer curves. In calcareous soils, pH is buffered by the dissolution and precipitation of $CaCO_3(s)$, calcite, by CO_2 according to

$$CaCO_3(s) + CO_2(g) + H_2O(l) = 2HCO_3^- + Ca^{2+} \quad [7]$$

Under conditions of CO_2 weathering, the net reaction is dissolution of calcite and leaching of Ca^{2+} and HCO_3^- ions. The pH buffer range for soil systems in equilibrium with $CaCO_3(s)$ varies considerably. A $CaCO_3(s)$ suspension in distilled water in equilibrium with the ambient atmosphere will contain two bicarbonate ions in solution for every Ca^{2+} ion and the pH will be 8.3 (Stumm and Morgan, 1996). However, higher P_{CO2} in soils can result in pH values of <7.5. If a calcareous soil has a

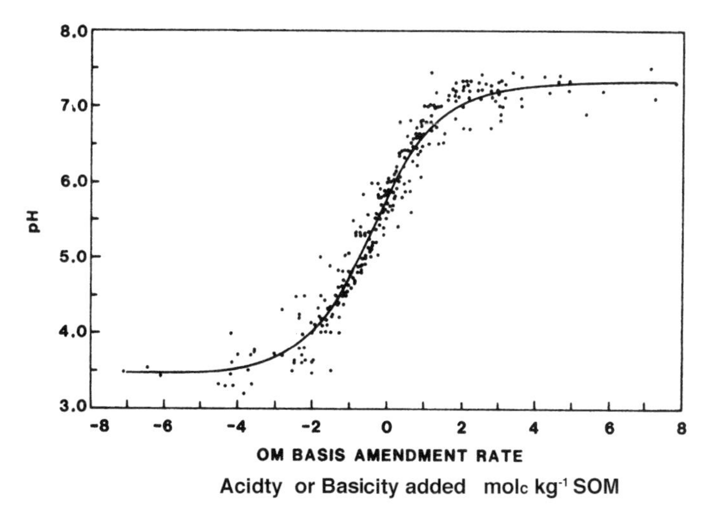

Fig. 8–2. pH_{CaCl2} after a 30-d reaction of 51 soils from Vermont, USA, with varying additions of H_2SO_4 and $CaCO_3$. The data for all of the soils were adjusted along the *x* axis to coincide with the zero addition pH for Soil 51 (see Fig. 1) and recalculated per kilogram SOM (Magdoff and Bartlett, 1985).

Table 8–2. pH-buffering reactions in soils with a pH range of 3.0 to 10.0.

pH buffer substance	pH range	Proton accepting or donating reactions
Limestone, $CaCO_3$	7–9.5	Dissolution and precipitation
Oxides and hydroxides of Fe and Al; silicate clay edges	whole pH range	H^+ adsorption and desorption on surface hydroxyl sites
H^+–SOM	whole pH range	Dissociation and protonation of carboxyl and phenol groups
Al-SOM(s) and $Al(OH)_3$(s)	5–8	Precipitation of organic bound Al^{3+} as $Al(OH)_3$ or dissolution of $Al(OH)_3$ by organic acids
Al^{3+}/H^+ exchange in SOM	<4.5	H^+ exchange with Al^{3+} on carboxyl and phenol groups
$Al(OH)_3$(s)	4–5.5	$Al(OH)_3$(s) dissolution and precipitation in soils with low SOM content
Silicate clay interlayer $Al(OH)_n^{3-n}$ in 2:1 clays	4.2–7	Hydrolysis and precipitation or dissolution of interlayer $Al(OH)_3$(s)
Permanent charge silicate clays	3.5–4.2	Ion exchange of H^+ and Al^{3+}
Irreversible dissolution of high activity 2:1 silicate clays and poorly ordered alumino-silicates	<3.5	Consumption of H^+ upon release of Al^{3+} from reactive silicates
Very slow irreversible weathering of primary silicate minerals	whole pH range	Consumption of H^+ upon dissolution of Ca^{2+}, Mg^{2+}, K^+, and Na^+ from primary minerals

source of HCO_3^- in addition to $CaCO_3$(s), such as Mg and Na bicarbonates, the pH of a calcareous soil can be >9.

Calcium carbonate is, per unit mass, the most effective pH buffer found in soils (Table 8–3). One kilogram of $CaCO_3$(s) can neutralize 2000 cmol of H^+ and buffer the pH at a value >7 until all of the $CaCO_3$(s) is dissolved. The buffer plot in Fig. 8–2 shows that when excess $CaCO_3$(s) was added the soils in the laboratory the buffering was infinite (slope of the plot = 0) at a pH value of 7.3.

Buffering by Weak Acid–Weak Base Systems

Reactions Involving H^+/Al^{3+} Exchange and Dissolution of Al in 2:1 Clays

Reactions involving exchangeable Al^{3+} and H^+ on permanent charge sites in silicate clays can be important in pH buffering. Hydrated protons, hydronium ions (H_3O^+), can occupy exchange sites on permanent charge clays: however, H^+–saturated high activity aluminosilicate clays—smectites, vermiculites, and illites—are not stable and readily decompose, producing Al^{3+} ions. The Al^{3+} ions have a much greater affinity for exchange sites than hydronium ions. When montmorillonite, a smectite, is saturated with H^+ the pH is <3.5 (Coleman and Craig, 1961). After 24 h of aging, acid Al^{3+} replaces the exchangeable H^+ ions (Coleman and Craig, 1961). Extraction of very acid mineral soils with neutral salts solutions (e.g., 1 *M* KCl) yields mostly Al^{3+} and not H^+ (Thomas and Hargrove, 1984).

If hydronium ions were the only acidic cations on permanent charge sites the role of permanent charge clays in pH buffering in acid soils could be described using ion exchange equations (Evangelou and Phillips, 2005, this publication). The chemistry of Al clays, however, is more complex. Aluminum clays are acidic, yielding

Table 8–3. Approximate maximum reversible proton donation or adsorption capacity of soil mineral phases. The estimations are based on either the base needed to raise the pH from 3.5 to 8.3 or the acid needed to lower the pH from 8.3 to 3.5. The silicate clays are assumed to be Al^{3+} saturated at pH 3.5.

Soil material	Capacity	References
	$cmol_c\ kg^{-1}$	
$CaCO_3$	2000	Stumm and Morgan, 1996
Silicate clay ion exchange reactions		
Smectites	80–150	McBride, 1994
Vermiculite	150–200	McBride, 1994
Illite	20–40	McBride, 1964
Kaolinite	1–5	Thomas and Hargrove, 1984
Allophane and immogolite ion exchange reactions	20–50	Wada, 1989
Proton adsorption and desorption on hydroxides and oxides of Fe and Al	4–40	Borggaard, 1983

H^+ ions when Al ions are displaced from clays by Ca^{2+}, K^+, or other cations. The production of H^+ ions is due to hydrolysis of Al^{3+} in solution and/or the precipitation of $Al(OH)_3(s)$. The hydrolysis reactions of Al^{3+} are (Nordstrom and May, 1996)

$$Al^{3+} + H_2O(l) = AlOH^{2+} + H^+ \qquad \log K_1 = -5.00 \qquad [8]$$

$$AlOH^{2+} + H_2O(l) = Al(OH)_2^+ + H^+ \qquad \log K_2 = -5.1 \qquad [9]$$

$$Al(OH)_2^+ + H_2O(l) = Al(OH)_3^0 + H^+ \qquad \log K_3 = -6.7 \qquad [10]$$

$$Al(OH)_3^0 + H_2O(l) = Al(OH)_4^- + H^+ \qquad \log K_4 = -6.2 \qquad [11]$$

Soluble Al^{3+} is a weak acid, and the pH in a suspension of Al^{3+} saturated clay is greater than in a suspension of H^+ saturated clay. The reaction for the precipitation of $Al(OH)_3(s)$ yields protons according to the equilibrium solubility reaction for $Al(OH)_3(s)$, which given by

$$Al(OH)_3(s) + 3H^+ = Al^{3+} + 3H_2O(l) \qquad \log K_{soil\ Al(OH)3} = 8.3 \qquad [12]$$

The log$K$ for Eq. [12] ranges from 7.74 for well crystalline gibbsite at 25°C (Palmer and Wesolowski, 1992) to 9.5 for poorly ordered amorphous $Al(OH)_3(s)$ (Bruggenwert et al., 1991). Gustafsson et al. (1998) reported a value for "soil $Al(OH)_3(s)$" of 8.3. Combining the solubility equilibrium and the hydrolysis reactions yields log $Al(OH)_x^{3-x}$ vs. pH activity plots that show the effect of pH on Al solubility in equilibrium with $Al(OH)_3(s)$ (Fig. 8–3).

In addition to the monomeric hydrolysis species shown by Eq. [8] to [11], Al can form soluble polymeric hydroxy species. Polymeric species are produced when NaOH is added slowly to an Al^{3+} solution (Bertsch and Parker, 1996). The predominant polymer is $AlO_4Al_{12}(OH)_{24}(H_2O)_{12}^{7+}$, which is often referred to as Al_{13}. To produce Al_{13}, sufficient OH^- must be added to yield metastable solutions that are highly oversaturated with respect to precipitation of $Al(OH)_3(s)$ (Bertsch and Parker, 1996). Although there is one report of the of Al_{13} in a soil solution of for-

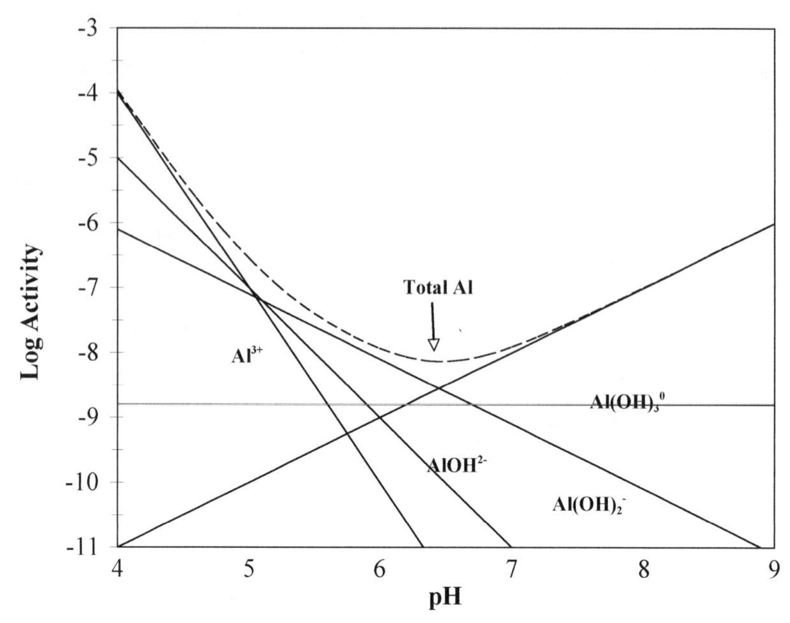

Fig. 8–3. Activities of dissolved Al^{3+} and Al monomeric hydrolysis species in equilibrium with a soil $Al(OH)_3(s)$ phase having a log K of 8.0 for dissolution.

est soil (Hunter and Ross, 1991), it is unclear whether conditions for formation of these species commonly occurs in soils.

The capacity of permanent charge clays to buffer pH varies widely because of the large difference in the CEC of clays (Table 8–3). Kaolinite, a low activity clay, has little or no permanent charge and has little capacity to retain Al^{3+}. Illite has a much greater CEC than kaolinite and is a much better buffer of pH. Smectites have an even greater capacity to buffer pH.

Precipitation of $Al(OH)_3(s)$ in 2:1 Layer Silicate Clay Suspensions

Most of the buffering that occurs during the addition of a strong base to Al-smectite clay suspensions can be explained by the precipitation of $Al(OH)_3(s)$ in clay interlayers (Bruggenwert et al., 1991). When Bruggenwert et al. (1991) hydrolyzed Al^{3+} solutions by adding NaOH to produce solutions with polymeric hydroxy-Al and then added Na-montmorillonite, they found that the pH increased due to the adsorption of Al^{3+} on the clay exchange sites and depolymerizaiton of Al hydroxy polymers. They found that in the presence of clay a hydrolysis model involving only the precipitation of $Al(OH)_3(s)$, ion exchange of Al^{3+} with Na^+, and the solution hydrolysis of Al^{3+} to $AlOH^{2+}$ (Eq. [8]) was sufficient to explain their observations. The solubility of $Al(OH)_3(s)$ precipitated in smectite interlayers at room temperature (22°C) is defined by a log$K$ of 8.7 (Bloom et al., 1977). Correcting this value to 25°C yields a log$K$ of about 8.5 (Gustafsson et al., 2001), similar to the soil $Al(OH)_3(s)$ solubility in Eq. [12].

The model of Bruggenwert et al. (1991) shows that clay exchange sites adsorb Al^{3+}, decreasing the potential for the formation of $Al(OH)_3(s)$. Titration of $AlCl_3$

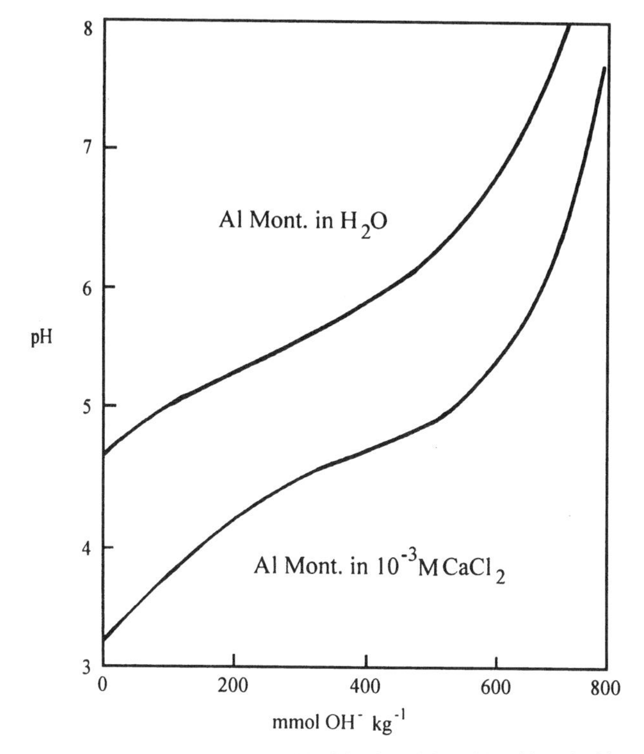

Fig. 8–4. Titration of an Al saturated montmorillonitic clay with NaOH with and without the addition 0.001 *M* $CaCl_2$ (Turner and Nichol, 1962).

with strong base in the presence of a base cation saturated clay will result in a titration curve that is buffered by $Al(OH)_3$ precipitation at a higher pH than if the clay were not present. In addition to adsorption of Al^{3+}, clay provides surface sites for $Al(OH)_3$ to precipitate when base is added. In Al-clay suspensions, addition of base cation salts increases displacement of Al^{3+} into solution and increases the potential for precipitation when strong base is added. This is clearly shown by the data of Turner and Nichol (1962) for titration of an Al^{3+} saturated montmorillonite clay with NaOH, with and without the addition of Ca^{2+} (Fig. 8–4). Without Ca^{2+} maximal buffering is at pH 5.5 and in 0.001 *M* $CaCl_2$ the maximal buffering is at pH 4.5. For Al-clay with only partial saturation with Al^{3+} the pH of maximal buffering would be greater than shown in Fig. 8–4.

Hydroxy-interlayered smectite (HIS) and hydroxy-interlayered vermiculite (HIV) are commonly found in acid soils, and under the proper conditions of OH/Al molar ratios, well-ordered $Al(OH)_3(s)$ interlayered smectite and vermiculite can readily be formed in the laboratory (Jardine and Zelazny, 1996). While hydroxy interlayers can have an impact on pH buffering over a wide pH range (Fig. 8–4 and Table 8–2), the native pH values in soils with well-developed HIV and HIS are restricted to values in the range of about 4.2 to 5.5. Karathanasis et al. (1983) reported soil solution pH values of 4.15 to 4.99 for 10 Ultisol Ap horizons with 38 to 57% of the clay fraction consisting of HIV. Lee et al. (1985) reported pH values of 4.5

to 5.0 in1.0 *M* Ba, Mg, K, and ammonium chloride leachates from AE, E, Bh, and Bw horizons of three Inceptisols and one Spodosol containing 25 to 69% HIV.

The formation and dissolution of $Al(OH)_3(s)$, independent of whether $Al(OH)_3$ is formed in clay interlayers or as a separate phase, is slow compared with the rates of Al^{3+} ion exchange. Therefore equilibrium in Al-clay or acid soil suspensions may not be reached during laboratory determinations of pH buffering. The dissolution of well crystalline gibbsite is very slow (Bloom and Erich, 1987), and the precipitation of the $Al(OH)_3(s)$ in the presence of clay is also slow. Schwertmann and Jackson (1963) reported that the titration of aged H^+ saturated montmorillonite (actually Al^{3+} saturated) has three different pH buffer ranges. The reactions at pH values of 5.5 to 7.6, which involves interlayer $Al(OH)_3(s)$ precipitation, are slow.

Reversible Charge Mineral Surfaces

Aluminosilicate clay edges and the surfaces of oxides and hydroxides of Al and Fe(III) provide reversible charge sites for the adsorption and desorption protons; reactions that can buffer over a wide range of pH values. These surface reactions are

$$S\text{-OH(s)} + H^+ = S\text{-OH}_2^+\text{(s)} \qquad K_1 \qquad [13]$$

$$S\text{-OH(s)} = S\text{-O}^-\text{(s)} + H^+ \qquad K_2 \qquad [14]$$

where *S*-represents Al or Fe on the outermost layer of a crystal (Fig. 8–4). The resulting surface charge is reversible. For oxides and hydroxides of Fe(III) $\log K_1 \approx 6.55$ and $\log K_2 \approx -9$ (McBride, 1994). An increase in pH results in a more negative net charge and a decrease in pH in a more positive net charge, as shown for the titration of hematite in Fig. 8–5. These surfaces are amphoteric and can be both acidic or basic, depending on pH. The proton adsorption and desorption reactions are rapid (Zettlemoyer and McCafferty, 1973). The most important reversible charged surfaces in soils are Al- and Fe oxides and hydroxides, kaolinite and short-range ordered Al-silicates. The latter includes allophane, imogolite, and protoimogolite.

Equations [13] and [14] suggest that it might be possible to model proton adsorption and desorption using a simple weak acid equilibrium model. However, the accumulation of charge on a surface results in an electrical potential difference between the surface and the solution, and this can greatly affect the surface ionization. This can be described by models that include a surface potential term (Goldberg, 2005, this publication). Because salt solutions can provide a shielding of the surface and decrease the surface potential, the desorption to produce a negative charge site at high pH or the protonation to form positive charge sites at low pH is increased with increasing salt concentrations. The titration of hematite $Fe_2O_3(s)$ in KNO_3 (Fig. 8–5) shows that the effect of higher ionic strength is to increase the buffer capacity. Note that the axes for pH and acid or base addition in Fig. 8–4 are reversed compared with the soil buffering plots shown in Fig. 8–1 and 8–2.

The point of zero net proton charge (pznpc) is an important pH value in the titration of reversible charge surfaces (Table 8–4). This is the pH at which proton

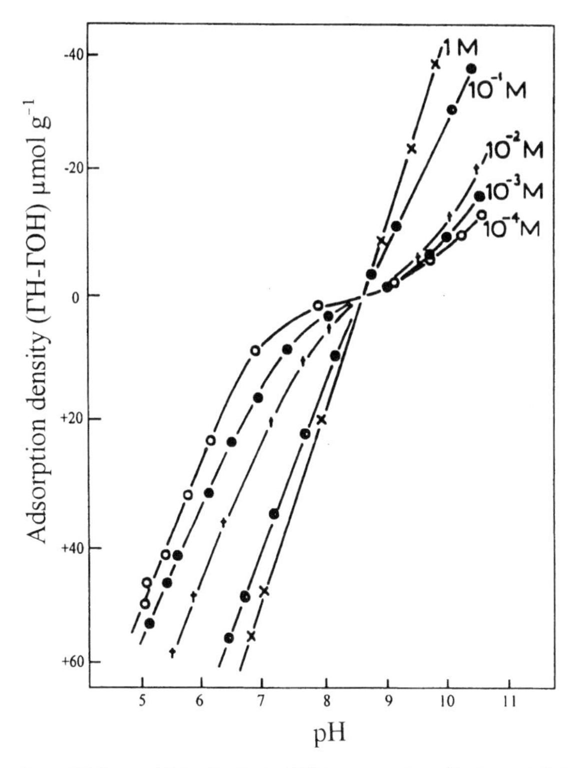

Fig. 8–5. Adsorption of OH^- and H^+ by Fe_2O_3 in KNO_3 suspensions (Parks and Debryun, 1962).

adsorption equals proton desorption, and there is no net surface proton charge. In Fig. 8–5 this point is represented by the point of intersection of titration plots in at varying ion strength of an indifferent electrolyte.[1] At this pH value the salt concentration has no effect on pH if surface charge is due to proton adsorption and desorption alone. The pznpc is the mean of the acidity constants pK_{a1} and pK_{a2} of Reactions [13] and [14], respectively.

$$pH_{pznpc} = 0.5(\log K_1 - \log K_2) \quad [15]$$

Specifically adsorbed cations (e.g., transition metal ions and Al^{3+}) and anions (e.g., phosphate and organic weak acids) can greatly affect proton adsorption and desorption on reversible charge surfaces. Strongly held cations shift the apparent pznpc to higher pH values, whereas specifically bound anions shift the apparent pznpc to a lower pH (Parks, 1967).

Titration of soil suspensions in a range of concentrations of NaCl or KCl, as shown in Fig. 8–5 for hematite, has been used to determine both the variation in surface charge with pH and pznpc values of soils that contain pH-dependent charge minerals (Sposito, 1992). For soils it is expected that SOM and permanent charge clays will shift the pznpc to lower pH values than for the titration of pure variable

[1] Salt ions that do not strongly adsorb on an oxide or hydroxide surface.

Table 8–4. Estimates of pznpc, cation exchange capacity (CEC), and anion exchange capacity (AEC) for reversible charge surfaces.

Mineral	pznpc	CEC (pH 7.0)/AEC (pH 4.0)	References‡
		$cmol_c\ kg^{-1}$	
$Al(OH)_3$ and am-$Al(OH)_3$	8.2–9.5	na†	1, 2, 3, 4, 5
$Fe(OH)_3$	7–9	na	6
Allophane	5.5–6.9	10–40/5–30	7, 8, 9, 10
Imogolite	na	19/37	9
Kaolinite	4.0–5.0	0–10	11, 12

† not available

‡ 1, Rakotonarivo et al. (1988); 2, Anderson et al. (1976); 3, Anderson and Malotky (1979); 4, Goldberg and Glaubig (1987); 5, Anderson and Benjamin (1990); 6, Borggaard (1983); 7, Perrott (1977); 8, Gonzales-Batista et al. (1982); 9, Theng et al. (1982); 10, Clark and McBride (1984); 11, Lim et al. (1980); 12, Sposito (1989).

charge minerals. (The contribution of SOM to soil proton charges will be discussed below in H^+ and Al^{3+} Adsorbed to Soil Organic Matter.) The shift is expected because SOM has only pH-dependent negative charge and little or no potential to develop positive charge, and permanent charge clays have high negative charge with only a small amount of pH-dependent positive charge on the edges. Determination of pznpc and magnitude of charge variation with pH by acid–base titration of soil suspensions is complicated at low pH by the release of Al^{3+} from SOM sites without generating positive surface charges. Three protons can displace one Al^{3+} without resulting in the development of positive charge.

Bloom (1979) titrated Al-substituted SOM in 0.01, 0.10, and 1.0 *M* NaCl suspensions and found an apparent pznpc of 3.9. At pH <3.9, Al^{3+} release increases with increased Na^+ concentration, and at pH >3.9, the ionization of weak acid protons and precipitation of organic Al to form $Al(OH)_3$ is also increased with increasing Na^+ concentration. This caused the titration plots to intersect at pH 3.9 and resemble the plots in Fig. 8–5, even though the intersection point does not represent a real point of zero charge. Skyllberg and Borggaard (1998) determined an apparent pznpc of 4.1 in a Spodosol Bs with 1.1% SOM using titration data. However, when they corrected for the displacement of organically adsorbed Al^{3+} by H^+, the calculated pznpc was 2.9. A pznpc of 2.9 suggests that the Bs soil material has little potential to develop positive charge and the cation exchange properties are almost completely dominated by SOM. The role of Al-SOM in pH buffering is discussed in more detail below.

H^+ and Al^{3+} Adsorbed to Soil Organic Matter

Organic matter can be the dominant buffering material in surface soils, even in mineral soils. This is illustrated by the buffer plots produced by Magdoff and Bartlett (1985) for the addition of $CaCO_3$ and H_2SO_4 to agricultural surface soils (Fig. 8–1 and 8–2). For the 51 surface mineral soils in this study, a plot of buffering based on SOM content accounted for buffering in the pH range of 4 to 7 for all of the soils (Fig. 8–2). At high pH, the buffering is due to $CaCO_3$ and at low pH is due to dissolution of aluminous minerals. These soils have a mean organic matter content of 2.17%, with a range 0.44 to 10.49%. Organic matter is dominant even though the mean clay content was 38%. These soils are from a glaciated region and are low in Al and Fe oxides and hydroxides, which would contribute to buffering in more highly weathered soils.

pH Buffer Capacity of Soil Organic Matter

The main pH buffering components in SOM are the carboxylic and phenolic groups. These groups account for the contribution of SOM to pH-dependent charge in soils. Based on chemical analysis of humic and fulvic acids extracted from SOM (Clapp et al., 2005, this publication) various organic acids have been proposed to represent the acidity of carboxyl and phenolic sites in SOM. Much of the acidity is thought to arise from carboxyl groups bound to aromatic ring structures. The simplest aromatic model compounds are benzene carboxylic acid, which has a pK_a of 5.0, and phenol, which has a pK_a of 10.0. Soil organic matter, however, is a very complex mixture with high contents of oxygen functional groups (Stevenson 1994; Hayes and Malcolm, 2001). With a greater density of O functional groups in SOM the average pK_a will be reduced (McBride, 1994). Any effort to model the details of the titration behavior of natural organic matter from soils or surface waters requires more than two pK_a values for weak acid protons (Perdue, 2001). Ionization of weak acid generates negative charges that accumulate on the macromolecular structures in SOM, and this surface charge reduces the proton donation ability of humic and fulvic acids. Thus, soil organic matter buffers pH over a wider range than predicted by a simple mixture of benzoic acid and phenol, and the titration plots do not show any clearly defined inflections that might define titration endpoints (Fig. 8–6).

The ionization of carboxyl groups is generally complete at pH 8, and the ionization of phenolic groups is complete at pH 11. Thus, the common methods for determination of carboxyl acidity involve titrating to pH 8 and titration to pH >11 for the determination of the phenol (Stevenson, 1994; Fig. 8–6).

The buffering capacity of SOM is high compared with most mineral components. The CEC values at pH 8.2 ($BaCl_2$–TEA acidity plus exchangeable base cations) reported for organic horizons in forest soils is in the range 42 to192 $cmol_c$ kg^{-1} (Table 8–5). Soil organic matter with greater degree of decomposition generally has a greater pH buffering capacity. Therefore, the buffer capacity of SOM (per dry mass of OM) is greater in mineral soil horizons and in agricultural soils than in most forest soil organic horizons. Helling et al. (1964) separated the pH-dependent CEC due to SOM from CEC due to clays by a multiple regression analysis for 60 mineral soils from Wisconsin. The general equation for the contribution of SOM to CEC in these soils is CEC ($cmol_c$ kg^{-1} of SOM) = −34.7 + 29.7pH. Thus, between pH 4 and 8 the total potential buffering capacity of SOM in mineral soils is about 240 $cmol_c$ kg^{-1}.

Importance of Al-SOM Associations for pH Buffering in Acid Soils

In acid soils many of the SOM oxygen-containing functional groups (i.e., potential CEC) are bound to Al^{3+}. The carboxyl and phenolic sites in SOM have a higher affinity for Al^{3+} than the permanent charge sites in clays; therefore, Al^{3+} re-

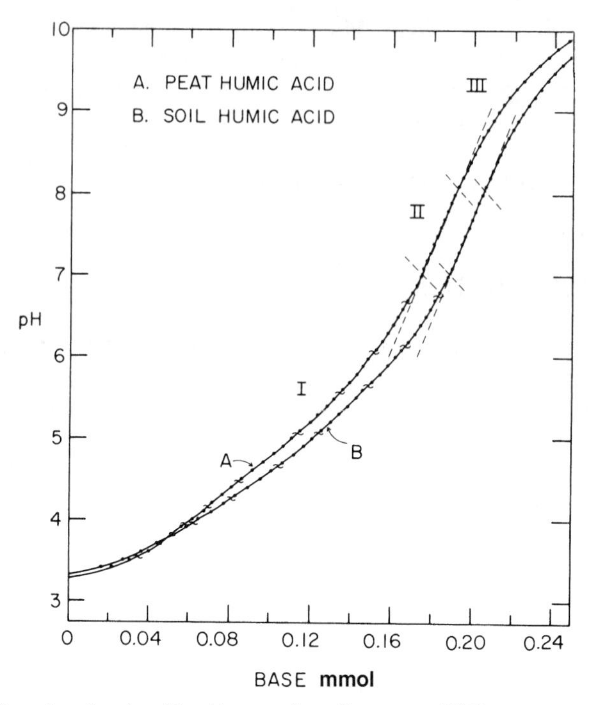

Fig. 8–6. Titration of two humic acids with strong base (Stevenson, 1994).

leased by mineral weathering is preferentially adsorbed by SOM. As in clays, Al-substituted SOM is less acid than the H^+-saturated form (Martin and Reeve, 1958; Hargrove and Thomas, 1982; Nätscher and Schwertmann, 1991; Skyllberg, 1994; Skyllberg et al., 2001). The saturation of oxygen-containing acid sites of SOM with Al is important for the buffering of pH, as well as for the solubility of Al and organic carbon in acid forest soils (Tipping and Woof, 1991; Skyllberg and Magnusson, 1995; Lofts et al., 2001a, 2001b). The Al saturation of SOM increases with greater depth in mineral soil as SOM concentrations decrease and the availability of Al from mineral weathering increases (Walker et al., 1990; Skyllberg, 1999; Skyllberg et al., 2001).

The neutralization of OH added to SOM partially substituted with Al^{3+} involves both the reaction of weak acid protons and bound Al^{3+} cations. At lower pH values and lower saturation of organic sites with Al^{3+}, titration with base results in the ionization of carboxyl groups and replacement of H^+ with the cation of the base.

Table 8–5. The CEC at pH 7.0 or 8.2 in organic horizons of forest soils.

Soils	CEC pH 7.0 or 8.2 $cmol_c\ kg^{-1}$ (SOM)	Reference
L, F and H layers	42–108	Wells and Davey, 1966
F layers	106–123	Van Cleve and Noonan, 1971
H layers	120–143	Van Cleve and Noonan, 1971
F and H layers	145–192	Skyllberg, 1993

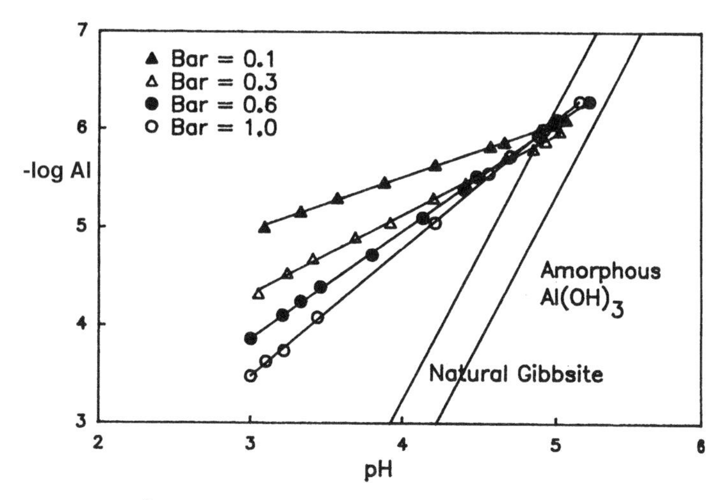

Fig. 8–7. Variation in Al^{3+} activity in equilibrium with O horizon soil organic matter saturated with different levels of Al. BAR = ratio of bound Al to titratable carboxyl groups (Walker et al., 1990).

Also the Al^{3+} in solution is adsorbed by organic sites as pH is raised. At higher pH values (i.e., 5–8) precipitation of $Al(OH)_3(s)$ can occur according to the reaction:

$$(R\text{-}COO)_3Al(s) + 3OH^- = 3R\text{-}COO^-(s) + Al(OH)_3(s) \qquad [16]$$

This is illustrated by the data of Walker et al. (1990), who saturated O-horizon soils with a range of Al contents (Fig. 8–7). They calculated a bound Al ratio (BAR), which is the ratio of $3Al^{3+}$ to the carboxyl content determined by titration of H^+ saturated SOM to pH 8. The pH was adjusted with $Ca(OH)_2$ in 0.01 *M* $CaCl_2$. Plots of pAl [$-\log(Al^{3+})$] vs. pH show that even when the quantity of Al in SOM is equivalent to the titratable carboxyl sites the equilibrium solutions are undersaturated with respect to the precipitation of microcrystalline gibbsite, and soil $Al(OH)_3(s)$, at pH <5.3. Data in Fig. 8–7 also show that with lower Al saturation the undersaturation with respect to precipitation is greater. At higher pH values solutions become oversaturation with respect to $Al(OH)_3(s)$, and precipitation can occur.

The titration of Al organic matter to pH 8.2 by strong base should yield a complete neutralization of Al-SOM sites if Eq. [16] goes to completion. However, reactions of Al^{3+} bound to SOM are not instantaneous. Hargrove and Thomas (1982) titrated Al-SOM in 0.1 *M* KCl and found that with increasing Al in the SOM the titration to pH 7.5 showed less consumption of NaOH. They allowed only 1 d for equilibration. With higher salt concentration and the use of divalent salts like $BaCl_2$ the displacement of Al from SOM to form $Al(OH)_3(s)$ is more rapid.

The inhibition of $Al(OH)_3(s)$ precipitation by organic ligands has been observed for the titration of Al salts in the presence of soluble organic ligands. Even at pH values above 7.0 oxalate and salicylate inhibit precipitation and ^{27}Al nuclear magnetic resonance (NMR) spectra and small angle X-ray scattering show that Al is not condensed but occurs as Al-monomers, some of which are both hydrolyzed and complexed by the organic ligands (Maison et al.,1994a, 1994b). A complete review of research on how organic complexation of Al affects (inhibits) the hydrolyses and the polymerization process during formation of $Al(OH)_3$ is given by Bertsch and Parker (1996).

Buffering by Dissolution of Aluminosilicates

At low pH in soils, high activity clays can dissolve, consuming protons and releasing Al^{3+} into solution. Equation [17] shows the rapid congruent dissolution of a Ca-saturated montmorillonite that takes place at pH <3.5:

$$Ca_{0.165}(Al_{1.67}Mg_{0.333})Si_4O_{10}(OH)_2(s) + 6H^+ + 4H_2O(l)$$
$$\rightarrow 0.165\ Ca^{2+} + 0.333\ Mg^{2+} + 1.67Al^{3+} + 4Si(OH)_4 \qquad [17]$$

where five of the six protons consumed in the dissolution are due to the release of Al^{3+}. This reaction likely contributes to the very strong buffering at pH 3.5 observed by Magdoff and Bartlett (1985) (Fig. 8–1 and 8–2).

SOLUBILITY AND EXTRACTABILITY OF ALUMINUM IN SOILS

The solubility of Al is important in acid soil because the concentration of soluble Al is linked to several important properties of acid soils. The Al^{3+} ion and its hydrolysis species are toxic to plants and can severely inhibit growth. The solution activity of Al^{3+} determines whether aluminous mineral phases are stable or will dissolve and degrade. Also, the maximum possible concentration of salt replaceable (exchangeable) Al is fixed by the availability of soluble $Al(OH)_3(s)$.

Al^{3+} Activity in Soil Solutions

The activity of Al^{3+} [(Al^{3+})] in soil solutions can be difficult to determine with accuracy because of the soluble complexes Al forms with dissolved organic matter (DOM). The products of hydrolysis in solution, Eq. [8] through [11], and the concentration of Al complexes with inorganic ions (i.e., sulfate and fluoride) can be calculated if pH and the concentration of Al^{3+} not associated with DOM [$Al_{inorganic}$] are known. However, the Al-DOM complexation constants are not well known, making it very difficult to calculate the binding of Al by DOM. Thus, $Al_{inorganic}$ is often determined empirically using methods that rely on the fact that the rate of reaction of Al-organic complexes with a cation exchange resin or an organic ligand (e.g., ferron or 8-hydroxyquinoline) is slow compared with rates of reaction with the various components of $Al_{inorganic}$ (Bloom and Erich, 1996). More recently, Göttlein (1998) presented a method to determine noncomplexed Al^{3+} directly using capillary electrophoresis. In general this method gives lower concentrations of free Al^{3+} than competitive complexation methods. The analytical concentrations of $Al_{inorganic}$ and inorganic ligands are used for calculation of activity and concentration of Al^{3+} and monomeric hydrolysis with a chemical speciation computer program like MINTEQA2 (Allison et al., 1991).

In the lower parts of the B horizon and deeper horizons in strongly acid northern temperate region forest soils, (Al^{3+}) data show that the solubility of Al is defined by soil $Al(OH)_3$(s) solubility (Eq. [12]). This is illustrated in Fig. 8–8 for Bs2 horizons from Spodosols in Sweden in 1:1 suspensions in 0.002 *M* NaCl at 8, 21, and 25°C (Gustafsson et al., 2001). A small amount of $AlCl_3$ was added to some suspensions to test the approach to equilibrium from oversaturation. All the pH vs. log(Al^{3+}) plots have the expected slope of 3.0 for $Al(OH)_3$(s) solubility, and at 25°C the solubility was defined by log*K* of 8.39. Log*K* values at 8 and 21°C and 9.4 and 8.54 can be predicted from the solubility at 25°C using a ΔH_r° value of −105.0 kJ mol^{-1} for the dissolution reaction of $Al(OH)_3$(s) (Palmer and Wesolowski, 1992). Gustafsson et al. (2001) compared their results with those of other investigators and concluded that, in general, in Bs2 horizons at soil solution pH >4.2, an $Al(OH)_3$(s) phase with a log*K* = 8.3 controls the solubility of Al^{3+}. This is the same value suggested by Gustafsson et al. (1998) for soil $Al(OH)_3$ solubility (Eq. [12]).

Soluble Si was also determined in the soils studied by Gustafsson et al. (2001), and the data were tested to see if imogolite might be another phase controlling the Al^{3+} activity. The soils did contain imogolite, but in the short time of the experiment imogolite did not attain equilibrium. The authors concluded that, given enough time, imogolite equilibrium is possible and that simultaneous equilibrium between imogolite and $Al(OH)_3$(s) can occur.

In surface horizons of acid temperate region forest soils, soil solutions are typically undersaturated with respect to $Al(OH)_3$(s). In these horizons organic matter binding of Al controls Al^{3+} activity. This is illustrated by suction lysimetry solution data from Spodosols in the Hubbard Brook Experimental Forest in New Hampshire, USA (Table 8–6). Driscoll et al. (1985) compared the variation in potential

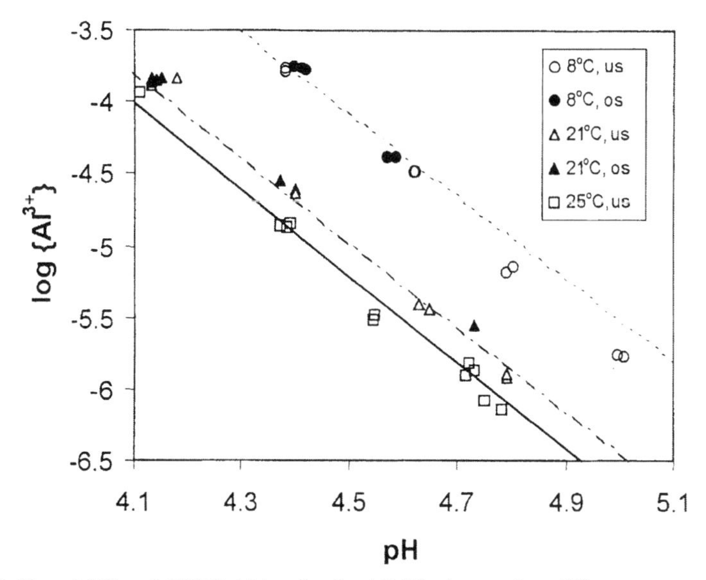

Fig. 8–8. The solubility of $Al(OH)_3$(s) in a Spodosol Bs2 horizon at three different temperatures obtained from undersaturation (us) and oversaturation (os). The dashed lines through the 8 and 21°C data are the calculated solubility using the ΔH value for the dissolution of $Al(OH)_3$(s) (Gustafsson et al., 2001).

Table 8–6. Mean and standard deviation of saturation index for equilibrium with $Al(OH)_3(s)$ in suction lysimeter solution obtained from a Spodosol in the Hubbard Brook Experimental Forest, New Hampshire, USA (Driscoll et al., 1985).

Horizon	Solution pH	Total soluble Al^{3+}	DOC	Saturation index
		μM		
O2	4.49 ± 0.59	7.0 ± 4.1	2400 ± 1100	−2.57 ± 1.21
E	4.93 ± 0.43	12.6 ± 9.3	1000 ± 670	−3.11 ± 1.73
Bhs1	5.19 ± 0.16	10.4 ± 3.3	400 ± 120	−1.48 ± 1.45
Bhs2	5.22 ± 0.26	10.0 ± 3.7	520 ± 180	−0.87 ± 0.64

for formation of $Al(OH)_3(s)$ with depth in Spodosols by calculation of the saturation index (SI) values where

$$SI = \log(IAP/K) \quad [18]$$

and IAP is the ion activity product in the soil solution:

$$IAP = (Al^{3+})/(H^+)^3 \quad [19]$$

The SI values in Table 8–6 are for equilibrium with microcrystalline gibbsite with a $\log K = 8.77$ (May et al., 1979). Soluble $Al_{inorganic}$ was determined using an ion exchange method (Driscoll, 1984), where $Al_{inorganic}$ is defined as the Al retained on a strong acid (sulfonate) ion exchange resin that has been adjusted to the pH of the soil solution using NaOH. No correction was made for the effect of field sampling temperatures <25°C. The data show more than a two order of magnitude undersaturation with respect to $Al(OH)_3(s)$ in the O and E horizons and less undersaturation with depth. The solution data in Table 8–6 also indicate that more than 85% of the soluble Al is organically complexed. This illustrates the crucial importance of good estimates of $Al_{inorganic}$ in determination of (Al^{3+}).

The solubility of Al^{3+} in organic-rich surface horizons, as well as in some organic-rich mineral soil horizons, is controlled by the binding of Al^{3+} to SOM (Walker et al., 1990; Berggren and Mulder, 1995; Skyllberg, 1999; Lofts et al., 2001a). In acid surface soils with sufficient organic matter a general equation for the pH-dependent Al^{3+} solubility is given by

$$\log(Al^{3+}) = a + b\text{pH} \quad [20]$$

where $b < 3$ and $a < 8.3$ (Bloom et al., 1979a; Walker et al., 1990). The equation describes the response (Al^{3+}) in 0.01 M $CaCl_2$ suspensions of highly organic surface soils to the addition of strong acid or strong base. Plots of $\log(Al^{3+})$ vs. pH in suspensions of O horizons with different levels of bound Al are shown in Fig. 8–7. With decreasing Al saturation of carboxyl sites the value of the slope term, b, decreases. In northern temperate forests the Al saturation of SOM generally corresponds to <10% of CEC_8 in the top of the O horizon and saturation increases with greater depth to more than 60% of CEC_8 in the E and SOM-enriched B horizons (Walker et al., 1990; Skyllberg, 1999; Skyllberg et al., 2001).

Few data on soluble Al^{3+} are available for acid soils from other than northern temperate region forest soils. However, the existing data suggest that in surface soils undersaturation with respect to $Al(OH)_3(s)$ is typical at very low pH. Bloom et al. (1979a) found that for an A horizon from an Inceptisol of New York the solubility of Al in 0.01 *M* $CaCl_2$ was defined by $\log(Al^{3+}) = 3.5 - 2.0pH$. The relationship did not change with time of reaction up to 30 d. The intercept increased to >3.5 with greater Ca^{2+} concentration in the solution and decreased with addition of organic matter. This suggests displacement of Al^{3+} from organic matter binding sites by both Ca^{2+} and H^+ was controlling (Al^{3+}). The value of $\log(Al^{3+})$ defined by $\log(Al^{3+}) = 3.5 - 2.0pH$ intersects at pH = 4.8 with the $\log(Al^{3+})$ defined by the solubility equation for soil $Al(OH)_3(s)$ (Eq.[12]). This means that at pH >4.8 oversaturation with respect to soil $Al(OH)_3(s)$ is expected while at lower pH the soil suspensions are undersaturated with respect to soil $Al(OH)_3(s)$. In the B horizon of the same Inceptisol the solubility was defined by $\log(Al^{3+}) = 5.8 - 2.4$ pH. This relationship intersects with soil $Al(OH)_3(s)$ solubility at pH 4.1 showing greater solubility of Al^{3+} at greater depth. The data for the plow layer of an agricultural Oxisol from Brazil gave results between the Inceptisol A and B horizons.

Salt Extraction of Al and H^+

Monovalent neutral salts, typically 1.0 *M* KCl or 1.0 *M* NH_4Cl, are used to extract salt-extractable or exchangeable acidly (see Quantity of Acidity in Acid Soils above). Other salts of alkaline earth or alkali metals have also been used, and 0.1 *M* $BaCl_2$ has been shown to give similar results to 1.0 *M* KCl (Khanna et al., 1986). In organic soil horizons (e.g., peats and Spodosol O and A horizons) the quantity (moles of charge) of extractable H^+ typically exceeds Al^{3+}, whereas for Spodosol B and C horizons and most other mineral soil horizons the opposite is true (e.g., Skyllberg, 1995). These salts extract only monomeric Al from acidic soils, and not hydroxy polymers (Lee, et al., 1985).

With repeated extraction the quantity of Al extracted depends on the number of extractions. As shown in Fig. 8–9 each successive extraction results in increased pH and decreased quantity of Al extracted (Bhumbla and McLean, 1965; Skeen and Sumner, 1967a, 1967b; Amedee and Peech, 1976; Bache and Sharp, 1976; Juo and Kamprath, 1979; Bloom et al., 1979b; Oates and Kamprath, 1983; Lee et al., 1985). Some investigators have attempted to separate exchangeable Al from nonexchangeable Al using sequential extraction data. Sivasubramanian and Talibudeen (1972) suggested that extrapolating the apparently linear phase of the cumulative extraction plot that is observed after about three or more extractions (see Fig. 8–9) to the *y* axis can be used to give an estimate of the exchangeable Al. In this interpretation the slope of the linear phase would represent the release of Al from nonexchangeable forms. However, different salts or different concentrations of the same salt yield different quantities of exchangeable Al (Khanna et al., 1986).

The quantity of Al extracted from surface soils depends on the ability of the extracting cations to replace both Al^{3+} bound to clay exchange sites and to organic sites (Skeen and Sumner,1967a, 1967b). Oates and Kamprath (1983) showed that the release of salt-extractable Al increased for chloride salts in the order: $K^+ < La^{3+} < Cu^{2+}$ in soils with substantial amount of organic matter, whereas there were no

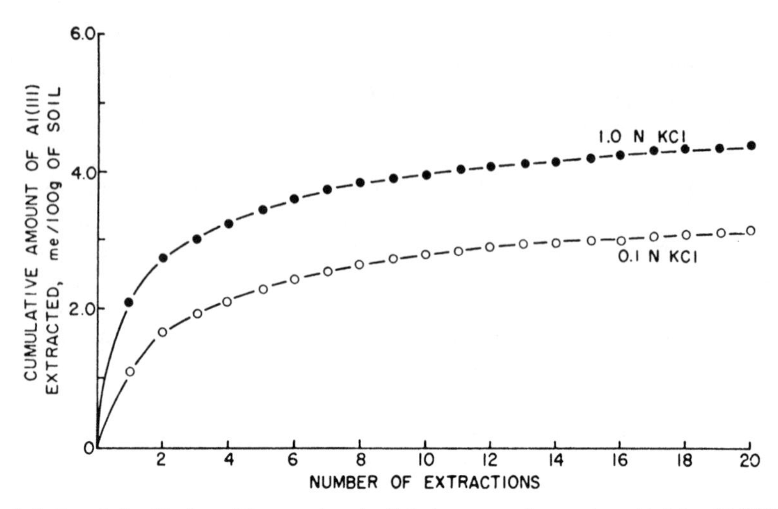

Fig. 8–9. Cumulative Al released from a mineral soil during repeated extraction with 0.1 and 1.0 *M* KCl extraction (Amedee and Peech, 1976).

differences in extractable Al among these cations in mineral subsoils low in organic matter. In the subsoils nearly all the extractable Al^{3+} was bound to clays and was readily extractable. The relative extractability of Al from SOM in the surface soils is defined by the relative affinity of the extracting cation for organic matter binding sites. Copper has a greater affinity for binding to organic matter weak acid sites than lanthanum, and La^{3+} binds much more strongly than K^+ (Bloom et al., 1979a, 1979b). Because of the ability of Cu^{2+} to displace both Al^{3+} and protons from SOM, 0.5 *M* $CuCl_2$ extracts of organic soils are much lower in pH than 1 *M* KCl extracts, and more much Al^{3+} is extracted from SOM by $CuCl_2$. In studies of artificially Al substituted peats Hargrove and Thomas (1981) showed that 0.5 *M* $CuCl_2$ extracted the same quantity of Al^{3+} as 2 *M* HCl, while KCl extracted only a fraction of the organic bound Al.

Al Extracted by $CuCl_2$

The Cu^{2+} ion strongly replaces organically bound Al, and $CuCl_2$ is a good extractant for quantitative displacement of both exchangeable Al on mineral exchange sites and organically bound Al (Juo and Kamprath, 1979; Hargrove and Thomas, 1984; Walker et al., 1990). In organic soil horizons $CuCl_2$ provides a much better estimate of organically bound Al than the traditional method of extraction with 0.1 *M* Na-pyrophosphate ($Na_4P_2O_7$) at pH 10 (McKeuge, 1967; Parfitt and Childs, 1988). The pyrophosphate method extracts significant fractions of Al from hydroxy interlayers and amorphous hydrated Al oxides (McKeuge et al. 1971; Kaiser and Zech,1996). The quantities of Al extracted from Spodosol Bs horizons using the pyrophosphate method exceeds the binding capacity of SOM (Higashi et al.,1981; Dahlgren and Walker, 1993; Berggren and Mulder,1995; Lofts et al., 2001a).

Extraction with $CuCl_2$ provides a method to separate titratable soil acidity attributed to organic and clay bound Al^{3+} from acidity due to weak acid protons in

SOM or on mineral surfaces. Skyllberg (1999) and Skyllberg et al. (2001) separated Al^{3+} acidity from proton acidity using Al^{3+} obtained by sequential 0.5 *M* $CuCl_2$ extraction and the total titratable acidity at pH 8.2. The contribution of weak acid protons is then calculated as

$$\text{Proton acidity} = \text{Total acidity}_{\text{pH 8.2}} - 3\text{Al}_{\text{CuCl2}} \quad [21]$$

For two Spodosols and one Inceptisol in Denmark, Skyllberg et al. (2001) found the contribution of proton acidity to total acidity varied between 75 to 85% in O, A, and E horizons at soil solution (centrifugation) pH of 3.3 to 4.0 and 35 to 38% in C horizons at soil solution pH of 4.2 to 4.4.

BASE SATURATION

Relationship between Base Saturation and pH

Because the quantity of acidity in soil is a function of the quantity of acid sites on oxides, clay edges, and SOM, soil acidity is not directly related to soil pH. In order to relate acidity to pH, soil chemists developed the concept of base saturation, which is the ratio of the quantity of exchange sites bound to base cations (Na^+, K^+, Mg^{2+} and Ca^{2+}) to the total quantity of cation exchange sites (Peech, 1941; Clark and Hill, 1964; Blosser and Jenny, 1971). If the CEC measured at pH 8.2 (CEC_8) is taken as the total quantity of potentially negative charge sites, the percentage base saturation (BS%) is

$$\text{BS\%} = 100[(\text{Na}^+ + \text{K}^+ + 2\text{Mg}^{2+} + 2\text{Ca}^{2+})/\text{CEC}_8] \quad [22]$$

An alternative definition of BS% that some authors have used is given when ECEC is used in the denominator of Eq. [22] (e.g., Reuss and Johnson, 1985).

The relationship between BS% and pH provides a way to model (explain) the effect of an addition of bases as carbonates, bicarbonates, oxides, or hydroxides to soil, or the effect of leaching of base cations from soils by acids.

A generally linear relationship is seen between pH and BS% (CEC_8) in agricultural soils in the pH_{CaCl2} range of 4 to 7 (Peech, 1941; Clark and Hill, 1964; Blosser and Jenny, 1971). Essentially linear pH buffering behavior in this pH range is also shown by the Magdoff and Bartlett (1985) data for H_2SO_4 and $CaCO_3$ additions to soil (Fig. 8–1 and 8–2). As discussed above in Overview of pH Buffering in Soils, Magdoff and Bartlett (1985) found that for surface soils from a northern temperate region (Vermont, USA) buffering by organic matter explains most of the titration behavior at pH values of 4 to 7. Titration plots of H saturated soil organic matter with strong base also show an approximately linear behavior at pH 4 to 7 (Fig 8–6). It is interesting to note the titration plots of oxides are also essentially linear for this pH range (Fig. 8–5).

Reuss (1983) used Eq. [22] to explain pH buffering data in Spodosols and Ultisols of the southeastern USA, but the BS% vs. pH plot was curvilinear at low pH for these soils that contain higher contents of hydrous oxides. A more linear fit was achieved when the effective ECEC rather than total CEC was used in the denominator of Eq. [22].

Bloom and Grigal (1985) developed an alternative model to explain the relationship between the BS% and pH based on the titration behavior of SOM. They used the modified Henderson–Hasselbach equation for the titration of humic acids (Posner, 1964):

$$pH = pK_a + n\log([A^-]/(HA) \quad [23]$$

where $[A^-]$ is the concentration of weak anionic sites, which is equivalent to the quantity of base that has reacted with weak acid protons; HA is the concentration of protonated sites; K_a is the acidity constant, and n is an empirical constant that accounts for the variation in K_a values for the weak acid sites (Bloom and McBride, 1979). When $n = 1$, Eq. [23] is the Henderson–Hasselbach equation for the titration of soluble monomeric–monoprotic weak acids. The modified equation with $n \neq 1$ was developed to fit titration data for synthetic carboxylate polymers (Katchalsky and Spitnik, 1947). The n value for titration of organic matter with Na or KOH is usually about 2, but it varies for different sources of SOM and for different cations in solution. Bloom and McBride (1979) found that in solution with 1/3 mol of Al for each titratable carboxylate group n was 1.2. Bloom and Grigal (1985) assumed that the sum of sites adsorbing Ca, Mg, K, and Na could be taken as an estimate of nonprotonated sites (A^-) whereas sites adsorbing H and Al were taken to represent protonated sites, yielding

$$pH_{CaCl2} = pK_a + n\log[BS\%/(100 - BS\%)] \quad [24]$$

A least square regression fit of equation to pH and BS% data for the top 25 cm of 57 forest soils of Minnesota, yielded a pK_a of 4.96, $n = 0.80$, and $R^2 = 0.63$. The low value of n that is much less than 2 may be a consequence of high Al saturation of SOM in these mineral soils. The equation is quite linear in the pH range of 4 to 6. To explain the very strong buffering seen by Magdoff and Bartlett (1985) below pH_{CaCl2} 4 (see Fig. 8–2) they added a term for the solubility of Al^{3+}. Bloom and Grigal (1985) studied soils from a northern glaciated region where the soils are low in hydrous oxides. Equation [24] may not work as well for soils high in hydrous oxides.

The relation of base saturation to pH shown in Eq. [24] relies on treating Al^{3+} and H^+ as equivalent acid cations and Na^+, K^+, Mg^{2+}, and Ca^{2+} as equivalent base cations. In Eq. [24] Al^{3+} is equivalent to three protons. At pH values <4.0, however, Al^{3+} shows little acid behavior in solution and <15% of soluble Al^{3+} ions hydrolyze to produce H^+. As discussed above, Al saturated clays and Al saturated SOM are weaker acids than H^+ saturated clays and SOM.

Because of differing acidic properties of Al^{3+} and H^+, the relationship between pH and BS%, calculated using Eq. [24], is not consistent in soils with pH <5.0 and can even be negative (Skyllberg, 1994, 1999; Ross et al., 1991, 1996; Skyllberg et al., 2001; Johnson, 2002). In Spodosols and Inceptisols of northern temperate forests it is normal for the soil solution pH value to increase with depth from about 3.5 in O to 4.5 to 5.5 in B/C horizons, while BS% decreases (Giesler and Lundström, 1993; Giesler et al., 1996; Skyllberg et al., 2001). This pattern is best understood if Al^{3+} bound to soil is treated more as a base cation than as an acid cation. The increase in pH with depth in these soils is associated with an increase in Al saturation of SOM (determined using $CuCl_2$) from <10% in O horizons to >45% in E and B horizons. The decrease in the acidity of the organic matter with increasing Al saturation is more than sufficient to overcome the decrease in BS% of 10 to 20% in the O horizons to <1.0% in the B horizons (Skyllberg, 1999; Skyllberg et al., 2001).

Biocycling of base cations results in seasonal inputs in the O horizon and a maintenance of an enhanced BS% in litter and in the surface of O horizons. Deeper into the O horizon, and in E and B horizons, the Al saturation is maintained at a higher level due to weathering of alumino-silicate minerals and Al hydroxides (Skyllberg, 1994; Rustad and Cronan, 1995). Hydrogen ions bound to oxygen containing functional groups of SOM (RCOOH) are consumed by the silicate (or hydroxide) dissolution, and the released Al^{3+} is complexed by SOM carboxylate and phenolate groups. Equation [25] illustrates one possible reaction between protonated SOM and orthoclase feldspar to yield Al-SOM.

$$R(COOH)_3(s) + KAlSiO_4(s) + H^+ \rightarrow (RCOO)_3Al(s) + H_4SiO_4(q) + K^+ \quad [25]$$

The increased Al saturation of SOM results in an increased pH. Skyllberg (1999) showed that this process alone explained the pH increase by depth in E horizons of two Swedish Spodosols.

Skyllberg et al. (2001) found that for the 60 samples taken from O, E, A, Bh, Bs, Bw, and C horizons of two acid Spodosols and an Inceptisol from Denmark the pH in soil solution (extracted by centrifugation) down to the 1-m depth could be quite well described by the model:

$$\mathrm{pH} = 6.78 - 1.21\log I - 1.45\log(100\mathrm{H}_{\mathrm{pH}\,8.1}/\mathrm{CEC}_{8.1}) \quad [26]$$

where I is ionic strength (mM) and $H_{pH\,8.1}$ is the proton acidity calculated by Eq. [21] at pH 8.1. The equation given in Fig. 8–10 is the best fit to data representing

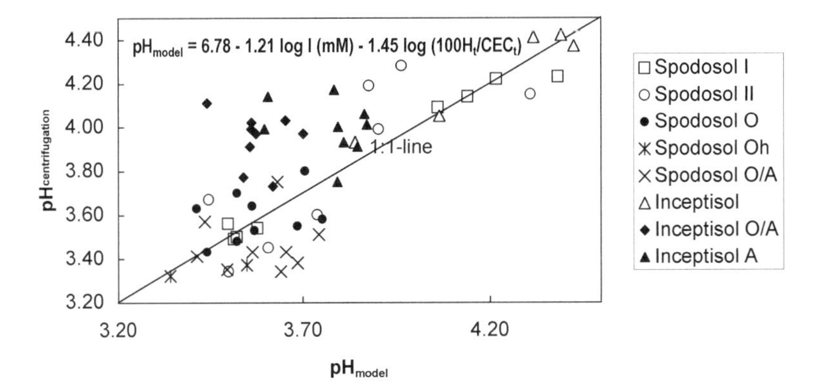

Fig. 8–10. Measured pH in soil solution (centrifugate) vs. the predicted pH using Eq. [26] for 60 soil samples (36 O and O/A horizon samples and 24 samples from complete soil profiles of O, A, B, and C horizons) obtained from Spodosols and Inceptisols and Danish forests (adapted from Skyllberg et al., 2001).

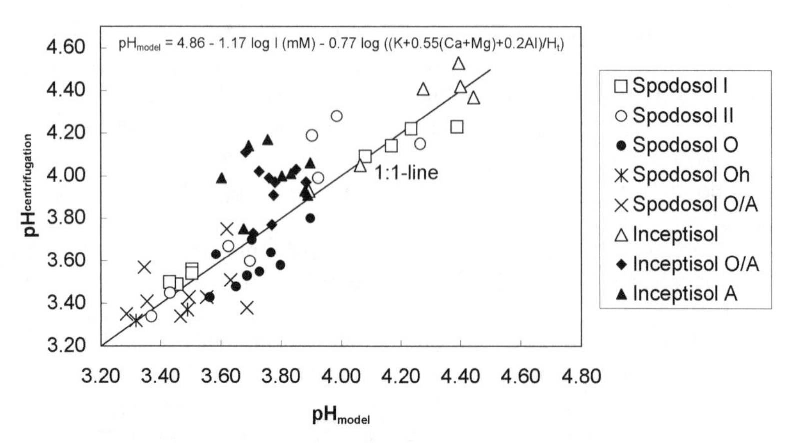

Fig. 8–11. Measured pH in soil solution (centrifugate) vs. the predicted pH using Eq, [27] for 60 soil samples obtained from Spodosols and Inceptisols and Danish forests (adapted from Skyllberg et al., 2001).

one sample for each horizon of the three soil profiles (n = 24). In this equation Al is included as a base cation, not an acid cation.

A further improvement of the fit is obtained by excluding some adsorbed Ca^{2+}, Mg^{2+}, and Al^{3+} that can be considered to be bound with sufficient strength not to participate in exchange reactions (Fig. 8–11). These cations are considered to be "specifically bound". This model is based on the extended Henderson–Hasselbach model (Eq. [23]). To determine the quantity of nonspecifically bound cations, Skyllberg et al. (2001) used independent, experimental data for five different O horizons. In these O horizons an estimated mean of 45% of the organically bound Ca^{2+} and Mg^{2+} is specifically bound at a pH of about 4.0 whereas the corresponding figure for Al^{3+} is approximately 80%. Sodium and K^+ are assumed to be 100% nonspecifically adsorbed to SOM. This means that 55% of Ca^{2+} + Mg^{2+} and 20% of organically bound Al^{3+} (determined by $CuCl_2$ extraction) are nonspecifically bound and can be considered to balance dissociated organic sites. The complete model for soil solution pH is given by

$$\mathrm{pH} = 4.86 - 1.17\log I + 0.77\log(\{[\mathrm{Na}] + [\mathrm{K}] + 0.55([\mathrm{Ca}] + [\mathrm{Mg}]) + 0.20\mathrm{Al}\}/\mathrm{H}) \qquad [27]$$

where [Na], [K], [Ca], [Mg], and [Al] are in units of charges per unit mass of soil and the ionic strength I is in millimolar units. When Eq. [27] is fitted to the complete data set of 60 soil samples the R^2 value is 0.76, as compared to a R^2 of 0.58 for Eq. [26]. It is interesting to note that the pK_a (4.86) and n (0.77) in Eq. [27] are similar to the values obtained by Bloom and Grigal (1985) for Eq. [24] for a set of soils that were generally not so acidic and in which Al was not included as a base cation. The assumption that some of Ca and Mg are "specifically bound" in this model does not mean that these cations are not easily exchangeable by 1 M KCl or other monovalent salt. Rather it means that these cations are bonded with enough strength to neutralize many of the surface charges from interaction with nearby ions in solution.

SOIL ACIDITY AND PLANT GROWTH

Toxicity of Al^{3+}, H^+, and Mn^{2+}

Reduced plant growth in acid soils commonly results from excessive dissolved Al^{3+}, H^+, and/or Mn^{2+}. In acid mineral soils the solution concentrations of Al^{3+}, H^+, and Mn^{2+} are inextricably linked because as H^+ concentration increases and pH falls below 5.5, the solution concentrations of Mn^{2+} and Al^{3+} increase. Also, in some acid soils low Ca^{2+} can be a problem (Runge and Rode, 1991), as well as low plant available P, Zn, or Mo (Sumner et al., 1991). It is exceedingly difficult to separate individual causative effects for poor plant growth in acid soils, but the toxicity of Mn^{2+} and Al^{3+} is generally important.

In laboratory experiments H^+ in nutrient solutions can negatively affect root growth. Also, in soils there is some evidence for the toxicity of H^+ (Alva et al., 1986; Cameron et al., 1986; Falkengren-Grerup and Tyler, 1993). More commonly Al is the toxic agent and both Al^{3+} and monomeric Al-hydroxy species strongly inhibit root growth, but soluble Al complexes with inorganic or organic anions are not as toxic (Bartlett and Riego, 1972; Suthipradit et al., 1990). In nutrient solutions the Al_{13} hydroxy polymer is more toxic than soluble monomers Al. However, as discussed above, there is uncertainty concerning whether Al_{13} occurs in soils. In addition to toxicity to root growth, Al^{3+} and H^+ negatively impact the growth and performance of rhizobia, the N-fixing bacteria associated with legumes (Foy, 1988).

In soils the toxicity due to Mn^{2+} is more complex than Al or H toxicity because the concentration of Mn^{2+} is related both to pH and to the redox chemistry of Mn(IV) oxides. Thus, Mn toxicity is related both to pH and Eh. The diagnosis of whether a soil might produce Mn toxicity is difficult because special precautions are necessary in taking and handling soil samples to avoid changes in redox status (Schlichting and Sparrow, 1988). Where Mn^{2+} toxicity is found it is seldom the only problem limiting plant growth, which further complicates the problem of assigning responsibility for poor plant growth to a particular element.

The negative response of plant species to acid soil conditions varies greatly across species, and within species variation among varieties can be very significant. Monocots are generally more tolerant of high H^+ and low Ca^{2+} than dicots (Runge and Rode, 1991). Some species are very tolerant to both H^+ and Al^{3+}. Some have tolerance only for H^+ but not Al^{3+}, whereas some species are intolerant to both.

The Role of Soluble Al Complexes in Al Toxicity

A number of soluble organic and inorganic ligands form complexes of varying stability with Al. These include soil organic components and the inorganic components phosphate, fluoride, carbonate, hydroxyl, and sulfate. Anions that form the stronger complexes with Al (e.g., dissolved humic and fulvic acids, monomeric polycarboxylic acids, phosphate, and fluoride) can greatly reduce Al toxicity. Sulfate, which forms a weaker complex, also can somewhat reduce toxicity. These complexes generally render the Al less toxic because they reduce the activity of Al^{3+} and monomeric hydroxy species in solution (Cameron et al., 1986; Pavan et al., 1982). Wong and Swift (1995) demonstrated that the activity of free Al^{3+} was decreased

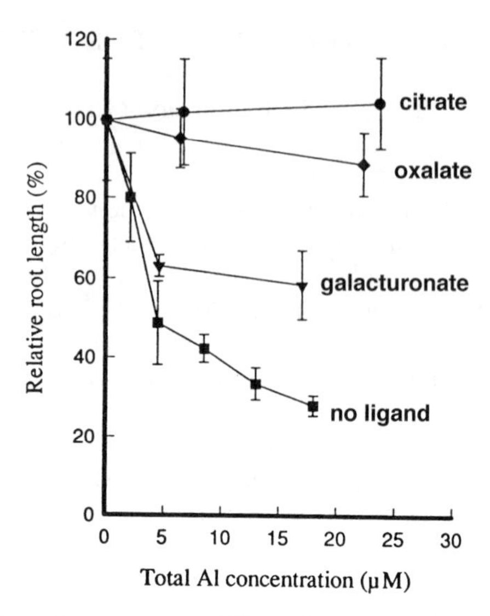

Fig. 8–12. The effect of organic ligands on the Al^{3+} toxicity root growth in mungbean. Mole ratio ligand/Al = 1/1 (adapted from Ostatek-Boczynski et al., 1995).

approximately fivefold by the addition of humic acid to Oxisols in the pH range 3.2 to 4.6, and they suggested that judicious management of soil organic matter could improve crop production. Ostatek-Boczynski et al. (1995) showed that monomeric organic ligands, such as citrate and oxalate, were highly effective in ameliorating Al toxicity to mungbean [*Vigna radiata* (L.) R. Wilczek] roots, whereas galacturonate was much less efficient (Fig. 8–12).

Plant Nutrient Assimilation and Soil Acidification

The cation–anion balances involved in nutrient uptake and in litter deposition and decomposition have important implications for acidification or alkalinity generation in soils. The internal charge balance in roots must be maintained, and to achieve balance, roots can excrete H^+ or HCO_3^- (Marschner, 1995). The predominant nutrient cations assimilated by plants are most often Ca^{2+}, Mg^{2+}, and K^+ but under certain conditions $NH_4{}^+$ can be significant and where Na^+ is high in soils it can be important. In accounting for the acid–base balance at root surfaces the efflux of one proton can be assigned to the uptake of one cationic charge. The predominant nutrient anion assimilated by plants is most often NO_3^-, but SO_4^{2-} must also be considered. Also, near the sea and in arid areas Cl^- can be important. In accounting for the acid–base balance at root surfaces the efflux of one bicarbonate can be assigned to the uptake of one anionic charge. The net acidification at a root surface is given by Eq. [28]. A negative value of acidification is net efflux of alkalinity.

Acidification by roots

$$= \text{Uptake of cation charge} - \text{Uptake of anionic charge} \qquad [28]$$

When some or all of the plant growth is recycled into soils the acid–base balance of the decomposition reactions must be considered. The cationic charges of base cations in plants are mostly balanced by carboxylic anions, and Eq. [29] illustrates for calcium gluconate, a plant component, that biological oxidation generates 1 mol of alkalinity for each mole of base cationic charge.

$$Ca(C_6H_{11}O_7)_2 \bullet H_2O + 11O_2 \rightarrow Ca^{2+} + 2HCO_3^- + 10CO_2 + 11H_2O(l) \quad [29]$$

During decomposition of the organic materials oxidation of plant amino-N to nitrate (Eq. [30]) and organic-thiol S to sulfuric acid produces acidity (Eq. [31]). The sulfur effect is much less than N because of the N content of plants is much greater than S.

$$H_2NCH_2CH_2CO_2H + 5O_2 \rightarrow HNO_3 + 3CO_2(g) + 3H_2O \quad [30]$$

$$\text{Organic-S} + 3/2O_2 + H_2O(l) \rightarrow H_2SO_4 \quad [31]$$

In a natural system with no harvest of biomass, little input of nutrients, and little loss by leaching, the net result of the processes in Eq. [28] through [31] is zero. However, if biomass is harvested Eq. [28] can be used to calculate the acidification effect of biomass removal. In modern agricultural management systems this effect is small compared with the acidification effects of N fertilizers, and farmers can compensate for acidification by liming. However, the effect of biomass removal can be significant for tree harvest on poorly buffered soils (Nilsson, 1993).

Soil Treatments

Liming

The most common approach for the amelioration of soil acidity in agriculture is the application of lime. However, there is no general agreement on the targets for pH, or extractable Al. The reasons for this lack of agreement stem from the multiple causes of poor growth under acid conditions and the differential tolerance of different varieties to acid soil problems. Addition of lime will add Ca and cause the release of adsorbed Mo and PO_4 while decreasing soluble Zn. Variation in soil properties, such as SOM content, also complicates the determination of liming needs. In certain circumstances, only topsoils are acid, and growing conditions in subsoils may not be adverse for root growth into subsoil and growth can be better than predicted from topsoil properties. This is particularly true were topsoil acidity results from the heavy use of ammoniacal fertilizers (see Acid-Producing Fertilizers above). In naturally acid soils that have been lined, subsoils may be too acid for vigorous root growth and crop grow can be less than predicted from topsoil properties.

The most common methods used in the USA for estimating liming needs in agricultural soils involve the estimation of the buffer capacity for neutralization of acidity to attain a desired pH value. Titration with a strong base could be used but this is slow and not adaptable for rapid routine analysis. Thus, methods have been developed that involve the reaction of mixed buffer solutions that have initial pH values near neutral. Mixed buffer solutions are used to give nearly linear pH response of the buffer with additions of acid. After a period of reaction of the buffer with a soil sample the pH decrease due to the reaction with the soil is measured. The Shoemaker–McLean–Pratt (SMP) method uses a pH 7.5 mixed buffer solution containing a mixture of acetate, triethanolamine, *p*-nitrophenol, and chromate that provides for a linear buffer response to acidity at pH values ranging from <5 to 7.5 (McLean et al., 1977). Target pH values for liming vs. the quantities of lime needed to raise the soil pH to a desired value have been tabulated (Sims, 1996). Other buffer methods for determination of lime requirement include the methods of Adams and Evans (1962), Woodruff (1948), and Nömmik (1983).

In areas of the world where highly weathered soils are common and the crop mix on farms is such that it is not necessary to raise the pH above a value of 6, the quantity of 1 *M* KCl extractable Al^{3+} can be used to determine the lime requirement (Sims, 1996). The quantity of Al^{3+} acidity is calculated as three times the molar quantity of extracted Al, and the lime requirement is calculated by multiplying this acidity by a factor of 1.5 or 2.0.

Treatment of Topsoils with Lime

When lime mixed into an acid soil, it dissolves and neutralizes the acid cations (Al^{3+}, Mn^{2+} and H^+) as follows:

$$2\text{Soil-Al}^{3+}(s) + 3CaCO_3(s) + 3H_2O \rightarrow 3\text{Exch-Ca}^{2+}(s) + 2Al(OH)_3(s) + 3CO_2(g) \quad [32]$$

$$2\text{Soil-Mn}^{2+}(s) + O_2(g) + 2CaCO_3(s) \rightarrow 2MnO_2(s) + 2\text{Exch-Ca}^{2+}(s) + 2CO_2(g) \quad [33]$$

$$2\text{Soil-H}^{+}(s) + CaCO_3(s) \rightarrow \text{Exch-Ca}^{2+}(s) + H_2O + CO_2(g) \quad [34]$$

The neutralization of acidity as shown in Eq. [32] and [34] was discussed above. The quantity of acidity due to Mn^{2+} is generally much lower than the acidity represented in Eq. [32] and [34], but Eq. [33] of is of interest in understanding the effect of pH on Mn^{2+} toxicity. In well-drained soils the response of Mn^{2+} to liming is due to the increase in the rate of oxidation of Mn^{2+} at higher pH values (Stumm and Morgan, 1996).

Where lime is relatively cheap, the tendency has been to adjust pH values to near neutrality (pH 6.5–7), which has proven to adequately overcome most of the Al^{3+} and H^+ acidity in relatively unweathered soils as found in the midwestern USA (McLean and Brown, 1984). However, in other regions, mainly in the tropics and subtropics where the soils are more highly weathered, liming to near neutrality has sometimes resulted in yield depressions due to reduction in plant available Zn and other micronutrients (Farina et al., 1982). In South Africa and the southeastern USA, where Mn toxicity is not common, liming to neutralize exchangeable Al^{3+} has proven to be generally effective (Kamprath, 1970; Reeve and Sumner, 1970). In other parts of the world, such as Australia, Brazil, Hawaii, and Nigeria, yield responses beyond the point where all Al is precipitated have been observed (van Raij et al., 1983; Fox et al., 1985). The need for extra lime due to an insufficient increase in pH to ade-

Table 8–7. Comparison of various invasive and noninvasive strategies for the amelioration of subsoil acidity in a South African Plinthic Paleudult as measured by the cumulative increase in corn (*Zea mays* L.) yield compared with conventional tillage for a period of 7 yr (adapted from Farina and Channon, 1988; updated by Dr. Farina).

Treatment	Increase in corn grain yield
	kg ha^{-1}
Limed topsoil + conventional tillage	0
Limed topsoil + modified subsoiler (0.9 m)	4 200
Limed topsoil + deep plow (Nardi)	6 800
Limed topsoil + 10 Mg lime ha^{-1} in furrow with Wye double digger	7 100
Limed topsoil + 10 Mg gypsum ha^{-1} surface applied	10 300

quately affect the rate of oxidation of Mn^{2+} is a problem in some acid soils in Australia (Moody et al., 1995). The common problem of subsoil acidity in highly weathered soils is not addressed by conventional liming of topsoils (van Raij, 1991). However, topsoil liming in some soils can have some effect on subsoils, as is discussed below.

Treatment of Subsoil Acidity

The mechanical incorporation of lime to depths beyond that of conventional agricultural tillage (0.5–1 m) has proven effective in neutralizing subsoil acidity (Bradford and Blanchar, 1977; Sumner et al., 1986), but it is usually impractical (Sumner, 1995). Practical strategies for acid subsoil amelioration fall into two categories, namely, those involving some disturbance or those involving no disturbance to the profile. In the former category, lime can be injected into the soil to depth (Doss et al., 1979; Gonzalez-Erico et al., 1979; Rechcigl et al., 1985; Coventry et al., 1987; Ellington et al., personal communication, 1992), or limed subsoil can be buried in vertical bands by various techniques (Farina and Channon, 1988; Coventry, 1991; Kirchhof et al., 1991). In all cases, significant yield responses can be obtained, often lasting for considerable periods of time as illustrated in Table 8–7.

Addition of lime to topsoil has been reported to reduce subsoil acidity, but Sumner (1995) reviewed the literature and found conflicting reports concerning lime movement down the profile in soils. He concluded that the controversy may stem from contamination of subsoil by topsoil during sampling, the use of only Ca as an indicator of lime movement, soil heterogeneity, mixing by soil fauna, or possibly by a mechanism proposed by Helyar (1991). Helyar (1991) demonstrated that liming a topsoil reduces acid leaching to the subsoil, but not necessarily the acid leaching from the subsoil. This results in a slight rise in subsoil pH.

If the role of plants and N fertilization is not considered, neutralization of subsoil acidity by lime applied in a topsoil can occur only by movement of lime particles or the leaching of soluble alkalinity from topsoil into the subsoils at a rate greater that acidity. In coarse-textured soils, undissolved lime particles can be transported physically through the pores to subsoil layers thereby increasing the pH. In addition, alkalinity in the form of soluble organic anions ($RCOO^-$) and HCO_3^- can move with water to lower horizons. If the leaching of alkalinity into subsoil exceeds acidity in the form of H^+, Al^{3+}, and Mn^{2+}, the pH can be increased. At the pH in agricultural soils, soluble H^+ is small, and it is safe to treat Al^{3+} as an acid.

Manganese is not an acid unless oxidation to Mn(IV) can take place in the subsoil but at pH values below 5.0 Mn^{2+} oxidation is very slow.

For the leaching of alkalinity to exceed the leaching of acidity the pH should exceed a value of about 5.5. At this pH Al^{3+} and Mn^{2+} are largely precipitated, the concentration of HCO_3^- is significant (see Eq. [2]), and organic anions can also contribute to leaching of alkalinity (Ritchie, 1989; Helyar, 1991). At pH <5.5 soluble alkalinity due to dissolved organic anions will exceed HCO_3^- and much of the organic alkalinity can be exceeded by Al^{3+} and H^+ acidity. The net reactions for soluble alkalinity with subsoil exchangeable Al^{3+} are given in Eq. [31] and [32].

$$2\text{Soil-Al}^{3+}(s) + 3\text{Ca(HCO}_3)_2 \rightarrow 3\text{Exch-Ca}^{2+}(s) + 2\text{Al(OH)}_3(s) + 6\text{CO}_2 \quad [35]$$

$$2\text{Soil-Al}^{3+}(s) + 3\text{Ca(ROO)}_2 \rightarrow 3\text{Exch-Ca}^{2+} + 2\text{Al(OH)}_3(s) + 6\text{RCOOH} + 6\text{CO}_2 \quad [36]$$

In addition the organic alkalinity can be converted by biological oxidation to bicarbonate (Eq. [29]), a stronger base.

With conventional liming of surface soils, alkalinity will move slowly. Porter and Helyar (1992) estimated that between 20 and 100 yr are required for the pH in the upper 0.1 m of a subsoil beneath a topsoil limed to the mid 5 pH range to be raised by 0.5 units. Raising the topsoil pH above 5.5 would promote more rapid movement of alkalinity.

Addition of acidifying ammoniacal fertilizers can acidify surface soils as shown in Eq. [3], but this reaction can also result in increased leaching of alkalinity to subsoil (Percival et al., 1955; Pearson et al., 1962; Abruna et al., 1964; Adams et al., 1967; Weir, 1975; Adams, 1981). The H^+ produced exchanges with soil Ca^{2+}, and $Ca(NO_3)_2$ can leach to the subsoil. This increases the Ca at depth, and if more NO_3^- than bases is taken up by subsoil roots alkalinity will be produced according to Eq. [28]. The overall reaction results in acidification of the surface soil and in alkalinity generation in the subsoil.

Adams et al. (1967) showed this effect very well. They found that in the presence of adequate lime, the pH of the entire profile is raised, whereas at lower rates, the pH of the upper part of the profile is increased but the pH in the lower portion is not raised as rapidly. In the absence of lime, acid penetration of the profile is much greater. The presence of roots in the subsoil is crucial to the generation of alkalinity it the subsoil by this strategy (Kotze and Deist, 1975). The efficacy of this strategy is substantiated by other reports involving the comparison of paired woodland and cultivated sites (Adams and Hathcock, 1984).

Addition of Organic Materials and Gypsum

Many investigators have reported large applications of animal manures or other organic materials can cause substantial increases in subsoil pH in agricultural soils (Long, 1979; Lund and Doss, 1980; Hern et al., 1982; Wright et al., 1985; Sharpley et al., 1991; Sweeten et al., 1995). However, other investigators suggest there is no effect (Sharpley et al., 1991) or that subsoil pH decreases (Kingery et al., 1994). The reactions taking place after the application of manure to soil are complex, and the resulting impact on subsoil pH will depend on the precise nature of the manure and the cropping system. Generally chicken manure has a greater effect in raising subsoil pH than other animal manures (Wood and Hattey, 1995). The transformation of NH_4^-–N to NO_3^- produces acidity in the topsoil as shown in Table 8–1. As described above, this can result in the leaching of $Ca(NO_3)_2$ and an increase in subsoil pH. Production of mobile organic ligands, which can move into the subsoil and are adsorbed on the surface Fe and Al hydrous oxides, can produce alkalinity by ligand displacement of OH^- ions (Hue, 1992). The leaching of organic alkalinity to the subsoil as described above can also contribute to the increase in subsoil pH. Complexation of Al^{3+} by the organic anions also renders Al less toxic.

Addition of various types of Ca-carboxylate materials to surface soils has been tried for the amelioration of subsoil acidity. Calcium-saturated coal-derived products have proven effective (van der Watt et al., 1991; Noble et al., 1995). These products move readily down the profile, decreasing exchangeable Al^{3+} and Mn^{2+} and increasing pH and exchangeable Ca. This can cause the direct precipitation of Al as shown in Eq. [36] or after oxidation of the organic anion to HCO_3^-. Organic anions can complex with Al^{3+} allowing for leaching of Al to greater depths (Noble et al., 1995). Similar results were obtained by Smith et al. (1995) using Ca citrate; they also found that Ca fulvate was ineffective, perhaps because, unlike citrate, the fulvate ion is not readily oxidized to bicarbonate (Eq. [36]).

Gypsum is also effective to ameliorate subsoil acidity by a process that was first demonstrated by Sumner (1970) and Reeve and Sumner (1972). Much the work on gypsum was summarized by Shainberg et al. (1989) and Sumner (1993, 1995). The comparative data for different treatments to alleviate subsoil acidity problems in Table 8–7 show that topsoil addition of lime plus gypsum is superior to mechanical incorporation of lime at depth. Gypsum is much more soluble than $CaCO_3$ and it moves down the profile, increasing subsoil Ca^{2+} levels, decreasing levels of exchangeable Al^{3+}, and reducing solution Al^{3+} activity. A number of mechanisms have been proposed to account for the observations of reduced levels of Al^{3+} in the subsoil. Originally, Reeve and Sumner (1972) proposed a "self-liming" effect in which sorption of SO_4^{2-} in by ligand on Fe and Al hydrous oxide surfaces produces OH^- . This ignores the possibility of precipitation of Al in one or more forms of basic Al sulfates (Adams and Rawajfih, 1977; Hue et al., 1985; Sposito, 1985) as follows:

$$3Al^{3+} + K^+ + 2SO_4^{2-} + 6H_2O(l) \rightarrow KAl_3(OH)_6(SO_4)_2(s) + 6H^+ \qquad [37]$$

$$4Al^{3+} + SO_4^{2-} + 10H_2O(l) \rightarrow Al_4(OH)_{10}SO_4(s) + 10H^+ \qquad [38]$$

$$Al^{3+} + H_2O(l) + SO_4^{2-} \rightarrow AlOHSO_4(s) + H^+ \qquad [39]$$

The formation of the $AlSO_4^+$ complex ion in solution also reduces Al activity and toxicity. However, in dynamic soil situations with frequent leaching events, it is unlikely that sufficiently high SO_4^{2-} concentrations can be supported in the soil solution to cause a significant decrease in Al toxicity due to formation of $AlSO_4^{2+}$.

Maintenance of Soil Organic Matter

A large body of work has demonstrated the beneficial effects of organic matter in ameliorating the toxic effects of Al in surface soils (Evans and Kamprath, 1970;

Thomas and Hargrove, 1984; Hue and Amien, 1989). Soil organic matter has a large capacity to complex Al^{3+}, as discussed in sections above. This greatly reduces the toxicity of Al and is one of the reasons that in northern temperate region forests plant rooting is very prolific in high organic surface horizons despite pH values of 4 or even less. However in agricultural soils the maintenance of high levels of organic matter is not always readily achieved in practice, especially under intensive row cropping. In certain areas, consideration has been given to concentrating crop residues from larger areas on small production plots as a strategy of counteracting soil acidity (Bessho and Bell, 1992).

Plants that take up large quantities of base cations from subsoils can pump subsoil alkalinity to the surface, and if the plant residues are not removed they can ameliorate surface soil acidity. In Australia under marginal farming conditions, Noble et al. (1996) demonstrated that certain species rich in base cations are effective in transferring alkalinity, via their leaf fall, from base-rich subsoils to anthropogenically acidified topsoils. This provides a potential means of restoring acidified soils that have become marginal for farming.

SOIL RESPONSE TO ACIDIC DEPOSITION

The acidification of soils by anthropogenic atmospheric deposition has been the topic of considerable research. The main components in acidic deposition are H^+ and NH_4^+ and the charge balancing anions, NO_3^- and SO_4^{2-} (Kennedy, 1992). Deposition very distant from sources can acidify poorly buffered lakes in a matter of decades (Kennedy, 1992). Soils are more strongly buffered than lakes, and acidification of soils is much slower than for lakes. However, the acidification of lakes is not independent of soil acidity because the transfer of alkalinity or acidity from soils to surface waters is important in determining the response of a lake to acidic deposition (Reuss and Johnson, 1985). The most sensitive soils and lakes are found in the areas in northern temperate forests where the soils are sandy and shallow over hard rock.

Deposition of strong mineral acids (HNO_3 and H_2SO_4) and NH_4 (oxidized in soils to HNO_3) eventually results in the leaching of base cations and soil acidification. In very poorly buffered soils this can result in a measurable reduction of pH and a reduction in plant available mineral nutrients within a few decades. With decreasing pH, Al and Mn may become toxic to some plant species. Several groups of investigators have developed biogeochemical models to predict the long-term effects of acidic deposition on forest soils and associated surface waters (Christophersen et al.,1982; Cosby et al., 1985; Gherini et al., 1985; Warfvinge and Sverdrup, 1992). These models have soil chemistry submodels that are based on the principles discussed in this chapter. Also Bloom and Grigal (1985) and Reuss et al. (1990) developed soil chemical models that may be used to predict response of soils to long-term inputs of strong acids. In this section we will use the numerical model of Bloom and Grigal (1985) to illustrate how concepts of soil acidity and pH buffering can be used to develop a rather simple model for the response of soils to long-term acid deposition. We will also discuss some of the problems in modeling soil responses to long-term inputs

The Bloom and Grigal Model

Bloom and Grigal (1985) based their model on the relationship between BS% and soil pH_{CaCl2} that is given by the extended Henderson–Hasselbach equation (Eq. [24]). This model was developed as a planning tool that uses existing data for prediction of the response of soils to long-term acidic deposition and data inputs are limited to readily available soil data and estimates that can be made from the data. Bloom and Grigal (1985) limited their modeling to the top 25 cm of mineral soil focusing on forest soils in northern Minnesota, USA. They considered the O horizon to be part of the biological system and excluded it from the modeling. The soils in northern Minnesota tend to have rather thin O horizons and very thin, poorly developed, E horizons. This contrasts to the more developed forest soils from the northeastern USA and northern Europe.

In the calculation of acidification, the quantity of protons adsorbed from the added acid, displacing base cations, is calculated as the difference between the input acidity and the output acidity in the soil leachate. Input acidity is the sum of H^+ and NH_4^+ minus NO_3^- in units of moles of charge per hectare per year. Ammonium and nitrate are included because of their effects on soil acidification. Because forests in Minnesota are N deficient it is assumed that no NH_4^+ or NO_3^- are leached from the top 25 cm. The volume of soil leachate is calculated as the difference between mean annual rainfall and evapotranspiration estimated from climatic data. Output acidity is calculated as the sum of H^+, $3Al^{3+}$, and $2AlOH^{2+}$. Soluble aluminum is significant only at low pH and at the pH values where Al is a significant component of leachate acidity, hydrolysis species other than $AlOH^{2+}$ are not significant. In the model the solubilization of Al^{3+} acidity in the leachate is calculated according to

$$\log(Al^{3+}) = 2.6 - 1.66pH \qquad [40]$$

This empirical Al solubilization equation was developed from data for the artificial acidification of a sandy forest Entisol from Minnesota. Equation [40] is an application of Eq. [20], and the coefficients are similar to equilibrium relationships Bloom et al. (1979a) measured in soil suspension for soils from other regions.

The initial pH and base saturation before the addition of acidic deposition are assumed to represent a steady state where the measured pH and BS% are due to the balance of the rate of natural generation of acidity, which is balanced by the rate of consumption of protons by weathering of bases from soil minerals. This assumption was justified by the fact that the soils of Minnesota have been little affected by acid deposition. Acidic deposition increases acid inputs and upsets the steady state by increasing the rate of leaching of bases (e.g., base cations with HCO_3^- and $RCOO^-$) to a value in excess of the rate of alkalinity generation by mineral weathering. Also, with a decrease in soil solution pH, the natural weathering due to $H_2CO_3{}^*$ is decreased because at lower pH the effectiveness of $H_2CO_3{}^*$ as a proton donor decreases (see Natural Acidification Processes above). The net effect on the loss of bases, S (kmol ha^{-1} yr^{-1}) is calculated in annual time steps as

$$S = I - A - C \qquad [41]$$

where I is the quantity of acid added in rain and dry deposition, A is the quantity of acidity in leachate (Al + H), and C is the decrease in bicarbonate weathering, compared with the initial condition.

The model uses pH_{CaCl2} as an input for the modified Henderson–Hasselbach equation (Eq. [24]). The ionic strength in soil solutions, however, is generally much lower than in 0.01 M $CaCl_2$, and the pH in soil solution leachates is much higher than the laboratory measurement of pH_{CaCl2}. The model therefore uses a pH value for the soil solution that is 0.8 units greater than pH_{CaCl2}. This value was chosen because it yielded the best fit of the model to laboratory data for sequential leaching of soils with 0.00025 M H_2SO_4, using a 1 d contact time for each leaching event.

Because different soils contain different mixtures of clay and organic matter it is not reasonable to expect the pK_a and n values in Eq. [24] should be the same for each soil. To overcome this difficulty Bloom and Grigal (1985) estimated a convergence point at BS% = 0.001 and $pH_{CaCl2} = 2.57$. The convergence condition was calculated from Eq. [24] using the pK_a and n values determined by a regression fit of pH_{CaCl2} vs. BS% for soils with pH_{CaCl2} values of 3.3 to 6 ($pK_a = 4.96$; $n = 0.797$). The model estimates unique pK_a and n values for each soil from the initial pH_{CaCl2} and BS% by simultaneous solution of Eq. [24] assuming both the convergence condition and the initial BS and pH_{CaCl2}. The model was shown not to be very sensitive to the pH at BS% = 0.001.

The plot in Fig. 8–13 shows a modeled response for a relatively poorly buffered forest soil (Fragiboralf with mean SOM = 2.0% and clay = 6% in the top 25 cm). The acid input was 0.3 kmol ha^{-1} yr^{-1} (acid rain pH = 4.65), a value typical of the area where the soil was obtained. In more strongly impacted areas the acidic deposition may easily exceed 1.0 kmol ha^{-1} yr^{-1} (Ulrich, 1991). The new steady-state pH_{CaCl2} after 500 yr is 3.4 (soil solution pH = 4.2). At this pH value more than 50% of the leachate acidity, as defined by the model, is to due to soluble Al. Aluminum buffering as defined by Eq. [40] is very significant because without this component the steady-state pH would be 0.3 units, or more, lower. A generally good fit was observed for the pH change in the laboratory leaching of four soils with 0.00025 M H_2SO_4 (Bloom and Grigal, 1985).

The model demonstrates that although the response to excess leaching caused by acidic deposition is very slow, significant changes can occur in the long term even under low levels of deposition. If the soils included in the Bloom and Grigal (1985) study were to be in a region of heavy acidic deposition, such as areas of central Europe with deposition exceeding 1.5 k mol ha^{-1} yr^{-1}, the time to maximal soil acidification would be <100 yr.

Forest harvest and the subsequent regrowth can also contribute to soil acidification by removal of base cations (removed in the biomass as metal RCOO-salts). Bloom and Grigal (1985) noted that repeated tree harvest and regrowth in Minnesota could contribute as much to soil acidification as the 0.3 $kmol_c$ ha^{-1} yr^{-1} deposition value used in their model simulations. In Swedish spruce (*Picea* spp.) forests soil acidification by stem biomass harvest is in the range of 0.1 to 0.4 $kmol_c$ ha^{-1} yr^{-1}, whereas whole tree harvest results in a soil acidification of 0.2 to 0.8 $kmol_c$ ha^{-1} yr^{-1} (Nilsson, 1993).

The model response shown in Fig. 8–13 did not consider the possible effect of lower pH on alkalinity generation by mineral weathering. Weathering rates in-

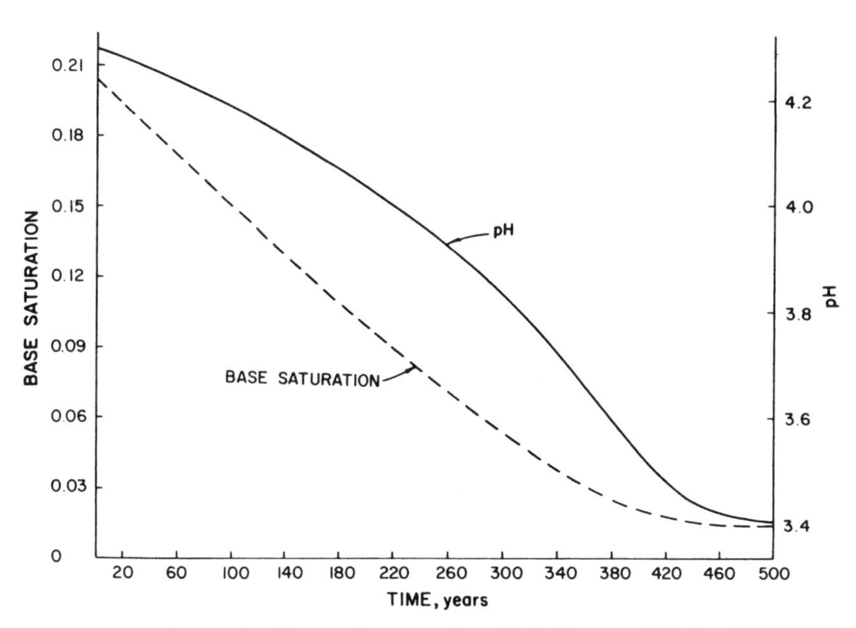

Fig. 8–13. Predicted change in pH_{CaCl_2} and base saturation with the Bloom and Grigal model (1985) for a Fragiboralf with mean SOM = 2.0% and clay = 6% in the top 25 cm with 500 yr of acid input at 0.3 kmol ha^{-1} yr^{-1} (acid rain pH = 4.65).

crease with decrease in pH and it is reasonable to assume that increases in weathering rates occur with long-term acidic deposition and this increase will to some extent; ameliorate the effect of acidic deposition on soils. To test the effects of pH on weathering Bloom and Grigal (1985) used a general expression for the effect of pH on weathering rate:

$$r = k(H^+)^x \tag{42}$$

where k is a rate constant and x is and empirical constant with a value typically about 0.5 (Lasaga, 1981). Transformation of this expression gives an equation that can be used to calculate increases in rate of weathering from a known initial rate:

$$r = r_0 10^{-0.5(pH-pH_0)} \tag{43}$$

where r_0 and pH_0 are the initial pH and rate of weathering in moles per hectare per year. When Eq. [43] was included, the final soil pH value after 500 yr was 0.2 units greater than shown in Fig. 8–13 and the final BS% was increased from 1 to 3. Thus, increases in weathering rates will have a significant effect, but not enough to change the general conclusion that the long-term impact of acidic deposition can result in very significant depletion of bases and lower pH.

Evaluation of the Bloom and Grigal Model

The model yielded a buffer curve that is similar in appearance to that obtained by Magdoff and Bartlett (1985) for acidification of soils in beakers (Fig. 8–1 and

8–2), except for some convexity at in the model response curve at pH_{CaCl2} 3.8 to 4.2. This is somewhat misleading because the final pH in the Bloom and Grigal (1985) model is a result of the balance of the input acidity vs. output acidity in leachate while the lowest pH value observed by Magdoff and Bartlett (1985) is due to the reaction of H_2SO_4 with soil in a closed vessel. In this static condition the low pH buffer limit is likely due to the dissolution of clay mineral Al^{3+} and precipitation of basic Al sulfates as shown in Eq. [37], [38], and [39]. The similarity in the two situations is that Al buffering reactions limit the low pH.

While the Bloom and Grigal (1985) model may be sufficient as a planning tool to compare and rank the response of different soils to acidic deposition, this model does not consider many of the complex interactions in soils that must be considered if a more accurate prediction is desired. For example, the model does not consider the differences in pH buffering for different soil horizons. A more complete model would include calculations for each horizon with the leachate from each horizon directed into the next horizon. Also, the model uses pH_{CaCl2} to predict solution pH and to calculate the response to decrease in BS%. Better results would be obtained using the approach of Skyllberg et al. (2001) that relates soil solution pH directly to the saturation of surface charges by base cations and Al. Soil solution pH values, however, are difficult to measure and vary with time of year (Skyllberg, 1991; Giesler and Lundström, 1993).

The Bloom and Grigal model estimates Al release using a simple equilibrium equation (Eq. [40]). The coefficients for this equation with short times for equilibration compared with the times calculated by the model. The discussion in Al^{3+} Activity in Soil Solutions suggests Al release calculated by Eq. [40] is due to the displacement of Al from SOM binding sites. However, the Bloom and Grigal model does not consider change in pH due to the replacement of Al in SOM with protons. Removal of Al^{3+} and replacement with H^+ results in a more acidic SOM. This process is complicated by the fact that protonated SOM can react with primary minerals, as shown in Eq. [25], resulting in the dissolution of Al followed by Al binding to SOM sites. The net result of Al removal from SOM by cation exchange and leaching and dissolution of Al from soil minerals Al will determine whether Al is stripped from SOM during long-term acidification or is accumulated. If the pH coefficient for the effect of pH on rate of weathering Al^{3+} from aluminous minerals is <1, as suggested in Eq. [43], and the pH coefficient for the equilibrium of Al with SOM is >1, as suggested by Eq. [40], the effect of a decrease in pH with time will be to decrease the Al saturation of SOM. If Al is stripped from SOM and SOM becomes more protonated, the drop in soil pH would be greater than calculated by the current model. This reaction has been proven to be of significance for pH decreases in sandy soils in The Netherlands (Mulder et al., 1989; Mulder and Stein, 1994; Wesselink et al., 1996).

The model also does not differentiate monovalent from divalent base cations in the BS% vs. pH relationship (Eq. [24]). This is contrary to the semiempirical model developed by Skyllberg et al. (Eq. [27]). Using Eq. [27] instead of Eq. [24] in the model would require a precise knowledge of the rates of weathering of Na, K, Ca, Mg, and Al from soil minerals and a consideration of inputs of base cations in rain and dustfall.

Another factor causing changes in pH in forest soils is changes in the quantity and quality of SOM in response to nutrient inputs in acidic deposition (Binkley et al., 1989; Binkley and Sollins, 1990; Binkley and Valentine, 1991; Skyllberg, 1996). The N in acidic deposition can result in increased productivity in N-limited northern forests (e.g., southern Sweden) (Elfving and Tenghammar, 1996). One effect of increasing growth is an increase in plant litter deposition resulting in increased SOM in both organic and mineral soil horizons. In addition, in Europe and North America large areas have reverted from agricultural use to dense forests with increases in SOM (Glatzel, 1991). Where fast growing coniferous tree species have deposited acid litter the increase in SOM has resulted in pH decrease (Binkley et al., 1989; Binkley and Sollins, 1990; Binkley and Valentine, 1991). Decreases in soil pH may also be caused by an increase in the soil solution ionic strength due to deposition of air pollutants, as well as possible increases in sea salt storm events. Thus, reported significant decreases in soil pH due to acidic deposition, such as in the report by Nilsson and Tyler (1995) for southern Sweden, may not primarily be due to acid rain, but rather to changes in quantity of SOM and soil solution ionic strength (Binkley and Högberg, 1997).

REFERENCES

Abruna, F., J. Vicente-Chandler, and R.W. Pearson. 1964. Effects of liming on yields and composition of heavily fertilized grasses and on soil properties under humid tropical conditions. Soil Sci. Soc. Am. Proc. 28:657–661.

Adams, F. 1981. Alleviating chemical toxicities: Liming acid soils. p. 269–301. *In* G.F. Arkin and H.M. Taylor (ed.) Modifying the root environment to reduce crop stress. ASAE, St. Joseph, MI.

Adams, F. 1984. Crop response to lime in the southern United States. p. 211–265. *In* F. Adams (ed.) Soil acidity and liming. 2nd ed. Agron. Monogr. 12. ASA, CSSA, and SSSA, Madison, WI.

Adams, F., and C.E. Evans. 1962. A rapid method for measuring of lime requirement of Red-Yellow Podzolic soils. Soil Sci. Soc. Am. Proc. 26:355–357.

Adams, F., and P.J. Hathcock. 1984. Aluminum toxicity and calcium deficiency in acid subsoil horizons of two coastal plains soil series. Soil Sci. Soc. Am. J. 48:1305–1309.

Adams, F., and Z. Rawajfih. 1977. Basaluminite and alunite: A possible cause of sulfate retention by acid soils. Soil Sci. Soc. Am. J. 41:686–692.

Adams, F., A.W. White, and R.N. Dawson. 1967. Influence of lime sources and rates on Coastal bermudagrass production, soil profile reaction, exchangeable Ca and Mg. Agron. J. 59:147–149.

Aitken, R.L., and P.W. Moody. 1991. Interrelations between soil pH measurements in various electrolytes and soils solution pH in acidic soils. Aust. J. Soil. Res. 29:483–491.

Allison, J.D., D.S. Brown, and K.J. Novo-Gradac. 1991. MINTEQA2/PRODEFA2: A geochemical assessment model for environmental systems. EPA/600/3-91/021. USEPA, Athens, GA.

Alva, A.K., D.G. Edwards, C.J. Asher, and F.P.C. Blamey. 1986. Effects of phosphorus/aluminum molar rations and calcium concentration on plant response to aluminum toxicity. Soil Sci. Soc. Am. J. 41:686–692.

Amedee, G., and M. Peech, 1976. The significance of KCl-extractable Al(III) as an index to lime requirement of soils of the humid tropics. Soil Sci. 121:227–233.

Anderson, M.A., and M.M. Benjamin. 1990. Modeling adsorption in aluminum–iron binary oxide suspensions. Environ. Sci. Technol. 24:1586.

Anderson, M.A., J.F. Ferguson, and J. Gavis. 1976. Arsenate adsorption on amorphous aluminum hydroxide. J. Colloid Interface Sci. 54:391

Anderson, M.A., and D.T. Malotky. 1979. The adsorption of protolyzable anions on hydrous oxides at the isoelectric pH. J. Colloid Interface Sci. 72:413

Bache, B.W., and G.S. Sharp. 1976. Characterization of mobile aluminum in acid soils. Geoderma 15:91–101.

Bartlett, R., and B. James. 1980. Studying dried stored soil samples—Some pitfalls. Soil Sci. Soc. Am. J. 44:721–724.

Bartlett, R.J., and D.C. Riego. 1972. Toxicity of hydroxy aluminum in relation to pH and phosphorus. Soil Sci. 114:194–200.

Berggren, D., and J. Mulder. 1995. The role of organic matter in controlling aluminum solubility in acidic mineral soil horizons. Geochim. Cosmochim. Acta 59:4167–4180.

Bertsch, P.M., and P.R. Bloom. 1996. Aluminum. *In* D.L. Sparks (ed.) Methods of soil analysis. Part 3. SSSA Book Ser. 5. SSSA, Madison, WI.

Bertsch, P.M., and D.R. Parker. 1996. Aqueous polynuclear aluminum species. p.117–168. *In* G. Sposito (ed.) The environmental chemistry of aluminum. Lewis Publ., Boca Raton, FL.

Bessho, T., and L.C. Bell. 1992. Soil solid and solution phase changes and mung bean response during amelioration of aluminum toxicity with organic matter. Plant Soil 140:183–196.

Bhumbla, D.R., and E.O. McLean. 1965. Aluminum in soils: VI. Changes in pH-dependent acidity, cation-exchange capacity, and extractable aluminum with additions of lime to acid surface soils. Soil Sci. Soc. Am. Proc. 29:370–374.

Binkley, D., and P. Högberg. 1997. Does atmospheric deposition of nitrogen threaten Swedish forests? For. Ecol. Manage. 92:119–152.

Binkley, D., and P. Sollins. 1990. factors determining differences in soil pH in adjacent conifer and alder–conifer stands. Soil Sci. Soc. Am. J. 54:1427–1433.

Binkley, D., and D. Valentine. 1991. Fifty-year biogeochemical effects of Norway spruce, white pine, and green ash in a replicated experiment. For. Ecol. Manage. 40:13–25.

Binkley, D., D. Valentine, C. Wells, and U. Valentine. 1989. An empirical analysis of the factors contributing to 20-year decrease in soil pH in an old-field plantation of loblolly pine. Biogeochemistry 8:39–54.

Bloom, P.R. 1979. Titration behavior of aluminum organic matter. Soil Sci. Soc. Am. J. 43:815–817.

Bloom, P.R., and M.S. Erich. 1987. The effect of soil solution composition on the rate and mechanism of gibbsite dissolution in acid solutions. Soil Sci. Soc. Am. J. 51:1131–1136.

Bloom, P.R., and M.S. Erich. 1996. The quantitation of aqueous aluminum. p. 1–38. *In* G. Sposito (ed.) The environmental chemistry of aluminum. Lewis Publ. Boca Raton, FL.

Bloom, P.R., and D.F. Grigal. 1985. Modeling soil response to acidic deposition in nonsulfate adsorbing soils. J. Environ. Qual. 14:489–494.

Bloom, P.R., and M.B. McBride. 1979. Metal ion binding and exchange with hydrogen ions in acid-washed peat. Soil Sci. Soc. Am. J. 43:687–692.

Bloom, P.R., M.B. McBride, and B. Chadbourne. 1977. Adsorption of aluminum by a smectite: I. Surface hydrolysis during Ca^{2+}–Al^{3+} exchange. Soil Sci. Soc. Am. J. 41:1068–1073.

Bloom, P.R., M.B. McBride, and R.M. Weaver. 1979a. Aluminum organic matter in acid soils: Buffering and solution aluminum activity. Soil Sci. Soc. Am. J. 43:488–493.

Bloom, P.R., M.B. McBride, and R.M. Weaver. 1979b. Aluminum organic matter in acid soils: Salt-extractable aluminum. Soil Sci. Soc. Am. J. 43:813–815.

Blosser, D.L., and H. Jenny. 1971. Correlations of pH and base saturation as influenced by soil-forming factors. Soil Sci. Soc. Am. Proc. 35:1017–1018.

Borggaard, O.K. 1983. Effect of surface area and mineralogy of iron oxides on their surface charge and anion-adsorption properties. Clays Clay Miner. 31:230–232.

Bradford, J.M., and R.W. Blanchar. 1977. Profile modification of a Fragiudalf to increase crop production. Soil Sci. Soc. Am. J. 41:127–131.

Brennan, R.F., and M.D.A. Bolland. 1998 Relationship between pH measured in water and calcium chloride for soils of southeastern Australia. Commun. Soil Sci. Plant Anal. 29:2683–2689.

Bruggenwert, M.G.M., T. Hiemstra, and G.H. Bolt. 1991. Proton sinks in soil controlling soil acidification. p. 8–27. *In* B. Ulrich and M.E. Sumner (ed.) Soil acidity. Springer Verlag, Berlin.

Cameron, R.S., G.S.P. Ritchie, and A.D. Robson. 1986. Relative toxicities of inorganic aluminum complexes to barley. Soil Sci. Soc. Am. J. 50:1231–1236.

Christophersen, N., H. Seip, and R.F. Wright. 1982. A model for streamwater chemistry at Birkenes, Norway. Water Resour. Res. 18:977–996.

Clapp, C.E., M.H.B. Hayes, A.J. Simpson, and W.L. Kingery. 2005. Chemistry of soil organic matter. p. 1–150. *In* M.A. Tabatabai and D.L. Sparks (ed.) Chemical processes in soils. SSSA Book Ser. 8. SSSA, Madison, WI.

Clark, J.S., and R.J. Hill. 1964. The pH–percent base saturation relationship of soils. Soil Sci. Soc. Am. Proc. 28:490–492.

Clark, C.J., and M.B. McBride. 1984. Cation and anion retention by natural and synthetic allophane and imogolite. Clays Clay Miner. 32:291–299.

Coleman, N.T., and D. Craig. 1961. The spontaneous alteration of hydrogen clay. Soil Sci. 91:14–18.

Cosby, B.J., G.M. Hornberger, and J.N. Galloway. 1985. Modeling the effects of acid deposition: Assessment of a lumped parameter model of soil water and stream water chemistry. Water Resour. Res. 21:51–63.

Coventry, D.R. 1991. The injection of slurries of lime, associated with deep tillage, to increase wheat production on soils with subsoil acidity. p. 437–445. *In* R.J. Wright et al. (ed.) Plant–soil interactions at low pH. Kluwer Academic Publishers, Dordrecht, The Netherlands.

Coventry, D.R., T.G. Reeves, H.D. Brooke, A. Ellington, and W.J. Slattery. 1987. Increasing wheat yields in north-eastern Victoria by liming and deep ripping. Aust. J. Exp. Agric. 27:679–685.

Dahlgren, R.A., and W.J. Walker. 1993. Aluminum release rates from selected Spodosol Bs horizons: Effect of pH and solid-phase aluminum pools. Geochim. Cosmochim. Acta. 57:57–66.

De Vries, W., and A. Breeuwsma. 1987. The relation between soil acidification and element cycling. Water Air Soil Pollut. 35:293–310.

Doss, B.D., W.T. Dumas, and Z.F. Lund. 1979. Depth of lime incorporation for correction of subsoil acidity. Agron. J. 41:541–544.

Driscoll, C.T. 1984. A procedure for the fractionation of aqueous aluminum in dilute acidic waters. Int. J. Environ. Anal. Chem. 16:267–284.

Driscoll, C.T., N. Van Breemen, and J. Mulder. 1985. Aluminum chemistry in a forested spodosol Soil Sci. Soc. Am. J. 437–443.

Elfving, B., and L. Tenghammar. 1996. Trends in tree growth in Swedish forests 1953–1992: An analysis based on sample trees from the National Forest Inventory. Scan. J. For. Res. 11:38–49.

Evangelou, V.P., and R.E. Phillips. 2005. Cation exchange in soils. p. 343–410. *In* M.A. Tabatabai and D.L. Sparks (ed.) Chemical processes in soils. SSSA Book Ser. 8. SSSA, Madison, WI.

Evans, C.E., and E.J. Kamprath. 1970. Lime response as related to percent Al saturation, solution Al, and organic matter content. Soil Sci. Soc. Am. Proc. 34:893–955.

Falkengren-Grerup, U., and G. Tyler. 1993. Soil chemical properties excluding field-layer species from beech forest mor. Plant Soil 148:185–191.

Farina, M.P.W. and P. Channon. 1988. Acid-subsoil amelioration: I. A comparison of several mechanical procedures. Soil Sci. Soc. Am. J. 52:169–175.

Farina, M.P.W., M.E. Sumner, and P. Channon. 1982. Lime induced yield depressions in maize (*Zea mays* L.) on highly weathered soils. Proc. 9th Int. Plant. Nutr. Coll. 1:162–168.

Fernandez, I.J., and P.A. Kosian. 1987. Soil air carbon dioxide concentrations in a New England spruce-fir forest. Soil Sci. Soc. Am. J. 51:261–263.

Fox, R.L., R.S. Yost, N.A. Saidy, and B.T. Kang. 1985. Nutritional complexities associated with pH variables in humid tropical soils. Soil Sci. Soc. Am. J. 49:1475–1480.

Foy, C.D. 1988. Plant adaptation to acid, aluminum-toxic soils. Commun. Soil Sci. Plant Anal. 19:959–987.

Gherini, S.A., L. Mok, R.J.M. Hudson, G.F. Davis, C.W. Chen, and R.A. Goldstein.1985. The ILWAS model: Formulation and application. Water Air Soil Pollut. 26:425–459.

Giesler, R., and U. Lundström. 1993. Soil solution chemistry: Effects of bulking soil samples. Soil Sci. Soc. Am. J. 57:1283–1288.

Giesler, R., U.S. Lundström, and H. Grip. 1996. Comparison of soil solution chemistry assessment using zero-tension lysimeters or centrifugation. Eur. J. Soil Sci. 47:395–405.

Glatzel, G. 1991. The impact of historic land-use and modern forestry on nutrient relations of central-European forest ecosystems. Fert. Res. 27:1–8.

Goldberg, S. 2005. Equations and models describing adsorption processes in soils. p. 489–518. *In* M.A. Tabatabai and D.L. Sparks (ed.) Chemical processes in soils. SSSA Book Ser. 8. SSSA, Madison, WI.

Goldberg, S., and R.A. Glaubig. 1987. Effect of saturating cation, pH, and aluminum and iron oxide on the flocculation of kaolinite and montmorillonite. Clays Clay Miner. 35:220.

Gonzales-Batista, A., J.M. Hernandez-Moreno, E. Fernandez-Caldas, and A.J. Herbillon. 1982. Influence of silica content on the surface charge characteristics of allophanic clays. Clays Clay Miner. 30:103–110.

Gonzalez-Erico, E., E.J. Kamprath, G.C. Naderman, and W.V. Soares. 1979. Effect of depth of lime incorporation on the growth of corn on an Oxisol of Central Brazil. Soil Sci. Soc. Am. J. 43:1155–1158.

Göttlein, A. 1998. Determination of free Al^{3+} in soil solution by capillary electrophoresis. Eur. J. Soil Sci. 49:107–112.

Gustafsson, J.P., D. Berggren, M. Simonsson, M. Zysset, and J. Mulder. 2001. Aluminum solubility mechanisms in moderately acid Bs horizons of podzolized soils. Eur. J. Soils Sci. 52:522–655.

Gustafsson, J.P., D.G. Lumsdon, and M. Simonsson. 1998. Aluminum solubility characteristics of spodic B horizons containing imogolite-type materials. Clay Miner. 33:77–86.

Hargrove, W.L., and G.W. Thomas. 1981. Extraction of aluminum from aluminum–organic matter complexes. Soil Sci. Soc. Am. J. 45:151–153.

Hargrove, W.L., and G.W. Thomas. 1982. Titration properties of Al-organic matter. Soil Sci. 134:216–225.

Hargrove, W.L., and G.W. Thomas. 1984. Extraction of aluminum from aluminum–organic matter in relation to titratable acidity. Soil Sci. Soc. Am. J. 48:1458–1460.

Hayes, M.H.B., and R.L. Malcom. 2001. Considerations of compositions and of aspects of the structures of humic substances. p. 3–40. *In* C.E. Clapp et al. (ed.) Humic substances and chemical contaminants. SSSA, Madison, WI.

Helling, C.S., G. Chesters, and R.B. Corey. 1964. Contribution of organic matter and clay to soil cation-exchange capacity as affected by the pH of the saturating solution. Soil Sci. Soc. Am. Proc. 28:517–520.

Helyar, K.R. 1991. The management of acid soils. p. 365–382. *In* R.J. Wright et al. (ed.) Plant–soil interactions at low pH. Kluwer Academic Publishers, Dordrecht, The Netherlands.

Hern, J.L., A. Menser, R. Sidle, R.L. Wright, and O.L. Bennett. 1982. The effects of organic complexing materials on ion movement and rooting environments in acid soils. p. 174. *In* Agronomy Abstracts. ASA, CSSA, and SSSA, Madison, WI.

Higashi, T., F. De Coninck, and F. Gelaude. 1981. Characterization of some spodic horizons of the Campine (Belgium) with dithionite-citrate, pyrophosphate and sodium hydroxide-tetraborate. Geoderma 25:131–142.

Hue, N.V. 1992. Correcting soil acidity of a highly weathered Ultisol with chicken manure and sewage sludge. Commun. Soil Sci. Plant Anal. 23:241–264.

Hue, N.V., and I. Amien. 1989. Aluminum detoxification with green manures. Commun. Soil Sci. Plant Anal. 20:1499–1511.

Hue, N.V., F. Adams, and C.E. Evans. 1985. Sulfate retention by an acid BE horizon of an Ultisol. Soil Sci. Soc. Am. J. 49:1196–1200.

Hunter, D., and D.S. Ross. 1991. Evidence for a phyto toxic hydroxy-aluminum polymer in organic soil horizons. Science 251:1056–1068.

Inskeep, W.P., and P.R. Bloom. 1986. Effects of soil moisture on soil pCO_2 soil solution bicarbonate and iron chlorosis in soybeans. Soil Sci. Soc. Am. J. 50:946–952.

Jardine, P.M., and L.W. Zelazny, 1996. Surface reactions of aqueous aluminum species. p. 221–270. *In* G. Sposito (ed.) The environmental chemistry of aluminum. Lewis Publ., Boca Raton, FL.

Johnson, C.E. 2002. Cation exchange properties of acid forest soils of the Northeastern USA. Eur. J. Soil Sci.53:271–282

Juo, A.S.R., and E.J. Kamprath. 1979. Copper chloride as an extractant for estimating the potentially reactive aluminum pool in acid soil. Soil Sci. Soc. Am. J. 43:35–38.

Kaiser, K., and W. Zech. 1996. Defects in estimation of aluminum in humus complexes of podzolic soils by pyrophosphate extraction. Soil Sci. 161:452–458.

Kamprath, E.J. 1970. Exchangeable aluminum as a criterion for liming leached mineral soils. Soil Sci. Soc. Am. Proc. 34:252–254.

Karathanasis, A.D., F. Adams, and B.F. Hajek. 1983. Stability relationships in kaolinite, gibbsite, and Al-hydroxyinterlayer vermiculite soil systems. Soil Sci. Soc. Am. J. 47:1247–1251.

Katchalsky, A., and Spitnik, O. 1947. Potentiometric titration of polymethacrylic acid. J. Polymer Sci. 2:432.

Kennedy, I.R. 1992. Acid soils and acid rain. John Wiley and Sons, New York.

Khanna, P.K., R.J. Raison, and R.A. Falkiner. 1986. Exchange characteristics of some acid organic-rich forest soils. Aust. J. Soil Res. 24:67–80.

Kingery, W.L., C.W. Wood, D.P. Delaney, J.C. Williams, and G.L. Mullins. 1994. Impact of long-term land application of broiler litter on environmentally related soil properties. J. Environ. Qual. 23:139–147.

Kirchhof, G., J. Blackwell, and R.E. Smart. 1991. Growth of vineyard roots into segmentally ameliorated acidic subsoils. p. 447–452. *In* R.J. Wright et al. (ed.) Plant–soil interactions at low pH. Kluwer Academic Publishers, Dordrecht, The Netherlands.

Kotze, W.A.G., and J. Deist. 1975. Amelioration of subsurface acidity by leaching of surface applied amendments. A laboratory study. Agrochemophysica 7:39–46.

Lasaga, A.C. 1981. Rate laws of chemical reactions. p. 1–68. *In* A.C. Lasaga and R.J. Kirkpatrick (ed.) Kinetics of geochemical processes. Mineralogical Soc. of Am., Washington, DC.

Lathwell, D.L., and M. Peech. 1964. Interpretation of chemical soil tests. Cornell Agric. Exp. Stn. Bull. 995. Cornell Univ., Ithaca, NY.

Lee, R., B.W. Bache, M.J. Wilson, and G.S. Sharp. 1985. Aluminum release in relation to the determination of cation exchange capacity of some podzolized New Zealand soils. J. Soil Sci. 36:239–253.

Lim, C.H., M.L. Jackson, R.D. Koons, and P.A. Helmke. 1980. Kaolins: Sources of differences in cation-exchange capacities and cesium retention. Clays Clay Miner. 28:223–229.

Lofts, S., B.M. Simon, E. Tipping, and C. Woof. 2001b. Modelling the solid–solution partitioning of organic matter in European forest soils. Eur. J Soil Sci. 52:215–226.

Lofts, S., C. Woof, E. Tipping, N. Clarke, and J. Mulder. 2001a. Modelling pH buffering and aluminum solubility in European forest soils. Eur. J Soil Sci. 52:189–204.

Long, F.L. 1979. Runoff water quality as affected by surface-applied dairy cattle manure. J. Environ. Qual. 8:215–218.

Lund, Z., and B. Doss. 1980. Coastal bermudagrass yield and soil properties as affected by surface-applied dairy manure and its residue. J. Environ. Qual. 9:157–162.

Magdoff, F.R., and R.J. Bartlett. 1985. Soil pH buffering revisited. Soil Sci. Soc. Am. J. 49:145–148.

Magnusson, T. 1992. Studies of the soil atmosphere and related physical site characteristics in mineral forest soils. J. Soil Sci. 43:767–790.

Magnusson, T. 1994. Studies of the soil atmosphere and related physical site characteristics in peat forest soils. For. Ecol. Manage. 67:203–224.

Maison, A., J.Y. Bottero, F. Thomas, and D. Tchoubar. 1994b. Chemistry and structure of Al(OH)/organic precipitates. A small-angle x-ray scattering study. 2. Speciation and structure of aggregates. Langmuir 10:4349–4352.

Maison, A., F. Thomas, J.-Y. Bottero, D. Tchoubar, and P. Tekley. 1994a. Formation of amorphous precipitates from aluminum-organic ligands solutions: Macroscopic and molecular study. J. Non-cryst. Solids 171:191–200.

Marschner, H. 1995. Mineral nutrition of higher plants. 2nd ed. Academic Press, London.

Martin, A.E., and R. Reeve. 1958. Chemical studies of podzolic illuvial horizons. III. Titration curves of organic-matter suspensions. J. Soil Sci. 9:89–100.

May, H.M., P. Helmke, and M.L. Jackson. 1979. Gibbsite solubility and thermodynamic properties of hydroxyaluminum ions in aqueous solution at 25°C. Geochim. Cosmochim. Acta 43:861–868.

McBride M.B. 1994. Environmental chemistry of soils. Oxford University Press, New York.

McKeuge, J.A. 1967. An evaluation of 0.1 *M* pyrophosphate and pyrophosphate-dithionite in comparison with oxalate as extractants of the accumulation products in podzols and some other soils. Can. J. Soil Sci. 47:95–99.

McKeuge, J.A., J.E. Brydon, and N.M. Miles. 1971. Differentiation of forms of extractable iron and aluminum in soils. Soil Sci. Soc. Am. Proc. 35:33–38.

McLean, E.O., and J.R. Brown. 1984. Crop response to lime in the Midwestern United States. p. 267–303. *In* F. Adams (ed.) Soil acidity and liming. 2nd ed. Agron. Monogr. 12. ASA, CSSA, and SSSA, Madison, WI.

McLean, E.O., J.F. Terwiiler, and D.J. Eckert. 1977. Improved SMP buffer method for determination of lime requirement of acid soils. Commun. Soil Sci. Plant Anal. 8:667–675.

Moody, P.W., R.L. Aitken, and T. Dickson. 1995. Diagnosis of maize yield response to lime in some weathered acidic soils. p. 537–541. *In* R.A. Date et al. (ed.) Plant–soil interactions at low pH. Kluwer Academic Publishers, Dordrecht, The Netherlands.

Mulder, J., and A. Stein, 1994. Solubility of aluminum in acidic forest soils: Long-term changes due to acid deposition. Geochim. Cosmochim. Acta. 58:85–94.

Mulder, J., N. van Breemen, and H.C. Eijck. 1989. Depletion of soil aluminum by acid deposition and implications for acid neutralization. Nature 337:247–249.

Nätscher, L., and U. Schwertmann. 1991. Proton buffering in organic horizons of acid forest soils. Geoderma 46:93–106.

Nilsson, S.I. 1993. Acidification of Swedish oligotrophic lakes—Interactions between deposition, forest growth and effects on lake-water quality. Ambio 22:272–276.

Nilsson, S.I., and G. Tyler, 1995. Acidification-induced chemical changes of forest soils during recent decades—A review. Ecol. Bull. 44:54–64.

Noble, A.D., P.J. Randall, and T.R. James. 1995. Evaluation of two coal-derived organic products in ameliorating surface and subsurface soil acidity. Eur. J. Soil Sci. 46:65–75.

Noble, A.D., I. Zenneck, and P.J. Randall. 1996. Leaf litter ash alkalinity and neutralization of soil acidity. Plant Soil 179:293–302.

Nômmik, H., 1883. A modified procedure for the rapid determination of titratable acidity and lime requirements in soils. Act. Agric. Scand. 33:337–348.

Nordstrom, D.K., and H.M. May. 1996. Aqueous equilibrium data for mononuclear aluminum species. p. 39–80. *In* G. Sposito (ed.) The environmental chemistry of aluminum. Lewis Publ., Boca Raton, FL.

Oates, K.M., and E.J. Kamprath. 1983. Soil acidity and liming: I. Effect of the extracting solution cation and pH on the removal of aluminum from acid soils. Soil. Sci. Soc. Am. J. 47:686–689.

Ostatek-Boczynski, Z., G.L. Kerven, and F.P.C. Blamey. 1995. Aluminum reactions with polygalacturonate and related organic ligands. p. 59–63. *In* R.A. Date et al. (ed.) Plant–soil interactions at low pH, principles and management. Kluwer Academic Publishers, Dordrecht, The Netherlands.

Palmer, D.A., and D.J. Wesolowski. 1992. Aluminum speciation and equilibria in solution. II. The solubility of gibbsite in acidic solutions from 30 to 70°C. Geochim. Cosmochim. Acta 56:1093–1111.

Parfitt, R.L., and C.W. Childs. 1988. Estimations of forms of Fe and Al: A review and analysis of contrasting soils by dissolution and Mössbauer methods. Aust. J. Soil Res. 26:121–144.

Parks, G.A. 1967. Aqueous surface chemistry of oxides and complex oxide minerals; Isoelectric point and zero point of charge. p. 121–160. *In* R.F. Gould (ed.) Equilibrium concepts in natural water systems. Adv. Chem. Ser. 67. ACS, Washington, DC.

Parks, G.A., and P.D. Debryun. 1962. The zero point of charge of oxides. J. Phys. Chem. 66:967–973.

Pavan, M.A., F.T. Bingham, and P.F. Pratt. 1982. Toxicity of aluminum to coffee in Ultisols and Oxisols amended with $CaCO_3$, $MgCO_3$, and $CaSO_4 \cdot H_2O$. Soil Sci. Soc. Am. J. 46:1201–1207.

Pearson, R.W., F. Abruna, and J. Vicente-Chandler. 1962. Effect of lime and nitrogen applications on downward movement of calcium and magnesium in two humid tropical soils of Puerto Rico. Soil Sci. 93:77–82.

Peech, M. 1941. Availability of ions in light sand soils as affected by soil reaction. Soil Sci. 51:473–486.

Percival, G.P., D. Josselyn, and K.C. Beeson. 1955. Factors affecting the micronutrient element content of some forages in New Hampshire. New Hamp. Agric. Exp. Stn. 93.

Perdue, E.M. 2001. Modeling concepts in metal–ion complexation. p. 305–316. *In* C.E. Clapp et al. (ed.) Humic substances and chemical contaminants. SSSA, Madison, WI.

Perrott, K.W. 1977. Surface charge characteristics of amorphous aluminosilicates. Clays Clay Miner. 25:417–421.

Porter, W.M., and K.R. Helyar. 1992. Subsurface acidity constraints to agricultural production. *In* Proc. Natl. Workshop on Subsoil Constraints to Root Growth and High Soil Water and Nutrient Use by Plants, Tanunda, South Australia.

Posner, A.M. 1964. Titration curves of humic acid. Int. Congr. Soil Sci., 8th, Vol. II. 17:161–174.

Rakotonarivo, E., J.Y. Bottero, F. Thomas, J.E. Poirier, and J.M. Cases. 1988. Electrochemical modeling of freshly precipitated aluminum hydroxide-electrolyte interface. Colloids Surf. 33:191

Rechcigl, J.E., R.B.J. Reneau, and D.E. Starner. 1985. Effect of subsurface amendments and irrigation on alfalfa growth. Agron. J. 77:72–75.

Reeve, N.G., and M.E. Sumner. 1970. Lime requirements of Natal Oxisols based on exchangeable aluminum. Soil Sci. Soc. Am. Proc. 34:595–598.

Reeve, N.G., and M.E. Sumner. 1972. Amelioration of subsoil acidity in Natal Oxisols by leaching of surface-applied amendments. Agrochemophysica 4:1–6.

Reuss, J.O. 1983. Implications of the calcium–aluminum exchange system for the effect of acid precipitation on soils. J. Environ. Qual. 12:591–595.

Reuss, J.O., and D.W. Johnson. 1985. Effect of soil processes on the acidification of water by acid deposition. J. Environ. Qual. 14:26–21.

Reuss, J.O., P.M. Walthall, E.C. Roswall, and R.W.E. Hopper. 1990. Aluminum solubility, calcium–aluminum exchange, and pH in acid forest soils. Soil Sci. Soc. Am. J. 54:374–380.

Ritchie, G.S.P. 1989. The chemical behaviour of aluminum, hydrogen and manganese in acid soils. p.1–60. *In* A.D. Robson (ed.) Soil acidity and plant growth. Academic Press, Sydney.

Ross, D.S., R.J. Bartlett, and F.R. Magdoff. 1991. Exchangeable cations and the pH-independent distribution of cation exchange capacities in Spodosols of a forested watershed. p. 81–92. In R.J. Wright et al. (ed.) Plant–soil interactions at low pH. Kluwer Academic Publishers, Dordrecht, The Netherlands.

Ross, D.S., M.B. David, G.B. Lawrence, and R.J. Bartlett. 1996. Exchangeable hydrogen explains the pH of Spodosol Oa horizons. Soil Sci. Soc. Am. J. 60:1926–1932.

Runge, M., and M.W. Rode. 1991. Effects of soil acidity on plant associations. p. 181–202. *In* B. Ulrich and M.E. Sumner (ed.) Soil acidity. Springer Verlag, Berlin.

Rustad, L., and C. Cronan. 1995. Biogeochemical controls on aluminum chemistry in the O horizon of a red spruce (*Picea rubens* Sarg.) stand in central Maine, U.S.A. Biogeochemistry 29:107–129.

Sanchez, P.A. 1976. Properties and management of soils in the tropics. Wiley, New York.

Schlichting, E., and L.A. Sparrow. 1988. Distribution and amelioration of manganese toxic soils. p. 277–292. *In* R.D. Graham et al. (ed.) Manganese in soils and plants. Kluwer Academic Publishers, Dordrecht, The Netherlands.

Schwertmann, U., and M.L. Jackson. 1963. Hydrogen-aluminum clays: A third buffer range appearing in potentiometric titration. Science 139:1052–1053.

Shainberg, I., M.E. Sumner, W.P. Miller, M.P.W. Farina, M.A. Pavan, and M.V. Fey. 1989. Use of gypsum on soils: A review. Adv. Soil Sci. 9:1–111.

Sharpley, A.N., B.J. Carter, B.J. Wagner, S.J. Smith, E.L. Cole, and G.A. Sample. 1991. Impact of long-term swine and poultry manure applications on soil and water resources in eastern Oklahoma. Oklahoma State Univ. Tech. Bull. T169. Oklahoma State Univ., Stillwater.

Sivasubramanian, S., and O. Talibudeen. 1972. Potassium–aluminum exchange in acid soils. 1. Kinetics. J. Soil Sci. 23:163.

Skeen, J.B., and M.E. Sumner. 1967a. Exchangeable aluminum: I. The efficiency of various electrolytes for extracting aluminum from acid soils. S. Afr. J. Agric. Sci. 10:3–10.

Skeen, J.B., and M.E. Sumner. 1967b. Exchangeable aluminum: II. The effect of concentration and pH value of the extractant on the extraction of aluminum from acid soils. S. Afr. J. Agric. Sci. 10:303–310.

Skyllberg, U. 1991. Seasonal variation of pH(H_2O) and pH($CaCl_2$) in centimeter layers of mor humus in a *Picea abies* (L.) Karst. stand. Scan. J. For. Res. 6:3–18.

Skyllberg, U. 1993. Acid-base properties of humus layers in northern coniferous forests. Ph.D. diss. Swedish University of Agricultural Sciences, Uppsala.

Skyllberg, U. 1994. Aluminum associated with a pH-increase in the humus layer of a boreal Haplic Podzol. Interciencia 19:356–365.

Skyllberg, U. 1995. Solution/soil ratio and release of cations and acidity from Spodosol horizons. Soil Sci. Soc. Am. J. 59:786–795.

Skyllberg, U. 1996. Small-scale pH buffering in organic horizons of two boreal coniferous forest stands. Plant Soil 179:99–107.

Skyllberg, U. 1999. pH and solubility of aluminum in acidic forest soils: A consequence of reactions between organic acidity and aluminium alkalinity. Eur. J. Soil Sci. 50:95–106.

Skyllberg, U., and O.K. Borggaard. 1998. Proton surface charges determination in Spodosol horizons with organically bound aluminum. Geochim. Cosmochim. Acta. 62:1677–1689.

Skyllberg, U., and T. Magnusson. 1995. Cations adsorbed to soil organic matter—A regulatory factor for the release of organic carbon and hydrogen ions from soils to waters. Water Air Soil Pollut. 85:1095–1100.

Skyllberg, U., K. Raulund-Rasmussen, and O.K. Borggaard. 2001. pH buffering in acidic soils developed under *Picea abies* and *Quercus robur*—Effects of soil organic matter, adsorbed cations and soil solution ionic strength. Biogeochemistry 56:51–74.

Slattery, W.J., and V.F. Burnett. 1992. Changes in soils pH due to long term storage. Aust. J. Soil. Res. 30:169–175.

Smith, C.J., K.M. Goh, W.J. Bond, and J.R. Freney. 1995. Effects of organic and inorganic calcium compounds on soil-solution pH and aluminum concentration. Eur. J. Soil Sci. 46:53–63.

Sposito, G. 1985. Chemical models of weathering in soils. *In* J.I. Drever (ed.) The chemistry of weathering. D. Reidel, New York.

Sposito, G. 1989. The chemistry of soils. Oxford University Press, New York.

Sposito, G. 1992. Characterization of particle surface charge. p. 291–314. *In* J. Buffle and H.P. van Leeuwen (ed.) Environmental particles. Lewis Publ., Boca Raton, FL.

SSSA. 2001. Glossary of soil science terms. SSSA, Madison WI.

Stevenson, F.J. 1994. Humus chemistry: Genesis, composition, reactions. 2nd ed. John Wiley and Sons, New York.

Strobel, B.W., I. Bernhoft, and O.K. Borggaard. 1999. Low-molecular-weight aliphatic carboxylic acids in soil solution under different vegetations determined by capillary zone electrophoresis. Plant Soil 212:115–121.

Stumm, W., and J.J. Morgan. 1996. Aquatic chemistry. John Wiley and Sons, New York.

Suarez, D.L. 1987, Prediction of pH errors in soil-water extractors due to degassing. Soil Sci. Soc. Am. J. 51:64–67.

Sumner, M.E. 1970. Aluminum toxicity—A growth limiting factor in some Natal sands. Proc. S. Afr. Sug. Tech. Assoc. 1–6.

Sumner, M.E. 1993. Gypsum and acid soils: The world scene. Adv. Agron. 51:1–32.

Sumner, M.E. 1995. Amelioration of subsoil acidity with minimum disturbance. p.147–185. *In* N.S. Jayawardane and B.A. Stewart (ed.) Subsoil management techniques. Lewis Publ., Boca Raton, FL.

Sumner, M.E., M.V. Fey, and A.D. Noble. 1991. Nutrient status and toxicity problems in acid soils. p. 149–182. *In* B. Ulrich and M.E. Sumner (ed.) Soil acidity. Springer Verlag, Berlin.

Sumner, M.E., and W.P. Miller. 1996. Cation exchange capacity and exchange coefficients. p. 1201–1229. *In* Methods of soil analysis. Part 3. SSSA Book Ser. 5. SSSA, Madison, WI.

Sumner, M.E., H. Shahandeh, J. Bouton, and J.E. Hammel. 1986. Amelioration of an acid soil profile though deep liming and surface application of gypsum. Soil Sci. Soc. Am. J. 50:1254–1258.

Suthipradit, S., D.G. Edwards, and C.T. Asher. 1990. Effects of aluminum on tap-root elongation of soybean (*Glycine max*), cowpea (*Vigna unguiculata*) and green gram (*Vigna radiata*) grown in the presence of organic acids. Plant Soil 124:233–237.

Sweeten, J.M., M.L. Wolfe, E.S. Chasteen, M. Sanderson, B.A. Auvermann, and G.D. Alston. 1995. Dairy lagoon effluent irrigation: Effects on runoff quality, soil chemistry, and forage yield. p. 99–106. *In* K. Steele (ed.) Animal wastes and the land–water interface. Lewis Publ., Boca Raton, FL.

Theng, B.K.G., M. Russel, G.J. Churchman, and R.L. Parfitt. 1982. Surface properties of allophane, halloysite, and imogolite. Clays Clay Miner. 30:143–149.

Thomas, G.W. 1982. Exchangeable acidity. p. 159–165. *In* A.L. Page et al. (ed.) Methods of soil analysis. Part 2. Agron. Monogr. 9. ASA and SSSA, Madison, WI.

Thomas, G.W. 1996. Soil pH and soil acidity. p. 475–490. *In* D.L. Sparks (ed.) Methods of soil analysis. SSSA Book Ser. 5. SSSA, Madison, WI.

Thomas, G.W., and W.L. Hargrove. 1984. The chemistry of soil acidity. p. 3–56. *In* F. Adams (ed.) Soil acidity and liming. 2nd ed. Agron. Monogr. 12. ASA, CSSA, and SSSA, Madison, WI.

Tipping, E., and C. Woof. 1991. The distribution of humic substances between the solid and aqueous phases of acid organic soils: A description of humic heterogeneity and charge dependent sorption equilibria. J. Soil Sci. 42:437 448

Tisdale, S.L., W.L. Nelson, J.D. Beaton. 1985. Soil fertility and fertilizers. 4th ed. Macmillan, New York.

Turner, R.C., and W.E. Nichol. 1962. A study of the lime potential: Relation between lime potential and percent base saturation of negatively clays in aqueous salt suspensions. Soil Sci. 94:58–63.

Ulrich, B. 1991, An ecosystem approach to soil acidification. p. 28–79. *In* B. Ulrich and M.E. Sumner (ed.) Soil acidity. Springer Verlag, Berlin.

van Breemen, N., J. Mulder, and C.T. Driscoll. 1983. Acidification and alkalinization of soils. Plant Soil 75:283–308.

van Cleve, K., and L.L. Noonan. 1971. Physical and chemical properties of forest floor materials. Soil Sci. Soc. Am. Proc. 35:356–360.

van der Watt, H.V.H., R.O. Barnard, I.J. Cronje, J. Dekker, G.J.B. Croft, and M.M. van der Walt. 1991. Amelioration of subsoil acidity by application of a coal-derived calcium fulvate to the soil surface. Nature 350:146–148.

Van Hees, P.A., J. Dahlen, U.S. Lundström, H. Borén, and B. Allard. 1999. Determination of low molecular weight organic acids in soil solution by HPLC. Talanta 48:173–179.

Van Lierop, W.M., and A.F. Mackenzie. 1977. Soil pH measurement and its application to organic soils. Can. J. Soil Sci. 57:55–64.

van Raij, B. 1991. Fertility of acid soils. *In* R.J. Wright et al. (ed.) Plant–soil interactions at low pH. Kluwer Academic Publishers, Dordrecht, The Netherlands.

van Raij, B., and M. Peech. 1972. Electrochemical properties of some Oxisols and Alfisols of the tropics. Soil Sci. Soc. Am. Proc. 36:587–593.

van Raij, B., A. Pereira de Camargo, H. Cantarella, and N. Machado da Silva. 1983. Aluminio trocavel e saturacao em bases como criterios para recomendacao de calagem. Bragantia 42:149–156.

Wada, K. 1989. Allophane and imogolite. p. 1051–1087. *In* J.B. Dixon and S.B. Weed (ed.) Minerals in soil environments. SSSA Book Ser. 1. SSSA, Madison, WI.

Walker, W.J., C.S. Cronan, and P.R. Bloom. 1990. Aluminum solubility in organic soil horizons from northern and southern forested watersheds. Soil Sci. Soc. Am. J. 54:369–374.

Warfvinge, P., and H. Sverdrup. 1992. Calculating critical loads of acid deposition with PROFILE—A steady state soil chemistry model. Water Air Soil Pollut. 63:119–143.

Weir, C.C. 1975. Effect of lime and nitrogen application on citrus yields and on the downward movement of calcium and magnesium in a soil. Trop. Agric. (Trinidad) 51:230–234.

Wells, C.G., and C.B. Davey. 1966. Cation-exchange characteristics of forest floor materials. Soil Sci. Soc. Am. Proc. 30:399–402.

Wesselink, L.G., N. van Breemen, J. Mulder, and P.H. Janssen. 1996. A simple model of soil organic matter complexation to predict the solubility of aluminum in acid forest soils. Eur. J. Soil Sci. 47:373–384.

Wong, M.T.F., and R.S. Swift. 1995. Amelioration of aluminum phytotoxicity with organic matter. p. 41–45. *In* R.A. Date et al. (ed.) Plant–soil interactions at low pH. Kluwer Academic Press, Dordrecht, The Netherlands.

Wood, C.W., and J.A. Hattey. 1995. Impacts of long-term manure applications on soil chemical, microbiological, and physical properties. p. 419–428. *In* K. Steele (ed.) Animal wastes and the land–water interface. Lewis Publishers, Boca Raton, FL.

Woodruff, C.M. 1948. Testing of lime requirement by means of a buffer solution and the glass electrode. Soil Sci. 66:53–63.

Wright, R.J., J.L. Hern, V.C. Baligar, and O.L. Bennett. 1985. The effect of surface applied soil amendments on barley root growth in an acid subsoil. Comm. Soil Sci. Plant Anal. 16:179–192.

Zettlemoyer, A.C., and E. McCafferty. 1973. Water on oxide surfaces. Croat. Chem. Acta 45:173–187.

Chapter 9

Chemistry of Redox Processes in Soils

RICHMOND J. BARTLETT AND DONALD S. ROSS, *University of Vermont, Burlington, Vermont, USA*

SOIL, perfect home for the actual and figurative roots of all life, including green plants, source of oxygen and food; synthesizer of itself, buffer of pH, pe, and temperature; supplier and recycler of water and carbon, manganese and iron, and all life-essential chemical elements; scavenger of free radicals, toxicity, and disease, cleanser of ecosystems; and lastly, essence of beauty and comfort from Mother Earth. Walk lightly.

The soil redox classification system (Table 9–1) divides soils into three categories with and three without oxidized Mn, and coincidentally into aerobic and oxygen-limited. This coarse separation is probably not capable of representing all soils but does establish a system that distinguishes degrees of redox status from strongly reducing to strongly oxidizing. Redox reactions in the more reduced soils, those containing Fe^{2+} or sulfides, appear to follow common scents and simple thermodynamically predicted reactions occur spontaneously. Although we will delve into wetland interfaces, redox chemistry in the more aerobic categories (I to III), being less straightforward, will be the primary focus of this chapter.

MAGIC OF MANGANESE, HUB OF SOIL REDOX PROCESSES

Almost perfectly poised in the earth's surface environment, Mn, the transcendental transition metal outshines even O_2 because in photosynthesis it is the provider of O_2. Manganese(III) is part of the key enzyme that unlocks oxygen from water in the green plant to become responsible for all the molecular O_2 in the atmosphere of Earth and in its soils and waters (Brudvig and Crabtree, 1989; Thorp and Brudvig, 1991; Schiller et al., 1998; Hanley et al., 2000). Dissolved Mn(III) is probably the strongest oxidant in the environment after O_2 (Laha and Luthy, 1990; Bartlett and James, 1993; Kostka et al., 1995).

Colloidal Mn oxide surfaces have unique properties that suit them for adsorption of both inorganic and organic substances, cationic or anionic. Because they form coatings over surfaces, colloidal Mn oxides exert chemical effects that are well out of proportion to their concentrations. The surfaces created act as unique protectors of life by scavenging heavy metals in soils and in waters by adsorption, complexation, and redox mechanisms (Murray, 1975; Davis and Leckie, 1978; Hem,

Chemical Processes in Soils. SSSA Book Series, no. 8.

Table 9–1. Soil redox classification system, abbreviated summary (Bartlett and James, 1995).

Soil redox category	Summary
Superoxic	Supermanganese free radical Mn^{3+} ion present Yellow with Tetramethylbenzidine
Manoxic	Contains reactive Mn oxides. Cr(III) oxidized
Suboxic	Contains Mn oxides, with reducing organics limiting their reactivity
Redoxic	Soils contain recently oxidized [easily reduced] Fe(III), or, if acidic, Fe(II)
Anoxic	Wetlands. Fe^{2+} present, nitrate absent
Sulfidic	Strongly reducing. Odoriferous, sulfides present

1978; Matocha et al., 2001a; Manceau et al., 2003; Nelson and Lion, 2003). These oxides also catalyze the destruction of oxygen free radicals formed at interfaces by oxidizing, reducing, or dismutating them (Bartlett and James, 1993). Manganese oxides also are involved in the oxidation of toxic organics (McBride, 1987; Ulrich and Stone, 1989), the synthesis of humus (Shindo and Huang, 1982, 1984), and evidence also suggests a role in the breakdown of humus (Sunda and Kieber, 1994). As a key to life and O_2 through photosynthesis, and the source of life, as the original electron parking place, and as protector of life, Mn would stand as the *ultimate element,* were it not for vital Fe, or O, or C, or N, etc.

Manganese oxides in soils are found as surface coatings and small concretions in aerobic surface soils, and nodules in subsurface zones that have fluctuating aeration. Identification of the specific mineral phase has been difficult because of poor crystallinity, the small size of the crystals (Dixon and White, 2002), and confusion over the structure and nomenclature of known Mn oxide minerals (Villalobos et al., 2004). Recent advances in the application of synchrotron-based spectroscopic techniques, especially by Manceau and coworkers (Drits et al., 1997; Friedl et al., 1997; Silvester et al., 1997; Manceau et al., 2003), have increased our knowledge of the structure of Mn oxides in the environment. The "higher" Mn oxides are arrangements of units of Mn(III,IV)O_6 octahedra in sheet or tunnel structures. These are usually associated with Fe oxides (Fe is often at higher concentrations) and nodules have been found to contain mixtures of lithiophorite and birnessite (both Mn oxides) and goethite (Manceau et al., 2003). The mineral phase of Mn oxide coatings (as opposed to nodules) in aerobic soils have yet to be identified but XANES spectroscopy has shown a high percentage of Mn(IV) in oxides of surface soils from Vermont (Ross et al., 2001b) and Indiana (Schulze et al., 1995; Guest et al., 2002). Recent work with lake sediments (Wehrli et al., 1992; Friedl et al., 1997) and biogenic oxides (Villalobos et al., 2004) have found hexagonal or "acid" birnessite, a phyllomanganate with relatively poor stacking and relatively low amounts of structural Mn(III). It is not unreasonable to predict a similar phase in aerobic soils and, for the purposes of this chapter, we will use MnO_2 to represent this mineral phase. Structural Mn(III) will make the actual formula MnO_x, where $1.75 < x < 2$. Reactive sites are thought to be at surface O that lie over a metal vacancy in the lattice structure or over structural Mn(III) (Manceau et al., 1997).

Because of their insolubility, Mn oxides are not directly available to plants. This indispensable oxygen carrier is not leached and it is a rare aerobic topsoil that is not well endowed with it. The typical range in soil is wide, from quite low to more than 0.3%, and some volcanic soils may contain greater than 10% Mn (Gambrell, 1996). We have found a number of high-pH Superoxic soils in Vermont with total Mn content between 1 and 2% (Ross et al., 2001b). Oxidation of Mn^{2+} to MnO_2 prevents its leaching, but Mn oxides are easily reduced by soil organics under wet or acid conditions. Depending on these soil conditions, plants may suffer either Mn deficiency or toxicity, somewhat independent of the total soil Mn concentration.

PROCESSES OF MANGANESE OXIDATION

It is axiomatic that living plants, animals, and microorganisms supply all of the soil organic substances that are redox reactive. Soil microorganisms lower pH by synthesizing and making available phenolic and aliphatic acids, and they increase pH near reactive surfaces by releasing cations through mineralizing organic substances and releasing base forming cations from the organics. Microorganisms also selectively "graze off" and metabolize the most available and electron rich reducing organics, those that would tend to reduce a Mn oxide or interfere with its oxidation.

Because the thermodynamically favored auto-oxidation of Mn(II) by atmospheric O_2 cannot be demonstrated unless the pH is above 8 (Diem and Stumm, 1984) and because rates of Mn(II) oxidation in aquatic systems are accelerated in the presence of microbial populations (Nealson et al., 1988), it has been commonly assumed that formation of Mn-oxides in the environment requires specific microbial Mn oxidizers, and soil microorganisms have been studied in this regard (Ehrlich, 1975; Silver et al., 1986; Sparrow and Uren, 1987). There is little doubt that microbial metabolites influence Mn oxidation and that a wide variety of specific bacteria can bring about the oxidation of Mn(II) to Mn(IV) (Tebo et al., 1997). Abiotic oxidation at high pH produces mixed Mn(III, IV) oxides (Junta and Hochella, 1994) while the products of microbial oxidation have usually been found to be primarily Mn(IV) (Mandernack et al., 1995; Villalobos et al., 2004). Unequivocal demonstration of microbial oxidation in soil has not been possible because the lack of Mn-oxide formation in the presence of a chemical microbial inhibitor cannot be used as evidence for dismissing the importance of abiotic mechanisms of Mn oxidation. Heat or radiation sterilization of soils or chemical inhibitors, e.g., chloroform or sodium azide, will reduce MnO_2 in any soil containing organic acids and will destroy a soil's natural Mn oxidizing mechanism (Ross and Bartlett, 1981). Autoclaving, metabolic inhibitors, even simple soil drying will increase Mn(II) concentrations (Goldberg and Smith, 1984; Berndt, 1988; Ross et al., 2001b) and destroy or greatly lower the ability of any soil to oxidize Cr(III) to Cr(VI) (Bartlett and James, 1980; Ross et al., 2001b). Such techniques might lead one to conclude that Cr oxidation is a biological process. Yet, in truth, such a conclusion would be false. It is also difficult to unequivocally demonstrate that Mn oxidation is an abiotic process in soil. Ross and Bartlett (1981) and Wada et al. (1978) both presented evidence for nonbiological oxidation in soils already containing Mn oxides. They hypothesized that the oxidation was autocatalytic and that fresh Mn oxides tended to form on old oxide surfaces. These reports are more than 20 years old but there have not been any additional findings either supporting or refuting the abiotic pathway in the soil system.

We could say that is not the nature of natural environments to provide a clear, or even a somewhat fuzzy, separation of purely abiotic from purely biologically mediated redox reactions. Sometimes moderately fuzzy is perfectly acceptable to a redox soil chemist who is mainly interested in understanding behavior of soils in the field, which is the only place where the soils really exist. At the molecular level, all redox reactions are chemical. But it is also correct to consider that most of the interacting molecules are biologically derived or grounded. Soil is a biological entity.

OXIDATION OF ORGANICS AND INORGANICS ON OXIDE SURFACES

The oxidation of Cr(III) by soil Mn oxides was first demonstrated by Bartlett and James (1979). Previous attempts to show this reaction had failed because of the accepted practice of using laboratory dirt (dried, sieved, and stored ex-soil) in these types of experiments (Bartlett and James, 1980). Much subsequent work has demonstrated that added Cr(III) salts are readily oxidized to Cr(VI) in a wide variety of soil types as long as they contain oxidized Mn (Bartlett, 1990, 1991; Kim et al., 2002; Ross et al., 2001b). The oxidized Cr(VI) is mobile and quite reactive, and therefore does not persist in most surface soils (Bartlett and James, 1988). Organically-complexed Cr(III) is not as reactive, which is why it is used instead of tannins in tanning leather; however, Cr(VI) has been found in groundwater near Cr(III) sludge waste sites and presumably arrived there after oxidation in the soil (Naidu et al., 2000). Oxidation by both synthetic and natural Mn oxides has been shown for contaminant metals such as Co(II) (Manceau et al., 1997) and Pu(III) (Amacher and Baker, 1982) and the nonmetals As(III) (Manning et al., 2002) and Se(IV). In some cases, the oxidized species is more soluble and mobile through oxyanion formation and, in some cases, the result is immobilization. Oxidation of hydroquinone (Stone and Morgan, 1984) and polyphenols (McBride, 1987) by MnO_2 was observed in the 1980s and subsequent work has demonstrated oxidative reactions with a wide variety of organics, both contaminant and natural. In most, but not all cases, Mn^{2+} is a product and may be released into solution. Reoxidation of the Mn(II) can be rapid and cloud the definition of the reaction, from the MnO_2 point of view, as either catalytic or reductive dissolution.

The mechanism for oxidation by MnO_2 of both inorganics and organics (e.g., polyphenols) appears to be sorption followed by electron transfer through surface oxygen bridges (Fendorf and Zasoski, 1992; Manceau et al., 1997; Matocha et al., 2001b). Many of the reactions involve more than a single e^- transfer and the reaction may proceed either stepwise or, if the sorption is multidentate, through simultaneous transfer. Questions remain as to the importance and occurrence of intermediate species of both Mn, i.e., Mn(III), and the sorbed substance, i.e., Cr(IV) and Cr(V). Nico and Zasoski (2000, 2001) presented evidence that surface Mn(III) sites in synthetic birnessite were the controlling factors in Cr(III) and phenol oxidation. Manceau and coworkers (1997) proposed two mechanisms for the oxidation of Co(II) by synthetic buserite (hydrated birnessite), both involving Mn(III) and leading to a change in the oxide towards the "acid" birnessite mentioned in the previous section. Other recent work (Banerjee and Nesbitt, 2001; Kim et al., 2002) has emphasized the importance of Mn(IV) in the reaction mechanism. These uncertainties are exacerbated by the fact that synthetic MnO_2 has been prepared by a variety of methods that result in different surface characteristics (Villalobos et al., 2004) and that there appear to be pronounced differences between the behavior of surface soil Mn oxides and synthetic or natural mineral oxides. Similar differences have been found between biogenic Mn oxides and synthetic oxides, especially in metal sorption behavior (Tebo et al., 1997; Nelson and Lion, 2003). Both soil and biogenic oxides appear to be less crystalline, resulting in more reactive surface area. Future work, especially with synchrotron-based spectroscopic methods, should clarify both the structure of non-nodule soil Mn oxides and the mechanisms of surface reactivity. As discussed in the next section, Mn(III) may play a central role in reactions taking place in mixed systems such as soils.

There are many possible immediate fates of the oxidized substance. The Cr(VI) oxyanion is released from the oxide surface, probably through electrostatic repulsion. Other metals, such as Co(III), are incorporated into the Mn oxide structure (Manceau et al., 1997). Products of reactions with organics range from gaseous CO_2 to polymers that sorb to the oxide surface (Majcher et al., 2000). Soluble organic intermediates may further react with the oxide surface, with each other, or migrate and react in other areas of the soil. Accumulation of products on the Mn oxide surface usually leads to a decrease in the reaction rate, presumably through blocking the reactive sites. Arsenic(V) may accumulate on the oxide surface (Manning et al., 2002) or precipitate with Mn(II) (Tournassat et al., 2002). The effect of soil drying on the decrease in oxidation capacity is probably related to desiccation-induced reduction of some surface Mn(III) or (IV) and blockage of reaction sites by Mn(II) (Ross et al., 2001b). Addition of Mn(II) to a moist oxidizing soil (Superoxic or Manoxic) will initially decrease the soil's ability to oxidize added Cr(III) but, after 1 or 2 d of incubation, result in a higher oxidation capacity as a result of newly formed, "fresh" oxide surfaces (Ross and Bartlett, 1981). The dynamic nature of soil Mn oxides probably leads to a restoration of a new oxide surface following reaction.

MANGANESE(III) AND REVERSE DISMUTATION

Ionic Mn^{3+} is unstable in soils, being readily dismutated, or disproportionated, to the two- and four-valent forms as shown in the thermodynamically spontaneous dismutation reaction in Eq. [1]. The reverse of this reaction can be easily demonstrated by combining one drop each of 0.1 *M* $MnSO_2$, 0.1 *M* amorphous MnO_2 suspension, and 0.2 *M* KH_2citrate. This reaction, shown in Eq. [2], produces a tawny-colored Mn(III)citrate solution.

Dismutation:

$$2Mn^{3+} + 2H_2O \rightarrow Mn^{2+} + MnO_2 + 4H^+ \quad [1]$$

$$\log K = +9.1$$

Reverse Dismutation:

$$MnO_2 + Mn(II)citrate^- + citrate^{3-} + 4H^+ \rightarrow 2Mn(III)\text{-}citrate + 2H_2O \quad [2]$$

$$\log K \sim +18 \text{ (Klewicki and Morgan, 1998)}$$

In the reverse dismutation reaction, Mn(II) gives up an electron to Mn(IV) as each becomes Mn(III). This reverse of dismutation is spontaneous only when it is driven by the energy of complex formation between Mn(III) and an organic ligand (Bartlett, 1988; Bartlett and James, 1993). Sorption of organics on MnO_2 surfaces creates local conditions favorable for this reaction. Thus, both dismutation and its reverse are thermodynamically spontaneous under soil conditions.

Apparently for reverse dismutation to occur, the organic acid must be oxidizable by the Mn(III) bound to it. Only acids with O attached to the C adjacent to a carboxyl are amenable to oxidation by Mn(III) (Waters and Littler, 1965) and can be shown to chelate Mn(III) and drive the reverse dismutation. Oxalic, citric, lactic, and tartaric acids are oxidized by Mn(III), and will drive reverse dismutation, but not succinic acid, even though it chelates both Fe(III) and Al(III). EDTA will complex Mn(III) but it is much less stable than oxidizable citrate (Klewicki and Morgan, 1998). Klewicki and Morgan (1998) suggested that the greater stability may be partly due to the ability of citrate, when in excess, to catalyze the oxidation of Mn(II) by O_2.

Most of the MnO_2 in a sample of field-moist soil can be converted temporarily to Mn(III)-citrate by simply substituting about 1 cm^3 of the soil for the MnO_2 suspension in the reverse dismutation demonstration (above). Testing the tawny solutions with tetramethylbenzidine (TMB) will produce the yellow color of a two electron oxidation, identifying Mn(III) (Bartlett, 1998). Dion and Mann (1946) extracted significant quantities of Mn(III) from soils with pyrophosphate, originally interpreting their results as evidence of stable Mn(III) in their soils. Both citrate and pyrophosphate form Mn(III) complexes through reverse dismutation reactions that have kinetic stability, especially with an excess of ligand (Klewicki and Morgan, 1998). In the case of citrate, and presumably other hydroxy-carboxylic acids, stability is limited because a portion of the Mn(III) is reduced while citrate is partially oxidized to form citrate free radicals (Senesi and Schnitzer, 1978). The citrate free radical converts the tawny solution from Eq. [2] into a powerful reducing solution. A small amount of this solution will readily reduce Cr(VI); however, the pH is increased as some of the newly-formed Mn(III) is reduced to Mn(II) and the citrate is slowly oxidized by it to CO_2 and reducing free radicals. The decomposition stops when the pH reaches 7.6, and the Mn(III)-citrate can remain stable almost indefinitely. Neutralizing a tawny pH 6 reducing solution to pH 7.6 by adding base will instantly convert the reducing solution to one that will oxidize added Cr(III) to Cr(VI). The change in pH induces the remaining Mn(III) that is still bound by citrate to form the free radical, Mn^{3+}, an extremely powerful oxidant (R.J. Bartlett, 1989–1994, unpublished data).

True or false? (i) A soil high in Mn oxides will readily oxidize added Cr(III). (ii) The same soil, high in Mn oxides, will readily reduce added Cr(VI). Both are true statements. The explanation is simple enough. The Mn(III) ion, Mn^{3+}, is a powerful oxidizer, when it is being reduced at low pH to Mn(II), and Mn(III) at higher pH is a powerful reducer when it is being oxidized to MnO_2.

Under laboratory conditions, Mn(III)-citrate or oxalate, formed by reverse dismutation in solutions or in soil samples, reduced Cr(VI) at pH 4–6. When citric acid or oxalic acid was added to a soil already containing Mn(II) and Mn(IV), Mn(III)-citrate was formed, which reduced Cr(VI) much faster than the organic acid alone (Bartlett, 1988). The Mn(III)-organic acids also effectively reduced nitrate. In other experiments (R.J. Bartlett, 1989–1994, unpublished data), we found that Mn(III)-citrate reduced methylene blue and tetrazolium blue and oxidized TMB two electron steps to the yellow color. Some reduction of Cr(VI) will occur when it is added to a soil already containing Mn(II), MnO_2 and a small amount of citrate. Likewise, some oxidation of Cr(III) will occur if soluble Cr(III) is added to a soil already containing Mn(II), MnO_2, and a small amount of citrate. Mn(III)-organic is especially effective as a reducer at low pH; MnO_2, without too much organic, effectively oxidizes Cr(III) at almost any soil pH, provided the Cr(III) is soluble, which, of course, soil Cr(III) rarely is. The balance between pH, reducing organics, complexing organics, Mn(II), Mn(III), and Mn(IV) is critical and redox reactions will vary as conditions change over time and over short distances.

Pyrophosphate drives the reverse dismutation of Mn(II) and Mn(IV) reaction by binding strongly to Mn(III) to form a violet-pink complex (Dion and Mann, 1946). Loss of complex color accompanying the oxidation of organic C compounds in solution by the Mn(III) serves as a simple colorimetric method for measuring dissolved organic C (Bartlett and Ross, 1988). Mn(III)-pyrophosphate is quite stable if kept in the dark, and it is a useful reagent for studying behavior of Mn(III).

The cycling of Mn(III) in a living soil is driven by the cohabitation of Mn(IV) oxides with Mn(II) (created by reductive dissolution primarily by reducing organics) and complexing organics from microbial or root exudation. In the sections that follow, applications of reverse dismutation reactions will be called upon to explain the mechanisms of redox processes in almost all soil redox processes involving Mn transformations, especially in compacted soils and root–soil and sediment–water interfaces.

HYPOTHESIZED MANGANESE–FREE RADICAL REDOX PROCESSES

$$(COO)_2\text{–}Mn(III)^+ \rightarrow C_2O_4^{\bullet -} + Mn^{2+} \quad [3]$$

$$C_2O_4^{\bullet -} + NO_3^- + 2H^+ \rightarrow NO_2^- + 2CO_2 + H_2O \quad [4]$$

$$C_2O_4^{\bullet -} + O_2 + H^+ \rightarrow HO_2 + 2CO_2 \quad [5]$$

$$HO_2^{\bullet} + H^+ + Mn^{2+} \rightarrow Mn^{3+} + H_2O_2 \quad [6]$$

$$2Mn^{3+} + 2H_2O_2 \rightarrow Mn^{2+} + MnO_2 + O_2 + 4H^+ \quad [7]$$

$$Mn(II) + 2H^+ + 2OH^{\bullet} \rightarrow Mn(IV) + 2H_2O \quad [8]$$

Free radicals are normally short-lived but can transfer e^- via numerous pathways, linking various soil processes. Equations [3] through [8] are arranged in a possible sequential order. In Eq. [3], Mn(III) oxidizes a single carboxyl group attached to it by a single electron step, becoming reduced to Mn(II). In giving up an odd electron from one of its carboxyls to Mn, the oxalate molecule becomes a reducing free radical with an extremely strong disposition to act as a reducing agent by giving up its remaining odd electron (Waters and Littler, 1965). In Eq. [4], the same oxalate free radical reduces nitrate to nitrogen dioxide, a gas free radical and, in Eq. [5], it reduces dioxygen to the protonated superoxide free radical, becoming oxidized to CO_2 in both reactions (Fridovich, 1975). In Eq. [6], the protonated superoxide reoxidizes Mn^{2+} to Mn^{3+} and is itself reduced to hydrogen peroxide in a thermodynamically spontaneous reaction (Halliwell, 1974). In Eq. [7], one of the hydrogen peroxide oxygens gives up a single electron to Mn^{3+} to become a biodestructive hydroxyl free radical (Pryor, 1976), toxic to all living cells. In Eq. [8], the reduced Mn(II) scavenges the hydroxyl free radical converting it to ordinary harmless water and is oxidized to Mn(IV) (Archibald and Fridovich, 1981). Equation [7] is the sum of Eq. [5] (× 2) and Eq. [8]. Dioxygen is formed in a thermodynamically spontaneous reaction, and a stable fresh Mn oxide surface with adsorbed Mn(II) is established:

$$\text{H-quinone} + MnO_2 + 3H^+ \rightarrow \text{quinone}^{\cdot} + {}^{\cdot}Mn^{3+} + 2H_2O \qquad [9]$$

Equation [9] shows the MnO_2 oxidation of hydroquinone by a single electron step to form two free radicals, the semiquinone free radical and the supermanganese free radical (Bartlett and James, 1993), Mn^{3+}. Free radicals could be the keys to the linking together of reducing phenolic compounds into humic polymers that are stable in the presence of O_2. These are soil reactions that occur in darkness, where Mn(IV) and Mn^{3+} seem to do mighty oxidative work, in the darkness of soil or in the "dark" reaction of photosynthesis. Mn(III) appears down in the dark soil again where it does powerful *reductive* redox work.

Because the hydroxyl ($OH^{\bullet}$) and the superoxide ($O_2^{\bullet -}$) free radicals are among the few species having the thermodynamic capability for oxidizing Mn(II), Mn is one of the few elements capable of scavenging these radicals and protecting life from destruction by them. By scavenging free radicals, Mn disrupts the tendency toward thermodynamic equilibrium between O_2 and soil organic matter or between soil organic matter and living roots or microbes. Thus, metastable humus and roots persist in an oxidative environment. The following equations demonstrate both oxidative and reductive reactions of the inorganic free radicals with Mn:

$$Mn^{2+} + 2OH^{\bullet} \rightarrow MnO_2 + 2H^+ \qquad \log K = +53.2 \qquad [10]$$

$$3Mn^{2+} + 2H_2O + 2HO_2^{\bullet} \rightarrow 3MnO_2 + 6H^+ \qquad \log K = +54.2 \qquad [11]$$

$$MnO_2 + 2OH^{\bullet} \rightarrow Mn^{2+} + 2O_2 + 2OH^- \qquad \log K = +28 \qquad [12]$$

$$MnO_2 + 2HO_2^{\bullet} + 2H^+ \rightarrow Mn^{2+} + 2O_2 + 2H_2O \qquad \log K = +62 \qquad [13]$$

$$Mn^{2+} + 2OH^{\bullet} + H_2O \rightarrow MnO_2 + OH^- + 3H^+ \qquad \log K = +38 \qquad [14]$$

In Eq. [12] and [13], Mn oxides scavenge the toxic hydroxyl and superoxide free radicals. This property of the oxides has been proposed as one of the competitive advantages for microbial oxidation of Mn (Tebo et al., 1997). Complexed Mn(III) also should react with the free radicals. Synthesized Mn(III)-porphyrins have been shown to mimic superoxide dismutase (Faulkner et al., 1994). Much of the work with free radicals referred to and discussed here originated in the medical and animal-related biochemical literature. Powerful reducing agents, including ascorbic acid, tocopherol, carotene, and bio-flavanoids, contained in natural fruits, vegetables, red grapes and cacao berries are extolled for their benefits to human health and are being extracted and synthesized and sold to us as "anti-oxidant" vitamins. It seems likely that in relatively unsophisticated soil systems, microorganisms and higher plants are rescued unpredictably from damage by free radicals and peroxides by a variety of reducing soil microbial metabolites. These naturally occurring reduced organic and inorganic substances, the "anti-oxidants" of the soil, render the toxic compounds harmless by donating electrons to them. Successful agriculture down through the ages has depended upon the additions to soils of various organic plant, animal, and human waste materials that rather indiscriminately donate electrons wherever there is a presence of reducible moieties. Without doubt, sustainable survival of life on earth is fundamentally dependent on this seemingly haphazard electron donating system that prevails in the soils of our earth.

One could argue that this is why we need manure and organic farming—to get electrons to get MnO_2 reduction so we can get active redox—that is, the main ingredients of a soil are lime and Mn^{2+} to get MnO_2. Also needed is manure or other organic additions so that we can accumulate OM. We need the OM to get some Mn^{2+} and start reverse dismutation; then we get Mn^{3+} and an active redox setup and from that we get oxidative polymerization and humus. Then we get a soil that will grow things to eat.

CATALYTIC REDOX PROCESSES

Given: Simple organic compounds including physically disrupted plant residues, various plant and animal parts, garbage, feces, and other organic wastes, are all ready electron donors finding themselves in close proximity to the matrix of a well-endowed soil. Visualize these substances as oozing electrons. They are ripe for plucking, that is, oxidation. The electron acceptors will come. Oxygen at the soil surface and in soil pores will capture the most available of the electrons. But for the moment, let us consider electron acceptors that are in and of the soil. There is an innumerable abundance of enzymes from myriads of bacteria, molds, and fungi. There are metals, especially Fe(III), Cu(II), and Mn(III) and Mn(IV), and frequently the metals are involved with the enzymes.

Because it is so readily oxidized by O_2, Fe can catalyze the oxidation of simple electron donors that also complex Fe(III). Phenolic compounds are especially easy to oxidize. A phenolic becomes oxidized to a quinone when it reduces the Fe(III) to Fe(II). The Fe then is reoxidized by O_2 (Brudvig and Crabtree, 1989), and the Fe(III) formed is complexed and reduced by the excess phenolic compound, and so on. Other simple electron-rich organics from plant residues are oxidized and decayed by reacting with Mn, Fe, or Cu by similar mechanisms.

Even at temperatures too low for most bacteria, trace levels of Fe, Mn, or Cu can catalyze the oxidation and spoilage of foods by comparable mechanisms. Citric acid is added to prepared foods to bind the Fe(III) and prevent its complexation and then reduction by easily oxidized food organics; however, in light Fe complexed by citrate may still promote the decomposition of foods, although more slowly. When light is present, oxidation of both citrate and easily oxidized food will be catalyzed by Fe. Thus it *is* important to know that the light in the refrigerator goes out when you close the door.

In soils and other biological systems, Fe appears to be more important than Mn in filling certain catalytic roles. Because its oxidation is so much easier than that of Mn, Fe is more effective than Mn in catalyzing complete oxidation of organics. In animals, Fe is sovereign as the chief carrier of oxygen in blood. It is a carrier of electrons in cytochrome and peroxidase enzymes in both plants and animals. In the heme structure of cytochrome C, Fe is reversibly oxidized and reduced, but in catalases and peroxidases, also heme compounds, the Fe remains trivalent. The Fe in hemoglobin must be in the ferrous form in order for it to combine reversibly with O_2 and act as the all important carrier of oxygen in blood (Fruton and Simmonds, 1961). If the Fe is oxidized to Fe(III), methemoglobin is formed, which does not combine with O_2. The danger of nitrate in drinking water, spinach, or forage plants is that nitrite, formed by reduction of the nitrate in an infant's intestine, or in the rumen of a cow, oxidizes hemoglobin to methemoglobin causing methemoglobinemia, the inability of blood to carry oxygen. Iron in hemoglobin with attached oxygen is a powerful oxidant. It is conceivable that related compounds could be important in soil oxidation processes, but such have not been reported.

Manganese, although essential in animals, shows some toxicity tendencies that may be partly caused by its free radical prowess. Manganese is more likely than Fe to partially oxidize organic residues, thereby creating free radicals, and setting them up for polymerization or further microbial breakdown. Wang and coworkers (1999) demonstrated that in the absence of O_2, MnO_2 catalyzes the dealkylation of atrazine. Manganese oxides act as catalysts in dismutating H_2O_2 in near neutral or alkaline soils (Eq. [15]). In more acid soils or in the presence of chelating ligands, MnO_2 oxidizes H_2O_2 (Eq. [16]). In order to be considered a "true" catalyst how fast must MnO_2 reform in its original state by oxidation of the resulting Mn(II)? Manganese(III) can act as an enzyme-like reductase, or oxidase:

$$2H_2O_2 \xrightarrow{MnO_2} 2H_2O + O_2 \qquad [15]$$

$$H_2O_2 + MnO_2 + 2H^+ \rightarrow 2H_2O + O_2 + Mn^{2+} \qquad [16]$$

Metal chelating ligands occurring naturally among soil organic residues, root exudates, and microbial by-products commonly form enzyme-like compounds with soil metals and metal oxides, and these substances appear to have major effects on N transformations and important soil redox reactions, although these effects have received little study. Unless we decide that to be considered a true enzyme, a catalytic substance must do its job within the confines of a living cell, there can be no clear boundaries between these enzyme-like compounds and so-called true enzymes. Manganese peroxidases are formed by some species of wood-decaying fungi (Tebo et al., 1997). This Mn(III) complex is an extracellular enzyme that diffuses into the soil and depolymerizes lignin, apparently working in concert with a non-Mn enzyme, lignin peroxidase (Perez and Jeffries, 1992). Lignin peroxidase will also catalyze the oxidization Mn(II). Laccase, another extracellular oxidase with Cu cofactors, again will catalyze the oxidization of Mn(II) and Schlosser and Höfer (2002) showed that the formation of Mn(III) complexes with oxalate and malonate (found in nature in the soil) results in a pathway to peroxide formation, which in turn is required for the lignolytic peroxidases. Oxalate and MnO_2, both common in wood decay in soil, will abiotically oxidize lignin (Hames et al., 1998), and is one of the first well documented Mn(III)-organic oxidation reactions in soil. The somewhat complex system for lignin degradation is an example of microbial enzymes working with soil transition metals and complexing lignins. Our nomenclature may not be adequate to categorize the components of this system.

Nitrogen redox couples also function as catalysts in soils. As nitrate accepts an electron from an organic compound, the nitrate is reduced to nitrite and the organic is oxidized. When nitrite is oxidized again by autotrophic *Nitrobacter* or by MnO_2, atmospheric O_2 becomes the terminal electron acceptor. When nitrite is oxidized to nitrate by MnO_2, the Mn(II) formed, on reoxidation, is acting as an electron carrier by transferring two electrons from nitrite to O_2.

REDOX PROCESSES IN THE SYNTHESIS AND STABILIZATION OF HUMUS

A fresh, field moist soil of good tilth has organic matter being synthesized. A worn out field soil in poor tilth has organic matter in the process of depletion. An extreme case of organic matter depletion is lab dirt—soil that is dried, stored, and stale.

Redox processes of manganese and iron, together with plant roots and organic residues, are responsible for the synthesis of humus, the essence of the soil A horizon, known generically everywhere as topsoil. Topsoil, the humified organic–mineral mantle covering a portion of the surface of the planet Earth, is the biological foundation of life. Generic topsoil serves as the nurturing home for the roots of all plants and C recycling microorganisms. It is the ultimate provider of food for all of the inhabitants of Earth. Clays have been called the backbone of topsoil; humus is its brain and spinal cord.

Although oxidized Fe is an essential link in the process of humus stabilization, the role of Mn redox is singularly preeminent in the domain of darkness beneath the soil surface where the oxidative synthesis of humus appears to take place. Manganese oxidizes and catalyzes the transformations of organics. In acid soils and in poorly drained soils, both Fe and Mn are important in forming A horizons, but in well drained non-acid soils, soils that rarely reduce it, Fe is secondary to Mn.

Oxidative Polymerization by Manganese

A browning polymerization reaction takes place when a sugar and an amino acid (e.g., glucose and glycine) are boiled together (the Maillard reaction) (Bartlett,

1988; Jokic et al., 2001). A polymer of similar appearance forms at room temperature when amorphous MnO_2 is added to the same solutions. Iron oxides produce little browning effect. Amorphous Mn(IV) oxides also will rapidly cause browning of various phenolic compounds (Bartlett, 1988) and may result in the synthesis of humic acids (Shindo, 1990). The browning product of hydroquinone by MnO_2 has spectral characteristics of polymerization (Shindo and Huang, 1982, 1984). Numerous studies have shown polymeric products formed in the presence of Mn oxides (Pohlman and McColl, 1988; Naidja et al., 1998; Majcher et al., 2000). The effects of Fe and Mn oxides, and other soil minerals, on abiotic catalysis of oxidation and browning and polymerization of phenolic compounds are discussed in a review by Wang et al. (1986).

Our observations indicate that very fresh recently formed MnO_2 is most effective in bringing about oxidative polymerization reactions (R.J. Bartlett, 1989–1994, unpublished data). Conspicuously least effective are crystalline manganese oxides, laboratory reagent oxides, oxides in old and dried stored soil samples (often labeled "lab dirt" to emphasize their relative uselessness for investigating soil redox processes). When Mn(III) is added to a glucose–glycine solution, polymerization is weak and the product is lighter in color than that formed from fresh MnO_2. The freshening of MnO_2 for maximum reactivity is a stepwise process beginning with seasoned and relatively old and stable soil Mn(IV) oxides and charged Mn(II) species. The rhizosphere where microbial life and organic acid exudates from live roots thrive and flourish is a likely locale for the process of reverse dismutation [discussed in Manganese(III) and Reverse Dismutation], the immediate result of which is Mn(III). Microorganisms oxidize the organics that drive the reverse dismutation and also produce by oxidation the first harvest of freshly formed and reactive MnO_2. The second harvest arrives when the surplus Mn(III) formed by reverse dismutation dismutates back to fresh MnO_2 and Mn(II), which is handily adsorbed onto the MnO_2, and which then is in good position to be oxidized to constitute the third harvest of fresh and highly reactive MnO_2 (Fig. 9–1).

Manganese(III) has the potential for creating semiquinone and organic acid free radicals from fulvic acids, tannins, and simpler phenolic acids (Eq. [3]). Like Mn(III) itself, an organic free radical can act as either a reducing agent or an oxidizing agent and help induce polymerization of humic substances to form stable humus (Schnitzer and Khan, 1972; Senesi and Schnitzer, 1978; Bartlett, 1998). These organic free radicals in humus serve as "electron pathways" in transporting electrons to oxidized species. Schnitzer and Khan (1972) proposed a model in which unstable free radicals induce humification, and presence of stable free radicals in organic matter denotes that humification already has taken place. Sulflita et al. (1981) hypothesized that a hydroxyl free radical dehydrogenates a phenolic substrate forming phenoxy free radicals, which should spontaneously couple among themselves or attach to the polymeric structure of soil humus. Regardless of the pathway, Mn(III) probably plays a central role.

A and E Horizons in Spodosols

We became aware of the Mn–humus connection in comparing profiles of acid forest soils in northeastern USA and southeastern Canada during a period of 20 or

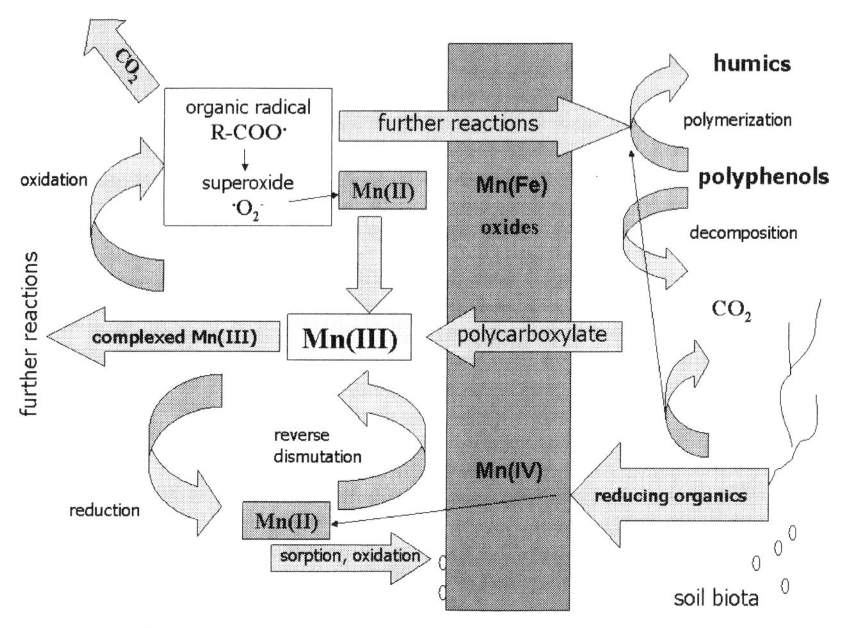

Fig. 9–1. Simplified Mn cycle in the rhizosphere of an aerobic soil showing how inputs of organics provide reductive power to create Mn(II), complexing ligands to temporarily stabilize Mn(III), and building blocks of humic substances. Possible free radical reactions are too numerous to mention.

more years. Some Spodosols had well developed leached, bleached, low-organic E horizons underlying acid black organic O horizons. Others had mull-type A horizons on the surface consisting of stabilized intermixed mineral and humified organic material, without E horizons beneath them. Profiles that had humified A horizons always tested positive for Mn-oxides by the TMD test throughout their profiles. Spodosol profiles with well developed E horizons generally were lacking in Mn-oxides by the TMB test. Subsequent chemical analyses of 41 pedons that had spodic horizons directly under E horizons and 37 pedons with spodic horizons directly beneath A horizons, pointed to the conclusion that Mn-oxides were associated with absence of developed E horizons and with presence of A horizons (Bartlett, 1990).

Two Mn related redox processes appear to be involved in the formation of A horizons in lieu of E horizons. (i) By processes discussed in Redox Processes within Rhizospheres, freshly formed reactive MnO_2 catalyzes the synthesis of humus in rhizospheres of the forest floor and (ii) Mn(III) formed by reverse dismutation of Mn(II) and MnO_2 [discussed in Mn(III) and Reverse Dismutation] catalyzes the decomposition of the organic acids that drove the reverse dismutation reaction. Without catalytic decomposition, these same organic acids would be available to bind Fe and Al and move it from an incipient A horizon, thereby causing it to develop as an E horizon, depleted in both humus and metal cations. Majcher and coworkers (2000) showed that catechol oxidation on synthetic birnessite at pH 4 resulted in both polymerization products and CO_2 evolution. The competing reactions mimicked respiration and biosynthesis but were abiotic.

Manganese will tend to be reduced and then leached from acid soils developed from parent materials low in Mn minerals. In the absence of Mn redox in such soils, most of the accumulated humified organic matter consists of fulvic acids bound by Al in the B horizon beneath an E horizon that is depleted of Mn, Al, Fe, and organic matter. Podzolized soils long cropped to potatoes (*Solanum tuberosum* L.) and fertilized with N, but never limed, are practically devoid of Mn. Liming exacerbates already severe Mn deficiencies in these soils. In the extreme case of a Aquod, developed from poorly drained sandy parent material in a warm climate, both Fe and Mn become reduced and migrate through an E horizon, which frequently is more than a meter thick. An A horizon will not form in such a profile, almost totally lacking Mn. This is an example of the white sand, black water phenomenon described by Jenny (1988), i.e., a pale mineral E horizon disassociated from organic matter that has been mobilized from the soil profile, giving organic color to the leaching waters.

Not only does Mn have a direct role in humus synthesis, it may have an indirect role in acid soils by oxidizing Fe (Bartlett and James, 1993). The Fe(III) thus formed should be more reactive with complexing organics than Fe(II) oxidized by O_2. A variety of soluble phenolic compounds, e.g., tannic acid, gallic acid, ordinary black tea brewed from partially humified leaves of *Thea sinencis* or tea-like infusions made from dried leaves of maple, oak, or other trees, all form insoluble black precipitates when Fe(III) salts are added to them. It seems likely that such resistant substances become part of metastable humus. Besides oxidizing Fe, the Mn oxides partially oxidize various phenolic acids to free radicals, which are now raw materials for oxidative polymerization.

Stabilization of Organic Matter in Near-Neutral Soils

Manganese and humus are intimately associated in A horizons, and these horizons are best expressed in near neutral high-Ca parent materials. Freshly formed Mn oxides are amorphous with measurable CECs and have strong tendencies to adsorb Mn(II) and other metal cations. Electrophoretic mobility demonstrates that adsorption of Mn(II) reverses the charge on a Mn oxide from negative to positive (Bartlett, 1988). Thus, initial sorption of negatively charged humus may be electrostatic. Stable complexes may be formed across nonoxidizable cation bridges and through other types of bonding.

In well drained parent materials containing quantities of base forming elements sufficient to maintain surface horizon pHs above about 5.5, reduction of Fe by organics usually does not occur within the stable soil matrix. Following its formation by reverse dismutation (Eq. [2]), Mn(III) is reduced by the organic acid ligands that drove the reverse dismutation reaction. The organic ligands are partially oxidized by single electron steps to form organic free radicals (Eq. [3]). At high pH, the Mn(II) formed is quickly reoxidized by one or more processes [see Processes of Manganese Oxidation]. The Mn(IV) oxides freshly formed by these processes catalyze (along with the organic free radicals) polymerization of organic substances to form humic materials. The organic acids either are decomposed to CO_2 and H_2O by the Mn oxides or they are incorporated into the humus being formed.

These well developed mull surface horizons are dark in color, high in organic C, and have strongly expressed structure related to their high humus content (Lutz and Chandler, 1946). It is easy to see how Mn-oxides would be maintained in a high state of oxidation in such soils and would have maximum contribution to development of A horizons in them. The rapid rates of reoxidation of Mn(II) following free radical forming oxidation of organics by Mn(III) and Mn-oxides lead to enhanced oxidative polymerization and stable OM. Thus, the organic acids are not available to mobilize Al from the organically enriched horizon to form an Al-depleted E horizon. The result is a strongly developed mull type of A horizon under the litter layer. The developing soil will not be a Spodosol.

In these strongly developed A horizons, leaching losses of Mn(II) are low. But Mn can become deficient in plants growing at neutral or higher pH in such soils, especially if tillage or erosion has led to loss of available "succulent" organic matter. In an unpublished experiment, we took a series of 100-g soil samples that were practically devoid of manganese (including pure quartz sand, Spodosol E horizon material, acid soils long cropped to potatoes receiving high N fertilizer) and divided them so that Mn was added to one half. Treatments consisted of Mn-free organics, cellulose and amino acids, with CO_2 collected during incubation. Without exception, the high-Mn treatments evolved less CO_2 than the no-Mn treatments, supporting the hypotheses that Mn promotes the binding of C in organic forms rather than having it become CO_2. High-Mn soils have been found to have high levels of OM (Ross et al., 2001b), although the cause and effect have not yet been conclusively demonstrated.

Cycling of Organic Matter and Nutrients

Controlled oxidation of organic residues tends to be slow and only partial. Free radical steps may be involved. The result is accumulation of low solubility stable organic by-products that eventually become humus. Fast oxidation of residues tends to be complete, turning them into CO_2, H_2O, and minerals with little organic buildup.

The rate of undoing of photosynthesis in soils is likely to be highest in a soil with the lowest reduced C in it. Factors that prevent thermodynamic equilibrium between soil humus and O_2 tend to take precedence over those responsible for the production of the organic matter. A drained clay soil contains more reduced C than a drained sandy soil because of organic bonding by clay surfaces coupled with lower O_2-containing porosity.

There can be a shortage of electrons available for redox in a soil that contains most of its reduced C in the form of stable humus, having already lost its more easily decomposed organic residues. Merkle (1955) offered an important insight into soil fertility when he observed that availability to plants of Fe, Mn, and Cu is adequate only if a soil is receiving additions each year of fresh organic residues. "Active," or oxidizable C (Weil et al., 2003), is necessary to maintain nutrient cycling. Weil and coworkers (2003) have recently developed a colorimetric test to estimate active C by its reaction with Mn(VII). Cyclic reduction and reoxidation, appear to be important in maintaining nutrient metals in hydrous, amorphous, readily reducible forms, freely available to plants.

Cycling of N in forest soil surface horizons has traditionally been assumed a microbial process of mineralization and oxidation. Recent evidence has suggested abiotic immobilization of NO_3^- and Davidson and coworkers (2003) proposed a novel redox cycle to explain the phenomenon—the “ferrous wheel” hypothesis. Nitrite is quite reactive with dissolved organic matter (DOM) and will incorporate N into DOM through nitration and nitrosation of aromatic rings. The “ferrous wheel” hypothesizes that reduced C, readily available in the rhizosphere, will reduce Fe(III) and the Fe(II) will reduce NO_3^- to NO_2^-, cycling back to Fe(III). Manganese(II) might substitute for Fe(II) but was not thought to be present in sufficient quantities in acid forest soils. Although thermodynamically feasible, the reduction of NO_3^- by Fe(II) is not always assumed because of the relatively low reactivity of NO_3^-. Future refinement of the mechanism may include organic free radicals and/or Mn(III).

REDOX PROCESSES WITHIN RHIZOSPHERES

Soil: Perfect Home for Roots

Roots are a major soil forming factor. And soils are a major root forming factor. Soils are the designing architects of roots, and roots are the builders of soils. Roots and soils grow together, mutually creating one another, both from an evolutionary standpoint and on a daily basis. There are many characteristics of soil and of roots in the realm of redox that reflect their adaptation one to the other. It seems reasonable to assume that the shape, the surface structure, the internal anatomical, physiological, and biochemical organization—all are perfection in terms of adaptation to life in the soil environment. We should be able to learn much from their intertwining redox.

It is a guiding principal in our management of soils for agriculture that vital roots need oxygen for growth, health, and function. In fact, we have tended to extend this axiom beyond stating that roots need oxygen, and our guiding principle in draining, tilling, and amending organic matter has become that roots need more oxygen. Dozens of studies have shown an overall increase in root growth in response to increased aeration in soils up to the 21% level of O_2 that is in the atmosphere (Stolzy, 1974); although, in hydrophytes, poor, compared with good, aeration appears to result in production of many fine branched high surface area/volume roots (Weaver and Himmel, 1930). If roots need more, why don’t they grow upward and out of the confining soil to where there is air in abundance? Earthworms (Subclass *Oligochaeta*) come right out on top when their O_2 is cut off down in the soil. Why don’t the roots of paddy rice (*Oryza sativa* L.) and cattails (*Typha latifolia* L.) grow upward from the oxygen deprived soil into the oxygen saturated water above the soil interface? In order to understand that there are not contradictions here, we must appreciate that most plant roots are adapted to living and growing in health in what could appear to be on first analysis a hostile redox environment.

In the soil, roots are surrounded by electron donors in a reduced system, and these same roots require electron sinks (oxidants) in order to carry out life-essential oxidative metabolic reactions. The roots depend on humified soil surrounding them for their buffered supply of essential air, water, and nutrients. The same humic environment that shelters and nourishes the roots was formed through processes involving free radicals in the synthesis of humic substances. Yet these same free radicals can be destructive to the living roots. The soil humus and the roots are in the same boat in that they are both in metastable equilibrium with each other and with their redox environments, but of course they are in thermodynamic disequilibrium with the O_2 of the atmosphere making then vulnerable to the oxidative factors that are basic to the tendency for increasing entropy. There is another obvious effect of soil on the roots living within it. It's dark down where they are. The soil provides a plant root with the absence of light and the presence of available Mn and Fe. The darkness prevents light induced reduction of Fe and concomitant formation of free radicals. Reduced Mn is a scavenger for oxygen free radicals, and oxidized forms of Mn together with Fe are responsible for accepting electrons from highly reactive reduced toxic or allelopathic organic compounds and then using these compounds for the synthesis of benign, nurturing, and stable humic substances. Manganese functions in the dark. As roots grow and develop in a soil, they find themselves benefitting from the humified materials surrounding them with the right balance between O_2 and H_2O, while at the same time the roots are being protected by redox metals from toxicity of oxygen free radicals.

Despite synthesis of reductants by rhizosphere microorganisms and secretion of various reducing compounds by plant roots, the net effect of most healthy growing roots is the formation of an oxidative rhizosphere. The soil of the root zones is oxidized, compared with that of the soil surrounding roots. In most plants, O_2 somehow gets down to the root surfaces, either through aerenchyma, in plants genetically adapted to wetland soils, or through open soil pores in drained soils with plants not so adapted. Cattails, sedges (*Carex* species), reeds (Genus: *Phragmites adans*), and wetland rice literally pipe in O_2 from the atmosphere through aerenchyma, the continuous pores that permeate leaves, stems, and roots of these and other hydrophytic plants. Paddy rice roots develop systems of interconnected internal air spaces through which O_2 diffuses from the atmosphere down to the roots and into the rhizosphere soil. Even though they are entirely in the reducing layer of a waterlogged soil, healthy rice roots are surrounded by sheathes of dark brown soil, reflecting conditions sufficiently oxidizing for the formation of Fe(III)-hydroxide and oxides of Mn. Oxygen has been shown to be produced near root surfaces from the enzymatic decomposition of glycollic acid to CO_2, through glyoxalate, oxalate, and formate, with the liberation of H_2O_2 (Russell, 1973).

Newly formed and fresh Mn-oxides often are found in rhizospheres at soil redox interfaces where atmospheric O_2 is in somewhat short supply. Oxygen free radicals are being formed there and Mn(II) is oxidized in scavenging them. Abounding microorganisms are part of the rhizosphere, and oxygen free radicals usually directly or indirectly result from microbial activity. The hypothetical "manganese oxidase enzyme" of Silver et al. (1986), rather than being an enzyme, may be a hydroxyl free radical [$OH^\bullet$], a protonated superoxide free radical [$HO_2^\bullet$] or a "supermanganese Mn^{3+} free radical" (Bartlett and James, 1993). Microbial processes favor Mn oxidation as they increase pH and p*e* in microsites where they decompose organic residues and release base forming cations from them.

There are other plants not considered hydrophytes that do not have aerenchyma, which nevertheless display healthy roots and oxidative rhizospheres under hydric soil conditions without ample supplies of molecular O_2. The roots of these plants appear to be blessed, usually for short periods of time, by the process of reverse dismutation including the formation of Mn^{3+} free radicals. In an experiment with maize (*Zea mays* L.) seedlings in flooded soil, the Mn(III) was shown to form black nonexchangeable Mn oxides on the seedling, probably by dismutation since O_2 was in short supply(R.J. Bartlett, 1989–1994, unpublished data). Thus it appears that Mn(III) can form by reverse dismutation in a rhizosphere, without O_2, and then serve as an oxidative reservoir of MnO_2 on the roots after dismutation. The Mn(III) had to give up electrons in order to accomplish this feat and to "appear to manufacture oxygen." It may have reduced a citrate or oxalate free radical that it "created" in the first place.

In 1961, long before we were aware of the supermanganese free radical, we tested adaptation of different forage crops to wet soil conditions and found that the root zones of forage crops that were adapted to somewhat poorly drained soils, e.g., birdsfoot trefoil (*Lotus corniculatus* L.) and Reed canary grass (*Phalaria arundinacea* L.), were shown to have rhizospheres that were considerably more oxidized than surrounding soil. The evidence for this was the formation of oxidized Fe coating the surfaces of the roots (Bartlett, 1961). Roots of alfalfa (*Medicago sativa* L.) and bromegrass (*Bromus secalinus* L.), which were not adapted to wet soils, were unhealthy and stunted and did not have Fe(III) coating the roots (Zelazny, 1966). Generally, the Fe(III) precipitated on a live root was found to be very easily reducible, as demonstrated using dipyridyl, oxalate, and light. Presence of such Fe(III) coatings on plant roots are indicative of a *Redoxic* soil redox category (Bartlett and James, 1995).

Redox Oxymoron: Oxidative Polymerization Increased by Restriction of Aeration

Compacted tractor and wagon wheel tracks and foot paths of cattle or joggers are seen frequently to stand out as darker green stripes against the background sward color of pastures, meadows, or lawns. Chemical analysis of the vegetation shows that the extra green of the grasses growing on the compaction stripes is associated with higher N content than the paler background vegetation. We hypothesize that the darker green, higher N is evidence of increased oxidative mineralization of soil organic matter in the root zones of the plants where the soil was compacted. Compaction causes partial closing of air-filled soil pores, lowering the soil O_2 content. This might be expected to cause a sharp decline in oxidative activity behavior in the rhizosphere. But it doesn't do that. All evidence, including that of green stripes and N mineralization, points to improved oxidation with compaction. The evidence indicates that the partial shutting out of O_2 by soil compaction results in greater oxidation. How can this be? The immediate answer is as follows: O_2 is not always the paramount oxidant in the rhizosphere. Our hypothesis suggests that the supermanganese free radical, Mn^{3+}, is the primary oxidant in the rhizosphere, a more effective oxidant than is O_2 itself. The activity of the Mn^{3+} free radical in the rhizosphere is kindled by moderate compaction? We suggest that rolling wheels or weighted hoofs or sneakers cause varying amounts of damage, by crushing, to below-ground roots and stems, and the result is some leakage of organic cell contents, exposing reactive functional groups on root surfaces. Leaked carboxyl groups, for example, drive reverse dismutation of Mn(II) and Mn(IV) to form Mn(III)-carboxylates (Eq. [2]). Hypothetical rhizosphere microbes metabolize the organics binding Mn(III) and thereby release oxidizing supermanganese Mn^{3+} free radicals. A Mn^{3+} free radical is a powerful oxidant that readily oxidizes nitrite to nitrate (Bartlett, 1981), Cr(III) to Cr(VI), and Mn(II) to Mn(IV). Two molecules of Mn^{3+} also readily dismutate to Mn^{2+} and MnO_2 (Eq. [1]), a more oxidizing potion than Mn(III)-carboxalate but less powerful than Mn^{3+}. Further proof of increased oxidation by soil Mn oxides under compacted green stripes is shown by higher Cr oxidation tests in the compacted soil than under a pale green or brownish stripe (R.J. Bartlett, 1989–1994, unpublished data).

Redox poise between reactivities of oxidants and reductants is not critical so long as the key Mn oxides are present. The activity of oxygen is less important than that of oxidized Mn. In the case of compacted grassland or turf, it is the organic acids that drive reverse dismutation and affiliated oxidative reactions. Germinating seeds exude organic acids and other labile organics that promote and drive reverse dismutations. In the absence of live roots, organic acids formed by fermentation processes in soils where O_2 is somewhat limiting are another possible source of organic acids that drive reverse dismutation and result in the formation of highly oxidative Mn^{3+}, and MnO_2, and provide oxidation by Mn species in the virtual absence of atmospheric O_2. For example, lactic acid forms under limited aeration, in a silo or in a crock containing cabbage salted and compacted for sauerkraut. There are many intermediate stopovers for "loose" electrons as they wend their ways toward the ultimate electron parking place, namely atmospheric O_2.

REDOX PROCESSES AT WATER–SEDIMENT INTERFACES

Redox transformations are rampant wherever oxidative and reductive soil environments share interfaces, in peds, paddies, sewage lagoons, manure storage pits, landfills, septic tank leach fields, sediments, humus or clay surfaces, and in rhizosphere root–soil boundary regions. A redox interface forms wherever two unlike redox environments have come together and distinctive redox transformations occur across the interface. At a redox interface there are free radicals and wondrous and important redox chemical happenings resulting from the formation and dissipation of the free radicals. Redox interfaces occur in wetlands between colloidal soil surfaces in flooded soils and also between roots and moist soil in rhizospheres of the many plants in both the very wet and the less wet wetlands. All are major sites of for humus synthesis and preservation. And it is the interface that defines the wetland.

A common soil redox interface occurs between a porous colloidal solid containing oriented water and a layer of somewhat freer water that is oxygenated by contact with atmospheric air. When soil in a beaker is flooded and left to stand without disturbing, three redox entities form in the beaker. There is saturated soil, water above it, and a thin interface between them, at the colloidal solid surface. The interface consists of the thin soil surface immediately underneath partially oxygenated free water in contact with the atmosphere. Immediately below the interface is the rest of the saturated soil that is practically devoid of free O_2. The soil below will become Anoxic or even Sulfidic, once dissolved O_2 in the interstitial water is used up. The soil without O_2 will serve as a source of electron donors to the Manoxic soil surface just beneath the oxygenated water. It is the electron sink zone. Reduced Fe and Mn, N, and organic substances are abundant in the bulk soil on the anaerobic side of the interface. Soluble C compounds, NH_4^+, Fe^{2+}, and Mn^{2+} diffuse to the aerobic surface of the interface zone where they become oxidized to nitrate and Fe and Mn oxides. Soluble organic compounds are oxidized at the interface. CO_2 escapes. Powerful superoxidizing free radicals, $OH^{\bullet}$, $HO_2^{\bullet}$, and Mn^{3+}, along with organic free radicals accumulate at the interface. The interface is where the action is.

Oxides of Mn and Fe are inclined to persist at the interface in spite of the giant pool of electron donors on the anaerobic side. The metal oxides meet with only very small portions of the diffusing reduced substances during any given time period so that the Mn and Fe are reoxidized as fast as they are reduced. When Fe^{2+} escapes into water in large quantities, it sometimes oxidizes at the air–water interface, and shiny and fragile iron oxide films are commonly seen floating on still waters. Because these films refract light, producing beautiful rainbow colors, they often are mistaken for oil (Goodman and Petry, 1991) or even for polluting aromatic hydrocarbons. At a touch of a finger, however, the Fe film breaks and scatters into tiny fragments across the water surface. Oil climbs up your finger.

Some nitrate diffuses free into the anaerobic zone and is reduced. The interface provides a mechanism for denitrification and also is an optimum environment for N fixation by nonsymbiotic microorganisms. Organic products of anaerobic respiration diffuse upward and provide readily available reduced C energy needed by the N fixing organisms living on the aerobic side of the interface. Using cellulose as an energy source in the anaerobic zone, Magdoff and Bouldin (1970) showed that N fixation was directly proportional to the interfacial areas in soil containers. When unavailable forms of Cr(III) (amorphous $Cr(OH)_3$, shredded Cr-tanned leather, and high-Cr tannery sludge) were incubated with a flooded high-Mn Typic Eutrochrept soil sample, varying amounts of Cr(VI) were found in the supernatant above the redox interface (Bartlett, 1996). The interface provides the redox gradient necessary to "free" and oxidize the strongly complexed Cr(III). Again somewhat counterintuitive, the lack of oxygen creates strongly oxidizing conditions.

Phosphate, strongly bound in insoluble Fe(III) compounds, can become soluble and mobile upon reduction of the Fe holding it in the anaerobic zone (Russell, 1973; Reddy et al., 1999). An adjoining wetland can be a vehicle for increasing the availability of P to a lake as the soluble phosphates diffuse into the water from an anaerobic zone below an interface. On the other hand, if these phosphates diffuse or are carried through or across the oxidized side of the interface, they are reprecipitated or adsorbed on the aerobic side by oxidized Fe and polymerized hydroxy Al. In this sense, an interface could act as a filter. On the third hand, sunlight shining on the topside of an interface can cause reduction of Fe and temporary increase in solubility of Fe-bound phosphate (see next section). For a wetland system to be effective in removing phosphate from the overlying water, at least two conditions are necessary. First, the overlying water must be sufficiently oxygenated in order to reoxidze the diffusing Fe(II) (Moore and Reddy, 1994). Secondly, the release of P from the anaerobic side of the interface must be slow enough to not saturate the reoxiding Fe on the aerobic side. Agricultural soils with a history of P fertilization may release large amounts of phosphate upon flooding (Young and Ross, 2001), overwhelming the reprecipitation rate.

Preserving Wetlands (Slowing Redox Equilibrium)

Because of very high accumulation of humus, the wetland soil might be considered an exaggeration of a well drained soil. Draining a wetland soil will bring about oxidation of humus, release of $CO_{2,}$, and an increase in the earth's overall entropy. Disturbing a well drained soil will do the same, but to a lesser extent because it is less of a storehouse of sequestered C. Preserving any soil, any biological system is preserving life. Since life on earth depends on the thermodynamic nonequilibrium between organic and inorganic, preventing the decomposition of a dead plant or animal preserves life in the generic sense. Making mummies of relatives won't save the family, but it does preserve reduced C, the evidence of life. All living systems, expressly soil humus, although temporary, are balanced in a remarkably stable state of nonequilibrium, wonderfully resistant to degradation.

All soils, and expressly wetlands, are very efficient systems for temporarily preserving generic life, slowing the earth's roll down the entropy hill. We must do the best we can to manage our soils sustainably in ways that are consistent with the survival in health of as many species as possible without neglecting any members of our own so-called *Homo sapiens*. Like life itself, humus is low entropy. Humus is the key to tilth of topsoil. Tilth provides the best of possible environments for healthy roots. Humus, tilth, roots, water, oxygen, and food—let's not try to live without them.

PHOTOCHEMICAL REDOX PROCESSES

Practically speaking, when we are trying metaphorically to shed light on colloidal solid–water–air interfaces, we find ourselves out in the sunshine digging or sampling, or else working with soil samples in a well-lit laboratory. Metaphorically, however, we are placing ourselves in the dark when we do these things because we don't consider that light may have effects on our observations. Plant roots, of course, and the majority of soil peds, are always in authentic darkness, except when we are studying them. The soil chemistry literature is filled with data, gathered in a variety of intensities and qualities of light, without any regard concerning influence of the light on such chemistry. Light is life. It is so overwhelmingly fundamental and obviously critical to all of the biology and chemistry within us and around us that we rarely deal with it or even think about it, being quite content to get by in our metaphorical semi-darkness.

We know that certain soil extractions must be done in the dark to avoid photochemical reactions. Tamm's acid (acid NH_4-oxalate) will extract greater amounts of Fe in the light due to photoinduced redox reactions of Fe(III) and Fe(III)-oxalate complexes (Bartlett and James, 1993). The effects of soil drying by sunlight are different from those by drying in the dark (Bartlett and James, 1980). Drying in the light has been found to lower the Cr oxidation test and increase extractable Mn relative to air drying (independent of soil moisture content) (Ross et al., 2001b).

Numerous soil constituents are photoactive and their chemistry will differ greatly in the sunlight. They include Fe(III) species, especially $Fe(OH)_2^+$ and Fe(III) polycarboxylates, humic acids, and MnO_2 (Bertino and Zepp, 1991; Jokic et al., 2001). Light will generally not penetrate the soil surface >2 mm but, on exposed soil, this depth will be sufficient to create a redox interface. Upward diffusion may occur and extend the effective depth of sunlight. Probably the most prevalent reactions in soil are photoredox transformations of Fe(III) and associated organic ligands:

$$Fe(OH)_2^+ + H_2O \; h\nu \rightarrow OH^{\bullet} + H^+ + Fe(OH)_2 \; \log K = -56.9 \qquad [17]$$

$$3Fe(OH)_2^+ + 2H_2O \; h\nu \rightarrow HO_2^{\bullet} + 3H^+ + 3Fe(OH)_2 \; \log K = -75 \qquad [18]$$

$$Fe(III)\text{-}oxalate_2^- \text{-}h\nu \rightarrow Fe(II)\text{-}oxalate^0 + oxalate^{\bullet -} \qquad [19]$$

(Faust and Zepp, 1993)

Log K values calculated from standard free energies of formation of the reactants are negative for these reactions, indicating that the equations do not represent thermodynamically spontaneous reactions. Nevertheless these reactions do take place without violating the second law of thermodynamics. They can be driven by the input of light energy, hν, or Planck's constant times the wavelength. In the presence of polycarboxylates, organic radicals are formed (Faust and Zepp, 1993). Numerous side reactions are possible, including oxidation of the organic to CO_2, formation of inorganic radicals and H_2O_2, reoxidation of Fe(II), and redox reactions with other nearby soil constituents. Changes in pH, along with the oxidation of complexing ligands, have the potential to affect other soil chemical processes, such as the polymerization of Al (Ross et al., 2001a).

Most recent developments in knowledge of photochemistry have been in the areas of aquatic and atmospheric chemistry (Waite, 1986; Sulzberger, 1990; 1992; Miller, 1994). The Fe(II) formed by photoreduction in aquatic systems can be reoxidized to Fe(III) by dissolved O_2 in the dark and a diurnal cycle is often found. In Colorado acid stream mine drainage, daytime production of Fe^{2+} by light processes was almost four times its night time reoxidation (McKnight et al., 1988). Kleber and Helz (1992) showed daytime reduction of Cr(VI) in a shallow estuary, and in the laboratory Cr(VI) was reduced in the light but not in the dark and not in water samples pre-filtered to remove ferric hydroxides and particulate organic matter. Collienne (1983) concluded, as a result of field observations in two lakes in Belgium and also as a result of laboratory experiments, that photochemical processes promoted by presence of humic acids explained increases in Fe(II) concomitant with lowering in Eh and of Fe(III). Degradation of aromatic hydrocarbons (Baxendale and Magee, 1955) also have been observed, caused by the hydroxyl free radical produced in the classical Haber-Weiss reaction between Fe(II) and H_2O_2. Iron and light will also degrade pyrene (R.J. Bartlett, 1989–1994, unpublished data). It seems possible that hydroxyl free radicals produced in photochemical reactions in wetlands might also degrade polluting polyaromatic hydrocarbons, such as those found in the Burlington, VT, Barge Canal Coal Tar Superfund site. With the full effects of sunlight on an unforested wetland, there is an opportunity for the photooxidation of a variety of toxic, allelopathic, and benign organic compounds on the solid surface or in the water phase.

Photoredox reactions in soil have not been as thoroughly researched. There are applications in the treatment of contaminants and in the cycling of Hg in the environment. Kieatiwong and Miller (1992), and Kieatiwong et al. (1990) showed photodegredation in soils of aryl ketones and of dioxins. Zhang and Lindberg (1999) outlined a number of possible mechanisms for Hg(II) reduction and Hg° release from soils covering a broad spectrum of photochemical reactions. Direct photolysis of $Hg(OH)_2$ and Hg(III)-organic acids may cause reduction to Hg°. Dissolved organic C and humic acids are photoactive and may directly reduce Hg(II) or indirectly through reduction of dissolved O_2, which in turn acts as a reductant. Both soluble and oxide Fe(III) organic acid complexes will react with light and lead to free radicals that can reduce solution Hg(II). Certain metal oxides (e.g., TiO_2 and ZnO_2) are photoactive and may produce e^- for Hg(II) reduction. The possible combination of mechanisms and numerous side reactions make it difficult to conceptually model the overall process of photoreduction on soil surfaces. It may simpler to be kept in the dark.

REFERENCES

Amacher, M.C., and D.E. Baker. 1982. Redox reactions involving chromium, plutonium, and manganese in soils. Final Rep. DOE/DP/04515-1. Pennsylvania State Univ., University Park.

Archibald, F.S., and I. Fridovich. 1981. Manganese, superoxide dismutase, and oxygen tolerance in some lactic acid bacteria. J. Bacteriol. 146:928–936.

Banerjee, D., and H.W. Nesbitt. 2001. XPS study of dissolution of birnessite by humate with constraints on reaction mechanism. Geochim. Cosmochim. Acta 65:1703–1714.

Bartlett, R.J. 1961. Iron oxidation proximate to plant roots. Soil Sci. 92:372–379.

Bartlett, R.J. 1981. Nonmicrobial nitrite-to-nitrate transformation in soils. Soil Sci. Soc. Am. J. 45:1054–1058.

Bartlett, R.J. 1988. Manganese redox reactions and organic interactions in soils. p. 59–73. *In* R.D. Graham et al. (ed.) Manganese in soils and plants. Kluwer Academic Publ., Dordrecht, the Netherlands.

Bartlett, R.J. 1990. An A or an E: Which will it be? p. 7–18. *In* J.M. Kimble and R.D. Yeck (ed.) Proc. 5th Int. Soil Correlation Meeting (VISCCOM). Characterization, Classification, and Utilization of Spodosols, Oct. 1–14, 1987. USDA Soil Conserv. Serv., Lincoln, NE.

Bartlett, R.J. 1991. Chromium cycling in soils and water: Links, gaps, and methods. Environ. Health Perspectives 92:17–24.

Bartlett, R.J. 1996. Chromium redox mechanisms in soils: Should we worry about Cr(VI)? p. 1–20. *In* S. Canali et al. (ed.) Chromium environmental issues. FrancoAngeli, Rome.

Bartlett, R.J. 1998. Characterizing soil redox behavior. p. 371–397. *In* D.L. Sparks (ed.) Soil physical chemistry. 2nd ed. CRC Press, Boca Raton, FL.

Bartlett, R.J., and B.R. James. 1979. Behavior of chromium in soils: III. Oxidation. J. Environ. Qual. 8:31–35.

Bartlett, R.J., and B.R. James. 1980. Studying air-dried, stored soil samples-some pitfalls. Soil Sci. Soc. Am. J. 44:721–724.

Bartlett, R.J., and B.R. James. 1988. Mobility and bioavailability of chromium in soils. p. 267–283. In J. Nriagu (ed.) Advances in environmental science and technology. John Wiley and Sons, New York.

Bartlett, R.J., and B.R. James. 1993. Redox chemistry of soils. p. 151–208 *In* D.L. Sparks (ed.) Advances in agronomy. Academic Press, New York.

Bartlett, R.J., and B.R. James. 1995. System for categorizing soil redox status by chemical field testing. Geoderma 68:211–218.

Bartlett, R.J., and D.S. Ross. 1988. Colorimetric determination of oxidizable carbon in acid soil solutions. Soil Sci. Soc. Am. J. 52:1191–1192.

Baxendale, J.H., and J. Magee. 1955. The photochemical oxidation of benzene in aqueous solution by ferric ion. Trans. Faraday Soc. 51:205–213.

Berndt, G.F. 1988. Effect of drying and storage conditions upon extractable soil manganese. J. Sci. Food Agric. 45:119–130.

Bertino, D.J., and R.G. Zepp. 1991. Effects of solar radiation on manganese oxide reactions with selected organic compounds. Environ. Sci. Technol. 25:1267–1273.

Brudvig, G.W., and R.H. Crabtree. 1989. Bioinorganic chemistry of manganese related to photosynthetic oxygen evolution. Prog. Inorg. Chem. 37:99–142.

Collienne, R.H. 1983. Photoreduction of iron in the epilimnion of acidic lakes. Limnol. Oceanogr. 28, 83, 1983.

Davidson, E.A., J. Chorover, and D.B. Dail. 2003. A mechanism of abiotic immobilization of nitrate in forest ecosystems: The ferrous wheel hypothesis. Global Change Biol. 9:228–236.

Davis III, J.A., and J.O. Leckie. 1978. The effect of complexing ligands on trace metal adsorption at the sediment–water interface. p. 1009–1024. *In* W.E. Krumbein (ed.) Environmental biogeochemistry and geomicrobiology. Vol. 3. Ann Arbor Sci. Publ., Ann Arbor, MI.

Diem, D., and W. Stumm. 1984. Is dissolved Mn^{2+} being oxidized by O_2 in absence of Mn-bacteria or surface catalysts. Geochem. Cosmochim. Acta 48:1571–1573.

Dion, H.G., and P.J.G. Mann. 1946. Three-valent Mn in soils. J. Agric. Sci. 36:239–245.

Dixon, J.B., and G.N. White. 2002. Manganese oxides. p. 367–388. *In* J.B. Dixon and D.G. Schulze (ed.) Soil mineralogy with environmental applications. SSSA Book Series no. 7. SSSA, Madison, WI.

Drits, V.A., E. Silvester, A.I. Gorshkov, and A. Manceau. 1997. Structure of synthetic monoclinic Na-rich birnessite and hexagonal birnessite: I. results from x-ray diffraction and selected-area electron diffraction. Am. Mineral. 82:946–961.

Ehrlich, H.L. 1976. Manganese as an energy source for bacteria. *In* J.O. Nriagu (ed.) Environmental biogeochemistry and geomicrobiology. Vol. 2. Ann Arbor Sci. Publ., Ann Arbor, MI.

Faulkner, K.M., S.I. Liochev, and I. Fridovich. 1994. Stable Mn(III) porphyrins mimic superoxide dismutase in vitro and substitute for it in vivo. J. Biol. Chem. 269:23471–23476.

Faust, B.C., and R.G. Zepp. 1993. Photochemistry of aqueous iron(III)-polycarboxylate complexes: Roles in the chemistry of atmospheric and surface waters. Environ. Sci. Technol. 27:2517–2522.

Fendorf, S.E., and R.J. Zasoski. 1992. Chromium(III) oxidation by gamma-MnO_2: I. Characterization. Environ. Sci. Technol. 26:79–85.

Fridovich, I. 1975. Superoxide dismutases. Annu. Rev. Biochem. 44:147–159.

Friedl, G.F., B. Wehrli, and A. Manceau. 1997. Solid phases in the cycling of manganese in eutrophic lakes: New insights from EXAFS spectroscopy. Geochim. Cosmochim. Acta 61:275–290.

Fruton, J.S., and S. Simmonds. 1961. General biochemistry. John Wiley and Sons, New York.

Gambrell, R.P. 1996. Manganese. p 665–82. *In* D.L. Sparks (ed.) Methods of soil analysis. Part 3. Chemical methods. SSSA, Madison, WI.

Goldberg, S.P., and K.A. Smith. 1984. Effect of drying and storage conditions upon extractable soil manganese. Soil Sci. Soc. Am. J. 48:559–564.

Goodman, B., and F. Petry. 1991. Oil spills: Nature's own. Audubon, Nov.–Dec. 82–87.

Guest, C.A., D.G. Schulze, I.A. Thompson, and D.M. Huber. 2002. Correlating manganese XANES spectra with extractable soil manganese. Soil Sci. Soc. Am. J. 66:1172–1181.

Halliwell, B. 1974. Manganese ions, oxidation reactions and the superoxide radical. Neurotoxicol. 5:113–118.

Hames, B.R., B. Kurek, B. Pollet, C. Lapierre, and B. Monties. 1998. Interaction between MnO_2 and oxalate: Formation of a natural and abiotic lignin oxidizing system. J. Agric. Food Chem. 46:5362–5367.

Hanley, J., J. Sarrou, and V. Petrouleas. 2000. Orientation of the Mn(II)–Mn(III) dimer which results from the reduction of the oxygen-evolving complex of photosystem II by NO: An electron paramagnetic resonance study. Biochemistry 39:15441–15445.

Hem, J.D. 1978. Redox processes at surfaces of manganese oxide and their effects on aqueous metal ions. Chem. Geol. 21:199–218.

Jenny, H. 1988. The soil resource. Springer-Verlag, New York.

Jokic, A., A.I. Frenkel, and P.M. Huang. 2001. Effect of light on birnessite catalysis of the Maillard reaction and its implication in humification. Can. J. Soil Sci. 81:277–283.

Junta, J.L., and M.F. Hochella. 1994. Manganese(II) oxidation at mineral surfaces: A microscopic and spectroscopic study. Geochim. Cosmochim. Acta 58:4985–4999.

Kieatiwong, S., and G.C. Miller. 1992. Photolysis of aryl ketones with varying vapor pressures on soil. Environ. Toxicol. Chem. 11:173–179.

Kieatiwong. S., L.V. Nguyen, V.R. Hebert, M. Hackett, G.C. Miller, M.J. Miille, and R. Mitzel. 1990. Photolysis of chlorinated dioxins in organic solvents and on soils. Environ. Sci. Technol. 24:1575–1580.

Kim, J.G., J.B. Dixon, C.C. Chusuei, and Y. Deng. 2002. Oxidation of chromium(III) to (VI) by manganese oxides. Soil Sci. Soc. Am. J. 66:306–315.

Kleber, R.J., and G.R. Helz. 1992. Indirect photoreduction of aqueous chromium(VI). Environ. Sci. Technol. 26:307–312.

Klewicki, J.K., and J.J. Morgan. 1998. Kinetic behavior of Mn(III) complexes of pyrophosphate, EDTA, and citrate. Environ. Sci. Technol. 32:2916–2922.

Kostka, J.E., G.W. Luther, and K.H. Nealson. 1995. Chemical and biological reduction of Mn(III)-pyrophosphate complexes: Potential importance of dissolved Mn(III) as an environmental oxidant. Geochim. Cosmochim. Acta 59:885–894.

Laha, S., and R.G. Luthy. 1990. Oxidation of aniline and other primary aromatic amines by manganese dioxide. Environ. Sci. Technol. 24:363–373.

Lutz, H.J., and R.F. Chandler. 1946. Forest soils. John Wiley and Sons, New York.

Magdoff, F.R., and D.R. Bouldin. 1970. Nitrogen fixation in submerged soil-sand-energy material media and the aerobic-anaerobic interface. Plant Soil 33:49–61.

Majcher, E.H., J. Chorover, J.M. Bollag, and P.M. Huang. 2000. Evolution of CO_2 during birnessite-induced oxidation of ^{14}C-labeled catechol. Soil Sci. Soc. Am. J. 64:157–163.

Manceau, A., V.A. Drits, E. Silvester, C. Bartoli, and B. Lanson. 1997. Structural mechanism of Co^{2+} oxidation by the phyllomanganate buserite. Am. Mineral. 82:1150–1175.

Manceau, A., N. Tamura, R.S. Celestre, A.A. MacDowell, N. Geoffroy, G. Sposito, and H.A. Padmore. 2003. Molecular-scale speciation of Zn and Ni in soil ferromanganese nodules from loess soils of the Mississippi Basin. Environ. Sci. Technol. 37:75–80.

Mandernack, K.W., J. Post, and B.M. Tebo. 1995. Manganese mineral formation by bacterial spores of the marine Bacillus, strain SG-1: Evidence for the direct oxidation of Mn(II) to Mn(IV). Geochim. Cosmochim. Acta 59:4393–4408.

Manning, B.A., S.E. Fendorf, B. Bostick, and D.L. Suarez. 2002. Arsenic(III) oxidation and arsenic(V) adsorption reactions on synthetic birnessite. Environ. Sci. Technol. 36:976–981.

Matocha, C.J., E.J. Elzinga, and D.L. Sparks. 2001a. Reactivity of Pb(II) at the Mn(III,IV) (Oxyhydr)Oxide–Water Interface. Environ. Sci. Technol. 35:2967–2972.

Matocha, C.J., D.L. Sparks, J.E. Amonette, and R.K. Kukkadapu. 2001b. Kinetics and mechanism of birnessite reduction by catechol. Soil Sci. Soc. Am. J. 65:58–66.

McBride, M.B. 1987. Adsorption and oxidation of phenolic compounds by iron and manganese oxides. Soil Sci. Soc. Am. J. 51:1466–1472.

McKnight, D.M., B.A. Kimball, and K.E. Bencala. 1988. Iron photoreduction and oxidation in an acidic mountain stream. Science 240:637–640.

Merkle, F.G. 1955. Oxidation–reduction processes in soils. p. 200. *In* F.E. Bear (ed.) Chemistry of the soil. VanNostrand Reinhold, New York.

Miller, W.L. 1994. Recent advances in the photochemistry of natural dissolved organic matter. p. 111–128. *In* G.R. Helz and R.G. Zepp (ed.) Aquatic and surface photochemistry. CRC Press, Boca Raton, FL.

Moore, P.A., and K.R. Reddy. 1994. Role of Eh and pH on phosphorus geochemistry in sediments of Lake Okeechobee, Florida. J. Environ. Qual. 23:955–964.

Murray, J.W. 1975. The interaction of metal ions at the manganese dioxide–solution interface. Geochim. Cosmochim. Acta 39:505–519.

Naidja, A., P.M. Huang, and J.M. Bollag. 1998. Comparison of reaction products from the transformation of catechol catalyzed by birnessite or tyrosinase. Soil Sci. Soc. Am. J. 62:188–195.

Naidu, R., R.S. Kookana, L.H. Smith, and D. Mowat. 2000. Fate and dynamics of chromium at long-term tannery waste contaminated site: I. Distribution of chromium. p. 400. *In* Agronomy Abstracts. ASA, Madison, WI.

Nealson, K.H., B.M. Tebo, and R.A. Rosson. 1988. Occurrence and mechanisms of microbial oxidation of manganese. Adv. Appl. Micro. 33:279–318.

Nelson, Y.M., and L.W. Lion. 2003. Formation of biogenic manganese oxides and their influence on the scavenging of toxic trace metals. p. 169–186. *In* H.M. Selim and W.L. Kingerly (ed.) Geochemical and hydrological reactivity of heavy metals in soils. CRC Press, Boca Raton, FL.

Nico, P.S., and R.J. Zasoski. 2000. Importance of Mn(III) availability on the rate of Cr(III) oxidation on d-MnO_2. Environ. Sci. Technol. 34:3363–3367.

Nico, P.S., and R.J. Zasoski. 2001. Mn(III) center availability as a rate controlling factor in the oxidation of phenol and sulfide on d-MnO_2 Environ. Sci. Technol. 35:3338–43.

Perez, J., and T.W. Jeffries. 1992. Roles of manganese and organic acid chelators in regulating lignin degradation and biosynthesis of peroxidases by Phanerochaete chrysosporium. Appl. Environ. Microbiol. 58:2402–2409.

Pohlman, A.A., and J.G. McColl. 1988. Organic oxidation and metal dissolution in forest soils. Soil Sci. Soc. Am. J. 52:265–274.

Pryor, W.A. 1976. The role of free radical reactions in biological systems. p. 1–49. *In* W.A. Pryor (ed.) Free radicals in biology. Vol. 1. Academic Press, New York.

Reddy, K.R., R.H. Kadlec, E. Flaig, and P.M. Gale. 1999. Phosphorus assimilation in streams and wetlands: Critical reviews. Environ. Sci. Technol. 29:1–64.

Ross, D.S., and R.J. Bartlett. 1981. Evidence for nonmicrobial oxidation of manganese in soil. Soil Sci. 132:153–160.

Ross, D.S., R.J. Bartlett, and H. Zhang. 2001a. Photochemically induced formation of the "Al13" tridecameric polycation in the presence of Fe(III) and organic acids. Chemosphere 44:827–832.

Ross, D.S., H.C. Hales, G.C. Shea-McCarthy, and A. Lanzirotti. 2001b. Sensitivity of soil manganese oxides: Drying and storage cause reduction. Soil Sci. Soc. Am. J. 65:736–743.

Russell, E.W. 1973. Soil conditions and plant growth. 10th ed. Longman, London.

Schiller, H., J. Dittmer, L. Iuzzolino, W. Dorner, W. Meyer-Klaucke, V.A. Sole, H.F. Nolting, and H. Dau. 1998. Structure and orientation of the oxygen-evolving manganese complex of green algae and higher plants investigated by x-ray linear dichroism spectroscopy on oriented photosystem II membrane particles. Biochem. 37:7340–784.

Schlosser, D., and C. Höfer. 2002. Laccase-catalyzed oxidation of Mn^{2+} in the presence of natural Mn^{3+} chelators as a novel source of extracellular H_2O_2 production and its impact on manganese peroxidase. Appl. Environ. Microbiol. 68:3514–3521.

Schnitzer, M., and S.V. Khan. 1972. Humic substances in the environment. M. Dekker, New York.

Schulze, D.G., S.R. Sutton, and S. Bajt. 1995. Determining manganese oxidation state in soils using x-ray absorption near-edge structure (XANES) spectroscopy. Soil Sci. Soc. Am. J. 59:1540–1548.

Senesi N., and M. Schnitzer. 1978. Free radicals in humic substances. p. 467–480. *In* W.E. Krumbein (ed.) Environmental biogeochemistry and geomicrobiology. Vol. 2. Ann Arbor Sci. Publ., MI.

Shindo, H. 1990. Catalytic synthesis of humic acids from phenolic compounds by Mn(IV) oxide. Soil Sci. Plant Nutr. 36:679–682.

Shindo, H., and P.M. Huang. 1982. Role of Mn(IV) oxide in abiotic formation of humic substances in the environment. Nature (London) 298:363–365.

Shindo, H., and P.M. Huang. 1984. Significance of Mn(IV) oxide in abiotic formation of organic nitrogen complexes in natural environments. Nature (London) 308:57–58.

Silver, M., H.L. Ehrlich, and K.C. Ivarson. 1986. Soil mineral transformations by soil microbes. p. 497–519. *In* P.M. Huang and M. Schnitzer (ed.) Interactions of soil minerals with natural organics and microbes. SSSA, Madison WI.

Silvester, E., A. Manceau, and V.A. Drits. 1997. Structure of synthetic monoclinic Na-rich birnessite and hexagonal birnessite: II. Results from chemical studies and EXAFS spectroscopy. Am. Mineral. 82:962–978.

Sparrow, L.A., and N.C. Uren. 1987. Oxidation and reduction of Mn in acidic soils, effect of temperature and soil pH. Soil Biol. Biochem. 19:143–148.

Stolzy, L.H. 1974. Soil atmosphere. p. 335–361. *In* E.W. Carson (ed.) The plant root and its environment. Univ. Press of Virginia, Charlottesville.

Stone, A. T., and J.J Morgan. 1984. Reduction and dissolution of manganese(III) and manganese(IV) oxides by organics: l. reaction with hydroquinone. Environ. Sci. Technol. 18:450–456.

Sulzberger, B. 1990. Photoredox reactions at hydrous metal oxide surfaces: A surface coordination chemistry approach. p. 401–429. *In* W. Stumm (ed.) Aquatic chemical kinetics. John Wiley and Sons, New York.

Sulzberger, B. 1992. Heterogeneous photochemistry. p. 337–367. *In* W. Stumm (ed.) Chemistry of the solid–water interface. John Wiley and Sons, New York.

Sunda, W.G., and D.J. Kieber. 1994. Oxidation of humic substances by manganese oxides yields low-molecular-weight organic substrates. Nature (London) 367:62–64.

Tebo, B.M., W.C. Ghiorse, L.G. vanWaasbergen, P.L. Siering, and R. Caspi. 1997. Bacterially mediated mineral formation: Insights into manganese(II) oxidation from molecular genetic and biochemical studies. p. 225–266. *In* J. F. Banfield and K. H. Nealson (ed.) Geomicrobiology: Interactions between microbes and minerals. Mineral. Soc. Am., Washington, DC.

Thorp, H.H., and G.W. Brudvig. 1991. The physical inorganic chemistry of manganese relevant to photosynthetic oxygen evolution. New J. Chem. 15:479–490.

Tournassat, C., L. Charlet, D. Bosbach, and A. Manceau. 2002. Arsenic (III) oxidation by birnessite and precipitation of manganese (II) arsenate. Environ. Sci. Technol. 36:493–500.

Ulrich, H., and A.T. Stone. 1989. Oxidation of chlorophenols adsorbed to manganese oxide surfaces. Environ. Sci. Technol. 23:421–428.

Villalobos, M., J. Bargar, B. Toner, and G. Sposito. 2004. Characterization of the manganese oxide produced by *Pseudomonas putida* strain MnB1. Geochim. Cosmochim. Acta (in press).

Wada, H., A. Seirayosakol, M. Kimura, and Y. Takai. 1978. The process of manganese deposition in paddy soils: 1. A hypothesis and its verification. Soil Sci. Plant Nutr. 24:55–62.

Waite, T.D. 1986. Photoredox chemistry of colloidal metal oxides. p. 426–445. *In* J.A. Davis and K.F. Hayes (ed.) Geochemical processes at mineral surfaces. ACS Symp. Ser. 323. Am. Chem. Soc., Washington, DC.

Wang, D., J.Y. Shin, M.A. Cheney, G. Sposito, and T.G. Spiro. 1999. Manganese dioxide at a catalyst for oxygen-independent atrazine dealkylation. Environ. Sci. Technol. 33:3160–3165.

Wang, T.S.C., P.M. Huang, C.H. Chou, and J.H. Chen. 1986. The role of soil minerals in the abiotic polymerization of phenolic compounds and formation of humic substances. p. 251–281. *In* P.M. Huang and M. Schnitzer (ed.) Interactions of soil minerals with natural organics and microbes. SSSA Spec. Publ. 17. SSSA, Madison, WI.

Waters, W.A., and J.L. Littler. 1965. Oxidation by vanadium(V), cobalt(III) and manganese(III). p. 186–241. *In* K.B. Wiberg (ed.) Organic chemistry, a series of monographs. Vol. 5-A. Academic Press, New York.

Weaver, J.E., and W.J. Himmel. 1930. Relation of increased water content and decreased aeration to root development in hydrophytes. Plant Physiol. 5:69–92.

Wehrli, B., G. Friedl, and A. Manceau. 1992. Reaction rates and products of manganese oxidation at the sediment–water interface. p. 111–134. *In* C.P. Huang et al. (ed.) Advances in chemistry. Series 244. Aquatic chemistry: Interfacial and interspecies processes. Am. Chem. Soc., Washington, DC.

Weil, R.R., K.R. Islam, M.A. Stine, J.B. Gruver, and S.E. Sampson-Liebig. 2003. Estimating active carbon for soil quality assessment: A simplified method for laboratory and field use. Am. J. Altern. Agric. 18:1–15.

Young, E.O., and D.S. Ross. 2001. Phosphate release from seasonally flooded soils: A laboratory microcosm study. J. Environ. Qual. 30:91–101.

Zelazny, L.W. 1966. Chemical characterization of rhizospheres of selected Vermont soils. M.S. thesis. Univ. of Vermont, Burlington.

Zhang, H., and S.E. Lindberg. 1999. Processes influencing the emission of mercury from soils: A conceptual model. J. Geophys. Res. 104:21889–21896.

Chapter 10

Equations and Models Describing Adsorption Processes in Soils

SABINE GOLDBERG, *USDA-ARS, George E. Brown, Jr., Salinity Laboratory, Riverside, California, USA*

Adsorption is defined as the net accumulation of a chemical species at the interface between a solid phase and an aqueous solution phase (Sposito, 1989). In adsorption this accumulation is restricted to a two-dimensional molecular structure on the surface. The chemical species that accumulates at the interface is called the adsorbate, while the surface where the accumulation takes place is called the adsorbent. Adsorption processes in soils can be described by a variety of models. Empirical adsorption models provide descriptions of experimental data without theoretical basis. Chemical adsorption models, on the other hand, provide a molecular description of the adsorption process using an equilibrium approach.

The purpose of this chapter is to review various equations and models used to describe ion adsorption in soils. First, the common empirical models used in soil chemistry will be described and their limitations evaluated. Second, the common chemical models used to describe adsorption on soil minerals will be presented and their advantages over the empirical approaches discussed. Last, limitations and approximations in the use of these chemical models in soil systems will be explained.

EMPIRICAL MODELS

Adsorption processes in soils have historically been described using empirical adsorption isotherm equations. Typically, such equations are excellent at describing experimental data despite their lack of theoretical basis. Popularity of the isotherm equations results in part from their simplicity and from the ease of estimation of their adjustable parameters.

Distribution Coefficient

A linear function is the simplest and most widely used adsorption isotherm equation. Such an adsorption isotherm equation is conventionally expressed in terms of the distribution coefficient, K_d:

$$x = K_d c \quad [1]$$

Chemical Processes in Soils. SSSA Book Series, no. 8.

where x is the amount of ion adsorbed per unit mass and c is the equilibrium solution ion concentration. Distribution coefficients have found wide use in describing contaminant adsorption in flowing systems (Reardon, 1981) and in transport models (Travis and Etnier, 1981). Because of the linear assumption, the distribution coefficient usually describes ion adsorption data only across a very restricted solution ion concentration range. Figure 10–1 shows the ability of the distribution coefficient to describe linear adsorption of the pesticide parathion from hexane onto a partially hydrated Israeli soil (Yaron and Saltzman, 1972). In this example adsorption is linear across the solution concentration range investigated.

Langmuir Isotherm Equation

The Langmuir adsorption isotherm equation was initially developed to describe the adsorption of gases onto clean solids. This equation can be derived theoretically based on evaporation and condensation rates. The Langmuir adsorption isotherm equation is:

$$x = \frac{x_m K c}{1 + K c} \qquad [2]$$

where x_m is the maximum adsorption per unit mass and K is an affinity parameter related to the bonding energy of the surface. Use of the Langmuir isotherm implies a finite number of uniform adsorption sites and absence of lateral interactions. Despite the fact that these assumptions are violated in soils, the Langmuir equation has often been used to describe ion adsorption on soil materials. In many studies the Langmuir isotherm equation only described adsorption for low solution ion concentrations. Such an example is presented in Fig. 10–2a for phosphate adsorption on two Australian soils (Mead, 1981).

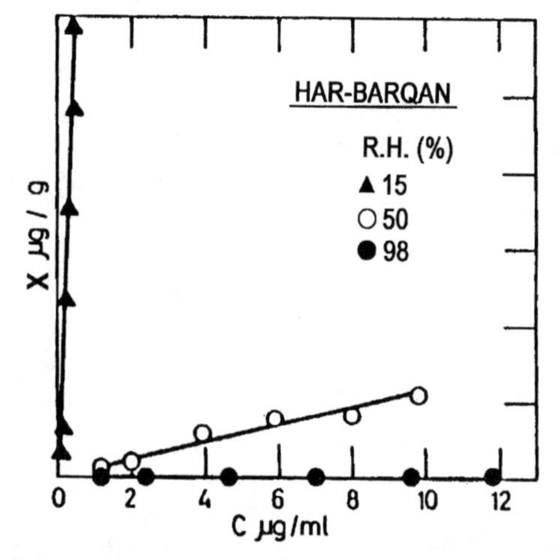

Fig. 10–1. Linear adsorption of parathion from hexane onto a partially hydrated Israeli soil at various relative humidities (from Yaron and Saltzman, 1972).

Multi-Surface Langmuir Adsorption Isotherm Equation

The Langmuir adsorption isotherm equation also has been formulated for the simultaneous adsorption of a gas by more than one surface. The multi-surface Langmuir isotherm is:

$$x = \sum_{i=1}^{n} \frac{x_{m_i} K_i c}{1 + K_i c} \quad [3]$$

where n is the number of sets of surface sites. Because of the increase in the number of adjustable parameters, fit to ion adsorption data with the multi-surface Langmuir isotherm equation is usually excellent. This improvement in fit can be observed in Fig. 10–2a for phosphate adsorption on two Australian soils (Mead, 1981).

Freundlich Adsorption Isotherm Equation

The Freundlich adsorption isotherm equation is the oldest of the nonlinear isotherms and its use implies heterogeneity of adsorption sites. The Freundlich isotherm equation is:

$$x = Kc^{\beta} \quad [4]$$

where β is a heterogeneity parameter, th e smaller β the greater the expected heterogeneity (Kinniburgh, 1985). This expression reduces to a linear adsorption isotherm when $\beta = 1$. Although the Freundlich equation is strictly valid only for

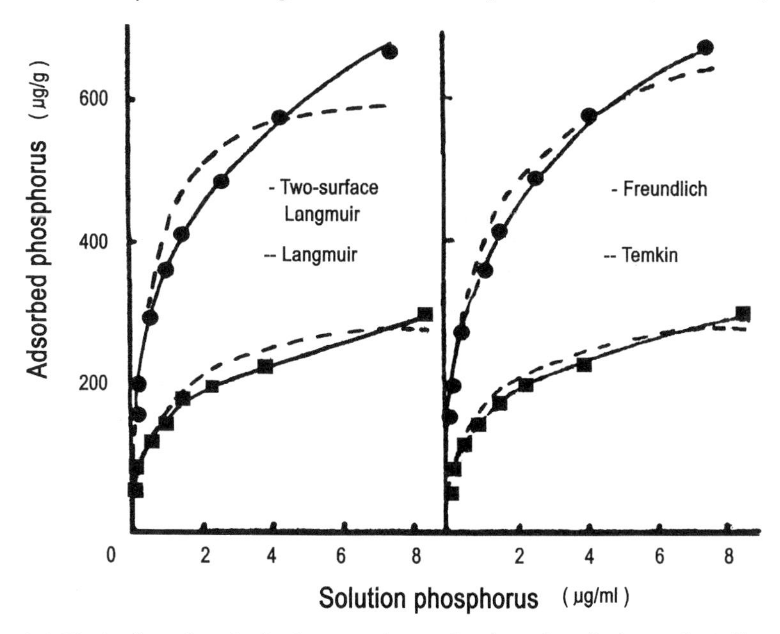

Fig. 10–2. Fit of various adsorption isotherm equations to phosphate adsorption by two Australian soils: (a) Euchrozem and (b) Black Earth (from Mead, 1981).

ion adsorption at low solution ion concentration (Sposito, 1984), it has often been used to describe ion adsorption by soils across the entire ion concentration range investigated. The ability of the Freundlich adsorption isotherm to describe data is indicated in Fig. 10–2b and 10–3 for phosphate adsorption by soils (Mead, 1981). The Freundlich adsorption isotherm does not obey Henry's law at low ion concentration nor does it reach an adsorption maximum at high ion concentration (Kinniburgh, 1985).

Temkin Adsorption Isotherm Equation

For the Temkin adsorption isotherm equation, the energy of adsorption is a linear function of the surface coverage (Travis and Etnier, 1981). The Temkin isotherm equation is:

$$x = a + b \log c \qquad [5]$$

where a and b are parameters. The Temkin isotherm is valid only for an intermediate range of ion concentrations (Kinniburgh, 1985). Figure 10–2b indicates that the ability of the Temkin adsorption isotherm to describe phosphate adsorption is reduced at higher solution concentration (Mead, 1981).

Toth Adsorption Isotherm Equation

The Toth adsorption isotherm equation obeys Henry's law at low ion concentration and reaches an adsorption maximum at high ion concentration (Kinniburgh, 1985). The Toth isotherm equation is:

$$x = \frac{x_m K c}{[1 + (Kc)^{\beta}]^{1/\beta}} \qquad [6]$$

This expression reduces to the Langmuir adsorption isotherm when $\beta = 1$ (Kinniburgh, 1985). Figure 10–3 shows the fit of the Toth equation to phosphate adsorption on a Scottish soil (Kinniburgh, 1986).

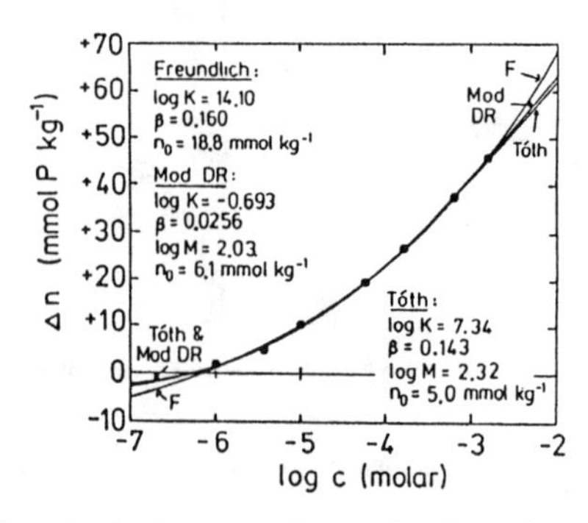

Fig. 10–3. Fit of various adsorption isotherm equations to phosphate adsorption on a Scottish soil (from Kinniburgh, 1986).

Dubinin-Radushkevich Adsorption Isotherm Equation

The Dubinin-Radushkevich adsorption isotherm equation is:

$$\log x = -\beta[\log(Kc)]^2 + \log x_m \quad [7]$$

This isotherm is suitable only for an intermediate range of ion concentrations because it exhibits unrealistic asymptotic behavior (Kinniburgh, 1985). Such a fit is indicated in Fig. 10–3 for phosphate adsorption (Kinniburgh, 1986).

Limitations of the Empirical Approach

Operationally the Langmuir adsorption isotherm equation can describe experimental data from a precipitation reaction, although theoretically this is impossible (Veith and Sposito, 1977). Two theorems have been developed for the Langmuir adsorption isotherm equation (Sposito, 1982). The mechanism theorem states that the adherence of experimental data to the Langmuir adsorption isotherm equation provides no information about the chemical reaction mechanism. The interpolation theorem states that any reaction process for which the distribution coefficient, K_d, is a finite, decreasing function of the amount adsorbed, x, and extrapolates to zero at a finite value of x, can be represented mathematically using a two-surface Langmuir adsorption isotherm equation.

Although all of the above adsorption isotherm equations are often excellent at describing ion adsorption they must be considered simply as numerical relationships used to fit data. Independent experimental evidence of an adsorption process must be present before any chemical meaning can be assigned to isotherm equation parameters. Since the use of the adsorption isotherm equations constitutes essentially a curve-fitting procedure, isotherm parameters are valid only for the chemical conditions under which the experiment was conducted. Use of these equations for prediction of ion adsorption behavior under changing conditions of solution pH, ionic strength, and solution ion concentration is impossible.

Linear Transformations

The Langmuir and the Freundlich adsorption isotherm equations can be transformed to linear form thereby allowing the parameters to be estimated graphically or with linear regression. Various linear transformations of the Langmuir adsorption isotherm equation result in:

1. the reciprocal Langmuir plot:

$$\frac{c}{x} = \frac{1}{x_m K} + \frac{c}{x_m} \quad [8]$$

2. the distribution coefficient or Scatchard plot:

$$\frac{x}{c} = x_m K - xK \quad [9]$$

3. the Eadie-Hofstee plot:

$$x = x_{\rm m} - \frac{x}{Kc} \qquad [10]$$

4. the double reciprocal or Lineweaver-Burk plot:

$$\frac{1}{x} = \frac{1}{x_{\rm m}} + \frac{1}{x_{\rm m}Kc} \qquad [11]$$

Each of these linear transformations produces changes in the original error distribution by giving greater weighting to low adsorption values than to high adsorption values (Kinniburgh, 1986). The reciprocal Langmuir plot is the linearization that has been used most commonly by soil scientists. An example of the reciprocal Langmuir plot is indicated in Fig. 10–4 for sulfate adsorption by four Oregon soils (Chao et al., 1962). The reciprocal Langmuir plot is less sensitive in detecting deviations from linearity than the Eadie-Hofstee plot (Dowd and Riggs, 1965). An additional disadvantage of Eadie-Hofstee and Scatchard plots is that they use the amount adsorbed, x, usually assumed to contain all of the measurement error, as the independent variable assumed to be error free in conventional regression analysis (Kinniburgh, 1986). The best linear transformation is not necessarily the one that gives the highest correlation coefficient but rather it is the one having an error distribution most closely matching the true error distribution (Kinniburgh, 1986).

The linear form of the Freundlich adsorption isotherm equation is:

$$\log x = \log K + \beta \log c \qquad [12]$$

The frequent good fit of data to the Freundlich adsorption isotherm equation is at least partially due to the insensitivity of the linear form. Log-log plots are well known for their insensitivity. Figure 10–5 shows fit of the linear Freundlich plot to sulfate adsorption data (Chao et al., 1962). An additional disadvantage of the linear form

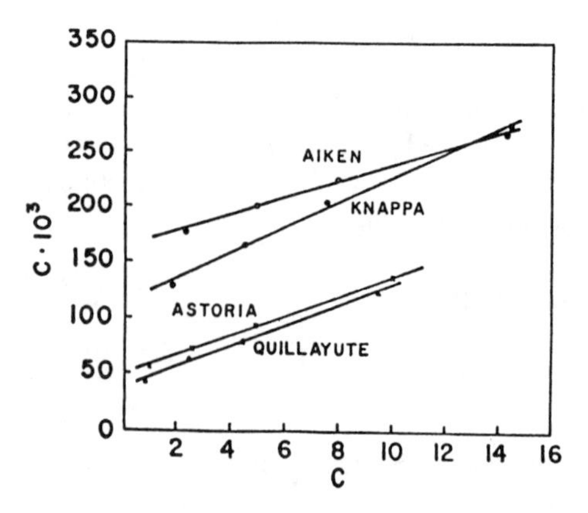

Fig. 10–4. Fit of the reciprocal Langmuir linearization to sulfate adsorption on four Oregon soils (from Chao et al., 1962).

is that all data points are not given equal weightings in the analysis of log-transformed values (Barrow, 1978).

Nonlinear Least Squares Methods

For equations such as the Temkin, Toth, and Dubinin-Radushkevich isotherms, containing three parameters, it is not possible to estimate parameters with linear regression or any reliable graphical method. Nonlinear regression is required to fit these isotherms. Nonlinear regression usually involves the minimization of residual sums of squares. This operation is no longer computationally difficult because of the wide availability of computer algorithms (Kinniburgh, 1986). Direct fitting of adsorption data using nonlinear least squares methods avoids the problems of changing error distribution and biased parameters associated with linear transformations. Such fitting is indicated in Fig. 10–3 for phosphate adsorption. Use of linear regression of any linear form with proper weighting, however, can provide parameter estimates close to those obtained using nonlinear least squares methods (Kinniburgh, 1986).

Nonlinear least squares methods involve finding the set of parameters that minimizes the weighted residual sum of squares, WRSS:

$$\mathrm{WRSS} = \sum_{i=1}^{m} w_{\mathrm{i}}(n_{\mathrm{i}} - \hat{n}_{\mathrm{i}})^2 \qquad [13]$$

where $\hat{n}_{\mathrm{i}}$ is the fitted value for observation i, w_{i} is the weighting factor, and m is the number of observations (Kinniburgh, 1986). The computer program ISOTHERM (Kinniburgh, 1985) contains nonlinear least squares routines for fitting numerous adsorption isotherm equations including those listed above. The principal criterion for comparing the goodness-of-fit of different isotherms to the same data set is the coefficient of determination, R^2 (Kinniburgh, 1985):

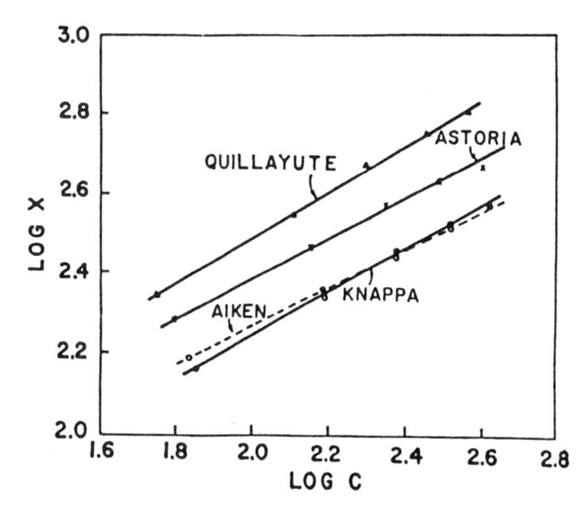

Fig. 10–5. Fit of the linear Freundlich isotherm to sulfate adsorption on four Oregon soils (from Chao et al., 1962).

$$R^2 = \frac{\text{RSS}}{\text{TSS}} \quad [14]$$

where RSS is the residual sum of squares:

$$\text{RSS} = \sum_{i=1}^{m} (n_i - \hat{n}_i)^2 \quad [15]$$

and TSS is the corrected total sum of squares:

$$\text{TSS} = \sum_{i=1}^{m} (n_i - \overline{n})^2 \quad [16]$$

where $\overline{n}$ is the mean value of n_i. Nonlinear fitting of adsorption data can produce values of the coefficient of determination greatly improved over linear transformations (Goldberg and Forster, 1991).

CHEMICAL MODELS

Various chemical surface complexation models have been developed to describe ion adsorption and potentiometric titration data at the oxide-mineral solution interface. Surface complexation models provide molecular descriptions of ion adsorption using an equilibrium approach that defines surface species, chemical reactions, mass balances, and charge balance. Thermodynamic properties such as solid phase activity coefficients and equilibrium constant expressions are calculated mathematically. The major advancement of the chemical surface complexation models is consideration of charge on both the adsorbate ion and the adsorbent surface. Application of these models to reference oxide minerals has been extensive but their use in describing ion adsorption in soils has been limited. Three chemical surface complexation models have been applied to soil systems and will be discussed.

Balance of Surface Charge

The balance of surface charge on a soil particle in aqueous solution is defined by the following expression (Sposito, 1984):

$$\sigma_s + \sigma_H + \sigma_{is} + \sigma_{os} + \sigma_d = 0 \quad [17]$$

where σ_s is the permanent charge due to isomorphous substitution in soil minerals, σ_H is the net proton charge due to formation of inner-sphere surface complexes between protons and hydroxyl ions and surface functional groups, σ_{is} is the charge due to formation of inner-sphere complexes between ions, other than protons and hydroxyls, and surface functional groups, σ_{os} is the charge due to formation of outer-sphere complexes with surface functional groups or with ions in inner-sphere complexes, and σ_d is the dissociated charge equal to minus the surface charge neutralized by background electrolyte ions in solution.

An inner-sphere surface complex contains no water between the adsorbate ion and the surface functional group. An outer-sphere complex, however, contains at least one water molecule between the adsorbate ion and the surface functional group. Inner-sphere surface complexation of ions is called specific adsorption, or ligand exchange in the case of surface complexation of anions. Outer-sphere surface complexation is called nonspecific adsorption. The surface functional group is defined as XOH, where X represents a metal ion in the oxide mineral that is bound to a reactive surface hydroxyl group. In the application of surface complexation models to clay and soil systems, the surface functional group can also represent an aluminol or silanol group at the edge of a clay mineral particle.

Constant Capacitance Model

The constant capacitance model of the oxide–solution interface was developed by two research groups in Switzerland (Schindler and Gamsjäger, 1972; Hohl and Stumm, 1976; Schindler et al., 1976; Stumm et al., 1976, 1980). This model contains the following assumptions:

1. All surface complexes are inner-sphere complexes.
2. No surface complexes are formed with ions from the background electrolyte.
3. The Constant Ionic Medium Reference State determines the activity coefficients of the aqueous species.
4. Surface complexes exist in a chargeless environment in the Standard State.
5. The relationship between surface charge, σ, and surface potential, ψ, is linear and given by:

$$\sigma = \frac{CSa\,\psi}{F} \quad [18]$$

where C is the capacitance (F m^{-2}), S is the surface area (m^2 g^{-1}), a is the suspension density (g L^{-1}), σ has units of mol_c L^{-1}, F is the Faraday constant (C mol_c^{-1}), and ψ has units of V.

The protonation and dissociation reactions of the surface functional group are:

$$XOH + H^+ \rightleftharpoons XOH_2^+ \quad [19]$$

$$XOH \rightleftharpoons XO^- + H^+ \quad [20]$$

The surface complexation reactions for ion adsorption are:

$$XOH + M^{m+} \rightleftharpoons XOM^{(m-1)} + H^+ \quad [21]$$

$$2\,XOH + M^{m+} \rightleftharpoons (XO)_2\,M^{(m-2)} + 2\,H^+ \quad [22]$$

$$XOH + H_xL \rightleftharpoons XH_{(x-i)}\,L^{(1-i)} + H_2O + (i-1)\,H^+ \quad [23]$$

$$2\,XOH + H_xL \rightleftharpoons X_2H_{(x-j)}\,L^{(2-j)} + 2\,H_2O + (j-2)\,H^+ \quad [24]$$

where M represents a metal ion, $m+$ is the charge on the metal ion, L represents a ligand, x is the number of protons present in the undissociated form of the ligand, $1 \leq i \leq n$, and $2 \leq j \leq n$ where n is the number of ligand surface complexes and is equal to the number of dissociations undergone by the ligand.

The equilibrium constant expressions describing the above reactions are:

$$K_+(\text{int}) = \frac{[\text{XOH}_2^+]}{[\text{XOH}][\text{H}^+]} \exp[F\psi/RT] \quad [25]$$

$$K_-(\text{int}) = \frac{[\text{XO}^-][\text{H}^+]}{[\text{XOH}]} \exp[-F\psi/RT] \quad [26]$$

$$K_{\text{M}}^1(\text{int}) = \frac{[\text{XOM}^{(\text{m}-1)}][\text{H}^+]}{[\text{XOH}][\text{M}^{\text{m}+}]} \exp[(m-1)F\psi/RT] \quad [27]$$

$$K_{\text{M}}^2(\text{int}) = \frac{[(\text{XO})_2\text{M}^{(\text{m}-2)}][\text{H}^+]^2}{[\text{XOH}]^2[\text{M}^{\text{m}+}]} \exp[(m-2)F\psi/RT] \quad [28]$$

$$K_{\text{L}}^{\text{i}}(\text{int}) = \frac{[\text{XH}_{(\text{x}-\text{i})}L^{(1-\text{i})}][\text{H}^+]^{(\text{i}-1)}}{[\text{XOH}][H_{\text{x}}L]} \exp[(1-i)F\psi/RT] \quad [29]$$

$$K_{\text{L}}^{\text{j}}(\text{int}) = \frac{[\text{X}_2\text{H}_{(\text{x}-\text{j})}L^{(2-\text{j})}][\text{H}^+]^{(\text{j}-2)}}{[\text{XOH}]^2[H_{\text{x}}L]} \exp[(2-j)F\psi/RT] \quad [30]$$

where R is the molar gas constant (J mol^{-1} K^{-1}), T is the absolute temperature (°K), and square brackets represent concentrations (mol L^{-1}). The electrostatic potential terms $\exp(-F\psi_i/\text{RT})$ are coulombic correction factors accounting for the effect of surface charge on surface complexation and can be considered as solid phase activity coefficients.

The mass balance for the surface functional group is:

$$[\text{XOH}]_{\text{T}} = [\text{XOH}] + [\text{XOH}_2^+] + [\text{XO}^-] + [\text{XOM}^{(\text{m}-1)}] + 2[(\text{XO})_2M^{(\text{m}-2)}] + \sum_{i=1}^{n} [XH_{(\text{x}-\text{i})}L^{(1-\text{i})}] + \sum_{j=2}^{n} 2[X_2H_{(\text{x}-\text{j})}L^{(2-\text{j})}] \quad [31]$$

and the charge balance is:

$$\sigma = [\text{XOH}_2^+] - [\text{XO}^-] + (m-1)[\text{XOM}^{(\text{m}-1)}] + (m-2)[(\text{XO})_2M^{(\text{m}-2)}] + \sum_{i=1}^{n} (1-\text{i})[XH_{(\text{x}-\text{i})}L^{(1-\text{i})}] + \sum_{j=2}^{n} (2-j)[X_2H_{(\text{x}-\text{j})}L^{(2-\text{j})}] \quad [32]$$

This set of equations can be solved by hand or with a computer program using the mathematical approach outlined in Westall (1980).

Triple Layer Model

The triple layer model of adsorption was developed at Stanford University (Davis et al., 1978; Davis and Leckie, 1978, 1980; Hayes and Leckie, 1986). The model contains the following assumptions:

1. Protons and hydroxyl ions form inner-sphere surface complexes.
2. Ion adsorption reactions produce either outer-sphere or inner-sphere surface complexes.
3. Ions from the background electrolyte form outer-sphere complexes.
4. The Infinite Dilution Reference State determines the activity coefficients of the aqueous species.
5. Three planes of charge represent the oxide surface.
6. The relationships between surface charges, σ_o and σ_d, and surface potentials, ψ_o, ψ_β, and ψ_d, are:

$$\sigma_d = -\frac{Sa}{F}(8\varepsilon_o DRTI)^{1/2}\sinh(F\psi_d/2RT) \quad [33]$$

$$\sigma_o = \frac{C_1 Sa}{F}(\psi_o - \psi_\beta) \quad [34]$$

$$\sigma_d = \frac{C_2 Sa}{F}(\psi_d - \psi_\beta) \quad [35]$$

where ε_o is the permittivity of vacuum, D is the dielectric constant of water, I is the ionic strength, and C_1 and C_2 are capacitance densities.

The equations for inner-sphere surface complexation are Eq. [19] through [24] as in the constant capacitance model, where ψ is replaced by ψ_o. The equations for outer-sphere surface complexation are (Davis et al., 1978; Davis and Leckie, 1978, 1980):

$$\mathrm{XOH} + M^{m+} \rightleftarrows \mathrm{XO^-} - M^{m+} + \mathrm{H^+} \quad [36]$$

$$\mathrm{XOH} + M^{m+} + \mathrm{H_2O} \rightleftarrows \mathrm{XO^-} - MOH^{(m-1)} + 2\,\mathrm{H^+} \quad [37]$$

$$\mathrm{XOH} + H^+ + L^{\ell-} \rightleftarrows \mathrm{XOH_2^+} - L^{\ell-} \quad [38]$$

$$\mathrm{XOH} + 2H^+ + L^{\ell-} \rightleftarrows \mathrm{XOH_2^+} - LH^{(\ell-1)-} \quad [39]$$

$$\mathrm{XOH} + C^+ \rightleftarrows \mathrm{XO^-} - C^+ + \mathrm{H^+} \quad [40]$$

$$\mathrm{XOH} + \mathrm{H^+} + A^- \rightleftarrows \mathrm{XOH_2^+} - A^- \quad [41]$$

where C^+ is the cation and A^- is the anion of the background electrolyte and outer-sphere complexes are indicated by splitting the surface complexes with dashes. In the triple layer model one of the inner-sphere metal surface complexes is represented as bidentate, Eq. [22], while one of the outer-sphere metal surface complexes is represented as a hydroxy-metal surface species, Eq. [37]. Outer-sphere hydroxy-metal complexation reactions were more consistent with experimental data (Davis and Leckie, 1978). For anion surface complexation, a protonated outer-sphere surface complex, Eq. [39], rather than a bidentate inner-sphere surface complex, Eq. [24], represented experimental data more consistently (Davis and Leckie, 1980).

The triple layer model equilibrium constants for inner-sphere surface complexation are Eq. [25] through [30] as in the constant capacitance model, where ψ is replaced by ψ_o. The equilibrium constant expressions for outer-sphere surface complexation are (Davis et al., 1978; Davis and Leckie, 1978, 1980):

$$K^1_M(\text{int}) = \frac{[XO^- - M^{m+}][H^+]}{[XOH][M^{m+}]} \exp[F(m\psi_\beta - \psi_o)/RT] \quad [42]$$

$$K^2_M(\text{int}) = \frac{[XO^- - MOH^{(m-1)}][H^+]^2}{[XOH][M^{m+}]} \exp[F((m-1)\psi_\beta - \psi_o)/RT] \quad [43]$$

$$K^1_L(\text{int}) = \frac{[XOH_2^+ - L^{\ell-}]}{[XOH][H^+][L^{\ell-}]} \exp[F(\psi_o - \ell\psi_\beta)/RT] \quad [44]$$

$$K^2_L(\text{int}) = \frac{[XOH_2^+ - LH^{(\ell-1)}]^-}{[XOH][H^+]^2[L^{\ell-}]} \exp[F(\psi_o - (\ell-1)\psi_\beta)/RT] \quad [45]$$

$$K_{C^+}(\text{int}) = \frac{[XO^- - C^+][H^+]}{[XOH][C^+]} \exp[F(\psi_\beta - \psi_o)/RT] \quad [46]$$

$$K_{A^-}(\text{int}) = \frac{[XOH_2^+ - A^-]}{[XOH][H^+][A^-]} \exp[F(\psi_o - \psi_\beta)/RT] \quad [47]$$

The mass balance for the surface functional group is:

$$\begin{aligned}[XOH]_T = {} & [XOH] + [XOH_2^+] + [XO^-] + [XOM^{(m-1)}] + 2[(XO)_2 M^{(m-2)}] \\ & + [XL^{(\ell-1)-}] + 2[X_2L^{(\ell-2)-}] + [XO^- - M^{m+}] + [XO^- - MOH^{(m-1)}] \\ & + [XOH_2^+ - L^{\ell-}] + [XOH_2^+ - LH^{(\ell-1)-}] + [XO^- - C^+] + XOH_2^+ - A^-] \end{aligned} \quad [48]$$

and the charge balances are:

$$\sigma_o + \sigma_\beta + \sigma_d = 0 \quad [49]$$

$$\begin{aligned}\sigma_o = {} & [XOH_2^+] + [XOH_2^+ - L^{\ell-}] + [XOH_2^+ - LH^{(\ell-1)-}] + (m-1)[XOM^{(m-1)}] \\ & + (m-2)[(XO)_2M^{(m-2)}] + [XOH_2^+ - A^-] - [XO^-] - [XO^- - M^{m+}] \\ & - [XO^- - MOH^{(m-1)}] - (\ell-1)[XL^{(\ell-1)-}] - (\ell-2)[X_2L^{(\ell-2)-}] - [XO^- - C^+] \end{aligned} \quad [50]$$

$$\begin{aligned}\sigma_\beta = {} & m[XO^- - M^{m+}] + (m-1)[XO^- - MOH^{(m-1)}] + [XO^- - C^+] \\ & -\ell[XOH_2^+ - L^{\ell-}] - (\ell-1)[XOH_2^+ - LH^{(\ell-1)-}] - [XOH_2^+ - A^-] \end{aligned} \quad [51]$$

The above set of equations can be solved with a computer program using the mathematical approach described in Westall (1980).

Stern Variable-Surface Charge Variable-Surface-Potential Model

The Stern VSC-VSP model of adsorption was developed in Australia (Bowden et al., 1977, 1980; Barrow et al., 1980, 1981; Barrow, 1987). This model contains the following assumptions:

1. Protons, hydroxyl ions, and "strongly adsorbed" oxyanions and metals form inner-sphere surface complexes.
2. Protons and hydroxyl ions reside in the o-plane close to the surface.
3. "Strongly adsorbed" ions reside in an a-plane a short distance away from the surface o-plane.
4. Major cations and anions form outer-sphere surface complexes and reside in the β-plane a short distance away from the a-plane.
5. The surface functional group is defined as OH-X-OH_2 allowing only one protonation or dissociation to occur for every two surface hydroxyl groups
6. The relationships between surfaces charges, σ_o, σ_a, σ_β, and σ_d, and surface potentials, ψ_o, ψ_a, ψ_β, and ψ_d, are Eq. [33] and:

$$\psi_o - \psi_a = \sigma_o/C_{oa} \qquad [52]$$

$$\psi_a - \psi_\beta = (\sigma_o + \sigma_a)/C_{a\beta} \qquad [53]$$

$$\psi_\beta - \psi_d = -\sigma_d/C_{\beta d} \qquad [54]$$

where σ_o, σ_a, σ_β, and σ_d have units of $mol_c\ m^{-2}$ and C_{oa}, $C_{a\beta}$, and $C_{\beta d}$ have units of $mol_c\ V^{-1}\ m^{-2}$. The diffuse layer charge, σ_d, Eq. [33] has units of $mol_c\ L^{-1}$ and must be divided by the surface area and the suspension density.

The Stern VSC-VSP model emphasizes parameter optimization. Therefore it does not define specific surface species and surface reactions and does not provide equilibrium constant expressions and mass balance. The charge balance expressions are:

$$\sigma_o + \sigma_a + \sigma_\beta + \sigma_d = 0 \qquad [55]$$

$$\sigma_o = \frac{N_S\{K_H[H^+]\exp(-F\psi_o/RT) - K_{OH}[OH^-]\exp(F\psi_o/RT)\}}{1 + K_H[H^+]\exp(-F\psi_o/RT) + K_{OH}[OH^-]\exp(F\psi_o/RT)} \qquad [56]$$

$$\sigma_a = \frac{N_T\sum_i Z_i K_i a_i \exp(-Z_i F\psi_a/RT)}{1 + \sum_i K_i a_i \exp(-Z_i F\psi_a/RT)} \qquad [57]$$

$$\sigma_\beta = \frac{N_S\{K_{cat}[C^+]\exp(-F\psi_\beta/RT) - K_{an}[A^-]\exp(F\psi_\beta/RT)\}}{1 + K_{cat}[C^+]\exp(-F\psi_\beta/RT) + K_{an}[A^-]\exp(F\psi_\beta/RT)} \qquad [58]$$

where N_S is the maximum surface charge density ($mol_c\ m^{-2}$), N_T is the maximum adsorption of specifically adsorbed ions ($mol_c\ m^{-2}$), K_i, a_i, and Z_i are the binding constant, the activity, and the charge of the ith specifically adsorbed ion, respectively.

To solve this set of equations, values of N_S, N_T, K_i, and C_i, are chosen to optimize model fit to the data. Subsequently, the charge densities, σ_i, and the electrostatic potentials, ψ_i, are calculated with a computer program (Barrow, 1979).

The Stern VSC-VSP model has been extended to describe ion adsorption processes by soil materials (Barrow, 1983) including the rate of the adsorption reaction (Barrow, 1986a). This extended Stern VSC-VSP model contains the following assumptions:

1. Individual sites react with adsorbing ions as with sites on variable charge oxides.
2. A range of sites exists whose summed adsorption behavior can be modeled using a distribution of parameters of the Stern VSC-VSP model.
3. The initial adsorption reaction induces a diffusion gradient into the particle interior that begins a solid-state diffusion process.

The equations in the extended Stern VSC-VSP model describe (Barrow, 1986a):

A. Heterogeneity of the surface:

$$P_j = 1/(\sigma/\sqrt{2\pi}) \exp[-0.5(\psi_{a0j} - \bar{\psi}_{a0}/\sigma)^2] \qquad [59]$$

where P_j is the probability that a particle has initial potential ψ_{a0j}, $\bar{\psi}_{a0}$ is the mean value of ψ_{a0j}, and σ is the standard deviation of ψ_{a0j}.

B. Adsorption on each component of the surface:

1. at equilibrium:

$$\theta_j = \frac{K_i \alpha \gamma c \exp(-Z_i F \psi_{aj}/RT)}{1 + K_i \alpha \gamma c \exp(-Z_i F \psi_{aj}/RT)} \qquad [60]$$

where θ_j is the proportion of the jth component occupied by the ith adsorbed ion, ψ_{aj} is the potential of the jth component, α is the fraction of adsorbate present as the ith ion, γ is the activity coefficient, and c is the total concentration of adsorbate.

2. rate of adsorption:

$$\theta_{jt} = \frac{K_1^* c(1 - \theta_j) - k_2^* \theta_j}{k_1^* c + k_2^*} \{1 - \exp[-t(k_1^* c + k_2^*)]\} \qquad [61]$$

θ_{jt} is the increment in θ_j over time interval t, and

$$k_1^* = k_1 \alpha \gamma \exp(\overleftarrow{\alpha} F \psi_{aj}/RT) \qquad [62]$$

$$k_2^* = k_2 \alpha \gamma \exp(\overrightarrow{\alpha} F \psi_{aj}/RT) \qquad [63]$$

where k_1 and k_2 are rate coefficients and $\overleftarrow{\alpha}$ and $\overrightarrow{\alpha}$ are transfer coefficients.

C. Diffusive penetration:

$$M_j = 2/\sqrt{\pi}\{C_{0j}\sqrt{(Dft)} + \sum_{0}^{k}(C_{kj} - C_{kj-1})\sqrt{[\tilde{D}f_k(t - t_k)]}\} \qquad [64]$$

where M_j is the amount of material transferred to the interior of the jth component on an area basis, C_{0j} is the surface concentration of the adsorbed ion at time t, C_{kj} is the value of C_{0j} at time t_k, $\tilde{D}$ is the coefficient related to the diffusion coefficient via the thickness of the adsorbed layer, and f is the thermodynamic factor.

D. Feedback effects on potential:

1. for a single period of measurement:

$$\psi_{aj} = \psi_{a0j} - m_1\theta_j \qquad [65]$$

where ψ_{aj} is the potential of the jth component after reaction and m_1 is a parameter.

2. for measurement through time:

$$\psi_{aj} = \psi_{a0j} - m_1\theta_j - m_2M_j/N_{mj} \qquad [66]$$

where N_{mj} is the maximum adsorption on component j and m_2 is a parameter.

E. Effects of temperature:

$$D = A\exp(-E/RT) \qquad [67]$$

where E is an activation energy and A is a parameter.

$$K_i = \exp(B/RT) \qquad [68]$$

where B represents potentials in specified standard states plus an interaction term (Bowden et al., 1977).

To describe phosphate adsorption in soil the continuous distribution of Eq. [59] was divided into 30 discrete elements (Barrow, 1983). These 30 sets of equations were solved iteratively with a computer program using the criterion of goodness-of-fit to experimental sorption data (Barrow, 1983). Despite its foundation in chemical principles, the Stern VSC-VSP model should be regarded as a curve-fitting procedure because of its very large number of adjustable parameters.

Advantages of Chemical Models

The major advantage of the chemical surface complexation models is that they consider surface charge arising from the protonation-dissociation reactions and from the ion surface complexation reactions. These models are descriptions of adsorption processes whose molecular features can be given thermodynamic significance

(Sposito, 1983); however, goodness-of-fit to experimental adsorption data cannot be used as evidence for the presence of any of the surface complexes postulated in the surface complexation models.

As the complexity of a model increases, the number of adjustable parameters also increases, and this improves the model's curve-fitting ability. Chemical significance of a model application suffers when model parameters whose value are available experimentally are adjusted mathematically. This is the case for the maximum surface charge density and the maximum ion adsorption density parameters in the Stern VSC-VSP model. The Stern VSC-VSP model also is compromised chemically because mass balance is not carried out for the surface functional group. Until such time as independent experimental evidence allows the determination of the exact structure of adsorbed surface complexes, models having chemical simplicity and a small number of adjustable parameters are preferable.

Applications of Chemical Models to Soil Systems

Constant Capacitance Model

Applications of the constant capacitance model to soil systems have been restricted to the description of anion adsorption. This model has been used to describe phosphate (Goldberg and Sposito, 1984), borate (Goldberg and Glaubig, 1986; Goldberg, 1999; Goldberg et al., 2000), selenite (Sposito et al., 1988; Goldberg and Glaubig, 1988a), arsenate (Goldberg and Glaubig, 1988b), sulfate (Kooner et al., 1995), and molybdate (Goldberg et al., 1998, 2002) adsorption on soils. For all studies except that of Kooner et al. (1995) values of the protonation and dissociation constants were averages obtained from a literature compilation of $\log K_{+}(\text{int})$ and $\log K_{-}(\text{int})$ values for aluminum and iron oxide minerals. Kooner et al. (1995) obtained $\log K_{+}(\text{int})$ and $\log K_{-}(\text{int})$ values by optimizing potentiometric titration data with the computer program FITEQL (Westall, 1982). FITEQL is an iterative nonlinear least squares optimization program designed to fit equilibrium constants to experimental data and contains surface complexation models including the constant capacitance model and the triple layer model. Capacitance density values for these soil studies ranged from 1.06 F m^{-2} (Goldberg and Sposito, 1984) to 2.7 F m^{-2} (Sposito et al., 1988).

Monodentate anion surface species, Eq. [23], were defined in all of the above studies. Sposito et al. (1988) assumed that in addition to monodentate selenite species formed on one set of surface sites bidentate selenite species formed on another set of surface sites. Fit of the constant capacitance model to selenite adsorption on a California soil is depicted in Fig. 10–6. Anion surface complexation constants for all studies were optimized with the FITEQL computer program. To describe borate (Goldberg and Glaubig, 1986) and selenite (Goldberg and Glaubig, 1988a) adsorption, values of $\log K_{+}(\text{int})$ and $\log K_{-}(\text{int})$ were optimized together with the anion surface complexation constants. The optimized value of $\log K_{-}(\text{int})$ for some of the soils was insignificantly small producing a chemically unrealistic situation that reduced the particular application of the model to a curve-fitting procedure. A reanalysis using both trigonal and tetrahedral surface species was well able to describe boron adsorption without this chemically unrealistic behavior (Goldberg, 1999).

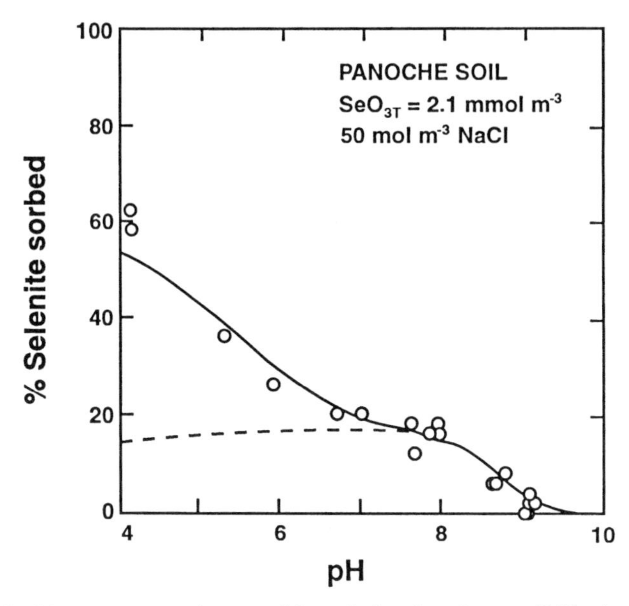

Fig. 10–6. Fit of the constant capacitance model to selenite adsorption on a California soil. The solid line indicates the model fit (from Sposito et al., 1988).

The predictive capability of the constant capacitance model to describe ion adsorption has been tested for phosphate (Goldberg and Sposito, 1984), borate (Goldberg and Glaubig, 1986; Goldberg et al., 2000), selenite (Sposito et al., 1988), and molybdate (Goldberg et al., 2002). Qualitative prediction of selenite adsorption on four California soils was possible using the selenite surface complexation constants obtained for one other California soil as indicated in Fig. 10–7 (Sposito et al., 1988). Using an average set of anion surface complexation constants obtained from numerous soils, the model qualitatively predicted phosphate (Goldberg and Sposito, 1984) and borate (Goldberg and Glaubig, 1986) adsorption on individual soils. A new approach for predicting boron adsorption used a general regression model to predict model parameters from the easily measured soil chemical properties surface area, organic and inorganic carbon and aluminum oxide content (Goldberg et al., 2000). This approach has provided a completely independent model evaluation and was well able to predict boron adsorption on numerous soils of diverse soil orders having a wide range of chemical properties. This predictive capability is presented in Fig. 10–8. A similar modeling approach was used to predict molybdate adsorption. The general regression equations predict model parameters from the chemical properties: cation exchange capacity, organic and inorganic carbon content, and iron oxide content (Goldberg et al., 2002) The predictive capability of this approach is shown in Fig. 10–9 for both monodentate and bidentate molybdenum surface configurations.

Triple Layer Model

The triple layer model has been applied to the adsorption of calcium, magnesium, sulfate (Charlet, 1986; Charlet and Sposito, 1987, 1989; Charlet et al., 1993),

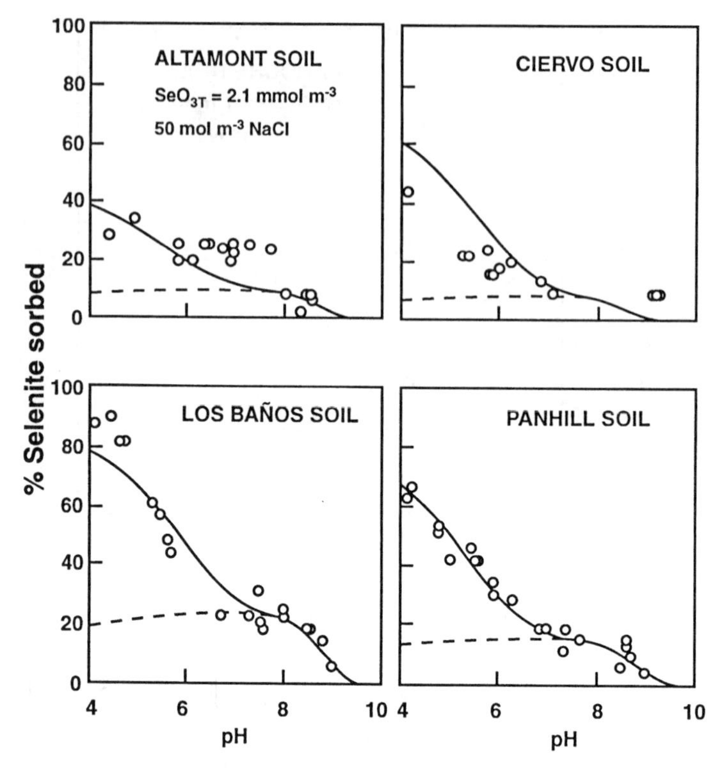

Fig. 10–7. Constant capacitance model predictions of selenite adsorption by California soils. Solid lines indicate the model fits (from Sposito et al., 1988).

chromate (Zachara et al., 1989), and molybdate (Goldberg et al., 1998, 2002) on soils. Potentiometric titrations and measurements of background electrolyte adsorption on a Brazilian oxisol were used to extrapolate values of the protonation and dissociation constants and the surface complexation constants for the background electrolyte (Charlet, 1986; Charlet and Sposito, 1987).

The triple layer model was well able to fit calcium, magnesium, and sulfate adsorption on a Brazilian oxisol using monodentate inner-sphere surface complexes and experimentally determined values of $\log K_+$(int), $\log K_-$(int), $\text{Log}K_{C+}$(int), and $\log K_{A-}$(int) (Charlet, 1986; Charlet and Sposito, 1989). The ability of the triple layer model to describe calcium and magnesium adsorption on a Brazilian oxisol is indicated in Fig. 10–10 and 10–11, respectively. On the other hand, Charlet et al. (1993) described sulfate adsorption on an acidic forest soil as a monodentate outer-sphere surface complex by using literature values of $\log K_+$(int), $\text{Log}K_-$(int), $\log K_{C+}$(int), and $\log K_{A-}$(int) previously determined for goethite (Hsi and Langmuir, 1985) with the justification that the clay components of the soil consisted of phyllosilicates and iron oxide minerals. In a similar approach, Zachara et al. (1989) used literature values of $\log K_+$(int), $\log K_-$(int), $\text{Log}K_{C+}$(int), and $\log K_{A-}$(int) previously determined for an aluminum-substituted goethite (Ainsworth et al., 1989) to describe chromate adsorption as a monodentate outer-sphere surface com-

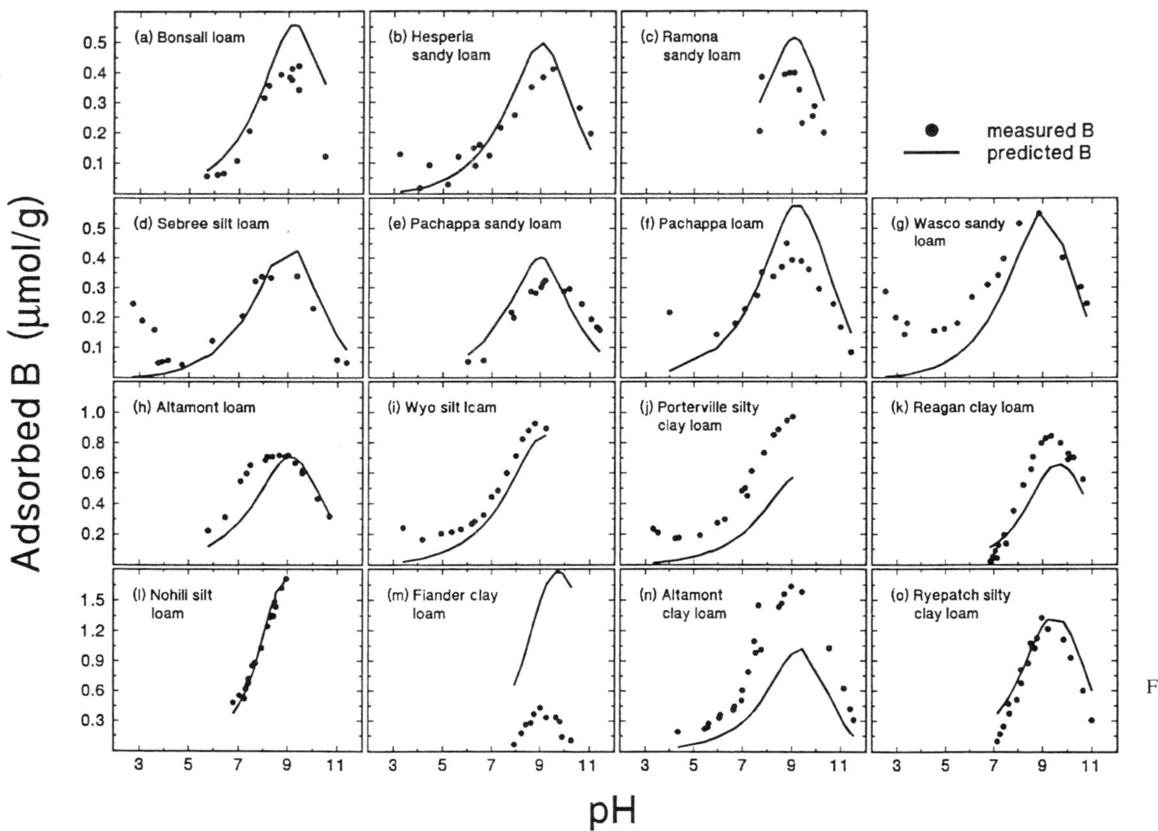

Fig. 10–8. Constant capacitance model predictions of boron adsorption by soils of various soil chemical properties and diverse soil orders. Solid lines indicate model fits (from Goldberg et al., 2000).

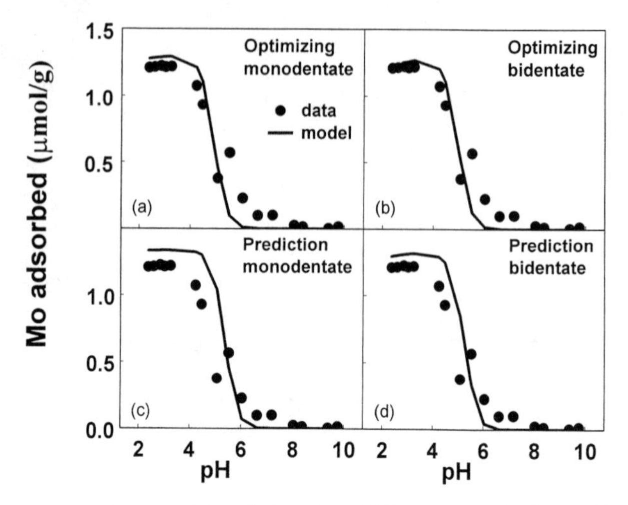

Fig. 10–9. Constant capacitance model predictions of molybdenum adsorption by an acid soil. Solid lines indicate model fits (from Goldberg et al., 2002).

plex by two soils assuming that chromate adsorbed only on the iron sites. The ability of this modeling approach to describe chromate adsorption is depicted in Fig. 10–12. On the other hand, Goldberg et al. (1998) used literature values of logK_+(int), logK_-(int), LogK_{C+}(int), and logK_{A-}(int) obtained for γ-Al_2O_3 (Sprycha, 1989a,b) to describe molybdate adsorption on two arid-zone soils using both inner-sphere and outer-sphere adsorption mechanisms. The assumptions were made that the aluminol group is the molybdate reactive functional group in these soils and that surface complexation reactions of aluminols can be described with constants for the reactive surface hydroxyls of γ-Al_2O_3. The ability of this modeling approach to describe molybdate adsorption on a California soil is depicted in Fig. 10–13.

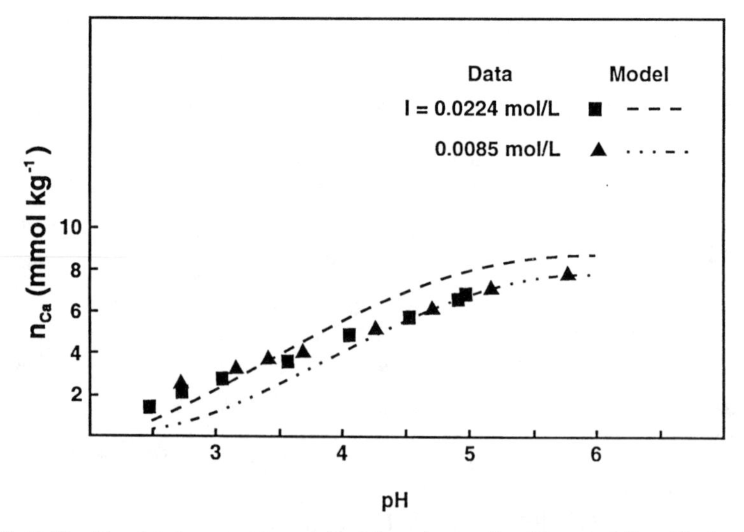

Fig. 10–10. Fit of the triple layer model to calcium adsorption on a Brazilian oxisol (from Charlet, 1986).

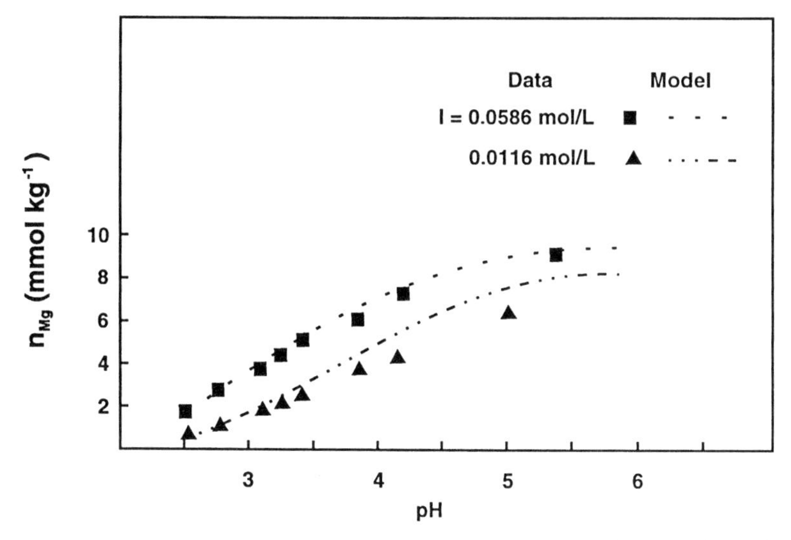

Fig. 10–11. Fit of the triple layer model to magnesium adsorption on a Brazilian oxisol (from Charlet, 1986).

Stern Variable-Surface Charge Variable-Surface Potential Model

The extended Stern VSC-VSP model of adsorption has been used to describe the effect of time and temperature on fluoride, molybdate (Barrow, 1986a), cadmium, cobalt, nickel, zinc (Barrow, 1986b, 1998), selenite, and selenate (Barrow and Whelan, 1989b) adsorption, the effect of pH on phosphate (Barrow, 1984,

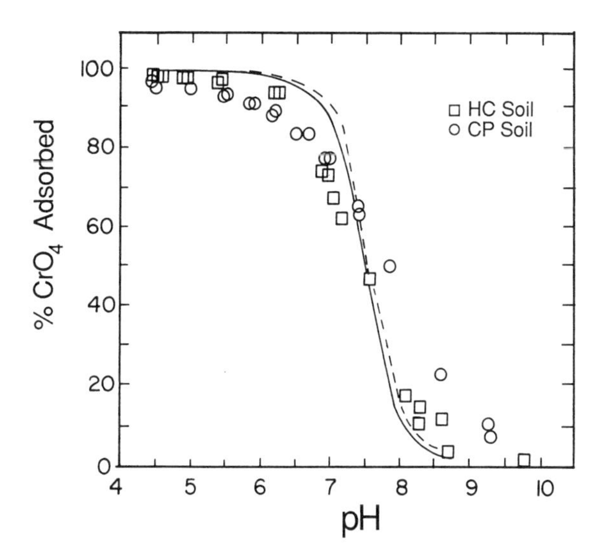

Fig. 10–12. Fit of the triple layer model to chromate adsorption on a North Carolina (CP) and a Tennessee (HC) soil. The solid line indicates the fit to the HC data. The dashed line indicates the fit to the CP data (from Zachara et al., 1989).

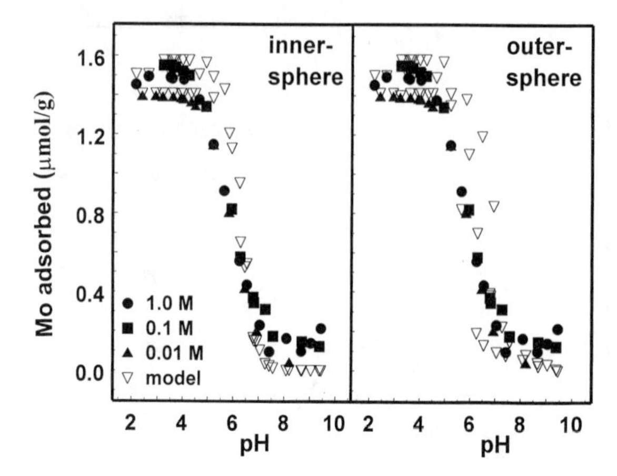

Fig. 10–13. Fit of the triple layer model to molybdate adsorption on a California soil as a function of solution pH and ionic strength. Model results are represented by open triangles (from Goldberg et al., 1998).

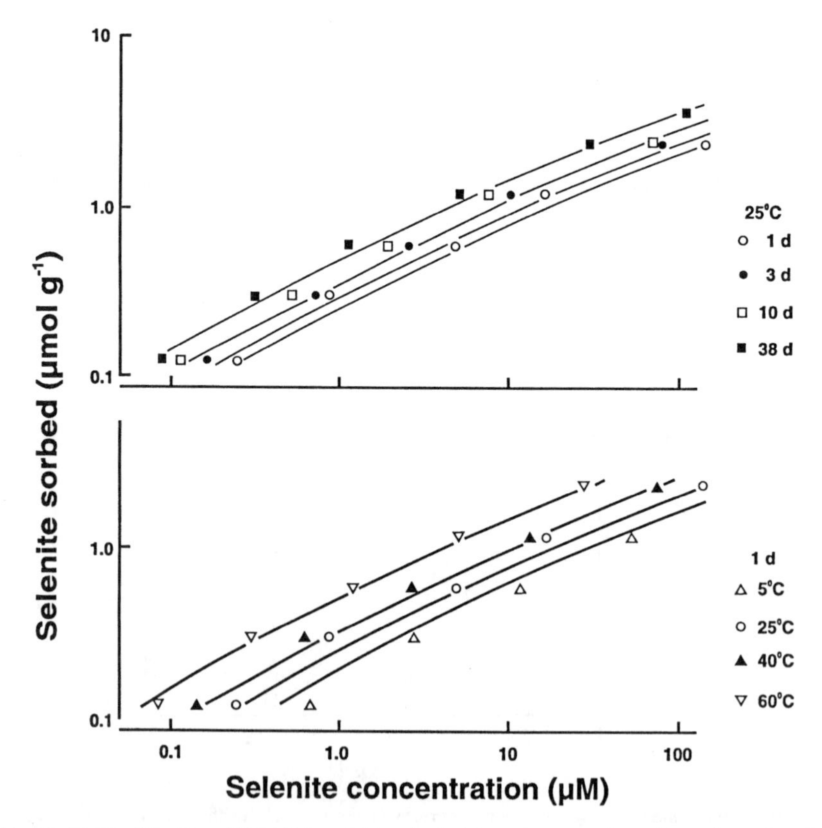

Fig. 10–14. Fit of the Stern VSC-VSP model to selenite adsorption on an Australian soil as a function of time and temperature (from Barrow and Whelan, 1989b).

1986d), zinc (Barrow, 1986c, d), fluoride (Barrow and Ellis, 1986), selenite, selenate (Barrow and Whelan, 1989a), and borate (Barrow, 1989) adsorption, and the effect of ionic strength on selenite, selenate (Barrow and Whelan, 1989a), and borate (Barrow, 1989) adsorption on Australian soils. The Stern VSC-VSP model was able to provide a good fit to the experimental data for all of the above cases because of its very large number of adjustable parameters as indicated in Fig. 10–14 through 10–17. In the description of the effect of pH on phosphate (Barrow, 1984) and zinc (Barrow, 1986c) adsorption on soils, binding constants obtained previously for the iron oxide goethite were used.

Approximations in the Use of Chemical Models in Soil Systems

The chemical surface complexation models described above contain the assumption that ion adsorption takes place on one or at most on two sets of reactive surface sites. This is clearly a gross simplification since soils are complex multisite mixtures having a variety of reactive surface functional groups. Thus the equilibrium constants determined for soils represent average composite values for all sets of reactive functional groups present in soils.

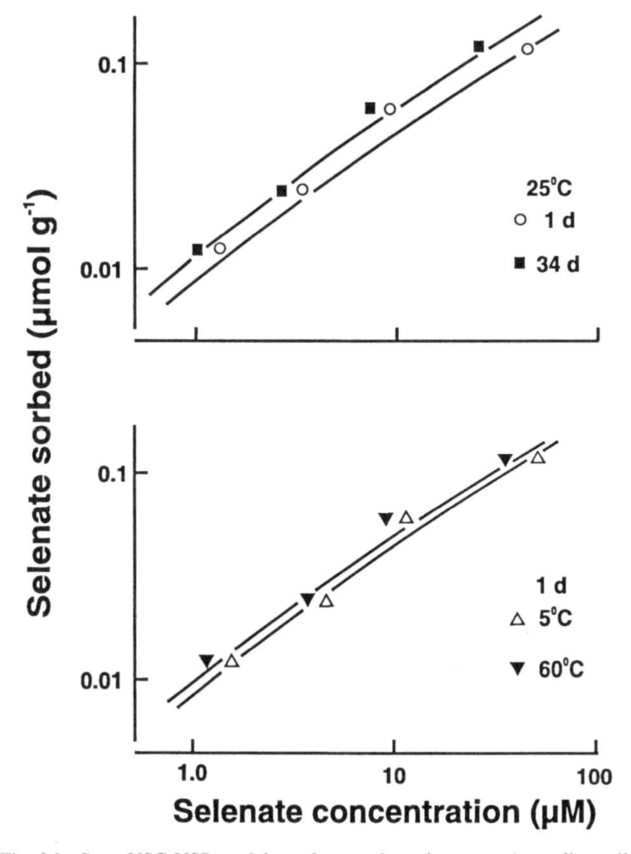

Fig. 10–15. Fit of the Stern VSC-VSP model to selenate adsorption on an Australian soil as a function of time and temperature (from Barrow and Whelan, 1989b).

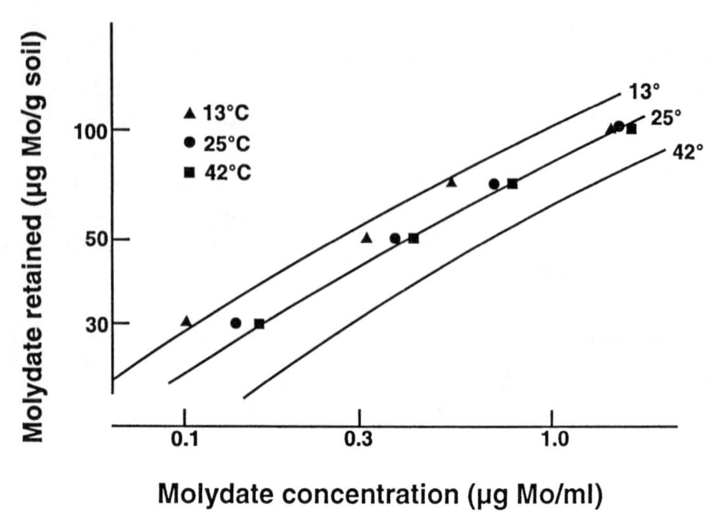

Fig. 10–16. Fit of the Stern VSC-VSP model to molybdate adsorption on an Australian soil as a function of temperature (from Barrow, 1986a).

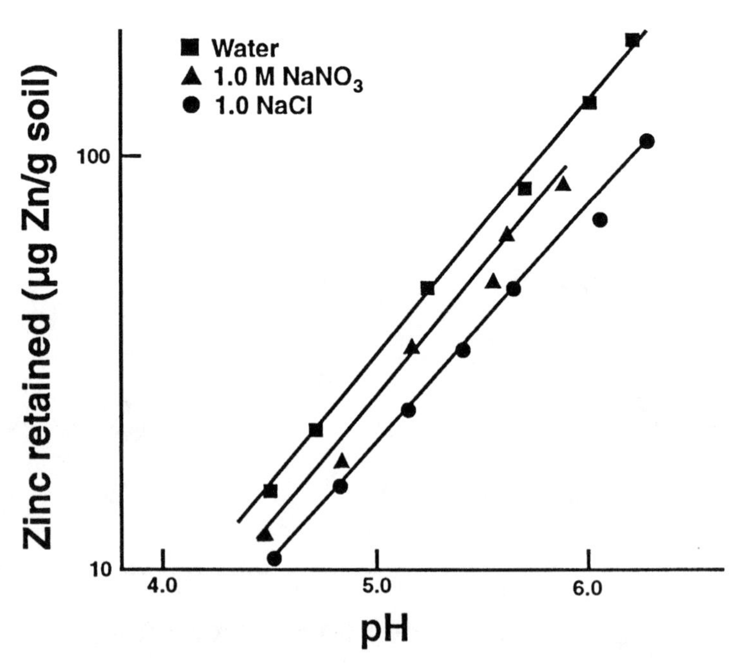

Fig. 10–17. Fit of the Stern VSC-VSP model to zinc adsorption on an Australian soil as a function of background electrolyte (from Barrow and Ellis, 1986).

The total number of reactive surface functional groups, $[SOH]_T$ is an important parameter in the chemical surface complexation models and is related to the surface site density, N_S, by the following expression:

$$[SOH]_T = \frac{Sa10^{18}N_S}{N_A} \qquad [69]$$

where N_A is Avogadro's number and N_S has units of sites nm^{-2}. In the application of chemical surface complexation models to soils, surface site density values have been obtained from maximum ion adsorption (Goldberg and Sposito, 1984) or optimized to fit the experimental adsorption data (Barrow, 1983). To allow standardization of surface complexation modeling approaches and the development of self-consistent thermodynamic databases, Davis and Kent (1990) recommended use of a surface site density value of 2.31 sites nm^{-2} for all natural adsorbents including soil materials. This site density values was used in the prediction of borate (Goldberg et al., 2000) and molybdate (Goldberg et al., 2002) adsorption.

In the application of the surface complexation models to soils dominant in clay minerals, the assumption is made that ion adsorption occurs on the aluminol and silanol groups of clay edges. The effect of the permanent charge sites on this adsorption process is not considered. This simplification may be inappropriate, particularly for anions, since repulsive electrostatic forces emanating from particle faces may spill over and affect the ion adsorption process on clay edges (Secor and Radke, 1985).

APPENDIX

Example: Boron Adsorption on a California Soil

In this section, the application of various equations and modeling approaches to one set of adsorption data will be provided to demonstrate their utility and limitations. Boron adsorption was determined as an isotherm (amount adsorbed as a function of equilibrium solution ion concentration) and as an envelope (amount adsorbed as a function of solution pH per fixed total ion concentration) on the Hanford soil from California.

The ISOTHERM program (Kinniburgh, 1986) was used to fit the Langmuir, Freundlich, two-surface Langmuir, Temkin, Toth, and Dubinin-Radushkevich equations to the B adsorption isotherm data. Figure 10–A1 shows the ability of the isotherm equations to fit the data. With the exception of the Temkin equation, fit of the models to the data is excellent with each having a highly statistically significant coefficient of determination, $R^2 = 0.98^{**}$. Although the fit of the Temkin equation ($R^2 = 0.84^{**}$, $a = 1.12$, $b = 2.97$) was also statistically significant, it is clearly inferior. Parameter values for the other isotherms are as follows: Freundlich: $K = 0.628$, $\beta = 0.761$, Langmuir: $K = 0.0302$, $x_m = 16.1$, two-surface Langmuir: $K_1 = 0.00524$, $x_{m1} = 53.5$, $K_2 = 0.413$, $x_{m2} = 1.18$, Toth: $K = 0.00000164$, $\beta = 0.124$, $x_m = 630625$, Dubinin-Radushkevich: $K = 3070000$, $\beta = -0.0518$, $x_m = 0.00441$. For the Toth and the Dubinin-Radushkevich equations, the fit results in highly unrealistic values of K and maximum adsorption, x_m. These values must therefore be regarded solely as empirical fitting parameters and not true measures of maximum adsorption.

The constant capacitance model was applied to boron adsorption isotherm and envelope data simultaneously. The ability to consider both of these variables is one of the ad-

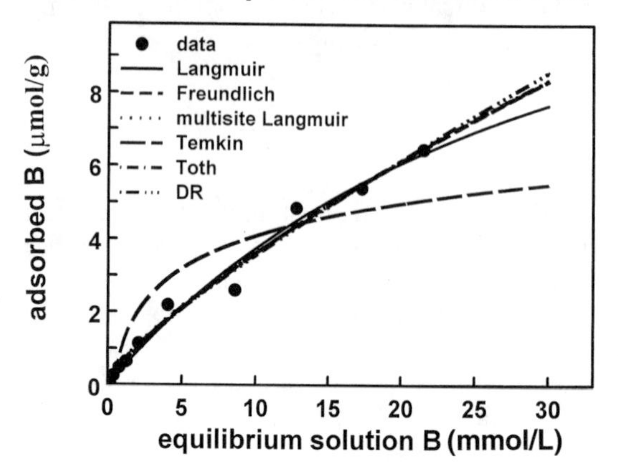

Fig. 10–A1.Fit of various adsorption isotherm equations to boron adsorption on a California soil.

vantages of surface complexation models over the empirical isotherm equations. In addition to Eq. [19] and [20], the surface complexation reaction for boron adsorption is defined as:

$$XOH + H_3BO_3 = XH_3BO_4^- + H^+ \quad [A1]$$

and the equilibrium constant expression for this reaction is:

$$K_{B-} = \frac{[XH_3BO_4^-][H^+]}{[XOH][H_3BO_3]} \exp(-F\psi/RT) \quad [A2]$$

The mass balance equation for the surface functional group is:

$$[XOH_T] = [XOH] + [XOH_2^+] + [XO^-]\ [XH_3BO_4^-] \quad [A3]$$

and the charge balance equation is:

$$\sigma = [XOH_2^+] - [XO^-] - [XH_3BO_4^-] \quad [A4]$$

The solid line in Fig. 10–A2 indicates the ability of the constant capacitance model to fit boron adsorption simultaneously as a function of solution boron concentration and solution pH. While the fit is not quite as good as it is when each type of adsorption curve is fit individually (not shown), the ability of the model to predict adsorption as a function of both of these variables is impressive. Parameter values are $\log K_{B-} = -7.14$, $\log K_+ = 7.06$, and $\log K_- = -10.77$. The dashed line in Fig. 10–A2 shows the boron adsorption predicted by the constant capacitance model when the generalized regression equations of Goldberg et al. (2002) are used to predict $\log K_{B-}$, $\log K_+$, and $\log K_-$. Using these equations the predicted values are $\log K_{B-} = -7.48$, $\log K_+ = 7.71$, and $\log K_- = -11.14$. The predicted boron adsorption is obtained independent of any experimental measurement. The model prediction for the adsorption envelope is similar in quality to the model fit. For the adsorption isotherm, the predicted adsorption is only about 10% different from the fitted adsorption, indicating the utility of this approach.

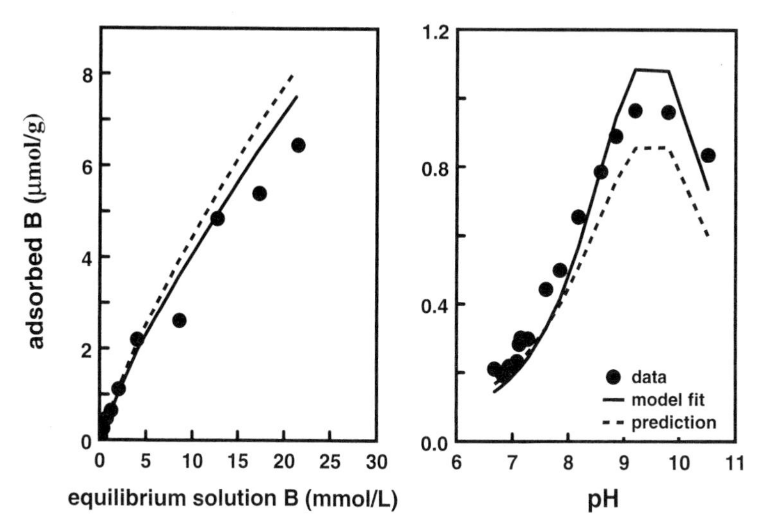

Fig. 10–A2.Constant capacitance model predictions of boron adsorption on a California soil as a function of solution B concentration and solution pH.

REFERENCES

Ainsworth, C.C., D.C. Girvin, J.M. Zachara, and S.C Smith. 1989. Chromate adsorption on goethite: Effects of aluminum substitution. Soil Sci. Soc. Am. J. 53:411–418.

Barrow, N.J. 1978. The description of phosphate adsorption curves. J. Soil Sci. 29:447–462.

Barrow, N.J. 1979. Computer programmes for calculating charge and adsorption of ions in variable charge surfaces. Technical Memorandum 79/3. CSIRO. Division of Land Resources Management, Perth, Australia.

Barrow, N.J. 1983. A mechanistic model for describing the sorption and desorption of phosphate by soil. J. Soil Sci. 34:733–750.

Barrow, N.J. 1984. Modelling the effects of pH on phosphate sorption by soils. J. Soil Sci. 35:283–297.

Barrow, N.J. 1986a. Testing a mechanistic model: I. The effects of time and temperature on the reaction of fluoride and molybdate with a soil. J. Soil Sci. 37:267–275.

Barrow, N.J. 1986b. Testing a mechanistic model: II. The effects of time and temperature on the reaction of zinc with a soil. J. Soil Sci. 37:277–286.

Barrow, N.J. 1986c. Testing a mechanistic model: IV. Describing the effects of pH on zinc retention by soils. J. Soil Sci. 37:295–302.

Barrow, N.J. 1986d. Testing a mechanistic model: VI. Molecular modelling of the effects of pH on phosphate and on zinc retention by soils. J. Soil Sci. 37:311–318.

Barrow, N.J. 1987. Reactions with variable-charge soils. Martinus Nijhoff, Dordrecht, the Netherlands.

Barrow, N.J. 1989. Testing a mechanistic model: X. The effect of pH and electrolyte concentration on borate sorption by a soil. J. Soil Sci. 40:427–435.

Barrow, N.J. 1998. Effects of time and temperature on the sorption of cadmium, zinc, cobalt, and nickel by a soil. Aust. J. Soil Res. 36:941–950.

Barrow, N.J., J.W. Bowden, A.M. Posner, and J.P. Quirk. 1980. An objective method for fitting models of ion adsorption on variable charge surfaces. Aust. J. Soil Res. 18:37–47.

Barrow, N.J., J.W. Bowden, A.M. Posner, and J.P. Quirk. 1981. Describing the adsorption of copper, zinc and lead on a variable charge mineral surface. Aust. J. Soil Res. 19:309–321.

Barrow, N.J., and A.S. Ellis. 1986. Testing a mechanistic model: V. The points of zero salt effect for phosphate retention, for zinc retention and for acid/alkali titration of a soil. J. Soil Sci. 37:303–310.

Barrow, N.J., and B.R. Whelan. 1989a. Testing a mechanistic model: VII. The effects of pH and of electrolyte on the reaction of selenite and selenate with a soil. J. Soil Sci. 40:17–28.

Barrow, N.J., and B.R. Whelan. 1989b. Testing a mechanistic model: VIII. The effects of time and temperature of incubation on the sorption and subsequent desorption of selenite and selenate by a soil. J. Soil Sci. 40:29–37.

Bowden, J.W., A.M. Posner, and J.P. Quirk. 1977. Ionic adsorption on variable charge mineral surfaces. Theoretical-charge development and titration curves. Aust. J. Soil Res. 15:121–136.

Bowden, J.W., S. Nagarajah, N.J. Barrow, A.M. Posner, and J.P. Quirk. 1980. Describing the adsorption of phosphate, citrate and selenite on a variable charge mineral surface. Aust. J. Soil Res. 18:49–60.

Chao, T.T., M.E. Harward, and S.C. Fang. 1962. Adsorption and desorption phenomena of sulfate ions in soils. Soil Sci. Soc. Am. Proc. 26:234–237.

Charlet, L. 1986. Adsorption of some macronutrient ions on an oxisol. An application of the triple layer model. Ph.D. diss. Univ. of California, Riverside.

Charlet, L., N. Dise, and W. Stumm. 1993. Sulfate adsorption on a variable charge soil and on reference minerals. Agric. Ecosyst. Environ. 47:87–102.

Charlet, L., and G. Sposito. 1987. Monovalent ion adsorption by an oxisol. Soil Sci. Soc. Am. J. 51:1155–1160.

Charlet, L., and G. Sposito. 1989. Bivalent ion adsorption by an oxisol. Soil Sci. Soc. Am. J. 53:691–695.

Davis, J.A., R.O. James, and J.O. Leckie. 1978. Surface ionization and complexation at the oxide/water interface. I. Computation of electrical double layer properties in simple electrolytes. J. Colloid Interface Sci. 63:480–499.

Davis, J.A., and D.B. Kent. 1990. Surface complexation modeling in aqueous geochemistry. Rev. Mineral. 23:117–260.

Davis, J.A., and J.O. Leckie. 1978. Surface ionization and complexation at the oxide/water interface: II. Surface properties of amorphous iron oxyhydroxide and adsorption of metal ions. J. Colloid Interface Sci. 67:90–107.

Davis, J.A., and J.O. Leckie. 1980. Surface ionization and complexation at the oxide/water interface: 3. Adsorption of anions. J. Colloid Interface Sci. 74:32–43.

Dowd, J.E., and D.S. Riggs. 1965. A comparison of estimates of Michaelis-Menton kinetic constants from various linear transformations. J. Biol. Chem. 240:863–869.

Goldberg, S. 1999. Reanalysis of boron adsorption on soils and soil minerals using the constant capacitance model. Soil Sci. Soc. Am. J. 63:823–829.

Goldberg, S., and H.S. Forster. 1991. Boron sorption on calcareous soils and reference calcites. Soil Sci. 152:304–310.

Goldberg, S., and R.A. Glaubig. 1986. Boron adsorption on California soils. Soil Sci. Soc. Am. J. 50:1173–1176.

Goldberg, S., and R.A. Glaubig. 1988a. Anion sorption on a calcareous, montmorillonitic soil—Selenium. Soil Sci. Soc. Am. J. 52:954–958.

Goldberg, S., and R.A. Glaubig. 1988b. Anion sorption on a calcareous, montmorillonitic soil—Arsenic. Soil Sci. Soc. Am. J. 52:1297–1300.

Goldberg, S., S.M. Lesch, and D.L. Suarez. 2000. Predicting boron adsorption by soils using soil chemical parameters in the constant capacitance model. Soil Sci. Soc. Am. J. 64:1356–1363.

Goldberg, S., S.M. Lesch, and D.L. Suarez. 2002. Predicting molybdenum adsorption by soils using soil chemical parameters in the constant capacitance model. Soil Sci. Soc. Am. J. 66:1836–1842.

Goldberg, S., and G. Sposito. 1984. A chemical model of phosphate adsorption by soils: II. Noncalcareous soils. Soil Sci. Soc. Am. J. 48:779–783.

Goldberg, S., C. Su, and H.S. Forster. 1998. Sorption of molybdenum on oxides, clay minerals, and soils: Mechanisms and Models. p. 401–426. *In* E.A. Jenne (ed.) Adsorption of metals by geomedia: Variables, mechanisms, and model applications. Proc. Am. Chem. Soc. Symp. Academic Press, San Diego.

Hayes, K.F., and J.O. Leckie. 1986. Mechanism of lead ion adsorption at the goethite–water interface. Am. Chem. Soc. Symp. Ser. 323:114–141.

Hohl, H., and W. Stumm. 1976. Interaction of Pb^{2+} with hydrous γ-Al_2O_3. J. Colloid Interface Sci. 55:281–288.

Hsi, C.-K.D., and D. Langmuir. 1985. Adsorption of uranyl onto ferric oxyhydroxides: Application of the surface complexation site-binding model. Geochim. Cosmochim. Acta 49:1931–1941.

Kinniburgh, D.G. 1985. ISOTHERM. A computer program for analyzing adsorption data. Report WD/ST/85/02. Version 2.2. British Geological Survey, Wallingford, England.

Kinniburgh, D.G. 1986. General purpose adsorption isotherms. Environ. Sci. Technol. 20:895–094.

Kooner, Z.S., P.M. Jardine, and S. Feldman. 1995. Competitive surface complexation modeling of sulfate and natural organic carbon on soil. J. Environ. Qual. 24:656–662.

Mead, J.A. 1981. A comparison of the Langmuir, Freundlich and Temkin equations to describe phosphate adsorption properties of soils. Aust. J. Soil Res. 19:333–342.

Reardon, E.J. 1981. K_ds — Can they be used to describe reversible ion sorption reactions in contaminant migration? Groundwater 19:279–286.

Schindler, P.W., B. Fürst, R. Dick, and P.U. Wolf. 1976. Ligand properties of surface silanol groups: I. Surface complex formation with Fe^{3+}, Cu^{2+}, Cd^{2+}, and Pb^{2+}. J. Colloid Interface Sci. 55:469–475.

Schindler, P.W., and H. Gamsjäger. 1972. Acid-base reactions of the TiO_2 (anatase)–water interface and the point of zero charge of TiO_2 suspensions. Kolloid-Z. u. Z. Polymere 250:759–763.

Secor, R.B., and C.J. Radke. 1985. Spillover of the diffuse double layer on montmorillonite particles. J. Colloid Interface Sci. 103:237–244.

Sposito, G. 1982. On the use of the Langmuir equation in the interpretation of "adsorption" phenomena: II. The "two-surface" Langmuir equation. Soil Sci. Soc. Am. J. 46:1147–1152.

Sposito, G. 1983. Foundations of surface complexation models of the oxide–aqueous solution interface. J. Colloid Interface Sci. 91:329–340.

Sposito, G. 1984. The surface chemistry of soils. Oxford Univ. Press, Oxford, England.

Sposito, G. 1989. The chemistry of soils. Oxford Univ. Press, Oxford, England.

Sposito, G., J.C.M. deWit, and R.H. Neal. 1988. Selenite adsorption on alluvial soils: III. Chemical modeling. Soil Sci. Soc. Am. J. 52:947–950.

Sprycha, R. 1989a. Electrical double layer at alumina/electrolyte interface: I. Surface charge and zeta potential. J. Colloid Interface Sci. 127:1–11.

Sprycha, R. 1989b. Electrical double layer at alumina/electrolyte interface: II. Adsorption of supporting electrolytes. J. Colloid Interface Sci. 127:12–25.

Stumm, W., H. Hohl, and F. Dalang. 1976. Interaction of metal ions with hydrous oxide surfaces. Croatica Chem. Acta 48:491–504.

Stumm, W., R. Kummert, and L. Sigg. 1980. A ligand exchange model for the adsorption of inorganic and organic ligands at hydrous oxide interfaces. Croatica Chem. Acta 53:291–312.

Travis, C.C., and E.L. Etnier. 1981. A survey of sorption relationships for reactive solutes in soil. J. Environ. Qual. 10:8–17.

Veith, J.A., and G. Sposito. 1977. On the use of the Langmuir equation in the interpretation of "adsorption" phenomena. Soil Sci. Soc. Am. J. 41:697–702.

Westall, J. 1980. Chemical equilibrium including adsorption on charged surfaces. Am. Chem. Soc. Adv. Chem. Ser. 189:33–44.

Westall, J.C. 1982. FITEQL: A computer program for determination of chemical equilibrium constants from experimental data. Rep. 82–01. Oregon State Univ., Corvallis.

Westall, J.C. 1986. Reactions at the oxide–solution interface: Chemical and electrostatic models. Am. Chem. Soc. Symp. Ser. 323:54–78.

Yaron, B., and S. Saltzman. 1972. Influence of water and temperature on adsorption of parathion by soils. Soil Sci. Soc. Am. Proc. 36:583–586.

Zachara, J.M., C.C. Ainsworth, C.E. Cowan, and C.T. Resch. 1989. Adsorption of chromate by subsurface soil horizons. Soil Sci. Soc. Am. J. 53:418–428.

Chapter 11

Sorption and Desorption Rates for Neutral Organic Compounds in Soils

THOMAS M. YOUNG, *University of California, Davis, California, USA*

WALTER J. WEBER, JR., *University of Michigan, Ann Arbor, Michigan, USA*

Organic chemicals are introduced into and/or transported through the soil environment for a wide variety of reasons, both accidental and purposeful and in many different forms, including: dissolved in water, as a separate liquid phase, as a wettable powder, as an emulsion, or bound to colloidal particles. Common contamination scenarios encompass a wide range of initial concentrations of the chemicals in soil, mixtures of chemicals released and contact times between the chemical and soil. The complexity of contamination events adds to the existing complexity of the soil environment, which features a minimum of three phases (air, water, solid) with a solid phase that is itself a complex assemblage of minerals and natural organic matter. Numerous physical-chemical and biological processes act to determine the transport and ultimate fate of the released chemicals. Because of the high soil to solution ratios present in both saturated and unsaturated soils, sorption to soil particles and subsequent desorption from them is one of the most important of such processes, even for chemicals with low affinities for the solid phase. Sorption processes have been the focus of significant research attention because of their potential to moderate the risks posed by organic chemicals to ecological and human health during direct contact or via negative impacts on the water quality of underlying aquifers or nearby surface water resources.

In many cases the distribution of a contaminant among solid, liquid and vapor phases is assumed to be at equilibrium for purposes of predicting the potential impacts of a chemical release. For example, risk-based approaches for making remediation decisions at contaminated sites commonly rely on equilibrium sorption models (American Society for Testing and Materials, 1995); however, in many real situations chemical distributions among these compartments may be far from equilibrium, and the mobility or retention of a chemical may therefore be misestimated in ways that result in inappropriate decisions about the application or removal of a particular chemical (Alexander, 1995; Cornelissen et al., 1998b; Stroo et al., 2000).

Rates of sorption processes in soils are highly variable, but in some instances times required to attain an equilibrium distribution of a contaminant between solid

Chemical Processes in Soils. SSSA Book Series, no. 8.

and solution phases have been as high as several years (Ball and Roberts, 1991b; Huang and Weber, 1998; Rügner et al., 1999). Determining whether the rates of the sorption process need to be taken into account in a particular soil-organic chemical system requires a comparison between the residence time of soil vapor or soil pore water within a source zone and the time scales required to attain local equilibria in the system. Recent research has shown that virtually all of the features of a contamination event outlined above can have significant impacts on the rate and extent of organic chemical sorption and desorption in soils.

The primary goals of this chapter are to:

- develop a conceptual model of the mass transfer process that occurs when organic chemicals are released to the soil environment;
- review current knowledge about physical-chemical processes governing the rates of uptake and release of organic chemicals in soils, with particular emphasis on how these rates are related to properties of the soil, the chemical, contact time, initial concentration and the presence of competing solutes;
- review and evaluate potential approaches for modeling the uptake and release processes, and;
- summarize unanswered research questions in this area and discuss ways to obtain the necessary information.

Several reviews of organic chemical sorption rates and mechanisms have been published over the last decade (Pignatello, 1989; Weber et al., 1991; Pignatello and Xing, 1996; Luthy et al., 1997; Pignatello, 2000). Consequently, this chapter focuses most heavily on research published within the last 10 yr, and on implications of that research that have not received as much attention in previous review articles. For more extensive discussion of research conducted more than 10 yr ago, readers should refer to the above-cited articles.

Because of the substantial body of research accumulated in this field, even over the limited time horizon under consideration, we have chosen to limit the coverage of the review to sorption and desorption rates of nonionic organic chemicals in soils. Although only soils and subsoils are covered in detail, much of the discussion also may relate to organic chemicals in sediments found in rivers, estuaries or oceans. Comparisons between soils and sediments must be viewed with caution, however, because the microstructure and organic matter composition of marine, estuarine and riverine sediments are typically substantially different from those of soils (Hedges and Oades, 1997). A few noteworthy studies on sediments, especially those with potential methodological implications for future studies with soils, are reviewed in some detail. Studies that explore rate-limiting mechanisms in sorption and desorption processes in soils by reference to experiments with model solids (e.g., synthetic polymers or organoclays) or subsurface materials with extreme properties (e.g., shale, coal, or peat) also are reviewed because they provide insights not always available from direct experimentation on bulk soils or aquifer materials. The discussion encompasses many examples of non-ionizable chemicals of environmental concern, including polychlorinated biphenyls, petroleum hydrocarbons, chlorinated solvents, and organochlorine insecticides. It also includes ionizable chemicals, but only with respect to their neutral form.

SORPTION DYNAMICS AND POTENTIAL RATE LIMITING STEPS

Overview of Soil Structure and Mass Transfer Processes

Soil is often envisioned as individual soil grains comprising clay mineral platelets, rock fragments, plant debris, and humic materials held together by physical forces, metal ion bridging, and microbially produced exopolymers. While this conceptual model is a useful one, it is important to remember that it is a simplification of a much more complicated reality. Scanning electron micrograph (SEM) images of two California surface soils, one from an agricultural plot near the University of California Davis campus and one from a forest in the foothills of the Sierra Mountains, are shown in Fig. 11–1. These pictures make it clear that the distinction between soil grain and soil aggregate is a rather arbitrary one and that the path taken by a sorbate molecule to reach any particular site within the soil matrix is likely to be complicated. When the potential for intraparticle porosity on the <10 nm scale

Fig. 11–1. Scanning electron micrographs of (a) a forest soil (Forbes, 4.30% organic C) and (b) an agricultural soil (Yolo surface, 1.20% organic C). In both cases the 53 to 425 μm sieve fractions are shown.

(which is not visible at the resolution of typical SEM images) is considered, it becomes obvious that accurately describing sorption rates in such a complex and heterogeneous medium is a challenging prospect. Aquifer materials are no less complex than surface soils. Organic matter in subsurface materials, which often have low total organic C contents, can be quite diverse in structure and intimately associated with the mineral matrix. Organic facies within sandstone and limestone aquifer materials have been shown to have regions of highly transformed organic C (vitrinite, inertinite, phytoclasts) with coal-like structures and very high affinity for organic contaminants (Kleineidam et al., 1999b; Fig. 11–2). Predicting sorption and desorption rates from independently measured soil and chemical properties is an even more daunting task. The goal of this section is to describe relationships between soil microstructure and potential rate limiting steps in the sorption–desorption process.

The soil environment is an extremely dynamic and heterogeneous system that is rarely in equilibrium with respect to any of its components. Mineral phases are constantly dissolving or precipitating. Moisture levels change as water infiltrates and evaporates, producing corresponding changes in redox states as the ability of oxygen to penetrate to a particular depth is reduced or increased. Natural organic compounds are continually added to the subsurface as vascular plants and microorganisms degrade, and as root exudates enter the soil solution. Soil microorganisms in turn transform the chemical states of these compounds in order to supply their metabolic and catabolic needs; fragments of these transformation products may eventually become intimately associated with the soil matrix as humic materials. Microbial activity at the soil solution interface alters the surface chemistry of the soil, and promotes aggregation. The nature and size of these microbial communities depends in turn on moisture levels, redox states, and organic matter inputs.

The reactive domain for uptake of organic chemicals is soil organic matter (SOM) unless the sorbent has extremely low (<0.1 wt%) organic C content (Schwarzenbach and Westall, 1981), and this limit may be even lower depending on the solute's hydrophobicity and the nature of the organic matter and mineral surfaces. Therefore, in most soils and subsurface materials, the dynamics of the sorp-

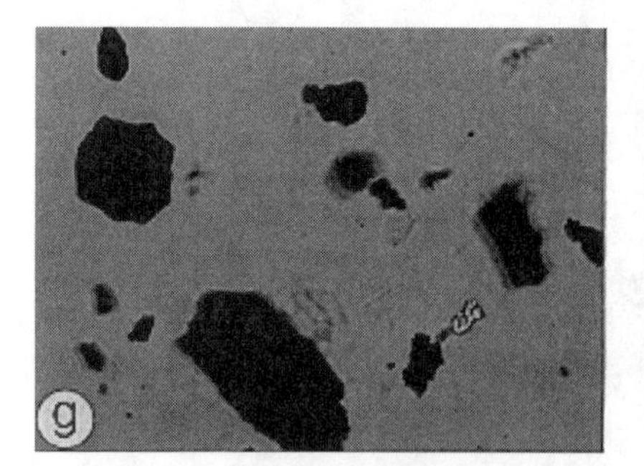

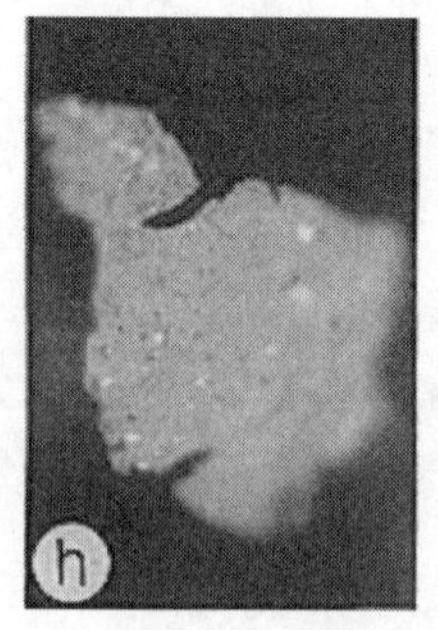

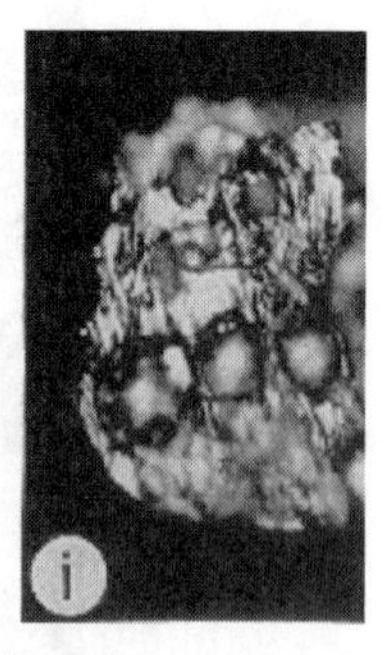

Fig. 11–2. Micrographs of organic facies isolated from subsurface sandstone or limestone particles. A combination of transmitted and reflected light sources were used to identify the organic facies within the mineral matrix as (g) coal fragments and vitrinitic plant remains, (h) a high reflecting vitrinite particle, and (i) a highly reflecting phytoclast (charcoal) showing the open cellular structure of wood tissue (Kleineidam et al., 1999b).

tion process are determined by how quickly the chemical can reach the SOM and/or be transported through it, and/or by how quickly the sorption or desorption reaction occurs once the chemical reaches the active site. Potential rate limiting steps in the uptake process from bulk soil water include:

- transport through stagnant water films near solid surfaces (film diffusion)
- diffusion through pores of varying dimensions, either between or within individual soil grains (pore diffusion)
- diffusion through soil organic matter to reach "internal" surface sites (intraorganic matter diffusion)
- reaction with the active site to create the sorbed form of the chemical

In the case of nonpolar organic chemicals in soil, the reaction step is typically a physical sorption process accompanied by low heats of sorption (Young and Weber, 1995). In the case of partitioning-type sorption processes the active site is ill-defined and the reaction process is thought to be very rapid. Desorption rates are dependent on the reverse set of processes, although the energy barrier for desorption is higher than that for adsorption in the case of an exothermic sorption process (Pignatello, 2000). This issue is discussed in greater detail later in the chapter. The remainder of this section discusses each of the above steps in greater detail and concludes with a review of current understanding of the location of sorbed chemicals within the soil matrix and their local chemical microenvironment within the soil organic matter.

Transport Processes

Soil Aggregates

Numerous mass transfer resistances are operative at the aggregate scale, including film diffusion and diffusion in interparticle void spaces (Fig. 11–1). Collectively these resistances are expected to result in times to reach sorptive equilibrium of hours to days in systems featuring advective flux of air or water. Diffusion coefficients will be relatively high and the void or interparticle "pore" sizes involved are large compared with molecular dimensions; however, relatively few studies have measured rates of organic chemical sorption or desorption in intact soil aggregates, largely because of experimental difficulties. Typical aggregates break apart during well-mixed batch reactor experiments, and the appreciable clay fractions of most aggregate forming soils present problems with respect to performing accurate column studies. Batch experiments without mixing pose difficulties related to obtaining representative samples of the water film in contact with the matrix surfaces, which may be too thin to sample in unsaturated soils and too heterogeneous to obtain representative results in saturated systems. As a consequence of these experimental limitations, most research that has sought to identify rate-limiting steps in the sorption–desorption process have focused on disaggregated mineral or organic matter as the domains of resistance. Focus on mass transfer resistance at the subaggregate scale also is justified by common observations of very slow sorption and desorption processes (years to attain equilibrium), even in completely mixed batch systems that include only relatively pure mineral or organic phases (described below). The remainder of this section examines our current knowledge about mass transport processes in mineral solids, within soil organic matter, and how organic matter-mineral associations affect the independent processes.

Mineral Phases

A number of studies have examined sorption or desorption rates for porous and non-porous inorganic solids to determine whether these types of materials might serve as the locus of significant mass transfer resistance. In one report it was noted that up to 5% of the initial mass of trichloroethene (TCE) sorbed on silica gels remained after 30 d of purging with humidified N even though >90% of the TCE desorbed within 10 min (Farrell and Reinhard, 1994b). Montmorillonite samples had very low residual TCE contents and were largely TCE-free within 10 min of purging suggesting that limited TCE penetration of the clay interlayer spacing occurred. Subsequent investigation suggested that the residual TCE in the silica gels might have been trapped in intragranular micropores by mineral precipitation because the activation energy for removing the TCE was similar to the dissolution enthalpy of silica (Farrell et al., 1999). Sorption of phenanthrene by α-Al_2O_3, and three silica gels with average pore sizes between 4 and 11.9 nm were reported by other investigators to reach apparent equilibrium within 10 min, suggesting that little or no penetration of the intraparticle porosity of these materials occurred during the 21 to 60 d observation period (Huang et al., 1996). Sorption rates on bentonite were only slightly slower, attaining equilibrium within 1 d. The slower rates and higher ultimate capacity of the bentonite on a N_2 surface area basis suggest that phenanthrene molecules were able to penetrate the expandable interlayer space of the bentonite (Huang et al., 1996) or that capillary condensation of phenanthrene occurred within bentonite aggregates (Hundal et al., 2001). Experimental TCE desorption isotherm and rate measurements at varying temperatures combined with modeling results have been used by other researchers to infer that slow desorption is controlled by activated diffusion in mineral pores, particularly at low solid loading levels (Werth and Reinhard, 1997a; Werth and Reinhard, 1997b); however, the similarity between the silica gel desorption rate results and those for the natural solids were cited as the basis for believing that the rate-controlling micropores were mineral, and these pores might have been blocked by mineral precipitates (Farrell et al., 1999). Mechanisms in the two studies might have differed since the activation energy for TCE desorption from silica gels determined in the two studies differed by a factor of two. Micropore diffusion was also implicated as a rate controlling process by experiments with isotopically labeled TCE in which the rate of uptake of deuterated TCE was slower in sorbents preloaded with TCE than TCE uptake in the same sorbents before preloading (Werth and Reinhard, 1999). Fitted counter diffusion coefficients were up to six times larger than co-diffusion coefficients with the magnitude of the difference between the constants depending on the apparent micorpore dimensions (McMillan and Werth, 1999). Further evidence that diffusion within micropores controls the rate of TCE desorption was provided by temperature stepped desorption measurements and modeling results with silica gel and natural materials (Werth et al., 2000; Castilla et al., 2000).

It is interesting to note that the results with TCE and silica gels suggesting the importance of mineral solids in controlling desorption resistance have all been obtained in unsaturated column systems near 100% relative humidity (Farrell and Reinhard, 1994b; Werth and Reinhard, 1997a; Werth and Reinhard, 1997b; Farrell et al., 1999; Castilla et al., 2000), while the phenanthrene desorption data that suggests this process is less important was obtained in fully saturated batch reactors (Huang et al., 1996). From the existing experimental database it is difficult to tell whether these differences are related to the nature of the experiments or the different solutes employed. One possibility is that TCE is accumulating as a neat phase within pores of dimensions of less than 20 molecular diameters because solvation forces act to promote water displacement by organics in such pores, even when the pore surface is hydrophilic (Farrell and Reinhard, 1994b; Corley et al., 1996). Free energy changes calculated for a pore filling process for a series of substituted benzenes on glass beads were quite similar to each compound's free energy change for condensation from water (Corley et al., 1996).

Organic Matter Domains

Slow uptake or release of organic chemicals in soils with higher organic matter contents are typically controlled by rates of transport through natural organic matter pores or matrices, where the critical organic carbon content for organic matter transport control appears to be lower than 0.1%. Transport rates through SOM appear to be strong functions of the structure of the organic matrices, as evidenced by greater hysteresis for more reduced and more aromatic organic matter (Huang and Weber, 1997; Schultz et al., 1999) and slower uptake or release rates observed for ancient organic matter phases such as kerogen or coal (Karapanagioti et al., 2000; Johnson et al., 2001b). Although the primary role of organic matter quantity in determining the equilibrium uptake of organic chemicals has long been recognized (Lambert, 1967; Chiou et al., 1983), the secondary role of organic matter structure in determining isotherm capacity has been the subject of more recent investigations (Garbarini and Lion, 1986; Grathwohl, 1990; Xing, 1997; Schultz et al., 1999; Ahmad et al., 2001; Kulikova and Perminova, 2002; Salloum et al., 2002). In some cases a relatively small fraction of highly transformed organic matter appears to account for the bulk of the sorption capacity of a soil, aquifer or sediment sample (Weber et al., 1992; Kleineidam et al., 1999b; Gustafsson et al., 1997). There is growing evidence that the importance of organic matter structure in determining sorption and desorption rates and the size of any desorption resistant fraction may be even more significant than its role in establishing sorption capacities.

Two important types of highly transformed organic matter found in soils and subsoils are (i) ancient sedimentary materials that have been diagenetically transformed under reducing conditions to such substances as kerogens and coals (Grathwohl, 1990; Weber et al., 1992) and (ii) black carbon (e.g., soot, charcoal) derived from anthropogenic combustion or forest fires (Baldock and Skjemstad, 2000; Gelinas et al., 2001). Despite significant differences in the origin and chemical nature of these materials, they appear to have similar impacts on sorption processes (Luthy et al., 1997). Organic matter buried in reducing sediments is converted over time to a more reduced C structure with a higher degree of aromaticity and a more highly condensed overall structure (Weber et al., 1999). Such diagenetically altered organic matter has been repeatedly shown to have higher organic C normalized sorption capacities, more nonlinear sorption isotherms, more heterogeneous heats of adsorption, greater degrees of sorption hysteresis, and slower sorption and desorption rates in comparison to more typical soil organic matter (Grathwohl, 1990; Weber et al., 1992; Young and Weber, 1995; Huang and Weber, 1997; Huang and Weber, 1998; Weber et al., 1998; Kleineidam et al., 1999a; Johnson et al., 2001b). Black carbon is potentially another important reactive domain in soils. In marine sediments it has been shown to greatly increase the capacity for sorption of polycyclic aromatic hydrocarbons, with soot carbon distribution coefficients exceeding those predicted from log K_{OC}–log K_{OW} correlations by about three orders of magnitude (Gustafsson et al., 1997). Although black carbon levels in soils appear to be lower than those in sediments based on a limited number of samples (Gelinas et al., 2001), the measured concentrations (~0.05 to 0.1%) may still be significant in driving sorption processes. The impact of black carbon on sorption and desorption rates remains to be explored.

Several research groups have drawn analogies between the structure of soil organic matter and that of synthetic macromolecules (Weber et al., 2001). This analogy grew out of an observation that many sorption phenomena, including sorption isotherms and rates, could be adequately described using two-compartment models (Huang et al., 1997; Cornelissen et al., 1997a) combined with the realization that the changes in sorption behavior observed for natural materials with increased diagenesis paralleled those for synthetic polymers above and below their glass transition temperatures (Pignatello, 1989; Carroll et al., 1994; Young and Weber, 1995). These parallels have been highlighted in a number of experimental studies comparing sorption behavior in polymers and soils. Isotherms for sorption of chlorobenzenes on polyvinyl chloride and a peat soil both became increasingly linear at higher temperatures as the fraction of glassy character in the materials is expected to decline, and both materials exhibited similar decreases in the distribution coefficient of 1,3-dichlorobenzene as the concentration of a competitor, chlorobenzene, was increased (Xing and Pignatello, 1997). The degree of sorption nonlinearity was observed to increase with time for 2,4-dichlorophenol sorption in a synthetic polymer (polyvinyl chloride) and a peat soil suggesting that in both systems high capacity "glassy" domains were filled more slowly than the less condensed "amorphous" domains (Xing and Pignatello, 1996). Similar increases in the nonlinearity of phase distribution relationships with increasing soil-solute contact time were observed for phenanthrene on a set of four EPA reference materials including three river sediment samples and a soil sample (Weber and Huang, 1996).

Significant support for the idea that synthetic polymer conceptual models can be applied to soil organic matter was provided by the observation of a glass transition temperature, T_g, in four different humic and fulvic acid samples from varied sources (LeBoeuf and Weber, 1997; LeBoeuf and Weber, 2000a; Young and LeBoeuf, 2000). The sorption isotherm of phenanthrene from water on one of the humic acid samples was approximately linear (Freundlich isotherm exponent 0.95) at a temperature above the water wet glass transition temperature and nonlinear (Freundlich isotherm exponent 0.76) below it (LeBoeuf and Weber, 1997). Similar changes in isotherm linearity upon changing temperature were observed for poly(isobutyl methacrylate), a synthetic polymer with a water wet T_g of 50°C, while nonlinear isotherms were observed at both temperatures for a coal sample with $T_g \cong 355$°C. Sorption–desorption hysteresis of phenanthrene on a range of natural and synthetic organic matrices was related to the measured glass transition temperature for these substances and was greater for each sorbent at temperatures below the material's T_g, further confirming that many "nonideal" sorption phenomena may be related to the glassy or rubbery state of natural organic matter (LeBoeuf and Weber, 2000b). Although the glass transition temperatures measured for humic substances in these studies are relatively low (36–73°C), they suggest that soil organic matter is likely to have appreciable glassy character at environmentally relevant temperatures. The lower glass transition observed for the fulvic acid sample compared with the humic acid samples (36 vs. 62–73°C) would be expected based on the lower molecular weight and lower aromaticity of the fulvic acid sample (Young and LeBoeuf, 2000). These findings are significant because they provide a possible explanation for the very slow sorption and desorption rates measured for soils in previous studies, since diffusion coefficients of organic contaminants in glassy polymers are 6 to 12 orders of magnitude slower than in corresponding amorphous polymers, with the exact ratio of diffusion coefficients depending on contaminant molecular size (Weber et al., 2001). Other investigators have failed to observe glass transitions for Aldrich humic acid or for lake sediments (Cornelissen et al., 2000b). Differences in results among these studies may be attributed to differences in sample purification (particularly for the Aldrich humic acid), sample thermal conditioning prior to analysis, mass of sample analyzed, and nature of the purge gas (N_2 vs. O_2).

Spectroscopic techniques have produced further evidence regarding the existence of varied structural domains within soil organic matter. The heterogeneity of the organic matrix makes unambiguous delineation of distinct structural features impossible; however, a number of different advanced solid-state nuclear magnetic resonance (NMR) techniques provide evidence consistent with the domain heterogeneity indicated by glass transition temperature measurements. Techniques applied include x-ray diffraction (Xing and Chen, 1999) and wide angle x-ray scattering (Hu et al., 2000) to identify condensed regions, NMR relaxation time experiments to differentiate between more and less mobile domains (Preston and Newman, 1992; Cook and Langford, 1998), variable temperature NMR to show the "melting" of crystallites (Hu et al., 2000), and two-dimensional wideline separation to demonstrate heterogeneous environments for particular types of functional groups (Mao et al., 2002). Taken together these investigations lend support to a model of soil organic matter that includes domains of distinctly different molecular mobility, polarity and packing density, each of which is likely to substantially alter organic contaminant transport rates within the respective domains.

Effects of Organic-Mineral Associations on Sorption Processes

The vast majority of studies examining sorption equilibria and rates for organic chemicals in soil, particularly those with a goal of identifying underlying mechanisms, have focused on either pure mineral phases or organic phases including isolated humic materials or synthetic polymers as reference materials. Rate-limiting mechanisms in whole soils or sediments have then been interpreted in light of the results for the chosen (relatively pure) reference materials. Although this approach has been used because it is experimentally tractable, it neglects the intimate association between mineral and organic phases in soils and subsurface materials and the diversity of their potential interrelationships (e.g., Fig. 11–1). Changes in sorption capacity on a carbon normalized basis have been observed for humic materials sorbed to clay surfaces suggesting that the conformation of the natural organic matter on a soil surface affects its affinity for uptake of organic chemicals, although the direction of the changes observed have not been consistent (Murphy et al., 1990; Jones and Tiller, 1999). The impact of clay mineral associations on rates has not been systematically explored, with the exception of synthesized organoclay materials to be used in pollutant containment structures (Deitsch et al., 1998). Slow desorption of a tetrachlorobiphenyl and naphthalene were observed on anionic surfactant coated anatase particles and the behavior was qualitatively similar to that observed for a whole soil (Hunter et al., 1996). The mechanism for retention of the target solutes by both systems was suggested to be rearrangement of the organic coating following sorption, effectively trapping some of the solute molecules in the sorbed state despite numerous subsequent washing steps with solute free solutions (Kan et al., 1997).

In many natural systems, organic matter exists in close association with mineral phases, with hydrous aluminum and iron phases showing particularly high affinity for accumulating organic matter (Kaiser and Guggenberger, 2000). The strong sorption capacity of these phases acts to selectively preserve hydrophobic organic matter containing carboxylic and aromatic functionalities and appears to be a critical step in the preservation of potentially labile organic materials in an otherwise microbially active subsurface environment (Baldock and Skjemstad, 2000). Therefore, there is undoubtedly a strong correlation between the type of mineral matter present in a soil and the quantity and composition of the associated natural organic matter. This cross-correlation is not considered in most studies that attempt to relate soil structure to sorption properties and could result in erroneously assigning the source of desorption resistance to aromatic groups within the organic matter when it is the organic-mineral interaction rather than the organic structure that is responsible for the observed phenomena. Evidence for such a view was provided in studies of polycyclic aromatic hydrocarbon sorption to three aquifer materials (Holmén and Gschwend, 1997). Fine-grained portions of the aquifer materials included a disproportionate share of both Fe and organic matter and the Fe–organic associations also appeared to be present as coatings on larger grains. Presuming that some fraction of the organic matter within the Fe oxide cements or coatings was kinetically unavailable for sorption of the target compounds based on the difference between column and batch reactor results allowed a first-order rate constant to be predicted from independently measured properties of the sorbent and aquifer materials (Holmén and Gschwend, 1997). Further evidence that overall sorption rates may be controlled by slow transport through inorganic phases to reach reactive organic matter was provided by studying the equilibrium and rate of phenanthrene sorption on various lithocomponents (Kleineidam et al., 1999a,b). In these systems <10% of equilibrium sorption was attained for 2 to 4 mm particles after 500 d in completely mixed batch reactors, largely because of the low intraparticle porosities of the associated inorganic phases. Residual or "desorption resistant" contaminants have been shown to be much more rapidly released following pulverization or acidification (HCl or H_2SO_4) suggesting that mineral phases that are broken up or dissolved by these treatments are largely responsible for the observed rate limitations (Pignatello, 1990a). Systematic study of the influence of organic matter-mineral associations on sorption and desorption rates is an area that deserves additional research given its demonstrated importance and the relatively limited amount of study afforded to it.

Sorbate Location and Environment

As the preceding discussion makes clear, a general consensus exists that the rate limiting step for sorption and desorption of nonpolar organic chemicals in soils is intraparticle mass transport through microporous minerals (in cases where such materials surround an organic matter phase or when soil organic matter contents are very low) and/or through expanded or condensed organic matter. A mechanistic description of the rate of the overall process depends on an understanding of the path that a sorbate molecule must take to reach a sorption site so that the appropriate diffusion coefficients and length scales can be specified. Until recently, little information about the location of sorbed molecules within organic matter or the distribution of organic matter within the soil matrix was available. Current research focused on these topics is reviewed in the remainder of this section.

Sorbate Locations

Direct observation of organic chemicals within natural particles has been made possible by application of advanced analytical techniques including infrared or nuclear magnetic resonance spectroscopy to characterize the local chemical environment of sorbed compounds and laser desorption-laser ionization mass spectrometry to identify physical locations within particles. Infrared laser desorption followed by ultraviolet laser ionization of the target compounds was shown to be capable of mapping surface concentrations of polycyclic aromatic hydrocarbons on soils and sediments at a resolution of 40 μm (Gillette et al., 1999). Results of the initial investigation showed variations in PAH concentrations by factors of between 5 and 29 across 500 μm particles in both lab-spiked and field-contaminated materials. These subparticle heterogeneities in PAH distribution are consistent with the heterogeneous distribution of hydrophobic and hydrophilic moieties predicted to occur in humic substances by conformational modeling (Sein et al., 1999). PAHs were found to be primarily associated with coal- and wood-derived particles in a harbor sediment, and to be concentrated within a few μm of the surface (Ghosh et al., 2000). PAH concentrations in silica particles from the same sample were several orders of magnitude lower and were localized within patchy organic matter coatings on the silica surface.

A variety of indirect evidence suggests that sorption nonlinearity, competition, hysteresis, and rate limitations may be primarily associated with a fraction of sorbed chemical that occupies matrix "holes" or microvoids within less flexible natural organic matter domains associated with many different types of soils and other geosorbents. Correlation of these effects with the "hardness" of the organic matter domains was initially noted by Weber et al. (1992). Subsequently, more nonlinear isotherms and a greater degree of competition was observed for sorption of 1,2– and 1,3-dichlorobenzene on a peat soil and three humic acids with increasing volume of microvoids measured by CO_2 adsorption at 77 K (Xing and Pignatello, 1997). More extensive investigation of CO_2 adsorption on soil materials has confirmed its utility in estimating micropore volumes and has suggested that diffusion of solutes within soil organic matter primarily occurs within micropores with maximum restrictions of around 0.5 nm (de Jonge and Mittelmeijer-Hazeleger, 1996). Further evidence for localization of sorbates within SOM microvoids was provided by the large negative enthalpic contributions calculated from Flory-Huggins interaction parameters on sorbents with relatively high T_g values, while constant, positive interaction parameters were calculated for materials studied above their T_g (LeBoeuf and Weber, 1999). Finally, the concept of the hole-filling mechanism is supported by consistent hole capacity values determined from Polanyi-Manes isotherms fitted to a range of nonpolar liquids on an aquifer material and the ability to interpret competitive effects between solutes on this basis (Xia and Ball, 1999, 2000). Results for a high organic matter peat soil, however, were not well described by Polanyi-Manes theory indicating that, contrary to the aquifer material examined in the studies referenced above, that the microvoids within the peat were not fixed but could be altered in their configuration by sorption history (Xia and Pignatello, 2001). Solutes differed in their plasticizing ability, with aromatic compounds (chlorobenzenes, benzene, and chloronitrobenzene) producing lower estimates of hole capacities than aliphatic compounds except for dichlorophenol, which behaved more similarly to the aliphatic compounds (Xia and Pignatello, 2001).

Spectroscopic methods have been used to identify the local chemical microenvironment of organic chemicals within natural organic matter, with a particular emphasis on differentiating between the populations of chemicals that sorb either rapidly or slowly. The sorption–desorption behavior of 1,2-dichloroethane to dry humic and fulvic acid samples was investigated using diffuse reflectance infrared spectroscopy to determine whether sorbed species showed a preference for either the anti or gauche conformer (Aochi and Farmer, 1997). Two distinct populations of sorbate were identified in the study. The first set of absorption bands was similar to that of DCA in the vapor state and the absorption bands rapidly (<24 h) disappeared when desorption was initiated. A second absorption band is not present in the vapor phase spectrum and it continues to grow even after 72 h of desorption suggesting that this species is more tightly bound and experiences significant mass transport resistance as it travels to its sorption site. A preference for the nonpolar anti conformer in sorbed DCA indicated that DCA was preferentially accumulating within nonpolar humic domains. This contrasts with results for DCA sorption on clay minerals in which sorption of DCA in the gauche conformation was favored suggesting sorption was occurring on clay surfaces with high effective dielectric fields within a network of pores that restricted transport. Sorption of the vapor-like species was postulated to occur in pockets within the expanded portion of natural organic matter while the second species was attributed to sorbate localization within microvoids in condensed and/or glassy regions; the presence of such microporosity was demonstrated using CO_2 adsorption (Aochi and Farmer, 1997). The generality of these results is somewhat questionable because humic material configurations change upon drying to reveal hydrophobic outer surfaces and hydrophilic core regions.

Several investigators have used solution- and solid-state nuclear magnetic resonance (NMR) spectroscopy to distinguish between kinetically different sorbate populations and to probe the binding environment experienced by sorbed molecules. Solid-state ^{19}F-NMR of sorbed hexafluorobenzene has been used to distinguish two chemically different compartments that appear to correspond to HFB that is either rapidly or slowly released (Kohl et al., 2000; Cornelissen et al., 2000a). Both studies identified at least two distinct peaks with chemical shifts downfield of that for liquid HFB but their assignment of these peaks to rapidly and slowly desorbing HFB differed. NMR spectra were collected at various times from 13 min to 24 h after the addition of liquid HFB to two peat samples and extracts from which lipids had been removed. Peaks at −167.0 and −167.4 increased rapidly and reached a plateau after about 4 h of sorption and were attributed to loosely bound, mobile HFB. A broader peak further downfield (−158.7 ppm) from liquid HFB increased more slowly, continuing to increase in size after 24 h, and was attributed to an immobile form of sorbed HFB. These peak assignments were further confirmed by running the same samples in a static mode in which the "mobile" peak remained sharp but the "immobile" peak became much broader (Kohl et al., 2000). In a second study ^{19}F-NMR spectra were collected for HFB sorbed to lake sediments and model organic materials including activated carbon and rubbery (polyacetal) and glassy (polystyrene) polymers (Cornelissen et al., 2000a). Longer contact times were allowed (1 wk to 2 mon) and spiking was performed from aqueous HFB solutions. Material that was presumed to contain only slowly desorbing HFB was prepared by subsequently removing the rapidly desorbing fraction using Tenax beads as an infinite sink for HFB. NMR experiments were not performed as a function of loading time in this study so peak assignments to rapid and slow fractions were based on some assumptions about the sorption rates. Rapidly desorbing HFB was attributed to peaks with chemical shifts of around −126 ppm while slowly desorbing HFB was identified with a peak at approximately −165 ppm. Two peaks with similar shifts were observed for HFB sorbed in the glassy polystyrene sample, which was taken as evidence that HFB in natural organic matter was divided between glassy and rubbery domains as in the synthetic polymer sample. No signal could be detected in static experiments to allow differentiation between mobile and immobile HFB. Differences in chemical shifts and peak assignments between the studies may relate to the contact time used prior to NMR analysis or differences in the spiking procedure, or may simply relate to errors in peak assignment. Neither study was able to detect evidence that liquid-like HFB was sorbed by a "hole-filling" mechanism, but the HFB loading levels in both studies were quite high (15–24 g kg^{-1} OC) in order to obtain reasonable NMR signals and therefore the small volume of HFB that might be present in sorbent micropores might not have been detectable (Kohl et al., 2000; Cornelissen et al., 2000a).

EXPERIMENTAL ASSESSMENT OF SORPTION DYNAMICS

Experimental Protocols

An excellent review of the range of experimental techniques that can be used in conducting sorption and desorption rate studies and how data from each can be used to generate sorption rate parameters has been recently published (Pignatello, 2000). Consequently, this section briefly summarizes the advantages and disadvantages of various methods for investigating sorption and desorption rates and reviews a limited number of recent studies with implications for experimental methods that were not covered in the earlier review by Pignatello. The major experimental methods for investigating sorption and desorption rates can be classified by the type of ideal reactor configuration they most closely resemble (Weber and DiGiano, 1996):

- Completely mixed batch reactors (CMBRs). Soil and solution are placed in a sealed container and mixed by rotary shaking, end-over-end tumbling, or stirring. Periodically the extent of sorption or desorption is determined by separating the phases (e.g., by centrifugation) and analyzing either the solution phase (typically) and/or the solid phase (less common).
- Completely mixed flow reactors (CMFRs). A soil–solution mixture is placed in a reactor and is mixed by stirring or shaking while a solution containing organic chemical (sorption) or background electrolyte (desorption) is pumped through the reactor while the effluent concentration is monitored.
- Plug flow reactors (PFRs). Soil is packed into a column (glass or stainless steel) between two porous endplates and solution containing organic chemical (sorption) or background electrolyte (desorption) is pumped through the reactor while the effluent concentration is monitored.

In choosing an experimental approach or evaluating a method chosen by others it is important to remember that no experimental method is ideal for all combinations of soils and solutes. The advantages of each method often directly create potential disadvantages. For example, completely mixed systems can eliminate rate limitations caused by slow transfer of chemicals across the boundary layer surrounding the particles by their intensive mixing but the mixing can cause changes in the soil's particle size distribution as particles are abraded during the experiment, especially if mixing is accomplished by paddle or stir bar. Other types of problems can affect any of the methods. Highly hydrophobic chemicals can display significant sorptive losses to system components (glass, Teflon, stainless steel). The remainder of this section reviews issues to consider in design and analysis of sorption and desorption rate studies using the reactor configurations listed above.

Extent and Nature of Mixing

Experimental sorption rate studies span a range of mixing intensity from none (ideal PFR) to intense mixing in some CMFR systems. The primary advantage of a column system (PFR) is that studies with intact or minimally disturbed soil aggregates can be performed, although uniform or representative column packing is always an experimental challenge (Pignatello, 2000). Column studies may therefore be most representative of organic chemical transport in soil systems where aggregate structure may serve as an important control on the rates of the transport process. Systems with minimal mixing are more complicated to analyze, though, since it is not certain that the rate controlling processes are at the intraparticle scale. Certain soil or solute characteristics may preclude the use of column-based experiments, however. For example, soils with high clay content are subject to plugging within the frits of columns resulting in unacceptably high back pressures and/or leakage while highly hydrophobic solutes (log $K_{OW} > 5$) may have retention times that are too long to be experimentally practical. Completely mixed systems simplify experimental interpretation by (presumably) eliminating boundary layer mass transfer as a significant process but, as noted above, can reduce average particle sizes and expose additional surface area for sorption or desorption during the experiment (Wu and Gschwend, 1986). The resulting change in surface area for interfacial mass transfer can result in unpredictable variation in the overall mass transfer coefficient. Particle abrasion appears to be worst for systems that are mixed by paddles or stir bars and least for systems that employ slow end-over-end tumbling. The potentially high solution to soil ratios in completely mixed experiments make these systems well suited for assessing sorption dynamics of moderately or highly hydrophobic organic chemicals. Less hydrophobic chemicals may be difficult to study using these techniques because obtaining sufficient extents of adsorption (~50%) may require solution to soil ratios that cannot be adequately mixed. Completely mixed flow reactors suffer from some of the same problems as both CMBRs and PFRs. Fine soils can clog the frits of flow cells, removing some of the particles from a completely mixed environment. Small particles that clog the frits may also contain high fractions of organic matter and consequently be significant in the overall sorption behavior of the bulk soil given the high correlation often observed between the content of clay and organic matter. Clogging problems can be alleviated to some extent within flow cells by periodically reversing the flow direction through the cell to backwash the frits. Although this technique can reduce clogging, (Heyse et al., 1997) found that the duration of experiments was still limited by clogging for some materials, and that the calculated sorption rate coefficients were uniformly higher than those estimated by column methods, possibly because of newly exposed surface area.

System Losses

Typically sorption data is analyzed to a first approximation by assuming that the added chemical is either in the solution phase or on the soil particles within the reactor. This assumption allows the sorbed phase concentrations to be calculated from the amount of chemical initially added and knowledge of the current solution phase concentration; however, the probe chemical also might be sorbed to system components, partitioned to reactor headspace, or transformed to another species by biological or abiotic processes. Investigators typically correct for such losses by testing losses in a control system that is soil-free and correcting their data accordingly. Important differences between the test and control reactors must be appreciated for such an approach to be valid, however. Transformation reactions may be greatly accelerated in the presence of soil, for example by soil bacteria that are not present in the control reactors or by indirect photolysis facilitated by soil organic matter. Biodegradation losses can be minimized by addition of an antimicrobial agent (e.g., NaN_3, $HgCl_2$, formaldehyde) or sterilization of the soil (γ-irradiation or autoclaving). Each of these treatments may alter soil organic matter structures and lead to changed sorption behavior and must be carefully assessed prior to application. Sorptive losses to system components are typically around 5% when all system components are glass or stainless steel, with greater losses when soft materials such as Teflon are employed. Losses might be greater in control reactors if the same initial concentration is used in both the test and control reactors because the equilibrium solute concentration will be lower in the soil-containing reactors. This effect can be corrected for by constructing a control isotherm that quantifies sorptive losses as a function of final aqueous concentration.

In many recent investigations of sorption and desorption rates close attention has been paid to the slow component of sorption and desorption reactions. Total uptake and release during these stages of the reaction may be only a few percent of the initially added solute and therefore even small system losses are important because of their potential to mask (desorption) or magnify (sorption) the process under study (Huang et al., 1998). As a general rule, sorptive loss corrections that are greater than about 25% of the slow sorption component (during the period over which the protracted sorption occurs) call into question conclusions about the magnitude of protracted sorption and desorption reactions. A number of investigators have recommended flame sealing of glass bottles as a means of minimizing both volatile and sorptive losses over long reaction periods (Ball and Roberts, 1991a; Huang et al., 1998). In all cases at least a fraction of the reactors employed in a sorption–desorption rate study should be exhaustively analyzed to verify that the added chemical can be accounted for and that calculated sorbed or solution phase concentrations agree with the extraction results.

Phase Separation and Colloidal Particles

A major difficulty in conducting sorption experiments in completely mixed soil slurries is obtaining a good separation of the solid and water phases. Settling is typically facilitated by the addition of background electrolyte (e.g., 0.01 *M* $CaCl_2$) but mechanical separation is usually still required. The preceding discussion emphasized the need for glass components, which effectively limits centrifugation speeds to ~8000 rpm. Filtration of soil slurries is made difficult by the substantial sorptive losses of hydrophobic chemicals on most membrane filter materials. Whether solutions are filtered or centrifuged, the supernatant will contain colloidal organic material that may have sorbed significant amounts of the solute of interest. The colloid bound material will be counted as dissolved by most measurement techniques artificially inflating the solution phase concentration and reducing the distribution coefficient (Gschwend and Wu, 1985). This effect is most important for highly hydrophobic organic compounds (log K_{OW} >~5) that show the greatest extent of sorption to colloidal material (Schwarzenbach et al., 1993). Since the source of the organic colloids is soil organic matter the concentration of colloidal material, and hence its impact, on a particular sorption experiment will depend on the soil to solution ratio (Voice et al., 1983). It is therefore important to keep a constant soil to solution ratio in all sorption and desorption experiments for a particular soil–solute system or to demonstrate that colloidal matter is not affecting the results. Other experimental ways to avoid problems due to nonsettling colloidal material are to keep the bulk solution and solid separate using selectively permeable materials such as dialysis tubing (Allen-King et al., 1995) or to purge the solution with gas to strip out the target chemical thereby measuring the aqueous activity of the dissolved chemical that is colloid independent (Coates and Elzerman, 1986; Brusseau et al., 1990). In each case an additional mass transfer resistance is introduced into the system complicating the interpretation of sorption rate phenomena and demanding careful attention to measurement of the characteristic response time of the system to changes in concentrations or flow rates in the absence of soil.

Colloidal material is particularly important in the study of sorption dynamics using decant and refill procedures. In a typical decant and refill experiment the batch reactor is centrifuged following an equilibration period and 60 to 90% of the solution is removed and replaced with water that contains solute or is solute-free for adsorption or desorption reactions, respectively (Kan et al., 1994; Hunter et al., 1996; Kan et al., 1997; Kan et al., 1998). The process is then repeated successively to promote adsorption or desorption by keeping the reaction driving force high. Colloid concentrations in supernatant would be expected to decline (to an unknown degree) in successive decant and refill steps as water extractable soil organic matter is removed. The removal of the colloidal material would produce an apparent increase in the distribution coefficient implying the presence of sorption–desorption hysteresis. A potential solution to this problem is to replace the withdrawn supernatant with an aqueous solution that has been pre-equilibrated with solute-free soil at the same solid to liquid ratio as in the experimental reactors, presumably keeping the colloidal organic concentration constant throughout the experiment. The importance of this phenomenon was assessed by comparing the extent of sorption hysteresis on three soils using the decant-refill method where the supernatant was either clean electrolyte solution or electrolyte solution pre-equilibrated with one of the three soils (Huang et al., 1998). The effect of colloidal material was found to be relatively insignificant for phenanthrene but might be more important for more hydrophobic chemicals. The decant-refill technique as typically practiced (<3 d equilibration times for each cycle) was found to significantly overestimate hysteresis relative to methods that employed longer equilibration times during the uptake and release segments of the experiment. When sorptive equilibrium is not attained, subsequent desorption steps will be incomplete since some of the added compound is still being adsorbed. Nonlinear isotherms compound the problems with the decant-refill technique because distribution coefficients will increase with decreasing solution phase concentration and, if the underlying nonlinearity is not recognized, will create the appearance of desorption hysteresis (Huang et al., 1998). A two compartment model in which a portion of the sites are at equilibrium with the solution phase following a Freundlich isotherm and the remaining sites are accessed via a first-order mass transfer process was capable of effectively representing data from decant and refill experiments with short contact times (Streck et al., 1995; Altfelder et al., 1999, 2000).

Driving Forces and Boundary Conditions

One of the critical problems in sorption and desorption rate studies is determining when the system has attained equilibrium. This is particularly true in the cases of protracted slow uptake and release observed for many geosorbents. In a finite-bath (CMBR) sorption or desorption rate study the driving force for mass transfer declines as the system approaches equilibrium. Identifying the point at which equilibrium has been attained against a background of experimental uncertainty and small system losses can be challenging. Minimizing system losses (e.g., by flame sealing vials) and working to improve experimental precision (e.g., by improving analytical methods) is therefore a critical component of designing a long-term rate study.

A second approach is to study sorption and desorption rates in a system that approximates an infinite bath by maintaining nearly constant aqueous concentrations throughout the course of the reaction. This is most common for desorption rate experiments in which various methods of stripping the desorbed contaminants from the aqueous system have been employed, most notably gas purging (Karickhoff, 1980; Coates and Elzerman, 1986) or the addition of another sorbent phase such as Tenax or XAD-4 polymeric beads (Pignatello, 1990b; Carroll et al., 1994; Cornelissen et al., 1997a). Obviously the former technique is best suited for studying volatile or semivolatile compounds while the sorbent approach is better suited to more hydrophobic compounds. Both approaches have disadvantages, however. The gas purge technique may alter the sediment characteristics by promoting changes in chemical composition (e.g., by oxidation) or aggregation state. Using Tenax to create an infinite-sink condition requires successive separations of the Tenax from the water and soil particles, and this separation may cause the loss of solids that become physically associated with the Tenax or the loss of dissolved organic matter that sorbs to the Tenax; however, the potential artifacts from both methods seem less severe than those associated with the decant-refill technique for studying desorption rates.

Experimental Duration and Sampling Strategy

Many early studies of sorption rate processes examined the process for 1 to 7 d, assuming that these periods were adequate for the systems involved to attain equilibrium. More recent studies have shown that much longer times are often required, particularly for materials that have been contaminated for long periods of time (i.e., "aged" samples). The anticipated duration of a sorption rate study has important implications for the experimental technique to be employed. For example, it is difficult to acquire early sorption or desorption time data (<10 min) using CMBRs because of the requirement for centrifugation to separate the phases. Early data must generally be obtained using column or flow cell techniques. In most realistic environmental situations, either natural or engineered, particle-fluid contact times are on the order of hours to days so detailed early time data for sorption or desorption reactions are not always required. Long desorption times often cannot be studied using CMFR techniques because of backpressure buildup, even when flow reversal is employed (Heyse et al., 1997). In batch studies with long equilibration times (months to years) flame sealing is recommended to reduce experimental variability and to lower total solute losses (Huang et al., 1998). The vast majority of studies on slow uptake and release of organic chemicals from soils over long periods of time have employed CMBR techniques for the reasons noted above, although some investigators have used column studies to good effect (Farrell and Reinhard, 1994a; Werth and Reinhard, 1997a).

When batch experiments are employed to obtain rate parameters, the timing and frequency of sampling must be decided upon. An organized method of determining desorption rate sampling times and experimental durations was developed by Opdyke and Loehr (1999) using Monte Carlo simulation and assuming that desorption followed a dual domain first-order profile, as had been shown experimentally by a number of previous researchers (e.g., Karickhoff, 1980). Most common experimental protocols appear to be sufficient for determining the fraction of sorbed chemical that is released rapidly and the rate constant associated with the fast fraction because these parameters require samples at early times, which are typically well represented in laboratory experiments. Estimating the release rate in the slow desorption compartment requires that samples be taken at times around $1/k_{slow}$, which would require experiments to be conducted for periods beyond 1000 d given values of k_{slow} reported in previous studies, while experimental durations of 14 to 300 d are more common in such experiments (Cornelissen et al., 1997a; Johnson et al., 2001b). Few investigators have followed sorption reactions for as long as 1000 d (Rügner et al., 1999). Estimating k_{slow} from a 250 d experiment yielded a coefficient of variation of >15% when the true value of k_{slow} was 10^{-3} d^{-1} (Opdyke and Loehr, 1999), probably a tolerable level of error given other uncertainties in fate and transport predictions. Other researchers have suggested shortening the time of the measurement of desorption rate parameters by conducting experiments at elevated temperatures (Johnson and Weber, 2001) or by conducting the desorption experiments in alternative solvents such as supercritical carbon dioxide (Hawthorne et al., 1999, 2001); however, these techniques have not been widely applied and there are some indications that the supercritical fluid extraction technique may not be generally useful for inferring aqueous sorption parameters (Weber and Young, 1997; Young and Weber, 1997).

Research Findings

Sorption and desorption rates are influenced by numerous factors including the chemical properties of the organic compound and its concentration, the chemical and physical characteristics of the soil, the amount of time the soil and solute have been in contact, the temperature of the system, the presence of competing solutes, the moisture level of the soil and the soil's particle size distribution. The majority of experimental investigations into sorption rates has combined experimental observations with speculation about controlling mechanisms and/or has constructed mathematical models of the data. Some of the mechanistic inferences of these studies are discussed in the previous section. Before delving into the diverse approaches used to model the process it seems useful to summarize the experimental observations regarding the extent to which these variables impact sorption and desorption behavior.

Sorbate Dimensions and Properties

Larger, more hydrophobic compounds are consistently observed to be sorbed and desorbed more slowly on soils and sediments. This is an expected result both because the larger compounds have smaller aqueous diffusion coefficients and because they exhibit more significant extents of sorption as reflected by their higher K_{OC} values. Exactly how sensitive sorption rate coefficients are to solute characteristics remains a source of debate, however. In a column study of 12 organic chemicals including alkyl- and chloro-substituted benzenes and polycyclic aromatic hydrocarbons, fitted first-order mass transfer coefficients ranged from 0.04 to 24 h^{-1} (Piatt and Brusseau, 1998). Diffusion coefficients estimated from the mass transfer coefficients spanned a similar range of almost three orders of magnitude while the aqueous diffusivities of the test compounds varied by less than a factor of three. An index of molecular connectivity that accounts for solute size and shape was more strongly correlated with the sorption rate coefficient than was the octanol water partition coefficient, suggesting that a solute's shape may be an important secondary factor in controlling rate processes (Piatt and Brusseau, 1998). Sorption rates of tetrachloroethene and 1,2,4,5-tetrachlorobenzene in size fractionated aquifer material differed by factors of around 40, more consistent with the differences in sorptive distribution coefficient than with the ~25% difference in aqueous diffusivities (Ball and Roberts, 1991b). Batch studies of desorption rates have revealed far smaller dependence of rate coefficients on solute characteristics. Fitted dual-domain first-order rate constants for slow desorption of nine chlorinated benzenes, polychlorinated biphenyls (PCBs) and PAHs from a lake sediment were negatively correlated with molecular volume, but the parameters varied by less than a factor of five (Cornelissen et al., 1997a). Additional study with eight chlorinated benzenes and PCBs on eight different soils and sediments resulted in a similarly narrow range in k_{slow} of 1.5 to 6×10^{-3} h^{-1} (Cornelissen et al., 2000b). It is unclear to what extent these different findings reflect true differences between the experimental systems studied and to what extent they arise from differences in experimental technique or the model used to fit the data. The two types of findings lead to very different conclusions about the nature of the rate limiting step in sorption processes, with the larger dependence on molecular properties suggesting that rates are controlled by diffusion through constricted pores or voids of molecular dimensions. Limited variation in rate constants across a wide range of soils suggests that a molecular rearrangement of the soil organic matter may serve as the mechanism "trapping" organic compounds, no matter what their size.

Concentration Effects

When sorption rate limitations arise from diffusive processes and sorption equilibria display decreasing distribution coefficients with increasing solution phase concentrations, as has been widely observed for hydrophobic organic chemical uptake by soils, the apparent diffusion coefficient of the solute within the soil matrix will increase with increasing concentration (Braida et al., 2001). This suggests that sorption equilibrium will be reached more quickly when the applied solute concentration is high. A second reason for increasing rates of sorption at higher aqueous concentrations may lie in the similarities between sorption processes in soils and those in synthetic polymers (Pignatello and Xing, 1996; Weber et al., 2001). When a sufficiently high concentration of a small organic solute is sorbed within a glassy polymer matrix the polymer in some cases swells enough that long-range segmental motion of the polymer chains, which had previously been prevented, is facilitated. This transition is referred to as a glass transition concentration. Above this level the polymer matrix becomes rubbery and exhibits significantly increased diffusion coefficients relative to those observed when the polymer is glassy.

Phase distribution relationships (PDRs) under non-equilibrium conditions have been employed by two groups of investigators to develop rate of approach to equilibrium data consistent with the notion that higher solution phase concentrations promote more rapid sorption reactions. One line of evidence is found in the increasing nonlinearity of phase-distribution relationships with increasing experimental contact times (Xing and Pignatello, 1996; Weber and Huang, 1996; Huang and Weber, 1998). Phase distribution relationships are similar to adsorption isotherms in that they describe the relationship between solid and solution phase concentrations of a solute, but the experiments are conducted at a series of different "equilibration times" designed to be shorter than those required to attain actual or even apparent equilibrium. For soils these relationships are nearly linear at short contact times (<1 d) but exhibit successively lower Freundlich exponents (n) and higher Freundlich capacity factors (K_F) as contact time increases, until both values stabilize as the system reaches an apparent equilibrium as illustrated, for example, in Fig. 11–3. The increasing nonlinearity is primarily manifested as increased solid phase concentrations with time at low solid phase loadings with far smaller changes (on a percentage basis) in the uptake at higher solid phase concentrations. For example, operationally defined "slow fractions" of dichlorophenol uptake in peat were 36% at aqueous concentrations of 0.02 mg L^{-1} while they were only 13% at 6 mg L^{-1} (Xing and Pignatello, 1996).

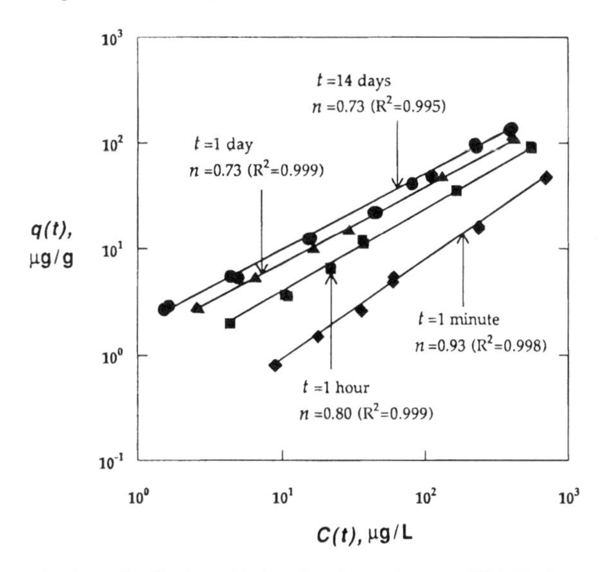

Fig. 11–3. Change in phase distribution with time for phenanthrene on EPA-23 river sediment (Weber and Huang, 1996).

As noted by Weber and Huang (1996), their observations that the time variable Freundlich exponent, $n(t)$, in a series of PDR experiments approaches equilibrium values more rapidly at high concentration values than at lower concentration values necessarily imposes a concentration dependence on the coefficient of any simple first-order (reaction or diffusion) rate model applied to interpretation of such data. For diffusion-controlled processes this situation is generally referred to as non-Fickian diffusion (Weber and DiGiano, 1996).

Batch spiking of phenanthrene from aqueous solution on soils and shales at low applied concentrations has been reported to produce more significant desorption resistant fractions (Johnson et al., 2001b) even though the absolute concentrations of desorption resistant species have been observed to increase with increasing concentrations for tetrachloroethene and 1,2-dibromo-3-chloropropane (Pignatello, 1990a). No differences in desorption rates of pentachlorobenzene from soil were observed in experiments involving different aging times (zero and six weeks) when spiking was performed at high applied concentrations, but a small effect was observed at lower applied concentrations (Schlebaum et al., 1999). In additional studies involving different concentrations, phenanthrene sorption on a range of solid materials including soils, river sediments, shale particles and kerogen extracts consistently exhibited faster approach to equilibrium for higher applied solute concentrations as illustrated in Fig. 11–4, although the absolute times required to attain equilibrium varied depending on sorbent characteristics (Huang and Weber, 1998). Pentachlorophenol uptake on a soil sample using a multiple spike methodology also exhibited more rapid uptake at higher total applied solute concentrations (Divincenzo and Sparks, 1997).

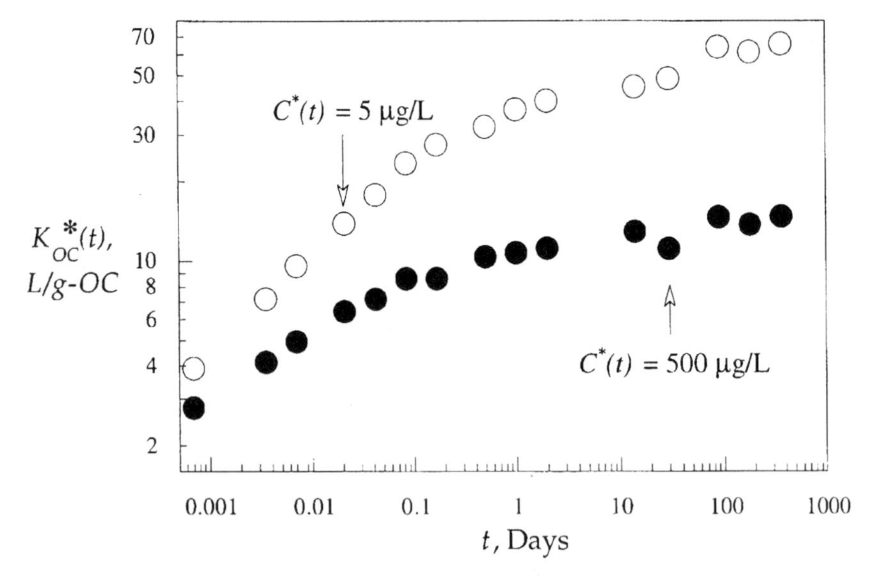

Fig. 11–4. Different rate of approach to equilibrium for two different applied phenanthrene concentrations (5 and 500 μg L^{-1}) on EPA-23 sediment. K_{OC}^* represents the organic C normalized distribution coefficient calculated at the index concentration $C^*(t)$ from the experimentally-derived time-dependent $K_F(t)$ and $n(t)$ values (Huang and Weber, 1998).

Although the preceding studies support the notion of concentration dependent rates of attaining sorptive equlibrium, they were not designed to directly address this question. One reason this effect may not have been more widely noted lies in the method used to conduct typical sorption rate experiments in CMBRs. Usually a constant solution-to-solid ratio is employed at varying initial aqueous concentrations of solute. As referenced earlier, if the isotherm for the system is nonlinear with a Freundlich exponent < 1 (concave-down), this will result in a smaller fraction of the added solute being adsorbed at equilibrium in the high concentration reactors than in the low concentration ones (Braida et al., 2001). For a diffusion controlled process the fractional rate of uptake declines as the proportion of the solute ultimately sorbed declines, leading to apparently slower sorbate uptake in the high concentration reactors, other things being equal. Common experimental investigations of concentration effects on uptake rates are therefore confounded because higher applied concentrations will tend to increase rates as a result of both isotherm nonlinearity and sorbent plasticization and simultaneously act to decrease rates due to declining gradient effects. The overall observed effect of solute concentration on uptake rate will depend on which of these several factors prevails. Characteristic sorption times for phenanthrene uptake on seven different soils generally declined with increasing initial concentration based on an analysis of experimental results using two models expected to bound the true time required to reach equilibrium (Braida et al., 2001). The maximum change in sorption times observed were about a factor of two as the solution phase concentration increased by two orders of magnitude.

Data consistent with the existence of a glass transition concentration was obtained in a study of trichloromethane (TCM) sorption and desorption on a peat soil (Xia and Pignatello, 2001). Significant hysteresis was observed after 21-d contact times for solution concentrations <3000 mg L^{-1} TCM but the hysteresis was greatly reduced at concentrations above that level. Further evidence that the soil organic matter was deformed by the presence of TCM was found in the increased capacity of the sorbent for TCM after previous uptake and release of TCM, referred to as a "conditioning effect" (Xia and Pignatello, 2001). In other studies the fraction of sorbate that desorbs slowly from soils, sediments and surfactant coated TiO_2 surfaces was found to increase when the sorbents were loaded at higher applied solute concentrations (Kan et al., 1994, 1997, 1998; Hunter et al., 1996; Chen et al., 2000).

Soil Organic Matter Composition Effects

The controlling role of organic matter in establishing slow desorption rates was demonstrated by the far smaller degree of desorption resistance of chlorinated benzenes and PCBs sorbed on model mineral solids (montmorillonite, zeolite, and organic matter free sediment) compared with whole sediment or model organic matter (XAD-8, polystyrene, polyacetal) phases (Cornelissen et al., 1998a). Organic matter was estimated to determine slow desorption behavior for samples with more than about 0.1% organic content.

The increase in sorption *capacity* of soils for hydrophobic organic compounds as the SOM increases in condensation and aromaticity and decreases in polarity, as measured by a wide variety of parameters (e.g., H/C or O/C atomic ratios, pyrolysis GC-MS peak areas, ^{13}C-NMR peak areas) has become well-established (Garbarini and Lion, 1986; Grathwohl, 1990; Rutherford et al., 1992; Xing, 1997; Young and Weber, 1997; Schultz et al., 1999; Kile et al., 1999; Chefetz et al., 2000; Ahmad et al., 2001; Kulikova and Perminova, 2002; Salloum et al., 2002). The influence of organic matter composition on sorption and desorption *rates* has been studied less, but some tentative conclusions are beginning to emerge. A variety of experimental evidence suggests that the widely observed "slow fraction" of organic compound uptake and release is more significant in natural organic matter that is richer in chemically reduced organic domains that are probably less subject to swelling by water under typical ambient conditions. One measure of sorptive reversibility used in several studies is the "hysteresis index" (Huang et al., 1998).

$$\text{Hysteresis Index} = \left. \frac{q_e^d - q_e^s}{q_e^s} \right|_{T,C_e}$$

where q_e^s and q_e^d are solid-phase solute concentrations for sorption and desorption experiments, respectively, that result in the same equilibrium aqueous solute concentration C_e and are conducted at a constant temperature T. The values are concentration dependent because the adsorption and desorption isotherms may be nonlinear to different degrees. Hysteresis indices for phenanthrene sorption reactions on soils and subsurface materials including shale and kerogen samples were negatively correlated with the atomic O/C ratio of the organic matter (Huang and Weber, 1997) and were negatively correlated with O and N containing pyrolysis fragments such as indole, methyl pyrrole, phenol, methyl phenol and methyl cyclopentenone (Schultz et al., 1999). Hysteresis indices were largest for shale and kerogen samples that have been subjected to extensive diagenetic alteration under reducing conditions (Huang and Weber, 1997; Weber et al., 1998; Huang et al., 1998).

Superheated water (water at temperatures exceeding the boiling point but at elevated pressures sufficient to maintain a liquid state) has been used to extract polar functionalities within natural organic matter, simulating in the laboratory diagenetic transformations that occur on geologic time scales (Johnson et al., 1999, 2001a). The extraction process has been shown to remove carboxylic, aliphatic and carbohydrate organic materials and may result in creation of additional aromaticity. The resulting materials were in many respects similar to geologically mature kerogen samples (Johnson et al., 2001a). The subcritical water extracted soil samples exhibited higher organic C normalized sorption capacities, greater degrees of sorption nonlinearity and an increase in the hysteresis index (Johnson et al., 1999, 2001a).

Hysteresis indices also were observed to be greater for soils and synthetic polymers when the samples were glassy (below their T_g) than when the same samples were rubbery, although the correlation between T_g and the hysteresis index was far from exact, further implicating condensed or glassy organic matter domains as a primary cause of slow adsorption and desorption behavior (LeBoeuf and Weber, 2000b). The importance of highly altered natural organic matter in controlling sorptive uptake of phenanthrene also was illustrated by examining the reactivity of subsamples of an alluvial sediment that contained different types of organic matter and different organic matter domain dimensions (Karapanagioti et al., 2000). Subsamples with coal derived particulate organic matter exhibited the slowest sorption rates, requiring 100 d to reach 10% of their ultimate sorption capacity. The coaly particulate matter also had the highest organic C normalized sorption capacities and isotherm nonlinearity. Sorption rates were much faster in amorphous organic coatings, reaching apparent sorptive equilibrium within several days (Karapanagioti et al., 2000).

The precise characteristics of organic matter that establish the rate of slow desorption remain elusive, however. Attempts to identify relationships between organic matter characteristics (C, N, or O content, C/(N + O) or aromaticity measured by ^{13}C NMR) and the desorption resistant fraction or first-order rate constants for slow desorption have been less than completely successful (Cornelissen et al., 2000b). Several investigators have in fact gone so far as to suggest that desorption resistant fractions and their associated desorption rate constants are relatively invariant, and independent of the exact nature of the SOM (Kan et al., 1998; Chen et al., 2000; Cornelissen et al., 2000b). This seems unlikely, however, in view of the substantial evidence that the nature of SOM controls swelling and deformation phenomena in response to temperature effects and sorbate concentration. In any event, it is clear that further research is required before the role of organic matter chemical structure in determining sorption rates can be unequivocally defined.

Soil Moisture

Changes in humic substance conformation (e.g., from a linear to a coiled configuration), with changes in solution phase composition are generally acknowledged, although the exact mechanisms of such transformations remain the subject of debate (Conte and Piccolo, 1999; Hosse and Wilkinson, 2001). Although changes in conformation of soil bound humic materials are less well investigated, primarily because of experimental difficulties, wetting and drying cycles have been hypothesized to cause such changes in soil organic matter structure. Chlortoluron (a urea herbicide) had higher distribution coefficients in saturated systems on soil that had been air-dried than on the same soil that had been maintained at field moisture, and the rate of approach to equilibrium was faster on the sample that had been air-dried (Altfelder et al., 1999). Changes in sorption equilibria and rates were found to be fully reversible upon rewetting of the soil samples, but the rewetting time required to restore the original activity was 80 and 500 h for a loess soil and a sandy soil respectively. Higher capacities and more rapid uptake by air-dried soil samples is consistent with a rearrangement of soil organic matter upon drying that leaves more hydrophobic regions at the soil-air interface and more hydrophilic moieties nearer the interior of the aggregate (Shelton et al., 1995). This conformational change is also likely to slow the penetration of water into the SOM interior, limiting competition of water for sorption sites and providing additional sorption capacity for nonpolar organic chemicals (Rutherford and Chiou, 1992).

Particle Size Distribution

A range of conflicting results have been obtained regarding the influence of bulk particle size on the rate of sorption or desorption, probably due to the wide range of soils, sediments and subsurface materials studied. Several studies have suggested that the appropriate length scale controlling sorption mass transfer is much less than the typical particle radius. This hypothesis has been supported by the failure of milling operations (Carroll et al., 1994) or particle size segregation (Farrell and Reinhard, 1994b; Cornelissen et al., 1999a) to produce significantly different sorption–desorption rates. In other investigations, smaller particles have exhibited faster rates of approach to equilibrium, indicating that the grain radius may be the appropriate length scale for describing mass transfer within soils and sediments. Milling was found to enhance the removal of aged organic contaminants (Steinberg et al., 1987; Pignatello, 1990a) and to speed the uptake of freshly added chemicals (Ball and Roberts, 1991b). Investigators who observed a particle size dependence of sorption–desorption rates have generally been able to describe their data with radial diffusion models and use of grain dimensions as diffusive length scales. One study found that the radial diffusion model worked well for subsurface fractions >63 to 500 μm, depending on the material, and that for smaller radii sorption rates were substantially slower than would be predicted based on particle radius and intraparticle porosity and was almost independent of pore size (Rügner et al., 1999). Photomicroscopic investigation of the smaller particles revealed particulate organic matter phases that were within one order of magnitude of the dimensions of the smaller particles and supported intraorganic matter diffusion as the rate limiting step in the smaller sieve fractions (Kleineidam et al., 1999b). The importance of accurately accounting for the distribution of particle sizes in sorption rate models has been stressed in studies that show the improved performance of multiple particle class models in comparison to single particle class models employing an average grain dimension (Pedit and Miller, 1994, 1995; Kleineidam et al., 1999a). These conflicting results provide additional support for the idea that more than one rate controlling process is probably operative for sorption reactions in soils and sediments and that these processes operate on widely varying length scales.

Aging Effects

The existence of an "aging effect" in which hydrophobic organic chemicals become increasingly resistant to removal by water, mild organic solvents or biological receptors with increased residence time is now widely documented, although the exact mechanisms remain subject to debate (Alexander, 1995; Pignatello and Xing, 1996; Luthy et al., 1997). Although one possible mechanism for the aging effect is the formation of covalent bonds between SOM and the introduced organic chemicals (Thorn et al., 1996), that mechanism is not considered further in this chapter since hot solvent extractions were found to be capable of recovering untransformed chemicals in most of the studies reviewed here. Organic chemicals with diverse structures and molecular sizes, including herbicides, chlorinated solvents, polycyclic aromatic hydrocarbons and polychlorinated biphenyls, have been shown to exhibit the aging effect to different degrees on different soils. Pesticides (atrazine, metolachlor, ethylene dibromide, simazine, picloram) have frequently been found to exhibit an increase in the slowly desorbing fraction, an increase in soil-water distribution coefficients and a decrease in the rate of desorption as the period of field aging increases (McCall and Agin, 1985; Steinberg et al., 1987; Pignatello and

Huang, 1991; Scribner et al., 1992). The phenomenon has been successfully replicated in the laboratory under controlled aging conditions. In laboratory spiking experiments that compared contact times between 2 and 34 d, an increase in (operationally defined) desorption resistant contaminant was observed with increased aging time for halogenated hydrocarbons (Pignatello, 1990a), chlorobenzenes, polycyclic aromatic hydrocarbons, and polychlorinated biphenyls (Cornelissen et al., 1997a). More significant changes in extractability were observed for trichloroethene at contact times that extended up to 15.5 mon (Pavlostathis and Mathavan, 1992). Although it is clear that the more significant aging effect sometimes observed for field contaminated soils compared with laboratory spiked materials is at least partly due to the far longer contact times in the field that permit slow diffusive transport to less accessible domains (Steinberg et al., 1987; ten Hulscher et al., 1999), other factors such as temperature fluctuations, redox chemistry changes, wetting–drying cycles or organic matter diagenesis cannot presently be ruled out as contributing to the phenomenon.

Temperature Effects–Activation Energy

Elevated temperatures are usually observed to promote more rapid desorption of organic contaminants from soils, and investigators have attempted to use the exact nature of the temperature dependence to draw inferences about rate controlling mechanisms. Desorption from porous materials and polymer matrices is commonly viewed as an "activated" process (i.e., one that requires energy input to overcome energy barriers to solute release from surface sites and transport to the bulk fluid). Such energy barriers might result from the strength of solid–solute interactions, diffusion through micropores or alterations of the solid phase such as mineral dissolution or SOM rearrangement. Wide ranges of desorption activation energies have been measured for different natural and model solids. Aged 1,2-dibromoethane contaminated soils displayed desorption activation energies of 66 kJ mol^{-1} (Steinberg et al., 1987). Activation energies for TCE desorption from silica gel, aquifer material and soil samples ranged from 47 to 94 kJ mol^{-1} increasing with longer purging times for the natural materials (Castilla et al., 2000). Chlorobenzene and PCB desorption from field contaminated sediment had activation energies of 57 to 70 kJ mol^{-1} for slow desorption while a laboratory contaminated sediment displayed statistically indistinguishable values of 60 to 74 kJ mol^{-1} (Cornelissen et al., 1997b). PAH desorption from coal-derived particles from a harbor sediment sample exhibited activation energies of 133 to 138 kJ mol^{-1} (depending on the desorption model applied to the data) while the clay–silt fraction of the sediment had far lower activation energies of 37 to 41 kJ mol^{-1} (Ghosh et al., 2001). Differences in activation energies were also observed between phenanthrene desorption from a shale sample (83–86 kJ mol^{-1}) and a surface soil (41–69 kJ mol^{-1}) although the values were not statistically distinguishable, suggesting that solute transport through more reduced and diagenetically altered organic matter is subject to more significant energy barriers than that from the more flexible and swollen organic matter in humic-dominated soils (Johnson and Weber, 2001). While reported activation energies for desorption vary widely, they are for the most part higher than those expected for hindered diffusion through mineral micropores, and more consistent with values expected for diffusion within organic polymers (ten Hulscher and Cornelissen, 1996; Cornelissen et al., 1997b; Johnson and Weber, 2001). For example, activation energies for trichloroethene desorption from organic matter-free silica gels with defined pore sizes are in the range of 15 to 57 kJ mol^{-1} (Werth and Reinhard, 1997a; Farrell et al., 1999; Castilla et al., 2000) generally lower than values observed for natural geosorbent materials. Because desorption activation energy ranges for glassy polymers and zeolite micropores overlap, firm distinctions about mechanism are impossible based on activation energies alone (Castilla et al., 2000). The danger of over-interpreting these values is stressed by the observation that, in at least one instance, a precipitation process rather than a diffusive one was found to be the rate-controlling step in desorption from a silica gel (Farrell et al., 1999). An important practical use of the temperature dependence of desorption processes is that experiments at elevated temperatures can be used to gather information about the rates of release at lower temperatures greatly shortening experimental times (Cornelissen et al., 1997b; Johnson and Weber, 2001).

Competing Solutes

As for many other soil sorption phenomena, the effect of competing solutes on equilibrium soil-water distributions is relatively well documented while the associated rate phenomena are only beginning to be explored. Competing solutes, including those that are structurally different than the primary solute, can suppress primary solute sorption even at competitor concentrations comparable to those of the primary solute (McGinley et al., 1993; Xing et al., 1996; Xing and Pignatello, 1997, 1998; White and Pignatello, 1999; Schaefer et al., 2000; Li and Werth, 2001). In the presence of competitors primary solute isotherms become more linear while the competitive effect is observed to be the strongest for sorbents that initially display the most nonlinear sorption for the target compound (McGinley et al., 1993). These observations suggest that the solutes are competing for a relatively limited number of holes or microvoids within the soil organic matter and that these sites are energetically preferred compared with the remaining, more flexible, organic matter domains. The limited evidence on the impact of competitors on sorption and desorption rates has confirmed this view. The addition of pyrene to a sandy loam soil and a peat significantly increased the rate and extent of phenanthrene desorption from the soils at both short (2 d) and long (42–76 d) contact times (White and Pignatello, 1999). In tests with field-contaminated sediments, the addition of large excesses of 1,2,4-trichlorobenzene increased the extent of desorption of aged PCBs by up to a factor of 2.7 (Cornelissen et al., 1999b). Aged PAH and mineral oil residues in another sediment were shown to block the access of a PCB to desorption resistant sites thereby reducing the extent of the slow desorbing fraction of the added chemical by a factors of between 9 and 14 compared with the same sediment from which the PAH/mineral oil mixture had been removed. These findings suggest that the slow desorption domain is not necessarily far removed from the soil surface since competitor induced desorption of the target compound can be seen soon after the competitor's addition. Uptake and release rate limitations are therefore more related to the solute's movement being held up by sorption within organic matter holes, and that transport is faster if these holes have previously been "plugged" by another solute (White and Pignatello, 1999).

MODELING APPROACHES

Although widely used fate and transport models and risk estimation methodologies still rely almost exclusively on linear equilibrium partitioning of organic compounds between solid and aqueous phases (American Society for Testing and Materials, 1995), the need to incorporate rate limitations into these models is critically apparent to those who have studied the sorption phenomena described in the preceding sections. Three main modeling approaches have emerged for describing rate limited uptake and release of hydrophobic organic chemicals in soils and sediments at the grain scale: radial diffusion models, discrete multi-site models, and stochastic models featuring a continuous distribution of site types. Radial diffusion models are the most mechanistically satisfying of the three model classes since the parameters of these models can mostly be compared with measurable, physically well-defined characteristics of the sorbent and the solute. They have been limited, however, primarily by the lack of information on the relevant diffusive length scale and the heterogeneity of soil and sediment particles. Multi-site models offer the simplest mathematical treatment of sorption rate processes, but there is no current consensus about how to relate the model parameters to independently measurable quantities. Suggestions for how this might be done have been advanced, however. Finally, the stochastic models that include a continuous distribution of desorption rates seem mechanistically reasonable since the heterogeneity of soils and sediments suggests that more than two or three distinct types of sorption "sites" are to be found within a soil particle and because interparticle variability in properties is the rule rather than the exception; however, fitted parameters in these models are the most difficult to relate to underlying mechanisms and therefore the least likely to be independently measurable. This section reviews major applications of each of the three model types, focusing on the most recent literature in this area. Comprehensive and more general treatments of sorption and desorption rate modeling in heterogeneous systems are available (Weber and DiGiano, 1996; Pignatello, 2000).

Radial Diffusion Models

Intraparticle diffusion models have been the most widely applied approach for describing sorption and desorption rates in laboratory-scale reactors. Typically the spherical form of the model is used:

$$\frac{\partial C}{\partial t} = \frac{D_{\mathrm{a}}}{r^2}\frac{\partial}{\partial r}\left[r^2\left(\frac{\partial C}{\partial r}\right)\right]$$

where C is the solution phase concentration, D_{a} is the apparent diffusion coefficient, and r is the radial distance from the particle center. If retarded diffusion through solvent filled pores is the rate limiting mechanism and a nonlinear Freundlich isotherm model can describe sorption within the pores, the apparent diffusivity can be related to the solute's diffusion coefficient in water (D_{aq}) by

$$D_a = \frac{D_{aq}}{\left(1 + K_F\, nC^{n-1}\, \frac{\rho}{\varepsilon}\right)\tau_f}$$

where ε is the intraparticle porosity, ρ is the sorbent's bulk density, K_F and n are the Freundlich isotherm capacity and exponent, and τ_f is the tortuosity factor. The ratio of the aqueous diffusivity to the tortuosity factor is often referred to as the effective diffusivity, D_e. This model was found to effectively describe nonlinear and hysteretic sorption and compared favorably to a coupled equilibrium first-order model and a second order model for different soils and organic solutes over time periods <10 d (Miller and Weber, 1984, 1986; Weber and Miller, 1988). Other investigators used the pore diffusion model to successfully describe short term (<2 d) adsorption and desorption data using τ_f (or D_e) as the only fitting parameter (Wu and Gschwend, 1986). Radial diffusion model fits to longer term data (weeks to years) are often improved by the addition of a second parameter that allows for a fraction of the solute to sorb instantaneously following the bulk isotherm, with the uptake or release of the remaining material governed by the radial diffusion model (Ball and Roberts, 1991b; Pedit and Miller, 1994; Johnson et al., 2001b); however, the instantaneously sorbing fraction in these models does not appear to be mechanistically meaningful because the parameter value is sensitive to whether earlier or later time rate data are used to derive it (Pedit and Miller, 1995).

Examining the history of applying radial diffusion models to describe environmental sorption reactions permits a number of generalizations about situations in which they will be successful. The models have been found most useful over extended time periods (>10 d) only for describing hydrophobic chemical uptake on narrow size fractions of aquifer material (Ball and Roberts, 1991b; Pedit and Miller, 1994, 1995; Holmén and Gschwend, 1997; Rügner et al., 1999). In each of the cited studies, the sorbent was a low organic C material (<0.004–0.11%) and in all of the cases there was evidence (microscopic examination or results on pulverized samples) that a portion of the organic matter was protected or encapsulated within mineral grains. In these situations it might be expected that diffusive transport of the solute through mineral micropores would be the rate-limiting step and that the grain radius might be the relevant diffusive length scale. The importance of conducting experiments and deriving model parameters from narrow size fractions was stressed by the significant improvements in model performance observed for multiple particle class radial diffusion models compared with those using a single average particle radius to predict uptake (Pedit and Miller, 1994, 1995). Success of the model for low organic carbon materials probably relates to the increased likelihood that the sorbing domain is present as a thin coating on a mineral surface or as relatively small inclusions. D_e/a^2 values were inversely related to particle radius squared for particle radii above 63 and 500 μm for two limestone fractions with maximum organic matter domain sizes of 5 and 50 μm, respectively (Rügner et al., 1999). These findings suggest that intraparticle diffusion is a rate controlling process as long as organic matter domain dimensions are small in comparison to the grain scale. Tortuosity factors for a variety of relatively large (>1 mm) low organic carbon rock fragments and aquifer material size fractions were inversely related to intraparticle porosity, allowing the radial diffusion model to be applied to such materials with no adjustable parameters (Rügner et al., 1999). Such an application, however, requires that pore sizes be large relative to solute dimensions so that hindered diffusion does not dominate the transport process (Farrell and Reinhard, 1994b).

Successful application of the radial diffusion model to higher organic C content materials have either involved shorter (<10 d) equilibration times (Weber and Miller, 1988) or have involved extensive efforts to document the size and location of the reactive organic matter domains (Karapanagioti et al., 2000). By using coal petrography techniques to visualize and classify the dimensions and types of organic matter within an alluvium sample, the different rate processes governing solute uptake on the various subfractions were elucidated. Coal derived particulate organic matter exhibited the highest organic normalized sorption capacity, the most nonlinear sorption isotherms and the slowest sorption rates of any of the subsamples, requiring 5700 d to reach 75% of its equilibrium solute uptake while the bulk sample took just 12 d to reach a similar level (Karapanagioti et al., 2000).

Description of long-term desorption data using radial diffusion models has been uniformly problematic, particularly for the fraction of chemical that is slowly released. Radial diffusion models, either with or without an instantaneously sorbing fraction, were the least successful of six models considered for describing phenanthrene desorption from a soil, an aquifer material and a shale (Johnson et al., 2001b) although the other models all had an equal or greater number of adjustable parameters. The particular form of the radial diffusion model employed in that study did not account for changes in the effective diffusivity with changing solid phase concentration such as are expected for systems described by a nonlinear isotherm. Further, in the case of the soil sample the model was applied to a very wide range of particle sizes without accounting for the particle size distribution (Johnson et al., 2001b). The radial diffusion model could not describe both fast and slow desorption of TCE from model inorganic materials (silica gels of varying particle and pore sizes) or natural solids (soil and aquifer material) with a single effective tortuosity, suggesting that the model was not mechanistically correct for these sorbents (Farrell and Reinhard, 1994b). One reason for this failure may be the heterogeneity in pore sizes, shapes, and constrictions in natural materials. Early time TCE desorption data at two temperatures was attributed to release from mesopores (2–50 nm diameter) and was successfully predicted using a pore diffusion model without adjustable parameters, while the later time data (t >~10 min) was successfully fitted using a micropore diffusion model (Werth and Reinhard, 1997a). A similar dual porosity interpretation was used to describe the two phases of adsorption of toluene on oven-dried montmorillonite and a loam soil (Arocha et al., 1996). A different dual diffusion coefficient model, originally developed to describe sorption–desorption rates of solutes in non-uniform polymer powders, was applied to analyze desorption of PCBs from Hudson River sediment (Carroll et al., 1994). The difference in diffusivities between fast and slow fractions was presumed to arise from chemicals sorbed within glassy and rubbery domains of sediment organic matter. Diffusion coefficients were therefore predicted from presumed humic substance permachor values and the model fits were then used to derive apparent length scales for the organic matter domains responsible for uptake, which were calculated to be on the order of 10 to 30 nm.

Independent measurement of SOM domain dimensions and sorbed compound distributions within them are clearly desirable to provide mechanistically meaningful model input parameters. Information on the distribution of PAHs within sediment particles obtained by two-step laser desorption-laser ionization mass spectrometry and has been recently used to construct a radial diffusion model that accounts for the initial distribution of the PAHs within the outer 3 to 5 μm of the coal derived particles having the highest contaminant concentrations (Ghosh et al., 2001). The radial diffusion model for these particles assuming all PAHs were initially within 3 μm of the particle surface successfully fit the thermal programmed desorption profiles of particular PAHs proving the utility of the independently derived information; however, there was no attempt to fit the radial diffusion model to the long-term aqueous PAH desorption profiles and it is not immediately clear if such fitting would be successful.

Multi-Site Models

The common observation that organic contaminant desorption from soils is "biphasic" with a rapid and slowly released fraction of chemical prompted the development of models that attempt to derive rate parameters for the fast and slow fractions separately. A diverse set of models has been developed to accomplish this goal. The two most widely used models in this class are a dual-domain first-order model and the equilibrium–first-order model, the latter of which assumes the "fast" sites are always at equilibrium while solute sorbs to "slow" sites at a rate that can be described by a first-order relationship. These two models will be the primary focus of this discussion, while some attention will also be given to other multi-site model formulations.

The dual-domain first-order model has been widely used in major part because of its mathematical simplicity, but it has been shown to be comparable to more sophisticated models for empirical description of sorption–desorption data (Johnson et al., 2001b). This model has the form:

$$\frac{q(t)}{q_0} = \phi_F e^{-k_F t} + (1 - \phi_F) e^{-k_S t}$$

where $q(t)$ is the solid phase concentration at time t, q_0 is the solid state concentration when desorption commences ($t = 0$), ϕ_F is the fraction of solute sorbed in rapidly desorbing domains and k_F and k_S are the first-order rate constants for rapid and slow chemical release, respectively. This model has been shown to successfully fit desorption data for a wide range of chemicals, either laboratory spiked or field-contaminated, on varied sorbents (Cornelissen et al., 1997a,b, 1998a, 2000b; Johnson et al., 2001b; Hawthorne et al., 2001). Although most of these investigations have involved desorption time periods of approximately 10 to 20 d, a few have involved desorption experiments >100 d (Hawthorne et al., 2001; Johnson et al., 2001b). The dual-domain first-order model performed better than the radial diffusion model, a polymer diffusion model, and a stochastic first-order model employing the γ distribution in fitting data for phenanthrene desorption from a surface soil, an aquifer material and a shale sample (Johnson et al., 2001b). Example fits from the six models tested for the Wagner soil are shown in Fig. 11–5. In addition to its mathematical tractability, the dual-domain first-order model has the advantage that its parameters can be visually related to a cumulative removal curve such as those in Fig. 11–5. The slope of the early and later portions of the graph are related to k_F and k_S, respectively, while the "break" between these two curve segment occurs approximately at ϕ_F, representing the fast fraction of the desorbing contaminant. Despite this appeal, however, the model parameters depend in complex ways on the nature of the sorbent, the solute, contact time, spiking concentration, and adsorption or desorption conditions. The model does not attempt to describe the underlying mechanisms controlling desorption rates and efforts to relate the model parameters to characteristics of the sorbent or the solute have been generally unsuccessful (Cornelissen et al., 2000b). Extension of the dual-domain first-order model by adding a third first-order term to describe release from "very slow" sites has been attempted (ten Hulscher et al., 1999), but such models suffer from their dependence on too many adjustable parameters, which can result in non-unique parameter values during calibration and optimization (Johnson et al., 2001b).

A second common discrete multi-site modeling approach is the equilibrium–first-order formulation, which presumes that a fraction of the sites come to equilibrium with the fluid phase instantaneously while uptake in the others is con-

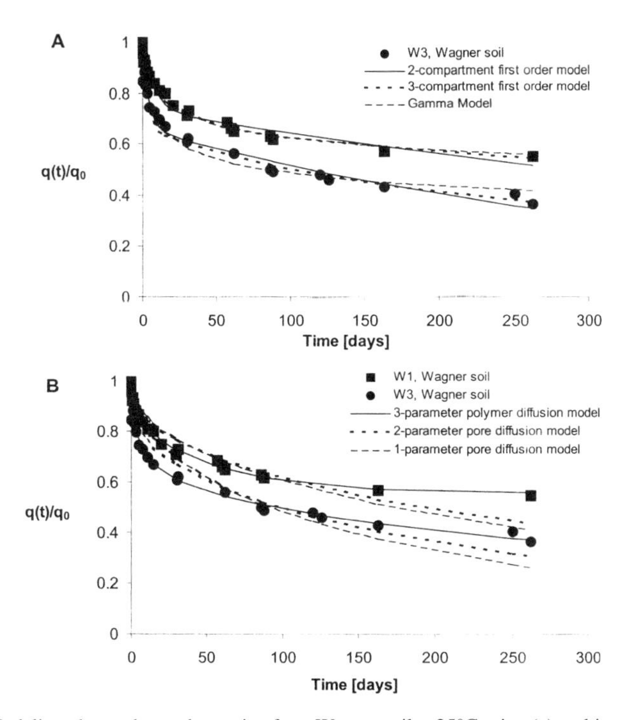

Fig. 11–5. Modeling phenanthrene desorption from Wagner soil at 25°C using (a) multicomponent first-order desorption models and (b) diffusion models (Johnson et al., 2001).

trolled by a first-order mass transfer process (Cameron and Klute, 1977). The two different site types could differ either in their chemical composition and reactivity toward the target solute or in their physical location, with transport to the "slow" sites being diffusion limited. If the distinction is based on chemical heterogeneity, we may presume that the sorption behavior of each compartment differs and that the overall rate of uptake is

$$\frac{dq}{dt} = n_F K_{F_F} C^{n_F - 1} \frac{dC}{dt} + \alpha \left(K_{F_S} C^{n_S} - q_S\right)$$

where $K_{F,F}$ and $K_{F,S}$ are the Freundlich capacity factors and n_F and n_S are the Freundlich exponents for the fast and slow fractions, respectively and α is the first-order rate constant for sorption within the slow compartment (Weber and Miller, 1988). Although the parameters in the equilibrium–first-order model can be ascribed physical meaning, it has not proven widely successful for fitting either adsorption or desorption data. For example, adsorption data for lindane and diuron on aquifer materials were described significantly better by radial diffusion models than by an equilibrium–first-order modeling approach, regardless of whether single or multiple particle classes were employed (Weber and Miller, 1988; Pedit and Miller, 1994). The equilibrium–first-order model also was less successful than stochastic first-order models in fitting sorption and desorption data for carbon tetrachloride and 1,2-dichlorobenzene on modified clays and a peat soil (Deitsch et al., 1998).

Various other discrete multi-site modeling approaches have been proposed in recent studies but have been applied to a more limited degree (Schlebaum et al., 1999; Kan et al., 1998; Chen et al., 2000). One approach extends the notion of different sorption isotherms within different soil compartments by postulating first-order uptake rates within the linear uptake domain and *n*th order rates within the nonlinear (Freundlich) isotherm domain (Schlebaum et al., 1999). When independent isotherms are not measured, this model has seven fitting parameters, the most of any of the models discussed in this chapter, and this increase in complexity does not seem justified given the success of other models in fitting sorption and desorption data with fewer adjustable parameters. Other investigators have drawn on experimental observations of prolonged, nearly constant leaching rates from the slow fraction to presume that the slow and fast fractions are governed by two separate linear sorption coefficients with the fast fraction following conventional hydrophobic partitioning theory and the slow fraction exhibiting organic C normalized sorption capacities several orders of magnitude larger (Kan et al., 1998; Chen et al., 2000). This approach has been shown to produce adequate descriptions of some laboratory and field data, but it fails to account for widespread observations of sorption isotherm nonlinearity and, like several of the models described in this chapter, it lacks a mechanistic basis.

Stochastic Descriptions of Mass Transfer Processes

The intrinsic heterogeneity of soils, sediments, and aquifer materials at both the intraparticle and interparticle scales has caused a number of researchers to conclude that deterministic models applying single-valued parameters to the description of sorption–desorption rates are bound to poorly characterize the underlying processes. Instead, a class of stochastic models has emerged in which mass transfer rates are described using continuous parameter distributions. Continuous distributions have been applied to diffusion coefficients in radial diffusion models (Werth and Reinhard, 1999; McMillan and Werth, 1999; Werth et al., 2000) and to first-order mass transfer coefficients in both single-site and two-site models (Connaughton et al., 1993; Culver et al., 1997). Both lognormal and γ distributions have been applied to describe the probabilistic variation in model parameters. The most common approach has been to use the two-parameter γ distribution to describe variations in the first-order mass transfer rates. One implementation of the model, which is a generalization of the equilibrium–first-order model described above, uses a linear isotherm to describe the equilibrium compartment and a discretized γ distribution to describe mass transfer rate coefficients for the first-order sites (Culver et al., 1997):

$$\frac{dq}{dt} = \sum_{i=1}^{N} \alpha_i \, (FK_D C - q)$$

where α_i is the first-order rate constant governing uptake or release at slow sites for the ith compartment of the γ distribution according to

$$p(\alpha) = \frac{\beta^{-\eta}\,\alpha^{\eta-1}}{\int_0^\infty x^{\eta-1} e^{-x} dx}\, e^{-(\alpha/\beta)}$$

where η represents the shape factor and β represents the scale factor for the γ distribution, and x is a dummy variable of integration. The model outlined above has been shown to successfully describe sorption and desorption of 1,2-dichlorobenzene, trichloroethene and carbon tetrachloride on a variety of soils and organoclays, generally outperforming the equilibrium–first-order model (Culver et al., 1997; Deitsch et al., 1998; Culver et al., 2000; Deitsch et al., 2000). Other similar first-order model formulations based on the γ distribution also have been shown to fit a wide range of sorption and desorption rate data (Connaughton et al., 1993; Pedit and Miller, 1994; Chen and Wagenet, 1995). A version of the gamma model was inferior to the dual-domain first-order model in describing phenanthrene desorption from a surface soil although it performed well on similar data for a shale and aquifer material sample (Johnson et al., 2001b). For a two (or three) parameter model the γ distribution approach has been highly successful in describing both sorption and desorption data with consistent model parameters; however, the physical significance of the parameters and the mechanistic insight they provide is limited.

CONCLUSIONS AND FUTURE DIRECTIONS

Heterogeneities in soil mineral phases, organic matter composition, and the aggregation of these diverse phases pose significant challenges for describing, and ultimately predicting, rates of sorption and desorption of organic chemicals in soils. Progress in this area during the last 15 yr has largely come through combinations of extended uptake and release rate measurements in bulk systems and detailed analyses of the structure and composition of soil matrices using microscopic and/or spectroscopic techniques. Numerous models have been shown capable of adequately describing particular sets of uptake or release data, although some are inevitably better than others under various conditions. In the absence of other supporting information, the ability of a particular model to fit a particular data set should not be viewed as imparting useful information about underlying mechanisms. Future advances in this area will probably occur as clever new ways are devised to independently measure parameters required in various sorption and desorption rate models, the ultimate goal being the development of models that are free of adjustable parameters.

The two leading candidates for rate controlling sorption mechanisms in soil remain intraparticle diffusion and intraorganic matter diffusion, as postulated by investigators over a decade ago (Brusseau and Rao, 1989; Pignatello, 1989). In the intervening years significant progress has been made in refining and applying these models, developing particular sub-cases of each model, and in proving their applicability by means other than model fitting. Intraparticle diffusion models have proven their utility in describing a wide range of sorption rate data, especially in combination with careful documentation of the location and dimensions of the reactive organic matter phases within low organic C content aquifer materials. Application of this modeling approach requires a detailed particle size distribution and microscopic evidence that the organic matter domains are within the mineral matrix. Crushing a portion of the sample to prove that the relevant diffusional length scale is at or near the grain scale also should be a routine part of applying this model. When the model is applied to describe contaminant desorption the initial intraparticle distribution of the sorbed organic chemical is also required. The intraparticle diffusion model is currently the best developed of the mechanistic models with each of the steps described having already been undertaken successfully by at least one research team. Although there are still many opportunities for future advances such as improved determination of intraparticle pore structures, application of the model in a predictive manner using existing methods seems within reach for low organic carbon content materials.

Intraorganic matter diffusion appears to be the most common rate-limiting step for higher organic matter soils and sediments since the relevant diffusional length scale is significantly less than the particle scale in many cases. In these cases the fast and slow portions of sorption–desorption behavior are attributed to migration through glassy and rubbery organic matter domains. Making such models predictive requires quantifying fractions of organic matter present within glassy and rubbery domains by determining the glass transition temperatures (e.g., by differential scanning calorimetry) or assessing the mobility of various organic matter domains (e.g., using NMR relaxation time measurements). Furthermore, methods for quantifying the populations of fast and slow desorbing species within a sorbent at a particular aging time are required to provide independently measured initial conditions for desorption studies. Although the existence of glassy and rubbery domains within soil organic matter and the existence of two different populations of sorbed molecules has been proven in particular cases, there has been no direct linkage of these findings to sorption–desorption rate modeling. All of the experimental tools required to construct an intraorganic matter diffusion model using independently derived parameters appear to be in place; assembling them in a comprehensive manner awaits future research. A remaining challenge in this area is the hypothesis advanced by a number of researchers that the organic matter matrix rearranges following contaminant uptake. If true, this makes it unlikely that model parameters obtained from uptake studies will ever effectively describe desorption behavior. This hypothesis is difficult to independently confirm. Evidence to date relies largely on some reasonable suppositions and interpretation of model fitting results. Many of the observations cited to prove the organic matter rearrangement hypothesis also could be caused by failure to obtain equilibrium during the uptake step, a situation that is difficult to conclusively prove or refute. Significant further work is required to document the existence of the rearrangement phenomenon and to consider how it might be incorporated into rate models.

The mechanistic models described above offer hope of predictive capability for sorption and desorption rate processes, but require significant investment of time and experimental resources to be implemented completely. This investment is certainly worthwhile if research is intended to prove rate-controlling mechanisms or requires a model that can be extrapolated beyond the range of available data. In many cases all that is required is a useful mathematical description of the available rate data for use in a fate and transport model that will operate within the range of the experimental observations. In such a case a more empirical fitting model may be desirable. Both the dual-domain first-order and γ distributed mass transfer rate models have proven to be rugged and capable of describing long-term desorption data. The dual-domain first-order model is mathematically simpler to apply and offers the advantage of parameters that can be obviously related to the desorption rate curve, however, it is only useful for describing desorption data. The γ distribution models implicitly account for heterogeneities in soils, do not require the postulation of a limited number of site types, and can employ the same parameters to describe both uptake and release of organic chemicals from soils. The proven utility of both models makes them attractive for data fitting applications; however, neither should be used for extrapolation beyond the available data since their mechanistic basis is unclear.

Significant advances in our understanding of sorption–desorption rate processes in soils have occurred during the last decade of intensive research on this subject. In our view, the most important finding to emerge from this body of research is the need to consider the complexity of the entire soil system, including mineral and organic phases and their interactions with organic contaminants and each other, in constructing both conceptual and mathematical models of sorption processes. Such considerations have shown that some early differences in experimental findings and modeling approaches regarding sorption–desorption rates arose out of important physical and chemical differences between the sorbent–solute systems being studied. A realization that multiple causes of slow sorption and desorption behavior exist and may be operative in any particular system is critically important to future progress in this field.

REFERENCES

Ahmad, R., R.S. Kookana, A.M. Alston, and J.O. Skjemstad. 2001. The nature of soil organic matter affects sorption of pesticides: 1. Relationships with carbon chemistry as determined by ^{13}C CPMAS NMR spectroscopy. Environ. Sci. Technol. 35:878–884.

Alexander, M. 1995. How toxic are toxic chemicals in soil? Environ. Sci. Technol. 29:2713–2717.

Allen-King, R.M., H. Groenevelt, and D.M. Mackay. 1995. Analytical method for the sorption of hydrophobic organic pollutants in clay-rich materials. Environ. Sci. Technol. 29:148–153.

Altfelder, S., T. Streck, and J. Richter. 1999. Effect of air-drying on sorption kinetics of the herbicide chlortoluron in soil. J. Environ. Qual. 28:1154–1161.

Altfelder, S., T. Streck, and J. Richter. 2000. Nonsingular sorption of organic compounds in soil: The role of slow kinetics. J. Environ. Qual. 29:917–925.

American Society for Testing and Materials. 1995. Standard guide for risk-based corrective action applied at petroleum release sites. Vol. E1739-95. ASTM, West Conshohocken, PA.

Aochi, Y.O., and W.J. Farmer. 1997. Role of microstructural properties in the time-dependent sorption–desorption behavior of 1,2-dichloroethane on humic substances. Environ. Sci. Technol. 31:2520–2526.

Arocha, M.A., A.P. Jackman, and B.J. McCoy. 1996. Adsorption kinetics of toluene on soil agglomerates: Soil as a biporous sorbent. Environ. Sci. Technol. 30:1500–1507.

Baldock, J.A., and J.O. Skjemstad. 2000. Role of the soil matrix and minerals in protecting natural organic materials against biological attack. Organ. Geochem. 31:697–710.

Ball, W.P., and P.V. Roberts. 1991a. Long-term sorption of halogenated organic chemicals by aquifer material: 1. Equilibrium. Environ. Sci. Technol. 25:1223–1237.

Ball, W.P., and P.V. Roberts. 1991b. Long-term sorption of halogenated organic chemicals by aquifer material: 2. Intraparticle diffusion. Environ. Sci. Technol. 25:1237–1249.

Braida, W.J., J.C. White, F.J. Ferrandino, and J.J. Pignatello. 2001. Effect of solute concentration on sorption of polyaromatic hydrocarbons in soil: Uptake rates. Environ. Sci. Technol. 35:2765–2772.

Brusseau, M.L., and P.S.C. Rao. 1989. Sorption nonideality during organic contaminant transport in porous media. Crit. Rev. Environ. Control 19:33–99.

Brusseau, M.L., R.E. Jessup, and P.S.C. Rao. 1990. Sorption kinetics of organic chemicals: Evaluation of gas-purge and miscible-displacement techniques. Environ. Sci. Technol. 24:727–735.

Cameron, D.R., and A. Klute. 1977. Convective–dispersive solute transport with a combined equilibrium and kinetic adsorption model. Water Resour. Res. 13:183–188.

Carroll, K.M., M.R. Harkness, A.A. Bracco, and R.R. Balcarcel. 1994. Application of a permeant–polymer diffusional model to the desorption of polychlorinated biphenyls from Hudson River sediments. Environ. Sci. Technol. 28:253–258.

Castilla, H.J., C.J. Werth, and S.A. McMillan. 2000. Structural evaluation of slow desorbing sites in model and natural solids using temperature stepped desorption profiles: 2. Column results. Environ. Sci. Technol. 34:2966–2972.

Chefetz, B., A.P. Deshmukh, P.G. Hatcher, and E.A. Guthrie. 2000. Pyrene sorption by natural organic matter. Environ. Sci. Technol. 34:2925–2930.

Chen, W., A.T. Kan, and M.B. Tomson. 2000. Irreversible adsorption of chlorinated benzenes to natural sediments: Implications for sediment quality criteria. Environ. Sci. Technol. 34:385–392.

Chen, W., and R.J. Wagenet. 1995. Solute transport in porous media with sorption-site heterogeneity. Environ. Sci. Technol. 29:2725–2734.

Chiou, C.T., P.E. Porter, and D.W. Schmedding. 1983. Partition equilibria of nonionic organic compounds between soil organic matter and water. Environ. Sci. Technol. 17:227–231.

Coates, J.T., and A.W. Elzerman. 1986. Desorption kinetics for selected PCB congeners from river sediments. J. Contam. Hydrol. 1:191–210.

Connaughton, D.F., J.R. Stedinger, L.W. Lion, and M.L. Shuler. 1993. Description of time-varying desorption kinetics: Release of naphthalene from contaminated soils. Environ. Sci. Technol. 27:2397–2403.

Conte, P., and A. Piccolo. 1999. Conformational arrangement of dissolved humic substances. Influence of solution composition on association of humic molecules. Environ. Sci. Technol. 33:1682–1690.

Cook, R.L., and C.H. Langford. 1998. Structural characterization of a fulvic acid and a humic acid using solid-state ramp-CP-MAS ^{13}C nuclear magnetic resonance. Environ. Sci. Technol. 32:719–725.

Corley, T.L., J. Farrell, B. Hong, and M.H. Conklin. 1996. VOC accumulation and pore filling in unsaturated porous media. Environ. Sci. Technol. 30:2884–2891.

Cornelissen, G., K.A. Hassell, P.C.M. Van Noort, R. Kraaij, P.J. van Ekeren, C. Dijkema, P.A. de Jager, and H.A.J. Govers. 2000b. Slow desorption of PCBs and chlorobenzenes from soils and sediments: Relations with sorbent and sorbate characteristics. Environ. Pollut. 108:69–80.

Cornelissen, G., H. Rigterink, M.M.A. Ferdinandy, and P.C.M. Van Noort. 1998b. Rapidly desorbing fractions of PAHs in contaminated sediments as a predictor of the extent of bioremediation. Environ. Sci. Technol. 32:966–970.

Cornelissen, G., M. van der Pal, P.C.M. van Noort, and H.A.J. Govers. 1999b. Competitive effects on the slow desorption of organic compounds from sediments. Chemosphere 39:1971–1981.

Cornelissen, G., P.C.M. Van Noort, and H.A.J. Govers. 1997a. Desorption kinetics of chlorobenzenes, polycyclic aromatic hydrocarbons, and polychlorinated biphenyls: Sediment extraction with Tenax and effects of contact time and solute hydrophobicity. Environ. Toxicol. Chem. 16:1351–1357.

Cornelissen, G., P.C.M. Van Noort, and H.A.J. Govers. 1998a. Mechanism of slow desorption of organic compounds from sediments: A study using model sorbents. Environ. Sci. Technol. 32:100–115.

Cornelissen, G., P.C.M. Van Noort, G. Nachtegaal, and A.P.M. Kentgens. 2000a. A solid-state fluorine NMR study on hexafluorobenzene sorbed by sediments, polymers, and active carbon. Environ. Sci. Technol. 34:645–649.

Cornelissen, G., P.C.M. Van Noort, J.R. Parsons, and H.A.J. Govers. 1997b. Temperature dependence of slow adsorption and desorption kinetics of organic compounds in sediments. Environ. Sci. Technol. 31:454–460.

Cornelissen, G., H. van Zuilen, and P.C.M. Van Noort. 1999a. Particle size dependence of slow desorption of in situ PAHs from sediments. Chemosphere 38:2369–2380.

Culver, T.B., R.A. Brown, and J.A. Smith. 2000. Rate-limited sorption and desorption of 1,2-dichlorobenzene to a natural sand soil column. Environ. Sci. Technol. 34:2446–2452.

Culver, T.B., S.P. Hallisey, D. Sahoo, J.J. Deitsch, and J.A. Smith. 1997. Modeling the desorption of organic contaminants from long-term contaminated soils using distributed mass transfer rates. Environ. Sci. Technol. 31:1581–1588.

de Jonge, H., and M.C. Mittelmeijer-Hazeleger. 1996. Adsorption of CO_2 and N_2 on soil organic matter: Nature of porosity, surface area and diffusion mechanisms. Environ. Sci. Technol. 30:408–413.

Deitsch, J.J., J.A. Smith, M.B. Arnold, and J. Bolus. 1998. Sorption and desorption rates of carbon tetrachloride and 1,2-dichlorobenzene to three organobentonites and a natural peat soil. Environ. Sci. Technol. 32:3169–3177.

Deitsch, J.J., J.A. Smith, T.B. Culver, R.A. Brown, and S.A. Riddle. 2000. Distributed-rate model analysis of 1,2-dichlorobenzene batch sorption and desorption rates for five natural sorbents. Environ. Sci. Technol. 34:1469–1476.

Divincenzo, J.P., and D.L. Sparks. 1997. Slow sorption kinetics of pentachlorophenol on soil: Concentration effects. Environ. Sci. Technol. 31:977–983.

Farrell, J., D. Grassian, and M. Jones. 1999. Investigation of mechanisms contributing to slow desorption of hydrophobic organic compounds from mineral solids. Environ. Sci. Technol. 33:1237–1243.

Farrell, J., and M. Reinhard. 1994a. Desorption of halogenated organics from model solids, sediments and soils under unsaturated conditions: 1. Isotherms. Environ. Sci. Technol. 28:53–62.

Farrell, J., and M. Reinhard. 1994b. Desorption of halogenated organics from model solids, sediments, and soil under unsaturated conditions: 2. Kinetics. Environ. Sci. Technol. 28:63–72.

Garbarini, D.R., and L.W. Lion. 1986. Influence of the nature of soil organics on the sorption of toluene and trichloroethylene. Environ. Sci. Technol. 20:1263–1269.

Gelinas, Y., K.M. Prentice, J.A. Baldock, and J.I. Hedges. 2001. An improved thermal oxidation method for the quantification of soot/graphitic black carbon in sediments and soils. Environ. Sci. Technol. 35:3519–3525.

Ghosh, U., J.S. Gillette, R.G. Luthy, and R.N. Zare. 2000. Microscale location, characterization, and association of polycyclic aromatic hydrocarbons on harbor sediment particles. Environ. Sci. Technol. 34:1729–1736.

Ghosh, U., J.W. Talley, and R.G. Luthy. 2001. Particle-scale investigation of PAH desorption kinetics and thermodynamics from sediment. Environ. Sci. Technol. 35:3468–3475.

Gillette, J.S., R.G. Luthy, S.J. Clemett, and R.N. Zare. 1999. Direct observation of polycyclic aromatic hydrocarbons on geosorbents at the subparticle scale. Environ. Sci. Technol. 33:1185–1192.

Grathwohl, P. 1990. Influence of organic matter from soils and sediments from various origins on the sorption of some chlorinated aliphatic hydrocarbons: Implications on K_{oc} correlations. Environ. Sci. Technol. 24:1687–1693.

Gschwend, P.M., and S. Wu. 1985. On the constancy of sediment–water partition coefficients of hydrophobic organic pollutants. Environ. Sci. Technol. 19:90–96.

Gustafsson, O., F. Haghseta, C. Chan, J. Macfarlane, and P.M. Gschwend. 1997. Quantification of the dilute sedimentary soot phase: Implications for PAH speciation and bioavailability. Environ. Sci. Technol. 31:203–209.

Hawthorne, S.B., E. Bjorklund, S. Bowadt, and L. Mathiasson. 1999. Determining PCB sorption–desorption behavior on sediments using selective supercritical fluid extraction: 3. Sorption from water. Environ. Sci. Technol. 33:3152–3159.

Hawthorne, S.B., D.G. Poppendieck, C.B. Grabanski, and R.C. Loehr. 2001. PAH release during water desorption, supercritical carbon dioxide extraction, and field bioremediation. Environ. Sci. Technol. 35:311–317..

Hedges, J.I., and J.M. Oades. 1997. Comparative organic geochemistries of soils and marine sediments. Organ. Geochem. 27:318–361.

Heyse, E., D. Dai, P.S.C. Rao, and J.J. Delfino. 1997. Development of a continuously stirred flow cell for investigating sorption mass transfer. J. Contam. Hydrol. 25:337–355.

Holmén, B.A., and P.M. Gschwend. 1997. Estimating sorption rates of hydrophobic organic compounds in iron oxide- and aluminosilicate clay-coated aquifer sands. Environ. Sci. Technol. 31:105–113.

Hosse, M., and K.J. Wilkinson. 2001. Determination of electrophoretic mobilities and hydrodynamic radii of three humic substances as a function of pH and ionic strength. Environ. Sci. Technol. 35:4301–4306.

Hu, W.-G., J. Mao, B. Xing, and K. Schmidt-Rohr. 2000. Poly(methylene) crystallites in humic substances detected by nuclear magnetic resonance. Environ. Sci. Technol. 34:530–534.

Huang, W., M.A. Schlautman, and W.J. Weber, Jr. 1996. A distributed reactivity model for sorption by soils and sediments: 5. The influence of near-surface characteristics in mineral domains. Environ. Sci. Technol. 30:2993–3000.

Huang, W., and W.J. Weber, Jr. 1997. A distributed reactivity model for sorption by soils and sediments: 10. Relationships between desorption, hysteresis, and the chemical characteristics of organic domains. Environ. Sci. Technol. 31:2562–2569.

Huang, W., and W.J. Weber, Jr. 1998. A distributed reactivity model for sorption by soils and sediments: 11. Slow concentration dependent sorption rates. Environ. Sci. Technol. 32:3549–3555.

Huang, W., H. Yu, and W.J. Weber, Jr. 1998. Hysteresis in the sorption and desorption of hydrophobic organic contaminants by soils and sediments: 1. A comparative analysis of experimental protocols. J. Contam. Hydrol. 31:129–148.

Huang, W., T.M. Young, M.A. Schlautman, H. Yu, and W.J. Weber, Jr. 1997. A distributed reactivity model for sorption by soils and sediments: 9. General isotherm nonlinearity and applicability of the dual reactive domain model. Environ. Sci. Technol. 31:1703–1710.

Hundal, L.S., M.L. Thompson, D.A. Laird, and A.M. Carmo. 2001. Sorption of phenanthrene by reference smectites. Environ. Sci. Technol. 35:3456–3461.

Hunter, M.A., A.T. Kan, and M.B. Tomson. 1996. Development of a surrogate sediment to study the mechanisms responsible for adsorption–desorption hysteresis. Environ. Sci. Technol. 30:2278–2285.

Johnson, M.D., W. Huang, and W.J. Weber, Jr. 2001a. A distributed reactivity model for sorption by soils and sediments: 13. Simulated diagenesis of natural sediment organic matter and its impact on sorption–desorption equilibria. Environ. Sci. Technol. 35:1680–1687.

Johnson, M.D., W. Huang, Z. Dang, and W.J. Weber, Jr. 1999. A distributed reactivity model for sorption by soils and sediments: 12. Effects of subcritical water extraction and alterations of soil organic matter on sorption equilibria. Environ. Sci. Technol. 33:1657–1663.

Johnson, M.D., T.M. Keinath II, and W.J. Weber, Jr. 2001b. A distributed reactivity model for sorption by soils and sediments: 14. Characterization and modeling of phenanthrene desorption rates. Environ. Sci. Technol. 35:1688–1695.

Johnson, M.D., and W.J. Weber, Jr. 2001. Rapid prediction of long-term rates of contaminant desorption from soils and sediments. Environ. Sci. Technol. 35:427–433.

Jones, K.D., and C.L. Tiller. 1999. Effect of solution chemistry on the extent of binding of phenanthrene by a soil humic acid: A comparison of dissolved and clay bound humic. Environ. Sci. Technol. 33:580–587.

Kaiser, K., and G. Guggenberger. 2000. The role of DOM sorption to mineral surfaces in the preservation of organic matter in soils. Organ. Geochem. 31:711–725.

Kan, A.T., G. Fu, M. Hunter, W. Chen, C.H. Ward, and M.B. Tomson. 1998. Irreversible sorption of neutral hydrocarbons to sediments: Experimental observations and model predictions. Environ. Sci. Technol. 32:892–902.

Kan, A.T., G. Fu, M.A. Hunter, and M.B. Tomson. 1997. Irreversible adsorption of naphthalene and tetrachlorobiphenyl to Lula and surrogate sediments. Environ. Sci. Technol. 31:2176–2185.

Kan, A.T., G. Fu, and M.B. Tomson. 1994. Adsorption–desorption hysteresis in organic pollutant soil–sediment interaction. Environ. Sci. Technol. 28:859–867.

Karapanagioti, H.K., S. Kleineidam, D.A. Sabatini, P. Grathwohl, and B. Ligouis. 2000. Impacts of heterogeneous organic matter on phenanthrene sorption: Equilibrium and kinetic studies with aquifer material. Environ. Sci. Technol. 34:406–414.

Karickhoff, S.W. 1980. Sorption kinetics of hydrophobic pollutants in natural sediments. p. 193–205. *In* R.A. Baker (ed.) Contaminants and sediments. Vol. 2. Ann Arbor Press, Ann Arbor, MI.

Kile, D.E., R.L. Wershaw, and C.T. Chiou. 1999. Correlation of soil and sediment organic matter polarity to aqueous sorption of nonionic compounds. Environ. Sci. Technol. 33:2053–2056.

Kleineidam, S., H. Rugner, and P. Grathwohl. 1999a. Impact of grain scale heterogeneity on slow sorption kinetics. Environ. Toxicol. Chem. 18:1673–1678.

Kleineidam, S., H. Rugner, B. Ligouis, and P. Grathwohl. 1999b. Organic matter facies and equilibrium sorption of phenanthrene. Environ. Sci. Technol. 33:1637–1644.

Kohl, S.D., P.J. Toscano, W. Hou, and J.A. Rice. 2000. Solid–state ^{19}F NMR investigation of hexafluorobenzene sorption to soil organic matter. Environ. Sci. Technol. 34:204–210.

Kulikova, N.A., and I.V. Perminova. 2002. Binding of atrazine to humic substances from soil, peat, and coal related to their structure. Environ. Sci. Technol. 36:3720–3724.

Lambert, S.M. 1967. Functional relationship between sorption in soil and chemical structure. J. Agric. Food Chem. 15:572–576.

LeBoeuf, E.J., and W.J. Weber, Jr. 1997. A distributed reactivity model for sorption by soils and sediments: 8. Sorbent organic domains: Discovery of a humic acid glass transition and an argument for a polymer-based model. Environ. Sci. Technol. 31:1697–1702.

LeBoeuf, E.J., and W.J. Weber, Jr. 1999. Reevaluation of general partitioning model for sorption of hydrophobic organic contaminants by soil and sediment organic matter. Environ. Toxicol. Chem. 18:1617–1626.

LeBoeuf, E.J., and W.J. Weber, Jr. 2000a. Macromolecular characteristics of natural organic matter: 1. Insights from glass transition and enthalpic relaxation behavior. Environ. Sci. Technol. 34:3623–3631.

LeBoeuf, E.J., and W.J. Weber, Jr. 2000b. Macromolecular characteristics of natural organic matter: 2. Sorption and desorption behavior. Environ. Sci. Technol. 34:3632–3640.

Li, J., and C.J. Werth. 2001. Evaluating competitive sorption mechanisms of volatile organic compounds in soils and sediments using polymers and zeolites. Environ. Sci. Technol. 35:568–574.

Luthy, R.G., G.R. Aiken, M.L. Brusseau, S.D. Cunningham, P.M. Gschwend, J.J. Pignatello, M. Reinhard, S.J. Traina, W.J. Weber, Jr., and J.C. Westall. 1997. Sequestration of hydrophobic organic contaminants by geosorbents. Environ. Sci. Technol. 31:3341–3347.

Mao, J., G. Ding, and B. Xing. 2002. Domain mobility of humic acids investigated with one- and two-dimensional nuclear magnetic resonance: Support for dual-mode model. Commun. Soil Sci. Plant Anal. 33:1679–1688.

McCall, P.J., and G.L. Agin. 1985. Desorption kinetics of picloram as affected by residence time in the soil. Environ. Toxicol. Chem. 4:37–44.

McGinley, P.M., L.E. Katz, and W.J. Weber, Jr. 1993. A distributed reactivity model for sorption by soils and sediments: 2. Multicomponent systems and competitive effects. Environ. Sci. Technol. 27:1524–1531.

McMillan, S.A., and C.J. Werth. 1999. Counter-diffusion of isotopically labeled trichloroethylene in silica gel and geosorbent micropores: Model development. Environ. Sci. Technol. 33:2178–2185.

Miller, C.T., and W.J. Weber, Jr. 1984. Modeling organic contaminant partitioning in ground water systems. Ground Water 22:584–592.

Miller, C.T., and W.J. Weber, Jr. 1986. Sorption of hydrophobic organic pollutants in saturated soil systems. J. Contam. Hydrol. 1:243–261.

Murphy, E.M., J.M. Zachara, and S.C. Smith. 1990. Influence of mineral-bound humic substances on the sorption of hydrophobic organic compounds. Environ. Sci. Technol. 24:1507–1516.

Opdyke, D.R., and R.C. Loehr. 1999. Determination of chemical release rates from soils: Experimental design. Environ. Sci. Technol. 33:1193–1199.

Pavlostathis, S.G., and G.N. Mathavan. 1992. Desorption kinetics of selected volatile organic compounds from field contaminated soils. Environ. Sci. Technol. 26:532–538.

Pedit, J.A., and C.T. Miller. 1994. Heterogeneous sorption processes in subsurface systems: 1. Model formulations and applications. Environ. Sci. Technol. 28:2094–2104.

Pedit, J.A., and C.T. Miller. 1995. Heterogeneous sorption processes in subsurface systems: 2. Diffusion modeling approaches. Environ. Sci. Technol. 29:1766–1772.

Piatt, J.J., and M.L. Brusseau. 1998. Rate-limited sorption of hydrophobic organic compounds by soils with well-characterized organic matter. Environ. Sci. Technol. 32:1604–1608.

Pignatello, J.J. 1989. Sorption dynamics of organic compounds in soils and sediments. p. 45–80. *In* Reactions and movement of organic chemicals in soils. Vol. 22. SSSA, Madison, WI.

Pignatello, J.J. 1990a. Slowly reversible sorption of aliphatic halocarbons in soils: 2. Mechanistic aspects. Environ. Toxicol. Chem. 9:1117–1126.

Pignatello, J.J. 1990b. Slowly reversible sorption of aliphatic halocarbons in soils: 1. Formation of residual fractions. Environ. Toxicol. Chem. 9:1107–1115.

Pignatello, J.J. 2000. The measurement and interpretation of sorption and desorption rates for organic compounds in soil media. p. 1–73. Advances in agronomy. Vol. 69.

Pignatello, J.J., and L.Q. Huang. 1991. Sorptive reversibility of atrazine and metolachlor residues in field soil samples. J. Environ. Qual. 20:222–228.

Pignatello, J.J., and B. Xing. 1996. Mechanisms of slow sorption of organic chemicals to natural particles. Environ. Sci. Technol. 30:1–11.

Preston, C.M., and R.H. Newman. 1992. Demonstration of spatial heterogeneity in the organic matter of de-ashed humin samples by solid-state ^{13}C CPMAS NMR. Can. J. Soil Sci. 72:13–19.

Rügner, H., S. Kleineidam, and P. Grathwohl. 1999. Long-term sorption kinetics of phenanthrene in aquifer materials. Environ. Sci. Technol. 33:1645–1651.

Rutherford, D.W., and C.T. Chiou. 1992. Effect of water saturation in soil organic matter on the partition of organic compounds. Environ. Sci. Technol. 26:965–970.

Rutherford, D.W., C.T. Chiou, and D.E. Kile. 1992. Influence of soil organic matter composition on the partition of organic compounds. Environ. Sci. Technol. 26:336–340.

Salloum, M.J., B. Chefetz, and P.G. Hatcher. 2002. Phenanthrene sorption by aliphatic-rich natural organic matter. Environ. Sci. Technol. 36:1953–1958.

Schaefer, C.E., C. Schuth, C.J. Werth, and M. Reinhard. 2000. Binary desorption isotherms of TCE and PCE from silica gel and natural solids. Environ. Sci. Technol. 34:4341–4347.

Schlebaum, W., G. Schraa, and W.H. Van Riemsdijk. 1999. Influence of nonlinear sorption kinetics on the slow-desorbing organic contaminant fraction in soil. Environ. Sci. Technol. 33:1413–1417.

Schultz, L.F., T.M. Young, and R.M. Higashi. 1999. Sorption–desorption behavior of phenanthrene elucidated by pyrolysis-gas chromatography-mass spectrometry studies of soil organic matter. Environ. Toxicol. Chem. 18:1710–1719.

Schwarzenbach, R.P., P.M. Gschwend, and D.M. Imboden. 1993. Environmental organic chemistry. John Wiley and Sons, New York.

Schwarzenbach, R.P., and J.C. Westall. 1981. Transport of nonpolar organic compounds from surface water to groundwater: Laboratory sorption studies. Environ. Sci. Technol. 15:1360–1367.

Scribner, S.L., T.R. Benzing, S. Sun, and S.A. Boyd. 1992. Desorption and biovailability of aged simazine residues in soil from a continuous corn field. J. Environ. Qual. 21:115–120.

Sein, L.T.J., J.M. Varnum, and S.A. Jansen. 1999. Conformational modeling of a new building block of humic acid: Approaches to the lowest energy conformer. Environ. Sci. Technol. 33:546–552.

Shelton, D.R., A.M. Sadeghi, J.S. Karns, and C.J. Hapeman. 1995. Effect of wetting and drying of soil on sorption and biodegradation of atrazine. Weed Sci. 43:298–305.

Steinberg, S.M., J.J. Pignatello, and B.L. Sawhney. 1987. Persistence of 1,2-dibromoethane in soils: Entrapment in intraparticle micropores. Environ. Sci. Technol. 21:1201–1208.

Streck, T., N.N. Poletika, W.A. Jury, and W.J. Farmer. 1995. Description of simazine transport with rate-limited, two-stage, linear and nonlinear sorption. Water Resour. Res. 31:811–822.

Stroo, H.F., R. Jensen, R.C. Loehr, D.V. Nakles, A. Fairbrother, and C.B. Liban. 2000. Environmentally acceptable endpoints for PAHs at a manufactured gas plant site. Environ. Sci. Technol. 34:3831–3836.

ten Hulscher, T.E.M., and G. Cornelissen. 1996. Effect of temperature on sorption equilibrium and sorption kinetics of organic micropollutants: A review. Chemosphere 32:609–626.

ten Hulscher, T.E.M., B.A. Vrind, H. van den Heuvel, L.W. van der Velde, P.C.M. Van Noort, J.E.M. Beurskens, and H.A.J. Govers. 1999. Triphasic desorption of highly resistant chlorobenzenes, polychlorinated biphenyls, and polycyclic aromatic hydrocarbons in field contaminated sediments. Environ. Sci. Technol. 33:126–132.

Thorn, K.A., P.J. Pettigrew, and W.S. Goldenberg. 1996. Covalent binding of aniline to humic substances: 2. N-15 NMR studies of nucleophilic addition reactions. Environ. Sci. Technol. 30:2764–2775.

Voice, T.C., C.P. Rice, and W.J. Weber, Jr. 1983. Effect of solids concentration on the sorptive partitioning of hydrophobic pollutants in aquatic systems. Environ. Sci. Technol. 17:513–518.

Weber, W.J., Jr., and F.A. DiGiano. 1996. Process dynamics in environmental systems. John Wiley and Sons, New York.

Weber, W.J., Jr., and W. Huang. 1996. A distributed reactivity model for sorption by soils and sediments: 4. Intraparticle heterogeneity and phase-distribution relationships under nonequilibrium conditions. Environ. Sci. Technol. 30:881–888.

Weber, W.J., Jr., W. Huang, and E.J. LeBoeuf. 1999. Geosorbent organic matter and its relationship to the binding and sequestration of organic contaminants. Coll. Surf. A: Physicochem. Eng. Aspects 151:167–179.

Weber, W.J., Jr., W. Huang, and H. Yu. 1998. Hysteresis in the sorption and desorption of hydrophobic organic contaminants by soils and sediments: 2. Effects of soil organic matter heterogeneity. J. Contam. Hydrol. 31:149–165.

Weber, W.J., Jr., E.J. LeBoeuf, T.M. Young, and W. Huang. 2001. Contaminant interactions with geosorbent organic matter: Insights drawn from polymer sciences. Water Res. 35:853–868.

Weber, W.J., Jr., P.M. McGinley, and L.E. Katz. 1991. Sorption phenomena in subsurface systems: Concepts, models and effects on contaminant fate and transport. Water Res. 25:499–528.

Weber, W.J., Jr., P.M. McGinley, and L.E. Katz. 1992. A distributed reactivity model for sorption by soils and sediments: 1. Conceptual basis and equilibrium assessments. Environ. Sci. Technol. 26:1955–1962.

Weber, W.J., Jr., and C.T. Miller. 1988. Modeling the sorption of hydrophobic contaminants by aquifer materials: I. Rates and equilibria. Water Res. 22:457–464.

Weber, W.J., Jr., and T.M. Young. 1997. A distributed reactivity model for sorption by soils and sediments: 6. Mechanistic implications of desorption under supercritical fluid conditions. Environ. Sci. Technol. 31:1686–1691.

Werth, C.J., S.A. McMillan, and H.J. Castilla. 2000. Structural evaluation of slow desorbing sites in model and natural solids using temperature stepped desorption profiles: 1. Model development. Environ. Sci. Technol. 34:2959–2965.

Werth, C.J., and M. Reinhard. 1997a. Effects of temperature on trichloroethylene desorption from silica gel and natural sediments: 2. Kinetics. Environ. Sci. Technol. 31:697–703.

Werth, C.J., and M. Reinhard. 1997b. Effects of temperature on trichloroethylene desorption from silica gel and natural sediments: 1. Isotherms. Environ. Sci. Technol. 31:689–696.

Werth, C.J., and M. Reinhard. 1999. Counter-diffusion of isotopically labeled trichloroethylene in silica gel and geosorbent micropores: Column results. Environ. Sci. Technol. 33:730–736.

White, J.C., and J.J. Pignatello. 1999. Influence of bisolute competition on the desorption kinetics of polycyclic aromatic hydrocarbons in soil. Environ. Sci. Technol. 33:4292–4298.

Wu, S.-C., and P.M. Gschwend. 1986. Sorption kinetics of hydrophobic organic compounds to natural sediments and soils. Environ. Sci. Technol. 20:717–725.

Xia, G., and W.P. Ball. 1999. Adsorption-partitioning uptake of nine low-polarity organic chemicals on a natural sorbent. Environ. Sci. Technol. 33:262–269.

Xia, G., and W.P. Ball. 2000. Polanyi-based models for the competitive sorption of low-polarity organic contaminants on a natural sorbent. Environ. Sci. Technol. 34:1246–1253.

Xia, G., and J.J. Pignatello. 2001. Detailed sorption isotherms of polar and apolar compounds in a high-organic soil. Environ. Sci. Technol. 35:84–94.

Xing, B. 1997. The effect of the quality of soil organic matter on sorption of naphthalene. Chemosphere 35:633–642.

Xing, B., and Z. Chen. 1999. Spectroscopic evidence for condensed domains in soil organic matter. Soil Sci. 164:40–47.

Xing, B., and J.J. Pignatello. 1996. Time-dependent isotherm shape of organic compounds in soil organic matter: Implications for sorption mechanism. Environ. Toxicol. Chem. 15:1282–1288.

Xing, B., and J.J. Pignatello. 1997. Dual-mode sorption of low polarity compounds in glassy poly(vinyl chloride) and soil organic matter. Environ. Sci. Technol. 31:792–799.

Xing, B., and J.J. Pignatello. 1998. Competitive sorption between 1,3-dichlorobenzene or 2,4-dichlorophenol and natural aromatic acids in soil organic matter. Environ. Sci. Technol. 32:614–619.

Xing, B., J.J. Pignatello, and B. Gigliotti. 1996. Competitive sorption between atrazine and other organic compounds in soils and model sorbents. Environ. Sci. Technol. 30:2432–2440.

Young, K.D., and E.J. LeBoeuf. 2000. Glass transition behavior in a peat humic acid and an aquatic fulvic acid. Environ. Sci. Technol. 34:4549–4553.

Young, T.M., and W.J. Weber, Jr. 1995. A distributed reactivity model for sorption by soils and sediments: 3. Effects of diagenetic processes on sorption energetics. Environ. Sci. Technol. 29:92–97.

Young, T.M., and W.J. Weber, Jr. 1997. A distributed reactivity model for sorption by soils and sediments: 7. Enthalpy and polarity effects on desorption under supercritical fluid conditions. Environ. Sci. Technol. 31:1692–1696.

Chapter 12

Metal Ion Complexation by Soil Humic Substances

NICOLA SENESI AND ELISABETTA LOFFREDO, *Università di Bari, Bari, Italy*

The behavior of metals in soil depends not only on the total soluble and insoluble metal concentrations, but also on their relative distribution and speciation in the soil liquid and solid phases. The knowledge of the type and concentration of individual species present in the system is, therefore, essential in assessing the impact of metals on the global ecosystem. Metal ion speciation is important in determining the general biological and physico-chemical behavior of metal ions. This, in turn, influences bioavailability of metal nutrients to plants and soil microorganisms, toxicity hazard of potentially toxic metals, migration-accumulation phenomena of metals in the soil–water system, pedogenic processes, geochemical transfers, and mobility pathways (Stevenson, 1986).

The distribution of metal ions in soils is extremely complex and is governed by a variety of reactions that include complexation with organic and inorganic ligands, ion exchange, adsorption and desorption processes, precipitation and dissolution of solids, and acid–base equilibria. In particular, complexation reactions involving natural organic matter play a key role in establishing the behavior of metal ions, especially in trace concentrations. Soil organic compounds that may form complexes with metal ions include: (i) humic substances (HS), which represent the major (between 70 and 80%) natural organic fraction in soils; (ii) organic substances of defined molecular structure and chemical properties such as aliphatic acids, polysaccharides, amino acids, polyphenols, etc; and (iii) xenobiotic organic chemicals applied to soil accidentally or by purpose.

In particular, HS are able to interact with metal ions to form water-soluble, colloidal, and water-insoluble complexes of varying properties and widely differing chemical and biological stabilities (Schnitzer, 1978; Stevenson, 1994). Accumulated evidence suggests that most processes in which metals are involved in soils, including mobility and transport, fixation and accumulation, chemical reactivity and bioavailability, are affected by their interaction with HS. The topic is of considerable practical interest also because of the continuous and increasing release of various heavy metals to soil by numerous modern agricultural practices.

Metal–HS complexation in soil is of practical and theoretical significance for several reasons. For example, chemical weathering of rocks and minerals, and related soil genesis and evolution are known to involve metal complexation reactions with HS. The bioavailability of several metal ions, especially trace elements, is

Chemical Processes in Soils. SSSA Book Series, no. 8.

strongly influenced by complexation with the soluble and insoluble fractions of HS. For instance, at pH values commonly found in soils, nutrient metal ions that would ordinarily be converted to insoluble forms may be maintained in solution by complexation with HS, thus increasing their bioavailability. On the contrary, the concentration of a toxic metal ion may be reduced to a nontoxic level through complexation to insoluble fractions of soil HS, whereas soluble HS complexes can function as metal-carrier in the transport to ground- and surface-water bodies, thereby rendering the water unfit for several uses. Further, complexation of metal ions by HS may represent a challenging problem for their analytical determination in soil samples.

METAL REACTIVITY, AND COMPLEXATION STOICHIOMETRY AND BINDING SITES OF HUMIC SUBSTANCES

Metal Ion Reactivity with Soil Organic Complexants

Inorganic cations may be subdivided into three classes according to their reactivity with organic ligands (Buffle, 1988). "Hard" cations such as alkaline-earth metals $Ca^{2+}$and Mg^{2+} are mainly involved in electrostatic interactions and may form rather weak, outer-sphere complexes only with hard oxygen ligands. "Soft" cations such as Cd^{2+}, Pb^{2+}, and Hg^{2+} possess a strong affinity for and tend to form covalent bonds with intermediate (N) and soft (S) ligands. "Borderline" cations, including most transition metals such as Fe^{3+}, Cu^{2+}, Zn^{2+}, and Mn^{2+}, have a character intermediate between hard and soft metals, and possess appreciable affinity for both hard and soft ligands.

Complexation of "hard" cations by soil organic ligands have little influence on speciation of these metals, since their concentration in soil is much higher than that of organic ligands. "Hard" cations, however, may have two important indirect effects on the complexation of other metals: (i) a competitive effect for oxygen ligands, with respect to "soft" and "borderline" cations, and (ii) a counterion effect for negatively charged polyelectrolyte complexing agents, thus influencing their reactivity by modifying their charge and/or conformation, as well as their degree of aggregation and dispersion–coagulation. "Soft" metal ions are generally found in very low concentration in soil, and can be easily complexed to N and S sites, despite the low concentration of these sites in soil organic matter. The concentration of "borderline" cations in soil is generally a little lower or similar to that of the organic complexants. These metals can, therefore, compete for hard ligands with "hard" metals, which are less strongly bound but at higher concentration in soil, and with "soft" metals, which are more strongly bound but at lower concentration.

The most important organic complexing functional groups present in soil organic matter, classified according to their affinity for hard, borderline, and soft metals, are listed in Table 12–1 (Buffle, 1988). For soft metals the order of donor atom affinity is: $O < N < S$, whereas a reverse order is observed for hard cations. For bidentate ligand sites, affinity for a given soft metal increases with the overall softness of the donor atoms, in the order: $(O,O) < (O,N) < (N,N) < (N,S)$, whereas this order is reversed for hard metals (Buffle, 1988). In general, the competitive reactions for

Table 12–1. Most important metal organic ligands of soil organic matter classified according to their preference for hard, borderline, and soft metals.

Ligands preferred by hard cations (hard bases)	Ligands preferred by borderline cations	Ligands preferred by soft cations (soft bases)
–C(=O)O$^{\ominus}$ (Carboxylate), –C(=O)O– (ester)	–NH2, =NH, ≡N (primary, sec., tert. amino groups)	R$^{\ominus}$ (alchil anion)
–OH (alcoholic and phenolic)	–NH–C(=O)– (amide)	–SH, –S$^{\ominus}$ (Sulfydril, sulfide)
>C=O, –O– (Carbonyl) (ether)		–S–S–S, S (disulfide) (thioether)

a given ligand essentially involve "hard" and "borderline" metals for O sites, and "borderline" and "soft" metals for N and S sites, with competition between "hard" and "soft" metals being weak (Buffle, 1988).

The typical affinity sequence of soil organic matter for divalent metal ions (at pH 5) generally parallels the metal electronegativity values by Pauling (Stevenson and Ardakani, 1972); however, the type, source, and concentration of organic matter in soil can affect metal binding affinity, and the relative affinities are often dependent on the method used to measure metal bonding and pH (Stevenson and Ardakani, 1972). Further, selectivity coefficients for metal binding vary with the amount of metal bound. For example, the strength of binding of Zn^{2+} and Cu^{2+} by soil HS increases with a decrease in the metal amount available (Davies et al., 1969). At low ligand concentrations, the relatively soft cations Cd^{2+} and Pb^{2+} are highly preferred by "soft", S-containing ligands in HS, and could compete successfully with "hard" cations, even with the very abundant Ca^{2+} ions. The reverse is true, however, for the Cd^{2+}/Ca^{2+} competition at high concentration where carboxylate ligands become dominant.

Metal Complexation Stoichiometry and Binding Sites in Humic Substances

Humic substances, i.e., humic acids (HA) and fulvic acids (FA), contain a large number of complexing sites per molecule, and thus behave as other natural "multiligand" complexing agents like proteins and metal oxides, which are distinguished from "simple" ligands such as inorganic anions and amino acids (Buffle, 1988). The principal molecular characteristics that govern the complexing ability of HS are: polyfunctionality, polyelectrolyte character, hydrophilicity, and the capacity to form intermolecular associations and change molecular conformation (Buffle, 1988).

The major functional groups in HA and FA that can bind metal ions are O-containing groups, including carboxylic, phenolic, alcoholic and enolic hydroxyl groups as well as carbonyl functionalities of various types (Stevenson and Fitch, 1986; Senesi, 1992). Aminogroups and S- and P-containing groups are also involved in metal binding.

Several types of binding reactions can be visualized between metal ions and HS, the simplest case being the 1:1 binding. Formation also can be expected of mononuclear complexes with the central group being either the HS macromolecule or the metal ion, and of polynuclear complexes.

The most common stoichiometries assessed are HS:metal = 1:1 and 2:1, often involving the formation of metal chelates. Aromatic carboxyls and phenolic groups play a prominent role in the 1:1 binding of metal ions by soil HA and FA, by forming chelates that involve two COOH groups in a phtalic-type site (Fig.12–1, Eq. [1]) and both phenolic and COOH groups in a salicylate-type site (Fig. 12–1, Eq. [2]). The most stable complexes involve the more strongly acidic COOH groups, whereas the less stable complexes are believed to be associated with weakly acidic COOH

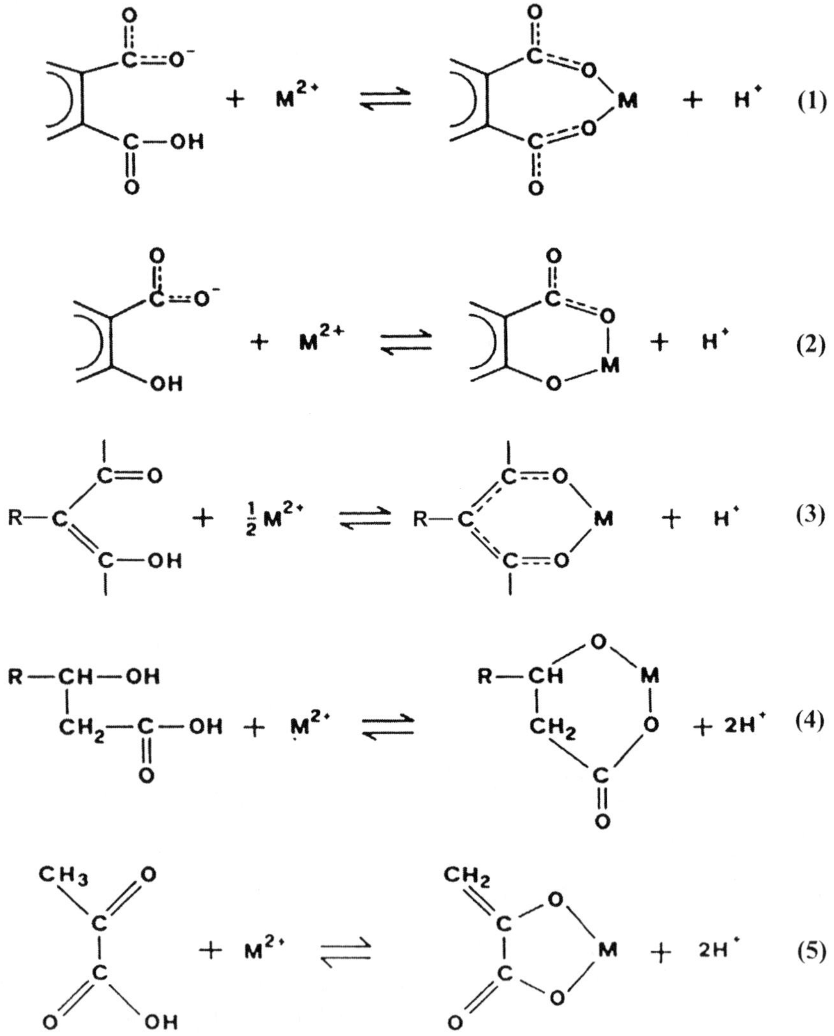

$$+ M^{2+} \rightleftharpoons \quad + H^{+} \quad (1)$$

$$+ M^{2+} \rightleftharpoons \quad + H^{+} \quad (2)$$

$$+ \tfrac{1}{2}M^{2+} \rightleftharpoons \quad + H^{+} \quad (3)$$

$$R-CH(OH)-CH_2-C(=O)-OH + M^{2+} \rightleftharpoons \quad + 2H^{+} \quad (4)$$

$$+ M^{2+} \rightleftharpoons \quad + 2H^{+} \quad (5)$$

Fig. 12–1. Most common binding sites for metal ions in humic substances and stoichiometries of complexation.

and phenolic OH groups. Other possible combinations involve two phenolic OH, quinone, NH_2, and sulphydryl groups, and conjugated ketonic structures (Fig. 12–1, Eq. [3]) (Stevenson, 1976a; Chen and Stevenson, 1986).

Nonaromatic carboxyl and hydroxyl sites also may be involved in metal ion binding by soil HA and FA. The monomeric analogues of some of these sites, e.g., pyruvic and glycolic acid (Fig. 12–1, Eq. [4] and [5]), have binding constants similar to phtalic and salicylic acids.

Metal ions also may coordinate with ligands belonging to two (or more) HS molecules, forming 2:1 complexes (Fig. 12–1, Eq. [6] and [7]) and/or chelates (Fig. 12–1, Eq. [8] and [9]), and eventually producing a chain structure (Fig. 12–1, Eq. [10]) that may result in the aggregation and precipitation as the chain grows at high metal to HS ratios (Stevenson, 1976a).

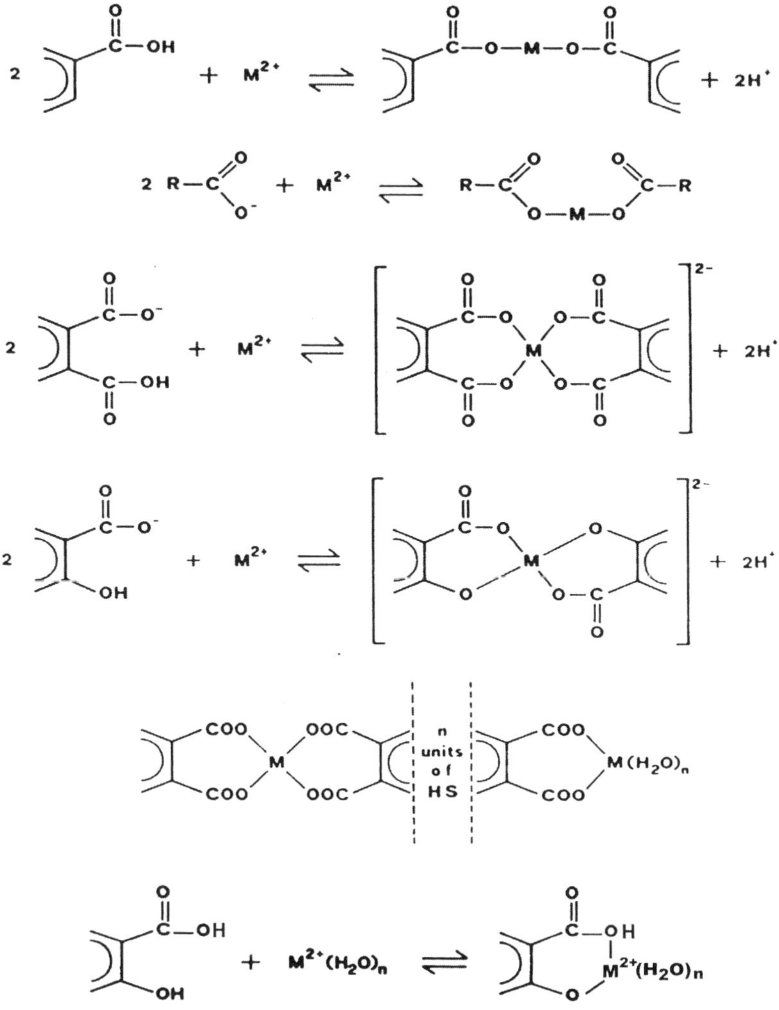

Fig. 12–1. Continued.

Two main types of complexes may be formed between metal ions and HS, that are: (i) inner-sphere complexes, resulting in the formation of bonds with some covalent character between the ligand atom(s) and the metal ion, both completely or partially dehydrated; and (ii) outer-sphere complexes that result in the electrostatic attraction between the ligand(s) and the metal ion that remains completely hydrated. For simplicity, all reaction schemes described in Fig. 12–1, Eq. [1] to [10], show formation of inner-sphere complexes, but they may represent outer-sphere HS complexes if the cation is solvated (e.g., Fig. 12–1, Eq. [11]).

The electronic and steric environment of the ligand site, such as its chemical network, geometry and conformation, and the physical and chemical characteristics of the surrounding medium, including pH, ionic strength and metal concentration, can exert a marked influence on the overall interaction process (Buffle, 1988). In a given HS macromolecule, identical coordinating groups can be bound to different types of aliphatic chains and aromatic rings of various structure, which can exert differing electronic effects. The steric microenvironment of the binding site, and particularly its size, will depend on the geometry, steric conformation, and flexibility of the whole complexant molecule. This is influenced by the formation–disruption of hydrogen bonds and metal bridges, which can vary with pH, ionic strength, and concentration of the metal to be complexed. The hydration of hydrophilic sites and electrostatic effects, i.e., the electric field determined by the extent of ionization of major acidic complexing groups, also can influence the formation process and stability of complexes (Buffle, 1988). The relative importance of these different effects varies with the degree of site occupation by metal cations and represents the fundamental difference between HS complexants and "simple" ligands.

QUANTITATIVE PARAMETERS OF METAL ION COMPLEXATION BY SOIL HUMIC SUBSTANCES

Metal Complexation Capacity of Humic Substances

In general, the complexation or binding capacity of HS is defined as the maximum capacity of HS for binding metal ions (Stevenson, 1994). According to Perdue (1989), the complexation capacity (CC) of a multiligand system such as HS may be viewed as a compositional rather than thermodynamic property. To a good approximation, the CC of HS may thus be considered as the weighted average of the complexation capacities of the individual ligands $(CC)_i$ in the system, according to (Perdue, 1989):

$$CC = \frac{\Sigma(CC)_i[\text{weight}]_i}{\Sigma[\text{weight}]_i} \qquad [1]$$

where $[\text{weight}]_i$ is a weighting factor related to the relative abundance of the ligand in the multiligand system. Given the relative abundance of acidic functional group ligands, primarily COOH, and the low concentrations of other O-containing and N- and S-containing ligands in HS, the maximum binding capacity of HS results approximately equal to the total functional acidic group content (Stevenson, 1994). Even though total acidity and complexation capacity are not equal, the former represents a useful surrogate parameter and provides an upper limit for the latter (Perdue, 1989). Further, total acidity can be measured by well-established methods (Perdue, 1985), thus avoiding much of the ambiguities and misunderstandings that can arise over complexation capacities and their reported dependence on experimental conditions (Perdue, 1989).

The complexation capacity of HS, however, is a function of various factors that include pH, ionic strength, HS concentration and properties, and nature of the metal ion (e.g., Stevenson, 1994; Perdue, 1989; MacCarthy and Perdue, 1991). The effect of pH on metal–HS complexation results from both changes in the extent of ionization of COOH groups and hydrolysis reactions of the metal ion involving the formation of oxyhydroxides (Stevenson, 1994). In general, the complexation capacity of HS has been reported to increase with increasing pH from acidity up to about neutrality, to decrease at higher ionic strength, to increase at higher HS concentrations, and to vary with the nature of the metal ion (in Perdue, 1989). For any given pH and ionic strength, trivalent cations are bound in greater amounts than divalent ones. Among the latter ones those forming the strongest coordination complexes (e.g., Cu^{2+}) are bound to a greater extent and at the stronger binding sites than those that form weak complexes (e.g., Mn^{2+}) (Stevenson, 1994).

Complexation capacity of HS should, however, be regarded as a constant quantity that is characteristic of the HS sample, rather than a variable parameter (MacCarthy and Perdue, 1991). According to these authors, the apparent variations of complexation capacity of HS would, in fact, result from the inability to saturate all the ligand sites under some experimental conditions, and this effect would arise from the variability in the conditional concentration quotient (see later), and the influence of dilution on complex formation.

Stability Constants and Related Parameters of Metal–Humic Substance Complexes

The stability constant is probably the most important quantitative parameter for the characterization of a metal–ligand complex in that it provides a numerical index of the affinity of the metal cation for the ligand, and allows the development of quantitative models able to predict the speciation of metal ions in the system studied. Several different theoretical and experimental approaches have been attempted for the determination of stability constants of metal–HS complexes and modeling metal–HS complexation reactions. Data analysis and interpretation is, however, still controversial due to the intrinsically complex and ill-defined nature of HS. The multiligand, polyelectrolitic nature of HS macromolecules results in the inability to describe quantitatively the tipes, concentrations and strengths of the several non-identical binding sites in HS, and in the impossibility to ascertain and measure the stoichiometry of metal–HS complexation (MacCarthy and Perdue, 1991). Further, comparison and use of literature data for modeling metal–HS complexation are generally of limited value because of the bewildering assortment of fundamentally different conceptual and experimental approaches used to obtain the data. Because of the extreme complexity and wide controversy existing on the subject, only a general overview of the theoretical and experimental approaches is provided in this chapter. For a more detailed information, the reader is referred to the several reviews and book chapters available on the topics of stability constants and conceptual and mathematical models of metal–HS complexation (e.g., Perdue and Lytle, 1983; Stevenson and Fitch, 1986; Sposito, 1986; Perdue, 1989; MacCarthy and Perdue, 1991; Tipping, 1998; Perdue, 2001).

The overall reaction of a single metal ion (M) and a single ligand (L) to form the complex M_mL

$$\mathrm{m}M + \mathrm{n}L \leftrightarrows M_mL_n \qquad [2]$$

can be described by (MacCarthy and Perdue, 1991):

$$K = \frac{\{M_mL_n\}}{\{M\}^m\{L\}^n} = \frac{[M_mL_n]}{[M]^m[L]^n} \bullet \frac{\gamma_{M_mL_n}}{(\gamma_M)^m(\gamma_L)^n} = K_c \bullet \Gamma \qquad [3]$$

where braces { } and square brackets [] denote activities and concentrations, respectively, γ values are activity coefficients, K is the true overall thermodynamic stability (equilibrium) constant, K_c is the concentration quotient, and Γ is the activity coefficient ratio. For convenience, the charges on the metal, ligand, and complex are generally not shown explicitely, even if their values are extremely important.

Most experimental investigations of metal complexation reactions yield concentrations, rather than activities, of reactants and products. Thus, overall formation constants are not measured directly but must be indirectly obtained by calculations, or by extrapolation techniques from the concentration coefficients whose values are not constant but depend, as well as those of the activity coefficient ratios, on the ionic concentration (ionic strength) of the solution. Basic electrostatic considerations (Debye-Hückel theory) indicate that Γ equals 1, and thus K_c values equal K values, at zero ionic strength, and Γ increases, and K_c tend to decrease, with increasing ionic strength. The concentration of the complex M_mL_n is thus expected to decrease upon addition of a background electrolyte to the solution of metal and ligand.

Competition from side reactions, especially the acid–base chemistry of both the metal ion and the ligand, i.e., the hydrolysis of the metal ion to produce hydroxy complexes and the protonation of the ligand, is another factor that affects the extent of complexation of M by L, and must be taken in due consideration. In general, protonation of ligands occurs at low pH, and hydrolysis of metals at high pH. Thus, the most favorable condition for the complexation of M by L is at intermediate pH values.

Metal–ligand complexation reactions are usually investigated at constant pH and using experimental techniques that are able to distinguish the metal–ligand complex from the uncomplexed metal, i.e., from all forms of the ligand that are not bound to the metal ion. Thus, a "conditional stability constant", or more precisely a "conditional concentration quotient", K_c^*, can be written (MacCarthy and Perdue, 1991):

$$K_c^* = \frac{[M_mL_n]}{\frac{[M]^m}{(\alpha_M)^m} \bullet \frac{[L]^n}{(\alpha_L)^n}} = (\alpha_M)^m(\alpha_L)^n \frac{[M_mL_n]}{[M]^m[L]^n} = (\alpha_M)^m \bullet (\alpha_L)^n \bullet K_c \qquad [4]$$

where α_M and α_L are the side reaction coefficients of the metal ion and the ligand, respectively. Eq. [4] thus describes the formation and stability of the complex relative to all other forms of the metal and ligand that may exist under the actual experimental conditions. Further, Eq. [4] shows that the conditional concentration quotient, K_c^*, includes a pH-invariant but ionic strength-dependent term, K_c, and pH-dependent terms, α_M and α_L.

Assuming for simplicity a 1:1 stoichiometry for all ML_i complexes, the complexation reaction of a single metal ion M with a multiligand system such as HS containing a number of ligands L_i can be described by an average stability constant, or, more appropriately, an average concentration quotient (MacCarthy and Perdue, 1991):

$$\bar{K}_c = \frac{\Sigma[ML_i]}{[M]\Sigma[L_i]} \qquad [5]$$

where $\Sigma[ML_i]$ and $\Sigma[L_i]$ are, respectively, the sum of the concentrations of all 1:1 complexes and all uncomplexed ligands in the mixture. In practice, the metal ion will distribute itself among the various ligands on the basis of its stability constant with each ligand and of the relative concentration of that ligand. The metal ion will thus tend to complex preferentially with ligands having the highest stability constant and/or highest concentration.

Similar to single-ligand systems, an "average conditional concentration quotient," also called "stability function," $\bar{K}_c^*$, that takes pH effects into account, can be defined for multiligand systems such as HS (MacCarthy and Perdue, 1991). This quantity is a measure of the overall binding of a single metal ion to a multiligand system; however, it cannot be calculated rigorously for HS because there is not sufficient information available to enable the full description of side reactions and the calculation of side-reaction coefficients of HS for which individual components are unknown. Thus the function $\bar{K}_c^*$ is often calculated directly from experimental data as (Perdue, 1989):

$$\bar{K}_c^* = \frac{C_M - [M]}{[M](C_L - C_M + [M])} \qquad [6]$$

where C_M and C_L are the total stoichiometric concentrations of the metal and ligand in the system examined and $[M]$ is the concentration of the free metal ion. The value of $(C_M - [M])$ corresponds to the sum of the concentrations of all complexes formed between M and the multiligand system, and is calculated by neglecting the presence of inorganic metal complexes. The quantity $(C_L - C_M + [M])$ represents the sum of the concentrations of all binding sites that are not associated with M, and is calculated assuming an average 1:1 metal/ligand stoichiometry for the total of binding sites. Stability functions can be defined also for complexes of stoichiometries other than 1:1 but this aspect will not be addressed here.

Average $\bar{K}_c^*$ values are ultimately functions of ionic strength, pH, and degree of saturation of the multiligand system with metal ion. At a given pH and ionic strength, $\bar{K}_c^*$ will decrease steadily as the total metal/ligand ratio (C_M/C_L) increases. At low metal/ligand ratios, the occurrence of preferential reactions of stronger ligands will determine the functional nature of $\bar{K}_c^*$ (Perdue, 1989). Thus, it is not possible to obtain true stability "constants" for metal–HS complexation, and reported values are not actually constant because the functional dependence is a fundamental feature of systems such as HS, which cannot be eliminated by any experimental method unless total fractionation be achieved into pure ligands that can be studied separately.

In conclusion, the ill-defined nature of HS and their component ligands, and the elusive nature of metal–HS interactions prevent metal–HS complexation reactions to be described in rigorous mathematical terms; however, important information can be obtained on these reactions by application of general principles and methods of coordination chemistry. In particular, the stability function concept provides an important overview and useful insights into the nature of metal binding by HS, and also may provide useful indications for most appropriate experimental methods in investigating metal–HS interactions.

Modeling and Models

The large body of field and laboratory data accumulated over the last four decades on the binding of metal ions and protons by HS has prompted efforts for the development of chemical speciation models that could encapsulate the ample knowledge and information acquired, and allow applications to "real-world" situations. Models formulated have been applied, however, with variable success. Standing the complexity and amplitude of the subject, it is not the objective of this chapter to attempt to provide a detailed discussion of metal–HS binding models, for which the reader may refer to the literature cited in the following text where a brief hystorical survey of various proposed and used models is provided.

The prevalent effort of any modeling approach is the quantitative description of the relative concentrations and strengths of the many nonidentical binding sites contained in HS. The various models applied to describe metal–HS complexation can be classified as "discrete" ligand models and "continuous" multiligand models.

In the discrete ligand approach only a few ligands or binding sites are required to fit the experimental data. Since early modeling work the competition for HS binding sites between metal ions and protons has been recognized, and metal binding has been estimated from displacements of pH titration curves (Van Dijk, 1971). In subsequent models this approach has been extended by also taking into account electrostatic effects associated with variation of coulombic charge (Stevenson, 1976b). Conditional stability constants have been used by several investigators (e.g., Schnitzer and Skinner, 1966, 1967; Bresnahan et al., 1978; Saar and Weber, 1980) to describe metal–HS interactions at a single pH and ionic strength. Modified and refined discrete models also incorporated site heterogeneity effects, i.e., variability in the intrinsic binding strengths among sites, and electrostatic effects related to the net negative charge of HS (Marinsky and Ephraim, 1986; Ephraim and Marinsky, 1986).

A simplified discrete ligand model denoted as Model V has then been developed (Tipping and Hurley, 1992), which included: (i) site heterogeneity, i.e., discrete sites with a range of affinities, together with the formation of bidentate sites; (ii) electrostatic effects; and (iii) competition among protons and metal ions. Model V, and its predecessor Model IV, have been used to describe published data sets on a range of metals with some success. More recently, the inadequacies of Model V have been demonstrated and discussed, and an improved model (Model VI) also based on the discrete-site–electrostatic formulation but moved towards a more distributional approach in describing binding sites for metals, has been presented and applied to those data sets analyzed previously with Model V and to additional data sets (Tipping, 1998).

The continuous multiligand distribution models are based on the assumption that a large number of heterogeneous ligands with a range of binding affinities are involved in metal binding with formation of complexes characterized by a continuous (e.g., Gaussian) distribution of the stability constants (Gamble et al., 1980; Perdue and Lytle, 1983; Altman and Buffle, 1988). The initial, noncompetitive Gaussian distribution models that did not account for proton–metal competition were then extended in the competitive Gaussian distribution model (Dobbs et al., 1989). Later, this model was chosen to describe cation binding by HS in MINTEQA2, a computer program for performing chemical equilibrium calculations (e.g., Allison and Perdue, 1995).

A continuous distribution model of the various ligands in HS, in which individual ligand concentrations are normally distributed with respect to the individual stability constants for metal binding, has been used successfully in conjunction with two fluorescence techniques to study metal–HS complexation (Susetyo et al., 1991). The model includes the effects of pH, ionic strength, and competing metal ions. The parameters of the model are estimated by fitting the spectral titration data to the calculated titration plot.

A series of models have been developed that characterize HS binding site heterogeneity by using a continuous distribution of equilibrium constants, together with electrostatic submodels, and by assuming monodentate binding of metals to proton binding sites (e.g., Milne et al., 1995; Benedetti et al., 1995). The most recent of these models is the NICA (non ideal competitive adsorption)—Donnan model (Kinniburgh et al., 1996) that fits very well extensive experimental data sets obtained for a wide range of metal ion concentrations. The incorporation of multidentate sites into the NICA-Donnan model is under consideration (Kinniburgh et al., 1996).

Different types of models have been compared in several reviews (e.g., Perdue and Lytle, 1983; Buffle, 1988; Dzombak et al., 1986; Nifant'eva et al., 1999). Sposito (1986) provided a detailed review of a variety of models, and illustrated the advantages of conceptual simplicity and calculational convenience of the "quasi-particle" model. This model includes previous concepts and models, and describes mathematically an aqueous system containing HS by replacing it with a set of hypothetical, average, noninteracting molecules whose behavior mimics that of the actual HS system. In a recent review (Perdue, 2001) three most modern and successful models for the description of competitive proton and metal binding by HS have been discussed comparatively focusing on similarities and differences in their conceptual approaches. These are the competitive Gaussian distribution model (Dobbs et al., 1989), Model V (Tipping and Hurley, 1992), and the NICA-Donnan model (Koopal et al., 1994; Kinniburgh et al., 1996).

In conclusion, although the continuous models have some distinct advantages, the literature survey indicates that the most popular are still the discrete models that offer different kinds of approaches that enable the explanation of complexation of metal ions by HS. It should be noted, however, that all the obtained parameters (e.g., constants and binding site concentrations) should be regarded just as mathematical curve-fitting parameters, without any chemical meaning, their principal utility being their ability to accurately predict overall metal complexation by HS within the experimental conditions used; however, since the primary purpose of metal–HS binding models is to perform chemical speciation calculations for field situations, this requires validation of the model on field data. More recent models reproduce trends quite well but still lack of knowledge exists on some key factors, and empirical adjustment of these factors is necessary to fit the data. Thus, more extensive field-testing is required to establish the usefulness or otherwise of any model of metal–HS binding.

Experimental Methods Other Than Spectroscopic

A variety of separation and nonseparation methods have been used to speciate metal ions in the presence of HS, assess complexing capacity of HS and calculate conditional stability constants/quotients of metal–HS complexes. Because of the extreme amplitude of the topic, only a brief overview of the methods most commonly applied to soil HS will be provided here. A number of reviews and book chapters have been published on the topic, to which the reader can refer for details (Saar and Weber, 1982; Stevenson and Fitch, 1986; Weber, 1988; Dabek-Zlotorzynska et al., 1998; Nifant'eva et al., 1999).

Most commonly used separation approaches include: (i) proton-release titration (Van Dijk, 1971; Stevenson, 1976b, 1977; Takamatsu and Yoshida, 1978); (ii) cation exchange with synthetic resins (e.g., Zunino et al., 1972; Crosser and Allen, 1977); (iii) gel filtration or permeation chromatography (GPC) (e.g., Mantoura and Riley, 1975; Mantoura et al., 1978; Krajnc et al., 1995); (iv) high performance cation exchange chromatography (HP-CEC) (Sutheimer and Cabaniss, 1997); (v) conventional dialysis (e.g., Zunino and Martin, 1977; Weber, 1983); (vi) equilibrium dialysis-ligand exchange (EDLE) (Van Loon et al., 1992; Glaus et al., 2000); (vii) ultrafiltration (UF) (Nifant'eva et al., 1999); and (viii) capillary electrophoresis (CE) (Dabek-Zlotorzynska et al., 1998).

Most important nonseparation, other than spectroscopic methods include: (i) ion-selective electrode (ISE) potentiometry (e.g., Saar and Weber, 1979, 1980; Bresnahan et al., 1978; Buffle et al., 1980; Bhat et al., 1981; Langford et al., 1983); and (ii) voltammetric techniques such as anodic stripping voltammetry (ASV) (e.g., Buffle and Greter, 1979; Bhat et al., 1981; Weber, 1983; Turner et al., 1987; Van den Hoop et al., 1995).

Separation approaches generally present the great advantage of allowing measurement of nearly any metal ion by sensitive techniques such as atomic absorption and plasma spectrometries; however, dialysis and chromatographic methods suffer of two major disadvantages that are the always existing possibility of disturbing complexation equilibria and of adsorption of interacting species, i.e., the metal ion, the HS and their complexes, on the membrane or chromatographic material. Further, equilibrium dialysis and cation exchange suffer from uncertainties due to lack of definition of chemical forms actually measured by the method. An important advantage of the UF methods is that they do not take so long experimental time as dialysis, and do not disturb the complexation equilibria.

Some nonseparation methods, such as ISE and ASV, although being the most sensitive and advanced, also suffer of the problems of adsorption and shifts in equilibria. Further, these methods are applicable adequately only to a limited typology of metal ions, e.g., Cu^{2+}, Cd^{2+}, Pb^{2+}, and Zn^{2+}.

In conclusion, the main problem is that the comparability of results obtained by different groups of researchers is hindered by the different nature of methodological approaches and instrumental techniques employed and parameters measured, and by the variability of system conditions used to obtain reported data.

THE SPECTROSCOPIC APPROACH TO METAL–HUMIC SUBSTANCE COMPLEXATION

The knowledge of the molecular structure and the binding mechanisms of metal–HS complexes is relatively limited. Progress in ecotoxicology of trace metals in environmental systems including soil requires a more precise and extended conceptual knowledge of molecular and mechanistic aspects of trace metal–HS reactions. A greater emphasis on basic information about chemical aspects is needed compared with quantitative monitoring studies (Stumm et al., 1983).

Spectroscopic techniques, such as ultraviolet-visible (UV-VIS), fluorescence, infrared (IR), electron spin resonance (ESR), or electron paramagnetic resonance (EPR), x-ray absorption spectroscopy (XAS), nuclear magnetic resonance (NMR) and Mössbauer, offer a great potential to provide a number of important insights at the molecular and mechanistic level of metal complexation reactions with HS, as well as unique information about the chemical and physical nature of the binding sites. Some of the mentioned spectroscopic techniques are also able to provide information on quantitative aspects of metal–HS complexation.

Table 12–2 shows an overview of the spectroscopic techniques that will be considered in the following text together with the indication of the metal ions of application, the level of specificity of each technique for metal ion measurement, and its capability for qualitative and/or quantitative application. Only results obtained with soil HS will be shortly reviewed, whereas data on HS from sources other than soil will not be examined, with some exceptions.

Ultraviolet-Visible Spectroscopy

Principle and Methodology

The absorption bands in the UV-Vis region arise from electronic transitions from bound states (outer valence orbitals) to excited states (McCoustra, 1990). The wavelength limits of these regions of the electromagnetic spectrum are 200 to 400 nm for the UV and 400 to 800 nm for the visible. In organic molecules such as HS, these exceptionally low energy transitions are associated with the presence of chromophores, i.e., conjugated double-bonds and aromatic and related molecules with

Table 12–2. Spectroscopic techniques used for metal ion studies.

Spectroscopic technique	Relevantmetal ions of application	Specificity level of the technique	Qualitative technique	Quantitative technique
Ultraviolet-visible	Cu^{2+}, Fe^{3+}, Al^{3+}	Medium	YES	YES
Fluorescence	Cu^{2+}, Fe^{3+}, Fe^{2+}, Co^{2+}, Ni^{2+}, Mn^{2+}, Cr^{3+}, VO^{2+}, Pb^{2+}, Al^{3+}, K^{+}, Na^{+}, Ca^{2+}, Ba^{2+}, Eu^{3+}	Medium	YES	YES
Infrared	Cu^{2+}, Mn^{2+}, Zn^{2+}, Pb^{2+}, Co^{2+}, Ni^{2+}, Ca^{2+}, Mg^{2+}, Sr^{2+}, Fe^{3+}, Al^{3+}	Low	YES	NO
Electron spin resonance	Cu^{2+}, Mn^{2+}, Fe^{3+}, VO^{2+}, Mo(V), Mo(III)	High	YES	NO
Nuclear magnetic resonance	Cd^{2+}, Al^{3+}, Pb^{2+}, Mn^{2+}, Fe^{3+}, Cu^{2+}, VO^{2+}, Zn^{2+}, Eu^{3+}	High	YES	YES
Mössbauer	Fe^{3+}, Fe^{2+}	High	YES	NO
X-ray absorption	Cu^{2+}, Pb^{2+}, Zn^{2+}, Ni^{2+}, Co^{2+}, Mn^{2+}, Fe^{3+}, Hg^{2+}	High	YES	NO

delocalized electronic orbitals. The strongest UV-Vis absorption bands are associated with $\pi \rightarrow \pi^*$ transitions, and weaker ones with $n \rightarrow \pi^*$ transitions, where π and π^* refer to bonding and antibonding *p*-type orbitals and *n* refers to lone-pair orbitals (McCoustra, 1990). Electronic transitions can occur within the molecular orbitals of chromophores or involve the transfer of an electron from one chromophore to another chromophore or to a nonchromophore (electron- or charge-transfer excitation). For metal–organic complexes, absorption can occur in the UV-Vis region from $d \rightarrow d$ orbital transitions associated with transition metals.

More information on the theory and methodology of UV-Vis spectroscopy can be found, among several others, in Brown (1980) and McCoustra (1990). Comprehensive reviews on the application of UV-Vis spectroscopy to the study of HS and metal–HS complexes were provided, e.g., by Bloom and Leenheer (1989) and Stevenson (1994).

The two most important parameters determined in UV-Vis spectroscopy are the wavelength(s) of maximum absorptions and the absorptivity, ε. A change in both parameters of the chromophores may result by ionization of carboxyl and hydroxyl groups (Scott, 1964). Interaction of some metal ions such as Cu^{2+} with polycarboxylates in dilute aqueous solution leads to typical spectral perturbations in the UV region. They occur: (i) in the 195 to 200 nm range, corresponding to the internal ligand transition band typical of bound (di)carboxylic groups; and (ii) between 245 and 265 nm, that is the strong electron transfer band due to electron transition between the central metal ion and the electronic systems of the ligands (charge-transfer process). With increasing the ratio of Cu^{2+} ions to polyelectrolyte ligand, a redshift and a strengthening is observed for the charge transfer band (Campanella et al., 1990).

Stoichiometry of Metal–Humic Substance Complexes

The UV-Vis spectra of HS show little structure, appearing largely as broad, coarsely structured absorption bands. This is because of heterogeneous substitution that results in chromophores with overlapping bands and spectral shifts due to slight differences in the macromolecular structures.

UV-Vis spectroscopy, however, has been successfully applied for the evaluation of the stoichiometry of metal–HS complex formation based on the Job's method. This method consists in measuring the variation of optical densities in the visible range (400 to 800 nm) of aqueous solutions containing different ratios of metal ion to complexing agent, while simultaneously maintaining a constant total concentration of both reactants. For example, the ions Cu^{2+}, Fe^{3+}, and Al^{3+} are shown to form 1:1 complexes with a soil FA at pH 3, whereas at pH 5 Cu^{2+} and Fe^{3+} form 2:1 molar complexes with the FA, while Al^{3+}–FA complex composition remains at 1:1 (Schnitzer and Skinner, 1963).

Fluorescence Spectroscopy

Basic Principles and Methodology

The absorption of visible and ultraviolet radiation raises a molecule from the ground electronic and vibrational state to excited states. The most important ways in which excited electronic states decay are radiative processes, i.e., fluorescence and phosphorescence, which occur when excited electrons return to ground state, and nonradiative processes, i.e., internal conversion and intersystem crossing (Creaser and Sodeau, 1990; Senesi, 1990a). In particular, fluorescence is a radiative photoprocess that occurs between two energy levels of the same multiplicity and consists in the emission of less energetic (lower wavelength) photons than the photons absorbed to produce the excited state.

Two types of conventional fluorescence spectra, emission and excitation spectra, are useful for studies of metal–organic complexation. The emission spectrum is recorded by measuring the intensity of radiation emitted as a function of wavelength for a fixed excitation wavelength. The excitation spectrum is obtained by measuring the emission intensity at a fixed wavelength while varying the excitation wavelength.

The intensity of fluorescence, I_f, depends on the concentration of absorbing species in solution and the efficiency of the fluorescence process according to:

$$I_f = \phi_f I_o [1 - \exp(-\varepsilon bC)] \qquad [7]$$

where ϕ_f is the fluorescence efficiency, or quantum yields, that measures the efficiency with which the absorbed energy is re-emitted, i.e., the ratio of total energy emitted as fluorescence per total of energy absorbed, I_o is the intensity of incident radiation, ε is the molar absorptivity at the excitation wavelength, b is the path length of the cell, and C is the molar concentration. For very dilute solutions, where εbC is sufficiently small, Eq. [7] reduces to a linear relationship between measured fluorescence intensity and concentration:

$$I_f = \phi_f I_o \varepsilon bC \qquad [8]$$

When εbC values are low the fluorescence intensity is essentially homogeneous throughout the sample.

The fluorescence behavior of a molecule, especially in the solution state, is affected to various extent by several molecular properties and environmental factors. In particular, an electronically excited molecule can lose its energy by interacting with another solute (a quencher) and thus its fluorescence is reduced. Metal ions, especially paramagnetic ions, are generally able to quench the fluorescence of organic ligands by enhancing the rate of some nonradiative processes that compete with fluorescence, such as intersystem crossing. Unlike most other methods, fluorescence spectroscopy allows the direct measurement of the complexing capacity of the ligand through the determination of the concentration of free ligands, thus differentiating free and bound ligands.

Detailed information on the theory and methodology of fluorescence spectroscopy are available in several books and reviews (e.g., Schulman, 1985; Lakowicz, 1986; Sharma and Schulman, 1999; Valeur et al., 2001; Kraayenhof et al., 2002). A review on application of fluorescence to HS and metal–HS studies was provided by Senesi (1990a).

Quenching of Humic Substance Fluorescence by Metal Ions

HS of any source and nature fluoresce due to the presence of conjugated double bonds and aromatic rings bearing various functional groups (Senesi, 1990a); however, only a small fraction of HS molecules that absorb radiation actually undergo fluorescence.

Paramagnetic transition metal ions such as Cu^{2+}, Fe^{3+}, Fe^{2+}, Co^{2+}, Ni^{2+}, Mn^{2+}, Cr^{3+}, and VO^{2+}, may effectively quench the fluorescence of HS ligands, whereas the quenching effect for diamagnetic metal ions such as Pb^{2+} and Al^{3+} is much less pronounced, and Cd^{2+}, which forms much weaker complexes, and cations like K^{+}, Na^{+}, Ca^{2+}, and Ba^{2+} show no quenching effect (Saar and Weber, 1980; Weber, 1988; Senesi, 1990a). The greater quenching ability of Cu^{2+} is attributed to its capacity to form strong inner-sphere complexes that involve in binding several weak sites causing conformational changes making available additional, internal binding sites of FA (Ryan and Weber, 1982). Differently, Mn^{2+}, and likely Co^{2+}, would form outer-sphere complexes with FA not involving weakly acidic phenolic sites, where the ion is further from the fluorophore, and thus exhibits low quenching ability (Ryan et al., 1983). Fluorescence quenching by Mn^{2+} also is shown to decrease as the molecular weight of soil FA increases, as a function of the FA nature (Shestakov et al., 1987).

Two binding mechanisms are proposed to explain adsorption of Cu^{2+} by a soil FA at pH 3.5, 5, and 7 and various Cu/FA ratio (Fig. 12–2) (Bartoli et al., 1987): (i) a proteolytic reaction (strictly complexation), occurring up to a Cu/FA ratio of 0.6; and (ii) a charge neutralization, from a Cu:FA ratio of 0.6 to 1.1. Above a 1:1 ratio, the almost constant fluorescence suggests Cu^{2+} is no longer adsorbed on FA. Additional data indicate that the most strongly acidic carboxyl groups react first with Cu^{2+}, as would occur in bidentate chelation sites with salicylic- or phthalic-type structures (Underdown et al., 1981); however, a comparative ISE, ASV, and fluorescence quenching study show that no single ligand could adequately describe Cu^{2+}–FA complexing across the entire pH range (Gregor et al., 1989). Thus, a mixed mode of coordination is proposed, with the dominant binding sites varying with pH

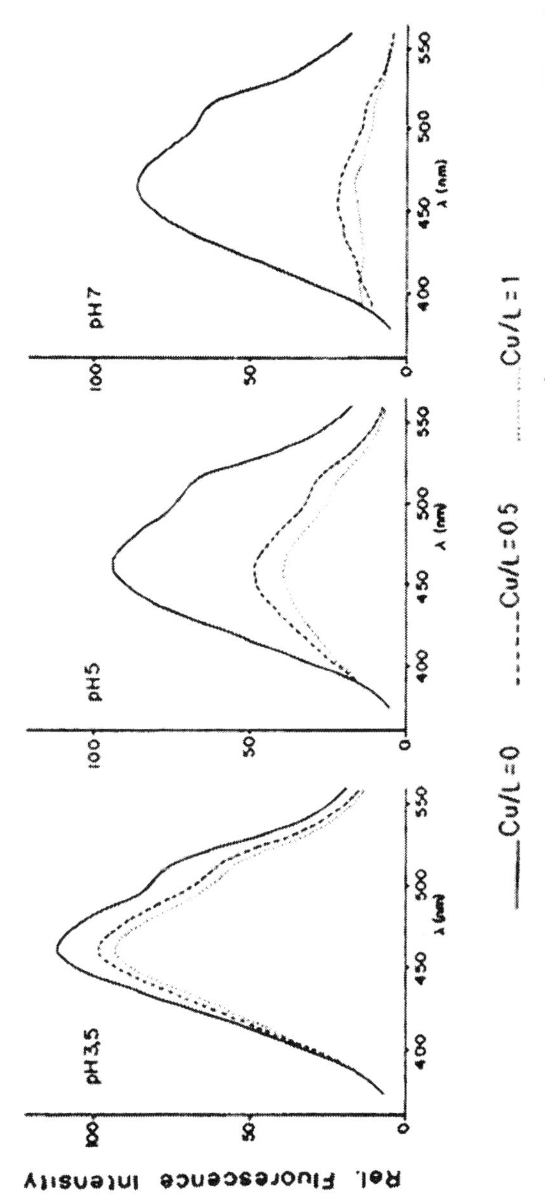

Fig. 12–2. Fluorescence emission spectra (λ_{exc} = 355 nm) of a soil fulvic acid (FA) and its complexes with Cu^{2+} obtained at three pHs and two different Cu^{2+}: FA ligand (L) ratios (from Bartoli et al., 1987).

and metal to ligand ratio. Comparison of FA–metal binding curves with those for model ligands indicates that potentially fluorescing groups such as salicylate or phthalates seem unlikely to be directly involved in Cu^{2+} complexing at low pH. Aromatic amino acid groups, such as tyrosine and phenylamine moieties, and citrate and malonate moieties in FA are suggested to provide important binding sites for Cu^{2+} at pH 3 to 7. At high pH, either polydentate fluorescing moieties are directly involved in complexing, or weakly complexing fluorescent centers become involved because of their close proximity to a strong donor site, i.e., within one or two carbon atoms of the complexing moieties (Gregor et al., 1989).

Wavelength shifts of fluorescence emission maximum and/or excitation peaks are also often observed upon interaction of HS with some paramagnetic metal ions (Senesi, 1990a). For example, the position and intensity of the peak at higher wavelength (465 nm) of a soil FA remain unchanged as more of either Cu^{2+} or Fe^{3+} is complexed, whereas the intensity of the lower wavelength peak (360 nm) decreases and shifts to longer wavelength (390 nm) at both pH 4 and 6 (Fig. 12–3) (Ghosh and Schnitzer, 1981).

Titration curves of HS fluorescence quenching vs. concentration of added metal quencher have been used to obtain the complexing capacity of HS ligands and the stability constants of HS–metal complexes (Saar and Weber, 1980, 1982; Underdown et al., 1981; Weber, 1983; Ryan et al., 1983; Dobbs et al., 1989; Grimm et al., 1991; Cook and Langford, 1995). Two fluorescence techniques, i.e., the lanthanide ion probe spectroscopy (LIPS) and fluorescence quenching of HS by Cu^{2+}, have been used in conjunction with a continuous distribution model to study metal–HS complexation (Susetyo et al., 1991). In the LIPS technique, the HS samples are titrated by Eu^{3+}ions, and the titration plot of the ratio of the intensities of two emission lines of Eu^{3+} is used to estimate the amount of bound and free species of the probe ion. In the other technique, titration curves of fluorescence intensity quenched by Cu^{2+} vs. the logarithm of total added Cu^{2+} are used.

Quantitative results, however, obtained by different researchers using different procedures are generally not comparable, i.e., complexing capacities of HS appear to be dependent on the method of measurement. Also the source of HS and the procedure used for its isolation, in addition to many experimental factors, including concentration of HS, ionic strength of solution, pH, temperature, and the method of data manipulation for the computation of stability constants can influence the results (Saar and Weber, 1982). For example, fluorescence quenching of FA and HA from different sources by various metal ions, including Cu^{2+}, Pb^{2+}, Co^{2+}, Ni^{2+}, and Mn^{2+}, increases with increasing pH (Saar and Weber, 1980).

Conclusions

Fluorescence spectroscopy has several advantages over most other methods for studying metal–HS complexation in aqueous media. The method is relatively rapid since no separation is required between bound and free metal ion, thus errors associated with the separation step in most speciation methods are avoided. Unlike most other methods, it allows the direct measurement of the complexing capacity of the ligand through the determination of the concentration of free ligands, thus differentiating free and bound ligands. Neither supporting electrolyte nor buffer nor adsorbing material are required to be added to the samples. The method is even more

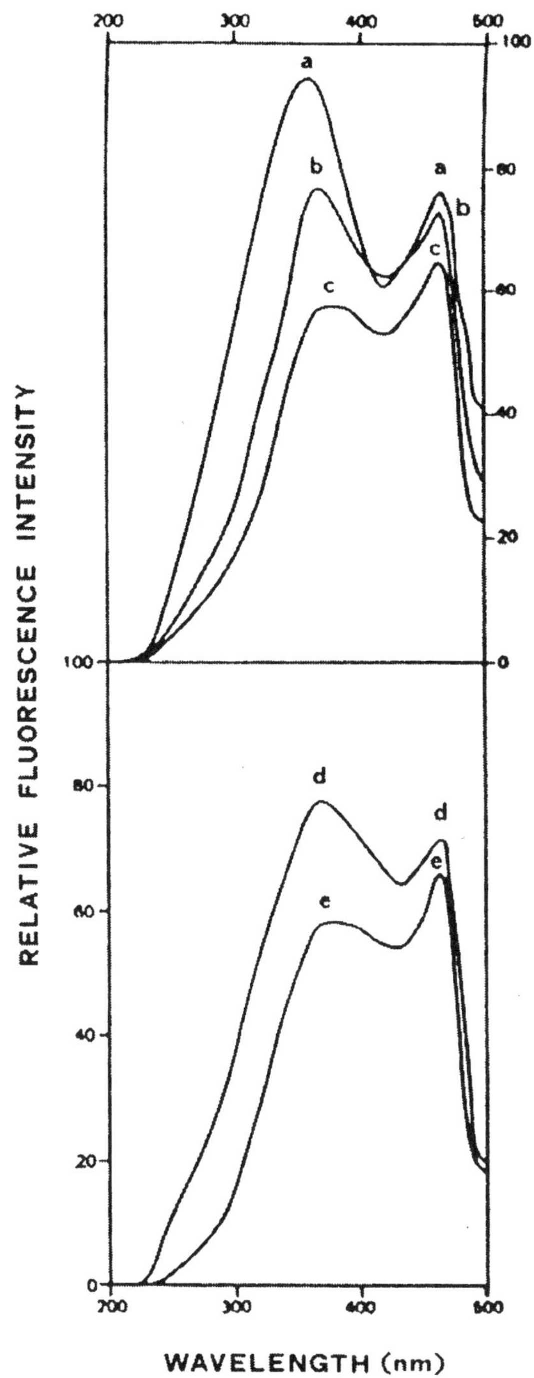

Fig. 12–3. Fluorescence excitation spectra at (a) pH 6 of a soil fulvic acid and its complexes with Cu^{2+} (%Cu: (b) 3.51 and (c) 7.22) and Fe^{3+} (%Fe: (d) 3.47 and (e) 6.27) (adapted from Ghosh and Schnitzer, 1981).

sensitive than ASV and ISE potentiometry and is sensitive enough for application to unmodified, natural organic ligands without preconcentration. It represents, therefore, an excellent complement to other, indirect complexing capacity measurement techniques based on the determination of free metal ion concentration with respect to the bound metal fraction.

The major disadvantage of fluorescence spectroscopy is, however, that it is very effective only with strongly binding, paramagnetic metal ions such as Cu^{2+}. This limitation can, however, be overcome by the use of fluorescent probes, such as the lanthanide ion probe that shows particularly promising for studies of metal binding by HS in environmental conditions, and at natural concentrations of both metal and HS.

Infrared Spectroscopy

Basic Principles and Methodology

The most interesting portion of the IR spectrum for the structural and analytical study of molecules is the medium IR region, between 4000 and 400 cm^{-1} (2.5 and 25 nm). Energy absorbed by an organic molecule in this region is converted into energy of molecular vibration, and the IR spectrum of the organic molecule consists of vibrational bands. There are two main types of bond vibration modes in simple molecules: stretching, that involves changes in the bond length between the atoms along the bond axis, and bending or deformation, which involves a change of bond angles.

The characteristic wavenumber (or frequency) ranges for the vibrations of particular functional groups, such as CH, NH, OH, C=O, C=C, aromatic rings, etc., depend on the vibrational mode, the strength of the bonds involved, and the masses of the atoms (Bellamy, 1975). Intensities in the IR spectra depend on dipole-changes occurring during the vibrations and so the polar bonds and groups frequently give the strongest absorptions. Since any change in the strength of the bonds and masses in a given system alters the vibrational frequency, the formation of metal–organic complexes may be studied by the shifts observed for vibrational frequencies. These can be used to identify the absorbing functional groups involved in the complexation and, possibly, provide information on the type of interaction occurring between the metal and the ligand in the complex formed (Nakamoto, 1986).

The greatest improvement in performance of IR spectroscopy was obtained when dispersive, diffraction grating spectrometers were replaced by interferometers, together with the necessary computing facilities for Fourier-transform (FT) conversion of the interferogram into a normal (intensity vs. wavenumber) spectrum. The very high sensitivity introduced by FT methods also determined the development and use of a number of additional nontransmission techniques including diffuse reflectance (Willis et al., 1988). Besides its high sensitivity, the great practical advantage of diffuse reflectance infrared Fourier-transform (DRIFT) technique is that powdered samples can be studied with little or no preparation. Compared with transmission spectra, the strongest IR absorption features of organic molecules are relatively reduced in intensity by selective reflection processes, thus weaker IR absorptions can be very well recorded with little distortion by DRIFT, and the result is generally adequate for structure and identification purposes.

Table 12–3. Main IR adsorption bands and assignments for humic substances.

Frequency (cm^{-1})	Assignment
3450–3300	O-H stretching, N-H stretching (trace), hydrogen-bonded OH
3080–3030	Aromatic C-H stretching
2950–2840	Aliphatic C-H stretching
1725–1710	C=O stretching of COOH, aldheydes and ketones
1660–1630	C=O stretching of amide groups (amide I band), quinone C=O and/or C=O of H-bonded conjugated ketones
1620–1600	Aromatic C=C stretching, COO^- symmetric stretching
1540–1510	N-H deformation and C=N stretching (amide II band), Aromatic C=C stretching
1460–1440	Aliphatic C-H deformation
1400–1380	OH deformation and C-O stretching of phenolic OH, C-H deformation of CH_2 and CH_3 groups, COO^- antisymmetric stretching
1260–1200	C-O stretching and OH deformation of COOH, C-O stretching of aryl ethers and phenols
1170	C-OH stretching of aliphatic O-H
1080–1030	C-O stretching of polysaccharide or polysaccharide-like substances, Si-O of silicate impurities
975–775	Out-of-plane bending of aromatic C-H

A wide number of books and reviews are available on general principles and methodology of IR techniques (e.g., Rao, 1963; Bellamy, 1975; Nakamoto, 1986; Willis et al., 1988), and their application to HS and metal–HS studies (MacCarthy and Rice, 1985; Bloom and Leenheer, 1989; Senesi, 1992; Stevenson, 1994).

Metal–Humic Substance Complexes

Despite the molecular complexity of HS, the most striking features of IR spectra of HS are their overall simplicity and similarity (Schnitzer, 1978; Stevenson, 1994), but these are more apparent than real. Simplicity, i.e., band broadness, generally results from the extended overlapping of very similar absorptions arising from individual functional groups of the same type but with different chemical environments. Similarity is in part due to the fact that most groups of atoms vibrate with almost the same frequency irrespective of the molecule to which they are attached. Thus, HS displaying similar spectra do not actually have the same overall molecular structure but only similar functional groups and structural units (MacCarthy and Rice, 1985).

A list of the most important and common IR band frequencies occurring in HS, and their corresponding assignment to specific functional groups and structural units are shown in Table 12–3. This information is very useful for the interpretation of IR spectra and for the possible identification of absorbing functional groups and mechanisms involved in metal–HS complexation.

Carboxylate Ligands. Results of IR spectroscopy have provided ample evidence of the prominent role played by COOH groups in metal ion complexation by HA and FA (reviewed in Senesi, 1992; Stevenson, 1994). The C=O stretching absorption band at about 1710 cm^{-1} and the C–O stretching and O–H deformation absorption at about 1200 cm^{-1} are strongly reduced in intensity or disappear upon ionization of the COOH groups following reaction of HA and FA with several divalent and trivalent metal ions including Cu^{2+}, Mn^{2+}, Zn^{2+}, Pb^{2+}, Co^{2+}, Ni^{2+}, Ca^{2+}, Mg^{2+}, Sr^{2+}, Fe^{3+}, and Al^{3+} (Fig. 12–4). Simultaneously, bands near 1600 and 1380 cm^{-1} arising, respectively, from the asymmetric and symmetric stretching vibrations of the COO^- groups are reinforced or appear (Fig. 12–4). The COO^- to the COOH absorption ratio in the IR spectra of a number of metal–HA complexes prepared at the same HA:metal ratio is shown to be dependent on the nature of the metal ion complexed, and to vary in the decreasing order: $Fe^{3+}>Cu^{2+}>Al^{3+}>Ca^{2+}>Mg^{2+}$. These results indicate a decreasing conversion of COOH groups to COO^- groups involved in metal complexation.

The frequency of the asymmetric and symmetric streching vibrations of COO^- may provide information on the ionic vs. covalent character of carboxylate–metal bonding. With increasing covalency, the COO^- asymmetrical stretching band shifts to a higher frequency and the symmetrical stretching band to a lower frequency (Nakamoto, 1986). In a study of the influence of several metal cations on the COO^- asymmetrical stretching in HA complexes, absorption bands ranging from 1585 cm^{-1} for Ca^{2+} to 1625 cm^{-1} for Al^{3+} are measured which suggest a different degree of bonding covalency for the various metals examined (Vinkler et al., 1976). Bonds with high covalent character for Cu^{2+} in HA and FA are preferentially formed at low levels of metal ion, whereas bonding becomes increasingly ionic as the system is saturated with the metal (Piccolo and Stevenson, 1982). Frequency shifts in the 1600 cm^{-1} region are, however, variable and slight, and interpretation are complicated by interference from other groups.

The nature of the carboxylate binding site in HS can be evaluated by measuring the separation between the two frequencies of the antisymmetric and sym-

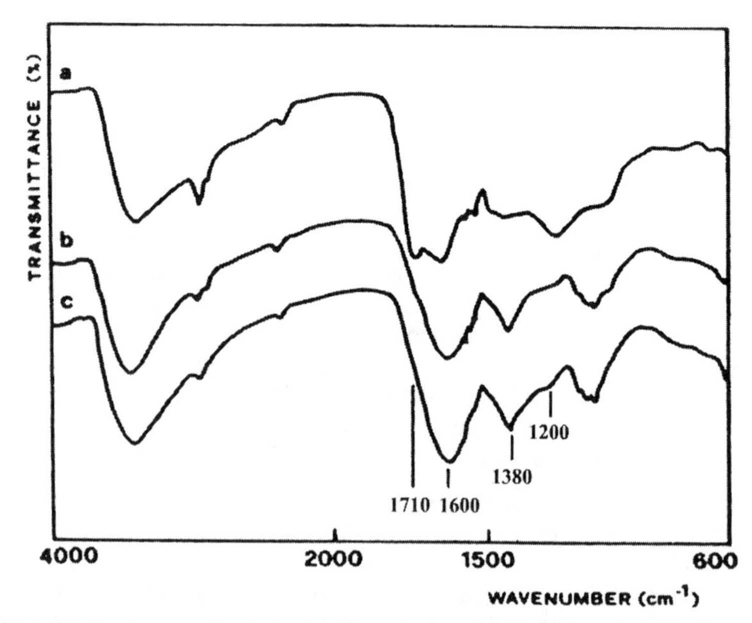

Fig. 12–4. Infrared (IR) spectra of a (a) soil humic acid (HA) and its complexes with (b) Cu^{2+} (Cu^{2+}:HA = 0.04), and (c) Cu^{2+} and Fe^{3+} [(Cu^{2+} + Fe^{3+}): HA = 0.08] (adapted from Senesi et al., 1986).

metric stretching vibrations for the metal-complexed COO^- group (near 1600 and 1380 cm^{-1}, respectively), with respect to the uncomplexed carboxylate ion. This separation is larger in unidentate complexes (Structure I), smaller in bidentate (chelate) complexes (Structure II) and comparable in bridging complexes (Structure III) (Nakamoto, 1986) (Fig. 12–5). A large separation is measured for Cu^{2+}, Fe^{3+}, Co^{2+}, and Zn^{2+} complexes of HA and FA, thus suggesting the formation of unidentate metal complexes (Boyd et al., 1981a; Prasad et al., 1987). These results also are consistent with metal chelation sites involving either two adjacent COOH groups (phtalate-type sites) or a COOH and an adjacent phenolic OH group (salicylate-type sites) (Boyd et al., 1981a; Prasad et al., 1987).

Phenolic and/or Alcoholic Hydroxyl Ligands. A decrease in intensity and/or a shift of the IR absorption near 3440 cm^{-1} is expected to occur where oxygen of various OH groups are involved in metal binding, alone or together with COOH groups. The net IR shift observed in the OH band, from 3500–3400 cm^{-1} in HA and FA, to 3300–3200 cm^{-1} in HA and FA complexes with Zn^{2+}, Fe^{3+}, Cu^{2+}, Pb^{2+}, and Mn^{2+} is ascribed to phenolic and/or alcoholic OH groups that bind the metal ion (Banerjee and Mukherjee, 1972; Tan, 1978; Piccolo and Stevenson, 1982). The extent of the OH stretching shift toward lower frequencies depends on the type of metal and follows the order: $Mn < Co < Cu < Fe$ (Banerjee and Mukherjee, 1972). Further, modifications of the typical band at 1070 cm^{-1} ascribed to polysaccharide components of FA suggest that OH groups of these structures also could be involved in metal complexation (Prasad and Sinha, 1983).

Conjugated Ketonic Complexing Sites. Although frequency shifts in the 1660 to 1600 cm^{-1} region are hardly identifiable due to the broad absorption, the shift to a lower frequency measured for the band at 1610 cm^{-1} in Cu^{2+}–FA complexes is attributed to the C=O group vibration in conjugated ketones weakened by resonance between C–O–Cu and C=O–Cu in the complexes (Piccolo and Stevenson, 1982).

Nitrogen and Sulphur Ligands. IR evidence also is provided for metal binding sites involving amide N (and possibly amide C=O) and sulfonic groups (SO_3H) in N- and S-rich HS (Sposito et al., 1976; Prasad et al., 1987).

Complexes of Hydroxylated and/or Hydrated Metals. The very sharp IR bands observed in the region 1130 to 1080 cm^{-1} and at between 890 and 697 cm^{-1} for HA and FA complexes with Cu^{2+}, Zn^{2+}, Al^{3+}, or Fe^{3+} are attributed to the metal–oxygen vibration of bound hydroxylated and/or hydrated metal ions (Juste and Delas, 1967, 1970; Tan, 1978; Piccolo and Stevenson, 1982).

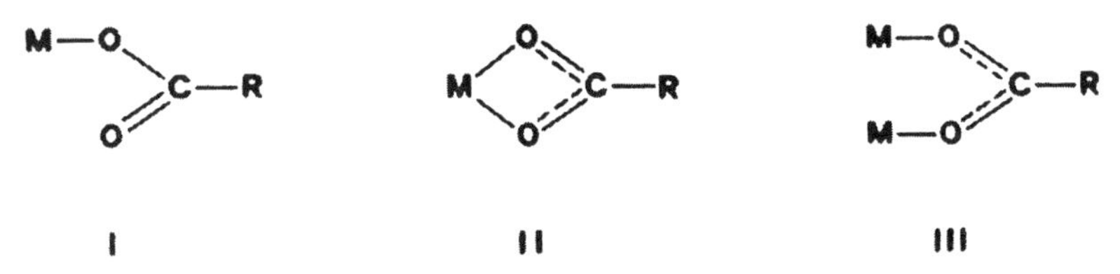

Fig. 12–5. Unidentate (I), bidentate (II), and bridging (III) metal–carboxylate complexes.

Conclusions

In conclusion, the complexity and heterogeneity of HS is still cause of ambiguity and uncertainty in the interpretation of IR spectra of their metal complexes. IR spectrometry can, however, provide useful information about the nature and reactivity of HS structural components that interact with metal ions and on the molecular arrangement of the binding sites in HS involved in metal complexation. The use of IR spectroscopy in conjunction with selective group blocking techniques and, especially, together with chemical derivatization methods has the potential for considerably enhancing the quality of IR spectra and facilitating their interpretation, thus providing IR data more informative on metal–HS complexation.

Application of IR spectroscopy to the study of HS–metal complexes in deuterated water solution and of FT-IR technique in common aqueous solution shows very promising in providing unique information on the interacting species that may be observed in the equilibrium state in low perturbed system and in close-to-environmental conditions. Further, the use of DRIFT spectroscopy applied to metal–HS complex powders will surely facilitate future studies of metal ion binding by HS.

Electron Spin Resonance Spectroscopy

Basic Principles and Methodology

The electron is a charged particle possessing an angular momentum and, in consequence, a magnetic moment that can be detected by its interaction with an applied magnetic field that produces the splitting of unpaired electrons possibly present in the system into two energy levels. This phenomenon is referred to as the Zeeman effect that is the basic physical phenomenon underlying ESR (or EPR) spectroscopy.

If an incident electromagnetic radiation is supplied to a sample containing unpaired electrons distributed into two energy levels in thermodynamic equilibrium in a magnetic field of value H_o, e.g., by applying an alternating magnetic field of frequency ν_o perpendicular to the static magnetic field H_o, absorption occurs provided that the energy of each incident quantum equals the difference in energy, ΔE, between the two electron states, that is

$$h\nu_o = \Delta E = g\beta H_o \qquad [9]$$

where h is the Planck's constant, g is the Landè or spectroscopic splitting factor (or magnetogyric ratio, i.e., the ratio of the magnetic moment to the angular momentum of the electron), and β is the Bohr magneton. This is known as the "resonance condition."

The ESR signal is highly dependent on the nature of the local environment about the absorbing electron, that is, the position of the ESR signal, and the overall ESR spectral pattern depend on the environment conditions in the vicinity of the electron. The most important types of interactions in the spin system that affect the position and pattern of the ESR spectrum are the "electron Zeeman," "nuclear hyperfine," and "ligand superhyperfine" interactions.

The "electron Zeeman" effect arises from the interaction of unpaired electrons with the external magnetic field and determines the position at which resonance occurs, i.e., the deviation of the g-factor from the free electron value ($g = 2.00232$). Species with "axial symmetry" such as Cu^{2+} and V^{4+}, i.e., with one principal axis of symmetry, conventionally the z-axis, and equivalent x- and y-axes, exhibit two g-values, usually labeled $g_{||}$ (= g_{zz}, i.e., the g-value along the z or symmetry axis) and $g_{\perp}$ (= $g_{xx} = g_{yy}$, i.e., the g-value perpendicular to the z-axis in the x–y plane). Anisotropy of the g-tensor is often averaged for paramagnetic species in solution by rapid rotation of the metal ion, and a single isotropic g-value is exhibited, g_{iso} or $g_o = 1/3(g_{xx} + g_{yy} + g_{zz})$.

A "nuclear hyperfine" interaction arises from the magnetic moments of the unpaired electron and its nucleus, if this is magnetic, i.e., it has a nonzero spin ($I \neq 0$), such as the nuclei of Cu ($I = 3/2$), Mn ($I = 5/2$), and V ($I = 7/2$). Nuclear spin causes a splitting of the ESR signal into $2I + 1$ components, i.e., four lines for Cu, six for Mn, and eight for V. The splitting of the hyperfine components is, in general, approximated by $A/g\beta$, where A is the magnitude of the nuclear hyperfine interaction, the so-called "hyperfine coupling constant," which, like g, exhibits an orientation dependence.

The "ligand superhyperfine" interaction can occur if the ligand atoms have nuclear spin, such as ^{14}N ($I = 1$). A number of components may thus result for each ligand nucleus, which leads to very complex ESR patterns, particularly if the ligands are not identical.

Three types of spectral parameters can thus be obtained from analysis of paramagnetic metal spectra: (i) the g-value(s) of the metal(s) present in the sample; (ii) the hyperfine coupling constant, A, if the metal nucleus has a nonzero spin and a nuclear hyperfine structure is apparent; and (iii) the ligand superhyperfine splitting, if the unpaired electron of the metal ion is delocalized by partial covalence onto magnetic nuclei of surrounding ligands and if the related ligand nuclear superhyperfine structure is observed. Rarely are resonance lines due to "forbidden transitions" observed, e.g., for Mn^{2+}.

In order to determine accurately and rigorously these parameters, the experimental ESR spectrum should be compared with a computer-simulated spectrum calculated using trial parameters, and a convenient mathematical representation and description of the ESR spectrum should be provided by use of the operator "spin Hamiltonian" (Wertz and Bolton, 1972).

In practice, the g-values and hyperfine and superhyperfine constants, A, can be obtained relatively simply, though not rigorously, by direct computation from data accurately derived from the experimental ESR spectrum, and from spectrometer setting values used in the measurement, according to the standard equations (adapted by Wertz and Bolton, 1972):

$$g = h\nu_0/\beta H_0 = 0.714484\ \nu_0/H_0 \qquad [10]$$

and

$$A\ (\mathrm{cm^{-1}}) = A(\mathrm{MHz})/c = 2.80247(ag)/(cg_e) = 0.469766 \times 10^{-4} ag \qquad [11]$$

where ν_0 (MHz) is the accurately measured value of the microwave frequency at the resonance condition, H_0 is the value of the magnetic field (in 10^{-4} T) at which the resonance is centered (on calibrated chart paper), g_e (= 2.00232) is the *g*-value for the spin of the free electron, *a* is the hyperfine splitting measured as the peak-to-peak separation (in 10^{-4} T) between the hyperfine lines in the experimental spectrum recorded on calibrated chart paper, and *c* is the speed of light in a vacuum. Accuracy of magnetic field calibration of chart paper should be checked by using a suitable standard.

More details on principles and practices of ESR spectroscopy and related techniques can be found in literature (e.g., Abragam and Bleaney, 1970; Wertz and Bolton, 1972; Kevan and Kispert, 1976; Kevan and Schwarz, 1979; Thomson, 1990). Reviews on application of ESR to HS and metal–HS studies were provided by Senesi (1990a, 1990b, 1992, 1996).

Metal–Humic Substance Complexes

The ESR technique has been extensively applied to elucidate the chemical and geometrical properties of naturally occurring and laboratory-prepared complexes formed by HA and FA of various origin and nature with paramagnetic transition metal ions of major chemical and biological importance to agriculture and environment, including Fe, Cu, Mn, V, and Mo (reviewed by Senesi, 1990b, 1992, 1996). The ESR analysis of metal–HS associations can provide useful and, in some cases, unique information about oxidation states of metals bound, symmetry and type of coordination sites in HS, binding mechanisms of metals to HS, and identity of ligand atoms and groups involved in metal complexing. The method of paramagnetic metal "probe" addition also has been widely applied to the study of the "residual" binding capacity of natural HA and FA, in order to ascertain additional molecular and quantitative aspects of metal complexation by HA and FA. These include the nature of binding sites involved in various experimental conditions, the degree of mobility of bound metals, and the stability of complexes formed towards competitive physical and chemical treatments, including proton and metal ion exchange (Senesi, 1990b, 1992, 1996).

Ferric Iron Complexes. The ESR spectra of soil HA and FA generally exhibit an asymmetrical, isotropic resonance line with a value of about $g = 4.2$ consistent with high spin (five unpaired *d*-electrons) Fe^{3+} ions held in tetrahedral or octahedral sites of low-symmetry (rhombic) ligand field (Fig. 12–6) (Senesi et al., 1977, 1986, 1989a, 1991; McBride, 1978; Schnitzer and Ghosh, 1982; Goodman and Cheshire, 1987). The ESR signal of these Fe^{3+} complexes does not specify the chemical nature of the ligands, although *g*-values near 4.2 are consistent with Fe^{3+} complexes with O functional groups, possibly carboxylic acids and/or polyphenols. This form of Fe exhibits considerable resistance to proton and metal exchange, and to chemical reduction, thus suggesting that Fe^{3+} is strongly bound and protected in inner-sphere complexes in HS (Senesi et al., 1977).

A very broad signal with a value of about $g = 2$ is often exhibited by soil HS (Fig. 12–6a) (Senesi et al., 1977; McBride, 1978; Goodman and Cheshire, 1987). Very probably, this signal consists of an envelope of several resonances arising from extended spin-spin coupling of various neighboring paramagnetic metal ions, most

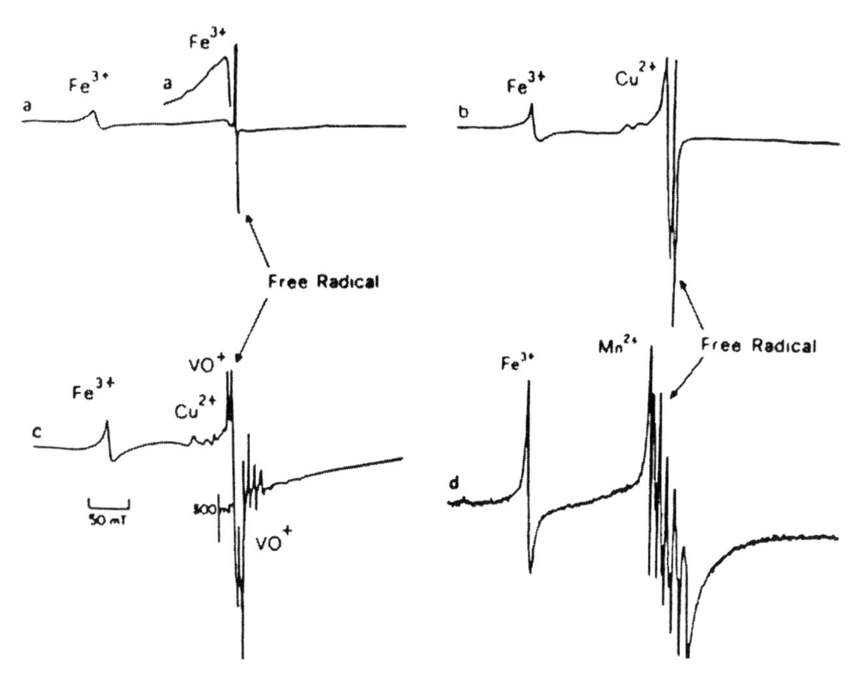

Fig. 12–6. Representative wide scan range (800 mT) ESR spectra at 77 K of a Mollisol humic acid (HA) from IHSS Reference and Standard collection (a; a′, higher gain), (b) Paleosol HA, (c) loam soil HA, and (d) decomposing leaf litter aqueous extract from a forest soil (from Senesi, 1996).

likely high-spin Fe^{3+} ions in octahedral sites with no or only small axial distortion from cubic symmetry. Iron in such sites is easily reduced by chemical agents and easily extracted by complexing agents, thus suggesting that it is weakly bound on external surfaces of HS (Senesi et al., 1977).

Two weak resonances at low-field positions (g-values of about 9 and 6) have been sometimes observed in ESR spectra of HS (Senesi et al., 1989a, 1991). The former resonance probably arises from Fe^{3+} in sites with near orthorhombic symmetry, and the latter from high-spin Fe^{3+} in largely distorted, axially symmetric crystal fields. No direct ESR evidence of Fe^{2+} species has been obtained in soil HS.

HS also possess a high residual binding capacity toward Fe^{3+} ion that can form complexes stable against various physical and chemical treatments. The intensity of the resonance at $g = 2$ relative to that at $g = 4.2$ increases with increasing Fe^{3+} addition to a soil FA, indicating that most of the added Fe is bound to surface octahedral sites (Senesi et al., 1977).

Divalent Copper Complexes. HS often exhibit an anisotropic, rigid-limit spectrum of the "axial" type in the region corresponding to about $g = 2$, with partially resolved $g_{||}$ and $g_{\perp}$ components of the g-value (Fig. 12–6b,c, and 12–7b,c), which is ascribed to the presence of Cu^{2+} ions (Goodman and Cheshire, 1976; Cheshire et al., 1977; Lakatos et al., 1977a; McBride, 1978; Boyd et al., 1981b; Schnitzer and Ghosh, 1982; Senesi and Sposito, 1984; Senesi et al., 1985, 1986, 1989b). Since the nuclear spin of both Cu isotopes, ^{63}Cu (69.2%) and ^{65}Cu (30.8%), has a value of $I = 3/2$, the ESR spectrum should be split into four (i.e., $2I + 1$) features both at $g_{||}$ and $g_{\perp}$; however, only the component at $g_{||}$ is generally resolved partially into a quadruplet, while the splitting of the $g_{\perp}$ component is generally too small to be resolved.

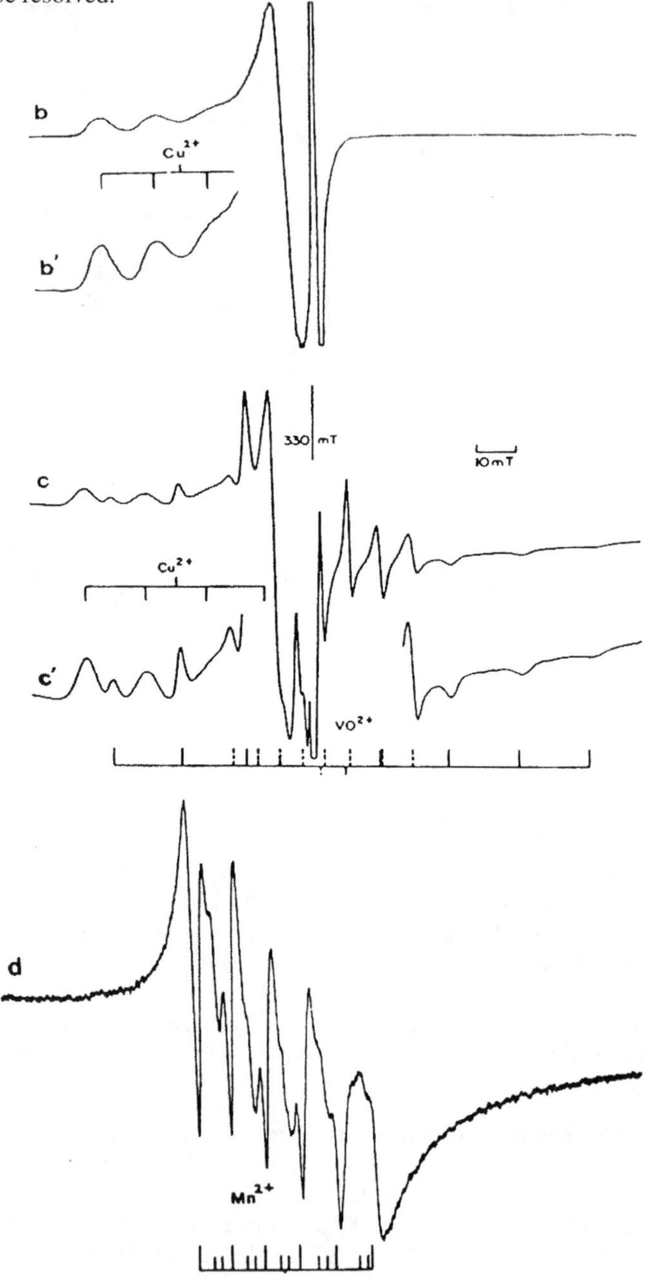

Fig. 12–7. Same ESR spectra as in Fig. 12–6, recorded on an enlarged scan range (200 mT) (from Senesi, 1996).

The ESR parameters of Cu^{2+}–HS complexes, either naturally occurring or obtained synthetically by Cu^{2+} ion-doping of HS samples, are in any case consistent with a d_{x2-y2} groundstate for Cu^{2+} held in inner-sphere complexes in the HS matrix, with ligands arranged in a square planar (distorted octahedral) coordination site (tetragonal symmetry). The experimental values of these parameters generally indicate binding sites for Cu^{2+} in HS involving either only O functional groups (carboxyls, phenolic hydroxyls, carbonyls and, often, water molecules), or both O and N ligands, or even only N atoms (i.e., a tetraporphyrin site) (McBride, 1978; Boyd et al., 1981b; Schnitzer and Ghosh, 1982; Senesi and Sposito, 1984; Senesi et al., 1985, 1986, 1989b, 1991). The participation of N in the binding of Cu^{2+} by HS also is supported by the resolved pattern observed in some cases at $g_{\perp}$, deriving from superhyperfine coupling of the Cu unpaired electron to N ligand nuclei ($I = 1$) (Goodman and Cheshire, 1976).

Measured ESR spectral parameters also provide evidence of a high covalent bond contribution (i.e., delocalization of the unpaired electron toward the ligands) for Cu^{2+} in HS. The combined effects of the crystal field splittings and covalency lead to the following empirical order of decreasing $g_{\parallel}$ and $g_{\perp}$ factors for different donor atoms: O > N > S. Similarly, the values of the hyperfine coupling constants, $A_{\parallel}$ and $A_{\perp}$, that are related directly to the degree of covalency and g-factors, increase and, obviously, their absolute values, $|A_{\parallel}|$ and $|A_{\perp}|$, decrease with increasing electronegativity of the ligand atoms (O < N < S) (Jameson, 1981); however, small deviations from square planar coordination will shift $A_{\parallel}$ to smaller values, while larger deviations will shift $g_{\parallel}$ to smaller values (closer to $g = 2$) (Peisach and Blumberg, 1974).

A good correlation exists between the $g_{\parallel}$ and $A_{\parallel}$ values and either the Cu^{2+} loading or the coordination environment of model and natural organic Cu complexes (McBride, 1989). There is evidence that covalent bonding, as indicated by low g-values and large hyperfine constants, is favored by low Cu^{2+} loading in HS, where complexation to amine-type N groups is preferred to O-containing ligands. In contrast, ESR parameters measured for Cu^{2+} at high loadings in HS indicate not very covalent, although rigid binding of Cu^{2+}, largely to O-containing ligands, probably as inner-sphere complexes with a higher degree of mobility of Cu^{2+}.

An accurate analysis of the Cu^{2+} ESR pattern at $g_{\parallel}$, possibly with the aid of a computer simulated spectrum, reveals that in some cases it consists of two (or more) superimposed quadruplets, often very difficult to be differentiated, each component arising from a different type of local environment for Cu^{2+}, i.e., from different classes of binding sites for Cu^{2+} in HS (Senesi, 1990b, 1992, 1996).

Large ESR evidence has been obtained of the capacity of Cu^{2+} ions to replace Mn^{2+}, VO^{2+}, and Fe^{3+} from FA and HA when these materials are treated with excess Cu^{2+} ions, and to form complexes of various stability toward protons or competing cations (Senesi et al., 1977, 1985, 1986, 1991).

Vanadyl Ion Complexes. More complicated, but relatively well-resolved spectra exhibiting a richly structured pattern with a value of about $g = 2$ are observed for some soil HA or FA (Fig. 12–6c) and/or their fractions. Analysis made on an enlarged spectrum of the region (Fig. 12–7c) indicates the presence of two distinct, overlapping rigid-limit spectra of the "axial" type. One comprises the typical anisotropic pattern of complexed Cu^{2+} previously discussed, the other consists of the partial superimposition of two hyperfine octuplets, corresponding to the parallel and perpendicular components of V^{4+} (nuclear spin, $I = 7/2$) (Goodman and Cheshire, 1975; Cheshire et al., 1977; Lakatos et al., 1977a; McBride, 1978; Senesi et al., 1989b).

The complexity of the spectrum may require computer simulation for the accurate determination of the ESR parameters, which are consistent with a VO^{2+} ion rigidly bound in HS as an inner-sphere complex in a square planar coordination site. Vanadyl complexes naturally occurring in soil HS are characterized by relatively strong ligand fields and high covalency, being consistent with oxygen ligands, mostly phenolate or possibly water molecules, and, more rarely, N ligands. This renders VO^{2+} ions difficult to remove from these sites, even with acid leaching. In contrast, complexes obtained by VO^{2+}-doping of HS involve weaker ligand fields and lower covalency indicating that VO^{2+} is primarily bound to surface carboxylate groups while it remains partially hydrated, thus resulting in relatively labile and exchangeable forms (McBride, 1978; Templeton and Chasteen, 1980; Senesi et al., 1986, 1989b, 1991).

ESR spectroscopy has been applied to study the dynamics of motion, molecular conformation, aggregation properties, stability constants, and stoichiometries in aqueous solution of VO^{2+} complexes with two gel-filtrated FA fractions of different molecular weight (MW) (FA-I and FA-II) (Templeton and Chasteen, 1980). ESR results indicate that both FA fractions involve similar binding sites of low symmetry consisting of four O ligands bound in the first coordination sphere (inner-sphere) of VO^{2+} ions. The higher MW fraction forms a complex approximated as $(VO)_2$ $(FA\text{-}I)_6$, whereas the lower MW fraction forms a simpler complex, VO-FA-II. Comparative evaluation of ESR parameters suggests that the ligand fields existing about the metal can be modeled by the complexes bis(phtalato)(salicylato)–oxovanadium (IV) and mono(salicylato)–oxovanadium (IV), respectively (Templeton and Chasteen, 1980).

Divalent Manganese Complexes. The ESR spectra of some soil HA (Fig. 12–6d, and 12–7d) and fractionated soil FA feature a well-resolved isotropic pattern with a *g*-value around 2 consisting of six almost equally spaced principal lines and, possibly, 10 secondary lines (corresponding to forbidden transitions) of lesser intensity. The ESR parameters of such spectra are consistent with high-spin hexahydrated Mn^{2+} ($I = 5/2$) bound in outer-sphere complexes by electrostatic forces to six O atoms of negatively charged carboxylate and phenolate groups in a distorted octahedral environment (Gamble et al., 1977; Cheshire et al., 1977; McBride, 1978; Senesi et al., 1991).

At low pH most Mn^{2+} is adsorbed by a soil FA as outer-sphere complexes, but at pH > 8, or $T > 50°C$, Mn^{2+} can enter inner-sphere multiligand complexation sites (Mc Bride, 1982). These results indicate that the type and stability of Mn^{2+}–HS complexes and, in turn, its ease of exchangeability and bioavailability in natural systems are strongly dependent on pH and temperature.

Further, HS isolated from various sources exhibit a high residual complexing capacity for added Mn^{2+} that can be bound in water-stable forms, but, unlike

Fe^{3+} and Cu^{2+}, it may be completely displaced by protons or strongly complexed metal ions (Senesi et al., 1991).

Weighted average equilibrium constants of water soluble Mn^{2+}–FA complexes were determined by ESR spectroscopy on the basis of the linear functionality existing between the height of hyperfine peaks of Mn^{2+} and the concentration of free Mn^{2+} (Gamble et al., 1977). The increasing relaxation line broadening observed in the ESR spectrum of Mn^{2+} with increasing addition of complexing ligand could be directly related to increasing complexation of Mn^{2+} by FA. The K_c values measured by ESR are in excellent agreement with K_c values determined by an ion-exchange method, but the ESR method is faster, more sensitive, and more convenient than the ion-exchange procedure.

Molybdenum Complexes. ESR evidence is obtained suggesting that HA can reduce molybdate to Mo(V) and complex the Mo(V) species in strongly acidic media (Lakatos et al., 1977a; Goodman and Cheshire, 1982). The ESR spectrum of a peat soil HA complex with Mo(V) enriched in ^{95}Mo (nuclear spin, $I = 5/2$) features two distinct components, each split into two six-line hyperfine patterns at $g_{\parallel}$ and $g_{\perp}$ consistent with two different axially symmetric Mo(V)–HA complexes.

Treatment with 0.1 M HCl of the ^{95}Mo(V)-enriched-HA produces a low-intensity six-line ESR spectrum probably arising from a Mo(III) species (Goodman and Cheshire, 1982). This result suggests that Mo(III) species can be formed and remain stable in the solid state even in aerobic conditions when protected in HS complexes.

Conclusions

Major advantages of ESR spectroscopy are its high sensitivity and the ability to measure spectra directly with minimal or no sample pretreatment. ESR provides evidence that small quantities of metal ions can bond selectively in inner-sphere complexes at the most preferred sites for the metal, whereas in the presence of high amounts of metal added the high degree of site occupation generally results in a loss of relative selectivity. ESR analysis also show that high pH values that generate a greater availability of negatively charged O ligands favors inner-sphere complexation for metals that are retained as hydrated ions at lower pH. Inner-sphere coordination also is preferred when competing water ligands are removed by dehydration, thus forcing the metal to enter into direct bonding with HS ligands. ESR data confirm that the more electronegative the metal ion, the stronger the metal bound to HS, the higher the degree of bond covalency. The ESR approach also allows, in principle, the determination of the free ion concentration and, therefore, of the degree of complexation of the metal ion.

The intrinsic limitation to the ESR technique is its applicability only to paramagnetic metal ions that give a detectable ESR signal such as Cu, Fe, Mn, V, and Mo. The major limitation of the ESR experiment is the inability to resolve signal component lines that may overlap to such an extent to result in merging of individual resonant lines or spin packets into a single overall broad line or envelope with a loss of information. Line broadening is determined by either an homogeneous or inhomogeneous mechanism. The first effect arises from microwave-power saturation that produces broad spectra for some metal ions such as Fe^{3+}, thus the choice of power is critical to avoid saturation. The second effect is caused by interactions with neighboring paramagnetic species of the same or different type, or neighboring nuclei. Temperature also is a critical parameter in the ESR experiment. Since sensitivity for paramagnetic species increases with lowering sample temperature, according to Curie's law, ESR measurements are often made at either liquid N (77 K) or liquid He (4.2 K) temperature, which may reduce some type of broadening.

ESR related spectroscopies that hold the potential to overcome some resolution limitations and yield more information than the classical ESR approach about the chemical environment of paramagnetic metal ions are the electron-nuclear double resonance (ENDOR) (Kevan and Kispert, 1976) and electron-spin echo envelope modulation (ESEEM) (Kevan and Schwartz, 1979) spectroscopies. Either ENDOR or ESEEM represent by principle a useful tool in extending the resolution of the ESR experiment; however, the sensitivity of ENDOR and ESEEM is much lower than that of ESR, and interpretation of ENDOR and ESEEM spectra is not a simple matter, especially if ligands are not well characterized, as it is the case for HS. Both ENDOR and ESEEM techniques have not yet been applied to strictly metal–HS complexes, but the sensitivity and ease of carrying out experiments are improving rapidly so that a major scientific activity may be expected to occur in this area of ESR spectroscopy.

Mössbauer Spectroscopy

Basic Principles

Mössbauer spectroscopy measures the resonant absorption of nuclear gamma rays involved with transitions between the ground and excited state of atomic nuclei with nonzero angular momenta. The precise energy of such transitions is influenced by the chemical environment of the nuclei and any external magnetic field and electric field gradient. The Mössbauer effect is highly isotopic-specific, and only approximately 30 isotopes are Mössbauer-active, including iron, nickel, zinc, and mercury, but the most easily studied metal is iron (Gutlich et al., 1978).

The principal energy-dependent parameters that can be obtained from a Mössbauer spectrum are the "isomer shift," the "quadrupole coupling constants," the "magnetic field," and the "peak width." The magnitude of the "isomer shift" is proportional to the difference in electron density at the metal ion nucleus in the sample and in reference, usually the metallic form. Factors influencing the isomer shift include also the total population of 3d orbitals. The "quadrupole splitting," or magnetic hyperfine interaction, is determined by the interaction between the quadrupole moment of the excited state and the electric field gradient at the metal nucleus. The latter mostly derives by a combination of charges originating from the electronic environment (valence electrons) of the metal ion and from surrounding atoms. In the case of iron, this effect gives rise to a doublet in the Mössbauer spectrum. The "magnetic hyperfine field" is proportional to any magnetic field experienced by the nucleus. In the case of iron, the interaction between the nuclear magnetic dipole moment and any magnetic field at the nucleus removes completely the degeneracy of nuclear energy levels, and produces six peaks in the spectrum with intensity ratios 3:2:1:1:2:3, if the magnetic domains are randomly oriented. Finally, the value of the "peak width" can provide useful information on the possible presence of unresolved components in the Mössbauer spectrum.

High-spin ferric ($3d^5$ configuration) and ferrous ($3d^6$ configuration) ions can be successfully studied by Mössbauer spectroscopy. The high-spin Fe^{3+} ion has a spherical and symmetrical electronic charge distribution in either octahedral or tetrahedral coordination that results in a zero contribution from valence electrons to the quadrupole splitting; however, an electric field gradient can arise, for example, in the presence of nonidentical ligands that distort the cubic symmetry. The additional electron present in the d orbitals of the high-spin Fe^{2+} ion determines an asymmetrical electron distribution due to the distortion of the first coordination sphere of ligands holding the ferrous ion in octahedral or tetrahedral arrangements in an ionic complex. This effect generally results in much larger values of both the quadrupole splitting and isomer shift for the ferrous iron, with respect to the ferric iron. Low spin states of both ferrous and ferric ions are much more difficult to be studied by Mössbauer spectroscopy; however, little evidence exists of the occurrence in nature of low-spin forms of iron ions.

The experimental values of Mössbauer parameters may provide information on the type of coordination, symmetry, and chemical nature of the groups bound to the high-spin iron ions. More information on general Mössbauer theory and methodology can be found in Gutlich et al. (1978), whereas a review on its application to Fe–HS complexes was provided by Senesi (1992).

Humic Substance Complexes with Iron Ions

All Mössbauer investigations so far reported on metal–HS complexes have been with the isotope ^{57}Fe (reviewed by Senesi, 1992). The Mössbauer spectrum obtained on a natural soil HA containing 0.57% Fe could be computer-fitted to three doublets exhibiting parameters suggesting the presence of three sites for Fe^{3+} ions in the HA, two with octahedral and one with tetrahedral coordination (Fig. 12–8a) (Senesi et al., 1977). Chemical reduction with hydrazine of the HA sample results in a Mössbauer spectrum in which one of the two doublets assigned to Fe^{3+} ions in octahedral sites disappears (3a in Fig. 12–8a), and a new doublet appears featuring much higher values of the isomer shift and quadrupole splitting, typical of Fe^{2+} ions in the same type of sites (3b in Fig. 12–8b) (Senesi et al., 1977).

Results of an extended Mössbauer study of several Fe^{2+}–HA complexes suggest the presence of partially hydrated high spin Fe^{2+} ions held in elongated octahedral arrangements in ionic, inner-sphere chelate-type complexes with carboxylate and phenolic and alcoholic hydroxyl groups and N-containing groups (Lakatos et al., 1977b); however, Mössbauer evidence of partial oxidation of Fe^{2+} to Fe^{3+} is obtained on exposure to air of some Fe^{2+}–HA complexes. On the contrary, Mössbauer evidence is obtained that about one half of the iron in a $^{57}Fe^{3+}$–HA slurry at pH 1 is present in two components with quadrupole splitting values consistent with Fe^{2+} ions partly in completely hydrated, outer-sphere complexes with HA, and partly associated directly with O ligands of HA in inner-sphere complexes (Goodman and Cheshire, 1979).

In other Mössbauer studies conducted on various naturally occurring and laboratory-prepared Fe^{3+}–HA and Fe^{3+}–FA complexes at various pH values and recording temperatures, one or two Fe^{3+} doublets, a weak Fe^{2+} doublet, and one or two magnetically split sextets are obtained (reviewed in Senesi, 1992). These results suggest the occurrence of iron in these materials partly as ferric and ferrous HA and FA complexes, and partly as inorganic species.

Conclusive Comments

The Mössbauer effect has been used successfully for studying iron complexes with HS, commonly in conjunction with other spectroscopic techniques, particu-

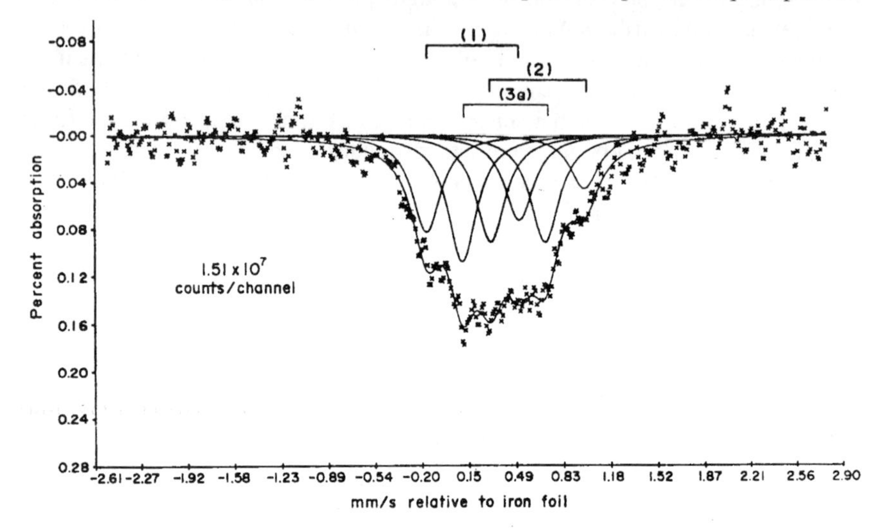

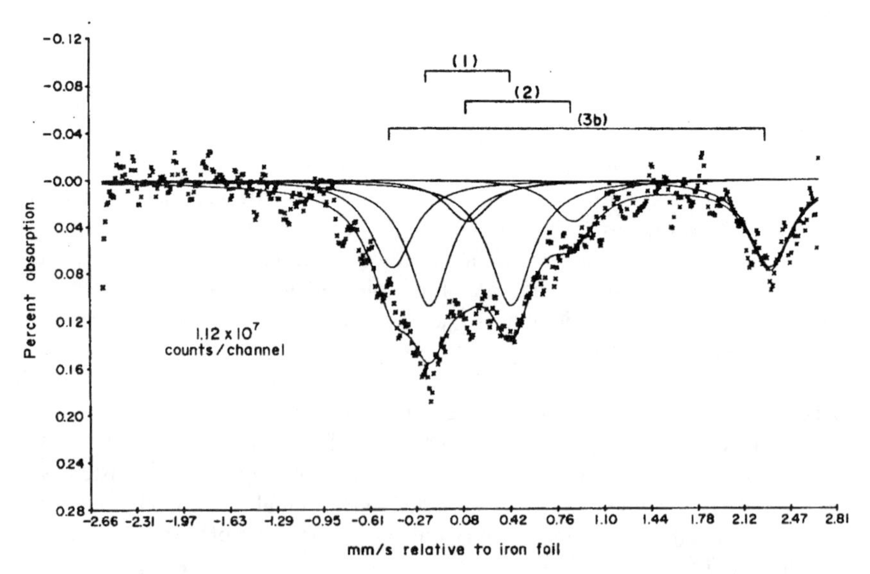

Fig. 12–8. Mössbauer spectra of (a) natural soil humic acid (HA) and (b) the same HA after reduction with hydrazine. Doublets (1), (2), (3) refer to three different sites for iron in HA. Doublets (3a) and (3b) refer, respectively, to Fe^{3+} and Fe^{2+} ions in similar octahedral sites (from Senesi et al., 1977).

larly ESR spectroscopy. Much Mössbauer evidence suggest that both ferrous and ferric ions occur mostly in high-spin forms in combination with HS. On these bases, Mössbauer spectroscopy is able to distinguish unambiguously between high-spin Fe^{3+} and Fe^{2+} ions on the basis of major differences observed in both the isomer shift and quadrupole splitting of these ions.

The intrinsic limitation of the technique is its high specificity for the ^{57}Fe isotope. Although the low natural abundance of this iron isotope (2.19%), it is possible to obtain reasonable Mössbauer spectra also with unenriched, natural Fe–HS complexes. Caution must be exercised, however, in interpretation of Mössbauer data, and model compounds should be investigated for comparison, as several binding sites exist that may give rise to similar spectra.

Nuclear Magnetic Resonance

Basic Principles

Nuclear magnetic resonance (NMR) spectroscopy has been applied to elucidate metal binding mechanisms to organic ligands mainly by two approaches, i.e., by measuring the effects of metal complexation either (i) on the relaxation times of ^1H of water molecules solvating the metal cation; or (ii) on the chemical shifts of NMR-active metal ions (e.g., Cd, Al, and Pb). A wide range of books and reviews have been published on NMR theory and methodology including some that are more related to NMR-active metals (e.g., Connors, 1987; Wilson, 1989; Macomber, 1998).

Solvation Water (^{1}H) Relaxation. In NMR experiments, relaxation is the process by which nuclear spin states subjected to an external magnetic field establish equilibrium. The relaxation times of ^1H in water molecules coordinated to a paramagnetic ion are altered relative to those of water molecules in the bulk solution according to the following relation of proportionality (Deczky and Langford, 1978):

$$(1/T_{1,2(obs)} - 1/T_{1,2A}) \propto qC_M/T_{1,2M} \qquad [12]$$

where $T_{1,2(obs)}$ is the observed relaxation time for either longitudinal ($T_{1\,(obs)}$, spin-lattice), or transverse ($T_{2\,(obs)}$, spin-spin) relaxation; $T_{1,2A}$ refers to relaxation times of protons in the bulk solvent, and $T_{1,2M}$ to protons in the solvation shell of the paramagnetic ion; C_M is the concentration of paramagnetic ions; and q is the number of water molecules coordinated to the paramagnetic ion.

According to Eq. [12], the coordination of the metal ion to a nonsolvent ligand may thus lead to either a decrease or an increase of observed relaxation rates ($1/T_{1,2(obs)}$). A decrease may result from a reduction of q due to replacement of coordinated water by the ligand and/or an increase of electronic relaxation times ($T_{1,2M}$), whereas an increase of $1/T_{1,2\,(obs)}$ may be due to a reduction of $T_{1,2M}$. Since the width at half height of the NMR peak of water is proportional to the value of $1/T_{1,2(obs)}$ addition of an organic ligand to a water solution of a paramagnetic metal ion may produce either line narrowing, which corresponds to increased $T_{1,2\,(obs)}$, or line broadening, corresponding to a reduction of $T_{1,2(obs)}$.

NMR-Active Metal Ions. Chemical shifts of NMR-active metal ions, such as Cd, Al, and Pb, are affected by their local environments, and can thus provide useful information on the metal coordination number and type and number and geometric arrangement of ligands complexing the metal, also as a function of metal/ligand molar ratios, pH, and other system properties.

Since metal chemical shifts are determined by the interaction between the electron cloud surrounding each nucleus and the external magnetic field, nuclei shielded by electron clouds will experience an effective field of a magnitude lower than that of deshielded nuclei (Macomber, 1998). When metal nuclei bind to ligand groups containing electronegative O, N, and S atoms, electrons are withdrawn, that is, nuclei are deshielded, and the resonance frequency, or chemical shift, increases consequently.

Stability Constants of Metal Complexes. The ability of NMR to discriminate between free metal ions and their complexes with organic ligands renders this technique a potentially powerful analytical tool for measuring metal complexation. In the NMR experiment, a nucleus that can partition between two magnetically nonequivalent sites is said to undergo "chemical exchange" between the sites. The features of the observable resonance signal thus depends on the rate of the exchange process (Connors, 1987).

For example, in the simple case of 1:1 complex formation, several possibilities of chemical exchange may exist (Connors, 1987). The first possibility is "very slow" exchange, where the lifetimes of the nuclear states in the free ligand site, S, and in the metal–ligand site, S–M, are very long. In this case resonance peaks, ν_s and $\nu_{S\text{-}M}$, appear at both resonance frequencies of the target nucleus in sites S and S–M, respectively, and peak areas are proportional to the fractional occupancy of the sites. The second possibility is "moderately slow" exchange, where resonances appear at (or near) ν_s and $\nu_{S\text{-}M}$ but the peaks are broader than in the very slow exchange. Third, "very fast" exchange can occur relative to the NMR time scale, so that the nucleus appears to be essentially stationary. In this case only a single resonance peak will be observed at ν_0, which is the weighted average of the site frequencies ν_s and $\nu_{S\text{-}M}$.

Slow exchange possibly involves strongly bound complexes, and fast exchange generally refers to weak binding. For a system in slow exchange conditions, the ratio of NMR spectral peak areas of nuclei in S and S–M sites is calculated, and the stability constant of the complex can be obtained knowing the concentration of M (Connors, 1987).

Metal–Humic Substance Complexes

Relatively few and sparse studies have been performed by NMR on metal–HS complexes. A comprehensive and updated review has been recently provided by Kingery et al. (2001) on the various applications of NMR spectroscopy to the study of metal–HS interactions.

Solvation Water (^{1}H) Relaxation. Only two studies are available in the literature on the effects of HS complexation on relaxation times of ^{1}H in the solution shells of paramagnetic ions. A slight but definite decrease in line widths of the water NMR signal is observed upon addition of a soil FA to a Mn^{2+} water solution (Gamble et al., 1976). Comparison made with NMR data of Mn^{2+} complexes with model simple ligands is in favor of outer-sphere binding of hexahydrated Mn^{2+} ion by FA in a slightly distorted octahedral symmetry. In a later more detailed study, the outer-sphere complexation of hexahydrated Mn^{2+} by FA is confirmed by ascribing the ^{1}H relaxation mechanism mainly to distortion of the octahedral hydrated-metal symmetry by collisions with water molecules outside the complex (Deczky and Langford, 1978).

A different behavior is suggested by NMR data for Fe^{3+} complexation by the same soil FA (Gamble et al., 1976). The great reduction of the line width of water NMR signal measured upon addition of FA to Fe^{3+} solution is more likely attributed both to a profound change of the q parameter in Eq. [12], which implies inner-sphere coordination of Fe^{3+}, and to a substantial alteration of electron relaxation time by the FA ligands, consequent to reduction of symmetry of the high-spin d^5-center.

In the presence of FA, a marked increase is observed of the longitudinal relaxation rate dependence on Cu^{2+} ion concentration, in comparison to Cu^{2+}-aquo ion and Cu^{2+}-bipyridine complex (Deczky and Langford, 1978). This result is consistent with an inner-sphere complex of Cu^{2+} with FA.

Active-Metals-NMR. Cadmium-113 NMR offers a useful means for characterizing Cd^{2+}–HS complexes. Evidence by ^{113}Cd-NMR is provided of two kinds of Cd^{2+}–HA complexes involving either 1 or 10 Cd ions per HA molecule (Pommery et al., 1988). The involvement of N-containing groups, as well as of O-containing groups, is suggested by ^{113}Cd-NMR in Cd complexation by a soil FA in acid conditions (Chung et al., 1996).

Most recent studies on ^{113}Cd –NMR of Cd^{2+}–HS complexes have been performed on aquatic HS. A comparative ^{113}Cd-NMR study of Cd^{2+} complexation to the Suwannee river FA and to ethylendiaminotetracetic acid (EDTA) at various FA/Cd ratios and in acidic conditions is indicative of predominant O-functionalities of FA in binding Cd (Boersma et al., 1997). On the contrary, experiments conducted with Suwannee river FA at basic pH values provide strong evidence of Cd^{2+} binding by N atoms, but not by S atoms (Li et al., 1998), although binding of Cd^{2+} by S ligands is suggested for a commercial HA.

The effects of HS concentration and pH on Al^{3+} complexation by a swamp water HS were studied by ^{27}Al-NMR spectroscopy (Howe et al., 1997). At low HS concentration, few water molecules surrounding the metal ion are replaced by HS ligands to form mixed Al–H_2O–HS complexes. Increasing HS concentration determines the involvement of more HS ligands in Al^{3+} complexation, leading to a complete displacement of water molecules by HS ligands at very high HS concentrations. Increasing the pH of the system increases proton dissociation from HS, and thus favors the formation of Al^{3+}–HS complexes.

Binding and reduction of V (V) to the paramagnetic V (IV) ions (VO^{2+}) upon addition of a swamp water HS at acidic pH were studied by ^{51}V-NMR spectroscopy in conjunction with ESR spectroscopy (Lu et al., 1998). The V (V) NMR signal quickly disappears after the addition of HS, and the formation of V (IV) is confirmed by ESR spectroscopy, thus implying the direct involvement of HS in the reduction of V (V) and consequent complexation of the produced VO^{2+} ion.

Structure-Binding Relationships. Few NMR experiments have been conducted to relate structure of HS to their metal-binding capacity. A positive correlation is reported between aromaticity of several HAs determined by solution-state ^{13}C-NMR and stability constants of their complexes with Cd^{2+} and Zn^{2+} ions (Ashley, 1996). Even though the carboxylic contents are similar, HAs with higher aromaticities have higher stability constants than those with lower aromaticity, thus suggesting the involvement of phenolic groups in the complexation of these metals.

Analysis of Distortion Enhancement by Polarization Transfer (DEPT)-^{13}C-NMR and Quaternary-C only (QUAT) subspectra of a soil HA and a soil FA suggest that carboxylate groups, of which the FA is richer than the HA, play a prominent role in the binding of Eu^{3+} to HS (Shin et al., 1996). This result is consistent with the higher binding constant measured for Eu^{3+} with FA than with HA that is richer in aromatic carbon and aliphatic methine and quaternary carbons than FA.

Dipolar dephasing CPMAS ^{13}C-NMR has been applied to relate the structure of a soil HA and a sediment HA to the stability constant of Cd^{2+}–HA complexes (Sohn and Rajski, 1990). The stability constant of the Cd^{2+}-sediment HA is greater than that of Cd^{2+}-soil HA, probably due to the higher content and stronger binding of amino acid N groups, in addition to carboxylic groups, of the former HA with respect to the latter HA. Further, greater metal stability constants of Zn^{2+}–HA and Cd^{2+}–HA complexes are found to be associated with a higher degree of substituted aromatic C groups, possibly due to chelation effects (Cameron and Sohn, 1992).

Metal–Humic Stability Constants. Only two studies concerning aquatic HS are available in the literature on the use of NMR to obtain the stability constants of metal–HS complexes. Conditional stability constants for Al^{3+} complexation with groundwater HS in a range of HS concentrations were determined by using ^{27}Al-NMR (Lambert et al., 1995). Binding of Cd^{2+} to Suwannee river FA was studied using ^{113}Cd-NMR (Larive et al., 1996). Evidence of fast exchange conditions is obtained in both cases, which would involve weak metal-HS binding.

Conclusions

To date, few, if any, of the NMR studies on metal–HS complexation have attempted an examination of metal-N spin couplings, due primarily to the low sensitivity and low natural abundance of ^{15}N nuclei; however, with high-resolution instrumentation, natural abundance ^{15}N-NMR may provide an additional useful technique for examining metal–HS complexation.

Two-dimensional NMR methods that yield a two-dimensional frequency spectrum have not yet been attempted successfully to study metal interactions with HS, although instances of the successful applications of these approaches can be found in the study of metalloproteins (Kingery et al., 2001). Mononuclear (^{1}H) two-dimensional NMR experiments, such as Total Correlation Spectroscopy (TOCSY), can show ^{1}H–^{1}H coupling throughout the complete spin system, and exchange protons can provide information on sites to which metal attach. For example, N-containing units in HS that bind the metal can be identified since the amido protons from these structure will exchange and disappear from the spectrum.

Through-space Nuclear Overhauser Effect (NOE) can provide information on the sites where metals interact even if the metals do not form stable bonds through which spin coupling can be transferred. For instance, metal–HS interactions can be studied by NOE spectroscopy (NOESY) by measuring interactions between the protons within the HS molecules before and after the addition of metals to understand the conformational changes occurred within the molecules (Kingery et al., 2001). An alternative approach is to directly measure the Heteronuclear Overhauser Effect (HOE) between the metal ion and HS proton in close proximity by HOE spectroscopy (HOES), as demonstrated for organo-Li complexes (Bauer, 1995).

The inverse detection Heteronuclear Multiple Quantum Coherence (HMQC) experiment is another approach of two-dimensional-NMR techniques, which consists in a transfer of chemical shift and coupling information from relatively insensitive nuclei, as ^{13}C and some metals, to more sensitive nuclei such as ^{1}H. The advantage of this method is a substantial increase in the sensitivity obtained, due to the greater natural abundance of ^{1}H (Kingery et al., 2001).

In conclusion, the powerful two-dimensional-NMR techniques are expected to evolve in very promising tools for the study of both HS structures and their interactions with metals.

Synchroton-Based X-Ray Absorption Spectroscopy

Basic Theory and Methodology

The intense x-ray beams produced by synchroton sources has allowed for the development of nondestructive x-ray absorption spectroscopy (XAS) techniques that can provide information on oxidation states and local chemical environments of elements in a variety of earth and environmental materials (Fendorf and Sparks, 1996; Fendorf, 1999). The high intensity x-rays required for XAS are produced by electrons/positrons moving in a vacuum ring, the synchroton, at velocities approaching the speed of light, i.e., at relativistic energies (1 to 6 GeV), in paths curved by a magnetic field. At a synchroton experimental station, a collimated portion of the x-ray beam radiated by these charged particles tangentially to their curved path will enter a series of chambers in their way to the sample and detector.

Monochromatized x-rays, falling on the sample may or may not be absorbed depending on the elements in the sample and the energy range scanned by the monochrometer. Most elements are XAS-active and can absorb monochromatized x-rays specifically. Thus, XAS consists of recording the absorption by a sample of x-rays as a function of the wavelength. The spectral scan is performed in the vicinity of an x-ray absorption edge (K, L, or M) of the chosen target element. Allowed electron transitions for XAS are: at the K-edge, 1s $\rightarrow$ np; at the L-edge, 2p $\rightarrow$ nd; and at the M-edge, 3d $\rightarrow$ nf. Thus, spectral features of different elements possibly present in the sample do not overlap since K edges are separated by several hundred eV, and the actual position of the absorption edge depends first and foremost on the element absorbing the x-rays and second on its oxidation state. The XAS technique can therefore be used to investigate complex materials by successively tuning the absorption edge of each spectroscopically active element present in the sample. Theory of XAS is discussed in detail in several books (e.g., Teo, 1986; Lytle, 1988; Koningsberger and Prins, 1988).

By convention the XAS technique is distinguished into two methods, the x-ray absorption near-edge structure (XANES) spectroscopy and the extended x-ray absorption fine structure (EXAFS) spectroscopy. The XANES analysis is performed in the range from about 10 eV below to about 60 eV above the absorption edge. The XANES region of a sample is isolated from each merged scan, the pre-edge background is then subtracted, the absorption coefficient is normalized relative to the intensity of the "white-line" peak in order to plot all the spectra on the same scale, and the first derivative of the spectrum is finally calculated. The XANES spectra can provide qualitative information about the oxidation state of an excited atom, its coordination geometry, and its bonding environment through comparison with model compounds (e.g., Bianconi, 1988; Fendorf, 1999).

The EXAFS analyzes the measured oscillatory structure that appears at higher energies, from about 50 to about 1000 eV above the absorption edge, and can reveal the local atomic environment surrounding the excited atom (Fendorf, 1999; Bloom et al., 2001). In particular, the structural information provided by EXAFS includes average interatomic distances and number and chemical identities of the atoms within a 5 Å radius of the atom that absorbs the x-ray photon. A number of steps are generally required to process and analyze an EXAFS spectrum. Briefly, these consist in: (i) merging of scans; (ii) isolation of the fine-structure scattering curve, $\chi(E)$, from the absorption edge by background removal; (iii) conversion from a photon energy (eV) scale, $\chi(E)$, to a photoelectron kinetic energy or wavevector ($Å^{-1}$) scale, k; (iv) isolation of the scattering curve, $\chi(k)$, which requires a number of operations that can be accomplished by several computer software packages; (v) Fourier transformation of the weighted scattering curve, $k^3\,\chi(k)$, to obtain the experimental "radial structure function" (RSF); and, finally, (vi) comparison through a curve-fitting process by an adequate EXAFS analysis program, of the RSF and Fourier-filtered scattering curves to computed spectra derived from model structures. A common and highly regarded computer software package used for computing theoretical scattering curves, χ (k), and for RSF calculation for the model structure is FEFF (Rehr, 1993; Rehr et al., 1994; Zabinsky et al., 1994).

Metal–Humic Substance Complexes

Only recently XANES and EXAFS techniques have been applied to obtain more detailed and additional information on the chemical structure of the binding sites in soil metal–HS complexes, and a comprehensive review on this subject has been provided very recently (Bloom et al., 2001).

Copper Complexes. Binding of Cu^{2+} by soil HS at various pHs, and soil and peat HAs has been studied in detail by XANES and EXAFS spectroscopies by several authors (Hersterberg et al., 1997; Xia et al., 1997a,b; Davies et al., 1997). The similar pattern obtained for K-edge XANES spectra of Cu^{2+}–HS complexes, and their first derivatives, at pH 4, 5, and 6 (Fig. 12–9) suggest that Cu^{2+} binding sites in soil HS are similar at these pH values (Xia et al., 1997a). The small pre-edge bump for the 1s $\rightarrow$ 3d transition and the splitting in α and β peaks for the 1s $\rightarrow$ 4p transition, which are apparent especially in the first derivative XANES spectra of Cu^{2+} in HS at all three pH values (Fig. 12–9) suggest a reduced symmetry, i.e., a tetrag-

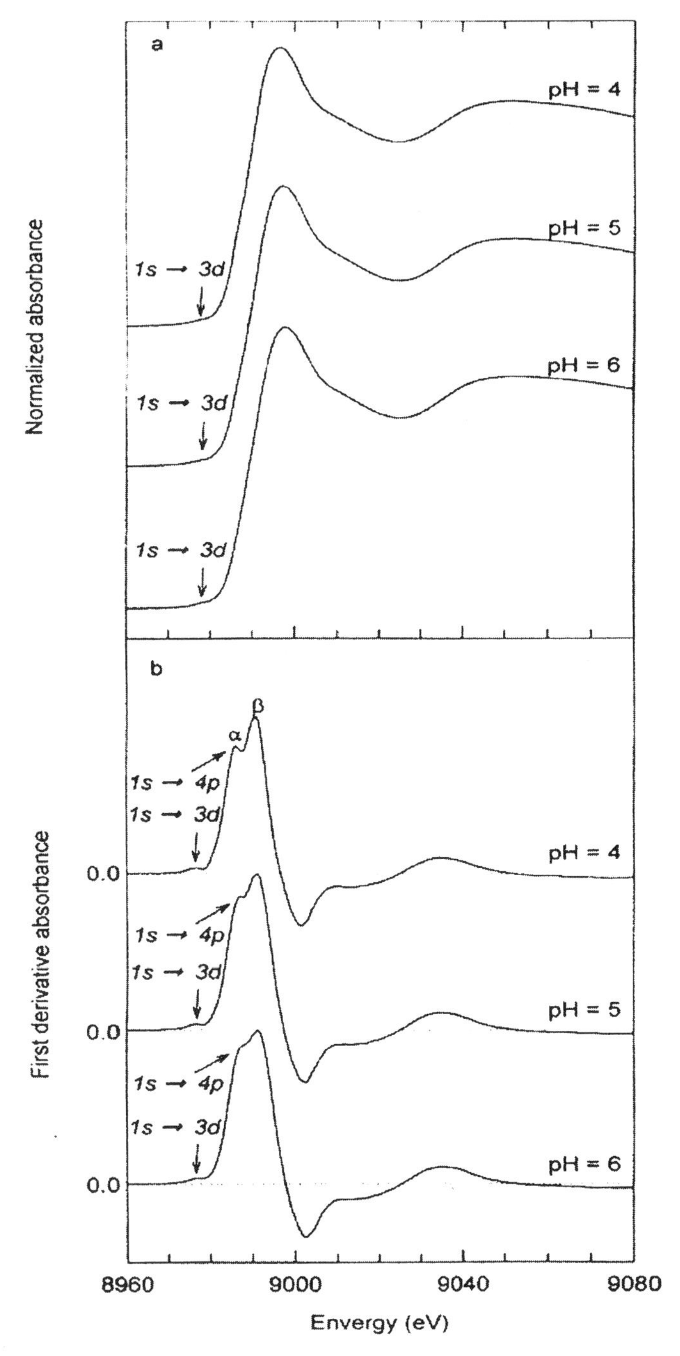

Fig. 12–9. (a) XANES spectra and (b) their first derivatives of Cu^{2+}–humic substance complexes (from Xia et al., 1997a).

onal distortion of the octahedral binding site of Cu^{2+} in HS, which is consistent with previous ESR results (Senesi, 1990b, 1992, 1996).

The RSF plots derived from EXAFS spectra and the corresponding FEFF simulations for Cu^{2+} bound to HS at pH 4, 5, and 6 show no change with pH (Fig. 12–10) (Xia et al., 1997a). All RSF plots show a major peak centered at 1.5 Å, which arise from atoms in the first coordination shell, and a minor peak centered at about 2.2 Å, which represents scattering from atoms in the second coordination shell.

Given the similarities of the scattering of photons and photelectrons by O and N atoms, it is not possible to distinguish bonding to O from bonding to N; however, being the amount of N in soil HS too small, O atoms from either water molecules or acidic HS functional ligands are considered to be the primary first-shell atoms. The position of the minor peak in the RSF plots suggests that C atoms rather than O atoms are more likely to appear in the second coordination shell of Cu^{2+} bound to HS. The presence of C atoms in the second shell is proof of inner-sphere complexation, while the number of C atoms in this shell can be an indication of the average number of acidic HS functional groups coordinating the Cu^{2+} ion; however, caution should be used in the evaluation of the second shell coordination numbers since experimental data usually have a large margin of error (>20%). This is especially true for the C atom that has a weak backscattering amplitude, particularly if the bond lengths are variable and/or heterogeneous such as with HS ligands. Further, the average second-shell coordination number calculated by EXAFS can mean either that it is valid for all metal atoms bound or that it is an average of a wide range of coordination numbers.

Analysis of experimental results and model calculations leads to the conclusion that at pH 4 to 6 the average environment of Cu^{2+} bound to soil HS is a tetragonally distorted octahedral binding site involving four equatorial O atoms at an average distance of 1.94 Å and two axial O atoms at an average distance of 2.02 Å in the first coordination shell, and possibly four C atoms at an average distance of 3.13 Å in the second coordination shell (Xia et al., 1997a). Average first shell Cu–O bond lengths of 1.92 to 1.93 are measured for Cu^{2+}–soil HA suspensions at various Cu/HA ratios and at pH 5.6 and 7.3 (Hersterberg et al., 1997).

Almost similar results are obtained in a XANES and EXAFS study by Davies et al. (1997, 2001) on Cu^{2+} complexes with three soil and peat HAs. No different metal binding sites are revealed at different metal-loading levels, and no evidence of metal clustering is obtained even in heavily metal-loaded HAs, thus suggesting that metal ions are widely separated. None of the Fourier transformed EXAFS spectra of Cu^{2+}–HS complexes show features expected for second shell atoms, even at 77 K. The authors conclude that Cu^{2+} in HAs has four nearest neighbor O or N atoms at 1.97 ± 0.02 Å, in good agreement with the slightly shortest distance estimates of Xia et al. (1997a).

Similar to ESR results, no evidence for Cu^{2+} bonding to reduced organic S in a reduced S-rich HA was obtained by EXAFS (Alcacio et al., 2001). Based on XANES data, these authors found that Cu^{2+} was mostly bound to O atoms of HA, and not to N atoms. Further, some differences found in Cu^{2+}–O (axial) bond lengths suggested that the local molecular configuration of Cu^{2+}–HS complexes may differ depending on the source and type of HS.

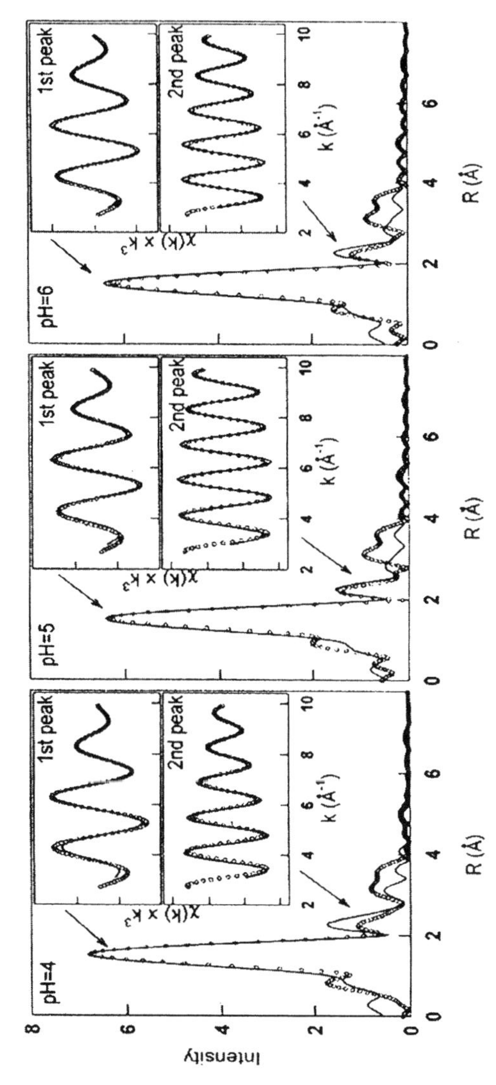

Fig. 12–10. Experimental radial structure function (RSF) for Cu^{2+}–humic substance complexes at pH 4, 5, and 6 (dotted line) and FEFF simulations (solid line) for an adjusted model of the coordination site derived from bond network analysis. The inset shows plots of experimental (dotted line) and fitted (solid line) inverse Fourier-transformed scattering curves for the first atomic shell (Cu–O) and second atomic shell (Cu–C) (from Xia et al., 1997a).

Lead Complexes. In an EXAFS study of a soil contaminated by alkyl-tetravalent Pb compounds, Pb is found in the divalent state and complexed to salicylate- and catechol-type functional groups of HS (Manceau et al., 1996). The best spectral resemblance to reference compounds is obtained for a mixture of 60% salicylate and 40% catechol functional groups. Varying the proportion of salicylate and catechol forms within 10 to 20% has minor effects on the general shape of the combined EXAFS spectrum (Manceau et al., 1996).

Binding of Pb^{2+} by a soil HS at pH 4, 5, and 6 has been studied in detail in the freeze-dried state by XANES and EXAFS spectroscopies (Xia et al., 1997a). The similarity of L_{III}-edge XANES spectra of Pb^{2+}–HS complexes, and of their first derivatives (Fig. 12–11), suggest that Pb^{2+} binding sites in soil HS are similar at these pH values. The experimental RSF plots and FEFF simulations for Pb^{2+} bound to HS at pH 4, 5, and 6 show small changes with varying the pH (Fig. 12–12). As in the case of Cu^{2+}–HS complexes, all RSF plots show a major peak arising from atoms in the first shell and centered at 1.9 Å (at low pH) and 1.7 Å (at higher pH), and a minor peak centered at 2.7 Å arising from atoms in the second shell.

The best simulations for Pb^{2+}–HS complexes are obtained for Pb^{2+} bound to four O atoms in the first shell at average distances decreasing from 2.46 Å to 2.32 Å with increasing pH from 4 to 6, and two C atoms at an average distance of 3.26 Å in the second shell. These result is an indication of inner-sphere bonding of Pb^{2+} to two O-containing functional groups of HS (Xia et al., 1997a). The greater width of the first-shell O peaks for Pb^{2+}–HS complexes, with respect to Cu^{2+}–HS complexes, indicates a higher degree of either chemical heterogeneity or distortion in the Pb binding sites compared with Cu binding sites.

Zinc, Nichel, and Cobalt Complexes. Similar to Cu^{2+}, the XANES spectra, and their first derivatives, of Ni^{2+} and Co^{2+} complexed by a soil HS at pH 4 feature an absorption in the pre-edge region normally due to 1s $\rightarrow$ 3d transitions, but no splitting of peaks α and β (Xia et al., 1997b). For Zn^{2+} complexes with the same HS sample the pre-edge features cannot be distinguished because of the inadequate number of data points collected in this region (Xia et al., 1997b). In any case, however, the experimental results indicate an average octahedral binding environment for these metal ions in soil HS.

The experimental RFS plots for Zn, Ni, and Co all have a major peak centered at 1.6 Å to 1.8 Å arising from atoms in the first coordination shell, and a much smaller peak centered at about 2.3 Å to 2.7 Å representing scattering from atoms in the second coordination shell (Xia et al., 1997b). Model fitting with FEFF indicates inner-sphere octahedral coordination with six O atoms in the first shell for Ni– and Co–HS complexes, whereas for Zn^{2+} the best fit is a coordination with 4 O and 2 S atoms. The bond distances for Ni–O and Co–O are 2.10 Å and 2.04 Å, respectively. The distances for Zn–O and Zn–S bonds are 2.13 Å and 2.33 Å, respectively.

Based on the position of the second peak of the RFS, Xia et al. (1997b) conclude that, similar to Cu^{2+} complexes, C atoms rather than O atoms are more likely to be present in the second coordination shell of Zn^{2+}, Ni^{2+}, and Co^{2+} bound to soil HS, thus indicating innersphere complexation. The number of C atoms in the second shell, which may represent the average number of acidic HS functional ligands

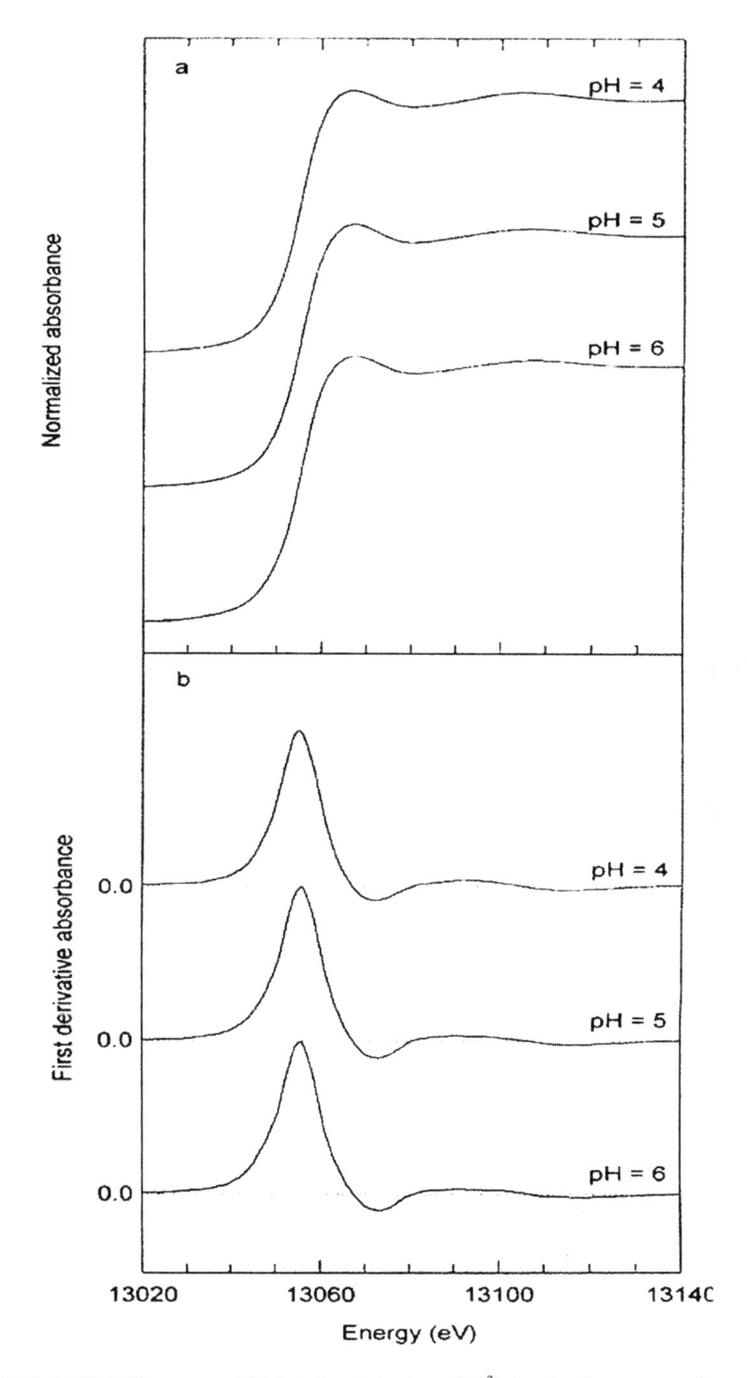

Fig. 12–11. (a) XANES spectra and (b) their first derivatives of Pb^{2+}–humic substance complexes (from Xia et al., 1997a).

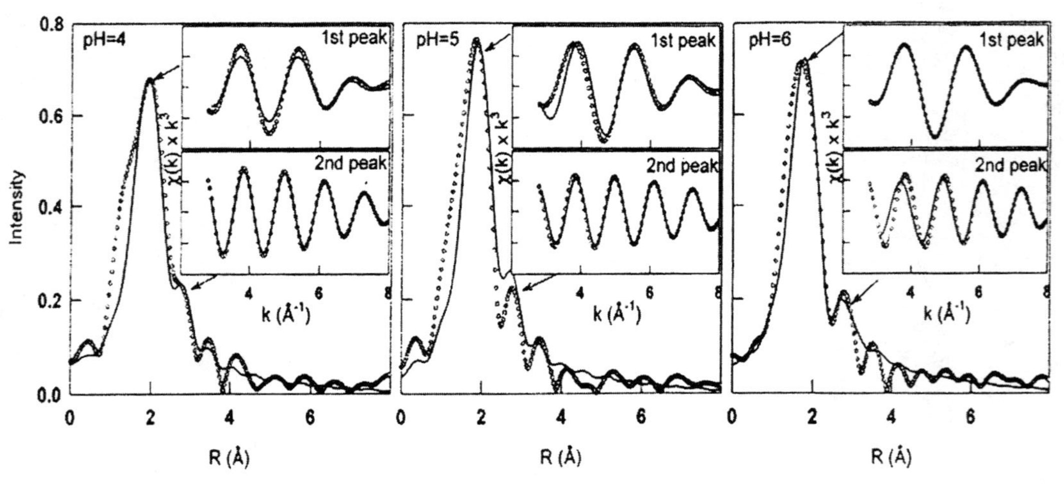

Fig. 12–12. Experimental radial structure function (RSF) for Pb^{2+}–HS complexes at pH 4, 5, and 6 (dotted line) and FEFF simulations (solid line) for an adjusted model of the coordination site derived from bond network analysis. The inset shows plots of experimental (dotted line) and fitted (solid line) inverse Fourier-transformed scattering curves for the first atomic shell (Pb–O) and second atomic shell (Pb–C) (from Xia et al., 1997a).

coordinating the metal, is: (i) one for Co, at a bond distance of 2.87 Å; and (ii) two for Ni and Zn, at a bond distance of 2.94 Å and 3.29 Å, respectively.

EXAFS spectra at the Zn–K edge and RSFs of Zn^{2+}–soil HA complexes obtained at various Zn^{2+} concentrations show that at low concentrations Zn^{2+} forms inner-sphere complexes of sixfold coordination to O ligands of HA (Sarret et al., 1997). As Zn^{2+} concentration is raised, the number of different ligand sites increases, the coordination being exclusively octahedral. At very high metal concentration, however, the major part of Zn (80–90%) is bound as an outer-sphere complex (Sarret et al., 1997).

Ferric Iron and Manganese (II) Complexes. Complexes of Fe^{3+} and Mn^{2+} ions with three soil and peat HAs have also been studied by XANES and EXAFS spectroscopies (Davies et al., 1997, 2001). No second shell peaks are observed in EXAFS spectra of Fe^{3+} and Mn^{2+}–HA complexes. Further, no evidence in EXAFS spectra is observed for oxidation number changes in these redox-active metals as a function of varying metal loading.

Both Fe^{3+} and Mn^{2+} ions show six nearest neighbor O or N atoms at a distance of 2.01 ± 0.05 Å and 2.20 ± 0.04 Å, respectively (Davies et al., 1997, 2001). The authors suggest that the highest uncertainty of interatomic distance for Fe^{3+} would indicate either that Fe^{3+} occupies more than one binding site with different nearest neighbor atom distances or that Fe^{3+} occupies a single distorted-octahedral site.

Mercury (II) Complexes. The Hg–L_{III} edge XAS spectrum obtained by Wang et al. (1997) after addition of a small quantity of Hg^{2+} (5 mmol Kg^{-1}) to a HA isolated from a soil contaminated with Hg yields a RSF plot showing a broad peak that is FEFF-fitted to an Hg-ligand bond length of 2.05 Å interpreted as an Hg–O bond; however, the EXAFS scan is not continued to higher energies and the spectral noise is high.

In another study (Skyllberg et al., 1997) of Hg^{2+} added to a soil HA at a ratio of 1 mol of Hg^{2+} for each mole of total organic S, the Hg–L_{III} edge spectrum yields an RSF plot with two peaks. The most prominent peak at 2.1 Å is consistent with the length for Hg–O or Hg–N bonds, whereas the minor peak at 2.5 Å matches exactly an Hg–S bond length. Using variable ratios of Hg to reduced S, only the Hg–O peak is visible when Hg^{2+} greatly exceeds reduced S, whereas both Hg–O and Hg–S peaks appear at ratios of less than one (Bloom et al., 2001).

A detailed EXAFS study of Hg^{2+} complexation by the same soil HA at a ratio of about 3 mol of Hg per mole of reduced S in the HA confirms the participation of reduced S in the binding of Hg^{2+} (Xia et al., 1999). The experimental RSF plot shows the same two peaks observed by Skyllberg et al. (1997) but with greater relative intensity for the Hg–S peak and better peak separation. A two-coordinate binding environment with one O atom and one S atom at distances of 2.02 and 2.38 A, respectively, in the first coordination shell is suggested for Hg^{2+} complexed by HA. Model calculations show that a second shell could contain one C atom and a second S atom at 2.78 and 2.93 A, respectively. Thus, in addition to thiol S, the possible contribution of disulfide/disulfane S is suggested for the bonding of Hg to soil HS (Xia et al., 1999). The appearance of C atom in the second shell suggests that one O-containing ligand such as COOH or phenol OH, rather than H_2O, is bound to Hg^{2+}.

Recently, Hersterberg et al. (2001) showed that the shift from Hg–O/N bonding to Hg–S bonding in HA was a continuous function of either the total S/Hg ratio or reduced S/Hg ratio.

Conclusions

X-ray absorption spectrocopy in the XANES and EXAFS regimes is a powerful, nondestructive and noninvasive tool for studying specifically the chemistry of several elements in a complex matrix like HS. In particular, these techniques allow the study of trace metal ion complexation by HS without any limitation due to the type of metal species, which needs not to be paramagnetic as in ESR spectroscopy. Further, additional details can be obtained on dentality of the central ions and bond lengths and distances, which cannot be revealed by ESR that lacks of a scattering component.

Further, the strong specific bonding sites involved in binding of some metals at low metal/C ratios in soil HS can be studied at more realistic, naturally occurring metal loading; however, synchroton light sources at large synchroton facilities are necessary to produce the intense x-ray beams needed for contemporary XAS.

REFERENCES

Abragam, A., and B. Bleaney. 1970. Electron paramagnetic resonance of transition ions. Clarendon Press, Oxford, England.

Alcacio, T.E., D. Hesterberg, J.W. Chou, J.D. Martin, S. Beauchemin, and D.E. Sayers. 2001. Molecular scale characteristics of Cu(II) bonding in goethite-humate complexes. Geochim. Cosmochim. Acta 65:1355–1366.

Allison, J.D., and E.M. Perdue. 1995. Modeling metal-humic interactions with MINTEQA2. p. 927–942. *In* N. Senesi and T.M. Miano (ed.) Humic substances in the global environment and implications on human health. Elsevier, Amsterdam.

Altmann, R.S., and J. Buffle. 1988. The use of different equilibrium functions for interpretation of metal binding in complex ligand systems: Its relation to site occupancy and site affinity distribution. Geochim. Cosmochim. Acta 52:1505–1519.

Ashley, J.T.F. 1996. Adsorption of Cu(II) and Zn(II) by estuarine, riverine, and terrestrial humic acids. Chemosphere 33:2175–2187.

Banerjee, S.K., and S.K. Mukherjee. 1972. Studies of the infrared spectra of some divalent transitional metal humates. J. Indian Soc. Soil Sci. 20:91–94.

Bartoli, F., A. Hatira, J.C. Andre, and J.M. Portal. 1987. Proprietes fluorescentes et colloidales d'une solution organique de podzol au cours du processus de complexation par le cuivre. Soil Biol. Biochem. 19:355–362.

Bauer, W. 1995. NMR of organolithium compounds: General aspects and application of two-dimensional heteronuclear Overhauser effect spectroscopy (HOESY). p. 125–172. *In* A.-M. Sapse and P.V.R. Schleyer (ed.) Lithium chemistry. Wiley Interscience, New York.

Bellamy, L.J. 1975. The infrared spectra of complex molecules. Chapman and Hall, London.

Benedetti, M.F., C.J. Milne, D.G. Kinniburgh, W.H. van Riemsdijk, and L.K. Kopal. 1995. Metal-ion binding to humic substances—application of the nonideal competitive adsorption model. Environ. Sci. Technol. 29:446–457.

Bhat, G.A., R.A. Saar, R.B. Smart, and J.H. Weber. 1981. Titration of soil-derived fulvic acid by copper(II) and measurement of free copper(II) by anodic stripping voltammetry and copper(II) selective electrode. Anal. Chem. 53:2275–2280.

Bianconi, A. 1988. XANES spectroscopy. p. 573–662. *In* D.C. Koningsberger and R. Prins (ed.) X-ray absorption: Principles, applications, techniques of EXAFS, SEXAFS and XANES. Wiley Interscience, New York.

Bloom, P.R., and J.A. Lenheer. 1989. Vibrational electronic and high-energy spectroscopic methods for characterizing humic substances. p. 409–446. *In* M.H.B. Hayes et al. (ed.) Humic substances. II: In search of structure. John Wiley and Sons, Chichester, England.

Bloom, P.R., W.F. Bleam, and K. Xia. 2001. X-ray spectroscopy applications for the study of humic substances. p. 317–350. *In* C.E. Clapp et al. (ed.) Humic substances and chemical contaminants. SSSA, Madison, WI.

Boersma, R.E., W.L. Kingery, M.H.B. Hayes, and R.P. Hicks. 1997. Characterization of humic substances complexation of cadmium by solution-state ^{113}Cd-NMR spectroscopy. p. 137–138. *In* I.K. Iskandar et al. (ed.) Proc. Int. Conf. Biogeochemistry of Trace Elements. Berkeley, CA. 22–27 June 1997. U.S. Army Cold Regions Res. And Eng. Lab., Hanover, NH.

Boyd, S.A., L.E. Sommers, and D.W. Nelson. 1981a. Copper(II) and iron(III) complexation by the carboxylate group of humic acid. Soil Sci. Soc. Am. J. 45:1241–1242.

Boyd, S.A., L.E. Sommers, D.W. Nelson, and D.X. West. 1981b. The mechanism of copper (II) binding by humic acid. An electron spin resonance study of a copper (II)-humic acid complex and some adducts with nitrogen donors. Soil. Sci. Soc. Am. J. 45:745–749.

Bresnahan, W.T., C.L. Grant, and J.H. Weber. 1978. Stability constants for the complexation of copper (II) ions with water and soil fulvic acid measured by an ion selective electrode. Anal. Chem. 50:1675–1679.

Brown, S.B. 1980. Ultraviolet and visible spectroscopy. p. 1–15. *In* S.B. Brown (ed.) An introduction to spectroscopy for biochemists. Academic Press, New York.

Buffle, J. 1988. Complexation reactions in aquatic systems: An analytical approach. Ellis Horwood, Chichester, England.

Buffle, J., P. Deladoey, F.L. Greter, and W. Haerdi. 1980. Study of the complex formation of copper(II) by humic and fulvic substances. Anal. Chim. Acta 116:255–274.

Buffle, J., and F.L. Greter. 1979. Voltammetric study of humic and fulvic substances: II. Mechanism of reaction of the Pb-fulvic complexes on the mercury electrode. J. Electroanal. Chem. 101:231–251.

Cameron, D.F., and N.L. Sohn. 1992. Functional group content of soil and sedimentary humic acids determined by CP/MAS ^{13}C NMR related to conditional Zn^{2+} and Cd^{2+} formation constants. Sci. Total Environ. 113:121–132.

Campanella, L., V. Crescenzi, M. Dentini, C. Fabiani, F. Mazzei, and A.I. Nero Scheffino. 1990. Polyelectrolytic metal ion sequestrants. p. 359–75. *In* J.W. Patterson and R. Passino (ed.) Metal speciation, separation, recovery. Vol. II. Lewis, Chelsea, MI.

Chen, Y., and F.J. Stevenson. 1986. Soil organic matter interactions with trace elements. p. 73–116. *In* Y. Chen and Y. Avnimelech (ed.) The role of organic matter in modern agriculture. Nijhoff, Dordrecht, the Netherlands.

Cheshire, M.V., M.L. Berrow, B.A. Goodman, and C.M. Mundie. 1977. Metal distribution and nature of some Cu, Mn and V complexes in humic and fulvic acid fractions of soil organic matter. Geochim. Cosmochim. Acta 41:1131 1138.

Chung, K.H., S.W. Rhee, H.S. Shin, and C.H. Moon. 1996. Probe of cadmium(II) binding on soil fulvic acid investigated by ^{113}Cd NMR spectroscopy. Can. J. Chem. 74:1360–1365.

Connors, K.A. 1987. Binding constants: the measurement of molecular complex stability. Wiley Interscience, New York.

Cook, R.L., and C.H. Langford. 1995. Metal-ion quenching of fulvic-acid fluorescence intensities and lifetimes: Nonlinearities and a possible 3-component model. Anal. Chem. 67:174–180.

Creaser, C.S., and J.R. Sodeau. 1990. Luminescence spectroscopy. p. 103–136. *In* D.L. Andrews (ed.) Perspective in modern vibrational spectroscopy. Springer-Verlag, Berlin.

Crosser, M.L., and H.E. Allen. 1977. Determination of complexation capacity of soluble ligands by ion exchange equilibrium. Soil Sci. 123:176–181.

Dabek-Zlotorzynska, E., E.P.C. Lai, and A.R. Timerbaev. 1998. Capillary electrophoresis: The state-of-art in metal speciation studies. Anal. Chim. Acta 359:1–26.

Davies, G., A. Fataftah, A. Cherkasskiy, E.A. Ghabbour, A. Radwan, S.A. Jansen, S. Kolla, M.D. Paciolla, L.T. Sein, Jr., W. Buermann, M. Balasubramanian, J. Budnick, and B. Xing. 1997. Tight metal binding by humic acids and its role in biomineralisation. J. Chem. Soc. Dalton Trans.1997:4047–4060.

Davies, G., E.A. Ghabbour, A. Cherkasskiy, and A. Fataftah. 2001. Tight metal binding by solid phase peat and soil humic acids. p. 371–396. *In* C.E. Clapp et al. (ed.) Humic substances and chemical contaminants. SSSA, Madison, WI.

Davies, R.I., M.V. Cheshire, and I.J. Graham-Bryce. 1969. Retention of low level of copper by humic acids. J. Soil Sci. 20:65–71.

Deczky, K., and H. Langford. 1978. Application of water nuclear magnetic resonance relaxation times to study of metal complexes of the soluble soil organic matter fraction fulvic acid. Can. J. Chem. 56:1947–1951.

Dobbs, J.C., W. Susetyo, L.A. Carreira, and L.V. Azarraga, 1989. Competitive binding of protons and metal ions in humic substances by lanthanide ion probe spectroscopy. Anal. Chem. 61:1519–1524.

Dzombak, D.A., W. Fish, and M.M. Morel. 1986. Metal-humate interaction: 1. Discrete ligands and continuous distribution models. Environ. Sci. Technol. 20:669–675.

Ephraim, J., and J.A. Marinsky. 1986. A unified physicochemical description of the protonation and metal ion complexation equilibria of natural organic acids (humic and fulvic acids): 3. Influence of polyelectrolyte properties and functional heterogeneity on the copper ion binding equilibria in Armadale horizons Bh fulvic acid sample. Environ. Sci. Technol. 20:367–376.

Fendorf, S.E. 1999. Fundamental aspects and applications of x-ray absorption spectroscopy in clay and soil science. p. 20–67. *In* D.G. Schulze et al. (ed.) Synchrotron x-ray methods in clay science. CMS Workshop Lectures. Vol. 9. Clay Minerals Soc., Boulder, CO.

Fendorf, S.E., and D.L. Sparks. 1996. X-ray absorption fine structure spectroscopy. p. 377–416. *In* D.L. Sparks et al. (ed) Methods of soil analysis. Part 3. Chemical methods. SSSA Book Series 5. SSSA, Madison, WI.

Gamble, D.S., C.H. Langford, and J.P.K. Tong. 1976. The structure and equilibria of a manganese (II) complex of fulvic acid studied by ion exchange and nuclear magnetic resonance. Can. J. Chem. 54:1239–1245.

Gamble, D.S., M. Schnitzer, and D.S. Skinner. 1977. Mn(II)-fulvic acid complexing equilibrium measurements by electron spin resonance spectrometry. Can. J. Soil Sci. 57:47–53.

Gamble, D.S., A.W. Underdown, and C.H. Langford. 1980. Copper(II) titration of fulvic acid ligand sites with theoretical, potentiometric, and spectrophotometric analysis. Anal. Chem. 52:1901–1908.

Ghosh, K., and M. Schnitzer. 1981. Fluorescence excitation spectra and viscosity behavior of a fulvic acid and its copper and iron complexes. Soil Sci. Soc. Am. J. 45:25–29.

Glaus, M.A., W. Hummel, and L.R. Van Loon. 2000. Trace metal-humate interactions: I. Experimental determination of conditional stability constants. Appl. Geochem. 15:953–973.

Goodman, B.A., and M.V. Cheshire. 1975. The bonding of vanadium in complexes with humic acid: An electron paramagnetic resonance study. Geochim. Cosmochim. Acta 39:1711–1713.

Goodman, B.A., and M.V. Cheshire. 1976. The occurrence of copper–porphyrin complexes in soil humic acids. J. Soil. Sci. 27:337–347.

Goodman, B.A., and M.V. Cheshire. 1979. A Mössbauer spectroscopic study of the effect of pH on the reaction between iron and humic acid in aqueous media. J. Soil Sci. 30:85–91.

Goodman, B.A., and M.V. Cheshire. 1982. Reduction of molybdate by soil organic matter: EPR evidence for formation of both Mo(V) and Mo(III). Nature (London) 299:618–620.

Goodman, B.A., and M.V. Cheshire. 1987. Characterization of iron–fulvic acid complexes using Mössbauer and EPR spectroscopy. Sci. Total Environ. 62:229–240.

Gregor, J.E., H.K.J. Powell, and R.M. Town. 1989. Evidence for aliphatic mixed mode coordination in copper(II)–fulvic acid complexes. J. Soil Sci. 40:661–673.

Grimm, D.M., L.V. Azarraga, L.A. Carreira, and W. Susetyo. 1991. Continuous multiligand distribution model used to predict the stability constant of Cu(II) metal complexation with humic material from fluorescence quenching data. Environ. Sci. Technol. 25:1427–1431.

Gutlich, P., R. Link, and A. Trautwein. 1978. Mössbauer spectroscopy and transition metal chemistry. Springer-Verlag, New York.

Hersterberg, D., J.W. Chou, K.J. Hutchinson, and D.E. Sayers. 2001. Bonding of Hg(II) to reduced organic sulfur in humic acid as affected by S/Hg ratio. Environ. Sci. Technol. 35:2741–2745.

Hersterberg, D., D.E. Sayers, W. Zhou, W.P. Robarge, and G.M. Plummer. 1997. XAFS characterization of copper in model aqueous systems of humic acid and illite. J. Phys. IV France 7(C2):833–834.

Howe, R.F., X. Lu, J: Hook, and W.D. Johnson. 1997. Reaction of aquatic humic substances with aluminum: A ^{27}Al NMR study. Mar. Freshwater Res. 48:377–383.

Jameson, R.F. 1981. Coordination chemistry of copper with regard to biological systems. p. 1–30. *In* H. Siegel (ed.) Metal ions in biological systems. Vol. 12. Dekker, New York.

Juste, C., and J. Delas. 1967. Influence de l'addition d'aluminium, de fer, de calcium, de magnésium ou de cuivre sur la mobilité électrophorétique, le spectre d'absorption infrarouge et la solubilité d'un composé humique. Ann. Agron. 18:403–427.

Juste, C., and J. Delas. 1970. Etude de quelques propriètés des complexes formés par les acides humiques et les cations. Bull. Assoc. Fr. Etude Sol. 4:39–49.

Kevan, L., and L. Kispert. 1976. Electron spin double resonance spectroscopy. Wiley-Interscience, New York.

Kevan, L., and R.N. Schwartz. 1979. Time domain electron spin resonance. Wiley-Interscience, New York.

Kingery, W.L., A.J. Simpson, F. Han, and B. Xing. 2001. Nuclear magnetic resonance studies of metal interactions with humic substances. p. 397–426. *In* C.E. Clapp et al. (ed.) Humic substances and chemical contaminants. SSSA, Madison, WI.

Kinniburgh, D.G., C.J. Milne, M.F. Benedetti, J.P. Pinheiro, J. Filius, L. Koopal, and W.H. van Riemsdijk. 1996. Metal ion binding by humic acid: Application of the NICA-Donnan model. Environ. Sci. Technol. 30:1687–1698.

Koningsberger, D., and R. Prins (ed). 1988. Principles, applications and techniques of EXAFS, SEXAFS and XANES spectroscopy. Wiley Interscience, New York.

Koopal, L.K., W.H. Van Riemsdijk, J.C.M. de Wit, and M.F. Benedetti. 1994. Analytical isotherm equations for multicomponent adsorption to heterogeous surfaces. J. Colloid Interface Sci. 166:51–60.

Kraayenhof, R., A.J.W.G. Visser, and O.S. Wolfbeis. 2002. Fluorescence spectroscopy, imaging and probes. New tools in chemical, physical and life sciences. Springer Verlag, Heidelberg, Germany.

Krajnc, M., J. Štupar, and S. Miliev. 1995. Characterization of chromium and copper complexes with fulvic acids isolated from soils in Slovenia. Sci. Total Environ. 159:23–31.

Lakatos, B., L. Korecz, and J. Meisel. 1977b. Comparative studies on the Mössbauer parameters of iron humates and polyuronates. Geoderma 19:149–157.

Lakatos, B., T. Tibai, and J. Meisel. 1977a. ESR spectra of humic acids and their metal complexes. Geoderma 19:319–338.

Lacowicz, J.R. 1986. Principles of fluorescence spectroscopy. Plenum Press, New York.

Lambert, J., J. Buddrus, and P. Burba. 1995. Evaluation of conditional stability constants of dissolved aluminum/humic substance complexes by means of ^{27}Al nuclear magnetic resonance. Fresenius J. Anal. Chem. 351:83–87.

Langford, C.H., D.S. Gamble, A.W. Underdown, and S. Lee. 1983. Interaction of metal ions with a well characterized fulvic acid. p. 219–237. *In* R.F. Christman and E.T. Gjessing (ed.) Aquatic and terrestrial humic materials. Ann Arbor Sci., Ann Arbor, MI.

Larive, C.K., A. Rogers, M. Morton, and W.R. Carper. 1996. ^{113}Cd NMR binding studies of Cd–fulvic acid complexes: Evidence of fast exchange. Environ. Sci. Technol. 30:2828–2831.

Li, J., E. M. Perdue, and L.T. Gelbaum. 1998. Using cadmium-113 NMR spectrometry to study metal complexation by natural organic matter. Environ. Sci. Technol. 32:483–487.

Lu, X., W.D. Johnson, and J. Hook. 1998. Reaction of vanadate with aquatic humic substances: An ESR and ^{51}V NMR study. Environ. Sci. Technol. 32:2257–2263.

Lytle, F.W. 1988. Experimental x-ray absorption spectroscopy. p. 135–223. *In* H. Winick et al. (ed.) Applications of synchrotron radiation. Gordam and Breach Sci.

MacCarthy, P., and E.M. Perdue. 1991. Complexation of metal ions by humic substances: Fundamental considerations. p. 469–492. *In* G.H. Bolt et al. (ed.) Interactions at the soil colloid-soil solution interface. NATO ASI Series. Vol. 190. Kluwer, Dordrecht, the Netherlands.

MacCarthy, P., and J.A. Rice. 1985. Spectroscopic methods (other than NMR) for determining functionalities in humic substances. p. 527–559. *In* G.R. Aiken et al. (ed) Humic substances in soil, sediment and water. Geochemistry, isolation, and characterization. Wiley-Interscience, New York.

Macomber, R.S. 1998. A complete introduction to modern NMR spectroscopy. Wiley Interscience, New York.

Manceau, A., M.C. Boisset, G. Sarret, J.L. Hazemann, M. Mench, P. Cambier, and R. Prost. 1996. Direct determination of lead speciation in contaminated soils by EXAFS spectroscopy. Environ. Sci. Technol.30:1540–1552.

Mantoura, R.F.C., and J.P. Riley. 1975. The use of gel filtration in the study of metal binding by humic acids and related compounds. Anal. Chim. Acta 78:193–200.

Mantoura, R.F.C., A. Dickson, and J.P. Riley. 1978. The complexation of metals with humic materials in natural waters. Estuar. Coastal Mar. Sci. 6:387–408.

Marinsky, J.A., and J. Ephraim. 1986. A unified physicochemical description of the protonation and metal ion complexation equilibria of natural organic acids (humic and fulvic acids): 1. Analysis of the influence of polyelectrolyte properties on protonation equilibria in ionic media: Fundamental concepts. Environ. Sci. Technol. 20:349–354.

McBride, M.B.1978. Transition metal binding in humic acids: An ESR study. Soil Sci. 126:200–209.

McBride, M.B. 1982. Electron spin resonance investigation of Mn^{2+} complexation in natural and synthetic organics. Soil Sci. Soc. Am. J. 46:1137–1143.

McBride, M.B. 1989. Reactions controlling heavy metal solubility in soils. p. 1–56. *In* B.A. Stewart (ed.) Advances in soil science. Vol. 14. Springer, New York.

McCoustra, M.R.S. 1990. Electronic absorption spectroscopy: Theory and practice. p. 88–101. *In* D.L. Andrews (ed.) Perspective in modern chemical spectroscopy. Springer-Verlag, Berlin.

Milne, C.J., D.G. Kinniburgh, J.C.M. De Wit, W.H. van Riemsdijk, and L.K. Koopal. 1995. Analysis of metal ion binding by a peat humic acid using a simple electrostatic model. Geochim. Cosmochim. Acta 59:1101–1112.

Nakamoto, K. 1986. Infrared and Raman spectra of inorganic and coordination compounds. Wiley-Interscience, New York.

Nifant'eva, T.I., V.M. Shkinev, B.Ya. Spivakov, and P. Burba. 1999. Membrane filtration studies of aquatic humic substances and their metal species: a concise overview: Part 2. Evaluation of conditional stability constants by using ultrafiltration. Talanta 48:257–267.

Peisach, J., and W.E. Blumberg. 1974. Structural implications derived from the analysis of the electron paramagnetic resonance spectra of natural and artificial copper proteins. Arch. Biochem. Biophys. 165:691–708.

Perdue, E.M. 1985. Acidic functional groups of humic substances. p. 493–526. *In* G.R. Aiken et al. (ed) Humic substances in soil, sediment, and water. Geochemistry, isolation, and characterization. Wiley-Interscience, New York.

Perdue, E.M. 1989. Effects of humic substances on metal speciation. p. 282–295. *In* I.H. Suffet and P. MacCarthy (ed.) Aquatic humic substances influence on fate and treatment of pollutants. Advances in Chemistry Series. Vol. 219. ACS, Washington, DC.

Perdue, E.M. 2001. Modeling concepts in metal-humic complexation. p. 305–316. *In* C.E. Clapp et al. (ed.) Humic substances and chemical contaminants. SSSA, Madison, WI.

Perdue, E.M., and C.R. Lytle. 1983. A critical examination of metal-ligand complexation models: Application to defined multiligand mixtures. p. 295–313. *In* R.F. Christman and E.T. Gjessing (ed.) Aquatic and terrestrial humic materials. Ann Arbor Sci., Ann Arbor, MI.

Piccolo, A., and F.J. Stevenson. 1982. Infrared spectra of Cu^{2+}, Pb^{2+}, and Ca^{2+} complexes of soil humic substances. Geoderma 27:195–208.

Pommery, J., J.P. Ebenga, M. Imbenotte, G. Palavitt, and F. Erb. 1988. Determination of the complexing ability of a standard humic acid with cadmium ions. Water Res. 22:185–189.

Prasad, B., and M.K. Sinha. 1983. Physical and chemical characterization of molecularly homogeneous fulvic acid fractions and their metal complexes. J. Indian Soc. Soil Sci. 31:187–191.

Prasad, B., G.D. Dkhar, and A.P. Singh. 1987. Cobalt(II), iron(III) and zinc(II) complexation by fulvic acids isolated from North-Eastern Himalayan forest and cultivated soils. J. Indian Soc. Soil Sci. 35:194–197.

Rao, C.N.R. 1963. Chemical applications of infrared spectroscopy. Academic Press, New York.

Rehr, J.J. 1993. Recent development in multiple-scattering calculations of XAFS and XANES. Japanese J. Appl. Phys. Suppl. 32-2:8–12.

Rehr, J.J., C.H. Booth, F. Bridges, and S.I. Zabinsky. 1994. X-ray absorption fine-structure in embedded atoms. Phys. Rev. B. 49:12347–12350.

Ryan, D.K., C.P. Thompson, and J.H. Weber. 1983. Comparison of Mn^{2+}, Co^{2+} and Cu^{2+} binding to fulvic acid as measured by fluorescence quenching. Can. J. Chem. 61:1505–1509.

Ryan, D.K., and J.H. Weber. 1982. Fluorescence quenching titration for determination of complexing capacities and stability constants of fulvic acid. Anal. Chem. 54:986–990.

Saar, R.A., and J.H. Weber. 1979. Complexation of cadmium(II) with water- and soil-derived fulvic acids: Effect of pH and fulvic acid concentration. Can. J. Chem. 57:1263–1268.

Saar, R.A., and J.H. Weber. 1980. Comparison of spectrofluorometry and ion-selective electrode potentiometry for determination of complexes between fulvic acid and heavy-metal ions. Anal. Chem. 52:2093–2100.

Saar, R.A., and J.H. Weber. 1982. Fulvic acid: Modifier of metal-ion chemistry. Environ. Sci. Technol. 16:510A–517A.

Sarret, G:, A. Manceau, J.L. Hazemann, A. Gomez, and M. Mench. 1997. EXAFS study of the nature of zinc complexation sites in humic substances as a function of Zn concentration. J. Phys. IV France 7 (C2):799–802.

Schnitzer, M. 1978. Humic substances: Chemistry and reactions. p. 1–64. *In* M. Schnitzer and S.U. Khan (ed.) Soil organic matter. Elsevier, Amsterdam.

Schnitzer, M., and K. Ghosh. 1982. Characteristics of water soluble fulvic acid-copper and fulvic acid–iron complexes. Soil Sci. 134:354–363.

Schnitzer, M., and S.I.M. Skinner. 1963. Organo-metallic interactions in soils: I. Reactions between a number of metal ions and the organic matter of a podzol Bh horizon. Soil Sci. 96:86–93.

Schnitzer, M., and S.I.M. Skinner. 1966. Organo-metallic interactions in soils: 5. Stability constants of Cu^{++}-, Fe^{++}-, and Zn^{++}-fulvic acid complexes. Soil Sci. 102:361–365.

Schnitzer, M., and S.I.M. Skinner. 1967. Organo-metallic interactions in soils: 7. Stability constants of Pb^{++}-, Ni^{++}-, Mn^{++}-, Co^{++}-, Ca^{++}-, and Mg^{++}-fulvic acid complexes. Soil Sci. 103:247–252.

Schulman, S.G. 1985. Molecular luminescence spectroscopy. Methods and applications. Wiley, Somerset, NY.

Scott, A.I. 1964. Interpretation of ultraviolet spectra of natural products. Pergamon Press, New York.

Senesi, N. 1990a. Molecular and quantitative aspects of the chemistry of fulvic acid and its interaction with metal ions and organic chemicals: Part II. The fluorescence spectroscopy approach. Anal. Chim. Acta 232:77–106.

Senesi, N. 1990b. Application of electron spin resonance (ESR) spectroscopy in soil chemistry. p. 77–130. *In* B.A. Stewart (ed.) Advances in soil science. Vol. 14. Springer-Verlag, New York.

Senesi, N. 1992. Metal-humic substance complexes in the environment. Molecular and mechanistic aspects by multiple spectroscopic approach. p. 429–496. *In* D.C. Adriano (ed.) Biogeochemistry of trace metals. Lewis, Boca Raton, FL.

Senesi, N. 1996. Electron spin (or paramagnetic) resonance spectroscopy. Methods of soil analysis: Part 3. Chemical methods. SSSA Book Series no.5. SSSA and ASA, Madison, WI.

Senesi, N., D.F. Bocian, and G. Sposito. 1985. Electron spin resonance investigation of copper (II) complexation by soil fulvic acid. Soil Sci. Soc. Am. J. 49:114–119.

Senesi, N., S.M. Griffith, M. Schnitzer, and M.G. Townsend. 1977. Binding of Fe^{3+} by humic materials. Geochim. Cosmochim. Acta 41:969–976.

Senesi, N., and E. Loffredo. 1999. The chemistry of soil organic matter. p. 239–370. *In* D.L. Sparks (ed) Soil physical chemistry. 2nd ed. CRC Press, Boca Raton, FL.

Senesi, N., T.M. Miano, M.R. Provenzano, and G. Brunetti. 1989a. Spectroscopic and compositional comparative characterization of some IHSS reference and standard fulvic and humic acids of various origin. Sci. Total Environ. 81/82:143–156.

Senesi, N., and G. Sposito. 1984. Residual copper (II) complexes in purified soil and sewage sludge fulvic acids: An electron spin resonance study. Soil Sci. Soc. Am. J. 48:1247–1253.

Senesi, N., G. Sposito, G.R. Bradford, and K.M. Holtzclaw. 1991. Residual metal reactivity of humic acids extracted from soil amended with sewage sludge. Water Air Soil Pollut. 55:409–425.

Senesi, N., G. Sposito, K.M. Holtzclaw, and G.R. Bradford. 1989b. Chemical properties of metal-humic acid fractions of a sewage-sludge amended aridisol. J. Environ. Qual. 18:186–194.

Senesi, N., G. Sposito, and J.P. Martin. 1986. Copper (II) and iron (III) complexation by soil humic acids: An IR and ESR study. Sci. Total Environ. 55:351–362.

Sharma, A., and S.G. Schulman. 1999. Introduction to fluorescence spectroscopy. Wiley, New York.

Shestakov, E.I., A.I. Karpukhin, V.V. Fadeev, and V.V. Chubarov. 1987. Fluorescence intensity of organic compounds containing manganese in podzolic soils. Izv. Timiryazevsk. 2:82–85.

Shin, H.S., S.W. Rhee, B.H. Lee, and C.H. Moon. 1996. Metal binding sites and partial structures of soil fulvic and humic acids compared: Aided by Eu(III) luminescence spectroscopy and DEPT/QUAT ^{13}C NMR pulse techniques. Org. Geochem. 24:523–529.

Skyllberg, U.L., P.R. Bloom, E.A. Nater, K. Xia, and W.F. Bleam. 1997. Binding of mercury(II) by reduced sulfur in soil organic matter. p. 285–286. *In* Extended Abstracts. 4th Int. Conf. on the Biogeochemistry of Trace Elements. 23–26 June 1997, Berkeley, CA., U.S. Army Cold Regions Res. and Eng. Lab., Hanover, NH.

Sohn, M., and S. Rajski. 1990. The adsorption of Cd(II) from seawater by humic acids of various sources of origin. Org. Geochem. 15:439–447.

Sposito, G. 1986. Sorption of trace metals by humic materials in soils and natural waters. CRC Crit. Rev. Environ. Control 16:193–229.

Sposito, G., K.M. Holtzclaw, and J. Baham. 1976. Analytical properties of the soluble metal-complexing frantions in sludge-soil mixtures: II. Comparative structural chemistry of fulvic acid. Soil Sci. Soc. Am. J. 40:691–698.

Stevenson F.J. 1976a. Binding of metal ions by humic acids. p. 519–540. *In* J.O. Nriagu (ed.) Environmental biogeochemistry. Vol. 2. Metals transfer and ecological mass balances. Ann Arbor Sci., Ann Arbor, MI.

Stevenson, F.J. 1976b. Stability constants of Cu^{2+}, Pb^{2+}, and Cd^{2+} complexes with humic acids. Soil Sci. Soc. Am. J. 40:665–672.

Stevenson, F.J. 1977. Nature of divalent transition metal complexes of humic acids as revealed by a modified potentiometric titration method. Soil Sci. 123:10–17.

Stevenson, F.J. 1986. Cycles of soil. Carbon, nitrogen, phosphorus, sulfur, micronutrients. Wiley–Interscience, New York.

Stevenson, F.J. 1994. Humus chemistry. Genesis, composition, reactions. 2nd ed. Wiley–Interscience, New York.

Stevenson, F.J., and M.S. Ardakani. 1972. Organic matter reactions involving micronutrients in soils. p. 79–114. *In* J.J. Mortvedt et al. (ed.) Micronutrients in agriculture. SSSA, Madison, WI.

Stevenson, F.J., and A. Fitch. 1986. Chemistry of complexation of metal ions with soil solution organics. p. 2–58. *In* P.M. Huang and M. Schnitzer (ed.) Interactions of soil minerals with natural organic microbes. SSSA, Madison, WI.

Stumm, W., R. Schwarzenbach, and L. Sigg. 1983. From environmental analytical chemistry to ecotoxicology: A plea for more concepts and less monitoring and testing. Angew. Chem. Int. Ed. Engl. 22:380–389.

Susetyo, W., L.A. Carreira, L.V. Azarraga, and D.M. Grimm. 1991. Fluorescence techniques for metal-humic interactions. Fresenius J. Anal. Chem. 339:624–635.

Sutheimer, S.H., and S.E. Cabaniss. 1997. Aluminum binding to humic substances determined by high performance cation exchange chromatography. Geochim. Cosmochim. Acta 61:1–9.

Takamatsu, T., and T. Yoshida. 1978. Determination of stability constants of metal-humic acid complexes by potentiometric titration and ion-selective electrodes. Soil Sci. 125:377–386.

Tan, K.H. 1978. Formation of metal–humic acid complexes by titration and their characterization by differential thermal analysis and infrared spectroscopy. Soil Biol. Biochem. 10:123–129.

Teo, B.K. 1986. EXAFS: Basic principles and data analysis. Springer-Verlag, Berlin.

Templeton, G.D., and N.D. Chasteen. 1980. Vanadium-fulvic acid chemistry: conformational and binding studies by electron spin probe techniques. Geochim. Cosmochim. Acta 44:741–752.

Thomson, A.J. 1990. Electron paramagnetic resonance and electron nuclear double resonance spectroscopy. p. 295–320. *In* D.L. Andrew (ed.) Perspectives in modern chemical spectroscopy. Springer Verlag, Berlin.

Tipping, E. 1998. Humic ion-binding Model VI: an improved description of the interactions of protons and metal ions with humic substances. Aquatic Geochem. 4:3–48.

Tipping, E., and M.A. Hurley. 1992. A unifying model of cation binding by humic substances. Geochim. Cosmochim. Acta 56:3627–3641.

Turner, D.R., M.S. Varney, M. Whitfield, R.F.C. Mantoura, and J.P. Riley. 1987. Electrochemical studies of copper and lead complexation by fulvic acid: II. A critical comparison of potentiometric and polarographic measurements. Sci. Total Environ. 60:17–34.

Underdown, A.W., C.H. Langford, and D.S. Gamble. 1981. The fluorescence and visible absorbance of Cu(II) and Mn(II) complexes of fulvic acid: the effect of metal ion loading. Can. J. Soil Sci. 61:469–474.

Valeur, B., J-C Brochon, and O.S. Wolfbeis. 2001. New trends in fluorescence spectroscopy: Applications to chemical and life sciences. Springer series on fluorescence, methods and Applications. Springer-Verlag, Heidelberg, Germany.

Van den Hoop, M.A.G.T., H.P. van Leeuwen, J.P. Pinheiro, A.M. Mota, and M. de L. Simoes Goncalves. 1995. Voltammetric analysis of the competition between calcium and heavy metals for complexation by humic material. Colloids Surfaces A 95:305–313.

Van Dijk, H. 1971. Cation binding of humic acids. Geoderma 5:53–67.

Van Loon, L.R., S. Granacher, and H. Harduf. 1992. Equilibrium dialysis-ligand exchange: A novel method for determining conditional stability constants of radionuclide-humic acid complexes. Anal. Chim. Acta 268:235–246.

Vinkler, P., B. Lakatos, and J. Meisel. 1976. Infrared spectroscopic investigations of humic substances and their metal complexes. Geoderma 15:231–242.

Wang, Z., D. Hesterberg, W. Zhou, D.E. Sayers, and W.P. Robarge. 1997. Extended x-ray absorption fine structure study of Hg speciation in a flood plain soil. *In* 3rd Int. Conf. on the Biogeochemistry of Trace Elements. Paris, 15–19 May 1995. Les Colloques no. 85. CD-ROM. INRA, Paris, France.

Weber, J.H. 1983. Metal ion speciation studies in the presence of humic materials. p. 315–331. *In* R.F. Christman and E.T. Gjessing (ed.) Aquatic and terrestrial humic materials. Ann Arbor Sci., Ann Arbor, MI.

Weber, J.H. 1988. Binding and transport of metals by humic materials. p. 165–178. *In* F.H. Frimmel and R.F. Christman (ed.) Humic substances and their role in the environment. Wiley-Interscience, Chichester, England.

Wertz, J.E., and J.R. Bolton. 1972. Electron spin resonance: elementary, theory and practical applications. McGraw-Hill, New York.

Willis, H.A., J.H. Van Der Maas, and R.G.J. Miller. 1988. Laboratory methods in vibrational spectroscopy. Wiley, Chichester, England.

Wilson, M.A. 1989. Solid-state nuclear magnetic resonance spectroscopy of humic substances: Basic concepts and techniques. p. 309–338. *In* M.H.B. Hayes et al. (ed.) Humic substances II. In search of structure. Wiley, Chichester, England.

Xia, K., U.L. Skyllberg, W.F. Bleam, P.R. Bloom, E.A. Nater, and P.A. Helmke. 1999. X-ray absorption spectroscopic evidence for the complexation of Hg(II) by reduced sulfur in soil humic substances. Environ. Sci. Technol. 33:257–261.

Xia, K., W.F. Bleam, and P.A. Helmke. 1997a. Studies of the nature of Cu^{2+} and Pb^{2+} binding sites of soil humic substances using x-ray absorption spectroscopy. Geochim. Cosmochim. Acta 61:2211–2221.

Xia, K., W.F. Bleam, and P.A. Helmke. 1997b. Studies of the nature of binding sites of first row transition elements bound to aquatic and soil humic substances using x-ray absorption spectroscopy. Geochim. Cosmochim. Acta 61:2223–2235.

Zabinsky, S.I., J.J. Rehr, A. Ankudinov, R.C. Albers, and M.J. Eller. 1994. Multiple-scattering calculations of x-ray absorption spectra. Phys. Rev. B. 52:2995–3009.

Zunino, H., G. Galindo, P. Peirano, and M. Aguilera. 1972. Use of the resin exchange method for the determination of stability constants of the metal-soil organic matter complexes. Soil Sci. 114:229–233.

Zunino, H., and J.P. Martin. 1977. Metal-binding organic macromolecules in soil: 2. Soil Sci. 123:188–202.

Chapter 13

Speciation of Metals in Soils

DARRYL ROBERTS, *Water and Earth Science Associates, Limited, Ottawa, Ontario, Canada*

MAARTEN NACHTEGAAL, *Paul Scherrer Institut, Villigen, Switzerland*

DONALD L. SPARKS, *University of Delaware, Newark, Delaware, USA*

The chemical, physical, and biological behavior of trace and heavy metals in soils control their movement and fate in soils. While organisms have evolved in the presence of metals in the natural environment for thousands and millions of years, it is only since industrialization that high metal concentrations are consistently being introduced to soil environments globally. This increased exposure of organisms to metals underscores the need to identify and quantify those species in soils that pose the greatest potential threat to organisms. In addition, metal species identification has use for researchers studying soil fertility (e.g., micronutrient availability to crops), land-use planning (e.g., application of metal-bearing biosolids), water quality (e.g., wastewater treatment), soil genesis and geomorphology (e.g., redoxamorphic features of Fe oxides), environmental quality (e.g., mine tailings), soil ecology (e.g., metal toxicity to microorganisms), and soil remediation (e.g., liming of smelter-impacted soils). The ubiquity of metals combined with the complexity of soils makes the study of metals one of the most important disciplines of soil chemistry.

Potentially toxic metals can be introduced to many natural systems, and evidence for this introduction can be found in freshwater bodies, marine and lacustrine sediments, soils, ice, vegetation, and animal populations. Once introduced into a particular environment, metals are not necessarily restricted to its initial host matrix as there is a dynamic cycle between all of the aforementioned phases. For instance, metals introduced to soils from industrial processes may be taken up by plants, which then can be consumed by animals, which may be consumed by other animals. Or, soils may undergo erosion, introducing metals to rivers and lakes and eventually marine environments. The speciation of metals in all of these environmentally important materials is beyond the scope of this chapter, but given the potential of soils to cycle metals between the various phases, metal speciation in soils can be used to assess regional and global metal cycling in many environmentally relevant materials.

Metals are present in soils as a result of both natural and anthropogenic processes, and separating out the two sources is often not a trivial task. Figure 13–1

Chemical Processes in Soils. SSSA Book Series, no. 8.

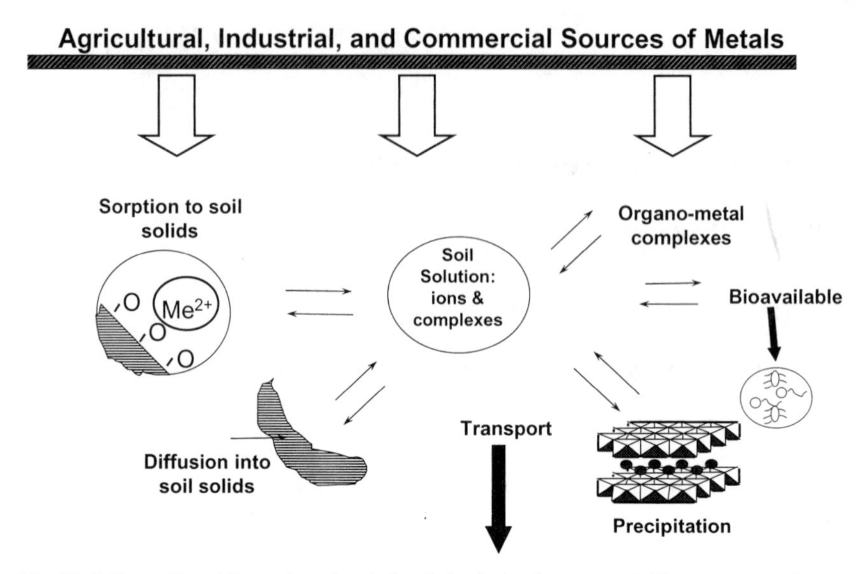

Fig. 13–1. Illustration of the various chemical and physical pathways a metal ion may encounter once introduced into the soil environment.

illustrates the fate of metals once introduced into soil environments from both natural and anthropogenic sources. Once the primary phase, be it naturally occurring or anthropogenic, is dissolved, metal ions may enter the soil solution and be subject to numerous pathways, all potentially overlapping. Each of the general pathways shown in the figure can be further divided into many more complex reactions, all with different kinetics and mechanisms. For instance, the role of colloids on metal partitioning and mobility in soils is a current research area that could keep a soil chemist occupied for decades. The soil solution may host the metal as a free ion or complexed to inorganic or organic ligands. Both the free ion and the metal-ligand complex can be exposed to one of several pathways, including: uptake by plants, mineral surfaces, and organic matter; transport through the vadose zone; precipitation as a solid phase; and diffusion into porous material. Reverse reactions also occur, making metal behavior in soils a truly dynamic process influenced by numerous physical and chemical processes. The three main pools a metal can be found in soils are (i) the soil solution, (ii) sorbed to solid phases, and (iii) as part of the structure of solid phases. The speciation of metals and trace elements in soil solutions is presented in Sauvé and Parker (2005, this publication), so the majority of this chapter will emphasize the latter two pools. Following an introduction to soil speciation and the various parameters influencing metal speciation, the various approaches and techniques that have been developed to determine metal speciation will be presented.

ORIGINS, INPUTS, AND SOURCES OF METALS IN SOILS

Before proceeding, it would be constructive to provide a definition for the broad term "metal" with respect to environmental soil chemistry and to present the metals that are of interest to researchers in this field. Metals have traditionally been classified based on categories such as light, heavy, semimetal (metalloid), toxic, and trace, depending on several chemical and physical criteria. Density, weight, atomic number, and degree of toxicity have all been used to classify metals. The term heavy metal is commonly used to encompass those metals found in soils and sediments that are associated with contamination and toxicity, but a definition of heavy metal is not universally agreed upon and a list of metals or metalloids considered to be a heavy metal will vary between researchers (Duffus, 2002). Metals can be further classified according to their hard and soft characteristics, based on the principle of hard and soft acids and bases (Sparks, 1995). All metal ions or atoms and most cations are Lewis acids and are capable of accepting a pair of electrons from a Lewis base (anion). Further categorizing metals, most are soft or transition acids, meaning they have low positive charge and large size and form covalent bonds with ligands.

Trace metal and micronutrient are terms that are often used in soil science and agronomy as those species found in low concentrations in soils that are essential for plant growth; however, a trace metal also may be found in elevated concentrations in soils and sediments due to both natural and anthropogenic processes, thereby negating the term "trace." For the purposes of this chapter, metals will include both heavy and moderately heavy (based on atomic mass) metals found in soils in both trace amounts and elevated concentrations to the point of plant and animal toxicity in some cases. Based on these criteria, the metals that can be considered important for the remainder of this chapter include: As, Cd, Co, Cr, Cu, Ni, Pb, Se, and Zn. In addition to high concentrations of certain metals causing plant and organism (both microbial and human) toxicity, deficiency problems are an issue with elements that are considered essential nutrients. Therefore, a concentration regime exists for several metals with respect to organism health and going below or above this regime can result in deficiency or toxicity symptoms, respectively.

Establishing a metal concentration range for a normal soil (i.e., background level) compared with a contaminated soil is very difficult. The parent material and geochemical history of a soil can result in metal concentrations that would be considered polluted compared with soils with so-called "normal" background metal concentration levels. Typical concentrations for some metals found in soils that are not considered contaminated and do not have parent material that is high in these metals are: 20 mg Cu kg^{-1} soil, 1 mg Cd kg^{-1} soil, 50 mg Ni kg^{-1} soil, 25 mg Pb kg^{-1} soil, and 50 mg Zn kg^{-1} soil (Förstner, 1995). That is not to say that metal concentrations that exceed these values can be considered polluted and pose a serious threat to organisms. As will be discussed in this chapter, it is not the total metal concentration that dictates the risk of toxicity, but rather the form the metal is in which is dependent on many chemical, physical, and biological parameters. This last statement is one of the fundamental reasons metal speciation is determined, although its acceptance is by no means universal.

One of the first things to consider when determining metal speciation in soils is the original source of the metal. Is the metal from natural weathering, industrial processing, use of metal components in commercial processes, aerial deposition of smelting materials, leaching from garbage and solid waste dumps, application of animal products to land, or some other source? These processes and more can introduce all of the metals previously mentioned, often concurrently. Physical and chemical alterations to the species are unavoidable in terrestrial and geochemical settings such as soils, so identifying the source will not guarantee that the species in the soil will be identified (Förstner, 1995). In the absence of pollutants, natural levels of metals in soils are dictated by the types of elements in rocks, weathering rates, organic matter content, soil texture, and soil depth. Most of the materials added to soils for agricultural purposes such as lime, inorganic fertilizers, and manure have low trace element levels and when applied at normal rates do not affect overall concentrations of trace metals in soil. Other sources of input with respect to agriculture include herbicides, fungicides, irrigation waters, biosolid application, dredged materials, fly ash, and municipal composts (Förstner, 1995).

Often times in the soil environment, the metal of concern in terms of potential toxicity is only toxic due to the characteristics of the soil environment it is in. For example, at circum- neutral pH values most metals are in a form that makes them unavailable for direct plant uptake in the aqueous form; however, this can be disputed since the environment in the immediate vicinity of plant roots, which is not necessarily measured when the bulk pH is determined, can be several pH units lower and solubilize metals that are considered stable. Often times the source of metal pollution is also responsible for the co-contamination by addition of chemical species that are capable of altering the soil environment. For instance, acid rain deposition, acid mine drainage, and deposition of sulfate from smelting activities can all cause acidic pH values in soils and lead to mobilization of metals that were once stable as adsorbed or precipitated complexes. As will be discussed, most metals are more readily available for plant uptake at low pH values, hence the onset of severe phytoxicity to plants at low pH values: the pH solublizes metal ions that are normally bound in a form not available for plant uptake. Clearly, the pH of the soil is one of the most important parameters in assessing metal speciation and many consider it the master variable when it comes to many environmental processes.

METAL SPECIATION AND BIOAVAILABILITY

The term *speciation*, just like "metal," is a multi-faceted term and difficult to assign a single definition. Metal speciation includes the chemical form of the metal in the soil solution, either as a free ion or complexed to a ligand, in the gaseous phase, and distributed amongst solid phases within the soil. Therefore, for a comprehensive description and understanding of metal speciation in soils, one would address all the various phases a metal may inhabit; however, the solid phase contains the majority of metals in soils and supplies the other two phases accordingly. This chapter will primarily deal with metal speciation in the solid phase, while the speciation of metals in the solution phase is addressed in Bartlett and Ross (2005, this publication). One should note that the separation of these topics into different chapters does not indicate they are separate phenomena in the soil. To the contrary, the long-term bioavailibility of metals to humans and other organisms is determined by the re-supply of the metal to the mobile pool (soil solution) from more stable phases (metals in and associated with solid species).

The quantitative speciation of metals as well as their variation with time is an important concept in environmental soil chemistry. In order to develop models capable of predicting the fate of nutrients and contaminants in soils an accurate description of the partitihoning of these constituents between the solid and solution phases is necessary (Schulze and Bertsch, 1995). Before any remediation strategy is attempted it is wise to determine and understand the nature of the metal species in soils. According to Mattigod et al. (1981) "positive identification of various solid phases of a trace metal in a soil, along with knowledge of their solubility and their kinetics of dissolution and precipitation, would provide sufficient information to make reliable predictions of trace metal activities in soil solutions." Speciation encompasses both the chemical and physical form an element takes in any geochemical setting. A detailed definition of speciation includes the following components: (i) the identity of the contaminant of concern or interest, (ii) the oxidation state of the contaminant, (iii) associations and complexes to solids and dissolved species (surface complexes, metal-ligand bonds, surface precipitates), and (iv) the molecular geometry and coordination environment of the metal (Brown et al., 1999). All of the above components can be interrelated and often difficult to separate. Moreover, they all have chemical, biological, and physical considerations and for this reason metal speciation is truly a multidisciplinary endeavor. The more of these parameters that can be identified the better one can predict the potential risk of toxicity to organisms by heavy metal contaminants.

Another vague term, albeit used often, used with respect to speciation of contaminants in soils is bioavailability. There is no universally accepted definition for bioavailability and it is usually a non-quantitative concept. An essential or toxic element is bioavailable if it is in a chemical form that plants can absorb readily and if, once absorbed, it affects the life cycle of the plants (Sposito, 1989). This definition is confined to the case of plants, but the same general definition can be used for humans and soil organisms (both micro- and macro-). Moreover, bioavailability can be desirous in the case of plant uptake of the required amounts of essential nutrients, or detrimental in the case of uptake of non-essential elements or essential elements at elevated levels. Despite the uncertainty associated with the term, there is a general consensus that the form the metal takes is correlated to the bioavailability of the metal. In the case of Pb, for example, the pathway for humans is often direct ingestion into the body and dissolution of phases in gastric acid. By simulating gastric conditions (pH 1–3; T = 278–328 K) it was found that the nature of the Pb phases in contaminated soils influenced the Pb release and, therefore, bioavailability (Gasser et al., 1996). While bioavailability may be considered a form in which the metal can pass through a living-cell membrane, this does not necessarily mean an organism will remove it from solution. For example, bacteria and fungi are known to have mechanisms to tolerate high metal concentrations in soils including binding the metal with proteins or extracellular polymers, formation of insoluble metal sulphides, and decreased uptake (Giller et al., 1998). Considering the fact that the presence of a metal, even in a potentially available form, does not automatically mean an organism will take it up, a better definition for bioavailability is the amount or concentration of a chemical that can be absorbed by an organism thereby creating the potential for toxicity or the necessary concentration for survival (Siegel, 2002).

Determining metal speciation in soils can be quite complex as thermodynamic models may give suggestions as to the possible species to expect in a system, but metal species are usually controlled by kinetics of the reactions. In addition, many techniques used to determine speciation directly are disruptive or destructive to a sample and may alter the chemical speciation (Förstner, 1987). To alleviate some of the difficulties in determining metal speciation in soils, laboratory-based approaches are often used since most parameters can be controlled and monitored; however, simulating the conditions found in the field is quite difficult, leading to questionable findings when trying to apply to field situations. In most laboratory experiments large quantities of metals are added as soluble salts during a short period of time, rather than gradually added during a long period of time, which is more indicative of what one may find in natural settings. In addition to possible changes in metal sorption mechanisms due to this discrepancy, there also is a difference to biological communities. If added gradually over time, a metal can constitute a constant stress that can be endured, avoided or overcome whereas an immediate addition of metal leads to a drastic and sudden disturbance that does not allow for adaptation (Giller et al., 1998). Despite the shortcomings in laboratory approaches to simulate field conditions, these types of studies have paved the way to understand metal speciation in soils and, therefore, will be referred to in upcoming sections of this chapter.

METAL SPECIES AND REACTIONS IN SOILS

The general category of reactions considered the most important with respect to metal speciation in soils is sorption. Sorption is a general term that encompasses many different mechanisms and refers to the general removal of a metal ion from solution and its subsequent association with the soil solid fraction. The reverse of this process, or the removal of a metal from a solid material and introduction into the soil solution, is termed desorption. The various mechanisms of metal sorption that occur in soils are illustrated in Fig. 13–2. These mechanisms will be discussed in the following subsections. Sorption reactions of metals in soils to a large extent dictate their mobility, fate, and bioavailability and are therefore vital to understand when attempting to understand metal speciation. The removal of metals from soil solutions by inorganic and organic phases is a process by which toxic metals can be sequestered, potentially alleviating deleterious environmental effects. It is possible that several mechanisms may contribute to the removal of a metal ion from solution concurrently. One way of considering the relationship between sorption of metals on soil components and metal speciation is to think of sorption as the reaction that involves the metal ion and the speciation as the end product of this reaction; however it is important to note that just as the speciation of a metal changes over time, sorption is also a dynamic process. The speciation of a metal at any one time is merely a snapshot and it is subject to changes as the sorption mechanisms change. Figure 13–3 illustrates the time scales of many metal sorption processes in soils and demonstrates the fact that, with respect to both time and metal concentration, several mechanisms of sorption may overlap with one another.

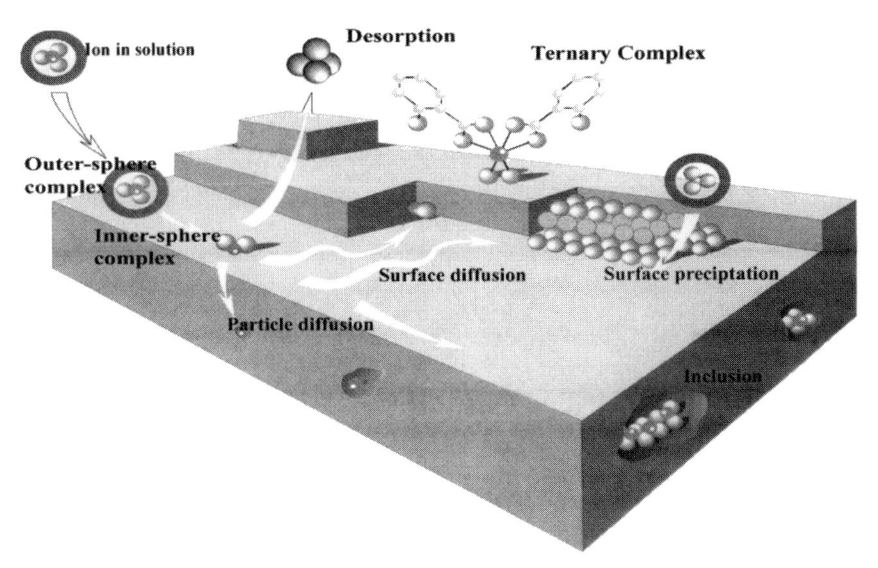

Fig. 13–2. Conceptual drawing of the various metal sorption processes that can occur on mineral surfaces (after Manceau et al., 2002).

In addition to time being a significant factor in determining metal speciation, the presence of crystalline and amorphous inorganic phases and organic material plays an important role in metal sorption and speciation. The solid fraction of a soil is a collection of non-living, living, and previously living material all capable of re-

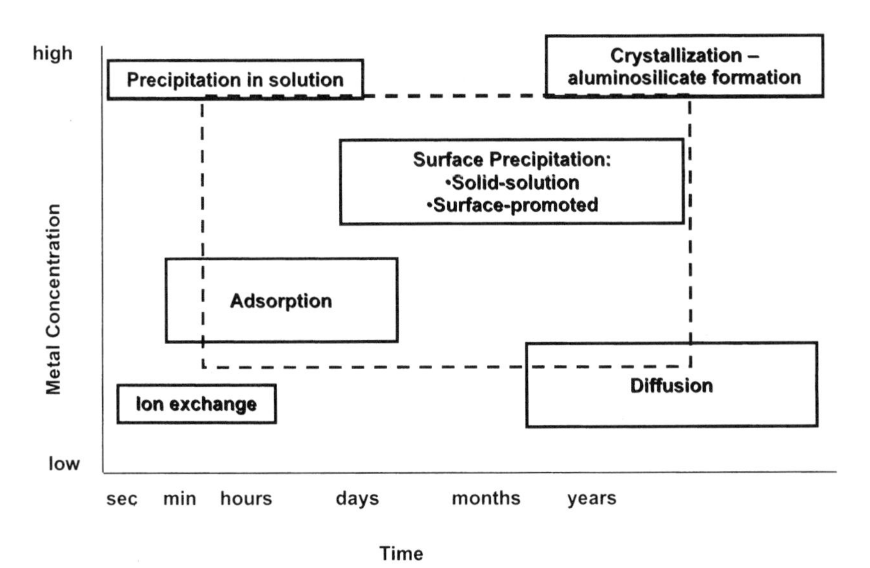

Fig. 13–3. The effect of time scales and metal concentration on various metal sorption mechanisms. The dotted line serves to connect many of the various mechanisms due to the potential continuum of processes (adapted from McBride, 1994).

acting with metal ions. Primary minerals supplied by the soil's parent material weather into secondary minerals while also releasing trace metals that may have been incorporated during formation. Secondary soil minerals include phyllosilicates (clay minerals), metal oxides, carbonate minerals, and sulfates. The oxides, hydroxides and oxyhydroxides of Al (gibbsite, bohemite), Mn (birnessite, pyrolusite), and Fe (goethite, hematite, ferrihydrite) are not the most abundant phases in soils, but they typically possess high surface reactivity and have large surface areas, so they are quite important in the sequestration of metals in soils and in many cases can be the primary reactive phases with respect to metal sorption (Manceau et al., 2002). The two types of surface sites on inorganic solids include permanent charge sites arising from vacancies or isomorphous substitution and sites originating from broken mineral lattice, known as variable charge sites (Charlet and Manceau, 1993). Aluminol and silanol groups occur on the edges of clay minerals and are variable charge sites. The variable charge functional groups found on organic molecules, include COOH, phenolic, alcoholic, and enolic groups. The reactions discussed in the following sections may take place at the various sites on both inorganic and organic phases.

Exchangeable Metal Ions (Outer-Sphere Complexes)

Both inorganic and organic solids in soils possess permanent charge sites that are often negatively charged, depending on pH. Positively charged metal cations that come into contact with these sites may form an electrostatic bond of low energy often referred to as cation exchange. In soil science, the term "cation exchange" is used to characterize the replacement of one adsorbed, readily exchangeable cation by another (Sposito, 1989). The metal cation in the soil solution that exchanges with one on the surface (or Ca^{2+}, Na^{+}, etc.) forms an outer sphere complex. For clay minerals, this type of reaction occurs mainly at planar sites of permanent structural charge and is therefore pH independent. Only for clay minerals with low structural charge (e.g., pyrophyllite) does significant electrostatic bonding at the clay edge sites of variable charge take place. In these systems electrostatic bonding is pH dependent (Stumm, 1992). Similar pH-dependent outer sphere complexation occurs between metals and organic matter. Indeed, organic matter is crucial in metal speciation in soils with its variable, and often high, cation exchange capacity. This is realized if one considers that the cation exchange capacity (CEC) of organic matter incrementally increased from 36 cmol kg^{-1} at pH 2.5 to 215 cmol kg^{-1} at pH 8.0, or 45% of the total CEC of the soil in a study on 60 Wisconsin soils (Helling et al., 1964).

The process of ion exchange and the formation of an outer sphere complex on a clay surface can be illustrated as follows (McBride, 1994):

$$Me^{n+} + nNa^{+} - \text{clay} \leftrightarrow Me^{n+} - \text{clay} + nNa^{+} \qquad [1]$$

where Me^{n+} is a metal cation with valence n.

In general, multivalent cations effectively displace monovalent cations from clay exchange sites when the monovalent cation concentration is low. Studies on Na^{+}–Me^{2+} exchange reactions on Na^{+}-saturated montmorillonite have shown that

Na^+–Me^{2+} is pH-independent below pH 6 (Inskeep and Baham, 1983). If, however, the monovalent cation concentration is high (e.g., high ionic strength), the competition for exchange sites may induce formation of adsorption complexes between the metal ion and surface. The dependence on ionic strength is one of the characteristic features of ion exchange–outer sphere complexation and is often used as a macroscopic assessment to determine if this sorption mechanism is operational.

Ion exchange reactions at surface sites exposed to solution are extremely fast and are difficult to measure by conventional methods. Cation exchange on 1:1 clays without interlayer regions (e.g., kaolinite) and 2:1 clays with expanded interlayer regions (e.g., montmorillonite) appears to be instantaneous (McBride, 1994). Kinetics of metal exchange is much slower on 2:1 interlayered minerals that may have K^+ within the interlayer region (e.g., vermiculite), since many of the exchange sites inaccessible for exchange with metal ions and the exchange process is diffusion limited (Sparks, 1995). In addition to having rapid kinetics of formation, outer sphere complexes are typically fully reversible and therefore do not represent a significantly stable metal sequestration pathway in most soil environments; however, the fact that this process is fairly rapid and energetically favorable is crucial if one considers the importance of micronutrient availability to plant roots, as ion exchange is the primary mechanism for this process. Just as the formation is fairly easy and fast, so is the reverse process. During long time scales, outer sphere complexes are not stable and will most likely convert to more stable sorption complexes.

Specifically Adsorbed Metal Species (Inner Sphere Complexes)

If a metal ion forms an ionic or covalent bond directly with a surface functional group, a stable molecular entity termed an inner sphere complex forms, otherwise known as a specific adsorption complex. These types of complexes do not have a water molecule present between the surface group and metal ion, resulting in a stronger bond compared to the electrostatic interaction of an outer sphere complex. Inner sphere complexes can be further categorized as monodentate if the metal is bound to one surface oxygen and bidentate if it is bonded to two (Sparks, 1995). Inner sphere adsorption complexes have been directly observed and established as quantitatively important species in soils contaminated with Pb (Morin et al., 1999) and Zn (Roberts et al., 2002). Adsorption complexes are two-dimensional molecular arrangements and do not include the formation of three-dimensional phases or diffusion phenomena, as will be discussed further in this section.

The adsorption of a metal ion, M, on an octahedral aluminol site has the following generalized reaction (McBride, 1994):

$$>\mathrm{Al{-}OH]}^{-1/2} + M(\mathrm{H_2O})_6^{\mathrm{n+}} \rightarrow\ >\mathrm{Al{-}O{-}M(H_2O)_5}^{(\mathrm{n}-3/2)+} + \mathrm{H^+} \qquad [2]$$

Reaction [2] is an example of a monodentate adsorption complex since only one oxygen group has participated in the reaction. A bidentate inner sphere complex has the generalized form:

$$2 > \mathrm{S} - \mathrm{OH} + M^{n+} = > (\mathrm{S} - \mathrm{O})_2\, M^{(n-2)+} + 2\mathrm{H}^+ \qquad [3]$$

where S–OH may be either a silanol or aluminal group.

Reactions [2] and [3] have at least four features that distinguish them from cation exchange (McBride, 1994):

1. Release of H^+ ions as Me^{n+} cations are adsorbed.
2. A high degree of specificity.
3. A desorption rate that is orders of magnitude slower than the adsorption rate.
4. A change in the measured surface charge toward a more positive value.

In addition to the above observations, further generalizations regarding metal adsorption via inner sphere complexes can be made:

1. As adsorption proceeds, it will affect the speciation of trace metals and ligands that remain in solution.
2. In general, >Al–OH groups are more effective at adsorbing metal ions than >Si–OH groups on mineral surfaces.
3. The more electronegative a metal, the higher its preference for adsorption on a reactive site on a mineral.

Abrupt increases in divalent metal ion adsorption in soils occur over a critical pH range, often less that one unit, termed the pH edge. This tends to correspond to the point where metal ions hydrolyze to form MOH^+ (Jones and Jarvis, 1981); however, changes in reactive sites on the sorbent phase as a result of pH changes also can play a role in this observed edge. Figure 13–4 shows a typical pH edge in the case of Ni on a soil clay fraction. To obtain this edge, an experiment was conducted in which all parameters (time, temperature, metal concentration) were held constant except for pH.

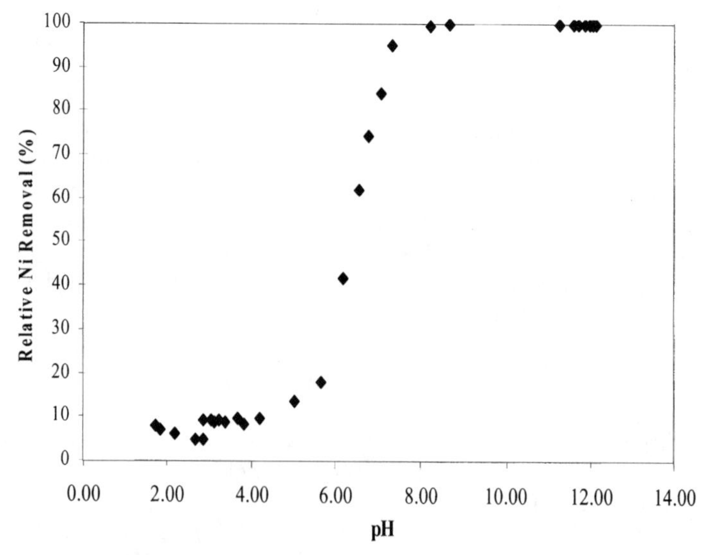

Fig. 13–4. Experimental Ni pH edge for Ni reacted with a soil clay fraction (from Roberts et al., 1999).

The methods of determining the mechanisms of metal adsorption (including inner sphere and outer sphere complexation) in soils include direct identification, macroscopic approaches, and one or both of the latter two combined with adsorption models. Direct identification methods will be discussed in the section on determining metal speciation. Macroscopically, adsorption isotherm experiments are often used to describe the relation between the equilibrium concentration of a metal ion in solution and the quantity of metal adsorbed to a solid surface. There are classically four types of isotherms based on the curve shape one gets when plotting amount of metal in solution vs. amount of metal adsorbed: S (s shaped), L (Langmuir), H (high affinity with steep slope), and C (linear) (Sparks, 1995). Often times experimentalists have assigned a particular mechanism to metal removal from solution by a solid phase based on the shape of the adsorption isotherm. Caution should be taken in making this type of assessment as these isotherms are based purely on macroscopic observations and in no way reveal any mechanistic information. Surface complexation models are further used to describe metal adsorption on soil surfaces, but again do not provide direct identification of metal species. A thorough review of surface complexation models as applied to soil chemistry can be found elsewhere (Goldberg, 1992).

The potentially strong inner sphere complex formed between a metal cation and a sorbent phase can provide an effective way to immobilize metals in soil environments. These phases may be quite stable over time and therefore should be considered in any metal speciation assessment. It is also important to realize that outer sphere and inner sphere complexes may not be mutually exclusive, and typically one may find a continuum between the two mechanisms exists. This is the case with any of the mechanisms that remove metal ions out of the soil solution. It is merely for the sake of simplicity of explanation of concepts that the various reactions of metals in soils are separated into separate sections in this chapter. The continuum phenomenon is especially evident in the case of adsorption and precipitation, as will be discussed in a later section.

Ternary Adsorption Complexes

In soils, metals are rarely the only species found in the soil solution and are often found complexed to both organic and inorganic ligands. For this reason metal adsorption may be different from the fairly "clean" description described above. Ligands are classified as atoms or molecules capable of donating electrons in a bond. By this definition, the oxygen atoms associated with silanol and aluminol groups on soil minerals are ligands. In solution, ligands can be inorganic, such as Cl^-, CO_3^{2-}, and SO_4^{2-}, or organic such as carboxyl and phenolic sites associated dissolved organic matter (DOM) (Sparks, 1995). Most metals discussed in this chapter are capable of reacting with both types of ligands in soils. The possible scenarios encountered include metal-ligand complexes that remain in solution, precipitated metal-ligand compounds, and metal-ligand complexes that adsorb on the sorbent phase (ternary complexes). The general effects of metal-complexing ligands in the soil solution on the adsorption of metal cations to soil minerals can be classified as follows (Sposito, 1989):

1. The ligand has a low affinity for the metal and for the adsorbent.
2. The ligand has a high affinity for the metal and forms a soluble complex with it, and this complex has a low affinity for the adsorbent.
3. The ligand has a high affinity for the metal and forms a soluble complex with it, and this complex has a high affinity for the adsorbent.
4. The ligand has a high affinity for the adsorbent, and the adsorbed ligand has a low affinity for the metal.
5. The ligand has a high affinity for the adsorbent, and the adsorbed ligand has a high affinity for the metal.
6. The metal has a high affinity for the adsorbent, and the adsorbed metal has a high affinity for the ligand.

Categories 3 and 5 result directly in enhanced metal adsorption from the presence of ligands by forming metal-ligand ternary complexes. Ternary complex formation can be represented by the following equations:

$$>\text{S–OH} + \text{M} + \text{L} = >\text{S–O–M–L} + \text{H}^+ \quad [4]$$

or

$$>\text{S–OH} + \text{M} + \text{L} = >\text{S–L–M} + \text{OH}^+ \quad [5]$$

where S–OH is the surface functional group, M is the metal, and L is the ligand.

In Reaction [4], the metal bonds to the surface functional group, and the ligand to the metal. In Reaction [5], the ligand is between the surface functional group and the metal. Due to ternary complex formation, solubilities of metals and anions in soils are lowered below those expected from either adsorption or precipitation.

Examples of ternary complex formation are quite difficult to demonstrate directly in actual soils, but in simulated laboratory systems recent advances in characterizing these reactions have been made. It has been demonstrated that in the case of U(VI) complexation to hematite in the presence of carbonate, a hematite-U(VI)-carbonato structure formed, similar to Reaction [4] (Bargar et al., 2000). Elzinga et al. (2001) demonstrated the formation of Pb–sulfate ternary complexes on the surface of goethite by probing both the sulfate and Pb with infrared spectroscopy and x-ray absorbtion spectroscopy, respectively. Details of these techniques will be discussed in the next section. In addition to ternary complex formation, the Pb promoted the adsorption of sulfate to goethite as a result of Pb changing the surface charge. The complexes formed in this system are presented in Fig. 13–5. In Fig. 13–5a, Pb and SO_4^{2-} complex directly to the goethite surface independently, but some interaction was still observed. In Fig. 13–5b, a similar complex as Reaction [4] above is seen. Whether the complex forms in solution prior to complexation or if the sulfate first adsorbs to the surface followed by complexation of the Pb to the sulfate was not determined. In soil environments, such ternary complexes may be more the rule than the exception, but few studies have successfully identified these complexes given the complexity of soils. The presence of ligands in an ion-sorbent complex has been shown to influence the atomic coordination environment of the ion and, therefore, may lead to differences in the stability of metal sorption complexes. One should keep in mind that both the solid and solution phases in soils are extremely

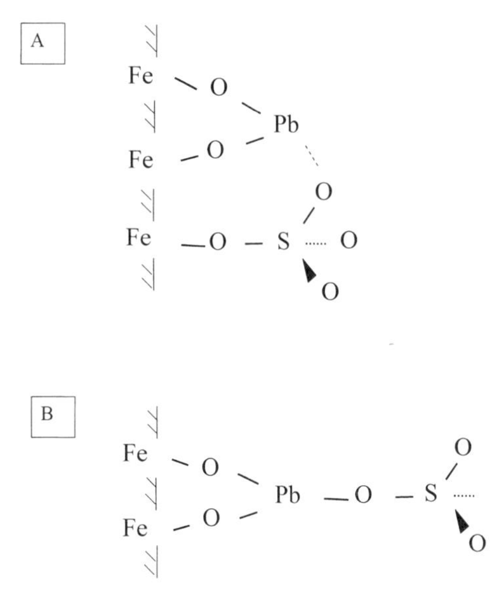

Fig. 13–5. Deduced surface complex formation for Pb and sulfate co-adsorbed on the surface of goethite. (A) Independent adsorption with some electrostatic interaction and (B) ternary complex (from Elzinga et al., 2001).

heterogeneous and competition for metals between sorption sites, ligands, and precipitated solids is commonplace.

Precipitated Metal Species

The previous two subsections have described metal species that were potentially easily bioavailable, especially if the sorption complexes had not aged and transformed to more stable entities and/or a perturbation of equilibrium induced metal release. Metal precipitates are potentially much more stable sinks for metals in soil environments and there are several pathways for their formation, both in the soil solution and on or near surfaces of solid phases in soils. In recent years, soil chemists have discovered that the occurrence of these phases in soils is probably more common than previously expected, thanks in part to the application of advanced analytical techniques (Ford et al., 1999; O'Day et al., 1994a; Roberts et al., 1999; Scheidegger and Sparks, 1996b). Prior to the application of techniques capable of directly probing metal species in soils and on soil minerals, macroscopic approaches were the main tool to decide whether or not a precipitated phase was a viable metal form.

One classical method for determining whether or not a solid precipitate is controlling the metal ions in the soil solution is the thermodynamic solubility product approach. In this approach, one compares the ion activity product (IAP) in soil solutions (assuming equilibrium) with the equilibrium ion activity products for various solid phases. The activity of any solid phase is defined as 1 if it exists in a pure form and is at standard pressure and temperature.

For example, in the case of a solid zinc hydroxide phase:

$$Zn(OH)_2 + 2H^+ = Zn^{2+} + 2H_2O \quad [6]$$

$$K_{dis} = (Zn^{2+})/(H^+)^2 = K^\circ_{so} \quad [7]$$

where K°_{so} = the thermodynamic solubility product constant and K°_{dis} = the thermodynamic dissolution constant. K°_{so} is numerically equal to K°_{dis} when the solid phase is pure. The right side of Reaction [6] is the ion activity product (IAP) and together with the K°_{so} is an index of whether the soil solution is in equilibrium with a given inorganic mineral component. If IAP/K°so = 1; it is in equilibrium. If IAP/K°so> 1 it is supersaturated; and if IAP/K°so < 1 the solution is undersaturated (Sparks, 1995).

The log K°_{dis} value can be calculated from the standard free energy accompanying the reactions (ΔG°_r):

$$\Delta G^\circ_r = \Sigma \Delta G^\circ_f \text{products} - \Sigma \Delta G^\circ_f \text{reactants} \quad [8]$$

where ΔG°_f is the standard free energy of formation.

Published values can be found for many solid phases for which standard free energy data are known (Lindsay, 1979). The minerals whose solubility products are equal to the measured IAP are assumed to be present and hence control metal activities in the soil solution. Unfortunately, this approach has many limitations. First, thermodynamic solubility data must be known for all potential precipitates, as well as their solid-solutions. In addition to equilibrium data not being known for solid phases, the dissolution–precipitation kinetic data also is limited. Often the most less-stable, more-soluble phase will precipitate out of solution faster than a more-stable, less-soluble solid phase. Also, this approach is most successful for elements with moderate to high total concentration in soils (Al, Fe, Ca) and it does not work as well for trace elements (Cu, Zn) unless the soil is grossly contaminated with the element in question (McBride, 1994).

A major shortcoming of using the equilibrium solubility approach is that it considers the precipitation of known phases from solution without consideration of a solid surface (clay minerals, oxides, organic matter). If one considers solid surfaces when investigating the formation of metal precipitates, it becomes clear that precipitation can occur under conditions in which bulk precipitation is not anticipated. According to Ford et al. (2001), the solid surface may promote metal precipitation by (i) the sorbent changing sorbate properties to drive precipitation and (ii) the sorbent modifying the solution composition near the mineral–water interface which induces precipitation. Their review article describes four scenarios under the general term of "surface precipitation": (i) increased metal ion activity at a mineral surface leading to precipitation, (ii) increased population of metal ions near the solid surface due to a net attractive force, (iii) a two-dimensional adsorption complex incorporating into the mineral structure as it continues to grow, and (iv) an unstable mineral surface dissolving as metal ions are sorbed, yielding a mixed metal precipitate phase. The last scenario does not necessarily require that the newly formed precipitate have a structural link to the substrate. Another term for this mechanism of surface precipitation is dissolution-induced homogeneous precipitation (Manceau et al., 2002). Considering the fact that soils are under a continuous state of weathering and are never truly at equilibrium, it stands to reason that Scenario 4 may be quite common in contaminated soils. For example, a mixed Ni–Al layered double hydroxide phase formed when Ni was reacted with a soil at pH 7.5 that contained many Al-bearing minerals (kaolinite, vermiculite, gibbsite) (Roberts et al., 1999). The identity of this phase in a soil sample would not have been possible without earlier experiments with Ni and reference Al-bearing minerals (Scheckel and Sparks, 2000; Scheidegger et al., 1997). The general formula for these phases can be written as:

$$\{Me^{2+}_{1-x}Me^{3+}_{x}(OH)_2\}^{+x} \bullet (x/n)A^{-n} \bullet mH_2O \qquad [9]$$

where Me^{2+} could be Co(II), Fe(II), Mg(II), Mn(II), Ni(II), or Zn(II), and Me^{3+} is Al(III), Cr(III), or Fe(III). Interlayer anions, A^{-n}, can be represented as Br^-, Cl^-, ClO_4^-, I^-, NO_3^-, or OH^-. The net positive charge, x, is counterbalanced by an equal negative charge, n. The remaining interlayer space is occupied by water molecules, m. The divalent and trivalent cations are distributed within the brucite-like octahedral hydroxide structure.

Several key observations have been made regarding the formation of solid solutions or mixed-metal surface precipitates in soil environments. In the past it was thought that most surface precipitates formed only after hours or days, but it has been seen that they can form in a matter of minutes (Elzinga and Sparks, 1999; Scheidegger et al., 1996). Over time these precipitates may become more resistant to dissolution since initial precipitates are often amorphous and have a higher free energy than crystalline phases of similar composition. The increased stability of these phases with time has been observed experimentally and in the case of Ni–Al layered double hydroxides has been attributed the gradual transformation to a precursor Ni phyllosilicate phase (Ford et al., 1999; Scheckel et al., 2000). In most cases the ionic radius of the metal must be small enough to allow it to enter octahedral sites. For this reason the relatively large Pb^{2+} ion can strongly sorb on aluminum hydroxide but it cannot substitute into the hydroxide during coprecipitation. This is a major consideration given the high toxicity level associated with Pb in soils. On the other hand, Cr^{3+} and Mn^{3+} can replace Fe^{3+} and Al^{3+} in precipitating oxides and hydroxides due to the similar ionic radii (McBride, 1994). Mn and Ni were found to be incorporated into both goethite and hematite, while Cd and Pb were not (Ford et al., 1997).

The pH of the soil also is a factor that may control the onset of surface precipitation. In experiments conducted from pH 6.0 to 7.5 it was found that Ni–Al layered double hydroxide phases only formed on a soil clay fraction above pH 6.8, regardless of reaction time. Figure 13–6 illustrates this finding, with sorption data at pH 6.0, 6.8, and 7.5 shown along with corresponding spectra from x-ray absorption fine structure (XAFS) spectroscopy in the right panel. At pH 6.0, even after several days, no sign of precipitation was found because the necessary threshold pH value was not met. The effect of this difference in sorption mechanisms influenced the release of the Ni, with the pH 6.0 sorption system releasing the greatest relative amount of Ni (Fig. 13–7).

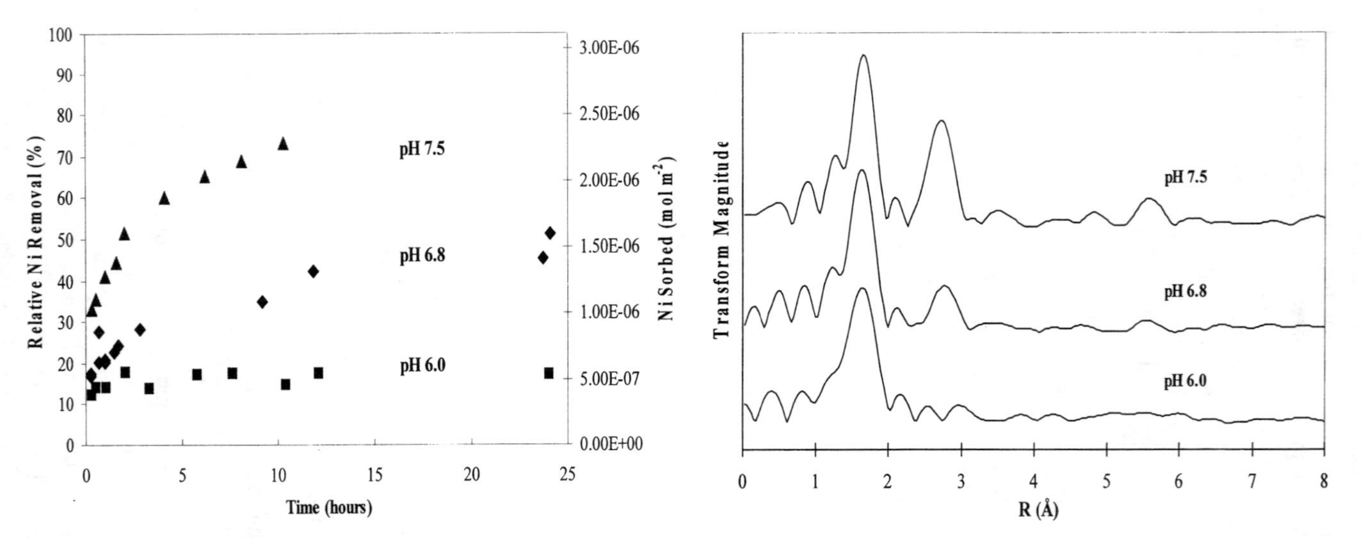

Fig. 13–6. Ni sorption kinetics on soil clay fraction at pH 6.0, 6.8, and 7.5. In the right hand panel, EXAFS radial structure functions demonstrate Ni–Al layered double hydroxide formation at pH 6.8 and 7.5, but not at pH 6.0 (from Roberts et al., 1999).

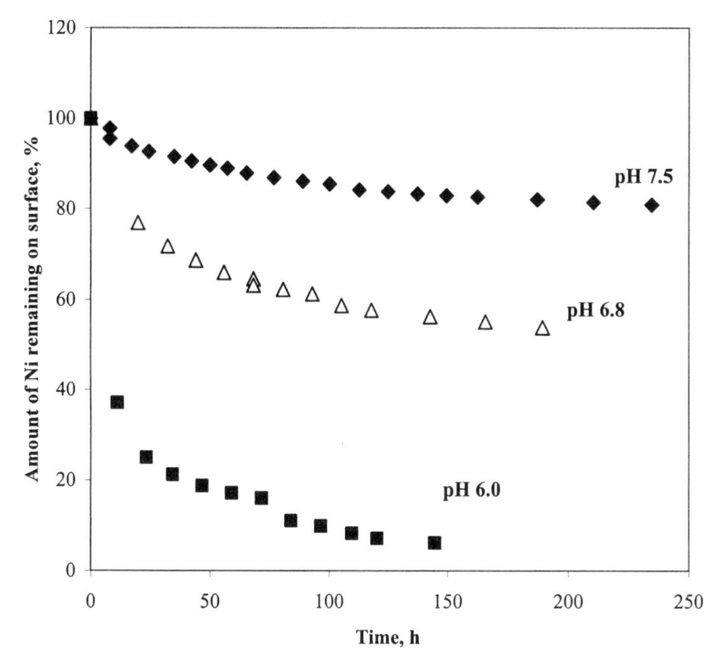

Fig. 13–7. Ni desorption–dissolution from reacted soil clay in Fig. 13–6 after reacting for several hundred hours.

Diffused Metal Species

Diffusion is a physical process whereby a substance can be removed from the bathing solution around a mineral phase and therefore falls under the broad classification of sorption. Soils are porous materials consisting of both macropores (>2 nm) and micropores (<2 nm) making diffusion a major mechanism of metal sorption, especially over long time scales found in natural settings (Sparks et al., 1999). For a metal ion to reach all potential sorption sites it must be transported through the bulk solution, travel through the liquid film on the solid surface (film diffusion), traverse pores within an individual particle (intra-particle pore diffusion) or pores between particles (inter-particle pore diffusion), and penetrate into the solid matrix (Pignatello and Xing, 1996; Sparks et al., 1999). Pore diffusion and matrix diffusion are considered as probably transport-limiting processes (Sparks et al., 1999). The sorbed metal ion diffuses into the sorbent, filling vacancies or substituting for sorbent atoms. This phenomenon may be responsible for the progressive decrease of metal mobility in soils.

There are several examples in the literature that ascribe slow metal sorption and hysteretic desorption behavior to diffusion processes. Brümmer et al. (1988) monitored the kinetics of Ni, Zn, and Cd adsorption and desorption by goethite. A diffusion-dependent adsorption and fixation process of the metal cations inside the goethite structure was proposed as the reason for the observed desorption hysteresis. For Cd and Se adsorption on porous aluminum oxide, XPS results verified that adsorbate intra-particle diffusion followed by sorption was the predominant Cd and

Se uptake mechanism (Papelis, 1995). For Mn, Co, Ni, and Cu cations, nonreversible adsorption was observed after sorption and aging on goethite and was attributed to slow pore diffusion (Coughlin and Stone, 1995). Both diffusion and precipitation can yield effective means of natural attenuation of metal contaminants; however, if extreme changes were to occur, such as a drop in pH due to acid rain deposition, these sequestered metals could provide a major source of metal contamination. This may require the assurance that the metal contaminated soil is stabilized by lime or some other amendment that can buffer the pH effectively.

Redox Reactions

In soils, abiotic and biotic reactions involving metals and many other species are continuously taking place and usually involve proton and electron transfers. A reaction in which a loss of electrons occurs is termed an oxidation reaction, while reduction is the gain of electrons (Sparks, 1995). The two reactions are not exclusive since the loss and gain of e^- must be balanced. Moreover, the electrons produced in a reaction are not species in solution, so oxidants (the species accepting electrons) and reductants (the species donating electrons) must be in immediate contact for the transfer to occur. In well-drained surface soils the main redox couple reaction that occurs is the production of e^- due to the biological oxidation of organic matter (simplified as CH_2O):

$$1/4CH_2O + 1/2H_2O = 1/4CO_2 + e^- + H^+ \tag{10}$$

$$O_2^- + e^- + H^+ = 1/2O_2 + 1/2H_2O \tag{11}$$

The reason O_2 is the favored oxidant is that its log $K°$ value is greater than those of other potential oxidants. The sequence of oxidants in soils are usually considered (from most energetically favorable to least): O_2 > NO_3^- > Mn(III/IV) > Fe(III) > SO_4^{2-}. A soil under water- logged or flooded conditions may see reductive dissolution of solid Mn (IV) and Fe (III) phases, and eventually reduce SO_4^{2-} to produce sulfide minerals, such as FeS_2 (pyrite). The extent to which a soil is reduced or oxidized is generally assessed by the values Eh and pe. Eh is a redox potential and is expressed in terms of electrochemical energy (millivolts) and assumes a system is at thermodynamic equilibrium. Oxidized soils have values ranging from +400 to +700 mV while reduced soils may have values from −250 to −300 mV (Sparks, 1995). The pe is the $-\log(e^-)$ and is an index of the electron free energy level per mole of electrons. In a redox couple, the oxidant will be reduced if the pe value of its half reaction is greater than the pe value of the reductant half reaction at a particular pH. A more thorough treatment of redox behavior in soils can be found in Chapter 9 of this text or in the review chapter by Bartlett (1998).

The redox status of a soil is an important factor in considering metal speciation in soils for both direct and indirect reasons. Metals that are directly sensitive to redox changes in soils include As, Cr, Hg, and Se. They all have reduced and oxidized forms and in many cases the redox status influences their toxicity and mobility. For example, reduced Cr(III) has a low hazard associated with it while oxidized Cr(VI) can be quite toxic to organisms (Fendorf and Sparks, 1994). The opposite holds true in the case of arsenic, with As(III), aresenite, being much more toxic to humans than As(IV), arsenate.

Perhaps the most important metals when it comes to redox reactions in soils are Fe and Mn. While we have not considered these elements as metals of interest for toxicity or deficiency purposes, the oxides and oxyhydroxides of Fe and Mn are well recognized as being indirectly involved in metal speciation due to adsorption, precipitation, and redox reactions. In oxidizing conditions, Fe(III) and Mn(III/IV) oxides are extremely common in soils and are known to retain metals given their reactivity and high surface areas. In reduced conditions, however, Fe and Mn oxides are subject to reductive dissolution and metals associated with these phases such as Cd, Ni, Pb, and Zn can have dramatic changes in their solubility and mobility in soils; however, if a soil is flooded for a long period of time, sulfide formation may ensue and immobilize the metal ions in the structure. This is the reasoning behind the construction of wetlands to sequester metals that may be toxic or mobile if not present in the sulfide fraction.

In soils, minerals such as δ–MnO_2 are present and are thermodynamically capable of oxidizing Cr(III) to Cr(VI) by the following reaction:

$$Cr(OH)^{+} + 1.5\ \delta\text{–}MnO_2 \leftrightarrow HCrO_4^{-} + 1.5\ Mn^{2+} \qquad [12]$$

While one might be concerned that any Cr(III) introduced to soils containing Mn oxides would transform into the more toxic species, another reaction may compete with this one at pH values >3:

$$Cr(OH)_{3(s)} + 1.5\ \delta\text{–}MnO_2 + 2H^{+} \leftrightarrow HCrO_4^{-} + 1.5\ Mn^{2+}\ 2H_2O \qquad [13]$$

Fendorf and Zasoski (1992) used several advanced analytical techniques to show that Reaction [13] occurred at elevated pH and Cr(III) concentrations, thereby covering the δ–MnO_2 surface with $Cr(OH)_{3(s)}$ and limiting the further oxidation of Cr(III) to Cr(VI). Given the seasonal variability of redox conditions in soils, it is important to consider redox reactions on the mobilization and sequestration of metals (Bostick et al., 2001; O'Day et al., 1998).

TECHNIQUES TO DETERMINE METAL SPECIATION IN SOILS

Metal speciation in soil continues to be a dynamic topic and of interest to soil scientists, engineers, toxicologists, and geochemists alike, as all of these disciplines do require knowledge of soil speciation to varying degrees. The accuracy and precision that is required will often dictate the approach one takes in determining metal speciation. In the case of molecular environmental science, one is often interesting in obtaining the most detailed information possible, with the result being the use of advanced analytical techniques. Often, these analytical techniques have been developed in other disciplines such as surface science, solid-state physics and medical research. Soil chemistry has benefited dramatically from the application of techniques from other disciplines to study metal speciation. The combination of several techniques still remains the most prudent, practical, and comprehensive approach to metal speciation in soils since there is no single method that is a panacea. While certain methods may have good detection limits and can speciate metals in trace amounts, they may not be ideally suited to provide structural or spatial information. Therefore, the most thorough and appropriate treatment of metal speciation methods should include a wide range of approaches, from classical methods to the cutting edge of technology.

Analytical Methods to Determine Metal Concentrations

While the majority of soil chemists agree that the total metal concentration does little to reveal information about metal speciation and gives no indication of bioavailability, it is still a necessary analytical step. For many of the methods that will be discussed, determination of metal concentration in either the solid soil matrix or in the soil solution (or extraction solution) is required. The total metal concentration of a soil has been used as a rough estimate to the degree of contamination of a soil. For instance, a partitioning coefficient, K_d, is defined as follows:

$$K_d = \frac{\text{Total Soil Metal}}{\text{Dissolved Metal}} \quad [14]$$

Total soil metal has a concentration of mg kg^{-1} and can include metals in minerals, as precipitates and as adsorption complexes. The dissolved metal is expressed as mg L^{-1}, so the units for K_d are L kg^{-1}. While a simple concept and equation, the actual relationship is controlled by many factors. This parameter is frequently used to estimate the potential availability of a metal to a plant or other organism; however, the metals in mineral phases typically are less available than the other two contributors to soil metals, so K_d values can often be overestimated. This determination alone, however, does little to reveal anything about the metal species and on its own is not a useful measure of the risk of metal bioavailability.

The soil solution contains the most mobile and potentially available metal species and are often found at very low concentrations and require sensitive analytical techniques to be measured (Walther, 1996). By measuring the metal concentration in a pristine soil solution, one would already be able to make some good predictions on the likelihood a soil poses a risk. The concentration of a particular ion in the soil solution (intensity factor) and the ability of solid components in soils to resupply an ion that is depleted from the soil solution (capacity factor) are both important properties of a given metal contaminated soil (Sparks, 1995). Water extraction and salt extractions can be used to determine the intensity factor, but the latter method may induce complexation of metals with inorganic anions. Tension-cup lysimiters can be used to collect soil solutions from the field. Metal ion concentrations of the soil solution can be done by spectrometry, chromatography, and colorimetry. To speciate the soil solution, it is necessary to apply ion association of speciation models as direct determination of all individual species is not possible.

There are several analytical techniques that can be used to measure metal concentrations in the soil solution or any solution in which the metal might be present (e.g., in extracting solutions). For soils, the solids are normally dissolved prior to measurement in a solution strong enough to cause total dissolution of all solid phases. Solutions of this type include microwave- HNO_3 digestion, hot-plate reflux digestion with HNO_3–HCl–H_2O_2, aqua regia-hydrofluoric acid mixtures, and digestion in hydrofluoric acid. The solutions can then be analyzed using one of several analytical techniques including atomic absorption spectrometry (AAS) and inductively coupled plasma-mass spectrometry (ICP-MS) spectrometry. X-ray fluorescence (XRF) also is a means of determining total metal concentration in soils and benefits from not having to use harsh chemicals for total dissolution because the analysis is done using the whole soil, but metal concentrations normally have to be upwards of 5 mg kg^{-1}. In addition, metal concentrations in solid slurries may be gleaned using ICP-MS.

Single Chemical Extraction Techniques

The use of single (one-step) extraction techniques finds the most application in soil fertility assessment in order to predict deficiencies or toxicities of trace elements. This approach to metal speciation considers several chemical pools of trace metals that share a common function (Walther, 1996). These definitions include plant available form, exchangeable cation, and labile species. Examples of solutions used to extract these chemical pools include ethylenediaminetetraacetic acid (EDTA), diethylenetriaminepentaacetic acid (DTPA), acetic acid, salt solutions, and water. While water is clearly the least expensive and simplest extracting solution for soils, it is not necessarily the best extracting solution to use, since salt-free water is rarely found in natural soil waters. Dilute salt solutions with a concentration of at least 0.0001 *M* such as $CaCl_2$, $Ca(NO_3)_2$, KNO_3, and $NaNO_3$ are routinely used. The routine use of these single extractions in soil fertility laboratories worldwide underscores their importance. This importance has led to several studies aimed at assessing extraction validity using plant uptake and crop yield studies. Alterations and adjustments to the extracting solutions have been implemented depending on the type of soils in the region of interest. Clearly, single extractions cannot estimate the amount of slowly-available metal that is released over time since extractions are carried out during a period of several hours. Moreover, the exact speciation of the metal is not gleaned using this type of approach. Despite these shortcomings, single extraction techniques will continue to be useful for both soil fertility and soil quality investigations.

Selective Sequential Extraction Techniques

A more rigorous and species-specific alternative to determining metal speciation than total metal concentration and one-step extractions is the use of selective sequential extractions. The purpose of sequential extractions is to provide detailed information on metal origin, biological and physicochemical availability, mobilization, and transport (Tessier et al., 1979). This approach to metal speciation conceptualizes soil as having several fractions that metals can be associated with, and these specific fractions can be attacked by chemicals specific to each individual fraction. Previous advances in soil chemical analysis that aimed at characterizing these fractions without the immediate goal of metal speciation were used to develop sequential extraction methods. Typically, the first step in a sequential extraction process extracts the most labile metals, and each successive step increases in strength until the most non-labile fraction is left. The most widely cited procedure for sequential extraction is in the article by Tessier et al. (1979). This procedure was developed to speciate metals in contaminated river sediments and defined five fractions of metals: (i) exchangeable, (ii) bound to carbonates, (iii) bound to iron and manganese oxides, (iv) bound to organic matter, and (v) residual. After many studies and refinements, chemical extractions steps also are designed to selectively extract physically and chemically sorbed metal ions, metal sulfides, and metals in other fractions. Researchers have altered the technique to account for variations in soil pH, soil texture, metal concentration, redox status, and other parameters that can vary from soil to soil and regionally. The resulting extract is operationally defined based on the proposed chemical association between the extracted species and solid phases in which it is associated. Given that the extraction is operationally defined, the extracted metal may or may not represent true chemical species that it is given, so care must be taken to report the step in which in was removed rather than the phases it is associated with.

The use of sequential extractions for metal speciation is not without its limitations and pitfalls. These include (i) the incomplete dissolution of target phases, (ii) the removal of a non- target species, (iii) the incomplete removal of a dissolved species due to re-adsorption on remaining soil components or due to re-precipitation with the added reagent, and (iv) change of the valence of redox-sensitive elements (Brümmer et al., 1983; Calmano et al., 2001; Gruebel et al., 1988; La Force and Fendorf, 2000; Ostergren et al., 1999). These limitations are becoming more evident as research coupling sequential extractions with analytical techniques capable of directly determining metal speciation in soils and sediments is performed (Adamo et al., 1996; Brümmer et al., 1983; Calmano et al., 2001; Gruebel et al., 1988; Henderson et al., 1998; La Force and Fendorf, 2000; Ostergren et al., 1999).

Given the fact that sequential extractions are the most common means of determining metal speciation in soils and sediments and other geomedia, refinement and improvements of this procedure are desired. The coupling of direct speciation procedures will enable extractions to become more complete and universal, significantly improving our understanding of metal partitioning and mobility in soils. Despite the limitations of sequential extraction procedures, they will continue to be valuable for relative comparisons between contaminated sites. Moreover, these techniques are readily available, economically practical, and provide quantitative results rapidly. Combined with other speciation techniques and separation of physical phases in soils based on particle size, magnetic separation, and density gradient separation, extraction techniques can be rather robust at revealing metal speciation in soils.

Fundamentals of Spectroscopy and Microscopy

Both spectroscopy and microscopy rely on harnessing various wavelength regions on the electromagnetic spectrum and bombarding a sample in order to glean chemical and physical details. The energy that is directed at a sample results in several potentially complicated processes, but can be generalized as a transition of an atom from a ground state to an excited state. This transition occurs at a very particular wavelength because atomic processes are quantized (O'Day, 1999). Spectroscopy and microscopy deal with the interaction of electromagnetic radiation with matter (Bertsch and Hunter, 1998). The broad energy range of the electromagnetic spectrum and the various means of measuring the various excitation phenomena that occur from the radiation bombarding the sample yield a large array of spectroscopic and microscopic techniques. Several sources provide overviews of these techniques and provide lists of acronyms (Calas and Hawthorne, 1988; Bertsch and Hunter, 1998; O'Day, 1999). The techniques useful for metal speciation in soils narrows this list down, and it can be further shortened, depending on specific interests such as in-situ requirements, cost limitations, and availability.

The two main categories of spectroscopy are defined by the interaction between the applied radiation and the sample. If the incident radiation of a particular frequency excites an internal process, this is absorption spectroscopy. Spectroscopies that fall under this category include nuclear, electronic and vibrational spectroscopies. In the other type of spectroscopy, the incoming radiation induces the emission of radiation from the sample but with a different frequency. Spectroscopies that fall in this category include energy loss spectroscopy, elastic scattering and luminescence spectroscopy (Calas and Hawthorne, 1988). The former category of spectroscopy will be the main one discussed for the remainder of this chapter. The frequency, υ of the radiation is related to the wavelength, λ by the following relation:

$$c = \upsilon\lambda \quad [15]$$

Where c is the velocity of propagation in vacuum (Calas and Hawthorne, 1988). The regions of the electromagnetic spectrum each have an unique associated phenomena once it comes into contact with a sample. The shorter the wavelength of the radiation, the smaller the size of the object that can be detected. For this reason, γ-rays and x-rays can provide atomic-scale resolution, while nuclear magnetic resonance (NMR) has spatial resolution corresponding to the size of a large animal. While lower energy radiation (longer wavelength) such as infrared (IR) cannot elucidate processes at the atomic level, it is able to cause vibration of bonds between atoms, giving rise to absorption spectra that can reveal information about molecules.

Electrons, like photons, are absorbed, scattered, and diffracted by matter, yielding the desired chemical and structural information (Manceau et al., 2002). Also, electrons can be focused with magnets down to the angstrom scale, like in the case of transmission electron microscopy (TEM); however, electron microscopy cannot identify structural forms of metals associated with minerals. Electron diffraction would have the best of both: good resolution and the ability to glean structures; however, electron microprobes are not both element specific and sensitive to the type and distance of neighboring atoms. With any technique one should be aware of influences the probing energy has on the sample. This is especially the case in microscopy where strong interaction of electrons with matter can induce a change in oxidation state, especially for moist soil samples that are hydrated and potential meta-stable with respect electron beams (Manceau et al., 2002).

Spectroscopic Techniques

Several analytical tools prevalent in characterization of materials in the surface sciences, chemistry, physics, and geology have been applied to direct speciation of heavy metals in soils and sediments for a number of years. The clear advantage in using direct techniques over chemical extractions is the lower risk of sample alteration and transformations of metal species from using extracting solutions. When selecting an analytical technique to speciate and quantify the form of metals in complex heterogeneous materials such as soils and sediments, a selective and non-destructive one is favorable (Manceau et al., 1996). Non-invasive, in-situ spectroscopies are those that can collect a spectrum from a sample with little alteration to the sample relative to its original state. In the case of soils this is extremely useful as soils always have some solution present, and exposing the sample to drying, heating, or pressure can substantially alter metal species (e.g., As(III) may transform to As(IV)). For laboratory-based studies designed to study single sorbent–sorbate interactions this also is necessary since experiments are performed in the hydrated state and altering the sample may alter the experimental outcome.

Spectroscopies capable of collecting data in situ include fluorescence, ultraviolet-visible (UV-vis), IR, NMR, electron spin resonance (ESR), Mössbauer, and XAFS. With the exception of the last technique that will be detailed in the next section, all of these spectroscopic methods can be found in most analytical laboratories in the field of chemistry, environmental science, surface science or solid-state physics. A good review of these techniques and others for determining metal sorption mechanisms can be found in Scheidegger and Sparks (1996a). In general, these techniques have traditionally been developed to characterize relatively clean systems, free of many of the organic and inorganic phases found in soil. Since many of the original types of experiments performed using these tools did not change significantly when dried or manipulated, sample alterations have been common in order to optimize the signal of the measurement. Unfortunately, this approach has been used for soil samples that are much more sensitive to sample alterations. In addition, the traditional experiments using these techniques rarely suffered from low element concentrations so the techniques were not optimized for metal concentrations found in most soils. This has resulted in experiments using unrealistic elevated levels of metals or other ions in order to have high surface loadings and collect high quality spectra. Fortunately, advances have been made in order to make these techniques more applicable to speciation studies of trace elements and contaminants in soils or on soil minerals. The Fourier Transform approach to IR (ATR-FTIR) and Raman (FT- Raman) spectroscopies has resulted in many experiments investigating speciation of elements at the water–mineral interface.

The principal invasive non-in-situ techniques used for soil and aquatic systems are x-ray photoelectron spectroscopy (XPS), auger electron spectroscopy (AES), electron energy loss spectroscopy (EELS) and secondary mass spectroscopy (SIMS). Each of these techniques yields detailed information about the structure and binding of minerals and bonding of minerals and the chemical species present on the mineral surfaces. The disadvantage of these techniques is that they have to be performed under ultra high vacuum, where dehydration of the sample and the particle bombardment might lead to misleading data due to experimental artifacts

(Scheidegger and Sparks, 1996a). This might especially be the case for a hydrated surface complex. Stable phases in soils that are not sensitive to invasive techniques may be well-suited for these techniques, but in the case of adsorbed complexes, amorphous precipitates or redox sensitive species, they should be avoided if possible (Bertsch and Hunter, 1998).

XPS is a surface analytical technique devised at the end of 1960s to provide chemical analyses of surfaces. X-ray spectroscopy uses x-rays as the stimulator, and photoelectrons are detected in response. For Auger spectroscopy, either x-rays or electrons can be used to generate Auger electrons whose energies are typically used for elemental identification only. Since electrons can only travel extremely short distances through solids without energy loss, these techniques are only sensitive to the near surface. The most important reason for using XPS and AES to study sorption reactions is that they are surface sensitive to most geochemically important elements. Another major advantage is that it can provide important information on the chemical state of the substrate surface before reaction, and both the substrate and chemical state of the sorbed species after reaction. Applications of XPS for determining metal speciation in soils include the study of Cr(III), Ni(II) and Cu(II) on chlorite, illite, kaolinite, and smectites; AsO_4^{3-}, CrO_4^{2-}, Zn^{2+} and Pb^{2+} sorption on ferrihydrite; and Cd and Se on corundum (Scheidegger and Sparks, 1996a).

Microscopic Techniques

Given the myriad of reactive phases in soils and their complex distribution in the soil matrix, a technique capable of providing spatial and morphological information on heavy metal speciation is desired. Microscopic techniques may resolve the different reactive sites in soil at the micron and submicron level, thus allowing for a more selective approach to speciation. Examples of these techniques include scanning electron microscopy (SEM), electron probe micro analysis (EPMA), and transmission electron microscopy (TEM). In order to glean elemental information and ratios, all the above techniques are often coupled with an energy dispersive spectrometer (EDS). While the above techniques have given insight into elemental associations and metal distributions in contaminated soils and sediments, they do have a few drawbacks. The most notable drawbacks are that EDS is only sensitive to >0.1% elemental concentration, it is insensitive to oxidation states of target elements, and it does not provide crystallographic data (La Force and Fendorf, 2000; Ostergren et al., 1999; Webb et al., 2000). A study investigating Zn speciation in contaminated sediments found that SEM coupled with x-ray EDS only provided elemental concentrations, but discerning between Zn sulfate and Zn sulfide was not possible (Webb et al., 2000). Similarly, electron microprobe analysis was unable to locate Hg grains within an Hg-contaminated sample and was unable to distinguish between polymorphs of Hg- bearing phases (cinnabar and metacinnabar) (Kim et al., 2000).

The application of scanning probe microscopy (SPM) has greatly advanced the understanding of the interaction of metals with solid phases. SPM represents a class of microscopic techniques that provide high resolution, multidimensional images of solid surfaces by monitoring the interactions between sharp tips and the surface (Bertsch and Hunter, 1998). SPM includes scanning tunneling microscopy (STM), atomic force microscopy (AFM), magnetic force microscopy (MFM) and chemical force microscopy (CFM). AFM, CFM, and MFM are all types of scanning force microscopies (SFM). SFM techniques are advantageous in that they are relatively low in cost and the samples may be run in-situ. SFM has been used to study the kinetics of Cr(III) sorption reactions on goethite and silica using a flow cell mounted in a SFM (Fendorf et al., 1996). Scanning force micrographs revealed the formation of a Cr(III) precipitate that was distributed across the entire surface of goethite while discrete surface clusters were observed on the SiO_2 surface. High resolution tunneling electron microscopy (HRTEM) was used to observe a kaolinite surface reacted with Co for several hours (Thompson, 1998). Co was spatially associated with Al and Si, suggesting an association with kaolinite. Reactions of metals at solid–solution interfaces have become possible to monitor thanks to the development of tapping-mode SFM and the fluid cell apparatus. In tapping-mode, the tip contacts the surface at a known frequency and is useful for using on fragile surfaces and limits artifacts caused by traditional SFM. With a fluid cell, the reaction of a metal with a surface and changes that may result due to dissolution or precipitation can be monitored. For example, AFM has been used to observe the fairly rapid growth of Ni-Al precipitates on pyrophyllite (Scheckel et al., 2000).

Synchrotron-Based Methods

The use of synchrotron light sources to address environmental issues has provided insight into the reaction mechanisms of heavy metals at interfaces between sorbent phases found in soils and the soil solution. Several synchrotron facilities are operational in the USA and more exist worldwide, most of which have beamlines dedicated to environmental research. In the last decade and a half, soil chemists and other environmental and earth scientist interested in determining the atomic coordination environment of target metals in geomedia have used such facilities. To date, the most widely used technique used at synchrotron facilities by these scientists is XAFS. XAFS results from the attenuation of x-rays by atoms of a given element yielding an absorption across a narrow energy range. This narrow energy range is the absorption edge and corresponds to the production of photoelectrons due to excitation of inner-core electrons by the x-ray photons. This occurs when the incident x-ray energy, E, is approximately equal to the binding energy of the core level electron, E_b and is the reason XAFS is element- specific (Schulze and Bertsch, 1995). The term XAFS is a general term encompassing a range of energies around an absorption edge for a specific element: the pre-edge, near-edge (XANES) and extended portion (EXAFS). Each region provides specific information on an element depending on the selected energy range, making XAFS an element specific technique. Several articles provide excellent overviews on the use of this technique in environmental samples (Fendorf et al., 1994; Schulze and Bertsch, 1995). Briefly, the first region is the low-energy side of the main absorption feature and is termed the preedge region. Preedge features are common for first row transition metals and can provide information to the oxidation state of the metal. In the XANES region, electron transitions lead to absorption edge from which chemical information of the target element, such as oxidation state, can be deduced. This region is often used for fingerprinting, or comparing known compounds to unknown samples. In addition, interatomic distances from the central absorbed to surrounding atoms can be estimated (Schulze and Bertsch, 1995). The extended region can provide the identity of the ligands surrounding the target element, specific bond distances, and coordination numbers of first and second shell ligands (Schulze and Bertsch, 1995). This information is extremely useful in speciation of metals in soils and sediments as it provides quantitative information on the geometry, composition and mode of attachment of a metal ion at a sorbent interface (Brown et al., 1999).

The type of spectra one collects during an XAFS experiment is displayed in Fig. 13–8. As one sees, the spectra require analysis in order to get detailed atomic information, such as the identity of the first and second neighboring atoms, coordination numbers, and atomic distance (in this case, for Zn). Theoretically, one can use this information to distinguish between outer sphere complexes, inner sphere complexes, and surface precipitates. This has certainly been successfully achieved in the case of studies involving one metal ion adsorbed onto one mineral surface. While the conditions of these types of studies may not be indicative of the conditions one may encounter in a field situation, they certainly have made the transition to more complex systems possible. As a result, several recent studies have used XAFS to quantitatively speciate metals in contaminated soils (O'Day et al., 1998; Manceau et al., 2000; Calmano et al., 2001; Roberts et al., 2002).

One of the major shortcomings of several analytical techniques used to speciate metals is the limited detection limit. Given the intensity of synchrotron facilities, this technique has a detection limit down to 50 ppm. Moreover, a specific metal of interest can be targeted, potentially with little interference from other elements in the complex matrix in which it is located. Gleaning molecular scale information in-situ is not possible with any other technique. Features that have dramatically increased the use of XAFS in environmental studies include; more synchrotron facilities are becoming available, more routine data analysis due to computer- based packages, and word of mouth via professional meetings and journal articles. Although not used by the majority of scientist working to speciate metals in soils, XAFS certainly has changed the way we think about metal speciation in soils and has revealed so many things that we were otherwise unaware of.

To date, standard, bulk XAFS has been the most widely used synchrotron-based technique used to characterize heavy metals in environmental samples; however, in soils and sediments, there exist microenvironments having isolated phases in higher concentrations relative to the average of the total matrix (Schulze and Bertsch, 1995). For example, the microenvironment of oxides, minerals and microorganisms in the soil rhizosphere has been shown to have a quite different chemical environment compared to the bulk soil (Wang et al., 2002). Often these phases may be very reactive and of significance in the partitioning of heavy metals, but quantitatively they are minority phases in the overall makeup of the soil and are therefore overlooked. As previously mentioned, electron microscopy can be used to provide micro-scale speciation of a metal in soil matrices, but they cannot provide all of the atomic and structural information of XAFS. The average x-ray beam size in an XAFS experiment is several centimeters. With focusing mirrors and other devices, the x-ray beam bombarding a sample may go down to a few square microns in area, nearing the size of the most reactive species in soils, enabling one to distinguish between individual species in a heterogeneous system. These same principles can be applied to XRD and one can employ μ-XRD to attain crystallo-

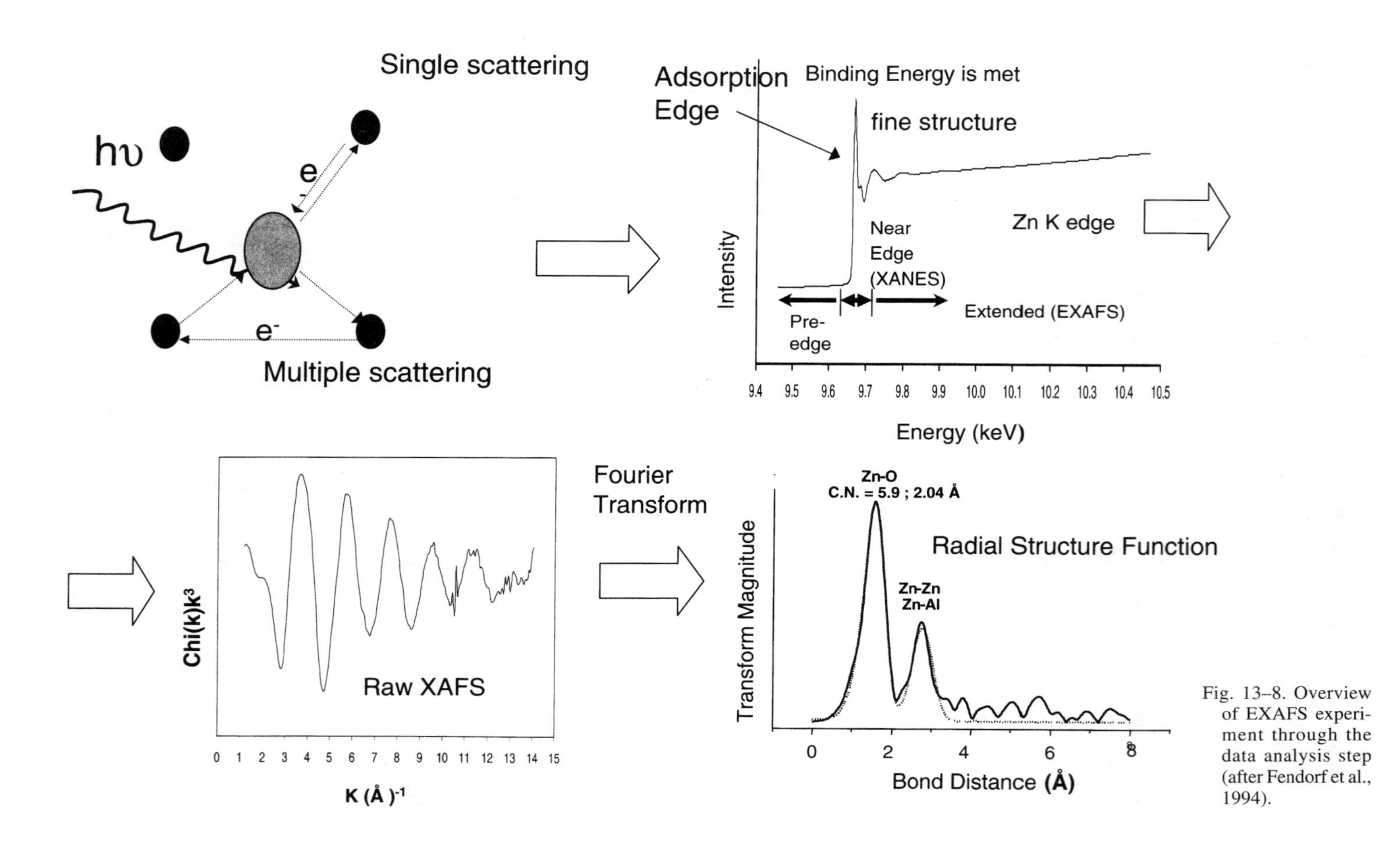

Fig. 13–8. Overview of EXAFS experiment through the data analysis step (after Fendorf et al., 1994).

graphic information on minerals that may occur in micrometer aggregates. In order to determine the exact location to place the focused x-ray beam on the sample, μ-XAFS is often combined with micro synchrotron-based XRF (μ-SXRF), allowing elemental maps to be obtained prior to analysis. While electron microprobe is often not sensitive enough to detect trace metals in soil, μ-SXRF offers sufficient sensitivity to investigate the spatial distribution of trace metals and their spatial correlation with other elements. Until recently, most studies have employed μ-XANES to determine the oxidation state of target elements in environmentally relevant samples since first and second generation light sources were not bright enough to achieve decent results for μ- EXAFS (Duff et al., 2001, 1999; Hunter and Bertsch, 1998; Manceau et al., 2000).

With the advent of brighter, third generation sources, μ-EXAFS has been used to speciate metals in soils and sediments (Isaure et al., 2002; Manceau et al., 2000; Roberts et al., 2002; Strawn et al., 2002). For example, Zn contaminated soils due to aerial deposition of smelter materials was probed using bulk EXAFS, μ-SXRF, and μ-EXAFS. The soil in question sits just below the soil surface and any Zn present is from the dissolution of Zn-bearing solid phases identified in the surface soil. Figure 13–9 presents the results of using all of these techniques. As one sees from the XRF elemental maps, Zn is spatially associated with both Fe and Mn in the sample, and also is concentrated in a region where neither element is present. The bulk XAFS reveal Zn has many second neighbor atoms, but distinguishing between them is difficult given the similar bond distances Zn shares with Fe and Mn atoms. With μ-XAFS, individual contributions from second neighbor Al, Fe, and Mn atoms can be observed and demonstrates that over an area of only a few millimeters, Zn can be adsorbed to three different phases. The results from a stirred-flow desorption experiment for the surface and subsurface soil demonstrated how a difference in speciation for Zn between the two soils influenced its release back into the soil solution. For the surface soil, Zn was in a fairly stable phase(s) and not easily dissolved. For the subsurface soil, Zn was more readily released into solution since adsorption complexes made up the majority of Zn species. A similar study by Manceau et al. (2000) use XRD, XAFS and μ-XAFS to demonstrate that upon weathering of Zn-mineral phases in soils, Zn was taken up by the formation of Zn-containing phyllosilicates and, to a lesser extent, by adsorption to Fe and Mn (oxyhydr)oxides. The major difference between the two Zn systems was the soil pH. In the former experiment, acidic pH values were operational, while in the latter study the soils were closer to neutral values. This demonstrates the influence pH has on metal speciation.

With XAFS, in order to discriminate between species and quantify them in a multi-species system, the species must have different oxidation states, or vary in atomic distances by ≥0.1 Å and/or coordination numbers by =1 (O'Day et al., 1994b). Using a nonlinear-least square fit of the raw data or a shell-fitting approach of Fourier-transformed data, typically only two species may be detected within a given sample and there is a tendency to overlook soluble species with weak or missing second-shell backscattering in the presence of minerals with strong second-shell backscattering (Manceau et al., 2000). This latter point often leads to the inability to successfully detect minor metal bearing phases, even though they may be the most reactive or significant in the metal speciation. Discrimination between species has

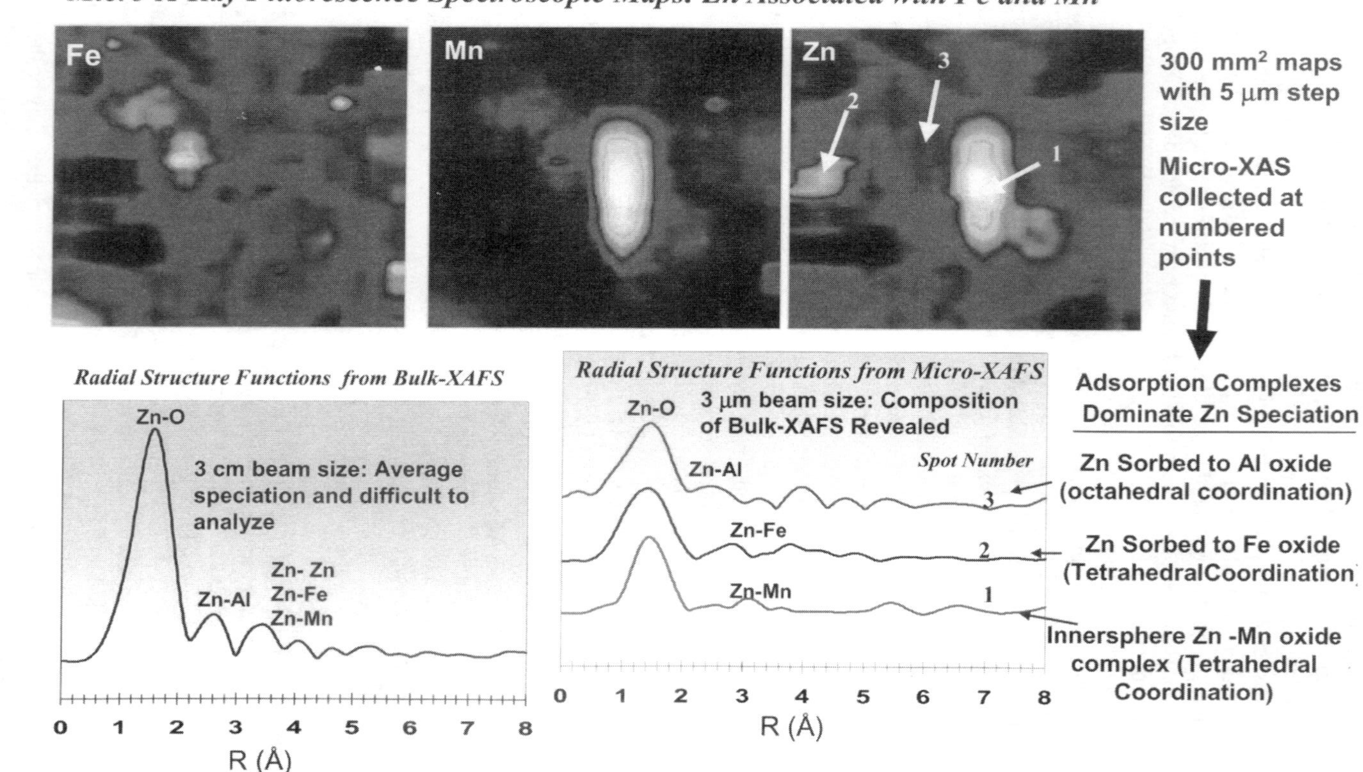

Fig. 13–9. Results from a synchrotron-based study for Zn contaminated soils combining several analytical techniques (from Roberts et al., 2001).

also been achieved using the linear combination fit (LCF) technique, where spectra of known reference species are fitted to the spectrum of the unknown sample. LCF has been successfully employed to identify and quantify up to three major species, including minerals and sorption complexes (Morin et al., 1999; Ostergren et al., 1999). The success of the speciation depends critically on a spectral database containing all the major species coexisting in the unknown sample, underscoring the need to have a thorough database of reference spectra.

Logistical drawbacks to doing synchrotron-based studies include the availability of synchrotron light sources, the increased demand for beamtime at these facilities, and the difficulty in analyzing data. Clearly, the number of metal-impacted sites requiring metal speciation information far exceeds the amount of time available at synchrotron facilities. Therefore, the combination of XAFS with more routine speciation techniques, such as sequential extractions, is important as the former technique has been able to detect artifacts and other shortcomings of the latter technique and may eventually lead to more specific and defined extraction procedures (Calmano et al., 2001; La Force and Fendorf, 2000). By combining sequential extraction techniques with XAFS, the number of species may be reduced by chemical separation prior to attempting their identification by XAFS. Moreover, the use of two independent methods for determining metal speciation in soils may provide a more reliable result than each of the methods alone.

Another shortcoming of using XAFS in metal speciation studies in soils is the requirement for sufficiently high metal concentrations, depending on the beamline conditions. This requirement has resulted in many studies that rely on high pH values (> pH 7) and high metal concentrations in order to ensure adequate surface loading on the soil minerals. Voegelin et al. (2002) attempted to circumvent this issue by performing flow-through column experiments in combination with XAFS measurements. This approach relied on relatively low initial solution concentrations of Zn, Ni, Cd, and Co reacted with soil for a period of 42 d, leading to an accumulation of metal in the soil. Their studies showed that Zn–Al LDH precipitates formed in the soils, as well as in the case of Ni and Co sorption, though the latter two metals were relatively low in concentration and spectral analysis was difficult.

Additional Techniques for Determining Metal Speciation

For characterization of crystalline phases and minerals, XRD is extremely useful; however, metal-contaminated soils and sediments often contain the metal in a form such that it is a minority phase below the detection limit of the instrument, or the important reactive phase is amorphous and only produces a large background in the diffractogram.

Even if metal concentrations are below a reasonable level, one is still able to characterize the mineralogy of a sample that is crucial to understanding metal speciation in soils. A misconception is that only information on crystalline material can be gleaned using this technique; however, amorphous Fe and Mn oxyhydroxides and organic matter yields broad features in a diffractogram that can be exploited to reveal differences between samples. A popular method of using x-ray diffraction in metal speciation studies is to combine it with SSE, a term often called differential XRD (dXRD). While the extracting solution may induce changes in the speciation of the metal or alter phases in an unnatural way, this approach is still useful. As stated earlier, synchrotron XRD has been developed and is being used to overcome some of the shortcomings of standard XRD instruments, namely flux and spatial resolution. Even more powerful is to combine μ-XRD with μ-XRF and XAFS so that one can glean information on both the metal species and the sorbent phase the metal may be associated with (Manceau et al., 2002).

High-resolution thermogravimetric analysis (HR-TGA) is not the first method one considers in performing metal speciation studies, but has recently been demonstrated to be quite powerful and discerning metal speciation. Using this technique a sample is gradually heated and while heated the weight loss is determined. Weight loss events are often specific to minerals in soils, and even to the surface functional groups on mineral surfaces. For example, Ford et al. (1997) was able to identify and quantify individual populations of surface OH groups on the Fe oxyhydroxide mineral goethite (FeOOH). While not directly determining metal speciation, studies of this type are still applicable to speciation studies since these surface sites can control effectively adsorb metal ions such as Pb and Ni.

Another useful spectroscopic technique used in metal-mineral studies is diffuse reflectance spectroscopy (DRS). DRS is the study of light as a function of wavelength that has been reflected or scattered from a solid, liquid, or gas (Clark et al., 1990). This technique is sensitive to elements in minerals that have unfilled electron shells. This makes DRS useful to investigate Ni precipitates since Ni^{2+} has an unfilled d orbital. This technique would not be useful in the case of identifying Zn^{2+} precipitates since this element has no unfilled electron orbital. The usefulness in using DRS to identify Ni hydroxide precipitates is due to the sensitivity of this technique to OH absorption bands. DRS is capable of distinguishing kaolinite from halloysite and montmorillonite from illite that is difficult using XRD alone (Clark et al., 1990). Scheinost et al. (1999) demonstrated the usefulness of DRS in differentiating $Ni(OH)_2$ from a Ni–Al layered double hydroxide phases. This task would be difficult using XAFS alone since Al is a fairly weak backscattering atom and would most likely be drowned out by the Ni signal.

Many computer models have been developed to calculate speciation of metals in soils. The advantage to this approach is its ease in execution if the proper parameters required are known. The drawbacks include, lack of proper thermodynamic data, overlooking reaction kinetics, no consideration for hysteresis, unknown identity of sorbent phases, inadequacies in describing ternary sorption systems, and overlooking the role of precipitates and solid-solutions. Regression models have been used in the literature and need to have inputs of pH, organic matter content, oxide content, CEC, metal concentration, competing ions, etc.

The more species are added to the system, the more complex the mathematics becomes. Several speciation programs with which one can calculate the speciation of a certain element in aqueous solutions are available. A comprehensive list of these programs can be found in (Sparks, 1995). Parameters in these kinds of programs are temperature, pH, I, initial concentrations, and solid phases. One should be careful not to put too much emphasis on these results. Data in the thermodynamic databases of these programs are often from different sources and sometimes insufficient.

The purpose of chemical modeling of soil systems is to obtain information on the distribution of elements within a soil between solid, aqueous, and gaseous phases at a given point in time. The modeling should be capable of predicting the types and quantities of various solids, the concentrations and distribution of exchangeable and/or adsorbed ions, the metal and ligand speciation in the aqueous phase, and the composition of the gas phase.

REFERENCES

Adamo, P., S. Dudka, M. Wilson, and W. McHardy. 1996. Chemical and mineralogical forms of Cu and Ni in contaminated soils from the Sudbury mining region and smelting region. Can. Environ. Pollut. 91:11–19.

Bargar, J.R., R. Reitmeyer, J.J. Lenhar, and J.A. Davis. 2000. Characterization of U(VI)-carbonato ternary complexes on hematite: EXAFS and electrophoretic mobility measurements. Geochim. Cosmochim. Acta 64:2737–2749.

Bartlett, R.J. 1998. Characterizing soil redox behavior. p. 371–397. *In* D.L. Sparks (ed.) Soil physical chemistry. 2nd ed. CRC Press, Boca Raton, FL.

Bartlett, R.J., and D.S. Ross. 2005. Chemistry of redox processes in soils. p. 461–488. *In* M.A. Tabatabai and D.L. Sparks (ed.) Chemical processes in soils. SSSA Book Ser. 8. SSSA, Madison, WI.

Bertsch, P.M., and D.B. Hunter. 1998. Elucidating fundamental mechanisms in soil and environmental chemistry: The role of advanced analytical spectroscopic, and microscopic methods. p. 103–122. *In* P.M. Huang et al. (ed.) Future of soil chemistry. SSSA Spec. Publ. 55. SSSA, Madison, WI.

Bostick, B.C., C.M. Hansel, M.J. La Force, and S. Fendorf. 2001. Seasonal fluctuations in zinc speciation within a contaminated wetland. Environ. Sci. Technol. 35:3823–3829.

Brown, G.E.J., A.L. Foster, and J.D. Ostergren. 1999. Mineral surfaces and bioavailability of heavy metals: A molecular scale perspective. Proc. Natl. Acad. Sci. USA 96:3388–3395.

Brown, G.E., Jr. G.A. Parks, J.R. Bargar, and S.E. Towle. 1998. Use of x-ray absorption spectroscopy to study reaction mechanisms at metal oxide–water interfaces. p. 14–37. *In* D.L. Sparks and T.J. Grundl (ed.) Mineral–water interfacial reactions: Kinetics and mechanisms. Vol. 715. Am. Chem. Soc., Columbus, OH.

Brümmer, G.W., J. Gerth, and K.G. Tiller. 1988. Reaction kinetics of the adsorption and desorption of nickel, zinc and cadmium by goethite: I. Adsorption and diffusion of metals. J. Soil Sci. 39:37–52.

Brümmer, G.W., K.G. Tiller, U. Herms, and P.M. Clayton. 1983. Adsorption–desorption and/or precipitation-dissolution processes of Zn in soils. Geoderma 31:337–354.

Calas, G., and F.C. Hawthorne. 1988. Introduction to spectroscopic methods. p. 1–9. *In* F.C. Hawthorne (ed.) Spectroscopic methods in mineralogy and geology. Vol. 18. Mineral. Soc. of Am., Washington, DC.

Calmano, W., S. Mangold, and E.F. Welter. 2001. An XAFS investigation of the artefacts caused by sequential extraction analyses of Pb-contaminated soils. J. Anal. Chem. 371:823–830.

Charlet, L., and A. Manceau. 1993. Structure, formation, and reactivity of hydrous oxide particles: Insights from x-ray absorption spectroscopy. p. 117–164. *In* J. Buffle and H.P. v. Leeuwen (ed.) Environmental particles. Lewis Publ., Boca Raton, FL.

Clark, R.N., T.V.V. King, M. Klejwa, G. Swayze, and N. Vergo. 1990. High spectral resolution reflectance spectroscopy of minerals. J. Geophys. Res. 95:12653–12680.

Coughlin, B.R., and A.T. Stone. 1995. Nonreversible adsorption of divalent ions (Mn^{II}, Co^{II}, Ni^{II}, Cu^{II}, and Pb^{II}) onto goethite: Effects of acidification, Fe^{II} addition and picolinic acid addition. Environ. Sci. Technol. 29:2445–2455.

Duff, M., D. Hunter, I. Triay, P. Bertsch, J. Kitten, and D. Vaniman. 2001. Comparison of two microanalytical methods for detecting the spatial distribution of sorbed Pu on geological materials. J. Contam. Hydrol. 47:211–218.

Duff, M.C., D.B. Hunter, I.R. Triay, P.M. Bertsch, D.T. Reed, S.R. Sutton, G. Shea-McCarthy, J. Kitten, P. Eng, S.J. Chipera, and D.T. Vaniman. 1999. Mineral associations and average oxidation states of sorbed Pu on tuff. Environ. Sci. Technol. 33:2169–2163.

Duffus, J.H. 2002. "Heavy metals" A meaningless term? Pure Appl. Chem. 74:793–807.

Elzinga, E.J., and D.L. Sparks. 1999. Nickel sorption mechanisms in a pyrophyllite-montmorillonite mixture. J. Colloid Interface Sci. 213:506–512.

Elzinga, E.J., D. Peak, and D.L. Sparks. 2001. Spectroscopic studies of Pb(II)-sulfate interactions at the goethite–water interface. Geochim. Cosmochim. Acta 65:2219–2230.

Fendorf, S.E., G. Li, and M.E. Gunter. 1996. Micromorphologies and stabilities of chromium (III) surface precipitates elucidated by scanning force microscopy. Soil Sci. Soc. Am. J. 60:99–106.

Fendorf, S.E., and D.L. Sparks. 1994. Mechanisms of chromium (III) sorption on silica: 2. Effect of reaction conditions. Environ. Sci. Technol. 28:290–297.

Fendorf, S.E., D.L. Sparks, G.M. Lamble, and M.J. Kelley. 1994. Applications of x-ray absorption fine structure spectroscopy to soils. Soil Sci. Soc. Am. 58:1583–1595.

Fendorf, S.E., and R.J. Zasoski. 1992. Chromium (III) oxidation by γ-MnO_2: I. Characterization. Environ. Sci. Technol. 26:79–85.

Ford, R.G., P.M. Bertsch, and K.J. Farley. 1997. Changes in transition and heavy metal partioning during hydrous iron oxide aging. Environ. Sci. Technol. 31:2028–2033.

Ford, R.G., A.C. Scheinost, and D.L. Sparks. 2001. Frontiers in metal sorption–precipitation mechanisms on soil mineral surfaces. Adv. Agron. 74:42–62.

Ford, R.G., A.C. Scheinost, K.G. Scheckel, and D.L. Sparks. 1999. The link between clay mineral weathering and structural transformation in Ni surface precipitates. Environ. Sci. Technol. 33:3140–3144.

Förstner, U. 1987. Metal speciation in solid wastes: Factors affecting mobility. p. 237–256. *In* S. Bhattacharji et al. (ed.) Speciation of metals in water, sediment, and soil systems. Springer-Verlag, Berlin.

Förstner, U. 1995. Land contamination by metals: Global scope and magnitude of problem. p. 1–24. *In* H.E. Allen et al. (ed.) Metal speciation and contamination of soil. CRC Press, Boca Raton, FL.

Gasser, U.G., W.J. Walker, R.A. Dahlgren, R.S. Borch, and R.G. Burau. 1996. Lead release from smelter and mine waste impacted materials under simulated gastric conditions and relation to speciation. Environ. Sci. Technol. 30:761–769.

Gerth, J., G.W. Brümmer, and K.G. Tiller. 1992. Retention of Ni, Zn, and Cd by Si-associated goethite. Z. Pflanzenernähr. Bodenk. 156:123–129.

Giller, K.E., E. Witter, and S.P. McGrath. 1998. Toxicity of heavy metals to microorganisms and microbial processes in agricultural soils: A review. Soil Biol. Biochem. 30:1389–1414.

Goldberg, S. 1992. Use of surface complexation models in soil chemical systems. Adv. Agron. 41:233–329.

Gruebel, K.A., J.A. Davis, and J.O. Leckie. 1988. The feasibility of using sequential extraction techniques for arsenic and selenium in soils and sediments. Soil Sci. Soc. Am. J. 52:390–397.

Helling, C.S., G. Chester, and R.B. Corey. 1964. Contribution of organic matter and clay to soil cation-exchange capacity as affected by the pH of the saturation solutions. Soil Sci. Soc. Am. Proc. 28:517–520.

Henderson, P.J., I. McMartin, G.E. Hall, J.B. Percival, and D.A. Walker. 1998. The chemical and physical characteristics of heavy metals in humus and till in the vicinity of the base metal smelter at Flin Flon, Manitoba, Canada. Environ. Geol. 34:39–58.

Hunter, D.B., and P.M. Bertsch. 1998. In situ examination of uranium contaminated soil particles by micro-x-ray absorption and micro-fluorescence spectroscopies. J. Radioanal. Nucl. Chem. 234:237–242.

Inskeep, W.P., and J. Baham. 1983. Adsorption of Cd(II) and Cu(II) by Na-montmorillonite at low surface coverage. Soil Sci. Soc. Am. J. 47:660–665.

Isaure, M.-P., A. Laboudigue, A. Manceau, G. Sarret, C. Tiffreau, P. Trocellier, G. Lamble, J.-L. Hazemann, and D. Chateigner. 2002. Quantitative Zn speciation in a contaminated dredged sediment by μ-PIXE, μ-SXRF, EXAFS spectroscopy and principal component analysis. Geochim. Cosmochim. Acta 66:1549–1567.

Jones, L.H.P., and S.C. Jarvis. 1981. The fate of heavy metals. p. 593–620. *In* D.J. Greenlan and M.H.B. Hayes (ed.) Chemistry of soil processes. John Wiley and Sons Ltd., Boca Raton, FL.

Kim, C.S., G.E. Brown Jr., and J.J. Rytuba. 2000. Characterization and speciation of mercury-bearing mine wastes using x-ray absorption spectroscopy. Sci. Total Environ. 261:157–168.

La Force, M.J., and S. Fendorf. 2000. Solid-phase iron characterization during common selective sequential extractions. Soil Sci. Soc. Am. J. 64:1608–1615.

Lindsay, W.L. 1979. Chemical equilibria in soils John Wiley and Sons, New York.

Manceau, A., M.A. Marcus, and N. Tamura. 2002. Quantitative speciation of heavy metals in soils and sediments by synchrotron x-ray techniques. p. 579. *In* P.A. Fenter et al. (ed.) Applications of synchrotron radiation in low-temperature geochemistry and environmental science. Vol. 49. Mineral. Soc. of Am., Washington DC.

Manceau, A., M.C. Boisset, G. Sarbet, J. Hazemann, M. Mench, P. Cambier, and R. Prost. 1996. Direct determination of lead speciation in contaminated soils by EXAFS spectroscopy. Environ. Sci. Technol. 30:1540–1552.

Manceau, A., B. Lanson, M.L. Schlegel, J.C. Hargé, M. Musso, L. Eybert-Bérard, J.-L. Hazemann, D. Chateigner, and G.M. Lamble. 2000. Quantitative Zn speciation in smelter-contaminated soils by EXAFS spectroscopy. Am. J. Sci. 300:289–343.

Mattigod, S.V., G. Sposito, and A.L. Page. 1981. Factors affecting the solubilities of trace metals in soils. p. 203–221. *In* M. Stelly (ed.) Chemistry in the soil environment. Spec. Publ. 40. ASA, Madison, WI.

McBride, M.B. 1994. Environmental chemistry of soils. Oxford University Press, New York.

Morin, G., J.D. Ostergren, F. Juillot, P. Ildefonse, G. Calas, and G.E. Brown, Jr. 1999. XAFS determination of the chemical form of lead in smelter-contaminated soils and mine tailings: Importance of adsorption processes. Am. Mineral. 84:420–434.

O'Day, P.A. 1999. Molecular environmental geochemistry. Rev. Geophysics 37:249–274.

O'Day, P.A., S.A. Carroll, and G.A. Waychunas. 1998. Rock–water interactions controlling zinc, cadmium, and lead concentration in surface waters and sediments, U.S. tri-state mining district: 1. Molecular identification using x-ray absorption spectroscopy. Environ. Sci. Technol. 32:943–955.

O'Day, P.A., G.A. Parks, and G.E. Brown, Jr. 1994a. Molecular structure and binding sites of cobalt(II) surface complexes on kaolinite from x-ray absorption spectroscopy. Clays Clay Mineral. 42:337–355.

O'Day, P.A., J.J. Rehr, S.I. Zabinsky, and G.E. Brown, Jr. 1994b. Extended x-ray absorption fine structure (EXAFS) analysis of disorder and multiple-scattering in complex crystalline solids. J. Am. Chem. Soc. 116:2938–2949.

Ostergren, J.D., G.E. Brown, Jr., G.A. Parks, and T.N. Tingle. 1999. Quantitative speciation of lead in selected mine tailings from Leadville, CO. Environ. Sci. Technol. 33:1627–1636.

Papelis, C. 1995. X-ray photoelectron spectroscopic studies of cadmium and selenite adsorption on aluminum oxide. Environ. Sci. Technol. 29:1526–1533.

Pignatello, J.J., and B. Xing. 1996. Mechanisms of slow sorption of organic chemicals to natural particles. Environ. Sci. Technol. 30:1–11.

Roberts, D.R., A.M. Scheidegger, and D.L. Sparks. 1999. Kinetics of mixed Ni–Al precipitate formation on a soil clay fraction. Environ. Sci. Technol. 33:3749–3754.

Roberts, D.R., A.C. Scheinost, and D.L. Sparks. 2002. Zinc speciation in a smelter-contaminated soil profile using bulk and microspectroscopic techniques. Environ. Sci. Technol. 36:1742–1750.

Sauvé, S., and D.R. Parker. Chemical speciation of trace elements in soil solution. p. 655–688. *In* M.A. Tabatabai and D.L. Sparks (ed.) Chemical processes in soils. SSSA Book Series, no. 8. SSSA, Madison, WI.

Scheckel, K.G., A.C. Scheinost, R.G. Ford, and D.L. Sparks. 2000. Stability of layered Ni hydroxide surface precipitates: A dissolution kinetics study. Geochim. Cosmochim. Acta 64:2727–2735.

Scheckel, K.G., and D.L. Sparks. 2000. Kinetics of the formation and dissolution of Ni precipitates in a gibbsite–amorphous silica mixture. J. Colloid Interface Sci. 229:222–229.

Scheidegger, A.M., G.M. Lamble, and D.L. Sparks. 1996. The kinetics of nickel sorption on pyrophyllite as monitored by x-ray absorption fine structure (XAFS) spectroscopy. J. Phys. IV 4:773–775.

Scheidegger, A.M., G.M. Lamble, and D.L. Sparks. 1997. Spectroscopic evidence for the formation of mixed-cation hydroxide phases upon metal sorption on clays and aluminum oxides. J. Colloid Interface Sci. 186:118–128.

Scheidegger, A.M., and D.L. Sparks. 1996a. A critical assessment of sorption–desorption mechanisms at the soil–mineral interface. Soil Sci. 161:813–831.

Scheidegger, A.M., and D.L. Sparks. 1996b. Kinetics of the formation and the dissolution of nickel surface precipitates on pyrophyllite. Chem. Geol. 132:157–164.

Scheinost, A.C., R.G. Ford, and D.L. Sparks. 1999. The role of Al in the formation of secondary Ni precipitates on pyrophyllite, gibbsite, talc, and amorphous silica: A DRS study. Geochim. Cosmochim. Acta 63:3193–3203.

Schulze, D.G., and P.M. Bertsch. 1995. Synchrotron x-ray techniques in soil, plant, and environmental research. p. 1–66. *In* D.L. Sparks (ed.) Adv. Agron. Vol. 55. Academic Press, New York.

Siegel, F.R. 2002. Environmental geochemistry of potentially toxic metals. Springer-Verlag, Berlin.

Sparks, D.L. 1995. Environmental soil chemistry. Academic Press, San Diego.

Sparks, D.L., A.M. Scheidegger, D.G. Strawn, and K.G. Scheckel. 1999. Kinetics and mechanisms of metal sorption at the mineral–water interface. p. 108–135. *In* D.L. Sparks and T.J. Grundl (ed.) Mineral–water interfacial reactions. Kinetics and mechanisms. Am. Chem. Soc., Washington, DC.

Sposito, G. 1989. The chemistry of soils. Oxford University Press, New York.

Strawn, D., H. Doner, M. Zavarin, and S. McHugo. 2002. Microscale investigation into the geochemistry of arsenic, selenium, and iron in soil developed in pyritic shale materials. Geoderma 108:237–257.

Stumm, W. 1992. Chemistry of the solid–water interface, John Wiley and Sons, New York.

Tessier, A., P.G.C. Campbell, and M. Bisson. 1979. Sequential extraction procedure for the speciation of particulate trace metals. Anal. Chem. 51:844–851.

Thompson, H.A. 1998. Dynamic ion partioning among dissolved, adsorbed, and precipitated phases in aging cobalt(II)/kaolinite/water systems. Ph.D. diss. Stanford Univ., Palo Alto, CA.

Voegelin, A., A.C. Scheinost, K. Bühlmann, K. Barmettler, and R. Kretzschmar. 2002. Slow formation and dissolution of Zn precipitates in soil: A combined column-transport and XAFS study. Environ. Sci. Technol. 36:3749–3754.

Walther, J.V. 1996. Relation between rates of aluminosilicate mineral dissolution, pH, temperature, and surface charge. Am. J. Sci. 296:693–728.

Wang, Z.W., X.Q. Shan, and S.Z. Zhang. 2002. Comparison between fractionation and bioavailability of trace elements in rhizosphere and bulk soils. Chemosphere 46:1163–1171.

Webb, S.M., G.G. Leppard, and J.-F. Gaillard. 2000. Zinc speciation in a contaminated aquatic environment: Characterization of environmental particles by analytical electron microscopy. Environ. Sci. Technol. 34:1926–1933.

Chapter 14

Chemical Speciation of Trace Elements in Soil Solution

SÉBASTIEN SAUVÉ, *University of Montreal, Montreal, Canada*

DAVID R. PARKER, *University of California, Riverside, California, USA*

The chemical speciation of various elements in soils, sediments, and surface waters has been the subject of innumerable research studies undertaken in the last two to three decades. Environmental chemists have used an extremely wide array of empirical tools and modeling approaches to better understand the chemical behavior of both major and minor elements. There has been some controversy over the use and meaning of the term "speciation." Originally, the term was strictly biological; it described the evolutionary differentiation of distinct and unique biological species, and reflects the human need to categorize groups of like individuals within a larger group of somewhat dissimilar organisms. Likewise, chemical speciation attempts to differentiate among the different chemical forms of a given substance.

The International Union for Pure and Applied Chemistry (IUPAC) has recommended that chemical species be defined as "the specific form of an element defined as to isotopic composition, electronic or oxidation state, and/or complex or molecular structure" (Templeton et al., 2000). The speciation of an element is in turn defined as "the distribution of an element amongst defined chemical species in a system." In contrast, "fractionation" was defined as "the process of classification of an analyte or a group of analytes from a certain sample according to physical (e.g., size, solubility) or chemical (e.g., bonding, reactivity) properties." These IUPAC definitions were recommended with the specific aim of reducing the confusion surrounding the use of these terms by different researchers and by different scientific disciplines (Templeton et al., 2000). Accordingly, much of what has been published as "chemical speciation" would not meet this strict definition, and should perhaps be termed fractionation instead. On the other hand, some people may have unknowingly been doing chemical speciation.

At this stage, there should be no value judgment implicit in the use of speciation versus fractionation. Chemical speciation may seem more precise in a chemical sense, but both approaches can rely on a number of assumptions that are not always readily verifiable. Both have their importance and usefulness, and there are a number of research methodologies that probably represent a "gray area" in between the two definitions. Nonetheless, the IUPAC definitions do allow a common

Chemical Processes in Soils. SSSA Book Series, no. 8.

language to be used, and should facilitate more accurate communication within this very active scientific field.

Speciation is critical to the understanding of trace-element behavior because different chemical species may have very different properties. An excellent example is chromium; trivalent Cr is a micronutrient beneficially involved in glucose metabolism of mammals, while and hexavalent Cr is toxic and carcinogenic. Trivalent chromium (Cr^{3+}) is cationic and has a very low solubility in soils and readily sorbs to mineral and organic surfaces, but hexavalent chromium ($Cr_2O_7^{2-}$) is anionic and has a much lower tendency to sorb to soil surfaces, and it is hence much more mobile and prone to leaching into groundwater. Clearly, a given chemical element can be either beneficial or detrimental, depending on its chemical speciation.

The role of speciation has been extensively researched in aquatic toxicology and, in general, the free metal ions in solution are more biologically active than other forms (complexed, colloidal or sorbed)—the so-called "free-ion model" of metal bioavailability. Occasional exceptions to this generalization have been documented (Campbell, 1995), and its application to soil system is more complex (Parker et al., 2001). Soils are quite unlike a water column: they are not well-mixed; they are spatially heterogeneous and dynamic; the soil solution phase is volumetrically small compared with the solid phase with which it is in contact; and the solid phase is in more-or-less intimate contact with the soil biota. Also, very different transport processes prevail (e.g., diffusion, transpirational water flux), that may locally affect chemical conditions, and the soil biota are often capable of modifying their immediate chemical environment (e.g., the rhizosphere). This emphasizes that, although speciation may exert significant control over biological responses (such as uptake and/or toxicity), it also is a very a dynamic parameter that demands a thorough understanding and careful experimental work in order to fully assess its importance to the bioavailability of trace elements in the environment.

An accurate description of trace-element speciation also is a requisite for accurate modeling of their (bio)geochemical behavior. Soil chemical reactions, such as mineral-phase solubility, are properly described in thermodynamic terms using the chemical activities of the free, uncomplexed metals and ligands in solution. Such solubility is independent of any "side reactions" involving other metals or ligands. Thus, proper modeling of the geochemical "controls" on solubility, mobility, and bioavailability must fully account for the various ion associations that can occur in the soil solution.

SOIL SOLUTION

Trace Elements in Soil Solution

Identifying "typical" soil solution concentrations for most trace elements is problematic. What is "typical" for a certain area (e.g., an ultramafic or serpentinitic soil) could well be abnormally high in another region. To further complicate things, different laboratories will use different means to obtain–extract–define what is the soil solution. Nevertheless, Table 14–1 gives a range of values representative of what is expected for uncontaminated system, along with upper limits that can be expected in contaminated soils.

Table 14–1. Range of soil solution concentrations observed for certain trace elements as estimated from the data compiled in Sauvé et al. (2000a).

Element	Soil solution
	$\mu g\ L^{-1}$
As	0.5–60 000
Cd	0.01–5 000
Cr	2–500
Cu	5–10 000
Ni	0.5–5 000
Pb	0.5–500
Se	1–100 000
Zn	1–100 000

When evaluating or comparing soil solution dissolved trace element concentrations, it is important to keep in mind that soil properties have a drastic influence on metal solubility. For example, in many instances pH is more influential than the actual level of contamination (i.e., total metal concentration). It is often observed that soils spiked with metal salt generally exhibit show much higher soil-solution concentrations than do field-collected, "natural" contaminated soils.

Important Metal-Complexing Ligands in Soil Solution

The inorganic ligands in soil solution that are commonly present in "macro" (i.e., millimolar) concentrations are nitrate, sulfate, chloride, and (bi)carbonate. As a group, these are relatively weak ligands that often exert minimal influence on trace-metal speciation. Some important exceptions occur, however. In alkaline soils, HCO_3^- or CO_3^{2-} can be a significant complexor of transition metals such as Cu^{2+}, and carbonate complexes dominate the aqueous speciation of the uranyl ion, UO_2^{2+}. Cadmium is complexed by Cl^- (log K = 2.0 for formation of $CdCl^+$) and SO_4^{2-} (log K = 2.5 for $CdSO_4^0$), and these ion pairs can be especially significant in saline soils. Important ligands that typically occur at micromolar levels in soil solution include phosphate and fluoride; both have high affinities for trivalent Fe and Al, but are of lesser importance to the speciation of the bivalent transition metals.

Organic compounds that have been identified in soil solution (including the rhizosphere) are many and diverse. They include sugars, organic acids, amino acids, phenolic compounds, lipids, and several other types. Some of these (e.g., the sugars) are not sufficiently strong ligands to warrant consideration in a discussion of metal complexation. The common amino acids are a relatively common component of root exudates, but they seem to lack sufficient metal-complexing power to significantly alter the speciation of soil metals, even in the rhizosphere (Jones et al., 1994). Low-molecular weight organic acids such as citric and malic acids have been extensively studied with respect to their role in metal speciation in soils. They are of primary significance here because of their (hydroxy)carboxylic functionality and consequent affinity for metals, and because they are excreted by both plant roots and soil microflora. Strictly speaking, the term "organic acid" should often be replaced with organic-acid *anion*, to reflect the dissociated state of the ligand at many soil pH values, but many authors use "organic acids" in the interests of brevity.

Various phenolic acids such as salicylic acid, caffeic acid, and gallic acid have been identified in soil solution and have, at times, been implicated in the solubilization and mobility of Al in soils. In general, however, the ring hydroxyl of these compounds does not deprotonate except at very alkaline pH values; thus their affinity for metals at typical pH values is usually rather low. In comparison to the aliphatic organic acids, the phenolics are probably of minor significance to the speciation of metals in most situations.

"Siderophore" is the general name for a diverse group of amine compounds that have been implicated in the chelation and biological acquisition of iron. The "phytosiderophores" are a relatively small group of related compounds bearing chemical similarity to the ubiquitous plant compound nicotianamine (Ma and Nomoto, 1996). Root secretion of phytosiderophores in response to physiological Fe deficiency seems to be universal within the grass family (the so-called "Strategy II" group of plants), and is thus of worldwide nutritional significance. Microbially produced siderophores are a much larger and more diverse group of Fe-chelating compounds that are produced a variety of terrestrial and aquatic habitats (Roosenberg et al., 2000).

All soil solutions contain naturally produced, higher molecular-weight compounds, although the concentrations vary widely, and may be vanishingly small in some cases. The most ubiquitous class of compounds is "fulvic acid" (FA), a term used to describe a diverse group that is water-soluble at near-neutral pH values (Hayes and Malcom, 2001). One of the stable end-products of microbial processing of soil carbon, the FAs have highly variable and complex macromolecular structures, and contain a number of functional groups of varying affinity for trace metals. Because of the affinity of carboxyl groups (and to a lesser degree acidic hydroxyls and amines), FA is potentially a very significant determinant of trace-metal speciation. Attempts to predictively model the binding between metals and FA have ranged from rather simple to quite elaborate conceptual and mathematical representations (see below).

Modulation of the Soil Solution by the Solid Phase

The solid phase is in intimate contact with the soil solution and has ample opportunities to influence it. One might hypothesize that, given a sufficiently high level of soil contamination, mineral solubility reactions would control soil solution concentrations of dissolved metals. That is, at high metal levels, discrete mineral phases may form that contain trace metals in equilibrium with the soil solution, and that would precipitate or dissolve in response to additions or withdrawals of metals from the system, respectively. That said, if such conditions did prevail, the speed at which those reactions occur becomes critical. A good example of this concept occurs in the case of soil contamination with lead, where Pb-phosphate minerals such as pyromorphite have such a low mineral solubility and reasonable kinetics of formation that they can actually control Pb solubility in many contaminated environments.

Most of the other trace metals, however, do not generally have a mineral phase with a solubility low enough to be the determinant of solution concentrations under realistic environmental conditions. Instead, the solid phase still controls soil solution concentrations, but via surface sorption reactions, as opposed to mineral solubility equilibrium. The resulting soil solution concentrations can be predicted or modeled using different variety of surface complexation models, or through simple partitioning coefficients (K_d) (e.g., Sauvé et al., 2000a).

CHEMICAL EQUILIBRIUM THEORY AND NOMENCLATURE

General Ion Association Theory

Although factors such as redox number are considered part of chemical speciation, the speciation of the soil solution is dominated by the tendency for metals and ligands to interact via both electrostatic (ionic) attraction and covalent bonding. The equilibrium distribution of species can be generically written as

$$M + L \Leftrightarrow ML \quad [1]$$

with its associated equilibrium quotient

$$K_{eq} = [ML]/[M][L] \quad [2]$$

The types of association formed between metal and ligand can be (roughly) classified as an ion pair, a complex, or a chelate, depending on the strength of the association (Table 14–2). From a molecular standpoint, we envision ion pairs as weak associations between ionic metals and ligands that are largely or completely based on electrostatic attraction. The hydration shells of the interacting ions remain largely intact, and ion pairs are thus viewed as *outer-sphere complexes*. Metal-ligand complexes are, to varying degrees, *inner-sphere* in nature, such that ligand partly or wholly replaces the hydration shell of the metal, and there is no water molecule between the metal and ligand centers. Complexation usually arises from a combination of ionic attraction and covalent bonding between the metal and ligand. The formation of metal-ligand chelates is sometimes used to describe complexes of unusually high stability (Table 14–2 and below). With a chelate, the ligand is typically a polyfunctional molecule that can fold itself around a metal center, completely replace the metal's hydration shell, and achieve a fully stable coordination number (e.g., 6) with the metal. For example, EDTA is an effective chelator of metals like Fe(III) and Cu(II) because it has four carboxyl groups and two amines, thus facilitating an extremely stable six-fold coordination.

Table 14–2. Approximate classification of metal–ligand associations on the basis of binding strength.

Type of association	Effective stability constant (K)
Ion pair	~10^0 to 10^4
Complex	~10^4 to 10^8
Chelate	~10^8 to 10^{20+}

Fundamental Thermodynamics

The tendency for metals and ligands in aqueous solution to associate, sometimes quite strongly, can be described using classical thermodynamic concepts of enthalpy, entropy, and free energy. Briefly, enthalpy (H) is a state function predicated on the first law of thermodynamics (energy conservation) that, in classical thermodynamics, describes pressure-volume work. In the aqueous chemistry of the soil solution, pressure is constant, so that enthalpy is the heat of reaction that reflects the comparative bond strength of the products vs. the reactants. Entropy (S) is a state function predicated on the second law (entropy law) that is best viewed as describing the level of "disorderedness" when the products are more "disordered" than the reactants, there is an *increase* in entropy. The Gibbs free energy (G) is defined as the combined effect of enthalpy and entropy:

$$G = H - TS \qquad [3]$$

and if the temperature does not change during the reaction:

$$\Delta G = \Delta H - T\Delta S \qquad [4]$$

The corresponding *free energy of formation* and *enthalpy of formation* for compounds derived from their constituent elements under standard conditions ($P = 1$ atmosphere; $T = 298.2$ K) are usually denoted as G_f^0 and H_f^0, respectively.

Based on definitions of *chemical potential*, μ (see Stumm and Morgan, 1996), it can be shown that the overall free energy change for a chemical reaction is

$$\Delta G = \Sigma G_{\text{prod}} - \Sigma G_{\text{reac}} = \sum_i v_i \mu_i \qquad [5]$$

where v_i is the stoichiometric coefficient (positive for products, negative for reactants) for the ith component in the reaction. Moreover, the free energy change can be related to the chemical potential of each participant in their *standard state*, μ^0, by

$$\Delta G = \sum_i v_i \mu_i^0 + RT \sum_i v_i \ln X_i \qquad [6]$$

where X_i is the mole fraction of the ith component. For aqueous systems where solutes are the focus, this equation usually is expressed as

$$\Delta G = \Delta G^0 + RT \sum_i v_i \ln C_i \qquad [7]$$

where C_i is the molar concentration of the ith component. Because $v\ln C = \ln C^v$, the summation can be replaced by the familiar mass action quotient Q:

$$Q = \frac{\Pi \text{ products}}{\Pi \text{ reactants}} \qquad [8]$$

where the quantities in the right-hand side are the molar concentrations (or more properly activities; see below) raised to the appropriate stoichiometric coefficient. Thus:

$$\Delta G = \Delta G^0 + RT\ln Q \qquad [9]$$

At equilibrium, ΔG is by definition zero (the total G of the system is minimized), so that

$$\Delta G^0 = -RT\ln Q = -RT\ln K_{eq} \qquad [10]$$

where K_{eq} is the familiar equilibrium constant, i.e., the unique mass-action quotient at the equilibrium condition. This is the fundamental equality between the Gibbs free energy change for a reaction and the equilibrium constant. Using common logarithmic notation:

$$\Delta G^0 = -2.303\ RT\log K_{eq} \qquad [11]$$

and when G^0 is in kilojoules per mole:

$$\Delta G^0 = -5.70\log K_{eq} \qquad [12]$$

Example 1. Dissociation of acetic acid in water:

$$\text{HAc}\ \ \text{Ac}^- + \text{H}^+$$

$$K_a = \frac{[\text{Ac}^-][\text{H}^+]}{[\text{HAc}]} = 10^{-4.75}$$

$$\Delta G^0 = -5.70(-4.75) = 27.1\ \text{kJ mol}^{-1} \qquad [13]$$

Nonideality

Although aqueous concentrations are often employed as a convenient simplification, the chemical behavior of solutes is more properly described in terms of the *ionic activity,* denoted with braces or parentheses:

$$\{X^n\} = (X^n)\ {}^X\gamma \qquad [14]$$

where γ is the single-ion activity coefficient. In very dilute solutions γ approaches 1, and the solute's concentration and activity are identical. With increasing ionic strength, however, several ion–ion and ion–solute interactions lead to nonideal behavior that alters the free energy state of the solute, the solvent, or both. At moderate ionic strengths (say 0.01 M), the dominant interaction is long-range coulombic attraction–repulsion between charged species that effectively lowers the free energy of the system, and the activity coefficients are <1 (activities are always < concentrations). This interaction is adequately explained by Debye-Hückel theory (see Computational Methods) up to ionic strengths of ~0.1 *M*, which spans the range of interest for most soil solutions. In highly saline solutions, however, additional nonideality is introduced by dipole interactions between solutes, and by volume exclusion effects wherein the solute molecules interact due to physical "crowding" (i.e., they no longer behave as isolated point charges). Debye-Hückel theory cannot explain the observed changes in γ values (which often increase to >1) at these high ionic strengths, and other theories or empirical relations must be invoked to account for the effect of ionic strength on activities (see Computational Methods).

Acids and Bases

The pH of the soil solution is perhaps the most fundamental chemical characterization of soil that we routinely make. The soil pH is buffered, largely by reaction with the soil solid phase, and this characteristic pH value leads to "speciation" of certain dissolved constituents—weak acids or bases that tend to gain or lose aquated protons (denoted here simply as H^+) at the pH values prevalent in soils. In many soil solutions, these protolytic components include (bi)carbonate, phosphate, silicate, ammonia, borate, sulfide, as well as a host of dissolved organic molecules that contain functional groups such as carboxyls and amines.

The distribution of these weak-acid constituents of the soil solution can be described using the classical concepts of proton dissociation and buffering. For example, an *undissociated* monoprotic acid reversibly loses a proton to yield its *conjugate base* (often shortened to "base"):

$$HA \Leftrightarrow A^- + H^+ \qquad [15]$$

and the equilibrium quotient (on a concentration scale, not activity) is

$$K_a = [A^-][H^+]/[HA] \qquad [16]$$

When the acid is exactly 50% dissociated, $[A^-] = [HA]$, and this expression reduces to the familiar characterization of acid strength on a negative log scale (for convenient comparison to pH):

$$pK_a = -\log K_a = 4.8 \qquad [17]$$

The pK_a corresponds to the region of maximum buffering, i.e., where the affinity of A^- for H^+ results in the greatest resistance to change in solution pH as strong base or acid is added to the system.

Multiprotic acids are computational more complex, but conceptually no different than the simple case outlined above. For a thorough treatment, any classic quantitative analysis text will usually provide the needed detail.

Metal Hydrolysis

The hydrolytic nature of cationic metals (Baes and Mesmer, 1976) also can be viewed as a special type of metal–ligand interaction, as well as a specialized protolytic reaction. Because of their high charge ($2.00) and small ionic radius, many metals effectively destabilize the water molecules in their inner hydration shell, resulting in the release of one or more protons:

$$M(H_2O)_6^{n+} \rightleftharpoons M(H_2O)_5OH^{(n-1)+} + H^+ \qquad [18]$$

Additional deprotonation steps can follow and, based on mass action considerations, metal hydrolysis is clearly favored by a higher prevailing pH in soil solution. Conversely, low pH favors a predominance of the aquated free metal ion at the expense of the hydrolysis products. In addition to the mononuclear complexes with a single metal center, soluble polynuclear complexes also may form, especially with the common trivalent metals Al(III) and Fe(III). Note that in chemical equilibrium models (see below), hydrolysis reactions are most often described as an *association* reaction between a metal and the hydroxyl ion (e.g., $M^{2+} + OH^- \rightleftharpoons MOH^+$). Mechanistically, this is not quite correct, but it yields accurate and correct stoichiometry with respect to H^+, OH^-, and H_2O.

Properties and Classification of Trace Elements

With the exception of Group VIIA elements (halogens), the noble gases, and H, B, C, N, O, P, and S, all elements in the periodic table are classified as metals or metalloids (semimetals). When we speak of trace-element speciation in soil solution, it is implicit that our emphasis is on many of these metallic elements (see Soil Solution). The ligands that in turn complex the metals and metalloids most commonly involve N, O, or S, but can include other electronegative elements such as P, Cl, and F.

Several classification schemes for metals have been proposed over the years, all relating to the hard and soft acid-base (HSAB) properties of Lewis acids and bases. Traditionally, "A-type" or "hard sphere" cations [such as Al(III)] have been so described due to their low polarizability and relatively nondeformable electron sheaths. These metals have noble gas electron configurations (d^0), form strong complexes with F- and with O-containing ligands, and they hydrolyze readily. Conversely, "B-type" or "soft sphere" metals (such as Hg[II]) are highly polarizability and preferentially complex with reduced forms of S and with halides such as Br^- and Cl^- (Stumm and Morgan, 1996). Of particular relevance here are the bivalent, first-row transition metals that are usually viewed as intermediate in character (Table 14–3). Note that the classification of Nieboer and Richardson (1980) is similar, but is based primarily on a metal's tendency to form covalent bonds (regard-

Table 14–3. Classification of selected metal cations important in the speciation of soil solutions (adapted from Stumm & Morgan, 1996).

A-Type Metals: "Hard"	"Borderline"	B-Type Metals: "Soft"
Al^{3+}, Cr^{3+}, Fe^{3+}, UO^{2+}	Mn^{2+}, Fe^{2+}, Co^{2+}, Cu^{2+}, Ni^{2+}, Pb^{2+}, Zn^{2+}	Ag^+, Cd^{2+}, Hg^{2+}, Sn^{2+}
Major ligands: F > O > N = Cl > S		Major ligands: S > Cl = N > O > F

less of ligand), and results in several metals (Fe^{3+}, Cd^{2+}, Sn^{2+}) being shifted into the borderline or intermediate grouping.

Strong complexes, and particularly chelates, are particularly likely to form between multidendate ligands and transition metals. This strength of association can be explained by large increases in entropy, as the enthalpy of bond formation is often about the same for mono- and multidentate ligands; the increase in entropy arises largely from the displacement of the highly coordinated water molecules in the metal's hydration shell (Morel and Hering, 1993). More mechanistically, the stability and reactivity of metal-ligand can be explored using *ligand-field theory* (LFT) or *molecular orbital theory* (MOT), both of which are beyond the scope of this chapter. These mechanistic models may help to explain the largely empirical observation that the strength of complexation by a variety of organic ligands conforms to the sequence:

$$Mn^{2+} < Fe^{2+} < Co^{2+} < Ni^{2+} < Cu^{2+} > Zn^{2+} \quad [19]$$

This is the widely know Irving-Williams series, and is observed with a very large number of N-, O-, and S-containing ligands of varying strength; the same order is observed in the solubility products of metal-sulfide solid phases. Thus, Cu tends to form the most stable organic complexes among the first-row transition metals in their +II oxidation state, while the weaker complexation behavior of Zn is usually comparable to that of Co. Note, however, that weak electrostatic interactions (e.g., ion-pairing with SO_4^{2-}) exhibit no such variation.

SPECIATION METHODS

Chemical speciation can either be estimated using computational methods, or it can be empirically measured using a variety of analytical techniques. It is not uncommon to use an analytical and a computational component together to assess the free-ion activities or other chemical details. The route taken to evaluate metal speciation, and/or the actual methodology chosen, will often depend on the chemical–biological system under study, the analytical resources and expertise available, and the objectives of the study. As we stressed at the outset, given the historical disparities in the definitions of speciation, care must be taken when surveying the literature since much of what is published as *speciation*, should more properly be considered as *fractionation* (Templeton et al., 2000).

Computational Methods

Computational methods, as the name suggests, are attempts at estimating the chemical speciation in a solution from ion-association datasets of diverse origin, scope, and complexity through mathematical (typically computerized) models of similarly variable complexity. Limitations to such modeling approaches include completeness and the quality of the analytical input data, the accuracy of the needed stability constants, and the overall appropriateness of the model to the actual environmental system under study.

General Approaches to Equilibrium Problems

The aqueous speciation of a soil solution can, and often is, estimated using chemical equilibrium models of varying complexity. Such models can account for ion association (the formation of pairs, complexes, and chelates), pH (as it affects acid-base equilibria and hydrolysis), the oxidation-reduction status of the system, and/or the formation–dissolution of solid phases. Sometimes, the model also will account for exchange of solutes with the soil solid phase, e.g., by modeling adsorption or ion-exchange reactions.

All chemical equilibrium models, whether simple or complex, include two fundamental constituents: the simultaneous consideration of all pertinent equilibrium reactions (and their associated thermochemical constants), and the constraint of overall mass (or mole) balance on all of the elements (actually components, see below) under consideration. Occasionally, charge balance (electroneutrality) or its close relative, the *proton condition* (Stumm and Morgan, 1996), is also invoked to facilitate solution of the problem. The following simple examples illustrate the fundamental methodology, without consideration of ionic strength effects, followed by a discussion of modern computerized algorithms.

Approximate Method. The simplest speciation problems can sometimes be solved by making suitable assumptions that allow the mathematics to be greatly simplified.

Example: What is the pH of a 0.01 M solution of the weak acid HA (pK_a = 4.8)?

$$HA \Leftrightarrow A^- + H^+ \qquad [20]$$

If we assume that [HA] >> $[A^-]$ (i.e., [HA] $\simeq A_T$), then because the dissociation of HA is the only source of H^+, $[A^-] = [H^+]$. Thus:

$$K_a = 10^{-4.8} = [A^-][H^+]/[HA] = [H^+]^2/A_T$$

$$[H+] = \sqrt{10^{-4.8} \times 0.01} = 3.98 \times 10^{-4}\ M; \quad pH \sim 3.40$$

$$\text{Check assumption: } [A^-] = [H^+] = 3.98 \times 10^{-4}\ M = 4\% \text{ of } A_T \qquad [21]$$

Exact Method. Example: What is the extent of complexation of Cd^{2+} by Cl^- in a solution containing 0.001 M $CdCl_2$? (Assume only the 1:1 complex is significant). The problem can be completely described by three equations, one for the equilibrium quotient, and two that define the mole balance for the two participants (Cd and Cl):

$$K = [CdCl^+]/[Cd^{2+}][Cl^-] = 10^2$$

$$Cd_T = [Cd^{2+}] + [CdCl^+] = 0.001M$$

$$Cl_T = [Cl^-] + [CdCl^+] = 0.002M \qquad [22]$$

There are thus three equations and three unknowns and, after suitable substitutions, the problem reduces to a single quadratic expression:

$$0 = K[Cd^{2+}]^2 + [Cd^{2+}]\,(1 + KL_T - KM_T) - M_T \qquad [23]$$

After substituting appropriate values for K, Cd_T, and Cd_T, we find that the free Cd^{2+} concentration is 8.4×10^{-4} M and, by difference, the concentration of the $CdCl^+$ ion pair must be 1.6×10^{-4} M.

Of course most speciation problems of interest to a soil chemist involve numerous dissolved components and many possible complexation reactions—many more than three equations in three unknowns! Occasionally, these can be simplified by making appropriate assumptions (as above), and by using successive approximation techniques. Today, however, almost all speciation computations are performed with the convenience of a computerized equilibrium model (see below). It is important to keep in mind, however, that the basic ingredients of the computations are identical to the preceding example: a set of equilibrium quotients in conjunction with a set of mole balance expressions for every component under consideration.

Ionic Strength Corrections in Chemical Speciation Models

In general, the chemical speciation of natural waters can be modeled in one of two ways. The first, the so-called "specific interaction" (or ion interaction) model, uses mean ionic activity coefficients for neutral salts, which are measurable, macroscopic properties that confer strict thermodynamic significance to the model calculations (Sposito, 1984). The second and much more common approach is the "ion association" (or ion pairing) model that relies on single-ion activity coefficients calculated from the Debye-Hückel relation (or one of its variants) and is thus dependent on some extrathermodynamic assumptions. The specific interaction model is advantageous because it can be extended to solutions of high ionic strength (i.e., >1 M) that are poorly described by conventional ion association models; drawbacks include the need for experimentally determined activity coefficients and the resulting ion interaction parameters (i.e., the so-called Pitzer coefficients or related parameters) (Bassett and Melchior, 1990). The latter has tended to limit the application of the specific interaction model to major ions (i.e., in seawater or saline soils). Within limits, both approaches can accurately account for the ionic strength-dependent nonideality of solutes described in Chemical Equilibrium Theory and Nomenclature, Nonideality.

In ion-association models, the effect of ionic medium is almost universally handled via single-ion activity coefficients, the generalized expression of which can be written as

$$\log \gamma = \frac{-Az^2\sqrt{I}}{1 + aB\sqrt{I}} + B^0 I \qquad [24]$$

where γ is the activity coefficient, A is the Debye-Hückel limiting parameter = 0.5116 $L^{1/2}\ mol^{-1/2}$, z is the ion valence, I = the ionic strength ($mol\ L^{-1}$), a is an ion size parameter (cm), $B = 0.3284 \times 10^8\ L^{1/2}\ mol^{-1/2}\ cm^{-1}$, and B° is a correction term ($L\ mol^{-1}$)

When all of these parameters are known (including a and B^0), the equation can be solved and the resulting coefficient is said to be based on the *extended Debye-*

Hückel convention. In many cases, however, not all of the terms are exactly known and some simplifying assumptions are made: a is assumed to be numerically equal to $1/B$, and $B°$ is taken as $0.3Az^2$. Thus, Eq. [1] reduces to the widely-used Davies (1962) equation for use at ionic strengths less than about 0.1 to 0.7 M:

$$\log\gamma = -0.5116z^2\left(\frac{\sqrt{I}}{1+\sqrt{I}} - 0.3I\right) \quad [25]$$

The effect of ionic strength on single-ion activity coefficients computed using the Davies equation for mono-, di-, and trivalent ions is illustrated in Fig. 14–1.

For ionic strengths greater than about 0.5 M, some models such as GEOCHEM-PC (Parker et al., 1995) use the Helgeson (1969) equation:

$$\log\gamma = \frac{-0.5116z^2\sqrt{I}}{1 + a0.3284\sqrt{I}} + 0.041I \quad [26]$$

where $a = 4$ for a monovalent ion, 5 for a divalent ion, or 6 for an ion of trivalent or greater charge.

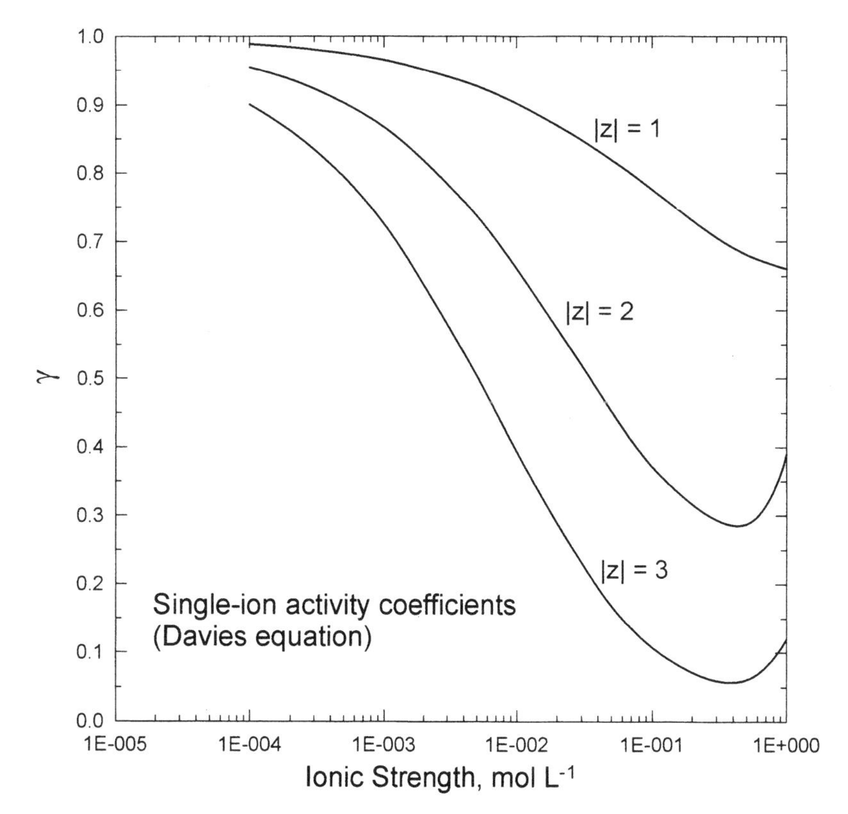

Fig. 14–1. Activity coefficients for ions of valence ±1, 2 or 3 as computed by the Davies equations for varying ionic strength.

The activity of neutral species is very weakly affected by ionic strength, and this interaction is often approximated using a simple relation such as that described by Harned and Owen, (1958):

$$\log \gamma = 0.1\, I \qquad [27]$$

Computer Algorithms

Computerized equilibrium models have become universal tools in soil and environmental chemistry, and they are routinely used to estimate or approximate the speciation of soil solutions or aqueous soil extracts. A few of the more common programs in use today include GEOCHEM-PC (Parker et al., 1995), MINEQL+ (Schecher and McAvoy, 1991), and MINTEQA2 (Allison and Brown, 1995), among others. The use of such programs always involves the same fundamental four steps:

1. The review and management of a thermodynamic database containing the needed stability constants.
2. The specification of the solution components and their concentrations, along with other needed chemical parameters (pH, temperature, etc).
3. The processing of the computations.
4. The manipulation and analysis of the program output to arrive at the distribution of species of interest.

Each computer model has its own list of *components* that are used to define the composition of the solution of interest. These components typically include uncombined cationic metals in a specified oxidation state (e.g., there are two components for Fe: Fe[II] and Fe[III]), as well as a few metals that normally combine with oxygen (e.g., U[VI] is defined as UO_2^{2+}). The list of component ligands includes simple anions such as Cl^-, oxyanions such as SO_4^{2-} and PO_4^{3-}, and an assortment of organic ligands (organic acids, amino acids, chelators, etc.). Note that the way in which many weak acid-base components are specified is largely up to the individual who creates or maintains the program's database. For example, the ammonium cation can be handled by including NH_3^0 as a *ligand* and then considering the appropriate protonation reaction ($NH_3^0 + H^+ = NH_4^+$), in which case the cation that prevails at pH << 9 would appear in the program output as an HL complex, not as an uncomplexed species. Alternatively, however, the user could define NH_4^+ as the fundamental component that then dissociates to NH_3^0 at high pH. [Note: in many programs such as GEOCHEM-PC, only *association* reactions can be handled computationally. Thus, the latter convention would treated as $NH_4^+ + OH^- = NH_3^0 + H_2O$.] Similarly, the final pK_a for vanadate is so high that the VO_4^{3-} species would seldom be of interest, so HVO_4^{2-} can be arbitrarily selected as the component species for use in the computer model (Parker et al., 1995). Note that any consideration of redox equilibria requires that at least two components for the same element are provided as input, for example sulfide (S[–II]) and sulfate (S[VI]).

From the input list of components, the computer algorithm generates an output list of *species* that includes pairwise combinations of metal and ligand components (e.g., ion pairs and complexes), along with associated protons or hydroxyls

Table 14–4. Representative examples of complexation and precipitation reactions considered by GEOCHEM-PC (Parker et al., 1995b), illustrating the conventions used.

	Stoichiometric coefficients
I. Acid-base equilibria	
$CH_3COO^- + H^+ = CH_3COOH$ (acetic acid)	01 1
NH_3 (aq) $+ H^+ = NH_4^+$	01 1
II. Complexation reactions	
$Ca^{2+} + SO_4^{2-} = CaSO_4^0$	11 0
$Ca^{2+} + PO_4^{3-} + 2H^+ = CaH_2PO_4^+$	11 2
$Fe^{3+} + DTPA^{5-} + H_2O = Fe(OH)DTPA^{3-} + H^+$	11-1
III. Metal hydrolysis	
$Al^{3+} + H_2O = AlOH^{2+} + H^+$	10-1
$Al^{3+} + 2H_2O = Al(OH)_2^+ + 2H^+$	10-2
IV. Precipitation of simple solids	
$Ca^{2+} + SO_4^{2-} + 2H_20 = CaSO_4{\bullet}2H_2O$ (gypsum)	11 0
$Ca^{2+} + PO_4^{3-} + H^+ + 2H_20 = CaHPO_4{\bullet}2H_20$ (brushite)	11 1
$Mg^{2+} + 3CO_3^{2-} + 5H_2O = Mg_4(CO_3)_3(OH)_2{\bullet}3H_2O + 2H^+$ (hydromagnesite)	43-2
V. Precipitation of oxides and hydroxides	
$Fe^{3+} + 3H_2O = Fe(OH)_3$ (amorphous) $+ 3H^+$	10-3
$Fe^{3+} + 2H_2O = \alpha$-FeOOH (goethite) $+ 3H^+$	10-3
$Fe^{3+} + 3H_2O = 1/2\ \alpha$-$Fe_2O_3$ (hematite) $+ 3H^+$	10-3

(e.g., species such as MLH or M_2L_2OH). Note that most speciation programs do not allow for mixed complexes containing, for example, two different metals in combination with a ligand. Depending on user-selected options, sparingly soluble solid phases may be predicted to precipitate from solution based on solubility data contained in the program's database. Table 14–4 provides a list of example reactions for the GEOCHEM-PC program. Note that all complexes and solids are defined in terms of association reactions between metals, ligands, and H^+ or OH^- as needed. The stoichiometric coefficients used in the thermodynamic database and program output tables indicate the number of metals, ligands, and protons (positive) or hydroxlys (negative) in a given complex or solid (Table 14–4).

Most of the commonly used speciation programs are computationally similar. Briefly, the requirement for mole balance for all components is coupled with the appropriate thermodynamic data for all reactions under consideration. These requirements provide a set of nonlinear equations linking the molar concentrations of the components with the conditional (ionic-strength adjusted) stability constants for the complexes and solids that may form from these components. These equations are solved simultaneously using an efficient successive approximation method (e.g., the Newton-Raphson method or a variant), until the sum of the concentrations of all metal and ligand species agree with the input concentrations of components to within a user-selected degree of precision (often 0.01%). Most programs do *not* impose or require electroneutrality, nor the closely related proton condition (Stumm and Morgan, 1996), which greatly minimizes computational difficulties (i.e., convergence problems, floating-point errors) when using input data that are incomplete or exhibit poor charge balance, but prevents *direct* solution of certain types of problems (most notably systems with open exchange of CO_2 to the atmosphere).

Chemical equilibrium modeling is entirely predicated on the availability of published stability (or formation) constants for the types of reactions shown in Table 14–4. Some researchers simply use the values provided with computer algorithms such as GEOCHEM-PC (Parker et al., 1995) or MINEQL+ (Schecher and McAvoy, 1991), while others obtain these values from published compendia such as the classic and widely cited volumes of Martell and Smith (1976–1989), now also available as a database for personal computers (NIST,1998). It is relatively uncommon for authors to consult the primary literature from which these thermodynamic databases are derived. This is not necessarily a flawed procedure, because these compilations are usually assembled as a critically reviewed dataset, but some caution is still warranted. Reactions schemes for some metal-ligand combinations have changed over various versions of the NIST database, and there is evidence of occasional errors (see, e.g., Parker et al., 2001).

Accurate and appropriate stability constants for the complexation of metals such as Al and Fe(III) are particularly problematic. Most constants are determined using potentiometric titrations, even though these metals are known to be extremely hydrolytic and very insoluble at circumneutral pH values. Such determinations can be quite accurate with very strong ligands (e.g., EDTA) but, with ligands of more moderate strength (e.g., the organic acids), these hydrolytic "side reactions" are critically important in the fitting of the titration data, and must be modeled using some assumed reaction. But, it is well known that polynuclear Al–OH complexes and solids form in a very unpredictable fashion that is dependent on the reaction conditions and methodology (Bertsch and Parker, 1996). These complications are probably responsible for the disparate reaction schemes and stability constants often reported the interaction of Al and organic acids such as citrate. As a consequence, we are presently inclined to view all speciation calculations for Al- and Fe(III)-complexation by the organic acids with a certain amount of caution.

Modeling Metal Interactions with Humic Substances

For at least three decades, environmental chemists have attempted to model the binding of metals (and protons) by dissolved humic substances (fulvic and humic acids). The models have ranged from rather simplistic (e.g., Sposito et al., 1982) to quite elaborate. In all cases, the models require intrinsic binding constants that are derived by statistically fitting model parameters to experimental data sets for metal binding by the humic substances. Consequently, it is important to recognize that the validity of such models generally depends on the quality of the data derived from on or more of the empirical speciation methods described next.

Probably the best-known such model is the series of humic ion-binding models developed by Tipping and coworkers as part of the comprehensive Windemere Humic Aqueous Model (WHAM) (Tipping, 1994, 1998; Tipping et al., 2002). In its current iteration (Tipping, 1998), this model is a discrete site–electrostatic model that accounts for competitive binding between metals and protons by both HA and FA, and is linked to an inorganic speciation model that accounts for metal complexation and hydrolysis, as well as ionic strength and temperature (Tipping et al., 2002). Proton binding is accounted for by two median proton dissociation constants along with two "range factors" that account for the "spread" around the median values. Similarly, metal binding is defined by two median intrinsic constants that are highly correlated, along with another "spread" term. An additional strong binding term for low-abundance sites is included based on a given metal's affinity for aqueous NH_3 (Tipping, 1998). In many cases, excellent agreement can be obtained between experimental measurements of metal binding and predictions made using this model. Importantly, the model also suggests that both Al and Fe(III), which are comparatively abundant geochemically, can compete effectively with trace metals such as Cu and Zn; binding of the latter by HA or FA may thus be overestimated in laboratory studies using purified humic substances (Tipping et al., 2002).

Another well-studied metal-ion binding model is the NICA-Donnan Model (NICA = non ideal competitive adsorption) model developed by Van Riemsdijk and coworkers (see Milne et al., 2003, and references therein). The NICA portion of the model describes specific binding of metals by HA–FA, and uses continuous distributions of binding constants to account for site heterogeneity. It is coupled to a Donnan sub-model that accounts for nonspecific electrostatic interactions, the importance of which diminishes with increasing metal "loading" as it the net negative charge of the humic substance. The two model components are linked and must be solved simultaneously (Milne et al., 2003). The NICA-Donnan also has been shown to adequately predict metal-binding in a large number of empirical datasets (Milne et al., 2003).

In the main, these modeling approaches have emphasized humic substances found in surface waters, especially freshwaters, although a few studies have explicitly addressed metal binding by soil organics in situ (as opposed to humic substances that have been extracted from soils and used in in vitro experiments). At this point, the utility of these models, for example in predicting trace-metal speciation (and perhaps biological response) in soil solution or the rhizosphere, remains largely unexplored.

Empirical and Analytical Speciation Methods

For more exhaustive literature reviews of analytical speciation techniques, the reader is referred to Apte and Batley (1995) for resins and ligand exchange methods; Florence (1986) and Mota and Correia dos Santos (1995) for electrochemical methods; Gulens (1987) for ion selective electrodes; Dabek-Zlotorzynska et al. (1998) for capillary electrophoresis; and Marshall and Momplaisir (1995), Sarzanini and Bruzzoniti (2001), and Szpunar et al. (2000) for chromatographic methods. Also, it must be emphasized that many speciation methods involve the pairing of a separation process with a detector; the basic principles underlining the functioning of different instruments are well explained in general textbooks (e.g., Skoog et al., 1998).

Size Separation

One of the simplest approaches to analytical speciation of trace metals is to effect a separation based on size (or, equivalently, molecular weight), for example using dialysis membranes (e.g., Berggren, 1989) or ultrafiltration membranes (e.g., Staub et al., 1984; Angehrn-Bettinazi, 1990). Florence (1982) suggested some nominal size cut-offs. Particulates larger than 450 nm are not considered part of the dissolved fraction, and this particulate fraction is easily segregated using common membrane filters (0.45 μm). Then come inorganic and organic colloids (between 10 and 200 nm in diameter), organic complexes (2–5 nm), and hydrated metal ions and inorganic complexes (~1 nm). Alternatively, Buffle and Leppard (1995a) defined colloids (= macromolecules) as all entities with nominal diameters between 10 and 1000 nm. Membranes are available with nominal molecular weight cutoffs as low as 500 to 1000 Da (= g mol^{-1}), and these may be used to separate colloidal metals and macromolecular complexes from metals complexed with inorganic and/or low-molecular-weight organic ligand. Such an approach may be particularly useful for separating metals associated with humic substances.

Originally, ultrafiltration (UF) consisted of a membrane filter sitting on porous support plate that comprised part of an air-tight reaction cell. Compressed air or an inert gas is used to pressurize the cell, pushing water through the membrane filter; constituents larger than the nominal cutoff are excluded and do not appear in the filtrate. Typically, a magnetically driven stirrer is used to minimize both membrane clogging and induced colloid self-coagulation in the diffusion layer adjacent to the filter surface (Buffle and Leppard, 1995b). In the last decade or so, much of the emphasis has been on cross-flow filtration (CFF, also known as tangential-flow filtration), wherein the aqueous sample flows parallel to the membrane with hydrostatic pressure driving water and small solutes through the membrane (the "permeate"). The "retantate," which becomes enriched in colloids larger than the membrane's nominal cutoff is continuously recycled through a large reservoir (Buessler, 1996). The method has been used extensively in attempts to better characterize colloidal organic material in marine systems (e.g., Larsson et al., 2002).

These methods have not enjoyed widespread usage in the speciation of trace metals, especially in soil solution. The CCF methods require very large volumes of sample (liters), seldom achievable with soil solution collection. A number of problems have been documented arising from interactions of the metals and/or organic macromolecules with the UF membrane itself (Buffle et al., 1995b). The separations represent a type for fractionation more than speciation per se, as the size cutoffs are approximate and very much operationally defined. Moreover, very little detailed information concerning speciation (e.g., free metal ion activities) is directly obtained.

In an application closer to fractionation than speciation, ultracentrifugation also can be used to study the metal-promoted aggregation of humic substances (Bryan et al., 2001). This shows that different cations vary in their ability to cause the aggregation or coagulation of dissolved organic matter (Römkens and Dolfin, 1998).

Ion-Selective Electrodes

Ion-selective electrodes (ISEs) are probably the most direct and simplest means of determining free metal activities. In principle, their functioning is much like that of a pH electrode responds directly to the activity of free H^+ ions. The surface of an ISE is similarly sensitive to free metal ions, and the electrode response can be thus be linked to overall metal speciation. Ion selective electrodes, like pH electrodes, measure a potential difference. That is, the reference electrode provided a constant reference potential that is independent of the composition of the solution it is inserted into, and the indicator electrode gives a response quantitatively related to the activity of the ion it is built to measure. The electrode response is usually given in millivolts, and comparison with standards of know ion activities allow calibration and interpolation of unknowns (much as we do when we calibrate a pH-meter with pH buffers).

Ion-selective electrodes are prone to interference by other ions, i.e., they are not fully selective. So care must be taken to insure that it is used within the range of solution properties where its response is valid. The Cu ion selective electrode is not prone to excessive interference (except very high chloride or mercury levels), and has been used to speciate soil solutions of various origins (Sauvé et al., 1995). The other electrodes (Pb, Cd) seem prone to excessive interferences from other metallic cations, organics, and iron in solution, and their use is limited to higher concentrations and/or purified–synthetic systems.

Nonetheless, it is important to emphasize that the detection limits reported by ISE manufacturers only apply to dilute solutions of metal salts. In well-buffered systems, where much of the metal of interest is bound to complexes in solution, the detection limits can be much lower (Gulens, 1987). For example, the reported limits for dilute $Cu(NO_3)_2$ salts is about 10^{-7} *M* free Cu^{2+}. In buffered synthetic systems, the same electrode has shown a linear response down to as low as 10^{-19} *M* free Cu^{2+} (Avdeef et al., 1983). Given that soil solutions are often rich in organic acids that can act as complexing ligands, the ion selective electrodes can be used for wider applications than often perceived, i.e., down to free Cu^{2+} and Pb^{2+} activities as low as 10^{-13} *M* (Sauvé et al., 1995). The activities of free Cu^{2+} in "clean," uncontaminated soils fall within the linear range of the Cu^{2+} electrode, making it a very useful tool for studying Cu speciation in soil solutions.

An example of the use of a Cu ISE is presented in Fig. 14–2. The variation in soil solution free Cu^{2+} is clearly a function of pH, in part. The free Cu^{2+} in this figure is shown as pCu^{2+} units, where $pCu^{2+} = -\log_{10}(\text{free } Cu^{2+} \text{ activity})$; as with pH, small numbers represent high concentrations, and vice versa. In this experiment, a Guatemalan soil used for banana culture was found to have elevated soil total Cu due to repeated pesticide application. The soil originally contained 76 g C kg^{-1}, and some of the samples were treated with H_2O_2 to remove organic matter level. Fig-

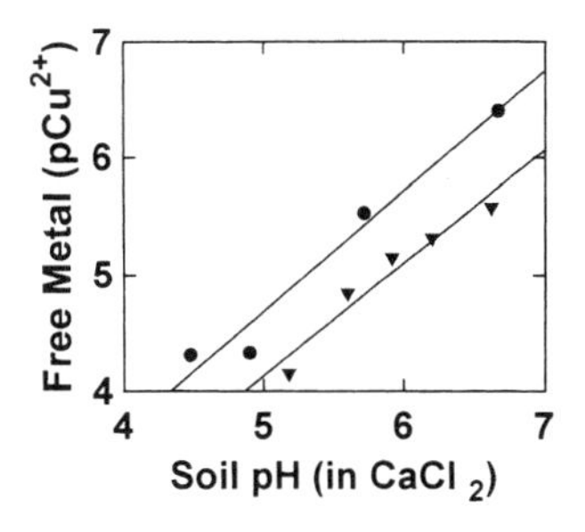

Fig. 14–2. Free Cu^{2+} activity in a Guatemalan banana soil containing 1020 mg Cu kg^{-1}. The pH was varied by adding small aliquots of NaOH or HNO_3. The circles represent the untreated soil and the triangles represent the H_2O_2-treated soil (lower organic matter content). Linear regressions for both soils are also included.

ure 14–2 clearly shows that pH is an important modulator of the soil solution free Cu^{2+}, and it also shows that the soil with reduced organic matter levels had a higher concentration of free Cu^{2+} in solution. This serves to illustrate that, at a given total metal loading level, overall solubility and chemical speciation are usually critically influenced by other soil characteristics.

Voltammetry

Various polarographic and voltammetric techniques have been used to quantify the labile or reactive forms of trace metals in solution. With, for example, anodic stripping voltammetry (ASV), the labile metal corresponds to that fraction that is reduced or deposited on the electrode. The ASV electrode often consists of a mercury droplet, but rotating disks of various compositions in combination with electrode coatings can also be used.

One limitation of ASV (or polarography) is that the technique is operationally defined through the electrochemical parameters used (electrode stirring or mercury drop diameter, deposition potential, pulse frequency, sample pH, temperature, sample composition, etc.). It also is important to note that standard additions or metal spikes should not be used for calibration. Such metal additions might increase total rather than labile metals levels, as a significant and variable amount of time might be required before they equilibrate with ligands in the sample. The peak heights obtained using ionic metal in solutions of similar inorganic composition should generally be used to standardize this method (Florence, 1986). If one is interested in the speciation under natural conditions, then any sample modifications or disturbances should be kept minimal, especially changes in pH. Even with minimal changes, voltammetric determinations are dynamic, and the measurement itself disturbs the original speciation in the sample to some degree. An accurate reporting of all methodological details is thus critical for proper evaluation of results.

Analytical ASV determinations can be optimized for electrochemical conditions, permitting the distinction of labile from nonlabile fractions. It is often assumed that the ASV-labile metals correspond to the sum of free and inorganically bound metals (Florence, 1986; Shuman, 1988), but the presence of any labile organic complexes could invalidate this assumption. Once the distinction is made between the free plus inorganically complexed and organically complexed forms, chemical equilibrium models are used to provide the exact distribution of metal species within the former fraction. An example of this approach, and the resulting values for free Cd^{2+} in soil solution as a function of pH and soil total Cd, are presented in Fig. 14–3.

Purging the sample with an inert gas before the analysis reduces the voltammetric interference from dissolved oxygen. But, CO_2 will probably degas, and any (bi)carbonate complexes that could be important in alkaline soil solutions would largely dissociate with the metal re-equilibrating with other solution ligands. It also is possible that organic matter adsorbs on the electrode, thus diminishing the signal. Except through comparison and validation with different techniques, it is problematic to distinguish physical-chemical hindrance due to sorption on the electrode from real speciation changes through the formation of unreactive (non ASV-labile) complexes (Florence, 1986). Such interferences are probably best evaluated using comparative validations with alternate speciation methods (Sauvé, 2002).

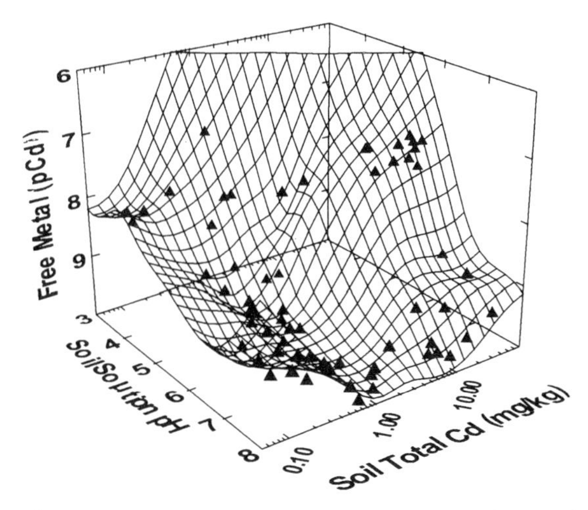

Fig. 14–3. Free Cd^{2+} as a function of soil solution pH and soil total metal content (mg Cd kg^{-1}). The graphic surface is obtained using a distance-weighted least-square smoothing algorithm (reproduced from Sauvé et al., 2002).

Ion Exchange Resins

Ion exchange resins can be used for many purposes in environmental analytical chemistry. It is important to recognize that not all applications of resins are compatible with the determination of trace-metal speciation. To ensure accurate speciation, it is important to use the resin to differentiate distinct metal species, and not just an operationally defined group. For example, if a soil-solution extract is reacted with a cationic resin, the free metal in solution should preferentially sorb to the resin, and this process could be used to separate free metal ions from the supernatant solution where the remaining metals are complexed by inorganic and organic ligands. Unfortunately such a simple scheme is complicated by the fact that the solution to resin ratio needs to be properly adjusted (see below), the affinity of the resin needs to be determined for the specific system conditions, the complexed metals in solution may dissociate and therefore bind to the resin, and the time of contact between the solution and the resin needs to be carefully evaluated. It also is necessary to evaluate whether the resin truly segregates only one complex species of a single metal with a single ligand (e.g., MeL), or if more elaborate complexes exist that might also bind to the resin (e.g., ML^+, MeL_2, MeHL, etc.).

The proper use of exchange resins to differentiate free metal ions depends on a number of key assumptions: (i) only free M^{n+} binds with resin; (ii) the proportion of macro-element cations (e.g., Na, Ca, Mg, K) on the binding sites on the resin are pre-equilibrated to that of the solution to be studied; (iii) the quantity of resin is very small relative to the volume of solution such that the removal of free metal from the solution does not induce a significant shift in the speciation. It is not trivial to fully achieve these three conditions for real environmental solutions.

Ion exchange resins have been used for the speciation of free metal in soil and environmental solutions (Holm et al., 1995; Fortin and Campbell, 1998). In such systems, the affinity of the cationic resin for the metal of interest is established in a preliminary experiment reproducing the characteristics of the studied solution with respect to ionic strength, chemical composition, and pH. Once the resin affinity is delineated, the original sample solution, the supernatant that has reacted with the resin, and the resin-adsorbed metals are analyzed using standard analytical techniques. The data are then treated to account for equilibrium exchange reactions and mass balance, and yield estimates of free metal-ion concentrations. Sample calculations are illustrated in Holm et al. (1995) and Fortin and Campbell (1998).

For ion exchange resins, the actual detection limits are dependent upon the instrumentation used for metal analysis. With some trace metals, notably Cu and Pb, the soil solution pool is dominantly complexed (>80 to 99%), and the very low concentrations of free metal can be very difficult to quantify, but given the advances and increased availability of inductively-coupled plasma-mass spectrometry (ICP-MS), this is becoming less of an issue. Still, the practical analytical detection limits available to a given laboratory will dictate the applicability of this methodology. Thus, speciation by ion exchange is often most useful with Cd or Zn that are more soluble, less complexed, and thus somewhat easier to quantify in soils solution.

In addition, an older alternative method similar to ion exchange resins uses manganese dioxide, MnO_2 (Apte and Batley, 1995). A small quantity of MnO_2 suspension is added to the sample, and acts as an ion-exchange medium, sorbing free metal only. After an equilibration period, the sample is filtered and the trace element concentration remaining in solution is determined. To be valid, the sorption properties of the MnO_2 need to be thoroughly characterized under conditions that are nearly identical to those that prevail during the assays. There are problems with variations in the properties of the MnO_2 suspensions, and the presence of colloidal particles in the sample also can interfere with the method. Here also, the MnO_2 spike itself must not alter the chemistry (and the speciation) of the sample, or else the speciation is not representative of the in situ condition.

Liquid Chromatography

In chromatography, a sample is eluted through a column in which a resin, gel, or other binding material effectively retards certain molecules based on properties such as charge, mass, polarity, hydrophobicity, or size. The aim is to segregate the solutes according to these properties, and thus allow a quantification of the successive waves (peaks) of analytes. The chromatographic separation is coupled to a detection method, preferably an on-line monitoring system. There is a plethora of potential combinations for separation and detection, and we do not wish to extensively review those here (see Marshall and Momplaisir, 1995; Szpunar et al., 2000; Sarzanini, 2002; He et al., 2002). Many chromatographic separations involving trace elements are geared toward separation of very nonlabile organic compounds [e.g., alkyllead and alkyltin species, metallothioneins, various proteins, etc. See examples for As and Cr in Ali and Aboul-Enein (2002), or for Sn and, As in Szpunar et al. (2000)]. It is less clear whether more labile organic complexes can be accurately separated without shifts in speciation due to eluent chemistry and/or column interaction.

Another issue emphasized by Marshall and Momplaisir (1995) is that many of the reported methods have been developed without validation using "real" environmental or biological samples; however, this trend is changing and some practical applications are starting to appear in the literature (e.g., Szpunar et al., 2000; He et al., 2002). Some recent applications of ion chromatography coupled with electrospray mass spectrometry for the determination of metal-EDTA complexes in soil solution and barley (*Hordeum vulgare* L.) xylem is pertinent to environmental issues pertaining to various EDTA usages (Collins et al., 2001). Other issues for the speciation of metals in biological materials include the means used to solubilize some recalcitrant fractions, i.e., it is critical that whatever dissolution technique is used prior to the chromatographic separation, it does not change or modify the chemical speciation (He et al., 2002).

A sample chromatogram for the speciation of arsenic in a standard solution illustrates the peak separation obtained for the inorganic forms of arsenic: As(III) and As(V) as well as for two organic forms: monomethylarsenic (MMA) and dimethylarsenic (DMA) (Fig. 14–4). This was done using anion-exchange liquid chromatography coupled with hydride-generation atomic fluorescence spectrometry.

Capillary Electrophoresis

The underlying principle of capillary electrophoresis (CE) is similar to that of liquid chromatography. Separation in electrophoresis is effected using the differential migration of ionic species in an electrical field. A background electrolyte, possibly containing an indicator needed for analytical quantification, is chosen for the analysis. One end of a capillary (typically of 25–100 μm internal diameter) is inserted in the sample, the other end is inserted in the background electrolyte, and a high voltage source is used to generate an electrical field between the ends of the capillary. According to their charge and reactions with the buffers, the ions in solution will migrate along the capillary. Different buffers can be used to better segregate the ions according to their affinity for complexation or reaction. A detector is placed at the end of the capillary and used to record the evolution in the solution concentration, with the most widely used detection methods being UV/VIS absorbance, fluorescence, atomic fluorescence spectrometry, laser-induced fluorescence, mass spectrometry, conductivity, amperometry, radiometry, and refractive index (Barker, 1995). One of the difficulties to apply CE to trace metal analysis is that most CE instruments are equipped with UV detectors, and most metal species of interest do not absorb in the UV. But, the detection method can be direct or indirect, i.e., it can directly measure the presence of a peak due to increased absorbance, or it can measure a decrease in the absorbance of the buffer due to the presence of a nonabsorbing species at the detector (i.e., a negative peak or depression). In this case, a highly UV-absorbing species is used in the electrolyte, and the UV-detector will measure a decrease in this species when the non-absorbing chemical species of interest reaches the detector.

Some difficulties may arise in segregating species with similar ionic conductivities and concomitant electrophoretic mobilities. It is possible to modify the electrolyte composition of the solution to attempt a separation using a weak complexing ligand. Each element has an individual complexation affinity for the medium and the respective mobility of each element will hopefully be altered to different extents (Weston et al., 1992). Depending on the actual sample conditions,

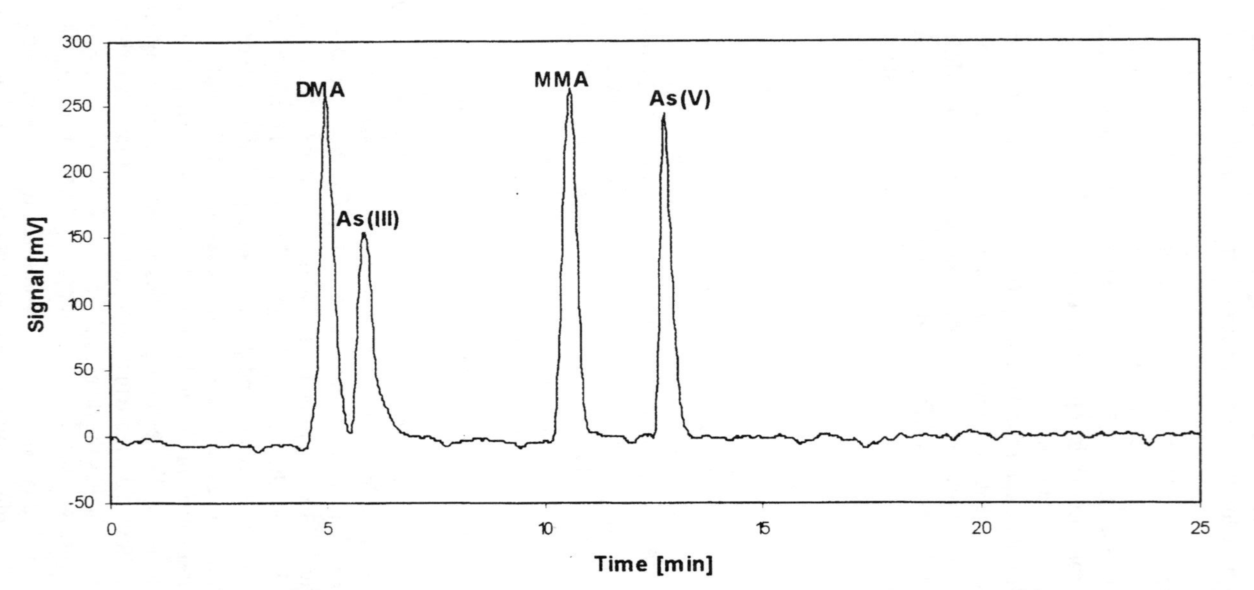

Fig. 14–4. Speciation of arsenic in a standard solution containing 5 µg As L^{-1} of each of dimethlyarsinic acid (DMA), monomethylarsonic acid (MMA) as well as arsenite (As(III)) and arsenate (As(V)). This was done using anion exchange chromatography coupled with hydride-generation-atomic fluorescence spectrometry (chromatogram graciously provided by Dirk Wallschlaeger from Trent University).

the possible presence of large concentrations of certain elements may reduce the sensitivity of the technique, and the detection limits are not always sufficient for relevant analyses in environmental samples (Dabek-Zlotorzynska et al., 1998). Capillary electrophoresis can be used for inorganic ions as well as complexes. For example, Owens et al. (2000) have used CE with UV detection to distinguish among the complexes of EDTA with different metals, and report detection limits between 2 and 7 μM for various complexes. There are few applications of this technique to soil solutions, and the background electrolyte needs to be minimized and checked for its tendency to alter the chemical speciation of the sample.

Donnan Dialysis

In Donnan exchange (or dialysis) methods, the aqueous sample is dialyzed across an ion-exchange membrane (Minnich and McBride, 1987; Fitch and Helmke, 1989; Fitch et al., 1986). The separation membrane can, in principled, be an anion- or cation-exchange membrane, but soil solution speciation has only been attempted for cationic metals (e.g., Cd^{2+}). A specially constructed Teflon chamber is used to dialyze a large donor (sample) solution (ca. 50 mL) across a Dupont Nafion-117 cation exchange membrane. The acceptor solution is quite small (e.g., 50 μL) and contains $Sr(NO_3)$ at the same ionic strength as the donor solution; the cationic metals in the donor solution exchange with Sr^{2+} in the acceptor solution. Because of the respective sizes of the two solutions, there is negligible mass transfer of metal from the original donor solution, and its equilibrium chemistry is thus unperturbed (Minnich and McBride, 1987). Upon equilibration, the free metal activity should be identical on both sides, provided that pH and ionic strength gradients between both sides of the membranes are negligible or are accounted for. Small differences in ionic strength can be accounted for by spiking a small quantity of ^{22}Na into the donor solution, and measuring its activity with a gamma counter after equilibration (Fitch and Helmke, 1989; Fitch et al., 1986). Typically, the total metal concentration of the acceptor solution is determined using GF-AAS, equated to $[Cu^{2+}]$, and converted to the free Cu^{2+} activity using the Davies equation.

Note that small cationic ions, including cationic complexes such as $CdCl^+$ *will* cross the membranes used in Donnan dialysis, and their activities should be similarly similar in both donor and acceptor solutions. If the concentrations of such ligands are all known, along with the relevant metal-binding constants, then it is relatively straightforward to compute the free metal activity in the acceptor solution. Provided that dialysis equilibration has not significantly altered the original speciation, the free metal speciation in the donor solution should be equivalent.

Because of the paucity of studies comparing simultaneously different metal speciation methods, we refer to an indirect comparison of the chemical speciation of soil solution Cd by Donnan exchange (Salam and Helmke 1998) and differential pulse anodic stripping voltammetry (Sauvé et al., 2000b). This comparison showed a convincing agreement between the two techniques (Sauvé et al., 2000b). On the other hand, a similar comparison of free Cu^{2+} activity by Donnan exchange (Salam and Helmke 1998) and ion-selective electrode (Sauvé et al., 1997) showed significant discrepancies, and an apparent lack of pH sensitivity with the Donnan technique (Sauvé, 2002). Similarly, Minnich and McBride (1987) noted that although the Cu ion selective electrode determinations were correlated with Donnan membrane measurements, the latter were about an order of magnitude lower. On the other hand, Wu et al. (2000) compared free Cu^{2+} by the Donnan method, and found a reasonable agreement with free Cu^{2+} based on a regression prediction derived from an ion-selective electrode dataset. A similar comparison for Pb showed that the results for Donnan membrane speciation was underpredicting free Pb^{2+} by one to two orders of magnitude relative another predictive equation based on stripping voltammetry determinations (Wu et al., 2000). It is apparent that Donnan membrane techniques are quite sensitive to the actual methodological setup, and in some cases tiny amounts of acceptor solutions are used. A standardized approach would help to compare its efficiency and its application to environmental samples.

Ligand Competition

In a few cases, ligand exchange is quite similar in principle to resin-exchange techniques, in that a ligand of known affinity for the trace metal of interest is equilibrated with the sample solution. The resulting chemical equilibrium is resolved to calculate free metal activity, and perhaps to infer the strength and complexation capacity of the ligands present in the solution. Ligand exchange or ligand competition can be combined with a solvent extraction (Adams and Kramer, 1999). There are various ligand-solvent extraction combinations available for different metals. The combinations of ligand and solvent compounds will vary according to the system and trace elements of interest, and also may vary according to the element concentration range. Some recent applications to Ag^{+} speciation are worth noting (Adams and Kramer, 1999; Herrin et al., 2001), but the methodological details of those techniques will not be elaborated upon further.

Ligands have been used on a number of occasions in competitive chelation method applied exclusively to soils (Lindsay, 1979). Here, a ternary system is created where a soil sample is equilibrated for a few days with an EDTA solution in the presence of an added solid mineral phase of a trace metal different than the one being analyzed. For example, to determine free Pb^{2+} in soil solution, Kalbasi et al. (1995) added 20 mg of solid $CdCO_3$ to 10 g of soil and 20 mL of solution of variable EDTA concentrations. At equilibrium, given the very high affinity of EDTA for trace elements over major cations, the solution ratio of Pb-bound to Cd-bound EDTA varies as a function of the EDTA concentration. That ratio can be compared with what is obtained with EDTA solutions exposed to $CdCO_3$ alone (without soil) and the ratio where control and soil solution crossover corresponds to the equilibrium conditions that prevail in the soil. Given that (i) the EDTA concentration is known (or measured), (ii) the free Cd^{2+} is fixed through mineral equilibrium with $CdCO_3$, and (iii) standard chemical parameters can be measured (pH, ionic strength, chloride, etc,), then the only unknown is the free Pb^{2+} activity. Although this method is elegant, it is potentially fraught with many artifacts; competition from major or other trace elements and the fate of the added EDTA need to be carefully evaluated (Kalbasi et al., 1995). Most importantly, it is questionable whether the speciation of a soil solution shaken 4 d with EDTA and $CdCO_3$ is representative of the original, in situ speciation. Indirect comparison of this technique with voltammetric or resin speciation determinations highlights some severe differences (Sauvé, 2002).

Ligand exchange can also be combined with voltammetry wherein an organic ligand (e.g., citric acid, EDTA, fulvic acid, etc.) is used as a competing ligand in conjunction with dimethylglyoxime. This is a potentially powerful tool to evaluate the dissociation kinetics of natural ligands and their mechanistic pathway (Celo et al., 2001; Xue et al., 2001). Although that method can be used for chemical speciation, the theory from which it is based is more complex than with most other techniques, and seems better suited to the exploration of the kinetics of metal-ligand exchange. It can, however, be applied to environmental samples, and becomes a polarography-dependent technique with fewer uncertainties than straight anodic stripping voltammetry (e.g., Aldrich et al., 2002).

Solid Phase Microextraction

Solid phase microextraction (SPME) is a well-known technique in the field of organic contaminants, but relatively new for the analysis of trace elements. This equilibrium technique uses a coated silica fiber which is inserted into a sample, either directly within the stirred solution or in the closed headspace above. The technique focuses on organic, volatile, or hydrophobic organic compounds. The silica fiber is coated with a polydimethylsiloxane, divinylbenzene, or another coating that will act to accumulate the organic compounds to the point where it is in equilibrium with either the solution or the air above that solution. The fiber is then retracted into the device and then deployed into a gas chromatograph, liquid chromatograph, or other instrument to desorb the accumulated contaminant and quantify it. Because the method targets volatile or hydrophobic compounds, its use for trace elements is limited to the speciation of organometallic species of lead, mercury, and tin (Mester et al., 2001). One of the main advantages of the technique is that, because it is based on equilibrium between the medium and the fiber and very little of the total mass of the compound is thus extracted; one can expect that the extraction itself does not modify the speciation.

Other Techniques

An early "speciation" technique used in aquatic chemistry was an algal bioassay that presumed that the organisms are sensitive only to free metal concentrations (Rueter and McCarthy, 1979; Petersen, 1982). Such biological studies have been refined and are still used (e.g., Reinfelder et al., 2000). There have been some similar approaches to soils using leaf litter extracts and relative algal uptake of $^{14}CO_2$ (Blaser et al., 1980). Although bioassays are useful ecotoxicological tools, they cannot be viewed as a reliable analytical method for quantifying speciation.

Nuclear magnetic resonance of ^{113}Cd has been used by Otto et al. (2001) to non-invasively probe metal binding. They have used calcium titration and competitive binding to characterize different fulvic acids. While this technique is restricted to diamagnetic metals it may prove most useful in certain applications in the future.

"Hybrid" Approaches

Few speciation techniques do not combine some form of chemical equilibrium computation with the empirical measurements, or borrow from more than one technique. A major distinction between computational or analytical speciation methods lies in how organic matter complexation is considered. Given the comparatively high level of confidence and consensus for inorganic speciation, calculations of inorganic ion-pairs and complexes are often integrated into other speciation schemes (e.g., voltammetry); however, even if the equilibrium calculations for inorganic chemical species are straightforward, the modeling of complexation by dissolved organic ligands is more complex and elusive. To date, no single model yields a consensus among the various users and each model depends upon a number of parameters that are metal-dependent, and that also vary according to the nature and source of the organic matter. It should thus be emphasized that, while the chemical equilibrium calculations for the inorganic species is relatively easy, the equivalent calculations for the association with dissolved organic matter is much more dependant on the actual model used, the assumptions made, and the analytical data used as input into the model.

Many empirical, some techniques are also "hybrid" methods in one or more ways. For example, some radiotracer analytical determinations are used within a competitive ligand exchange-solvent extraction (van Ginneken and Blust, 2000). In general, the majority of speciation methods entail a separation or segregation step, which is distinct and separate from an analytical step. For example, a resin or membrane can physically separate free from complexed metals, but atomic spectrometry, voltammetry, or spectrophotometric colorimetric methods are then used in the quantification of metal concentrations in the segregated solutions. In the same sense, all chromatography-based speciation methods are coupled or "hyphenated," since the detection method is distinct from the chromatographic separation. It is for this reason that chemical speciation of soil solutions are rarely, if ever, probed "directly" using wet chemical means.

Obstacles to Accurate Speciation of Soil Solutions

Collection of Soil Solution in situ. It is not trivial to collect soil solutions in situ. Soil solutions can be obtained using displacement, centrifugation, and lysimeters. Water displacement and centrifugation are do not yield truly in situ samples; the best they can achieve is a laboratory sample obtained immediately after soil collection. Even then, it is difficult to obtain enough solution from soils at low moisture levels by simple centrifugation, and it is often necessary to treat large amounts of soils to obtain small quantities of solution. It also has been suggested that, when looking at rhizosphere soils, centrifugation yields a solution that is closer to that of the bulk soil solution; Lorenz et al. (1994) attributed this to buffering by soil colloids.

Lysimeters are often made with a porous ceramic cup to sample soil solutions in field or pot experiments, and this is a reasonable approach for examining nutrients and macro-elements (e.g., Bowman et al., 2002). On the other hand, it is not appropriate for trace element analysis, as ceramic cups have been shown to reduce the concentrations or almost completely remove metals from soil solutions (Wenzel et al., 1997). Comparisons of passive capillary samplers and zero-tension samplers showed that capillary samplers are not suitable, the weathering of the fibreglass wicking seems to increase pH, Ca, Na, and Si while it reduces the concentrations of Al (Goyne et al., 2000).

Sampling Errors. Soils are heterogeneous, especially contaminated soils, and trace element analysis is sometimes problematic even with relatively homogeneous matrices. Thus, the first step in any environmental analysis is ensuring the representativeness of the sample, and providing some measure of the errors associated with the various steps involved. Errors associated with sample preparation and treatment for "total" content determinations (see review by Hoenig, 2001) are equally important with respect to chemical speciation. The problem is somewhat different, but speciation analyses similarly depend on measured concentrations in different media or analytical steps. Air drying and storage also can influence soil chemistry, especially redox sensitive metals (Ross et al., 2001).

Since pH is a dominant parameter modulating the chemical speciation of metals in soils, any factors having an impact on pH need to be accounted for. This includes factors that would not impact total metal analysis, but could influence speciation, such as the need for thorough rinsing of residual cleaning acids, and the effects of CO_2 degassing, soil drying and sieving, and remoistening air-dried soils after prolonged (Franzluebbers et al., 2000). Also experimental treatments may impact speciation, e.g., acidification with hydrochloric acid would generate two sources of variation, one for pH and one for chloride levels.

Obtaining a representative, uncontaminated sample of soil solution is not a trivial matter. Various research groups favor different methods, and each may have different strengths and weaknesses (Sauvé, 2002). Most importantly, chemical speciation is dynamic; speciation analysis reflects an instantaneous picture of the distribution of a given element at the moment of the analysis. Storage of extracted solutions is therefore problematic: acidification is unacceptable; cold storage will slow but not eliminate microbial transformation of organic matter and shifts in metal oxidation state; flocculation of organic matter also may occur and might be promoted if the sample is frozen. Consequently, it is necessary to make speciation determinations of soil solution samples as quickly as possible. Although this implies that the extracted solutions are more sensitive to transformation than solids, soil drying and storage nevertheless has important effects on soil properties, especially with regards to oxidation state and dehydration of surfaces (Bartlett and James, 1980).

Finally, whatever speciation technique is employed must minimize any perturbations in the equilibrium chemistry of the soil solution. While evaluating the specific methodological details of using each respective speciation techniques, one must keep in mind the potential perturbation and interferences. The method used to obtain the soil solution may itself result in perturbation of the equilibrium conditions of the soil sample.

Validation of Speciation Methods. It is not trivial to unequivocally validate soil solution speciation techniques. Known organic or inorganic ligands can be used to create "synthetic" samples with known chemical speciation (based on equilibrium modeling). These in turn can be used to show that a given empirical method yields "right" answer, at least for solutions containing model compounds (but probably not natural DOC). But even here a true validation is difficult; the mathematical models themselves are really based on earlier speciation studies (e.g., the potentiometric determination of stability constants) that have associated sources of error and uncertainty. Nevertheless, there is a large body of literature covering various computational and analytical speciation techniques, and a validation of sorts can be achieved through comparison of results obtained with various techniques. Unfortunately, when the results of two or more methods fail to agree, there is seldom an *a priori* set of criteria for determining which method(s) is/are "correct." This is an area of potential improvement for among environmental and soil chemists: objective criteria, for example concerning quantifiable parameters such as analytical accuracy and precision, need to be developed and used more consistently.

SUMMARY AND CONCLUSIONS

Although factors such as oxidation number are considered part of chemical speciation, the chemistry of trace elements in soil solution is dominated by the tendency for metals to form stable complexes with an assortment of inorganic and organic ligands. During the last three decades, the speciation of metals in natural waters, including soil solutions, has been one of the most active areas of research in environmental chemistry. Understanding the speciation of trace metals is essential to explaining and predicting their biological activity, including their toxicity, as only certain aqueous species seem to elicit any toxic response. An accurate description of metal speciation also is necessary for proper modeling of geochemical behavior, including mineral-phase solubility.

In this chapter, we have reviewed the fundamental definition of chemical speciation, and outlined the main physico-chemical parameters and theoretical concepts needed to understand and apply chemical speciation in soil systems. The range of available methods—computational as well as empirical/analytical—have been surveyed, while highlighting some of their advantages and weaknesses. The challenges inherent in accurate sampling and characterization of soil solutions also are discussed. Despite the enormous surge in interest in speciation of trace metals in environmental media during the last decade or so, the methods remain somewhat crude and often lack rigorous validation. This is especially so with soil solutions, which are chemically complex, spatially heterogeneous, and very dynamic. In the future, improved methods, as well as more assiduous application of existing methods, will be needed to develop unambiguous portraits of trace element speciation in situ. Only then will we be able to evaluate the true utility of such information to issues such trace-metal toxicity to terrestrial biota.

REFERENCES

Adams, N.W.H., and J.R. Kramer. 1999. Determination of silver speciation in wastewater and receiving waters by competitive ligand equilibration/solvent extraction. Environ. Toxicol. Chem. 18:2674–2680.

Aldrich, A.P., D. Kistler, and L. Sigg. 2002. Speciation of Cu and Zn in drainage water from agricultural soils. Environ. Sci. Technol. 36:4824–4830.

Ali, I., and H.Y. Aboul-Enein. 2002. Speciation of arsenic and chromium metal ions by reversed phase high performance liquid chromatography. Chemosphere 48:275–278.

Allison, J.D., and D.S. Brown. 1995. MINTEQA2/PRODEFA2: A geochemical speciation model and interactive preprocessor. p. 241–252. *In* R.H. Loeppert et al. (ed.) Soil chemical equilibrium and reaction models. SSSA Spec. Publ. no. 42. ASA and SSSA, Madison, WI.

Angehrn-Bettinazi, C. 1990. Factors affecting the investigation of heavy metal speciation in forest soils using thin-channel ultrafiltration. Int. J. Environ. Anal. Chem. 39:81–89.

Apte, S.C., and G.E. Batley. 1995. Trace metal speciation of labile chemical species in natural waters and sediments: Non-electrochemical approaches. p. 259–306. *In* A. Tessier and D.R. Turner (ed.) Metal speciation and bioavailability in aquatic systems. John Wiley and Sons, New York.

Avdeef, A., J. Zabronsky, and H. Stuting. 1983. Calibration of copper ion selective electrode response to pCu 19. Anal. Chem. 55:298–304.

Baes, C.F., Jr., and R.E. Mesmer. 1976. The hydrolysis of cations. John Wiley and Sons, New York.

Barker, D. 1995. Capillary electrophoresis, techniques in analytical chemistry. Wiley Interscience, New York.

Bartlett, R., and B. James. 1980. Studying dried, stored soil samples: Some pitfalls. Soil Sci. Soc. Am. J. 44:721–724.

Bassett, R.L., and D.C. Melchior. 1990. Chemical modeling of aqueous systems. p. 1–12. *In* D.C. Melchior and R.L. Bassett (ed.) Chemical modeling of aqueous systems II. ACS Symp. Ser. 416. Am. Chem. Soc., Washington, DC.

Berggren, D. 1989. Speciation of aluminum, cadmium, copper, and lead in humic soil solutions: A comparison of the ion exchange column procedure and equilibrium dialysis. Int J. Environ. Anal. Chem. 35:1–15.

Bertsch, P.M., and D.R. Parker. 1996. Aqueous polynuclear aluminum species. p. 117–168. *In* G. Sposito (ed.) The environmental chemistry of aluminum. 2nd ed. CRC Press, Boca Raton, FL.

Blaser, P., M. Landolt, and H. Flühler H. 1980. Metal binding properties of leaf litter extracts: II. A bioassay technique. Soil Sci. Soc. Am. J. 44:717–720.

Bowman, M.S., T.S Clune, and B.G. Sutton. 2002. A modified ceramic sampler and lysimeter design for improved monitoring of soil leachates. Water Res. 36:799–804.

Bryan, N.D., M.N. Jones, J. Birkett, and F.R. Livens. 2001. Aggregation of humic substances by metal ions measured by ultracentrifugation. Anal. Chim. Acta 437:291–308.

Buesseler, K.O. (ed.) 1996. The use of cross-flow filtration (CFF) for the isolation of marine colloids. Mar. Chem. 55:1–204 (special issue).

Buffle, J., and G.G. Leppard. 1995a. Characterization of aquatic colloids and macromolecules: 1. Structure and behavior of colloidal material. Environ. Sci. Technol. 29:2169–2175.

Buffle, J., and G.G. Leppard. 1995b. Characterization of aquatic colloids and macromolecules: 2. Key role of physical structures on analytical results. Environ. Sci. Technol. 29:2176–2184.

Campbell, P.G.C. 1995. Interactions between trace metals and aquatic organisms: A critique of the free-ion activity model. p. 45–102. *In* A. Tessier and D.R. Turner (ed.) Metal speciation and bioavailability in aquatic systems. John Wiley and Sons, New York.

Celo, V., J. Murimboh, M. Salam, and C. Chakrabarti. 2001. A kinetic study of nickel complexation in model systems by adsorptive cathodic stripping voltammetry. Environ. Sci. Technol. 35:1084–1089.

Collins, R., B. Onisko, M. McLaughlin, and G. Merrington. 2001. Determination of metal-EDTA complexes in soil solution and plant xylem by ion chromatography-electrospray mass spectrometry. Environ. Sci. Technol. 35:2589–2593.

Dabek-Zlotorzynska, E., E.P.C. Lai, and A.R. Timerbaev. 1998. Capillary electrophoresis: The state of the art in metal speciation studies. Anal. Chim. Acta 359:1–26.

Davies, C.W. 1962. Ion association. Butterworths, London.

Fitch, A., and P.A. Helmke. 1989. Donnan equilibrium/graphite furnace atomic absorption estimates of soil extract complexation capacities. Anal. Chem. 61:1295–1298.

Florence, T.M. 1982. The speciation of trace elements in waters. Talanta 29:345–364.

Florence, T.M. 1986. Electrochemical approaches to trace element speciation in waters. A review. Analyst 111:489–505.

Fortin, C., and P.G.C. Campbell. 1998. An ion-exchange technique for free-metal ion measurements (Cd2+, Zn2+): Applications to complex aqueous media. Int.. J. Environ. Anal. Chem. 72:173–194.

Franzluebbers, A.J., R.L. Haney, C.W. Honeycutt, H.H. Schomberg, and F.M. Hons. 2000. Flush of carbon dioxide following rewetting of dried soil relates to active organic pools. Soil Sci. Soc. Am.. J. 64:613–623.

Goyne, K.W., R.L. Day, and J. Chorover. 2000. Artifacts caused by collection of soil solution with passive capillary samplers. Soil Sci. Soc. Am. J. 64:1330–1336.

Gulens, J. 1987. Assessment of research on the preparation, response, and application of solid-state copper ion-selective electrode. Ion-Selective Electrode Rev. 9:127–171.

Harned, H.S., and B.B. Owen. 1958. The physical chemistry of electrolytic solutions. Reinhold, New York.

Hayes, M.H.B., and R.L. Malcom. 2001. Considerations of compositions and of aspects of the structures of humic substances. p. 3–40. *In* C.E. Clappet al. (ed.) Humic substances and chemical contaminants. SSSA, Madison, WI.

He, B., Y. Fang, G. Jiang, and Z. Ni. 2002. Optimization of the extraction for the determination of arsenic species in plant materials by high-performance liquid chromatography coupled with hydride generation atomic fluorescence spectrometry. Spectrochim. Acta B 57:1705–1711.

Helgeson, H.C. 1969. Thermodynamics of hydrothermal systems at elevated temperatures and pressures. Am. J. Sci. 267:729–804.

Herrin, R., A. Andre, M. Shafer, and D. Armstrong. 2001. Determination of silver speciation in natural waters: 2. Binding strength of silver ligands in surface freshwaters. Environ. Sci. Technol. 35:1959–1966.

Hoenig, M. 2001. Preparation steps in environmental trace element analysis: Facts and traps. Talanta 54:1021–1038.

Holm, P.E., T.H. Christensen, J.C. Tjell, and S.P. McGrath. 1995. Speciation of cadmium and zinc with applications of soil solutions. J. Environ. Qual. 24:183–190.

Jones, D.L., A.C. Edwards, K. Donachie, and P.R. Darrah. 1994. Role of proteinaceous amino acids released in root exudates in nutrient acquisition from the rhizosphere. Plant Soil 158:183–192.

Kalbasi, M., F.J. Peryea, W.L. Lindsay, and S.R. Drake. 1995. Measurement of divalent lead activity in lead arsenate contaminated soils. Soil Sci. Soc. Am. J. 59:1274–1280.

Larsson, J., O. Gustafsson, and J. Ingri. 2002. Evaluation and optimization of two complementary crossflow ultrafiltration systems toward isolation of coastal surface water colloids. Environ. Sci. Technol. 36:2236–2241.

Lindsay, W.L. 1979. Chemical equilibria in soils. John Wiley and Sons, New York.

Lorenz, S.E., R.E. Hamon, and S.P. McGrath. 1994. Differences between soil solutions obtained from rhizosphere and non-rhizosphere soils by water displacement and soil centrifugation. Eur. J. Soil Sci. 45:431–438.

Ma, J.F., and K. Nomoto. 1996. Effective regulation of iron acquisition in graminaceous plants. The role of mugineic acids as phytosiderophores. Physiol. Plant. 97:609–617.

Marshall, W., and G.-M. Momplaisir. 1995. Chromatographic approaches to trace element speciation of non-labile chemical species. p. 307–362. *In* A. Tessier and D. Turner (ed) Metal speciation and bioavailability in aquatic systems. John Wiley and Sons, New York.

Martell, A.E., and R.M. Smith. 1976–1989. Critical stability constants. 6 Vol. Plenum Press, New York.

Mester, Z., R. Sturgeon, and J. Pawliszyn. 2001. Solid phase microextraction as a tool for trace element speciation. Spectrochim. Acta B 56:233–260.

Milne, C.J., D.G. Kinniburgh, W.H. van Riemsdijk, and E. Tipping E. 2003. Generic NICA-Donnan model parameters for metal-ion binding by humic substances. Environ. Sci. Technol. 37:958–971.

Minnich, M.M., and M.B. McBride. 1987. Copper activity in soil solution: I. Measurement by ion-selective electrode and Donnan dialysis. Soil Sci. Soc. Am. J. 51:568–572.

Morel, F.M.M., and J.G. Hering. 1993. Principles and applications of aquatic chemistry. John Wiley and Sons, New York.

Mota, A.M., and M.M. Correia dos Santos. 1995. Trace metal speciation of labile chemical species in natural waters: Electrochemical methods. p. 205–257. In A. Tessier and D.R. Turner (ed.) Metal speciation and bioavailability in aquatic systems. John Wiley and Sons, New York.

National Institute of Standards and Technology. 1998. NIST Critically Selected Stability Constants of Metal Complexes Database. Version 5.0. U.S. Dep. of Commerce, Gaithersburg, MD.

Nieboer, E., and D.H.S. Richardson. 1980. The replacement of the nondescriptive term heavy metals by a biologically and chemically significant classification of metal ions. Environ. Pollut. Ser. B 1:3–26.

Otto, W., W. Carper, and C. Larive. 2001. Measurements of cadmium(II) and calcium(II) complexation by fulvic acids using ^{113}Cd NMR. Environ. Sci. Technol. 35:1463–1468.

Owens, G., V. Ferguson, M. McLaughlin, I. Singleton, R. Reid, and F. Smith. 2000. Determination of NTA and EDTA and speciation of their metal complexes in aqueous solution by capillary electrophoresis. Environ. Sci. Technol. 34:885–891.

Parker, D.R., R.L. Chaney, and W.A. Norvell. 1995. Chemical equilibrium models: Applications to plant nutrition research. p. 163–200. *In* R.H. Loeppert et al. (ed.) Soil chemical equilibrium and reaction models. SSSA Spec. Pub. no. 42. ASA and SSSA, Madison, WI.

Parker, D.R., W.A. Norvell, and R.L. Chaney. 1995. GEOCHEM-PC: A chemical speciation program for IBM and compatible personal computers. p. 253–269. *In* R.H. Loeppert et al. (ed.) Soil chemical equilibrium and reaction models. SSSA Spec. Pub. no. 42. ASA and SSSA, Madison, WI.

Parker, D.R., J.F. Pedler, Z.S. Ahnstrom, and M. Resketo. 2001. Reevaluating the free-ion activity model of trace metal toxicity toward higher plants: Experimental evidence with copper and zinc. Environ. Toxicol. Chem. 20:899–906.

Petersen, R. 1982. Influence of copper and zinc on the growth of a freshwater alga, *Scenedesmus quadricauda*: The significance of chemical speciation. Environ. Sci. Technol. 16:443.

Reinfelder, J.R., R.E. Jablonka, and M. Cheney. 2000. Metabolic responses to subacute toxicity of trace metals in a marine microalga (*Thalassiosira weissflogii*) measured by calorespirometry. Environ. Toxicol. Chem. 19:448–453.

Römkens, P.F.A.M., and J. Dolfing. 1998. Effect of Ca on the solubility and molecular size of DOC and Cu binding in soil solution samples. Environ. Sci. Technol. 32:363–369.

Roosenberg, J.M., Y.-M. Lin, Y.L, and M.J. Miller. 2000. Studies and syntheses of siderophores, microbial iron chelators, and analogs as potential drug delivery agents. Curr. Med. Chem. 7:159–197.

Ross, D.S., H.C. Hales, G.C. Shea-MCarthy, and A. Lanzirotti. 2001. Sensitivity of soil manganese oxides: Drying and storage cause reduction. Soil Sci. Soc. Am. J. 65:736–743.

Rueter, J.G., and J.J. McCarthy. 1979. The toxic effect of copper on *Oscillatoria* (*Trichodesmium*) *theibautti*. Limnol. Oceanogr. 24:558–562.

Salam, A.K., and P.A. Helmke. 1998. The pH dependence of free ionic activities and total dissolved concentrations of copper and cadmium in soil solution. Geoderma 83:281–291.

Sarzanini, C. 2002. Recent developments in ion chromatography. J. Chromat. A. 956:3–13.

Sarzanini, C., and M.C. Bruzzoniti. 2001. Metal species determination by ion chromatography. Trends Anal. Chem. 20:304–310.

Sauvé, S. 2002. Speciation of metals in soils. p. 7–58. *In* H.E. Allen (ed.) Bioavailability of metals in terrestrial ecosystems: Importance of partitioning for bioavailability to invertebrates, microbes and plants. Soc. for Environ. Toxicol. and Chem., Pensacola, FL.

Sauvé, S., W. Hendershot, and H.E. Allen. 2000a. Solid-solution partitioning of metals in contaminated soils: Dependence on pH, total metal and organic matter. Environ. Sci. Technol. 34:1125–1131.

Sauvé, S., M.B. McBride, and W.H. Hendershot. 1995. Ion-selective electrode measurements of copper(II) activity in contaminated soils. Arch. Environ. Contam. Toxicol. 29:373–379.

Sauvé, S., M. McBride, W.A. Norvell, and W. Hendershot. 1997. Copper solubility and speciation of *in situ* contaminated soils: Effects of copper level, pH and organic matter. Water Air Soil Pollut. 100:133–149.

Sauvé, S., W.A. Norvell, M. McBride, and W. Hendershot. 2000b. Speciation and complexation of cadmium in extracted soil solutions. Environ. Sci. Technol. 34:291–296.

Schecher, W.D., and D.C. McAvoy. 1991. MINEQL+: A chemical equilibrium program for personal computers. Version 2.1 User's Manual. Environmental Research Software, Edgewater, MD.

Shuman, M.S. 1988. Comparison of anodic stripping voltammetry speciation data with empirical model predictions of pCu. p. 125–133. *In* J.R. Kramer and H.E. Allen (ed.) Metal speciation: Theory, analysis and application. Lewis, Chelsea, MI.

Skoog, D.A., F. Holler, T.A. Nieman. 1998. Principles of instrumental analysis. 5th ed. Harcourt Brace and Company, Philadelphia.

Sposito, G. 1984. Chemical models of inorganic pollutants in soils. CRC Crit. Rev. Environ. Control. 15:1–24.

Sposito, G., F.T. Bingham, S.S. Yadav, and C.A. Inouye. 1982. Trace metal complexation by fulvic acid extracted from sewage sludge: II. Development of chemical models. Soil Sci. Soc. Am. J. 46:51–56.

Staub, C., J. Duffle, and W. Haerdi. 1984. Measurement of complexation properties of metal ions in natural conditions by ultrafiltration: Influence of various factors on the retention of metals and ligands by neutral and negatively charged membranes. Anal. Chem. 56:2843–2849.

Stumm, W., and J.J. Morgan. 1996. Aquatic chemistry. 3rd ed. John Wiley and Sons, New York.

Szpunar, J., S. McSheehy, K. Polec, V. Vacchina, S. Mounicou, I. Rodriguez, and R. Lobinski. 2000. Gas and liquid chromatography with inductively coupled plasma mass spectrometry detection for environmental speciation analysis: Advances and limitations. Spectrochim. Acta B 55:779–793.

Templeton, D., F. Ariese, R. Cornelis, L.-G. Danielson, H. Muntau, H. Van Leeuwen, and R. Lobinski. 2000. Guidelines for terms related to chemical speciation and fractionation of elements. Definitions, structural aspects and methodological approaches (IUPAC Recommendations 2000). Pure Appl. Chem. 72:1453–1470.

Tipping, E. 1994. WHAM: A chemical equilibrium model and computer code for waters, sediments, and soils incorporating a discrete site electrostatic model of ion-binding by humic substances. Comp. Geosci. 20:973–1023.

Tipping, E. 1998. Humic ion-binding model VI: An improved description of the interactions of protons and metal ions with humic substances. Aquatic Geochem. 4:3–48.

Tipping, E., C. Rey-Castro, S.E. Bryan, and J. Hamilton-Taylor. 2002. Al(III) and Fe(III) binding by humic substances in freshwaters, and implications for trace metal speciation. Geochim. Cosmochim. Acta. 66:3211–3224.

van Ginneken, L., and R. Blust. 2002. Determination of conditional stability constants of cadmium-humic acid complexes in freshwater by use of a competitive ligand equilibration-solvent extraction technique. Environ. Toxicol. Chem. 19:283–292.

Wenzel, W.W., R.S. Sletten, A. Brandstetter, G. Wieshammer, and G. Stingeder. 1997. Adsorption of trace metals by tension lysimeters: Nylon membrane vs. porous cup. J. Environ. Qual. 26:1430–1434.

Weston, A., P. Brown, A. Heckenberg, P. Jandik, and W. Jones. 1992. Effect of electrolyte composition on the separation of inorganic metal cations by capillary ion electrophoresis. J. Chromatogr. A 602:249–256.

Wu, Q., W. Hendershot, W. Marshall, and Y. Ge. 2000. Speciation of cadmium, copper, lead, and zinc in contaminated soils. Commun. Soil Sci. Plant Anal. 31:1129–1144.

Xue, H., S. Jansen, A. Prasch, and L. Sigg. 2001. Nickel speciation and complexation kinetics in freshwater by ligand exchange and DPCSV. Environ. Sci. Technol. 35:539–546.

Chapter 15

Chemistry of Salt-Affected Soils

DONALD L. SUAREZ, *USDA-ARS, George E. Brown, Jr. Salinity Laboratory, Riverside, California, USA*

Saline environments occur predominantly in arid and semiarid regions of the world. Saline soils occur less frequently in humid environments, primarily due to marine intrusions and anthropogenic inputs. The chemistry of these soils differs significantly from those present in non-saline, predominantly humid environments. Among the important differences are mineralogy (such as predominance of smectites over kaolinite), typically neutral to elevated pH, and base saturation of cation exchange sites. Salt affected soils usually contain calcite and less frequently gypsum. These soils typically contain high concentrations of exchangeable Na, consistent with the principles governing ion exchange as well as the relative solubility of Na as compared to Ca salts. It is necessary to measure all major dissolved species when examining solubility controls or constraints of any ion in solution.

ORIGIN OF SALTS

Various processes either singularly or in combination lead to the development of saline soils. A necessary condition is of course limited drainage. Ultimately the source of salts is directly linked to the development of liquid phase water on earth; the volcanic outgassing of water and other gases over geologic time and the corresponding weathering of silicate rocks. Since the early development of the earth surface, salts have been recycled into sedimentary rocks and redistributed via the hydrologic cycle.

Weathering

Mineral weathering is an important aspect of the development of saline soils. Primary aluminosilicate minerals, formed under high temperature and pressure, are almost all unstable under earth surface conditions. Weathering may consist of selective leaching of ions, such as base cations and resultant alteration into a secondary weathered mineral, such as feldspar alteration to smectite or formation of kaolinite and x-ray amorphous silica. These dissolution or weathering rates are very slow, thus soluble reaction products (salts) accumulate only under conditions that com-

Chemical Processes in Soils. SSSA Book Series, no. 8.

bine high evapotranspiration, and or low rainfall with limited drainage. Low rainfall, moderate pH and presence of soluble salts all reduce the weathering process.

It is rare that saline soils develop in situ from weathering of igneous or metamorphic rock. More commonly, weathering processes in high precipitation areas (and in lower pH and salinity environments) result in dilute stream and ground water that carry soluble ions to more arid environments in low-lying areas. In this manner the transported dissolved salts become concentrated in areas with little or no drainage. A well documented case is the development of saline, high pH salt deposits in depressions east of the Sierra Nevada Mountains in California. Silicate weathering along the eastern slope of the Sierra range results in dilute solutions in stream flow. Deposition of this stream water into the depressions in the very arid eastern valleys, combined with the hot summer climate, results in very concentrated brines. Depending on the ion ratios present in the initial dilute water, distinctly different water compositions result upon concentration and precipitation. The evolutionary paths of waters undergoing concentration are discussed later in this chapter.

In contrast to igneous and metamorphic rock, sedimentary rocks form under ambient or slightly elevated temperature. Most of the minerals in the rock are either primary unweathered minerals from parent materials or altered minerals such as clays, formed as a result of weathering. If the sediments were deposited in marine or shallow saline inland seas, reactive saline minerals may be present, primarily, calcite, dolomite, and gypsum. In some instances marine shales may have other evaporite minerals, such as halite or mirabilite, but this is relatively less common. In some instances the bulk rock may appear relatively unreactive.

Mancos Shale formed under shallow inland seas during the Cretaceous Age, apparently under hot, dry conditions. Fresh rock samples contain calcite, sulfide minerals and/or elevated concentrations of organic matter, and high exchangeable Na and Mg concentrations on the clay minerals. When reacted overnight in water these rock samples generate a relatively non-saline solution. Long term weathering results in release of acid from sulfide mineral oxidation, dissolution of calcite, exchange of the released Ca with Na and Mg and ultimately a high pH, high sulfate, Mg and Na, saline solution.

Atmospheric Deposition

Atmospheric deposition of salts consists of both inputs from rain and dry deposition–redistribution from other land surfaces during wind storms. Salt deposition in rain is greatest in coastal areas. Rain from high energy storms that develop over the oceans, such as hurricanes, can be quite saline but rain is generally in the range of electrical conductivity (EC) of 0.01 to 0.05 dS m^{-1}. The salinity of rain generally decreases as a storm moves inland, with a more rapid decrease in the concentrations of the larger ions, Na and Cl. Average annual deposition of salts ranges from 10–20 kg ha^{-1} in continental interiors to 100–200 kg ha^{-1} in coastal environments (Mason, 1964).

During recent geologic time, large portions of the Australian continent were covered by deep rooted vegetation, such as eucalypts. These trees were able to use almost all of the rainfall. The lack of deep percolation resulted in concentration of salts in the unsaturated zone (at and below the lower portion of the root zone). The highly weathered Australian landscape lacks salt-containing sedimentary rocks, in contrast to other saline regions of the world. Based on chemical and hydrological studies, these subsurface dissolved salts in Australia are attributed to concentration of salts present in rain (predominately Na and Cl).

Deposition of wind blown salts is also another important input of salts. Among these is the transport of calcite-containing dust into soils free of sedimentary pedogenic carbonates. These soil carbonates exert a major influence on the overall chemistry of the soil, inasmuch as it elevates soil pH and increases the base cation saturation of the soil. Most arid zone soils, even if classified as noncalareous, contain trace quantities of calcite that exert a large effect on the soil solution chemistry.

Anthropogenic Factors

The areal extent of saline soils has expanded as a result of human land use. Among the human land uses, the most influential has been introduction of irrigation into areas with insufficient drainage. Under high water table conditions salts are not leached downward and can accumulate due to surface evaporation. Most irrigated areas throughout the world contain saline soils, many of these saline regions developed as a result of irrigation.

Salinity also may develop in non-irrigated areas as a result of land use practices, such as conversion of eucalypt forests to wheat fields or conversion of prairie land into wheat fields with fallow rotations. In Australia the conversion to wheat increased subsurface recharge, displacing saline subsurface water into receiving streams, thus resulting in increased salinity in the rivers and in downstream irrigation projects. In the Northern Plains states and southern Canadian plains, fallowing had the desired effect of increasing soil moisture in years when wheat was planted, thereby decreasing crop failures, but it also had the effect of increasing recharge. Increased subsurface flow, combined with poor deep drainage resulted in lateral movement of naturally saline subsurface water and saline drainage water into low-lying areas, resulting in saline seeps, which are surface discharges. Continued discharge into low-lying areas, coupled to further salt concentration by evaporation, results in moist, high salinity soils.

Other anthropogenic inputs that cause soil salinization include surface discharge of brines from oil and natural gas operations, overgrazing soils underlain by saline formations, resulting in erosion and surface exposure of the saline materials, surface stockpiling of tailings from mining operations, and discharge of waste water from domestic water softeners.

Measurement of Soil Salinity

Various methods are used for reporting of salinity data. Total dissolved solids (TDS) refers to the dissolved material that remains upon evaporating a fixed volume of water. The values are commonly reported as mg L^{-1} TDS, but also may be reported as mg kg^{-1}, in this instance weight per kg of solution (corresponding to the molal scale). Under saline conditions the assumption that the molal concentration scale (moles per kg of water) can be approximated by concentrations in the molar scale (moles L^{-1} of solution) is not valid. As TDS measurements are time consuming, requiring sample filtration and evaporation of the solutions, salinity is commonly reported in terms of the solution's specific electrical conductivity. Total soluble salts (TSS) is generally reported as the sum of the concentrations of the soluble ions, expressed as $mmol_c L^{-1}$.

The conductance of a solution is proportional (non linearly) to the concentration of ionizing salts, and is measured in units of reciprocal ohms (mhos), or siemens (S) in the SI system. Measurement is made by generating a potential across two electrode surfaces (generally Pt) and determining the resistance. After correction for the cell constant (related to the cell geometry used) the data are reported as Specific Conductance (EC or SpC) per unit volume of solution (most commonly in units of $dS\ m^{-1}$). As a rough approximation TSS can be estimated from EC by using the relationship

$$\text{TSS}\ (\text{mmol}_c\ \text{L}^{-1}) = 10 \times \text{EC}\ (dSm^{-1}) \qquad [1]$$

At elevated salinity, the approximate conversion value decreases (8.2 at a TSS value of 100 $mmol_c L^{-1}$ and 6.8 at TSS = 1000), based on saturation extracts of soils analyzed at the Salinity Laboratory (U.S. Salinity Laboratory Staff, 1954). The following empirical relationship (Marion and Babcock, 1976)

$$\log C = 0.955 + 1.039 \log \text{EC} \qquad [2]$$

provides better accuracy but is limited in application since C in this instance was calculated based not on total concentration as earlier but on concentrations of individual species (corrected for ion pairs), thus requiring use of a chemical speciation program. If the chemical analysis of the sample is known very accurate calculations of EC can be done using the published data on EC relationships for either various single salt solutions and ion speciation, or calculations of individual ion conductances. The error in all of the approximation equations increases with increasing salinity, and is increasingly dependent on the specific ion composition of the water.

The detrimental effects of salinity on plant growth are primarily attributed to osmotic effects. Expression of salinity in terms of osmotic pressure is thus preferred to the alternative electrical conductivity or concentration expressions. Osmotic pressure can be estimated from electrical conductivity by the approximate relationship

$$\Pi\ (kPa) = 40\ \text{EC}\ (dSm^{-1}) \qquad [3]$$

where Π is the osmotic pressure in kPa and EC is at 25°C. A more accurate representation is presented in the thermodynamics section.

Measurement of soil salinity is commonly made using the analysis of the extract from a saturated paste. This method provides a reference condition that minimizes sample dilution (there is no free standing water in the paste) and relates to water content in the field (usually a 1.5 to 1.8 dilution relative to "field capacity" water content). This provides a general reference state suitable for salinity assessment in irrigated lands. A disadvantage of this method, in addition to dilution, is that there is not a fixed soil/water ratio in the extract. Other extraction methods such as 1:2 and 1:5 soil/water extracts have fixed soil water ratios, but result in greater dilution than saturation paste extracts, and thus the analyzed solutions deviate even further from the in situ soil solution composition.

Addition of water to a soil sample beyond the amount present under field conditions makes quantitative interpretation of the analysis difficult. Addition of water is undesirable because for saline soils, the assumption that concentrations can be simply corrected back to field water content is not correct, due to the processes of cation exchange, desorption, and dissolution reactions. The greater the dilution factor the greater the chemical changes from the in situ water composition. Dilution in the presence of exchangeable cations, results in release of Na to solution and loss of Ca and Mg from solution. If calcite or gypsum is present mineral dissolution will occur upon dilution. Use of computer models are useful in refining the estimated composition of in situ water at a given water content based on an extract analysis and the field water content and the extract water content.

The SAR (sodium adsorption ratio) is often reported in analyses of waters used for irrigation and soil water extracts. This ratio defined as

$$\mathrm{SAR} = \frac{(\mathrm{Na}^+)}{\left[\frac{(\mathrm{Ca}^{2+} + \mathrm{Mg}^{2+})}{2}\right]^{0.5}} \qquad [4]$$

where concentrations are expressed in $\mathrm{mmol_c\ L^{-1}}$. The SAR is a useful water quality parameter since it relates to the exchangeable sodium percentage (ESP), which is the percentage of the cation exchange capacity (CEC) occupied by Na (where CEC is expressed as $\mathrm{mmol_c\ kg^{-1}}$.

The relationship between SAR and ESP can be estimated from the relationship

$$\mathrm{ESP} = \frac{100(-0.0126 + 0.01475\ \mathrm{SAR})}{1 + (-0.0126 + 0.01475\ \mathrm{SAR})} \qquad [5]$$

developed by the Salinity Laboratory Staff (1954), from a series of analyses of saturation extracts and exchangeable cations. The equation can be further simplified without much loss of accuracy to the relation

$$\mathrm{ESP} = \frac{100\ (0.1475\ \mathrm{SAR})}{(1 + 0.01475\ \mathrm{SAR})} \qquad [6]$$

Adjustment to the SAR is sometimes made to correct for the increase in SAR associated with degassing of a ground water when exposed to atmospheric conditions and the resultant decrease in Ca caused by precipitation of calcite. The use of SAR adjustment using $\mathrm{pH_c}$ is to be avoided as it results in large errors, exceeding the errors associated with no adjustment, however other adjustment methods are suitable (Suarez, 1981).

Thermodynamics of Soil Solution

Thermodynamic calculations are essential when considering soil chemistry but as often stated, these calculations provide us with knowledge of what reactions can occur, and not, what reactions are occurring. Thermodynamic calculations enable us to determine the solute composition of water in equilibrium with selected solid phases. These calculations also can be used to enable a comparison between water analyses and the equilibrium concentrations, to determine saturation status and the potential for selected phases to be controlling solution composition. Thus the equilibrium model is a useful reference, indicating which processes or specific reactions are possible and merit further consideration. In some instances these calculations also may serve to predict natural water compositions. This later application is frequently not satisfactory due to kinetic considerations. Models such as WATEQ, PREEQEE, MINTEQ, and GEOCHEM. provide the ability to calculate saturation status. Some of these models also have the capability to predict the solution composition that a given water would have in equilibrium with specified solid phases.

The reasons for kinetic controls are complex and varied; however for highly soluble chloride and sulfate salts, equilibrium calculations may serve to provide upper boundaries for solution concentrations. These calculations are generally not suitable for predictions of solution concentrations of heavy metals and many sparingly soluble salts, where supersaturation is prevalent, nor for silicate minerals whose reactivity is low at 25°C, thus almost always undersaturated. Specific examples will be presented in later sections.

For dilute solutions it is reasonable to assume that the commonly used molarity (moles of solute L^{-1} of solution), can be substituted for the thermodynamically required molality (moles of solute kg^{-1} of water). For saline waters this is clearly not a reasonable assumption, and molarity must be converted to molality, the following equation can be used:

$$m = M \frac{W_{\text{solution}}}{(W_{\text{solution}} - W_{\text{solute}})\, d} \quad [7]$$

where W is the weight (kg), and d is the density (kg dm^{-3}).

Calculation of Ion Activities

Activity Coefficients

Non-ideal behavior of soluble ions is accounted for in calculation of activities from the concentrations and activity coefficients. Under non saline to moderate saline conditions the single ion activity approach is commonly used. Using this approach the free ion activity is calculated from the ion concentration and the ionic strength. The model assumption is that there are no specific interactions among free ions in solution and the activity can be predicted from the ionic strength. The ionic strength, I is defined by,

$$I = 1/2 \sum_{i=1}^{c} C_i Z_i^2 \quad [8]$$

where C is the molal ion concentration and z is the charge on the ion. Two approaches are most common, the Davis equation, and various versions of the Debye-Huckel equation. The Debye-Huckel limiting law is given as

$$\log \gamma = -Az^2 \sqrt{I} \quad [9]$$

where $A = 1.82 \times 10^6 (\varepsilon T)^{-3/2}$, equal to 0.509 at 25°C, and z is the charge of the ion, and ε is the dielectric constant. This equation is suitable only for dilute solutions(<0.005 M). The Davis equation, is given as (Stumm and Morgan, 1995)

$$\log \gamma = -Az^2\left(\frac{\sqrt{I}}{1+\sqrt{I}} - 0.2I\right) \quad [10]$$

This equation is listed by Stumm and Morgan (1995) as approximately applicable to 0.5 M, but generates significant errors above 0.1 M and is overly simplified in that it does not consider ion specific parameters, i.e., all ions of the same charge are treated equally. The second term in the equation is sometimes given as 0.2 instead of 0.3. An extended version of the Debye-Huckel equation is given by

$$\log \gamma = -Az^2 \frac{\sqrt{I}}{1 + Ba\sqrt{I}} \quad [11]$$

where $B = 50.3 (\varepsilon T)^{-3/2}$, or 0.33 at 25°C, and a is an adjustable corresponding to the hydrated ion radius, calculated from experimental data. This equation appears useful from 0.1 to 0.2 M, depending on the electrolyte composition.

A further extension of the Debye Huckel equation was developed by Truesdell and Jones, 1974,

$$\log \gamma = -\frac{Az^2\sqrt{I}}{1 + BA\sqrt{I}} + bI \quad [12]$$

where, a and b are empirical adjustable parameters. Truesdell and Jones (1974) fit the parameters to salt solutions. These fits were made using the experimental data available from a series of mean molal activity coefficients of numerous salts. Single ion activity coefficients were obtained by first assuming that $\gamma_{K^+} = \gamma_{Cl^-}$ and calculating γ_{K^+} and γ_{Cl^-} from mean salt data for KCL up to 4.0 M. From these data and mean salt activity coefficients for other salts (e.g., $CaCl_2$), activity coefficients are obtained for all other ions (e.g., Ca^{2+}).

Application of these fits to mixed salt solutions is based on the assumption that there are no specific ion–ion interactions, other than the ion pairs used. Excellent fits were made to 4 M, indicating that these activity coefficients are accurate for the salts that were fit (predominantly Cl salts) but this does not ensure that they can be used under mixed salt environments, especially those with high concentrations of SO_4 or alkalinity. Equation [12] is thus useable up to 0.3 M and as high as 4.0 M, depending on ion composition of the solution. Use of different activity calculation models results in significant discrepancies in the prediction of single ion activities (Suarez, 1999).

Activity of water cannot be assumed equal to one for concentrated solutions. An approximate value is obtained using the empirical correction (Garrels and Christ, 1965)

$$a_{H_2O} = 1 - 0.017\sum_{i=1}^{m} m_i \qquad [13]$$

where a is the activity, Σm_i is the sum of the molalities of dissolved anions, cations, and neutral species. The equation yields reasonable values if Σm_i is less than one molal. At higher concentrations it is necessary to consider osmotic coefficients and the expression given by Felmy and Weare (1986) can be used.

The osmotic pressure can be determined using the following equation from Robinson and Stokes (1965)

$$\Pi = \frac{-\nu RTW_A}{1000V_A}\,\phi m \qquad [14]$$

where Π is the osmotic pressure, R is the gas constant, T is absolute temperature, m is the molality, W is the molecular weight of the solvent (18 g $mole^{-1}$ for water), V_a is the partial molal volume of the water and ϕ is the osmotic coefficient.

Pitzer Expressions

Under high ionic strength (>0.3 M) activity coefficients cannot be represented solely by ionic strength expressions and ion–ion interactions must be considered. Pitzer (1979) described activity coefficients using a viral-type expansion

$$\ln\gamma_i = \ln\gamma_i^{DH} + \sum_j B_{ij}(I)\,m_j + \sum_j\sum_k C_{ijk}\,m_j m_k + \ldots \qquad [15]$$

where γ^{DH} is a modified Debye-Huckel expression and B_{ij} and C_{ij} are specific coefficients for each ion interaction. Felmy and Weare (1986) have reported the coefficients for the major ions and numerous minor species (cations, anions, and neutral species). Use of this model has required consideration of only a few complex species. Application of this model requires use of a thermodynamic database consistent with the activity calculations and selection of ion complexes. Under saline conditions ions exhibit non-ideal behavior.

Complexation, Ion Pairs, and Equilibrium Constants

Some ion interactions are sufficiently strong as to form bonds that are detectable with spectroscopic methods. The concentrations of these complexes are subtracted from the total ion concentrations before calculating individual ion concentrations. In some instances it has been observed that the calculated ionic activities do not correctly predict the solubility of a salt under increasing ionic strength. Using these solubility calculations and specific conductance data it has been inferred that other species, ion pairs, must exist, reducing the free ion concentration and activity and reducing the specific conductance of a solution. The Pitzer model does not include these ion pairs as separate chemical entities but rather represents the ion activity data by consideration of the ion–ion interactions.

Stability constants have been developed for a large number of ion pairs. In most instances there is no physical evidence for these species, and they are constructs to enable accurate predictions with the single ion model. The free ion activity is thus calculated from the total dissolved concentration of the element, activity coefficient of the single ion, association constants for the ion pairs and the activity coefficients of the individual ion pairs. In the case of Ca the total Ca concentration in solution is given by

$$Ca_T = Ca^{2+} + CaSO_4^0 + CaHCO_3^+ + CaCO_3^0 + CaOH^+ \ldots \quad [16]$$

In some models $CaCl^+$ is also considered. Each species also has an associated single ion activity coefficient. A chemical speciation computer program is essential for calculation of individual ion activities and distribution of the concentrations of individual species.

Calculation of stability constants require accurate calculation of ion activities, thus of activity coefficients of both the major ions and of the complex. The stability constants based on solubility studies are thus dependent on the model used for activity coefficients and stability of ion complexes in solution. Several models, including the thermodynamic model PREEQEEC (Parkhurst, 1995), and the transport model UNSATCHEM (Suarez and Simunek, 1997) have the option of using these activity calculation routines.

Solubility products for various minerals have been calculated from both calorimetric and solubility studies. In the case of solubility studies it is important to note that these calculations depend on the accurate calculation of ion activity. Use of different sets of activity coefficients and, ion pair species and stability constants for the ion pair species, all result in discrepancies in the activity of the free ion species, thus in the reported solubility product. As a result it is important that these solubility constants be used only with the activity coefficient model, ion pair species and ion pair dissociation constants with which the constant was determined. These errors increase in importance under saline conditions, as the activity coefficient values of the different models diverge and the complexes and ion pairs become an increasing percentage of the total dissolved salts. There are important differences in ion calculations and in solubility predictions among the most commonly used models, PREEQEC (Parkhurst, 1995), GEOCHEM (Sposito and Mattigod, 1977), and MINTEQA2 (Allison et al., 1990).

Mineral Solubility

General Concepts

For a given solid phase of composition AB that dissolves, the equilibrium expression is given by

$$K = \frac{[A^+]\,[B^-]}{[AB]} \quad [17]$$

where brackets represent activity of a species. Using the convention that the activity of the pure thermodynamically stable well crystallized solid phase equals one then the solubility product, K_{sp} at specified temperature is

$$K_{sp} = [A^+][B^-] \quad [18]$$

In this case the activities represent the activities of the species that would be in equilibrium with the solid phase. An evaluation of the saturation status of a water with respect to a mineral phase can be made by calculating the activities of the ions of interest from the water analysis and ion speciation program and then comparing the ion activity product wit the K_{sp} value. The IAP, given as

$$\text{IAP} = [A^+][B^-] \quad [19]$$

is based on the calculated activities from measured concentrations.

If temperature dependent data are available then the dependence on temperature can be fit with a power function (Truesdell and Jones, 1974) In the absence of this information and for moderate changes in temperature the Van't Hoff equation is often used

$$\log K_T = \log K_{Tr} - \frac{\Delta H_{Tr}}{2.3R}\left(\frac{1}{T} - \frac{1}{Tr}\right) \quad [20]$$

where T is the absolute temperature of interest, T_r is the reference temperature (298 K), R is the gas constant and ΔH_{Tr} is the enthalpy of reaction and K_{tr} is the equilibrium constant at the reference temperature. This equation is generally suitable for the small temperature range encountered by aqueous solutions at ambient pressure.

Processes Affecting the Chemistry of Saline Waters

Solid Phases

Calcite, $CaCO_3$, is almost always present in saline soils, the exception being in acid, saline soils. It is considered to be relatively insoluble, however its solubility is enhanced by the presence of elevated concentrations of CO_2, as occurs in most soils. It is generally the first mineral to precipitate in an evaporating sequence, resulting in loss from solution of Ca and alkalinity (in most instances HCO_3^- and CO_3^{2-}). The solubility expression is given by

$$CaCO_3 \Leftrightarrow Ca^{2+} + CO_3^{2-} \quad [21]$$

however, CO_3^{2-} is rarely the dominant carbonate species in solution. The carbonate species are not conservative since they depend on CO_2 partial pressure, as seen by the following set of equations. The partial pressure of CO_2 is related to the concentration of carbonic acid by the Henry's Law expression

$$[H_2CO_3^*] = P_{CO_2} K_H \qquad [22]$$

where $H_2CO_3^*$ denotes the sum of dissolved CO_2 gas and the aqueous species H_2CO_3. The summed expression $H_2CO_3^*$, decreases with increasing temperature since dissolved CO_2 gas is the major component. The concentration of $H_2CO_3^*$ depends only on the CO_2 partial pressure, temperature and to a minor extent, the ionic strength of the solution. The dissociation of carbonic acid is given by

$$[HCO_3^-] = \frac{H_2CO_3^* K_{a1}}{[H^+]} \qquad [23]$$

with a K_{a1} value of $10^{-6.35}$ at 25°C. As shown by Eq. [22] and [23], increasing the CO_2 partial pressure results in a decrease in pH and an increase in HCO_3^-. Carbonate ion is related to bicarbonate by the second dissociation expression

$$[CO_3^{2-}] = \frac{[HCO_3^-] K_{a2}}{[H^+]} \qquad [24]$$

with the constant K_{a2} equal to $10^{-10.33}$ at 25°C. In the absence of any additional reactions the change in CO_2 has no change on the net alkalinity since the change in H^+ (ΔH^+) = $\Delta HCO_3^- + \Delta CO_3^{2-} + \Delta OH^-$, expressed in $mmol_c L^{-1}$.

Under most commonly encountered conditions (pH 6.5–9.5) HCO_3^- is the dominant solution carbonate species. Based on the above reasoning it is more realistic to consider the following overall reaction for calcite dissolution and precipitation.

$$CaCO_3 + CO_2(g) + H_2O \Leftrightarrow Ca^{2+} + 2HCO_3^- \qquad [25]$$

and thus the overall equilibrium expression is given by

$$[Ca^{2+}][HCO_3^-]^2 = K_{SP}^C \frac{K_{CO_2} K_{a_1} P_{CO_2} (H_2O)}{K_{a_2}} \qquad [26]$$

Where K_{SP}^C is the solubility constant for calcite. Analysis of this equation indicates that an increase in CO_2 results in an increase in the solubility of calcite.

Under acid, saline conditions dissolution of calcite is represented by the following equation

$$CaCO_3 + 2H^+ = Ca^{2+} + H_2CO_3 \qquad [27]$$

Calcite is thus useful to neutralize acidity in acid sulfate soils, without further increasing the salinity.

Magnesium substitutes into the calcite structure, generally in the range of 2 to 5%, providing a minor sink term for Mg. Under high Mg/Ca ratios this substitution may increase to 8 to 12%; however, this is not usually observed in pedogenic calcite.

Soil solutions and shallow ground waters in arid environments are supersaturated with respect to calcite if calcite is present and concentrating processes such as evaporation and plant water uptake occur. This supersaturation, on average three fold (Suarez, 1977, Suarez et al., 1992), corresponds to the supersaturation level in the presence of dissolved organic matter at which nucleation stops and only crystal growth occurs (Lebron and Suarez, 1996). Soil solutions in arid zones thus contain substantially higher concentrations of calcium than predicted by equilibrium speciation models.

Dolomite, $CaMg(CO_3)_2$, is often present in saline soils but is derived from the soil parent materials rather than formed pedogenically. Dissolution of dolomite under non acid conditions is generally very slow, nonetheless over time it provides a source of Mg to solution. Dolomite dissolution is represented by the following reaction

$$CaMg(CO_3)_2 + 2CO_{2(g)} + 2H_2O \rightarrow Ca^{2+} + Mg^{2+} + 4HCO_3^- \quad [28]$$

Dolomite solubility is roughly comparable to that for calcite $K_{sp} = 10^{-17}$ and as with calcite almost all arid soils are supersaturated. Kinetic constraints prevent precipitation under earth surface temperature and pressures (even under clean and controlled laboratory conditions). Under high degrees of supersaturation a mixture of hydrated magnesium carbonate and calcite may form, sometimes called protodolomite. Since it is a mixture of two phases, it is preferable to consider the solubility and formation of the phases as separate entities.

Gypsum, ($CaSO_4 \bullet 2\,H_2O$), is generally the second most common precipitated phase in arid soils, and may occur in acid as well as alkaline soils. Dissolution and precipitation of gypsum is represented by

$$CaSO_4 \bullet 2H_2O \Leftrightarrow Ca^{2+} + SO_4^{2-} + 2H_2O \quad [29]$$

Saline acid soils are derived from the oxidation of sulfide minerals, generating sulphuric acid that reacts with the existing soil minerals. Gypsum will precipitate in these soils if small amounts of carbonates and other calcium source minerals are present. These soils typically occur in coastal areas, for example resulting from draining coastal wetland areas that were formerly under anaerobic conditions.

Gypsum readily precipitates and dissolves, thus gypsiferous soils can usually be assumed to be at equilibrium with respect to gypsum. The following solubility expression is applicable.

$$K_{SP}^{G} = [Ca^{2+}][SO_4^{2-}][H_2O]^2 \quad [30]$$

where K_{SP}^{G} is the solubilty constant for gypsum. Due to high solubility, most gypsiferous soils are high in dissolved Ca, however, in some instances more soluble sulfate salts may be present and $SO_4^{2-} >> Ca^{2+}$.

Magnesium Carbonates

Magnesite, ($MgCO_3$) is often considered the thermodynamically most stable magnesium carbonate mineral with a solubility comparable to that for calcite;

however, magnesite can generally be ignored as a mineral of interest in saline soils as it is not formed under earth surface temperature and pressure conditions and is rare, existing primarily as a hydrothermal mineral. The mineral is kinetically slow to dissolve.

The carbonate minerals potentially limiting Mg in solution are hydromagnesite and nesquehonite, both highly soluble and both able to readily precipitate when the solubility is exceeded. Use of these minerals in predictive models provides for an upper limit to magnesium in solution, and in combination with calcite, provides a mixed mineral phase material that is called protodolomite, as discussed above. Nonetheless these minerals, as well as the mixed phase protodolomite, are rare in saline soils.

The predicted ratio of Mg/Ca based on simultaneous equilibrium with magnesium carbonate and calcite, is roughly 1,000/1, well above values commonly seen in natural waters. More typically, saline waters have Mg/Ca ratios that range from 1 to 10, suggesting that other minerals or processes either control or limit Mg in solution. Seawater has a Mg/Ca ratio of 5, also suggesting other Mg removal processes.

As waters concentrate, the SAR increases proportionally with the square root of the concentration, thus in the absence of other reactions, Ca and Mg would be released from clays exchange sites into solution; however, precipitation of Ca results in additional removal of Ca from exchange sites and is often sufficient to result in Mg loss from solution and adsorption onto the clay exchange sites despite the increasing SAR. In the case of freshwater sediments, upon reaction with seawater there is a depletion of Mg from seawater (Sayles and Manglesdorf, 1977).

Loss of Mg from solution during the process of concentration is attributed to cation exchange and minor substitution of Mg into precipitated calcite, as discussed with seawater chemistry above. Magnesium carbonate minerals such as nesquehonite, $MgCO_3$ $3H_2O$, and hydromagnesite, rarely occur but if the solution concentration is above the saturation level, they will readily precipitate. Their rare occurrence indicates that other less soluble phases may control Mg in solution. The reaction of nesquehonite and hydromagnesite are represented by,

$$MgCO_3 \bullet 3H_2O + CO_{2_{(g)}} \Leftrightarrow Mg^{2+} + 2HCO_3^- + 2H_2O \qquad [31]$$

and,

$$Mg_5(CO_3)_4(OH)_2 \bullet 4H_2O + 6CO_{2_{(g)}} \Leftrightarrow 5Mg^{2+} + 10HCO_3^- \qquad [32]$$

respectively. Additional unaccounted for sinks may include Mg uptake into silicate minerals, including precipitation of sepiolite, formation of palygorskite or Mg diffusion into silicate minerals. These processes also impact terrestial saline waters.

Magnesium Silicates

Precipitation of sepiolite is possible under earth surface conditions and serves to put an upper limit on Mg concentrations in solution. Precipitation–dissolution of amorphous sepiolite can be represented by,

$$Mg_2Si_3O_{7.5}(OH) \bullet 3H_2O) + 4.5H_2O + 4CO_{2(g)}$$
$$\Leftrightarrow 2Mg^{2+} + 3H_4SiO_4 + 4HCO_3^- \quad [33]$$

Mg silicates may provide a sink for Mg, however in soil with sulfate dominant solution conditions, Mg in surface salt crusts is found in mixed Na–Mg sulfate hydrated salts, bloedite, and konyaite (Kohut and Dudas, 1993).

Cation Exchange

Cation exchange is a major process affecting the chemistry of saline soils. Cation exchange can be represented by the following reaction for $Ca^{2+} - Na^+$,

$$2NaX + Ca^{2+} \rightarrow 2Ca_{0.5}X + 2Na^+ \quad [34]$$

where the exchanger site can be expressed in the units of mol_c, in this case generating an expression similar to the commonly applied Gapon equation. The Gapon equation considers concentrations rather than activities of both the solid and solution phase,

$$k_G = \frac{(Ca^{2+})^{0.5}(NaX)}{(Na^+)(Ca_{0.5}X)} \quad [35]$$

This equation is often applied due to the relative stability of the selectivity value. A true thermodynamic expression is rarely used because the solid phase activity coefficients are not independently determined. Alternatively, if we write the exchange reaction in terms of moles of species rather than moles of charge we obtain an expression such as

$$2NaX + Ca^{2+} \rightarrow CaX_2 + 2Na^+ \quad [36]$$

The exchange expression for this reaction,

$$k_V = \frac{(Na^+)^2(CaX)}{(NaX)^2(Ca^{2+})} \quad [37]$$

is called the Vanselow equation. This equation is not as commonly used, probably due to the greater variation in the selectivity constant.

Evolution of Saline Waters

A representation of the chemical evolutionary path of saline waters upon concentration is shown in Fig. 15–1. This figure has been modified from representations of closed basin evaporation shown by Eugster and Jones (1979), among others. These processes and changes in water composition are directly applicable to changes in saline soil waters, with several exceptions.

As shown in Fig. 15–1, the most important determinant of the change in a water composition upon concentration is the ratio of Ca +Mg to alkalinity (expressed

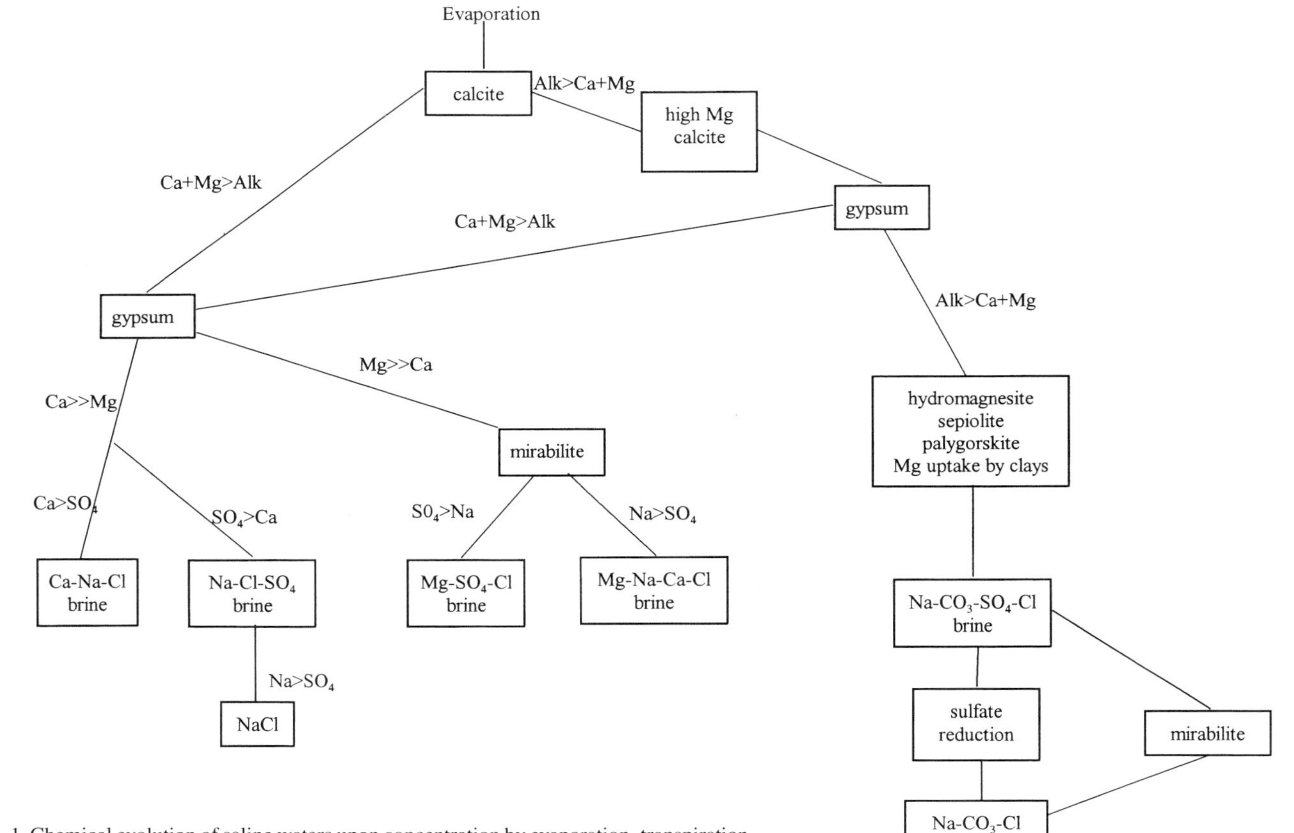

Fig. 15–1. Chemical evolution of saline waters upon concentration by evaporation–transpiration.

in $mmol_c\ L^{-1}$). Typically the origin of the dissolved salts in the soil is not primarily from silicate weathering, thus Ca+ Mg usually exceeds alkalinity. For example dissolution of silicate minerals can be represented by release of cations (Ca, Mg, Na, and K) precipitation of a Al oxide and release of silica, in the form of silicic acid, with the counter anion being HCO_3^-. A representation for dissolution of Na feldspar (albite) is given by

$$NaAlSi_3O_8 + 8H_2O + CO_2 \rightarrow Na^+ + Al(OH)_3 + 3H_4SiO_4 + HCO_3^- \quad [38]$$

In many instances a more realistic reaction includes the formation of kaolinite as a weathering product such as

$$2NaAlSi_3O_8 + 11H_2O + 2CO_2 \rightarrow Al_2Si_2O_5(OH)_4 + 4H_4SiO_4 + 2Na^+ + 2HCO_3^- \quad [39]$$

From these reactions it follows that for the release of every base cation (Ca^{2+}, Mg^{2+}, Na^+, and K^+) there is a formation of one or more HCO_3^- ion, thus (in $mol_c\ L^{-1}$) $Ca^{2+} + Mg^{2+} + Na^+ + K^+ = HCO_3^-$ and $Ca < HCO_3$.

When the (Ca+Mg)/alkalinity ratio is less than one the water evolves upon evaporation and/or transpiration to a high pH, high alkalinity and Ca and Mg depleted water. Waters of this composition are observed in the saline soils of the eastern Sierra Mountains. They are generally low in sulfate and the dissolved salts are derived from granitic weathering. Among the suite of minerals expected are calcite, which is almost always the first mineral to precipitate. With increasing concentration and continued calcite precipitation, Ca becomes increasing depleted in solution.

The cation exchange process exerts an important effect on the soil solution composition. As the water increases in salinity upon concentration, there is a decreasing preference for divalent as opposed to monovalent ions (see Eq. [35]. Upon concentration of the water this process serves to release Ca and Mg to solution and remove Na from solution. This process thus serves to buffer changes in the cation composition of the soil water, since Ca and Mg are being removed by precipitation. In some instances the Ca and Mg release may be sufficient to change the ratio of Ca+Mg to alkalinity to a value greater than one, thus moving the evolutionay path to the alkalinity depleted left side of Fig. 15–1. Cation exchange prevents the use of Na ratios as an indicator of the degree to which a water has been concentrated.

Upon further concentration of these waters where Ca+Mg exceeds alkalinity, gypsum may precipitate if sulfate is sufficiently elevated. Further concentration and precipitation of Mg minerals is often observed, however, the process may be more prevalent than reported as these minerals are relatively x-ray amorphous. With increasing concentration exchange sites become almost completely Na saturated with release of Mg to solution (at this stage Ca concentrations in solution and on exchange sites are very low). The waters are now enriched in Na sulfate and alkalinity, and mirabilite. a hydrated Na sulfate mineral, may also precipitate. Over long time scales there also is Mg substitution into clay mineral structures.

When Ca+Mg exceeds alkalinity, the waters evolve to a low alkalinity and reduced pH condition. Calcite and then gypsum will be the precipitating minerals. In almost all instances Mg will exceed Ca and if the waters are further concentrated, will evolve to either a Mg SO_4–Cl or Mg Na Ca Cl brine. With depleted alkalinity Mg minerals remain highly soluble.

REFERENCES

Allison, J.D., D.S. Brown, and K.J. Novo-Gradac. 1990. MINTEQA2/PRODEFA2. A geochemical assessment model for environmental systems. Version 23.0. Office Res. Dev. USEPA, Athens, GA.

Eugster, H.P., and B.F. Jones. 1979. Behavior of major solutes during closed-basin brine evolution. Am. J. Sci. 609–631.

Felmy, A., and J. Weare. 1986. The prediction of borate mineral equilibria in natural waters: Application to Searles Lake, California. Geochim. Cosmochim. Acta. 50:2771–2783.

Garrels, R.M., and C.H. Christ. 1965. Solutions, minerals, and equilibria. Harpur and Row, New York.

Kohut, C.K., and M.J. Dudas. 1993. Evaporite mineralogy and trace element content of salt-affected soils in Alberta. Can. J. Soil Sci. 73:399–409.

Lebron, I., and D.L. Suarez. 1996. Calcite nucleation and precipitation kinetics as affected by dissolved organic matter at 25°C and pH.7.5. Geochim. Cosmochim. Acta 60:2765–2776.

Marion, G.M., and K.L. Babcock. 1976. Predicting specific conductance and salt concentration of dilute aqueous solutions. Soil Sci. 106:393–398.

Mason, B. 1964. Principles of geochemistry. 2nd ed. John Wiley and Sons, New York.

Parkhurst, D.L. 1995. User's guide to PHREEQC-A computer program for the speciation, reaction path, advective transport and inverse geochemical calculations. U.S. Geol. Surv. Water Resources Invest. Rep. 95-4227. U.S. Geol. Survey. Washington, DC.

Pitzer, K.S. 1979. Activity coefficients in electrolyte solutions. p. 157–208. CRC Press, Boca Raton, FL.

Robinson, R.A., and R.H. Stokes. 1965. Electrolyte solutions. 2nd ed. Butterworths, London.

Sayles, F.L., and P.C. Manglesdorf. 1977. The equilibration of clay minerals with seawater: Exchange reactions. Geochim. Cosmochim. Acta. 41:951–960.

Sposito, G., and S. Mattigod. 1977. GEOCHEM: A computer program for the calculation of chemical equilibria in soil solutions and other natural water systems. Dep. Soil and Environ. Sci., Univ. of California, Riverside.

Stumm, W., and J. Morgan. 1995. Aquatic chemistry. 3rd ed. John Wiley and Sons, New York.

Suarez, D.L. 1977. Ion activity products of calcium carbonate in waters below the root zone. Soil Sci. Soc. Am. J. 41:310–315.

Suarez, D.L. 1981. Relationship between pH_c and SAR and an alternative method of estimating SAR of soil or drainage water. Soil Sci. Soc. Am. J. 45:469–475.

Suarez, D.L., and J. Simunek. 1997. UNSATCHEM: Unsaturated water and solute transport model with equilibrium and kinetic chemistry. Soil Sci. Soc. Am. J. 61:1633–1646.

Suarez, D.L., J.D. Wood, and I. Ibrahim. 1992. Reevaluation of calcite supersaturation in soils. Soil Sci. Soc. Am. J. 56:1776–1784.

Suarez, D.L. 1999. Thermodynamics of the soil solution. p. 97–134. *In* D.L. Sparks (ed.) Soil physical chemistry. 2nd ed. CRC Press, Boca Raton, FL.

Truesdell, A.H., and B.F. Jones. 1974. WATEQ, a computer program for calculating chemical equilibria of natural waters. J. Res. U.S. Geol. Surv. 2:234–248.

U.S. Salinity Laboratory Staff. 1954. Diagnosis and improvement of saline and sodic soils. Agric. Handb. 60. U.S. Dep. of Agric. U.S. Gov. Print. Office, Washington, DC.

SUBJECT INDEX